5th International Conference on Agricultural and Biological Sciences (ABS 2019)

AA006736

IOP Conference Series: Earth and Environmental Science
Volume 346

Macao, China
21 - 24 July 2019

ISBN: 978-1-5108-9670-3
ISSN: 1755-1307

Printed from e-media with permission by:

Curran Associates, Inc.
57 Morehouse Lane
Red Hook, NY 12571

proceedings.com

Some format issues inherent in the e-media version may also appear in this print version.

This work is licensed under a Creative Commons Attribution 3.0 International Licence.
Licence details: http://creativecommons.org/licenses/by/3.0/.

No changes have been made to the content of these proceedings. There may be changes to pagination and minor adjustments for aesthetics.

Printed with permission by Curran Associates, Inc. (2020)

For permission requests, please contact the Institute of Physics at the address below.

Institute of Physics
Dirac House, Temple Back
Bristol BS1 6BE UK

Phone: 44 1 17 929 7481
Fax: 44 1 17 920 0979

techtracking@iop.org

Additional copies of this publication are available from:

Curran Associates, Inc.
57 Morehouse Lane
Red Hook, NY 12571 USA
Phone: 845-758-0400
Fax: 845-758-2633
Email: curran@proceedings.com
Web: www.proceedings.com

TABLE OF CONTENTS

The 5th International Conference on Agricultural and Biological Sciences (ABS 2019)

Preface

Thanks to the support and contributions from more than 140 participants of more than 25 countries, *the 5th International Conference on Agricultural and Biological Sciences (ABS 2019)* held successfully from July 21st to July 24th in Macau University of Science and Technology, Macau, China.

International Conference on Agricultural and Biological Sciences (ABS) is convened annually to gather related experts to exchange the newest ideas and experiences in the Agricultural and Biological Sciences. Previous four ABS Conferences brought together a forum of renown personalities from the fields of agriculture and biological sciences from the leading research institutes and universities making it successful international gatherings. Based on ABS2018, the main topics of ABS2019 consist of five parts on Animal Sciences: Behavior, Reproduction and Biodiversity; Microbial Application and Applied Biochemistry; Agricultural Economics, Ecology and Soil Sciences; Food Science and Modern Technology; and Botanical Science: Physiology and Genetic Analysis.

More than 300 papers were submitted to ABS2019, and 120 submitted manuscripts have met the scope of IOP Conference Series: Earth and Environmental Science (EES). After a pre-review on the originality and language, peer review process was arranged by the Editorial Committee and 92 best manuscripts were selected for publication in IOP Conference Series: Earth and Environmental Science (EES). The Editorial Board was led by the Guest Editor Prof. Corina Carranca and Reviewers like Prof. Izabela Michalak, Prof. Claudio Frezza, Prof. Eduardo Abel Rienzi, Prof. Job Nmadu, Prof. Vítor Martinho, Prof. Aurelija Rudzianskaite, and Prof. Hisayoshi Hayashi, and we believe those selected 92 papers will provide some insights in the field of soil, environmental and biological sciences.

On behalf of the Conference Organizers, we would like taking this opportunity to express our sincere thanks to the Guest Editor and all the Reviewers for their tremendous efforts and dedication to the conference, to all the authors for their relevant contributions to the conference, as well as all the colleagues from IOP publisher for their support and their endless efforts towards the publication of the Conference Proceedings. We believe that with their earnest support and contributions, future ABS Conferences would scale new heights.

Below are some interesting pictures from ABS2019:

Content from this work may be used under the terms of the Creative Commons Attribution 3.0 licence. Any further distribution of this work must maintain attribution to the author(s) and the title of the work, journal citation and DOI.
Published under licence by IOP Publishing Ltd

澳門科技大學
MACAU UNIVERSITY OF SCIENCE AND TECHNOLOGY

IOP Conf. Series: Earth and Environmental Science **346** (2019) 011002 doi:10.1088/1755-1315/346/1/011002

Peer review statement

All papers published in this volume of *IOP Conference Series: Earth and Environmental Science* have been peer reviewed through processes administered by the proceedings Editors. Reviews were conducted by expert referees to the professional and scientific standards expected of a proceedings journal published by IOP Publishing.

Content from this work may be used under the terms of the Creative Commons Attribution 3.0 licence. Any further distribution of this work must maintain attribution to the author(s) and the title of the work, journal citation and DOI.
Published under licence by IOP Publishing Ltd

IOP Conf. Series: Earth and Environmental Science **346** (2019) 012001 doi:10.1088/1755-1315/346/1/012001

Temporal and spatial characteristics of soil moisture dynamics in fixed dune of the Horqin sandy land

S X Yao and C C Zhao[1]

Lanzhou City University, Lanzhou 730070, China

E-mail: zhao_chch1978@163.com

Abstract. The spatio-temporal variability of soil moisture (SM) in the fixed dune was analyzed based on the long-term monitoring data of SM at a depth of 1.5m in the fixed dune of the Horqin Sandy Land from May to September (that is growing season) for the 6th consecutive year (2009~2014) and the synchronize rainfall data. The results indicated that: (1) The changes of SM with the month generally showed a "W-shaped" curve in 2009, 2010 and 2011, and higher SM appeared in July. In 2012, the SM was lower in May and higher in June, and decreased slightly in the following three months. In 2013 and 2014, the changes of SM with the month showed a continuous downward trend. (2) Except for 2013, the SM was lower in the surface layer (0~10 cm), and the SM increased first, then decreased, and finally increased with soil depth. (3) As the soil depth increases, both SD and CV of SM showed a decreasing trend. According to the SD and CV method, the vertical variation layers of SM in fixed dunes of the study area were initially divided into the fast changing layer of 0~50 cm, the active layer of 50~90 cm and the second active layer of 90~150 cm.

1. Introduction

SM is an important controlling factor of many hydrological, ecological, and biological processes, and it is usually affected by rainfall, groundwater, terrain and vegetation, etc [1]. And hence, the SM has received important attention from various related research areas [2]. Although SM accounts for only 0.001% of the total water storage on Earth, it is a key factor for the climate system [3-4]. SM in semi-arid regions is the limiting factor for the development of ecosystems, and the spatio-temporal variation rule of SM is the main reference factor for ecological construction and planning in these regions [5].

The spatio-temporal variation of SM refers to the significant differentiation and diversity of SM characteristics in different times, sites and soil layers within a certain landscape [6]. The spatio-temporal variation of SM is the result of the comprehensive effect of various factors in multiple scales, such as land use (vegetation), weather (rainfall), terrain, soil and human activities. In recent years, the study on the spatio-temporal variation of SM has become one of the research hotspots at home and abroad [7]. The temporal variation of SM decreases as soil depth increases [8]. In arid regions, the spatial variation of SM is the result of the combined effect of moisture input (rainfall) and output (evapotranspiration and leakage). Generally, the spatial distribution of SM is divided into three types, namely decreasing type, fluctuating type and increasing type, which vary with place and time [7]. For example, Qiu Yang et al. studied the SM dynamics of the Loess Plateau in China and concluded that there is a significant difference in the average SM at different soil depths, and the SM content increases significantly as soil depth increases [9]. Yang Xinmin studied the dynamics of SM in the loess hilly region of China, and

Content from this work may be used under the terms of the Creative Commons Attribution 3.0 licence. Any further distribution of this work must maintain attribution to the author(s) and the title of the work, journal citation and DOI.
Published under licence by IOP Publishing Ltd

considered that the SM tends to decrease as soil depth increases [10]. Yang Yongdong's study on SM of different vegetation types in the Loess Plateau found that the SM changes slowly in the soil profile at the early stage of vegetation growth, while the SM shows a decreasing trend in the profile at the middle stage of vegetation growth [11].

Horqin Sandy Land lies in a semiarid area of southeast Inner Mongolia. Due to the long-term influence of extensive fuelwood gathering, heavy grazing and reclamation, it has become one of the most severe desertification regions in recent years [12]. Thus, discussing the variation of SM in fixed dunes of this region can not only help us understand the temporal and spatial distribution characteristics of SM in sandy land, but also have a important effect on the next study of SM in sandy land under different environmental conditions.

2. Methods

2.1. Study area

The Horqin Sandy Land located in the agro-pastoral transitional zone between the Inner Mongolian Plateau and the Northeast Plains (120°42′ E, 42°55′ N). The climate is temperate, semiarid continental monsoonal in this area, and with an average annual precipitation of 360 mm. The annual mean temperature is about 6.4℃, and the annual mean open-pan evaporation is around 1935 mm. The annual mean wind velocity is in the range of 3.2~4.1 $m \cdot s^{-1}$. The topography is characterized by dunes and interdune lowlands [13].

2.2. Data obtaining

In this research, a special fixed dune was selected, and we installed three neutron probes (CNC503DR) with equal distances to record the soil moisture (%) of eight soil layers, the lower limit of soil was10 cm, 30 cm, 50 cm, 70 cm, 90 cm, 110 cm, 130 cm and 150 cm. The three neutron probes were installed to measure the SM, and taken the average value for analysis. From 2009 to 2014, the SM measured once every seven days and increased measurement after precipitation. For each soil layer, the total measured times was all 101. The synchronize precipitation data was received from nearby weather station.

2.3. Classification of precipitation year

We used the domestic precipitation year classification method [14] to determine the types of precipitation years: If $P-0.33\delta>P_i$, it is dry year; if $P+0.33\delta<P_i$, it is wet year. Where P_i denote the current year precipitation (mm), P and δ represent the annual mean precipitation (mm) and its square error (mm), respectively.

2.4. Soil division

The commonly method was depended on CV and SD to divide the soil layer into four levels [15-16]: If SD<2, CV<0.1, it is relatively steady layer (RS); If SD: 2~3, CV: 0.1~0.2, it is second active layer (SA); If SD: 3~4, CV: 0.2~0.3, it is active layer (AL); And if SD>4, CV>0.3, it is fast changing layer (FC).

2.5. Data analysis

The statistical characteristics of SM in each month and each soil layer was obtained by Micro-Excel, and the graph that shows the dynamic of SM with month and soil depth was compiled by the Micro-Origin 8.0 software.

3. Results and discussion

3.1. Precipitation characteristics

During the experimental period, the statistical characteristics of precipitation in the study area (Table 1) showed that, in the past 6 years (2009~2014), the mean annual precipitation of the study area was 280.7 mm, with the mean square error of 78.12 mm, P+0.33 δ =306.48 mm, P-0.33 δ =254.92 mm. According to the classification standard for precipitation years [14], the rainfall was the highest (>306.48 mm) in 2012; at the lower levels (<254.92 mm) in 2009, 2011 and 2013, which were dry years; and at normal levels in 2010 and 2014, which were normal years. The annual rainfall was mainly imploded from May to September, more than 70% of the annual rainfall imploded in this period and even reaching up to 92.8% in 2014. In the meantime, the mean rainfall of different months was the highest (82.5%) in July and the lowest in January. The rainfall was scarce in January, February, November and December, and the total rainfall in the four months accounted for no more than 3% of the annual rainfall. The monthly rainfall was the highest in September 2012 at 144.8mm, which exceeded half of the annual rainfall in all years except 2012 and accounted for 33.6% of the annual rainfall in that year.

Table 1. Rainfall characteristics of experimental year (2009~2014).

Year	2009	2010	2011	2012	2013	2014	Mean
Jan.	0.0	0.0	0.0	0.0	0.0	0.0	0.0
Feb.	3.5	0.0	0.2	3.4	2.0	0.0	1.5
Mar	0.0	5.4	0.2	4.2	3.8	11.8	4.2
Apr.	64.4	11.2	3.0	17.0	30.0	2.2	21.3
May.	23.2	51.0	34.4	28.6	10.0	79.0	37.7
Jun.	54.4	13.4	33.4	87.4	36.0	69.2	49.0
Jul.	86.6	82.0	90.8	93.8	85.2	56.8	82.5
Aug.	9.8	26.8	24.8	27.8	59.2	7.8	26.0
Sep.	4.4	29.2	2.0	144.8	7.6	36.8	37.5
Oct.	1.6	58.2	7.8	17.4	17.4	5.2	17.9
Nov.	1.7	1.8	5.2	6.2	0.4	0.2	2.6
Dec.	2.2	0.4	0.0	0.0	0.0	0.0	0.4
Sum	251.8	279.4	201.8	430.6	251.6	269.0	280.7
Yearly rainfall pattern	Dry year	Normal year	Dry year	Rainy year	Dry year	Normal year	/

3.2. SM changes with month

The changes of the average SM in different soil layers with the month from 2009 to 2014 (Figure. 1) illustrated that, the varies of SM with the month generally showed a "W-shaped" curve in 2009, 2010 and 2011. Higher SM appeared in July and higher rainfall also occurred in July of these three years, indicating that heavy rainfall can greatly replenish SM at the middle stage of vegetation growth. In these three years, the SM was relatively lower in June and August. In 2012, the SM was lower in May and higher in June, and decreased slightly in the following three months, to the extent that the heavy rainfall (144.8 mm) in September failed to significantly increase the SM content. It shows that the recharge effect of heavy rainfall on SM at the later stage of vegetation growth is not as obvious as that at the middle stage of vegetation growth. In 2013, although higher rainfall also appeared in July (85.2 mm), the changes of SM with the month didn't show a "W-shaped" curve like that in the previous three years, but showed a continuous decline. In 2014, the highest rainfall (79 mm) occurred in May, when the SM was not the highest in all months. This is because May belongs to the early stage for vegetation growth in the study area, and the initial SM was at a low level. Combined with strong soil evaporation, the SM was not effectively recharged from heavy rainfall. The SM continued to decline since June, which was similar to that in 2013.

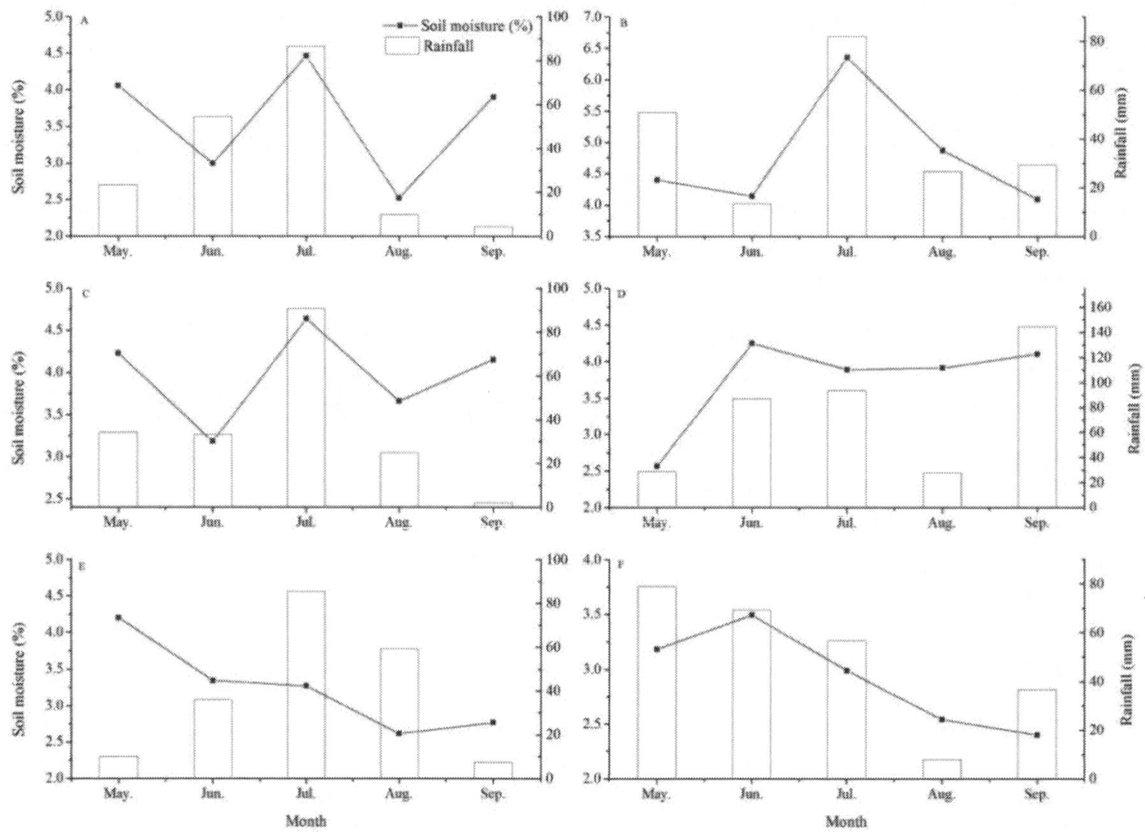

Figure 1. Variation of SM and rainfall with month.
A, B, C, D, E, F denote the year of 2009, 2010, 2011, 2012, 2013 and 2014, respectively.

3.3. SM changes with soil depth

The varies of SM along with soil depth (Figure. 2) displayed that, except for 2013, the average SM in each month of the growing season was lower in the surface layer (0~10 cm), mainly because the strong evaporation of the surface soil results in lower SM. On the whole, the SM increased first, then decreased, and finally increased with soil depth. Higher SM occurred in the soil layer of 10~30 cm from 2009 to 2011 and 30~50 cm in 2012 and 2014, and then decreased to the second low content in 90-110 cm, followed by an increasing trend again. In 2013, the SM basically continued to increase as the soil depth increased, which was different from that of the other five years. In the analysis of SM changes with the month, there was also a great difference between 2013 and other years. The reason remains to be further explored.

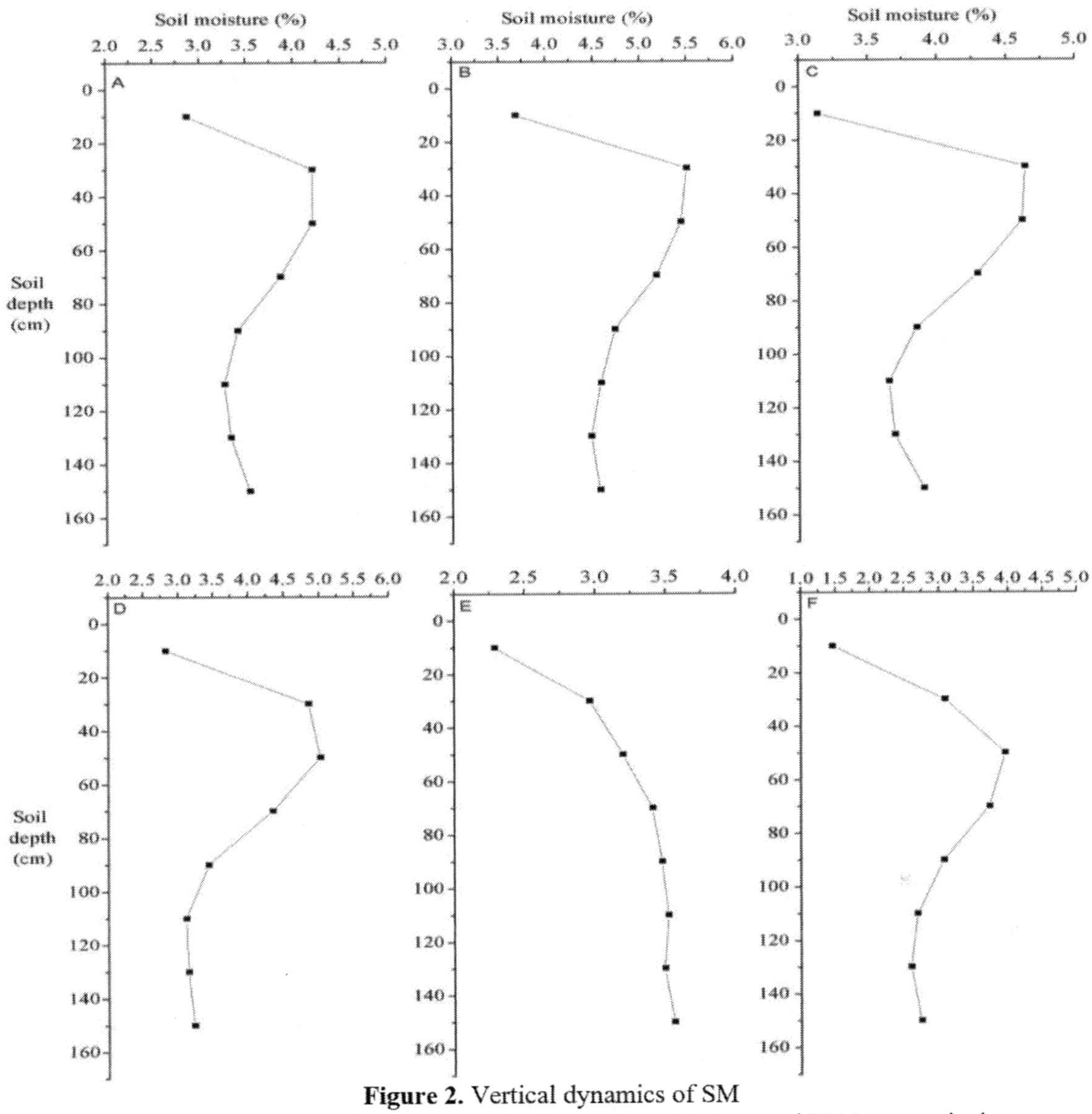

Figure 2. Vertical dynamics of SM
A, B, C, D, E, F denote the year of 2009,2010,2011,2012,2013 and 2014, respectively.

3.4. SM changes with soil layer

The statistical results of SD and CV of SM in each soil layer from 2009 to 2014 (Table 2) showed that, the SD of SM was relatively small, not exceeding 2.5. As the soil depth increases, both SD and CV showed a decreasing trend, which indicates that the variability of SM decreases with the increase of soil depth.

We divided the soil layer into four levels depending on the CV and SD of SM [15-16], the result indicated that, in 2009, 0~30 cm was FC, and the soil layer >30 cm was AL; In 2010, 0~70 cm was FC, 70~90cm was AL, and the soil layer >90 cm was SA; In 2011, only 0~10 cm was AL, and the soil layer >10 cm was SA; In 2012, 0~50 cm was FC, 50~70 cm was AL, and the soil layer >70 cm was SA; In 2013, 0~10 cm was FC, 10~110 cm was AL, and the soil layer >110 cm was SA; And in 2014, 0~50 cm was FC, 50~70 cm was AL, 70~90 cm was SA, and the soil layer >90 cm was RS. Based on the above analysis results, the vertical variation layers of SM in fixed dunes can be generally divided into the fast changing layer of 0~50 cm, the active layer of 50~90 cm and the second active layer of 90~150 cm.

Table 2. Vertical division of SM.

Year	Soil depth (cm)	0~10	10~30	30~50	50~70	70~90	90~110	110~130	130~150
	SD	1.12	1.36	1.10	0.87	0.77	0.77	0.80	0.79
2009	CV	0.39	0.32	0.26	0.23	0.23	0.23	0.24	0.22
	Layers	FC	FC	AL	AL	AL	AL	AL	AL
	SD	1.20	1.66	1.74	1.67	1.12	0.85	0.70	0.66
2010	CV	0.33	0.31	0.32	0.32	0.24	0.19	0.16	0.14
	Layers	FC	FC	FC	FC	AL	SA	SA	SA
	SD	0.70	0.90	0.70	0.74	0.54	0.42	0.45	0.58
2011	CV	0.22	0.19	0.15	0.17	0.14	0.12	0.12	0.15
	Layers	AL	SA	SA	SA	SA	SA	SA	SA
	SD	0.91	1.51	1.57	1.14	0.53	0.58	0.54	0.50
2012	CV	0.32	0.31	0.31	0.26	0.15	0.19	0.17	0.15
	Layers	FC	FC	FC	AL	SA	SA	SA	SA
	SD	1.08	0.65	0.68	0.64	0.73	0.75	0.68	0.43
2013	CV	0.47	0.22	0.21	0.21	0.21	0.21	0.19	0.12
	Layers	FC	AL	AL	AL	AL	AL	SA	SA
	SD	2.40	1.08	1.20	0.99	0.39	0.14	0.08	0.15
2014	CV	0.31	0.35	0.30	0.26	0.13	0.05	0.03	0.05
	Layers	FC	FC	FC	AL	SA	RS	RS	RS

4. Conclusions

(1) The changes of SM with the month generally differently at different month, which showed a "W-shaped" curve in 2009, 2010 and 2011, a continuous downward trend in 2013 and 2014, and it was lower in May and higher in June, and decreased slightly in the following three months in 2012.
(2) Except for 2013, the SM was lower in the surface layer (0~10 cm), and the SM increased first, then decreased, and finally increased with soil depth.
(3) As the soil depth increases, both SD and CV of SM showed a decreasing trend. The vertical variation layers of SM in fixed dunes of the study area were initially divided into the fast changing layer of 0~50 cm, the active layer of 50~90 cm and the second active layer of 90~150 cm.

References

[1] Li Y B, Gao M, Wei C, et al 2003 Differences of soil water characteristics under different land use patterns in karst mountains *J. Soil Water Conserv.* **17**(5) pp 63-66
[2] Liu S X, Bo X, Yuan G F, et al 2013 Relationship analysis between soil moisture in root zone and top-most layer in China. *Chin. J. Plant Ecol.* **37**(1) pp 1-17
[3] Wang J, Qi L, He J H, et al 2016 Relationship between spring soil moisture in the Tibetan Plateau and summer precipitation in the Yangtze river basin and its possible mechanism *Chin. J. Geophys.* **59**(11) pp 3985-3995
[4] Li D F, Liu T X, Hu Y Y, et al 2012 The dynamic simulation of soil water in the Korqin sandy land with sand-meadow land features. *J. Inner Mongolia Agricul. Univ.* **33**(2) pp 78-82
[5] Yao X L, Fu B J, Lü Y H 2012 Spatial patterns of soil moisture at transect scale in the Loess

IOP Conf. Series: Earth and Environmental Science **346** (2019) 012001 doi:10.1088/1755-1315/346/1/012001

Plateau of China *Acta Ecol. Sinica* **32**(16) pp 4961-4968
[6] Western A W and Bloschl G 1999 On the spatial scaling of soil moisture *J. Hydrol.* **217** pp 203-224
[7] Qiu Y, Fu B, Wang J, et al 2010 Soil moisture variation in relation to topography and land use in a hill slope catchment of the Loess Plateau, China *J. Hydrol.* **240** pp 243-263
[8] Qiu Y, Fu B, Wang J, et al 2001 Spatial variability of soil moisture content and its relation to environmental in dices in a semi-arid gully catchment of the Loess Plateau, China *J. Arid Environ.* **49** pp 723-750
[9] Yang X 2001 Study on the characteristics of water environmental in shrubbery land of loess plateau *Arid Zone Res.* **18**(1) pp 8-13
[10] Jia Z Q 2006 Soil moisture dynamic variation law of typical shrub-grass vegetation in loess hill area of Northwestern Shanxi *Bull. Soil Water Conserv.* **26**(1) pp 10-15
[11] Yang Y D, Zhang J S, Cai G J, et al 2008 Soil water dynamics of different vegetation in gully and hilly regions of the loess plateau *Res. Soil Water Conser.* **15**(4) pp 149-150
[12] Li Y H, Zhao Y, Hu M Y, et al 2003 Land capacity in agriculture grazing ecotone of east Keerqin lands- the case of Keerqimuoyihouqi county *Chin. J. Ecol.* **22**(3) pp 23- 28
[13] Zuo X, Zhao H, Zhao X, et al 2008 Spatial pattern and heterogeneity of soil properties in sand dunes under grazing and restoration in Horqin Sandy Land, Northern China *Soil Tillage Res.* **99**(2) pp 202-212
[14] Guo A, Chang J, Wang Y, et al 2015 Variation characteristics of rainfall-runoff relationship and driving factors analysis in Jinghe river basin in nearly 50 years *Trans. Chin. Soc. Agric. Eng.* **31**(14) pp 165-171
[15] Li H, Wang M and Chai B 2003 Spatial and temporal characteristics of soil moisture dynamics in Loess Plateau *J. Appl. Ecol. Res. Shenyang* **14**(4) pp 515-519
[16] Chen H and Shao M 2003 Review on hillslope soil water movement and transformation mechanism on the loess plateau *Adv. Water Sci.* **14**(4) pp 513-520

IOP Conf. Series: Earth and Environmental Science **346** (2019) 012002 doi:10.1088/1755-1315/346/1/012002

Expression and antibody preparation of Small Ubiquitin-like Modifier (SUMO) from *Aspergillus flavus*

W Z Wu[1, #], S Ahmad[1, #], S Wang[1], Y F Zhang[1], H Yang[1], S H Wang[1, 2] and Y Wang[1, 2]

[1]Key Laboratory of Mycotoxins and Pathogenic Fungi of Fujian Province Fuzhou, Key Laboratory of Chemical Biology and Biopesticide of Education Ministry, School of Life Sciences, Fujian Agriculture and Forestry University, Fuzhou, 350002 China
[#] These two authors equally contributed in this work.

E-mail: wshyyl@sina.com (S H Wang); wangyu@fafu.edu.cn (Y Wang)

Abstract. *Aspergillus flavus* is a ubiquitous filamentous fungus, which contaminates crops, also causes severe disease in animals. The well conserved representative of the ubiquitin-like family SUMO (small ubiquitin-like modifier) is associated with the superfamily of ubiquitin-like polypeptides. In this study, the recombinant protein SUMO from *A. flavus* was cloned and expressed in *E. coli* BL21 (DE3). The protein was purified by nickel column affinity chromatography method and identified by mass spectrometry with molar mass was 10296 Da. The yield of recombinant protein was > 90% pure. The SUMO protein was immunized into BALB/c mice, and its potency was measured by western blot. The predicted superposition model of SUMO protein was analyzed by ITASEER server. These results indicated that SUMO proteins and its polyclonal antibody might provide scientific basis to study the SUMOylation in *A. flavus*.

1. Introduction

Aspergillus flavus is a widespread pathogen which is mainly found in tropical and subtropical regions for numerous economically important crops, as well as maize, cotton peanut is sources of yield losses all over the world [1, 2]. This kind of pathogenic fungi can also produce mycotoxins-aflatoxins, which is a class of strong carcinogen [3], Aflatoxins B1 is the most toxic toxins among these metabolites [4]. Moreover, it is lethal for human and animals that could cause myocardial, kidney tissue damage and liver cancer [5, 6].

Ubiquitination of proteins for proteasomal degradation is a common mechanism to ensure degradation of key factors for cellular regulation in eukaryotic organisms, which is necessary for removing misfolded proteins [7]. SUMOylation is directed by an enzymatic cascade similar to that elaborated in ubiquitination [8]. Furthermore, SUMOylation is not employed to tag proteins for degradation comparison with ubiquitination [9]. Depending on their function, the SUMOylating of proteins has several consequences for the target substrates [10]. In higher eukaryotes, it was shown that SUMOylating played an important role in signal transduction, transcription regulation, stability of genome , repair of DNA, and progression of cell cycle [11].

SUMO (Small Ubiquitin-like Modifier) proteins are associated with the superfamily of ubiquitin-like polypeptides which covalently attached to several intracellular target proteins to controls proteins sub-cellular localization and is involved in protein-protein interactions [12]. Moreover, the SUMOylating of proteins was also shown to be elaborated in ubiquitin-dependent breakdown by the

Content from this work may be used under the terms of the Creative Commons Attribution 3.0 licence. Any further distribution of this work must maintain attribution to the author(s) and the title of the work, journal citation and DOI.
Published under licence by IOP Publishing Ltd

proteasome [13]. SUMO could block the access for further putative collaboration partners, which SUMO could help as a binding interface to alleviate the relationship with further proteins or DNA and could influence the conformation of target protein's through binding to a another non-covalent site of SUMO [14, 15]. Furthermore, numerous controlling mechanisms for SUMOylating were reported, such as phosphorylation capability act as positive as well as negative signal for sumoylation [16]. In this study, the purified SUMO protein from *A. flavus* (NRRL3357), which was expressed by a prokaryotic expression system, molar mass was identificated by mass spectrometry and the predicted superposition model of SUMO protein was analyzed. The polyclonal antibody against SUMO was prepared for further detection of the potential SUMOylating sites.

2. Material and methods

2.1. Multiple sequence alignment

Comparison of SUMO protein with other known fungal species. Sources of SUMO (OQ_E24410.1), *Penicillium steckii* (XP_020056676.1), *Aspergillus aculeatus*, (OX_S03525.1), *Aspergillus thermomutatus* (GA_O88816.1), *Aspergillus udagawae* (XP_015405215.1), and *Aspergillus nomius* (OX_N33420.1), were obtained from NCBI. The multiple alignments was done by DNAMAN software [17].

2.2. Molecular cloning and protein expression

The SUMO coding gene (accession NO. XM_002380391.1) found in NCBI was used for the construction of the recombinant plasmid, and the primers (upstream primer: 5'-ATGGCCGATGCAGGACTC-3 ', downstream primer: 5'-CTAACCACCGATCTGCTCCT-3') were used for PCR amplification from cDNA of *A. flavus*. The PCR products were cloned into pET28a containing a 6×His tag monitored by a PreScission protease cleavage at the site of N-terminus of SUMO, and the recombinant plasmid was moved into *E. coli* BL21 (DE3). The transformed cells were cultured in 10 ml LB medium which contains 50mg/L kanamycin at 37°C, 0.3 mM isopropyl-*β*- D-thiogalactopyranoside (IPTG) was added. After that cells were incubated on ice for 10 min then for overnight at 37°C. Then the cells were stored at 4°C with centrifugation of 6000 rpm for 5 min [18].

2.3. Protein purification

The cells were resuspended and sonicated on ice sage of binding buffer A (50mM Tris-HCl, 500 mM NaCl, 20mM imidazole pH 8.0). The suspension was shattered for 3 seconds interval (total 40 min). The lysate of the cell was centrifuged 3 times at15000 rpm/min for 20 minutes at temperature 4°C then the supernatant was loaded into the Ni-NTA column (nickel-charged resin). The protein was eluted from the column with Buffer A, B (imidazole concentration 20 mM, 50 mM, 100 mM and 300 mM), Buffer C (500 mM NaCl, 50 mM Tris-HCl, 300 mM imidazole pH 8.0), then the 6×His tag (polyhistidine-tag) was removed and incubation of fusion protein with buffer C at temperature 4°C for 16 h. After that mixed sample was dialyzed against buffer D (500 mM NaCl, 50mM Tris) for the purpose to separate the untage SUMO (cleaved by PreScission protease) from the cleaved mixture, After that loaded into another Ni-NTA column[18, 19]. Totally fractions that contain the purified SUMO were analyzed by SDS-PAGE.

2.4. Mass spectrometry

Mass spectrometry examination of SUMO was done by Beijing Protein Innovation. The SDS gel band was erased, and digested into decolorizing reagent contains acetonitrile (50%) and 25 mM ammonium bicarbonate. For the whole absorption of colloidal elements, DTT (dithiothreitol) was used and incubated at 50°C for 1 hr. After DTT was detached, 55mM IAM (iodoacetamide) was added after that incubated at 24°C for 45 min. Then excess IAM was removed and twice wash away with 25 mM ammonium bicarbonate for 10 min. Furthermore, enzyme was diluted as 1μg/μL up to 15 folds with 25mM ammonium hydrogen carbonate then addition with the dehydrated colloidal particles. The colloidal particles were immersed with 25mM ammonium hydrogen carbonate and incubated into a water bath at 37°C overnight for digestion purpose. The reaction of digestion was completed with 0.1% concentrated FA (formic acid), in addition sample was identified through Q-TOF (Quadrupole Time-of-Flight). The identification protein was performed with the Mascot software [20].

2.5. Immunization and iELISA assay of anti-serum

In the initial immunization, old female BALB/c mice have age six months (Wushi animal laboratory Shanghai, China) was administrated primarily by purified SUMO protein (60 µg/mL) slightly modified and added with the similar volume of Freund's complete adjuvant at interparatoneal injection of the final water-in-oil emulsion. Booster injections of purified protein (60µg/mL) additional with the same volume of Freund's incomplete adjuvant were injected, intervals at fourteen days. After three times injection, the anti-serum was collected from the tails of the two mice and assayed by ELISA. The ELISA plate contains 96-well was coated with coating antigen (purified SUMO protein, 100 µL/well), then incubated overnight at 4°C. The plate was wash-down away by PBS (phosphate buffered saline), jammed with PBSM (skim milk powder 200 µL/well), then incubated at 37°C for 1 h. Then plate was wash away by PBS and PBST (0.5% Tween 20 in PBS) three times. The serum of the mice was included to the plate, and then incubated at 37°C for 2h. Anti-mouse of goat IgG-HRP (1:8000, 100µl/well) was added after washing then incubated 37°C for 1h. These chemicals were purchased from (Sigma-Aldrich Chemical St. Louis, USA) after washing, TMB (100 µL/well) was added then incubated 37°C for 15 mint. Then 2M H_2SO_4 (50 µL/well) was added, then yellow colour was formed for measuring at 450nm by microplate reader [21, 22].

2.6. Western blot

After SDS-PAGE electrophoresis, the proteins on the gel were shifted to a PVDF (Polyvinylidene difluoride) and the membrane was placed in a closed solution (5% skimmed milk powder) with shaking at room temperature for 1h. Primary antibody incubation, with 5% nonfat dry milk diluted each film is about 4ml; closed PVDF membrane after hybridization carefully placed. Then added with a good antibody solution, slow shaker at room temperature and incubated with shaking 1.5h; the membrane washed with TBST (Tris-buffered saline-Tween) three times shaking every 5min, to remove residual antibody. The second antibody incubation: with TBS diluted with the instructions by the ratio of the secondary antibody (typically 1: 4000). The PVDF film is carefully placed in hybridization box. Added 4ml of each film with a good II antibody solution, hybridization with a mouth sealed then incubated shaking at room temperature for 1.5h. The film was taken after incubation; the membrane washed with TBST shaking three times 10min. After the secondary antibody incubation membranes finished membrane washed with TBST, a luminescent solution A, B in the darkroom exposure [23].

2.7. Predicted model

To choose the final models, I-TASSER uses the program (SPICKER) which was built on the pair-wise structure resemblance. The assurance of the model was calculated by C-score which was created on the importance of threading template alignments and the merging endowments of the structure gathering simulations. After assembly of structure simulation, I-TASSER was used the program TM-align structural alignment for the purpose to match the model to the entire structures in the PDB [24]. The structural analogue identified in PDB i.d IwzOA, so we had combined the predicted model and structure analogue using 1.12.Chimera software [25].

3. Results and discussion

3.1. Molecular cloning

The cDNA coding gene for SUMO was successfully amplified by PCR as in (Figure 1A), The PCR fragment was consistent with the size of the target gene (279 bp). The PCR product was cloned into the pET28a and transformed into the *E. coli* BL21 (DE3). The sequencing of DNA was confirmed that

IOP Conf. Series: Earth and Environmental Science **346** (2019) 012002 doi:10.1088/1755-1315/346/1/012002

recombinant plasmid was constructed successfully. The SUMO open reading frame encodes a predicted protein of 92 amino acids, which belongs to ubiquitin superfamily. Comparisons of amino acid sequences, it revealed that SUMO showed the highest similarity to other Aspergillus members (Figure 1B).

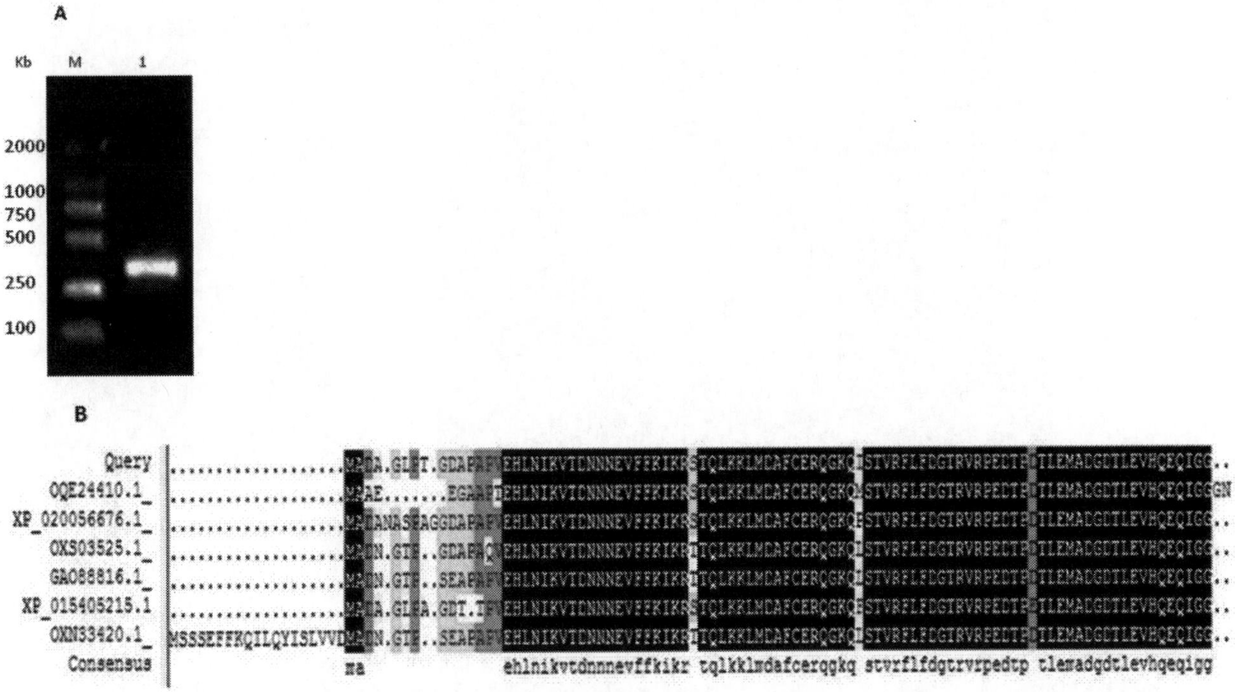

Figure 1. Cloning and sequence alignment (A) Cloning of SUMO gene from cDNA of *A. flavus*. M: DNA marker, Lane 1: cloned PCR fragment of 279 bp. (B) Comparison of SUMO protein with other known fungal species. The identical residues are in black colour.

3.2. Protein purification

The yield of SUMO which expresses in *E. coli* expression system was very high, and the protein showed excellent solubility. The recombinant SUMO was initially purified by Ni-NTA column successfully (Figure 2A). The 6×His-tag of the recombinant protein was removed by PreScission protease at 4°C, about 90% His-tag was removed from SUMO protein (Figure 2B). The untagged SUMO protein from the proteolytic digestion was isolated with the Ni-NTA column by collecting imidazole-free, and low concentration of imidazole (10 mM imidazole) wash fractions. Our results were mostly similar as the previous study about the purity of nucleoside diphosphate kinase (NDK) proteins nearly >98% [18].

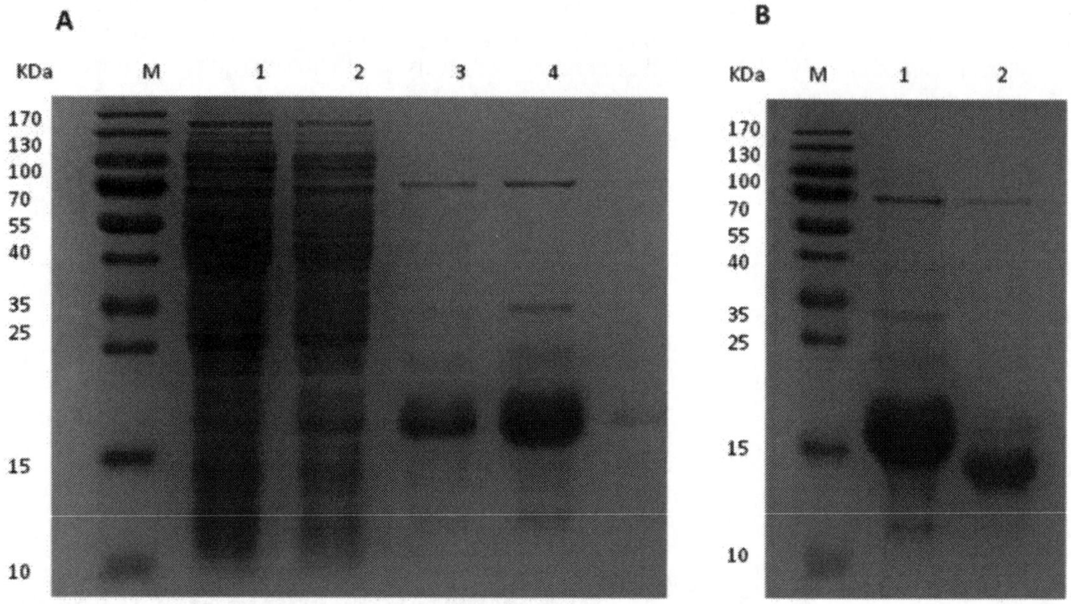

Figure 2. SUMO protein expression and purification. (A) SUMO protein purification with 6×his tag after cell extract by column chromatography (Ni–NTA). Where M showed marker Lane 1: supernatant; Lane 2: fractions of elution by Ni-NTA Column; Lane 3: fractions of elution with 50 mM imidazole concentration and Lane 4 fractions of elution with 100 mM imidazole concentration. (B) Purification of untagged SUMO by a proteolytic reaction of preScission protease. Where M indicated marker, Lane 1 SUMO protein (control): Lane 2 fraction flow by Ni-NTA Column (imidazole 10 mM).

3.3. The verification of peptides through Mass spectrometry

The recombinant target protein was purified and analyzed by liquid chromatography-mass spectrometry. Overall 8 peptides (FLFDGTR, LMDAFCER, LMDAFCER, KLMDAFCER, KLMDAFCER, VTDNNNEVFFK, VRPEDTPDTLEMADGDTLEVHQEQIGG, VRPEDTPDTLEMADGDTLEVH-QEQIGG) was obtained, and the two peptides as in (Figure 3A/B) showed single domain polypeptides. The target protein's molar mass was 10296 Da. The score was 12.84 emPAI (Exponentially modified protein abundance index). The identified peptides were matched with the creative sequence of SUMO from NCBI, which represented, the protein identification was accurate. The theoretical molar mass of recombinant protein was 10245.52 Da by Expasy's ProtParam prediction server, and experimentally mass spectrometry showed 10296 Da, which confirmed that the molar mass was very closely related. In the previous study to identify conjugation sites of ubiquitin-like protein by MS methods [26], but we analysis the mass spectrometry, to achived the molar mass, and peptides matched the creative sequence from NCBI, representing the identification of the protein.

IOP Conf. Series: Earth and Environmental Science **346** (2019) 012002 doi:10.1088/1755-1315/346/1/012002

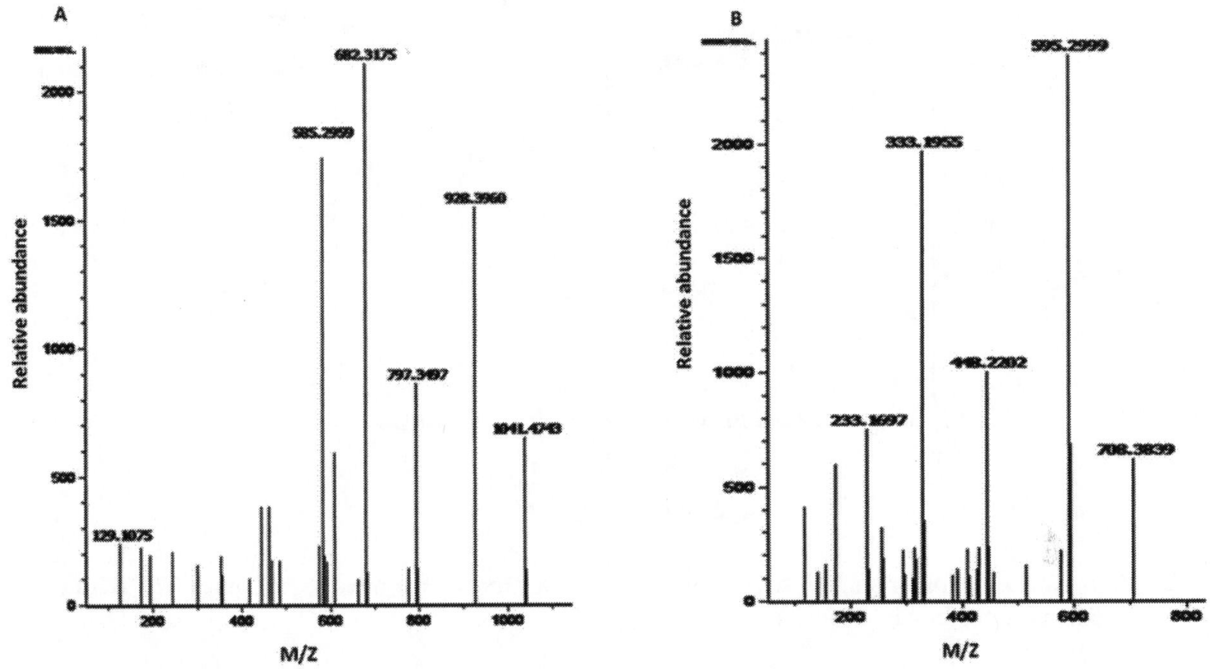

Figure 3. SUMO peptide determination by mass spectrometry (A) Determination of peptide by MS-MS spectrum FLFDGTR (B) Analysis of peptide by MS-MS spectrum VRPEDTPDTLEMADGDTLEVHQEQIGG m/z = ratio of mass to charge

3.4. Immunization and antiserum analysis

The mice were injected with purified SUMO protein, and booster injections were assumed intervals at fourteen days. After three times injection, the anti-serum was collected from the tails of the two mice, and the anti-serum titer of SUMO injected mice was tested through indirect noncompetitive ELISA. In the previous study higher antiserum titer was 1:8000, which indicated that the immunized mice had high Interferon gamma antibody titer [22]. Our result showed that two mice showed high titer, and mice 1 confirmed the highest anti-serum titer (1:64000). The result of the data was processed using Microsoft Excel to make a graph (Figure 4). Which indicated that immunogenic response was fully induced by SUMO protein, and the anti-SUMO antibody was successfully obtained.

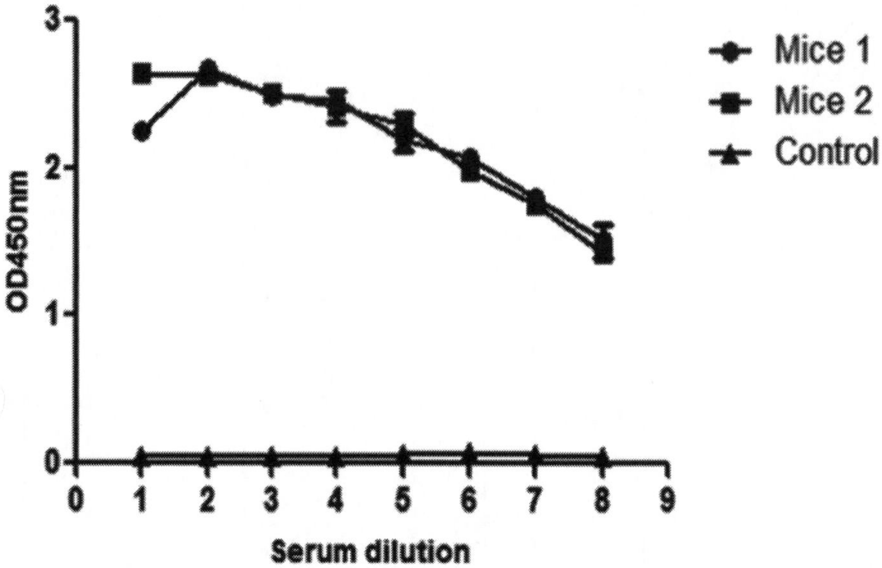

Figure 4. Detection of Sumo antiserum titer by ELISA, Blue colour mice 1, Pink mice 2 antiserum titer and the black is for control mice.

3.5. Antibody detection by western blott

To further confirm the identity of the above anti-SUMO antibody, western blot experiments were implemented by anti-SUMO antibody. In the previous study SUMO-1 proteins analyzed by western blotting by C21 antibody [27].Our result in (Figure 5), (Lane 2) showed that the SUMO protein was successfully recognized by this anti-SUMO antibody from western blot, while the negative control protein (Lane 1) was not bind to the anti-SUMO antibody. This result indicated that this anti-SUMO antibody bind to SUMO protein with high specificity.

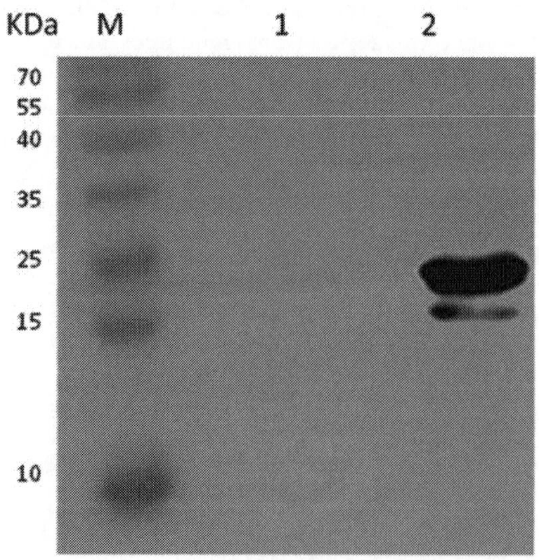

Figure 5. Detection of antiSUMO antibody by western blot. M: marker, Lane 1 control, Lane 2 the untagged SUMO protein.

3.6. Predicted superposition of SUMO

SUMO protein from the PDB has the closest structural similarities, and the maximum TM-score to the I-TASSER predicted model. The comparison of predicted model and known protein model (Figure 6).3D structure of the template (known protein) showed in pink colour. The predicted model at N-terminus contains the long loop. Protein loops take on specific conformations that are important to protein function [28]. The C-score was usually in the range of -5, 2. When a C-score showed a higher value that signifies a model among high confidence and C Score was 0.53. When a C-score shows a higher value that means a model with higher certainty. To measure the similarity between template and the structural query of the protein represent 0.625 TM score.

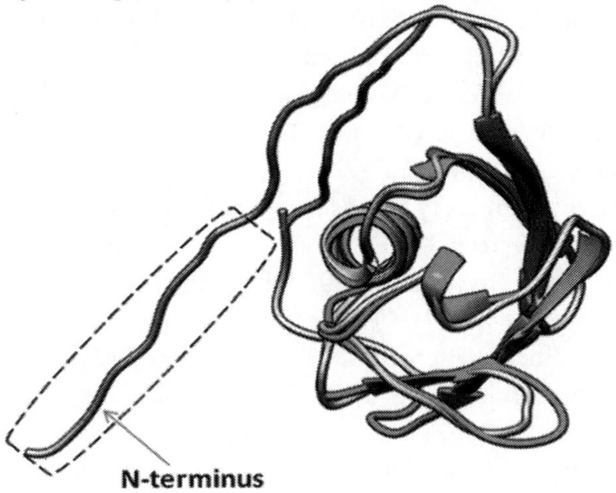

Figure 6. Superposition of SUMO protein. Superposition model pink colour showed known protein and blue colour demonstrated the predicted model consist of an extra loop nearby N-terminus (black circle).

4. Conclusions

The sumoylation of proteins has numerous consequences aimed at the target substrates, which depend on the function. In this study, to understand the SUMO biological function in the SUMOylated and non-SUMOylated prospective, the SUMO gene was cloned then expressed in the *E. coli* BL21 (DE3), and the protein was purified through nickel column affinity chromatography and analyzed by mass spectrometry. The BALB/c mice was immunized by SUMO protein, and anti-SUMO antibody was successfully prepared with high specificity. The prediction of superposition model of SUMO protein was analyzed by ITASEER server. It could be an available tool by using antibody produced from this study to determine the SUMOylation of *A. flavus*.

Conflicts of Interest

We state that, there is no conflict of interest between any authors.

Acknowledgements

The research was supported by the National Program 973 (No. 2013CB127802) from the Ministry of Science and Technology of China and (NSFC) (No. 2017J0102).

References

[1] Razzaghi-Abyaneh M, Shams-Ghahfarokhi M, Allameh A, Kazeroon-Shiri A, Ranjbar-Bahadori S, Mirzahoseini H and Rezaee M B 2006 A survey on distribution of Aspergillus section Flavi in corn field soils in Iran: population patterns based on aflatoxins, cyclopiazonic acid and sclerotia production. *Mycopathologia*, **161**(3), 183-92.

[2] Abbas H K, Zablotowicz R M and Locke M A 2004 Spatial variability of *Aspergillus flavus* soil populations under different crops and corn grain colonization and aflatoxins. *Can. J. Bot.*, **82**(12), 1768-75.

[3] Richard J L 2007 Some major mycotoxins and their mycotoxicoses An overview. I*nt. J. Food Microbiol.*, **119**(1-2), 3-10.

[4] Giorni P, Magan N, Pietri A, Bertuzzi T and Battilani P 2007 Studies on Aspergillus section Flavi isolated from maize in northern Italy. *Int. J. Food Microbiol.*, **113**(3), 330-8.

[5] Abdel-Wahhab M A and Aly S E 2003 Antioxidants and radical scavenging properties of vegetable extracts in rats fed aflatoxin-contaminated diet. *J. Agri. Food Chem.*, **51**(8), 2409-14.

[6] Hamid A S, Tesfamariam I G, Zhang Y and Zhang Z G 2013 Aflatoxin B1-induced hepatocellular carcinoma in developing countries: Geographical distribution, mechanism of action and prevention. *Oncol. Lett.*, **5**(4), 1087-92.

[7] Lodish H, Baltimore D, Berk A, Zipursky S L, Matsudaira P and Darnell J 1995 In Molecular Cell Biology. New York: WH Freeman.

[8] Welchman R L, Gordon C and Mayer R J 2005 Ubiquitin and ubiquitin-like proteins as multifunctional signals. *Nature Revi. Mole. Cell Biol.*, **6**(8), 599.

[9] Tatham M H, Geoffroy M C, Shen L, Plechanovova A, Hattersley N, Jaffray E G and Hay R T 2008 RNF4 is a poly-SUMO-specific E3 ubiquitin ligase required for arsenic-induced PML degradation. *Nature Cell Biol.*, **10**(5), 538.

[10] Gill G 2005 Something about SUMO inhibits transcription. *Curr. Opin. Genet. Devel.*, **15**(5), 536-41.

[11] Branzei D and Foiani M 2008 Regulation of DNA repair throughout the cell cycle. *Nature Revi. Mole. Cell Biol.*, **9**(4), 297.

[12] Sha T I A N, Poukka H, Palvimo J J and Jänne O A 2002 Small ubiquitin-related modifier-1 (SUMO-1) modification of the glucocorticoid receptor. *Biochem. J.*, **367**(3), 907-11.

[13] Geoffroy M C and Hay R T 2009 An additional role for SUMO in ubiquitin-mediated proteolysis. *Nature Revi. Mole. Cell Biol.*, **10**(8), 564.

[14] Verger A, Perdomo J and Crossley M 2003 Modification with SUMO. *EMBO Reports*, **4**(2), 137-42.

[15] Meulmeester E and Melchior F 2008 Cell biology: SUMO. Nature, **452**(7188), 709.

[16] Hicke L 2001. Ubiquitin and proteasomes: protein regulation by monoubiquitin. *Nature Revi. Mole. Cell Biol.*, **2**(3), 195.

[17] Han F and Zhang X 2007 Characterization of a ras-related nuclear protein (Ran protein) up-regulated in shrimp antiviral immunity. *Fish Shellfish Immunol.*, **23**(5), 937-44.

[18] Wang S, Li W B, Jiang L G, Wang S H, Zhang Y F and Wang Y 2016 Expression, purification, crystallization and preliminary crystallographic analysis of nucleoside diphosphate kinase (NDK) from *Aspergillus Flavus*. *Chin. J. Struct. Chem.*, **35**(11), 1708-13..

[19] Ramirez M A and Chen Z 2014 Synthesis of an intein-mediated artificial protein hydrogel. *J. Visual. Experim.* (83), e51202.

[20] Shilov I V, Seymour S L, Patel A A, Loboda A, Tang W H, Keating S P and Schaeffer D A 2007 The Paragon Algorithm, a next generation search engine that uses sequence temperature values and feature probabilities to identify peptides from tandem mass spectra. *Mole. Cell. Proteom.*, **6**(9), 1638-55.

[21] Saeed A, Ling S, Yuan J and Wang S 2017 The preparation and identification of a monoclonal antibody against domoic acid and establishment of detection by indirect competitive ELISA. *Toxins*, **9**(8), 250.

[22] Yang H, Zhong Y, Wang J, Zhang Q, Li X, Ling S and Wang R 2018 Screening of a ScFv antibody with high affinity for application in human IFN-γ immunoassay. *Front. Microb.*, **9**, 261.

[23] Wang R, Xiang S, Feng Y, Srinivas S, Zhang Y, Lin M and Wang S 2013 Engineering production of functional scFv antibody in E. coli by co-expressing the molecule chaperone Skp. *Front. Cell. Infect. Microb.*, **3**, 72.

[24] Roy A, Kucukural A and Zhang Y 2010 I-TASSER: a unified platform for automated protein structure and function prediction. *Nature protocols*, **5**(4), 725.

[25] Huang C C, Meng E C, Morris J H, Pettersen E F and Ferrin T E 2014 Enhancing UCSF Chimera through web services. *Nucleic Acids Res.*, **42**(W1), W478-W484.

[26] Jeram S M, Srikumar T, Pedrioli P G and Raught B 2009 Using mass spectrometry to identify ubiquitin and ubiquitin - like protein conjugation sites. *Proteomics*, **9**(4), 922-34.

[27] Desterro J M, Rodriguez M S and Hay R T 1998 SUMO-1 modification of IκBα inhibits NF-κB activation. *Molecul. Cell*, **2**(2), 233-9.

[28] Hu X, Wang H, Ke H. and Kuhlman B 2007 High-resolution design of a protein loop. *Proceedings of the National Academy of Sciences*, **104**(45), 17668-73.

Phytochemical Screening, antimicrobial and antioxidant efficacy of some plant extracts and their mixtures

M H Mahmood, A K Osama, E A Makky[1], M H Rahim, N H M Ali and N D Hazrudin

Faculty of Industrial Sciences & Technology, Universiti Malaysia Pahang, Gambang, 26300 Kuantan, Pahang, Malaysia

E mail: essam22001@gmail.com

Abstract. Various phytochemicals and bioactive compounds are sourced from medicinal plants. This study focused on finding the most potent plant extracts whose combination could exhibit better antioxidant and antimicrobial activities compared to the efficacy of the individual extracts. The combined plant extracts were *O. stamineus, E. longifolia, A. bilimbi, P. granatum,* and *M. nigra.* Ultrasound extraction technique was used for the extraction of the plant material before screening the extracts for the presence of several phytochemical groups. After the screening for the presence of phytochemicals, the extracts were evaluated for antimicrobial activity against five microorganisms (*S. aureus, B. subtilis, P. vulgaris, E. coli, and C. albicans*) at 3 different concentrations (600, 400 and 200 mg/mL) using disc diffusion method. The aim of this antimicrobial screening was to identify and select the most potent extracts for the combination study. From the screening result, *P. ganatum* and *M. nigra* had the highest antimicrobial activity against all the tested organisms, as well as the highest DPPH radical scavenging activity of 95.40 % (IC_{50} = 120.2 µg/mL) and 90.20 % (IC_{50} = 330.0 µg/mL), respectively. Hence, both extracts were combined and screened for antimicrobial and antioxidant activities at the same concentration range but at different combination ratios of 1:1, 1:2 and 2:1 (v/v). The results showed the extract mixtures at different combination ratios of 1:1, 1:2 and 2:1 (v/v) to exhibit good antioxidant and antimicrobial activities. However, the combination ratio of 1:1 gave the best antioxidant activity as it achieved an IC_{50} value of 82.0 µg/mL. Therefore, it is concluded that the mixture of the selected extracts exhibited strong antioxidant and antimicrobial activities, coupled with a significant level of antifungal activity.

1. Introduction

One of the ancient forms of healthcare known to man is herbal medicine. More than 50% of the modern drugs used in healthcare are sourced from natural sources; hence, the role of natural products in the development of the pharmaceutical sector is paramount. Phytochemicals are plant chemicals which are secreted by plants in response to environmental constraints or diseases. They are not required for plant metabolism but are beneficial to man in managing several diseases. To encourage the use of plant materials as the possible sources of antimicrobial agents, there must be a careful evaluation of their composition and biological activity prior to use [1]. The emergence of antibiotic-resistant microorganisms has propelled the discovery of novel antibacterial drugs. The increase in the rate of multiple drug-resistant microorganisms is attributable to the rashness use or malmanagement of antibiotics which has resulted in many antibiotics losing their effectiveness against certain microorganisms [2].

The unbalanced generation of free radicals creates abnormal physiological conditions that lead oxidative damage to cells through lipids, proteins and nucleic acid biomolecules degradation [3]. Oxidative damage

Content from this work may be used under the terms of the Creative Commons Attribution 3.0 licence. Any further distribution of this work must maintain attribution to the author(s) and the title of the work, journal citation and DOI.
Published under licence by IOP Publishing Ltd

caused by reactive oxygen species is repaired by the protective action of antioxidant which continuously converts the generated free radicals into less harmful molecules by intercepting radical chained reactions [4]. Plant-derived bioactive compounds have a protective role in minimizing oxidative stress. Several crude plant extracts possess high oxidative capacity and a remarkable concentration of phenolic compounds [5]. This study was designed to examine the phytochemicals content, antimicrobial, and antioxidant activities of selected medicinal plant extracts. The biological activity of the plant extracts was screened individually before combining the most potent extracts for the combination studies.

2. Methodology

2.1. Plant materials

Five plants (*Averrhoa bilimbi, Orthosiphon stamineus, Punica granatum, Morus nigra,* and *Eurycoma longifolia*) were collected from a herbal market in Kuantan, Pahang, Malaysia and transported to the laboratory unit of the Faculty of Industrial Sciences & Technology (FIST), Universiti Malaysia Pahang, Gambang campus for further preparation and investigations. The plant materials were manually cut into small pieces to increase the surface area during the extraction process. For the drying process, the plants were placed in a conventional oven at 40-45°C for 3 days [6]. Table 1 presents the plant materials used in this study.

Table 1: The English name, scientific name, code, family, and part of all plants used in this study.

NO.	English name	Scientific name	Code	Family	Part used
1	Tree sorrel	*Averrhoa bilimbi*	Ab	Oxalidaceae	Leaves
2	King of bitters	*Andrographis paniculata*	Ap	Acanthaceae	Whole plant
3	Cat's whiskers	*Orthosiphon stamineus*	Os	Lamiaceae	Leaves
4	Longjack	*Eurycoma longifolia*	El	Simaroubaceae	Roots
5	Pomegranate	*Punica granatum*	Pg	Lythraceae	Peels

2.2. Sample preparation

After drying the plant materials for 3 days at 40-50°C [6], they were pulverized in a dry mill blender (Panasonic MX-GM1011 H) and stored in airtight plastic bags in desiccators at room temperature until further use [7].

2.3. Ultrasound-assisted extraction (UAE)

A method modified by [8] was used during the extraction process. Five grams of the pulverized plant materials were placed in 100 mL of 80% ethanol and sonicated at the sonication power of 20 kHz. The sonication probe immersed into the extraction vessel during the process to ensure maximum power transfer. The sonication was set for 15 minutes at pulse-on pulse-off times of 20 and 5 seconds, respectively. After the sonication time, the mixture was filtered, and the filtrate was concentrated in a rotary evaporator to remove the solvent. The concentrated filtrate was later placed in a fume hood to ensure the evaporation of the remaining solvent in the extract.

2.4. Screening for phytochemicals content

2.4.1. Alkaline reagent test for flavonoid. About 50 mg of each extract was mixed with 5 mL of dilute ammonia solution and filtered. Then, concentrated H_2SO_4 was added to the filtrate and monitored for the formation of a yellow colour which will indicate the presence of flavonoids in the extracts [9].

2.4.2. Wagner's test for alkaloids. The plant extract (50 mg) was dissolved in a small volume of dilute HCl and filtered. Then was followed by a gentle addition of few drops of Wagner's reagent to the filtrate along the side of the test tube. The solution was monitored for the development of reddish-brown precipitates which indicates the presence of alkaloids in the extract [10].

2.4.3. Salkowski's test for glycoside. The plant extract (2 mL) was diluted in chloroform (2 mL), followed by the addition of a few drops of concentrated H_2SO_4. The mixture was observed for the formation of a reddish-brown color to indicate the presence of glycosides in the extract [11].

2.4.4. Tannin test. The plant extract (50 mg) was weighed and dissolved in distilled water (d.H_2O) before adding a few drops of $FeCl_3$ solution. The development of green precipitates after mixing indicates the presence of tannins in the extract [12].

2.4.5. Steroid test. The plant extract (50 mg) was dissolved in 2 mL each of chloroform and concentrated H_2SO_4 and allowed for some seconds. The presence of steroids is manifested by the formation of red colour in the lower layer of the mixture [11].

2.4.6. Froth test for Saponin. The plant extract (50 mg) was dissolved in d.H_2O and later made up the volume to 20 mL with d.H_2O. Then, the mixture was vigorously shaken for 15 min and observed for the formation of stable foam of at least 2 cm length to indicate the presence of saponins [10].

2.4.7. Ferric chloride test for phenolics. The plant extract (50 mg) was dissolved in d.H_2O (5 mL), followed by the addition of a few drops of ferric chloride (5%). The mixture was monitored for the formation of a bluish-black colour to indicate the presence of phenolics in the extract [10].

2.5. Microbial culture preparation

The five plant extracts were evaluated for antimicrobial activities against pure cultures of *S. aureus*, *B. subtilis*, *E. coli*, *P. vulgaris*, and *Candida albicans*. The microbes were sourced from the Microbiology Unit of FIST laboratory, UMP, Malaysia. The bacteria were cultured in Nutrient broth while the fungi were cultured in potato dextrose broth media (PDB). Both cultures were incubated for 24 h at 37 and 30 °C, respectively. After the incubation period, the optical density (OD) of microbial suspension was determined at 600 nm. An OD reading of 0.5 McFarland standard was considered to represent a microbial concentration of 10^8 CFU/mL [10].

2.6. Determination of antimicrobial activity

The extracts were evaluated for antimicrobial activity at different concentrations of 200, 400 and 600 mg/mL. The desired concentrations of the extracts were prepared in 80% ethanol and stored in the chiller to avoid evaporation. The disc diffusion method was used to screen the extracts for antimicrobial activity. A drop of each microbial suspension containing about 10^8 CFU/mL (determined by hemocytometer) was gently spread on the Mueller-Hinton agar (MHA) and PDA using a sterile cotton bud. Then, 6 mm diameter discs were sterilized at 121°C for 15 min and loaded with the extracts at different concentrations. Chloramphenicol (30 mg) and fluconazole (25 mg) were used as positive controls. The discs were dried and placed in contact with the surface of the plates containing the organisms, labeled, and incubated for 24 h (bacteria) and 48 h (fungi) at 37°C. The experiments were done in replicates and the results were reported as the diameter (mm) of the inhibition zones [10].

2.7. Determination of antioxidant activity

DPPH (2,2-diphenyl-1picrylhydazyl) scavenging assay: The antioxidant capability of the extracts was determined based on their DPPH free radical scavenging power [13]. The stock DPPH solution was prepared by dissolving DPPH (24 mg) in 100 mL of absolute methanol. The solution was covered with foil and stored at 20°C. For the working solution, the stock solution was further diluted with absolute methanol to the absorbance reading of 0.950 ± 0.02 at 517 nm. The extract solution was prepared by dissolved the crude extracts in absolute methanol at the concentration of 1mg/mL and further diluted for the analysis.

The test was performed by adding 1 mL of DPPH working solution into 100 μL of the extract solution prepared in different concentrations (10, 40, 100, 200, 500, 600 and 700 μg/mL). The mixtures were shaken and placed in the dark for 10 min at room temperature. The absorbance (Ab) of the mixture was read at 517 nm against a methanol blank. The decrease in the absorbance reading was correlated with the radical scavenging potential of extracts. The IC_{50} values of the extracts were calculated as the concentration of the extracts required to stabilize 50% of the radical population. Ascorbic acid was used as the positive control while the DPPH radical scavenging activity of the extracts was calculated using the relation:

$$\text{DPPH (\%)} = \frac{\text{Ab (control)} - \text{Ab (sample)}}{\text{Ab (control)}} \text{X } 100$$

Where Ab (control) is the absorbance of the control, and Ab (sample) is the absorbance of the test sample. The antioxidant activity of the combined extracts was also determined. The mixture prepared by mixing the extract of *P. granatum* and *O. stamineus* in 3 different ratios of 1:1, 2:1, and 1:2. The Absorbance of the different combination ratios was also read at 517 nm and the IC_{50} value of each combination was calculated as earlier described.

2.8. Data analysis
The data obtained from the experiments were analyzed carefully by calculating their mean and standard deviations. The IC_{50} values were calculated via linear regression analysis.

3. Results and discussion

3.1. Phytochemical screening
From the results of the phytochemicals screening, the presence of medically important phytochemicals was detected in the ethanolic extract of the studied plants. Glycosides were present in all the extracts while alkaloids were absent from all the extracts except in *E. longifolia.* Phenolics and tannins were present in all the samples except *E. longifolia*. Steroids were not detected in the ethanolic extract of *O. stamineus* and *A. bilimbi* while saponin was absent from both *E. longifolia* and *M. nigra* extracts (Table 2). Belkacem *et al.* [14] reported the extraction of phytochemicals from *P. granatum* fruit using ethanol as the extraction solvent. The absence of alkaloids and flavonoids in some of the extracts might be due to the non-suitability of ethanol as the extraction solvent for these group of compounds since Sharma *et al.* [15] reported the successful extraction of alkaloids and flavonoids using methanol and water. The ethanol extract of *P. granatum* contained most of the screened phytochemicals; however, *O. stamineus* extract contained all the phytochemical groups except alkaloids, steroid, and flavonoid [16].

Table 2: The preliminary phytochemical screening on the five plant extracts under test.

Phytochemical test	The plant extract				
	P. granatum	*O. stamineus*	*E. longifolia*	*M. nigra*	*A. bilimbi*
Phenolic	+	+	-	+	+
Saponin	+	+	-	-	+
Alkaloids	-	-	+	-	-
Glycoside	+	+	+	+	+
Tannin	+	+	-	+	+
Steroid	+	-	+	+	-
Flavonoids	-	-	+	+	+

(+): indicates presence; (-): indicates absence.

The extracts of *E. longifolia* revealed the presence of alkaloids, glycosides, steroids and flavanoids compounds while phenolic, saponins, and tannins were absent. Different types of chemicals were used to extract the plant, yet, all the extracts showed negative saponin tests, indicating that *E. longifolia* might contain little or no saponin content. Some studies have explained that tannins and phenolic compounds were present in *E. longifolia* but certain factors such as the plant source can cause some differences in their phytochemicals content [10]. *M. nigra* has been reported to contain alkaloids [17]. According to Hussain *et al.* [18], *M. nigra* does not contain saponin but the outcome of this study showed the presence of phenolics, saponin, glycoside, tannins, and flavonoids in *M. nigra* while alkaloids and steroids were not detected. Makky *et al.* [19] reported the presence of alkaloids and steroids in ethanolic extract of *A. bilimbi*.

3.2. Antimicrobial activity

To evaluate the antimicrobial efficacy of the extracts, discs containing ethanolic extract of the five plants at three different concentrations were placed onto the media inoculated with five different microbial pathogens. The varying degrees of the antimicrobial potentiality of the five plant extracts were presented in Table 3.

Chloramphenicol (30 mg) was used as the standard positive control and it showed inhibition zones of diameter ranging from 20-24.67mm against all the tested organisms. The result showed the most potent antibacterial activity against *S. aureus, B. subtilis* and *P. vulgaris* to be attributed to *P. granatum* extract at 400 mg/mL (diameter of inhibition zone = 12 ± 2.5 mm, 7.7± 0.5 mm, and 9.3± 1.3 mm at 400, 600, and 800 mg/mL, respectively). *M. nigra* extract was effective against only *S. aureus* (inhibitory zone of 8.0 ± 0.8 mm at 400 mg/mL), followed by *P. vulgaris* (inhibitory zone of 3.3 ± 4.7 mm at 400 mg/mL). *E. longifolia, A. bilimbi,* and *O. stamineus* extracts showed the lowest inhibitory activities to most of the tested microorganisms. *E. coli* was resistant to all the plant extracts except *E. longifolia* which was effective at the concentration of 200 mg/mL (inhibition zone = 3.0 ± 4.2 mm). The resistance of the Gram-negative organisms to the extract could be attributed to the outer membrane in the cell wall of the bacteria which prevented the penetration of hydrophilic compounds into the cells [20]. Tannins are one of the phytochemicals in *P. granatum* which have been revealed to prevent microbial growth [21]. The phenolic compounds for antimicrobial mechanisms in *P. granatum* are involved in reactions with microbial cell membrane proteins which will stimulate protein precipitation-mediated cell death and enzymes inhibition [22]. As per Souza *et al.* [23], the compounds present in *M. nigra* that inhibits bacterial growth are phenolics and flavonoids. As *M. nigra* has high steroids, flavonoids and alkaloids content, it is commonly used in herbal medicine [24] [18-19]. *O. stamineus* exhibits good antimicrobial properties [20]. Flavonoid and phenolic compounds present in *O. stamineus* can successfully inhibit bacterial growth. *S. aureus* has been reported to be sensitive to plant phytochemicals [16]. *E. logifolia* is also believed to be an effective antimicrobial agent even when extracted with different solvents [25]. According to Das *et al.* [26], *A. bilimbi* has effective antimicrobial activity especially against *E. coli* and *C. albicans*. The results of this study showed less effectiveness of *E. longifolia* possibly due to the extraction solvent and technique used. Some of the plant extracts gave the best yields when polar solvents are used for the extraction [10]. The antibiotic used as a control was effective against all the bacteria except *B. subtilis* and *C. albicans* which showed resistance to all the plant extracts and fluconazole. Similarly, chloramphenicol produced no zones of inhibition against *E. coli* and *B. subtilis* (Figure 1).

Table 3: Antimicrobial activity of five plant extracts at different concentrations against different microorganisms. The data were express in term of mean ± standard deviation.

Microbial isolates	Extract Conc. (mg/mL)	Inhibition zone diameter (mm±SD)					
		E. longifolia	*P. granatum*	*A. bilimbi*	*M. nigra*	*O. stamineus*	Chloramphenicol (30 mg)
S. aureus	600	0.0	12± 0.82	0.0	7.33± 0.47	7.67± 0.47	24.67± 1.25
	400	0.0	12± 2.45	0.0	8±0.82	2.67± 3.77	24± 1.41
	200	0.0	11± 0.82	3±4.24	5.33±4.71	0.0	21± 1.41
B. subtilis	600	0.0	7.67± 0.47	0.0	0.0	0.0	0.0
	400	0.0	5.67± 4.20	0.0	0.0	0.0	0.0
	200	0.0	5.67± 4.20	0.0	0.0	0.0	0.0
P. vulgaris	600	0.0	9.33± 1.25	0.0	0.0	3± 4.24	21± 1.41
	400	0.0	6.67± 4.99	0.0	3.33± 4.71	0.0	21.67± 0.47
	200	0.0	9± 2.16	0.0	3.33± 4.71	0.0	20± 1.63
E. coli	600	0.0	0.0	0.0	0.0	0.0	21.67± 0.47
	400	0.0	0.0	0.0	0.0	0.0	21.33± 0.47
	200	3 ± 4.24	0.0	0.0	0.0	0.0	20.67±0.94
							Fluconazole (25 mg)
C. albicans	600	0.0	0.0	0.0	0.0	0.0	0.0
	400	0.0	0.0	0.0	0.0	0.0	0.0
	200	0.0	0.0	0.0	0.0	0.0	0.0

(a) (b)

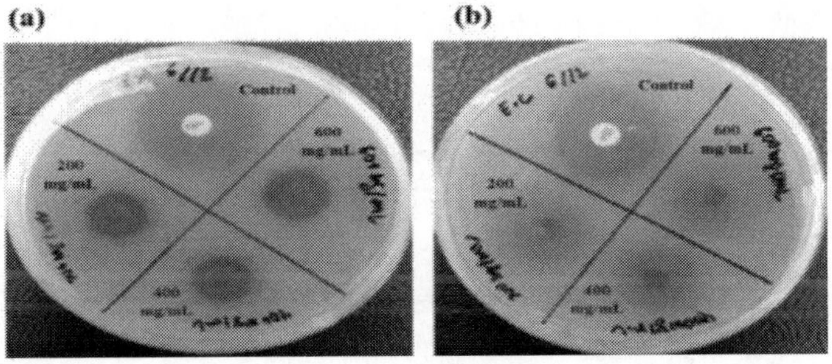

Figure 1: Growth inhibition zone of *P. granatum* against (a) *S. aureus* and (b) *E. coli*.

(a) (b)

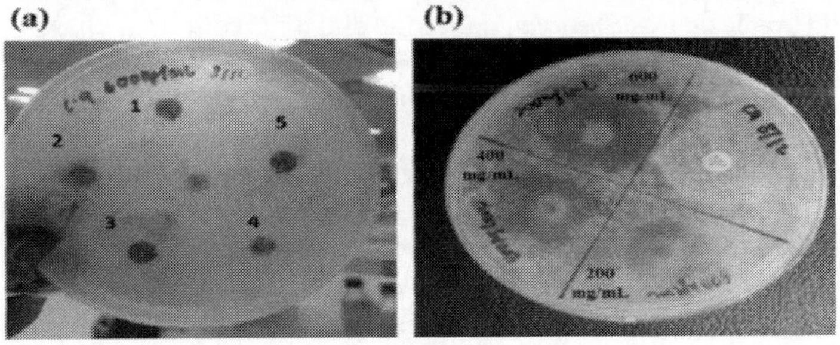

Figure 2: Growth inhibition zone of plant sample against *C. albicans* of all five plant extracts (a); and the mixture of *E. longifolia* and *M. nigra* (b) at different concentrations. 1: *E. longifolia*; 2: *P. granatum*; 3: *A. bilimbi*; 4: *M. nigra*; 5: *O. stamineus*.

The antimicrobial efficacy of the combined extracts of *P. granatum* and *M. nigra* at different concentrations was shown in Table 4. All the combination ratios of the plant extracts showed activity against all the microbial isolates except *E. coli* possibly due to nature of the bacterial cell wall. *S. aureus* showed the highest sensitivity at 600 mg/mL (inhibition zone = 12.0 ± 1.4mm) while the lowest inhibition zone of 5.7 ± 4.0 mm was recorded against *P. vulgaris* at 200 mg/mL. However, *C. albicans* was successfully inhibited at all the concentrations despite its resistance to the individual extracts. Fluconazole did not show any activity as it is a fungistatic agent rather than a fungicide. So, treatment with this agent can result in the development of acquired resistance. The impact of different Upc2p mutations on *ERG11* expression in *C. albicans* and its fluconazole-resistant phenotype and reported that the sustained production of the azole target annuls the activity of fluconazole, resulting in resistance [27] (Figure 2).

Table 4: Antimicrobial activity of plant extract mixture of *P. granatum* and *M. nigra* at different concentrations.

Microbial isolate	Extract conc. (mg/mL)	Inhibition zone diameter (mm±SD)	
		Extract mixture of *P. granatum/M. nigra*	Chloramphenicol (30 mg)
	600	12± 1.4	22.7± 1.2
S. aureus	400	11.7± 0.5	NA*
	200	7.7± 5.4	NA
	600	9.7± 0.5	0.0
B. subtilis	400	9.0± 0.8	NA
	200	7.7± 0.5	NA
	600	11.3± 0.5	21.7± 2.6
P. vulgaris	400	9.7± 0.5	NA
	200	5.7± 4.0	NA
	600	0.0	21.3±1.7
E. coli	400	0.0	NA
	200	0.0	NA
			Fluconazole (25 mg)
	600	10.0± 0.8	NA
C. albicans	400	8.7± 0.5	0.0
	200	8.3±0.5	NA

NA*: Not applicable.

3.3. Antioxidant assay

The free radical scavenging activity of the five plant extracts was assessed on DPPH radicals. The disappearance of DPPH radicals was observed due to the effect of the antioxidants in the extracts. Different concentrations of the extract caused the fading of the purple color of the DPPH solution to yellow. DPPH radicals are usually used to study the antioxidant activity of extracts as it can accept electrons and get stabilized [28]. DPPH is a purple stable free radical with an adsorption band at 517 nm and can be reduced to yellow by accepting electrons from antioxidants [29]. Figure 3 depicts the ability of different concentrations (10-1000 µg/mL) of the plant extracts to scavenge DPPH free radicals compared to the ability of ascorbic acid. From the table. Ascorbic acid presented the lowest IC_{50} value (37.5 µg/mL). Ascorbic acid is one of the potent naturally-occurring antioxidants in the biological system [30]. As per Arrigoni & Tullio, [31], efficient biological antioxidants are supposed to: (a) exist in good quantities in the cell, (b) react with several free radicals, (c) be ideal for regeneration other than just reacting with free radicals. These properties are typical for ascorbic acid, making it a good antioxidant for the cells of almost all aerobic species. Among the five plant extracts, the radical scavenging activity of *P. granatum* was almost equivalent to that of ascorbic acid at the concentration of 500 µg/mL and above. This proved that *P. granatum* extract possesses antioxidant properties like ascorbic acid at high concentration. On the other hand, *O. stamineus* extract only showed good scavenging activity at 700 µg/mL. Statistically, *P. granatum* and *O. stamineus* extracts were able to inhibit the formation of DPPH free radicals at the inhibition percentages of 95.40 and 90.20 %, respectively at 700 μg/mL. These results showed *P. granatum* and *O. stamineus* extracts to contain higher inhibitory compounds which may act as primary antioxidants and react with free radicals. Some studies reported that tannins possessed antioxidants and antiradical activities. Tannins are antinutrient of plant origin due to their ability to cause protein precipitation, lower the availability of vitamins & minerals, and inhibit digestive enzymes. It can be considered as the "health-promoting" components in foods derived from plants and beverages. Tannins do not function only as primary antioxidants, it also functions as a secondary antioxidant [32].

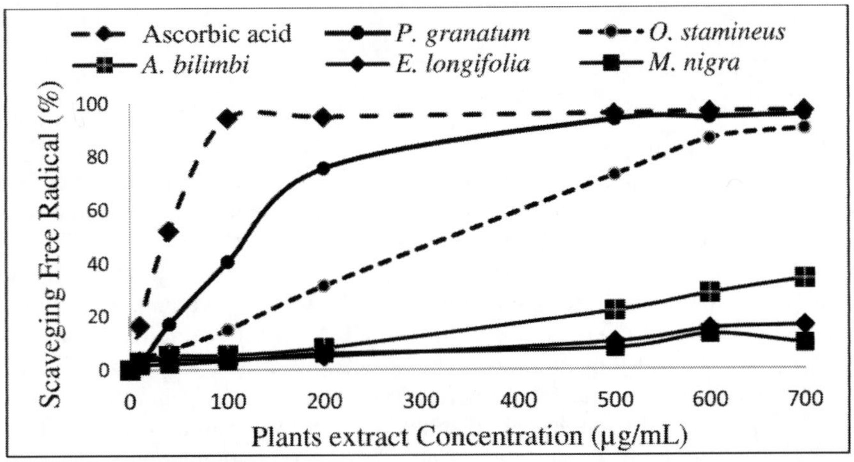

Figure 3: The DPPH free radical scavenging activity of the five plant extracts at different concentrations.

As the plants extract with the highest scavenging activity, *P. granatum* must contain phytochemicals that exhibit antioxidant properties, such as flavonoid and tannin. *P. granatum* extract does not contain flavonoid but rich in the other phytochemicals; hence, it presented the highest scavenging activity. This result can be explained by certain circumstances, such as the dependence of antioxidant activities on the type of extraction solvent used, the presence of other antioxidants of different chemical properties, as well as the compound polarity in the extraction solvent. To enhance the extraction of phytochemicals, the mixture of methanol and ethanol has been reported in several studies [33]. Contrarily, the scavenging activity of *A. bilimbi, E. longifolia,* and *M. nigra* was observed to be 33.78 %, 16.24 %, and 9.51 %, respectively at 700 μg/mL. These activities are weak to be considered antioxidant activities. Comparing with the positive control, the scavenging activity of the three extracts showed significant differences to that of ascorbic acid (more than 62.98%), indicating that these extracts have weak antioxidant activities. The scavenging activity of plant extracts has been reported to increase with the extract concentrations. The IC_{50} values of all the samples were calculated and reported as the concentration of the extracts required to scavenge 50% of the DPPH population [34]. Table 5 showed that *P. granatum* extract exhibited the highest antioxidant activity (IC_{50}= 120.2), followed by *O. stamineus* (IC_{50}= 330.0). Studies have shown that the IC_{50} values can decrease as the free radical scavenging activity of the extract increases. No IC_{50} values were recorded for *E. longifolia, A. bilimbi,* and *M. nigra* extracts as they exhibited poor antioxidant activity. This study also showed all the studied extracts to exhibit some levels of activity against DPPH radical even though *P. granatum* and *O. stamineus* extracts exhibited the best radical scavenging activity compared to *E. longifolia, A. bilimbi,* and *M. nigra*. From the observed antioxidant activities, it can be concluded that *P. granatum* and *O. stamineus* showed the best antioxidant activities and as such, were combined and investigated for improved antioxidant activity.

Table 5: IC_{50} values of individual plant extracts samples.

Samples	**IC_{50} values (µg/mL)**
Ascorbic acid	37.5
P. granatum	120.2
O. stamineus	330.0
E. longifolia	ND
A. bilimbi	ND
M. nigra	ND

ND: Not detected.

3.4. Antioxidant activities of the extract mixtures

The extracts of *P. granatum* and *O. stamineus* were combined at three different combination ratios (1:1, 2:1, 1:2) and studied for antioxidant activity. The results showed the extract combinations at the respective ratios to exhibit DPPH radical scavenging activities of 94.31, 91.80 and 87.79%, respectively at the concentration of 700 µg/mL. The combination ratio of 1:1 showed the highest radical scavenging activity while the other two combination ratios showed comparable radical scavenging activities to that of ratio 1:1. These results indicate that all the combination ratios of the extracts exhibited good antioxidant activities (Figure 4). The IC_{50} values of the extract combinations showed ratio 1:1 to exhibit the lowest IC_{50} value (82 µg/mL), followed by ratio 1:2 (196.0 µg/mL) and ratio 2:1 (255.0 µg/mL). The low IC_{50} value of combination ratio 1:1 showed that it has good antioxidant activity (Table 6). The good antioxidant activity of the extract combination at different ratios could be due to the interaction of different phytochemicals contained in both extracts.

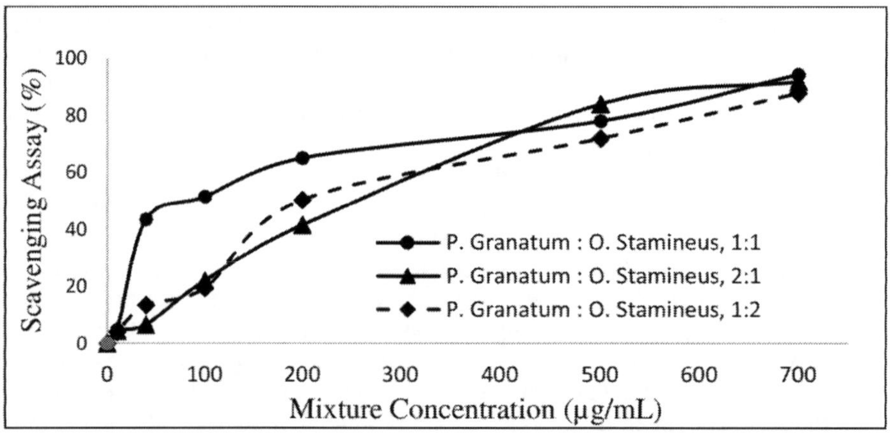

Figure 4: Scavenging activities of the different ratios of mixture *P. granatum* and *O. stamineus* extracts at different concentrations.

Table 6: IC_{50} values of *P. granatum* and *O. stamineus* mixture extracts at different ratios.

The ratio mixture of *P. granatum and O. stamineus* extracts	**IC_{50} value (µg/mL)**
1:1	82.0
1:2	196.0
2:1	255.0

4. Conclusion

In conclusion, the outcome of this study suggests that the five plant extracts have the potential to be used as antioxidant and antimicrobial compounds. The current report will provide preliminary findings for the selection of these plants for further antioxidant and antimicrobial studied which might be necessary to characterize, isolate, and purify the bioactive compounds they contain for onward use as complementary and alternative medicine in treating various diseases. The five extracts contained glycoside 100 %, phenolic and tannin compounds 80%, steroid 60%, flavonoids and alkaloids 20%. *P. granatum* extract showed better antimicrobial activity (60%) against all the microbial isolates, while, *M. nigra* and *O. stamineus* were effective at only 40%. *E. longifolia* and *A. bilimbi* showed only 20% antimicrobial activity while none of the extracts had antifungal activity. Regarding the combined extracts, the mixture exhibited 80% antimicrobial activity and some levels of antifungal activity.

Acknowledgments
The authors gratefully acknowledge Faculty of Industrial Sciences & Technology (FIST), Universiti Malaysia Pahang (UMP), for the financial support under grants (FRGS) RDU190163 and PGRS160392 that enabled the authors to accomplish this work.

Conflict of interest
The authors declare that there are no conflicts of interest.

References

[1] Nair R and Chanda S 2006. Activity of some medicinal plants against certain pathogenic bacteria strains. *Indian J. Pharmacol.*, **38**(2) 142-144

[2] Chowdhury R, Anupam C, and Costas D M 2015. Using Gene Essentiality and Synthetic Lethality Information to Correct Yeast and CHO Cell Genome-Scale Models. *Metabolites*, **5**(4) 536-570.

[3] Percival M 1998. Antioxidants. Clinical Nutrition Insights, Advanced Nutrition Publications, Inc., Revised.

[4] Wichi HP 1998. Enhanced tumor development of butylated hydroxyanisol (BHA) from the prospective of effect on fore stomach and oesophageal squamous epithelium. *Food Chem. Toxicol.*, **26** 717–723.

[5] Kähkönen M P, Hopia A I, Vuorela H J, Rauha J P, Pihlaja K, Kujala T S and Heinonen M 1999. Antioxidant activity of plant extracts containing phenolic compounds. *J. Agric. Food Chem.*, **47**(10) 3954-62.

[6] Ghazi F, Najmuldeen, Hariitaraan F, Ghasak G F and Zulkafli H 2017. Characterization of *Eurycoma longifolia* (Tongkat Ali) Essential Oils Extracted by Microwave Assisted Extraction. *J. Appl. Pharm. Sci.*, **7**(6). 062-068. ISSN 2231-3354

[7] Pang S F, Yusoff M M and Gimbun J 2014. Extraction and microencapsulation of polyphenols from *Orthosiphon stamineus* leaves. *J. Mech. Eng. Sci.*, **7**(1) 1033-1041.

[8] Saleh TA, Al-Shalalfeh M M and Al-saadi A A 2016. Graphene Dendrimer-stabilized silver nanoparticles for detection of methimazole using surface-enhanced Raman scattering with computational assignment. *Sci. Repor.*, **6** 32185.

[9] Mir M A, Parihar K, Tabasum U and Kumari E 2016. Estimation of alkaloid, saponin and flavonoid, content in various extracts of *Crocus sativa*. *J. Med. Plants. Stud.*, **4**(5) 171-174.

[10] Khanam Z, Wen C S and Bhat I U H 2015. Phytochemical screening and antimicrobial activity of root and stem extracts of wild *Eurycoma longifolia* Jack (Tongkat Ali). *J. King Saud. Univ. Sci.*, **27**(1) 23–30. https://doi.org/10.1016/J.JKSUS.2014.04.006

[11] Hengpratom T, Lowe G M, Thumanu K, Suknasang S, Tiamyom K and Eumkeb G 2018. *Oroxylum indicum* (L.) Kurz extract inhibits adipogenesis and lipase activity in vitro. *BMC Complem Altern Med.*, **18**(1), 177.

[12] Sajjad W, Sohail M, Ali B, Haq A, Din G, Hayat M, Khan S 2015. Antibacterial activity of *Punica granatum* peel extract. *Mycopath*, **13**(2) 105-111.

[13] Afsar T, Razak S, Khan M R, Mawash S, Almajwal A, Shabir M and Haq I U 2016. Evaluation of antioxidant, anti-hemolytic and anticancer activity of various solvent extracts of *Acacia hydaspica* R. Parker aerial parts. *BMC Complement Altern Med.*, **16**(1) 1–16. https://doi.org/10.1186/s12906-016-1240-8

[14] Belkacem N, Djaziri R, Lahfa F, El-Haci I A and Boucherit Z 2014. Phytochemical screening and *in vitro* antioxidant activity of various *Punica granatum* l. Peel extracts from Algeria: A comparative study. *Phytothérapie*, **12**(6), 372-379.

[15] Sharma B Lal Ratnakar S, Geetika G, Deen D S, Ashok K 2018. Comparative Evaluation of Antibacterial Activities of *Punica granatum* Leaves and *Acacia leucophloea* Roxb. Leaves. *World J. Pharm. Pharm. Sci.*, **7**(6) 790-802.

[16] Malahubban M, Alimon A R, Sazili A Q, Fakurazi S and Zakry, F A 2013. Phytochemical analysis of *Andrographis paniculata* and *Orthosiphon stamineus* leaf extracts for their antibacterial and antioxidant potential. *Trop. Biomed.*, **30**(3) 467–480.

[17] Wadood A, Ghufran M, Jamal S B, Naeem M, Khan A and Ghaffar R 2013. Asnad. Phytochemical analysis of medicinal plants occurring in local area of Mardan. *Biochem. Anal. Biochem.*, **2**(4) 1-4.

[18] Hussain F, Rana Z, Shafique H, Malik A and Hussain Z 2017. Phytopharmacological potential of different species of Morus alba and their bioactive phytochemicals: A review. *Asian. Pac. J. Trop. Biomed.*, **7**(10) 950-956.

[19] Makky E A, Mashitah M Y and Ibrahim M M 2012. Impact of medicinal plants phytocomponents against antibiotic resistant bacteria. *J. Chem. Pharm. Res.*, **4**(1) 881-893.

[20] Ashraf K, Sultan S and Adam A 2018. Orthosiphon stamineus Benth. is an outstanding food medicine: Review of phytochemical and pharmacological activities. *J. Pharm. Bioall. Sci.*, **10**(3) 109.

[21] Dastjerdi E V, Abdolazimi Z, Ghazanfarian M, Amdjadi P, Kamalinejad M and Mahboubi A 2014. Effect of *Punica granatum* L. flower water extract on five common oral bacteria and bacterial biofilm formation on orthodontic wire. *Iran. J. Public Health*, **43**(12) 1688.

[22] Albarri O M, Var I, Boushihassal A, Meral M, Önlen C, Mohamed M H and Köksal F 2017. The potential effects of Pomegranate on Bacteria and Viruses: A review. *J. Biotech. Sci. Res.*, **3**(6).

[23] Souza G R, Oliveira-Junior R G, Diniz T C, Branco A, Lima-Saraiva S R G, Guimarães A L, Oliveiraa A P, Pachecob A G M, Silvaa M G, Moraes-Filhod M O, Costad M P, Pessoad C Ó and Almeidaa J R G S 2018. Assessment of the antibacterial, cytotoxic and antioxidant activities of Morus nigra L.(Moraceae). *Braz. J. Biol.*, **78**(2) 248-254.

[24] Eman N A, Tan C S, Makky E A 2014. Impact of *Moringa oleifera* cake residue application on waste water treatment: a case study. *J. Water Resour. Prot.*, **6**(7) 677.

[25] Rehman S, Choe K and Yoo H 2016. Review on a traditional herbal medicine, *Eurycoma longifolia* Jack (Tongkat Ali): its traditional uses, chemistry, evidence-based pharmacology and toxicology. *Molecules*, **21**(3) 331.

[26] Das A, Rajkumar V and Dwivedi D 2011. Antioxidant effect of curry leaf (*Murraya koenigii*) powder on quality of ground and cooked goat meat. *Int Food Res. J.*, **18** 559-565.

[27] Berkow E L and Lockhart S R 2017. Fluconazole resistance in Candida species: a current perspective. Infect. *Drug Resist.*, **10** 237.

[28] Sumitra C, Nirupa A and Kalpna R 2013. Evaluation of Antioxidant Properties of Some Indian Vegetable and Fruit Peels by Decoction Extraction Method. *Am. J. Food Technol.*, **8** 173-182.

[29] Malviya S, Arvind A and Hettiarachchy N 2014. Antioxidant and antibacterial potential of pomegranate peel extracts. *J. Food Sci. Technol.*, **51**(12) 4132–4137. https://doi.org/10.1007/s13197-013-0956-4

[30] Oh B., Butow P, Mullan B, Clarke S, Beale P, Pavlakis N and Rosentha, D. 2010. Impact of Medical Qigong on quality of life, fatigue, mood and inflammation in cancer patients: A randomized controlled trial. *Ann. Oncol.*, **21**(3) 608–614. doi:10.1093/annonc/mdp479

[31] Arrigoni O, De Tullio M C 2002. Ascorbic acid, much more than just an antioxidant. *Biochim Biophys Acta* **15** 69:1–9.

[32] Amarowicz R 2007. Tannins: the new natural antioxidants? *Eur. J. Lipid Sci. Technol.*,**109** 549–551.
[33] Sultana B, Anwar F and Ashraf M 2009. Effect of extraction solvent/ technique on the antioxidant activity of selected medicinal plant extracts. *Molecules*, **14** 2167–2180.
[34] Suluvoy J K, Sakthivel K M, Guruvayoorappan C and Grace V M B 2017. Protective effect of *Averrhoa bilimbi* L. fruit extract on ulcerative colitis in wistar rats via regulation of inflammatory mediators and cytokines. *Biomed. Pharmacother.*, **91** 1113-1121.

IOP Conf. Series: Earth and Environmental Science **346** (2019) 012004 doi:10.1088/1755-1315/346/1/012004

Effect of organic acid pretreatment of water hyacinth on enzymatic hydrolysis and biogas and bioethanol production

P Tantayotai[1], P Mutrakulchareon[2], A Tawai[2], S Roddecha[3] and M Sriariyanun[2,4]

[1]Department of Microbiology, Faculty of Science, Srinakharinwirot University, Bangkok, Thailand
[2]Department Chemical and Process Engineering, Thai-German Graduate School of Engineering, King Mongkut's University of Technology North Bangkok (KMUTNB), Bangkok, Thailand
[3]Department of Chemical Engineering, Faculty of Engineering, Kasetsart University, Bangkok, Thailand

E-mail: macintous@gmail.com

Abstract. Due to rapid growth and uncontrolled spreading of water hyacinth in aquatic natural and urban niches, the conversion of this weed to value-added chemicals and biofuels leads to the reduction of pollution and greenhouse effect. The rate-limiting step of the conversion process is hydrolysis reaction of lignocellulosic biomass to sugars, which are subsequently converted to biochemicals and biofuels. To improve the enzymatic hydrolysis efficiency of water hyacinth, the organic acid pretreatments were conducted in this study by using two types of organic acids, including oxalic acid and citric acid. The pretreatment parameters, including pretreatment time, pretreatment temperature, and acid concentration were optimized by using Response Surface Methodology (RSM) based on Box-Behnken design. The highest reducing sugar concentration at 66.92 mg/100 mg-biomass was obtained when using oxalic acid pretreatment. The effects of organic acids on biogas and bioethanol production were observed compared to unpretreated water hyacinth. The biogas and bioethanol yields were increased for 75.61% and 23.26% when using citric acid and oxalic acid pretreatment, respectively. The improvement of biofuel production from water hyacinth by organic acid pretreatment suggested the possibility to applications and solutions in biorefinery industry and environmental situation.

1. Introduction

Lignocellulosic material is the most abundant biomass available worldwide due to its function in plant cell wall. Currently, it is targeted to be used as potential raw materials to produce valuable biochemicals and biofuels .Due to the awareness of greenhouse effect and environmental problem, renewable energy production has been promoted by governmental and private sectors .The main sources for lignocellulose biomass for renewable energy production are agricultural residues and dedicated energy crops . Additionally, aquatic weeds are included in the list for biofuel production because, in addition to biofuel production, the uses of aquatic weeds could reduce environmental problem as well [1-2].

Water hyacinth (*Eichhornia crassipes*) is one type of rapid-growing aquatic weed found in tropical areas, especially in South-east Asia. In Thailand, it was firstly introduced as an alien species, and thus it lacks of natural competitors making it uncontrolled spreaded in canals and rivers. Although water hyacinth growth causes environmental problems, in the bright side, it has high potential to convert to biogas due to its high total volatile solid for 80-86% of dried biomass. For 400 m^2-sized growing area

Content from this work may be used under the terms of the Creative Commons Attribution 3.0 licence. Any further distribution of this work must maintain attribution to the author(s) and the title of the work, journal citation and DOI.
Published under licence by IOP Publishing Ltd

of water hyacinth, it could be converted to biomethane for 4,200 m^3/yr, equivalent to 2,100 kg of LPG/yr [3]. In fact, the yields of biofuels from water hyacinth could be further increased by biomass pretreatment with various means, including chemical, physical and biological methods [4-8].

The production process of renewable energy from lignocellulosic biomass is composed of four steps as follows; namely, pretreatment, hydrolysis, fermentation and separation [1]. The bottlenecks of the whole process in terms of the operational cost are pretreatment and hydrolysis (~40-60% of total cost of the process) [8]. To enhance hydrolysis efficiency, the recalcitrant structures of cellulose fibrils should be modified and the inhibitors of hydrolysis and fermentation should be removed by an appropriated pretreatment method [1-2]. In general, diluted strong inorganic acid or alkaline, i.e. sulfuric acid and sodium hydroxide, are applied in pretreatment. Unlike organic acid pretreatment, those harsh inorganic acid pretreatment produces inhibitory by-products, i.e. 5-hydroxymethylfurfural (HMF) and furfural, to prevent enzymatic hydrolysis, fermentation as well as the growth of microorganism used in downstream process [9-12].

To reduce the downfall of strong inorganic acid pretreatment, organic acid pretreatment has been developed for various types of lignocelluloses [10-11]. In this study, the organic acid pretreatment on water hyacinth was optimized based on Response Surface Methodology (RSM) with Box-Behnken design by estimation of the level of pretreatment parameters to obtain the maximum sugar yields. In addition, the subsequent effects of the optimal pretreatment on biogas and bioethanol production were assessed to get the better knowledge on selection of an appropriate pretreatment method for water hyacinth.

2. Materials and methods

2.1. Materials and chemicals

Water hyacinth sample was collected from urban canal located in Nonthaburi province, central part of Thailand. Fresh water hyacinth was cut to approximately 1-cm-long pieces and dried in hot air oven at 60 °C for 24 h. Dried sample was milled, sieved through 20 mesh and stored in a sealed plastic bag at room temperature. CelluClast 1.5L, (produced by *Trichoderma reesei* ATCC 26921) and β-glucosidase (produced by *Aspergillus niger*) were purchased from Sigma-Aldrich (MO, USA) and Megazyme (Wicklow, Ireland), respectively. 3,5-dinitrosalicylic acid was purchased from Sigma-Aldrich (MO, USA). Other chemicals used in this study were supplied by Ajax Finechem (New South Wales, Australia).

2.2. Pretreatment and hydrolysis

The optimal pretreatment condition of water hyacinth was obtained by using Response Surface Method (RSM) with Box-Behnken design as described in previous studies [13-16]. A three levels, three factors of pretreatment parameters, including pretreatment time (X_1: 30 – 90 min), pretreatment temperature (X_2: 100 – 140 °C) and acid concentration (X_3: 0.5 – 2% for inorganic acid and 2 – 12% for organic acid), were assigned as independent variables, where reducing sugar yield (Y) were dependent variables (Table 1). A total 17 experimental runs were designed by using Design-Expert software version 7.0.0 (Stat-Ease, Inc., MN, USA)). The regression analysis of RSM and estimation of the model coefficient of determination (R^2) were conducted to assess the reliability of generated model. The second order polynomial regression model was calculated to maximize the dependent response by optimizing the three pretreatment parameters.

To pretreat biomass, dried samples were mixed with different acids, including oxalic acid, citric acid and hydrochloric acid (representing strong inorganic acid) in a glass screw-capped bottle containing different acid concentration (Table 1). Each pretreatment was carried out in a hot air oven using temperature and time conditions as designed in Table 1. After rapid cooling the pretreated sample to room temperature, solid fraction was separated by using a funnel and filter paper. The pretreated samples were washed with deionized water until neutral pH, then dried in hot-air oven at 60°C. The dried samples were subjected for downstream process and analysis. The pretreated samples were hydrolyzed by addition of a cellulase cocktail containing 35 µl of Celluclast® 1.5L (Sigma-aldrich, USA) and 10 µl of β-glucosidase (Megazyme, USA) per 100 mg of biomass. Each hydrolysis reaction was prepared in 20 mL of 50 mM sodium citrate buffer (pH 4.7) containing 200 µl of 2 M sodium azide with 2.5% loading ratio of biomass in a screw-capped plastic tube [9-10]. The mixture was incubated in a shaking incubator at 45 °C, 200 rpm for 72 h. The concentrations of reducing sugars were measured using the 3,5-dinitrosalicylic acid (DNS) assay and glucose was used for calculation in the standard curve [17].

Table 1. Box-behnken experimental design of RSM for water hyacinth pretreatment with hydrochloric acid, oxalic acid and citric acid.

Run	X_1 (time, min)	X_2 (temperature, °C)	X_3 (acid concentration, %w/v)		Y (reducing sugar, mg/ml)		
			Inorganic acid	Organic acid	Hydrochloric acid	Oxalic acid	Citric acid
1	60	100	0.5	2	9.37	12.19	2.60
2	30	100	1.25	7	11.52	8.32	1.55
3	90	100	1.25	7	15.92	15.76	3.74
4	60	100	2	12	13.66	14.54	3.12
5	30	120	0.5	2	1.71	14.17	3.04
6	90	120	0.5	2	2.20	16.02	3.66
7	60	120	1.25	7	13.74	16.44	3.86
8	60	120	1.25	7	11.87	16.51	4.00
9	60	120	1.25	7	11.00	15.75	3.74
10	60	120	1.25	7	15.54	15.71	3.82
11	60	120	1.25	7	14.60	15.32	3.82
12	30	120	2	12	11.22	15.35	3.60
13	90	120	2	12	15.39	16.73	4.03
14	60	140	0.5	2	7.92	12.89	2.70
15	30	140	1.25	7	11.70	13.90	2.90
16	90	140	1.25	7	6.20	14.11	3.02
17	60	140	2	12	12.60	15.24	3.82

2.3. Fermentation and composition analysis

To test the potential of pretreated water hyacinth in ethanol production, the hydrolysate were prepared without addition of sodium azide. One milliliter of yeast inoculum (*Saccharomyces cerevisiae* TISTR 5606, provided by Thailand Institute of Scientific and Technological Research (TISTR)) was added into 19 ml of culture media (containing 1% (w/v) glucose, 1% (w/v) yeast extract, and hydrolysate up to 19 ml, pH 5.0) [18]. The fermentation was set up in a shaking incubator at 32 °C, 100 rpm for 48 h. The culture supernatant was harvested by centrifugation at 8,000 rpm for 10 min. The ethanol concentration was analyzed using gas chromatography with FID detector (GC-2010, Shimadzu, Japan) equipped with DB-wax column (30 m x 0.25 mm, 0.25 µm). The oven temperature was operated at 40 °C for 4 min, ramping to 100 °C with 5 °C/min and ramping to 200 °C with 10 °C/min [18-19].

To test biogas production of pretreated water hyacinth, a set of batch anaerobic digester was set up. The fiber compositions of water hyacinth, including cellulose, hemicellulose and lignin were analyzed as described by Van Soest and Wine, 1967 [20]. The total solids (TS) and volatile solids (VS) were quantified according to the standard water and wastewater examination methods [21]. The seed inoculum was obtained from cattle manure (with %TS and %VS at 9.11% and 8.02%, respectively). After starving the seed inoculum for 48 h by incubating at 35 °C with 100 rpm agitation, each digester was set up with 200 ml of working volume containing 5% (TS) of seed inoculum, 5% of water hyacinth, 10 ml of Nutrient broth, and 15 ml of bicarbonate buffer (pH 6.0). Each digester was adjusted to pH 7.0, purged with nitrogen gas before start up, then put in waterbath at 38 °C. The production of biogas of each digester was recorded with 45-day period by using water displacement method.

3. Results and discussion

The optimal pretreatment conditions were obtained by using RSM with Box-behnken design by variation of the level of each pretreatment parameter, including pretreatment time (X_1), pretreatment temperature (X_2) and acid concentration (X_3). In this study, hydrochloric acid was selected as a strong inorganic acid, whereas oxalic acid and citric acid were chosen for organic acid pretreatment. The experimental values of each test parameters were designed to 17 runs, and the released reducing sugars from enzymatic hydrolysis were measured by DNS assay and filled in the Table 1.

Table 2. ANOVA of the predicted model representing the influence of pretreatment parameters on reducing sugar concentrations of pretreated water hyacinth.

Source	Sum of Squares	df	Mean Square	F Value	p-value Prob > F
Hydrochloric acid pretreatment					
Model	241.43	6	40.24	7.57	0.0029
A-Time	1.58	1	1.58	0.30	0.5977
B-Temp	18.10	1	18.10	3.41	0.0947
C-Conc	125.33	1	125.33	23.58	0.0007
AB	24.52	1	24.52	4.61	0.0573
A^2	28.64	1	28.64	5.93	0.0427
C^2	39.31	1	39.31	7.40	0.0216
Residual	53.15	10	5.31		
Lack of Fit	38.96	6	6.49	1.83	0.2907
Pure Error	14.19	4	3.55		
Cor Totol	294.58	16			
Predicted model :Y - =33.44876 +0.85720*Time +0.17236*Temp +18.83884*Conc –4.12622*10^{-3}*Time*Temp –2.89376*10^{-3}*Time^2 – 5.42455*Conc^2					
Oxalic acid pretreatment					
Model	61.28	5	12.26	20.36	< 0.0001
A-Time	14.75	1	14.75	24.50	0.0004
B-Temp	3.55	1	3.55	5.90	0.0334
C-Conc	5.42	1	5.42	9.01	0.0121
AB	13.5	1	13.05	21.68	0.0007
A^2	24.51	1	24.51	40.72	< 0.0001
Residual	6.62	11	0.60		
Lack of Fit	1.065.56	7	0.79	3.00	0.1527
Pure Error	1.06	4	0.26		
Cor Totol	67.90	16			
Predicted model :Y - =100.37160 +0.40648*Time +1.65740*Temp +0.16464*Conc –3.014020*10^{-3}*Time*Temp –6.01442*10^{-3}*Temp^2					
Citric acid pretreatment					
Model	5.07	7	0.72	3.53	0.0412
A-Time	0.13	1	0.13	0.63	0.4394
B-Temp	0.19	1	0.19	0.91	0.3643
C-Conc	0.43	1	0.43	2.08	0.1829
AB	1.09	1	1.09	5.33	0.0464
AC	1.31	1	1.31	6.36	0.0326
A^2	0.71	1	0.71	3.45	0.0961
C^2	1.11	1	1.11	5.42	0.0448
Residual	1.85	9	0.21		
Lack of Fit	1.75	5	0.35	14.33	0.0116
Pure Error	0.098	4	0.024		
Cor Totol	6.92	16			
Predicted model :Y + =4.31780 –0.018956*Time –0.044615*Temp +0.56213*Conc +8.71089*10^{-4}*Time*Temp –3.80837*10^{-3} *Time*Conc –4.54991*10^{-4}*Temp^2 – 0.020530*Conc^2					

The hydrolysis results showed that the concentration of reducing sugars were varied ranging from 1.71-15.92 mg/ml, 8.32-16.73 mg/ml and 1.55-4.03 mg/ml in hydrochloric, oxalic, and citric acid pretreatment, respectively. Then, dependent variables, e.g. reducing sugars concentration, and correspondent pretreatment parameters in each run were evaluated based on RSM to generate the model

IOP Conf. Series: Earth and Environmental Science **346** (2019) 012004 doi:10.1088/1755-1315/346/1/012004

representing the correlations between independent and dependent variables. Quadratic mathematic models were suggested to represent the influence of pretreatment parameters on reducing sugar concentrations with high coefficient of determination (R^2) values of 0.8366, 0.9211, and 0.7332 in the sets of hydrochloric, oxalic and citric acid pretreatment, respectively. The reliability and significance of each proposed model were determined using ANOVA and expressed as F value (Table 2). The significant model was indicated with the p-value (Prob. > F) of less than 0.05. The results showed that the models of three acid pretreatments were significant because the p-values were 0.0029, <0.0001 and 0.0412, in hydrochloric, oxalic and citric acid sets, respectively. These results of statistical analysis advocated the high reliability of generate model.

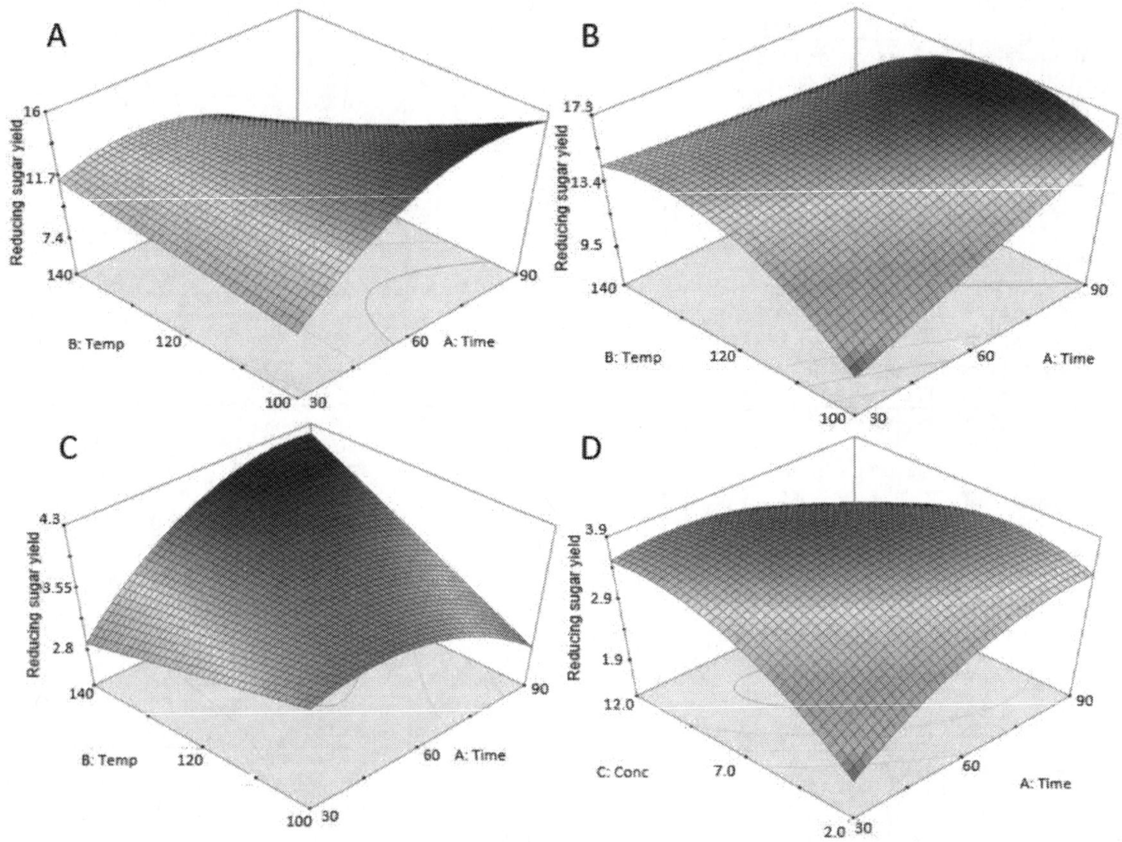

Figure 1. Response surface plots representing interaction effects of the independent variables on reducing sugar yield (mg/ml) obtained from water hyacinth pretreatments. (A) Hydrochloric acid, (B) Oxalic acid, (C-D) Citric acid.

Response surface contour plots were generated to visualize the interactive effects of the pretreatment parameters on the reducing sugar concentrations (Figure 1). As observed in the plots of hydrochloric, oxalic and citric acid pretreatment, pretreatment time and pretreatment temperature influenced on the saccharification yield. While, in citric acid pretreatment, acid concentration additionally affected the yield. Theoretically, long pretreatment time and high temperature lead to further hydrolysis and oxidation of the sugars, which decrease the overall yields. However, low level of pretreatment parameters are also not sufficient to get the maximum yield [10-11, 15]. It could be observed in the contour plots that the interactive effects of temperature and time were synergistic to increase or decrease sugar yields (Figure 1A-1C). Interestingly, the maximum sugar yield was obtained when using middle level of citric acid concentration and pretreatment time (Figure 1D). These observations suggested the differences in mechanisms of acid pretreatments by each type of acid and highlighted the importance of optimization experiments.

Based on the generated mathematical models, the predicted optimal sugar yields were 16.03 mg, 16.73 mg and 4.03 mg/ml for the hydrochloric, oxalic and citric acid pretreatment, respectively (Table 3). It could be seen that the predicted optimal reducing sugar yields obtained from hydrochloric and oxalic acid pretreatment were similar, although the hydrochloric acid concentration use here was much less compared to oxalic acid. Interestingly, when pretreated water hyacinth samples were subjected for biogas and bioethanol production, the yields of biogas and bioethanol obtained from oxalic (203.8 ml/g-TS and 3.55%) and citric pretreatments (221.1 ml/g-TS and 3.36%) were similar and both were clearly higher than that of hydrochloric pretreatment (136.8 ml/g-TS and 2.78%). Additionally, the yields of biogas obtained from oxalic and citric pretreatments were increased for 61.87% and 75.62%, respectively compared to unpretreated water hyacinth. Likewise, the ethanol productions were increased by the effect of oxalic and citric pretreatments for 23.26% and 16.67%, respectively. The improvement of bioethanol and biofuel production by using organic acid pretreatment could be the result of less production of fermentation inhibitors during the pretreatment compared to the inorganic acid pretreatment. Additionally, the remaining organic acid residues could be metabolized by microbes to produce biofuels. This result was agreed well with our previous researches on pretreatments of rice straw, oil palm shell and napier grass [10-11, 15, 18]. However, hydrochloric acid pretreatment showed high improvement on enzymatic saccharification, it could not improve production of biogas and bioethanol compared to unpretreated sample. This is possibly due to the presences of fermentation inhibitors causing the detrimental effects on fermentation.

Table 3. Predicted yields of optimal reducing sugar (mg/ml), biogas and bioethanol obtained from hydrochloric, oxalic and citric acid pretreatments

Acid pretreatment	Predicted optimal pretreatment parameters			Predicted optimal reducing sugar yield (mg/ml)	Experimental yield	
	Time (min)	Temp.(°C)	Acid concentration (%)		Accumulated biogas (ml/g-TS)	Bioethanol (%)
Hydrochloric	80.69	103.55	1.41	16.03	136.8	2.78
Oxalic	70.01	126.33	11.38	16.73	203.8	3.55
Citric	68.21	136.34	7.90	4.03	221.1	3.36
Unpretreated	n.d*.	n.d.	n.d.	n.d.	125.9	2.88

*n.d.: not determined

4. Conclusion

Water hyacinth is aquatic weed causing various environmental problems in Thailand because it is an alien species, thus lacking of natural competitors. Attempts to convert these lignocellulose wastes to biofuels have been made, yet the technology is not readily available for actual applications. This study focused on optimization of organic acid pretreatment of water hyacinth to improve the biofuel production yields by using oxalic acid and citric acid. The results obtained from enzymatic saccharification, biogas and bioethanol production suggested that the optimal pretreatment by organic acid improved process yields. The knowledge gained from this work could benefit to development of industrial processes and mechanisms of acid pretreatments on hydrolysis and fermentation.

Acknowledgements

The authors would like to thank Srinakharinwirot University (Research Grant contract No. 091/2562) for financial support of this work.

References

[1] Rodiahwati W and Sriariyanun M 2016 *KMUTNB. Int. J. Appl. Sci. Technol.* **9**(4) 289-298

[2] Akkharasinphonrat R, Douzou T and Sriariyanun 2017 *KMUTNB. Int. J. Appl. Sci. Technol.* **10**(2) 89-96

[3] Kammeng T, Srijaranai S and Viriyachai C 1987 *KKU Science J.* **15** 180-185

[4] Kobkam C, Tinoi J and Kittiwachana S 2018 *KMUTNB. Int. J. Appl. Sci. Technol.* **11**(4) 247-256

[5] Riansa-ngawong W, Suwansaard M and Prasertsan P 2015 *KMUTNB. Int. J. Appl. Sci. Technol.* **8(1)** 65-75

[6] Cheenkachorn K, Douzou T, Roddecha S, Tantayotai P and Sriariyanun M 2016 *Energy Procedia* **100**(2016) 160-165

[7] Sriariyanun M, Tantayotai P, Yasurin P, Pornwongthong P and Cheenkachorn K 2016 *Elec. J. Biotechnol.* **19** 23-28

[8] Tantayotai P, Rattanaporn K, Tepaamorndech S, Cheenkachorn K and Sriariyanun M 2018 *Waste. Biomass Valori* **2018**

[9] Hassan S S, Williams G A and Jaiswal A K 2018 *Bioresour. Technol.* **262 (2018)** 310-318

[10] Amnuaycheewa P, Hengaroonprasan R, Rattanaporn K, Kirdponpattara S, Cheenkachorn K and Sriariyanun M 2016 *Ind. Crops Prod.* **84(247)** 254

[11] Amnuaycheewa P, Rodiahwati W, Sanvarinda P, Cheenkachorn K, Tawai A and Sriariyanun M 2017 *KMUTNB. Int. J. Appl. Sci. Technol.* **10(2)** 107-117

[12] Rattanaporn K, Roddecha S, Sriariyanun and Cheenkachorn K 2017 *Energy Procedia* **141C(2017)** 146-149

[13] Sriariyanun M 2014 *KMUTNB. Int. J. Appl. Sci. Technol.* **7(4)** 47-52

[14] Sriariyanun M, Yan Q, Nowik I, Cheenkachorn K, Phusantisampan T and Modigell M 2015 *Kasetsart Journal (Natural Science)* **49(1)** 146-154.

[15] Junnienkul N, Douzou T, Yasurin P, Asavasanti S and Sriariyanun M 2018 *KMUTNB. Int. J. Appl. Sci. Technol.* **11(3)** 199-207

[16] Srisampao I, Pornwongthong P, Roddecha S, Rodiahwati W and Sriariyanun M 2018 *Proc. Int, Conf. on Informatics, Environment, Energy and Applications (IEEA2018)* 169-173

[17] Miller GL 1959 *Anal. Chem.* **31** 426–428

[18] Rattanaporn K, Tantayotai P, Phusantisampan T, Pornwongthong P and Sriariyanun M 2018 *Bioprocess Biosyst Eng.* **41** 467-477

[19] Sriariyanun M, Mutrakulcharoen P, Tepaamorndech S, Cheenkachorn K and Rattanaporn K 2019 *Orient. J. Chem.* **35**

[20] Van Soest P and Wine R 1967 *J. Assoc. Off. Anal. Chem.* **50** 50-55

[21] *APHA standard methods for the examination of water and wastewater* 21th ed 2005 In. American Public Health Association, Washington DC, USA

IOP Conf. Series: Earth and Environmental Science **346** (2019) 012005 doi:10.1088/1755-1315/346/1/012005

Characterizations of lignocellulose waxes and study of their effects on enzymatic saccharification for biofuel production

K Kitsubthawee[1], K Cheenkachorn[1], S Chuetor[1], K Rattanaporn[2] and M Sriariyanun[3, 4]

[1]Department of Chemical engineering, King Mongkut's University of Technology North Bangkok (KMUTNB), Bangkok, Thailand
[2]Department of Biotechnology, Faculty of Agro-Industry, Kasetsart University, Bangkok, Thailand
[3]Department of Chemical and Process Engineering, Thai-German Graduate School of Engineering, King Mongkut's University of Technology North Bangkok (KMUTNB), Bangkok, Thailand

E-mail: macintous@gmail.com

Abstract Wax is considered to be by-products of biofuel production from lignocellulosic biomass and agricultural processing plants. Although, attempts to find applications of waxes obtained from these sources have been made, yet, little proportions of cases were successfully practical in industrial applications due to lacking of wax characterization data. This research aims to characterize wax properties of three types of lignocellulose biomass; Rice straw (RS), Sugarcane bagasse (SB) and Napier grass (NG); obtained from Soxhlet extraction technique with hexane solvent. The chemical and physical properties of waxes were analysed by Fourier Transform Infrared analysis (FT-IR) analysis and Differential Scanning Calorimeter (DSC) suggesting the similarity of properties between NG and SB waxes. To test the effect of wax on biofuel production, both dewaxed and non-dewaxed biomass were pretreated with diluted alkali and enzymatically hydrolysed by cellulase. The results showed that the combination of dewaxing and alkali pretreatment increased the cellulose proportions in biomass compared to alkali pretreatment for 21.8%, 10.1% and 16.1% in RS, NG and SB, respectively. Additionally, the combined dewaxing and pretreatment improved the saccharification yields of RS, NG and SB for 5.84%, 35.03% and 26.10%, respectively. Altogether, the existence of wax clearly showed the negative effect on biofuel production, thus dewaxing process should be further considered to be included to improve the biorefining process.

1. Introduction

Due to awareness of greenhouse effect happening worldwide, Thai government and private sectors put efforts in development of renewable energy production from biomass, especially lignocellulose biomass obtained from agricultural processing plants and agricultural activities. Many types of agricultural wastes and by-products, i.e. rice straw, sugarcane bagasse, palm oil shell, cassava pulp, corn cobs, as well as dedicated energy crops, i.e. napier grass were used as raw materials for biofuel productions in a lab scale and a pilot scale in various forms, including biogas and bioethanol [1]. Based on several reports on these biofuel productions, wax that obtained as unused by-products caused the practical cleaning problem and reduced heat transfer efficiency of the reactors. Currently, the use of wax was overlooked due to its relatively low proportion in total biomass weight (0.5% - 4% w/w) [2]. Many discussions and works were brought out to find applications of these by-products, however, very few cases were success

Content from this work may be used under the terms of the Creative Commons Attribution 3.0 licence. Any further distribution of this work must maintain attribution to the author(s) and the title of the work, journal citation and DOI.
Published under licence by IOP Publishing Ltd

in industrial scale production due to lacking of economically feasible separation process, and actual potential properties of wax obtained from different biomass.

Natural wax found in plants and animals is categorized as an ester molecule consisting of long alkyl chains of fatty acids, alcohols, ketones and aldehydes. Wax and its derivatives have been applied as additives in value-added products, such as cosmetic ingredients for lip-sticks, mascara formulations, and many others .Additionally, based on an analysis of wax composition, a polyacrylamide (Policosanol) was also present, which this substance has antihypertensive properties, LDL fat reduction, triglycerides reduction, chlorate reduction in the bloodstream and heart disease prevention. Regarding to the current price of polycocanol as a therapeutic supplement at 620 USD/kg, it is very attractive to promote the use of wax in various applications [3]. In biofuel production side, wax has natural property as a coating agent and an anti-wetting agent. Previously, a study showed that the existence of wax reduced the ethanol yield in sorghum grain from 106.2 g/L to 86.1 g/L [4]. Thus, wax prohibits the accessibility of hydrolysis enzyme to interact with cellulose fibrils in biomass, resulting in reduction in hydrolysis efficiency, and biofuel yields ultimately.

In general, production of biofuels from lignocellulosic biomass consists of 4 main steps including pretreatment, hydrolysis, fermentation and product recovery. The critical steps in practical and economical aspects are pretreatment and hydrolysis that contribute to 40-60% of overall process cost. Many studies demonstrated that the efficiency of hydrolysis of lignocellulose could be improved by pretreatment to alter recalcitrant structure of biomass and remove hydrolysis inhibitors, such as lignin [5-9]. On the other hand, there are only 2 research works focused on the effect of wax on the biofuel production, and both works selected the sorghum plant as the model biomass [9-10]. The main purposes of this study were the chemical and physical characterization of waxes obtained from three types of lignocellulosic biomass, including Rice straw (RS), Napier grass (NG), and Sugarcane bagasse (SB). The effects of dewaxing on enzymatic hydrolysis were demonstrated by observing the released reducing sugars that subsequently converted to different forms of biofuels .The knowledge gained from this study could highlight the importance of wax in biorefining process and biofuel production.

2. Materials and methods

2.1. Extraction of plants wax

Rice straw (RS), Napier grass (NG), and Sugarcane bagasse (SB) were obtained from local farms located in the central part of Thailand. These fresh samples were dried at 80 °C in hot-air oven, then ground to pass a 1.0 mm sized-sieve and kept in the desiccators at room temperature prior to use. The wax was extracted from plants according to the procedure described by Athukorala methods [11] with minor modifications. Ground plants (five g) was extracted thoroughly with hexane (250 mL) in a Soxhlet extraction apparatus for six h [12-14]. Waxes compounds were collected after the solvent was evaporated in a vacuum oven for further analysis. Determinations of fiber compositions, including cellulose, hemicelluloses, and lignin were conducted using Van Soest methods [15].

2.2. Characterization of wax

2.2.1. Differential scanning calorimeter (DSC). Physical properties of crude plants waxes were analyzed by using Differential scanning calorimeter (TA Instruments DSC Model Q100) to measure their melting temperature (Tm). An amount of 3–4 mg of the sample was crimped in the aluminum pan and loaded along with reference pan without a sample. The heating rate was 10 o C/min, and heating temperature was carried from 30°C to 100°C.

2.2.2. Fourier transform infrared analysis (FT-IR). The structural constituents and chemical changes of functional groups of each biomass sample were analysed by using FT-IR instrument (Perkin Elmer Spectrum 1). Absorbance spectra of samples in chloroform were recorded using KBr windows. The spectra scanning ranged from 4000 to 400 cm-1 at resolution 4 cm-1.

2.3. Alkaline pretreatment

The pretreatment of ground plant samples and dewaxed plant samples were carried out in an autoclave machine with operating temperature at 121 °C for 30 min. Samples were pretreated with 2.0% NaOH solutions at a solid loading of 10% (w/v) in the screw-capped glass bottle. The pretreated sample was cooled down at room temperature, and the solid and the liquid fractions were separated using a funnel and a filter paper. The pretreated samples were washed several times with deionized water until neutral pH, and then the solid residue was dried in the oven at 65 °C, and was kept in dried form for further analysis and hydrolysis.

2.4. Enzymatic hydrolysis

The untreated, dewaxed and pretreated samples were enzymatically hydrolyzed with biomass loading at 5% w/v of dried material in 50 mM sodium citrate buffer at pH 4.8. Enzyme cocktails containing 400 µl of 2% sodium azide and 350 µl of CelluClast 1.5L® (Sigma-Aldrich, USA) and 100 µl of β-glucosidase (Megazyme, USA) per gram of tested biomass were added. The mixture was kept in a shaking incubator (150 rpm) at 45 °C for 72 h [16-18]. The total reducing sugar concentration was determined according to the 3,5-Dinitrosalicylic acid (DNS) assay using D-glucose as a standard.

3. Results and discussion

3.1. Effect of dewaxing on modification of fiber compositions

Naturally, wax deposits on plant surfaces and prevents water activities, a.k.a. aqueous alkaline solution for pretreatment in this study. NaOH pretreatment on lignocellulosic biomass was shown to remove lignin, a hydrolysis inhibitor, and thus to increases proportions of cellulose and hemicellulose contents of pretreated biomass [19]. To prove whether dewaxing has an impact on pretreatment efficiency and modification of fiber compositions, each type of biomass, including RS, NG and SB, was divided to four groups, including untreated, dewaxed, pretreated and dewaxed+ pretreated samples.

Table 1. The contents of cellulose, hemicellulose, and lignin in RS, NG and SB after dewaxing and alkaline pretreatment.

Samples	Treatment	Cellulose (%)	Hemicellulose (%)	Lignin (%)
RS	Untreated	47.062 ± 1.682	25.354 ± 1.908	6.003 ± 1.128
	Dewaxed	46.967 ± 0.637	23.331 ± 0.164	10.717 ± 0.238
	Pretreated	55.327 ± 0.270	16.125 ± 3.916	4.278 ±0.376
	Dewaxed + Pretreated	77.140 ± 3.139	8.028 ± 3.574	1.300 ± 0.521
NG	Untreated	40.416 ± 0.594	25.348 ±1.091	15.101 ± 0.608
	Dewaxed	38.402 ± 0.686	24.945 ± 1.658	14.751 ± 0.323
	Pretreated	64.780 ±2.208	13.396± 0.917	2.959±0.264
	Dewaxed + Pretreated	74.872± 0.708	11.693 ± 2.692	2.563 ± 0.376
SB	Untreated	57.669 ± 0.784	28.842 ± 0.665	12.381 ± 0.409
	Dewaxed	57.732 ± 1.461	25.421 ± 1.381	11.145 ± 0.241
	Pretreated	64.135 ± 0.728	12.501 ±1.862	3.122 ± 0.06
	Dewaxed + Pretreated	80.225 ± 0.669	11.324 ± 0.428	2.899 0.206

After dewaxing process by Soxhlet extraction, the wax yields obtained from RS, NG and SB were 0.564%, 1.678% and 0.577% w/w of dried biomass, respectively. The fiber composition analysis, including cellulose, hemicellulose and lignin was conducted for each sample (Table 1). The results from three types of plants showed that dewaxing did not really change the compositions of cellulose, hemicellulose and lignin compared to untreated sample. After alkaline pretreatment, the lignin content was removed and therefore cellulose was increased compared to untreated and dewaxed samples. For example, in NG, the cellulose content increased from 40.416% to 64.780%, while the lignin content decreased from 15.101% to 2.959%. This result supported the fact that alkaline pretreatment helps to remove lignin from the biomass. In case of dewaxed + pretreated sample, the cellulose contents of RS, NG and SB were further increased than pretreated samples, and the lignin contents were also more reduced (Table 1). For example, the cellulose contents of RS, NG and SB were increased to 77.140%, 74.872% and 80.225%, respectively. Altogether, the changes in fiber compositions suggested that dewaxing has the positive effect on pretreatment by uncovering the fiber complexes to interact with alkaline and water molecules.

3.2. Effects of dewaxing on enzymatic saccharification

In previous section, dewaxing was demonstrated to improve pretreatment efficiency. Likewise, in this section, the effect of dewaxing on enzymatic saccharification was assessed. To fully function, cellulase enzymes should be able to gain access to lignocellulose fibrils. The more hydrolysis interactions, the more reducing sugars were released. The contents of reducing sugars were quantitatively measured using DNS assay as described in the method section.

Comparing between pretreated and dewaxed + pretreated sample, all types of tested biomass in this study, including RS, NG and SB, showed improvement on saccharification in dewaxed + pretreated sample. The reducing sugars of dewaxed + pretreated samples of RS, NG and SB were increased for 5.83%, 35.03% and 26.10%, respectively compared to pretreated samples (Table 2). The increased yield of enzymatic saccharification could be expected to improve the biofuel yield in downstream process. This result could be explained that after wax was removed from the biomass, the accessibilities of alkaline and cellulase to cellulose fibrils were gained to remove lignin inhibitor and to hydrolyse cellulose subsequently.

Table 2. Measurements of released reducing sugars from enzymatic hydrolysed biomass.

Samples	Released reducing sugars (mg/ml)
Pretreated sample	
RS	17.111
NG	12.438
SB	15.262
Dewaxed + Pretreated sample	
RS	18.110
NG	16.795
SB	19.246

3.3. Physical and chemical properties of RS, NG and SB waxes

DSC was used to investigate phase transitions in the plant wax and to estimate its melting points, which is one of the most important physical property of wax. The phase transition of melting is endothermic and therefore reflects a negative heat flow. The decomposition temperature (T_d) was determined during DSC analysis, the maximum temperature should not exceed its decomposition point. All the waxes were found to have a T_d above 100 °C, so that a temperature of 80 °C was set to be the maximum tested temperature. The thermograms for the three biomass waxes, including RS, NG and SB, were shown in figure 1. The RS, NG and SB waxes possessed very similar phase transitions as their main peaks were positioned at 77 °C.

IOP Conf. Series: Earth and Environmental Science **346** (2019) 012005 doi:10.1088/1755-1315/346/1/012005

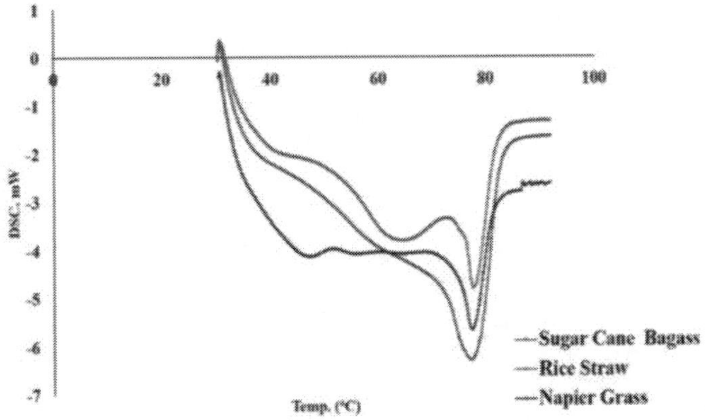

Figure 1. DSC thermograms of three biomass waxes, including RS, NG and SB.

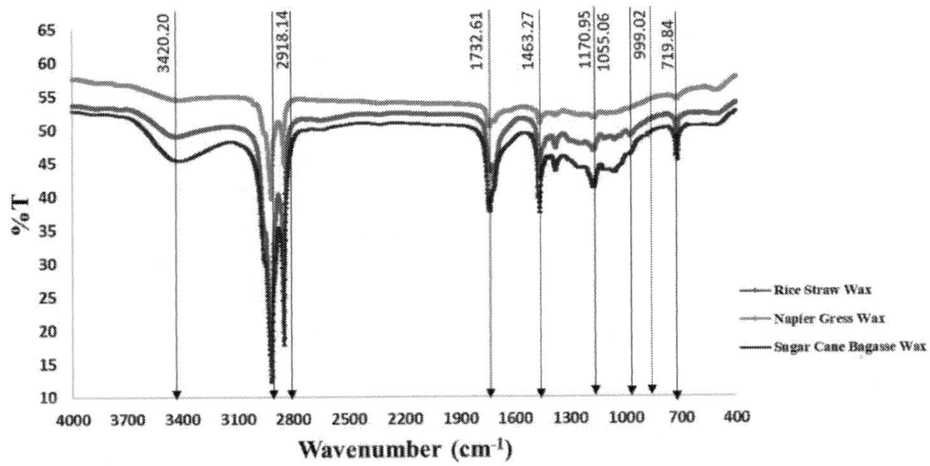

Figure 2. FT-IR spectra of three biomass waxes, including RS, NG and SB.

FT-IR analysis was conducted to monitor the alterations of chemical compositions and chemical interactions of substances. Extracted wax samples obtained from RS, NG and SB were analysed by using FT-IR and the absorption spectrums in the wavenumber ranging between 4000-400 cm^{-1} were observed (Figure 2). Similar to DSC analysis, three types of waxes also have similar spectral patterns, although the magnitudes of absorption peaks were differentiated. It could be seen that NG and SB waxes had similar magnitudes of peaks suggesting the similar chemical properties of these two waxes, which were in well agreement with saccharification results in Table 2. The bands at 2918 cm^{-1} and 2849 cm^{-1} were assigned to the asymmetric and symmetric CH_2 stretching vibrations, which were wax characteristic for an aliphatic fraction [20]. The band at 1713 cm^{-1} was represented to carbonyl stretching in fatty acids. The band at 1463 cm^{-1} was referred to the bending of CH_2 in lignin or wax. The bands at 1600 and 1510 cm^{-1} were characteristic for lignin fraction [21]. The bands at 1377, 1170, 1130, 1055 and 999 cm^{-1} were also matched to lignin [22]. These FT-IR peaks indicated the presence of fatty acid ester and lignin in wax.

3.4. Structural constituents and chemical changes

The chemical changes and structural constituents of A) untreated, B) dewaxed, C) pretreated and D) dewaxed+pretreated samples of RS, NG and SB were investigated by using FT-IR spectroscopy (Figure 3). Comparing between untreated and dewaxed samples in three types of biomass, the patterns and magnitudes of peaks were similar suggested that dewaxing did not engage in any chemical alteration. This observation was agreed well with the analysis of fiber composition in Table 1 that dewaxing did not cause any change in contents of cellulose, hemicellulose and lignin.

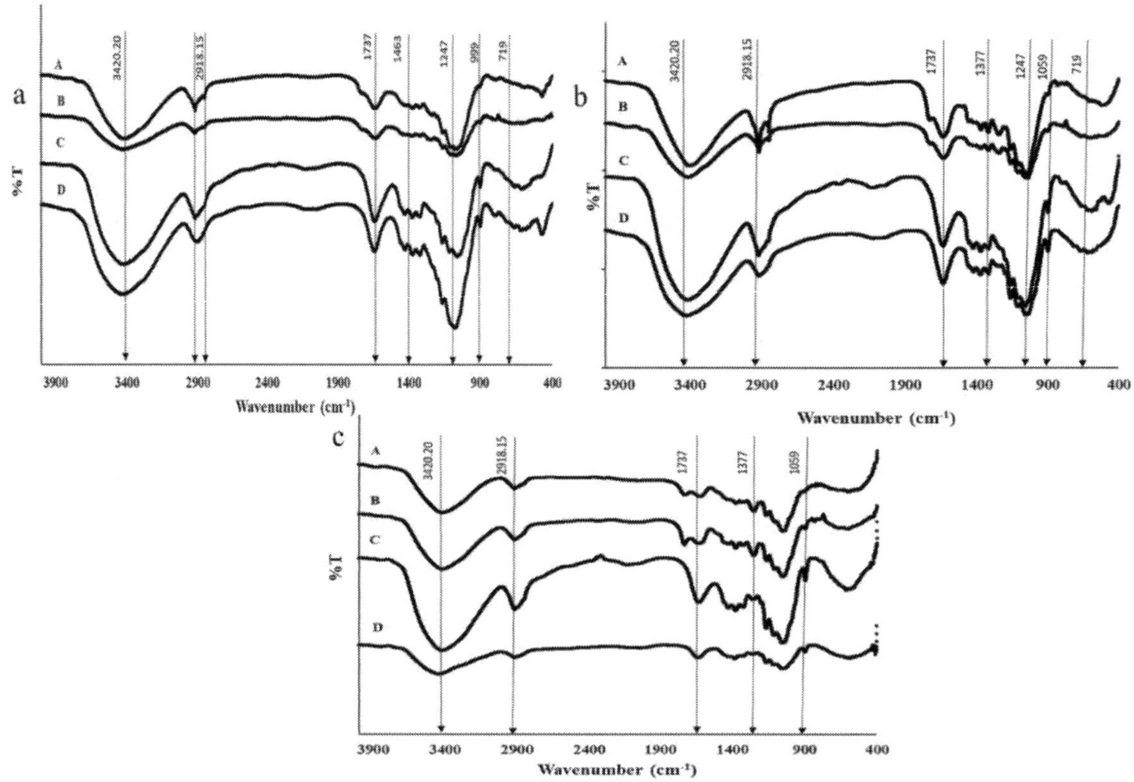

Figure 3. FT-IR spectra of (a) RS, (b) NG and (c) SB biomass, including A) Untreated, (B) Dewaxed, (C) Pretreated and (D) Dewaxed+Pretreated.

On the other hands, NaOH pretreatment caused the reductions of peak magnitudes at 1,734 cm^{-1}, 1,432 cm^{-1}, 1,376 cm^{-1}, 1,247 cm^{-1}, 1,162 cm^{-1}, 1,059 cm^{-1} and 719 cm^{-1} (Figure 3). It could be explained the presence of NaOH in biomass dissolve the cellulose, and split and reform of new inter- and intramolecular hydrogen bonds between cellulose fibrils [23]. Moreover, the magnitude of bands at 1,377 cm^{-1} and 1,463 cm^{-1}, represented CH_2 scissoring motion in crystalline cellulose [24] were reduced in NaOH-pretreatment sample and this reduction was even bigger in dewaxed+pretreated sample. This result indicates that NaOH pretreatment decreased the crystalline cellulose and exposed more surface areas of amorphous cellulose for enzymatic interactions that subsequently leaded to enhancement of reducing sugar productions. Furthermore, a band at 897 cm^{-1}, representing the β-glucoside linkage belonging to cellulose and hemicellulose, was more intense in dewaxed+pretreated sample than the others, which correlated to the fiber composition analysis in Table 1. Ether bonds at 1,250 cm^{-1} peak disappeared after pretreatment meaning that after pretreatment, these linkages between lignin and carbohydrates were disrupted, e.g. loss of lignin [25]. The peak at 1,739 cm^{-1}, representing aromatic C=C stretching of the aromatic ring of lignin [26], was diminished. It is firmly concluded that the disappearance of lignin peaks at 1,236 cm^{-1} and 1,650 cm^{-1} were due to removal of lignin coating on the surface of cellulose fibers of pretreated samples [27].

4. Conclusion

In this study, dewaxing process of RS, NG and SB showed clear improvement of pretreatment efficiency and enzymatic saccharification. Although the analysis of physical and chemical characters of waxes by using FT-IR and fiber composition analysis suggested the similarity of untreated and dewaxed samples, the combination of dewaxing and pretreatment showed clear modification of biomass as following mechanisms. First, combined dewaxing and pretreatment removed lignin inhibitor, and increased the proportion of cellulose in biomass. Second, it altered the inter- and intramolecular interactions of cellulose fibril complexes. Third, it transformed cellulose from the crystalline to amorphous structure. The synergistic effects of dewaxing and pretreatment in promotion of biofuel productions from different biomass demonstrated in this study could help to establish a feasible biorefining process in the future.

Acknowledgments
The authors would like to thank King Mongkut 's University of Technology North Bangkok (Research Grant contract No. KMUTNB-61-GOV-03-47) for financial support of this work.

Reference

[1] Rodiahwati W and Sriariyanun M 2016 *KMUTNB. Int. J. Appl. Sci. Technol.* **9**(4) 289-298
[2] Di C, Chang Z, Wang C, Ren W, Wang Z, Qin P and Tan T 2013 *Bioresour. Technol.* **149** 470–47
[3] Akkharasinphonrat R, Douzou T and Sriariyanun M 2017 *KMUTNB. Int. J. Appl. Sci. Technol.* **10**(2) 89-96
[4] Nhuan P, Nghiem J, Megan E and Hums 2018 *J. Ferment. Bioeng.* **4** 12
[5] Kobkam C, Tinoi J and Kittiwachana S 2018 *KMUTNB. Int. J. Appl. Sci. Technol.* **11**(4) 247-256
[6] Riansa-ngawong W, Suwansaard M and Prasertsan P 2015 *KMUTNB. Int. J. Appl. Sci. Technol.* **8**(1) 65-75
[7] Amnuaycheewa P, Hengaroonprasan R, Rattanaporn K, Kirdponpattara S, Cheenkachorn K and Sriariyanun M 2016 *Ind. Crops. Prod.* **84** 247-254.
[8] Cheenkachorn K, Douzou T, Roddecha S, Tantayotai P and Sriariyanun M 2016 *J. Energy Procedia.* **100**(2016) 160-165
[9] Amnuaycheewa P, Rodiahwati W, Sanvarinda P, Cheenkachorn K, Tawai A and Sriariyanun M 2017 *KMUTNB. Int. J. Appl. Sci. Technol.* **10**(2) 107-117
[10] Cai D, Chang Z, Wang C, Ren W, Wang Z, Qin P and Tan T 2013 *J. Bioresour. Technol.* **149** 470–473.
[11] Athukorala Y and Mazza G 2010 *Ind. Crops. Prod.* **31** 550–556
[12] Chaichoowong S, Bol JB, Bol P, Gamse T and Sriariyanun M 2017 *Oriental Journal of Chemistry.* **33**(1) 66-73
[13] Wongpanit B, Chotikamas C, Roddecha S, Tantayotai P and Sriariyanun M 2018 *Proc. Int. Conf. on Chemical Materials and Process* (ICCMP2018). **187** 01002
[14] Chotikamas S, Cheenkachorn K, Wongpanit B, Tantayotai P and Sriariyanun M 2018 *Proc. Int. Conf. on Chemical Materials and Process* (ICCMP2018). **187** 01001
[15] Athukorala Y and Mazza G 2010 *Ind. Crops. Prod.* **31** 550–556
[16] Sriariyanun M, Yan Q, Nowik I, Cheenkachorn K, Phusantisampan T and Modigell M 2015 *J. Nat. Appl. Sci.* **49**(1) 146-154.
[17] Junnienkul N, Douzou T, Yasurin P, Asavasanti S and Sriariyanun M 2018 *KMUTNB. Int. J. Appl. Sci. Technol.* **11**(3) 199-207
[18] Rattanaporn K, Tantayotai P, Phusantisampan T, Pornwongthong P and Sriariyanun M 2018 *Bioprocess. Biosyst. Eng.* **41** 467-477.

[19] Kim S J, Lee Y Y and Kim H T 2016 *Bioresour. Technol.* **199** 42–48
[20] Dubey P, Sharma P and Kumar V 2017 *Data. inBrief.* **15** 615–622
[21] Azizul M H, Barman N D, Kang H T and et al. 2012 *J. Microbiol. Biotechnol.* **12** 1681–1691
[22] Maryana R and Ma'rifatun D 2014 *Energy Procedia.* **47** 250 – 254
[23] Qi G. Peng F, Xiong L, Lin X, Huang C and et al. 2016 *Bioresour. Technol.*
[24] Mirahmadi K, Kabir M M, Jeihanipour A, Karimi K, Mohammad J and et al. 2010 *Bioresour Bioprocess.* **5**(2) 928-938.
[25] Athukorala Y, Mazza G and Oomah D B 2009 *J. Lipid Sci. Technol.* **11** 705–714
[26] DeswarteI E F, Clark H J, Jeffrey J E and Rose M P 2006 *Green Chem.* **8** 39–42
[27] Moreira P H S S, Freitas J C O, Braga R M, Araújo M R and Souza de F A M 2018 *Ind. Crops. Prod.* **116** 144–149

IOP Conf. Series: Earth and Environmental Science **346** (2019) 012006 doi:10.1088/1755-1315/346/1/012006

Removal of malachite green toxicity using water hyacinth (*Eichhornia crassipes*) biomass

R Pinontoan[1], M T Supandi, J Andhika, M I Sutanto and H Victor

Biology Department, Faculty of Science and Technology, Pelita Harapan University, Tangerang, Indonesia

E-mail: reinhard.pinontoan@uph.edu

Abstract. Malachite green (MG), a common synthetic dye that is used extensively in many industries including aquaculture, is highly toxic to cells. One of its metabolites, leucomalachite green, poses more toxic effects to cells compared to MG itself. Adsorption of this synthetic dye seems to be a promising method in removing dye without the formation of other harmful products. This study evaluated the potency of water hyacinth (*Eichhornia crassipes*) biomass of both root and leaf part of the plant in removing MG dye toxicity. Biomass from both root and leaf part of the plant showed high decolorization efficiency in less than 1 hour of incubation. Results from thin layer chromatography (TLC) analysis showed no formation of other degradation metabolites, implying the decolorization process that occurred was solely adsorption. Toxicity test towards *Saccharomyces cerevisiae* showed that water hyacinth biomass could reduce MG toxicity. Therefore, the root and leaf biomass collected from *E. crassipes* does have the potential to detoxify aqueous solution contaminated by MG.

1. Introduction

Synthetic dyes have been commonly used in many industries, such as textiles, food, pharmaceutical, cosmetic, aquaculture, pulp and paper industries to color products [1]. Once disposed of, the wastewater ends up in the environment, contaminating the soil and water bodies. Dyes are very detectable contaminants, because of their intense color in water bodies. Dyes are more stable and resistant to microbial degradation due to their complex aromatic molecular structures [2]. Moreover, colored dye effluents are known to be very toxic and are potential mutagenic and carcinogenic agents [3].

One of the most extensively used toxic synthetic dye is Basic Green 4, commonly known as malachite green (MG). MG, belonging to the triphenylmethane dye group, is highly soluble in water and is commonly used in textile industry (for dyeing silk, wool, jute, leather, ceramics, and cotton), in the aquaculture industry (as a fungicide, parasiticide, and disinfectant), as well as a cytochemical staining agent [4,5]. MG has been extensively used in aquaculture as a topical treatment by bath or flush methods to control skin and gill flukes [4]. However, the use of MG brings controversy due to the toxic side effects it produces. Though the use of MG as a biocide or coloring agent has been banned in several countries, many parts of the world still use this dye due to its low cost, ready availability and efficacy [6].

MG is highly toxic to mammalian cells, even in concentrations of less than 0.1 ppm. It is reported to cause carcinogenesis, mutagenesis, chromosomal aberration, and respiratory toxicity [7]. It is also

Content from this work may be used under the terms of the Creative Commons Attribution 3.0 licence. Any further distribution of this work must maintain attribution to the author(s) and the title of the work, journal citation and DOI.
Published under licence by IOP Publishing Ltd

IOP Conf. Series: Earth and Environmental Science **346** (2019) 012006 doi:10.1088/1755-1315/346/1/012006

known to increases liver tumor formation in rats and generates reproductive abnormalities in rabbits and fish [8]. One of MG degradation product, leucomalachite green (LMG) is known to be more toxic than MG itself and remains in cells for a longer time period [3,9]. Many methods have been tested for the removal of MG from aquatic environments. Current methods of detoxifying and decolorizing MG involve the use of boron-doped diamond (BDD) anodes and electro-Fenton systems, peroxi-coagulation, all of which are neither readily available nor effective when dealing with wastewater that has entered the environment [2]. Degradation of MG by various methods might not fully detoxify the products and might be even more toxic. There is therefore a need for an affordable and efficient method to remove these effluents from wastewaters.

One efficient method in removing dyes is adsorption. Adsorption is a promising approach for the removal of various pollutants from water and wastewater. Agricultural solid wastes and biosorbents, such as biopolymers and biomass adsorbents, have shown interesting adsorption abilities to remove MG [10]. One promising biosorbent is water hyacinth (*Eichhornia crassipes*) biomass. Water hyacinth is a wild aquatic plant that often grows in lakes, rivers, swamps, or ponds. It can easily grow in extensive amount that causes environmental problems and can be considered a weed by some, therefore, it has little to no economic value in public. Nevertheless, water hyacinth has the potential to remove pollutants, when used as a biological filtration system. It has also been reported that the water hyacinth has excellent tolerance towards toxic pollutants. The water hyacinth's biomass has large potential in decolorizing textile dyes and able to adsorb heavy metals [11,12]. Therefore, this study was aimed to evaluate the potential of water hyacinth biomass in removing MG dye toxicity through a yeast assay system.

2. Materials and methods

2.1. Materials

Malachite green (927.01 g/mol, dye content ≥ 90 %) and $CuSO_4.5H_2O$ were purchased from Merck. Potato dextrose agar (PDA) and broth (PDB) were obtained from Pronadisa. The solvents (acetic acid, ethyl acetate, n-propanol) were purchased from CV Anugrah. *Saccharomyces cerevisiae* was isolated from a commercial baker's yeast (Fermipan). Samples of *Eichhornia crassipes* were collected from Danau Kelapa Dua, Tangerang, Indonesia.

2.2. Biomass preparation

Water hyacinth were cleaned completely with tap water for several times to remove dirt and all the soil particles followed by separation of the leaf and root from the stem. Root and leaf were separated, and both samples were washed with distilled water. Samples were dried in the oven of 70 °C for 24 hours, then were put in desiccator for another 24 hours. The dried samples were grounded, and the powder was used in this study.

2.3. MG decolorization assay

MG decolorization assay was carried out by putting 0.1 g of both root and leaf biomass into a 100 ml solution of 50 ppm MG. The samples were incubated at 25 °C in shaker incubator at 120 rpm speed. Samples were withdrawn at time intervals (0, 5, 10, 15, 20, 25, 30, 45, 60, 90, 120, 180, 240, and 300 minutes) and the residual MG concentration in solution was measured immediately using UV/VIS spectrophotometer (BioDrop) in wavelength 617 nm. Absorbance spectra was measured at 300-800 nm wavelength to see the overall change. This experiment was repeated for four times [3,9,13]. The absorbance value was converted into concentration (ppm) using linear equation obtained from standard curve.

Based on the absorbance value using UV/VIS spectrophotometer, percentage of decolorization was measured [3]. The decolorization efficiency of the adsorption was calculated using the following equation:

$$Decolorization\ (\%) = \frac{(A_0 - A)}{A_0} \times 100 \tag{1}$$

In which A_0 is the initial absorbance and A is the absorbance of medium after dye decolorization at 617 nm. All the values shown were mean ± standard deviation of four replicates. Percentage of decolorization for each time interval between root and leaf sample were compared by two sample t-test in Microsoft Excel.

2.4. Thin layer chromatography (TLC) assay for metabolites
After complete adsorption, metabolites produced in decolorized dye solution were extracted with equal volume of ethyl acetate and a small amount of $CuSO_4.5H_2O$. The metabolites formation was examined by TLC using silica gel. The solvent mixture used was n-propanol: acetic acid: ethyl acetate: distilled water (6:1:1:2 v/v). The separated products were examined directly [14].

2.5. Toxicity test
Toxicity test of the decolorized product was done using *Saccharomyces cerevisiae*. *S. cerevisiae* was isolated from a commercial baker's yeast by adding a certain amount of the yeast into PDB media as a starter culture. The culture was incubated for 24 hours in room temperature. The culture was then streaked into PDA and incubated for 24 hours in 37 °C until pure culture was obtained.

The toxicity test was done using a well-diffusion assay. PDA was poured in petri dish; depth of agar was 3–4 mm. With sterile cotton swab, *S. cerevisiae* was spread evenly over the plate successively in two directions to obtain an even inoculum. The plate was allowed to dry for 5 min. Wells of 7 mm diameter were cut on the surface of the agar. Twenty microliters of each treatment (MG solution treated for 0, 5, 15, 30, 60, 180 and 300 minutes) were added in one well. This was done in four replicates for both root and leaf samples. In addition, a standard was made by adding untreated MG solution of different concentrations (0, 5, 7.5, 10, 25 and 50 ppm) to each well. All test plates were incubated aerobically at 37 °C for 24 hours and observed for the inhibition zones. The zone of inhibition was measured by a scale to the nearest mm excluding well diameter [15,16]. Means between groups were compared by One Way ANOVA followed by Tukey Honest Significant Difference (HSD) post-hoc analysis in Minitab 18.

3. Result and discussion

3.1 MG decolorization assay
MG decolorization assay was done to evaluate the ability of *E. crassipes* biomass in removing MG. Both root and leaf biomass of *E. crassipes* were dried, grounded and added to the MG solution. The initial concentration of the dye was 50 ppm. The test was carried out by examining the absorbance spectra between 300 - 800 nm. Figure 1 shows the graph of absorbance spectra.

The change in absorbance at 617 nm indicates reduction of MG concentration in solution. As time progressed, the peak of A617 decreased correspondingly with MG concentration in the solution (Table 1). Barapatre *et al.* [17] reported that untreated MG has three main absorbance peaks at 315 nm, 420 nm and 617 nm, while degraded MG by ligninolytic fungus *Aspergillus flavus* has increased absorption peaks at 220, 254 and 370 nm. Mukherjee & Das [18] also reported that the absorption peak of MG at 617 nm was completely lost while three new absorption peaks appeared at 612, 335, 321 nm due to biodegradation by enzymes of *Enterobacter asburiae* XJUHX-4TM. Meanwhile, if the decolorization is due to the adsorption, the original absorption peak would decrease proportionally. Therefore, the process of decolorization in this study indicates adsorption of the dye by *E. crassipes* biomass.

Table 1 showed that both root and leaf could decolorize MG up to 90% in less than 1 hour. In comparison between root and leaf part of the plant, the root section showed a significantly ($p<0.05$) higher adsorption ability up to 180 minutes of incubation. The root part also showed faster decolorization ability where it reached equilibrium at 20 minutes compared to the leaf which needed 45 minutes. In comparison with other biomass, Burca *et al.* [19] reported that *Camellia sinensis* leaves could reach equilibrium in 80 minutes for removal of MG with concentration of 75-140 ppm.

Meanwhile for other dyes, Tan *et al.* [20] reported that whole part of *E. crassipes* plant could remove methylene blue and methyl orange up to 98% and 67% respectively in 20 days in open air and exposed to sufficient sunlight. This indicates that the use of *E. crassipes* biomass in this study has a promising decolorization ability towards MG in a short period of time.

Overall, the root biomass of *E.crassipes* performed better than the leaf biomass in this study. This is likely due to the cellulose content of the plant parts. Cellulose fibres are hydrophilic due to the hydroxyl groups found on the surface of the structure and could therefore contribute to the effectiveness in which the root biomass absorbs the dye. Lara-Serrano *et al.* [21] showed that the roots of *E. crassipes* contained significantly higher amounts of cellulose, holocellulose and hemicellulose when compared to the plant's leaves. It is therefore likely that the root biomass has better absorption properties the leaf biomass.

A. Root

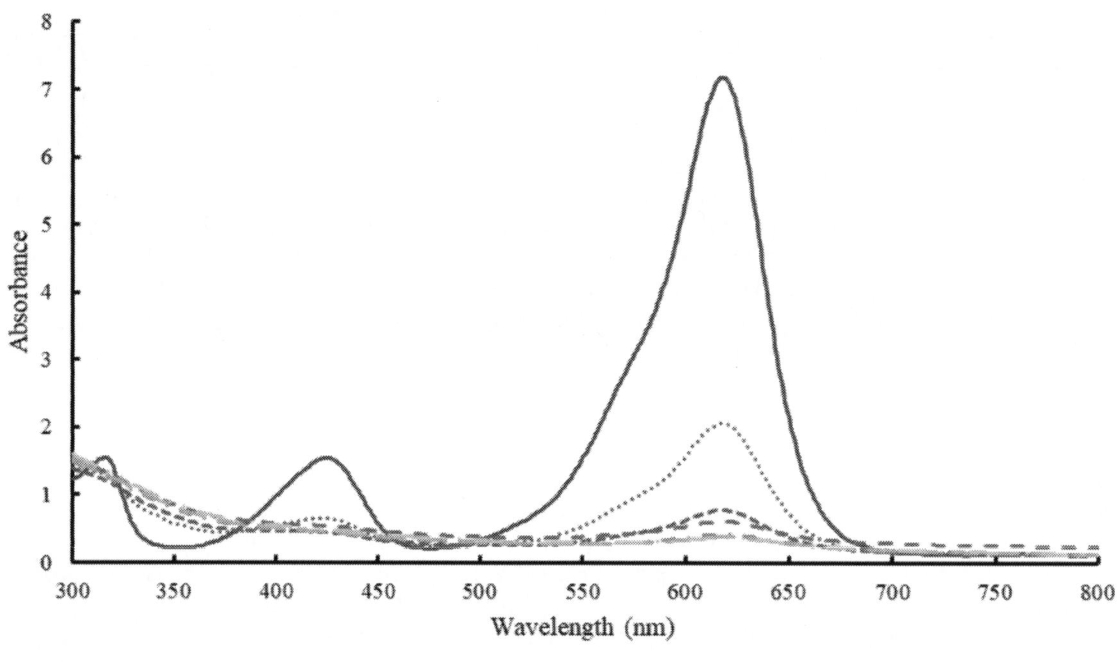

B. Leaf

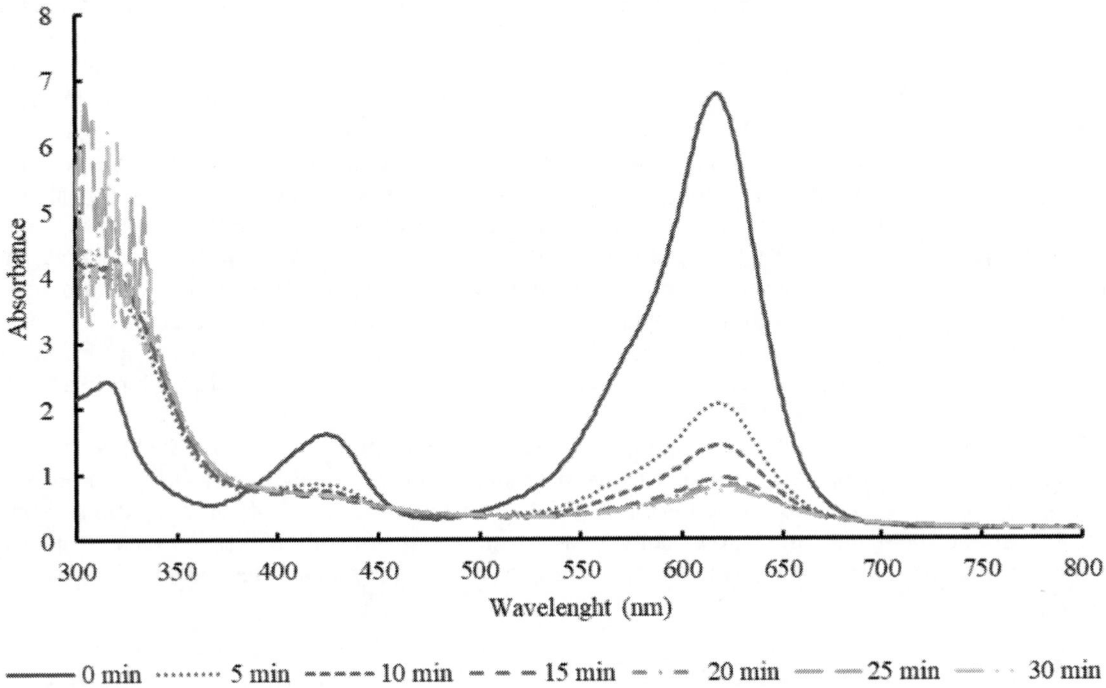

Figure 1. Absorbance spectra of MG solution after treated with *E. crassipes* biomass addition, taken every five minutes. The initial MG concentration was 50 ppm. The samples were incubated in a shaker incubator at 25 °C and 120 rpm speed. Data shown were mean of four replicates.

Table 1. Decolorization percentage of *E. crassipes* biomass towards MG and its concentration.

Incubation time (min)	Dye concentration (ppm)		Decolorization percentage (%)	
	Root	Leaf	Root	Leaf
0	40 ± 2.4	37 ± 2.5	0 ± 0.0	0 ± 0.0
5	12 ± 1.1	12 ± 0.7	71 ± 1.2[a]	70 ± 0.5
10	5 ± 0.2	9 ± 0.4	89 ± 0.4[a]	79 ± 0.6
15	4 ± 0.2	6 ± 0.2	93 ± 0.4[a]	86 ± 0.3
20	3 ± 0.2	5 ± 0.3	94 ± 0.6[a]	88 ± 0.6
25	3 ± 0.2	5 ± 0.4	95 ± 0.6[a]	88 ± 0.5
30	3 ± 0.2	5 ± 0.3	95 ± 0.6[a]	89 ± 0.4
45	3 ± 0.2	4 ± 0.3	95 ± 0.7[a]	91 ± 0.6
60	3 ± 0.2	4 ± 0.4	95 ± 0.7[a]	91 ± 0.7
90	3 ± 0.2	4 ± 0.4	95 ± 0.7[a]	92 ± 0.7
120	3 ± 0.2	4 ± 0.3	95 ± 0.7[a]	92 ± 0.5
180	3 ± 0.2	4 ± 0.4	94 ± 0.7[a]	92 ± 0.7
240	4 ± 0.2	4 ± 0.4	93 ± 0.8	92 ± 0.6
300	4 ± 0.2	4 ± 0.3	92 ± 0.9	92 ± 0.8

[a] Indicates significant difference of decolorization percentage ($p < 0.05$) between root and leaf during the same time.

Note: The values shown were mean ± standard deviation of four replicates. Means were compared using two sample t-test. Dye concentration was determined using standard curve of MG.

3.2 TLC analysis of metabolites

When metabolized, MG can form a colorless yet more toxic product called LMG, which is also known to retain in the muscles and cells of fish for longer periods of time [22]. A TLC analysis using the product of the adsorption process was therefore used to determine if the adsorption conducted by the biomass produces any other metabolites. The samples collected from different time points of the adsorption process (Figure 2Aa-g and 2Ba-g) were run through the TLC with Figure 2 A1-5 and B1-5 as controls. The controls, which contain MG on its own, display a single spot at the same retardation factor (Rf) value of around 0.7 with a decrease in intensity and size in proportion to the concentration of MG present in the sample. Similar to the controls, a large, single, vivid spot at the same Rf value as the control samples can be seen at time 0-300 minutes that decreased in intensity and size over time (Figure 2Aa-g and 2Ba-g). This is especially true for the samples incubated with the root biomass as the spot formed by the sample at 300 minutes is much more faded and smaller than the spot formed by the lowest concentration of the control samples, which therefore implies a concentration of MG of less than 5 ppm in the sample, measured as 4 ppm (Table 1). The decrease in intensity of the spots formed indicates that a process of decolorization and adsorption occurs when the leaf and root biomass of *E. crassipes* are incubated with the MG.

TLC analysis also showed that no other metabolites were present in the final product and that the biomass was able to adsorb the product without degrading MG. Wanyonyi *et al.* [14] reported that TLC of degraded MG by crude protease enzymes showed the appearance of three additional spots with Rf values of 0.96, 0.92 and 0.90 as compared to control Rf value of 0.87. This indicates that the process of decolorization demonstrated by the biomass in this study was indeed adsorption and not degradation.

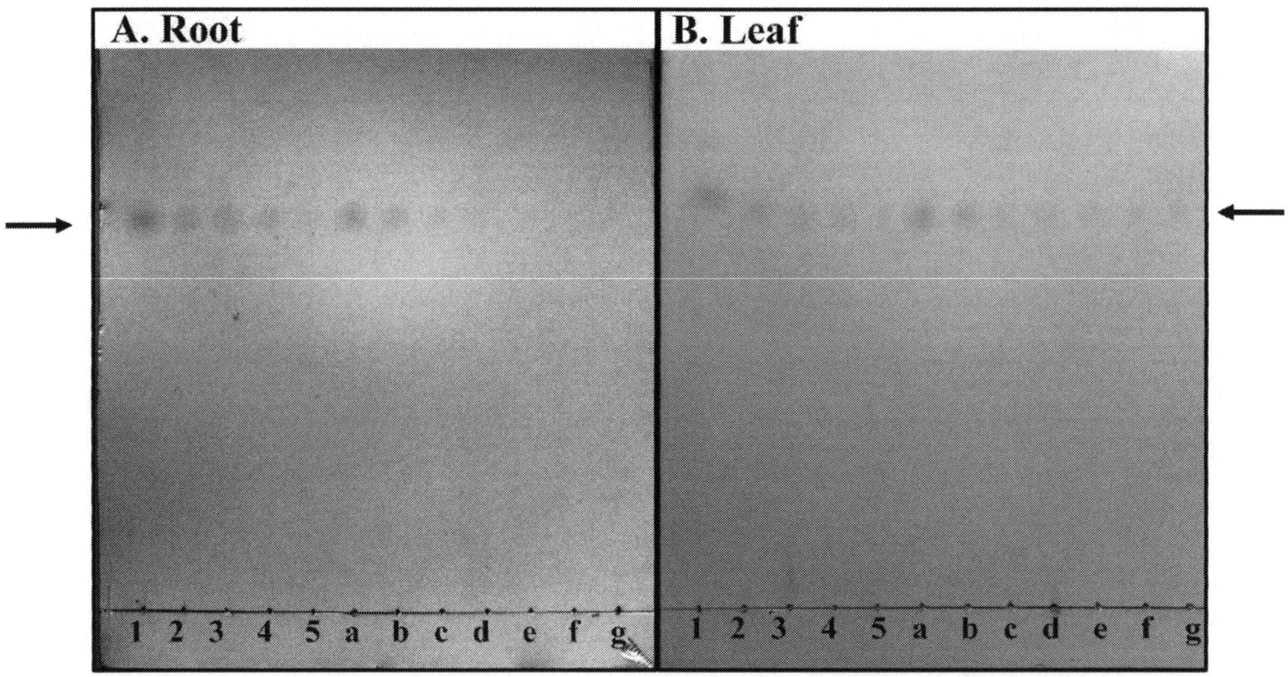

Figure 2. TLC analysis of metabolites produced. The arrows show separated dye compounds being extracted with Rf value of ± 0.7. 1: MG 50 ppm; 2: MG 25 ppm; 3: MG 10 ppm; 4: MG 7.5 ppm; 5: MG 5 ppm; a: 0 min; b: 5 mins; c: 15 mins; d: 30 mins; e: 60 mins; f: 180 mins; g: 300 mins.

3.3 Toxicity analysis of metabolites in yeast assay system

Despite evidence that adsorption has occurred, a decrease in the intensity of the color does not necessarily mean that the samples are detoxified and therefore a toxicity assay was conducted using baker's yeast (*S. cerevisiae*) to determine the toxicity of the samples incubated with the root and leaf biomass of *E. crassipes*.

Table 2. Toxicity test of root and leaf metabolites

Time (minutes)	Clear zone diameter (mm) Root	Leaf
0	10.3 ± 0.80^{a}	10.0 ± 0.70^{a}
5	2.0 ± 0.70^{c}	4.5 ± 1.50^{b}
15	0^{c}	1.3 ± 0.40^{c}
30	0^{c}	0^{c}
60	0^{c}	0^{c}
180	0^{c}	0^{c}
300	0^{c}	0^{c}

Note: Clear zone area was calculated by excluding diameter of wells. Means that do not share a letter are significantly different ($p < 0.05$).

The samples incubated with the root and leaf biomass were plated into the wells at different time points and the results can be seen in Table 2. As seen in Table 2, the diameter of the clear zone at 0 minute was similar to the diameter of that MG concentration 50 ppm (data not shown) and gradually decreases as time passes. After 5 minutes, the clear zone diameter was reduced significantly for both root and leaf sample. However, the diameter for the leaf samples was still significantly higher ($p<0.05$) than the root samples. This might be due to the slower adsorption by the leaf samples compared to the root samples at 5 minutes. After 15 minutes, both samples toxicity were comparable to MG 0 ppm (data not shown). This indicates a lower growth inhibition effect of the yeast colonies surrounding the wells and thus demonstrating a gradual decrease of the toxicity of the samples up until 300 minutes. The adsorption that occurs when the root and leaf biomass are incubated with MG can therefore decolor and detoxify MG contaminated areas. Based on the results, it can be concluded that the root and leaf biomass collected from *E. crassipes* does have the potential to detoxify aqueous solution contaminated by MG.

4. Conclusion

E. crassipes root and leaf biomass were proved to have potential in decolorizing MG solution by adsorption process. Both the root and leaf biomass could decolorize up to 90% MG in less than 1 hour. Moreover, TLC analysis showed that decolorization process occurred did not produced other metabolites. It was supported by yeast toxicity test which proved decreased toxicity of treated MG compared to untreated MG. In conclusion, *E. crassipes* biomass has potential to detoxify MG, therefore can be used in bioremediation of industrial waste water as an affordable source of biosorbent.

Acknowledgement

This work was financially supported by LPPM-UPH (P-030-FaST/V/2019).

References

[1] Wanyonyi W C, Onyari J M, Shiundu P M and Mulaa F J 2017 *Energy Procedia* **119** 38-51
[2] Ghanbari F and Moradi M 2016 Electrooxidation processes for dye degradation and colored wastewater treatment *Advanced Nanomaterials for Wastewater Remediation* ed R K Gautam and M C Chattopadhyaya (London: CRC Press LLC) chapter 3 p 62-3
[3] Ramezani S, Pourbabaee A A and Javaheri Daneshmand H 2013 *J Bioremed Biodeg* **4** 175

[4] Alderman D J 1985 *J Fish Dis* **8** 289-98
[5] Henderson A L, Schmitt T C, Heinze T M and Cerniglia C E 1997 *Appl Environ Microbiol* **63** 4099-101
[6] Srivastava S, Sinha R and Roy D 2004 *Aquat Toxicol* **66** 319-29
[7] Cleinmensen S, Jensen J C, Jensen N J, Meyer O, Olsen P and Würtzen G 1984 *Arch toxicol* **56** 43-5
[8] Rao K V K 1995 *Toxicol lett* **81** 107-13
[9] Culp S J, Mellick P W, Trotter R W, Greenlees K J, Kondell R L and Beland F A 2006 *Food Chem Toxicol* **44** 1204-12
[10] Raval N P, Shah P U and Shah N K 2017 *Appl Water Sci* **7** 3407-45
[11] Uddin Md N, Islam T Md and Das S 2014 *Int J Chem Eng* **2014** 1-13
[12] Murithi G, Onindo C O, Wambu E W and Muthakia G K 2014 *Bioresources* **9** 3613-31
[13] Dubey A, Mishra A and Singhal S 2014 *Int J Environ Sci Technol* **11** 1043-50
[14] Wanyonyi W C, Onyari J M, Shiundu P M and Mulaa F J 2014 *J Environ Sci Toxicol Food Technol* **8** 58-64
[15] Aneja K R, Joshi R, Sharma C and Aneja A 2010 *Braz J Oral Sci* **9** 421-6
[16] Yadav S, Trivedi N A and Bhatt J D 2015 *Ayu* **36** 203-7
[17] Barapatre A, Aadil K R and Jha H 2017 *CLEAN–Soil, Air, Water* **45** 1-22
[18] Mukherjee T and Das M 2014 *CLEAN–Soil, Air, Water* **42** 849-56
[19] Burca S, Maicaneanu A and Indolean C 2016 *Revue Roumaine de Chimie* **61** 541-7
[20] Tan K A, Morad N and Ooi J Q 2016 *Int J Environ Sci Dev* **7** 724-8
[21] Lara-Serrano J S, Rutiaga-Quiñones O M, López-Miranda J, Fileto-Pérez H A, Pedraza-Bucio F E, Rico-Cerda J L and Rutiaga-Quiñones J G 2016. *BioResources* **11** 7214-7223
[22] Radko L, Minta M and Stypula-Trebas S 2011 *Bull Vet Inst Pulawy* **55** 347-53

IOP Conf. Series: Earth and Environmental Science **346** (2019) 012007 doi:10.1088/1755-1315/346/1/012007

Genome Sequencing of *Pantoea agglomerans* strain NC, isolated from peanut bacterial wilt nursery

X Z Wang[1], Y Y Tang[1], Z W Wang[1], C Z Wu[2], J H Chen[3], Q Wu[1, 4] and C T Wang[1, 4]

[1]Shandong Peanut Research Institute (SPRI), Qingdao 266100, PR China
[2]Hezhou Branch, Guangxi Academy of Agricultural Science / Hezhou Agricultural Science Research Institute, Hezhou 542813, PR China
[3]Quanzhou Institute of Agricultural Science, Quanzhou 362212, PR China

E-mail: qi_wu@126.com (Q Wu); chinapeanut@126.com (C T Wang)

Abstract. *Pantoea agglomerans* was ubiquitous in nature and it has been identified as an effective microbial antagonist against fungal pathogens associated with several plants. Here, we report the draft genome sequence of *P. agglomerans* strain NC which was isolated from bacterial wilt infected peanut (*Arachis hypogaea*) in Nan Chang city of China.

1. Introduction

Pantoea agglomerans is ubiquitous frequently inhabiting plants, animals, soil and water [1-4]. It has been identified as an effective antagonist against fungal pathogens of several plants [5-8]. Here, we report the draft genome sequences of *P. agglomerans* strain NC which was isolated from bacterial wilt infected peanut (*Arachis hypogaea*) sample in Nan Chang city of China.

The *P. agglomerans* genome was obtained by Illumina MiSeq sequencing platform from a PE400 library (insert size 500 bp). Twenty three contigs (average length of 0.195 Mbp) and twenty one scaffolds (average length of 0.214 Mbp) were assembled *de novo* from 595.3M clean data (1,247,589 trimmed reads with average length of 239 bp) by using Newbler v2.8 (20110517_1502), providing a 132 - fold coverage and gaps were filled by GapCloser software. The size of assembled final sequence was 4.50 Mbp.

Putative protein coding sequences were predicted by Glimmer v3.0 software [9]. Non coding RNA genes were predicted by tRNAscan-SE, RNAmmer v1.2 and Rfam [10-12]. CRISPR recognition tool (CRT) was used to scan CRISPR (Clustered regularly interspaced short palindromic repeats) [13]. Protein coding genes were also analyzed by Swiss-Prot database (https://www.uniprot.org/) and blastall software. GO, COG and KEGG functional annotation were based on BLAST2GO, eggNOG and KEGG databases, respectively.

The G+C content of overall genome sequence is 55.3%. It consists of 4,179 protein - coding genes covering whole genome 85.8% with average length of 924.9 bp and mean G+C content of 56.7%. 64 tRNAs, 3 rRNAs and 85 other ncRNA (summary 152) were predicted comprising 0.48% of the genome with mean length of 144 bp.

2. Nucleotide sequence accession number

Content from this work may be used under the terms of the Creative Commons Attribution 3.0 licence. Any further distribution of this work must maintain attribution to the author(s) and the title of the work, journal citation and DOI.
Published under licence by IOP Publishing Ltd

This whole-genome shotgun project has been deposited at GenBank with the accession number QIBS00000000. The sequencing reads have been deposited under the accession number SRR8449985. All reads have been deposited to SRA and are associated with BioProject number PRJNA471388.

Acknowledgments
This research was supported by Natural Science Funds for Young Scholar of Shangdong Academy of Agricultural Sciences (2015YQN13), Natural Science Foundation of Shangdong province (ZR2015YL064), Qingdao science and technology plan of basic research project (12-1-4-11-(1)-jch), China Agricultural Research System (CARS-13), Agricultural Scientific and Technological Innovation Project of Shandong Academy of Agricultural Sciences (CXGC2018E21).

References

[1] De Araujo W L et al 2001 *Can. J. Microbiol.* **47** 229
[2] Dillon R J, Vennard C T and Charnley A K 2001 *Nature* **403** 851
[3] Yeung K F, Lee K M and Woodard R W 1998 *J. Nat. Prod.* **61** 207
[4] Mosso M A, De la Rosa M C, Vivar C and Medina M R 1994 *J. Appl. Bacteriol.* **77** 370
[5] Nunes C, Usall J, Teixido N and Vinas I 2001 *Int. J. Food. Microbiol.* **70** 53
[6] Nunes C, Usall J, Teixido N, Fons E and Vinas I 2002 *J. Appl. Microbiol.* **92** 247
[7] Wright S A, Zumoff C H, Schneider L and Beer S V 2001 *Appl. Environ. Microbiol.* **67** 284
[8] Feng Y, Shen D and Song W 2006 *J. Appl. Microbiol.* **100** 938
[9] Delcher A L, Harmon D, Kasif S, White O and Salzberg S L 1999 *Nucleic Acids Res.* **27** 4636
[10] Lowe T M and Eddy S R 1997 *Nucleic Acids Res.* **25** 955
[11] Lagesen K et al 2007 *Nucleic Acids Res.* **35** 3100
[12] Burge S W et al 2013 *Nucleic Acids Res.* **41** D226
[13] Bland C et al 2007 *BMC Bioinformatics* **8** 209

Soil phosphorus changes in a grey desert soil in Xinjiang, China, using ^{31}P nuclear magnetic resonance spectroscopy

B Wang[1, 4], X Zhang[2], H Fu[2], H Liu[1], X H Wang[1] and Y B Ma[3, 4]

[1]Institute of Soil Fertilizer and Agricultural Water Saving, Xinjiang Academy of Agricultural Sciences, Urumqi 830091, China
[2]College of Chemistry and Molecular Engineering, Peking University, Beijing 100871, China
[3]National Soil Fertility and Fertilizer Effects Long–term Monitoring Network, Institute of Agricultural Resources and Regional Planning, Chinese Academy of Agricultural Sciences, Beijing 100081, China

E–mail: wbx_wm@126.com (B Wang), mayibing@caas.cn (Y B Ma)

Abstract. This study investigated the effects of different long-term fertilization regimes on the total phosphorus (TP) content and the phosphorus (P) composition of grey desert soil in Xinjiang, China, and clarified the variation in P content under different long-term fertilization regimes. Solution ^{31}P nuclear magnetic resonance spectroscopy (^{31}P NMR) was used to analyze the variation in TP and P composition under different long-term fertilization regimes. The results showed that phosphate monoester was the major form of P and that no other forms of soil organic P (OP) were detected in grey desert soil, Xinjiang, China.

1. Introduction

^{31}P NMR (^{31}P nuclear magnetic resonance spectroscopy) has been widely used to determine the phosphorus (P) forms in soil samples since the early 1980s [1-5]. Compared with traditional sequential extractions [6,7], ^{31}P NMR is able to determine the P forms in soil. Due to this advantage, the technique is widely used. A long-term experiment was carried out in a grey desert soil at the monitoring station in Xinjiang, China. The soil samples in the experiments were used to study organic P (OP) composition in a grey desert soil under different long-term fertilization regimes by solution ^{31}P-NMR. This study evaluated the effects of long-term fertilization (including organic and inorganic fertilizers) on total OP

Content from this work may be used under the terms of the Creative Commons Attribution 3.0 licence. Any further distribution of this work must maintain attribution to the author(s) and the title of the work, journal citation and DOI.
Published under licence by IOP Publishing Ltd

and its composition content, and provides the basis for a reasonable fertilization regime combined with P application in grey desert soil in China.

2. Materials and methods

2.1. Experimental site conditions

The experiment site of the grey desert soil is a part of "the National Long-term Monitoring Network of Soil Fertility and Fertilizer Effects Program", which was established in 1989 to investigate the effects of soil fertility, crop rotation, climate, and the application of organic and inorganic fertilizers on crop yield in the major agricultural areas of China. The experiment site (43°58′23"N, 87°25′58"E) is located on the outskirts of Urumqi, Xinjiang, China, which has a typical continental arid climate. Its climate characteristics are as follows: the effective accumulated temperature, altitude, frost-free period, precipitation, average annual temperature and sunshine hours of the location are 1734°C, 600 m, 156 days, 242 mm, 7.6°C and 2454 hours, respectively. Grey desert soil is Chinese classification system (Calcaric Cambisol by FAO) [8,9]. The experimental plot covers 468 m^2 and all the experimental plots were separated by 1.2 m deep cement slabs. The main soil properties of the 0–20 cm soil layer prior to the experiment were soil organic matter 8.8 $g \cdot kg^{-1}$, pH (soil : water = 1 : 2.5) 8.1, total N (Nitrogen) 0.87 $g \cdot kg^{-1}$, total P (Phosphorus) 0.67 $g \cdot kg^{-1}$, alkaline hydrolysable N (available N) 55.2 $mg \cdot kg^{-1}$, Olsen-P (available P) 3.9 $mg \cdot kg^{-1}$, total K (Potassium) 23.0 $g \cdot kg^{-1}$ and NH_4OAc–K (available K) 288.0 $mg \cdot kg^{-1}$. For decrease the experimental field soil variability prior to the long-term fertilization experiments in 1990, a local wheat was grown for two years (1988–1989 year) without fertilizer. From 1990, wheat maize or cotton in rotation was planted each year at the local experimental site.

2.2. Experimental design

This experiment uses a randomized complete block design. The treatments were: (1) CK (unfertilized), (2) NPK (nitrogen, phosphorus and potassium), (3) NPKM (NPK and manure) and (4) NPKS (NPK and crop straw returned to the field). The NPK fertilizer application rates are listed in Table 1. The 60% of the N, all the P and K fertilizers as basal fertilizers were applied prior to seeding. In crop growing season, the remaining N fertilizer was applied as a top dressing by drip irrigation.

Table 1. Nutrient input between 1990 and 2012.

Treatments	Inorganic fertilizer ($kg \cdot ha^{-1}$)			Organic fertilizer ($kg \cdot ha^{-1}$)	
	N	P	K	Manure (M)	Straw (S)
CK	0.00	0.00	0.00	0.00	0.00
NPK	99.40–241.50	29.20–60.20	18.80–50.40	0.00	0.00
NPKM	29.80–84.90	8.70–22.40	6.70–10.10	30 000	0.00
NPKS	89.40–216.70	24.50–50.80	16.90–42.30	0.00	4 500–9 000

Notes: CK means no fertilizer applied to the crops. The application rates for N, P and K were 99.40 $kg \cdot ha^{-1}$, 29.20 $kg \cdot ha^{-1}$ and 18.80 $kg \cdot ha^{-1}$, respectively, for 1990–1994, and 241.50 $kg \cdot ha^{-1}$, 60.20 $kg \cdot ha^{-1}$ and 50.40 $kg \cdot ha^{-1}$ between 1995 and 2011 in NPK treatment. In NPKS treatment, the contribution of straw to soil nutrients was calculated by using 10% of the current year's utilization rate. In NPKM treatment, organic manure is sheep

IOP Conf. Series: Earth and Environmental Science **346** (2019) 012008 doi:10.1088/1755-1315/346/1/012008

manure, and its nutrient contribution is calculated according to the 15% utilization rate of organic manure in that year. The nutrient utilization rate of the two treatments was the same as that of NPK treatment.

Except in NPKS, all harvested crop biomass was removed from the plots, no crop or with little residues returned to the field. The long-term location monitoring experiment was conducted from 1990 to 2012. Maize, wheat and cotton were be planted in long-term monitoring experimental plot, but only one crop was grown per year. Spring wheat and cotton were planted in 2002 and 2012 year, respectively.

2.3. Test items and methods

2.3.1. NMR analysis. Pretreatment method for the soil samples: A 3.00 g soil sample was filtered through a 2 mm sieve and then placed in a 100 mL centrifuge tube. The sample was extracted with 0.05 mol·L^{-1} Na_2EDTA (sodium ethylene diamine tetra acetic acid) and 0.25 mol·L^{-1} NaOH using a sediment: solution ratio of 1:20 (w: v). The extraction procedure involved shaking the sample for 16 h at 20°C [1,10], and then centrifuging (20°C, 10 000 g, 20 min) and filtering the supernatant with a 0.45 μm sieve before removing the supernatant. The 15 mL supernatant was immediately frozen and dried. The lyophilized extract was re-dissolved in 1 mL of 0.25 mol·L^{-1} NaOH and centrifuged (4 °C, 5 min, 10 000 g). The 0.6 mL supernatant was then mixed with 0.05 mL D_2O (heavy water) and transferred to a 5 mm NMR tube for ^{31}P NMR analysis. The TP (total soil P) content of the soil was determined by inductively coupled plasma-optical emission spectroscopy (ICP-OES) after 5 mL of the sieved supernatants were digested in a H_2SO_4–$HClO_4$ mixture. The chemical shifts were recorded relative to an 85% H_3PO_4 standard ($\delta = 0$ ppm) [4, 11-13]. ^{31}P-NMR analysis was undertaken by the Analytical Instrumentation Center, Peking University, Beijing, China, which used a Bruker AVANCE III 500 MHz NMR (Switzerland).

2.3.2. Conventional analysis. Total NPK, available NPK, soil organic matter, pH, soil bulk density and cation exchange capacity (CEC) were determined and analyzed using conventional agricultural analysis methods [14].

2.4. Data analysis

The test data were analyzed by SAS 9.1 and Excel 2010 software, and the LSD (least significant difference) method was used to compare the data.

3. Results and Analysis

3.1. NMR characteristics of organic phosphorus (OP) in grey desert soil

The ^{31}P NMR spectra show the effects of different fertilization times and fertilizer treatments on the OP of the grey desert soil under the long-term fertilization experiment (Figure 1) in Xinjiang, China. Compared to the results reported by Andrea (2012) [15], Margarita (2004) [16] and Turner (2003) [11,12], the chemical shifts (X axis) at around 6.44 and 6.38 PPM were the characteristic peaks for inorganic orthophosphate and the phosphate monoesters, respectively. The spectra showed in Figure 1 indicate that the phosphate monoesters were the main form of soil OP and other forms of OP were not found in CK, NPKS, NPKM and NPK treatments.

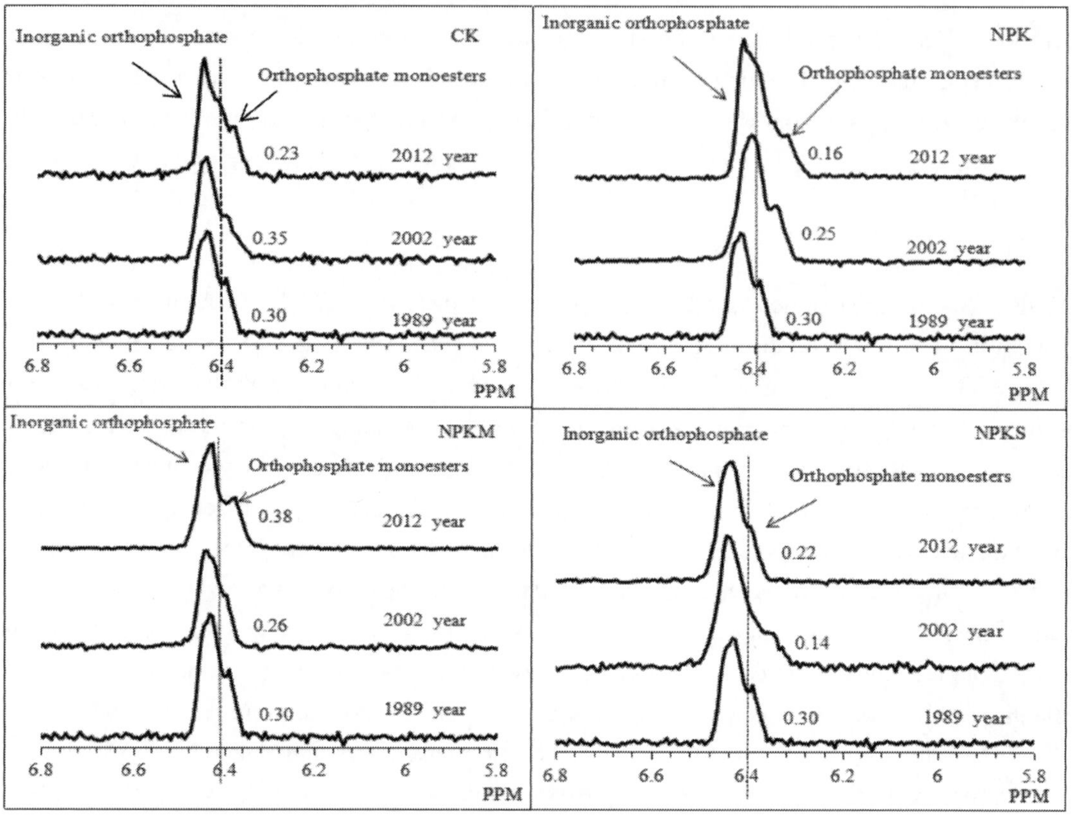

Figure 1. NMR characteristics of soil OP in different treatments and years.

Note: The X-axis is the chemical shift (PPM). The size of the area enclosed by the NMR spectra and the X-axis indicates the content of the corresponding P molecules. The dotted line is the boundary between inorganic orthophosphate and the phosphate monoester. The numbers on the right side of the dotted line represent the ratio of the phosphate monoester to inorganic orthophosphate.

The relative contents of the phosphate monoester varied across different treatments and years (Figure 1). The relative content of the phosphate monoester increased, but then decreased in CK treatment. It gradually reduced in NPK, whereas it decreased first and then increased in NPKS and NPKM treatments. To a certain extent, the change trends for phosphate monoester relative contents indicated that it gradually decreased in CK and NPK because there was no external supplementation with organic materials. In contrast, the relative contents of the phosphate monoester decreased at first, but then increased again in NPKM and NPKS when organic matter was gradually added.

The absolute contents of soil OP and inorganic phosphorus (IOP) can be obtained by calculating the ratio of the P compositions in NMR spectra. The soil P compositions after different fertilization times (1989, 2002 and 2012 years) and treatments were quantified and analyzed in order to further study the impact of different long-term fertilization regimes on the accumulation of various P compositions in grey desert soil across Xinjiang, China.

3.2 P content variations in grey desert soil

The results of a 23-year monitoring experiment showed that TP, OP and IOP contents in different treatments (CK, NPK, NPKM, NPKS) all changed significantly in grey desert soil under long-term fertilization (Figure 2 and Table 2).

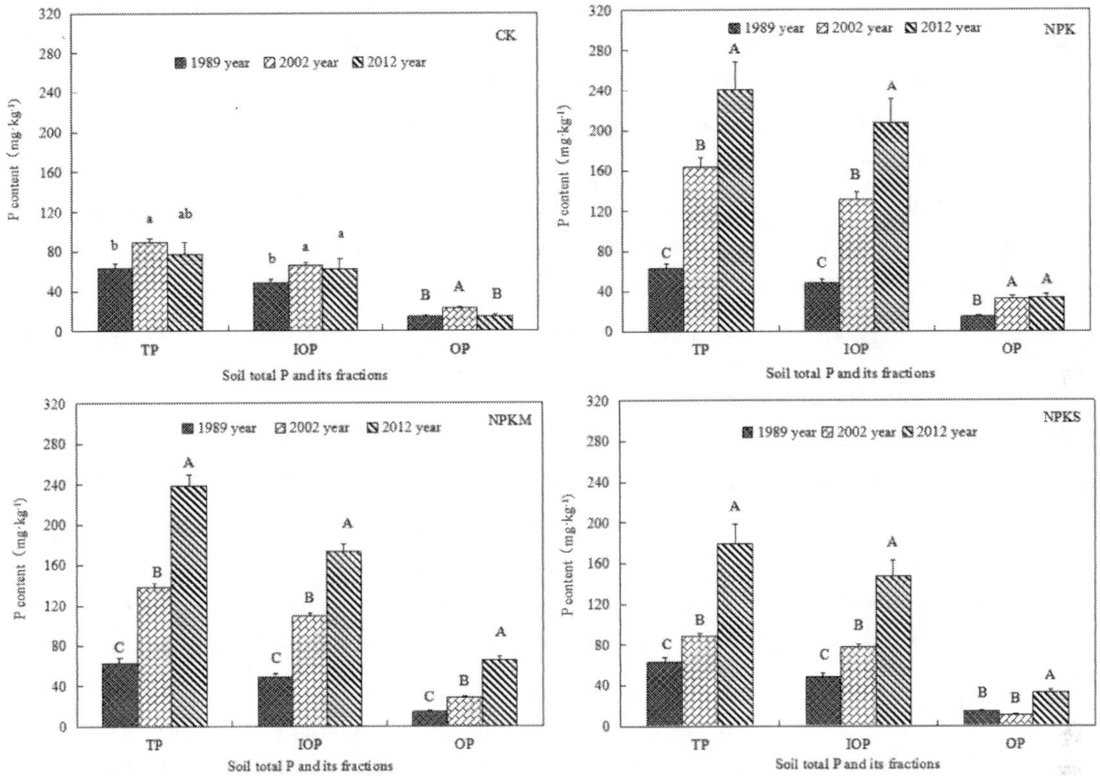

Figure 2. Total soil phosphorus (TP) and its fractions changed with time in different treatments.
Note: The upper and lower case letters above the column show significant differences at $P < 0.01$ and $P < 0.05$, respectively. Soil IOP and OP were inorganic orthophosphate and orthophosphate monoesters, respectively, which were measured by ^{31}P-NMR.

Table 2. Soil phosphorus contents for different years and treatments.

Treatments	Total phosphorus (g·kg^{-1})		Inorganic orthophosphate (mg·kg^{-1})		Orthophosphate monoesters (mg·kg^{-1})	
	2002	2012	2002	2012	2002	2012
IS	0.063±0.004D	0.063±0.004C	48.34±3.03E	48.34±3.03D	14.50±0.91D	14.50±0.91C
CK	0.089±0.003C	0.077±0.010C	66.10±2.22D	62.50±8.19D	23.14±0.78C	14.38±1.88C
NPK	0.164±0.008A	0.241±0.022A	130.96±6.34A	207.70±19.28A	32.74±1.58A	33.23±3.08B
NPKM	0.138±0.003B	0.238±0.009A	109.22±2.75B	172.54±6.65B	28.40±0.72B	65.57±2.53A
NPKS	0.089±0.002C	0.180±0.015B	77.94±1.66C	147.14±12.70C	10.91±0.23E	32.37±2.79B

Note: "IS" is the initial sample in 1989. Different capital letters after the number indicate that they are significantly different at $P < 0.01$.

3.2.1 Total P (TP) content variations in grey desert soil. Although the soil TP contents cannot directly reflect the soil P supply capacity, a low content indicates that the soil is incapable of supplying sufficient P for plant growth [17]. Therefore, the TP content can be used as a potential indicator of soil P fertility. In general, the TP content in farmland does not change greatly over the short term. However, the results of a 23-year old long-term located experiment (Figure 2) show that the TP contents in grey desert soil changed significantly under the different fertilizer regimes during the monitoring period ($P < 0.01$). The TP content increased significantly along the years after fertilization ($P < 0.01$) (Figure 2). The TP contents in NPK, NPKM and NPKS fertilization treatments after 13 (2002 year) and 23 (2012 year) years increased by 2.60 and 3.82 times, 2.18 and 3.78 times, and 1.41 and 2.85 times, respectively. And it compared with CK, the values increased by 1.84 and 3.13 times, and 1.55 and 3.09 times, and 1.00 and 2.33 times. The TP content in CK increased at first, but then decreased. After 13 years, the TP content in CK significantly increased by 1.42 times ($P < 0.05$). Over 23 years (2012 year), the TP content in CK decreased at first, but then increased by 1.22 times (Table 2) compared to the relative basal values in 1989.

3.2.2 IOP content variations in grey desert soil. Generally speaking, the soil IOP compounds are all orthophosphates [18] and the IOP concentration *accounts* for 50%–80% of the soil TP [17]. The P concentration level is directly related to the soil P supplying capacity, and the IOP concentration was significantly affected by fertilization. The results showed that the concentrations of IOP in grey desert soil after the different fertilization treatments changed significantly during the 23 years ($P < 0.01$). Figure 2 shows that the concentrations of IOP in grey desert soil treated with fertilizers increased significantly year by year ($P < 0.01$). However, the concentration of IOP in CK, which received no additional fertilizer, significantly increased at first ($P < 0.05$), but then slightly decreased ($P > 0.05$). After 13 (2002 year) and 23 (2012 year) years, the relative basal values for IOP in NPK, NPKM and NPKS treatments had increased by 2.71 and 4.30 times, 2.26 and 3.57 times, and 1.61 and 3.04 times, respectively. The three treatments increased P after 13 and 23 years by 1.98 and 3.32 times, 1.65 and 2.76 times, and 1.18 and 2.35 times, respectively, compared to CK. The IOP concentration in CK significantly increased at first by 1.37 times (2002 year) ($P < 0.05$), but then the concentration decreased by 1.29 times compared to the relative basal values (2012 year) (Table 2).

3.2.3 OP content variations in grey desert soil. The soil OP content generally accounts for 20%–50% of the soil TP [18]. The OP content and variation have an important effect on the P supplying capacity of a soil and P absorption efficiency of plants. The long-term (23 year) fertilization treatment had a considerable impact on OP in grey desert soil (Figure 2). The OP in NPKM treatment significantly increased each year ($P < 0.01$), and the change trend for OP in grey desert soil after NPKS treatment showed that it remained constant at first (2002 year) ($P > 0.05$), but then increased significantly (2012 year) ($P < 0.01$). The results for the NPK treatment were the opposite. However, the results for CK without fertilization showed that there was at first a significant increase (2002 year) ($P < 0.01$) followed by a significant decrease in OP (2012 year) ($P < 0.01$). The OP content in NPKM after 13 (2002 year) and 23 (2012 year) years increased by 1.96 and 4.52 times, respectively; the OP content in NPKS after

23 years increased by 2.23 times and in NPK after 13 years increased by 2.26 times, but remained almost constant over the following 10 years. The OP content in CK significantly increased by 1.6 times after 13 years, but then decreased significantly over the following 10 years. However, the OP content changes in CK were not significantly different to the base values (Figure 2, Table 2).

3.3 Variation in relative OP contents in grey desert soil

With the increase number of years, the relative soil OP content in CK significantly increased, but then significantly decreased (Figure 3). It decreased significantly year by year in NPK treatment. In NPKM and NPKS treatments, the relative contents of the soil OP decreased significantly, but then significantly increased. The relative OP content change trend in each treatment was consistent with the absolute OP content (Figure 1 and Figure 3). In NPK and CK treatments (Figure 3 left), the relative OP content decreased extremely significantly, but it increased significantly in NPKM and NPKS as the years increased. In 2002, the relative soil OP content was significantly highest in CK (Figure 3 right), but was lowest in NPKS treatment ($P < 0.01$). The relative content values for OP in NPK and NPKM treatments were between CK and NPKS, but the difference between them was not significant. The relative content in CK significantly increased because the OP content was significantly higher. In contrast, the significant increase in TP content meant that the relative OP content significantly decreased in NPKS, NPKM and NPK treatments.

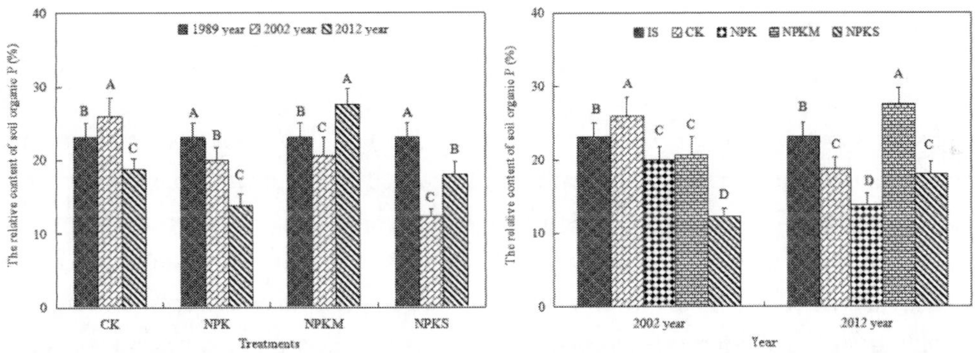

Figure 3. Changes in relative contents of soil OP in different treatments and years.

Note: The upper-case letters above the column show that the values are significantly different at $P < 0.01$. The relative contents of the soil OP (%) = OP/TP*100%.

In 2012, the relative OP content significantly increased in NPKM, but significantly decreased in CK, NPK and NPKS treatments. The main reason is that the OP content in NPKM significantly increased, and the ratio of OP to TP was greater than in CK, NPK and NPKS. Soil OP content was basically constant, but the TP content significantly increased in NPK, which meant that the relative OP contents in NPK and CK treatments significantly decreased. In 2012, compared with 2002, the relative OP content was significantly higher in NPKM and lower in NPK compared with other treatments. The relative contents of CK and NPKS treatments were between NPKM and NPKS, but there was no significant difference between CK and NPKS.

4. Discussion

Under the different long-term fertilization conditions in grey desert soil, Xinjiang, China, the phosphate monoester was the main form of OP, and no other forms of OP were detected, which may be due to the low soil organic P content. In this study, the content of total P and its composition in non-fertilization treatment increased from 1989 to 2002, and gradually decreased from 2002 to 2012. Maybe, the mainly reason was that P supplementation of crop stubble and other organics in early stage, but it gradually declined over time as crop production goes down.

In the same way, the IOP and OP contents in NPKS were lower than in NPKM and NPK treatments, which was the main reason why the TP content in NPKS was significantly lower than in NPK and NPKM. The IOP and OP contents in NPKM were between NPK and NPKS, which also meant that the TP content of NPKM was between the NPK and NPKS treatments.

In 2012, the OP content in NPKM was significantly higher than in NPKS and NPK, but there was not significantly difference between NPKS and NPK. This was mainly because the TP in NPK and NPKM were not significantly different, but were very significantly greater than in NPKS. The TP content in CK decreased, mainly due to the decreases in IOP and OP. The OP content significantly decreased, which had a great effect on TP content. However, the TP contents in NPKS, NPKM and NPK significantly increased. The increases in TP contents in NPK and NPKS were mainly due to the significant increases in IOP and OP contents. In particular, IOP significantly increased in NPK. This showed that over the long term monitoring experiment, applying fertilizer and non-fertilizer application regime had a considerable impact on the OP contents in the soil.

5. Conclusion

The ^{31}P NMR spectra showed that the phosphate monoester was the only form of soil organic phosphorus composition under the different fertilization treatments. The phosphate monoester organic phosphorus form in grey desert soil was not affected by long-term fertilization and the fertilizer regimes. Under long-term organic fertilization, the organic phosphorus content in grey desert soil treated with fertilizers significantly increased, and NPKS (nitrogen, phosphorus, potassium and crop straw returned to the field), NPKM (nitrogen, phosphorus, potassium and manure) by 2.23 and 4.52 times, respectively, compared to non-fertilization.

Acknowledgments

We thank the National Natural Science Foundation of China (41661075, 41161054), Tianshan Youth Project (2017Q006) and Xinjiang Key Research and Development Program of China (2016B02017-4, 2018B01006-1) for providing funds for this work.

References

[1] Cade-Menun B J and Preston C M 1996 A comparison of soil extraction procedures for P NMR spectroscopy *Soil Sci.* **161**(11) 770-785

[2] Hinedi Z R, Chang A C and Lee R W K 1988 Mineralization of phosphorus in sludge-amended soils monitored by phosphorus-31-nuclear magnetic resonance spectroscopy *Soil Sci. Soc. Amer. J.* **52** 1593-1596

[3] Koopmans G F, Chardon W J, Dolfing J, Oenema O P, Meer D V and Riemsdijk W. H 2003 Wet chemical and phosphorus-31 nuclear magnetic resonance analysis of phosphorus speciation in a sandy soil receiving long-term fertilizer or animal manure applications *J. Envir. Qual.* **32**(1) 287-295

[4] Peng X L, Fang H L, Zhan X H, Hao G J, Lv Z W, Ma G J and Gu B 2009 Transformation of phosphorus forms in soils with application of sewage sludge by using phosphorus-31-nuclear magnetic resonance spectroscopy (In Chinese) *J. Agro-Environ. Sci.* **28**(10) 2104-2110

[5] Wang H, Song S H, Zhang J Y and Liu Y X 2017 Research advance in soil phosphorus fractionations and their characterization by chemical sequential methods and ^{31}P-NMR techniques (In Chinese) *J. Plant Nutri. Fert.* **23**(02) 512-523

[6] Bowman R A and Cole C V 1978 An exploratory method for fractionation of organic phosphorus from grass land soils *Soil Sci.* **125** 95-101

[7] Fan Y K and Li S J 2004 A modification of Bowman-Cole' fractionation method of soil organic phosphorus (In Chinese) *Chin. J. Soil Sci.* **06**(35) 743-749

[8] FAO 2006 World reference base for soils resources *World Soil Resource Report No.* **103**. (Rome, Italy)

[9] Wang B, Sun J S, Liu H and Ma Y B 2017 The characteristics of phosphorus adsorption and desorption in grey desert soil of Xinjiang, China *IOP Conf. Series: Earth Envir. Sci.* **77** 012020

[10] Wang X J, Ding S M, Jiang H Q, Bai X L and Fan C X 2008 Characterizing phosphorus in environmental model samples by ^{31}P-NMR (In Chinese) *J. Lake Sci.* **01** 45-50

[11] Turner B L, Mahieu N and Condron L M 2003 Phosphorus-31nuclear magnetic resonance spectral assignments of phosphorus compounds in soil NaOH-EDTA extracts *Soil Sci Soc Am J.* **67** 497-510

[12] Turner B L, Mahieu N and Condron L M 2003 Quantification of myo-inositol hexakisphosphate in alkaline soil extracts by solution P-31 NMR spectroscopy and spectral deconvolution *Soil Sci.* **168**(7) 469-478

[13] McDowell R W, Stewart I and Cade-Menun B J 2006 An examination of spin-lattic relaxation times for analysis of soil and manure extracts by liquid state phosphorus-31 nuclear magnetic resonance spectroscopy *J. Environ. Qual.* **35** 293-302

[14] Bao S D 2007 *Analytical Methods for Soil and Agro-chemistry* (In Chinese) (Beijing: China Agriculture Press)

[15] Andrea G V, Ju¨rgen S, Gerhard G, Johan V, Persson P, Mats J and Reiner G 2012 Changes in organic phosphorus composition in boreal forest humus soils: the role of iron and aluminium *Biogeochemistry* **108** 485-499

[16] Margarita B, Mauricio E, Gerardo G, Dan B and Andrew C 2004 Characterization of chemical phosphorus forms in volcanic soils using ^{31}P-NMR spectroscopy *Commn. Soil Sci. Plant Anal.* **35** 1323-1337

[17] Lu R K 2000 *Analytical Methods of Soil Agricultural Chemistry* (In Chinese) (Beijing: China Agricultural Science and Technology Press)

[18] Lu R K 1998 *Priniciples of soil-plant nutrition and fertilization* (In Chinese) (Beijing: Chemical Industry Press)

Changes in farmers' income in Fuling District of Chongqing China from 2018 to 2028: A grey prediction method perspective

J Q Peng[1,2,3]

[1]Southwest Minzu University, Chengdu, Sichuan, 610041, China
[2]Yangtze Normal University, Fuling, Chongqing, 408100, China

E-mail: 158818697@qq.com

Abstract. Urbanization development can have a great impact on farmers income structure. Through empirical analysis, the results of this paper show that with the development of urbanization the per capita disposable income of farmers in Fuling District will reach70,890.69 yuan per person by 2028, nearly five times that of 2017. The influence of urbanization rate on various income sources of Fuling farmers are as follows: family business income, wage income, transfer income, and property income. The government should increase investment in rural education, promote the improvement of farmers' quality, accelerate infrastructure construction, increase support for agriculture, vigorously promote accurate poverty alleviation, promote the increase of farmers' family business income, and steadily increase farmers' wage income.

1. Introduction

In recent years, the development of urbanization in less developed areas is unprecedented. The incomes of urban workers in the second and tertiary industries grew rapidly, and the absolute incomes of rural residents increased significantly. However, compared with the growth rate of per capita income of urban residents, the growth rate of per capita net income of rural residents is slow. The increase of rural residents' income in underdeveloped areas is an important material prerequisite and foundation for solving the problems of agriculture, rural areas and farmers and building a moderately prosperous society in all respects. How to promote rural economic development, raise people's incomes, and optimize the income structure is very important.

Many scholars have done a lot of research on the relationship between urbanization and farmers' income: Brian C. Briggema thought government policies have a great influence on people's income [1] while Lerman, IddoKan, AyalKimhi and Zvi Lerma focused on affecting factors [2,3] and Liu Hongyuan, Liuke paied more attention on the impact of farmers' income structure in China [4]. Empirical studies also show that urbanization impact farmers' income [5-8]. Most of the literature analyzes the influencing factors of farmers' income growth in a general way. However, there are not many studies focusing on the changes in income structure and the actual development of regional development plans to promote farmers' income increase, and many suggestions are not operational.

This paper analyzes the changes in farmers' income and the impact of urbanization on rural residents' income structure in Fuling district, which has practical reference significance for less developed areas to promote urbanization and the improvement of rural residents' income level and the upgrading and optimization of income structure.

Content from this work may be used under the terms of the Creative Commons Attribution 3.0 licence. Any further distribution of this work must maintain attribution to the author(s) and the title of the work, journal citation and DOI.
Published under licence by IOP Publishing Ltd

2. Characteristics of changes in farmers' income in the process of urbanization in Fuling District

Since 2000, with the rapid development of urbanization Fuling District, the urbanization rate increased from 38.3% in 2000 to 67.18% in 2017. Rural per capita disposable income increased rapidly, from 1892 yuan to13,466 yuan. There are several characteristics:

2.1. The growth rate of urbanization rate is synchronized with the growth of the proportion of farmers' wages and incomes

In 2004, the urbanization rate of Fuling district was 47.46%, and the farmers' wage income was 789.66 yuan per person, accounting for 21.64% of the total income. While in 2017, these numbers are as follows: 67.18%, 3969 and 29.47%. Over the past decade, the urbanization rate has increased by 19.72 percent while farmers' income has increased (table 1).

Table 1. Urbanization rate and rural per capita disposable income (yuan per person).

year	Urbanization	Disposable income	Salary income	Household income	Transfer income	Property income	Income share (%)	Household operating income (%)
2004	47.46	2528.52	789.66	2687.15	164.91	7.54	21.64	73.64
2005	49.00	2779.79	904.16	2872.21	181.12	10.91	22.78	72.38
2006	50.36	2853.95	1119.79	2448.50	262.51	53.20	28.83	63.04
2007	51.80	3499.26	1413.76	2938.43	337.10	57.36	29.78	61.91
2008	53.30	4168.43	1627.13	3725.22	338.85	58.74	28.3	64.79
2009	54.20	4651.13	2051.31	3715.34	517.81	57.20	28.3	58.58
2010	55.80	5548.80	2431.40	4597.30	514.30	75.90	31.91	60.34
2011	57.56	6858.30	2822.27	5440.55	799.76	147.58	30.64	59.08
2012	59.28	7941.92	3398.37	5683.10	1082.24	266.60	32.58	54.49
2013	60.68	8817.00	2602.22	6428.40	2533.53	260.74	22.00	54.37
2014	62.18	9963.00	3011.82	8620.35	2889.59	335.02	20.27	58.02
2015	63.78	11089	3338	9115	3260	380	30.10	82.20
2016	65.45	12253	3655	9646	3788	404	29.83	78.72
2017	67.18	13466	3969	10455	4451	438	29.47	77.64

Data Source: "Fuling Statistical Yearbook 2018"

2.2. Urbanization has led to a slight increase in the proportion of household income

The family business income in Fuling District was 2,687.15 yuan per person in 2004, and in 2017, it was 10,455 yuan per person. The proportion of family business income to total income rose from 73.64% to 77.64%. In general, with the rapid development of urbanization, the total income of farmers' family business income has increased. Due to various entrepreneurship training, farmers have started their own businesses, which has led to an increase in family business income.

2.3. Increased transfer income and property income

The proportion of transfer income and property income in total income has increased. The property income was 164.91 yuan per person in 2004 and 4,451 yuan per person in 2017, accounting for 4.52% to 33.06%, an increase of nearly 30%. The property income accounted for 0.2% in 2004 to 3.30% in 2017.

3. Problems in the growth of farmers' income in the process of urbanization in Fuling District

3.1. Low growth rate of farmers' income

Although farmers' incomes have grown rapidly, the growth rate of farmers' income is still slow relative to the rate of economic growth. In 2017, the per capita disposable income of farmers in Fuling District reached 13,466 yuan, more than double the amount of 1,892 yuan in 2001. However, compared with the growth rate of GDP per capita, the growth rate of farmers is still low. In 2017, the per capita GDP of Fuling District was 85,968 yuan per person, which was 10.2 times of the per capita GDP (8425 yuan per person) in 2001, with an average annual growth rate of 57%.

3.2. Large income gap between urban and rural areas

The income of farmers has increased rapidly, and the income gap between urban and rural areas has also improved. However, the gap between urban and rural incomes is still large. In 2017, the per capita disposable income of urban residents was 33,709 yuan per person, ranking 12 in Chongqing. At the same time, the per capita disposable income of rural residents was only 13,466 yuan per person, and the city ranked 22nd.

3.3. Uneven income structure of farmers

In horizontal comparison, the proportion of farmers' property income is still at a very low level. According to foreign research, property income generally accounts for about 30% of the disposable income of residents, and the income from property of the US agricultural industry accounts for nearly 40% of disposable income. By contrast, the property income of rural residents is seriously low, which was 3.43% in 2016 and 3.30% in 2017.

4. Grey theory prediction of farmers' income changes in the process of urbanization in Fuling District

4.1. GM (1, 1) model

The grey system theory regards all random variables as the amount of gray that changes within a certain range. The method of data processing is applied to the gray quantity, and the unorganized raw data is sorted into a regular generation series and then a new model is established [9]. GM(1,1) represents a first-order, one-variable differential equation model , which is a commonly used gray prediction model in gray systems. It is a model consisting of a first-order differential equation containing only univariates. The test of the model generally adopts three kinds of test methods, namely the residual size test, the correlation test and the post-test difference test. In this study, the post-test difference method is used to test the accuracy of the model.

Assume $x^{(0)}$ For the original data sequence, there is

$$x^{(0)} = (x^{(0)}(1), x^{(0)}(2), \ldots, x^{(0)}(n)) \tag{1}$$

among them: $x^{(0)}(k) \geq 0 (k = 1,2,\ldots,n)$. and $x^{(1)}$ equals $x^{(0)}$ after a cumulative sequence of data, there is $x^{(1)} = (x^{(1)}(1), x^{(1)}(2),\ldots, x^{(1)}(n))$, among them $x^{(1)}(k) = \sum_{i=1}^{k} x^{(0)}(i)(k = 1,2,\ldots,n)$.

$z^{(1)}$ is immediate sequence of mean generation for $x^{(0)}$

$$z^{(1)} = (z^{(1)}(2), z^{(1)}(3),\ldots, z^{(1)}(n)) \tag{2}$$

Then get the GM(1,1) model as

$$x^{(0)}(k) + az^{(1)}(k) = b \tag{3}$$

Substituting a and b into the gray GM (1,1) model, and there is

$$\hat{x}^{(1)}(k+1) = (x^{(1)}(0) - \frac{b}{a})e^{-ak} + \frac{b}{a}, k = 1,2,\cdots,n \tag{4}$$

Finally ,get the gray prediction model

$$\hat{x}^{(0)}(k+1) = \hat{x}^{(1)}(k+1) - \hat{x}^{(1)}(k), k = 1,2,\cdots,n \tag{5}$$

When $k \leq n$, it is the original data simulation value; $k > n$ is the model prediction value.

In this paper the use of a difference test method for testing the accuracy of the model, which is

IOP Conf. Series: Earth and Environmental Science **346** (2019) 012009 doi:10.1088/1755-1315/346/1/012009

generally used after posterior error ratio c and a small error probability p comprehensive evaluation index. Standard deviation of the original sequence $S_1 = \sqrt{\frac{1}{n}\sum_{t=1}^{n}[x^{(0)}(t) - \overline{x^{(o)}}]^2}$, the standard deviation of the residual sequence $S_2 = \sqrt{\frac{1}{n}\sum_{t=1}^{n}[e^{(0)}(t) - \overline{e^{(o)}}]^2}$, Posterior variance ratio $c = \frac{s_2}{s_1}$, And small error probability $p = p(|\varepsilon(k) - \bar{\varepsilon}| < 0.6745 s_1)$. In general, the smaller c is, the better, the larger p is the better.

4.2. Forecast of farmers' income in the process of urbanization in Fuling District

According to the data from "Fuling Statistical Yearbook 2007-2017," GM (1,1) model can be established and related parameters are as follows: S1 = 2593.103, S2 = 206.5335, c = 0.0796, p =1. According to the model accuracy evaluation standard, the accuracy of the GM (1, 1) model is rated as one level, which indicates that the model has good prediction accuracy and can be predicted by the model.

The results are shown in table 2.

Table 2. GM(1,1) prediction value for farmers' per capita disposable income (yuan per person).

ear	2007	2008	2009	2010	2011	2012	2013	2014	2015	2016	2017
Actual value	3499.26	4168.43	4651.13	5548.8	6858.3	7941.92	8817	9963	11089	12253	13466
Forecast value	3588.74	4168.08	4856.93	5636.19	6541.36	7591.91	8811.17	10226.2	11029.2	12566.8	14296.1
Relative error(%)	0.0256	0.0001	0.0408	0.0133	-0.046	-0.04	-0.006	0.026	-0.005	0.024	0.062

According to formula (5), the model equation is:

$$x^{(1)}(k+1) = 16480.45276e^{0.149653k} - 13951.93276 \quad k = 1,2,...,n.$$

As shown in table 2, per capita disposable income model predictions and actual disposable income average error is relatively small, the maximum relative error, 0.0 62 (2017), and were below 0.1%, The prediction results are quite good. According to the same method, the predicted value of per capita disposable income from 2018-2028 is obtained（Table 3）. All the data can be described in Figure 1.

Table 3. Rural residents' per capita disposable income predictive value (yuan per person).

year	2018	2019	2020	2021	2022	2023	2024	2025	2026	2027	2028
Predictiv	1861	2162	2511	2916	3387	3934	4569	5306	5983	6108	7089
e value	5.30	0.33	0.51	4.12	2.11	0.11	0.82	6.73	1.20	1.08	0.69

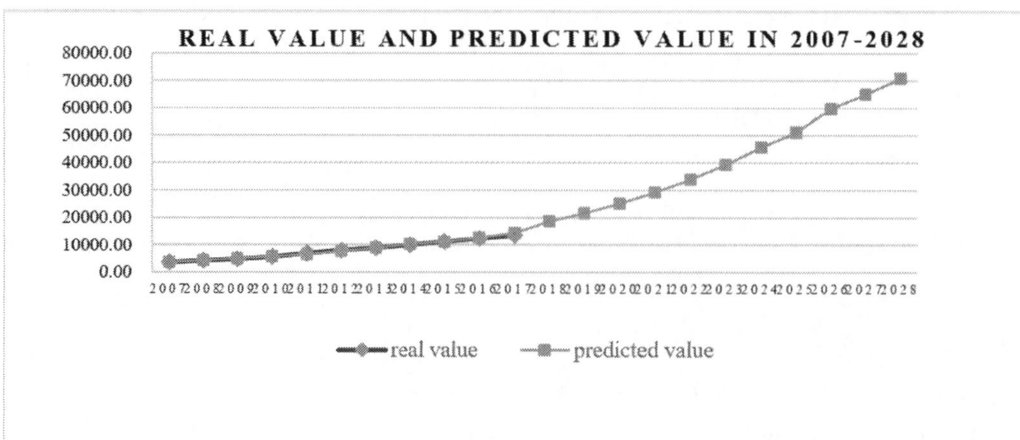

Figure 1. Comparison of predicted and actual value of per capita disposable income from 2007 to 2028 (yuan per person).

4.3. Grey correlation analysis of urbanization and farmers' income structure

Grey correlation theory is an analysis method to judge the degree of correlation between factors according to the similarity degree of development and change of various factors. Grey correlation analysis tries to seek the main relationship between various factors in the system, find out the important factors affecting the target value, and then master the main characteristics of things.

The urbanization rate of Fuling District from 2004 to 2017 is selected as the reference sequence: X0(k)= (X0(1),X0(2),...,X0(10)); The family business income, property income, and transfer income are comparative sequences as a comparison sequence: Xi(k)=(Xi(1),Xi(2),...,Xi(10)), where K=1, 2...10; i =1, 2, 3, 4. Then through the calculation of initial value standardization, absolute difference, maximum value and minimum value, the correlation coefficient and the correlation between farmers' income sources and urbanization rate are obtained (see table 4).

Table 4. Correlation coefficient table.

year	X_1 wage income	X_2 family business income	X_3 transfer income	X_4 property income
2004	1.00000	1.00000	1.00000	1.00000
2005	0.99604	0.99872	0.99768	0.98558
2006	0.98756	0.99474	0.98162	0.82539
2007	0.97593	0.99993	0.96747	0.81305
2008	0.96798	0.99080	0.96817	0.80953
2009	0.95114	0.99158	0.93414	0.81472
2010	0.93706	0.98147	0.93583	0.76119
2011	0.92308	0.97215	0.88626	0.60683
2012	0.90270	0.97035	0.84210	0.45379
2013	0.93356	0.96218	0.66799	0.45973
2014	0.91881	0.93723	0.63609	0.39655
2015	0.90765	0.93259	0.60599	0.36616
2016	0.89712	0.92763	0.56756	0.35185
2017	0.88698	0.91967	0.52562	0.33333
Correlation	0.87040	0.96993	0.82261	0.64126

The results are as follows: family business income, wage income, transfer income and property income. Among them, the relationship between urbanization rate and family business income, wage income, and transfer income is extremely strong, and the correlation with property income is strong.

From the above analysis, the following conclusions can be drawn: (1) In the process of urbanization, the sources of income of farmers have been developed to varying degrees. (2) The growth of wage income of farmers is basically synchronized with urbanization; the proportion of family business income to total income has decreased year by year. (3) The influence of urbanization rate on farmers' income was: family business income, wage income, transfer income, and property income.

5. Suggestions on improving farmers' income structure and promoting farmers' income increase

5.1. Promoting an increase in family business income

Increasing investment in rural education and promoting the quality of farmers. Since family business income is highly correlated with urbanization rate, family business income is still a major component of farmers' income; therefore, it is essential to improve the quality of farmers: First, training for rural adult education. Continue to promote the new farmers' poverty alleviation and entrepreneurship training in Fuling District, the Three Gorges immigrant training, the new professional farmers training, and the management of professional farmers. The second is the training of rural teachers, through training to improve the overall business quality; rural teachers can be regularly exchanged in urban areas to implement urban and rural educational resources sharing. Third, accelerate infrastructure construction and increase support for agriculture. Strengthen financial support for rural areas and provide strong guarantee for agricultural intensification;

5.2. Steadily increasing the wage income of farmers

Adjust and optimize the industrial structure and increase farmers' income. The proportion of farmers' wage income in the income of farmers is gradually increasing, and the role of wage income in farmers' income is becoming more and more important. Urbanization construction can solve the absorption of peasant surplus labor; it should promote the employment of peasants through the project, revolve around regional characteristic industries and products, vigorously develop the township economy, support township enterprises to absorb employment, and increase wage income.

5.3. Promoting property income and transfer income

First, Promoting precision poverty alleviation. Promoting the dynamic management mechanism of poor households, and use poverty alleviation and development as the main way to get rid of poverty. The second is to increase the financial transfer payments to rural public goods. Improve the rural production environment, promote the development of agricultural mechanization, increase the subsidies for the purchase of agricultural machinery, effectively reduce the burden on farmers and increase the income of farmers.

Acknowledgment

This research was financially supported by Southwest Minzu University Graduate Innovative Research Project (Project Name: Research on the factors affecting financial exclusion in Western China ethnic regions, Project No. CX2016BS03).

References

[1] Briggeman B C, Gray A W, Morehart M J *et al* 2007 A new US farm household typology: Implications for agricultural policy *Review Agri. Econ.* **29** 765-82
[2] Lerman Z, Csaki C and Feder G 2004 Evolving farm structures and land use patterns in former socialist countries *Q. J. Natl. Agr.* **43** 309-36
[3] Kan I, Kimhi A and Lerman Z 2006 Farm output, non-farm income, and commercialization in rural Georgia *Electr. J. Agr. Developm. Econ. 3* 276-86
[4] Liu H Y and Liu K 2015 The dilemma of farmers' property income growth and the solution to the problem *Econ. Sys. R.* **2015** 90-4
[5] Huang Z M 2016 An empirical study on the relationship between agricultural modernization, urbanization and farmers' income growth *Stats. Dec.* **2016** 106-9
[6] Li A L and Wang K 2018 Research on the impact of new urbanization on farmers' salary income-Taking Hubei Province as an example *CER* **2018** 153-6
[7] Qian K K and Mo W 2018 New urbanization and farmers' income increase-based on spatial econometric analysis of prefectural cities in China *Eco. Survey* **35** 30-6
[8] Tan F L and Peng Y W 2018 Analysis of the correlation between urbanization, economic growth and farmers' income *J. HAU* **19** 94-100
[9] Wang L 2013 Application of grey GM (1,1) model in population prediction of Xiamen City *JXU T* **2013** 35-38

IOP Conf. Series: Earth and Environmental Science **346** (2019) 012010 doi:10.1088/1755-1315/346/1/012010

First record of *Lyophyllum shimeji* mashrooms in Dazhou

D Xu[1], C Q Sun[1], B Li[1], J J Xu[1], H Zhao[1, 2], Z D Wang[1] and J Ma[1]

[1]Dazhou Academy of Agricultural Sciences, Dazhou, China
D Xu and C Q Sun contributed equally to this work

E-mail: 1207927707@qq.com

Abstract. a new record of *Lyophyllum shimeji* was collected during field trip to masson pine at Dazhou (31.35° N, 107.72° E; 1100 m altitude) district in 16th of October 2017. It was identified using morphological and molecular techniques. Complete description was performed for the collected fresh fruiting bodies and isolated pure culture. Radial growth rate of culture was estimated on media1, media 2, media 3, media 4, media 5 and media 6 (0.196, 0.192, 0.122, 0.092, 0.200, 0.160 cm/ day, respectively). *Lyophyllum* are very close in morphological characters; hence, the identification was confirmed by DNA sequence analysis of the ribosomal 5.8S rRNA gene including the flanking internal transcribed spacers (ITS). The results showed that it was *Lyophyllum shimeji* and this is the first record in other parts of China. The suitable culture medium component was potato 200g, $MgSO_4 \cdot 7H_2O$ 0.5g, glucose 20 g, KH_2PO_4 0.3 g, tryptone 0.5 g, VB_1 20 mg, ager 20 g, ddH_2O 1000ml, PH5-6. These results could provide a new reference for its artificial domestication and cultivation studies.

1. Introduction

Lyophyllum is a world widespread genus, with many edible species such as *Lyophyllum decastes* , *Lyophyllum ulmarium*, *Lyophyllum shimejie* et al, and widely exist in North-temperate and arctic areas, and display broad ecological strategies, ranging from plant decayers to parasite species [1]. *Lyophyllum shimejie* has been recognized as a delicacy for centuries with known distribution in East Asia, especially in Japan and Korea but rare in China except in Yunnan. This species was reported by Peigui Liu from Jinggu,Yunnan in *Saprosma ternatum* woods and *pyrus pyrifolia* woods [2]. Hitherto, there is no any record for *Lyophyllum shimejie* in other parts of China.

Remarkable evolution has been made to affirm phylogenetic relationships in the largest order; Agaricales [3]. However, continued assessment of evolutionary relationships within this order is necessary. The species of *Lyophyllum* are generally regarded as a species complex and a great deal of taxonomic confusion, and are somewhat difficult to be differentiated by morphological descriptions. Thus, molecular techniques such as ribosomal RNA gene sequencing had been developed for their identification [4].

Though the *L. decastes* have traditionally been characterized as saprotrophic, several of these species are later seen to ectomycorrhiza and are difficult to cultivate artificially without host plant. However, *L. shimeji* example, in 1994, Ohta successfully obtained the fruit body through the artificial cultivation using barley grain without the host plant [5]. And Watanabe as well as Yoshida reported the same results at the same time, although the yields have been relatively low [6]. In recent years, a huge infections by the pine wood nematode caused a rapidly decline in the fructification of *L. shimeji* in Dazhou areas and the wild-grown *L. shimeji* is available only in high-class restaurants. Little research has been conducted for its artificial domestication and cultivation in China. Thus, it is necessary to explore its artificial

Content from this work may be used under the terms of the Creative Commons Attribution 3.0 licence. Any further distribution of this work must maintain attribution to the author(s) and the title of the work, journal citation and DOI.
Published under licence by IOP Publishing Ltd

domestication and cultivation.

In order to provide theoretical and applied references for propagation and artificial domestication and cultivation of *L. shimeji*, the purposes of this study were: (1) to morphology and molecular analysis of *Lyophyllum* mushroom that grows in Dazhou, China; (2) to screen the culture medium of *L. shimeji* for its artificial domestication and cultivation studying.

2. Material and method

2.1. Studied material

The fruit-bodies used in this experiment were found at Dazhou (31.35° N, 107.72° E; 780 m altitude) district in 16th of October 2017. Prior to the molecular analysis, the samples for the present study were fully described by expert field mycologists and assigned to taxa according to the literature.

2.2. Medium compositions and culture conditions

Isolation into pure culture was carried out directly after collection from the field sites according to the method of El-Gharabawy *et al* [7]. The pure culture mycelium, obtained by separation and purification from wild *Lyophyllum* fruiting bodies were cultured on different culture mediums (medium1: $CaCl_2$ 0.05 g, $MgSO_4$ 0.15 g, NaCl 0.025 g, $FeCl_3$ (1%) 0.001, KH_2PO_4 0.5 g, $VB_1$100 ug, $(NH_4)_2HPO_4$ 0.25 g, malt extract 10 g, glucose 10 g, citric acid 0.2 g, ager 20 g, ddH_2O 1000 ml; medium2: potato 200 g, $MgSO_4 \cdot 7H_2O$ 0.5 g, glucose 20 g, KH_2PO_4 0.3 g, tryptone 0.5g, VB_1 20 mg, ager 20 g, ddH_2O 1000 ml, PH5-6; medium3: KH_2PO_4 1.0 g, $Na_2MO_4 \cdot 2H_2O$ 0.0027 g, $MgSO_4 \cdot 7H_2O$ 0.5 g, $CaCl \cdot 2H_2O$ 0.05 g, H_3BO_3 0.028 g, $VB_1$0.0001 g, $MnCl_2 \cdot 2H_2O$ 0.003 g, $ZnSO_4 \cdot 7H_2O$ 0.0023 g, EDTA 0.02 g, $CuCl_2 \cdot 2H_2O$ 0.0063 g, glucose 20.0 g, ager 15 g, ddH_2O 1000 ml, pH5.4±0.2 (25°C); medium4: malt extract 20 g, soy tryptone 1 g, glucose 20 g, ager 15 g, ddH_2O 1000 ml; medium5: yeast extract 3 g, glucose 20 g, KH_2PO_4 3 g, $MgSO_4 \cdot 7H_2O$ 5 g, tryptone 5 g, VB_1 20 mg, ager 20 g, potato 200 g+pine 80 g were soaked in distilled water at constant temperature 100°C for 30 min, filter and let stand for 1000 ml, PH 5.5-6.0; medium6: potato 200 g, ager 20 g, glucose 20 g, ddH_2O 1000 ml) under sterile conditions.

2.3. Determination of Radial growth rate (RGR)

Isolation plates were incubated at 25°C and pure cultures were maintained on potato dextrose agar (PDA) slopes at 4°C. Radial growth rate (RGR) was quantified on 9 cm petridishes of six different culture mediums using 1cm discs of actively growing cultures at 25°C.

2.4. DNA extraction and polymerase chain reaction amplification parameters

Genomic DNA was extracted from the fresh fruiting body by the cetyltrimethylammonium bromide (CTAB) method [8]. The oligonucleotide primers (ITS1:5'-TCCGTAGGTGAACCTGCGG-3' and ITS4: 5'-TCCTCCGCTTATTGATATGC-3') used for amplification and the PCR amplifications were performed in ABI 9700 thermal cyclers. The PCR mixture with a total volumes of 25 μl and contained 0.5 μl of each primer, 0.5μl template (20~50 ng), 2.5 μL 10 × Taq buffer (Mg^{2+}plus), 0.5 μl dNTP (2.5 mM), and 0.5 μl Taq DNA polymerase (Tiangen Biotech, Beijing Co., Ltd.).Thermal condition was performed with initial denaturation at 94°C for 4 min, followed by 35 cycles at 94°C for 45s, 55°C for 45s and 72°C for 1 min, then final extension at 72°C for 10 min. PCR products were analyzed and separated on 1.0–1.5% (w/v) agarose gels, stained with ethidium bromide, and visualized using UV light. The expected PCR products were purified using the SanPrep Column DNA Gel Extraction Kit (Shanghai Biological Technology Co., Ltd.). Obtained ITS nucleotide sequences were subjected to a BLAST search against the NCBI database (http://www.ncbi.nlm.nih.gov/) to match the best similarities with other related ITSs on database.

3. Results

3.1. Morphological characteristics

The taxonomic position: Fungi – Basidiomycota – Agaricomycetes – Agaricales –Tricholomataceae-Lyophyllum.

The samples were collected during field trip to masson pine at Dazhou (31.35° N, 107.72° E; 1100 m altitude) district in 16th of October 2017. The fruit bodies were gregarious with the base of a bulged waist and usually in areas where rotten branches and leaf litter collects; cap is 2 to 6 cm across; cap surface is smooth and the middle part slightly convex or concave; Gills sinuate; Stem is 2 to 8 cm long and 1.5 to 3.5 mm in diameter and white at base with no ring (figure 1).

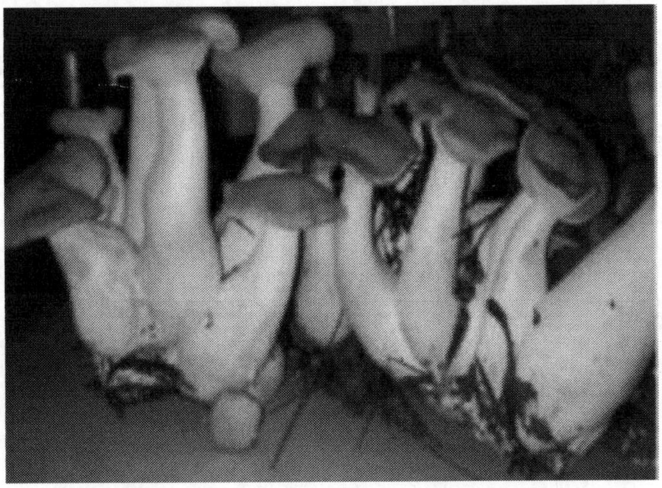

Figure 1. Morphology of *Lyophyllum shimeji.*

3.2. Molecular characterization of L. shimeji

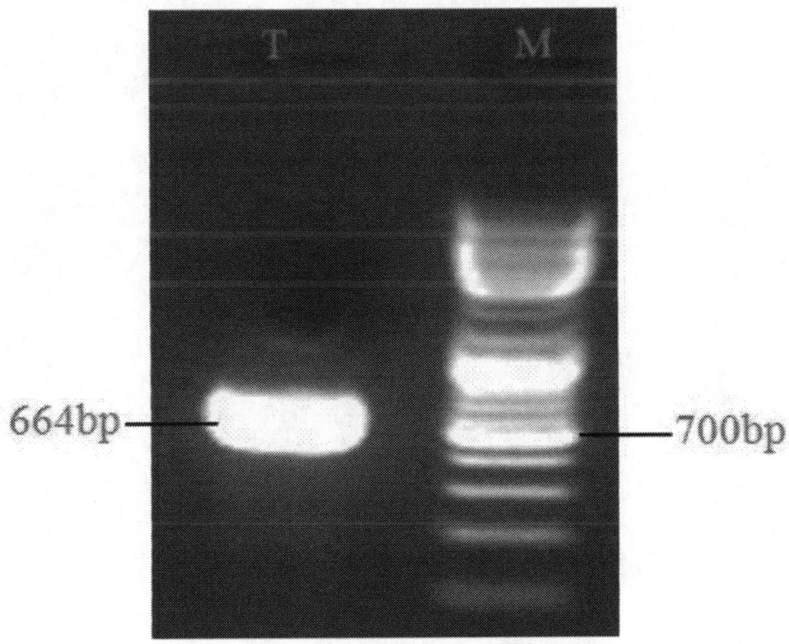

Note: M: marker; T: target ITS

Figure 2. PCR of the electrophoresis pattern.

The best DNA sequence similarities with our ITS nucleotide sequence (664 bp) (figure 2) were obtained from NCBI GenBank and aligned results (figure 3) showed that the sample which collected at Dazhou (31.35° N, 107.72° E; 1100 m altitude) district in 16th of October 2017 was *L. shimeji.*

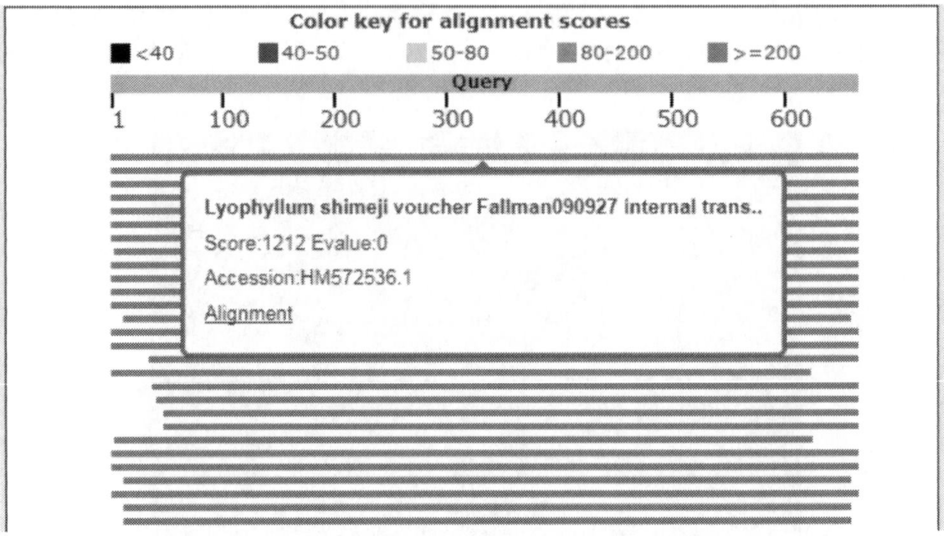

Figure 3. target ITS nucleotide sequence alignment results.

3.3. *The mycelium morphology and culture characteristics analysis*

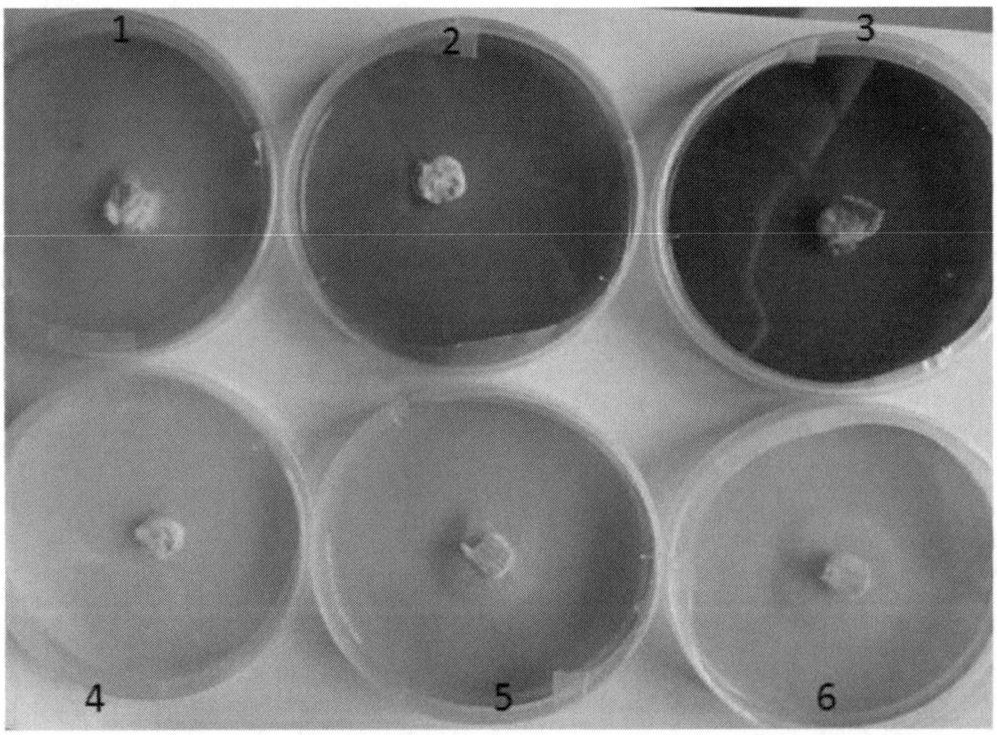

Figure 4. Mycelium morphology of 3 days later after inoculation in different mediums. 1 stands for medium 1, 2 stands for medium 2, 3 stands for medium 3, 4 stands for medium 4, 5 stands for medium 5, 6stands for medium 6.

The mycelium morphology of different culture mediums were shown in figures 4 and 5, and the culture characteristics analysis were shown in table 1. The mycelia began to germinate 2 days later on medium1, medium 5 and medium 6 while the mycelia began to germinate 3 days later on medium2, medium 3 and medium 4. The mycelia are all white except on medium 3, in which the mycelium is canary. The mycelia are denser on medium2, medium 3 and medium 5, dense on medium 1and medium 6, sparsely on medium 4. Culture showed more aggressive appearance on medium2 with thick growing tenacious mycelium.

1 2 3

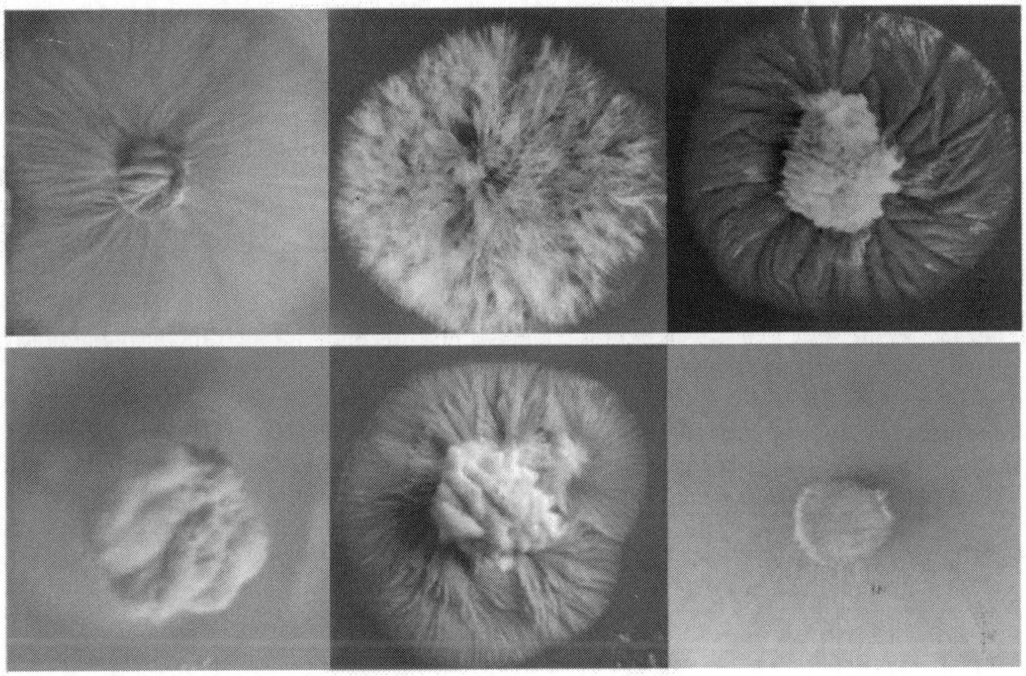

4 5 6

Figure 5. Mycelium morphology of 7 days later after inoculation in different mediums. 1 stands for medium 1, 2 stands for medium 2, 3 stands for medium 3, 4 stands for medium 4, 5 stands for medium 5, 6stands for medium 6.

Table 1. Comparison of characteristics of different parent medium.

mediums	mycelium germination time(d)	mycelium dense	mycelium color	colony marginal morphology
1	2	++	white	neatly
2	3	+++	white	neatly
3	3	+++	canary	neatly
4	3	+	white	irregular
5	2	+++	white	neatly
6	2	++	white	irregular

Note: +++ denser, ++ dense, + sparsely.

3.4. Radial growth rate (RGR) of mycelium

The radial growth rate (RGR) of mycelium were shown in table 2. Although the RGR were not significant different at 0.05 level on medium 2 and medium 5, the RGR on medium 2 and medium 5 were significantly different on medium 3. According to the RGR and the mycelium morphology and culture characteristics analysis, medium 2 is the best culture medium for *Lyophyllum shimeji*.

Table 2. Radial growth rate (RGR) of mycelium on different mediums.

mediums	Radial growth rate ($cm^{-1}d$)
1	0.196±0.001a
2	0.192±0.004a
3	0.122±0.002c
4	0.092±0.002d
5	0.200±0.005a
6	0.160±0.003b

Note: the data are the mean ± standard error of 10 replicates, and the different lowercase letters indicated significant differences at 0.05 level.

4. Discussion

The ectomycorrhizal fungus *Lyophyllum shimeji* (Kawam.) Hongo is one of the most valuable edible mushrooms in East Asia, especially in Japan but rare in China [9]. Hitherto, there is no any record for *Lyophyllum shimejie* in other parts of China except in Yunnan which reported by Peigui Liu in 1993. In this study, morphological characteristics and molecular techniques results showed that the samples fund at Dazhou (31.35° N, 107.72° E; 780 m altitude) district in 16th of October 2017 was *Lyophyllum shimejie*. The existence of *Lyophyllum shimejie* in other parts of China is reported for the first time.

The ectomycorrhizal fungi are difficult to cultivate artificially without the host plant. However, *L. shimeji* example, in 1994, Ohta successfully obtained the fruit body through the artificial cultivation using barley grain without the host plant, although the yields have been relatively low. In recent years, a huge infections by the pine wood nematode caused a rapidly decline in the fructification of *L. shimeji* in Dazhou areas and the wild-grown *L. shimeji* is available only in high-class restaurants. Little research has been conducted for its artificial domestication and cultivation in China. Therefore, in order to explore its artificial domestication and cultivation without the host plant, we screen the suitable medium for this fungus in this report. Results showed that the suitable medium component was potato 200g, $MgSO_4 \cdot 7H_2O$ 0.5g, glucose 20 g, KH_2PO_4 0.3 g, tryptone 0.5g, VB_1 20 mg, ager 20 g, ddH_2O 1000 ml, PH5-6. These results could provide a new reference for its artificial domestication and cultivation studies.

Acknowledgments

The authors would like to thank the key Research Projects of National modern agricultural industrial technology system Sichuan edible fungi team (Sichuan Agriculture [2009] No. 75) for providing financial support.

References

[1] Bellanger J M, Moreau P A, Corriol G, Bidaud A, Chalange R, Dudova Z *et al* 2015 Plunging hands into the mushroom jar: A phylogenetic framework for Lyophyllaceae (agaricales, basidiomycota) *Genetica* **143** 169-94

[2] Peigui L 1993 Introduction of a valuable edible fungas from Yunnan-*Lyophyllum shimeji* (kawam.) *Edible Fungi of China* **12** 2-29

[3] Binder M, Larsson K H, Matheny P B and Hibbett D S 2010 Amylocorticiales ord. nov. and Jaapiales ord. nov.: Early-diverging clades of Agaricomycetidae dominated by corticioid forms *Mycologia* **102** 865-80

[4] Stott K, Desmerger C and Holford P 2005 Relationship among Lepista species determined by CAPS and RAPD *Mycological Research* **109** 205-11

[5] Ohta A 1994 Production of fruit-bodies of a mycorrhizal fungus, *Lyophyllum shimeji*, in pure culture *Mycoscience* **35** 147-51

[6] Watanabe K, Kawai M and Obatake Y 1994 Fruiting body formation of *Lyophyllum shimeji* in pure culture *Mokuzai Gakkaishi* **40** 879-82

[7] El-Gharabawy H, Detheridge A, El-Fallal A, El-Sayed A and Griffith G 2016 Analysis of wood

decay and ligninolysis in Polyporales from the Nile Delta region of Egypt *Mycosphere* **7** 392-404

[8] Lee S B and Taylor J W 1990 Isolation of DNA from fungal mycelia and single spores *PCR Protocol. A Guide to Method and Applications*, M A Innis, D H Gelfand, J J Sninsky and T J White Ed *Genetics and Evolution* **34** 282-7

[9] Larsson E and Sundberg H 2011 Lyophyllum shimeji, a species associated with lichen pine forest in northern fennoscandia *Mycoscience* **52** 289-95

Design and test of tilling depth measurement device based on angle sensor

C Li[1,2], Y He[1,2], Q Feng[1,2], K Jiang[1,2], X Wang[1,2,3] and C Zhang[1,2]

[1]Beijing Research Center of Intelligent Equipment for Agriculture, Beijing 100097, China

[2]National Research Center of Intelligent Equipment for Agriculture, Beijing 100097, China

E-mail: wangx@nercita.org.cn

Abstract. Inter-tillage is one of the most important methods to increase crop yield, reasonable tillage depth can improve crop yield and economic benefits. In the research, an angle sensor-based tillage depth measurement device was designed and its function was tested. The tillage depth measurement device was mainly composed of a supporting palm, an angle sensor, a spring, shaft components, a connection block, a regulating damper and so on, through reasonably designing the geometric structure of the supporting palm in the tillage depth measurement device and combining with the angle sensor, the information of tillage depth was converted into electrical signal, and the linear relationship between tillage depth and electrical signal was obtained. In laboratory test, the calibration and accuracy verification experiments of the tillage depth measurement device were carried out. The correlation coefficient (r) of reference values and measured values using the tillage depth measurement device was 0.9996 with the standard error (SE) of 0.08 cm. In field test, the accuracy of the tillage depth measurement device was also verified with the maximum absolute error of 0.8cm. The research results showed that the measurement device based on angle sensor had high measurement accuracy, providing a simple and direct tillage depth measurement method.

1. Introduction

Tillage depth is an important parameter to measure the operating performance of farm machinery, reasonable tillage depth can improve crop yield and economic benefits [1,2]. Traditional manual measurement method of tillage depth has high labor intensity, low efficiency and large measurement error, can't obtain the tillage depth status in real time. With the applications of electronic, sensor and information technology in the tractor, the traditional manual measurement method of tillage depth has been replaced by various electronic measurement sensors, instruments or devices gradually.

Based on the literature about tillage depth measurement in China and abroad, the existing tillage depth measurement technology mainly includes the use of ultrasonic sensor to measure the tillage depth [3], which is vulnerable to the influences of uneven ground, crop branches and stubble, and such influences are difficult to be removed. The technology also contains the use of contour wheel (land wheel) or copying slip palm to measure the tillage depth [4-7], the contour wheel or the copying slip palm is installed on a plough frame, when the plough falls, the contour wheel or the copying slip palm still keeps contact with the cultivated land. After the plough is deep into the ground, the contour wheel or the copying slip palm is raised relatively, leading to rotation angle changes of the contour wheel or the copying slip palm, at last, the tillage depth was obtained by converting the rotation angle changes,

Content from this work may be used under the terms of the Creative Commons Attribution 3.0 licence. Any further distribution of this work must maintain attribution to the author(s) and the title of the work, journal citation and DOI.
Published under licence by IOP Publishing Ltd

this method is more practical, but needs to install a set of additional devices. The use of position sensor is also a main technology to automatically measure the tillage depth of agricultural tools [8-11], the tillage depth is calculated by measuring the angle change of the lifting arm of the suspension mechanism, this method is convenient for installation and integration, but relies on the geometric relationship of the suspension mechanism, and the calibration method is complex and tedious.

In the presented research, a tillage depth measurement device was constructed based on angle sensor and a corresponding method of tillage depth measurement was studied, and the measurement precision was tested to verify the feasibility of the device and method.

2. Design of the tillage depth measurement device

The tillage depth measurement device was mainly composed of a supporting palm, an angle sensor, a spring, shaft components, a connection block, a regulating damper, etc., whose three-dimensional structure is shown in figure 1. The measurement device was installed on a tillage machine with a certain horizontal distance from the tillage shovel of the tillage machine, and the supporting palm of the measuring device was in direct contact with the cultivated land surface. When the tillage shovel was deep into the ground, the supporting palm was in direct contact with the cultivated land surface throughout, it was raised relatively, and a certain height difference between the supporting palm and the tillage shovel was produced, through the angle sensor, the measurement device measured this height difference, namely the tillage depth, the schematic of the device in different working positions is shown in figure 2, *H0* is the initial position of the supporting palm when it was naturally suspended and in direct contact with the cultivated land surface, *H* is the working position of the supporting palm, *ΔH* is the height difference between the supporting palm and the tillage shovel, namely the tillage depth. By reasonably designing the geometric structure of the supporting palm in the tillage depth measurement device and using the angle sensor, the information of tillage depth was converted into electrical signal, and the linear relationship between tillage depth and electrical signal was obtained.

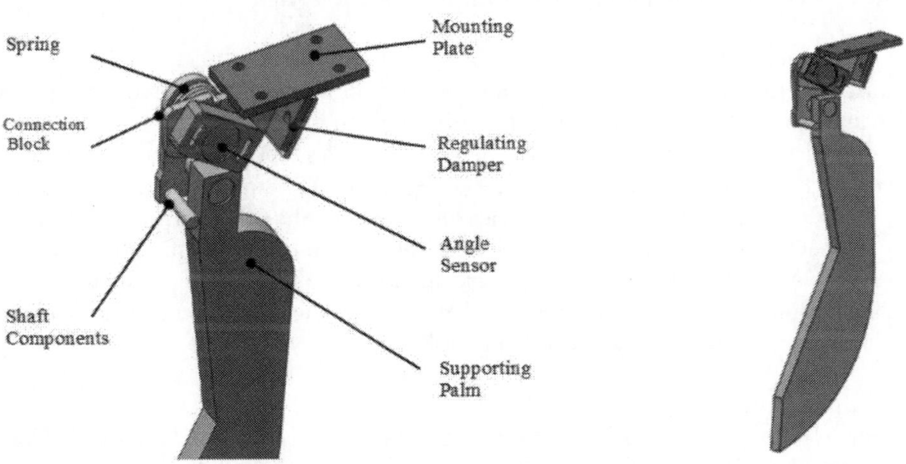

Figure 1. The three-dimensional structure of the tillage depth measurement device. (a) Drawing of partial enlargement and (b) Full thumbnail.

The angle sensor converted angle data into electrical signal, the angle data and the electric signal showed a linear relation. The geometric structure of the supporting palm in the tillage depth measurement device was reasonably designed, so that the vertical coordinate of the edge trajectory in contact between the supporting palm and the cultivated land surface had a linear relationship with the rotation angle of the supporting palm, and the rotation angle of the supporting palm was measured by the angle sensor accurately, therefore, the vertical coordinate of the edge trajectory in contact between the supporting palm and the cultivated land surface had a linear relationship with the electrical signals, and then, there appeared a linear relationship between the electrical signal and the tillage depth of the tillage shovel. The linear regression analysis of the electrical signal and the tillage depth measured by the tillage depth measurement device was carried out, and the linear regression equation was established.

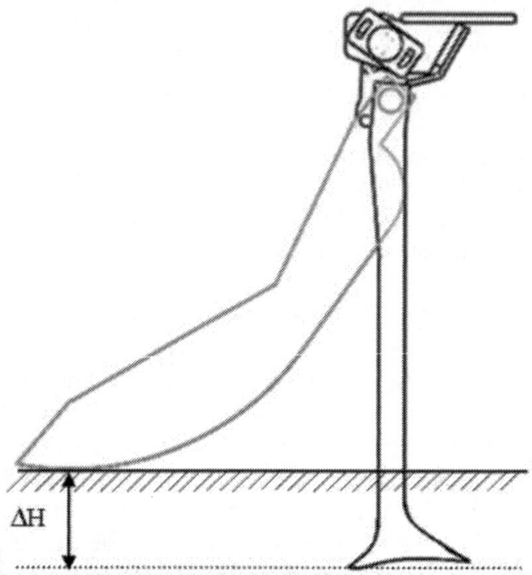

Figure 2. The schematic of the device in different working positions.

3. Calibration experiment of the tillage depth measurement device

3.1. Experimental materials

In the calibration experiment, the tillage depth measurement device, a lifting platform, a support frame, a ruler, a data collector and a computer were utilized. The lifting platform kept the supporting palm of the tillage depth measurement device at different heights, and the supporting palm was always in contact with the surface of the lifting platform during the experiment. The support frame was used to fix the tillage depth measurement device. The ruler measured the height of the lifting platform. The data collector was connected with the angle sensor of the tillage depth measurement device to obtain the digital voltage signals of the angle sensor in real time and send the digital voltage signal to the computer through wireless communication. The computer received the voltage data sent by the data collector in real time to fit the linear model of tillage depth and voltage.

3.2. Experimental methods

The initial height of the supporting palm when it was naturally suspended was taken as the height datum. In the course of working, the supporting palm was always in contact with the surface of the lifting platform, the lifting range of the lifting platform was 1.0～15.5 cm. When the lifting platform rose or fell with the interval of 0.5 cm, the voltage data of the tillage depth measurement device was recorded accordingly, there were 30 voltage data totally for one way which was repeated for 3 times.

3.3. Data analysis and modeling

Averaged 3 measured voltage values for each height, and regarded the average voltage value as the voltage value of the corresponding height, there were 30 groups of height and voltage values totally. Taking the height value as the tillage depth, the linear fitting of the tillage depth and the electric signal was carried out, and the linear model of the tillage depth and electric signal was established, the relation between the tillage depth and the electric signal was shown in figure 3, the fitting determination coefficient (r^2) was 0.9998 with the standard error (*SE*) of 0.07 cm, The linear equation of the relationship between the tillage depth and voltage is shown in equation (1),

$$y = -10.911 \cdot x + 20.655 \quad (1)$$

where, *y* represents the tillage depth with the unit of cm. *x* is the voltage, and the unit is V.

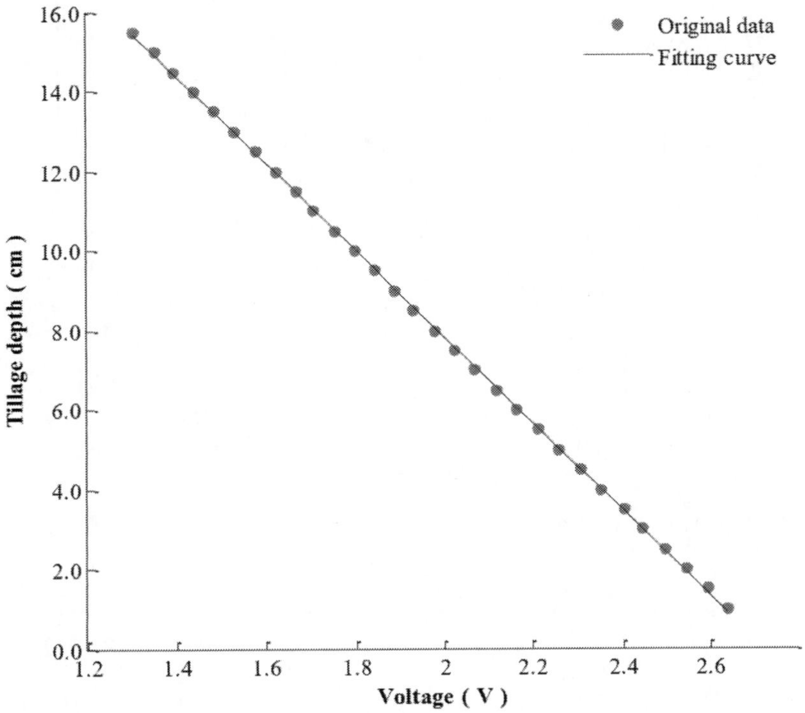

Figure 3. The fitting effect chart of the tillage depth and the electric signal.

4. Laboratory verification test

4.1. Experimental materials

The experimental materials were the same as the calibration experiment. The difference was that the computer brought the voltage data into the linear model of tillage depth and voltage, output and saved the tillage depth data.

4.2. Experimental method

Took the height of the supporting palm as the height datum when it is naturally suspended, the supporting palm was in contact with the surface of the lifting platform, the lifting range of the lifting platform was 0.5~12.5 cm. Once the lifting platform rose or fell, the current height *Hn* of the lifting platform (the support palm) and the tillage depth data *Dn* measured by the tillage depth measurement device were recorded, there were 25 groups of depth data, n was equal to 1, 2, 3..., 25.

4.3. Data analysis

Correlation analysis was conducted between the actual height *Hn* (reference value) of the supporting palm and the tillage depth data *Dn* (measured value) measured by the tillage depth measurement device, the correlation coefficient (*r*) was 0.9996 with the *SE* of 0.08 cm. The correlation analysis effect diagram of the reference value and the measured value is shown in figure 4. The experimental results showed that the tillage depth measurement device had high measuring accuracy under static condition.

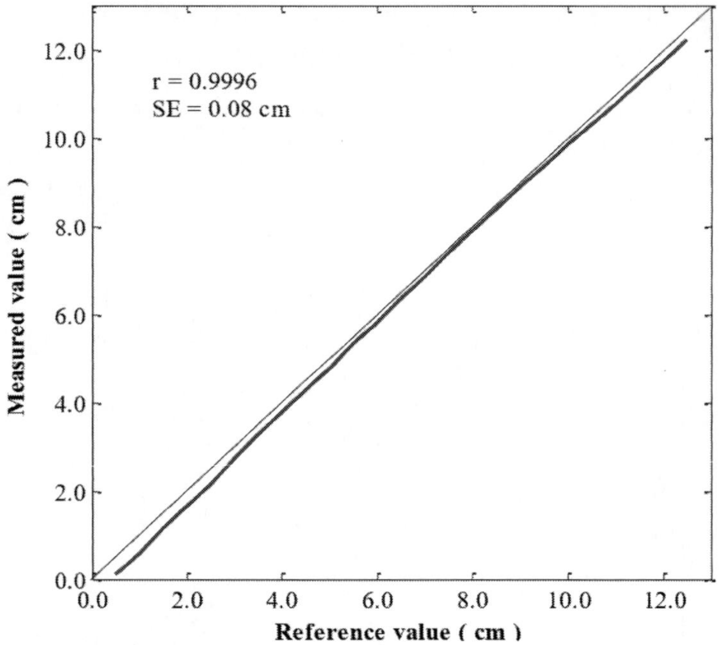

Figure 4. The correlation between the measured value and the real value in laboratory verification test.

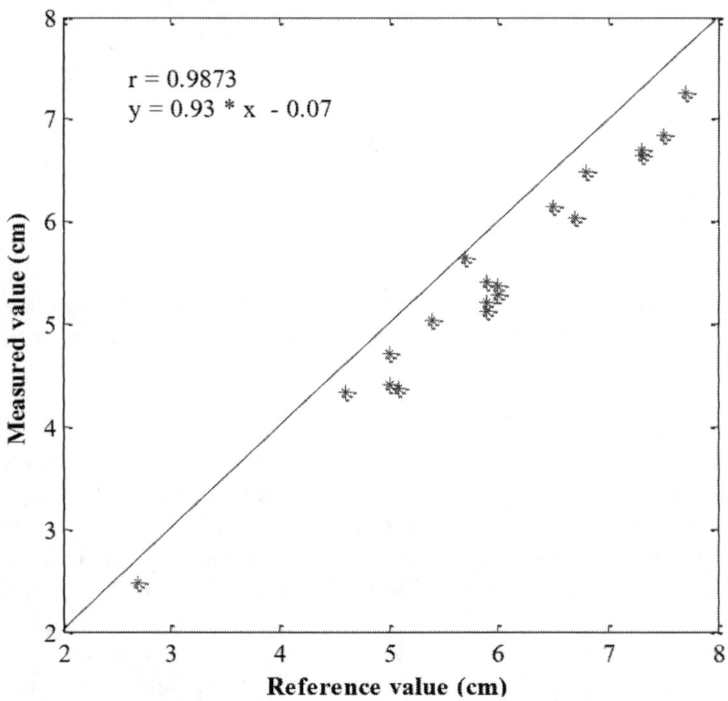

Figure 5. The correlation between the measured value and the real value in field validation test.

5. Field validation test

In the field condition, the measurement device was installed on the front side of a tillage machine and driven by a tractor. Totally, 20 measuring points were selected in the test area. Firstly, the tillage depth of each test point was measured with this measurement device as the measured value. Then, the tillage depth of each test point was measured manually as the reference value, a total of 20 groups of measured values and reference values were compared. Correlation analysis was conducted between reference value and measured value, the correlation coefficient (r) was 0.9873. The correlation analysis effect diagram of the reference value and the measured value is shown in figure 5, x means the reference value and y is the measured value. The experimental results showed that the measurement device had a high measuring accuracy under the field dynamic condition, and the measuring absolute error was no more than 0.8 cm.

6. Conclusion

In this study, a tillage depth measurement device based on angle sensor was designed. By reasonably designing the geometric structure of the supporting palm in the tillage depth measurement device, and using the angle sensor, the information of tillage depth was converted into electric signal, and the relationship between the tillage depth and electric signal was linear. According to the measurement principle analysis and calibration results, the fitting determination coefficient (r^2) was 0.9998 with the standard error (*SE*) of 0.07 cm.

Laboratory verification test results showed that there was a high correlation between the measured value and the reference value with the correlation coefficient (r) of 0.9996 and the standard error (*SE*) of 0.08cm under static conditions, the tillage depth measurement device had high measurement accuracy. Field verification test results showed that the measuring absolute error of the measurement device was no more than 0.8cm in dynamic condition, the measurement accuracy was high.

In the working process of the tillage depth measurement device, the supporting palm was in direct contact with the cultivated land surface, which was easy to be worn out in long-term use. Therefore, it was necessary to replace the support palm regularly and calibrate the tillage depth measurement device.

Acknowledgments
The authors gratefully acknowledge the National Key Research and Development Plan (2017YFD0700901-2), National Key Research and Development Plan (20170701605) and Scientific and Technological Innovation Team of Beijing Academy of Agricultural and Forestry Sciences (JNKST201619) for supporting this research.

References

[1] Jia H, Guo M, Yu H, Li Y, Feng X, Zhao J and Qi J 2016 An adaptable tillage depth monitoring system for tillage machine *Biosysems. Eng.* **151** 187-199
[2] Yao S, Teng X and Zhang B 2015 Effects of rice straw incorporation and tillage depth on soil puddlability and mechanical properties during rice growth period *Soil. Tillage. Res.* **146** 125-132
[3] Li Z 2000 Research and design of operating depth measurement apparatus for farm implement *Trans. Chin. Soc. Agric. Mach.* **3** 88-91
[4] Mouazen A M, Anthonis J, Saeys W and Ramon H 2004 An automatic depth control system for online measurement of spatial variation in soil compaction, part 1: Sensor design for measurement of frame height variation from soil surface *Biosysems. Eng.* **2** 139-150
[5] Wu W and Ma R 2007 Design of the automatic control of the plowing depth of the integrated plowing set *J. Agric. Mech. Res.* **9** 77-79
[6] Li L, Li J, Li X, Huang Y and Wang Y 2001 Design researches on tiling depth electron testing system *J. Shihezi Univ. Nat. Sci.* **3** 246-248
[7] Xie B, Mao E and Tan Y 2006 Development of electronic hydraulic hitch controller based on CAN bus *Mach. Tool Hydraul.* **8** 185-187
[8] Lee J, Yamazaki M, Oida A, Nakashima H and Shimizu H 2000 Field performance of proposed foresight tillage depth control system for rotary implements mounted on an agricultural tractor

J. Terramechanics **2** 99-111
[9] Bentaher H, Hamza E, Kantchev G, Maalej A and Arnold W 2008 Three-point hitch-mechanism instrumentation for tillage power optimization *Biosysems. Eng.* **1** 24-30
[10] Singh C D and Singh R C 2011 Computerized instrumentation system for monitoring the tractor performance in the field *J. Terramechanics* **5** 333-338
[11] Du Q, Xiong X and Wei J 2008 Design and experiment on the control system of electro-hydraulic plow depth of tractor hydraulic hitch mechanism *Trans. Chin. Soc. Agric. Mach.* **8** 62-65

Plant community structure and species diversity in Lingang wetland park in Tianjin China

H Q Liu[1, 4], Y Kan[1], Y Liu[1], W M Zhang[1], H Y Shan[1], Y H Wang[2], F H Liang[2] and X L Liu[3]

[1]College of Horticulture and Landscape, Tianjin Agricultural University, Tianjin 300384, China
[2]Tianjin Lvzhiben Biotechnology Co. LTD, Tianjin, 301804, China
[3]Tianjin Xingning Industrial Co. LTD, Tianjin, 301500, China

E-mail: wjxlhq@126.com

Abstract. Lingang wetland park is a newly built wetland park in Lingang district, Tianjin, China. The community composition, community structure, frequency of occurrence of arbor and shrub layer and dominant species of plant community in lingang wetland park were investigated. The results showed that the community of Lingang wetland park was mainly characterized by the structure of trees-shrubs-lawns, and the ratio of species and number of evergreen trees to deciduous trees was 2:15 and 1:23. Wetland park is rich in plant diversity, the dominance index at park's east gate and "qushuihuajing" were 0.8183 and 0.8152, which were far higher than other quadrants and the dominant species is golden elm. Among the tall trees, *Ulmus pumila*, *Salix babylonica*, *Aureus locustae* with the highest application frequency accounted for 23.8%. For shrubs, lonicera japonica, *Euonymus japonicus* and *Lonicera maackii* had the highest frequency of application. *Festuca elata* and *Hemerocallis fulva* had highest application frequency in lawns. They accounted for only 13.2% in herbs. It indicated that only several plants in the park have high frequencies.

1. Introduction

This park green space is the main part of urban green space, which can not only beautify the city, but also balance the urban weather changes, improve the urban air quality and maintain the urban ecological balance [1]. People's growing spiritual needs can be met in the appreciation and use of high-quality urban parks. With the development of social economy, the requirements for park construction are becoming higher and higher, and the ecological integrity, diversity and ecological interaction of park vegetation are gradually taken into account.

Lingang wetland lies in Tianjin Lingang industrial zone. It is located in the downstream of Hai river basin, the central area of the east of Tianjin, on the verge of the Bo Sea. It is located in the temperate semi-humid continental monsoon climate zone. Obviously, weather changes along the seasons, with a hot and humid summer and a dry and cold winter. Annual average temperature is 12.3℃ and annual average rainfall is 617 mm.

Wetland park in Lingang is predominantly a water treatment artificial ecological wetland park with both landscape and ornamental value. The park's water area is about 170000 square meters; all kinds of nursery stock are planted with more than 120 varieties in total. Among the plants, number of trees is 261200, number of individual shrubs is 384000, number of tufty shrubs is 193600, number of

Content from this work may be used under the terms of the Creative Commons Attribution 3.0 licence. Any further distribution of this work must maintain attribution to the author(s) and the title of the work, journal citation and DOI.
Published under licence by IOP Publishing Ltd

IOP Conf. Series: Earth and Environmental Science **346** (2019) 012012 doi:10.1088/1755-1315/346/1/012012

ground cover plants is 344500, lawn area is 493500 square meters. And wetland park is a combined artificial and natural park. Sewage are been treated and recycled through an international advanced ecological sewage treatment technology [2]. It not only beautifies the park landscape, but also reduces sewage pollution. The Binhai new area will become an outdoor leisure base and an ecological demonstration base for residents.

The plant community structure and diversity of the wetland park were investigated in the harbor economic zone in this paper. The purpose is to master the research of plant growth status and plant diversity, grasps the plant diversity and the natural law, establish scientific and suitable system mechanisms for the ecological construction and maintenance of the wetland park and improve the provision of basic data and theoretical background.

2. Materials and methods

2.1. Site selection

Quadrants were selected from seven typical scenic spots in wetland park, such as Yousiting, Tanghaixunyou, Qushuihuajing science center, Jinweichaosheng, Shuidanggutian and park's east gate, with a size of 100 ×100 m. In the research of plant diversity of wetland park, trees and shrubs in the quadrants were investigated and analyzed with the method of registering each tree, and the frequency and dominant species of plants and other important indicators were analyzed.

2.2. Calculation method

2.2.1. Calculation formula. The Dominance index is on the basis of the important value of each species in the community and reflects the concentration of dominant species in the community [3]. The calculation formula is as follows:

$$C = \sum_{i=1}^{s} \left(\frac{n_i}{N} \right)^2$$

2.2.2. Importance value. Importance value = (relative density + relative frequency + relative significance /300) [3]. Relative density (RD)=(the density of a plant/the total density of all plants) 100% =(the number of individual plants/the number of individuals of all plants) 100%; Relative frequency = (the frequency of this species/the total frequency of all species) 100%; Relative significance (relative Dominance) = (sum of individual chest area in the sample square and/total individual chest area in the sample square) 100%.

2.2.3. Species diversity index. Species diversity index is a comprehensive quantitative index that reflects the number, structure, individual number and evenness of species in a community [4]. The most commonly used ones are species richness index (S), Simpson index (D), Shannon-Weiner index (H) and Pielou's evenness index (J) [5], which measure community dominance and judge community diversity.

3. Results and analysis

3.1. Community composition

The quadrants were representative of the landscape areas in the wetland park. It can be seen that most local tree species are planted to assist the introduction of trees from other places, the natural succession of ground cover plants is utilized, and the planting of tree species is reasonably arranged. Due to its geographical location in saline-alkali land, the park attached great importance to the promotion and application of local native species, especially salt-tolerant species. From the perspective of engineering, the park was mainly represented by the classic structure of trees, shrubs and grasses. In the tree layer, poplar, willow, locust and jujube are used as the most important tree species. In the shrub layer, the easy pruning varieties were mostly chose such as spindle tree, virgin bulb and purple berberis. According to the geographical characteristics, the development and introduction of new species, large size of seedlings for embellishing and other ways to gradually enrich the variety of ornamental plants and reasonable dense planting, has contributed to the park being clearly structured, fascinating and with a harmonious landscape collocation.

3.1.1. Common native plants in Tianjin The local tree species in Tianjin are closely related to soil, water and climate conditions. 33 species of local tree species were identified there [6]. Common deciduous trees include weeping willow, golan tree, Chinese toon, Chinese pear, elm, national locust, apricot, persimmon, mulberry, coniferous tree, jujube tree, apple, dry willow, walnut, poplar, ginkgo, pear, etc. Commonly used deciduous shrubs are: midget crabapple, lilac, gold and silver wood, forsythia, begonia fruit, cypress, wild Martin incense, tamarix, Chinese rose. Comparing with the plant management account of Lingang wetland park, more than 80% of the plant species applied in Lingang wetland park coincide with the local tree species in Tianjin, which conformed to the assumption that the adaptability of local plants to the environment was much higher than that of alien species. On this basis, the plant diversity can be reasonably enriched through grafting, introduction and other means.

3.1.2. Composition of plant types. The relationship of plant type composition was an indispensable aspect to study the characteristics of plant species composition. Evergreen The percentage of evergreen species was the least, accounting for 2.7%, and that of deciduous tree species was the largest, accounting to 37.4% in the park (Figure 1). The species ratio of evergreen trees and deciduous trees in wetland park was 1:13.9 and the number ratio is 1:23, showing that the park was dominated by deciduous trees, and there were significant differences in the species and number of evergreen and deciduous trees. Shrub species accounted for about 30.8%; and ground covers and aquatic plants accounted for 29.1%. The both ratio were similar (Figure 1). The shrub layer was divided into solitary shrub and secondary shrub, with a species ratio of 2:1 and a number ratio of 1:1. Therefore, it can be seen that the design of solitary shrub and cluster shrub was more scientific and average. Under the condition of meeting the phenological conditions of Tianjin, the species and quantity of evergreen trees should be increased through introduction and cuttage, so as to improve the species richness and the ornamental value of the park.

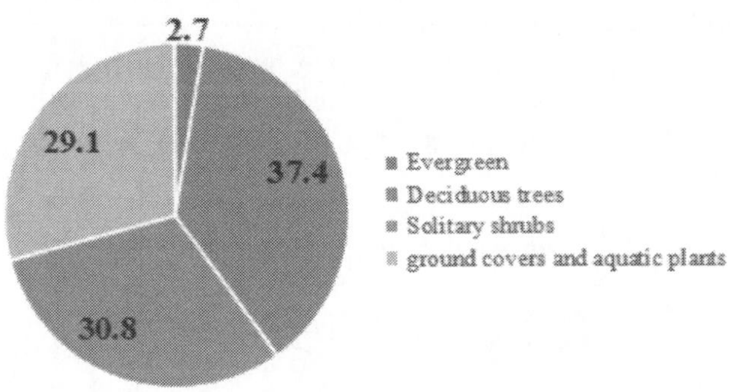

Figure 1. Percentage of species distribution in wetland park.

3.1.3. Composition of ornamental species. Flowering plants were mostly rosaceae, and other species of the genera were also planted in the shrub layer. Due to factors such as soil and climate conditions in wetland park, people were very fond of landscape areas formed by flowering plants. Therefore, that kind of plants, especially shrubs, were widely used in wetland park. The main flowering plants such as rose, begonia, plum, cherry blossom, jasper and other rose shrubs have a flowering period concentrated from April to June, the flower color being red, pink, white, purple and other colors. They have a very important garden landscape value.

In the foliage plants, except purple plum, gold privet and evergreen cypress, most autumn leaves are yellow, such as weeping willow, national locust, golden elm, gold and silver wood, goldenrod tree and so on. Because of the influence of Tianjin's geographical location and climate, deciduous plants were larger than others, and there were abundant ornamental effects in autumn. There were fewer spring leaf plant and evergreen.

3.2. Community structure

3.2.1. Dominant species of tree layer in the quadrants. Among the 7 quadrants, the dominance index of the meander path and the park's east gate was significantly higher than that of other landscape areas, reaching to 0.8, and the dominant species were gold leaf elm (table 1). It was shown that gold leaf elm had a prominent dominant position in wetland park. In addition, it had beautiful leaf color, developed root system, resistance to cold, drought and barren, and strong adaptability. Therefore, the planting of gold leaf elm in wetland park had great landscape significance and value.

Table 1. Dominance index and dominant species of tree layer in quadrants.

Quadrants	Dominance index	Dominant species
Yousi pavilion	0.7001	107 poplar
Haitangxunyou	0.7968	*Mmidget crabapple, Malus halliana*
Qushuihuajing	0.8152	Gold leaf elm
Science centre	0.6852	Elm
Jinweichaosheng	0.1953	Mmedlar
Park east gate	0.8183	gold leaf elm, *Acer negundo*
Shuidanggutian	0.7275	*Robinia idaho*

3.2.2. application frequency of plants in the sample plot. The commonly used trees in wetland park were poplar (*Populus L*), Pagoda tree (*Sophora japonica*), weeping willow (*Salix babylonica*), golden elm (*Ulmus pumila*), accounting for 23.8% in the total tree species in the park. The results showed that, except for a few trees with high presence rate, the frequency of other tree species in the vegetation community of wetland park was not high (table 2). The frequency of deciduous trees was relatively large. Due to the characteristics of northern weather, the frequency of evergreen trees in wetland park was relatively small, and it can be seen that there were more presence of foliage species, which was suitable for the design characteristics of northern gardens. Among them, species such as *Malus micromalus* and *Albizia julibrissin* can be planted on both sides of the road to form a landscape of flowers, which had high ornamental value but low appearance frequency. It was worth to be popularized and enlarged.

Table 2. Appearance frequency in common trees.

Frequency (%)	Plant names	Species
20≤f<25	*Populus L, Sophora japonica, Salix babylonica, Ulmus pumila*	4
15≤f<20	*Crataegus pinnatifida, Amygdalus persica, Robinia pseudoacacia, Sabina chinensis, Sophora japonica*	5
10≤f<15	*Malus micromalu,Malus halliana, Sabina chinensis, Salix matsudana*	4
f<10	*Rhus Typhina, Morus alba, Zelkova serrata, Albizia julibrissin, Acer negundo, Armeniaca vulgaris*	7

Among shrubs, *Rosa*, *Lonicera japonica* etc., were commonly used shrubs in gardens (table 3).

IOP Conf. Series: Earth and Environmental Science **346** (2019) 012012 doi:10.1088/1755-1315/346/1/012012

Buxus megistophylla was often trimmed into a low giant sphere in the park, or planted as a single plant, or symmetrically arranged in a regular way, forming a characteristic landscape with flowers and leaves set off against each other for years. Although winter jasmine (*Jasminum nudiflorum*) and Flowering plum (*Euonymus maackii) we*re both less than 15% in frequence, their flowers were blooming like broomfuls and sea of flowers, which had high ornamental value and can form beautiful and rich levels of garden landscape, so they should be popularized and favored.

Table 3. Application frequency of common shrubs.

Frequency (%)	Plant names	Species
20≤f<25	*Lonicera japonica, Buxus megistophylla, Lonicera maackii*	3
15≤f<20	*Rosa sp, Hibiscus syriacus, Ligustrum lucidum, Ribes mandshuricum*	4
10≤f<15	*Euonymus maackii, Jasminum nudiflorum*	2
f<10	*Berberis thunbergii, Forsythia suspensa, japonica, Prunus ceraifera, Amygdalus triloba, Sabina chinensis*	6

Table 4. Application frequency of common herb ground covers.

Frequency (%)	Plant names	Specials
20≤f<25	*Festuca elata, Hemerocallis fulva*	2
15≤f<20	*Iris lactea, Phalaris arundinacea, Echinacea purpurea, Gaillardia pulchella*	4
10≤f<15	*Coreopsis drummondii, Gomphrena globosa, Nepeta cataria*	3
f<10	*Physostegia virginiana, Salvia japonica, Aster novi-belgii, Sedum spectabile, Parthenocissus quinquefolia*	5

The most frequent herbs used for cover ground were *Festuca elata* and *hemerocallis* (*hemerocallis fulva*) (table 4). In the park, they are often planted under the trees to form a better landscape under trees with other shade tolerant plants, but it only accounted for 14.2% of all herbaceous species. This indicated that the repeated use of these herbaceous plants in wetland park green space was serious, while the appearance frequency of most ground covers with good ornamental effect was generally low. Single application of plants was easy to cause monotony, so the species of ground covers, such as compositae, should increase. Compositae aare the most commonly used flowers in the park in spring, they can be planted around in the spring, and help tothe landscape of the park. Compositae plants with good ornamental effect such as *Echinacea purpurea*, *Coreopsis drummondii*, *Gomphrena globosa*, and others. had beautiful shape and strong adaptability and, can be used to shape a beautiful landscape with richer park species.

4. Discussion

4.1. Common local species in the parks
In this paper, the species of common native plants in wetland parks were investigated, but the genera and families of these plants including exotic and native plants were not counted and classified. Data and proportions of all plants lacked data. The number, species and proportion of native plants and exotic plants in the park will have an important impact on the landscape construction of the park. Therefore, these aspects should be further studied so that more perfect guidance and suggestions can be put forward for the reconstruction of the park in the future.

4.2. Selection of evergreen and deciduous trees
In the selected quadrat, when tall trees are selected and configured, the species and number of

deciduous species in wetland park were significantly higher than those of evergreen species and the difference was obvious. However, evergreen and deciduous tree species can only play a good ecological function in a reasonable combination, so the presence rate of woody plants in the park should be increased, especially the proportion of evergreen trees. It can enrich the winter garden landscape, and bring into play important ecological, aesthetic and socio-economic benefits.

4.3. Waterfront plants in park

Lingang wetland park had both ecological and landscape construction uniqueness. Its land is surrounded by water, the park covered a large area, the water body was unevenly distributed and the coastal plant species were simple, so investigation quadrants were not divided into water samples separately. We did not have a thorough understanding of the waterfront plants in the wetland park, lack basic data support in the selection of their species and number and their distribution and configuration and cannot put forward scientific suggestions on the ecological construction, conservation and improvement of water bodies in the wetland park.

4.4. Ornamental plant configuration

After investigating the ornamental and seasonal features of plants in the park, it was found that there were still some problems in the design of the park. There are few varieties of plants flowering in autumn in the park, and the flowering period was mainly concentrated in spring and summer. Green form was simpler, lawn area was bigger and herbaceous plant chooses collocation simply. Therefore, wetland park should improve the structure of green space community on the basis of the original plant community, species selection and plant configuration. It would be important to enrich the winter plant landscape by focusing on the cultivation of flower plants; and bring more autumn flowers such as hollyhock to beautify the wetland park landscape effect and make the best of the characteristics of different plants.

Acknowledgments

This research was supported by the University Young and Middle-aged Backbone Innovation Talent Training Program in Tianjin (J01009030709), Postgraduate innovation training program in Tianjin (2017YPY001), Tianjin agricultural scientific and technological achievements transformation and promotion project (201703080) and Agricultural Science and Technology; Achievements Transformation and Extension Project in Tianjin (201603060).

References

[1] Dai X F Investigation and application of plant diversity in Changchun park 2016 (Jilin, China: Jilin Agricultural University)
[2] Wen T, Shu Z L and Ding Y *Chin. Water Wastewater* 2014 **30** 42-45
[3] Xing X H *Hubei Agr. Sci.* **2010** 27-30
[4] Xiong X F and Wu T H *Nor. Hort.* **2011** 19 87-89
[5] Wu Z L, Zhou X N and Zheng L F *J. Fore. Res.* **2011** 11-5
[6] Tan J Study on plant landscape in Binhai new area of Tianjin 2011 (Changsha, China: Central South University of Forestry and Technology)

Comparing straw checkerboards paving methods in deserts of China with sustainability indicators

T T Sui[1], J H Liu[1,3], T Huang[2], Q Q Huang[1], D Wang[1] and K Ma[1]

[1]School of Technology, Beijing Forestry University, Beijng 100083, Beijing, China
[2]School of Art and Design, Southern Illinois University, Carbondale, 62901, Carbondale, US

E-mail: liujinhao@bjfu.edu.cn

Abstract. The straw checkerboard sand barrier (SCSB) is widely used to fix sand and prevent desertification n China. In this paper, two methods of straw checkerboard (SC) paving are compared using sustainability indicators: the traditional method employing workers using straw and shovel, and the mechanized method using straw and a sand fixing vehicle. The paper describes the processes of the two different paving methods. Data regarding these two different methods of SC0020 paving were obtained from prior research experiments conducted by a sand-fixing vehicle prototype. The paper proceeds to use three sustainability indicators representing the three aspects of sustainable development. Using these indicators, data were compared to assess these two methods. The results show that the operational efficiency of the mechanized method is nearly 200 times higher than the traditional method, and that the total cost in the operational process can be reduced by about 7, 200 Yuan (about 1,145 USD) per hectare. The completion cycle of the break-even point (164 hectares) is 52.5 days in mechanized method, which is about one tenth of the traditional method. This not only indicates a clear advantage of performance and cost-effectiveness for the mechanized process over the traditional method, but also shows that this mechanized process can contribute to sustainable development for the region greatly because the preliminary work on degraded land and soil restoration can be completed in a significantly shorter period and the health conditions of a large number of manual workers can be improved.

1. Introduction

As of 2014, desert covers 172 million hectares of land in China, accounting for 17.93% of the country's total region. Desertification affects 920 counties and districts throughout the country, 93.95% of which is distributed in North and Northwest China [1]. In recent years, sandstorms caused by strong winds blowing over loose sand have become one of the most serious environmental problems in China. It does significant harm to the environment, transportation, communication, and infrastructures in the affected areas as well as the adjacent regions, undermining the government's efforts to promote sustainable development of the country [2]. To alleviate these problems, scholars identified the key factors that produce sandstorms and suggested potential solutions to reduce the frequency and intensity of sandstorms, and to prevent further erosion and suspension of loose silt and sand [3].

Measures to combat desertification have been developed in many countries affected by deserts. These methods include biological, mechanical, and chemical methods. In China, many such measures have been experimented with and implemented to reverse desertification and alleviate its influence on the environment [4]. One such method, straw checkerboards sand barrier (SCSB) has been proven to be one

Content from this work may be used under the terms of the Creative Commons Attribution 3.0 licence. Any further distribution of this work must maintain attribution to the author(s) and the title of the work, journal citation and DOI.
Published under licence by IOP Publishing Ltd

of the most successfully implemented sand stabilization methods [5-7]. There are two main methods to pave straw checkerboards (SC): the traditional method and the mechanized method. Traditionally, a SC is planted by manual labors using spades. Though a plate shaped shovel head specialized for paving SC has been designed to provide an easier way to press straw into sand [8], and a straw checkerboard reusing method was put forward to improve the utilization of materials [9], the paving efficiency of the traditional method remains low. An adult worker is reasonably estimated to complete about 100 to 200 m^2 SC in an eight-hour work day on average according to long-term observation (it is a wide range because the actual units of square meters of a worker's accomplishment vary greatly in reality, determined by combined factors related to the individuals, the weather, and so forth). However, in China, the area affected by desertification is so large that the cost in terms of man hours of labor to pave SC would be enormous, while the cost to the laborers' health is equally high. Not only is desert work physically demanding, but long-term exposure to large numbers of dust particles and strong UV radiation also significantly damages workers' health [10-12]. More problematically, the SCSB can only last about 3 to 5 years as a result of the erosion caused by wind, sunlight and rain. This affects the desert improvement negatively because in most cases, SC paving is the first step in the long process of restoring degraded land and soil, followed by vegetation recovery. The SC paved to protect infrastructure such as highways, railways, and mining area etc. are particularly vulnerable, as those areas are not suitable for the survival of vegetation. The rate of SC deterioration and sand movement is so fast that manual labor alone cannot replace SC quickly enough to maintain their effectiveness. To overcome these difficulties, a mechanized method has been developed to pave SC.

Several research and development projects on the mechanized SC paving equipment's functions have already been completed [13,14]. The first physical prototype of a SC paving robot has already been field-tested. The experimental data showed that in desert areas with gentle slopes, SC paving can be highly mechanized [15]. In order to further improve the equipment's performance, a new generation of vehicles for stabilizing sand in surface has been manufactured through repeated optimization.

Sustainable development is defined by the United Nations as development that is able "to meet the needs of the current generation without compromising the ability of future generations to meet their needs" [16]. Sustainability normally includes three aspects – economic, environmental, and social [17,18]. Sustainable development can be evaluated by Sustainability Assessment (SA), a process that aims to contribute to a better understanding of the meaning of sustainability. There are multiple frameworks for Sustainability Assessment Methods (SAM), including product-related assessment tools, indicators, indices, and integrated assessment methods. Among these, sustainability indicators have been determined to be the most useful assessment method suitable for different fields [19-21]. The extensive use of sustainability indicators is due to their following useful characteristics: easy summarization and the ability to focus and condense massively complex information to a series of valuable data capable of being managed. The results derived from the indicators simplify data comparison, allowing trends, hot-spots, and other informative insights to be more easily assessed [22,23]. Sustainability indicators developed by the United Nations are listed in the "Global indicator framework for the Sustainable Development Goals (SDGs) and targets of the 2030 Agenda for Sustainable Development" and are used in this paper.

In this paper, we focus on analyzing the positive impacts of a mechanized SC paving method on sustainable development. Our research question is: compared with the traditional method, is the mechanized method more advantageous than the traditional manual method to the economy, to the environment, and to the society in terms of sustainability?

2. Data and methods

2.1. Traditional (manual) straw checkerboards paving method

Straw checkerboards (SC) with the size of 1 m × 1 m has been shown to be the most effective for sand stabilization in most areas [7], so this size of checkerboard was used in this study. The process includes three steps [24], (i). On the soil surface, outline the straw's grid by pressing a groove with a width of about 1000 mm and at 1000 mm of soil depth. (ii). Orderly lay the straw on the sand surface vertical with the marked line with enough thickness to ensure the sand resistance of the SC, pave to a width of about 1,000 mm band, ensuring that the middle point of the straw is coincident with the middle point of the mark line. (iii). Put the flat spade in the middle of the straw laid in step 2, press the spade heavily with the foot, and draw out the spade after two ends of straw are erected, then push the sand on both sides of the straw root to make the straw stand firmly.

In order to achieve the purpose of a windbreak and sand stabilization, the SC should meet the following criteria [25]: (i) the space between adjacent straw sand barriers should be 1000 mm (shown as figure 1(b)), range from 800 mm to 1000 mm is reasonable; (ii) the depth of SCSB beneath the sand surface (h in figure 1(a)) should range from 150 mm to 200 mm; (ii) the effective height (H in figure 1(a)) of SCSB should range from 150 mm to 200 mm.

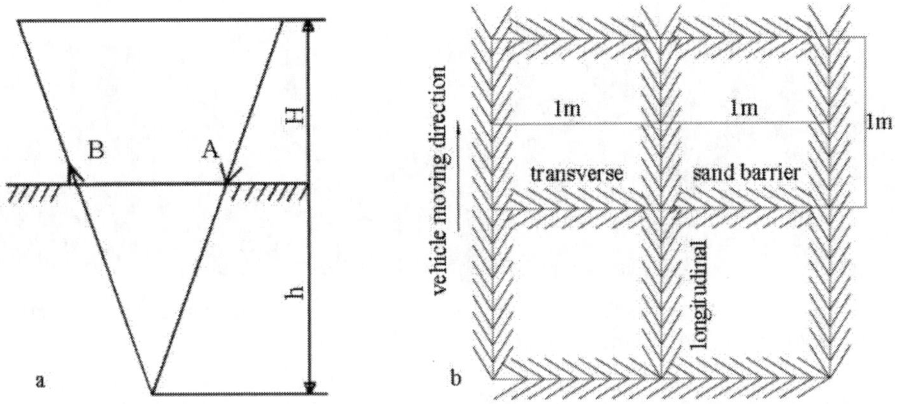

Figure 1. Characteristics of straw-checkerboard, (a) Plane structure; (b) Lateral section structure.

2.2. Mechanized straw checkerboard paving method

During the traditional SC paving process, the most time consuming and physically demanding steps for the workers are paving the straw with moderate thickness and then pressing the spade with their feet. Therefore, these two steps should be the first to be considered when proposing the mechanized paving method and designing the sand stabilization vehicle. The equipment should have the ability to pave mechanically, with low manual labor input and the flexibility to pave a variety of sizes.

To achieve the above purpose, a vehicle for sand stabilization was manufactured. We utilize a paving mechanism and a heavy-duty truck that has the ability to work in the special desert terrain using wheels as two main parts in the prototype, shown in figure 2. The tractor moves the paving mechanism and produces traction, while the paving mechanism paves the SC.

Figure 2. Sand fixing vehicle.

The structure of the paving mechanism determines the process of the mechanical SC paving method. The paving mechanism consists of two parts: transverse paving mechanism and longitudinal paving mechanism (we define the vehicle's moving direction as the longitudinal direction, and the direction perpendicular to the longitudinal direction as the transverse direction), as can be seen in figures 3(a) and 3(b).

When the sand fixing vehicle is working, in the longitudinal direction, straw is conveyed to the sand surface from the straw box (which stores a certain amount of straw) by an intermittent feeding mechanism. Then the transverse inserting tool inserts the straw into the sand to form the transverse straw sand barrier. In theory, adjacent straw sand barriers are spaced 1000 mm apart.

At the same time, in the longitudinal direction, the straw is conveyed and inserted into the sand continuously by a pressure wheel, paving a constant straw sand barrier. Together with the transverse straw sand barrier, the straw checkerboard sand barriers are formed, shown as figure 3(c).

Figure 3. Mechanized straw checkerboard paving method. (a) Transverse straw sand barrier insertion; (b) Longitudinal straw sand barrier insertion; (c) Straw checkerboards.

The parameters measured in the experiments on the mechanized SC paving method are in reasonable agreement with the SC technical criteria introduced in Section 2 (errors were caused by the vibration of the equipment and possibly the occasional imprecision of measurement). The depth of SC ranges from 165 mm to 240 mm, the height of SC ranges from 150 mm to 210 mm, the distance between two adjacent straw sand barriers ranges from 945 mm to 990 mm. According to the results, we conclude that the SC paved by the mechanical method on relatively flat terrain can achieve the sand stabilization effect.

2.3. Data collection

For the purpose of assessing the contribution of the mechanized SC paving method for sustainable development, a series of sustainability indicators were selected. We chose the traditional SC paving method, which has been successfully implemented by manual labor in China, as a baseline for comparison. The mechanized SC paving method with a sand stabilization vehicle was applied under comparable conditions. After, the data of the sustainability indicators of the two SC paving methods were assessed and compared.

In this paper, data for selected sustainability indicators by comparing experiments using both, manual and mechanical methods. The data are subsequently analyzed from the perspective of sustainable development. The key factor in the sustainability assessment process is to choose the most reasonable indicators related to sustainability [17, 26]. This is the first time that the sustainable effect of mechanized sand stabilizing procedure has been evaluated. The mechanized method will be compared to a baseline traditional manual method with three indicators, as shown in (table 1). Our selection of indicators and methods are listed below. (i) Government cash support: In China, SC is mainly paved for to protect agricultural land, communication facility and other infrastructures in desert areas, supported by government cash flow to assist development. Costs of a unit area determine the total area can be developed with the same total official flows. In this paper, both the costs of equipment and costs during the operation process are taken into consideration to calculate the official flows to a unit area of SC. For the mechanized method, this includes the costs of vehicle manufacturing, maintenance, diesel fuel, material, a driver and an assistant to the driver. As for the traditional method, it has been implemented in China for many years, the average costs of material and workers are based on the market prices. (ii) Frequency rates of non-fatal occupational injuries: Occupational health and safety ranked second and fifth respectively in social indicators for sustainable development [17]. The manpower employment (which is defined as the number of direct working hours needed to complete a hectare of SC (h/ha)) representing the physical labor volume of workers employed for SC paving, is related to the workers' health. The manpower employments of the two methods are derived from their operational efficiency. (iii) The time involved to complete a selected unit area was evaluated in both procedures: The complete cycle of a unit area of SC indicates the beginning of the subsequent land and soil restoration. In most cases, only by combining sand stabilization with ecological restoration can be achieved the long-term effect on restoring degraded land. We take the break-even point of the two methods as an example, and compare the completion cycle of the two methods, and use the Green Cover Index [27] to evaluate the state of land recovery.

Table 1. Indicators for sustainability assessment.

Category	Indicators	Unit	Data Collection
Economy	Government cash to a unit area of SC	Yuan/ha	Calculation
Society	Frequency rates of non-fatal occupational injuries		Interviewing/ Literature
Environment	Degraded land and soil restoration	Day	Recording/ Literature

2.4. Experiment site

The experiment was carried out at the experimental site, an agricultural experimental station for sand stabilization in Heerhongde (located at N49°11′, E118°27′, figure 4), in the summer of 2010. This base belongs to Prairie Chenbarhu banner located at the northern part of the Hulun Buir Sandy Land, Inner Mongolia, China. This land is characterized by fixed and semi fixed dunes with a loose soil layer and low humus content, which keeps moving under wind action. Only sparse sand plants such as Corispermum hyssopifolium and Agriophyllum squarrosum grow on quicksand and flora species of sheep firewood, wheatgrass and caragana grow on fixed and semi-fixed sand dunes [28]. The weather in spring is dry and windy, in the winter it is very cold, in the summer and autumn there is a great temperature difference between day and night. The research site belongs to the continental monsoon climate zone with a semi-humid and semi-arid transitional climate where the annual average temperature is −2.5 to 0°C and the annual sunshine duration is 2900 to 3000 h. The annual precipitation ranges from 280 mm to 320 mm, and this precipitation mostly concentrated in summer and autumn. The annual evaporation ranges from 1400 mm to 1900 mm, which is five times more than the annual precipitation rate. Mean annual wind velocity ranges from 3.2 to 4 m/s [29].

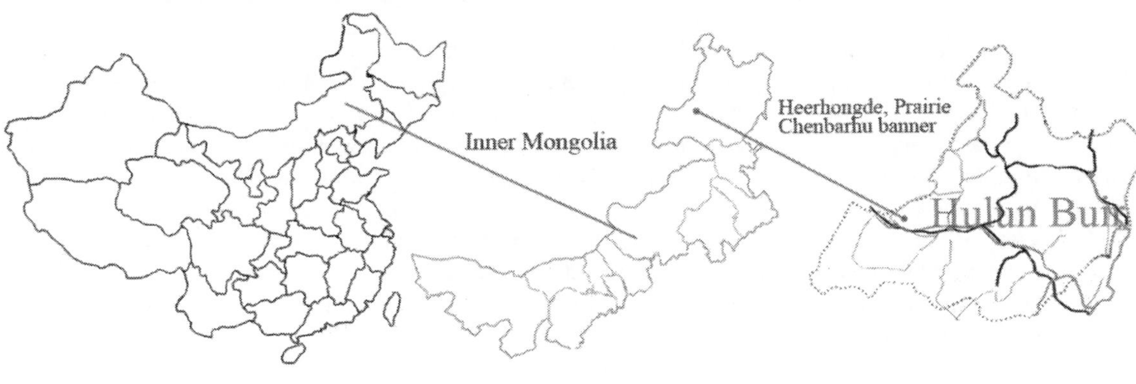

Figure 4. Experiment site.

The geography of the experimental base for wind and sand stabilization in Heerhongde is characterized by sand dunes alternating with gently undulating inter lowlands. It is object free, without stones and shrub stumps. Our SC paving experiment was conducted by the sand fixing vehicle prototype at the speed of 0.5 m/s (or 1.8 km/h, approximately), the distance between two adjacent straw sand barriers was set at 1000 mm. The straw used in the experiment was sorted and ranged from 600 mm to 750 mm in length. We measured the parameters of SC randomly with measurement tools, including tape, steel ruler and stopwatch, to test if the mechanical paving method fulfilled the SC paving criteria. Using this data, we were able to calculate the practical work efficiency of the sand fixing vehicle.

3. Results and discussion

3.1. Results

With a satisfactory SC paving effect, operation efficiency plays an important role in the following results and analysis of sustainability indicators. In this study, operation efficiency was considered as the area of SC paved in an hour (m^2/h). The data for the two methods is given in table 2. The SC paving experiment was carried out under the condition of keeping sand fixing vehicle moving at a constant speed of 0.5 m/s (or 1.8 km/h, approximately). At this speed, the vehicle paves 3 m^2 SC every 2s (because the transverse inserting tool is designed as 3 m in width and the longitudinal inserting tool includes three pressure wheels with 1,000 mm spacing), thus the theoretical operation efficiency is 5400 m^2/h. But in practice, 1 hectare SC can be completed in about 2.6 hours. Because the sand fixing vehicle needs to stop to resupply straw material, and occasionally the straw feeding mechanism may be stuck and require maintenance. Thus, about an average of 3,900 m^2 SC can be paved in 1 hour, namely the machine's actual efficiency is 3,900 m^2/h. Note that two workers are required for the operation of the sand fixing vehicle, a driver controls the truck and another worker completes the rest of the work. In the traditional method, an adult worker can complete about 100 ~ 200 m^2 SC in eight hours at an efficiency of 12.5 ~ 25 m^2/h.

Table 2. Operations of two methods.

	Traditional Method	Mechanized Method
Completed area (m^2)	100 ~ 200	10,000
Work hours(hr)	8	2.6
Efficiency (m^2/h)	12.5 ~ 25	3,900
Work speed(m/s)	/	0.5
Number of driver	0	1
Number of labor	1	1
Size of checkerboard (m)	1 × 1	1 × 1
Tool	Spade	Sand fixing vehicle

Government cash flows to a unit area of SC (as of 2017): In a regional project such as restoration of deserts, the government funds were assigned to cover costs incurred are funneled to cover costs incurred by many collaborating units over many years. The SC paving project is a small portion of this large effort and by reducing the cost of the project, we can effectively reduce the government cash support. Before commencing operations, the straw preparations are the same in both the traditional and the mechanized methods. Therefore, the costs of preparation can be omitted in our calculation. Here, we take 1 hectare SC as the study area in order to estimate the government cash flow needed to accomplish the unit area with both, manual and mechanical methods to a unit area of two methods.

In the SC paving project, operational cost accounts for the major proportion, which includes the cost of materials, tools, and labors. In the traditional method, 1 hectare SC needs five to six metric tons of straw [30]. The average price of straw is 600 Yuan/ton in 2017, so the official flows to straw for 1 hectare SC is around 3,000 to 3,600 Yuan (about 454 to 545 USD). As for the government cash support for labor, we calculate with an estimated average efficiency of 20 m^2/h (it takes 500 hours to finish 1

IOP Conf. Series: Earth and Environmental Science **346** (2019) 012013 doi:10.1088/1755-1315/346/1/012013

hectare), one worker costs on average 15 Yuan/h, hence the total government cash support per worker would be 7,500 Yuan/ha (about 1,165 USD). With the help of straw feeding mechanism in sand fixing vehicle, the straw is relatively evenly spread on sand surface, requiring 4.25 to 4.5 metric tons per hectare which cost 2,550 ~ 2,700 Yuan (about 395 ~ 420 USD). 1 hectare of SC can be completed in 2.6 hours with the efficiency of 3,900 m^2/h, the assistant's cost is 25 Yuan/h, and the driver's hourly wage is an average of 60 Yuan/h. The government support for labor is 221 Yuan/ha (about 35 USD). The theoretical diesel fuel consumption of 2.6 hours is about 148L according to power rating (249 KW) and the minimum fuel consumption at full load (≤192 g/kW.h) of the engine WD615.95C, and the density of diesel fuel (0.84 g/ml), thus the government support to fuel (average price of diesel fuel is 5.77 yuan/L in Sep, 2017) is about 854 yuan (135 USD). As for the price of sand fixing vehicle, we can reasonably estimate the production price instead of providing the accurate purchase price because the preliminary research, design and manufacturing of prototypes was a process of continuous optimization that incurred extra costs that might be recuperated by the sales of patents in the long run. In addition, this process was subsidized by the government in the forms of research grants, which might be difficult to justify to include in the final retail prices. To make relatively reasonable price estimation, we consider only the costs of parts in the paving mechanism manufacturing and the price of the heavy-duty truck (about 350,000 yuan), on these bases, the price of a sand fixing vehicle is estimated at about 900,000 yuan (142,100 USD). Clearly, the cost of production of a single prototype is far higher than that of mass production. Therefore, if the sand fixing vehicle is in mass production, its price may be drastically reduced. The maintenance cost of the sand fixing vehicle (with a life cycle of 10 years, which is the mandatory retirement standard for heavy duty vehicles in China) is estimated to be about 300,000 yuan based on its components' prices.

As can be seen in table 3, to complete 1 hectare SC, the total government cash support to labor employment and material is 11,100 yuan (about 1,743 USD) in traditional method and 3,775 yuan (about 598 USD) in mechanized method. After considering the mechanized method's costs of one-time purchase and ongoing maintenance of the vehicle, the break-even point for the two different methods is 164 hectares (1.64 km^2), when the total official flows of two methods is only a few thousand yuan in difference.

Table 3. Operation costs (as of 2017).

Category	Traditional Method (yuan/ha)	Mechanized Method (yuan/ha)
Worker/Assistant	7,500	65
Materials	3,000~3,600	2,550~2,700
Driver	/	156
Fuel	/	854
Vehicle	/	900,000
Maintence	/	300,000
Total for 164 ha	1,233,000~1,233,600	1,214,200~1,214,350

Frequency rates of non-fatal occupational injuries: The amount of worker's exposure to occupational environment can be estimated using the operational efficiency discussed earlier. The manpower employment of the traditional method is 500 hours/ha and that of mechanized method is about 5.2 hours/ha considering one driver and one assistant. The mechanized method can prevent more workers suffering from physical harms, such as dehydration, ultraviolet radiation and high levels of fine particulate matter (e.g. PM2.5 and PM10) [31], which can be analyzed.

Restore degraded land and soil: As mentioned earlier, degraded land and soil restoration process is affected by SCBC's paving operation cycle. In theory, to complete 164 hectares (the break-even point) SC, about 420 hours are needed by the mechanized method. The completion cycle is 52.5 days calculated on the basis of an 8-hour work day, with a driver and an assistant. In comparison, 82,000 hours are needed for the traditional method, the completion cycle is 5,125 days with two workers. In other words, the traditional method can complete the same work using the same number of days as the mechanized method, as long as about 390 workers are employed to do the work.

3.2. Discussion

It is reasonable to conclude that at the working speed at 0.5 m/s(1.8 km/h approximately), the operational efficiency of the mechanized method using a sand fixing vehicle is nearly 200 times more productive than that of the traditional method. The most important aspect brought about by the mechanized method's higher efficiency is the reduction of government subsidies to a unit area's development assistance. The result indicates that paving 164 hectare of SC is the break-even point of the two methods. If more SC were paved, the official flows to a unit area (1 hectare) saved by the mechanized method are more than 7,200 Yuan/ha (about 1,145 USD), which is multiplied with the number of hectares involved. This reduction is particularly important for this region because agriculture is the fundamental way of living for the poor and vulnerable groups in these rural and remote areas. Arable land is the foundation for agricultural production, food supply and therefore farmers' income. Large area of agricultural land is affected by moving sand in China, forcing the government to invest billions to protect the existing arable land and reform the desertification land for agriculture. Similarly, grassland used for livestock husbandry around the desert is also seriously affected. The grassland resources are indispensable to the livestock husbandry, which is a traditional trade in the region. Therefore, protection and utilization of grassland are the fundamental measures to sustain livestock husbandry and achieve stable and high yield, to raise the income level of herdsmen, and to promote the comprehensive economic development in the pastoral area. The mechanized method helps to pave larger area of SC with a reduced government cash flows, which has a positive impact on protecting local infrastructures and increasing the area of land that can be engaged in agricultural production. In other words, this method can complete an equal area of SC with less money. Taking the "13th Five-Year Project" of China as an example, the government planned to improve 100 thousand km^2 of sandy land during this period [1]. Calculating with the saving cost (7200yuan per hectare) after the break-even point (1.64 km^2), about 72 billion yuan (11.45 billion USD) can be saved if mechanical method were applied in this project. The saving investment could be possibly redirected to develop other facilities in rural area. This helps fulfill the requirements of the SDGs because it helps to enhance agricultural productive capacity and improve the local economic sustainable development in developing countries.

Another equally important aspect is that the higher efficiency is related to the social sustainable development from a view of decreasing frequency rates of non-fatal occupational injuries. Non-fatal injuries of short-term labor working in harmful environment such as mining, construction [32] and deserts (including labors paving SC) have not received enough attention. Extreme heat, drastic temperature fluctuation, rapid evaporation and strong reflection of sunlight characterize the desert environment. The maximum temperature of some deserts can reach 45°C, the perceived temperature is often even higher.Desert environment affects physical labor's health in various ways. Wilk et al. reported body weight losses of 0.65% following a 3-hour activity (intermittently at 35 +/- 1°C, 45-50% relative humidity) when individuals were allowed water ad libitum, accompanied by the loss of a large number of electrolytes. Another investigation made in warm environment (30°C dry bulb, 50% relative humidity) showed that when the individuals are dehydrated by 2% of their body weight, their physical performance will be impaired and a decrease by 30% of work capacity will occur when they are dehydrated by more than 5% of their body weight. If they bear high and sustained sweat rates of more than 1 L/hour, that is body weight loss of 1 Kg/hour, subsequent rapid body weight losses are possible in protective clothing [33]. Investigation carried out in simulated desert conditions showed that the water loss following prolonged exercise is up to 11 liters per day. The water loss is not shared by the whole body equally: the blood serum is the main source, which causes a decrease of the circulating blood volume, bringing server influences [33]. Secondly, it has been proven that there is a strong correlation between ultraviolet (UV) light and the risk of both non-melanoma skin cancer and malignant melanoma [10,34]. An investigation reported that erythemal radiant exposure of 65% outdoor workers exceeded 1000 J/m^2 (10 SED), corresponding to MED with adaptation for skin phototypes III and IV, based on which the occupational risk estimation are high [35,36]. It is reported that the annual total amount of

UV in the Taklimakan Desert reaches 305.64 MJ/ (m^2. A). May to August is a period of greater intensity of UV, and the cumulative value of UV for these 4 months accounted for 46.67% of the total annual UV. The maximum monthly average daily amount of UV is 1.23 MJ/m^2 in July, and the average annual amount is 0.84 MJ/m^2 [37]. Workers in desert from May to August suitable for SC paving are exposed to higher UV radiation. What's more, dust storm is a frequent phenomenon in sandy sites. PM2.5 and PM10 originating from the dust contain various pollutants and biological components, some of which have been shown to be positively related to increased mortality and hospitalizations of the people who have been exposed to airborne particles [10,38,39], causing respiratory disorders.

Labors for SC paving in traditional method bear a long-term exposure to the extreme hot and dry environment, intensive UV radiation and air-bone particles, resulting in harmful effects that may only reveal themselves years after exposure. This increases the frequency rates of non-fatal occupational injuries. Furthermore, these contract labors are usually employed from rural regions near the working site, including many women, who are largely uneducated and lack the legal knowledge to protect themselves from occupational injuries with these contracts. In these often tight-knit rural communities, once these workers are chronically injured in the desert occupational environment, they and their family will suffer enormous pressure, which restricts the sustainable development of local society. "Protect labor rights and promote safe and secure working environments for all workers, especially women and precarious employment" is suggested in SDGs [31]. The sand fixing vehicle is a method to reduce physical labor required for SC paving. And its higher efficiency reduces the length of exposure for the workers to the unhealthy occupational environment, helping to decrease the local people's frequency rates of non-fatal occupational injuries, coinciding with the SDGs.

The SDGs suggested that governments should implement resilient agricultural practices to help maintain ecosystems, strengthen capacity for adaption to extreme weather and drought and progressively improve land and soil quality [31]. SC combined with ecological restoration works well for these objectives. A great deal of research shows the SCSB has a positive link with sand stabilization, its contribution is to increase the sand surface roughness and reduce the wind speed, forming a stable aeolian environment where plant seeds can sprout and take root while water, soil organic carbon, and fine particle contents can accumulate to facilitate vegetation survival [40,41]. The most common degraded land and soil restoration method is to sow seeds and plant native vegetation suitable for survival in the desert with the SC. With the succession of plants, the land surface will form a biological soil crust. Gradually, the area will form a complex ecosystem with other species moving into the area, achieving ecological restoration [42,43]. However, the ecological recovery process requires a long period in the desert environment due to the extreme climate and water shortage. The crust is still in the initial stage and the soil nutrient content remains low even after 3 to 5 years [41]. Even in the Shapotou desert, one of the most successful sites in China using SCSB to fix sand and restore degraded land, the highest coverage of artificial vegetation planted there is a scant 40% even after 8 to 9 years of recovery [44].

Figure 5. Vegetation change. (a) Original; (b) 1 year later; (c) 3 year later.

Specific to our program, the status of degraded land and soil restoration after 3 years is shown in figure 5(c). Note that the original proportion of bare sand land in figure 5(a) is 62% with an annual vegetation coverage of 18%. The reason bare sand was not 100% and vegetation cover was not 0% is because the experiment base is not a total desert, it is the grassland eroded by wind and buried by sand in recent years. In other words, it is a semi-sandy and semi-arid area. The rate of moisture here is relatively higher than desert condition, bringing good results for land and soil restoration. The green cover index results show that soil wind erosion was reduced by 60%, the most obvious decrease of all. The other indicator (the proportion of bare sand to vegetated area) fell by almost 50% too. The proportion of annual vegetation coverage in our demonstrative plot continuously rose from 18% to 58% over the course of 3 years.

Since SC pavement is the preliminary step, its completion cycle plays a decisive role for subsequent vegetation restoration. Paving equal area of SC with the same number of labors, the completion cycle of the traditional method is nearly 100 times longer than that of the mechanized method. Due to the arid climate, the period suitable for SC paving in most desert areas is from late spring to early autumn. The mechanized method has a definite advantage in shortening the SC paving cycle to start the degraded land and soil restoration much sooner. This is especially beneficial in the areas with the most urgent need to restore land and recover green cover, helping local environmental sustainable development. In the past, large area of SC was paved by traditional method to restore degraded land and protect agricultural land, mining area, transportation and other infrastructures in China, contributing to the sustainable development of agriculture and animal husbandry in rural desert region. Due to the rapid rate of sand expansion and SC degradation, the mechanized method is developed to improve the operation efficiency and protect the manual workers' health. In addition, the restored land and recovered green belt is conducive to decrease the mean levels of finest particles, which is related to reduce mortality rate attributed to ambient air pollution.

4. Conclusion

The SDGs highlight that combating desertification, and halting and reversing land degradation and halting biodiversity loss are significant parts to achieve global sustainable development. Straw checkerboard sand barrier is one of the most effective methods to resist desertification. Both the traditional method and the mechanized method are used for SC paving, conducive to sustainable development. In comparison, the mechanized method can reduce the government cash support to development assistance, decrease frequency rates of non-fatal occupational injuries and greatly shorten the operation cycle, which helps to largely reduce the time involved in restoring degraded land To achieve the UN's SDGs and targets of the 2030 Agenda, we suggested that the local governments should organize these sand-fixing operations as a collective effort, and reasonable tax incentives and/or subsidies to be established by the central government to encourage the mechanized method since desertification negatively affects cities outside of the desert regions as well.

Sand fixation and afforest the desert area are essential parts of Chinese government's sustainable development policies [16]. This paper should be of great importance to the decision-makers as it provides an evaluation of the mechanized method relative to the baseline manual method in terms of sustainability. These results will help them determine whether the sand fixing vehicle should be implemented to further the sustainable development objectives of halting further desertification and facilitating revegetation in the region.

Acknowledgments

The study was partially funded by the Development of Key Equipment and Monitoring System for Vegetation Restoration (200-GK121901062) and The National Key Technology R & D Program during the Eleventh Five-year Plan Period (2006BAD11A15).

References

[1] State Forest Administration Available online: http://www.forestry.gov.cn/main/69/content-831684.html (accessed on 1 March 2018).
[2] Wang T, Chen G T and Qian Z A *et al* 2001 Situation of sand-dust storms and countermeasures

in North China *J. Desert. Res.* **21** 322-27

[3] Zhang R, Han Z and Wang M 2002 Dust storm weather in China: New characteristics and origins *Quaternary Sci.* **22** 374-80

[4] Liu J, Shi B and Lu Y *et al* 2012 Effectiveness of a new organic polymer sand-fixing agent on sand stabilization *Environm. Earth Sci.* **65** 589-95

[5] Li Y L, Cui J Y and Zhang T H *et al* 2009 Effectiveness of sand-fixing measures on desert land restoration in Kerqin Sandy Land, northern China *Ecolog. Engin.* **35** 118-27

[6] Zhang T H, Zhao H L, Li S G, Li F R, Shirato Y, Ohkuro T and Taniyama I 2004 A comparison of different measures for stabilizing moving sand dunes in the Horqin Sandy Land of InnerMongolia, China *J. Arid. Environ.* **58** 202-13

[7] Qiu G Y, Lee I B and Shimizu H *et al* 2004 Principles of sand dune fixation with straw checkerboard technology and its effects on the environment *J. Arid Environ.* **56** 449-64

[8] Zhou G Y *et al* 2015 A spade and the method of using the shovel to lay straw squares (C.N. Patent 105,155,604A)

[9] Li S Y and Xu X W *et al* 2014 A reused straw-checkerboard method for sand stabilization (C.N . Patent 103,541,341A)

[10] Stafoggia M, Zaulisajani S, Pey J, Sternberg T and Edwards M 2017 Desert dust and health: A central Asian review and steppe case study International *J. Environm. Res. Public Health* **14** 1342

[11] Goudie A S 2014 Desert dust and human health disorders *Environm. Int.* **63** 101

[12] Håkansson N, Floderus B and Gustavsson P *et al* 2001 Occupational sunlight exposure and cancer incidence among Swedish construction workers *Epidemiology* **12** 552

[13] Shu Q and Liu J 2007 Research on trafficability of straw-checkerboard sand barriers paving robot *Trans. Chin. Soc. Agricult. Machinery* **38** 99-103

[14] Lu H M, Guo X R and Zhao Z G *et al* 2006 Desert trafficability study of trellis grass paving robot *T. Chin. Soc. Agric. Mach.* **98** 99-102

[15] Liu J H, Pan H B and Shu Q 2009 Multi-body dynamic simulation and experiments of the straw-checkboard barriers paving robot *Trans. Chin. Soc. Agricult. Machinery* **40** 153-7

[16] WCED-World Commission on Environment and Development 1987 *Our Common Future* (Oxford: Oxford Univ. Press)

[17] Singh R K, Murty H R and Gupta S K *et al* 2009 An overview of sustainability assessment methodologies *Ecolog. Indicators* **9** 189-212

[18] Waas T, Hugé J, Verbruggen A and Wright T 2011 Sustainable development: A bird's eye view *Sustainability* **3** 1637-61

[19] Ness B, Urbel-Piirsalu E, Anderberg S *et al* 2007 Categorising tools for sustainability assessment *Ecolog. Econom.* **60** 498-508

[20] Paehlke R 2009 Methods for sustainability assessment: sustainability indicators *Ecolog. Indicators* **9** 189-212

[21] Mahsa M, Mohammad J J, Nabiollah M, Reza A, Reza A and Tomás B R 2017 Sustainability indicators for municipalities of megacities: Integrating health, safety and environmental performance *Ecolog. Indicators* **83** 271-91

[22] Waas T, Huge J, Block T *et al* 2014 Sustainability assessment and indicators: Tools in a decision-making strategy for sustainable development *Sustainability* **6** 5512-34

[23] Hunt D V, Lombardi D R, Jefferson I *et al* 2008 Application of sustainability indicators in decision-making processes for urban regeneration projects *Engineer. Sustainabil.* **161** 77-91

[24] Cheng G M, Sun D X and Xu L F 2009 Protecting transmission line foundation by paving sand checkerboard barriers of plant straws *Hubei Electric Power* **33** 35-37

[25] Wang Z T and Zheng X J 2002 A simple model for calculating measurements of straw checkerboard barriers *J. Desert Res.* **22** 229-32

[26] Liu G, Brown M T and Casazza M 2017 Enhancing the sustainability narrative through a deeper understanding of sustainable development indicators *Sustainability* **9** 1078

[27] Clap A J, Bishop P A, Smith J F *et al* 2002 A review of fluid replacement for workers in Hot Jobs *Aiha J.* **63** 190-8
[28] Feng F, Wang S, Hong X *et al* 1993 Desert control and revegetatlon experiment on typical steppe in hulunbeer grassland *Chin. J. Grassland* **4** 38-40
[29] Na Q U, Min D U, Yuan L M *et al* 2013 Effect of bio-engineering technology of sandy blowout in hulunbeier *J. Inner Mongolia Forest. Sci. Technol.* **39** 25-27
[30] Liu Y X 1987 The establishment and effect of protecting System along the Bautou-Lanzhou railway in Shapotou sandy area *J. Desert Res.***7** 1-11
[31] United Nations Available online: https://sustainabledevelopment.un.org/post2015/transformingourworld (accessed on 5 March 2018).
[32] Alfers L, Phumzile X and Richard D 2016 Promoting workplace health and safety in urban public space: Reflections from Durban, South Africa *Environ. Urbaniz.***28** 391-404
[33] Adolph E F 1947 *Physiology of Man in the Desert* (Interscience Pub:Newyork USA)
[34] Modenese A and Gobba F 2017 Occupational exposure to solar radiation at different latitudes and pterygium: A systematic review of the last 10 years of scientific literature *International J. Environm. Res. Public Health* **15** 37
[35] Vecchia P, Hietanen M, Stuck B E, van Deventer E and Niu S edi. 2007 *Protecting Workers from Ultraviolet Radiation* (Oberschleissheim, Germany: International Commission on Non-Ionizing Radiation Protection (ICNIRP))
[36] Wolska A 2013 Occupational exposure to solar ultraviolet radiation of Polish outdoor workers: risk estimation method and criterion *Int. J. Occupat. Safety Ergonomics Jose* **19** 107-16
[37] Qing H E, Jin L, Shuai L I *et al* 2010 Solar ultraviolet radiation characteristics of the Taklimakan Desert Hinterland *Res. Sci.* **32** 1462-70
[38] Trianti S M, Samoli E, Rodopoulou S *et al* 2017 Desert dust outbreaks and respiratory morbidity in Athens, Greece *Environ. Health* **16** 72
[39] W Zeng, L Lang, Y Li *et al* 2017 Estimating the excess mortality risk during two red alert periods in Beijing, China *Int. J. Environ. Res. Public Health* **15** 50
[40] Li X R, Xiao H L, He M Z *et al* 2006 Sand barriers of straw checkerboards for habitat restoration in extremely arid desert regions *Ecolog. Engin.* **28** 149-57
[41] Shirato Y, Zhang T H, Ohkuro T *et al* 2005 Changes in topographical features and soil properties after exclosure combined with sand-fixing measures in Horqin Sandy Land, Northern China *Soil Sci. Plant Nutrit.* **51** 61-8
[42] Li X R, Xiao H L, He M Z *et al* 2006 Sand barriers of straw checkerboards for habitat restoration in extremely arid desert regions *Ecolog. Engin.* **28** 149-57
[43] Wang N, Jiao J, Jia Y *et al* 2017 Influence of afforestation on the species diversity of the soil seed bank and understory vegetation in the Hill-Gullied Loess Plateau, China *Int. J. Environ. Res. Public Health* **14** 1285
[44] Yu Y, Lin Q, Shi Q *et al* 2002 Changes of habitat and vegetation in man-made vegetation area of Shapotou section along Baotou-Lanzhou railway *Acta Ecolog. Sin.* **22** 433-9

Coupling of urban green space system and sponge city construction

X J Pan[1,2]

[1]Wuxi City College of Vocational Technology, China

E-mail: 962482391@qq.com

Abstract. Through quantitatively analyzing the implementation of the sponge city concept in urban green space systems, this research explored the role of urban green spaces in collecting and using rainwater in a urban landscaping. By combining the ecological and recreational functions of the urban green spaces, as well as appropriate quantitative indexes, reasonable approaches for applying rainwater collection, infiltration and retention technologies in urban green space design and construction are suggested. Thus, the implementation of the sponge city concept in the initial planning stage of green spaces was considered. The Taihu New City in Wuxi was chosen as study case. Thus, this paper combined current planning status between green space and rainwater collection systems and analyzed them. As result, usable green spaces for the sponge city construction were obtained. Also, details on the construction characteristics of various usable green spaces were acquired. The role of these green spaces in the sponge city construction were classified and quantitatively assessed and compared with relevant standards. On the basis of guaranteeing green space intended purposes, this paper analyzed the results of coordinating the green space areas with rainwater collection systems and specified the role of various green spaces in the rainwater collection.

1. Introduction

Some consequences of modernity may threaten urban life. For instance, while some cities are often flooded by stormwater, some others are vulnerable to drought and pollution-induced water shortage. The concept of "sponge city" was officially proposed in 2013, and since the construction of this type of structures started, experts from architecture, water supply, drainage, and landscaping in China have done preliminary research on this topic in their own field.

Urban green spaces are expected to play a key role in sponge city design and construction. However, considering their intended functions for ecology, recreation, emergency sheltering and others, the focus of the experts is to explore: (a) how urban green spaces help in collecting stormwater under the premise of guaranteeing their original role; (b) how much rainwater can be carried through the total rainwater volume without negatively influencing ecological and recreational functions; and (c) what appropriate technologies can be applied.

1.1. Research background

1.1.1. Rainwater regulation system abroad. Severe environmental problems emerging as a result of urbanization include frequent waterlogging, aggravating runoff pollution, loss of water resources, and degradation of aqueous environments. In this regard, in the past 2 decades developed countries such as UK, USA, Australia, Germany and Japan have formulated the sustainable urban stormwater

Content from this work may be used under the terms of the Creative Commons Attribution 3.0 licence. Any further distribution of this work must maintain attribution to the author(s) and the title of the work, journal citation and DOI.
Published under licence by IOP Publishing Ltd

development and management system by simulating natural drainage. A number of relevant countermeasures and technologies have been largely developed and applied in practices [1]. The Council for Sustainable Development (USA) suggested the concept of green infrastructure based on the theories and practices of BMPs and LID, providing strategic instructions and technological support for the construction of the "sponge city" in the field of stormwater management [1]. Germany is one of the countries that better makes use of urban rainwater. German's MR system (Mulden-Rigolen-System) ensures infiltration efficiency by maintaining rainwater in low-lying grassland and in infiltration ditches for short and long term processes, respectively [2]. Japan has applied a rainwater retention and infiltration plan since 1980 by promoting infiltration pools, pipes and wells, penetrative pavements, infiltration side ditches, regulating reservoirs, and green spaces [2]. In the WSUD systems (water-sensitive urban design) of Australia, the rainwater bio-infiltration system (also known as stormwater bioretention, rain-gardens or biofilters) is widely applied since it displays flexible design scalability, configurations and structures, as well as the possibility of integrated beauty. Particularly, small-scale rain-gardens, large bio-retention basins, and low-lying lands can be flexibly designed with this system [3].

Other similar concepts include Sustainable Urban Drainage System (SUDS) in the UK, and the Active Beautiful and Clean Waters Program (ABC) in Singapore [4]. By integrating the theories of LID and WSUD, New Zealand established a Low-Impact Urban Design and Development System [1].

1.1.2. Problems of the sponge city construction in China. The sponge city construction in China shows potential problems related to understanding and orientation. Urban green spaces are often expected to be responsible for rainwater storage in the sponge city construction. Every facility of the rainwater garden, grass ditch and infiltration gallery, and stormwater garden type are taken as green space facilities. This situation neglects two key factors that might result in undesired consequences:

- City development conditions and population density determine the particular properties of the urban green space system in China. Almost all domestic green spaces are built with expected functions. Both, land use quality and property of urban green lines, take the ecological security and recreational needs into consideration, where the area should not be reduced and functions not be replaced, same as for the blue line water system. However, in less-populated countries, the urban green space system mainly focuses on the overall layout, not on regulatory details, and the construction of sponge city facilities does not influence its green space system layout. In the sponge city construction in China, the blind application of sponge facilities to green spaces may destroy the layout, influence basic functions, and even threaten the ecological security of these green areas.
- Urban green space system's capacity of carrying sponge facilities. Undoubtedly, the urban green space system is an ideal sponge because of its high infiltration rate and low runoff coefficient. However, the carrying capacity of a green space system should be comprehensively studied in the sponge city construction. Sponge facilities can only be built in urban green spaces under the premise of maintaining their ultimate ecological and recreational functions. Moreover, scientificity of the green spaces, appropriate amount and reasonable construction are also the preconditions for the construction of sponge facilities.

1.2. Literature review

China officially proposed the concept of sponge city and launched the construction of sponge city in 2013. From this time on, experts from planning, ecology, architecture, water supply, drainage, and landscaping have devoted great efforts in this type of endeavor. Yu Shaowu, Ding Nian, Ren Xinxi and Yangchen introduced many forms of sunken green spaces and explained their effect in *Exploration of Stormwater Storage, Infiltration and Utilization in Urban Sunken Green Spaces*. They stated that the sunken green spaces can use rainwater by properly regulating the green space structure. In addition, they barely need more construction investment or energy consumption, and help reducing the degradation of the environment [5]. Su Yijing, Wang Sisi, Che Wu, Wei Yizhe and Dong Yin in *Sunken*

Green Spaces Based on the Concept of "Sponge City" relate the construction of sunken green spaces by integrating "sponge city" and the latest domestic and international rainstorm management concepts, trying to achieve the multiple effect of sunken green spaces such as waterlogging control, water resource utilization, and landscape embellishment [6]. Li Yongfu and Wang Dongmei in *Research Progress on the Role of Sunken Green Spaces for the Storage and Utilization of Urban Rainwater* reviewed domestic cases on rainwater utilization in sunken green spaces [2]. Zhang Jinlong and Zhang Zhizheng in *Storage and Infiltration Capacity of Sunken Green Spaces and Its Influence Factors* referred to the sunken green spaces as types of ecological rainwater infiltration facilities that are important for the infiltration, storage and utilization of rainwater runoff. Thus, they are a key factor in urban rainstorm utilization [7]. Li Junqi, Che Wu, Chi Lian and Liu Song in *Key Parameters and Influence Factors for the Low-lying Green Spaces in Residential Areas,* discussed fundamental parameters such as the sunken depth of the low-lying green spaces, the loading rate of the low-lying green spaces and waterlogging time, and proposed the design concepts and methods for the low-lying green spaces [8]. Fan Qunjie in *Research on the Effect of Urban Green Space System in Regulating Stormwater Runoff and Reducing Pollution* described the regulating effect of urban green spaces on rainwater runoff. They also demonstrated the reducing effect of urban green space systems on rainwater runoff pollution, and established a number of actions and management strategies for urban green space systems with the aim of regulating rainwater runoff and reducing runoff pollution [9].

1.3. Research objectives
In this investigation, Taihu New City in Wuxi was chosen as case study to comprehensively explore the ecological functions of urban green spaces and scientific and reasonable construction of the sponge city, in order to provide suggestions on sponge functions and quantification indexes for the carrying capacity of various green spaces so as to their original ecological functions are guarranteed.

2. Methodology
Coupling of green space system and drainage system planning: a case study of Taihu New City.

2.1. Green space system of Taihu New City
The urban green space system layout of Wuxi City highlights the integration of river regulation and construction of riverfront belt parks. Based on the natural conditions and city development tendency, green spaces were designed along rivers, lakes, roads, and mountains. Taihu New City is included in the urban green space system layout of Wuxi, "One core and dual rings, three axes and four wedges, multiple patches and parks, green corridor and water network" (figure 1). The outer ring incorporates the ecological background ring of Taihu New City that consists of Taihu Lake shelter forest and landscape woods. The inner ring incorporates Gaolang Road and Liangtang River of Taihu New City. In addition, a 100-m-wide forest belt and 400-m-wide of ecological green spaces connect the main green areas and suburban parks into a "green necklace" for the city, showing the outstanding landscaping characteristics of Wuxi. Three axis incorporate the canal eco-landscape into Taihu New City. Multiple patches and parks, including various urban parks, public open spaces and ecological green spaces are distributed in the City. In addition, the construction of the Jinkui Park, the Shangxian River Wetland Park, the Changguang River Wetland Park, multiple belt parks along Gaolang Road and the canal, and others has resulted in a ratio of 18 m^2 of green spaces per capita. The main framework of "five horizontal and eight longitudinal roads" and the secondary framework of "five horizontal and nine longitudinal roads" (figure 2), and 25 river systems connect the parks and green spaces to create an open green network (figure 3).

The planned green spaces in Taihu New City (excluding affiliated green spaces) make a total of 3,542 hm^2, of which 2,071 hm^2 are parks, 1,237 hm^2 are road protective green spaces, and 234 hm^2 are river protective green spaces.

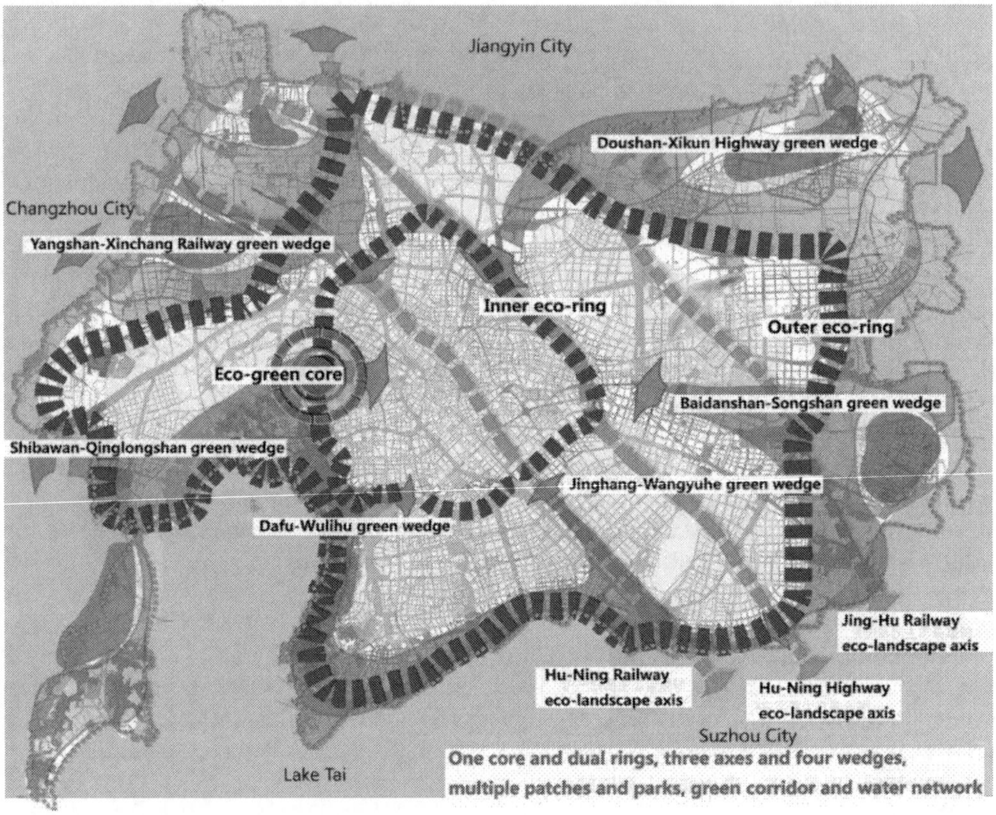

Figure 1. Urban green space system structure of Wuxi City [10].

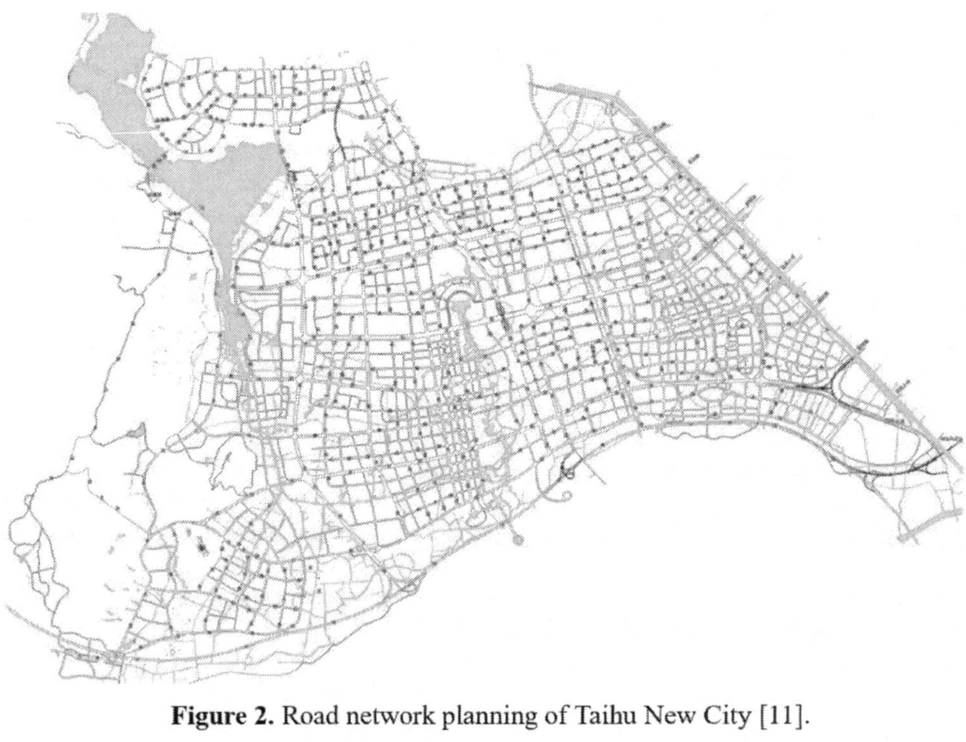

Figure 2. Road network planning of Taihu New City [11].

Figure 3. River planning of Taihu New City [11].

2.2. Rainwater planning of Taihu New City

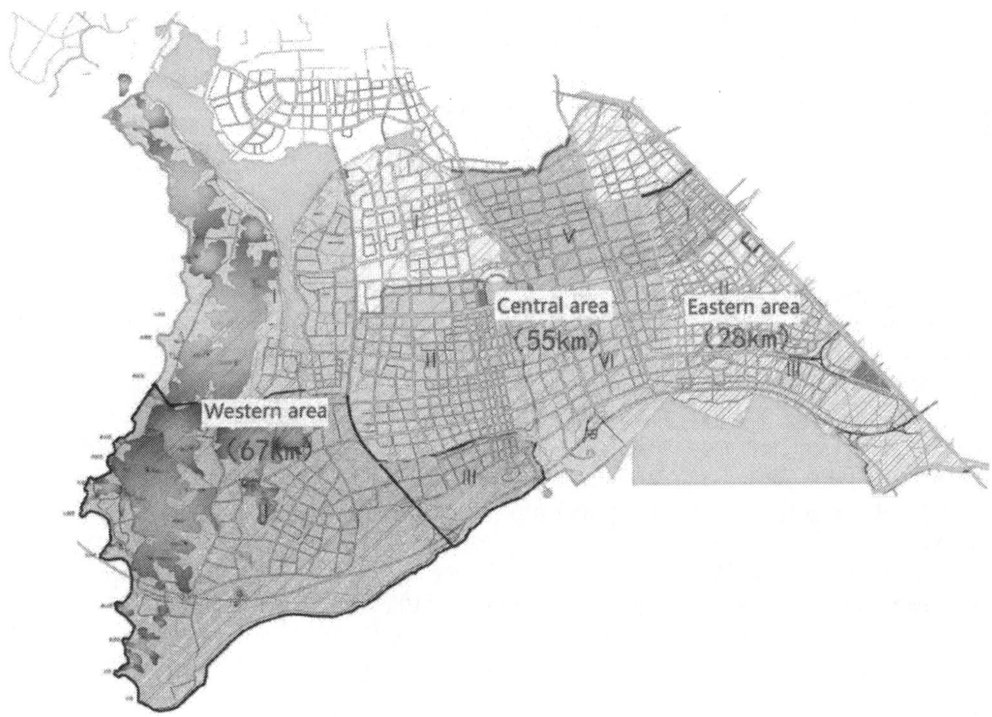

Figure 4. Rainwater planning of Taihu New City [11].

The rainwater pipes in Taihu New City add for a total of 150 km^2. The pipe network is divided into 3 major rainwater drainage parts according to the functional division of the new city: eastern, central, and western area. Each of these are designed with an independent drainage system, and are further divided into many smaller independent drainage units (figure 4).

The rainwater is discharged into the rivers that were planned to accept drainage. Main drainage rivers are given in table 1.

Table 1. Main rivers in Taihu New City [11].

No.	Name	Origin-destination	Remark
1	Changguang River	Dongli Lake-Henggang Riverlet	Main rivers
2	Miaoqiao Riverlet	Shangxian River-Changguang River	
3	Renzi Riverlet	Miaoqiao Riverlet-Henggang Riverlet	
4	Liangtang River	Dongli Lake-Jing Hang Canal	
5	Nanda Riverlet	Shangxian River-Li River	
6	Lihe River	Liangtang River-Lake Tai	
7	Daxigang	Jing-Hang Canal-Lake Tai	
8	Jing-Hang Canal	Liangtang River-Wangyu River	
9	Henggang Riverlet	Lake Tai-Renzi Riverlet	Secondary rivers
10	Wuwang Riverlet	Changguang River-Miaoqiao Riverlet	
11	Banqiao Riverlet	Changguang River-Miaoqiao Riverlet	
12	Fangqiao Riverlet	Shanxi River-Shangxian River	
13	Shanxi River	Miaoqiao Riverlet-Banqiao Riverlet	
14	Shajinggang	Liangtang River-Nanda Riverlet	
15	Guanjiaqiao River	Miaoqiao Riverlet-Qianlongqiao Riverlet	
16	Yangmuqiao River	Li River-Shangxian River	
17	Qianlongqiao Riverlet	Miaoqiao Riverlet-Shangxian Riverlet	
18	Mianzhang Riverlet	Jing-Hang Canal-Shajing Riverlet	
19	Bashiqiao Riverlet	Li River-Daxi Riverlet	
20	Luquqiao River	Li River-Jing Hang Canal	Tributaries
21	Kanjiali River	Li River-Jing Hang Canal	
22	Shangxian River	Fangqiao Riverlet-Lake Tai	
23	Rengangzhi Riverlet	Miaoqiao Riverlet-Renzi Riverlet	
24	Zhangzhuangxiang River	Dongli Lake-Shanxi River	

2.3. Coupling of the green space system and the sponge city construction in Taihu New City
Since Taihu New City is a newly-developing area, most construction related activities should apply sponge technologies to collect and use rainwater. This paper reports on coupling the rainwater collection with its utilization in public areas such as road squares and green spaces.

The planning of the rainwater collection and the urban green space systems of Taihu New City have to consider the road network and water system of the city. For the rainwater collection system, rainwater pipes are installed along the road network, and the collected rainwater is drained to the rivers nearby. On the other side, green spaces are planned along rivers, lakes and roads so they can take advantage of the pressure in the rainwater pipes guaranteeing water supply of green spaces and in consequence, the maintenance of their basic functions (figure 5).

- The planning of the rainwater collection systems considers that all the rainwater is to be collected into the pipe network. When combining the strategies, since road green spaces and public green spaces also collect rainwater, volume entering the pipe system is reduced. The water collection in these green spaces is properly controlled, and the overflow goes into the pipes.
- All the rainwater drainage into rivers should be properly planned. Riverfront protective forest belts may also share the rainwater, which may help in reducing flood pressure. This also ensures that the forests have proper supply of the vital liquid. In this type of systems and for

control purposes, no rainwater pipes are to be installed along rivers. This is to ensure that water goes into the forest belts. Moreover, Moreover, part of the rainwater from the pipes may also be available for them.

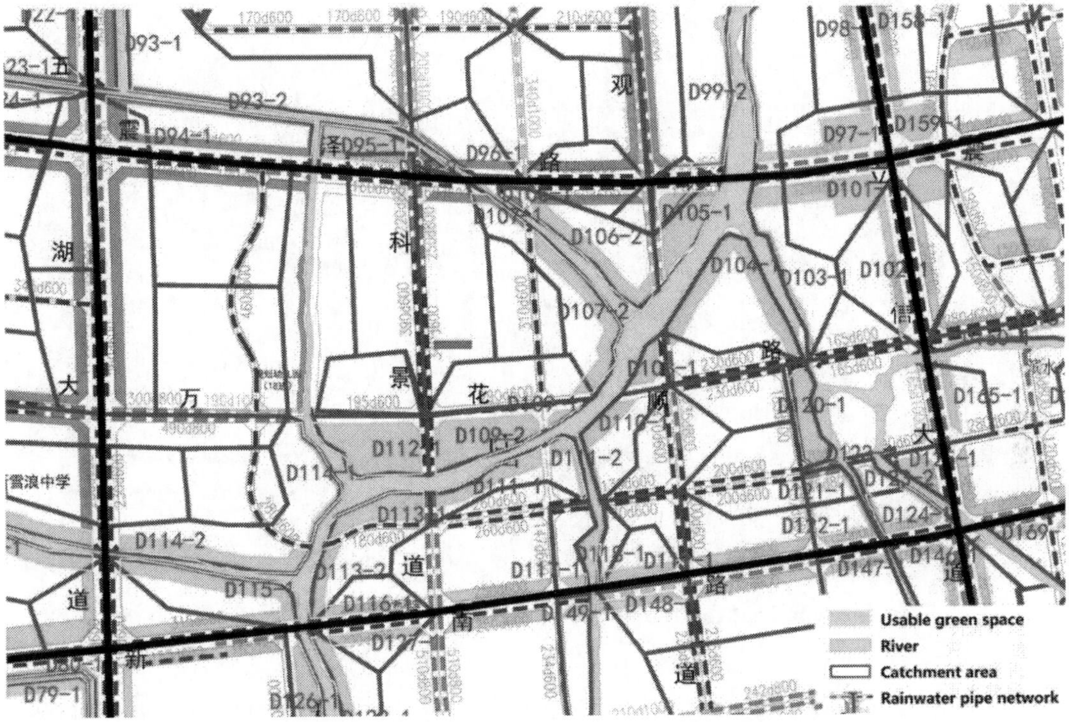

Figure 5. Green space layout for rainwater collection.

- Combining rainwater and green spaces in Taihu New City (table 2). The total size of the planned rainwater areas in Taihu New City is 2,579.46 hm^2, the annual runoff is 12.228 million m^3. Close to the collecting space for each subarea, the usable green space for the construction of the sponge city part (excluding affiliated green space) is of 882.96 hm^2 (figure 6), which may eventually share the rainwater from pipelines and rivers.

Table 2. Coupling of rainwater planning and green spaces in Taihu New City.

Rainwater collecting subareas	Area// hm^2	Runoff// m^3	Usable green space area// hm^2	Ratio of green space to rainwater subareas//%
Western I	216.09	899,044.93	100.18	46.36
Western II	293.12	1,390,200.69	100.20	34.18
Central I	169.29	761,253.01	66.58	39.33
Central II	457.53	2,240,110.89	141.74	30.98
Central III	65.37	249,292.81	35.04	53.60
Central V	363.85	1,852,859.49	97.80	26.88
Central VI	322.08	1,677,378.00	78.79	24.46
Eastern I	136.40	739,778.83	27.21	19.95
Eastern II	75.68	345,029.82	28.78	38.03
Eastern III	169.32	857,023.22	46.60	27.52
Eastern V	310.73	1,216,323.57	160.04	51.50
Total	2579.46	12,228,295.26	882.96	34.23

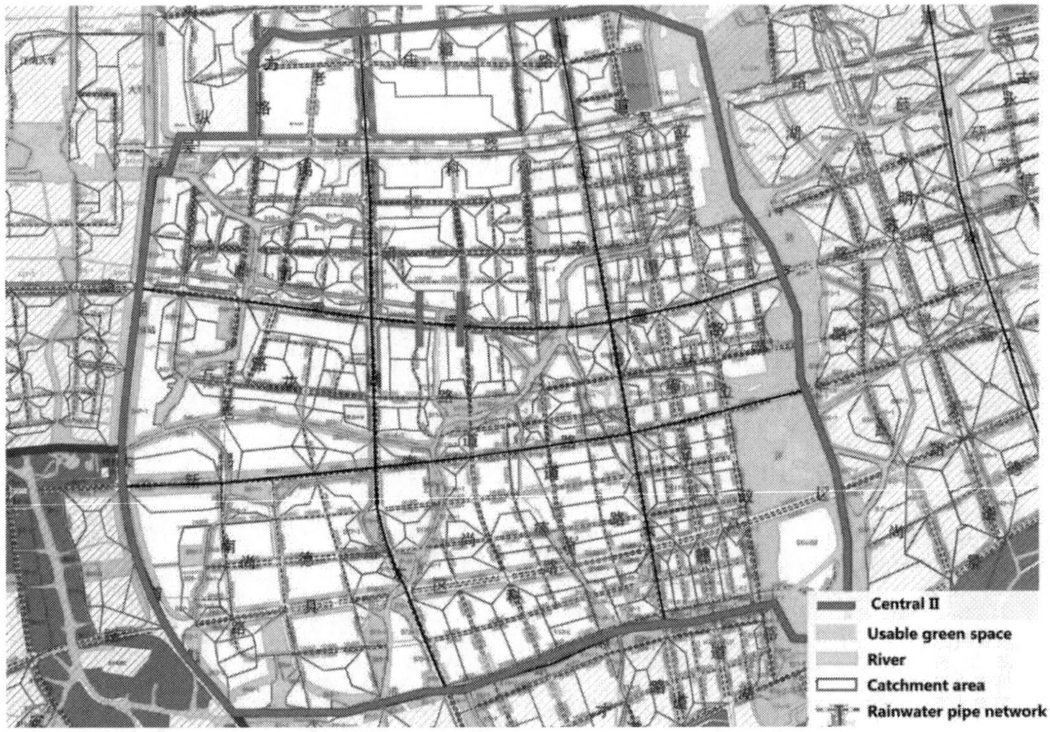

Figure 6. Distribution map of usable green spaces in Central II.

3. Results

Functional purpose of urban green spaces in coupled areas depends on rainwater interception and collection capacity in the beginning, middle and end of the rainwater collection season. In addition, sponge facilities capacities including rainwater retention, infiltration, and utilization are important. By adopting the measures shown below, runoff peak flow in Taihu New City may be delayed by 4.4 min, and the annual total runoff reduced in 15.6%.

3.1. Increasing the green space area to delay the runoff peak flow

Fan Qunjie's reported in *Research on the Effect of Urban Green Space System in Regulating Stormwater Runoff and Reducing Pollution*, that "as the ratio difference of green space area up to 10%, the drainage system with a larger green space ratio witnessed the runoff peak about 20 min later than the system with a smaller green space ratio did" [9]. Although the greening rate in the coupled area reaches the 34.23%, 2.2% more can be applied according to relevant standards of landscaping industry. This increment in green space would delay runoff peak by 4.4 min.

3.1.1. Increase green space in parks. According to the specifications for park green space service radius coverage included in *Evaluation Standards of Urban Landscaping* GB/T50563-2010, there should be a public park over 5,000 m^2 within 500 m from citizens' homes. Thus, 184 green spaces should be built within the planned rainwater collecting area of Taihu New City. This means that 114 more of these spaces should be constructed in addition to the currently existing 70. According to the previously mentioned standard of 5,000 m^2, an additional area of 57-hm^2 of park green spaces should be constructed. This would improve the greening rate by 0.4%, and delay the runoff peak by 0.8 min.

3.1.2. Broaden the protected green spaces. Within the rainwater planning area of Taihu New City, the 2 grade-II rivers display a total length of 11.71 km, while for the 6 grade-III rivers the number is of 24.57 km, the 49 grade-IV rivers sum 135.86 km, the 42 grade-V rivers 65.47 km, and the 84 grade-VI rivers add a total length of 51.36 km. The extent of protected green spaces along the rivers should comply with these numbers according to their grade as: (a) grade-I 50 m; (b) grade-II 30 m; (c) grade-III 20 m; (d) grade-IV 15 m; (e) grade-V 10 m; and (f) grade-VI 10 m (table 3). According to this standard, an area of about 404.73 hm^2 of riverfront protective green spaces should be considered. This means that at least 170.73 hm^2 more of these areas should be built, in addition to the existing 234 hm^2. This would improve the greening rate by 1.2% and delay the runoff peak by 2.4 min.

Table 3. Control regulations for riverfront protective green belts [12].

River grade	Extent of protective green spaces	Remark
Grade-I river	50 m outside from the river boundaries	For rivers with higher requirements, the greening control and construction scope can be upgraded according to the river grade, or designed according to higher planning standards.
Grade-II river	30 m outside from the river boundaries	
Grade-III river	20 m outside from the river boundaries	
Grade-IV river	15 m outside from the river boundaries	
Grade-V river	10 m outside from the river boundaries	
Grade-VI river	10 m outside from the river boundaries	

3.1.3. Increase roof greening. According to the *Ecological Water System Planning of Taihu New City*, the planned roof areas reaches a number of 9,218 hm^2. Thus, expanding the green roofs in 10% (92.18 hm^2) represents a rate increase of 0.6%, and may result in a delay of the runoff peak of 1.2 min.

3.2. Increase sunken squares and sunken green spaces to store rainwater

In the EXPO 2010 Shanghai, Rotterdam displayed its "water plaza system" provided a new option for preventing floods by optimizing the urban water circulation system. This system was built on natural terrains, and consisted of a series of pools of varying shapes, sizes and heights, connected by ditches. The place serves for recreation purposes during business hours, and in the event of a rainstorm works as a flood control system. This is because the rainwater flows to the low-level of the plaza, avoiding street flooding. All the pools conform a circulation network, where water is distributed to ditches during heavy rainfall, returning to the big pools during small rainfall. Rainwater in this system can both, flow among the pools, and be stored as a fresh water resource [5].

Table 4. Distribution of usable green spaces in the rainwater subareas.

Rainwater subareas	Usable green space//hm^2	Park green space// hm^2	Protective green space// hm^2	Others// hm^2
A	100.18	37.48	0.31	62.39
B	100.20	26.39	17.91	55.91
C	66.58	61.58	4.27	0.72
D	141.74	118.78	20.45	2.51
E	35.04	35.04	0	0
F	97.80	24.94	72.01	0.85
G	78.79	52.24	26.55	0
H	27.21	1.14	16.91	9.17
I	28.78	5.71	19.94	3.13
J	46.60	17.39	29.18	0.03
K	160.04	26.74	72.11	61.18
Total	882.96	407.43	279.64	195.89

By building sunken squares in street parks, water retention in the green space system can be better achieved. According to the *Park Design Standards*, street park roads and pavement account for 10%-30% of the land use. In this paper, a value of 20% was selected. In addition, 5% was set up as the fraction of total land that is used for paved squares. As Clause 4.3.2 of the *Standards* specifies, "water depth of the paved-bottom waters within 2.0 m away from the bank should not exceed 0.7 m". Thus, the sunken square depth should be 0.7 m to store rainwater. In the rainwater planning area of Taihu

New City, the usable park green space area is of 407.43 hm^2 (table 4), so the sunken square area (5% of the total) will be 20.37 hm^2, and the 0.7-m depth will store about 142,600 m^3 rainwater, reducing the total runoff by 1.2%.

For high soil dampness and long-term waterlogging, there are a few alternatives of plant species that can be grown in sunken green spaces. Those species suitable for both dry and wet lands are better options, such as perennial herbs. The most selected species for sunken green spaces are rainwater garden and grass ditch. According to Clause 4 of the *Technological Standards for Urban Landscaping of Wuxi City*, "urban planning by considering regional characteristics of the city should take trees as the keynote because of the largest ecological quantity, the application area of trees and big shrubs should be over 70% of the total landscaping area, and the rest is lawns not under canopies and other ground covers". Therefore, 30% of the green space can be selected as the sunken green space (i.e. 246.89 hm^2), which may result in an increase in rainwater storage of 864,100 m^3, reducing the total runoff by 7.1% (considering an average depth of 0.35 m.)

3.3. Increase sunken green space to enhance rainwater infiltration

In determination of the infiltration rate in the Ecological Water System Planning of Wuxi Taihu New City, the following spaces and percentages were considered: sunken green space 50%, common green space 16%, and permeable pavement 8% [13]. The infiltration rate in sunken green spaces is much higher than that of common green spaces. Therefore, by setting up a certain amount of sunken green space within the shelter forests and parks, the rainwater infiltration rate would increase, the surface runoff would be reduced without changing the properties of the green space, which may additionally enhance underground water conservation and reduce the drainage pressure of rainwater pipes.

In the event 30% of the green space was of the sunken type (i.e. building 246.89 hm^2 sunken green space in Taihu New City), the infiltration rate might increase in 34%, and runoff and total runoff might be reduced by 892,300 m^3, and 7.3%, respectively.

It is suggested that sunken green spaces are designed as grass ditches, which serve as both, drainage and irrigation facilities according to the season. As for parks, they can be structured as rainwater gardens to satisfy ornamental, water infiltration and retention needs.

3.4. Increase tree population for water retention

Original big trees should be maintained in place, but more tress should be planted in order to help retaining rainwater during the rainfall season. According to Clause 14 in the *Principle for the Construction of Wuxi Urban Landscaping*, minimum rates are: "trees in green patches such as parks, small amusement gardens and courtyards should be over 5 per 100 m^2, those in green stripes such as roads and rivers should be over 9 per 100 m^2". Diversity of tree species is also required, and the suggestion is to have over 40 species for green spaces below 3,000 m^2, over 60 species for green spaces between 3,000-10,000 m^2, over 80 species for green spaces between 10,000-20,000 m^2, and over 100 species for green spaces above 20,000 m^2. The standard for road green spaces and protected green spaces can be moderately reduced.

4. Discussion and recommendations

4.1. Planning and design requirements for the sponge facilities in the coupled green spaces

4.1.1. Control Planning. Urban green space areas should be guaranteed: "by increasing green space areas by 10%, the runoff peak may be delayed by 20 min". Thus, it is an essential premise for the sponge city construction to plan green lines strictly according to national standards. In view of the current conditions, Wuxi City should carry out this plan taking care of the following aspects: (a) First, the projected area of green park spaces should increase. According to the service radius coverage of a park green space indicated in *Evaluation Standards of Urban Landscaping* GB/T50563-2010. there should be a public park with at least 5,000 m^2 within 500 m of citizens' homes. (b) Second, protective forest belts (i.e. roads, riverfront and urban cluster green spaces) should be designed according to the standards, and a reasonable extension of the forest belt should be maintained. (c) Third, the index of roof greening should be added as construction project control. Roof greening control was extensively promoted in Taihu New City, and all roofs where people can stand up were to be designed with roof greening. However, effective application areas were extremely limited for various reasons. For example, in some cases, roof space was already occupied by external air conditioner units. On the other hand, difficulties in quantitative assessment and acceptance were also factors in the low success rate of these type of projects. It seems that, because of those difficulties, some construction units abandoned the roof greening projects. Thus, it is suggested that ratio of roof greening to total roof areas where people are able to stand should be reduced to 10%. This would help in increasing the possibility of reaching this number.

4.1.2. Design Key points. In designing landscape sponge facilities, considerations should include the fact that these type of structures should comply with the quantitative requirements of appropriate standards. Also, sponge facilities should be perfectly integrated into landscape design.

- For sunken squares, the elevation relationship among roads, rainwater collection pipes and green spaces must be properly handled, and vertical design of pathways and green spaces integrated for the rainwater collection and storage. Rainwater utilization after the runoff peak should also be taken into consideration. Most importantly, every point previously mentioned should be implemented under the premise of maintaining the recreational purposes of the square during dry seasons.
- Sunken green spaces also work as rainwater gardens and ecological grass ditches. Rainwater garden designs should first focus on the optimization of plant distribution. Plants suitable for both, dry and wet seasons are best choices for rainwater gardens. In addition, complete arrangement of trees, shrubs and herbs, determines the rainwater garden landscape. Moisture intolerant trees and big shrubs can be established on elevated dikes with thick soil layers and low moisture. Herbaceous, perennials, and even aqueous grasses and flowers can be planted in low-lying lands with thin soil layers and high moisture content.
- Ecological grass ditches are mainly applied in protected forest belts for rainwater storage purposes during rainy seasons, and the water can be used to irrigate the forest woods. This saves the cost of mobile watering. When designing ecological ditches, the whole forest should be considered. One or more rainwater pools could be built in the open forest, to a rate of no more than 10% of the total forest belt area.
- The key to roof greening design is to properly cover with soil. Considering the load bearing, a thin soil layer can be applied to increase the covering depth so water retention capacity and water permeability can be improved. Also, landform characteristics should be integrated with plant arrangements. Uniformity in covering depth is not necessary. For example, according to the *Jiangsu Provincial Technological Regulation for the Application of Urban Landscape Plants*, a 40-60 cm thick layer is appropriate for planting small trees; 40 cm for shrubs; 30 cm for ground covers and vines; and 20 cm for lawns. It is also recommended that mesophanerophytes be plantes as species tolerant to drought and wind are better choices.

4.2. Overall design of greening and rainwater-collecting areas in planning the urban green space system

4.2.1. Micro adjustments of urban green space systems for green space functions conservation. According to *Evaluation Standards of Urban Landscaping* (GB/T50563-2010), the evaluation of the service radius coverage indicated that an even distribution of park green spaces is required, and 57 hm^2 more of these areas should be created in Taihu New City.

According to *Water System Planning of Taihu New City*, the creation of 170.73 hm^2 more riverfront protected forest belts are needed in Taihu New City.

4.2.2. The Role of various green spaces in rainwater collection (table 5).

- Sunken green spaces and sunken squares serve as the sponge facilities for park green spaces and other green spaces. The ratio of sunken squares should be within values of 5% paved squares and 0.7 m sunken depth; 30% sunken green space, with an average depth of 0.35 m.
- Sponge facilities are not to be established in road green spaces by principle. However, for green belts over 5 m wide, sponge facilities for protected green spaces can be created.
- In residential and working areas, roof greening, sunken squares and sunken green spaces can be used for rainwater collection.
- In protected green spaces, grass ditches and rainwater pools can be used as sponge facilities. The ratio of grass ditches should be of 5% of the green space area. This should be first used as a drainage facility, and also as irrigation channel in dry seasons. One or more rainwater pools can be designed in open forests, and the size maintained to a 10% of the forest belt.

Table 5. Sponge facilities for various green spaces.

Category	Sponge facilities							Role						Area control
	Sunken green space	Sunken square	Grass ditch	Rainwater pool	Roof greening	iIncreasing planning area	Increasing tree application	Infiltration	Storage	Retention	Purification	Use	Drainage	
Park green space	√	√				√	√	√	√	√		√		Sunken square 5%, 0.7 m deep; sunken green space 30%, 0.35 m deep; grass ditch 5%, 0.35 m deep; rainwater pool 10%, 0.35 m deep; roof greening 10%.
Protective green space			√	√		√	√	√	√	√		√	√	
Green spaces for working and residential areas	√	√			√		√	√	√	√		√		
Road green spaces														
Others	√						√	√	√	√				

5. Conclusion

Urban green spaces are highly efficient sponges. It is very important that planning and construction of sponge cities comply with national standards. By building sponge facilities with the premise of guaranteeing the ecological conditions and recreational purposes of green spaces, and ensuring that quantitative indexes are not disregarded, the peak runoff of Taihu New City may be delayed by 4.4 min, and the annual total runoff reduced by 15.6%. Therefore, with strict planning and design control, urban green space systems based on the coupling of urban green spaces and sponge facilities will synergistically improve individual characteristics and may be promote the construction of both urban green space systems and sponge cities.

References

[1] Che S Q, Xie C K, Chen D *et al* 2015 Development and constructive approaches for theories and technologies of sponge city system *Chinese Landscp. Archit.* **2015(5)** 11-5

[2] Li Y F and Wang D M 2011 Research progress on the role of sunken green spaces for the storage and utilization of urban rainwater *South-to-North Water Transf. and Water Sci. &*

Tech. **2011(1)** 160-5

[3] Deletic A, Hatt B and McCarthy D *et al* 2014 Ecological landscape—stormwater biofiltration for delivering multiple benefits to cities *Chinese Landscp. Archit.* **2014(4)** 25-33

[4] Wong H F T and Wang J B 2014 Ecological landscapes, water sensitive urban design and green infrastructure *Chinese Landscp. Archit.* **2014(4)** 20-4

[5] Yu S W, Ding N, Ren X X *et al* 2010 Exploration of stormwater storage, infiltration and utilization in urban sunken green spaces *Water & Wastewater Eng.* **2010(S1)** 116-8

[6] Su Y J, Wang S S, Che W *et al* 2014 Sunken green spaces based on the concept of "sponge city" *South Archit.* **2014(3)** 39-43

[7] Zhang J L and Zhang Z Z 2012 Storage and penetration capacity of sunken green spaces and its influence factors *Water Saving Irri.* **2012(1)** 44-7

[8] Li J Q, Che W, Chi L *et al* 2004 Key parameters and influence factors for the low-lying green spaces in residential areas *Water & Wastewater Eng.* **2004(9)** 41-6

[9] Fan Q J 2006 Research on the effect of urban green space system in regulating stormwater runoff and reducing pollution (China National Knowledge Infrastructure)

[10] Wuxi Bureau of Planning, Wuxi Bureau of City Services and Landscaping 2010 *Planning of Wuxi Urban Green Space System*

[11] Wuxi Municipal Design 2007 *Regulatory Planning of Wuxi Taihu New City*

[12] Wuxi Bureau of Water Conservancy 2014 *Water System Planning of Wuxi Taihu New City*

[13] Shanghai Jiaotong University 2008 *Ecological Water System Planning of Wuxi Taihu New City*

IOP Conf. Series: Earth and Environmental Science **346** (2019) 012015 doi:10.1088/1755-1315/346/1/012015

Spatiotemporal variation of vegetation dynamics and correlations with climatic factors in the Tibetan Plateau, China

Q L Fang[1,2], L X Zhou[1,2], L P Zhu[1,2], X J Zhang[1,2], J Yang[1], J Zhou[1,2], W W Song[1,2], X J Yu[1,2] and Z H Luo[1,2,3]

[1]South China Institute of Environmental Sciences, Ministry of Environmental Protection, Guangzhou 510655, China
[2]State Environmental Protection Key Laboratory of Urban Ecological Simulation and Protection, Guangzhou 510655, China

E-mail: luozhaohui@scies.org

Abstract. The Tibetan Plateau has been recognized as one of the most sensitive areas responding to climate change, and has becomes a hotspot for coupled studies on terrestrial ecosystem variation and climate change. Leaf area index (LAI) is a key indicator that reflects vegetation dynamics and has been widely used to analyze the responses of vegetation to climate change. In this study, the spatiotemporal variation of LAI in the growing season and its correlations with climatic factors were analyzed. The results showed that the spatial pattern of LAI decreased from southeast to northwest. In terms of the temporal trend of LAI, 85% of the total study area experienced an increased trend. Additionally, 74% of the whole plateau will experience an improved vegetation growth in the future. Furthermore, temperature, precipitation and solar radiation all showed positive correlations with LAI for most of the study area. Our results effectively revealed the variation of LAI and its correlations with climatic factors. However, grassland in the plateau have been shown to have a greater and more rapid response to climatic fluctuations. Therefore, more managements should be made by local governments to improve the fragile environment, especially for areas with a decreasing LAI trend.

1. Introduction

As a key connectivity among atmosphere, pedosphere, and hydrosphere, vegetation plays an irreplaceable component in atmosphere adjustment, soil conservation, and the maintenance of climatic and ecosystem stability [1,2]. Due to the critical role of vegetation coverage in regulating the balance of regional ecosystems [1,3,4], the variation of vegetation dynamics has been recognized the basis of the protection of the ecological environment [3,5].

With wide coverage and long time series span, freely remote sensing data with high temporal resolution has recognized as the most important data source for monitoring the variation of vegetation dynamics over long time period at large scales [6-8]. Widely applied as the parameters of vegetation dynamics, vegetation index such as normalized difference vegetation index (NDVI) has been most widely used as an effective index for monitoring vegetation dynamics [1,9]. Based on the National Oceanic and Atmospheric Administration (NOAA) Advanced Very High-Resolution Radiometer (AVHRR) NDVI, Duan *et al* [10] found a significant increasing trend of NDVI for most part of Xinjiang Province during 1981-2006. Moreover, the relationships between NDVI and climatic factors have also

Content from this work may be used under the terms of the Creative Commons Attribution 3.0 licence. Any further distribution of this work must maintain attribution to the author(s) and the title of the work, journal citation and DOI.
Published under licence by IOP Publishing Ltd

been recognized to reflect the responses of vegetation dynamics to climate change [11-13].

Alpine grassland in the Tibetan Plateau covers 34.3% of the total grassland in China [14], and plays a key role in global carbon cycles [15]. Due to the fragile ecological environment and sensitivity to climate change [16], the Tibetan Plateau act as a suitable region for the study of vegetation dynamics and its correlation with climatic factors. Numerous research have been explored vegetation dynamics in the plateau by means of NDVI and net primary productivity (NPP) as an indicator. Fang *et al* [17] found that NDVI in the plateau displayed an increasing trend during 1982-1999. Peng *et al* [3] put forward that the plateau experienced a significant increasing trend of NDVI during 1982-2003 based on linear regression analysis. Pan *et al* [18] revealed that annual grassland NDVI slightly increases with fluctuation during 1982-2013. In contrast, Wang *et al* [19] observed that NPP in the plateau showed a decreasing magnitude of 0.16 Tg C a^{-1} during 2000-2012. In terms of methodology, linear regression analysis method was mostly applied to analyze vegetation dynamics due to its simplicity and robustness [20]. However, mutation would be smoothed in the method. Moreover, the trend of vegetation dynamics is usually not a linear trend at the background of climate change and intensive anthropogenic activities occurred simultaneously, therefore, a non-linear trend in this method is unable to detected [3]. In addition, the variation trends in the future seems more significant than that of study period, because the former could really respond to vegetation change in the future [3] under the background of climate change. Unfortunately, few studies have explored the variation trend in the future [3,21].

In the present study, we took the Tibetan Plateau as a study region with the objectives of (1) to analyze the variation of grassland LAI in growing season during period 1982-2013 and after 2013; and (2) to understand the relationship between LAI and climatic factors. Our results improved the understanding of physiological and biophysical processes further, and will motivate local government to focus ecological protection and/or restoration on the plateau. Moreover, our findings could contribute to ecological construction achievements in the plateau at the background of climate change and intensive anthropogenic activities.

2. Materials and methods

2.1. Study area

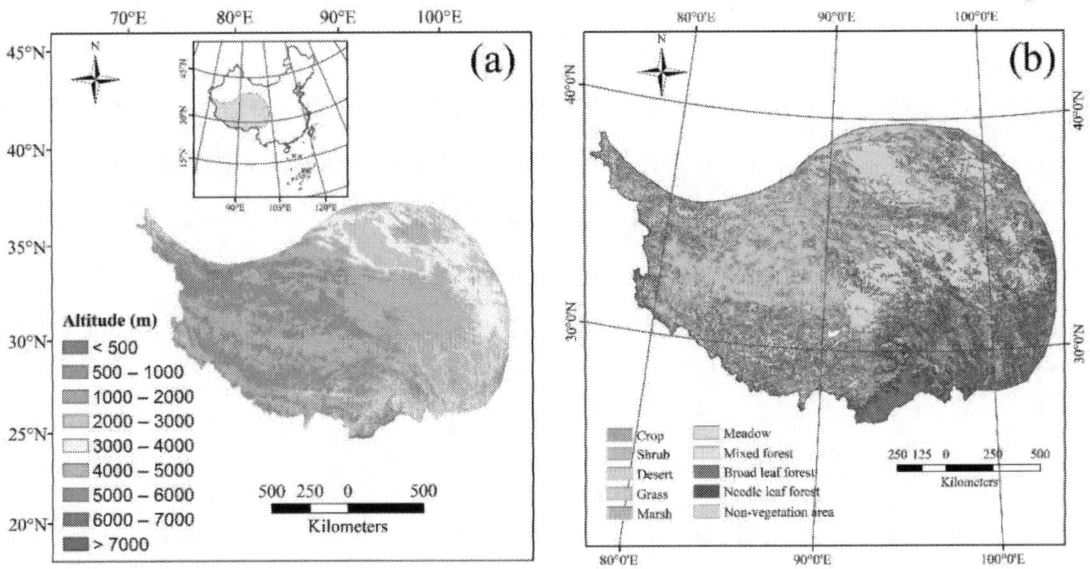

Figure 1. Location and elevation (a) as well as vegetation types (b) in the Tibetan Plateau.

The Tibetan Plateau (26°00′12″ N — 39°46′50″ N, 73°18′52″ E — 104°46′59″ E) lies in the southwest of China, covering nearly 2.5 million square kilometers with an average elevation of over 4000 m (figure 1(a)). It is the largest plateau and most extensive highland in the world, and known as roof of the world [16]. The plateau has profound thermal and dynamical effects on climate and atmospheric circulation because of its unique topography and geographical location. The major climate features in the plateau were characterized by low air temperature, strong solar radiation and clear demarcation [22]. Mean annual temperatures and mean annual precipitation ranging from −5°C to 12°C and more than 800 mm to less than 200 mm, respectively [23], and the vegetation types were dominantly by alpine meadows, alpine steppes, shrubs and forests (figure 1(b)).

2.2. Data sources

Datasets used in this study include Leaf area index (LAI), defined as one-half of the total leaf area per unit ground surface [24], climate data, and vegetation type data. Global LAnd Surface Satellite (GLASS) LAI product with a spatial and temporal resolution of 0.05° and 8 days respectively from 1982-2013 were applied in this study. The LAI values of the dataset were retrieved by general regression neural networks, and provides much better quality and accuracy than other LAI datasets such as CYCLOPES LAI dataset. To analyze the correlation between LAI and climatic factors, site observed temperature, precipitation and solar radiation from 1982 to 2013 were download from the China Meteorological Data Sharing Service System. It has been illustrated that the missing and false data in this dataset has been excluded [2]. Kriging spatial interpolation method was applied to climate data, and then a spatial resolution of 0.05° were then output to consistent with the LAI dataset. Because the dominant vegetation type in the plateau was grassland, and our purposes were to explore the variation of grassland LAI dynamics and its correlation with climatic factors, therefore, pixels with non-grassland (e.g., forests) in growing season were masked out based on vegetation type dataset in the plateau (figure 1(b)). Additionally, the growing season was defined from May to October according to our previous publish [25], then the mean LAI of each year for growing season were applied to further analysis.

2.3. Methods

The magnitude of LAI variation was calculated based on Theil-Sen Median analysis. It is a robust statistical method [26] to estimate noisy variable changes and/or variable changes with short time series [27]. The Theil-Sen Median trend for each pixel was calculated as follows:

$$slope_i = \text{Median}\left(\frac{LAI_m - LAI_n}{m-n}\right) \text{ for } i = 1, \ldots, N \quad (1)$$

where $slope_i$ is the magnitude of LAI variation at *i* pixel; LAI_m and LAI_n are the *LAI* values at times *m* and *n* ($m > n$), respectively.

To test the significance of the magnitude, the Mann–Kendall (M-K) method, a non-parametric statistical test [28], was applied. It has been widely applied in analyzing hydrological and meteorological time series [1]. The M-K was calculated as equations (2)-(5).

$$S_{LAI} = \sum_{q=1}^{m-1} \sum_{p=q+1}^{m} sgn(LAI_p - LAI_q) \quad (2)$$

where m is the number of data points, LAI_p and LAI_q are the data values at times p and q ($p > q$), respectively, and $sgn(LAI_p - LAI_q)$ is the sign function

$$sgn(LAI_p - LAI_q) = \begin{cases} +1, & if\ LAI_p - LAI_q > 0 \\ 0, & if\ LAI_p - LAI_q = 0 \\ -1, & if\ LAI_p - LAI_q < 0 \end{cases} \quad (3)$$

The test statistic *Z* is calculated using equation (4). It has been illustrated that the statistic *Z* is nearly normal distribution when the sample size m greater than 8 [28]. The sample size m in this study was equal to 32, and trends were tested at $\alpha = 0.05$.

$$Z=\begin{cases} \frac{S_{LAI}-1}{\sqrt{Var\ (s)}}, & if\ S_{LAI} > 0 \\ 0, & if\ S_{LAI} = 0 \\ \frac{S+1}{\sqrt{Var\ (s)}}, & if\ S_{LAI} < 0 \end{cases} \tag{4}$$

where $Var\ (s)$ is calculated as follows:

$$Var(S) = \frac{m(m-1)(2m+5)-\sum_{i=1}^{t} k_i(k_i-1)(2k_i+5)}{18} \tag{5}$$

where t is the number of tied groups and k_i denotes the number of ties of extent i.

To identify the consistency of LAI variation between the study period and after the study period, the Hurst exponent based on R/S analysis was applied. The exponent has been widely applied in many fields such as climatology, hydrology, economics, vegetation dynamics, geo-chemistry and geology [1,21,29]. The exponent was calculated as follows:

- Divide the time serious {LAI (i)} (i =1, 2, …, m) into i sub series $S_{LAI(n)}$, and for each series n = 1, …, i.
- Define the mean sequence of the time series,

$$\text{LAI}\ (i) = \frac{\sum_{n}^{i} LAI_n}{i} \quad i = 1,2, \ldots, \text{m} \tag{6}$$

- Calculate the range sequence,

$$S_{LAI(n,i)} = \sum_{n=1}^{n}(LAI_n - \overline{\text{LAI}}\ (i)) \quad 1 \le n \le i \tag{7}$$

- Create the range sequence,

$$R_i = \max_{1\le n\le i} S_{LAI(n,i)} - \min_{1\le n\le i} S_{LAI(n,?)} \quad i = 1,2, \ldots, \text{m} \tag{8}$$

- Create the standard deviation sequence,

$$S_i = \sqrt{\frac{\sum_{n=1}^{i}(LAI_n - LAI_i)^2}{i}} \quad i = 1,2, \ldots, \text{m} \tag{9}$$

- Calculate the Hurst exponent,

$$\frac{R_i}{S_i} = ci^H \tag{10}$$

where H is the Hurst exponent, and calculated based on the following fitting equation using the least squares method.

$$\text{Log}\ (\frac{R_i}{S_i})_m = \text{a} + \text{H} \times \log(m) \tag{11}$$

According to previous study [30,31], the range of H value is from 0 to 1. When the H value is equal to 0.5, it means the trend of LAI after the study period would be unrelated with that of the study period. However, when the H value is greater or lower than 0.5, it means that the LAI trend in the future is consistent or inconsistent with the study period, respectively.

To observe the relationship between LAI and climatic factors (e.g., temperature), partial correlation analysis, which excludes the effects of other variables and has been widely used to explore vegetation–climate interactions [32], was applied. Moreover, the relationship was tested at a signicicance level of α = 0.05.

3. Results

3.1. Spatial pattern of LAI

The spatial pattern of average LAI and its standard deviation for the 32 years from 1982 to 2013 are presented in figure 2. The spatial distribution of average LAI was regional difference, showing a decreasing trend from southeast to northwest. The highest value of LAI distributed in east of the plateau, with values higher than 4.46 m^2/m^2, accounted approximately 8.99% of the total area. In contrast, average LAI is usually lower than 1.00 m^2/m^2 was distributed in northwest of the plateau and accounted more than 21% of the total area. For the remaining areas, LAI varied from 1 to 4.46 m^2/m^2, and mostly occurred in middle and southwest of the plateau (figure 2(a)). The spatial distribution of standard deviation is similar to that of mean LAI (figure 2(b)). The lowest standard deviation with values lower than 0.42 m^2/m^2 were found in west of the plateau, while the highest values occurred in south of the plateau, with values mostly higher than 1.4 m^2/m^2. For the remaining areas, the standard deviation mostly varied between 0.42 and 1.40 m^2/m^2.

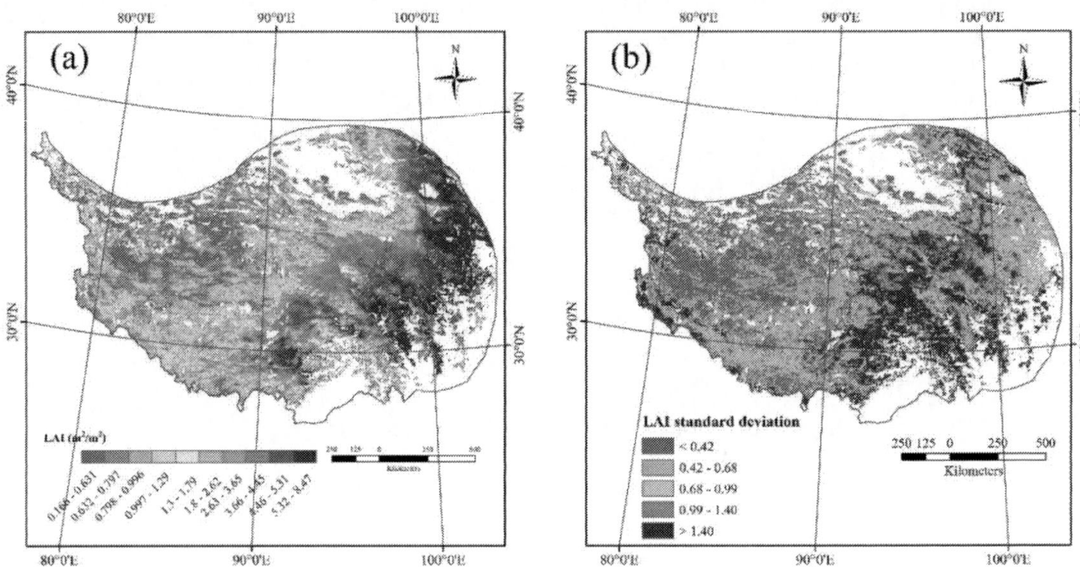

Figure 2. Spatial pattern of average LAI (a) and corresponding standard deviation (b) in the Tibetan Plateau over 1982-2013.

3.2. Temporal trend of LAI from 1982 to 2013

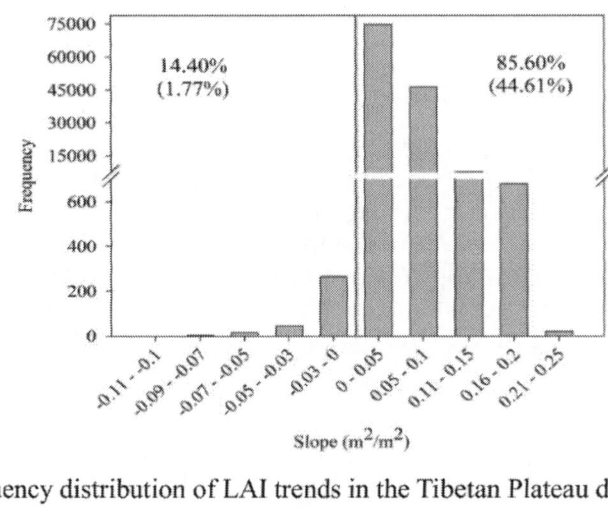

Figure 3. Frequency distribution of LAI trends in the Tibetan Plateau during 1982-2013.

Temporal trend of LAI in the growing season was illustrated in figures 3 and 4. During the study period (1982-2013), increasing trend of LAI accounted 85.60% of the total plateau, of which 44.61% displayed a significant positive trend ($p < 0.05$; figure 3). Significant positive trend were mainly occurred in northeast and southwest of the plateau (figure 4(b)). The magnitudes of increasing LAI mostly ranged from 0 to 0.1 $m^2/m^2/a$ (figure 3). In contrast, 14.40% of the plateau showed a negative trend, of which 1.77% showed a significant decreasing trend (figure 3). The decreasing magnitude of growing season LAI in the plateau mostly occurred at -0.03—0 $m^2/m^2/a$, and scattered in west and southeast of the plateau (figure 4(a)).

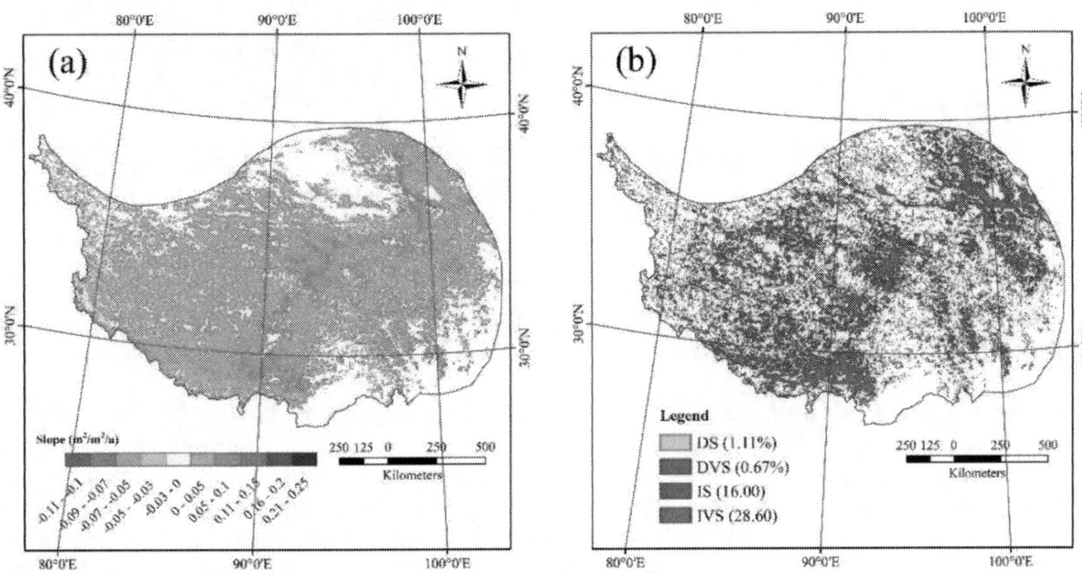

Figure 4. Trends of LAI in growing season within the plateau between 1982 and 2013 (a). Areas with significant ($p < 0.05$) and very significant ($p < 0.01$) variation are also displayed (b). Positive value means increased LAI and vice verses; DS and IS indicate significantly decreased and significantly increased LAI trend, respectively, while DVS and IVS indicate very significantly decreased and very significantly increased LAI trend, respectively.

3.3. Consistency of LAI trend analysis

As shown in table 1, areas with Hurst exponents higher than 0.5, accounted 77.06% of the whole plateau, and they were mostly found in southwest and middle of the plateau, suggesting a consistent trend of LAI variation between the study period and after the study period. Approximately 23% of the total pixels displayed that the Hurst exponents were lower than 0.5, and they mostly found in northwest and southeast of the plateau (figure 5(a)), suggesting an inconsistent trend of LAI variation between the study period and after the study period.

Table 1. Hurst exponent and sustainability of LAI in the growing season across the plateau during 1982 — 2013.

Hurst exponent	Percentage (%)	Variation types	Percentage (%)
0 - 0.1	0.35	SI	68.55%
0.1 - 0.3	3.41	USID	17.05%
0.3 - 0.5	19.18		
0.5 - 0.7	43.68	SD	8.68%
0.7 - 0.9	31.49		
0.9 - 1	1.89	USDI	5.72%

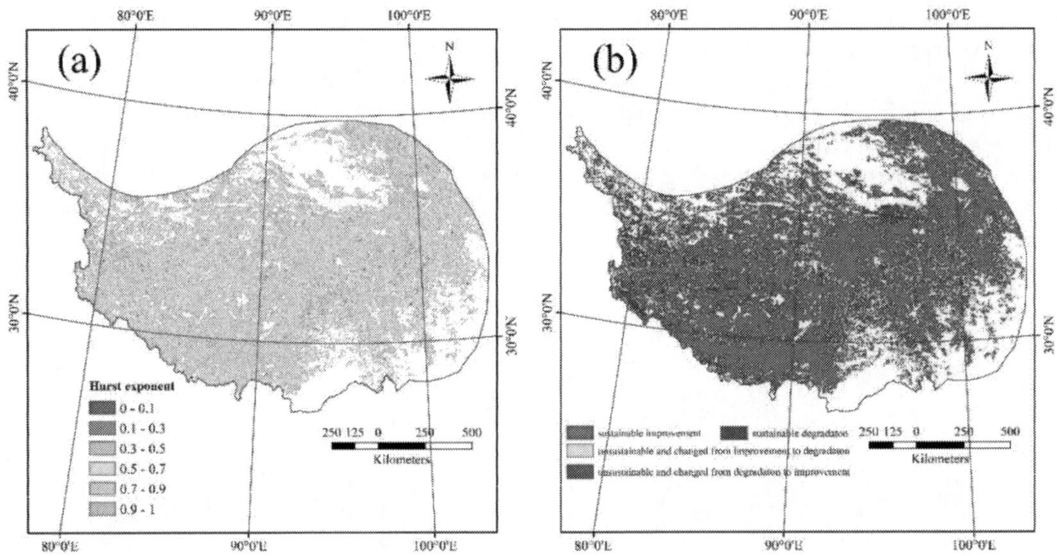

Figure 5. Spatial pattern of Hurst exponent (a) and sustainability (b) of LAI in the growing season across the plateau during 1982-2013.

Moreover, trends result from the Mann-Kendall method was superimposed with the Hurst exponent results to yield the coupled information of variation in trends sustainability. The coupling results were analyzed in four cases, namely (1) sustainable improvement (SI): a positive variation trend with the Hurst exponent higher than 0.5; (2) unsustainable and changed from improvement to degradation (USID): a positive variation trend with the Hurst exponent lower than 0.5; (3) sustainable degradation (SD): a negative variation trend with the Hurst exponent higher than 0.5; (4) unsustainable and changed from degradation to improvement (USDI): a negative variation trend with the Hurst exponent lower than 0.5.

The results from table 1 shows that the sustainable area accounted 77.23% of the total, of which 68.55% was SI, and 8.68% was SD. The SI of LAI was mostly found in southwest and east of the plateau, while the SD area was mostly located in northwest of the plateau (figure 5(b)). Approximately 22.77% of the total area showed unsustainable, of which the variation of LAI changed from improvement to degradation account for 17.05% (table 1), and they mainly scattered in north and southeast of the plateau. The variation of LAI changed from degradation to improvement account for 5.72%, and they mostly scattered in northwest and middle of the plateau. Our results indicate that approximately 74.27% of the study region displayed an improved LAI, while 25.73% of the study region undergoing a degradation trend in the future.

3.4. Correlations between LAI and climatic factors

Climatic factors displayed a positive relationship with LAI (figure 6). Approximately 76.28% and 70.73% of the total area exhibited positive relationships between LAI and temperature as well as precipitation, respectively. In addition, more than 11% of the total displayed significant positive correlations between LAI and temperature as well as precipitation ($p < 0.05$). The positive relationship between LAI and temperature mainly occurred in east and middle of the Tibetan Plateau, while for precipitation, they mostly found in middle and southwest of the plateau. The positive partial correlation coefficient between LAI and temperature and precipitation was mostly distributed between 0 and 0.5 (figures 6(c) and 6(f)). In terms of the correlation between LAI and solar radiation, about 65.34% of the total indicated positive, but only 5.76% of the total showed significant ($p < 0.05$; figure 6(i)). The positive relationships were mostly distributed in southwest of the plateau, while the negative relationships were mainly occurred in southeast of the plateau. The positive partial correlation coefficient between LAI and solar radiation was mostly distributed between 0 and 0.3, while the negative partial correlation coefficient was mainly ranges from -0.3 to 0 (figure 6(i)).

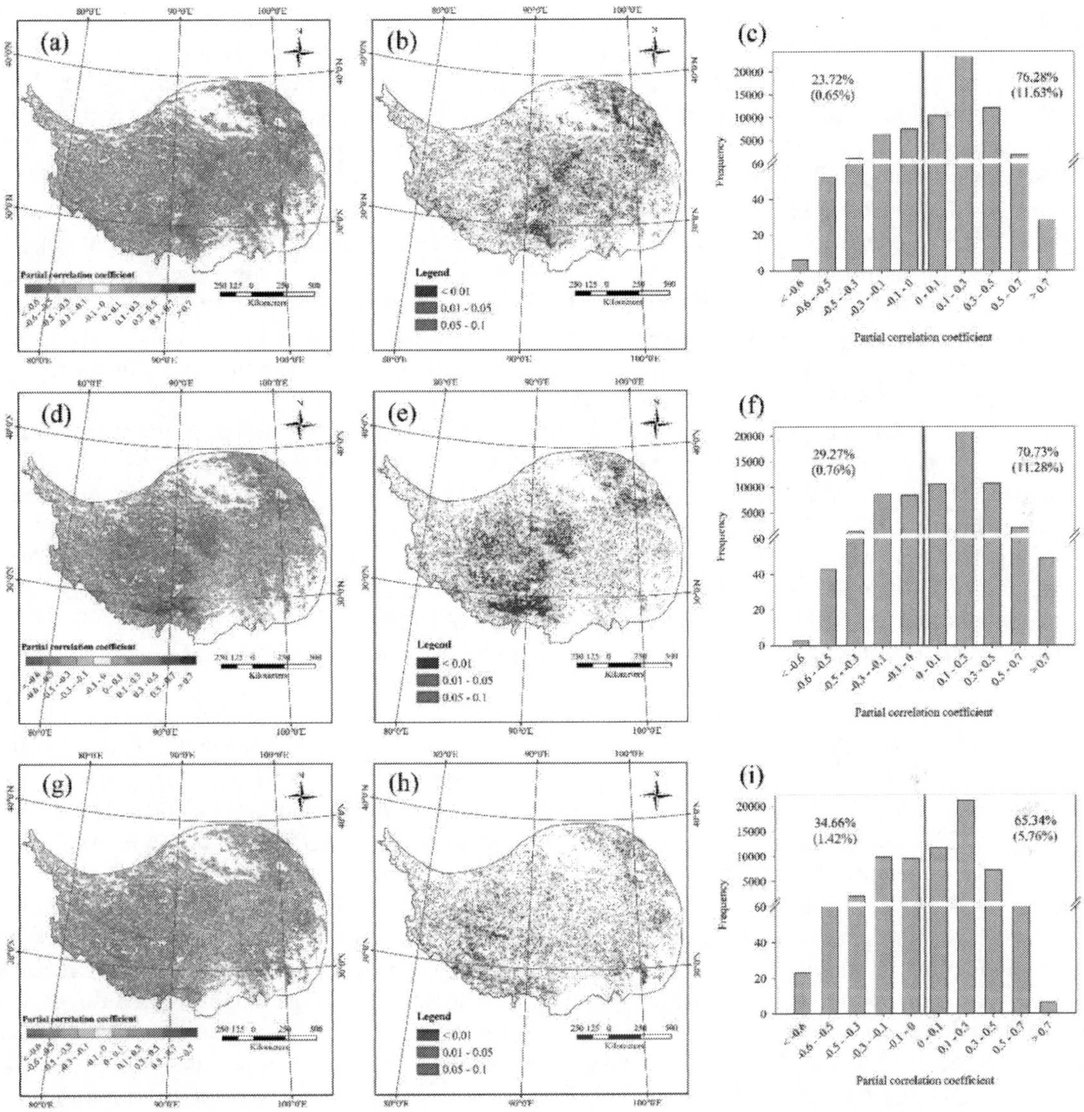

Figure 6. Partial correlation coefficients and the spatial distribution of significant relationship between LAI and (a,b) temperature, (d,e) precipitation and (g,h) solar radiation in the plateau from 1982 to 2013. The corresponding frequency distributions (c,f,i) were also displayed. Data listed within bracket were the ratio of significant pixels at the level of $p < 0.05$.

4. Discussion

4.1. Spatiotemporal variation of LAI

Generally, the spatial distribution of LAI in growing season on the plateau decreasing from southeast to northwest, which was consistence with the spatial pattern of the hydrothermal conditions in the plateau [33]. The highest LAI with value more than 4.46 m^2/m^2 was displayed in east and southeast of the plateau, and mainly due to monsoons that providing a favorable hydrothermal conditions in growing season for vegetation growth [16]. In contrast, LAI in northwest was mostly lower than 1 m^2/m^2, which may cause by low temperature and insufficient water supply because of the high elevation and inland location.

Previous studies show that environment conditions in the plateau have changed heavily [34-36]. For instance, climate changes in the plateau have been characterized by increasing temperature and decreasing precipitation [16], which bring a warmer and drier environment for hindering vegetation growth [34]. In addition, anthropogenic activities such as agriculture use, unsustainable practices (e.g., over-logging), urbanization and collection of hurb medicines [35-37] were also responsible for the harsh environment that hampered vegetation growth. However, our results revealed that more than 85% of the total displayed an increasing trend of the growing season LAI in the plateau during 1982 — 2013), suggesting little impacts of harsh environment conditions on vegetation growth. Alternatively, the increased LAI for most of the study area may also reflects that vegetation in these areas have their own strategies to adapt the current changing conditions. However, the trend of LAI in west and southeast of the plateau were decreasing during the study period. Therefore, more attentions should be paid and more managements such as local government in adopting policies or projections should be taken to resist harsh environment conditions and to promote vegetation growth in the future.

The Mann-Kendall analysis together with Hurst exponent revealed the future sustainability of the LAI trends in the plateau. Our results illustrated that the Hurst exponent was larger than 0.5 in most of the plateau, which indicated a high consistency of LAI variation after year 2013. Furthermore, among the areas with the consistency, 8.68% experienced consistency degradation, and mainly distributed in hinterland. This may cause by relatively stable harsh environment conditions such as warming-drying climate in the northwest of the plateau [3]. Approximately 22.77% of the total area displaying inconsistency of future LAI trend, which might be due to warming and drying after late 1990s [16,38] and the intensified anthropogenic activities such as overgrazing and urbanization [3].

4.2. Relationship between LAI and climatic factors

Previous studies have illustrated that climatic factors are the critical factors that affect vegetation dynamics [16,39]. Overall, temperature in growing season were positively correlated with LAI, accounted more than 76% of the total, which was consistence with previous studies [16,21]. This may be caused by high temperature that provides a warm environmental conditions for vegetation growth especially for vegetation that growing in high elevation. In addition, the activity of photosynthetic enzymes would be increased by a warmer temperature and then enhanced the capacity of vegetation for photosynthesis [40,41]. Alternatively, high temperature usually promotes the thaw of frozen soil and glacier in the plateau, which provide more water for vegetation growth. The correlations between LAI and precipitation supported this conclusion. More than 70% of the total area displayed that the growing season precipitation have a positive relationship with LAI. This is because grassland usually has a shallow root depth, and the water utilized by grasses usually comes from superficial soil profiles. As such, more precipitation usually results in increase of soil moisture, which can meet water requirements for vegetation growth, especially for areas with a decreasing trend of precipitation [16,19]. More than 65% of the whole area showed a positive relationship between the growing season LAI and solar radiation. This may because sufficient solar radiation supply enough energy for vegetation photosynthesis, which enhanced photosynthesis and promote vegetation growth. Previous studies have illustrated that alpine grassland in the plateau have been shown to have a greater and more rapid response to climatic fluctuations [42]. If grass in the plateau could adapted the fluctuations after responses, more protection and managements should be made for this kind of grassland to avoid unsustainable practices (e.g., over-logging). However, if they cannot, more managements should be made to help them to adapt or replaced them by adapted grassland. Therefore, more managements should be made by local governments to improve the fragile environment to promote vegetation growth.

4.3. Limitations and future research directions

Hurst exponent, served as an index to identify the consistency of variable variation between the study period and after the study period, has been widely applied in various fields, such as geology, hydrology, geochemistry and climatology to distinguish sustainability of time series data [29]. However, it could not analysis how long the sustainability or unsustainability of the predicted LAI trends would continue into the future, despite in the importance for forecasting future LAI variation. Therefore, it is urgent to determining the duration of LAI trends. Additionally, the variation of LAI was influenced by various factors, such as elevation, soil characteristics (e.g., temperature and water content), CO_2 concentrations, nitrogen enrichment and deposition, and local government policy and planning changes. However, only three climatic factors were included for our study. Therefore, the factors mention above should also be considered in future studies. Although it is acknowledged that the spatial resolution will get more and more coarse along with the enlarging of the study area, the spatial resolution of LAI used in this study made it difficult to validate the accuracy of its variation. Therefore, a new remote sensing dataset with more fine spatial and time resolution and long time series for LAI variation analysis need to develop in the future.

5. Conclusions

the growing season GLASS LAI dataset and climatic gridded data were applied to analysis the spatiotemporal change of LAI and its correlation with climatic factors in the plateau during 1982 — 2013. Our results indicated that the spatial distribution of average LAI was area difference, and showed gradients decreasing from southeast to northwest. In addition, more than 85% of the total study area experienced an increased LAI trend. However, 14.40% of the total area displayed a negative trend, and scattered in west and southeast of the plateau. According to the superimposed of Mann-Kendall analysis and Hurst exponent analysis, more than 77% of the whole area has high consistency of future LAI variation trends with the former trend during the period of 1982-2013, of which 68.55% was sustainable improvement. Additionally, nearly 6% of the total area displayed unsustainable and LAI trend changed from decrease to increase in the future. This meant that more than 74% of the whole plateau will experience an enhanced vegetation growth in the future. Furthermore, temperature, precipitation and solar radiation all showed positive relationship with LAI for most of the study area.

Acknowledgments

This study is funded by the Central Fund Supporting Nonprofit Scientific Institutes for Basic Research and Development (Grant No. PM-zx421-201803-056).

References

[1] Jiang W L, Yuan L H, Wang W J, Cao R, Zhang Y F and Shen W M 2015 *Ecol Indic* **51** 117-26
[2] Luo Z H, Song Q M, Wang R, Zeng H M, He T, Zhang H J and Wu W C 2018 *ISPRS Int J Geo-Inf* **7** 1-14
[3] Peng J Z, Liu Z H, Liu Y H, Wu J S and Han Y N 2012 *Ecol Indic* **14** 28-39
[4] Luo Z H, Tian D L, Ning C, Yan W D, Xiang W H and Peng C H 2015 *Environ Earth Sci* **74** 4549-59
[5] Zhang Z M, Coilie F V, De Clercq E M, Ou X K and De Wulf R 2013 *Ecol Indic* **31** 49-58
[6] Fu X Y, Tang C J, Zhang X X, Fu J Y and Jiang D 2014 *Ecol Indic* **40** 102-8
[7] Fensholt R and Proud S 2012 *Remote Sens Environ* **119** 131-47
[8] Zhang T T, Zeng S L, Gao Y, Ouyang Z T, Li B, Fang C M and Zhao B 2011 *Ecol Indic* **11** 1552-62
[9] Fensholt R T *et al* 2012 *Remote Sens Environ* **121** 144-58
[10] Duan H C, Yan C Z, Tsunekawa A, Song X, Li S and Xie J L 2011 *Environ Earth Sci* **64** 1011-20
[11] McPartland M Y, Kane E S, Falkowski M J, Kolka R, Turetsky M R, Palik B and Montgomery R A 2019 *Global Change Biol* **25** 93-107
[12] Li G, Sun S B, Han J C, Yan J W, Liu W B, Wei Y, Lu N and Sun Y Y 2019 *Sci Total Environ* **660** 177-87

[13] Zheng K, Zhou J Z, Pei J Y, Chen H, Zhang X L, Huang F Q, Li F M and Ye J S 2019 *Sci Total Environ* **660** 236-44
[14] Fan J H, Zhong H P, Harris W, Yu G R, Wang S Q, Hu Z M and Yue Y Z 2008 *Climatic Change* **86** 375-96
[15] Ni J 2002 *J Arid Environ* **50** 205-18
[16] Luo Z H, Wu W C, Yu X J, Song Q M, Yang J, Wu J H and Zhang H J 2018 *Remote Sens* **10** 1-22
[17] Fang J, Piao S L, He J S, Ma W H 2004 *Sci China Ser C* **47** 229-40
[18] Pan T, Zou X T, Liu Y J, Wu S H and He G M 2017 *Ecol Eng* **108** 307-17
[19] Wang J J, Dong J, Yi Y, Lu G, Oyler J, Smith W K, Zhao M Liu J and Running S 2017 *J Geophys Res-Biogeo* **122** 261-78
[20] Fensholt R, Rasmussen K, Nielsen T T and Mbow C 2009 *Remote Sens Environ* **113** 1886-98
[21] Jiapaer G, Liang S L, Yi Q X and Liu J P 2015 *Ecol Indic* **58** 64-76
[22] Zhong L, Ma Y M, Salama M S and Su Z B 2010 *Climatic Change* **103** 519-35
[23] Shen M Y, Tand Y H, Chen J, Zhu X L and Zheng Y H 2011 *Agr Forest Meteorol* **151** 1711-22
[24] Chen J M and Black T A 1992 *Plant Cell Environ* **15** 421-9
[25] Luo Z H and Yu S X 2014 *Remote Sens* **9** 1-19
[26] Sen P K 1968 *J Am Stat Assoc* **63** 1379-89
[27] Zhou J H *et al* 2016 *Sci Total Environ* **572** 119-28
[28] Neeti N and Eastman J R 2011 *Trans. GIS* **15** 599-611
[29] Hou X Y, Wu T, Yu L J and Song Q 2012 *Acta Ecologica Sinica* **32** 297-304
[30] Hurst H E 1951 *Trans Am Soc Civil Eng* **116** 770-99
[31] Mandelbrot B B and Wallis J R 1969 *Water Resour Res* **5** 967-88
[32] Wu D H, Zhao X, Liang S L, Zhou T, Huang K C, Tang B J and Zhao W Q 2015 *Global Change Biol* **21** 3520-31
[33] Gao Q Z, Guo Y Q, Xu H M, Ganjurjav H, Li Y, Wan Y F, Qin X B, Ma X and Liu S 2016 *Sci Total Environ* **554** 34-41
[34] Cui X F, Graf H F, Langmann B, Chen W and Huang R H 2006 *Global Planet Change* **54** 33-56
[35] Houghton R A and Hackler J L 2003 *Global Biogeochem Cy* **17** 1-19
[36] Liu J Y, Tian H Q, Liu M L, Zhuang D F, Melillo J M and Zhang Z X 2005 *Geophys Res Lett* **32** 1-5
[37] Zhang Y, Zhang C B, Wang Z Q, Chen Y Z, Gang C C, An R and Li J L 2016 *Sci Total Environ* **563** 210-20
[38] Wang Q C, Qin N S, Tang H Y, Chen F and Qi R Y 2007 *Arid Zone Res* **24** 234-9
[39] Richardson A D, Keenan T F, Migliavacca M, Ryu Y, Sonnentag O and Toomey M 2013 *Agr Forest Meteorol* **169** 156-73
[40] Shi C G, Sun G, Zhang H X, Xiao B X, Ze B, Zhang N N and Wu N 2014 *Plos One* **9** 1-10
[41] Liu Q, Fu Y S H, Zeng Z Z, Huang M T, Li X R and Piao S L 2016 *Global Change Biol* **22** 644-55
[42] Shen M G, Piao S L, Dorji T, Liu Q, Cong N, Chen X Q, An S, Wang S P, Wang T and Zhang G X 2015 *Natl Sci Rev* **2** 454-67

IOP Conf. Series: Earth and Environmental Science **346** (2019) 012016 doi:10.1088/1755-1315/346/1/012016

Identification of suitable culture media for stimulation of *Dendrobium* 'Earsakul' growth

S Hualsawat[1], K Ngoenyu[1], K Phanya[1], S Inthaisong[1], S Moranok[1], S Promaunand[1] and P A Tantasawat[1,2]

[1]School of Crop Production Technology, Suranaree University of Technology, Muang District, Nakhon Ratchasima, Thailand

E-mail: piyada@sut.ac.th

Abstract. *Dendrobium* is one of the most important economic cut-flowers in Thailand because of its attractive floral characteristics and popularity among consumers worldwide. However, its slow growth has prevented rapid mass micropropagation. The objective of this work was to stimulate *Dendrobium* growth *in vitro* using supplements, plant growth regulators and natural extracts. *Dendrobium* plantlets were cultured on VW0 media supplemented with 3 doses each of the energy drink Ready® (5, 10 and 15 mL/L), 6-benzylaminopurine (BAP; 1, 2 and 3 mg/L), V8® juice (100, 200 and 300 mL/L) and shrimp paste (2.5, 5 and 7.5 g/L) for 90 days. Plain VW0 media were used as control. The results showed that at 90 days the highest fresh weight (0.39 ± 0.05 g), plant height (3.99 ± 0.23 cm) as well as total root length (7.08 ± 1.30 cm) was obtained on VW0 media supplemented with 1 mg/L BAP. These were 1.8, 1.5 and 2.5-fold higher than those observed on plain VW0 media, respectively. However, the number of roots and number of leaves of plantlets cultured on this medium were not statistically significant. Results suggest that this culture medium is useful in promoting *D.* 'Earsakul' growth *in vitro* and may be applied for any future commercial micropropagation.

1. Introduction

Dendrobium is one of the largest genera of orchids and is widely used in commercial cut-flower production. It has become popular worldwide because of its floral characteristics i.e., color, size, and shape [1]. To enhance its production, improved techniques/methods of micropropagation are important because of their ability to produce a significant number of quality plants throughout the year. Additionally, micropropagated plants are uniform in age, size, growth and maturity which in turn increase their economic value. They can also be disease-free and genetically true to type [2,3]. However, micropropagation of orchids in general is limited due to slow growth and thus necessitates the development of culture media to stimulate orchid growth *in vitro*. The success of tissue culture depends on the components and state of culture media, namely, macronutrients, micronutrients, plant growth regulators, vitamins, carbon sources as well as organic compounds [4], which are necessary for the growth and development of *in vitro* cultured plantlets.

Plant growth regulators (PGRs) affect plant growth and cell differentiation. They have been supplemented in culture media to induce the physiological processes of plant tissues, allowing them to grow and develop *in vitro* [5]. Generally, cytokinin is the most common PGR group that is widely used in micropropagation. 6-Benzylaminopurine (BAP) is also largely used because of its high efficiency in stimulating cell elongation, shoot proliferation and growth of lateral buds. Previous studies showed that BAP was efficient for stimulation of growth and development of orchids, namely,

Content from this work may be used under the terms of the Creative Commons Attribution 3.0 licence. Any further distribution of this work must maintain attribution to the author(s) and the title of the work, journal citation and DOI.
Published under licence by IOP Publishing Ltd

Cymbidium finlaysonianum, *Vanda tessellate* and *Violet phalaenopsis* [6-8]. In addition to BAP, shrimp paste derived from either shrimp or krill contains kinetin which is the essential precursor of chitosan helps in increasing the cell division of plants [9,10]. There are a few reports describing the effectiveness of shrimp paste in accelerating the root formation of 'Nam Dok Mai' rose apple ex vitro [9]. However, it has not been used in the tissue culture of orchids as a supplement.

Vitamins are essential components for plant growth and development when they are directly or indirectly combined with other media components [11]. This study aimed to reduce the costs of micropropagation by using additives in the form of common products available on the market, e.g. the energy drink Ready® containing goji berry and mixed berry juices, V8® and shrimp paste. Ready® possesses vitamins, especially vitamin C (ascorbic acid), which has antioxidant effects on scavenging free radicals, i.e. hydroxyl radicals (•OH), hydrogen peroxide (H_2O_2), and singlet oxygen ($^1O_2$) [12-15]. Therefore, plant tissues are protected from high light intensity stress by vitamin C. In addition, it helps preserve nitrate which is an important nutrient for growth [16]. Chaikhiri and Chouychai [17] reported that 10 mL/L of Ready boot® was the most appropriate supplement for *in vitro* growth of *Cymbidium finlaysonianum,* especially for shoot and root length, number of leaves and roots as well as fresh weight. Vegetable juice 'V8®' is also considered appropriate because it contains water and the concentrated juices of tomatoes, carrots, celery, beets, parsley, lettuce, water cress and spinach including vitamin C, beta carotene as well as citric acid, which are precursors that participate in various biological processes in plants such as photosynthesis, photomorphogenesis, photoprotection, and development [18-20]. Both products are rich in vitamins and are cheaper than PGRs and therefore considered as alternative additives in the culture media of orchids. The objective of the present study is to investigate the effects of various culture media for the stimulation of *D.* 'Earsakul' growth, especially during a culture period of 90 days. The knowledge gained can be applied to provide maximum benefits to farmers and entrepreneurs for the micropropagation of orchids and other plant species for commercial purposes.

2. Materials and methods

2.1. Plant materials
Lengths of approximately 2.4 - 2.7-cm plantlets of *D.* 'Earsakul' were initially grown on Vacin and Went 0 medium (VW0); VW basal salts, 100 g/L potato (*Solanum tuberosum*), 200 mL/L coconut water, 50 g/L 'Hom Thong' banana, 2% (w/v) sucrose and 0.2% (w/v) activated charcoal [21] for the present experiment.

2.2. Culture media for growth stimulation
The VW0 medium was developed for the *in vitro* culture of many orchid species [21-23]. This medium was supplemented with the energy drink Ready® (5, 10 and 15 mL/L; T.C. Pharmaceutical Industries Company Limited, Thailand), 6-benzylaminopurine (BAP) (1, 2 and 3 mg/L; Acros Organics, New Jersey, USA), V8® juice (100, 200 and 300 mL/L; Campbell Soup Company, Camden, New Jersey, USA) and shrimp paste (2.5, 5.0 and 7.5 g/L; SGS (Thailand) Limited, Thailand). Plain VW0 medium was used as control. The pH of the culture media was adjusted to 5.0 before gelling with agar (0.7% w/v). All culture media were dispensed in volume of ca.15 ml into glass bottles (4 oz.). The media were autoclaved at 121°C for 20 min at 103.4 KPa pressure. Roots were cut from plantlets under aseptic conditions, and the plantlets were transferred to various media randomly. Cultures were maintained at 25±2°C during a 16/8 h photoperiod.

2.3. Measurement of growth parameters
The following growth parameters were measured and recorded at 0, 45 and 90 days: fresh weight (g), height (cm), number of leaves, number of roots and total root length (cm). To determine the fresh weight, the plantlets were removed from the media, thoroughly washed in distilled water and dried with paper before measuring with a 4 digit balance. The plant height was measured from the base to

the top of the pseudobulbs. The numbers of leaves and roots were manually counted. The total root length was measured from all of the roots from the base to the longest point.

2.4. *Statistical analysis*
The effects of culture media on stimulation of *D.* 'Earsakul' growth were statistically analyzed using the analysis of variance (ANOVA) and a completely randomized design (CRD) with 15 replications. The means among all treatments were compared by Duncan's new multiple range test (DMRT). All statistical analyses were performed using SPSS version 16.0 [24].

3. Results and discussion
When the effects of the culture media on the growth and development of *D.* 'Earsakul' plantlets were evaluated based on various growth parameters (fresh weight, plant height, number of leaves, number of roots and total root length) at 0, 45 and 90 days, it was found as expected that all growth parameters of the plantlets at the initial stage (0 day) were not significantly different among various culture media. After 45 days of the culture period, significant differences were observed on most growth parameters of the plantlets cultured on different media ($p<0.05$). The differences for all growth parameters were highly significant ($p<0.01$) at 90 days among the different media.

- *Fresh weight:* At 45 days, plantlets that were cultured on 1 and 2 mg/L BAP had the highest total fresh weight, however, no significant difference was observed among these media and those supplemented with 5 and 10 mL/L Ready® and VW0 control. After 90 days of culture on different media, this parameter also differed significantly ($p<0.01$) among the culture media. *Dendrobium* plantlets had a tendency to grow and develop best when they were cultured on the medium supplemented with 1 mg/L BAP. The total fresh weight of plantlets grown on this medium was highest (1.8-fold higher than VW0 control), but it was not significantly different from those supplemented with 2 mg/L BAP and 10 mL/L Ready® (figure 1A). The culture media supplemented with V8® and shrimp paste negatively affected the fresh weight of *D.* 'Earsakul' plantlets. Plantlets grown on these media were small and had yellowish-green pseudobulbs and leaves (figure 2: H-M).
- *Plant height:* At 45 days, plantlets grown on the culture media supplemented with 1 mg/L BAP and 10 mL/L Ready® showed highest plant height (1.2 and 1.3-fold) and this was significantly greater ($p<0.05$) than those plantlets which were cultured on VW0 (control). Nevertheless, no significant differences among these media and those supplemented with 2 mg/L BAP and 15 mL/L Ready® were observed. After 90 days of culture on different media, the highest plantlets were obtained in the medium supplemented with 1 mg/L BAP (1.5-fold higher than VW0 control), followed by 2 mg/L BAP, 10 and 15 mL/L Ready® respectively, which were significantly higher ($p<0.05$) than those grown on other media (1.2-fold higher than VW0 control) (figure 1B). On the other hand, plantlets cultured on the medium supplemented with 300 mL/L V8® had the lowest height and this medium also caused yellowish-green pseudobulbs and leaves (figure 2J).
- *Number of leaves:* At 45 days of the culture period, no significant differences among the culture media were observed for the number of leaves. However, *Dendrobium* plantlets cultured on different media began to exhibit significant differences in the number of leaves ($p<0.01$) after 90 days of culture. The highest number of leaves was achieved in plantlets cultured in the media containing 10 at ($p<0.05$) and 15 mL/L Ready®, but these were not significantly different from those grown on the media supplemented with 1 and 2 mg/L BAP, 5 g/L shrimp paste and VW0 as control (figure 1C). By contrast, plantlets cultured on the media containing 2.5 and 7.5 g/L shrimp paste had fewer leaves with the lower leaves turning brown (figure 2: K and M).
- *Number of roots:* At 45 days of the culture period, plantlets showed the highest number of roots when they were cultured in media supplemented with 1, 2 and 3 mg/L BAP. However, they were not significantly different from those cultured on VW0 control and many other

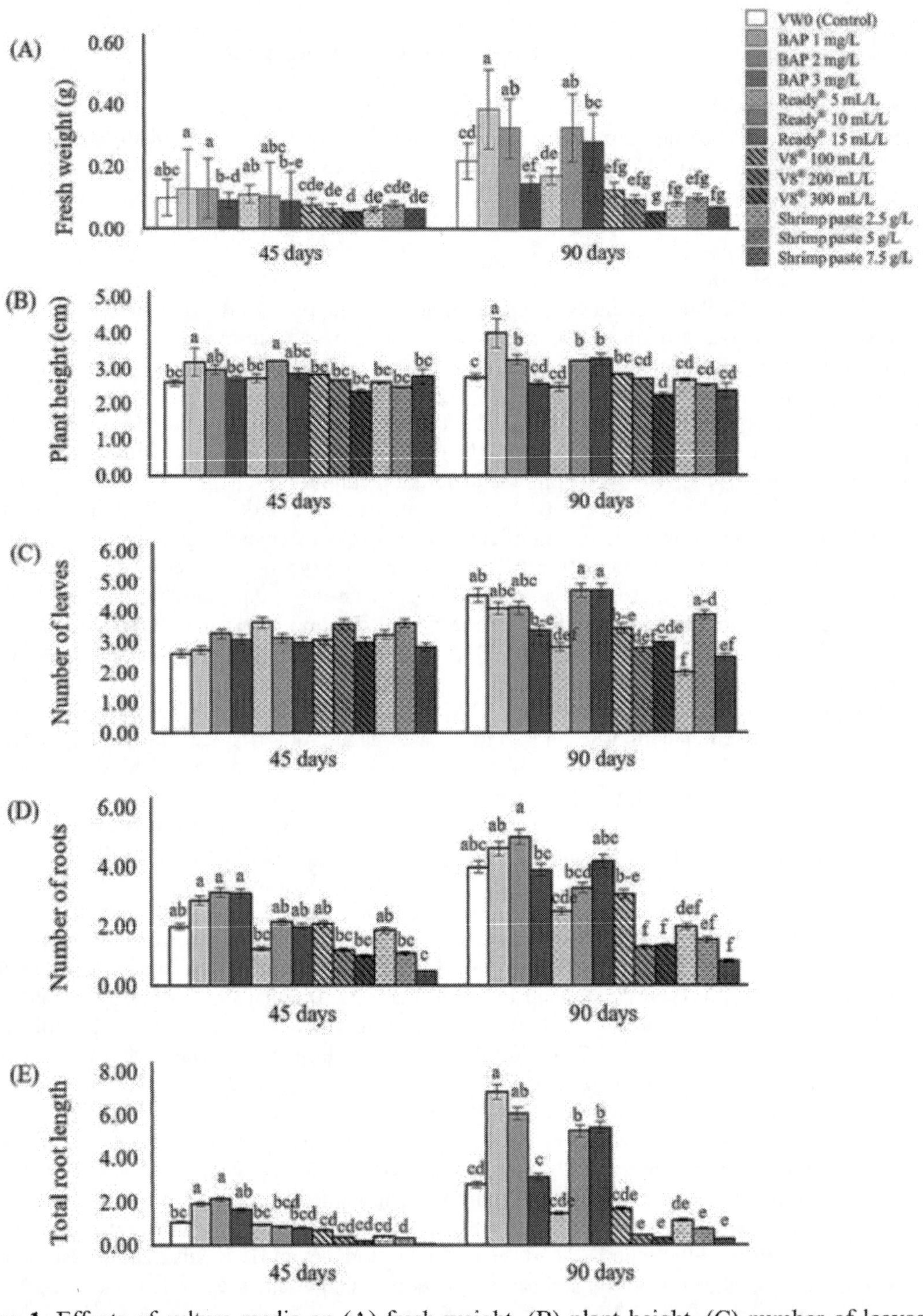

Figure 1. Effects of culture media on (A) fresh weight, (B) plant height, (C) number of leaves, (D) number of roots and (E) total root length of D. 'Earsakul' plantlets. Bars represent standard error (SE) of the mean. Different letters (a, b, c, d, e, f and g) show statistically significant differences among the treatments (Duncan's new multiple range test (DMRT)).

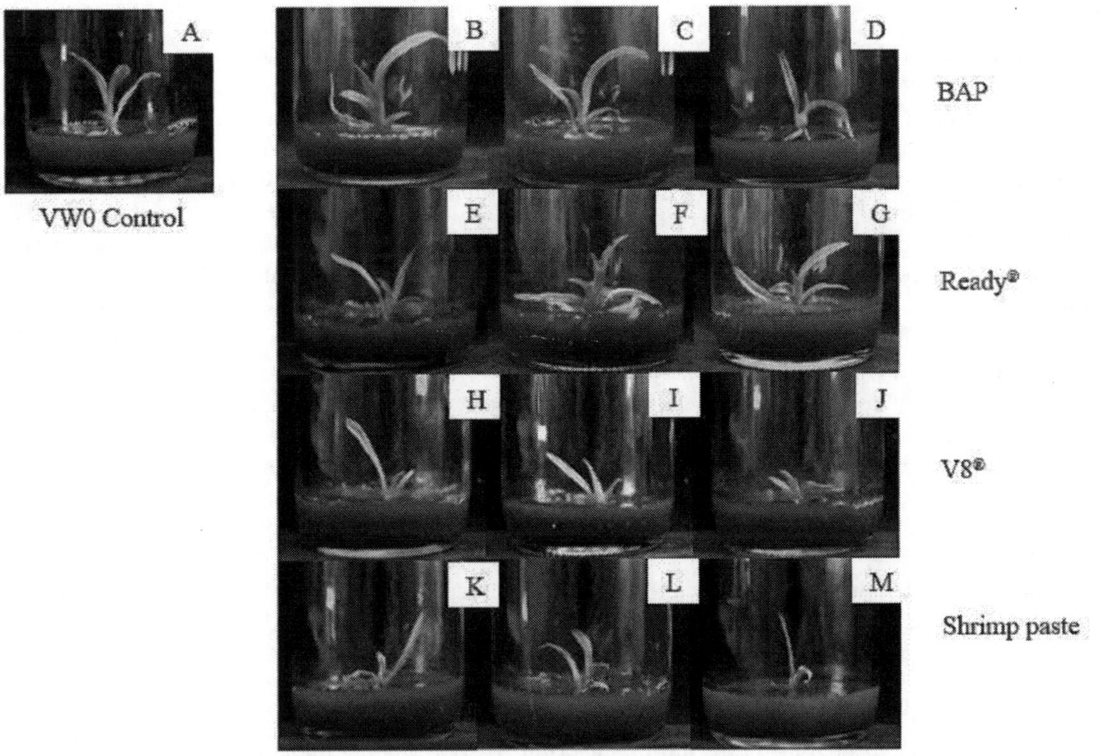

Figure 2. The effects of culture media on growth and development of *D.* 'Earsakul' *in vitro* at 90 days (A) VW0 control, (B) 1 mg/L BAP, (C) 2 mg/L BAP, (D) 3 mg/L BAP, (E) 5 mL/L Ready®, (F) 10 mL/L Ready®, (G) 15 mL/L Ready®, (H) 100 mL/L V8®, (I) 200 mL/L V8®, (J) 300 mL/L V8®, (K) 2.5 g/L shrimp paste, (L) 5 g/L shrimp paste and (M) 7.5 g/L shrimp paste.

media. After 90 days of culture on different media, plantlets cultured on the medium supplemented with 2 mg/L BAP tended to have the highest number of roots (figure 1D), but with no significant difference from those cultured on media supplemented with 1 mg/L BAP, 15 mL/L Ready® and VW0 control. By contrast, culture *media* supplemented with shrimp paste and high concentrations of V8® adversely influenced root numbers. Plantlets grown on these media had the lowest numbers of roots (significantly lower than those cultured on VW0 control; $p<0.05$).

- *Total root length:* At 45 days of the culture period, total root length of *D.* 'Earsakul' plantlets was highest in media supplemented with 1 and 2 mg/L BAP (1.8 and 2.0-fold significantly higher than VW0 control), but these were not significantly different from the medium supplemented with 3 mg/L BAP. After 90 days, the total root length of *D.* 'Earsakul' plantlets increased maximally when they were cultured on the medium supplemented with 1.0 mg/L BAP (2.5-fold significantly longer than VW0 control) but they were not significantly different from those cultured on 2 mg/L BAP (Figure 1E). Media containing 200 and 300 mL/L V8® (figure 2: I and J) as well as 5 and 7.5 g/L shrimp paste (Figure 2: L and M) had significantly inhibited the root length of *D.* 'Earsakul' plantlets and also induced yellowish-green pseudobulbs and leaves.

When the results were taken together, it was observed that *D.* 'Earsakul' plantlets cultured on media supplemented with 1 mg/L BAP had the highest growth and were healthier and more vigorous (figure 2B). These results are in agreement with previous studies in which it was reported that 1 mg/L BAP induced the growth of orchids (*Vanda tessellate*, *Dendrobium* and *Violet phalaenopsis*) [6-,25-27]. In addition, BAP promoted the expansion of leaves, roots and promoted chloroplast maturation

[28-30].

However, the increasing concentrations of BAP may induce programmed cell death (PCD) in plants by accelerating senescence, and causing slow and stunted growth [31,32]. In addition, the supplementation of 10 and 15 mL/L Ready® also produced relatively better growth of *D.* 'Earsakul' plantlets than those of V8® and shrimp paste and VW0 control. These may result from Ready® containing goji berry and mix berries juices which have high vitamin C and have acted as major redox buffers by regulating various physiological processes controlling growth and development [33]. Chaikhiri and Chouychai [17] reported that 10 mL/L Ready boot® was the most appropriate supplement for growth of *Cymbidium finlaysonianum in vitro*, especially with respect to shoot length, number of leaves, roots and fresh weight. By contrast, culture media supplemented with V8® and shrimp paste negatively affected *D.* 'Earsakul' plantlets *in vitro*, possibly due to the inappropriate concentrations used, or the possible toxicity of some components. Therefore, *D.* 'Earsakul' plantlets grown on these media were small and had yellowish-green pseudobulbs and leaves as well as root inhibition (Figure 2 H-M). However, previous reports showed that shrimp paste can accelerate ex vitro root formation in rose apple, *Ixorachinensis*, *Carmona retusa* (Vahl) Masam and *Duranta erecta* L. [9,10].

4. Conclusion

In summary, our results suggest that the medium supplemented with 1 mg/L BAP was the most appropriate culture medium for stimulation of *D.* 'Earsakul' plantlets growth *in vitro*, especially with regard to fresh weight, plant height and total root length.

Acknowledgement

We thank Mr. Peter Bint for editing the manuscript. This work was supported by the Office of the Higher Education Commission under the NRU project of Thailand and grants from Suranaree University of Technology (SUT), Thailand.

References

[1] Kuehnle A R 2006 Orchids: *Dendrobium Flower Breeding and Genetics* ed N O Anderson (Hawaii: University of Hawaii Springer) chapter 20 pp 539-60

[2] Pillay M and Tenkouano A 2011 *Banana Breeding: Progress and Challenges* (New York: CRC Press)

[3] Pragya, Singh S K, Misra R L and Ranjan J K 2012 *In vitro* shoot regeneration from cormel derived callus of gladiolus and bio-hardening of plantlets *Indian J. Biotechnol.* **11** 99-104

[4] Tantasawat P A and Woranyuwat A 2008 *Plant Tissue Culture Laboratory Manual* (Bangkok: Agentex) p 109

[5] Davies P J 2010 Their nature, occurrence, and functions *The Plant Hormones* ed P J Davies (Dordrecht: Springer) pp 1-15

[6] Rahman M S, Hasan M F, Das R, Hossain M S and Rahman M 2009 *In vitro* micropropagation of orchid (*Vanda tessellata* L.) from shoot tip explant *J. Bio-Sci.* **17** 139-44

[7] Pradhan S, Paudel Y P and Pant B 2013 Efficient regeneration of plant from shoot tip explants of *Dendrodium densiflorum* Lindl., a medicinal orchid *Afr. J. Biotechnol.* **12** 1378-83

[8] Poonpipat N, Soipetkasem P and Chaiprasart P 2018 Effect of NAA and BA on micropropagation of *Dendrobium dixathun* Lchb.f., *Dendrobium signatum* Rchb.f. and *Dendrobium tortile* Lindl *Agricultural Sci. J.* **49(S)** 230-3

[9] Chumpookam J, Weeratatprapa N and Pichakum N 2015 Effects of auxin and shrimp paste on layering of 'Nam Dok Mai' rose apple *Agricultural Sci. J.* **46(S)** 669-72

[10] Homjan T, Komen A, Keawprasong K and Nabhadalung N 2011 Study on effects of shrimp paste on root growth of cutting *Ixora chinensis*, *Carmona retusa* (Vahl) Masam and *Duranta erecta* L *Proc. Int. Conf. on Science and Agricultural Technology* 1st (Bankkok, Thailand) pp 1–6

[11] Abrahamian P and Kantharajah A 2011 Effect of vitamins on *in vitro* organogenesis of plant *AJPS.* **2** 669-74
[12] TCP 2019 Winning power momentum products [on-line] Available: https://www.tcp.com/en/product/products-ed/ready
[13] Hacışevki A 2009 An overview of ascorbic acid biochemistry *J. Fac. Pharm.* **38** 233-55
[14] Arrigoni Oand De Tullio M C 2002 Ascorbic acid: much more than just an antioxidant *Biochim. Biophys. Acta.* **1569** 1-9
[15] Pehlivan F E 2017 Vitamin C: An Antioxidant Agent ed Amal H Hamza *IntechOpen* [on-line] Available: https://www.intechopen.com/books/vitamin-c/vitamin-c-an-antioxidant-agent
[16] King Y 2019 The effects of vitamin C & folic acid on the growth of plants [on-line] Available: https://www.hunker.com/13429040/the-effects-of-vitamin-c-folic-acid-on-the-growth-of-plants
[17] Chaikhiri A and Chouychai W 2017 Effect of media on growth of *Cymbidium finlaysonianum in vitro* and effect of fertilizer on acclimatization *Khon Kaen Arg. J.* **45(S)** 1197-202
[18] Veggies for all 2016 V8® Vegetable juice original [on-line] Available: https://www.campbells.com/v8/vegetable-juice/v8-original/
[19] Borlongan J 2017 What effects does citric acid have on plant life [on-line] Available: https://www.gardenguides.com/92182-effects-citric-acid-plant-life.html
[20] Nisar N, Li L, Lu S, ChiKhin N and Pogson B J 2015 Carotenoid metabolism in plants *Cell Press.* **8** 68-82
[21] Tantasawat P A, Khairum A, Arsakit K, Poolsawat O, Pornbungkerd P and Kativat C 2015 Effects of different culture media on growth and proliferation of *Dendrobium* 'Earsakul' protocorm-like bodies *HortTechnology* **25** 681-6
[22] Sebastinraj J, BrittoS J, Kumar D V, Robinson J P and Thangavel P 2014 Rapid propagation of *Vanda testacea* (Lindl.) Rchb.F. – A highly medicinal value epiphytic orchid of India *WSRJ.* **10** 223-30
[23] Aktar S, Nasiruddin K M and Hossain K 2008 Effects of different media and organic additives interaction on *in vitro* regeneration of *Dendrobium* orchid *J. Agric. Rural. Dev.* **6** 69-74
[24] Levesque R and SPSS Inc 2006 SPSS *Programming and Data Management: A Guide for SPSS and SAS* 4rd ed SPSS (Chicago: SPSS Inc.) p 540
[25] Gansau J A, Indan H, Abdullah S N, David D, Marbawi H and JawanR 2016 Effects of organic additives and plant growth regulators on protocorm development of *Dendrobium lowii Transactions on Science and Technology* **3** 462-8
[26] Gnasekaran P, Rathinam X, Sinniah U R and Subramaniam S 2010 A study on the use of organic additives on the protocorm-like bodies (PLBs) growth of *Phalaenopsis violacea* orchid *J. Phytol.* **2** 29-33
[27] Nambiar N, Tee C S and Maziah M 2012 Effects of organic additives and different carbohydrate sources on proliferation of protocorm-like bodies in *Dendrobium Alya* Pink *Plant Omics J.* **5** 10-8
[28] Chory J, Aguilar N and Peto C A 1991 The phenotype of *Arabidopsis thaliana det1* mutants suggests a role for cytokinins in greening *Symp. Soc. Exp. Biol.* **45** 21-9
[29] Chory J, Reinecke D, Sim S, Washburn T and Brenner M 1994 A role for cytokinins in de-etiolation in *Arabidopsis*: *det* mutants have an altered response to cytokinins *Plant Physiol.* **104** 339-47
[30] Spiro M D, Torabi B and Cornell C N 2004 Cytokinins induce photomorphogenic development in dark-grown gametophytes of *Ceratopteris richardii Plant Cell Physiol.* **45** 1252-60
[31] Carimi F, Terzi M, Michele R D, Zottini M and Schiavo F L 2004 High levels of the cytokinin BAP induces PCD by accelerating senescence *Plant Sci.* **166** 963-9
[32] Otroshy M, Khalili Z, Ebrahimi M A, Nekoui M K and Moradi K 2013 Effect of growth regulators and explant on plant regeneration of *Solanum lycopersicum* L. var. *cerasiforme.*

Russ. Agr. Sci. **39** 226-35
[33] Hossain M A, Bosch S M, Vivancos P V, Burritt D, Fujita M and Lorence A 2018 *Ascorbic Acid in Plant Growth, Development and Stress Tolerance* (Switzerland: Springer International Publishing AG)

IOP Conf. Series: Earth and Environmental Science **346** (2019) 012017 doi:10.1088/1755-1315/346/1/012017

Improvement of potato cultivation technology

M Kalimullin[1, 4], R Abdrakhmanov[1], R Andreev[2], A Semenov[2], O Vasilyev[2], P Zaitsev[2] and S Arkhipov[3]

[1]Kazan State Power Engineering University, Russia
[2]Chuvash State Agricultural Academy, Russia
[3]Kazan State University of Architecture and Engineering, Russia

E-mail: marat-kmn@yandex.ru

Abstract. Existing methods for assessing the economic efficiency of the system of machines and technological methods contain numerous similar indicators. The multifaceted nature of the problem, its dynamism also cannot be covered by a single criterion – the value of the reduced costs. A method of systematic research and analysis of the technological efficiency of the production process has been developed to improve the technological processes and technical means for potato cultivation and to achieve higher economic indicators of agricultural production development. A generalized criterion characterizing the efficiency of the working process as a whole, which can be used to evaluate various technological operations in the production of potatoes, is obtained. Various technological processes performed by the working bodies of combined and serial machines were analyzed to verify and confirm the effectiveness of the developed provisions. At the same time, vector quantity was adopted as the main criterion for the comparative evaluation of machines and technologies. As a result of the implementation of this method, a scientifically based technological map for the cultivation of environmentally safe potatoes is proposed. The complex of agricultural practices for the implementation of the proposed technology is selected taking into account soil and climatic conditions and is designed for the cultivation of potatoes with minimal labor and money. The use of technological operations, such as the distribution of straw rolls on the field with its incorporation into the soil, germination of seed potatoes tubers in the peat-mineral shell, inter-row processing with sprinkling of plants with soil, simultaneous removal of the tops by crushing in the process of harvesting combines can significantly reduce costs in the production of potatoes.

1. Introduction

The peculiarity of agriculture, as an industry, is largely due to the need to preserve and improve soil fertility. As well as technological operations on a wide area, broken in time by the biological cycle of plant development. Traditional technologies, involving the execution of operations with a time gap, are inferior to new technological schemes of combining operations performed in one pass of the machine unit. This allows reducing the number of machine passes through the field, reduce labor and material costs by creating a more favorable regime for the development of plants and technological operations in the optimal time, to obtain higher yields of cultivated crops [1,2].

Reducing the cost of potato production is one of the main tasks of potato growers in Russia. In the cultivation of this crop, the cost largely depends on the technology of potato production [3,4].

The degree of perfection of the specific technology of potato cultivation is determined by the fact that the quality indicators of the machines used in the production process meet numerous

Content from this work may be used under the terms of the Creative Commons Attribution 3.0 licence. Any further distribution of this work must maintain attribution to the author(s) and the title of the work, journal citation and DOI.
Published under licence by IOP Publishing Ltd

agrotechnical requirements. In addition, technical and economic indicators are taken into account: productivity, labor and cost per unit of output [5,6].

Agricultural machines and tools operate in conditions determined by the interaction of processes of different nature, and therefore their design requires a comprehensive solution of technical, biological, social and technological problems [7,8].

The industrial technology of potato cultivation is based on the maximum mechanization of the processes of its planting, cultivation and harvesting, but has a number of disadvantages, which include the inability to extend the growing season and obtain physiologically full tubers with dense skins in the early period, the unsatisfactory operation of existing potato planters in the planting mode of sprouted tubers and the discrepancy between the known methods of germination and the requirements for the preparation of seed potatoes [9,10].

Comparative assessment of the level of technical means, analysis of options for mechanized technological processes and identification of the most effective among them is a complex task, which is associated with a variety of performance indicators. Existing methods for assessing the economic efficiency of the system of machines and technological methods contain numerous similar indicators. The multifaceted nature of the problem, its dynamism also cannot be covered by a single criterion – the value of the reduced costs.

2. Materials and methods

Theoretical studies were conducted on the basis of generally accepted laws of classical mechanics, mathematics and mathematical statistics. Experimental studies were conducted in accordance with the generally accepted methods of experiments, the current standards and regulations. Calculation and processing of the obtained results were performed by methods of mathematical statistics using a PC with Microsoft Excel 2010 and Statistica application programs.

A generalized labor productivity is taken in the form of an integral quality indicator, reflecting the ratio of the total positive effect of consumption to the total cost of total labor for the creation and use of products:

$$\varepsilon_{\kappa} = \frac{\Sigma Э_i}{(S_a + S_n)} = \frac{k_a \Sigma C_i}{(C_{np} + H_i)} \tag{1}$$

where ΣE_i - total useful effect; S_a - the cost of creating hardware; S_n - the cost of their operation (consumption); k_a is a generalized indicator of the quality of work performed (products); ΣC_i is the total cost of functioning of a machine or unit for the production of a unit of production (unit cost of a product of a given quality); C_{pr} - reduced costs in units of current expenses; H_i - the cost of unit expenses of living labor, referred to the unit of production or output.

According to the π-theorem of the similarity theory, the functional dependence between the quantities characterizing the process can be represented as a relationship between the similarity criteria made from them. Therefore, from the obtained set of similarity criteria, taking into account their logical relationship, it is possible to obtain one generalized criterion characterizing the efficiency of the workflow as a whole:

$$J = \frac{(Q_T G_T P_T k_c \tau_c)}{10^4 N_y M_y q_T W} \tag{2}$$

where W - hourly capacity of the unit, ha / h; G_T is the intensity of technological cargo flows, kg / m^2; M_y is the specific material consumption of the technological process, kg / ha; q_T - specific fuel consumption, kg / ha; k_c is an indicator of the speed of the technological process; τ_c is a measure of the intensity of total costs, characterized by the ratio of current costs to reduced ones; P_T - technological material consumption of the production process, kg / ha; Q_T is an indicator characterizing energy consumption for processing and distribution of technological material, N · m / s; N_y - total power consumption, kW.

The values of the complex indicators (1) and (2) can be represented either as a vector space or as a functional that takes into account their weight in order to comprehensively evaluate the effectiveness of the entire production process and obtain its quantitative characteristics. Thus, various technological processes carried out by the working bodies of combined and serial machines were analyzed to verify and confirm the effectiveness of the use of the developed provisions. At the same time, the vector value was adopted as the main criterion for the comparative evaluation of machines and technologies.

$$F^2 = \varepsilon_\kappa^2 + J^2 \tag{3}$$

A comprehensive indicator obtained to assess the effectiveness of new technology reflects the specifics and content of the process of increasing the productivity of social work. It characterizes the efficiency of the mobile unit, reflects its technical level and the rationality of the technological impact on the processed environment, as well as the intensity of the workflow.

3. Results and discussion

As a result of the calculations of the parameters of the system analysis, a graph of the dynamics of the complex indicator of the efficiency of tillage equipment, depending on the speed of translational motion and its width, which is shown in Figure 1, was built.

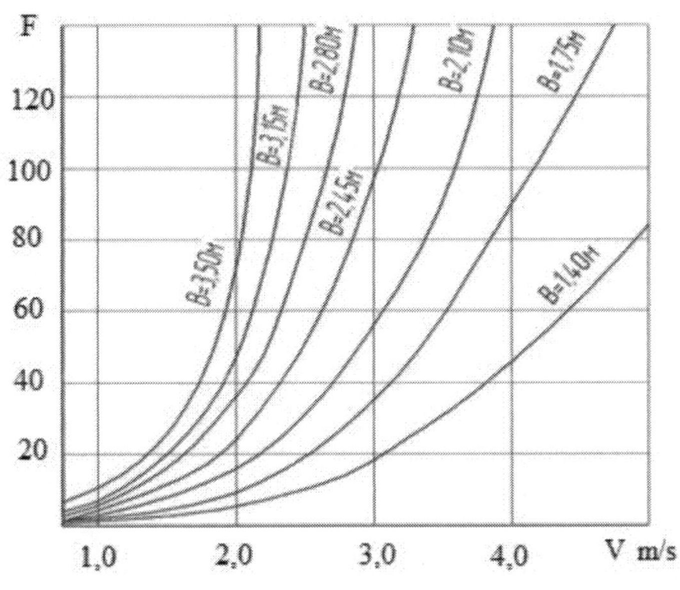

Figure 1. Dynamics of changes in the complex index of efficiency of tillage equipment depending on the speed of translational motion: F - a comprehensive indicator of the efficiency of tillage equipment; V - forward speed, m / s; B -width, m.

As follows from the graph, a rather intensive increase in the value of the complex index occurs with an increase in the speed of movement of the aggregate. Moreover, the intensity of its growth increases more sharply with the increase in the width of the working grip. If the plows with a width of 1.40 ... 1.75 m intensive growth of this indicator begins at a speed of 2.60...3.0 m / s, with a width of 2.8 m corresponding to its values are achieved at a speed of 1.40...1.5 m/s. Therefore, wide - angle plows with the number of buildings from 8 to 10 are most efficiently used at speeds of about 2 m/ s, while the four-and five-body plows similar indicators are achieved only at speeds of 3.7...4.0 m/s.

But the width of the capture cannot be increased indefinitely, since this growth is limited both by the power of the energy means that aggregates the combined machine, and by the impossibility of uniform execution of the technological process on the entire width of the work corresponding to agrotechnical requirements. This also applies to the number of types of weapons that can be aggregated at the same time.

Therefore, on the basis of many years of experience of advanced potato farms in our country, using some agrotechnical requirements of foreign technologies, based on the analysis of technological schemes of construction of the production process, assessment of its organizational and technical level, a comprehensive assessment of the effectiveness of technology and the intensity of mechanized technologies was developed scientifically based zonal technology of production of commercial potatoes with minimal labor and funds, the list of operations which is presented in the technological map (Table 1).

Table 1. Technological card on the cultivation of potatoes planting method – semi-ridge. Predecessor - winter rye.

Ser No.	List of operations	Agrotechnical indicators
1	The distribution of rolls of straw on the field with simultaneous incorporation into the soil	Processing depth 8 ... 12 cm.
2	Loading of fertilizers	-
3	Fertilization	50-60 t / ha
4	Autumn plowing with a turnover of formation	Processing depth 8 ... 12 cm.
5	Sorting seed potatoes into fractions	Weight of tubers 50 ... 80g.
6	Loading of seed potatoes	-
7	Transportation of seed potatoes	-
8	Germination of seed potatoes in the peat-mineral shell	The duration of 25...30 days
9	Surface treatment of soil	Processing depth 8 ... 12 cm.
10	Main processing of the soil	Processing depth 8 ... 12 cm.
11	Loading of seed potatoes	-
12	Transportation of seed potatoes	-
13	Transportation of chemicals for processing tubers	-
14	The cutting ridges with the application of mineral fertilizers	Processing depth 8 ... 12 cm.
15	Planting potatoes with sprinkling	Sprinkle with soil 1.5 ... 2 cm thick
16	Leveling the field surface, shelter tubers	Method of movement - diagonal
17	Pre-emergence treatment of fields	-
18	Treatment of the seedlings	-
19	The first inter-row treatment with sprinkling of plants with soil	The depth is 5-6 cm, the protective zone of 5 cm.
20	The second inter-row treatment with sprinkling of cultivated plants with soil	The depth is 5-6 cm, the protective zone of 5 cm.
21	Treatment of potato against fitoftoroza	Preparation
22	Potato hilling	-
23	Combine harvesting of potatoes with simultaneous crushing of a tops and destruction of an imago of the Colorado potato beetle	-
24	Transportation of potatoes	-

The use of technological operations, such as the distribution of straw rolls across the field with its incorporation into the soil, sprouting seed potato tubers in a peat-mineral shell, inter-row processing with plant sprinkling with soil, the simultaneous removal of the tops by crushing during harvesting with potato harvesters can significantly reduce the costs of potato production.
The developed technology of potato production provides for:

- rational use of organic fertilizers and balanced use of mineral fertilizers;
- creation of planned on-farm seed production on a virus-free basis;
- provides conditions for sustainable harvesting;
- organization of specialized contract and rental units for potato cultivation in farms, which allow obtaining high yields and high quality products;
- timely and correct performance of works on protection of plants from weeds, pests and diseases, using modern equipment, without polluting the environment.

4. Conclusions

The complex of technological operations for the implementation of the proposed technology is selected taking into account soil and climatic conditions and is designed for the cultivation of potatoes with minimal labor and money.

The introduction of the proposed zonal technology of cultivation and har-vesting of potatoes can improve productivity and reduce labor costs by 30...35%.

A comprehensive indicator obtained to assess the effectiveness of new technology reflects the specifics and content of the process of increasing the productivity of social work. It characterizes the efficiency of the mobile unit, reflects its technical level and the rationality of the technological impact on the processed environment, as well as the intensity of the workflow.

References

[1] Kalimullin M N, Abdrakhmanov R K and Arkhipov S M 2015 Rotary Haulm chopper parameters development and substantiation for root and tuber crops. *Intern. J. Appl. Engin. Res.* **10** 25691-7.

[2] Kalimullin M N, Abdrakhmanov R K and Galiev I G 2016 Improvement of potato cultivation technology *Magazine "Machinery and equipment for the village"* **4** 6-9.

[3] Semenov A V and Gavrilov V N 2017 A new method of germination of tubers of early potatoes *Potat. Veget.* **6** 35-6.

[4] Byshov N V, Borychev S N, Simdyankin A A, Kolotov A S, Kolupaev S V, Kiryushin I N, Uspenskii I A, Shemyakin A V and Yukhin I A 2017 Modern view on potato production *Polythematic network electronic scientific journal of the Kuban State Agrarian University* **128** 146-53

[5] Karpov M V, Shardina G E and Zhizdyuk A A 2018 Study of the effectiveness and economic evaluation of the application of the developed potato planting machine *Agricul. Sci. J.* **4**. (LLC Amirit, Saratov) pp 41 – 7.

[6] Simakov E A *et al.* 2009 Improving the efficiency of potato growing - an integrated approach *Potat. Veget.* (Institute of potato farming) **1** pp 2-4.

[7] Rembalovich G K, Byshov N V, Borychev S N, Uspenskii I A, Ryazanov N A, Beznosyuk R V and Bulatov E P 2013 Innovative solutions harvesting and transport technological processes and technical means in potato *Agricul. Machin. Tech.* **1** 23-5.

[8] Shardina G E and Karpov M V 2011 Justification kinematical parameters spoon-belt apparatus are planted *Scientific Rev.* (Saratov, LLC "APEX-94") **4** 117.

[9] S.S.Tubolev *et al.* 2010 Machine technologies and equipment for potato production (Moscow: Agrospas) 316

[10] Semenov A V and Novikov A M 2018 The effectiveness of seed germination in the cultivation of early potatoes *Bull. Chuvash State Agricul. Acad.***2** (5) pp 22-6.

IOP Conf. Series: Earth and Environmental Science **346** (2019) 012018 doi:10.1088/1755-1315/346/1/012018

Influence of soil-protective technologies on the characteristics of the soils of hop plants

M Smirnov[1, 2], P Smirnov[1], E Alexeev[1], I Maksimov[1], Y Kazakov[1] and E Prokopeva[1]

[1]Chuvash State Agricultural Academy, 29, K. Marx St., Cheboksary, 428003, Russia

E-mail: sttmo@yandex.ru

Abstract. It is known that one of the main and complex indicators of the soil is the density of its composition (volume mass). The widespread introduction of new technologies of tillage in our country only increases interest in the values of density in the new conditions. At the same time, soil hardness is also an important parameter characterizing agrophysical conditions of plant growth. We believe that research on the diagnosis of these soil parameters is not enough, so we have conducted research in this area. Traditional methods of tillage do not have a significant compacting effect in the surface layer. Most of the acquired parameters of the additional density are characterized by us as optimal. Only in some cases of ploughing the soil density decreased to 0.7...0.99 g / cm^3. This indicates an excessively friable state. Processing of fields by "no-till" (zero) type leads to an increase of values of density of addition of the soil to 0.94...1.25 g / cm^3. Hardness values also increased and reached 22.8 kgf / cm^2 (2.24 MPa). We see the reason for this phenomenon in the inevitability of the use of heavy-duty tractors. At the same time, the mass of the tractor itself almost levels the type of engine used – caterpillar or pneumatic wheels. Therefore, the objects of study were tractors of different classes, energy saturation, and weight. Soil samples were taken from arable and subsurface horizons from five conventional layers. The results showed that the impact of the drivers of the tractors increases the density of the composition of the soil in all conditional layers. However, the compact intensity on the layers differentiated depending on the weight of the tractor. The soil in the traces of tractors weighing up to 6-7 tons was compressed mainly in the Arab horizon (at a depth of 0.3 m), and in the subsurface (at a depth of 0.3-0.5 m) the compact was practically absent. Soil compaction by tractors weighing 8-11 tons occurred to a depth of 0.5 m. At the same time, the increase in volume mass in the layers of the arable horizon was from 0.25 to 0.35 g/cm^3. The most effective way to restore the subsurface layer is the use of various subsurface agents in the autumn. Application in the winter of snow retention techniques allows one to accumulate in the loosened subsurface layer of additional moisture needed in the spring presowing period.

1. Introduction

The last decades are characterized by a wide spread of technologies of minimum tillage in the cultivation of crops in our country. For the implementation of such methods of tillage, the industry has established a range of powerful tractors, soil-cultivating tools, etc. At the same time, the total mass of the aggregate inevitably increased, and the influence of propulsion systems of the undercarriage systems on the arable and subsurface soil layers increased [1, 2].

Content from this work may be used under the terms of the Creative Commons Attribution 3.0 licence. Any further distribution of this work must maintain attribution to the author(s) and the title of the work, journal citation and DOI.
Published under licence by IOP Publishing Ltd

It is known that the main functions of tillage are the creation of optimal addition of the arable layer, the formation of conditions for the preservation of potential and effective fertility, as well as the protection of arable land from erosion. At the same time, the following indicators of the physical properties of the soil are mandatory: mechanical and granulometric compositions, micro - and macro-structural formulations, the hardness, and density of the addition of soil (bulk density) and its moisture content [3, 4].

2. Purpose of research

Study of the influence of dump and minimum processing technologies on the change in the density of addition and hardness of gray forest soils of the Chuvash Republic.

3. The object of research

Studies on the effect of resource-saving methods of tillage on its agro-physical parameters were carried out in the current hop gardens of the Chuvash State Agricultural Academy, LLC. "Agrokhmel" Vurnary district and LLC "Agricultural Resources" of Urmar district of the Chuvash Republic. Garden hops and land with a total area of about 20 hectares were investigated in these farms [5, 6].

The topsoil is represented by a combination of gray and light gray forest soils. Their mechanical composition is medium-loamy and heavy-loamy. According to the agrochemical analysis in the fertile layer of 2.1 ... 2.5% humus. Humus content up to 2.8 ... 3.1% was recorded on separate garden hops. The depth of the fertile layer is 23 ... 28 cm. The reaction of the soil is average. The slope of all of the hop gardens – out of or within 1.5...5% [7, 8].

The following tractors were chosen as the objects of our research:

- wheel Agromash-30TK (22,1 kW, 2500 kg), MTZ-1025 (77 kW, 4500 kg), BTZ-243K (184 kW, 8660 kg), CLAAS Atles 926 (166 kW, 10200 kg);
- crawler Agromash-90TG (69.1 kW, 6900 kg).

4. Method of research

Chemical and physicochemical parameters were obtained by the generally accepted formulations of modern methods. The density of the soil was determined by the drilling method and was evaluated on a scale of N.A. Kachinsky; soil hardness-in the field using a hand penetrometer Wile soil; humidity - thermostatically-by weight method. Statistical data analysis was carried out using MS Excel software package [9,10].

5. Results and discussion

When studying the impact of basic tillage on crop yields, special attention should be paid to the soil layer below the plant roots. It is known that the ratio of soil moisture and density affects the absorption of mineral elements by plants. As a result of the use of tillage tools (or their complex) in the soil, the density of the arable layer characteristic for this processing is formed. Getting the maximum yield depends on the moisture regime. It is known that plowing provides the lowest soil density. Therefore, it is used in excess of precipitation. The lack of moisture transferred to the treatments, providing a denser structure of the plough layer.

Let's consider how the studied methods of processing changed the addition of soil at a depth of 20 cm. According to the estimates, the density was characterized as "negligible". In the conditions of dump plowing in all terms of observations of 2018, the density of the addition of the arable layer did not exceed the border of 1 g/cm3 and corresponded to the level of "loose" (Table 1).

The data obtained indicate the reduced values of the indicator and are consistent with a number of soil researchers in the Volga region. The loose structure of soils of the agricultural part of the Chuvash Republic is associated with their long stay in the frozen state, cracking as a result of periodic drying in the summer and increasing porosity. Research materials also indicate a significant unproductive loss of moisture on physical evaporation when the density of the soil is below optimal. It should be noted that all these characteristics were obtained for soils treated by the classical dump method.

Table 1. Statistical parameters of the density of addition of gray and light gray forest soils. (2018)

Option of tillage	Layer, cm	20.05.2018		29.06.2018		02.10.2018	
		The bulk density of the layers, g/cm^3	Humidity, %	The bulk density of the layers, g/cm^3	Humidity, %	The bulk density of the layers, g/cm^3	Humidity, %
Stoker	0-5	0.68±0.05	16	0.76±0.06	14	0.83±0.05	5
	5-20	0.70±0.06	11	0.86±0.04	9	0.82±0.04	5
Minimum	0-5	0.87±0.06	13	0.85±0.09	17	0.78±0.06	13
	5-20	0.87±0.06	13	0.95±0.05	10	0.91±0.06	11
Zero	0-5	1.08±0.07	12	1.07±0.06	10	0.94±0.06	11
	5-20	1.20±0.06	9	1.19±0.07	10	1.17±0.04	6

Replacement of the dump treatment with surface loosening (for example, disk coulters) was accompanied by a significant increase in the density of all the studied layers in the spring and summer of 2018 (Table 1). By the end of the growing season, there was a significant decrease in density from 0.88 to 0.78 g / cm3. Probably, one of the factors that influenced the formation of the loose structure is the impact of the core root grain crops and the formed mulching layer from the plant material of field crops of previous years.

Rejection of mechanical tillage found exceeding the level of its compaction in comparison with the dump and minimal processing, but still, it was within the optimum. It is interesting to note that the greatest changes affected only the top 0 ... 5 cm of the root layer of grain crops: from summer to autumn there was a significant decrease in density. In layer 5 ... 20 cm its parameters remained statistically equivalent. It is obvious that the preservation of plant residues of previous field crops on the soil surface, activating the accumulation of young organic substances, increases resistance to deformation and improves the macroporosity of the soil.

Immediately after treatment, the soil has agrophysical characteristics favorable for plant development. However, it is derived from its equilibrium state and, as a rule, its density is lower than the equilibrium state. Subsequently, rapid shrinkage of the soil occurs, the speed of which is determined by the amount and intensity of precipitation, by permeability.

During the growing season of 2018, the values of the density of addition in the conditions of dump plowing did not exceed the level of the first gradation but slightly exceeded the parameters of the previous year. Thus, the traditional method of tillage created its excessively loose structure, which, according to the principles of agrophysics, leads to poor seed contact with the soil, unproductive diffuse moisture loss, increased mineralization processes of organic matter and damage to root systems of field crops during subsidence.

The use of surface loosening revealed a significant decrease in density in a layer of 0 ... 5 cm from mid-summer to autumn. There were no significant changes with the depth, and the soil level varied in the range of 1.01...1.06 g/cm3, significantly exceeding the top layer.

Thus, the maximum density was reached by the soil in the layer of 5 ... 20 cm to the autumn period in the absence of machining and corresponded to the level of "dense". Whereas in the upper 0 ... 5 cm layer, "biological self-loosening" appears - the result of the accumulation of a layer of mulch on the soil surface along with the passages of dead root systems of crops. We believe that a slight differentiation of the arable layer by density can slow down the involuntary evaporation of moisture and contribute to the activation of the natural processes of soil formation.

Thus, the estimated methods of influence on the soil as a whole did not exert a compaction effect on critical values. Density parameters were characterized as "optimal", and as excessively loose in the variant with dump plowing, in some periods. When using soil-protective technologies, there was a tendency to differentiation of the studied layers of the root-inhabited stratum according to the density of structure.

Hardness is an integral indicator of the physical condition and indicates the measure of the strength of the soil against the destructive mechanical action of external forces: roots of plants, tillage implements, raindrops. It is functionally dependent on the granulometric composition, structure, density, and humidity of the soil.

In order to determine the depth of the spreading of the compacting effect of tractors in the soil, two-year (2017-18) studies were carried out by a model field method on experimental garden hops. The objects of study selected tractors, differing class, type of propulsion, engine power, and total weight. The evaluation criterion is the bulk density of the soil - the most comprehensive indicator of soil condition. The results of the study are presented in Table 2.

Table 2. The change in bulk density of the soil under the action of running systems of tractors.

Option	The bulk density of the layers, g/cm^3				
	0 ... 10 cm	10 ... 20 cm	20 ... 30 cm	30 ... 40 cm	40 ... 50 cm
Control	1.00	1.21	1.31	1.42	1.40
Agromash-30TK	1.21	1.29	1.36	1.42	1.40
MTZ-1025	1.26	1.40	1.39	1.41	1.42
Agromash-90TG	1.23	1.32	1.41	1.43	1.41
BTZ-243K	1.31	1.48	1.57	1.57	1.59
CLAAS Atles 926	1.35	1.46	1.58	1.59	1.60

Analyzing the values of the bulk density by layers, it can be noted that the density of the addition of the soil increases depending on the depth. For example, the original soil (control) was loose and had a bulk density in the topsoil up to 1.21 g / cm^3. In the subsurface horizon, this value increased from 1.31 (20 ... 30 cm) to 1.40 g / cm^3.

The mechanical effect of tractor thrusters increased the density of the soil in all layers, and the intensity of compaction attenuated with depth. So, in a layer of 0 ... 10 cm, the increase in bulk density was 0.23 ... 0.35 g / cm^3, in a layer of 20 ... 30 cm - 0.08 ... 0.27 g / cm^3, and at a depth of 50 cm, the soil compacted only 0.01 ... 0.20 g / cm^3.

After the use of various tractors, the intensity of soil compaction was different. The soil was compacted in traces of tractors up to 7,000 kg in the arable horizon, and the sealing effect of these tractors was almost the same. Note that the layers of 30 ... 40 cm and 40 ... 50 cm after their passage remained almost uncompacted. In other words, light tractors compact mainly the arable horizon of gray and light gray forest soils in layers up to 30 cm.

In traces of tractors weighing more than 7,000 kg, the soil compaction pattern is different. Heavy tractors compacted the soil to a depth of 50 cm and the degree of compaction was defined as"significant". The increase in volume mass in the layers of the arable horizon ranged from 0.25 to 0.35 g/cm^3. Moreover, the degree of compaction of such tractors was almost the same and depended mainly on the design features of tractors (twin wheels, caterpillar engines, etc.).

Thus, studies have shown that the soil is compacted not only in the arable horizon but also in the deeper layers of the soil under the action of the tractor's running systems. It should be noted that the compaction of the upper layers of the soil is affected by the specific pressure of machines on the soil, and the lower layers (up to 50 cm) are compacted mainly under the influence of the weight of the tractor.

It is necessary to pay attention to rather high values of the hardness of hop rows. In our opinion, there are several reasons:

first, during the low level of mechanization rows of hops were weeded and loosened manually and efficiently;

second, the aisle on the garden hop is normally 2.2...3.3 m; therefore, each pass of the tractor compacts the row of hops.

The results of the research allowed substantiating the ways to justify the methods of deep tillage, allowing neutralizing the negative impact of tractor running systems. A compacted arable horizon can be brought to an optimal state by conventional tillage implements, and new treatment methods are required to eliminate the negative effects of compaction of the subsoil layers of the soil. Carrying out deep loosening of the subsurface soil layers without inverting the horizon can significantly improve the drainage of the arable layer and aeration of the entire root layer, increases the accumulation of productive moisture in the lower loosened horizons.

Research materials show that soil over-compaction by tractor and agricultural machinery engines violates growth conditions and reduces crop yields. Pressure on the soil causes not only a decrease in yield but also a change in the microrelief of the field – this affects the reliability and performance of machines. In addition, re-compaction of the soil will increase the resistivity of tillage tools in subsequent treatments.

Table 3. Soil moisture in the experimental plots of the hop garden in the spring of 2018.

Option	Soil moisture, %				
	0 ... 10 cm	10 ... 20 cm	20 ... 30 cm	30 ... 40 cm	40 ... 50 cm
Section No. 1	18.8	23.1	25.1	25.7	27.8
Section No. 2	17.4	20.2	23.6	25.3	27.4
Section No. 3	37.6	37.1	31.7	24.2	23.0
Section No. 4	36.8	35.9	30.5	23.8	23.1

The next stage of our study is to study the method of increasing the moisture content in the subsurface layer. For this, we carried out the methods of snow retention in winter 2017-2018 on the garden hop of the Chuvash State Agricultural Academy. Garden hop was divided into four conditional plots of 1 ha each. The row spacing in the sections No.1 and No.2 after harvesting hops in the autumn of 2017 were treated with a deep-ripper to a depth of 45 ... 50 cm. Sections No.3 and No.4 were not exposed to tillage. Then, during the winter period, snow retention was conducted twice in sections No. 1 and No. 3. In the period from December to the end of February (3 months), measurements were made of the thickness of snow cover in all four sections. We found out that the height of the snow cover by the end of February at sections No.2 and No.4 (without snow retention) reached 45 ... 50 cm, and at sections No.1 and No.3 (with snow retention) was equal to 70 ... 80 cm. The results of soil moisture measurement in the spring pre-sowing period of 2018 are presented in Table 3.

It follows from it that the soil moisture in the arable and subsurface layers on the experimental section №1 is higher than on section No.2. The same pattern was observed in the comparison of sections No. 3 and No. 4. Thus, the use of snow retention in the winter period makes it possible to accumulate in the loosened subsurface layer the additional moisture required during the spring pre-sowing period.

It should be noted that observations in the spring of 2018 revealed another feature – the difference in soil ripeness in the experimental plots. For example, sections No. 3 and No. 4 were in the condition bogged by thawed waters till the middle of May. Row spacing on sections No. 1 and No. 2 in mid-April, was freed from meltwater. Thus, the autumn deep tillage to a depth of 50 cm contributed to the destruction of the subsurface layer in the aisle. This allowed the spring meltwater to penetrate into the subsurface horizon. At the same time, the top layer of soil 0...20 cm reached its physical ripeness much earlier.

References

[1] Chernoivanov V, Yezhevskiy A and Fedorenko V 2012 *World trends in machine-technological support of intellectual agriculture* (Moskow: Rosinformagrotech) pp 200-18

[2] Alekseev E P *Investigation of seed uniformity under field and laboratory conditions* et al 2018 IOP Conf. Ser.: Mater. Sci. Eng. **450** 062011

[3] Smirnov P 2018 *Kazan. Bull. Kazan State Agrarian Univer.***4 (51)** 124-9

[4] Medvedev V, Kazakov Yu, Pushkarenko N, Smirnov P and Vasilev A 2017 *St. Petersburg. Bull. Int. Academy Agrarian Education* **37** 27-31

[5] Vasiliev A, Andreev R, Pushkarenko N, Smirnov P and Ivanov E 2017 *Madrid. Proceedings of the 30th International Business Information Management Association Conference, IBIMA 2017* 5289-94

[6] Fadeev A 2016 *Tech. Cultiv. Hop* (Cheboksary: Chuvash NIISH) pp 22-7

[7] Milosta G and Lapa V 2010 *Agrobiological bases of hop growing in the Republic of Belarus: monograph* (Grodno: GGAU) 286 p

[8] Lozhkin A, Makushev A, Pushkarenko N, Kornilova L, Stepanova V and Fedorova O *Madrid. Proceedings of the 30th International Business Information Management Association Conference, IBIMA 2017* 5295-9

[9] Smirnov P and Maksimov I 2010 *The practical results of the influence of mulching on the hardness and intensity of soil tillage* (Yoshkar-Ola: Mari State Technical University) pp 223-5

[10] Medvedev V 2009 *Hardness Soil* (Kharkov) pp 98-110

The effectiveness of the introduction of promising varieties and new technology in the cultivation of common hops

A V Korotkov[1], N N Pushkarenko[2], A O Vasiliev[2, 5], V I Medvedev[3], A P Akimov[3] and G S Yunusov[4]

[1]Head of Scientific and Practical Center for Research in Hop Production, Chuvash State Agricultural Academy, 29 K/Marks Street, Cheboksary 428000, Russian Federation
[2]Technical Service Department, Chuvash State Agricultural Academy, 29 K/Marks Street, Cheboksary 428000, Russian Federation
[3]Department of Transport And Technological Machines and Complexes, Chuvash State Agricultural Academy, 29 K/Marks Street, Cheboksary 428000, Russian Federation
[4]Department Of Mechanization of Production and Processing of Agricultural Products, Mari State University, 1 pl. Lenin, Yoshkar-Ola 424000, Republic of Mari El, Russian Federation

E-mail: 3777222@bk.ru

Abstract. In the conditions of the middle zone of the Russian Federation, incl. the Chuvash Republic, it is possible to cultivate early, mid-early and mid-season hop varieties: Tsivil'skii, Druzhnyi; Sumer', Krylatskii, Podvyaznyi, Mikhailovskii, Favorit, Flagman, Feodal, Forvard and Fakir with a growing season of 95-120 days, with a certain ratio of bitter substances and essential oils, with the appropriate cultivation technology. On the newly restored plantations, the location of the pillar rows - 10 m, for the use of general purpose tractors, it is recommended to hold the tab of hops with 3.3 m spacing between rows, and 1.1 m in the rows between the plants. The use of autumn pruning of the main rhizomes of hops accelerates the emergence of sprouts and reduces the duration of the phases of growth and development in the spring time. The use of new technologies and innovative techniques allow reducing the cost of 123 person days in comparison with the traditional technology.

1. Introduction

National economic significance of hops is due to the fact that cones are a mandatory and indispensable raw material for brewing. They are also used in medicine, bakery and perfume industry.

Hop is a laborious culture. Labor costs for its cultivation are 120-180 people / days per 1 ha [1]. Currently, the Russian Federation is taking the next steps to revive the hop industry. The task is to provide Breweries of the country with high-quality domestic product. The solution of this issue will require a large - scale technical re-equipment of the industry through the introduction of modern and innovative technology, and the use of promising varieties of hops suitable for meteorological conditions of the middle zone of the Russian Federation. The proposed technology includes the use of a set of technological measures, the most important of which are the mechanization of labor-intensive processes, including pruning the main rhizomes, hanging supports, preparing holes for planting, laying of hop-

Content from this work may be used under the terms of the Creative Commons Attribution 3.0 licence. Any further distribution of this work must maintain attribution to the author(s) and the title of the work, journal citation and DOI.
Published under licence by IOP Publishing Ltd

gardens, combining technological operations in one pass of the unit [2-4]. In order to improve the existing technology, research has been conducted on the development of highly efficient innovative methods of hop cultivation based on the implementation of intensive varieties - Podvyaznyi, Krylatskii, Fakir, Feodal, Forvard, Favorit, Flagman, which allow minimizing costs and means to increase plant productivity. Work continues on the study of hop varieties with a combination of different characteristics, the main of which are belonging to different groups of ripeness, biochemical type, resistance to wintering conditions, phytophages and phytopathogens not only in Russia, but also in other countries of the world - in the Czech Republic, Ukraine, Poland and Germany [5-8].

2. Materials and methods

The component composition of bitter substances and essential oils was investigated by HPLC (High Pressure Liquid Chromatography) and Gasliquid chromatography. Extraction of bitter substances, according to the method, was carried out using methanol (methyl alcohol). Various elements of technology of cultivation of hop ordinary with a combination of an agronomic assessment of receptions were studied. The methods of direct accounting, visual assessments and field observations of plants were applied in the studies.

When selecting varieties for use in a production environment, the main economic and biological features of hops are essential: the length of the growing season, the color of the main and lateral sprouts, the shape and size of the cones and the bush, the degree of disclosure of scales, productivity, alpha-acid content, resistance to damage by pests and diseases, suitability of the variety for mechanized harvesting. Also, the responsiveness of the hop variety to various types of mineral fertilizers, its environmental sustainability and adaptability to wintering conditions, requirements for growth and development, attitude to soils and biochemical characteristics

The duration of the frost-free growing season in the central part of the Russian Federation, incl. Chuvash Republic, rarely exceeds 115-120 days, so in such conditions it is possible to cultivate early, medium early and mid-season hop varieties: Tsivil'skii, Druzhnyi; Sumer', Krylatskii, Podvyaznyi, Mikhailovskii, Favorit, Flagman, Feodal, Forvard and Fakir with a growing season of 95-120 days, with plantings on chernozem, gray forest, sod-podzolic soils, with a certain amount of effective temperatures and moisture. Late-ripening varieties in our region do not reach the technical ripeness phase, during wintering they suffer from high soil moisture, low temperatures, and are affected by various root rot and hop bacteria: Fusarium rot (Fusarium oxysporum Schl.), gray rot (Botrytis cinerea Pers.), sclerotic rot (Sslerotinias clerotiorum Lib.), penicillous rot (Penicillum sp.), rhizoctonosis (Rhizoctonia sp.), a diffuse form of downy mildew and mild bacterial rot (Erwinia carotovora (Jones.) Holl) [9-11].

The composition of hops has a decisive influence on the quality of beer. Bitter resin, alpha and beta acids, cohumulone, colupulone; essential oils myrcene, caryophyllene, farnesene, humulene, alpha and beta silinen give the beer special bitterness and aromatic properties [12].

It is noted that in order to reduce the annual cycle of hop development and to facilitate the work of hop growers in the spring, autumn pruning of the main rhizomes over the first pair of kidneys is carried out in the second decade of October until the top layer of soil freezes [13]. Various authors emphasize that it is recommended to use green manure crops on hop-gardens in order to improve the agrochemical and physical properties of the soil. Also, feeding or the first hilling series with ammonium nitrate at a dose of N_{60} or complex fertilizers $N_{45}P_{45}K_{45}$, and the second - to a depth of 15-20 cm with a dose of $N_{30}P_{45}K_{45}$ or $N_{45}P_{45}K_{45}$ with simultaneous harrowing. It is necessary to carry out lime treatment of hop-gardens once in 5 years.

As a result, the efficiency of the use of mineral fertilizers increases by 20-30 % [14].

3. Results and discussion

Agroecological features of hop varieties are different, so the Feodal variety is characterized by intensive spring regrowth, the Flagman variety has low winter hardiness, lunges make up to 10-15%. The productivity and quality of hops is reduced when damaged by aphids, mites and defective powdery mildew. Attitude to damage and damage to various types of phytophages and phytopathogens is average in the Druzhnyi, Mikhailovskii varieties. Krylatskii, Podvyaznyi, Sumer', Favorit, Flagman, Feodal varieties are more stable. A sum of effective temperatures from 1800 to 2000 °C and moisture of at least 250-300 mm is required (Table 1) for normal growth and development of hops in the conditions of the Russian Federation. All varieties are suitable for mechanized cleaning.

The biochemical composition of cones varieties of domestic selection is very diverse. The content of bitter substances, essential oils and their components depends on the variety, place of growth, mineral nutrition, weather conditions and the quality of primary post-harvest processing.

Table 1. The main characteristics of hop varieties recommended for cultivation in the conditions of the Russian Federation.

Varieties	Vegetation period (days)	Sum of effective temperatures, °C	Cold resistance	Resistance to major pathogens
Druzhnyi	103-110	1800-1900	high	average
Tsivil'skii	95-105	1800-1900	high	high
Krylatskii	112-120	1900-2000	high	high
Mikhailovskii	110-120	1900-2000	high	average
Podvyaznyi	110	1800-1900	high	high
Sumer'	106-117	1900-2000	high	high
Favorit	102	1800-1900	average	high
Fakir	106	1800-1900	high	high
Faraon	104	1800-1900	high	average
Feodal	103	1800-1900	high	high
Flagman	112	1800-1900	high	high
Forvard	104	1800-1900	average	high

Chernozem, gray forest and sod-podzolic soils have the most favorable effect on yield. Moisture - 250-300 mm. Suitability to the mechanized processing of hop-gardens is noted. Recommended planting scheme 2,5x1,2; 3,0x1,1 or 3,3x1,1 m.

Regionalized varieties contain cohumulone of 16.9-26.3 per cent, and the ratio alpha/beta of 0.69-1.51 shows the attitude of an aromatic or bitter-aromatic type (Table 2).

Table 2. Biochemical characteristics of common hop varieties.

Components	Podvyaznyi	Krylatskii	Sumer'	Favorit	Faraon	Feodal	Flagman	Forvard	Fakir
Alpha acids, %	7.9	5.7	6.6	3.9	3.8	4.0	5.5	2.9	5.2
Beta acids, %	5.5	3.8	4.6	5.4	3.6	5.7	7.9	5.4	4.5
Alpha/beta	1.4	1.5	1.4	0.7	1.0	0.8	0.7	0.5	1.4
Cohumulone, %	23.2	26.3	25.8	19.2	16.9	20.0	18.2	17.7	19.8
Colupulone, %	46.4	48.2	46.8	40.5	34.5	39.5	37.1	34.1	37.4
Total ml / 100g.	1.45	0.9	1.0	0.40	0.80	0.27	1.37	0.46	0.41
Myrcene, %	46.0	34.0	30.0	28.3	43.7	25.4	25.1	42.9	37.3
Caryophyllene, %	9.3	15.8	13.9	8.3	6.6	12.3	10.4	5.7	9.8
Farnesene, %	12.2	1.7	5.5	21.0	12.6	<0,01	22.4	16.9	14.4
Humulene %	17.8	42.8	43.8	23.0	16.0	35.4	26.8	13.7	17.0

The content of myrcene, which determines the aroma of beer wort, in varieties is 25.1-46.0%; caryophyllene - 5.7-15.8%. The greatest amount of essential oils is found in bitter and aromatic varieties Podvyaznyi, Krylatskii, Sumer' and Flagman.

According to the results of many years of research, innovative methods of cultivation of hops with minimal labor and money are proposed. Thus, it was recommended to plant new plantations into the planting furrow to a depth of 28 cm, both in free and in columnar rows by stem cuttings or annual hop saplings according to the 2.5x1.2 scheme; 3.0x1.1 m, and newly restored hop-gardens by placing long-pillar rows after 10 m, for the use of general purpose tractors, when laying, use a row spacing of 3.3 m, and plant the rows in the rows through 1.1 m. with the number of bushes 3, 3; 3.5; 2,7 thousand / ha, respectively. With timely and high-quality planting, the survival rate of stem cuttings and annual seedlings will be 75-80; 90-95%, and the first year of the growing season allows receiving 1-3; 3-5 c / ha, on the second - 5-8; 8-12 centners per hectare, and the costs for such planting are no more than 15 people. days / ha.

In order to reduce or completely eliminate the weed in the rows in early spring, the introduction of soil herbicides is recommended, followed by embedding with a boring unit. In order to reduce the annual cycle of hop development and to facilitate the work of hop growers in the spring, autumn pruning of the main rhizomes over the first pair of kidneys is carried out in the second decade of October until the top layer of soil freezes. As a result, due to the effective use of reserves of productive moisture and photosynthetic radiation, the phase of germination and technical ripeness of plants is accelerated in the next year. This reduces the period of growth and development of hops, which allows starting harvesting cones in the early stages.

After mechanized pruning, in order to improve the agrochemical and physical properties of the soil, regardless of the timing of pruning of the main rhizomes, to conduct continuous sowing of hop-gardens with green manure crops (mustard, rape), locally tape two-fold feeding of vegetative plants at a distance of 20-25 cm from the row to or at the first hilling with ammonium nitrate in a dose of N_{60} or complex fertilizers $N_{45}P_{45}K_{45}$, and in the second - to a depth of 15-20 cm with a dose of $N_{30}P_{45}K_{45}$ or $N_{45}P_{45}K_{45}$ with simultaneous harrowing and to carry out liming hop-gardens once every 5 years. As a result, the efficiency of the use of mineral fertilizers increases by 20-30 %.

Table 3. Labor costs with traditional and new technology, people-days per 1 ha.

No.	Name of works	Traditional technology	New technology	Labor saving	Increased productivity
1	Introduction of herbicides	-	2	-2	-
2	Horrowing of hop-gardens	0.2	0.2	-	-
3	Trimming the main rhizomes	33	2	31	10
4	Stretching the string over the row instead of pin pegs	25	2.5	22.5	10
4	Supports mounting	27	8	19	3.4
5	Fixing lower ends of the supports	25	2.5	22.5	10
6	Planting stalks for support	10	10	-	-
7	Frame	10	1	9	10
8	Removal of lower leaves, side branches, weeds	20	1	19	20
9	Interrow loosening, hilling with mineral fertilizers	2	2	-	-
10	Spraying with pesticides	2	2	-	-
11	Other mechanized transport operations	5	3	2	-
12	Harvesting	75	75	-	-
13	Drying of hops on PCB and pressing	4	4	-	-

14	Total	238.2	115.2	123.0	2

With mechanized pruning of the main rhizomes, productivity increases by 2 times, and with hanging supports, only 8 people / days are spent per 1 ha (Table 3).

The lowest leafiness and high productivity of varieties of domestic selection of hops contribute to an increase in the daily volume of gross collection of cones from 1.5 to 4.0 tons with mechanized harvesting.

4. Conclusions

In the conditions of the middle zone of the Russian Federation, incl. the Chuvash Republic, it is possible to cultivate early, mid-early and mid-season hop varieties: Tsivil'skii, Druzhnyi; Sumer', Krylatskii, Podvyaznyi, Mikhailovskii, Favorit, Flagman, Feodal, Forvard and Fakir with duration of the vegetation period of 95-120 days. With placement of plantings on chernozems, gray-forest and sod-podzolic soils.

As a result of biochemical evaluation of cones, it was found that Tsivil'skii, Mikhailovskii, Favorit, Faraon, Feodal, Forvard and Fakir varieties belong to the aromatic type. Sumer', Krylatskii, Podvyaznyi and Flagman varieties - to the bitter-aromatic type. Druzhniy Variety - to the bitter type.

It is advisable to plant the new hop-garden according to the scheme 2.5x1.2; 3.0 x1.1 m, and the newly restored plantations, the location of the columnar rows after 10 m, the tab - with row spacing of 3.3 m. The arrangement of plants in rows of 1.1 m.

The use of autumn pruning of the main rhizomes of hops accelerates the emergence of sprouts and reduces the duration of the phases of growth and development in the spring time.

The lowest leafiness and high productivity of varieties of domestic selection of hops contribute to an increase in the daily volume of gross collection of cones with mechanized harvesting.

The use of new technologies and innovative techniques allow reducing the cost of 123 person days in comparison with the traditional technology.

Typically, gray forest soils found in the extreme south-eastern part of forest - Glossic, Greyzemic, Luvic Phaeozems (Abruptic, Clayic, Albic) on Cretaceous clay eluvium, occupy an insignificant area of 2 hectares.

References

[1] Zakharov A I, Makushev A E , Zakharov D A and Dobrova A N 2017 Resource conservation in the production of hop raw materials (on the example of the Chuvash Republic) *Bull. Chuvash State Agricul. Acad.* **3** pp 105-10.

[2] Medvedev V I, Kazakov Y F, Pushkarenko N N, Smirnov P A and Vasil'ev A O 2017 The current level of mechanization of cultivation of hops in the Chuvash Republic: problems and direction of development *Proceedings Intern. Acad. Agrarian Educ.* **37** pp 27-31.

[3] Dmitriev Y P, Medvedev V I, Akimov A P, Dmitrieva O Y, Dmitriev S Y, Maksimov A N and Andreev V A 2018 Machine technology for the cultivation of hops *Bull. Kazan GAU* **2** (49) pp 86-92.

[4] Zakharov A I 2016 Factors of intensification in the hops cluster of Chuvashia *Intern. J. Environ. Sci. Educ.* **11**(17), 10651-9.

[5] Scomra U 2008 Genetic resources of hops in Poland: collection, evaluation and utilization in breeding *Proceeding of the International Scientific Meeting*. Chmelarsky Institute, Zatec. P. 13-8.

[6] Krofta I, Klapal J, Jezek J and Ticha J 2008 Hodnoceni kvalitativnich ukazatelu ceskych chmelu ze sklizne (Chmelarstvi) pp 1-8.

[7] Seigner E, Lutz A and Seefelder S 2008 Utilization of genetic resources in breeding programmes at the Hop Research Center Huell *Proceeding of the International Scientific Meeting.* (Chmelarsky Institute, Zatec) pp 8-12.

[8] Fadeev A A and Nikonova Z A 2015 The results of the study of hop varieties of different groups

IOP Conf. Series: Earth and Environmental Science **346** (2019) 012019 doi:10.1088/1755-1315/346/1/012019

of ripeness on economically important traits and resistance to the main diseases *Agrarian Sci. Euro-Northeast* 5 (48) pp 29-33.
[9] Nikonova Z A and Korotkova Z P 2017 Creating and preserving a collection of common hops as a gene pool for breeding *Niva Volga* **4** (45) pp 104-8.
[10] Fadeev A A and Nikonova Z A 2018 Assessment of variety samples of the collection of common hops according to phenological and morphological features *Bull. Russian Agricul. Sci.* **2** pp 40-2.
[11] Danilova Y S 2008 Types of hops depending on the biochemical composition *Agrarian Sci. Euro-Northeast* **11** pp 51-6.
[12] Dresel M, Dunkel A and Hofmann T 2015 Sensomics analysis of key bitter compounds in the hard resin of hops (Humulus lupulus L.) and their contribution to the bitter profile of Pilsner-type beer. *J. Agricul. Food Chem.* T63 **13** pp 3402-18.
[13] Efimov A D and Korotkov A V 2009 Productivity of hops depending on the timing and height of the cut stems when cutting the main rhizomes *Agricul. Sci. Euro-North-East* **3**(14) pp 45-7.
[14] Korotkov A V 2015 Influence of aftereffects of liming, fertilizers and green manure crops on the parameters of fertility of gray-forest heavy loamy soil and yield of hops *Bull. Mari State Uni.. Series Agricul. Sci. Econ.* **4** (4) pp 29-33.

IOP Conf. Series: Earth and Environmental Science **346** (2019) 012020 doi:10.1088/1755-1315/346/1/012020

Effects of rhus typhina and robinia pseudoacacia on physicochemical properties of saline soil in Dagang Area, Tianjin

Y Liu[1], H Q Liu[1,3], F H Liang[1], Y Y Shi[1] and X L Liu[2]

[1]College of Horticulture and Landscape, Tianjin Agricultural University, Tianjin 300384, China
[2]Tianjin Xingning Industrial Co. LTD, Tianjin, 301500, China

E-mail: wjxlhq@126.com

Abstract. In order to explore the effect of using two different afforestation tree species to improve soil in coastal saline-alkali land, the method of combining field sampling and indoor analysis and test was adopted. The physical and chemical properties of soils planted with *Rhus typhina* and *Robinia pseudoacacia* forest were determined. The results showed that afforestation could improve soil structure, decrease soil bulk density and increase soil fertility. To reduce soil salinization, the soil physical and chemical indexes of *Robinia pseudoacacia* plantation were better than those of growing *Rhus typhina* forest, so the planting area of *Robinia pseudoacacia* plantation in Dagang area could be increased. Planting *Rhus typhina* forest could also improve the soil in this area, but its effect was lower than planting *Robinia pseudoacacia* plantation.

1. Introduction

Soil salinization is a global problem, especially in developing countries, which affects the environment and agricultural productivity [1,2]. The total area of salinized land in the world is about 10 million km^2, and the total area of salinized land in China is about 1 million km^2. It is mainly distributed in the northeast, northwest, inland areas of north China and coastal areas north of the Yangtze river [3]. The area of salinized land in Tianjin is about 7800 km^2, accounting for 65.8% of the total soil area in Tianjin. The land with greater than 0.2% soil salinity is about 4700 km^2, accounting for 39.3% of the total soil area. The salinized land in Tianjin Binhai new area is about 1958.9 km^2, accounting for 86.3% of the total area of Binhai area. As a coastal city, the coastal area of Tianjin contains 25% of the total area of salinized land, among which the area of salinized soil accounts for 79.9% of the total area of salinized land, and the severely salinized land accounts for 32.1% of the total area of severely salinized land [4-8]. Soil salinization in Dagang area of Tianjin is serious, with high salt content, low organic matter content, poor soil fertility and alkaline soil [9,10].

Rhus typhina is a small deciduous tree belonging to the sumac family. It is native to Europe and the United States, often in the open sand or gravel soil growth, and more cultivated in the provinces north of the Yellow River basin to use mainly for barren hills greening as well as saline landscape forest species.

Robinia pseudoacacia is a deciduous leguminous tree, which require higher water condition, for example the growth was slower in places with high water table, was extremely easy to induce diseases, and had certain drought resistance [11].

At present, Wang, Li and Hou *et al* have deployed 6 representative soil salinization profiles in coastal

Content from this work may be used under the terms of the Creative Commons Attribution 3.0 licence. Any further distribution of this work must maintain attribution to the author(s) and the title of the work, journal citation and DOI.
Published under licence by IOP Publishing Ltd

areas of Tianjin, and studies had shown that soil surface vegetation was an important factor affecting soil salinization [12]. Land salinization was one of the major obstacles to the utilization of land resources. Wang, Sun, Li *et al* discussed the effects of different salt isolation measures on soil water and salt dynamics and photosynthetic characteristics of *Robinia pseudoacacia* in coastal saline-alkali land [13]. In the development of coastal new urbanization, soil quality problems were mainly concentrated on soil salinization, soil salinization and other problems. Ma studied the relationship between the physical and chemical characteristics of coastal soil and soil quality [14]. Y H Gao conducted research on soil water and salt transport and salt-tolerant plants' improvement of saline soil in the saline and alkaline land of Bohai bay [15]. Wang *et al* made various studies on soil physicochemical properties from soil depth, soil organic carbon, planting methods of different plants, microbial effects and other aspects [16,17]. Cao, Li, Qin *et al* studied the growth rules of *Robinia pseudoacacia* in pure and mixed forests of 2-3 years in coastal saline and alkaline lands [18]. Lu, Liu, Yang *et al* studied the physicochemical properties of soil in the artificial pure forest of Chinese scholartree and *Robinia pseudoacacia* planted in the coastal saline and alkaline land for six years and three years of seedling age [19]. In the past five years, there had been few studies on the effects of *Rhus typhina* on saline soil.

Because of the harsh soil conditions of saline-alkali land, there were few plant species that could grow, and it greatly restricted the local economic development, so the research on saline-alkali land improvement technology had become urgent. Test materials mainly were artificial pure forest planting acacia and torch for two sample, and physical and chemical properties of soil were tested to explore the acacia and torch influence on soil physical and chemical properties so as to provide theoretical basis for soil salinization, governance, and improvement and for the popularization and application of saline-alkali land improvement and landscaping.

2. Materials and methods

2.1. Test materials

The test soil was taken from the north of Dagang district of Tianjin in spring 2018. The soil was planted with *Robinia pseudoacacia* forest and torch forest respectively. The plant spacing of *Robinia pseudoacacia* was 2 m×2 m, and the plant spacing of the torch was 3 m×3 m. Take the wasteland soil of the same plot as the reference.

2.2. Sampling method

Select a square sample plot and draw two intersecting diagonals on each sample plot. Take 4 vertices of 2 diagonals and a center point as sampling points. Take 5 soil samples from each sample plot and repeat them for 3 times. A ring knife was used for sampling physical traits, and a shovel was used for sampling chemical traits.

2.3. Determination of soil physical properties

Soil bulk density using ring knife method; Soil porosity and soil moisture content were determined by drying method.

The percentage of pores in the soil per unit volume is called the total porosity of the soil. Can be calculated by the following formula:

$$\text{Total porosity }\% = (1 - \text{bulk density/specific gravity})\ 100$$

Single factor randomized block design was adopted in the experiment.

Data were analyzed using anova and significance test.

2.4. Determination method of soil chemical properties

Cations in soil water-soluble salts include Ca^{2+}, Mg^{2+}, K^+ and Na^+. Among them, Ca^{2+} and Mg^{2+} were commonly determined by EDTA titration. K^+ and Na^+ were determined by flame photometry. Carbonate ion and bicarbonate ion were determined by double indicator -- neutralization titration method, chloride ion by silver nitrate titration method, soluble salt by conductance method, organic matter content by potassium dichromate capacity method -- external heating method, and soil pH value by pHS-3B type acidity meter [20,21].

Single factor randomized block design was adopted in the experiment.

Data were analyzed using anova and significance test.

- Flame spectrophotometer Model: FP640 Manufacturer: Shanghai Hongji Instrument Equipment Co., LTD
- pHS-3B ph Meter Model: pHS-3B Manufacturer: Xi'an Yunyi Analytical Instrument Co., LTD
- Conductivity Meter Model: HI2316 Manufacturer: Beijing Zhongke Dilian Technology Development Co., LTD

3. Results and analysis

3.1. *Physical properties of soil*

The physical properties of soil planted with *Rhus typhina* forest and *Robinia pseudoacacia* forest were analyzed, and the results were shown in table 1: Soil physical properties of *Robinia pseudoacacia* plantation were significantly better than that of the soil planted with *Rhus typhina*, but there was no significant difference from the control.

Table 1. changes of soil moisture content, bulk density and porosity in planting *Rhus typhina* forest and *Robinia pseudoacacia* forest.

	water content (%)		Bulk density(gcm^{-3})			Porosity (%)		
CK	27.478 a	A	1.982	a	A	51.232	a	AB
Robinia pseudoacacia	25.712 ab	A	1.656	a	AB	53.319	a	A
Rhus typhina	22.766 b	A	1.293	b	B	43.432	b	B

3.2. *Soil chemical properties*

3.2.1. Content analysis of K^+, Ca^{2+} and Mg^{2+} in soil of planting Rhus typhina forest and Robinia pseudoacacia forest. The results of the determination of K^+, Ca^{2+} and Mg^{2+} in the soil planted with *Rhus typhina* forest and *Robinia pseudoacacia* forest and the soil used as the control were shown in table 2: In the K^+ content determination, there was no significant difference between the soil planted with plants and the control, while there was significant difference between the soil planted with *Robinia pseudoacacia* forest and the soil planted with *Rhus typhina* forest. In the content determination of Ca^{2+} and Mg^{2+}, the soil planted with *Robinia pseudoacacia* forest and *Rhus typhina* forest was significantly different from that of the control group, while there was no significant difference in the content of Ca^{2+} and Mg^{2+} between the two plants.

Table 2. Contents analysis of K^+, Ca^{2+} and Mg^{2+} in soil of planting *Rhus typhina* forest and *Robinia pseudoacacia* forest.

	K^+(mg/L)			Ca^{2+}(mg/L)			Mg^{2+}(mg/L)		
CK	0.0328	ab	A	0.0713	a	A	0.269	a	A
Robinia pseudoacacia	0.0410	a	A	0.0458	b	B	0.0363	b	B
Rhus typhina	0.0305	b	A	0.0365	b	B	0.0420	b	B

3.2.2. Content analysis of Cl^-, Na^- and HCO_3^- in the soil of planting Rhus typhina forest and Robinia pseudoacacia forest. The results of the determination of Cl^-, Na^- and HCO_3^- in the soil planted with *Rhus typhina*forest and *Robinia pseudoacacia* forest and the soil used as the control were shown in table 3: The content of Cl^- in the soil of *Robinia pseudoacacia* was significantly different from that of the control, and the difference was significant between the soil of planting *Robinia pseudoacacia* and the soil of planting *Rhus typhina*. However, there was a significant difference between the soil planted with the *Rhus typhina* and the control.

Table 3. Content Analysis of Cl^-, Na^- and HCO_3^- in soil of planting *Rhus typhina* forest and *Robinia pseudoacacia* forest.

	Cl^-(mg/L)			Na^+(mg/L)			HCO_3^-(mg/L)		
Robinia pseudoacacia	0.0413	a	AB	0.0400	a	A	0.0375	a	A
CK	0.0315	b	A	0.0335	a	A	0.0478	a	A
Rhus typhina	0.0205	c	B	0.0440	a	A	0.0384	a	A

There was no significant difference in Na^- and HCO_3^- contents between the soil planted with *Robinia pseudoacacia* forest and *Rhus typhina* forest and the control soil.

3.2.3. Content analysis of pH, organic matter and electrical conductivity of soil in planting Rhus typhina forest and Robinia pseudoacacia forest. The results of pH, conductivity and organic matter content of the soil planted with *Rhus typhina* forest and *Robinia pseudoacacia* forest and the control soil are shown in table 4: The pH differences of the three soils were not significant. In the determination of soil organic matter content, the difference of organic matter content between the soil planted with *Robinia pseudoacacia* and the other two kinds of soil was extremely significant, while the difference of organic matter content between the control soil and the soil planted with *Rhus typhina* was not significant. There was no significant difference between the electrical conductivity of the soil planted with *Robinia pseudoacacia* and the control soil. The electrical conductivity of the soil planted with *Robinia pseudoacacia* and the control soil was significantly higher than that of the soil planted with *Rhus typhina* .

Table 4. Content analysis of pH, organic matter and electrical conductivity in soil of planting *Rhus typhina* forest and *Robinia pseudoacacia* forest.

	pH			organic matter(g/kg)	Electrical conductivity (us/cm)		
Robinia pseudoacacia	6.80	a	A	0.508aA	0.173	a	A
CK	7.20	a	A	0.315bB	0.156	a	A
Rhus typhina	6.98	a	A	0.380bB	0.148	b	B

4. Discussions

4.1. Discussions on the physical properties of soil for growing Rhus typhina forest and Robinia pseudoacacia forest

Soil bulk density refers to the dry soil weight per unit volume of soil in the natural state of undamaged soil structure. Soil porosity was related to soil bulk density and water content. Bulk density was also an important index of soil porosity. The decrease of soil bulk density leads to the increase of soil porosity and water content. In this experiment, soil bulk density decreased, water content and total porosity increased after planting *Rhus typhina* forest and *Robinia pseudoacacia* forest, effectively improving soil physical properties, which was consistent with the research results of Liu *et al* [22]. The physical properties of soil planted with *Robinia pseudoacacia* forest were better than those of soil planted with *Rhus typhina.*

4.2. Effects of planting Rhus typhina forest and Robinia pseudoacacia forest on the content and electrical conductivity of soil ions

The conductivity of soil reflects the content of main ions in soil, and the higher the conductivity, the higher the content of ions [22]. The results showed that the content of water-soluble ions in the soil of different tree species was different in the same season, which might be the difference between different tree species or related to the growth and development characteristics of plants themselves. This was consistent with the research of Bi and Zhang [23]. In the experiment, the content of Cl^- and K^+ in the soil planted with *Robinia pseudoacacia* forest was significantly higher than that in the soil planted with *Rhus typhina*, and there was no significant difference in the content of HCO_3^-, Na^+, Mg^{2+} and Ca^{2+} between them.

4.3. Effects of planting Rhus typhina forest and Robinia pseudoacacia forest on soil pH and organic matter

A comprehensive evaluation was conducted on salinized soil by pH and organic matter. A high pH index value indicated that the higher salinization degree of soil, the greater the harm to plants. The higher the organic matter content was, the better the soil nutrients were. In the pH measurement, there was no significant change in the soil pH value of the two plants. In the determination of organic matter content, the organic matter content in the soil planted with *Robinia pseudoacacia* forest was higher than that in the soil planted with *Rhus typhina* forest, which was consistent with the study of Wang, Li, Dong *et al* [24,25].

5. Conclusion

In this experiment, according to the analysis of physical and chemical properties, the soil planted with *Robinia pseudoacacia* forest was superior to the soil planted with *Rhus typhina* forest in terms of soil moisture content, bulk density and porosity. In the determination of K^+ content and electrical conductivity of the soil planted with *Robinia pseudoacacia* forest, the soil was higher than that planted with *Rhus typhina* forest, and there was no difference between the two in the determination of pH, Na^+, Cl^-, HCO_3^-, Ca^{2+} and Mg^{2+}. The soil organic matter content of *Robinia pseudoacacia* plantation was higher than that of *Rhus typhina* plantation. The soil physical and chemical properties of *Robinia pseudoacacia* plantation were better than that of *Rhus typhina* plantation. Therefore, in Dagang district of Tianjin city, the proportion of planting *Robinia pseudoacacia* forest should be increased and the planting scale of *Rhus typhina* forest should be reduced.

Acknowledgments

This research was supported by the University Young and Middle-aged Backbone Innovation Talent Training Program in Tianjin (J01009030709), Agricultural Science and Technology Achievements Transformation and Extension Project in Tianjin (201603060), the Major Science and Technology Project of the Seed Industry of Tianjin Committee of Science and Technology (17ZXZYNC00070) and Postgraduate Innovation Training Program in Tianjin (2017YPY001) and Tianjin Applied Basic and Frontier Technology Research Plan (14JCYBJC30200).

References

[1] Liu G X 2018 *China High New Technol.* **8** 51-3
[2] Liu X and Chi C M 2014 *Anhui Agric. Sci.* **30** 95-7
[3] Yang Z, Shang J M and Chen Y L 2019 *Gansu Agri. Sci. Technol.* **01** 13-20
[4] Long Y X, Zhang X H and Tan Z H 2018 *J. Agronomy* **8** 39-43
[5] Zhang H 2019 *Communications on Forestry Sci. Technol.* **02** 57-8
[6] Li W 2018 *Shelterbelt Sci. Techno.* **06** 34-35+37
[7] Ao W M and Yin X G 2018 *Extension Basic Agri. Technol.* **6** 33-5
[8] He Q F, Peng Y H, Gou Z Y, Tan Y B, Zheng W and Shen W H 2018 *Guangxi Forestry Sci.* **47** 29-35
[9] Liu X and Chi C M 2014 *Anhui Agri. Sci.* **30** 95-7
[10] Liu X P 2015 *Resour. Environ. Sci. Soil Fertilizer* **1** 217-21
[11] Xue X 2013 *Modern Rural Sci. Technol.* **17** 47
[12] Wang W X, Li Q K, Hou J Y and Liu Y T 2015 *Geophysical Geochem. Exploration* **39** 172-9
[13] Wang L L, Li S Y, Sun X Y, Zhang T, Fu Y and Zhang H L 2015 *Acta Ecol. Sinica* **5** 84-94
[14] Ma W M 2011 Relationship between physical and chemical properties of coastal soil and soil

quality Northwest A & F Univ., Shanxi, China
[15] Gao Y H 2011 Study on soil water and salt dynamics and salt-tolerant plant improvement of saline soil in saline and alkaline land of Bohai Bay Chinese Academy of Forestry
[16] Huang Q P 2017 *Shelter Forest Sci. Technol.* **12** 35-7
[17] Du Y W, Zeng X F and Deng X Z 2017 *J. Forest Environ.* **37** 181-7
[18] Cao B H, Li F M, Qin Y J, Zhang Y J, Liu Y Y and Zhao T 2012 *J. Shandong Agri. Univ. (Natural Science Edition)* **43** 201-5
[19] Lu X X, Liu Y, Yang Y H, Wang P H, Yuan W H and Liu Y J 2016 *Northern Hortic.* **4** 72-5
[20] Ma J, Liu X D, Jin M, Zhao W J, Cheng C X, Jun M H and Rong W X 2019 *J. Water Soil Conser.* **33** 207-13
[21] Shao H Y, Zhang A F, Li Z Y, Liu D, Li Y F, Lu L, Wang X D, Du H Y and Zhang Y L 2019 *J. Northwest Agri. Forestry Univ. (Natural Science Edition)* **08** 1-8
[22] Liu L L, Zhu Q K, Zhao W J, Yao W J, Ma H, Chen W S and Bu Y J 2014 *Chin. Soil Water Conser. Bull.* **34** 311-5
[23] Du Y W, Deng X Z and Li S L 2017 *Hubei Forestry Sci. Technol.* **02** 9-13+42
[24] Zhang L, Xu P P, Li S Y, Sun X Y, Zhang T, Liu Z H and Yao L Y 2017 *Sci. Soil Water Conserv. China* **15** 92-9
[25] Wang H Y, Li H L, Dong Z and Chen X C 2016 *Soil Water Conserv. Res.* **23** 161-5

Biologically active additives, their effect on the egg production of geese, hatchability, and development of goslings

I A Alekseev[1], V G Semenov[1], T V Pastukhova[1], A V Obukhova[1], A F Kuznetsov, V G Sofronov[2], A Kh Volkov[3] and A I Dimitrieva[1, 4]

[1]Chuvash State Agricultural Academy, 28003, Cheboksary, K. Marx str., 29, Chuvash Republic, Russia
[2]Kazan State Academy of Veterinary Medicine named after N.E. Bauman, 420029, Kazan, Siberian Tract, d. 35, Rep. Tatarstan, Russia
[3]St. Petersburg State Academy of Veterinary Medicine, 196084, St. Petersburg, Chernihiv Str., 5, Russia

E-mail: dekanatFvmz@mail.ru

Abstract. The prospects of goosewatch development are determined by several factors: goose meat contains 17.5-18.5% protein, 22-23% of fat, 0.9-1.0% of ash. Insufficient reflection of these issues in the husbandry industry, peasant farms determined the direction of our study – the study of the impact of "Pozitron-2 Air 1" artificial ozonizer on the microclimate of the plant growing geese and biologically active additives "Bacell", "Larikarvit" and "Sporobacterin" on the body of geese and gooslings. When performing research work, zoohygienic, hematological, biochemical, zootechnical, veterinary and sanitary, economic research methods using modern methods and instruments were used. The conducted studies had shown that 3-hour sessions of ozonation with the specified device at a dose of 0.02 m^3 of ozonated air per 1 m^3 of purified air of the room significantly changed the state of the microclimate in it for the better. Thus, the relative humidity in the room for growing gooslings after ozonation significantly decreased from 78.64±0.16 to 71.12±0.14%, that is, on average by 7.52% ($P<0.05$), ammonia concentration - by 22.73% ($P<0.01$), hydrogen sulfide – by 32.27% ($P<0.01$), carbon dioxide – by 18.18% ($P<0.01$), dust – by 60.56% ($P<0.001$), microorganisms – by 46.94% ($P<0.01$) ($P < 0.001$). These data indicated that the tested artificial ozonizer Pozitron-3 Air-1 improved and normalized the microclimate in the room for growing gooslings. When testing new dietary supplements, geese egg laying, hibernation hatching and hatchability were important indicators. The results of the study showed that egg production in geese of experimental groups, against the background of the use of dietary supplements "Larikarvit", "Bacell", "Sporobacterin", compared with analogues in the control of February – June was higher by an average of 6.44%, 7.49%, 6.64% ($P<0.01$). Among these experimental groups of birds, in comparison with intact birds, the highest egg production was in the second experimental group of gooslings, where the biologically active additive "Bacell" was used as part of the main diet, within 7.49% ($P<0.01$). As it is known, the output of young animals in goose breeding is determined by the number of viable, healthy gooslings, expressed as a percentage of the number of eggs laid in the incubator. The results showed that this important indicator in the application of dietary supplements in the diet of geese in experimental batches, compared with the control batch was significantly higher by 7.71, 10.31, 8.85 ($P<0.05$) percent.

1. Introduction

Content from this work may be used under the terms of the Creative Commons Attribution 3.0 licence. Any further distribution of this work must maintain attribution to the author(s) and the title of the work, journal citation and DOI.
Published under licence by IOP Publishing Ltd

Geese are an important reserve for increasing of poultry production. Their meat is characterized by high caloric content. The experience of advanced geese farming shows that geese give an exceptional economic effect, especially when using intensive technology of growing geese for meat [1-10].

With intensive methods of geese farming at specialized poultry farms, large farms and peasant farms, the creation and maintenance of an optimal microclimate remains a serious problem. In this regard, the poultry yards microclimate quality indicators improvement should be accompanied not only by the regulation of the known complex of air environment physical and chemical factors in the modern conditions, but also by an increase in its biological activity, as well as air sanitation through its purification and disinfection. Currently, there are several ways of air sanitation: application of air filters, ozone-air mixture, bactericidal radiation, and air ionization, etc. [11-15].

Prospectivity of the geese farming development is determined by a number of factors. Geese are fast growing birds, body weight of gooslings increases by 45-50 times daily from to 9-10 weeks of age and reaches 4.6-6.0 kg, at a cost of no more than 3 kg of concentrated feed per 1 kg of body weight gain. Muscle tissue accounts for 37-39%, skin with subcutaneous fat – 14-16%, internal fat – 6,7% in carcasses of geese up to 9 -10 weeks of age. Goose meat contains 17.5-8.5% protein, 22-23% fat, 0.9-1.0% ash [16-26].

Biologically active additives are used as growth promoters of animals, birds and are becoming increasingly popular in modern poultry farming. After all, it is not only antibacterial agents that reduce the risk of diseases of the gastrointestinal tract but also stimulate the consumption of feed, due to the active secretion of saliva and digestive enzymes that increase its digestibility. In poultry, increasing attention is paid to improving bird health and productivity. In this regard, the role of the immune system is difficult to assess. On the one hand, it provides reliable protection against most infectious and invasive diseases, and, on the other hand, the effectiveness of immunization depends on immunocompetence. There are a lot of problems associated with both increased morbidity and mortality of birds, especially young, and with their poor growth and development with a low nonspecific and specific resistance of the body in birds [27-34].

The lack of attention to the problematic issues in geese farming industry determined the direction of our research. So, we focused on the study of the effect of biologically active feed additives as "Laricarvit", "Bacell" and "Sporobacterin" on the body of geese and goslings.

New biologically active additives and probiotic feed preparations were tested in production conditions and introduced into the practice of poultry farming in accordance with the "Regulation on the Procedure for Testing of New Veterinary Drugs". The correct use of various feed additives and biologically active substances, the number of which is constantly increasing, is one of the factors of increasing the natural resistance, survival rate and productivity of birds, reducing of feed consumption per unit of production and improving the efficiency of the poultry breeding [35-41].

The creators' applications of "Laricarvit", "Bacell" and "Sporobacterin" to the leadership of the Chuvash State Agricultural Academy to test these biologically active additives on different types of poultry farming in the conditions of farms of the Chuvash Republic were the basis for research on the testing of biologically active additives (LLC "Biotekhagro", the Krasnodar Territory, Timashevsk; LLC "Bakoren", Orenburg)

The aim of this work was the optimization of the microclimate in the goose growing stock plant using artificial ozonizer "Pozitron - 3 Air 1" and the study of influence of biologically active additives "Laricarvit", "Bacell" and "Sporobacterin" on the egg production of geese, morphological, biochemical status and development of goslings.

2. Material and methods

Testing of "Laricarvit", "Bacell" and "Sporobacterin", the biologically active additives, was conducted in geese farming farm of the Tsivilsky District of the Chuvash Republic on 200 geese and 480 goslings of "Linda" breed. Scientific and economic experiments on studying of the effect of above mentioned biologically active additives on egg production, hatching and hatching of goose eggs, as well as survival rate, quality, and productivity of goose growing stock were conducted in the period from 2017-2018.

Pozitron-3 Air 1 is a device of ozone disinfection of air, which is intended to clean air in rooms from dust, microorganisms, carcinogenic and strongly smelling substances. This is due to the fact that ozone has high penetrating power, has a bactericidal effect on many pathogenic and opportunistic bacteria. Also, ozone kills the virus, destroying its membrane, damaging the cell both outside and inside. If you compare ozone with the usual means against bacteria, it is 600 times stronger than chlorine, and it kills bacteria several times better than ultraviolet rays [13].

Air ozonation was carried out in the room for growing geese to improve the microclimate. In this case, the specified ozonizer was used. After the birds were cleared from the room, and its pre-cleaning, Pozitron-3 Air 1 was used daily for 6 days to ozone the room, with a three-hour exposure, at a dose of 0.02 m3 of the ozonated air per 1 m3 of air to be cleaned. The main parameters of the microclimate were estimated with the appropriate devices before and after this procedure.

Experiment on establishing the effect of the tested biologically active additives on the egg production of geese was carried out on 200 geese of "Linda" breed, which according to the principle of analogues were divided into 4 groups (control and three experimental), 50 heads in each. Adult geese daily got biologically active additives from February to June as part of the main diet during 150 days: the first experimental group got "Laricarvit" at the rate of 3.0 kg/t of feed, the second experimental group got "Bacell" in the amount of 2kg/t of feed, the third experimental group got "Sporobacterin" at a dose of 1.0 ml/100 kg of body weight. The control group birds were on the main diet without biologically active additives.

480 geese, which were divided in four groups (control and three experimental) 120 heads each, were selected in order to establish the influence of above mentioned biologically active additives to be tested on the growth, development and survival rate of birds on the principle of analogues. Gooslings got biologically active additives from the first days of life during 75 days as a part of a diet: the first experimental group got "Laricarvit" at the rate of 2.5 kg/t of feed, the second experimental group got "Bacell" in the amount of 1.5 kg/t of feed, the third experimental group got "Sporobacterin" at a dose of 0.5 ml/kg of body weight.

The birds of the control group were on the main diet. Zoohygienic conditions of keeping and feeding for all groups of birds were the same. The following research methods are used in the performance of research work:

1) zoohygienic methods - the air temperature was measured with a thermometer, the relative humidity of the air with the "TKA-PKM" universal device (model 42), the carbon dioxide concentration by the Subbotin-Nagorsky method, the content of ammonia and hydrogen sulfide by the universal gas analyzer UG -2, the speed of air movement was estimated with hot-wire anemometer; concentrations of dust and microorganisms in the rooms - with Krotov apparatus when assessing the microclimate in the growth room. Air ozonation was carried out with "Pozitron-3 Air 1", a modern artificial ozonizer, in accordance with the instructions for its use, in the growth room.

2) hematological methods - the number of erythrocytes, leukocytes, hemoglobin in the blood was determined using veterinary hematological analyzer "Vetsan HM-5". Blood from geese was usually taken in the morning before feeding from the axillary vein, 15 samples from each test group, every 10 days of the experiment.

3) biochemical methods – the total protein content in the blood serum of animals was determined by IRF-454B-2M Refractometer, individual protein fractions were determined by the nephelometric method.

4) immunological methods – lysozyme activity of blood serum was estimated using *M. lisodeiticus* sewage culture, phagocytic activity of neutrophils using daily agar culture *St. aureus*, bactericidal activity of blood serum using daily *E. Coli* culture.

5) zootechnical - the gain in live weight of goslings was determined by regular weighings of goslings on electronic scales on the 1st, 20th, 20th, 4th, 50th, 60th, 76th day of the experimental work;

6) veterinary and sanitary examination of meat was carried out according to organoleptic studies (appearance, consistency, smell, the degree of bleeding) and in accordance with the "Rules of Veterinary Inspection of Poultry and Veterinary and Sanitary Examination of Poultry Meat".

IOP Conf. Series: Earth and Environmental Science **346** (2019) 012021 doi:10.1088/1755-1315/346/1/012021

7) economic methods - economic efficiency of the use of biologically active additives in the cultivation of goose growing stock was calculated by the method of I. N. Nikitin (Nikitin, I. N. Organization and Economics of veterinary business: textbook. / I. N. Nikitin, M. H. Voskoboinik. M.: Humanitarian Publishing Center VLADOS, 1999. P. 209–251).

8) statistical methods - statistical processing of the results was carried out by the method of Oivin using Studen's tables.

"Laricarvit" is a biologically active additive produced by JSC "Petrokhim" (Belgorod City), which is intended to enrich diets of animals and birds with beta-carotene, chlorophyll, and flavonoids for the purpose of increase of survival rate, productive and reproductive qualities. The following active ingredients are contained in 1 kg of the preparation: chlorophyll - not less than 500 mg/kg, beta-carotene – not less than 1700 mg/kg, bioflavonoid complex of larch – not less than 700 mg/kg, as well as filler – silicon dioxide – up to 1 kg may be used for food purposes without restrictions [18].

"Bacell" is produced by LLC "Biotekhagro"(Registration Number PVR-1-4.7/02100 d. Timashevsk of the Krasnodar Territory), consists of microbial mass of spore-forming bacteria *Bacillus subtilis 945* (B - 5225); acidophilic bacteria *Lactobacillus acidofilus L917* (B – 4625); *Ruminococcus albus 37* (B-4292); sunflower meal, or products of grain or legumes. 1 g of biologically active additive contains, at least, 1 x 108 CFU (colony forming units) of bacteria of each species and does not contain GMO [39].

«Sporobacterin liquid» is a probiotic of a new generation of production Bakoren, LLC (Orenburg City, Registration Certificate PN 000792/01 of 14.03.2011), is a suspension of living bacteria of the strain *Bacillus subtilis 534* [38]. The preparation is allowed for use in medical practice, veterinary medicine, and livestock farming. The bacteria of the strain produce proteolytic enzymes (proteases, lipases, endogenous interferon), contributing to the improvement of protein digestibility by, on the average, 4.9%, fat by 6%, fiber by 10.7%. At the same time, there is an increase in the absorption of minerals by 7.3% and nitrogen by 9.3% [38].

3. Results.

The results of the experiment on the rehabilitation of rooms are shown in table 1.

Table 1. Indicators of the microclimate in the room for goosling farming on the background of air ozonation (M±m).

Indicator	Original air	Processed air
Air temperature, °C	22.16±0,18	21.94±0,24
Relative humidity, %	78.64±0.16	71.12± 014*
Air speed, m/s	0.19±0.01	0.20±0.01
Concentration: ammonia, mg/m3	7.92±0.04	6.12±0.02**
hydrogen sulfide, mg/m3	3.44±0.06	2.33±0.08**
carbon dioxide, %	0.22±0.02	0.18±0.01*
Dust, mg/m3	12.22±0.10	4.82±0.08**
Microorganisms, t/m3	62.72±0.22	33.28±0.19**

Note: * P<0.05; * * P<0.01
M-arithmetic mean value;
m - mean square error

The presented data indicate that sanitation for 6 days, with 3-hour sessions of ozonation at a dose of 0.02 m3 of ozonated air per 1m3 of the purified air of the room using an ozonizer, significantly changed the quality of the microclimate for the better.

Thus, the relative humidity in the room for growing of goose growing stock decreased from 78.64 ± 0.16% to 71.12±0.14% (P<0.05), ammonia concentration - from 7.92±0.04 mg/m3 to 6.12±0.02 mg/m3 (P<0.01), hydrogen sulfide – from 3.44±0.06 mg/m3 to 2.33±0.08 mg/m3 (P<0.05), carbon dioxide – from 0.22±0.02 to 0.18±0.01% (P<0.05), dust – 39.44% (P<0.001), microorganisms - 53.06% (P<0.01).

The ozonizer did not have a noticeable effect on the temperature and speed of the air in the room. Egg production of geese is presented in Table.2.

Table 2. Data of egg productivity of geese using biologically active additives (average value per one goose for 1 month and only per season, in pieces).

Period of a year	Group of geese			
	Control MD	1 expermental MD+"Laricarvit"	2 experimental MD+"Bacell"	3 experimental MD+"Sporobacterin"
February	2.26	2.29	2.35	2.30
March	12.82	13.52	13.74	13.64
April	13.55	14.31	14.50	14.24
May	12.28	13.21	13.28	13.23
June	2.66	2.82	2.88	2.85
Total	43.57	46.15	46.75	46.26

Note: MD - main diet

The egg production against the background of the use of "Laricarvit", a biologically active additive, in the ration, compared with the control analogues, during this period (February - June) was significantly higher, on the average, by 2.58 pcs./head in geese in the first experimental group, with the introduction of "Bacell", a biologically active additive, into a diet - by 3.18 pcs./head in the second experimental group of birds, with the use of "Sporobacterin", a biologically active additive - by 2.69 pcs/head, or by 5.92%, 7.29% (According to Table 2, P <0.01), 6.17% (P <0.01) in the third experimental group.

Hatching in the control group of eggs was 69.48%. This figure was characterized by 77.19%, 79.79%, 78.33% (P<0.01) in the experimental groups, against the background of the tested biologically active additives (Table 3).

Table 3. Indicators of egg incubation in "Universal", the incubation cabinets.

Indicators of incubations	Group of trays with goose eggs			
	Control, MD	1 experimental MD+ "Laricarvit"	2 experimental MD+"Bacell"	3 experimental MD+"Sporobacterin"
Laid trays, pieces.	16	16	16	16
Number of eggs in the tray, psc.	60	60	60	60
Total number of laid eggs.	960	960	960	960
Of them fertilized, pcs.	799	820	849	830
in per cent	83.23	85.42	88.44	86.46
Hatching, heads	667	741	766	752
in percent	69.48	77.19	79.79	78.33
Hatchability, percent	85.62	90.36	90.22	90.60

Note: MD - main diet

Hatchability compared with the control group of eggs was higher by 6.88%, 7.82%, 7.12% (P<0.01). This indicator in the second experimental group of birds, compared with the first and third groups was higher by 0.94% and 0.70%.

Bactericidal, lysozyme and phagocytic activity were determined in serum (Table 4) as indicators of nonspecific resistance against the background of the use of these biologically active additives in goosling.

The analysis of the numerical values of Table 4 shows that the activation of factors of nonspecific resistance of blood serum of gooslings was observed against the background of the use of biologically active additives. Growth of bactericidal activity amounted to 6,74%, lysozyme activity of 5.79%, phagocytic activity of 5.94% (as per table 4, P<0.05) in the first experimental group of birds, in applying the "Laricarvit", in relation to the control. In the second experimental group of birds, where «Bacell» was used in the diet, a significant increase in the indicated indices of nonspecific blood serum resistance of goslings was 6.87%, 5.93%, 6.10% (according to table 4 P <0.05). In the third experimental group of birds, against the background of the use of «Sporobacterin», the growth of these indicators was 8.31%, 6.72%, 6.96% (according to table 4 P <0.01).

Table 4. Indicators of nonspecific resistance of blood serum of gooslings in the application of dietary supplements (M±m).

Indicator	Groups of gooslings			
	Control without additives	1 experimental "Laricarvit"	2 experimental "Bacell"	3 experimental "Sporobacterin"
Bactericidal activity, %				
at the beginning of the experiment	44.62±0.82	43.91±0.78	44.36±0.80	43.89±0.76
at the end of the experiment	46.57±1.06	49.71±1.10**	49.77±1.21**	49.90±1.26**
Lysozyme activity, %				
at the beginning of the experiment	32.10±0.65	31.96±0.58	31.82±0.52	32.14±0.68
at the end of the experiment	34.18±0.72	36.16±0.76*	36.21±0.81*	36.48±0.93**
Phagocytic activity, %				
at the beginning of the experiment	48.07±0.85	47.83±0.79	48.16±0.86	47.97±0.91
at the end of the experiment	51.12±1.08	54.16±1.12**	54.24±1.34**	54.68±1.31**

Note: * P<0.05; **P<0.01.
M - arithmetic mean value;
m - mean square error

Table 5. Indicators of the safety of geese in the biologically active additives.

Indicator	Control (MD)	1 expermental MD+"Laricarvit"	2 experimental MD+ «Bacell»	3 experimental MD+"Sporobacterin"
Number of birds in the experiment, heads.	120	120	120	120
Mortality of birds: heads	6.00	4.00	4.00	3.00
%	5.00	3.33	3.33	2.5
The survival rate of birds, heads	114	116	116	117
%	95.00	96.67	96.67	97.50

Note: MD - main diet

The analysis of this indicator during the whole period of the experimental work showed that the introduction of these biologically active additives into the main diet of young geese has a positive effect on the viability of geese, this is evidenced by the increase in the safety of geese in the experimental groups against the background of the use of biologically active additives (table 5).

During the experiment in the control group where these biologically active additives were not used, 6 out of 120 goslings died from diseases of the gastrointestinal tract of non-infectious etiology, which is 5.00%, the viability of young birds in this group was 95.00%.

Survival rate was characterized by, the average, 96.67%, as compared with counterparts in the control, mortality decreased by, on the average, 2 heads, and birds' survival rate increased by 1.67% in the first and second experimental groups of gooslings, against the background of the use of biologically active additives for the specified period. The survival rate of birds was 97.50%, compared to the control it was higher by, on the average, 2.50% in the third experimental group, when "Sporobacterin" was used in the diet.

The number of red blood cells in the goslings of the control group by the age of 15 days decreased from 2.19 ± 0.14 to 2.10 ± 0.12 × 1012 / l, in the first experimental group - from 2.30 ± 0.18 to 2.20 ± 0.13x1012 / l, in the third experimental group - from 2.29 ± 0.15 to 2.18 ± 0.12x1012 / l, or 4.11%, 4.37%, 4.35%, 4, 81% By the age of 30 days, this figure in these groups decreased by 3,66%, 3,50%, 3,48%, 3,94%. At the same time, a regular and reliable increase in the number of red blood cells in the blood of experienced goslings is clearly observed, compared with control analogues, depending on the use of biologically active additives. So, in the blood in birds of the first experimental group, in relation to control analogues, a significant increase in this indicator by 15 days of the experiment averaged 4.28% (P <0.05), in the second experimental group - 4.76% (P <0.05), in the third experimental group of birds - 3.81% (P <0.05), and by the 30th day of the experiment this growth averaged 4.73% (P <0.05), 5, 21% (P <0.05), 4.26% (P <0.05), respectively.

The number of white blood cells in the blood of geese, both in the control and in the experimental groups also gradually decreased with the growth of birds. At the same time, there is a slight increase in their number against the background of the use of these tested additives, however, without statistical reliability.

At the beginning of the experiment, the level of hemoglobin in the blood of the control and experimental geese had no significant differences and ranged from 108.04±1.89 g/l to 108.89±2.18 g/l. By the 15th day of the experiment, the level of hemoglobin in the experimental goslings compared to the control analogues, on the background of the use of biologically active additives in the first experimental group was higher on average by 5.49%, in the second - by 6.17%, the third - by 5, 61% On the 30th day of the experiment, these indicators were characterized by a change in the direction of growth: in the first experimental group of geese on the background of the use of "Laricarvit" by 6.11%, in the second - on the background of "Bacell" – by 6.95%, in the third - when using "Sporobacterin" – by 6.67%, respectively.

The total protein content in blood serum was slightly higher in geese of experimental groups against the background of the use "Sporobacterin", the biologically active additive. So, this indicator in the first experimental group of birds, compared with the control analogs, was significantly higher by6.43, 5.83, 6.05%, in the second experimental group - by 7.94, 8.07, 6.91%, in the third experimental group - by 6.54, 6.19, 6.18% respectively.

Studies have shown that the level of γ-globulins in the serum of the experienced gooslings significantly exceeded the analogous indicator of the control group of birds, compared with the control analogues, against the background of the use of the tested biologically active additives. Thus, this value was higher, on the average, by 9.27, 9.53, 10.09 % (P<0.01) in the blood serum of the first experimental group of geese, compared with intact poultry on 14, 28, 56th day of the experiment, by 9.69, 10.33, 12.30% (P<0.01) was in the second experimental group, by 9.86, 11.32% (P<0.01) in the third experimental group.

Among the important indicators characterizing the growth and development of geese is the average daily increase in their body weight (table 6).

Daily weight gain exceeded that of control analogues by 6.00% - 6.24% (P<0.01) in the gooslings of the first experimental group in 10, 20, 30, 40 - day of the experiment with application of the "Laricarvit" average. More intensive growth of this indicator occurred on 50-th, 60-th, 75-th days of the experiment in birds of this experimental group, against the background of the use of this biologically active additive. So, the figure in this group of birds compared to the control, was greater, on the average, by 6.45 to 7.50% (P<0.01) within the time frame of work. This value in the gooslings was higher by 7.18, - 7.51% (P <0, 01) on 10, 20, 30, 40 days of the experiment, compared to the control, by 6.99-7.88% (P <0.01) on the 50th, 60th, 75th days of the experiment in the second experimental group, with the introduction of "Bacell" in the main diet.

Table 6. Dynamics of the average daily increase in body weight of geese when using biologically active additives, g (M±m).

Age, day	Control MD	1 expermental MD+"Laricarvit"	2 experimental MD+"Bacell"	3 experimental MD+"Sporobacterin"
10	32.14±0.50	34.07±0.46*	34.45±0.55**	34.16±0.46*
20	41.96±0.51	44.50±0.62*	44.94±0.66**	44.52±0.62*
30	71.12±0.67	75.03±0.74**	76.09±0.72**	75.53±0.74**
40	73.70±1.05	78.30±1.14**	79.24±0.91**	78.23±1.14**
50	70.81±1.22	75.38±1.10**	75.76±1.33**	75.20±1.04**
60	65.98±1.34	68.84±1.41**	69.73±1.48**	69.00±1.12**
75	64.80±1.30	69.66±1.69**	69.91±1.64**	69.75±0.69*

Note: * P<0.05; ** P<0.01; MD - main diet.
M –mean arithmetic value;
m – mean square error

The increase of body weight in geese of the 3-rd experimental group, relative to control, was higher by 6.28 – 6.14% (P<0.01) when applied "Sporobacterin", the biologically active additive, on the 10, 20, 30, 40-th day of the experiment. This indicator in this group exceeded the control analogues by 6.19-7.63% (P<0.01) on the 50th, 60th, 75th days of this biologically active additive use. From the presented analysis it follows that "Bacell", the biologically active additive, had a more effective influence on the body of experimental birds, which is expressed by an increase in their average daily weight gain on the 75th day of scientific and production experience by, on the average, of 7.88% (P<0.01). Results of control slaughter of birdsare given in table 7.

Table 7. Data of slaughter of geese of control and experimental groups in the application of biologically active additives (M±m).

Indicator		Group: Control (MD)	1 expermental (MD+ «Laricarvit»)	2 experimental (MD+"Bacell")	3 experimental (MD+"Sporobacterin")
Pre-slaughter weight, g.		4535.15±31.54	4767.80±32.41*	4808.16±34.11*	4782.23±33.41*
Carcass weight	uneviscerated. g	3628.12±27.14	3814.24 ±28.09*	3846.52±29.11*	3825.78±28.08*
	half-eviscerated. g	3541.01±26.10	3726.03±28.34**	3774.88±30.21**	3739.70±29.14**
	%	78.10±0.41	78.15±0.48**	78.51±0.51**	78.20±0.50**
	viscerated. g	2546.03±82.24	2695.71±86.07**	2722.38±92.16**	2704.35±88.10**
	%	56.14±0.81	56.54±0.98**	56.62±1.38**	56.55±1.12**

Note: * P<0.05; * * P<0.01; MD - main diet.
M–mean arithmetic value; m–mean square error

The numerical data of the table 7 show that the highest pre-slaughter weight of geese was in the experimental groups. This value in these groups of birds, compared with the control, was higher by 5.12%, 6.0%, 5.44%, respectively (according to Table 7 P<0.05).

The presented table shows that the weight of the uneviscerated carcass exceeded the same figure in the control analogues by 6.01%, 6.02%, 5.45% (P<0.01) in the experimental groups of geese. The yield of the mass exceeded the same indicator in the control by on the average 5.87%, 6.925, 6.21% (P<0.01) in the viscerated form in the geese of the experimental groups. The moisture content in the meat of gooslings of the first experimental group was less by 1.14%, compared with the control, by 1.18% in the second experimental group, by 1.24% (p<0.05) in the third experimental group, while the amount of crude protein, on the contrary, was more by 1.07%, 1.21%, 1.26% (P<0.05).

The content of dry substances in the samples of meat of the experimental groups, in relation to the intact group of birds, against the background of the tested biologically active additives was also higher, compared with the control analogues, by 1.14%, 1.18%, 1.24% (p<0.05).

The cost of the additional gross gain in body weight of gooslings were equal 6516.00 RUB, 7614.00 RUB, RUB 6678.00, respectively, at the sales price of 200.0 rubles per kg, in these experimental groups of gooslings.

The profit minus all expenses in the first experimental group of gooslings amounted to 5316.00 rubles, 6494.00 rubles in the second group and 5498.00 rubles in the third group. The conditional profit from one goosling in the first experimental group was 44.30 RUB, 54.11 RUB was in the second group, 45.81 in the third group, as result of the application of biologically active additives "Laricarvit", "Bacell" and "Sporobacterin" at growing of goose growing stock. Based on each ruble spent, when using biologically active additives, net profit was 4.44 rubles in the first experimental group, 5.79 rubles in the second experimental group and 4.66 rubles in the third experimental group.

Thus, the tested domestic biologically active additives not only contributed to the improvement of physiological, morphological, biochemical, qualitative indicators of poultry meat but also to the increase in the live weight of geese and allowed to get additional profit.

4. Discussions

The results of the study showed that the egg production was increased, on the average, by 5.92%, 7.29%, 6.17% (P <0.01) in the goose of the experimental groups, compared with the use of dietary supplements "Laricarvit", "Bacell" and "Sporobacterin" compared to their analogues in the control for February - June. The highest egg production was in the second experimental group of geese, where "Bacell", the biologically active additive, was included into the main diet, within 7.29% (P<0.01) among these experimental groups of birds, compared with intact birds.

As you know, the output of goose growing stock is determined by the number of hatched viable healthy gooslings, expressed as a percentage of the number of eggs laid in the incubator. The results showed that this indicator was significantly higher by 7.71, 10.31, 8.85 (P <0.01) percent in the experimental batches, as compared with the control batch.

Further studies have shown that 667 viable healthy gooslings were obtained from fertilized eggs in the control batch, the hatchability in this batch (the percentage of hatched geese from the number of fertilized eggs) was 85.62%. In the first experimental group, this indicator relative to the control analogues was significantly higher by 4.74% (P<0.01), in the second – by 4.60% (P<0.01), in the third experimental group – by 4.98% (P<0.01), respectively.

An increase in the bactericidal activity of blood serum was observed in the first experimental group of birds by 6.74% (P <0.01), lysozyme activity - by 5.79% (P <0.05), phagocytic activity - by 5.94% (P <0.01) against the background of the use of "Laricarvit". The increase in these indicators of nonspecific resistance of the blood serum of geese was 6.87% (P<0.01), 5.93% (P<0.05), 6.10% (P<0.01) in the second experimental group, where the biologically active additive "Bacell" was used. The growth of marked components of nonspecific resistance of the blood serum of gooslings in relation to the intact poultry was, on the average, 7.15% (P<0.01), 6.73% (P<0.01), 6.96% (P<0.01) in the third experimental group on the background of the application of "Sporobacterin".

As you know, total protein is one of the most stable biochemical parameters. The total protein content in the blood serum was slightly higher than that of the control bird in our studies when using biologically active additives in the diet of experimental birds. This indicator was significantly higher, on the average, by 6.05-8.07% (P<0.01) on the 14th, 28th, 56th days of the experiment in blood serum of the experimental geese, in relation to the control.

Albumin fraction of protein is used in the body goose growing stock for the synthesis of tissue proteins. The results of studies showed that the level of this protein fraction in the blood serum of experimental geese, compared with intact poultry, was slightly higher, in the range of 2.97 - 3.65% (P<0.5), but the biometric analysis of these values was not statistically reliable.

A regular increase in the serum of the experienced goslings was established, compared with the intact bird, the gamma - globulin fraction of the protein, was within 9.27% - 12.30% (P <0.01) against the background of the use of tested biologically active additives. These figures indicate that the tested biologically active additives "Laricarvit", "Bacell" and "Sporobacterin" contributed to the stimulation of the Central organs of the immune system of geese to some extent, as gamma globulins are carriers of immune bodies [3].

The preparations used contributed to a significant increase in serum bactericidal activity by 6.74% (P<0.01), lysozyme activity – by 5.79% (P<0.05), phagocytic activity – by 5.94% (P<0.01). The increase in these indicators of nonspecific resistance of the blood serum of geese was 6.87% (P<0.01), 5.93% (P<0.05), 6.10% (P<0.01) in the second experimental group, where the biologically active additive "Bacell" was used. The growth of marked components of nonspecific resistance of the blood serum of goslings in relation to the intact poultry was, on the average, 7.15% (P<0.01), 6.73% (P<0.01), 6.96% (P<0.01) in the third experimental group on the background of the application of "Sporobacterin", survival rate was higher by 3.67 - 5.50% (P<0.05) in the experimental groups of birds.

Studies conducted and the data obtained allow us to state that the tested biologically active additives "Laricarvit", "Bacell" and "Sporobacterin" help to stimulate the egg production of geese, increase the percentage of hatching and hatching of goose eggs, erythrocytes, hemoglobin, non-specific resistance and obtaining healthy, viable goslings.

As you know, the growth rate of geese is characterized by indicators of the average daily growth of their body weight. The increase in body weight of birds both in the control and in the experimental groups occurred according to the same pattern - as they grew. The growth of geese was more intense at the same time in the experimental groups. So, the average daily gain in body weight exceeded the same indicator of the intact group on the 10th, 20th, 30th, 40th days of the goslings of the experimental group of scientific and economic experiment by, on the average, 6.00 - 6.24% (P<0.05), by 6.45 - 7.50% (P <0.01) on the 50th, 60th, 75th days in the first experimental group, when using "Laricarvit", the biologically active additive.

A similar change in body weight of the goslings was observed in the second experimental group of goslings, where "Bacell", the biologically active additive was introduced into their diet. The average daily weight gain of goslings was, on the average, 64.80±1.30 g in the control group at the end of the scientific and economic experiment. This indicator exceeded the same indicator of the intact group by, on the average, 7.88% (P<0.01) in the second experimental group of birds.

The experimental work performed showed that "Sporobacterin", the biologically active probiotic additive liquid, also has a fairly high growth-stimulating effect at the completion phase of the experiment against the background of the use of this preparation in the diet of the average daily increase in body weight of goslings in the third experimental group of birds, on the average, was 69.75 g, which exceeds the same indicator in the control by 7.63% (P<0.01%).

From the presented analysis, it can be concluded that "Bacell", the biologically active additive, causes the highest growth-promoting effect on growing goslings of the three tested biologically active additives, which contains the microbial mass of the spore-forming bacteria *Bacillus subtilis 945* (B-5225), acidophilic bacteria *Lactobacillus acidophilus L917* (B-4625), *Ruminococcus albus 37* (B-4292), sunflower meal, beet molasses, nonfat milk and water in its composition. This indicator was higher by, on the average, 7.88% (P<0.01) in the group of goslings, where "Bacell" was used, in relation to the control.

The chemical composition of meat in goslings of experimental groups on separate indicators slightly differed from those of the control group. Thus, the moisture content in the meat of goslings of experimental groups, compared with control analogues, was less by 1.14-1.24% (P<0.05), the raw protein was, on the contrary, was more by 1.07 - 1.26% (P<0.05). The content of crude fat and ash in samples of meat of experimental and control goslings were, approximately, identical and had no reliable difference. At the same time, the dry matter content was higher by, on the average, 1.14-1.24% (P<0.05) in the samples of goslings' meat of the experimental groups, compared with those analogues of the control group.

5. Conclusion

Three-hour ozonation sessions at a dose of 0.02 m3 of ozonated air per 1 m3 of purified air improved the sanitary and hygienic condition and optimization of the microclimate in the room for growing s for 6 days. The content of the relative humidity of the air in the room is reduced by 8.52% (P <0.05), the concentration of ammonia - by 22.73% (P <0.01), hydrogen sulfide - by 32.27% (P <0.001), carbon dioxide - by 0.10% (P <0.05), dust - by 60.56% (P <0.001), microorganisms - by 46.34% (P <0.001). The egg production for the average goose in the experimental groups, with the introduction of the dietary supplements "Laricarvit", "Bacell" and "Sporobacterin" in the main ration, was 5.92, 7.29, 6.17% higher (P <0.01), output of goslings - by 4.59, 6.67, 5.73% (P <0.05), hatchability - by 8.32, 9.07, 7.98% (P <0.01), than in the control.

The increase in the blood content of the number of erythrocytes by 4.73 - 5.21 (P <0.05), leukocytes - by 2.91 - 2.95%, hemoglobin - by 6.11 - 6.95% (P <0.05), the level of total protein in blood serum - by 5.83% - 6.19% (P <0.01), gamma globulins - by 9.27 - 12.30% (P <0.01) under the influence of biologically active additives "Laricarvit", "Bacell" and "Sporobacterin" on the 30th day of the experiment occurred within the physiological norm.

6. Summary

1. The bactericidal activity of the blood serum increases by 6.74 - 7.15% (P <0.01), lysozyme activity increases by 5.74 - 6.75% (P <0.01), phagocytic activity - by 5.99 - 6.96% (P <0.01), which indicates the activation of nonspecific resistance in the body of birds, as evidenced by an increase in survival rate of goslings at 1.67 - 2.50% against the background of the use of "Laricarvit", "Bacell" and "Sporobacterin" as part of the main diet.
2. The use of biologically active additives "Laricarvit", "Bacell" and "Sporobacterin" stimulates the growth and development of goose growing stock. The average daily gain of body weight of geese was higher by an average of 7.50 - 7.88 (P<0.01) in the phase of completion of the experiment in the experimental groups of birds, compared with the control. Higher growth of this indicator was noted in the group of goslings, where "Bacell" biologically active additive was introduced into the main diet - 7.88% (P <0.01).
3. The use of biologically active additives "Laricarvit", "Bacell" and "Sporobacterin" increases the pre-slaughter weight of goslings by 5.13 – 6.47% (P<0.05), the weight of the viscerated carcass - by 6.83-7.64 % (P<0.01). These figures were higher in the experimental group of goslings, where "Bacell" was introduced to the main diet of goslings.
4. Organoleptic and biochemical parameters of meat of experienced geese were identical with those of the control group and met the requirements of SanPiN 2.3.2.1078-01 "Hygienic Requirements for Safety and Nutritional Value of Food Products". The amount of moisture in the meat of the gosling of the experimental groups was less by 1.14 - 1.24% (P <0.05), the content of crude protein was by 1.07 - 1.26% more (P <0.05), dry matter was by 1.14 - 1.24% (P <0.05) than in the control group of goslings.
5. Enrichment of the diet of goslings with biologically active additives “Laricarvit”, “Bacell” and “Sporobacterin” contributes to obtaining additional profit in the domestic economy, based on each spent ruble 4.44 - 5.79 rubles, respectively.

7. Recommendations

1. We recommend using an artificial ozonizer brand "Pozitron-3 Air 1", with sanitation for six days, daily three-hour sessions at a dose of 0.02 m3 of ozonated air per 1 m3 of purified air, when growing goslings to improve the microclimate of the rooms.

2. We recommend to enter into the diet biologically active additives "Laricarvit" at the rate of 3 kg/t, "Bacell" 2 kg/t feed and "Sporobacterin" at a dose of 1 ml per 100 kg of dose weight of birds to increase the egg production of geese, hatching of healthy gooslings .
3. We recommend to introduce the dietary supplements "Laricarvit" into the goslings at the rate of 2.5 kg/t, "Bacell" - 1.5 kg/t of feed, "Sporobacterin" in a dose of 0.5 ml per 100 kg of body weight of birds to increase the intensity of growth, meat productivity, natural resistance, survival rate, quality of meat.

References

[1] Mahalov A G, Sukhanova F S and Roiter I S 2014 Geese, breeds, technologies and even recipes, *Kurgan: Publishing house of the Kurgan State Agricultural Academy*. 280 – 4.
[2] Pimenov B and Smirnov B 1984 Breeding resources and opportunities of husbandry *Poultry* **5**. 30-1.
[3] Reuter J S 2007 Prospects for the development of geese farming in Russia *Poultry*. **6** C2-4.
[4] Reuter J S 2009 Ural white geese – a promising breed. *Poultry* **10**, 9-12.
[5] Reuther J S 2015 Highly productive goose with grey plumage *Poultry* **5** 21 – 3.
[6] Snitkin M 2005 Prospects of development of husbandry in Russia *Poultry*. **10** 4-6.
[7] Sukhanova S F and Asaubaev G 2001 Heredity of phagocytic reactions of young geese of different breeds *Poultry* **9**. 15-6.
[8] Sukhanova S F 2012 Feed additive "Stimulus" for gooslings *Poultry* **5**. 31-2.
[9] Chikov A G and Pyshmintsev N A 2010 Features of feeding geese when growing them for meat *Effective Farming* **7** (57) 46 – 8.
[10] Shtele A A 1987 Geese on the farm and farmstead *Agropromizdat* 110 p.;
[11] Alekseev I A, Pastukhova T V and Ishtudova E R 2015 Optimization of microclimate in the room for growing geese with ozonizer "Pozitron-3 Air 1", Proceedings of the international scientific - practical conference "Food security and sustainable development of the agricultural sector", Cheboksary. - ChGSHA 2015 393-6.
[12] Alekseev I A and Pastukhova 2012 Hygiene of raising geese with the use of probiotic and vitaminous feed additives "Bacell" and "Laricarvit", Proceedings of the All-Russian Scientific and Practical Conference of Young Scientists, Postgraduates and Students "Agrarian Science - the Basis for the Development of the Agro-Industrial Complex, Cheboksary: CHGSHA.- 2012. 172-4.
[13] Balyasnikov V E 2012 Installation of ozone disinfection of air: passport and instruction manual *PC "POSITRON"*. 3-7.
[14] Dedyukina A S 2010 Ozone - possibilities of practical application of ozone. mhtml: File: 57-9.
[15] Melnik V 2014 Microclimate and productivity of poultry *Anim. Husband. Russia*. **4**. p. 21 – 2.
[16] Azaubaev G S 2008 Biological and productive qualities of geese bred in Kurgan region *Vniitei Agroprom*, **26 sun**. 122-4.
[17] Alekseev F F, Aralov A V and Belyakova L S 2007 Technology of production of meat of geese *Meat poultry*, St. Petersburg.: 178-80.
[18] Temporary guidance on the use of carotene feed additives.-2010.-Belgorod, JSC "Petrokhim".- P. 1-2;
[19] Gadiyev R R and Galina C R 2014 Meat quality of crossbred geese *News of the Samara State Agricultural Academy*. **1**. 124-7;
[20] Gadiyev R R, Farrakhov A R and Galina C R 2013 Reproductive qualities of geese of different genotypes *Bulletin of Bashkir State Agricultural University*. **3** (27). 66 – 9.
[21] Gadiev R R, Farrakhov A R and Khaziev D D 2009 Breeding geese *Recommendations, Ufa: Bashkir State Agricultural University*. p. 40.
[22] Galimullin T R and Galina C R 2014 Phase feeding of geese of parent herd Materials of the second all-Russian. science.- Prak. Conf. dedicated to the 100th anniversary of Professor H. V. Ayupova "Modern achievements of veterinary medicine and biology - in agricultural

production", Ufa: Bashkir State Agricultural University 334-7;
[23] Galina C R and Gadiyev R R 2014 Productive qualities of geese in a phase-feeding *News of the Orenburg State Agricultural University*. **3** (47). 125-8;
[24] Ignatova V V 2002 Questions of geese farming were raised in Suzdal *Anim. Husband. Russia* **10**. 40-1.
[25] Kazantseva M and Toboev G 2013 Economic - useful signs geese of lindovskaya *Poultry* **5** 2-3.
[26] Kirillov N and Toboev G 2011 Formation of meat productivity of Lindovskaya breed of geese *Poultry* **2**. 44-6.
[27] Gerasimenko V V 2005 Morphokinetic effect of the microflora of the gastrointestinal tract on the body of geese *Bulletin of OGU* **2**. 133-7;
[28] Pastukhova T V, Alekseev I A and Martynov G N 2015 Growth and development of goose growing stock when probiotic Sporobacterin is administered in the diet of scientific notes of KGAVM. **24** (4). 170-4.
[29] Sukhanova S F 2014 The use of drugs series "Vetom" in the feed of young geese *Poultry* **10**. 25-7.
[30] Tarakanov B V and Gerasimenko V V 2008 Using of Microcikol when grown geese *Zootechnics*. **4**. 20-2.
[31] Tarakanov B V, Nikulin V N and Gerasimenko V V 2005 Nonspecific resistance and productivity of geese using lactoamilovorin *Veterinary* **2** 55-8.
[32] Shevchenko A I, Nozdrin G A and Smolovskaya O V *et al.* 2009 Influence of probiotics, Vetom 1.1, 1.2 Vetom, Selen and Sinbiotic complexes on their basis on the meat productivity of geese *Achievements of Science and Technology of Agriculture*. **4**. 48-50.
[33] Shilova E, Donnik E and Shkuratova I *et al.* 2011 Germavit to improve efficacy of vaccination against influenza in geese *Poultry* **7** 32-3.
[34] Bordello S P and Scand J 1984 Bacteria and gastrointestinal secretion and motility *Gastroenterology* **93** 34-6 (Sapl).
[35] Gebbers O 1984 Functional morphology of the mucosal barrier *Microecol. Therapy*. **14** 117-23.
[36] Meslin J 1984 Effect of microflora on the dimensions of enterocyte microvili in the rat. *Reprod Nutr. Dev.* **24**(3) 81-2.
[37] Meslin J 1987 Effect of amyloamize starch on intestinal mucosal morphometry and ileal epithelial renewal. Comparison between germfree and conventional rats In: Gnotobiol. and its applications. Versales, France. 1987. 94-5.
[38] Temporary guidance on the use of liquid Sporobacterin.-2011. Orenburg, "Bakoren", OOO. 1-2;
[39] Gorkovenko L K 2011 Instruction on the use of probiotic additives "Prolam", "Monosporin" and "Bacell" in animal husbandry *Krasnodar*. 1-3;
[40] Donnik I. Shkuratova I and Zaslonov A *et al.* 2011 Germavit impact on the health of young geese *Poultry* **2** 41-3.
[41] Fisinin V I, Gadiyev R R and Farrakhov A R 2012 Effect of heterosis in geese farming Proceedings of the international XV11. conf. "Innovative developments and their development in poultry industry". Sergiev Posad.-2012. 114-6.

IOP Conf. Series: Earth and Environmental Science **346** (2019) 012022 doi:10.1088/1755-1315/346/1/012022

Study on the effect of stress treatment on embryogenic callus of raspberry

F H Liang[1], Y J Liu[1, 3], J H Yang[1], Y Liu[1], Y Y Shi[1], H Q Liu[1] and C Zhang[2]

[1]College of Horticulture and Landscape, Tianjin Agricultural University, Tianjin 300384, China
[2]Tianjin Shuofang Green Technology Development Co. LTD, Tianjin, 300384, China

E-mail: 2485674628@qq.com

Abstract. In order to obtain the embryogenic callus of raspberry, the friable calluses of raspberry were used as materials to be treated by different time of starvation and desiccation, and were further cultivated and induced embryoid. By use of embryogenic observation, the embryogenic degree of callus could be determined. The results showed that the embryogenic degree of calluses which were treated by starvation for 6 days was the highest, and the desiccation treatment had no effect on embryogenic degree of calluses. Therefore, appropriate starvation treatment can facilitate the transformation of non-embryogenic callus into embryogenic callus. The results of this experiment could provide reference for callus embryogenesis induction of other plants.

1. Introduction

Raspberry, as a new functional fruit, is more and more popular among consumers due to its rich nutrition [1]. With the expansion of the planting area of raspberry, the requirement for the provenance of raspberry also becomes diversified [2-5], so it is urgent to accelerate the breeding research of raspberry.

Raspberries are mostly wild varieties, and cultivated varieties obtained through artificial breeding are relatively stable plant varieties with heredity. Through conventional hybridization, it is difficult to obtain new excellent characters, so the development of conventional breeding meets a bottleneck. However, gene transformation technology is restricted by the market and cannot be industrialized. Therefore, mutagenesis technology is basically adopted in the breeding of raspberry [6-10]. Conventional physical and chemical mutagenesis using seeds or plants has the lower mutagenesis efficiency. Mutation breeding with embryonic callus can greatly improve the mutagenesis efficiency [11-13].

Embryogenic callus is different from common callus and has strong advantages in somatic mutation breeding and gene transformation. Its main advantages are as follows: first, the embryonic callus is in the state of vigorous division, and the genetic material in the cell is very active, and it is easy to produce variation, which is more obvious under the condition of added mutagens; Second, the number of embryonic callus cells has an obvious advantage, and each cell can be regarded as a single mutagenesis object, thus increasing the mutagenesis efficiency. Third, embryogenic callus can be dispersed into multiple independent individuals, so that each mutant plant is formed by the division and differentiation of a single embryogenic mutant cell, thus avoiding the chimerism phenomenon under conventional mutagenesis. Fourthly, embryogenic callus can obtain regenerated plants at any

Content from this work may be used under the terms of the Creative Commons Attribution 3.0 licence. Any further distribution of this work must maintain attribution to the author(s) and the title of the work, journal citation and DOI.
Published under licence by IOP Publishing Ltd

time. Because embryogenic callus can regenerate directly on the culture medium without hormone, without hormone induction, which saves time for research and improves the efficiency of research [14,15].

Although the regeneration efficiency of embryogenic callus is high, it is still not ideal to conduct tissue culture studies on some plants due to the difficulty in inducing embryogenic callus. How to improve the proportion of embryogenic cells is a difficult point for many scholars Therefore, obtaining better callus induction conditions and in vitro regeneration techniques are the preconditions for the utilization of embryonic callus [16]. In this study, the callus of raspberry was treated by stress method in order to obtain embryonic callus, so as to provide good experimental materials for breeding research and lay a foundation for the cultivation and promotion of bramble.

2. Materials and methods

2.1. Test materials

The callus used in this study was the raspberry loose callus induced by the garden plant laboratory team of Tianjin agricultural university. The loose callus was easy to be separated and differentiated stably, which was a good material for loose embryo callus induction.

2.2. Test methods

2.2.1. Definition of embryonated callus. Generally, the morphological characteristics of loose embryogenic callus are loose texture, milky white or yellow color, and spherical particles on the surface. Nikon.P. s. XY64 somatoscope was used for observation. From the perspective of cytology, embryogenic callus is characterized by small cells, dense protoplasm, small vacuoles and large nuclei, while non-embryogenic callus is characterized by large cells, small nuclei and large vacuoles. The instrument used for observation was a LEICA DM 2000 microscope. The characteristics of the spherical embryo are the spherical structure composed of multiple cells. The spherical surface is smooth, the internal cell division is vigorous, and the cell individual is small, which will present the development and change process similar to the plant seed embryo by further cultivation.

2.2.2. drying treatment of callus of raspberry. Dissected the cultivated raspberry loose callus into small pieces of 2 mm size with the scalpel, and paid attention not to crush the callus in the process of transfer and segmentation. Take the sterile petri dish and transfer the separated callus of raspberry to the empty petri dish in the ultra-clean table. The petri dish must be dry without the presence of steam. Distribute the callus evenly in the petri dish and seal the petri dish with paraffin sealing film. Twenty callus groups were placed in each culture dish, and three replicates were set for each treatment. The treated calluses were cultured under the following conditions: 2000 lux light intensity, 24 h light. Temperature is 23°C, dry processing time respectively 1 d, 2 d, 3 d 4 d.

2.2.3. Hunger management of raspberry callus. The starvation treatment adopted in this experiment was very simple. The cultivated loose callus of raspberry was taken and cut into small pieces of 2 mm size for later use. Take the sterile petri dish, put the sterile filter paper in the petri dish in the ultra-clean table, add a small amount of sterile water to the petri dish, and note that the amount of water is subject to all the soaked filter paper. Do not add more filter paper or do not completely soak. The calluses were placed in a prepared dish. 15 callus groups were placed in each culture dish, and 5 replicates were set for each treatment. The treated calluses were cultured under the following conditions: 2000 lux light intensity, 24 h light. Temperature of 23°C, hungry processing time respectively for 3 d, 6 d, 9 d, 12 d.

2.2.4. Recovery culture of callus of raspberry on MS medium. The dried callus and starved callus were transferred from the petri dish to MS medium for embryoid culture, and the changes of callus were observed after 14 days.

3. Results and analysis

3.1. Experimental results of drying treatment of callus of raspberry

After drying for different days, the callus of raspberry changed significantly in appearance and under microscopic observation, and then transferred to MS medium. After about 14 days of culture, the callus basically did not show embryonic structure, as shown in table 1.

Table 1. Effects of drying treatment on embryogenesis of callus of raspberry.

Treatment day	Number of callus inoculated	Changes of callus after treatment	Microscopic changes of callus after treatment	The change of culture was restored on MS medium after 14 days
1	60	no change	The cells were parenchyma without obvious embryogenesis.	The callus grew normally, but no embryonic structure was produced.
2	60	A small amount of callus were dehydrated and dried	The cells on the surface lost water and died. The internal cells are non-embryonic.	Callus could not grow and all died with culture time.
3	60	Nearly half of callus lost water	Most cells lose water and die.	All callus died after Browning.
4	60	All callus lost water and atrophied.	All cells lost water and died.	All callus died after Browning.

As can be seen from table 1, with the increase of drying treatment time, the surface of the callus loses water and the callus shrinks. When the treatment time reaches 4d, the callus is basically dry and has no living condition. In terms of the anatomical structure, the callus was still non-embryonated parenchyma cells even after 1 day of treatment. With the increase of treatment time, the cells outside the callus lost water, atrophied and died, forming a structure similar to the cork layer, while the internal cells were normal, but there was no change of embryogenesis. The callus of raspberry treated for 4d was completely dehydrated, and the cell structure could not be found under the microscope, so embryogenic changes were more unlikely.

When these dried callus were transferred to MS medium, only the callus treated for 1d could grow, but there was no embryoid structure. The microscopic observation was non-embryonal callus. Other callus showed browning after 2-3 d on MS medium, and almost all of them turned black and died after 7d.Therefore, the drying treatment had no effect on the embryogenesis of the callus, and the drying treatment could not accelerate the embryogenesis of the callus, or even hinder the normal growth of the callus.

3.2. Effects of starvation treatment on embryogenesis of callus of raspberry

After starvation treatment for different days, the callus of raspberry changed significantly in appearance and under microscopic observation, and then was transferred to MS medium. After about 14 days of recovery culture, some callus were treated to spherical embryos, as shown figures 1-3. In figures 1-3, T3 represents the starvation treatment for 3 days, T6 represents the starvation treatment for 6 days, T9 represents the starvation treatment for 9 days, and T12 represents the starvation treatment for 12 days.

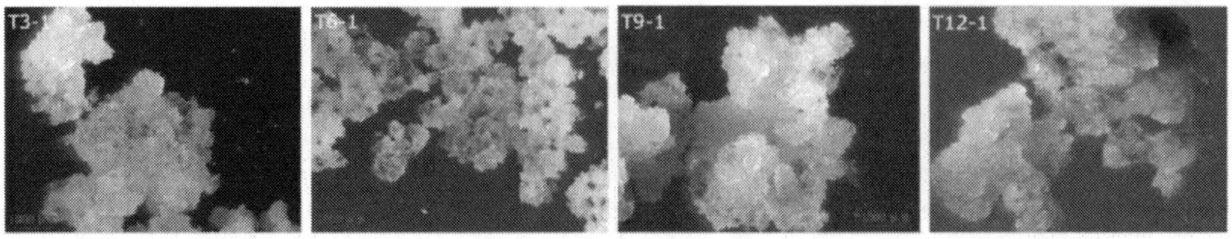

Figure 1. Effects of different starvation treatment days on callus morphology.

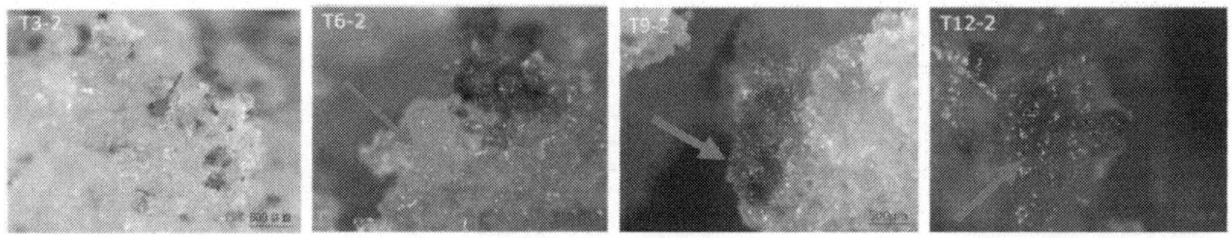

Figure 2. Effects of different starvation treatment days on granule structure formation of callus.

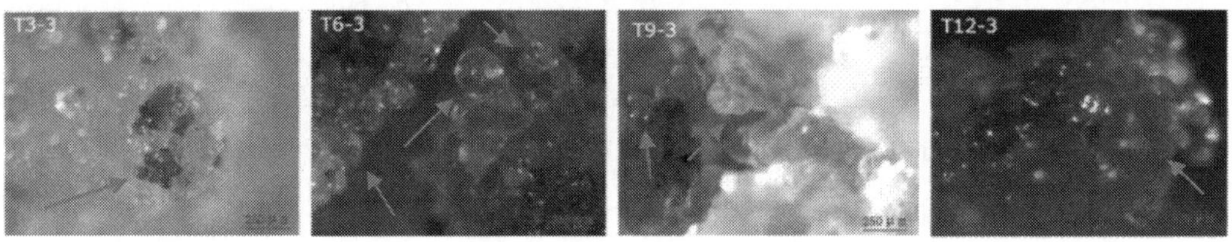

Figure 3. Spherical embryo structure at different starvation treatment time.

Table 2. Effects of starvation treatment on embryogenesis of callus of raspberry.

Treatment day	Number of callus inoculated	Changes of callus after treatment	Microscopic changes of callus after treatment	The change of culture was restored on MS medium after 14 days
3	75	no change	Most of the cells were parenchyma, and 5% of them showed embryogenesis.	The callus grew normally and the smooth spherical embryo structure appeared on the callus with embryogenic characteristics.
6	75	Callus are hard and granular	More than 95% of the cells showed embryogenesis and several cells clustered together.	Almost all callus have smooth spherical embryo structure.
9	75	Callus are hard and granular. Browning occurred in 15% of callus	All the living cells showed embryogenic characteristics, and multiple cells were clustered together. Individual cells die.	All the callus of the surviving cells had globular embryo structure with smooth surface. But soon 85% of callus died after Browning.
12	75	Callus are hard and granular. Nearly 50% of callus showed Browning.	All the living cells showed embryogenesis, but half of them died.	Soon all callus died after Browning.

As can be seen from table 2, the effect of starvation treatment on the number of spherical embryos generated after callus recovery culture was obvious. After 14 days of recovery culture, about 5% of the spherical embryos in the starved treated callus for 3 d grew normally, about 95% of the spherical embryos in the starved treated callus for 6 d grew normally, about 13% of the spherical embryos in the starved treated callus for 9 d grew normally, and almost all the callus in the starved treated callus for 12 d died. It can be seen that the proportion of callus to embryogenesis is the highest when the starvation time of callus is 6 d, which is most conducive to the embryogenesis of callus.

4. Discussion

4.1. The theoretical basis that starvation treatment is beneficial to the embryogenic transformation of callus

The mechanism of the effect of starvation treatment on the embryogenesis of callus can be summarized as the effect of stress on cell differentiation. Specifically, under the premise of nutrient deficiency, some cells grow slower, some cells disintegrate and die, and some competitive cells will absorb the nutrients of these cells to maintain their basic needs. These cells to resist adversity, nutrition lack necessary absorbs the nutrition of surrounding cells, mainly is the nutrition of disintegration of some cells. These cells in order to continue to resist adversity stress gradually reduce their metabolism, leading to splitting speed drops, causing cell excessive nutrient accumulation, cytoplasm concentration increases, the intracellular nutrients increases, leading cell toward embryonic development and callus toward the embryonic callus. This is similar to the performance of many plant cells under adversity, such as the budding reproduction of some plants under adversity. This mechanism is the self-protection function acquired in the process of plant evolution, and it is the potential ability of plants [17].

4.2. In this experiment, the reason why drought treatment had little effect on callus embryogenesis and the measures to change

Drought stress can also promote the callus to embryonic callus transformation, but the effect is not obvious in this experiment. The main reason is that the callus was inoculated directly to the triangle in the bottle without any medium. The callus surface water loss fast and died, so as to no prepare for embryonic structure transformation. Therefore, to transform callus into embryonic by drying stress it is necessary to make a relatively dry environment, allows cells to respond to the environment gradually, to adjust, Rather than trying to induce cells to adapt in this extreme way. In the future, some humidity can be added to the environment to control the degree of drying, such as adding a few drops of water in a tripod bottle to create a relatively dry environment. Or control the relative humidity in the culture environment to prevent the rapid disappearance of water in the tripod. Another method is to add polyethylene glycol into the culture medium to cause physiological drought, so as to change the water absorption environment of the callus, which can also receive the same treatment effect as the drought. In a word, drought has some effect on embryogenesis of callus, but the treatment method must be selected correctly to get the desired effect.

4.3. Effects of the timing and mode of hunger treatment on the embryogenic effect

some preliminary experiments were done to find the best time to deal with hunger in this study. In the process of callus culture, the physiological state of cells is different, and with the change of medium composition and culture environment, the callus will also change correspondingly, and there will be various changes in physiology and structure. Generally speaking, the time should be chosen to transform the callus into embryogenic tissue as far as possible, that is, the closer the callus is to the embryogenic state, the easier it is to be induced to produce embryogenic transformation by stress. At this time, the embryogenic transformation of the callus can be realized once and for all by simple stress.

5. Conclusion

In this experiment, the callus of bramble was embryogenic induced by starvation and drought stress. The results showed that the embryogenic effect obtained by starvation treatment was more ideal, and the embryogenic effect obtained by starvation treatment at 6d was the best. Although drought is not ideal for callus embryogenesis, it can be overcome by improving the experiment. In conclusion, adversity can promote the transformation of callus into embryogenic tissue, which is especially necessary for plant species whose embryogenic callus cannot be induced by common measures.

Acknowledgments

This research was supported by the Science and Technology Commissioner Project in Tianjin (18JCTPJC68000), Tianjin Modern Agricultural Industry Technology System Innovation Team Construction (ITTHRS2018002), The Major Science and Technology Project of the Seed Industry of Tianjin Committee of Science and Technology (17ZXZYNC00070), Agricultural Science and Technology Achievements Transformation and Extension Project in Tianjin (201603060) and Tianjin Applied Basic and Frontier Technology Research Plan (14JCYBJC30200).

References

[1] Shi X, Chen Q Q and Bi J F 2015 *Food Ind. Sci. Technol.* **36** 376-81
[2] Du Y Y, Liu H L and Li B G 2016 *Comm. Forestry Sci. Tech* **01** 76-9
[3] Ren J and Wei P 2013 *Henan Sci. Technol.* **12** 174-5
[4] Luo S Q, Zhou C Y and Tang J G 2012 *Guizhou Sci.* **30** 84-9
[5] Wu C Y 2015 *Northern Fruit Trees* **25** 49-50
[6] Li C Y and Liao W J 2007 *Anhui Agric. Sci.* **04** 998-9
[7] Xiao Y P, Wang Z and Zhang Z Q 2001 *Chin. Herbal Med.* **08** 69-72
[8] Li W L, Li H Y, Wang X M and Wu W L 2009 *Develop. Forestry Sci. Technol.***03** 9-14
[9] Miao B Y, Liu Y J and Yang J H 2014 *Shanxi Agric. Sci.* **42** 209-12
[10] Tang F Y, Fang L and Wang X M 2013 *Shandong Agric. Sci.* **45** 51-4
[11] Gao J, Wang Y Z and Huang H Y 2016 *Plant Res.* **36** 52-7
[12] Zhang H F 2016 *Sci. Technol. Forestry* **41** 9-12
[13] Wen F, Xiao S X and Nie Y M 2012 *Chin. Agric. Sci.* **45** 4050-6
[14] Li Q, Sun D and Li H B 2014 *Seeds* **33** 5-9
[15] Wang H Y 2010 *J. Liaocheng Univ. (Nat. Sci. Ed.)* **02** 51-3
[16] Wu J 2016 *Chin. Acad. Agric. Sci.* 25-28
[17] X Q Yang 2013 *Yangzhou Univ.* 44-49

IOP Conf. Series: Earth and Environmental Science 346 (2019) 012023 doi:10.1088/1755-1315/346/1/012023

Influence of spatial structure of road greening on particle diffusion

F H Liang[1, 3], Y Liu[1], Y Y Shi[1], H Q Liu[1] and C Zhang[2]

[1]College of Horticulture and Landscape, Tianjin Agricultural University, Tianjin 300384, China
[2]Tianjin Shuofang Green Technology Development Co. LTD, Tianjin, 300384, China

E-mail: 630923263@qq.com

Abstract. In order to study the influence of the transparency and crown density of road green belts on particle dispersion, the green belts along Tinjing highway in Tianjin, China, were selected to study in this study and the green space was divided into four types, which respectively were the high crown density-high transparency green space, the low crown density-low transparency green space, the high crown density-low transparency green space and the low crown density-high transparency green space. The particle pollution concentration in sidewalk and green space was researched in the four types of green space. It was concluded that: in the sidewalk, the high crown density-high transparency green space was the most adverse to the diffusion of automobile exhaust pollutants because those contents of pm2.5, pm10 and dust were all the highest; in the green space, the high crown density-high transparency green space and the low crown density-low transparency green space were the most adverse to the diffusion of automobile exhaust pollutants because those contents of pm2.5, pm10 and dust were all the highest. So among the four types of green space, the high crown density-high transparency green space was the most adverse to pollutant diffusion.

1. Introduction

With the continuous progress and rapid development of China's economy, more and more families had bought private cars, and the continuous increase of car ownership made the emission of automobile exhaust become the primary source of air pollution [1], which also indirectly led to the continuous deterioration of the environmental quality of urban roads. Tang [2] had shown that dust pollution caused by aggregation of solid particles had become the most serious problem in urban pollution. Meanwhile, a large amount of exhaust emission made the contents of pm2.5, pm10 and dust in the city increased sharply. However, urban roads were not only private vehicles, but also belong to a large number of non-motor vehicles and pedestrians. Air quality was closely related to our health and life [3], so now we were faced with how to solve the problem of road pollution.

Many scholars in China began to study the relationship between automobile exhaust diffusion and environmental quality. According to the study, a large number of road green plants could adsorb dust [4-7] and a certain amount of air pollutants [8-10], so the plant configuration according to the terrain and the surrounding environment was applied to purify air in the urban green space system [11,12]. Therefore, the number of plants in the road greenbelt increased year by year, and different species and forms as well as plant community layers [13,14] were increasingly abundant. However, few studies had been conducted on the relationship between the greenbelt space and the quality of road environment in woodland and sidewalk.

Content from this work may be used under the terms of the Creative Commons Attribution 3.0 licence. Any further distribution of this work must maintain attribution to the author(s) and the title of the work, journal citation and DOI.
Published under licence by IOP Publishing Ltd

IOP Conf. Series: Earth and Environmental Science **346** (2019) 012023 doi:10.1088/1755-1315/346/1/012023

At present, the scholars mainly focused on the absorption of exhaust pollution on plant leaves, and had done a lot of research. However, due to different cities and different allocation modes, conclusions of the same tree species were also different. Some scholars began to study the influence of the space structure of the whole green space on the dust retention capacity of the green space. However, no better configuration of the space structure of the green space was proposed. Therefore, the method of the real road environment pollutants measurement was adopted and according to the transparency and crown density of the road green space, four types of green space form were divided, which were the high crown density-high transparency green space, the low crown density-low transparency green space, the high crown density-low transparency green space and the low crown density-high transparency green space. The contents of pm2.5, pm10 and dust on the sidewalk and green space were researched in the four green space types. The feasible green space structure form was put forward, which could effectively improve the pedestrian traffic space environment and also provide the basis for perfecting the theory of green space system planning.

2. Materials and methods

2.1. The choice of road sample

Tianjin located in the north China plain is in the north latitude 38°34'-40°15', longitude 116°43'-118°04' between, is one of the four municipalities directly under the central government of China, which is a city with a population of more than 10 million and one of the areas with severe smog in China (figure 1).

Figure 1. Tianjin location.

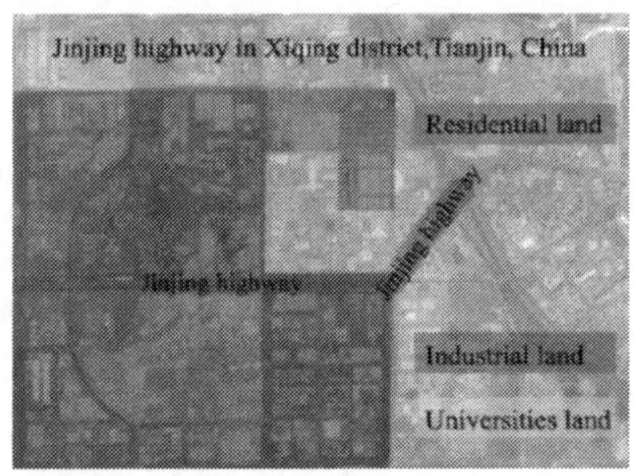

Figure 2. Jinjing highway location.

The experimental time was selected in autumn, mainly because the autumn climate in Tianjin is moderate in temperature with low wind speed and plant growth entered a stable stage. The climatic conditions during autumn are relatively stable, which is conducive to the consistency of experimental conditions. In contrast, spring in Tianjin was windy and plants were still in the growth stage, which greatly affected the changes of environment factors. Summer is hot, and rain affected greatly the exhaust diffusion is concentrated. The average annual precipitation in Tianjin is about 550～600 mm, and the number of precipitation days is 63-70 days. The precipitation in June, July and August accounts for about 75% of the whole year. Winter is cold, dry less snow and high wind speed and the most important reason is that plants lose their leaves in winter.

The test sample was selected as the Jinjing highway in Xiqing district, Tianjin. The road was two boards and three belts with 8 two-way lanes. There were factories, schools, research institutes, enterprises and institutions around (figure 2), per square kilometer 3,800 people on working days, so the traffic flow in the morning and evening was large, with 70-90 vehicles per minute. The greenbelt bandwidth on the north side of the road ranged from 15 to 20 m. The greenbelt was rich in plant forms and diverse in spatial structure, which ensured environmental consistency for studying the influence of greenbelt structure on pollutant diffusion.

According to the spatial structure of different greenbelts on the north side of Jinjing highway, four sample points were selected according to the difference in transparency towards the road direction and canopy density of road in green space, that was, the high crown density-high transparency green space, the low crown density-low transparency green space, the high crown density-low transparency green space and the low crown density-high transparency green space, as shown in table 1.

Table 1. Sample land types and plant species.

Sample type	crown density	transparency	Plant species
high crown density -low transparency	90%	10%	gold leaf privet, big leaf poplar, forsythia, purple Leaf Lee, golden tree, pagoda tree
low crown density -low transparency	10%	10%	gold leaf privet, big leaf poplar, forsythia, purple leaf lee, honeysuckle, golden tree
low crown density -high transparency	10%	60%	gold leaf privet, Chinese redbud, jujube, Canada yang
high crown density -high transparency	90%	60%	gold leaf privet, big leaf poplar, gold leaf yu, Canada yang

2.2. Method of measurement

Measuring time was September 28 and 30, 2017, 4 PM to 6 PM (at this time, it was the off-duty peak, and the traffic flow was large, with an average of 80 vehicles per minute), within 1.0 m/s wind speed, temperature from 16.3°C to 26.5°C, humidity of 33% to 37%. In each sample two positions were chosen, then one was in the middle of the sidewalk and the other was 10 meters inside the green space. On each selected position, using handheld pm2.5 tester, pm10 tester, dust meter and anemometer, the contents of pm2.5, pm10 and dust were measured in 0.5m, 1m, and 1.5m. Each measurement repeated three times and the average value at the same position averaged all measurement. The overall data of the test were collated, analyzed and plotted by SSPS and Excel2007.

3. Results and analysis

3.1. The relationship between pm2.5 concentration and spatial structure of green belt

As could be seen from figure 3, the variation rules of pm2.5 concentration in sidewalk and green space were consistent. Low crown density-low transparency green space and high crown density-high transparency green space had higher pm2.5 concentration and had significant differences from others, but the difference between them was not significant. However, the pm2.5 concentration of high crown density-low transparency green space and low crown density-high transparency green space was lower and the difference between them was not significant.

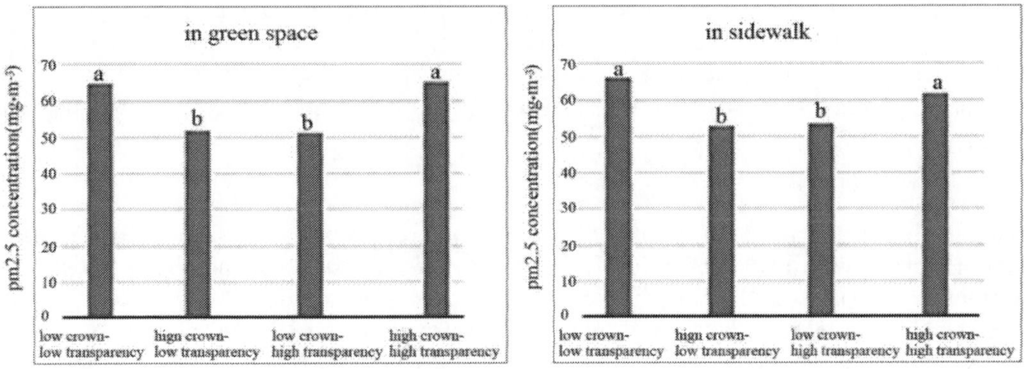

Figure 3. Pm2.5 concentration of four types of green space in sidewalk and green space.

Except the high crown density-high transparency green space, the pm2.5 concentration was higher than that in green space. Automobile exhaust was the main source of pm2.5. When vehicles passed by, pm2.5 concentration increased, while the spatial structure of different green Spaces affected the diffusion speed and scope of pm2.5. Transparency affected the horizontal diffusion of pm2.5, while canopy density affected the vertical diffusion of pm2.5. The two factors worked together and the situation was relatively complex. Within the sidewalk, the green space with low transparency hindered the horizontal diffusion. However, if the tree species was in the shape of a gentle slope, it was conducive to the vertical diffusion [8]. The green space with high transparency was beneficial to the horizontal diffusion, but it would hinder the vertical diffusion if the tree species was in a steep slope, as shown in figure 4.

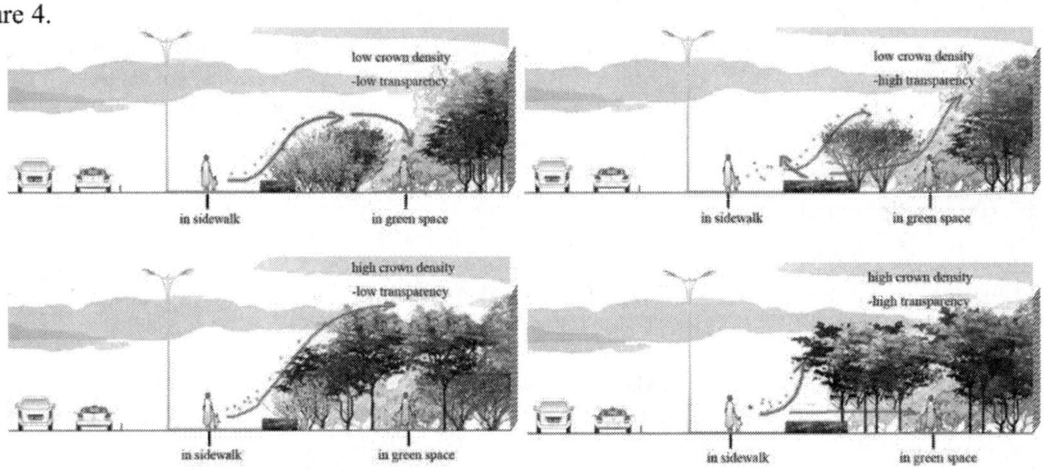

Figure 4. Four selected sample elevation and direction of the wind.

Within the green space, pm2.5 mainly took the form of sedimentation and retention. In low crown density-low transparency green space, pm2.5 settled from top to bottom in the green space, and spread slightly in the middle and lower layers of plants. In high crown density-high transparency green space, pm2.5 spread from outside to inside in the green space, and then spread slightly in the forest land. Both types of green space were dominated by pm2.5 aggregation and plant adsorption. Green land with low pm2.5 concentration had weak pm2.5 aggregation effect. Low crown density-high transparency green space, on the one hand, was affected by horizontal diffusion; on the other hand, it was affected by vertical settlement, so that there was turbulence in the green space, which could spread outward again. High crown density-low transparency green space is closed around, and relatively isolated space was formed inside, so pm2.5 concentration was also low.

Through the above analysis, it would be concluded that the four green space types that were conducive to pm2.5 concentration diffusion in sidewalk and in green space were also in the order of the high crown density-high transparency green space and the low crown density-low transparency green space > the high crown density-low transparency green space and the low crown density-low transparency green space.

3.2. The relationship between pm10 concentration and spatial structure of green belt

As could be seen from figure 5, the variation rules of pm10 concentration in sidewalk and green space were consistent. Low crown density-low transparency green space and high crown density-high transparency green space had higher pm10 concentration and had significant differences from others, but the difference between them was not significant. However, the pm10 concentration of high crown density-low transparency green space and low crown density-high transparency green space was lower and the difference between them was not significant.

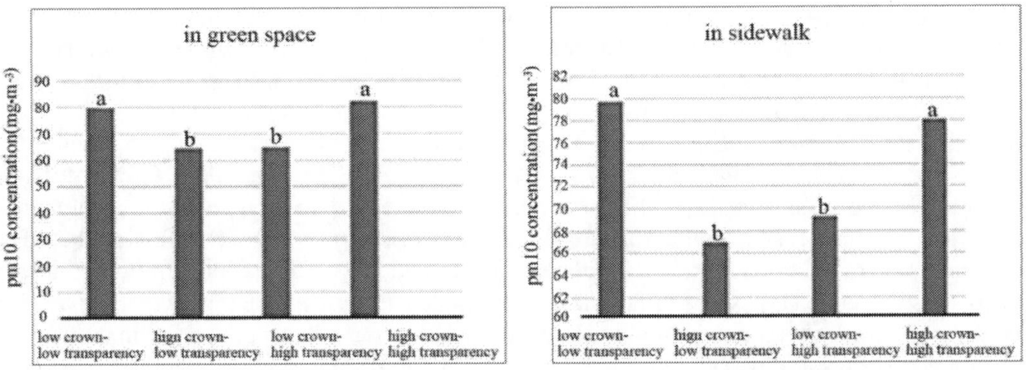

Figure 5. Pm10 concentration of four types of green space in sidewalk and green space.

In low crown density-low transparency green space and high crown density-high transparency green space, pm10 concentration of sidewalk were lower than that in green space. These two types of green space were conducive to the diffusion of pm10 in the sidewalk and the settlement in green space, while the other two types of green space were contrary. Pm10 particles were larger than pm2.5 particles and spread close to each other, so they could be directly concentrated in nearby woodlands after diffusion, which was different from pm2.5 particles, as shown in figure 4.

Through the above analysis, it would be concluded that the four green space types that were conducive to pm10 concentration diffusion in sidewalk and in green space were also in the order of the high crown density-high transparency green space and the low crown density-low transparency green space > the high crown density-low transparency green space and the low crown density-low transparency green space.

3.3. The relationship between dust concentration and spatial structure of green belt

As could be seen from figure 6, the dust concentration of low crown density-low transparency green space within the sidewalk was significantly lower than that of other green space types, and the difference of other green space types was not significant. Within the green space, the dust concentration of low crown density-high transparency green space was significantly lower than that of other green space types and the difference of other green space types was not significant.

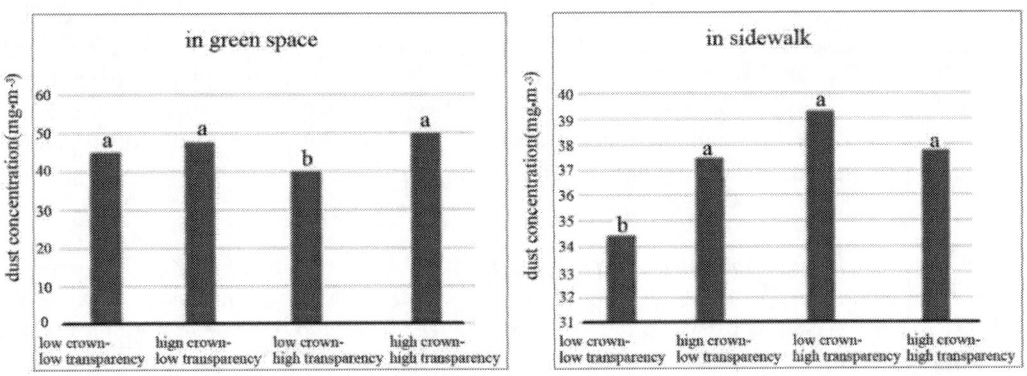

Figure 6. Dust concentration of four types of green space in sidewalk and green space.

Concentrations of dust in the sidewalk were lower than the concentrations of dust in the green space. The dust was the floorboard of all kinds of graded structure, not only came from automobile exhaust, but also from all kinds of dust in the air. The vehicle traffic mainly produced small fraction structure that couldn't increase dust concentration a lot in the overall, but impact on local wind direction and wind speed. The air circulation and the dust diffusion in the sidewalk were fast, while a large part of dust would be retained in the green space, so the dust concentration in the green space was much higher than that in the sidewalk, as shown in figure 4.

In the sidewalk, dust with high crown density-high transparency green space slowed down diffusion because the spread in the vertical direction could form vortex. Dust with high crown density-low transparency green space slowed down diffusion because the green space style could form higher slope. Dust with low crown density-high transparency green space slowed down diffusion because the horizontal direction of dust was hindered by plant branches and leaves. The concentration of dust in the sidewalk of these greenbelt types was high. The low crown density-low transparency green space formed a low gentle slope, which was very conducive to dust diffusion, so the dust concentration is significantly lower than other green space types.

In green space, dust could be adsorbed by a lot by plants, but also easy to cause the secondary dust [15]. High crown density-high transparency green space and high crown density-low transparency green space were relatively closed, so the dust easily stranded, and much dust was accumulated on the plant leaves. When environmental wind entered, dust could be kicked up, so dust concentration of the two green space types was higher. In low crown density-low transparency green space, the dust concentration was also higher, mainly due to settlement. In low crown density-high transparency green space, dust diffusion modes were more and air circulation was relatively active, so the dust concentration was significantly lower than other green space types.

Through the above analysis, it could be concluded that the sequence of the four green space types in the sidewalk that was conducive to the diffusion of dust concentration was the high crown density-high transparency green space and the low crown density-low transparency green space > the high crown density-low transparency green space and the low crown density-low transparency green space, and the sequence of the four green space types in the green space was the high crown density-high transparency green space, the low crown density-low transparency green space and the high crown density-low transparency green space > the low crown density-high transparency green space.

4. Discussion

4.1. Influence of green space structure on pm2.5 and pm10 concentration

Green space structure could change the diffusion velocity of pm2.5 and pm10 made from car tail gas, transparency affected horizontal diffusion velocity, crown density affected the vertical diffusion velocity. Green space type that was the most suitable for sidewalk environment should be low transparency and had the gentle slope shape. However the steep slope could prevent vertical diffusion. Green space type that was the most suitable for inside of green space environment should be high crown density-low transparency green space closed to a closed type, which could isolate the spread in all directions, so the overall pm2.5 and pm10 concentration were low. But the relatively open green space structure could produce the secondary dust of pm2.5 and pm10 because the internal air flow was not stable.

4.2. Influence of green space structure on dust concentration

Dust came from a wide range of sources, including automobile exhaust and all kinds of dust in the air. The mobility in the sidewalk was greater than that in the green space, so the concentration of dust in the sidewalk was lower. Excessive plants in green space would hinder dust diffusion, resulting in increased dust concentration. Green space structure with high canopy density was more likely to retain dust, while green space structure with low canopy density and high permeability had more air circulation, which was easier to accelerate dust diffusion rate, reduce dust accumulation and reduce concentration.

5. Conclusion

In this study, the influence of road green belts with different spatial structures on exhaust pollutant diffusion was studied from the two positions of green space and sidewalk, and the conclusion was drawn through analysis: among the four types of green space, the high crown density-high transparency green space and the low crown density-low transparency green space were the most adverse to the diffusion of pm2.5 and pm10, while the other two types were the most advantageous to the diffusion of pm2.5 and pm10; the low crown density-high transparency green space was the most advantageous to the diffusion of dust, while the other three green spaces were the most adverse to the diffusion of dust. In the sidewalk, the contents of pm2.5, pm10 and dust of the high crown density-high transparency green space were all the highest, so this type of green space was the most adverse to the diffusion of automobile exhaust pollutants; in the green space, the contents of pm2.5, pm10 and dust of the high crown density-high transparency green space and the low crown density-low transparency green space were all highest, so these two types of green space were the most adverse to the diffusion of automobile exhaust pollutants; while the contents of pm2.5, pm10 and dust of the low crown density-high transparency green space were all lowest, so this type of green space was the most advantageous to the diffusion of automobile exhaust pollutants. Therefore, in the road green space landscape design in the future, the influence of different green space structure on the pollutant diffusion of the road should be fully considered. From the perspective of human health, reasonable plant configuration should be planed and designed.

Acknowledgments
This research was supported by the Science and technology commissioner project in Tianjin (18JCTPJC68000), the University Young and Middle-aged Backbone Innovation Talent Training Program in Tianjin (J01009030709), Agricultural Science and Technology Achievements Transformation and Extension Project in Tianjin (201603060), Postgraduate Innovation Training Program in Tianjin (2017YPY001) and the major science and technology project of the seed industry of Tianjin committee of science and technology (17ZXZYNC00070).

References
[1] Du A K, Luo J and Li H 2016 *Resour. Conserv. Environ. Protec.* **10** 75
[2] Tang S Z, Zhang H R and Xu H 2000 *Forestry Sci.* **36** 20-7
[3] Shan L, Jiang F, He X J, Xie R and Wang H K 2008 *Chin Environ. Manag.* **10** 59-64
[4] Chen Y J and Liu Z B 2017 *Green Sci. Technol.* **15** 24-6
[5] Liu H R, Gao Y D and Wang W 2016 *Northern Hortic.* **12** 49-54
[6] Huang Y M and Wang B B 2016 *Modern Hortic.* **04** 170-1
[7] Liu Y, Li C W, Xing W Y, Zhu Y and Wei J F 2015 *Northern Hortic.* **03** 77-81
[8] Liu G R 2019 *J. Hengshui Univ.* **21** 21-4
[9] Yan J W and Liu T F 2018 *Northern Hortic.* **17** 110-4
[10] Zhao Y B, Yan W Z and Yan H W 2014 *J. Northwest Forestry Univ.* **29** 197-200
[11] Dong Y Q 2015 *Southern Agric.* **09** 65
[12] Xu X Y 2018 *Fujian Agric. Forestry Univ.* 60-65
[13] Yang L 2018 *Inner Mongolia Agric. Univ.* 34-39
[14] Wang J, Lu C D, Niu L W and Zhang F F 2008 *J. Agric. Eng.* **34** 225-32
[15] Fang Z G, Ouyang Z Y, Hu L F, Wang X K and Lin X Q 2005 *J. Applied Trans. Environ. Biol.* **11** 487-95

IOP Conf. Series: Earth and Environmental Science **346** (2019) 012024 doi:10.1088/1755-1315/346/1/012024

Effect of intercropping peonies under walnut trees for growth and development of roots in two kinds of soil

R Xu[1], Z J Wang[2], Y Luo[3], N Zhang[4], C Liang[2,5] and G P Zhu[3,5]

[1]Henan Forestry Vocational College, Luoyang, 471002, P. R. China
[2]Luoyang Academy of Agriculture and Forestry Sciences, Luoyang, 471023, P. R. China
[3]Non-timber Forest Research and Development Center, Chinese Academy of Forestry, Zhengzhou, 450003, P. R. China
[4]Henan Provincial Institute of Science of Technology Information, Zhengzhou, 450003, P. R. China

E-mail: lcjjlgg@sina.com (C Liang); 120695387@qq.com (G P Zhu)

Abstract. To increase the economic benefit and to reduce the cost at early stages of walnut (*Juglans regia* L.) cultivation, the interaction of roots between walnut cultivar Zhongsong No. 1 and peony (*Paeonia ostii* T. Hong & J. X. Zhang) was studied. The growth rhythm per year and the vertical and horizontal distribution of roots were measured. Statistically evaluated data for five consecutive years of study using a random sampling method at yellow-brown and red soil types were presented. The results showed that walnut had three root growth peaks per year, whereas peony had two peaks. At vertical distribution of roots, walnut was concentrated under 40 cm and peony under 20 cm of soil. At vertical distribution of roots, walnut was concentrated 40 -120 cm far from the tree, but peony was 20 -60 cm. We are confident that the root growth and development when intercropping of peony was carried out under walnut trees was alternating and the negative influence was low. So, the cultivation model of "intercropping peony under walnut trees" was suitable at the Luoyang area, Henan province, but the cultivation distance ought to be at 120 -180 cm.

1. Introduction

Walnut (*Juglans regia* L., Fam.: Juglandaceae) is one of the world's four famous tree nuts (hazelnut, walnut, almond, cashew). It is classified as deciduous tree. In cultivation, large space between rows and low-density were required to meet the needs of plant growth and development. In the People's Republic of China, the common planting distance of walnut cultivation is 4 - 6 m and the vacant land under walnut tree can be used to improve the land utilization efficiency by intercropping. This will increase the multiple cropping indexes and will also add farmers' income. According to a statistical source, in 2016 the cultivation area of walnut was more than 80 million 667 m^2 in China [1]. The huge vacant land resulted under the cultivation field of walnut tree could not be used at all for effective development and utilization.

In intercropping, the selection of crops under forests is important. Inappropriate plant species may cause the opposite effect due to allelopathy and competitive growth [2,3]. However, using suitable crops can promote tree growth [4] and bring economic benefit through this type of cultivation processes. It was found that intercropping of different crops beside walnuts had different effects [5]. Walnuts

Content from this work may be used under the terms of the Creative Commons Attribution 3.0 licence. Any further distribution of this work must maintain attribution to the author(s) and the title of the work, journal citation and DOI.
Published under licence by IOP Publishing Ltd

intercropped with peonies, woads and *Semen cassiae* were found beneficial for growth, development and yield for both the plants [6-10]. Otherwise, intercropping radish, cabbage and mung bean under walnut tree significantly hindered the growth of the latter [11-13]. Cropping maize and wheat together produced allelopathic stress which significantly affected the accumulation of walnut biomass [14,15]. Normally, intercropping affects overground biomass accumulation. But in case of walnuts even cropping with suitable kind of crops, the vertical and horizontal distribution of roots in soil are the main limiting factors for the growth and development for both the intercropped plants. Walnut is a deep-rooted plant, but the distribution of walnut roots is greatly affected by soil conditions and management procedures. Generally, fertile, loose and deep soils result in walnut roots developing much better. In walnut tree, the main root is vertically and deeply distributed while the lateral roots are distributed widely and horizontally. Under the condition of viscous and barren soil, the vertical distribution of walnut root is restricted by the bad soil characteristics. Whereby, the principal root grows shallower, but the lateral roots get distributed widely along horizontal direction. When irrigation is applied, the roots of walnut trees grew relatively shallow. On the other hand, under drought stress, the roots were deep in the vertical direction in order to ensure better soil water utilization [14,16,17]. Root growth of walnut is also species specific of the intercrops. For example, walnut trees intercropped with wheat, roots were seen to extend below the wheat roots in order to avoid competition with wheat for water and fertilizer. At this time, the roots of walnut were distributed in a deep longitudinal direction, but the density of roots in each layer of soil profile was lower than that of the purely cultivated walnut trees. The effect was positively correlated with the distance of intercropped wheat [18]. Therefore, it is necessary to select crops suitable for intercropping inside walnut forests.

The peony plant（*Paeonia ostii* and *P. suffruticosa*) could be used as ornamental and for pharmaceutical purposes, and as a source of oil. In the recent years, characteristic economic forests of peony were developed with high economic value [19]. Cultivation of this plant has a long history particularly in Yonglin county, Anhui province; Heze city, Shandong province; Luoyang city of Henan province and other places of China. In some previous studies it has been found that intercropping peony under walnut forest make reasonable use of the suitable shading and reduces the light inhibition of peonies, thereby promoting their growth and development [6]. In the meantime, intercropping peony and walnut significantly improved the quality of nuts and times of farming operation. This had ensured the peony and walnuts enough nutrient supply and shortened the growing period of the walnuts from juvenile to adult stage. At the same time, intercropping interactions stimulated the vegetative growth of underground parts of peony and the accumulation of biomass which increased yield [7]. The harvested peony seeds showed high quality with increased oil yield [8] and thus had brought remarkable economic benefits in cultivation processes. So, peony has been considered an ideal intercropping plant with walnut. However, the growth and underground distribution of roots under the intercropping system of both walnut and medicinal peony are not clear. Knowledge gap remains still regarding the selection of best models for such intercropping system. Information on suitable intercropping distance, time, density, etc. is not yet available. So, there is an urgent need of carrying out research to boost production of both walnut and peony. The objectives of the present research were to study the growth and distribution of roots in interplanting systems with peonies under walnut trees. So that a suitable cultivation mode of intercropping is developed and this will provide scientific guidance for obtaining higher economic yield in the walnut and peony cultivation processes.

2. Materials and methods

2.1. Growth condition of samples

According to the growth conditions of principal production areas of peony in Henan province, two different soil types were selected. Those were situated in Mengjin county and Luoyang city (table 1). The test site is located in the mountainous area but beset with shallow hills in the west of Yu River, near the Yellow River. Warm and continental monsoon (semi-humid) affects obviously the monsoon circulation. Four distinct seasons do prevail in the area. The spring is dry and windy, the summer enjoys heat abundant rainfall, autumn is cool and rainy and winter is also cold but with less snow. The annual average temperature is 13.7°C, and the highest mean occurs in July (26.2°C). The average minimum temperature, 0.5°C, is experienced in January. The annual average frost-free period is counted as 235 d. Among other climatic factors, sunshine hours are 2270.1 h/year. Annual average rainfall is 650.2 mm, with precipitations unevenly distributed: 50% are concentrated in the flood season, e.g., in July, August and September. The highest rainfall occurs in July (an average of 164.1 mm). In January, the average rainfall is 6.9 mm. The average annual evaporation capacity for a number of years approached 1599 mm which is generally strong from April to June mostly around 200 mm with a minimum of less than 100 mm. This condition is typical and representative of the shallow-hilly area between Yellow River and Huai River.

Table 1. Growth condition of samples in this study.

Number	Acreage (hm^2)	Agrotype	pH	Organic matter (g·kg-1)	Content of rapidly available nutrients (mg·kg-1)		
					nitrogen	phosphorus	potassium
Block1	7.5	Red- clay	7.3±0.2	9.8±0.5	61.5±3.2	12.9±0.8	149.5±11.5
Block2	10.0	Yellow-brown	7.5±0.1	11.7±0.4	59.1±1.9	13.2±0.6	151.3±9.8

2.2. Experimental design

In the middle of March 2011, walnut (cultivar: Zhongsong No. 1) seedlings were planted, with 'Xiangling' as the pollination variety following a density of 3×6m. In the middle of October, 2011, 2 year-old family seedlings of medicinal phoenix peony (Fengdan white, *Paeonia ostii* T. Hong & J. X. Zhang, Fam.: Paeoniaceae) hereinafter called peony, were planted with a planting distance of 2 m away from the walnut tree and a planting density of 0.4×0.5 m following the conventional field management technique. Each treatment area was 200 m^2 and prepared in 3 replicates. Two control treatments, one in each of walnut and peony planting gardens were used. The growth dynamics and distribution of walnut and peony roots in the experimental fields and the controls were analyzed by random sampling method for 5 consecutive years.

2.3. Sampling method

On the vertical section, the roots of walnuts were dug with a spade at the following depths: 0-20 cm, 21 -40 cm, 41 -60 cm, 61 -80 cm and 81 -100 cm. On the horizontal section, the root system of walnut was excavated at different distances ranging from 0 -40 cm, 41 -80 cm, 81-120 cm, 121 -160 cm and 161 - 200 cm away from the tree. The distribution of peony roots and its intersection with walnut roots were investigated by the whole plant random sampling method.

Five walnut trees and peonies were randomly selected for root biomass measurement. The sampling time started from the falling leaves of walnut and the root distribution and growth were measured once in a year. During the rooting season of walnuts and peonies, roots were dug up every 10 days for the observation of new roots. The changes of root weight ratio, the time of new rooting and quantity of dry roots in different soil layers were calculated.

2.4. Data analysis

The new roots were counted first in the field and then taken back to the laboratory for further processing. In the laboratory, the sample of roots were rinsed off with clear water and then dried in Kraft paper bags at 105°C to constant weight. Office Excel 2010 was used for data entry, preliminary processing and statistical analysis. DPS v6.50 was used for the measured sets of data to carry out significant variance analysis and Duncan multiple comparisons. Differences were considered statistically significant at $p \leq 0.05$.

3. Results

3.1. Annual trend of new root growth between walnut and peony intercropping

Walnuts exhibited evident growth peaks three times a year which, however, decreased in intensity with the growth stage (figure 1). Both intercropped and cultivated walnuts showed three significant growth peaks in a year, with the first and second growth peaks occurring in succession within a short period of time. New-root growth peaks appeared for the first time in the middle of May (May 12), then the new-root ratio gradually reduced in late June (June 22) (figure 1) to achieve the second growth peak; both growth peaks were nearly 40 days apart. This may be due to the heavy precipitation in the Yellow-Huai River basin in winter, which was conducive to the growth of the root system after the recovery of the ground temperature in spring. Later on, with the continuous rise of soil temperature, the amount of evaporation increases and the soil moisture content drops significantly, causing difficulties for growing the new root of walnut. By the end of June, the rainy season was advancing and precipitations increased, which was conducive to the growth of new walnut roots. Then, with the yearly peak in temperature, the occurrence of new roots was inhibited, resulting in a rapid decrease in the proportion of new roots. The third growth peak occurred in early September (about September 2), and by late October new roots gradually ceased to form. From the results, we are confident that the trend of growth of walnut new root is closely related to soil temperature and moisture. Therefore, water storage and conservation of walnut tree in hilly arid land is very important.

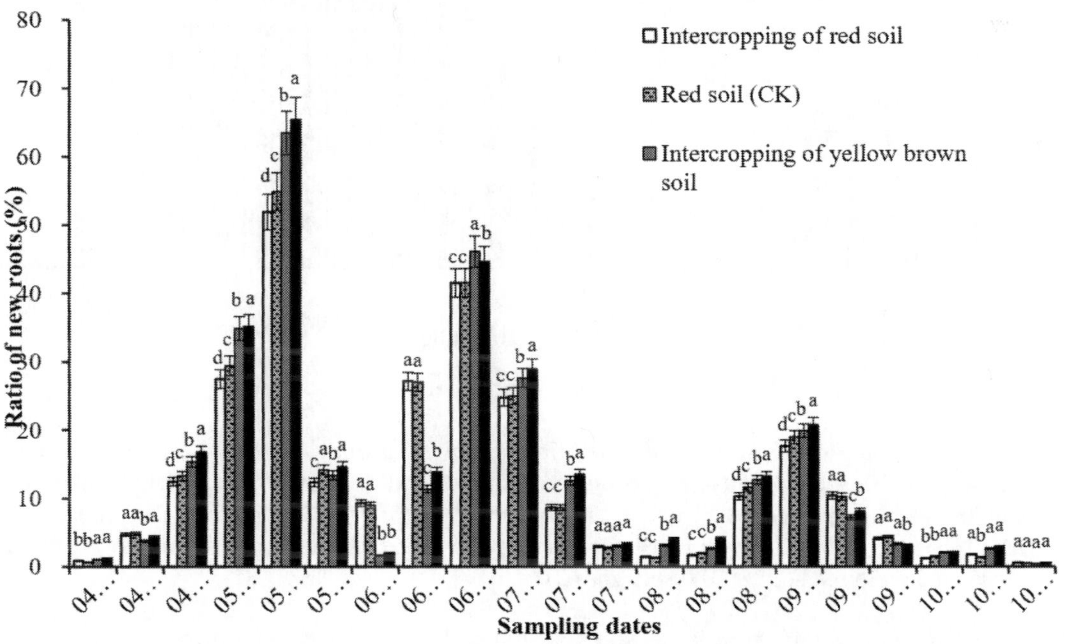

Figure 1. Ratio of new roots in walnut trees in two soil types.

In the walnut intercropping, different soil types had different effects on the generation of new roots (figure 1). High soil permeability was good for the development of walnut new roots. The proportion of new roots in yellow-brown soil was significantly higher than that in red soil (figure 1). Intercropping walnut significantly ($p \leq 0.05$) influenced the generation of new root. This trend in yellow-brown soil was significant ($p \leq 0.05$); the ratio of new roots in red soil was significantly lower than that in the yellow-brown soil in most cases.

In terms of biomass accumulation of walnut roots in different years, it has been seen that in the first year of seedlings, the roots of walnut were damaged in different degrees during transplanting. This had caused a later development of new root growth than usual in a typical year (data not shown). This was mainly related to the formation of callus tissue after root injury which caused new root development in the following period. With the increase in the age of walnut trees, the root system reached the growth peak in the third year, which is also the rapid expansion period of the crown of walnut trees [16]. After the fourth year, the tree crown began to reduce and root growth gradually weakened.

The annual growth peak of new roots of peony had two significant peaks, and they occur alternately with the growth peak of walnut roots (figure 2). The first growth peak of peony appeared in early April (about April 3rd), and gradually stopped in early May (about May 3rd), that is 40 days later than the growth peaks of walnuts. The second growth peak started in late September, until mid-November peak and this is about 20 days later than that of walnut trees. The growth continued until the next year early January (about January 3rd), then entered physiological dormancy. Thus, it can be seen that there was no overlap between the annual growth peak of new roots in the intercropping of walnut and peony in Yellow-Huai River basin.

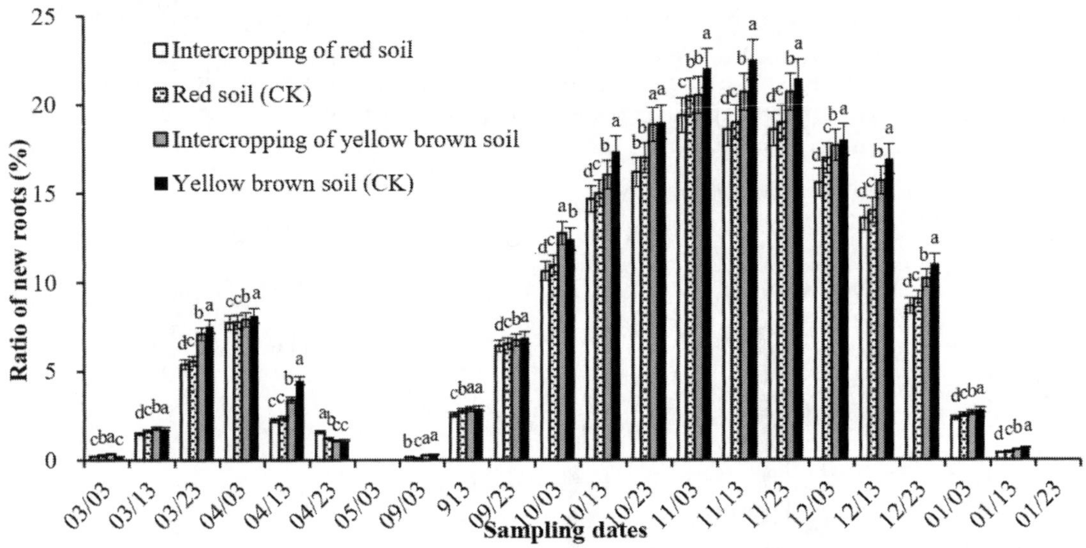

Figure 2. Ratio of new roots of medicinal peony in two soil types.

In the intercropping with walnuts, the generation of new roots of peony varied according to soil type (figure 2), whereas the generation of new roots in walnut was little affected. The yellow-brown soil with good soil permeability promoted the occurrence of new roots of peony. However, intercropped walnut significantly affects the occurrence of new roots of peony; this phenomenon was more significant in yellow-brown than in red soil ($p \leq 0.05$) (figure 2).

3.2. Vertical distribution of new roots between walnut and peony during intercropping

In the vertical distribution of root system, the roots of walnut showed the trend of "low - high - low". The ratio of new roots peaked at 40 cm depth from the ground and the difference between treatments was significant ($p \leq 0.05$) (figure 3). Different soil types and intercropped peony had significant effects on the vertical distribution of new roots in walnut ($p \leq 0.05$) (figure 3). The new roots in red soil were significantly lower ($p \leq 0.05$) than those in yellow-brown soil (figure 3), which may also be related to the clay of the red soil and the resistance of the new root growth. Intercropping walnut in red and yellow-brown soil showed that the ratio of new roots was less ($p \leq 0.05$) than not-intercropped walnuts (figure 3). This proves that the vertical distribution the of root system of walnut suffers certain degree of inhibition when it is intercropped with peony.

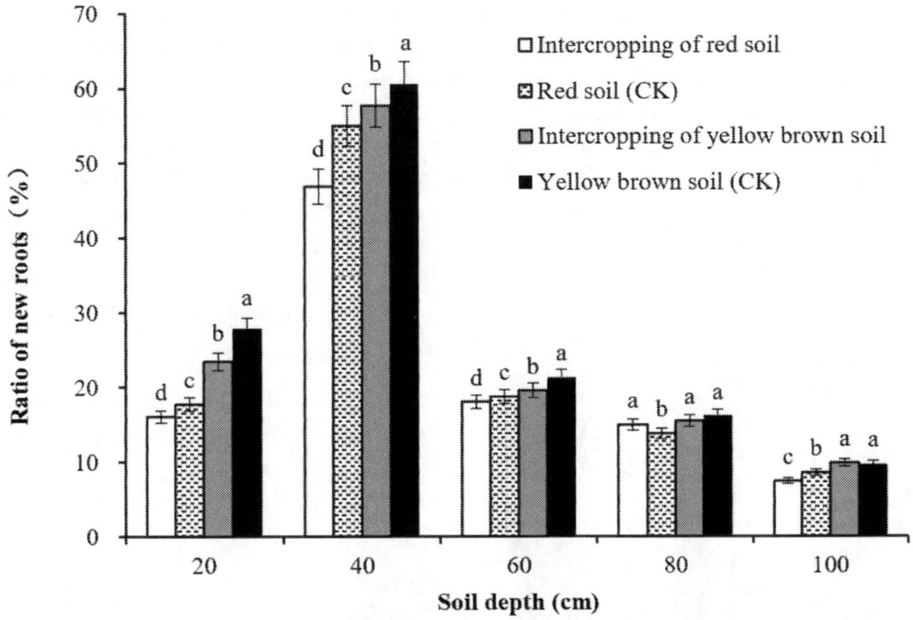

Figure 3. Vertical distribution of roots of walnut in two soil types.

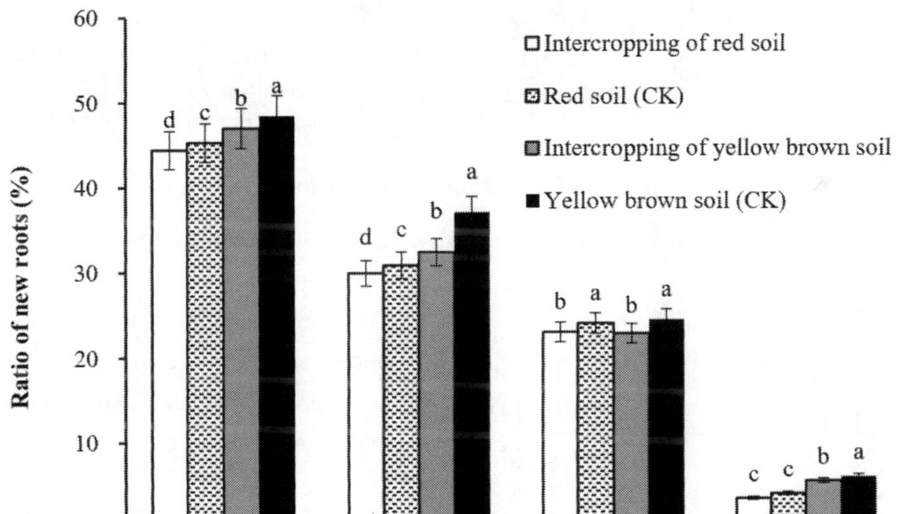

Figure 4. Vertical distribution of peony roots in two soil types.

Vertical distribution of peony roots showed a continuous declining trend. The proportion of new roots in the soil layer reached the peak at 20 cm depth which decreases further with an increasing depth of soil layer. The difference between treatments reached a significant level ($p\leq 0.05$) (figure 4). Vertical distribution of new roots of peony when intercropped showed significant effects with different soil types. The new roots in red soil were significantly lower than those in yellow-brown soil ($p\leq 0.05$) (figure 4). This may still be related to the clay of red soil. In the intercropping model, the new roots of peony were less than that of not-intercropped which indicated that intercropping can inhibit the peony new root generation through vertical soil profile (figure 4).

3.3. Horizontal distribution of new root between walnut and peony during intercropping

The roots of walnut showed a pattern of increasing abundance up to 40 cm from the trunk, and then a decreasing trend. The generation of new roots was still abundant at 80 cm distance from the main stem. Different soil types and intercropped peony had significant effects on the horizontal distribution of new walnut roots ($p \leq 0.05$) (figure 5). The number of new roots in the range of 0-80 cm in red soil was lower than that in yellow-brown soil ($p \leq 0.05$). In intercropping, walnut roots within 0 -120 cm in yellow brown soil were significantly ($p \leq 0.05$) more developed than those in red soil. However, in the 160 -200 cm range, an opposite trend was found.

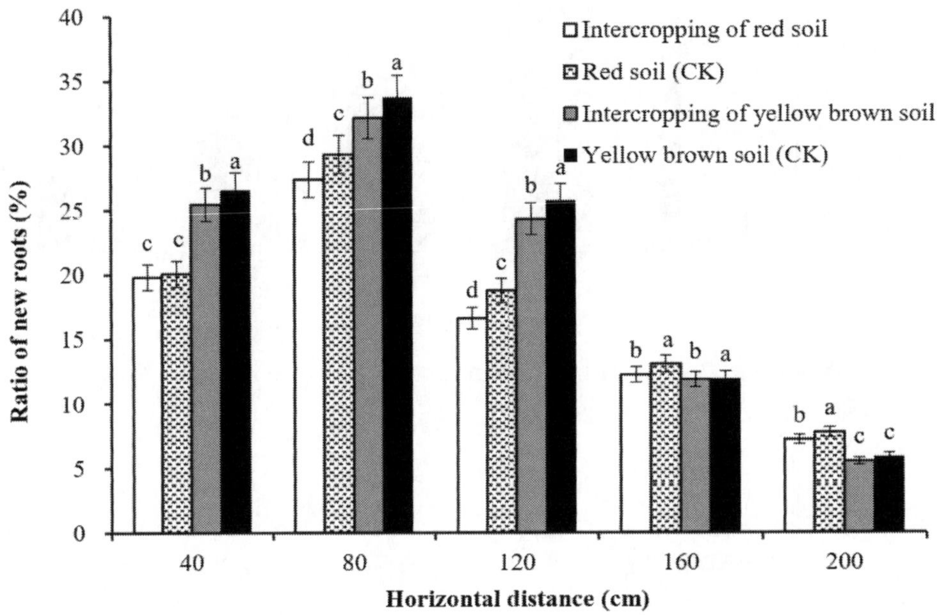

Figure 5. Horizontal distribution of walnut roots in two soil types.

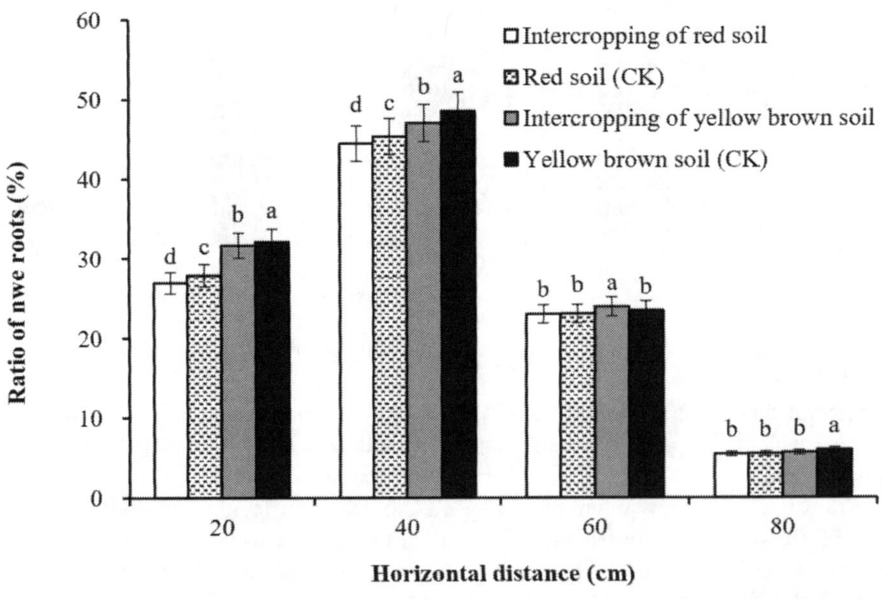

Figure 6. Horizontal distribution of peony roots in two soil types.

Regarding the horizontal distribution of peony roots, a trend similar to that found for walnut was found (figures 5 and 6). The distribution of new roots peaked at 40 cm from the main stem (figure 6). Different soil types and intercropping had significant effects on the horizontal distribution of new roots of peony. The number of new roots in yellow-brown soil in the range of 0-40 cm was significantly higher than that in red soil. However, it showed an opposite trend, i.e. the proportion of new roots in red soil was higher ($p \leq 0.05$) than that in yellow-brown soil. In the intercropping model, the roots of walnut within 0-40 cm deep in red soil and yellow-brown soil were significantly less than in CK ($p \leq 0.05$), but the difference was not obvious at 60-80 cm soil depth (figure 6).

4. Discussion

The trend of annual growth pattern of walnut roots in the Yellow-Huaihe River basin can be divided into three phases. In the first phase, the growth peak occurred between the middle of April and end of May. The second one took place from the middle of June to the beginning of July while the third growth peak occurred between the beginning of September and mid-July. All the three growth peaks followed a decline over time. However, growth peak in peony occurred twice in a year. The first peak occurred between late March and mid-April. The second and late growth peak started in late September and continued till December. This result was consistent with Liu *et al* [20]. Considering the time period, it could be said that the first growth peak of peony occurred 20 days or so earlier than that of walnut trees. The second peak of peony occurred at 40 days later than walnut's third growth peak. But in the last year of the experiment the growth peak of peony was significantly longer compared the growth peaks of walnut. This indicated that in the intercropping of walnut and peony the new roots growth peak of both occurred alternately, so the mutual influence was small. Unlike walnut, peony had two significant growth peaks, first growth peak occurred 40 days later than walnut, the second growth peak occurred 20 days later than walnut's third growth peak. So, basically peonies and walnut trees in intercropping systems did not overlap in their periods of new-root growth. From the perspective of soil type, both horizontal and vertical directions, the red soil contains high K^+ content which is conducive to the growth and development of nucleoli [16]. However, compared with the yellow-brown soil with good permeability, it is still not conducive to the development of new roots of walnuts and peonies. In terms of horizontal distribution of roots, new roots of walnut were mainly distributed at 40-120 cm far from the main trunk. But in peony, root distribution in the horizontal direction occurs at 20-60 cm distance. This indicates that peony actually takes a longer time for optimum root growth. This view is also supported by medicinal peony, in which 4 years from planting to harvesting are usually necessary to reach the medical specifications [19]. The intercropping distance between walnut and peony should be kept at 120-180 cm. Too close a distance will result in competitive stress growth and may have adverse effects on the occurrence of new roots of both the species. In summary, we are confident that the intercropping of walnuts and peony in the Yellow-Huaihe River basin was one of the effective modes to improve the multiple crop indexes. Last but not least; the planting distance between the two crops should be kept above 120-180 cm, so as to reduce the horizontal competition of water and nutrients. This will facilitate both the species to obtain a large growth space. In turn it will also be convenient for the effective implementation of field management measures to ensure good production. Planting peony among peach trees has been an important measure to compensate for the economic deficit in the early stage, as intercropping not only reduced the occurrence of photoinhibition [6], but also made full use of land resources. Unified field management measures reduced the input of repeated labour incentives in the walnut garden. Moreover, the results of peony planting in the third year have been getting high economic value [21]. Therefore, the intercropping pattern of "walnut + medicinal peony" may be regarded as an appropriate intercropping measure in the Yellow-Huaihe River basin. The present research provides some recommendations that could be popularized for rising production. In combination, it is necessary to study the mechanism of mutual growth and development to explore the optimal intercropping management method.

Acknowledgment

This work supported by "the Fundamental Research Funds for the Central Non-profit Research Institution of CAF (No. CAFYBB2017ZA005).

References

[1] Deng J L 2016 Current situation and development strategy of walnut industry in China *China Forest Prod. Ind.* **43**56-8
[2] Díaz-García A, García-Riaño J and Zapata-Narvaez J 2015 Improvement of sporulation conditions of a new strain of *Bacillus amyloliqiens* in liquid fermentation *Adv. Biosci. Biotech.* **6** 302-10
[3] Wang Y Q, Jiao Y J, Chen D M, Yuan L, Huang Y, Wu Y K and Du R W 2016 Effects of *eupatorium adenophorum* extracts on seed germination and seedling growth of pasture species *Acta Prataculturae Sin.* **25** 150-9
[4] Hu J, Zhu W, Xu N and Sun G 2013 Effects of mulberry-soybean intercropping on growth of mulberry and soybean, and photosynthesis responsive to light intensity *J. Cent. South Forest. Univ.* **33** 44-9
[5] Zhang Q, Fu S L, Yao X H, Teng J H, Shao W Z, Ren H D, Wang K L and Chang J 2015 Allelopathic effects of water extractions from leaves and husks of *Carya illinoensis* on three plant species *Forest Res.* **28** 674-80
[6] Tang Z, Shen Z, Ding X and Chen J 2017 Effects of shading on photosynthetic characteristics in *Paeonia ostii*cv. 'Phoenix White' seedlings *Econ. Forest Res.* **35** 84-9
[7] Yin H, Wei S, Liang C, Wang Z and Zhang X 2017 interaction effect of peony planting under walnut *North. Hortic.* **14** 74-8
[8] Xu S Q, Liu Q, Huang S W, Ni S and Hu G R 2017 Studies on general situation of oilseed peony intercropping with woods in China *Mod. Hortic.* **14** 28-30
[9] Li Y R, Zhang J Y, Sang X F, Teng Q, Wang Y and Guo Z R 2015 Analysis on the cultivation model and prospect of the double oil cultivation of peonies with thin shell hickory and oil *Mod. Agr. Sci. Tech.* **21** 172-2
[10] Chen P 2014 *Study on Interspecific Water Relations of Walnut-Woad/Cassia Agroforestry System* (Beijing: Chinses Academy of Forestry)
[11] Wang Q, Xu Z, Hu T, Hu H, Chen H and Shi H 2014 Effect of decomposing walnut leaf litter on the growth and physiology of radish *Acta Bot. Boreal. - Occident. Sin.* **34** 2475-82
[12] Yan T, Zhai M Z, Wang Y and Hao F 2012 Allelopathic effects of root extracts from walnut on seed germination and seedling growth of three plant types *J. Huazhong Agr. Univ.* **31** 713-9
[13] Xu Z, Hu T, Li Z, Chen H, Wang Q and Hu H 2015 Effect of secondary metabolites in *Juglans regia* leaf litter from different habitats on photosynthetic physiological characteristics of *Brassica chinensis* during early decomposing stage of leaf litter *Acta Eco. Sin.* **35** 5147-56
[14] Zhang W 2014 *Interspecific root intercations, nitrogen acquisition and utilization in fruit trees/crops intercropping in southern Xinjiang province* (Beijing:China Agriculture University)
[15] Zhou P 2014 *Research on intercropping pattern and technique in young walnut orchard* (Taiyuan, China: Shanxi Agriculture University)
[16] Liang C and Zhang X 2014 *High-Efficient Technology of Walnut* (Beijing: Jindun Press)
[17] Li D 2016 Study on impact of growth characteristics and soil water changes in root zone for walnut tree under drip irrigation (Urumchi, China: Xinjiang Agriculture University)
[18] Wang L 2011 Effects of agroforestry models on roots of some trees and crops in northern Wei River (Yanglin, China: Northwest Agriculture and Forestry University)
[19] Zhang Y F, Wang J, Fang C W, Li J and Wu J H 2015A preliminary study of specification and grade of cortex moutan *J. Anhui Tradit. Chinese Med. Coll.* **34** 80-3
[20] Liu Z, Kong D, Li Y and Yang Q 2008 Dynamic carbon and nitrogen metabolism in roots of potted and field *Paeonia suffruticosa* 'luoyanghong' *Sci. Silvae Sin.* **44** 162-4
[21] Gao L L, Li Y Q, Wang Z S, Sun G J and Mo H Z 2017 Physicochemical characteristics and

functionality of tree peony (*Paeonia suffruticosa* Andr.) seed protein *Food Chem.* **240** 980-8

The effect of salt stress on the growth of 3 species of *Nitraria* seedlings

Y Y Shi[1,4], J H Yang[1], T Liu[1], Z W Zhang[2], K Mu[3], F H Liang[1] and Y Liu[1]

[1]College of Horticulture and Landscape, Tianjin Agricultural University, Tianjin 300384, China
[2]Tianjin Plant Protection and Quarantine Station, Tianjin 300061, China
[3]Tianjin Yuanda Garden Engineering Co. LTD, Tianjin 301600, China

E-mail: syyzzw@126.com

Abstract. In order to compare the salt tolerance of the three species of Nitraria, NaCl with different concentrations was used to stress the Nitraria seedlings. The results showed that salt treatment had different effects on the growth of Nitraria; The inhibiting effect of salt stress on the growth of plant height was in the order of Nitraria tangutorum Bobr>Nitraria roborowskii Kom>Nitraria sibirica Pall; Salt stress caused a rapid decrease in the number of leaves of Nitraria tangutorum Bobr and the number of leaves of Nitraria sibirica Pal rapidly decreased when the salt concentration was greater than 6%, while the relative number of leaves of Nitraria roborowskii Kom significantly decreased when the salt concentration was 15%; With the increase of salt concentration, the order of relative reduction of stem diameter was Nitraria tangutorum Bobr>Nitraria roborowskii Kom>Nitraria sibirica Pall; With the increase of salt concentration, the relative leaf area decreased significantly in the order of Nitraria tangutorum Bobr>Nitraria sibirica Pall>Nitraria roborowskii Kom; With the increase of salt concentration, there was no significant difference between the dry weight and fresh weight of Nitraria roborowskii Kom, but there were significant changes between Nitraria tangutorum Bobr and Nitraria sibirica Pal. According to the comprehensive membership function analysis, the salt tolerance of Nitraria roborowskii Kom was the best, followed by that of Nitraria sibirica Pal, and that of Nitraria tangutorum Bobr was the worst.

1. Introduction

Nitraria was a xerophyte or super-xerophyte shrub or small shrub of the Zygophyllaceae, which was of great significance to improve and restore the ecological environment. *Nitraria* had strong adaptability and salt-alkali resistance, drought resistance, barren resistance, cold resistance, high temperature resistance, sand resistance and other characteristics [1], which was an excellent plant for improving saline-alkali land and preventing wind and sand fixation [2]. However, there were significant differences in resistance and reproduction characteristics between the *Nitraria* species [3-6]. In order to study the salt-tolerance characteristics of three species of *Nitraria*, the growth indexes of *Nitraria sibirica* Pall, *Nitraria roborowsk*ii Kom and *Nitraria tangutorum* Bobr under NaCl stress at different concentrations were analyzed to provide a basis for the better application of *Nitraria* in saline-alkali and even heavy saline-alkali lands [7-10].

Content from this work may be used under the terms of the Creative Commons Attribution 3.0 licence. Any further distribution of this work must maintain attribution to the author(s) and the title of the work, journal citation and DOI.
Published under licence by IOP Publishing Ltd

2. Materials and methods

The tested materials were Nitraria sibirica Palli and Nitraria tangutorum Bobr seeds purchased from Gansu Minqin Linquan Ecological seed Co., Ltd., and *Nitraria roborowskii* Kom seeds purchased from Ningxia Yuansheng Green Yang Grass Co., Ltd. (Yinchuan). The seeds stored in the sand for 3 months were washed with clean water and seeded on the vermiculite seedbed. After 2 true leaves were grown, the seeds were transferred into a 10x10 flowerpot with 9 plants in each pot. Three weeks later, the seedlings were stressed with saline of the corresponding concentration. Each pot was 80 mL and NaCl solution was used for salt stress. The treatment concentrations were 0 (control), 3%, 6%, 9%, 12% and 15%, and each was treated with 6 repetitions. After salt treatment for 4 weeks, plant height, leaf number, stem diameter, leaf area, fresh weight and dry weight of the plants were measured respectively. The membership function method was used to calculate the relative plant height, relative leaf number, relative stem diameter, relative leaf area, fresh weight and dry weight of the plant, and root to crown ratio. The salt tolerance of the three kinds of white thorn was compared through comprehensive analysis of various growth indexes [11-14].

3. Results and analysis

3.1. Effects of NaCl stress on 3 kinds of Nitraria plant height

The increase of relative plant height was the ratio of the height of plant increase at the end of salt stress treatment to that at the beginning of stress treatment. As could be seen from figure 1, the increase of relative plant height of 3 species of *Nitraria* showed a decreasing trend with the increase of NaCl concentration. *Nitraria tangutorum* Bobr relative height increment was significantly; The change of *Nitraria roborowskii* Kom was not significant when the NaCl concentration was 0-6%, but it was significant when the NaCl concentration was 9%; The change of *Nitraria sibirica* Pall was significant when the NaCl concentration was 0-9%, and it was not significant when the NaCl concentration was 12-15%. Compared with the control group, the three kinds of *Nitraria* decreased by 62.5% to 94.6% at most, with the least decrease in the number of *Nitraria roborowskii* Kom and the most decrease in the number of *Nitraria tangutorum* Bobr. The results indicated that NaCl stress had the greatest inhibitory effect on the growth of *Nitraria tangutorum* Bobr plant height.

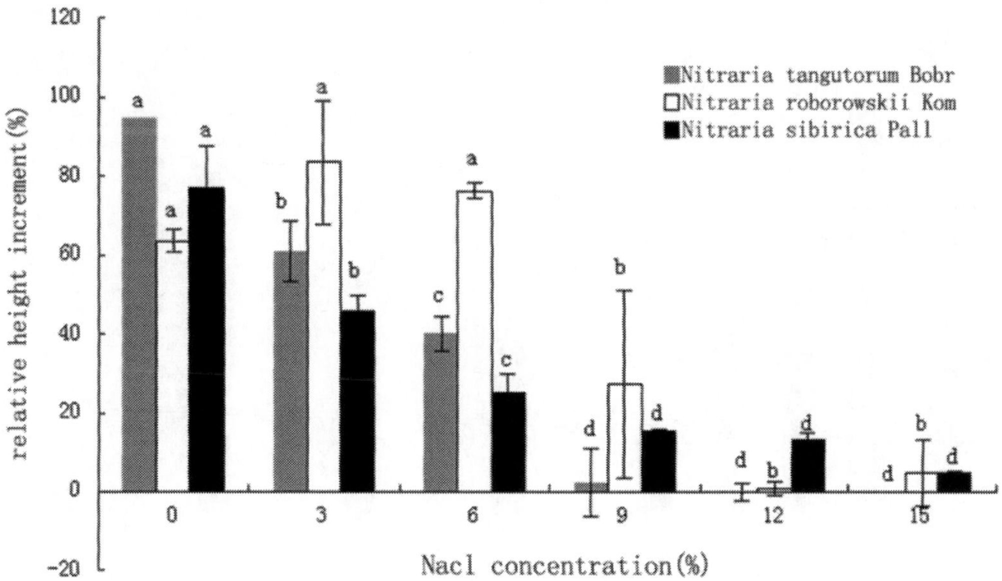

Figure 1. NaCl stress of 3 kinds of *Nitraria* relative height increment.

3.2. Effects of NaCl stress on 3 kinds of Nitraria leaf number

As could be seen from figure 2, the relative number of leaves of 3 species of *Nitraria* decreased after NaCl stress. The results showed that the leaf growth of *Nitraria tangutorum* Bobr was faster than that of *Nitraria roborowskii* Kom and *Nitraria sibirica* Pall under no-salt stress or low-concentration salt stress, and the leaf growth was significantly inhibited under high-concentration salt stress (> 6%); *Nitraria roborowskii* Kom was similar to *Nitraria tangutorum* Bobr, but the turning point was 9%; Although *Nitraria sibirica* Pall growth general under low concentration of salt, but high concentration of salt can be better growth, only at 15% of the salt concentration, growth will be significantly inhibited. As a result, the *Nitraria sibirica* Pall from its leaf growth, then the *Nitraria tangutorum* Bobr and the *Nitraria roborowskii* Kom more tolerant of salt.

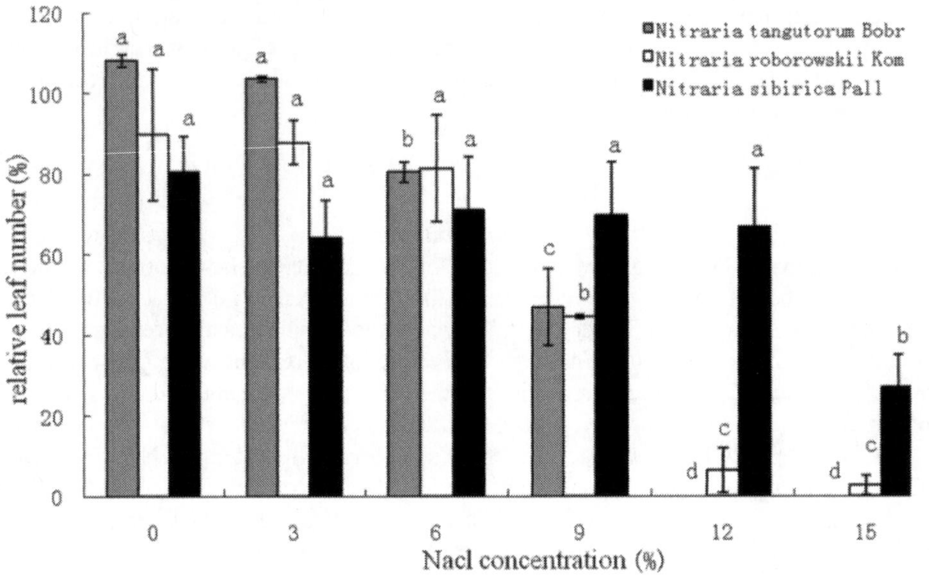

Figure 2. NaCl stress on 3 kinds of *Nitraria* seedling relative leaf number.

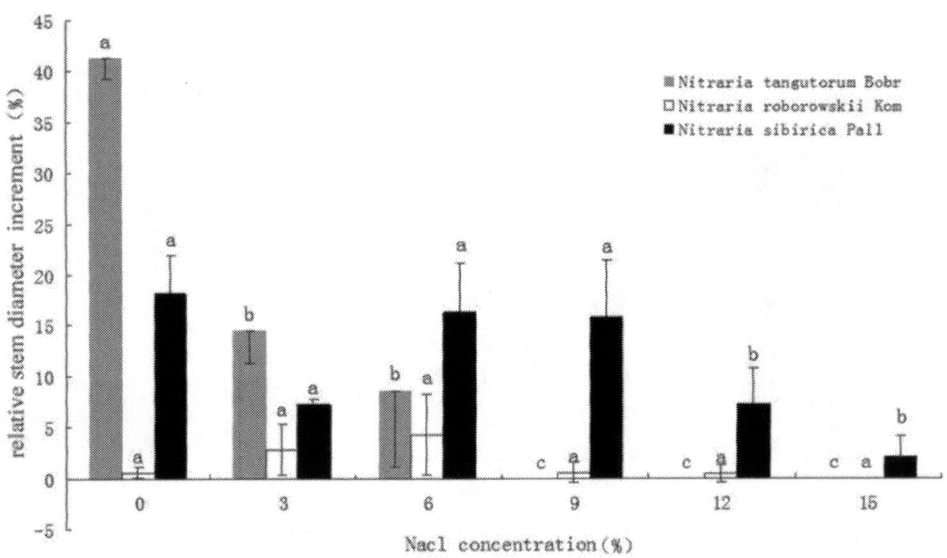

Figure 3. NaCl stress of 3 kinds of *Nitraria* seedling relative stem diameter increment.

3.3. Effects of NaCl stress on 3 kinds of Nitraria stem diameter

As could be seen from figure 3, the relative increase of stem diameter of *Nitraria tangutorum* Bobr and *Nitraria sibirica* Pall showed a decreasing trend with the increase of NaCl stress concentration, while there was no significant change in *Nitraria roborowskii* Kom. NaCl stress had a certain inhibitory effect on both *Nitraria tangutorum* Bobr and *Nitraria sibirica* Pall. *Nitraria tangutorum* Bobr was more sensitive to NaCl stress, and 3% NaCl concentration would be significantly reduced, while *Nitraria sibirica* Pall's sensitive concentration to salt was 12%, and the salt concentration had no significant effect on *Nitraria roborowskii* Kom. This indicated that the salt-tolerant concentration of *Nitraria sibirica* Pall was significantly higher than that of *Nitraria tangutorum* Bobr in terms of the relative increase of stem diameter.

3.4. NaCl stress on the influence of 3 kinds of Nitraria relative leaf area

As could be seen from figure 4, the relative leaf area of the three kinds of *Nitraria* seedlings decreased with the increase of NaCl stress concentration. With low NaCl concentration, *Nitraria tangutorum* Bobr did not change much, but decreased significantly with the increase of NaCl concentration. The relative leaf area of *Nitraria roborowskii* Kom increased significantly with 6%NaCl, but decreased rapidly with 9%NaCl and reached 0 with 15%NaCl. *Nitraria sibirica* Pall was no significant difference in NaCl between 0 and 9%, but NaCl decreased significantly in both 12% and 15%, 41.3 to 78.2% lower than the control group. This indicated that the leaf area of *Nitraria tangutorum* Bobr was sensitive to 6% NaCl, *Nitraria roborowskii* Kom was 9% NaCl and *Nitraria sibirica* Pall was 12% NaCl, so the salt-tolerance of *Nitraria sibirica* Pall was better than that of *Nitraria tangutorum* Bobr and *Nitraria roborowskii* Kom.

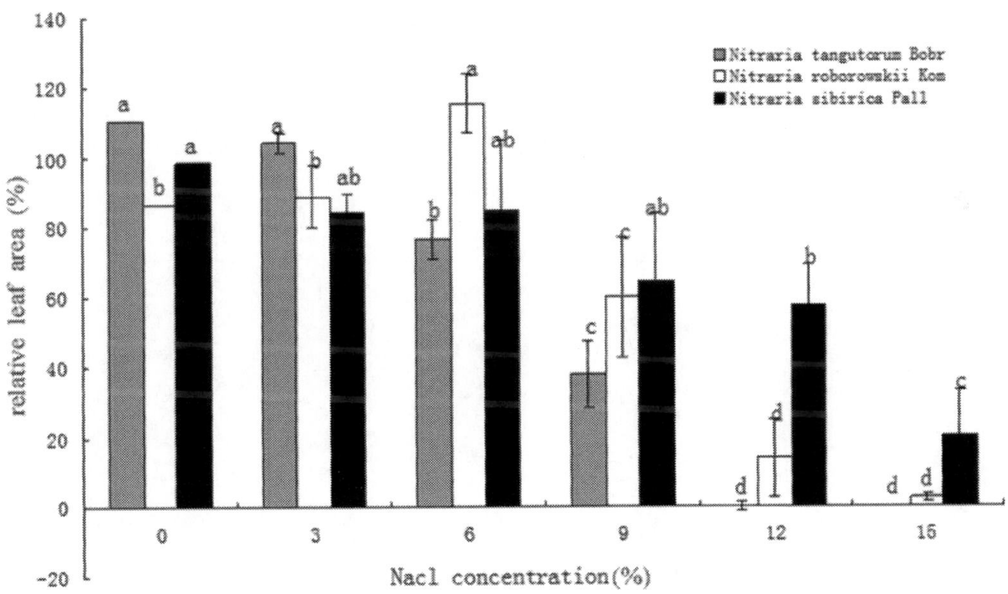

Figure 4. NaCl stress on 3 kinds of *Nitraria* seedling relative leaf area.

3.5. NaCl stress on 3 species of Nitraria plant fresh weight dry weight and root shoot ratio

As could be seen from figure 5, the fresh weight of *Nitraria tangutorum* Bobr decreased with the increase of NaCl stress concentration, while the fresh weight of *Nitraria roborowskii* Kom and *Nitraria sibirica* Pall first increased and then decreased. Under the stress of 0-6% NaCl, there was little change in *Nitraria tangutorum* Bobr and *Nitraria roborowskii* Kom, but significant increase in *Nitraria sibirica* Pall. Under the concentration of 3%NaCl, the above ground fresh weight of *Nitraria sibirica* Pall increased to the maximum, 1.4 times of that of the control group. When NaCl stress was higher than 6%, the above ground fresh weight of *Nitraria tangutorum* Bobr and *Nitraria roborowskii* Kom decreased rapidly, and reached the lowest level when NaCl reached 12%, which was 30.0% and 25.0% of the control group respectively. Although *Nitraria sibirica* Pall decreased, there was no significant difference from the control group. NaCl stress had the least effect on *Nitraria sibirica* Pall, and low concentration promoted the increase of fresh weight.

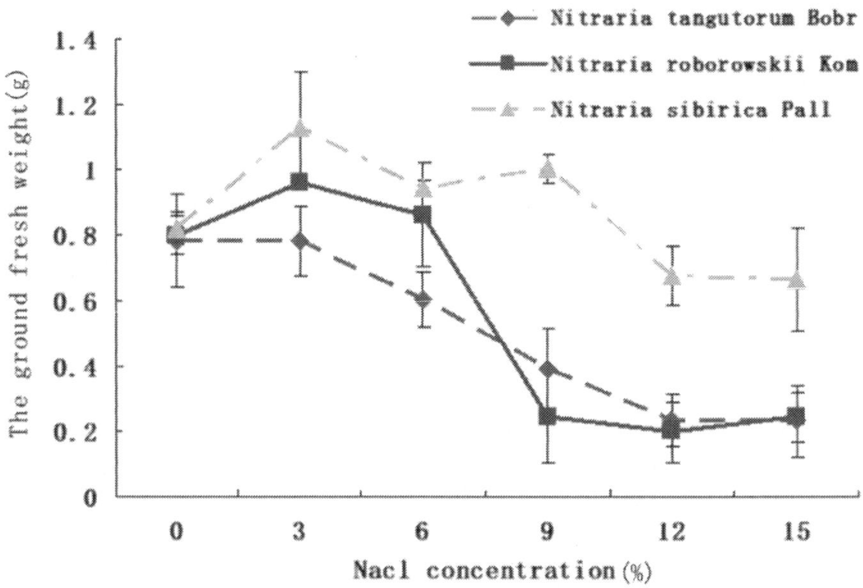

Figure 5. NaCl stress on 3 kinds of *Nitraria* relative fresh weight.

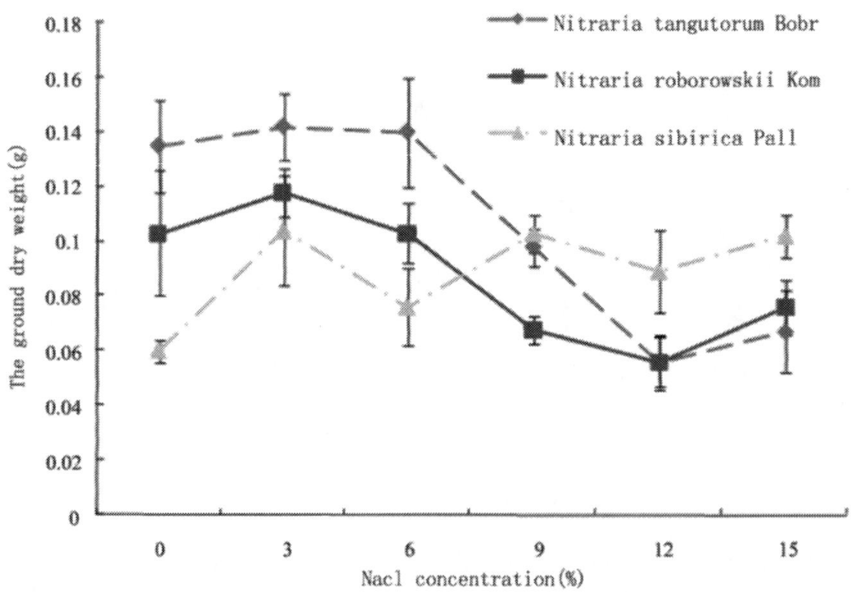

Figure 6. NaCl stress on 3 kinds of *Nitraria* relatively dry weight.

As could be seen from figure 6, with the increase of NaCl stress concentration, the aboveground dry weight of *Nitraria tangutorum* Bobr and *Nitraria roborowskii* Kom decreased, while the aboveground dry weight of *Nitraria sibirica* Pall fluctuated up and down, but did not increase or decrease significantly. Under the concentration of 0-6% NaCl, the above ground dry weight of the three species showed little change. With more than 6% NaCl concentration, the above-ground dry weight of *Nitraria tangutorum* Bobr and *Nitraria roborowskii* Kom decreased significantly with the increase of NaCl concentration. The dry weight of *Nitraria tangutorum* Bobr and *Nitraria roborowskii* Kom was 41.4% and 53.9% of the control at 12% NaCl concentration,respectively. At this time, the above-ground dry weight dropped to the lowest level. It was clear that the salt tolerance of thorn was very high. Only when the concentration of NaCl was higher than 6%, the aboveground dry weight of *Nitraria tangutorum* Bobr and *Nitraria roborowskii* Kom was significantly reduced, while the aboveground dry weight of *Nitraria sibirica* Pall was always unchanged. The salt tolerance of the *Nitraria sibirica* Pall was significantly better than that of the *Nitraria tangutorum* Bobr and the *Nitraria roborowskii* Kom.

It could be seen from figure 7 that the underground fresh weight of *Nitraria tangutorum* Bobr and *Nitraria roborowskii* Kom did not change much at the salt concentration of 0-6%. When the salt concentration was greater than 6%, the underground fresh weight of *Nitraria tangutorum* Bobr and *Nitraria roborowskii* Kom decreased significantly, with the lowest being 40.9% and 34.5%, respectively, of the control group. However, with the increase of NaCl stress concentration, the *Nitraria sibirica* Pall fresh underground weight first increased and then decreased. Under the 9% NaCl stress, the fresh underground weight increased to the maximum value, which was 1.6 times of that of the control group. With the increase of salt concentration, the fresh underground weight decreased but showed no difference from the control group. The results showed that NaCl concentration greater than 6% could significantly inhibit the accumulation of fresh substances in the underground of *Nitraria tangutorum* Bobr and *Nitraria roborowskii* Kom, and the inhibitory effect was enhanced with the increase of the stress concentration. NaCl stress had little effect on the fresh weight in the underground of *Nitraria sibirica* Pall, and low-concentration NaCl could significantly promote the accumulation of fresh substances in the underground of *Nitraria sibirica* Pall.

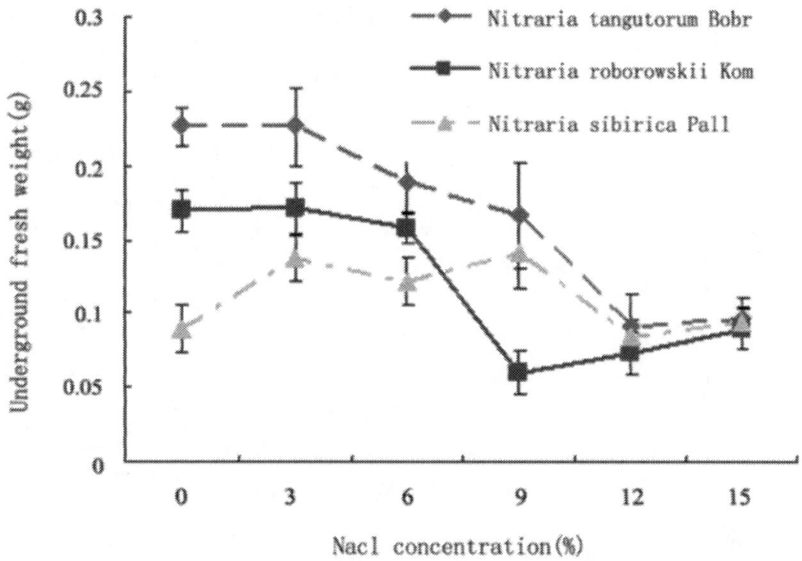

Figure 7. NaCl stress on 3 kinds of *Nitraria* underground fresh weight.

As could be seen from figure 8, the underground dry weight of *Nitraria tangutorum* Bobr and *Nitraria sibirica* Pall increased first and then decreased with the increase of NaCl stress concentration. Under the stress of 0-6% NaCl, the underground dry weight of *Nitraria tangutorum* Bobr and *Nitraria sibirica* Pall increased and reached the maximum at 6% and 3% NaCl respectively, which was 1.2

times and 2.4 times of the control group. Under the concentration of 9-15% NaCl, *Nitraria tangutorum* Bobr and *Nitraria roborowskii* Kom were reduced to 50.4% and 22.8% of the control group respectively, while *Nitraria sibirica* Pall was not different from the control group. This indicated that low concentration NaCl had a promoting effect on the accumulation of dry weight underground of *Nitraria tangutorum* Bobr and *Nitraria sibirica* Pall, and the promoting effect on *Nitraria sibirica* Pall was more obvious. NaCl had a strong inhibitory effect on the accumulation of dry matter of *Nitraria tangutorum* Bobr and *Nitraria roborowskii* Kom, and had a stronger inhibitory effect on *Nitraria roborowskii* Kom.

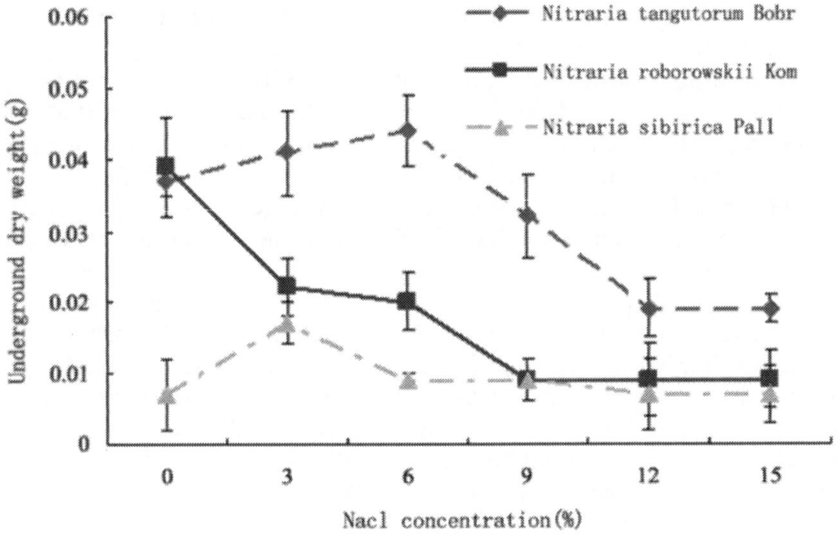

Figure 8. NaCl stress on 3 kinds of *Nitraria* underground dry weight.

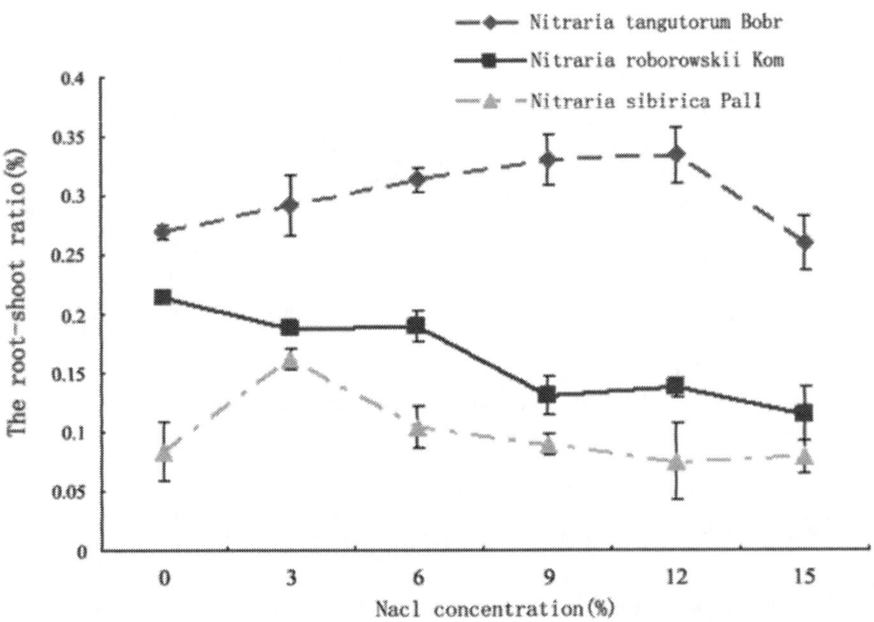

Figure 9. NaCl stress on 3 kinds of *Nitraria* relative root shoot ratio.

From the change trend, the plant weight of *Nitraria sibirica* Pall changed relatively little, the fresh weight on the ground decreased slightly, and the rest fluctuated slightly. The dry and fresh weight of the plants of *Nitraria tangutorum* Bobr and *Nitraria roborowskii* Kom decreased rapidly when the salt concentration was 3-9% NaCl. When NaCl concentration was 6-9%, the variation of plant weight of *Nitraria roborowskii* Kom was the largest, and 6% NaCl concentration could be the sensitive concentration. The dry weight variation of *Nitraria tangutorum* Bobr seems to have the same trend. Overall, the *Nitraria sibirica* Pall was better than the other two kinds of *Nitraria*.

As could be seen from figure 9, with the increase of stress concentration, *Nitraria tangutorum* Bobr was no significant difference in the ratio of roots to crowns, but the *Nitraria roborowskii* Kom kept decreasing, while the *Nitraria sibirica* Pall first increased and then decreased. Under the stress of 0-6% NaCl, there was no significant change in the *Nitraria roborowskii* Kom, while under the stress of 3% NaCl, the *Nitraria sibirica* Pall reached the maximum value, 1.9 times higher than the control group. When the concentration of NaCl was more than 6%, the *Nitraria roborowskii* Kom decreased significantly, the lowest was only 28.1% of the control, but the *Nitraria sibirica* Pall had no significant difference from that of the control. NaCl stress had a significant inhibitory effect on the ratio of root to crown of *Nitraria roborowskii* Kom, which inhibited root growth more. The root to crown ratio of *Nitraria tangutorum* Bobr was increased, which was related to the small effect of salt on the root growth and the large decrease of overground growth. The effect of low concentration on the ratio of root to crown of *Nitraria sibirica* Pall was greater than that of high concentration.

3.6. 3 species of Nitraria growth index comprehensive analysis of salt resistance

From the above results, it could be seen that NaCl stress had a certain impact on the changes of relative plant height, relative number of leaves, relative stem diameter and relative leaf area (except for the change of relative stem diameter of *Nitraria roborowskii* Kom) as well as the changes of relative weight of plants of *Nitraria tangutorum* Bobr, *Nitraria roborowskii* Kom and *Nitraria sibirica* Pall. 0-6% low concentration of NaCl stress except plant height and stem diameter index inhibition for the growth of other, smaller plant weight change of *Nitraria sibirica* Pall and more obvious role in promoting, The results showed that the growth indexes of *Nitraria tangutorum* Bobr and *Nitraria roborowskii* Kom were obviously inhibited when 6% NaCl was greater than that of *Nitraria sibirica* Pall, but the plant weight of small fruit was not affected obviously, and the relative weight of the plant changed little.

According to the relative changes in plant height, stem diameter, number of leaves, leaf area, average fresh weight of plants, average dry weight of plants and other growth indexes of the three *Nitraria* species under NaCl stress, the average value of their membership function under 5 NaCl concentrations of 3-15% was calculated and the salt-tolerance was analyzed [15-21]. All indexes were given equal weight. The results were as follows:

Table 1. 3 species of Nitraria under NaCl stress growth index average subordinate function value and salt resistance.

	plant height	stem diameter	leaf number	leaf area	fresh weight	dry weight	average value	Salt tolerant sorting
Nitraria tangutorum Bobr	0.137	0.230	0.562	0.200	0.067	0.302	0.250	3
Nitraria roborowskii Kom	0.816	0.022	0.305	0.467	0.292	0.035	0.323	2
Nitraria sibirica Pall	0.500	1.000	0.600	0.670	1.000	0.929	0.783	1

It could be seen from table 1 that the average value of membership function of *Nitraria sibirica* Pall was the highest, far higher than the other two species. Therefore, according to the growth condition of the plant, the *Nitraria sibirica* Pall was the most salt-tolerant, ranking the first, and the *Nitraria roborowskii* Kom was the second, ranking the second. *Nitraria tangutorum* Bobr was the worst, third in order.

4. Discussion

4.1. Single index analysis and comprehensive index analysis and salt tolerance of plants

In this experiment, there was no significant change in the increase of the relative stem diameter of *Nitraria roborowskii* Kom under NaCl stress, while *Nitraria sibirica* Pall and *Nitraria tangutorum* Bobr were more sensitive to NaCl stress. However, by considering the plant height, leaf number, leaf area, average fresh weight, average dry weight, and comprehensive membership function of 3 Species of *Nitrarial*. *Nitraria sibirica* Pall was more salt-tolerant than *Nitraria tangutorum* Bobr and *Nitraria roborowskii* Kom. Ha *et al*, Xu *et al*, after analyzed each growth index, all believed that the comprehensive analysis results of each index were more reliable.

4.2. Comparison of salt tolerance between three species of Nitraria and other species of Nitraria

Due to the time limitation, only the salt tolerance of *Nitraria tangutorum* Bobr, *Nitraria roborowskii* Kom and *Nitraria sibirica* Pall were analyzed in this study. According to the reports of other researchers, such as zhang li and wu xiang, etc., the comprehensive evaluation by the membership function method showed that the salt-tolerance of the three kinds of ***Nitraria***, from the largest to the smallest, the *Nitraria sibirica* Pall., the *Nitraria tangutorum* Bobr and the *Nitraria roborowskii* Kom. By referring to previous studies, it was believed that the salt-tolerance of *Nitraria sibirica* Pall was also stronger than that of *Nitraria roborowskii* Kom., which may be similar to or higher than that of *Nitraria sibirica* Pall.

4.3. Relationship between plant growth indexes and salt tolerance

In this study, six plant growth indexes were determined, and there were some differences among the indexes. Generally, the dry and fresh weight of the plant in the growth index could better reflect the growth characteristics or conditions of the plant. For example, Zhang *et al* concluded that under salt stress, fresh weight water content and dry weight water content of different varieties all decreased. Fu *et al* concluded that with the increase of salt stress intensity, plant height, fresh weight and dry weight of the two types of inbred lines presented a gradually declining trend. Ren concluded that the salt stress treatment had a great influence on the characters of wheat seedlings at different seedling stages, ranging from fresh weight and dry weight above ground. Therefore, in the comprehensive analysis, such as the calculation of the comprehensive membership function, the weight of indicators such as the dry matter weight of plants can be increased.

5. Conclusion

- NaCl stress had the greatest inhibitory effect on the plant height growth of *Nitraria tangutorum* Bobr, followed by *Nitraria sibirica* Pall, and finally *Nitraria roborowskii* Kom.
- With the increase of NaCl stress concentration, the number of leaves of *Nitraria tangutorum* Bobr decreased rapidly, and the number of leaves of *Nitraria roborowskii* Kom decreased rapidly when the salt concentration was greater than 6%. The relative number of leaves of *Nitraria sibirica* Pall was significantly reduced at 15% salt concentration.
- With the increase of salt concentration, the relative increase of stem diameter of *Nitraria tangutorum* Bobr was decreased in the first place, *Nitraria sibirica* Pall was decreased in the second place, but *Nitraria roborowskii* Kom was not significantly affected .
- With the increase of salt concentration, the relative leaf area of *Nitraria tangutorum* Bobr decreased significantly. The relative leaf area of *Nitraria roborowskii* Kom increased significantly at 6% and then decreased significantly. The relative leaf area of *Nitraria sibirica* Pall decreased significantly at 12%.

- With the increase of salt concentration, the dry weight and fresh weight of the plant, the weight of the *Nitraria sibirica* Pall decreased little, there was no significant difference, the rest fluctuated slightly up and down; The changes of *Nitraria tangutorum* Bobr and *Nitraria roborowskii* Kom were more consistent.
- NaCl stress had the least effect on the fresh weight of *Nitraria sibirica* Pall, and the changes of *Nitraria tangutorum* Bobr and *Nitraria roborowskii* Kom were basically the same. Under the salt stress of different concentration, the dry weight of *Nitraria sibirica* Pall had no obvious change.
- The comprehensive membership function analysis of the growth indexes showed that the salt tolerance order was successively *Nitraria sibirica* Pall > *Nitraria roborowskii* Kom > *Nitraria tangutorum* Bobr

Acknowledgments

This research was supported by the Major Science and Technology Project of the Seed Industry of Tianjin Committee of Science and Technology (17ZXZYNC00070) and University Young and Middle-aged Backbone Innovation Talent Training Program in Tianjin (J01009030709) and Agricultural Science and Technology Achievements Transformation and Extension Project in Tianjin (201603060) and Postgraduate Innovation Training Program in Tianjin (2017YPY001) and Tianjin Applied Basic and Frontier Technology Research Plan (14JCYBJC30200).

References

[1] Yang J H, Meng N and Zuo F Y 2013 *J. Tianjin Agricul. Uni.* **20** 5-8
[2] Wang Y G, Yang X H and Yu C T 2007 *Res. Soil Water Conserv.* **14** 74-9
[3] Zuo F Y, Hao X F and Chen Z F 2013 *J. Tianjin Agricult. Univ.* **20** 11-4
[4] Liu Y, Yang J H and Zuo F Y 2013 *Northern Horticult.* **11** 69-71
[5] Zhu J 2001 *Trends Plant Sci.* **6** 66-71
[6] Zhang J, Xing S and Sun Q 2004 *Chinese Forest. Sci. Technol.* **3** 12-6
[7] Li S F, Zhang Q C and Zhang Q C 2005 *J. North China Univ.: Natural Sci. Edition* **6** 78-81
[8] Dorothea B and Ramanjulu S 2005 *Critic. Rev. Plant Sci.* **1** 18-21
[9] Asish K P and Anath B D 2004 *Ecotoxicol. Environm. Safety* **3** 27-32
[10] Lutts S, Kinet J M and Bouharmont J 1996 *Plant Growth Regul.* **3** 43-6
[11] Zuo F Y 2013 The Effect of NaCl treatment on the seed germination, the seeding growth, the leaf physiology and biochemistry and the anatomical structures of leaf and stem of 3 *Nitraria* species, PhD Thesis of Southwest University, Chongqing, China
[12] Zhang K 2005 Study on standards and quality indexes of cucumber seedling from acupoint and dish, Doctor Thesis of China Agricultural University, Beijing, China
[13] Yu W J 2014 Mushroom residue in the application of melon seedling research, PhD Thesis of Shanghai Jiaotong University, Shanghai, China
[14] Wei M Z 2005 Effect of substrate physical properties on quality and later growth of cucumber seedlings, PhD Thesis of China Agricultural University, Beijing, China
[15] Ha L J, Xu Z and Yujisakai 2009 *Anhui Agricul. Sci.* **37** 5466-7
[16] Xu S S, Zhang G C and Ye J X 2011 *North Hor.* **8** 5-9
[17] Zhang L 2010 Beijing: Chinese Academy of Forestry, 3 species of *Nitraria* response to salt stress and study on the mechanism of salt tolerance
[18] Wu X, Ni J W and Zhang X H 2012 *J North Forest Univ.* **40** 44-7
[19] Zhang S H, Liu L X and Yang K J 2010 *North Rice* **40** 18-21
[20] Fu Y, Gao S R and Yang K J 2011 *J H Univ. Agricul.* **23** 13-5
[21] Ren Y Z 2012 *Hubei Agricul. Sci.* **51** 3702-5

Efficacy of herbicides on crabgrass in Golf Course, Southern China

L Fei[1,2,3] and P Jiang[1]

[1]Zhuhai College of Jilin University, Zhuhai City, China
[2]Shanghai Jiaotong University, Shanghai City, China

E-mail: 14270088@qq.com

Abstract. Crabgrass, a malignant weed on golf turf, affects the hitting effect and golf course landscape. In this study, three herbicides were used for crabgrass in driving range of the golf course in Southern China. Metsulfuron-Methyl, Flazasulfuron and Quinclorac were sprayed at three concentration gradients of 1, 2 and 3 times respectively. Results showed that three times the recommended amount of quinclorac had the best killing effect on crabgrass and the other two herbicides were still unsatisfactory even if the concentration reached three times of the recommended amount. The efficacy of quinclorac at recommended dose is 2.2 times higher than that of metsulfuron-methyl, and 1.3 times higher at three times recommended dose. Because this experiment was carried out in November, the leaf and sheath of crabgrass is old, cellulose content is high, lignification is serious, and it has high resistance to chemical herbicides. Manual removal is more environmentally friendly and efficient method.

1. Introduction

A standard 18-hole golf course generally covers an area of about 67 hm, and the turf planting area in the whole golf course accounts for about 2/3 of the total golf course area. The golf course turf grass is a kind of functional lawn which is planted at one time and used for many years, and needs to be highly managed artificially in the process of using. The growth of weeds is the main cause of turf degradation. Sun [1] pointed out that the weeds destroy the uniformity of the lawn, not only affect the beauty of the course, but also compete with the planted turfgrass. Among all kinds of harms faced by turfgrass planted on golf course, the harm of weeds is the greatest threat to lawn.

Crabgrass is the most serious invasive weed in the golf course. At present, crabgrass grows in the fairway and rough area of the whole golf course, the most serious area is the fairway. Crabgrass (*Digitaria Sanguinalis* (L.) Scop.), an annual herbaceous plant of Digitaria Scop in the grass family, is wet, fertile and light-loving. It does not require strict soil requirements and can grow well in weak acid and alkaline soils. It has the characteristics of fast propagation, strong fecundity, fast growth and many tillerings. Turner and Van Acker [2] observed that more than 700 m^2 of crabgrass seedlings emerged on some residential lawns in Canada in the last two weeks of July.

Especially in loose, wet and fertile areas, it is often the most prosperous breeding area of crabgrass. The planting layer of golf course turf is soft sandy soil, and for the healthy growth of turfgrass, it is necessary to keep the soil moist all the year round. At the same time, the greenkeeper will fertilize the turf from time to time. These characteristics of the soil that coincide with the life habits of crabgrass in the course.

Johnson [3] showed that the management of crabgrass was 99% with using ansar at 2.2 kg and

Content from this work may be used under the terms of the Creative Commons Attribution 3.0 licence. Any further distribution of this work must maintain attribution to the author(s) and the title of the work, journal citation and DOI.
Published under licence by IOP Publishing Ltd

dithiopyr at 0.3 kg. He also found that after 2.2 kg of ansar, the control effect of 1.1 kg of dinitrotoluidine and 1.1 kg of oxadiazon reached 95% and 94% respectively. However, if the ansar is not used first, the control effect is less than 79% when the dinitrotoluidine or oxadiazon is used alone. Smith [4] used norflurazon from 0.56 to 1.68 liters to control crabgrass in Bermuda grass lawn. The effect was over 99%. However, when the dosage of norflurazon exceeded 1.68 liters, Bermuda grass lawn showed toxicity. Jeffery [5] found that crabgrass of golf course in New Jersey resistant to oxazol and chlorpyrifos, while in the rough area, it was sensitive to oxaprop. When crabgrass emerged, Fenoxaprop-P consumption was 0.76 kg/ha, which was 4 times the maximum, the biomass of the resistant crabgrass could be reduced by 35%, when the dosage was 1.52 kg/ha, 8 times the maximum. The biomass of crabgrass was reduced by 64%. It is evident that the efficacy of herbicides in general dosage is limited. The amount of herbicide is less than the effect of weed control, and the amount of herbicide is more harmful to turfgrass. Therefore, for chemical weed control of turf weeds, not only the choice of herbicide species, but also the amount of precise control.

2. Materials and methods

2.1. Test subjects

Zhuhai City is located in north latitude 21°48'-22°27', longitude 113°03'-114°19', and has a typical subtropical monsoon maritime climate, which annual average temperature is 22.5°C, annual average relative humidity is 80%, and the average annual rainfall of 206.9 mm. Gulf golfcourse is located in Zhuhai City, Guangdong Province. It is the first high standard 27-hole championship course designed by Colin Montgomerie in mainland China in 2003. The total area of the course is about 2.5 km^2. The total length of the fairway area is 6532 m. The turf area is about 49000 m^2 and the planting layer are sand and coarse sand. The turfgrass planted in fairway area, rough area and driving range is Seashore Paspalum grass. This experiment was conducted in a real grass driving range, which has the same maintenance measures as the fairway and the rough in irrigation and fertilization.

2.2. Test design

Nine square areas with an area of 1×1 m^2 were selected in the test area of Gulf golf course. Three herbicides namely Metsulfuron-Methyl, Flazasulfuron and Quinclorac, were used to dissolve the herbicide solution according to the recommended dosage in the product instructions, and it was set to the standard concentration. Prepared herbicides liquids were sprayed by spot-spray in all the tested areas at using three concentrations of each herbicide. This experiment is divided into three concentration gradients: 1-, 2- and 3-fold of recommended dosage. The experimental area with 1-fold recommended dosage was sprayed with three herbicides once, the experimental area with 2-fold recommended dosage was sprayed three herbicides twice, and the experimental area with 3-fold recommended dosage was sprayed three herbicides three times. The experiment was repeated three times and the average value was determined.

Before spraying the tested herbicides, the shoots number of crabgrass and the leaf age in the experimental area was recorded. No water was allowed within 12 hours after spraying herbicides. The first observation was made 24 hours later. The yellowing degree and dead area of each plant were recorded. Then every 2 days, the observation was made and recorded for 30 days.

2.3. Data processing of test

The obtained data were analyzed using Excel 2013 according to Neil and Mack [6].

3. Results and discussion

3.1. Effect of metsulfuron-methyl on the control of crabgrass

Table 1 is based on the ratio of the number of dead shoots to the total number of shoots in the experimental area of after metsulfuron-methyl herbicide was sprayed. It can be seen from the table 1 that the first spraying started on November 17, and on November 18, the next day, no shoots death occurred in the test area using metsulfuron-methyl recommended dosage, while the same number of shoots died in the test area using 2 times recommended dosage and 3 times recommended dosage. Three days after spraying, the death of the shoots of crabgrass was found in the all of the test areas. From Nov. 18 to Nov. 26, that is, one week after spraying, the death peak period of the shoots of crabgrass was observed. The number of dead shoots in three testing dosages of the metsulfuron-methyl showed a straight upward trend. The results showed that metsulfuron-methyl had the best killing effect on crabgrass after 1 week of application. It can also be seen from the table 1 that there is almost no new increase in the number of dead shoots of crabgrass in the period from the eighth day to the half month after spraying. That is to say, the killing effect of metsulfuron-methyl on crabgrass is greatly reduced after sprayingone week. The killing effect of standard dose of metsulfuron-methyl on crabgrass is much lower than that of two- or three-times standard dose. It was also found that the killing effect of 3-fold standard concentration was better than that of 2-fold standard concentration. As far as metsulfuron-methyl is concerned, it was found that there was a certain degree of chlorosis of seashore paspalum grass in the test area with three times the standard concentration. When using herbicides to kill crabgrass, we should not only request high efficacy of herbicides, but also evaluate the damage of herbicides to turfgrass (mainly chlorosis of grass leaves).

Table 1. Number of dead shoots after spraying metsulfuron-methyl.

Dose	Days								
	0	1	2	4	6	8	10	12	14
1 recommended dose	0	0	0.3	1.3	1.6	2	2.3	2.3	2.3
2 times	0	0.3	1	1.6	2.3	3	4	4	4
3 times	0	0.3	1.6	2.3	3.6	4	5	5	5

On the 10th day after metsulfuron-methyl was applied, the number of dead shoots of crabgrass reached the peak, and there were no new dead leaves. One week later, on December 5th, new leaves of crabgrass were germinated in the experimental area using 1 and 2 times standard concentrations of metsulfuron-methyl. Only in the experimental area at applied this treatment 3 times, there were no new leaves of crabgrass. However, by December 6, the experiment area using 3 times of standard concentration also began to young leaf germinate. From Dec. 7 to Dec. 13, the number of the new leaf of crabgrass reached its peak. The results showed that after spraying by metsulfuron-methyl for 20 days, the killing efficacy was completely lost, and the tilllers of crabgrass which had not been completely killed would germinate into new leaf again. Generally speaking, the inhibitory effect of 3-fold of standard concentration on the growth of crabgrass was better than that of 2-fold of standard dose and at applied one time of standard dose.

Table 2. Number of germinating crabgrass leaves after spraying metsulfuron-methyl.

Dose	Days						
	17	19	21	23	25	27	29
1 recommended dose	1	1.5	2	2.5	3	3	3
2 times	1	1	1.3	1.6	2.3	2.3	2.3
3 times	0	1	1.3	1.6	2	2	2

3.2. Effect of flazasulfuron on the killing of crabgrass

Table 3 is based on the ratio of the number of dead shoots to the total number of shoots in the experimental area after the application of flazasulfuron herbicide. As can be seen from the table 3, the 10 days from the second day of flazasulfuron spraying, that is, from November 18 to November 28, are the best time for flazasulfuron to kill crabgrass. Using of flazasulfuron at standard concentration three times had the fastest effect and the best efficacy, however, even if it was used at 1-fold standard concentration, the killing effect was better, because the number of dead shoots after spraying the standard concentration of the herbicide one time was similar to those standard doses which applied two or three times. It can also be seen from the table 3 that the efficacy period of flazasulfuron controlling crabgrass is 10 days. Compared with metsulfuron-methyl, the efficacy period lasts longer. It can be seen that the control effect of flaza Sulfuron on crabgrass is better than metsulfuron-methyl. After 28 November, the number of dead shoots did not increase, indicating the end of the effective period of flazasulfuron.

Table 3. Number of dead shoots after spraying flazasulfuron.

Dose	Days								
	0	1	2	4	6	8	10	12	14
1 recommended dose	0	0.3	1	2.3	2.6	3	3.6	3.6	3.6
2 times	0	0.6	1.6	2.5	3	3.6	4	4	4
3 times	0	1	2	3.3	4	5	5.3	5.3	5.3

Table 4. Number of germinating crabgrass leaves after spraying flazasulfuron.

Dose	Days						
	17	19	21	23	25	27	29
1 recommended dose	1	1.3	1.6	2	2.5	2.5	2.5
2 times	1	1.3	1.6	1.8	2	2	2
3 times	0.3	0.5	1	1.3	1.6	1.6	1.6

According to table 4, on Nov. 28, 10 days after flazasulfuron spraying, the killing effect of crabgrass reached its peak, and the number of dead shoots of crabgrass did not increase in the next 7 days. However, since Dec.5, new leaves have germinated more or less in all the experimental areas, which lasted until Dec.13. New leaves have continued to germinate in succession, indicating that flazasulfuron, as an herbicide, has completely lost its efficacy in killing crabgrass after 17 days of application. However, compared with metsulfuron-methyl, flazasulfuron could inhibit the germination of crabgrass. There are only 1.6 germinated leaves one month after the application of 3 times dose flazasulfuron, however, 2 new leaves were germinated under the condition of metsulfuron-methyl. Because the time limit of the experiment to Dec.13 did not continue to increase the newly leaves, it was not entirely herbicide effect, to some extent, also affected by the weather.

3.3. Effectiveness of quinclorac in the control of crabgrass

Table 5 is based on the ratio of the number of dead shoots to the total number of shoots in the experimental area after spraying quinclorac herbicide. It can be seen that the efficacy of quinclorac is the best from Nov. 18 to Nov. 28 begaing from after sprayed the next day. Even if it is used according to the standard concentration dose, the killing effect of quinclorac is also good. Comparing the efficacy of the three herbicides, it can be seen clearly that quinclorac has the best killing effect on crabgrass. After spraying quinclorac, the number of dead shoots of crabgrass was larger, and in the process of herbicide efficacy observation, it was found that quinclorac was the least harmful to turfgrass, and chlorosis of seashore paspalum grass leaves was not obvious.

Table 5. Number of dead shoots after spraying the quinclorac.

Dose	Days								
	0	1	2	4	6	8	10	12	14
1 recommended dose	0	0.3	1.3	1.6	2.3	3.3	5	5	5

2 times	0	1.6	1.6	2.3	4	4.6	5.3	5.3	5.3
3 times	0	1.3	2.3	3	4.3	5	6.3	6.3	6.3

Table 6. Number of germinating crabgrass leaves after spraying quinclorac.

Dose	Days						
	17	19	21	23	25	27	29
1 recommended dose	0.6	1	1.3	1.6	2	2	2
2 times	0.3	0.6	1	1.3	1.6	1.6	1.6
3 times	0	0	0	0	0	0	0

According to table 6, from the spraying of 3 times the standard concentration of quinclorac on Nov. 16to the end of observation on Dec. 17, there was no young shoot germinated in the experimental area, which meant that crabgrass was completely killed and no new leaves germinated. This also proves that the efficacy of quinclorac is the best when it is used at 3 times the standard concentration. No leaf died since Dec. 11 and began to germinate newleaves in the test area at using the standard concentration of quinclorac one time. The test area with 2-fold standard concentration was later, new leaves germinating were observed for the first time on Dec. 13. Moreover, compared with the pharmacodynamic curves of metsulfuron-methyl and flazasulfuron, the controlling effect of quinclorac is still the best of the three.

3.4. Discussion

Echinochloa crus-galli L. is a major weed in rice fields in China, and quinclorac has been long used for its control [7]. Quinclorac was double effectiveness than metsulfuron-methyl to control crabgrass. Flazasulfuron also better than metsulfuron-methyl, but less than quinclorac. The optimum period of efficacy of these three herbicides was within 12 days after spraying. After that, the number of dead shoots of crabgrass showed a steady decline. Among them, the number of new leaf germinated after metsulfuron-methyl was nearly twice as much as that in the experimental area of quinclorac. By observing the turf biomass in the experimental area, it was also found that the efficacy of metsulfuron-methyl was worse than that of flazasulfuron and quinclorac, that is to say, the comprehensive efficacy of metsulfuron-methyl was worse than that of flazasulfuron and quinclorac. On the contrary, the number of shoots killed by quinclorac in the first 12 days was the highest, and the number of leaf regeneration in the later period was the lowest, and the comprehensive effect was the best.

Large crabgrass is a major grass weed widely distributed across China. Yu [8] collected 26 large crabgrass populations from maize field, found out 25 populations resistant to nicosulfuron. Li [9] reported that field applications of nicosulfuron have been ineffective in controlling crabgrass populations in Shandong Province, China and then found out Trp574Arg substitution was the main reason for crabgrass resistance to ALS-inhibiting herbicides. Zhu [10] summarized 30 weed species that have evolved resistance to 11 herbicide sites of action in recent years in China, in which crabgrass can resiste to the nicosulfuron, quizalofop-ethyl, and Sethoxydim. However, metsulfuron-methyl and flazasulfuron are similar with nicosulfuron,which are same kind of sulfonylureas systemic herbicides, maybe that is the one of reasons metsulfuron-methyl are weak to control crabgrass in this experiment.

4. Conclusion

In this experiment, quinclorac are better efficacy than metsulfuron-methyl and flazasulfuron. The efficacy of quinclorac at recommended dose is 2.2 times higher than that of metsulfuron-methyl, and 1.3 times higher at three times recommended dose. Using quinclorac is the better option when crabgrass have over 2 tillers.

In the study of weed survey [11,12], crabgrass was found to be one of the most abundant malignant weeds in landscape. For a golf course, chemical weed control of crabgrass is undoubtedly the best choice at present. Brian [13] pointed out there are no POST herbicides are registered for crabgrass control, because of high turfgrass injury and little residual control, there are not an effective substitute for using pre-emergence herbicides for southern crabgrass control. So, for crabgrass controlling, using pre-emergence herbicides are better than post-emergence herbicides. This strategy allows for the reduction of the number of herbicide applications, reducing costs and the environmental impact of pesticides [14].

References

[1] Sun J X 2008 *Grass Science* 3rd ed. (China Agricultural Publishing House, Beijing China)
[2] Turner F A, Van A and Rene C 2013 In situ emergence timing of large and small crabgrass in residential turfgrass of Southern Ontario *Can. J. Plant Sci.* **93** 503-9
[3] Johnso B J 1994 Tank-mixed herbicides on Large crabgrass (*digitaria-sanguinalis*) and goosegrass (*eleusine-indic*) control in common bermudagrass (*cynodon-dactylon*) Turf *Weed Sci.* **42** 216-21
[4] Smith A E 1994 Influence of norflurazon induced chlorosis on bermudagrass (*cynodon-dactylon*) hay production *Weed Technol.* **8** 508-11
[5] Jeffrey F D 2002 Detection of fenoxaprop resistant smooth crabgrass (*digitaria ischaemum*) in Turf *Weed Technol.* **16** 396-400
[6] Neil D, Mack S C and Fuqua R W 2015 Phase-change lines, scale breaks, and trend lines using excel 2013 *J. Appl. Behav. Anal.* **48** 478-93
[7] Peng Q, Han H P, Yang X *et al* 2019 Quinclorac resistance in echinochloa crus-galli from China *Rice Sci.*
Online First Article available http://www.ricescience.org/EN/abstract/abstract9823.shtml
[8] Yu M, Si C and Liu M J *et al* 2017 Investigation of resistance levels and mechanisms to nicosulfuron conferred by non-target-sitemechanisms in large crabgrass (*Digitaria sanguinalis* L.) from China *Pestic. Biochem. Phys.* **141** 84-9
[9] Li J, Li M and Gao X X *et al* 2017 A novel amino acid substitution Trp574Arg in acetolactate synthase (ALS) confers broad resistance to ALS-inhibiting herbicides in crabgrass (*digitaria sanguinalis*) *Pest Manag. Sci.* **73** 38-43
[10] Zhu J W, Wang J and DiTommaso A *et al* 2018 Weed research status, challenges, and opportunities in China *Crop Prot.*
[11] Saha D, Marble S C and Pearson B J *et al* 2019 Mulch type and depth, herbicide formulation, and post application irrigation volume influence on control of common landscape weed species *Horttechnology* **29** 65-77
[12] Parker E T, McElroy J S and Flessner M L 2015 Smooth crabgrass and goosegrass control with metamifop in creeping bentgrass *Horttechnology* **25** 757-61
[13] Brian D G, Barry J B and Unruh J B *et al* 2015 Evaluation of alternative herbicides for southern crabgrass (*digitaria ciliaris*) control in St. Augustine grass *Weed Technol.* **29** 536-43
[14] Agostinetto D, Fontana L C, Vargas L, Perboni L T, Polidoro E and Silva B M 2014 Competition periods of crabgrass with rice and soybean crops *Planta Daninha* **32** 31-8

Mechanization of the process of drying fruits and vegetables through the introduction of microwave drying unit UMS-2-10 in agricultural and food enterprises

E V Voronov[1,4], E A Shamin[1], V A Bocharov[1], O B Terekhova[2], L G Shashkarov[3], G A Larionov[3], P V Zaitsev[3] and N K Kirillov[3]

[1]Nizhny Novgorod Engineering and Economic University, Knyaginino, Russia
[2]Nizhny Novgorod State Agricultural Academy, Nizhny Novgorod, Russia
[3]Chuvash State Agricultural Academy, Cheboksary, Russia

E-mail: e_voronov@list.ru

Abstract. The article describes the installation of microwave drying UMS-2-10, its use in agricultural enterprises. Installation of combined drying (microwave and convective heating). The installation was developed by the research and production Association "Gamma" in Nizhny Novgorod. The principle of operation is based on the use of microwave energy in combination with the circulation of "warm" air. It is established that when choosing a drying unit for the food industry, it is necessary to keep in mind that the scientifically justified choice of the design of the dryer intended for drying specific types of plant substrates containing biologically active substances is more dependent on the structural properties and biological composition of the plant raw materials used in production. Drying plant UMS-2-10 installation provides for the use of microwave, convective and combined methods of heating the feedstock; characterized by small size and ease of operation; has optimal performance, focused on the orders of commercial enterprises, and not on the formation of large batches of finished products, the implementation of which may be in doubt; characterized by high maintainability, low cost and tariffs for delivery, and installation of the dryer.

1. Introduction

The results of the analysis of patent data of leading foreign firms (USA, UK, Germany, France, Finland, Canada, Sweden, Japan, etc.) show scientific and technological progress in the food and processing industries is carried out in two main areas: improvement of food production on the basis of traditional principles and radical change of production processes on the basis of the latest achievements of science and technology [1, 2].

Drying technologies are undergoing significant changes. Short-term drying processes using gentle modes for maximum preservation of physiologically valuable substances for the body - vitamins, dissolved carbohydrates, mineral substances and quick removal of moisture to increase shelf life - are becoming increasingly common. More and more importance is attached to fast food and therefore find their place in the human diet of the new century dried foods that are restored in a matter of minutes [3].

2. Materials and methods

The research was carried out at the Nizhny Novgorod State Agricultural Academy.

Content from this work may be used under the terms of the Creative Commons Attribution 3.0 licence. Any further distribution of this work must maintain attribution to the author(s) and the title of the work, journal citation and DOI.
Published under licence by IOP Publishing Ltd

The technological processes of drying samples of carrots, beets and onions using convection heating, microwave (heating in the electric field of microwave) heating and heating at the same time combining the effects on the product of convection of warm air and electric microwave energy were chosen as the objects of research.

The combined drying unit (microwave and convective heating) was used in the studies. The installation was developed by the research and production Association "Gamma" in Nizhny Novgorod.

3. Results and discussion

The principle of operation is based on the use of microwave energy in combination with the circulation of "warm" air.

The microwave drying unit has three modes: microwave, combined, combining microwave and electric processing methods and traditional electric heating.

Microwave processing in the installation is carried out by the method of volumetric heating, in which the field energy generated by magnetrons is converted into heat throughout the volume of the processed product. The air and product containers do not heat up because of the flow microwave field, without absorption.

The unit performs all the functions of microwave drying in the combined mode. In addition, in the operating mode, the warm air coming from the tubular electric heaters, with the help of a fan, is evenly distributed throughout the volume, carrying out additional surface drying of the product.

Under the conditions of convection (traditional electric heating), the heating of products is carried out with warm air coming from heating elements using a fan, microwave generators (magnetrons) are disabled in this mode. It is used not only for the main convective drying but also to remove moisture from the surface of the product after extracting it from the inside using microwave energy.

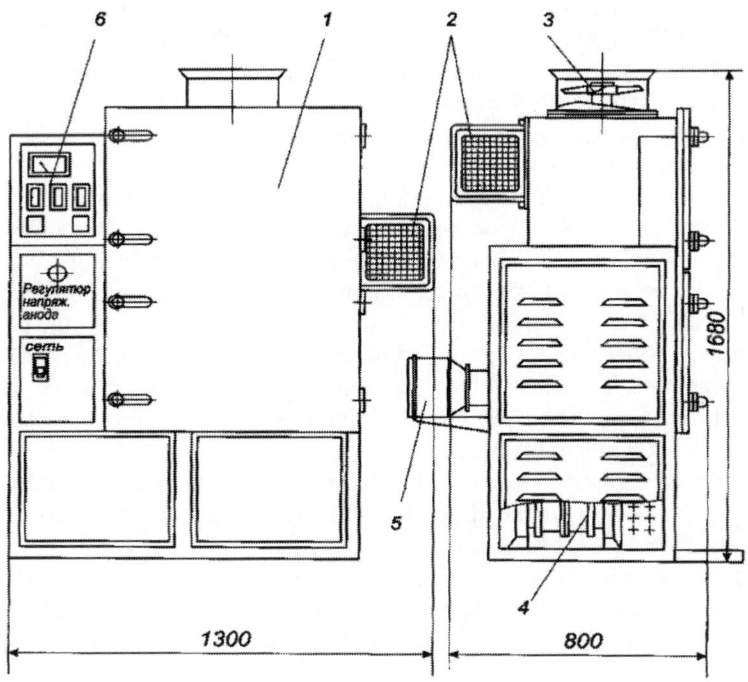

1 – drying (resonator) chamber; 2 – microwave generator unit; 3 – ventilator; 4 – power supply; 5 – heater; 6 – control panel

Figure 1. Installation of microwave drying UMS-2-10.

The installation consists of the following components (figure 1): drying (resonator) chamber with a door; three blocks of microwave generators that convert AC power into microwave power; fans, which removes moisture from the drying chamber; three power supplies that convert the mains voltage to the type required for the operation of magnetrons; heater, which is heated air entering the chamber from the bottom by means of a fan; remote control.

The speed of movement of the material in the dryer can vary by changing the angle of inclination of the springs, the amplitude, and frequency of oscillations, as well as the speed of the gaseous coolant, supplied to the sublattice of the dryer. Dryers of this type are produced both by the domestic industry and by some foreign companies. The grid area is 0.5...10 m, and the specific removal of moisture from the surface unit 30...70 kg/m2.

The door of the drying chamber with rubber seals, which are wearing a metal braid, fits snugly to the chamber and thus provides ideal protection against leakage of microwave energy. To control the tightness of the door to the drying chamber, a limit switch is installed between the case and the door. When the door is not tightly closed, the limit switch blocks the inclusion of microwave generators – which ensures the safety of the installation.

Each microwave generator unit includes magnetron, waveguide, fan, and high-voltage local transformer.

The power supply includes a single-phase power transformer, a high-voltage diode rectifier, and a voltage regulator.

The structure of the heater includes three tubular electric heaters and exhaust system.

On the control panel there are power buttons: ventilation chamber, heating and anode voltage of magnetrons, exhaust systems and tubular electric heaters; time relay; the switch of the control device for measuring the currents of magnetrons; microammeter; warning lights; Handles "Regulation of the anode current of magnetrons" toggle switch mode selection; lamp alarm magnetrons [4].

Rational design of the dryer can be selected only for a specific product or a small group of materials with similar physical and chemical characteristics.

Selecting the rational structures of the dryer was guided by the following requirements:

• providing high-quality indicators of the finished product (swelling, digestibility, preservation of attractive organoleptic characteristics, etc.);

• the minimum specific consumption of heat, steam, air, and electricity referred to 1 kg of evaporated moisture or 1 ton of finished product;

• the intensity of the process, ensuring the minimum overall dimensions of the installation, removal of moisture from 1 m2 of the capacity of the device or from 1 m2 of the production area;

• the possibility of maximum automation and mechanization of the drying process.

In all cases, it is necessary to carefully study the influence of the main parameters of the drying agent and the characteristics of the material on the dewatering process, namely: the possibility of using a higher temperature, air flow rate, its relative humidity, oscillating modes, combined methods of heat supply and drying methods, as well as pre-technological preparation of raw materials. At the same time, however, we should not forget about the chemical composition of the finished product (the safety of a group of vitamins, provitamins, proteins, sugars, etc.).

When choosing a drying unit for the food industry, it should be borne in mind that the scientifically justified choice of the design of the dryer intended for drying specific types of plant substrates containing biologically active substances is more dependent on the structural properties and biological composition of the plant raw materials used in production [5].

At the enterprises of vegetable-drying production, mainly convection drying lines are used using belt dryers, spray dryers (for drying juices and purees), tunnel dryers, cabinet boilers with boiling and vibratory boiling layer and lines using single-roll and double roll dryers. They are designed for drying a large volume of raw materials under specified conditions and uninterrupted supply of raw materials [6].

Currently, the use of large production capacities is inefficient, as there is a shortage of agricultural products, the constancy of supply of raw materials for enterprises is broken and vegetable drying plants are not fully loaded.

In such a situation, small enterprises with small-sized equipment designed for small batches of raw materials come to the first place. Installation dryers small performance gives a significant economic effect.

As a result of the analysis of literary sources as an installation for scientific research, the choice is made on the drying plant UMS-2-10, guided by the following:

- the installation provides for the possibility of using the microwave, convective and combined methods of heating the feedstock;
- small dimensions of the installation and ease of use;
- optimal performance focused on the orders of commercial enterprises, and not on the formation of large batches of finished products, the implementation of which may be in doubt;
- high maintainability, low cost, and tariffs for delivery and installation of the dryer.

Justification of parameters and modes of drying

The temperature of the drying agent – air – is one of the main factors affecting the drying process.

Experiments on drying, in which the relative humidity and air velocity were maintained constant, and only the temperature changed, it was found that at the beginning of drying, the increase in air temperature increases the speed of the process to a lesser extent than at a subsequent stage. when the thermal effects. associated with evaporation, can be neglected and the temperature of the material becomes close to the air temperature.

However, as the air temperature increases, heat losses increase, which are most significant at the final stage of drying the material with low humidity.

Relative humidity is the second important factor affecting the drying rate. In the case of constant temperature and air flow rate, the decrease in the drying rate at the first stage is directly proportional to the increase in the relative humidity of the air. Following this, the dependence of the process speed on the relative humidity decreases, and in the final section increases again.

The air flow rate at a constant drying rate section also affects the drying rate of the material (at a constant temperature and relative humidity). This effect is significant mainly at an air flow rate of less than 5 m/s. Further increase in the air flow rate has no significant effect on the drying rate of the material in the first section. The increase in the air flow rate after reaching this value is also limited by the fact that the jet begins to "tear" small pieces of the dried material from the drying surface. At the drying area of the material with low humidity, the air flow rate does not have a significant effect on the drying rate. In this area, it makes little sense to set the airspeed above 1 m/s.

The atmospheric phenomenon also affects the drying rate. Lowering the barometric pressure has a significant impact on the drying rate only at the first stage.

Grinding the material greatly reduces the drying time. This is the secret of the success of spray drying, which, if the material is well crushed and the particle size does not exceed a few microns, occurs in a few seconds.

The thickness of the material loading layer, or specific loading (in kg/m2), significantly affects the rate of drying. The loading weight of sliced vegetables is usually 10...15 kg/m2. in drying plants with a good organization of the process much attention is paid to the uniformity of the material loading. In particular, each pallet (baking tray) of the chamber dryer must be loaded with the same amount of material. In belt dryers, uniform material loading is ensured by special equipment. The increase in the thickness of the load reduces the speed of drying, mostly on the first initial part of the drying. As a result of the shrinkage of the material as it dries, the loading layer is more easily penetrated by the air flow, and its thickness decreases. Thus, the initial decrease in the drying rate when drying a thick layer stops as the material is dehydrated. This phenomenon makes it possible to install a lower speed of the lower tapes than the upper ones on the belt dryers. Material loading greatly affects the performance of the equipment, and its optimal value is established empirically.

All these laws are entirely fulfilled only in that case. if the material crushed into particles of the same size is spread out on a pallet in thin uniform layers and the whole is permeated by a stream of warm air. However, most of the established rules and laws applied to other methods of drying [7].

The optimal drying mode should provide a product of standard quality with high technical and economic indicators. When justifying and choosing the drying mode, it is necessary to proceed from the technological properties of the material that changes during the drying process, that is, it is necessary to choose such operating parameters (temperature, humidity, airspeed, etc.), the impact of which on certain characteristics of the material would provide its best technological properties.

When choosing the mode, it is necessary to take into account the thermal stability of the product, its biological nature and structural and mechanical properties, on which the warping of the samples and the formation of cracks depend.

In order to avoid warping and cracking it is necessary to strive for the most uniform drying over the cross-section of the sample. The maximum permissible moisture gradient is experimentally determined for each product [8].

The quality of dried vegetables is directly influenced by preparatory technological operations: a form of cutting, type and time of pre-heat treatment.

Vegetables in preparation for drying cut into pieces of different sizes and shapes: columns, circles, segments, chips, blocks, and records. The shape and size of the pieces has a great influence on the drying speed, and therefore on the performance of the drying plant. With a decrease in the thickness of the product pieces, the duration of dehydration is reduced and the recovery time of the dried product during its cooking is accelerated.

If the products are cut into small pieces, surface hardening occurs to a lesser extent [8].

Intensification of the drying process improves the quality of the dried product and reduces the loss of vitamins and other valuable nutrients. However, the thickness of the piece can be reduced to a certain value (2 mm for vegetables), as cutting into thinner pieces leads to the formation of a large number of crumbs.

Preference is given to dried vegetables, sliced in the form of cubes, plates, and chips, since such a product has a large bulk mass, evenly mixed in soup mixes, well dosed in soft packaging on machines and has a more attractive appearance in the finished dish.

The content of trifle in the cut raw material should not exceed 5...8%. The increased content of the crumb worsens the drying conditions and leads to unnecessary losses, as this reduces the yield of the standard dried product and increases the consumption of raw materials.

Uneven cutting across the width and thickness, the presence of agglomerated or incompletely cut particles is also unacceptable, as the correct drying mode is disturbed, the product is unevenly dehydrated, resulting in the expenditure of additional labor for sorting and finishing the large pieces leaving the dryer with high humidity. The surface of the cut should be flat, smooth. in this case, the cells of raw materials are destroyed less and the loss of vitamin C is also much less [9].

Currently, there are conflicting opinions among scientists and specialists in the field of drying vegetables about the advisability of using pre-heat treatment before drying raw materials [10].

On the one hand, steam or water blanching is considered to be a necessary condition for the preservation of color, taste, smell, vitamin activity, acceleration of recovery, as well as, mainly, the destruction of oxidative enzymes – oxidases and the prevention of hydrolysis or oxidation of lipids in order to prevent loss of the level of quality of products in the process of dehydration and especially subsequent storage. On the other hand, numerous studies show that blanching is not particularly necessary [11].

A negative factor of heat treatment of vegetables before drying is the partial leaching of soluble substances (sugars, minerals, acids, etc.) and the loss of water-soluble vitamins [12].

When blanching sugar, gelatinized starch, and gelling pectin penetrate into the intercellular space and clog pores, thereby making it difficult to remove moisture during subsequent drying. Also, in the process of pre-heat treatment, there are significant losses of coloring substances, which adversely affects the commercial properties of finished products [13].

Blanching reduces the activity of vitamin-destroying enzymes and preserves the taste, but vitamin losses can reach 30% [14].

As studies show, if combined heating is used, blanching is not particularly necessary. Volumetric and deep microwave heating at the first stage of drying creates a sufficiently high temperature (70...800C) inside the product particles. This temperature gives blanchardi effect. As the drying rate decreases, both the temperature inside the product and it dries to a characteristic drying crust.

4. Conclusion

Based on the analysis of literary sources, the following conclusions can be drawn:

1. Among the modern methods of drying, combined methods are increasingly important, which combine different forms of heating the product during drying.
2. In the development of new designs of drying units, the development of small-size chamber dryers, in which a combination of two or more drying methods is possible, comes to the fore.
3. The quality of dried products largely depends on the preparation of raw materials for drying, in which it is necessary to take into account the types of raw materials and the purpose of the finished product.

References

[1] Khan A and Smalley I J 1973 Observation of particle segregation in vibrated granular systems *Powder Tech.* **Vol. 8.** N 1-2 pp 69–75

[2] Anderson R E and Hardenburg R E 1997 *Results, and recommendations on controlled modified atmosphere storage and transport of stone fruits* (Mich. State Univ. Hort. Rpt.) **28** pp 235–241

[3] Ginzburg A S 1973 *Fundamentals of food drying techniques* (Moscow: Food industry) p 538

[4] *Installation of microwave drying UMS-2-10.* Passport MVUZ-00.00.0.00 PS

[5] Gorbatyuk V I 1999 *Processes, and apparatus of food production* (Moscow: Kolos) p 335

[6] Grishin M A, Tanasevich V I and Semenov Yu G 1989 *Food drying plants:* Handbook (Moscow: Agropromizdat) p 215

[7] Burich O and Berki F 1978 *Drying fruits, and vegetables* Hungarian translation (Budapest, 1974) (Moscow: Food industry) p 279

[8] Ginzburg A S 1960 *Drying of food products* (Moscow: Pishchepromizdat) p 683

[9] Katz Z A 1984 *Production of dried vegetables, potatoes, and fruits* (Moscow: light and food industry) p 216

[10] Tsapalova I E, Maurnikova L A, Poznyakovsky V M and Stepanova E N 2007 *Examination of products of processing fruits and vegetables. Quality and safety* (Novosibirsk: Siberian University publishing house) p 271

[11] Ginsburg A S 1987 *Fundamentals of theory and technology of food drying* (Moscow: Economics) p 218

[12] Voskoboinikov V A, Gulyaev V N, Kats Z A and Popov O A 1978 *Dried fruits, and vegetables* (Moscow: Food industry) p 279

[13] Flaumenbaum B L, Tanchev S S and Grishin M A 1986 *Basics of food preservation* (Moscow: Agropromizdat) p 494

[14] Volonchuk S K and Shornikova L P Nutrition and infrared drying of vegetable raw materials *Food Indust.* **5** pp 6-8

IOP Conf. Series: Earth and Environmental Science **346** (2019) 012028 doi:10.1088/1755-1315/346/1/012028

Formation of yield and commodity qualities of potatoes, depending on the varietal characteristics

E V Voronov[1, 4], O B Terekhova[2], L G Shashkarov[3], G A Mefodiev[3], L V Eliseeva[3], S V Filippova[3] and A A Samarkin[3]

[1]Nizhny Novgorod Engineering and Economic University, Knyaginino, Russia
[2]Nizhny Novgorod State Agricultural Academy, Nizhny Novgorod, Russia
[3]Chuvash State Agricultural Academy, Cheboksary, Russia

E-mail: e_voronov@list.ru

Abstract. The article presents data on the study of the effect of varietal characteristics of potatoes on the formation of yield and quality of tubers. The aim of the work was to identify the best varieties of potatoes for cultivation in Nizhny Novgorod region. For conducting research, the following methods were used: experiments at cultivation, storage, use of potatoes, laboratory studies, tasting evaluation, and ranking. The effect of tuber damage by common scab and rhizoctoniosis on the safety of potato tubers was revealed. It had been established that all varieties of potatoes in conditions of highly cultivated sod-podzolic soils form a high yield of tubers. The maximum yield provides cultivation varieties Crystel and Sandrine. Crystel and Sandrine Varieties were characterized by a maximum yield of marketable tubers and weight per tuber. They also had the high starch content in tubers, the maximum tasting and overall assessment of tubers. It was noted that tubers of all varieties were slightly affected by common scab and rhizoctoniosis, which had a positive effect on their ability to work. To a lesser extent, these diseases affected the tubers of the varieties Crystel, Sandrine, and Serafina. The quality of these varieties was at the level of the standard.

1. Introduction

Potatoes are a type of perennial tuberiferous herbaceous plants of the Solanum genus of the Solanaceae family. In our country it is cultivated annually, because the entire life cycle, starting with the germination of the tuber and ending with the formation and formation of mature tubers, took place during one growing season [1].

Potatoes are the most important food, technical and forage crops, which have great economic importance. The diverse food use of potatoes, the wide popularity which it enjoys among the population, gives reason to consider it one of the most important food products – the second is bread, - and for the food industry raw materials for the production of food alcohol, starch and molasses [2, 3].

In Russia potatoes are of great and versatile importance. It is used as a food, technical and forage culture. The tubers contain about 25% dry matter, including 12-22% starch, 1.4-3% protein and 0.8-1% ash. They are composed of various vitamins – C, B, PP, K, and carotenoids. Potatoes have the great importance as a technical culture. It serves as a raw material of starch-treacle, dextrin industry and the production of glucose and alcohol. Potatoes are widely used for forage purposes. It is especially valuable for pigs and dairy cattle. Animals are fed tubers, tops and products of industrial processing of potatoes (Bardu, pulp). As a tilled crop, potatoes serve as a good precursor to spring crops (spring wheat, corn, beet, barley, millet, etc.) [4-7].

Content from this work may be used under the terms of the Creative Commons Attribution 3.0 licence. Any further distribution of this work must maintain attribution to the author(s) and the title of the work, journal citation and DOI.
Published under licence by IOP Publishing Ltd

IOP Conf. Series: Earth and Environmental Science **346** (2019) 012028 doi:10.1088/1755-1315/346/1/012028

Currently, the world knows more than 1,000 varieties of potatoes, different from each other in terms of maturation, dry matter content, taste, and other consumer properties.

A necessary condition for the intensification of potato production is the constant improvement of the quality of the varietal composition. However, many of the zoned domestic potato varieties are not sufficiently adapted to these conditions, which lead to yield losses and wide variability over the years. No less important is the correct choice of the variety most adapted to the soil and climatic conditions of a particular zone of cultivation [8, 9].

In this regard, the issue of identification of new varieties with a high level of productivity and high adaptive capacity to local agroecological conditions is relevant. The optimal selection of assortment for each specific region is one of the main factors determining the increase in yield and quality of both food and seed potatoes. The introduction of new domestic table varieties that can withstand the impact of adverse abiotic and biotic environmental factors, will better meet the needs of the population in quality potatoes [10, 11].

Our research was conducted in Nizhny Novgorod region, one of the largest regions of the Volga Federal district. The most widespread varieties were chosen at evaluating potato tubers.

2. Materials and methods

In 2015-2016 an assessment was carried out on a set of economically valuable attributes of 6 varieties of potato samples: Nevsky, Gloria, Christel, Sandrine, Serafina, and Aspia.

The research was carried out on the experimental field of Gorodetsky variety site, located in the SEC "Kolkhoz Red KrasniyMayak" Gorodetsky district, Nizhny Novgorod region.

The soils of the experimental site were sod-podzolic, highly cultivated, light-loamy in granulometric composition, which are the most common in considered region. All agrotechnical requirements at cultivation of potatoes were fulfilled.

Both years of research were quite favorable for the growth and development of potato plants.

For conducting research, the following methods were used: experiments at cultivation, storage, use of potatoes, laboratory studies, tasting evaluation, and ranking.

3. Results and discussion

One of the main indicators of potato production efficiency is its yield [10]. Analysis of the studied varieties by yield showed significant differences in these indicators as among varieties (table 1).

The average yield on varieties for two years was 33.9 t/ha with variation in varieties within 31.2 – 44.3 t/ha 23.1%.

Table 1. The yield of potatoes depending on varietal characteristics.

Variety	2015	2016	Average for 2 years	± to control	
				t/ha	%
Nevsky (st)	45.3	25.0	35.1	-	-
Gloria	33.7	16.9	25.3	-9.8	-27
Crystel	57.0	31.6	44.3	9.2	26
Sandrine	53.8	30.0	41.9	6.8	19
Serafina	54.1	16.8	35.4	3.0	0.1
Aspia	42.6	19.8	31.2	- 3.9	-11
Average	47.8	23.4	33.9	- 1.2	- 3

The control sample was taken from Nevsky variety, as this variety had been a monosort for potato growing in Nizhny Novgorod region for the last decades.

According to the years of research, the maximum yield of 42.6 — 57.0 t / ha in all of varieties was formed in the conditions of 2015, and the minimum - 19.8 – 31.6 t / ha in the conditions of 2016.

In both years of research, the maximum yield was formed by the variety Crystel, and the lowest – by the variety Gloria.

The most unstable in terms of yield formation was the Serafina variety. So, in 2015, the yield (54.1 t/ha) was close to the most productive grade of Crystel (57.0 t/ha), but in 2016 it formed the lowest yield among the studied varieties (16.8 t/ha). And the difference in yields from 2016 was 37.3 t / ha lowers than in 2015 and higher than the same difference over the years, in other varieties.

During the study, the analysis of the structure of the crop showed that the average weight of one tuber changed depending on the varietal characteristics of potatoes (table 2).

Table 2. Tuber weight and yield of commodity potato tubers on average for considered period.

Variety	The weight of the tuber (g)	The yield of marketable tubers (%)
Nevsky (st)	73	68
Gloria	92	69
Crystel	96	81
Sandrine	81	80
Serafina.	55	52
Aspia	90	74
Average	81	71

One of the indicators of the structure of the crop is the mass of the tuber [11, 12]. In general, it varied in from 55 g for the Serafina variety to 96 g for the Crystel variety. All varieties except Serafina were characterized by a higher weight of one tuber compared to the standard – Nevsky variety. Compared to the control, the increase amounted to 8 g in the Sandrine variety, 17 g in the Aspia variety, 19 g in the Gloria variety and 23 g in the Crystel variety.

By weight of tubers, they were divided into large tubers with a mass of more than 80 g, medium tubers with a mass of 50 – 80 g and small tubers with a mass of less than 50 g.

Varieties Gloria Crystel, Sandrine and Aspia in size formed tubers belonging to the group – large, Nevsky variety and Serafina to the group – medium.

Potatoes tubers are subject to various diseases, some of which affect the commercial quality of tubers are a common scab and Rhizoctoniosis (black scab). Common scab is manifested in the form of ulcers on the surface of the tubers. Rhizoctonia blight (scab black) is manifested in the form of ulcers and cavities in the hot weather – lots with net necrosis. Conidia which are similar to the black lumps of the earth which stuck to a tuber peel [13].

Our studies (table 3) showed that all varieties were not equally affected by common scab and rhizoctoniosis. The affection of tubers with rhizoctoniosis varied in the range of 0 – 8.0%. It was found that the variety of Crystel was not affected by this disease (0 %). The maximum affection – 8 % was noted in the variety of Aspia. Gloria variety was affected at the level of 6 %, Nevsky variety at the level of 4 %, Sandrine and Serafina varieties at the level of 3 %.

Table 3. Potato's bed capacity depending on tuber disease on average for considered period.

Variety	Common scab	Rhizoctonia blight	Keeping quality
Nevsky (st)	4	6	94
Gloria	6	7	93
Crystel	0	3	94
Sandrine	3	3	92
Serafina	3	3	91
Aspia	8	0	95

The affection of tubers with rhizoctoniosis varied in the range of 0 – 7 %. Aspia variety was not affected by this disease. Potato varieties of Gloria varieties were the most affected by 7 %. Tubers of the Nevsky variety were affected by 6.0%, and tubers of the varieties Crystel, Sandrin and Serafina – by 3 %.

Keeping the quality of potatoes largely depends on the affection of tubers diseases [14].

In our studies, all varieties were characterized by high keeping quality of tubers which varied within 91 – 95 %. The maximum keeping quality – 95 % was characterized by the Aspia variety. The keeping quality was 94 % in the Nevsky and Crystel varieties, in the Gloria varieties – 93 %, Sandrine – 92 % and Serafina – 91 %.

Potatoes serve as a valuable raw material for various products – chips, potato grits, flakes, mashed potatoes, frozen semi-finished products, as well as for the production of starch, alcohol (one tonpotatoes contain 17.5 % starch content from which you can get up to 175 kg of pure starch or 112 liters of alcohol). Potato is also important as a forage crop [15].

The quality and yield of potatoes vary dramatically depending on climate and soil. Potatoes were grown on light sandy, but fertile soils in central Russia has excellent taste, large and clean tubers, mature starch grains, well-boiled.

With a lack of moisture (southern semi-deserts and steppes) or in areas with excessive moisture (Kola Peninsula) potatoes are tasteless, soap when cooked, suitable only for livestock feed and technical purposes.

Potatoes grown in the Northern regions have a lower starch content than in the Central and southern regions. In the conditions of our country in the transition from the middle latitudes (50-57°) to the North (57-67°) starch decreases by 0.5% for each degree of latitude. In the transition from the middle latitudes to the southern starch often increases, but is constrained by a lack of moisture and the spread of viral diseases. Sugar content increases from South to North [16, 17, 18].

Starch - the main carbohydrate, it is about 20 % crude and 80% dry weight. Distributed unevenly in tubers, at the apex, where there are more eyes, its content is 2-3 % less than at the base. In tubers of medium size (60-100 g), it is more than in large ones. The starch content is directly correlated with the dry matter content and also depends on the maturity of tubers, variety and weather [19].

Starch is deposited in tubers in the form of starch grains with a layered structure of 20-60 microns. The culinary qualities of potatoes depend on the sizes of starch grains; the larger they are, the higher the solubility and viscosity of starch, the variability of tubers.

The starch content determines the nutritional value and digestibility of potato tubers. Depending on the starch content, potato varieties with the low starch content (12 – 15 %), medium (16 – 20 %) and high (over 20%) are distinguished.

All varieties in our studies, all varieties of potatoes belonged to varieties with low starch content (table 4). The starch content in them ranged from 12.4 % in the Crystel variety to 14.7 % in the Sandrine variety. In other varieties, tuber starch content varied between 13.4 – 13.7 %.

Table 4. Quality of potatoes according to variety on average for considered period.

Variety	Starch content, %	Degustation evaluation, point	Overall assessment, point
Nevsky (st)	13.7	4.3	3.0
Gloria	14.6	4.0	3.0
Crystel	12.4	5.0	4.0
Sandrine	14.7	4.7	4.0
Serafina	13.6	4.7	3.0
Aspia	13.4	4.7	3.0
Average	13.7	4.6	3.3

Boiled tubers are evaluated by appearance, digestibility, consistency, taste, resistance to darkening [20]. Each indicator and overall assessment is characterized by appropriate quality levels, which are expressed in points.

On the basis of a set of individual indicators, the culinary qualities of potatoes as a whole are estimated as follows in table 5 [21].

Table 5. Culinary qualities of potatoes.

List	Description	Score
excellent	the appearance of boiled potatoes is excellent, the tubers are quite boiled, the consistency is mealy-related, the taste is pleasant, the flesh is resistant to darkening	5
good	good appearance, tubers moderately boiled or decayed to varying degrees, the consistency of mealy-bound or crumbly, the taste is characteristic, it is permissible weak darkening	4
satisfactory	appearance, good or mediocre, the potatoes slightly overcooked, the texture is dense or densely-bland, mediocre taste, acceptable explicit darkening of flour	3
bad and very bad	whole tubers, faded, consistency and taste with signs of watery and soapy, the flesh are very darkening	2-1

Of all the varieties we studied, only the Crystel variety received excellent marks, the other varieties received good marks.

The overall assessment of the varieties as a whole was also quite high. It had different varieties in the range of 3.0 – 4.0 points. Crystel and Sandrine varieties received the maximum overall rating of 4.0 points. Other varieties were estimated at 3.0 points.

4. Conclusion

Crystel variety forming the maximum yield of 41.9 t / and an increase to the standard of 9.2 t / ha, with a maximum mass of tuber 96 g and yield of commercial tubers 81 %, a slight defeat by scab and rhizoctoniosis, breeze 94.0 %, tasting assessment 5.0 points and an overall rating of 4.0 points is recommended for cultivation on highly cultivated sod-podzolic light loamy soils on the left bank of Nizhny Novgorod region.

References

[1] Alekseev V A 2003 The method of tillage, fertilizer and crop *Potatoes Veget.* **2** p 10
[2] Gusev M I 1994 About removal of nutrients from the soil by high yields of potatoes *Chemization* **11** p 57
[3] Zakharov V N 1991 Diagnosis of mineral nutrition *Potatoes Veget.* **1** pp 9–13
[4] Kosyanchuk V P 2000 Potato production in the Bryansk region - an integrated approach *Potatoes Veget.* **1** p 10
[5] Usanova Z I 2004 The productivity of potato varieties in the conditions of the upper Volga region *Achievements of agricultural science – development of the agro-industrial complex. Collection of scientific papers* (Tver: TGSH) pp 36–40
[6] Devyatkina L N 2018 Potato production: global and national discourse *Bullet. NGIEI* **5 (84)** pp 122–34
[7] Ivenin V V, Ivenin A V, Sakov A P, Bogomolov V N and Novosadov A A 2017 Optimization of the Dutch system of tillage for potatoes (Nizhny Novgorod: FSBEI of the Nizhny Novgorod State Agricultural Academy)
[8] Smirnov N A 2018 Strengthening of food independence of the region on the basis of increase of efficiency of branch potato *Bullet. NGIEI* **1(80)** pp 111–20
[9] Zaikin V P, Lisin A Yu, Rumyantsev F, Shamin A E and Borisova E E 2015 *Efficiency of green fertilizers in crop rotations on loamy soils of the Volga-Vyatka region (*Monograph. Knyaginino)

IOP Conf. Series: Earth and Environmental Science **346** (2019) 012028 doi:10.1088/1755-1315/346/1/012028

[10] Zaikin V P 2003 *Scientific bases of systems of agriculture of the Volga – Vyatka region* (N. Novgorod: study guide NGSKHA) p 288
[11] Savel'ev V A 2017 *Potato* (Monograph. Publ: Lan')
[12] Lorch A G 1948 *Dynamics of accumulation of potato crops* (Moscow: Selkhozgiz) p 191
[13] Maltsev V F 2002 *System of biologization of agriculture of non-Chernozem zone of Russia* vol 2 (Moscow: FGNU Rosinformagrotekh) p 574
[14] Pshechenkov K A 2007 *Storage technology of potato* (Publishing House "Potato") p 192
[15] Batsanova N S 1970 *Potato* (Moscow: Kolos) p 375
[16] The research reports. Stage 17.03.06.01 2005 *To develop zonal, resource-and energy-efficient, environmentally safe, cost-effective potato production technologies to ensure high sustainable yields, high-quality products* (Moscow) p 94
[17] Tabakov A G 2013 Potato yield, depending on the agrotechnical methods of cultivation ed Tabakov A G, Samarkina M A and Shashkarov L G *Bulletin of Kazan State Agrarian University* **T8 4(30)** pp143–145
[18] Efimov V M 2009 Yield of potatoes depending on mineral fertilizers, tripoli and its mixtures with mineral fertilizers ed Efimov V M and Shashkarov L G *Bulletin of the Kazan State Agrarian University* **T4 3(13)** pp 100–101
[19] Sokol P F 1963*Storage of potatoes and vegetables* (Moscow: Publishing house of agricultural literature, magazines and posters) p 256
[20] Davydenkova O N 2004 *Influence of growing and storage conditions of different potato varieties on consumer quality and processed products* Diss. of Cand.ofAgr. Sciences (Moscow) p 159
[21] *The method of state variety testing of crops (potatoes, vegetables and melons)* 2015 (Moscow: FGBU "Gossortkomissiya") p 61

IOP Conf. Series: Earth and Environmental Science **346** (2019) 012029 doi:10.1088/1755-1315/346/1/012029

Digital transformation of agricultural industry

A Shamin[1], O Frolova[1], V Makarychev[1], N Yashkova[1], L Kornilova[2] and A Akimov[2, 3]

[1]Nizhny Novgorod State Engineering and Economic University, Knyaginino, Russia
[2]Chuvash State Agricultural Academy, Knyaginino, Russia

E-mail: akimov_mechfak@mail.ru

Abstract. Introduction: Until recently, in our country, agriculture as a sector of the economy had low automation of business processes. At the present stage, digitalization and automation of the maximum number of agricultural processes is a conscious necessity. The task of it is the maximum automation of all stages of the production cycle to reduce losses, increase the productivity of agricultural business, optimal resource management. **Materials and methods:** This article use abstract logic and calculation and constructive methods. The historical experience of the formation of a multi-structure is considered, the digital stage is allocated. The system of comprehensive indicators of efficiency of agricultural production of various forms of management by types of efficiency is offered. **Discussion:** At the first stage agricultural producers on the basis of developed indicators of digital efficiency determine the level of digitalization for its further increase. At the second stage of implementation of elements of digitalization it is necessary for agricultural organizations to participate in the developed business model for the introduction of digital technologies, which will contribute to the development of agricultural production, growth of entrepreneurial activity and increase the sustainability of rural areas. **Results:** Based on the analysis of economic literature, foreign digital technologies used in advanced agricultural organizations of various forms of ownership in Russia are highlighted. It is concluded that the intensive introduction of digitalization and the Internet of things in agriculture will turn the industry less affected by it into a high-tech business due to the explosive growth of productivity and reduce unproductive costs and ensure food security of the regions and the country as a whole.

1. Introduction

We live in a special era when new generation technologies are being introduced in all spheres of human life. The widespread use of big data, robotics, virtual and augmented reality, the Internet of things, adaptive systems, bio, and Neurotechnology have long ceased to be innovative in the classical sense but rather moved into the category of those phenomena that we must learn to manage competently.

Digitalization gives hope for the continuation of the process of human social development, which has already led to a sharp improvement in the quality of life of a billion people.

To achieve this, multi-stakeholder collaboration is needed to achieve the three main objectives:

- distribute the benefits of technological breakthroughs fairly;
- to keep them from the inevitable negative effects;
- to ensure that new technologies will expand, not limit, the capabilities of all the inhabitants of the earth.

The technologies that are at the heart of digitalization are largely interrelated. Promising technologies should be considered neither as simple tools that are completely under our conscious control nor

Content from this work may be used under the terms of the Creative Commons Attribution 3.0 licence. Any further distribution of this work must maintain attribution to the author(s) and the title of the work, journal citation and DOI.
Published under licence by IOP Publishing Ltd

as external forces that cannot be controlled. Instead, we should try to understand: how and where human values are embedded in new technologies and how technologies can be used for the common good, the protection of the environment and human rights.

All interest groups should engage in a global dialogue on how technology is changing the systems around us and affecting the lives of every human being on the planet. In particular, discussions related to the management of advanced technologies and their role should be more broadly presented.

On the basis of the analysis of economic literature [1-9] the basic conditions for the development of the agricultural sector in the near future are identified:

- after 25 years, agriculture on the gross output index exceeded the level of 1991;
- after 25 years, agriculture on the livestock production index reached about 80% of the 1991 level;
- the average long-term level of grain production is 100 million tons – the level of the 80s;
- beef and milk production remain stagnant;
- opportunities for tangible growth of state support are exhausted;
- depending on the import of breeding material and seeds, technologies, equipment and machinery, finished products, including plant protection products, feed additives, etc.
- opportunities for expansion of agricultural exports are significantly limited;
- digitalization of all business processes as one of the elements of improving the efficiency of agricultural production.

2. Materials and methods

There is no consensus in the economic literature on the definition of specific periods and stages of formation of diversity in Russia, a rational combination of forms of management to improve the efficiency of agricultural production and the organization as a whole. The multi-layered structure of the Russian economy in the process of its historical development was characterized by alternating movement from simplification to complication and vice versa. Studying the evolution of the formation, development, and extinction of structures, the authors in their works previously identified 4 periods – pre-revolutionary, post-revolutionary, modern and new period of development of multiculturalism in the country (Table 1).

The pre-revolutionary period was represented only by small forms of management. In the course of structural transformations in the post-revolutionary period, a multi-structured agrarian structure was formed in agriculture, represented by various forms of management (agricultural organizations, peasant farms, households). The modern period is characterized by the formation of market relations and the emergence of a legislative framework that makes it possible to develop these forms of management. The new period is characterized by the strengthening of state support for all forms of management.

In 2006 and up to the present time (a new period of development) there was a strengthening of the role of the state in the development of small farms and agricultural organizations with the adoption of the national project "Development of agriculture" [10].

In 2017, both in agriculture and in the country's economy as a whole, the authors highlighted the digital stage. The potential for the introduction of advanced information technologies in domestic agriculture is huge. And the prospect for the development of the digital economy, including agriculture, is one of the topical topics for discussion both in the scientific sphere and at the state level. By Decree of the President of the Russian Federation No. 203 of May 9, 2017 "On the strategy of the development of the information society in the Russian Federation" [11] the main state priorities in this area were identified. In July 2017, the program "Digital economy in the Russian Federation" [12] was approved, the implementation of which should ensure the introduction of digital technologies in the daily life of society and this is especially important for agricultural residents. In the Strategy of scientific and technological development of the Russian Federation, approved by presidential decree of December 1, 2016 No. 642, stated that it is necessary the development of technologies, which, among other things, provide a transition to a highly productive and environmentally friendly agriculture, the development and implementation of systems for the rational use of chemical and biological protection of plants and animals [13].

According to the authors, the following structures function on the modern, new and digital periods:

- family consumer lifestyle;
- family business;
- private;
- the public way;
- the cooperative way;
- social order.

Table 1. Periodization of the formation and development of multi-structure in agricultural production in Russia

No.	The name of the period	Period	Ways
1	Prerevolutionary	from ancient times to 1917	the individual peasant, the state, landlords, community-based, tribal, capitalist, artisanal, craft
2	Post-revolutionary	1917-1990	state, co-operative, individual labor
3	Modern	1990-2006	family consumer, family business, private, state, cooperative, public
4	New	2006 – 2017	family consumer, family business, private, state, cooperative, public
5	Digital	2017 - present time	family consumer, family business, private, state, cooperative, public

Taking into account the different opinions [14,15,16, etc.] of the concept of efficiency and taking into account the allocation of the digital stage, the authors propose a system of indicators that allows a comprehensive description of the efficiency of agricultural production in different forms of management and proposes to highlight another type of efficiency – digital. Performance indicators are arranged by type of efficiency. First, the most significant types of efficiency and their indicators are presented (Table 2).

Table 2. Types and indicators of agricultural production efficiency of various forms of management.

Order of significance	Types of efficiency	Indicators of the efficiency of forms of management		
		Agricultural organizations	Peasant (farmer) economy	Households (HN)
1	Innovative	- the coefficient of commercial innovation efficiency - the coefficient of budget innovation efficiency - the coefficient of social innovation efficiency		
2	Digital	% automation of business processes in an organization, % -% of funds aimed at automating business processes in an organization (of the total amount of funds received) % -% annual update of licensed programs; - the share of costs allocated to training (retraining) employees to work with software products; - the proportion of employees who use software products for the computerization of production processes; - the proportion of internal workflow conducted through automation		

IOP Conf. Series: Earth and Environmental Science **346** (2019) 012029 doi:10.1088/1755-1315/346/1/012029

3	Budget-ary	*Economic* - budget subsidies attributable to the results of economic and financial activities per 1 ha of agricultural land. land, rub. - profit before tax on all activities on 1 hectare of agricultural land (including subsidies from the budget, without subsidies from the budget), RUB. - net profit of the organization on 1 hectare of agricultural land, RUB. - profit before tax for all activities, including subsidies from budgets, RUB. - profit before tax for all activities without subsidies from the budget, RUB. - the net financial result (profit minus loss of the organization), RUB. - increase in revenue received from the area (from livestock), processed (grown) by state support (acquisition of fuels and lubricants, fertilizers, leasing equipment, etc.), RUB. - savings in production costs by increasing material costs from the area (with livestock), processed (grown) at the expense of state support (acquisition of fuel, fertilizers, feed, leasing equipment, etc.), RUB. *Social* - subsidies for agricultural products per 1 agricultural worker, rubles. - the number of attracted workers in the agriculture sector, in which products were subsidized, pers. - the share of wage costs in total subsidies, %	
4	Techno-logical	*Natural* Production of main products (potatoes, vegetables, milk, meat in live weight, etc.) per 100 hectares of arable land (agricultural land), c *Monetary* Land allocation, RUB/ha Labor intensity, RUB/RUB Capital productivity, RUB/RUB Material consumption, RUB/RUB	*Natural* Production of main types of products (potatoes, vegetables, milk, meat in live weight, etc.) per 1 hundred square meters of agricultural land, kg. *Monetary* Earth productivity, rub.
5	Eco-nomic	*Natural* The productivity of C/ha Productivity, C *Monetary* Production profitability, % Profitability of sales, % Gross output per 1 RUB. costs, RUB. Gross income for 1 RUB costs, RUB. Net income per 1 RUB of expenses, RUB	*Natural* Productivity kg/hund Productivity, kg *Monetary* Gross output per 1 RUB. costs, RUB.

6	Social	*Natural* The birth rate in rural areas, % Level of availability of comfortable housing, sq. m Provision of children with places in kindergartens and educational institutions, places. Provision of health facilities in rural municipal areas., % % of rural population morbidity, % A number of students of universities, colleges of the agrarian profile, people. Unemployment rate, % Level of mechanization, automation of production processes, % The coefficient of the population in rural areas Average life expectancy in rural areas; The coefficient of aging of the rural population; *Monetary* Consumption Fund per employee, rub. Provision of medical, educational, commercial, cultural and living conditions for 1 person, rub. The level of paid services for 1 person, RUB. Standard of living in rural areas (average per capita income) The ratio of average wages in rural areas to the minimum subsistence level The ratio of average wages in rural areas to average wages in the region	*Natural* Level of availability of comfortable housing, sq. m Provision of children with places in kindergartens and educational institutions, places. Level of transport accessibility, seats for 1 person. Availability of paved roads, km *Monetary* Income level per 1 family member, RUB. Provision of medical, educational, commercial, cultural and living conditions for 1 person, rub. The level of paid services for 1 person, RUB.
7	Environmental	*Natural* Number of organic agricultural products obtained, C The ratio of the obtained organic products to the total volume of products (by types of products), % The level of environmental friendliness of the territory on emissions of pollutants into the atmosphere. The level of losses of agricultural products from man-made pollution. Earth preservation (to preserve the acreage) The volume of genetically modified raw materials used for food production in the region; *Monetary* The ecological and economic effect on 1 RUB. costs, adjusted for the amount of ecological and economic damage, RUB.	

These indicators can be used to assess the activities of various forms of management for specific types of efficiency, including digital. The introduction of new information technologies in agriculture will increase the efficiency and manageability of information flows, so this type of efficiency is in second place.

Calculation of these indicators by types of efficiency and by forms of management, their analysis and comparison will allow drawing conclusions, to determine trends in the development of agricultural producers.

3. Discussion

Digital tools and technologies help to address information asymmetries and provide users with more free and faster access to information. On the basis of the conducted research, the typology of digital technologies for agriculture is offered. It includes four groups:

1. Agricultural knowledge and information systems:

-collection of information among various forms of management;

-information management;

-providing relevant information to various forms of management.

2. Tools and services that provide market access and market opportunities.

3. Automation of business processes in various forms of management.

4. Automation of the process of ensuring food security in the region

Currently, in agriculture, the main factors of production are data in digital form. It is necessary that all information about the ongoing production processes should be collected in a single array of data, the analysis of which can take appropriate measures and significantly improve production efficiency. The question of creating a single information space of agriculture, and, accordingly, agricultural science, is still open [17 p. 282].

Currently, in Russia, support for the market of advanced agricultural technologies takes place at the Federal level. In the framework of projects such as the Internet Initiatives Development Fund (IIDF), the Skolkovo Foundation and the Skolkovo LLC, venture investments as well as government non-profit organizations, including the Innovation Promotion Foundation. Such assistance includes the establishment of incubators and programmes to attract mentors and industry experts to work with high-tech entrepreneurs on their products and business plans; the construction of office and laboratory infrastructure to promote business development and product development; the provision of tax incentives and access to a wide range of sources of funding through the organization of demonstration events and competitions of enterprises.

The rapid development of new technologies for the collection, management, transmission and use of information in digital format has led worldwide to a radical transformation of methods of agriculture and the functioning of the agro-industrial complex of the Nizhny Novgorod region. These changes come to the fore in agricultural holdings, where advanced digital technologies are used. However, many farmers in the region, especially small and medium-sized ones, do not even have permanent access to the Internet and do not have the skills to take advantage of digital technologies. The low level of demand for technology leads to missed opportunities and crop losses in the region. The authors propose in the Nizhny Novgorod region to create a model of interaction between the government, businessmen and the scientific community for the introduction of digital technologies in the regional agro-industrial complex and work in a single digital space. These models have already been created and are actively working in many regions of Russia (for example, Perm region).

Agricultural business community makes an order for the development of projects in the field of digitization the scientific community and IT companies, they are developing these projects, consider the effectiveness of the implementation based on the basis of expert selection (bodies of the state and the legislature jointly with the scientific community and the IT community) these projects are recommended for implementation.

This business model will allow to combine the competence of educational institutions in the region in the field of IT for more active development of this sphere of scientific and practical knowledge. The created structure will facilitate the interaction of participants and eliminate unnecessary competition. Together, members of business model will be able to reduce the time and material costs of innovative developments in the field of IT for the agro-industrial complex.

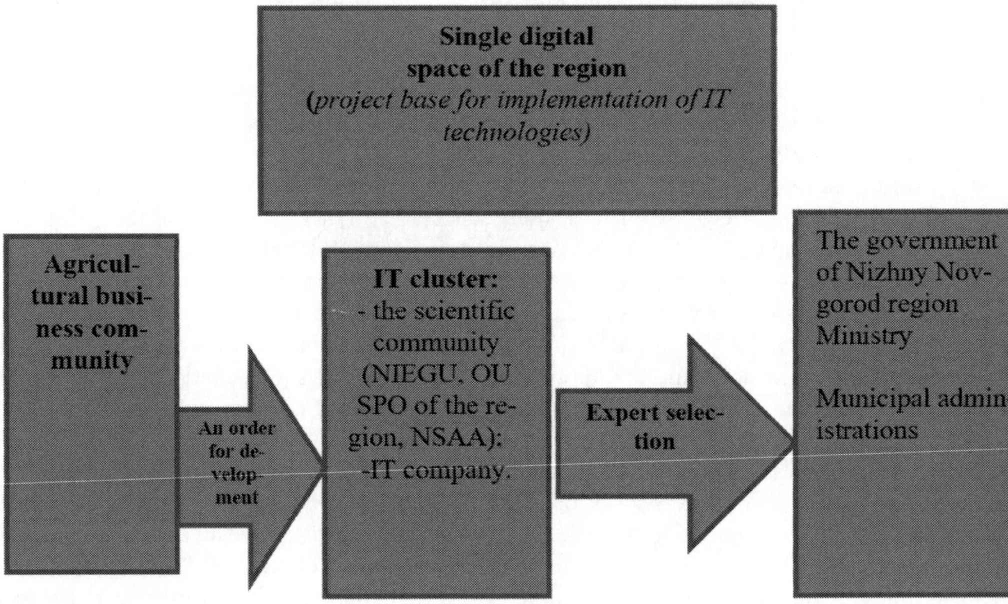

Figure 1. Business model for the introduction of digital technologies in Nizhny Novgorod region.

4. Results

The authors of this article based on the analysis of economic literature [17,18,19,20] identified foreign digital technologies used in advanced agricultural organizations of various forms of ownership in Russia:

In plant growing:

- **No-till** - zero tillage;
- **Strip-till** is a gentle tillage technology;
- precision farming;
- global positioning technology **(GPS);**
- geographic information systems **(GIS);**
- productivity assessment technologies **(YieldMonitorTechnologies)**;
- technology variable rate **(VariableRateTechnology);**
- Earth remote sensing technology.

In animal husbandry:

- renewal of breeding stock and improvement of genetic potential;
- robotic milking;
- automatic milking system (milking parlor);
- automatic animal feeding stations;
- computer control of the herd;
- automated monitoring of herd reproduction processes

The advantages of the main digital technologies were evaluated and determined by agricultural producers of Russia.

Nowadays global positioning systems provide a significant improvement in the accuracy of most agricultural operations performed by mobile units without the use of technological gauge or other devices in agricultural organizations.

The main advantages achieved by the use of geographical information systems are the reduction of the costs of controlling the use of sown areas by conducting remote monitoring based on satellite images, including the determination of the boundaries and areas of fields occupied by winter and spring crops under steam, as well as the assessment of harvested crop areas; the identification of locus of grain crops destruction, effective damage assessment, etc.

Based on the above, the authors determined the vector of development of the Russian agro-industrial complex in the medium and long term -the formation of a new domestic "breakthrough" digital and technological base, including the need to perform:

1. To develop and implement an effective mechanism of commercialization of innovations both in the field of technologies and in the field of digitalization and their transfer to agro-industrial production.
2. Neutralize the problem of dependence on imports of breeding material and seeds, technologies, equipment and machinery, finished products, including plant protection products, feed additives, etc.
   What awaits us in the future, from the transition to a digital AP to - radically changes will occur:
   -production technology;
   -organizational structure;
   - competence of personnel;
   - the system of relations with suppliers;
   -model of activities;
   -employee payment mechanisms;
   -financial, management and tax accounting and much more.

The digital economy plays an important role in assessing the food security of the region. We believe that digital technologies will allow:

- optimizing the logistics of supply of raw materials;
- optimizing the composition of the raw material zone of the region according to predetermined criteria;
- calculating the required domestic volume of food production based on the population of the region and its gender composition;
- determining the necessary volume of production of raw materials in the region, as well as the need for resources for its production (acreage, livestock, availability of livestock facilities, agricultural machinery, and labor), based on the planned productivity.

In order to realize economic benefits from digital transformation in agriculture, Russia must continuously carry out research and development aimed at building a global architecture, designed for simultaneous operation of different industries ; to achieve great maturity of the market, including such aspects as the way of thinking of users, and pricing structure; as well as increase public partnership in its framework to create a sustainable business model by promoting broad use of digital technologies at all levels of the agricultural sector.

References

[1] Order of the Government of the Russian Federation of 02.02.2015 N 151-p (as amended on January 13, 2017) “On approval of the Strategy for Sustainable Development of Rural Territories of the Russian Federation for the Period up to 2030.” Art. 1014.
[2] Korskaya M A 2016 Development of the methodology of an economic study of the problems of agriculture. *New Sci.: Theoret. Practical View*. 1**1-1C**: 97-9.
[3] Fedorova O V and Krutikov V K 2016 New approaches to the effective solution of problems of management of innovative development of agro-industrial complex. *APK: Econom., Manag.* **11** 40-5
[4] Polulyah Y G, Adadimova L Y, Oydup T M and Chupikova S A 2016 Sustainable development as a single category of problems of agriculture and rural areas. In the collection: food security, import substitution and socio-economic problems of agriculture. Materials of The International Scientific and Practical Conference, 2016. 335-9.
[5] Edrenkina N M 2008 Solutions to the problems of development of small forms of management

in the framework of the priority national project "agribusiness Development" in Siberia. *Achievem. Sci. Tech. Agricult.* No. **4** 11-3.

[6] Ryzhenkova N E and Chepik D A 2014 Development of innovative potential of agroindustrial complex: problems, directions *Econ. Agricul. Russia.* **6** 53-7.

[7] Berezina N N and Rakitina I S 2009 Innovative development of agriculture and problems of its financing. In the collection: *Econ. Managem.: Problems, Trends Prospects*. 5-9.

[8] Frolova O A 2011 *Development of economic forms in a multi-layered agricultural economy*. Monograph. Knyagino: NGIEI. 260 p.

[9] Chernova A A 2015 Agribusiness development: problems and solutions *Strat. Sustain. Develop. Russian Regions*. **29** 49-50.

[10] Priority national project "Development of agro-industrial complex" of September 5, 2005, approved by the Presidium of the Council under the President of the Russian Federation for the implementation of priority national projects.

[11] Decree of the President of the Russian Federation of 09.05.2017 N 203 "On the Strategy of development of the information society in the Russian Federation for 2017 - 2030"//"Assembly of legislation of the Russian Federation", 15.05.2017, N 20, Art. 2901.

[12] Order of the Government of the Russian Federation of 28.07.2017 No. 1632-p "On Approval of the Digital Economy of the Russian Federation Program" // "Collected Legislation of the Russian Federation", 07.08.2017, N 32, Art. 5138.

[13] Presidential Decree of 01.12.2016 No. 642 No. A of the Strategy for the Scientific and Technological Development of the Russian Federation//"Collection of Legislation of the Russian Federation", 05.12.2016, N 49, Art. 6887.

[14] Alexandrov G A 1984 What is the efficiency of the economy. M.: Economics, 65 p.

[15] Altukhov A I and Kundius V A 2006 *The Russian market APK: current state and mechanisms of development*. Monograph, 396.

[16] Altukhov I A and Shamin A E 1996 New agricultural system of Russia. – Nizhny Novgorod. GIPP "Nizhpoligraf", 281.

[17] Stenkina M V 2018 Problems of information support of agricultural and economic research in the digital economy. Agricultural Economics: state, tasks for the future. -M.: VIAPI named after A.A. Nikonov, 281-3.

[18] Larina T N and Zavodchikov N D 2018 Potential and prospects of development of "digital" agriculture in Russia. Agricultural Economics: state, tasks for the future. -M.: VIAPI named after A.A. Nikonov, 283-5.

[19] Balabanov V I, Tsvetkov I V and Zhogin I M 2018 Methods of assessing the effectiveness of the introduction of information technologies in agriculture. Agricultural Economics: state, tasks for the future. -M.: VIAPI named after A.A. Nikonov, 283-5.

[20] Nielson D, Meng Y T, Buivolova A and Kopyan A Unlocking the potential of digital technologies in Russian agriculture and finding prospects for small farms. www/worldbank.org.

[21] Schwab K 2018 *Technologies of the Fourth industrial revolution*:[translated from English]/Klaus Schwab, Nicholas Davis.- Moscow: Eksmo, 320c.

IOP Conf. Series: Earth and Environmental Science **346** (2019) 012030 doi:10.1088/1755-1315/346/1/012030

Enzyme preparations in compound feed for ducklings

V S Sherne[1, 2], A Y Lavrentev[1], N V Evdokimov[1], N S Petrov[1], E Y Nemtseva[1] and N V Danilova[1]

[1]Federal State Budgetary Educational Institution of Higher Education "Chuvash State Agricultural Academy", K. Marx street, 29, Cheboksary, 428003, Russia

E-mail: v.sherne@mail.ru

Abstract. The paper presents the results of the enzyme preparations in various combinations influence on the dynamics of growth, development and slaughter quality of ducklings. Scientific and economic experiment was held in the Chuvash Republic. For the experiments, the control and 2 experimental groups of ducklings (50 animals each), were selected and formed. Ducklings of the 1st experimental group received enzyme preparations amilosubtilin in the amount of 0,05 kg/t and protosubtilin in the amount of 0,05 kg/t during the entire period of cultivation in combination with the feed. Ducklings of the 2nd experimental group of feed got enzyme preparations amilosubtilin in the amount of 0,1 kg/t and celloviridin in the amount of 0,075 kg/t. It was established that ducklings, who received enzyme preparations in addition to animal feed had increased growth energy throughout the experiment. By the end of the experiment, the live weight of the ducklings control group was - 3,424 kg, the first experimental group, who received enzyme preparations amylosubtilin in the amount of 0,05 kg/t and protosubtilin in the amount of 0,05 kg / t in addition to feed – 3,712 kg. In the second experimental group, live weight by the end of the experiment was - 3,804 kg, where ducklings received enzyme preparations amylosubtilin in the amount of 0,1 kg/t and celloviridine in the amount of 0,075 kg/t in addition to feed. The average daily gain in live weight of animals in the control group was 53,52 g, and in the first experimental group – 58,10 g, in the second – 59,56 g. The inclusion of a mixture of enzyme preparations in animal feed in experimental animals contributed to an increase in body length compared with the control by an average of 0,9 cm and 1,6 cm, the tarsus length by 2,5 cm and 4,3 cm, and the chest girth by 1,8 cm and 2,5 cm, respectively. Based on the measurements, body indexes were calculated and had similar values, but with a slight increase in the experimental groups: in terms of massiveness - by 3,16 % in the first and 7,12 % - in the second, in the wide-body index - by 3,84 % in the first and 4,54 % - in the second and by erisomia, respectively - by 2,51 % and 2,66 % compared to the control. Increasing the productivity of ducklings during the cultivation has affected the reduction of feed costs and the unit cost of production, which contributed to an increase in profits in the experimental groups. Thus, the cost of 1 kg of gain in the second experimental group compared with the control group decreased by 7.9 rubles. The profit per 1 kg of gain in the second experimental group compared with the control was more by 7.9 rubles or 22,6%. Compared with the first experimental group at 2.88 rubles, or by 8,3 %.

Content from this work may be used under the terms of the Creative Commons Attribution 3.0 licence. Any further distribution of this work must maintain attribution to the author(s) and the title of the work, journal citation and DOI.
Published under licence by IOP Publishing Ltd

1. Introduction

Poultry farming is one of the most important branches of agriculture, designed to provide the population with dietary food: eggs and poultry meat, characterized by a high content of animal protein with low caloric content. Farmers and private farms can make a great contribution to the increase in poultry meat production. Depending on the available conditions in such farms, it is possible to contain from several birds to several hundred and even thousands. Birds, when feeding compound feed, give 3-5 times more gain in live weight per unit of consumed feed than other farm animals.

The main goal in the poultry industry at the present stage is the efficient production of the industry, with the lowest cost and high quality. This goal can be achieved by increasing the productivity of birds, which is now extremely important [1, 2, 3].

In the poultry industry, one of the early maturity sectors is duck breeding, which in recent years, like other branches of the poultry industry, have been gradually increasing production. Duck meat in its chemical composition is characterized by a high content of protein, mineral elements, and vitamins. Due to the comparative simplicity of ducks to the conditions of detention and feeding, it is easy to breed them in any climatic zone of the country. Production of duck meat is recognized as one of the promising areas. The current level of duck breeding in Russia and countries with market economies is characterized by a continuous process of concentration of production, deepening its specialization, improving the technology of keeping and feeding conditions of poultry, as well as mechanization and automation of labor-intensive processes. The issues of improving the technology of growing and maintaining poultry, rational use of feed, saving material and labor resources are also relevant [4, 5, 6].

In order to reduce the cost of its production, the poultry farmers are set the task of increasing the body weight at the end of cultivation with a relative reduction in their terms.

It is necessary to include components of local production to reduce the cost of production throughout the composition of animal feed, such as grain cereals, bagasse, and meal pellets and many others. However, due to the significant content of fiber and non-starchy polysaccharide they have low nutritional value and poor digestibility. Studies by a number of authors show that this problem is solved by incorporating biologically active substances (BAS) into the compound feed, and enzymes play a large role in this, which contributes to an increase in the efficiency of nutrient use of feed [7, 8].

In poultry farming, as in other livestock industries, there is some experience in the use of enzymes, but a number of issues on their use in duck breeding are currently insufficiently studied. Therefore, the need arises to study them to increase the gain in live weight and meat productivity, and it is an actual problem of modern zoo-technical and practice [9, 10].

The purpose of this work is to study the effect of the use of a mixture of enzyme preparations of domestic production in feed for ducklings on their productive qualities.

2. Material and methods

The experimental part of the work was carried out on true-breeding young ducks "Agidel" farm. The material was normally developed healthy ducklings. Three groups of young ducklings were formed for experiments, on the principle of groups of analogs.

The age of the ducklings was 1 day. The duration of the experiment was 60 days. The birds of the control group from day 1 to day 20 received compound feed PC 21-2, from 21 to 56 days – PC 22-2 and from 57 to 63 days – PC-23-1. The formula and nutritional value of feed for ducks are presented in Table 1.

The parameters of the microclimate in the poultry house during the scientific and economic experiments met the established zoohygienic standards. The experimental bird was fed with loose full-feed, twice a day (morning and evening).

Monitoring of productivity was carried out by studying live weight and average daily gains, by individual weighing them at the beginning, at the end and during the experiment at every certain interval.

In order to better understand the proportionality of physique, mutual development relative to each other different parts of the body, typical birds use the method of analysis and comparison of body indices, which are the ratio of one measurement to anatomically associated with it another measurement, expressed as a percentage.

Biometric processing of digital data obtained in the studies was carried out using the variational-statistical method described by N. A. Plokhinsky.

The ducklings of the 1st experimental group received amylosubtilin enzymes in the amount of 0,05 kg/t and protosubtilin in the amount of 0,05 kg/t during the entire growing period along with the compound feed. For the 2nd experimental group, the feed was enriched with amylosubtilin enzymes in the amount of 0,1 kg/t and celloviridin in the amount of 0,075 kg/t.

Table 1. Composition and nutritional value of feed for ducklings by age, in %.

Components	Ducklings age in weeks		
	1-3 (PC-21-2)	4-8 (PC-22-2)	9-22 (PC-23-1)
Corn	15,0	40,8	20,5
Wheat	45,0	30,0	15,0
Barley	17,45	9,50	25,0
Oat	-	-	4,0
Pea	-	-	3,0
Wheat bran	-	-	15,0
Meal pellets	7,0	5,0	3,6
Yeast, hydrolytic	3,0	3,0	2,0
Fish meal	7,0	5,0	1,0
Bone tankage	-	2,0	2,0
Herbal fertilizer	4,0	3,0	5,0
Alpha phosphate	-	-	0,8
Chalk	1,4	1,5	2,6
Table salt	0,15	0,2	0,5
To 1 ton of feed, the following is added, g:			
Lysine	1200	-	250
Meteonin	500	400	800
Antioxidants	150	150	150
Antibiotics	20	-	-
100 g of feed contains, %:			
Metabolizable energy (MJ)	1,197	1,241	1,240
Crude protein	18,09	16,47	14,62
Crude fiber	4,6	3,8	6.0
Calcium	1,17	1,16	1,44
Phosphorus	0,84	0,76	0,78
Sodium	0,39	0,35	0,36
Lysine without additives (mg)	888,7	776,8	628,7
Methoinine+cystine without supplements (mg)	685,0	567,9	450,7

Amylosubtilin G3h contained amylolytic enzymes (activity 600 u/g) and a small amount of proteolytic ones. The overall effect was associated with the combined effect of all its constituent enzymes, including beta-glucanase, xylanase and cellulase, catalyzing the cleavage of hard-to-digest polysaccharides of barley, wheat and rye.

Protosubtilin G3h (with proteolytic activity of 70 u/g) was characterized by proteolytic action, which provide neutral and alkaline proteases against proteins.

Celloviridin (cellolux) is a complex of cellulases (2000±200 u/g), xylanase (up to 8000 u/g) and glucanase (up to 1500 u/g). This drug catalyzes the breakdown of cellulose, xylans, beta-glucans of plant cells to readily available sugars.

3. Research results

Accounting for the consumption of specified feeds and their residues, the research showed that during the experimental period, the experimental ducks did not have a difference in the amount of feed eaten, they willingly ate the specified feed.

During the entire growing period, the safety of ducklings in the control group was 94,0 %, in the I experimental group — 96,0 %, and in the II experimental group — 98 %.

The dynamics of changes in live weight and average daily gains are shown in Figures 1 and 2. The live weight of the ducklings when fed up was almost the same and ranged from around 52 g. The highest gain in live weight in all study groups was observed at the 6th week of the experiment. In the control group, an average head gain of 680 g was obtained during this period. In 1 experimental group - 706 g, which is 103,8 % of the control group. In 2 experimental group, 714 g of gain for the same period, that in relation to control group makes 105% and respectively 101,1 % in comparison with 1 experimental group is received.

Upon the completion of the experiment, the live weight of ducks in the control group was 3,424±4,6 kg, the first experimental group was 3,712±4,9 kg, and the second experimental group was 3,804±4,9 kg (P<0,05). At the end of the growing, the live weight of the ducklings of the experimental groups was higher compared to the control by 8,4 and 11,27%.

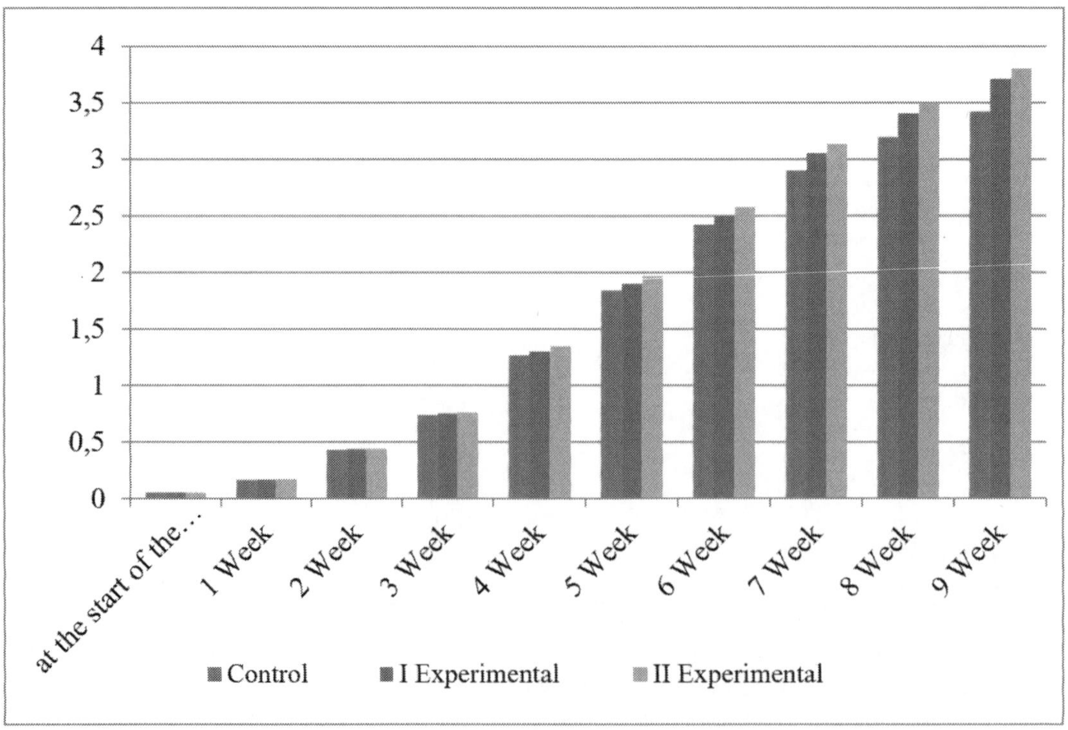

Figure 1. Dynamics of live weight changes of ducklings in groups, kg (P<0,05).

For the first week of farming (3,5 weeks of age), all groups of experimental birds showed a sharp increase in average daily weight gain, the next 2 weeks were more stable (a slight increase from 3,5 to 4,5 weeks and a slight decrease from 4,5 to 6, 5 weeks) (Figure 2).

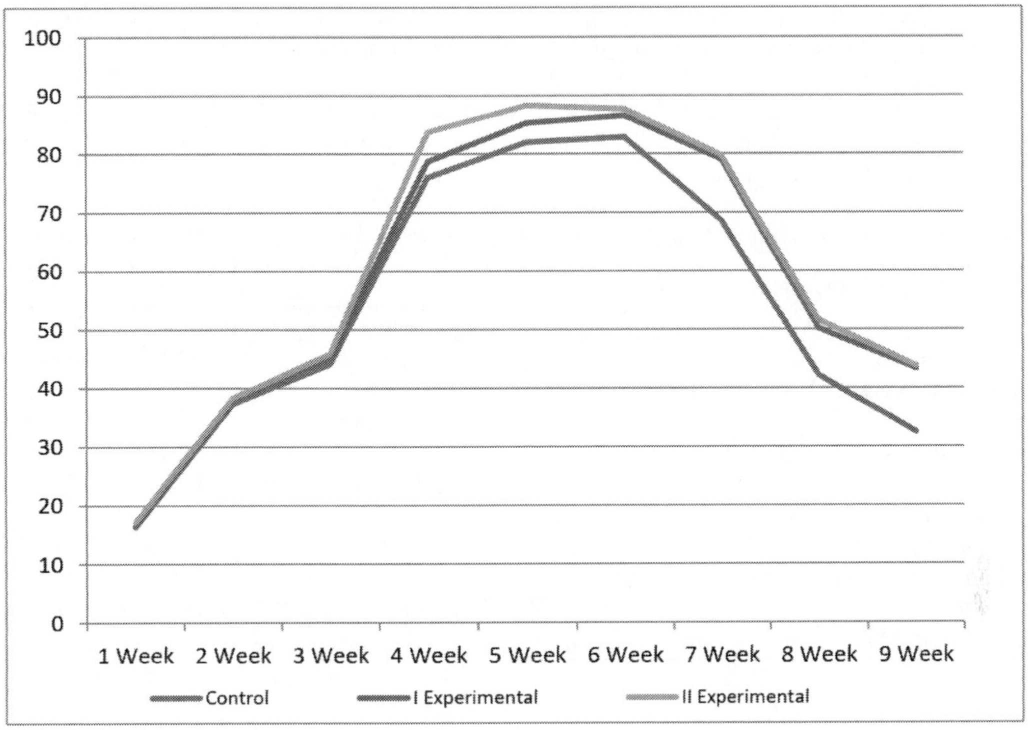

Figure 2. Dynamics of the average daily gain changes of ducklings in groups, g.

After that, there is a sharp decrease in productivity, which is clearly reflected in the diagram in Figure 2. In general, for the growing period, the average daily gain in live weight of animals in the control group was 53,52±4,6 g, and in the first test group 58,10±4,9 g, in the second 59,56±4,9 g ($P<0,05$).

In the scientific and economic experience, the body parameters development at the age of 63 days in the studied groups corresponded to the cross standard and were better in the ducks of the second experimental group with respect to control and the first experimental group. The results of exterior measurements of ducklings are presented in Table 2.

Table 2. Exterior measurements of ducklings (on average per 1 head in groups), cm.

Indicators	Group		
	Control	I experimental	II experimental
Body length	43,4±0,61	44,3±0,49**	45,0±0,52**
The width of the pelvis	9,7±0,32	11,6±0,44	12,1±0,50
Length of the keel bone	16,8±0,29	17,4±0,4**	17,7±0,44**
Tarsus length	11,9±0,21	14,4±0,33	16,2±0,34
Leg length	17,4±0,33	20,3±0,41	21,9±0,38
Chest Girth	38,8±0,51	40,6±0,51	41,3±0,49

at $**P \leq 0,01$

The inclusion of a mixture of enzyme preparations in animal feed in experimental animals contributed to an increase in body length, compared with the control by an average of 0,9 cm in the 1st experimental group and 1,6 cm in the 2nd experimental group, the length of the tarsus by 2,5 cm and 4,3 cm, chest girth 1,8 cm and 2,5 cm, respectively. A similar pattern occurred in the nature of

IOP Conf. Series: Earth and Environmental Science **346** (2019) 012030 doi:10.1088/1755-1315/346/1/012030

changes in the pelvis width and leg length of birds of the compared groups. Consequently, the introduction of enzyme preparations to animal feed of ducklings had positive impact on their growth and development.

Based on the measurements taken, body indexes were calculated. The results of the indexes of the constitution in ducks are presented in Table 3. Each of these indices allows us to estimate the particular productivity of a bird.

Table 3. Indexes of the constitution in ducklings (on average per 1 head in groups), %.

Indicators	Group		
	Control	I experimental	II experimental
Massiveness (The torso weight / Body length)	78,85±0,52	83,71±0,51*	84,47±0,53*
Breadth (The pelvis width x 100 / Body length)	22,35±0,33	26,19±0,31*	26,89±0,34*
Underpinning (Keel length x 100 / Body length)	38,70±0,42	39,28±0,43*	39,33±0,43*
Long-legged (The shank length, thigh (tibia) x 100)	68,39±0,55	70,93±0,54*	73,97±0,55*
Aresome (chest girth / body length x 100)	89,40±0,58	91,65±0,57*	91,78±0,56*

at *P≤0,01

The massiveness index characterizes compact build and fatness of the bird. Index of erysomomy - the development of the front of the torso. These indices give an idea of the poultry compactness, indirectly - the development of pectoral muscles in thickness, and had similar values, but with a slight increase in the experimental groups: in terms of massiveness - by 3,16 % in the first and 7,12 % - in the second, in the wide-body index - by 3,84 % in the first and 4,54% - in the second and by erisomia, respectively - by 2,51 % and 2,66 % compared to the control.

The research results of the ducklings slaughter qualities are given in Table 4. It was found that the use of a mixture of enzyme preparations in compound feed in experimental birds contributed to the increase in the killing qualities of ducks. Pre-slaughter live weight in the first experimental group was higher by 8,4 %, and in the second experimental group by 11,1 % than in the control group.

Table 4. Killing quality of ducklings, g.

Indicators	Group		
	Control	I experimental	II experimental
Pre-slaughter live weight	3424±4,6	3712±4,9*	3804±4,9*
The weight of the uneviscerated carcass	3111,0±4,4	3378,6±4,6*	3465,8±4,6*
Uneviscerated carcass yield, %	90,86	91,02	91,11
The weight of semi-eviscerated carcass	2968,9±4,3	3229,0±4,3*	3331,9±4,5**
Semi-uneviscerated carcass yield, %	86,71	87,26	87,59
The mass of the eviscerated carcass	2320,1±4,1	2671,5±4,2*	2751,4±4,3**
Eviscerated carcass yield, %	67,76	71,97	72,33
The yield of eviscerated carcasses by grade, %			
Grade 1	92	94	96
Grade 2	8	6	4

at * P≤0,05, ** P≤0,01

IOP Conf. Series: Earth and Environmental Science **346** (2019) 012030 doi:10.1088/1755-1315/346/1/012030

The mass of the uneviscerated carcass exceeded in the first experimental group, the amylosubtilin and protosubtilin enzymes with analogous indicators in the control by 267,6 g, i.e. by 8,6 %, in the second experimental group, treated with the enzymes amylosubtilin and celloviridin at 354,8 g, respectively, by 11,4 %. A similar pattern occurred in the nature of changes in the mass of viscerating carcass and the release of viscerating carcasses by grade.

As a result of carcass boning, it was found (table 5) that the absolute yield of muscle tissue of the carcass of ducklings of the 1st experimental group was higher than in the control group by 228,6 (114,7 %) and in the 2nd experimental group by 297,3 g (119,2%) (P<0,05-0,01), and the yield of skin with fat, respectively – by 35,0 and 45,3 g. the yield of edible parts in the second experimental group was higher than in the control group by 15,0 %, and in the first experimental group by 11.0 %. Output of non-edible parts (bones) in the control group made up 9,05 % in 1 experimental group of 9,31% and in the 2nd experimental group of 9,49%.

Table 5. The ratio of edible and inedible parts of the carcass.

Indicators	Group		
	Control	I experimental	II experimental
Edible parts:			
Muscles, g.	1547,6±14,7	1776,2±16,3*	1844,9±17,1**
%	45,20	47,85	48,50
skin with subcutaneous fat, g.	681,3±4,6	716,3±4,7	726,6±4,8
%	19,90	18,30	19,10
Kidneys,+fat+lungs	104,4±1.3	110,2±1,5	111,0±1,45
%	3,05	2,97	2,92
Total, g.	2335,2	2592,8	2685,6
%	68,20	69,85	70,60
Inedible parts:			
Bones, g.	309,8±3,2	345,6±3,4*	360,9±3,6**
%	9,05	9,31	9,49
Waste, g.	0	0	0
The ratio of edible parts to inedible parts	7,54	7,50	7,44
Muscle mass to bone mass ratio	4,99	5,14	5,11

* P<0,05, ** P<0,01

Due to the fact that the accounting of the given forages was carried out as a whole on group, instead of individually, and statistical processing of indicators on use of forages wasn't possible. Since the highest gains in live weight were observed in the first 6 weeks of cultivation, respectively, and in these periods of cultivation, the lowest feed costs per 1 kg of growth are observed (Figure 3). At the end of the growing period, when there was a decrease in the average daily growth, feed costs increase.

In total, over the period of the experiment, 9,835 kg of feed was spent in each group. In the first group, on 1 kg of gain, 2,91 kg of compound feed was spent in the control group, and in the first experimental group, 2,64 or 9,28 % less than in the control group and in the second experimental group, 2,62 kg or 9,97 % less than in the control group and by 2,26 % than in the first experimental group.

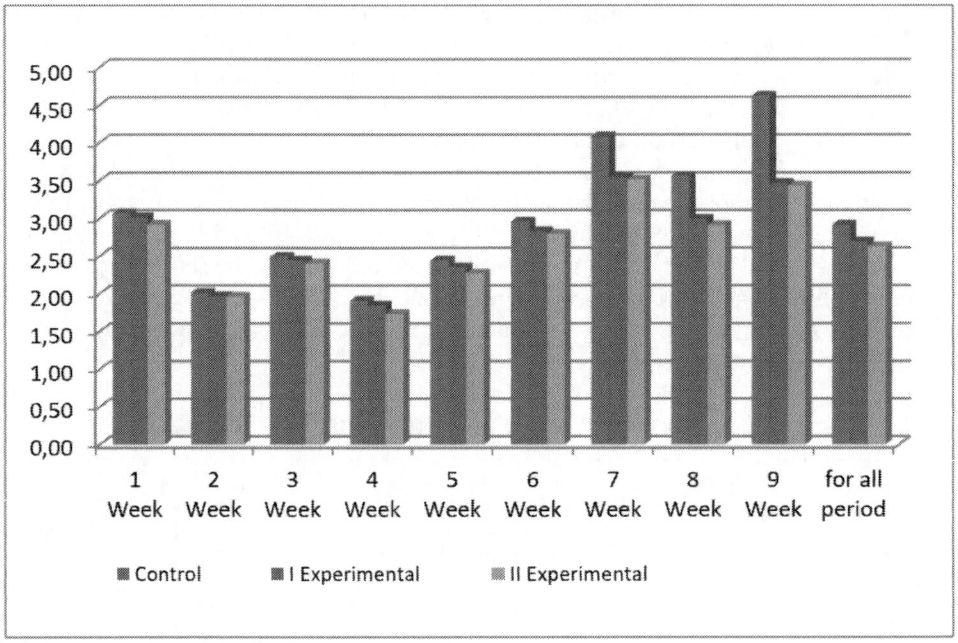

Figure 3. Feed costs by growing periods by groups, kg.

4. Conclusion

Thus, the analysis of the results of the study shows that the use of enzyme preparations in feed for ducklings contributes to the achievement of a higher absolute and average daily growth of live weight and increase the economic efficiency of its production. Since ducklings after 7 weeks of age reduces the intensity of growth, the optimal period of their cultivation, in our opinion, should not exceed 8 weeks. The results of studies suggest that the choice of enzyme preparations preference should be given to a mixture of amylosubtilin and celloviridine.

References

[1] Ivanova E Y and Lavrentiev A Y 2015 The effectiveness of the inclusion of enzyme preparations in feed for laying hens *Poult. Poult. Prod.* **2**. p.43-5.

[2] Zazhivikhina E I and Markova S A 2000 The use of biologically active substances based on terpenoids for farm animals and birds, Chemical-forest complex – scientific staffing in the XXI century. Problems and solutions. International scientific. practical. conference. Collection of articles on the materials of the conference, Krasnoyarsk: The Siberian State Technological University, 2000 287-9.

[3] Ageev V N, Egorov I A and Okolelova T M 1987 Feeding the bird: Handbook M.: Agropromizdat, 192

[4] Lavrentiev A Y, Sherne V S and Yakolev V I 2016 Influence of enzyme preparations on the productivity of goslings *Feed.* **7-8**. 78-9.

[5] Mikhailova L R and Nemtseva E Y 2018 Improving the feeding of broiler chickens on an industrial basis. Theory, and practice of modern agricultural science: materials of national (all-Russian) scientific. conference (20 February). – Novosibirsk: Novosibirsk State Agrarian University, 2018. 322 – 5.

[6] Nemtseva E Y and Mikhailova L R 2017 Analysis of growing broiler chickens on an industrial basis, Environmental management and socio-economic development of rural areas as the basis for the effective functioning of the regional agro-industrial complex: materials of the

All-Russian scientific-practical conference with international participation. - Cheboksary, Chuvash State Agricultural Academy. - 2017. 290 - 4.

[7] Petryankin F P, Semenov V G and Ivanov N G 2015 Immunostimulants in the practice of veterinary medicine: monograph Cheboksary: New Time, 272.

[8] Petryankin F P 2011 Feeding, metabolism and immunity in animals: monograph Cheboksary, 100.

[9] Sherne V S 2018 Characteristics of meat productivity of goslings that consumed feed with enzymes in various combinations *Agrarian Russia*. **8**. 21-4.

[10] Fisinin V I, Tishchenkov A N and Egorov I A 1998 Quality assessment of feed, organs, tissues, eggs, and poultry meat Methodical management / VNITIP. Sergiev Posad, 114.

IOP Conf. Series: Earth and Environmental Science **346** (2019) 012031 doi:10.1088/1755-1315/346/1/012031

Study on optimized combination and utilization model of agricultural and animal husbandry resources in mountain ecotone

H G Zhang[1] and X Y Tang[2, 3]

[1]Agriculture Resource and Environment Research Institute, Tibet Academy of Agriculture and Animal Science, Lhasa 850000, China

[2]Laboratory of Development and Application of Rural Renewable Energy, Ministry of Agriculture and Rural Areas / Biogas Institute of Ministry of Agriculture and Rural Areas, Chengdu 610041, China

E-mail: tangxiaoyu@caas.cn

Abstract: [Purpose] The development of agriculture and animal husbandry in Tibet is basically based on the interlaced development of agriculture and animal husbandry in mountainous areas. The interlaced agricultural and animal husbandry areas in mountainous areas are the unity of special, harmonious and harmonious regional small environment. The special attributes of mountainous areas determine the rich agricultural and animal husbandry resources, ecological development cycle areas, characteristic agricultural products areas and tourism. The division and regionalization of recreational zones and the optimal combination and utilization of agricultural and animal husbandry resources in mountainous ecotone have very important effects on promoting the benign cycle of regional agricultural and animal husbandry economic development, the development of urban integration and the protection of industrial ecological environment. How to develop and construct a new type of "environment-friendly and resource-saving" agricultural and animal husbandry areas is very important. It is of great practical significance to promote the development of regional agriculture and animal husbandry industry, to establish an efficient production system of agriculture and animal husbandry and to speed up the increase of farmers' and herdsmen's income. Rational and effective utilization of various beneficial resources can effectively solve the problems of material shortage, resource shortage and contradiction between supply and demand, and ensure the interlaced agriculture and animal husbandry economy in Tibet's mountainous areas. Sustainable development. [Method]On the basis of domestic and foreign literatures, this paper analyses and evaluates in detail the optimal combination and utilization

Content from this work may be used under the terms of the Creative Commons Attribution 3.0 licence. Any further distribution of this work must maintain attribution to the author(s) and the title of the work, journal citation and DOI.
Published under licence by IOP Publishing Ltd

mode of agricultural and animal husbandry resources in mountainous ecotone area by means of management induction, [Result]that is, the modern high-efficiency utilization mode of new energy, the green integration development mode of "chain fusion" in seed and culture, and the multi-level agricultural capital. Four basic modes, i.e. the mode of efficient utilization of resources and the mode of recreational sightseeing and sightseeing in mountainous agriculture and animal husbandry. [Conclusion] The above four modes of optimal combination and utilization of agricultural and animal husbandry resources are the basic macro-combination modes of agricultural and animal husbandry resources in the mountainous ecotone of Tibet. They are discussed in detail from the perspectives of natural resources utilization, agricultural products production and processing, agricultural and animal husbandry production and construction, and eco-tourism development in order to provide better services. Development and construction of agricultural and animal husbandry circular economy in Tibet. Finally, according to the characteristics of Tibet's own development, this paper puts forward to promote the integration of scientific concepts, the strength of scientific and technological support, and the fund guarantee system in the basic fields, and to strengthen the integration of farmer-herdsman cooperatives and market organizations in practical application. Reference and extension of the model, incentive and compensation security system, etc., for the future development of mountain-type interlaced agriculture and animal husbandry in Tibet circular agricultural and animal husbandry and resource optimization and combination of construction to provide a reference and research results.

1. Introduction

A total of 87 villages and towns in 31 counties and districts, covering an area of 335,000 square kilometers, accounting for 27.28% of the total area of Tibet, are important areas for the production, living and living of farmers and herdsmen in Tibet, as well as important bases for agricultural cultivation and livestock breeding in Tibet. The National Plan for the Protection of Ecologically Vulnerable Areas clearly points out that the ecotone between agriculture and animal husbandry in Tibet Autonomous Region is an economic development. The foundation and key point are that it has many vulnerabilities, such as weak anti-jamming ability, sensitivity to global climate change, strong fluctuation of time and space, and significant marginal effect. For the development of interlaced agriculture and animal husbandry in mountainous areas, there is a contradiction between the limited resources development space and progressive growth of agriculture and animal husbandry caused by the growth of agriculture and animal husbandry economy and the development of modern cities and towns. There are contradictions between environmental security and excessive waste and exploitation of resources, and between the effective utilization rate of agricultural and animal husbandry resources supply and the economic output rate of agriculture and animal husbandry, etc.[1-2] Faced with various problems, how to fully understand and scientifically guide the rational development and utilization of superior natural resources, change the mode of economic development and growth, adjust the industrial structure of agriculture and animal husbandry, improve the integration and complementarity of diverse resources, promote the development of agriculture and animal husbandry with Tibetan characteristics and increase the income of farmers and herdsmen are of great practical significance. It is necessary to speed up the development of modern agriculture and animal husbandry in the plateau. Actively exploring the mode of optimizing the combination and utilization of agricultural and animal husbandry resources, i.e. vigorously developing the ecological circulation agriculture and animal husbandry on the plateau and developing the circular agriculture and animal husbandry is the direction and fundamental way out for the future development of agricultural and animal husbandry economy in Tibet. Only in this way can the basic self-sufficiency rate of grain in Tibet be guaranteed, food safety and nutritional health be undertaken, and the benign, sustained and Development.

The model of optimum combination and utilization of agricultural and animal husbandry resources in mountainous ecotone studied in this paper is helpful for academia to have a deeper understanding of the efficient utilization of resources under the special natural environment and ecological characteristics of Tibet at present, and also provides ideas and methods for realizing sustainable development of agriculture and animal husbandry in plateau, protection of regional ecological environment and harmonious development of human and nature in the future. The factors such as the great difference of altitude gradient, the diversity of climate types, the complexity of natural conditions, the abnormal change of weather, the abundance and diversity of agricultural and animal husbandry resources in the mountain-type ecotone determine that there are more mountains and less land to change, the landscape is complex and diverse, and the per capita arable land area is relatively small; the vulnerability of natural ecology is obviously enhanced, the resistance ability is obviously low; the scale of agricultural and animal husbandry industrialization production is small, and the agricultural production is relatively low. Deep processing and high-quality brand development need to be improved. Therefore, in view of the mode of optimum combination and utilization of agricultural and animal husbandry resources in mountainous ecotone, we should combine theory with practice, innovate development ideas, tap resources potential, change resource allocation, develop and utilize core technologies, optimize the best combination mode, improve the productivity of agricultural and animal husbandry resources utilization, and promote the rational, full and effective utilization and development of agricultural and animal husbandry resources in Tibet to the maximum extent possible. Protecting the natural ecological environment, highlighting the comparative advantages and local characteristics of mountain-based agriculture and animal husbandry, and creating mountain-based circular agriculture and animal husbandry industrial clusters, in accordance with the principle of "superior regions, superior resources, superior industries, priority development", in-depth excavation and research to promote the efficient utilization of mountain-based circular agriculture and animal husbandry mode, which is different from conventional plains, basins and other agricultural and animal husbandry development modes, for the future West China. The practice of Tibet's modern circular agriculture and animal husbandry model on the plateau provides reference and reference.

2. Basic situation and function of the development of mountainous agriculture-animal husbandry ecotone in Tibet

2.1. General situation of the development of mountainous agro-pastoral ecotone in Tibet

The mountainous ecotone of agriculture and animal husbandry in Tibet Autonomous Region is the main agricultural and animal husbandry production and implementation area, as well as the important residential area and rural urban construction area, as well as the important production base of agricultural and animal products. It is also the birthplace of natural ecological environment protection in the plateau. The Agro-Animal husbandry in the ecotone basically maintains the original natural ecosystem. The overall climate characteristics are as follows: thin air, low air pressure and low oxygen content. More sunshine, strong radiation; lower temperature, large temperature difference. Mainly including the following different altitude gradients of mountains, low mountains, gentle slopes and valleys intersected by staggered rivers and corresponding farming and animal husbandry planting areas, breeding areas, etc., because of the particularity of the terrain, the different types of areas and levels are determined, the scope of concentration and dispersion are unified, according to the statistical yearbook of Tibet Autonomous Region, which occupies 30.78%, 26.56%, 7.53%, 2.01% respectively. 15.27% and 3.54%. Different regions have different landforms, and the structure and functional attributes of mountain agriculture and animal husbandry have their own characteristics. According to the altitude gradient section of planting and raising, there are not only barley planting areas and yak farming areas with relatively flat and vast elevation gradient between 4000 and 4800m, but also border counties such as Lazi County, Dingri County, Dingjie County and Gangba County. Highland barley planting areas and yak farming areas with relative concentration between 0 m are mainly along the Yarlung Zangbo River basin and the melting of alpine snow water, covering Qushui County, Linzhou County, Gacha County and Gongga County, and Linzhi County, Gongbu Jiangda County with an elevation gradient of less than 3 000 m are relatively rich in natural resources, agriculture and animal husbandry. The development of industry economy is relatively fast. Different regions have different development modes and characteristics of agriculture and animal husbandry. Although Tibet has incomparable advantages in climate, light, environment, resources, soil and other aspects, it is still in a state of extensive primitive agricultural and animal husbandry production in the development of agriculture and animal husbandry. The productivity benefits of farmers and herdsmen are low, the level of mechanization of Agriculture and animal husbandry is low, and the economic development of agricultural and animal husbandry areas is slow. The specific reasons are summarized as follows: the small scale, small quantity and weak strength of the development of agriculture and animal husbandry industry lead to the absence of a virtuous circle and value-added industrial chain of agriculture and animal husbandry, seriously affecting the enthusiasm of farmers and herdsmen in production; the special attributes of natural resources, high altitude, complex terrain, large temperature difference between day and night, and the restriction and influence of natural conditions on agricultural and animal husbandry production and construction. There are many risks and uncertainties; far away from the deep plateau of the interior, relatively advanced agricultural and animal husbandry production methods and methods, scientific concepts, information technology and so on lag far behind. The result of many reasons is that the development of Qinghai Province, Yunnan Province, Guizhou Province and Sichuan Province is extremely slow and low relative to the surrounding Northwest region [3].

2.2. Discussions on the functions of the mountainous agricultural-pastoral ecotone in Tibet

Tibet's natural attributes give special development environment to alpine, canyon, River and farming and animal husbandry planting areas, forming unique farming and animal husbandry cultivation and breeding methods. Mountain-based farming and animal husbandry has created a relatively dynamic and balanced internal environment for development according to the characteristics of different regions, effectively blocking the negative impact of external climate change and ecological disasters on regional small-scale farming and animal husbandry, and forming mountains. Natural safety barriers and protective measures in agro-pastoral ecotone, especially in maintaining natural ecological environment, sand control and soil consolidation, vegetation restoration and soil and water conservation, have played a natural attribute and coordination role; long-term agricultural and animal husbandry production and construction have basically formed a fixed development model, and complex and changeable topography and vegetation have provided abundant for agricultural and animal husbandry production. Biodiversity-vegetation resources and production conditions; the melting of alpine snow water and the important water resources in river valleys provide necessary conditions for the construction of agriculture and animal husbandry and the survival of human beings; the enrichment of solar energy and wind energy is not only an indispensable condition for human beings and even for the survival of species, but also an important precondition for promoting the regulation of regional agricultural and animal husbandry construction and maintaining the four-season alternation of agriculture and animal husbandry. The abundant forest and grass resources and livestock excrement around are good renewable resources for agriculture and animal husbandry, which are helpful to the understanding and utilization of farmers and herdsmen's livelihood in the micro-field. Superior, natural and unique ecological tourism resources, rich and colorful human landscape, characteristic culture, folk customs and other resources have laid the foundation for the development of tourism, the construction of ecological towns and the promotion of tourism brand. Basics. This paper mainly relies on the regional advantages of Tibet's mountainous agriculture and animal husbandry, fully taps and combines the potential available resources of agriculture and animal husbandry, takes "clean production" of agriculture and animal husbandry as the breakthrough point, effectively promotes the maximum utilization of resources to better serve the production and construction of agriculture and animal husbandry, promotes the harmonious coexistence of human and nature, and unifies economy and ecology [4][5].

3. optimized combination and utilization model of agricultural and animal husbandry resources in mountainous agricultural-pastoral ecotone of Tibet

Land resources utilization is limited, climate is unusually changeable, landforms are peculiar and diverse, resources and vegetation are abundant, and the optimal combination and utilization of agricultural and animal husbandry resources should depend closely on the development and construction of local agriculture and animal husbandry. First, the production and life construction of agriculture and animal husbandry should be brought into full play by maximizing the use of renewable energy in nature, with emphasis on the efficient utilization of modern new energy resources. The second is to combine the cultivation of agriculture and animal husbandry production, planting, breeding and post-processing with the development model of "chain melting" green integration, so as to make full use of solar energy resources, wind energy resources and water energy resources to carry out rational and effective comprehensive development and utilization. The second is to deepen the scale production and marketing of comprehensive agricultural products from green aquaculture, green planting, characteristic deep processing and brand marketing. Industrial development model; three is to focus on the separation and recycling of fuel, green manure, organic fertilizer, farm manure and agricultural and animal husbandry wastes; four is to support the multi-level agricultural and animal

IOP Conf. Series: Earth and Environmental Science **346** (2019) 012031 doi:10.1088/1755-1315/346/1/012031

husbandry resources efficient utilization model of the agricultural and animal husbandry system; four is to feature farm yards, tourist attractions, ecological leisure vacation and folk customs experience as one of the mountain agricultural and animal husbandry leisure sightseeing tour. Travel mode. Mountainous Agro-Animal husbandry ecotone incorporates planting, aquaculture, agricultural product processing, leisure agriculture and natural clean production energy into the development system and scope of circular agriculture and animal husbandry. Through equipments and devices, the natural energy resources are reasonably and effectively transformed and applied, and the waste resources generated by planting and breeding are correspondingly transformed into reuse and recycling. In order to achieve the goal of reducing agricultural and animal husbandry waste resources, effectively improving the production performance and utilization efficiency of waste, promoting the optimal combination and utilization of various superior agricultural and animal husbandry resources, meeting the needs of regional agricultural and animal husbandry development to the greatest extent, and improving the development of mountainous agriculture and animal husbandry. The stability, sustainability and cyclicity of the ecosystem have enhanced the lasting momentum and power of the development of mountain agriculture and animal husbandry, laid a reasonable, effective and stable ecosystem of agriculture and animal husbandry, and realized a virtuous circle of promoting the development of agricultural economy, increasing the income of farmers and herdsmen, and protecting the ecological environment of agriculture and animal husbandry.

3.1. Analysis and evaluation of efficient utilization model of modern new energy

The regeneration and utilization of efficient clean production energy is one of the important means and measures for the development of circular agriculture and animal husbandry in mountainous areas. During the production and construction of agriculture and animal husbandry in Tibet, the natural world has endowed abundant light, wind and water resources with clean energy. Since the formation of the earth, it has been closely related to the existence of the world, human beings, animals and agricultural production departments. There are no pollution, unlimited utilization, easy to collect and store, etc. The natural energy to guide the optimal combination and utilization of mountain agriculture and animal husbandry resources has the advantages of less investment, quick results, easy to understand and learn, easy to understand and simple. Through the use of modern science and technology and equipment, as well as many years of production and living practice of farmers and herdsmen, the following utilization modes are explored and discussed: solar energy In terms of utilization, the emphasis is on heating and lighting. Firstly, the solar integrated circuit board on the roof of the house is designed to store and transmit the heat for night lighting; secondly, the solar water heater on the roof is used to absorb the solar radiation heat, so as to promote the heating and heat preservation of the cool water, so as to facilitate the use and demand of hot water in the daily life of farmers; thirdly, it is connected. Through the establishment of solar street lamp, the absorption and storage of light energy during the day, and the release of solar energy through the light of street lamp at night; fourth, the use of solar cookers, cooking and boiling water and daily needs. In the aspect of wind energy utilization, the main body is power generation, power supply and lighting. By setting up wind energy utilization device to absorb the air flow circulating between mountains for energy storage, the wind energy can be converted into output current for power supply in agricultural and pastoral areas. In the aspect of water energy utilization, the emphasis is on agricultural production and animal husbandry, aquaculture and rough processing and utilization of agricultural products. Firstly, the top-down watershed of mountain-type Alpine water source can effectively irrigate different agricultural production and Cultivation in mountain-type, which provides convenience for rational and effective utilization of water resources. At the same time, agricultural cultivation along the river basin can effectively use water source for agricultural irrigation. Secondly, clean water resources provide safe drinking water sources for livestock breeding around; thirdly, good water sources provide broad space and development scope for regional agriculture and animal husbandry and aquaculture; fourthly, traditional farm water grinding turbines are established through small watershed rivers to make full use of water resources energy for daily roughing of agricultural products. (See Figure 1 for details).

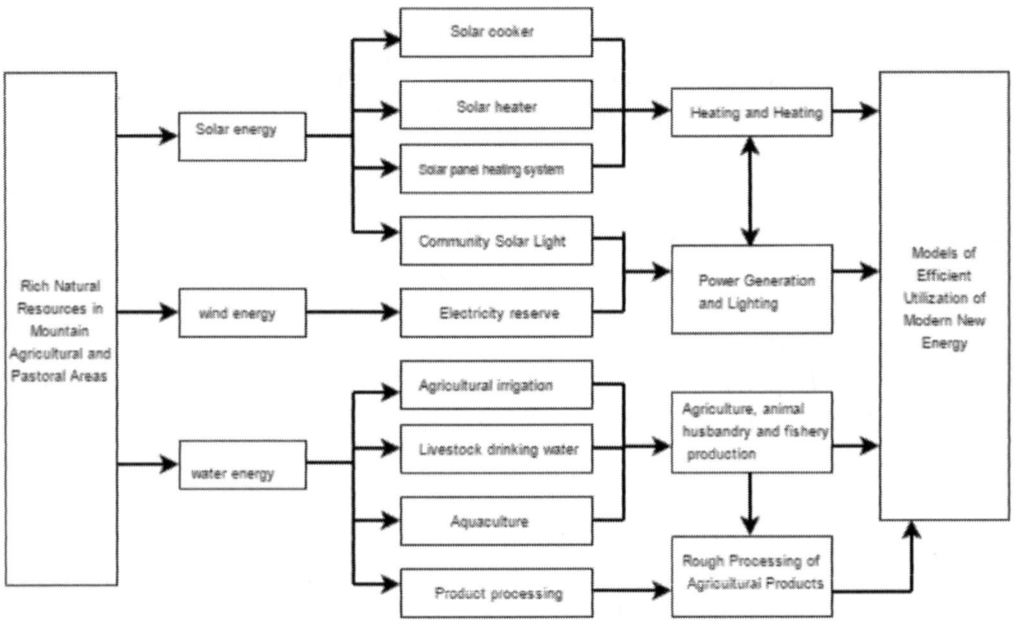

Figure 1: Schematic diagram of resource optimization combination for efficient utilization of modern new energy resources.

3.2. Analysis and evaluation of development modes of green fusion of nutrition chain melt

The green fusion development mode of "chain integration" of planting, breeding and processing is the management form of the industrialization chain of "chain integration" of mountain circular agriculture and animal husbandry. It is the production and life style closest to the local farmers and herdsmen. The concept of circular economy development runs through the pre-production, mid-production and post-production links of agricultural and animal husbandry development, and the production links give full play to the basis of agriculture and animal husbandry. This production function enables them to make full use of their resources and make rational use of their resources. Life links promote resource conservation and clean production, and give full play to the multi-function of agricultural and animal husbandry production. By relying on scientific and technological support, we firmly grasp the economic entity and mutual benefit as a link, improve the enthusiasm of farmers and herdsmen in

IOP Conf. Series: Earth and Environmental Science 346 (2019) 012031 doi:10.1088/1755-1315/346/1/012031

agricultural and animal husbandry construction, expand the space for agricultural and animal husbandry development, promote large-scale production and construction of agricultural and animal husbandry, and enhance agricultural production. Value-added resource utilization optimization combination model. Planting industry: first, advocating basic large-scale agricultural production in planting industry, increasing grain reserves laid the foundation for aquaculture; second, feeding maize and alfalfa were planted in the spare land to provide nutritious feed for livestock and make up for seasonal green storage feed; at the same time, legume crop planting and maize straw returning effectively could improve and enhance soil productivity; By effectively utilizing the idle resources of space, such as fruit trees planted on the ground, autumn fruiting can effectively increase the income of farmers and herdsmen, and also provide a forest habitat for free-range livestock. Breeding industry: first, stocking Tibetan chickens can effectively peck natural insects and effectively control the occurrence of local environmental diseases and insect pests. At the same time, all kinds of insects are also natural edible ingredients of Tibetan chickens. The two are mutually beneficial functions of forming resources. Two, Tibetan pigs possess the properties of arch soil, which play a regular protective role in the loose and ploughing of surface soil, and can effectively promote ground planting. Physical growth and labor saving; three, Tibetan Pig and Tibetan chicken can be returned to the field by raising the excrement of livestock, which provides a good source of fertilizer for the growth of ground crops; four, through livestock stocking, it can effectively promote harmonious coexistence and coexistence among different species, not only providing good outdoor activities for livestock, but also providing suitable housing for livestock. Realm, through field stocking to create green agricultural products, enhance the natural taste of livestock meat. Processing industry: first, through the deep processing of crop agricultural products-wheat crops, the intrinsic green value is enhanced; second, through the deep processing of livestock products, separated products and processed products are classified, which can effectively enhance the added value of products, through the development of market economy can effectively improve the value of enterprises and promote the income of farmers and herdsmen; third, through economic creation. The impact of income can effectively change farmers' ideas and management mode, promote the effective connection and development of planting, aquaculture and agricultural products processing industry, further re-investment, re-operation and re-development of funds, lay the foundation for expanding production scale and promoting economic development. (See Figure 2 for details).

IOP Conf. Series: Earth and Environmental Science **346** (2019) 012031 doi:10.1088/1755-1315/346/1/012031

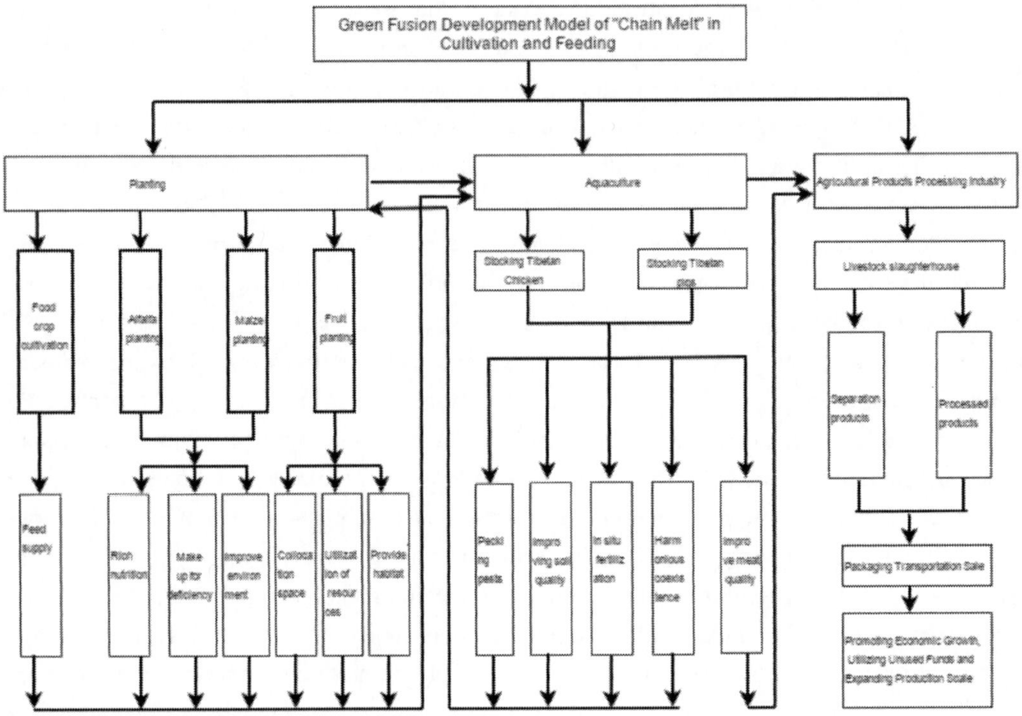

Figure 2: A sketch map of resource optimization and combination of "chain melt" green fusion development model for planting and feeding.

3.3. Multi-level high-efficiency utilization model of agricultural and animal husbandry resources

The multi-level mode of efficient utilization of agricultural and animal husbandry resources is that the development of mountain-type circular agriculture and animal husbandry understands and understands the optimal combination and utilization of resources from three angles of agricultural waste, animal husbandry waste and the production of living waste by farmers and herdsmen. The development of agriculture and animal husbandry and the livelihood of farmers and herdsmen will also produce a large amount of waste materials and variety of varieties. Different regions have different solutions. The utilization of multi-level agricultural resources in mountainous areas has both scientific reference and agricultural inheritance. The emphasis of this study is to make full use of various waste materials and optimize their combination and utilization. Agricultural waste: First, agricultural barley and rapeseed are the main crops of waste - straw, mainly by means of direct feeding livestock, auxiliary means - crushing and returning to the field and compressing to produce fuel, which is conducive to livestock fattening, improving land productivity and increasing economic output value; second, the autumn harvest of farmers and herdsmen will be residual in the field of withered grass, weeds and a large number of leaves to carry out. Land burning or returning to the field or collecting and retenting farm green manure. Livestock waste: First, chicken manure and pig manure produced by chicken farms and pig pens are usually fermented to produce organic fertilizer; second, sheep manure is the main fertilizer used by local farmers and herdsmen in crop cultivation, mainly through collecting

IOP Conf. Series: Earth and Environmental Science **346** (2019) 012031 doi:10.1088/1755-1315/346/1/012031

returned farmland and fermentation to produce organic fertilizer; third, cow manure produced in*farming*pastoral areas is used as fuel and market by air-drying. Trading is one of the sources of household heating in farming and pastoral areas, which has the advantages of burning odorless, pollution-free, high calorie and is well received by the market. This is the practice inherited by Tibetan farmers and herdsmen for a century. At the same time, the conditional farming and pastoral areas collect cow manure as the raw material of edible mushroom culture media centrally, which is the introduction of scientific and technological innovation. The waste materials produced by farmers and herdsmen's livelihood are as follows: first, the waste plastic mulch and shed film produced by planting industry; plastic bottles and glass bottles produced by agriculture; and waste household appliances, clothing and paper produced by farmers and herdsmen's livelihood are collected, classified, recycled, treated and reused centrally; second, the production and living of human beings in agricultural and pastoral areas. The construction and layout of biogas digesters for farming and animal husbandry are based on sewage, feces and other materials, and a small amount of agricultural and animal husbandry wastes. Because mountainous farming and animal husbandry areas are located in high altitude and low temperature areas, most of them are equipped with glass caps or plastic mulches to improve the heating effect, give full play to the biogas outgassing rate, serve the needs of heating and living in farming and animal husbandry areas, and effectively solve the waste pair of farming and animal husbandry areas. Pollution of soil resources, water resources, air and surrounding environment not only reduces the cost of treatment, but also realizes community cleaner production and living, and achieves the purpose of rational and effective utilization of waste resources. (See Figure 3 for details).

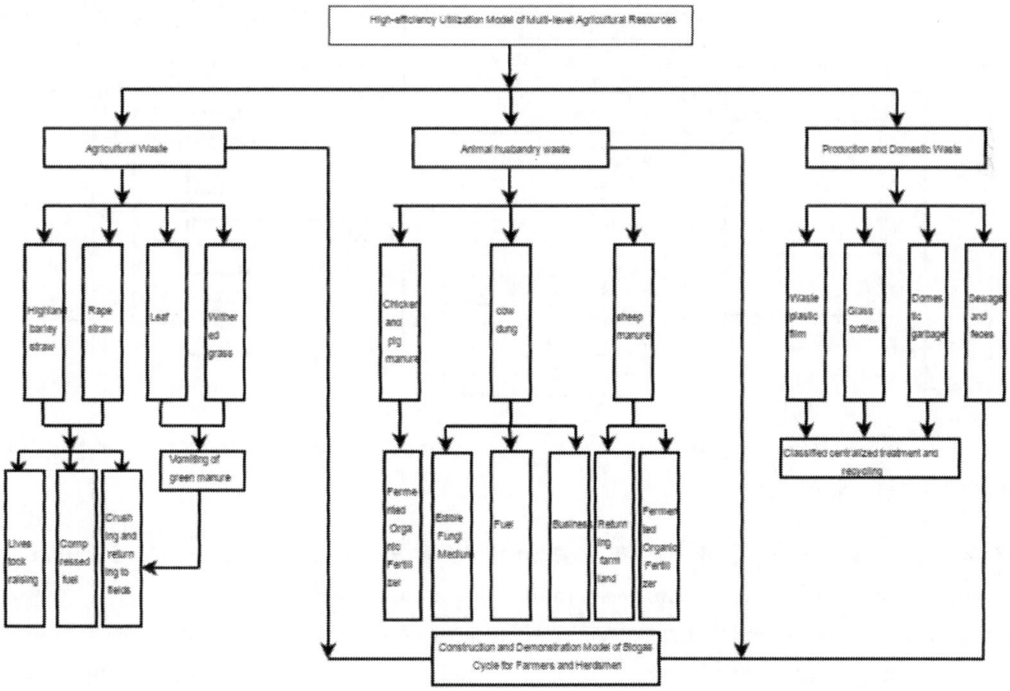

Figure 3: Schematic diagram of resource optimization and combination of multi-level efficient utilization model of agricultural and animal husbandry resources.

3.4. Mountain agriculture and animal husbandry leisure tourism model

Mountain agriculture and animal husbandry leisure tourism mode is an effective combination and utilization mode of mountain circular agriculture and animal husbandry development, which fully combines natural landscape resources, human history and culture resources, pastoral agriculture and animal husbandry life development resources, leisure and recreation vacation resources, folklore tourism resources, farmers' pleasure returning to rural resources, and provides people-oriented development concept in an all-round way, providing appreciation, entertainment and drinking. Food, accommodation, experience and other comprehensive services, integrated with economic, ecological, humanistic and social functions, belong to the rising industry of regional economic development in the future. They belong to the new industrial form and consumption mode of modern agriculture and animal husbandry. They are also important for promoting the development of regional agriculture and animal husbandry economy, optimizing the industrial structure of Agriculture and animal husbandry, stimulating the employment and income of farmers and herdsmen, and increasing production. It is of great practical significance to extricate agriculture from poverty and promote the structural reform of the supply side of agriculture and animal husbandry. Fully tapping and utilizing the rich types of resources in Tibet to blend and complement each other, and promoting the development of leisure tourism products in mountain areas of Tibet from low-level supply and demand to high-level supply and demand will become an important way for farmers and herdsmen to increase their income in the future[6][7]. (See Figure 4 for details).

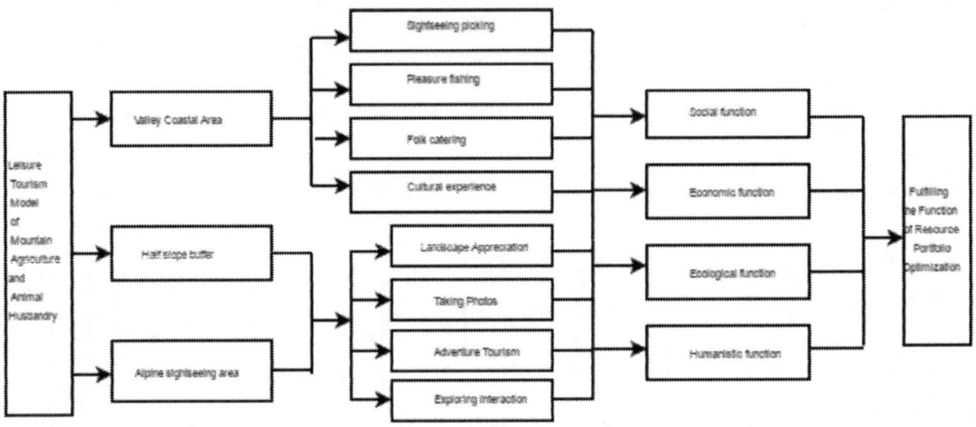

Figure 4: A sketch map of optimum combination of resources for mountain agriculture, animal husbandry, leisure and sightseeing tourism model.

4. Future prospect of optimized combination utilization model of agricultural and animal husbandry resources in 4 mountainous agricultural-pastoral ecotone

The optimum combination and utilization of circular agricultural and animal husbandry resources is a brand-new mode of agricultural and animal husbandry development. From the perspective of management, it is both a brand-new development concept and an advanced thought strategy. It uses the scientific concept of development, the concept of sustainable development, the theory of circular economy and the technology of optimum combination of resources, mainly through the optimum combination and utilization of resources, the Agro-Animal husbandry ecosystem. Improvement of quality and dynamic balanced development mechanism of agricultural and animal husbandry economic growth can effectively make full and rational use of various types of resources. The purpose is to achieve the best combination and utilization of agricultural and animal husbandry resources, reduce the input and use of harmful substances, minimize environmental pollution and ecological damage, promote the construction of "clean production" system of agriculture and animal husbandry, and promote population, resources and economic growth. The economic growth mode of agriculture and animal husbandry in the ecotone of agriculture and animal husbandry in mountainous areas of Tibet is coordinated development of environment and other living organisms. The future practice of optimizing combination and utilization mode of agricultural and animal husbandry resources in the ecotone of agriculture and animal husbandry in mountainous areas of Tibet is described below.

4.1. Integration of scientific ideas

This paper mainly elaborates on the following three aspects: taking the circular economy theory of various countries as the guidance, fully considering the natural attributes and characteristics of regional agricultural and animal husbandry development, taking the development of ecological circular economy as the goal, re-examining and evaluating the advantages and disadvantages of the traditional mode in the general development mode, evaluating the theory and prospects of the existing and future resource combination and optimum utilization mode, and further understanding and excavation. The internal relationship and coordination among human beings, resources and environment should be established, and an innovative development and efficient combination utilization model should be established to guide the development of circular economy of agriculture and animal husbandry in Tibet in the future, to change the excessive dependence of agricultural and animal husbandry production on resource development and consumption, and to promote the optimal combination and utilization model of ecological agricultural and animal husbandry resources in the plateau; to actively promote the ecological environmental protection consciousness and scientific development of the plateau farmers and herdsmen Understanding, raising the awareness of ecological environment protection and green consumption of farmers and herdsmen, abandoning the concept and thought of "great development, large production, large construction and large consumption", advocating a good situation of "saving resources, protecting ecology, improving environment and promoting development", actively promoting the scientific development concept of "reducing, reusing and recycling" in agricultural and animal husbandry production, and rationally and fully combining and utilizing oneself Natural resources and agricultural and animal husbandry waste resources in the world, to minimize the production of agricultural and animal husbandry production and living waste, to achieve the rational use of resources and reduce pollution emissions; design different sectors and industries of resource mix recycling industry chain, conducive to different levels and categories of waste become the premise of recycling economic development, and further realize orderly circulation of resources. Ring and rational development, to maximize the demand and supply of resources between different types of industries, to achieve a balance between the rational use of resources and sustainable development.

4.2. Science and technology support

The optimization and combination of agricultural and animal husbandry resources follow the concept of "science and technology is the first productive force". Science and technology are the basic guarantee for development. We should fully tap the regional agricultural and animal husbandry circulation technology demonstration system, carry out a series of research on technology, development and innovation, and give full play to the brand-new concept of production, values and agricultural economy of agriculture and animal husbandry. This paper mainly discusses from the following four aspects: setting up the concept of scientific training of talents, actively training and introducing professional and technical personnel of different disciplines, attaching importance to the status and role of professional and technical personnel in circular agriculture, giving full play to the enthusiasm, initiative and creativity of talents to integrate various disciplines into a comprehensive development; increasing investment in scientific research funds to support the cycle Development and development of agricultural and animal husbandry technology, promotion and application of new technology, new technology, new equipment and new methods, so as to promote the universal application and promotion of new technologies, methods and products, and give full play to the role of scientific and technological achievements in agricultural production; regular construction of farmer and herdsman training system, training seminars and practical guidance of professional technical experts, and promote understanding of agriculture and animal husbandry The frontier development of agriculture and animal husbandry is conducive to stimulating the enthusiasm, initiative and creativity of peasant households. Effective exploration and establishment of an integrated entity of production, education and research and a scientific and technological support platform for efficient utilization of agricultural and animal husbandry resources will promote scientific and technological breakthroughs in the technology of optimizing the combination of resources and utilization, give full play to the effect of multi-links, effectively link up traditional agriculture and animal husbandry with modern agriculture and animal husbandry, and Research has been fruitful, farmers and herdsmen have demonstrated, and the region has driven the reciprocal and mutually beneficial relationship, so as to realize the scientific and technological innovation platform of production, science and research to serve the development and construction of agricultural economy.

4.3. Agricultural and herdsman cooperatives and market operating organizations

Efficient utilization and combination of agricultural and animal husbandry resources and even the construction and development of agricultural and animal husbandry cycle are the economic integration that takes the family unit of farmers and herdsmen as the main body, the development organization that takes many individual individuals as the cooperative body, that is, the cooperative of farmers and herdsmen as the backing, and the development of market economy as the goal. This paper mainly discusses from the following three aspects: farmers and herdsmen are the main body of practice in the actual production and construction of agriculture and animal husbandry, as well as the entity that has the most say in the development of agriculture and animal husbandry. They have accumulated rich experience and practices in the long-term production and construction. These experiences and practices are the concise of scientific theory and methods, because small farmers with family as the unit disperse the main body, agricultural production and scale. Limited ability to resist natural risks and the contradiction between small-scale production and large-scale production have contributed to the development of agricultural and herdsman cooperatives. Farmers and herdsmen are the basic unit and indispensable component of the development of circular agriculture and animal husbandry. Cultivating new farmers and herdsmen who "learn knowledge, understand culture and know technology" is the main practitioner and operator of the development of circular economy of Tibetan agriculture and animal husbandry in the future. Society is not only a carrier organization for farmers and herdsmen, but also a good link for the development of market economy. It plays an important role as a link between the preceding and the following. It has its own service function, as well as the function of propaganda, creation and promotion of circulation mode. It also serves the economy and society, responds to existing problems to government departments and puts forward collective organizations for effective decision-making of optimizing the combination and utilization of resources. Weaving is a third-party social organization that enhances collective economy of farmers and herdsmen, opens up agricultural products market, develops marketing strategy and drives regional farmers'economic development. Based on market economic organization, individual farmers and herdsmen, farmer and herdsman cooperatives all have a unified and indivisible organic carrier, and the absence of any aspect affects the development of agricultural and animal husbandry economy.

4.4. Reference, promotion and application of resource optimized combination utilization model

Modern agriculture and animal husbandry and traditional agriculture and animal husbandry have different development directions, orientations and modes. Different regions have different resources, environments and layout of agriculture and animal husbandry. How to effectively tap, rationally utilize and match the resources of agriculture and animal husbandry varies. In terms of production and construction of agriculture and animal husbandry, processing and circulation, technology and mode, and consumption integration, the actual development situation should be considered. Further study is in line with its own development system. The successful models of circular economy development of agriculture and animal husbandry both at home and abroad have their own development attributes and characteristics. The model should be used for reference and introduced with full consideration of their own regional economic development, natural resources richness and the location conditions of development. The suitability, advantages and disadvantages of the model should be evaluated by referring to the effect of the model. Major innovation and digestion and reabsorption rationally design their own optimal combination and utilization mode of resources; the extension of the mode fully takes into account the scope of application of existing conditions, whether farmers and herdsmen accept and accept the degree, whether the input cost of the mode matches the economic benefits of output and other issues; the application of regional characteristics, reasonable, effective and give full play to various types of diversified resources Effective utilization of resources, improve the best utilization mode of various resources, adhere to the development concept of "innovation, coordination, green, open and sharing", and gradually realize the effective allocation of diversified resources, rational and effective utilization of waste resources and good sustainable development of natural ecological environment [8].

4.5. Financial security and other support systems

Everything's development is based on the guarantee of funds, which produces economic benefits and promotes the benign development of economic construction. Therefore, the development of circular agriculture and animal husbandry in Tibet should actively increase the financial investment in agriculture and animal husbandry, promote the construction of agricultural and animal husbandry infrastructure and the management of agricultural and animal husbandry environment, and provide a good positive guarantee system. The following three aspects should be done well: giving full play to the policy leading position of local government and fully mobilizing all parties to participate in the construction of circular economy through the investment and support of financial funds. The initiative of the establishment should be given preferential and incentive policies from tax, subsidy, finance and project fund construction, regulated and restrained by administrative and legal means, guided by propaganda and education to promote the rational and effective development of agricultural and animal husbandry resources, and lay a foundation for the efficient utilization of agricultural and animal husbandry resources. New construction and development, improvement of relevant laws and regulations, such as the Cleaner Production System Law of Agriculture and Animal Husbandry, Regulations on Pollution Prevention and Control of Agricultural and Animal Husbandry and the Law on Construction of Eco-cycle Agriculture, which are conducive to promoting the development of circular economy of agriculture and animal husbandry, standardize the sharing of implementation standards and results of production and construction, and actively encourage farmers, enterprises and companies to take safety in production as the basis, and put agriculture into practice. Efficient utilization and combination of animal husbandry resources should be brought into the track of standardization, institutionalization and clarification [9]. In compensation mechanism, the government and relevant departments should give distinct treatment to the implementation of efficient utilization of agricultural resources, and carry out ecological compensation, technological compensation and financial compensation for typical demonstration-driven circulation mode to enhance "key products and funds compensation". Based on the principles of key regions, key technologies and key development, an effective mechanism of technological compensation for circular agriculture and animal husbandry was explored and constructed [10].

5. Conclusion

At present, the construction and development of circular economy of agriculture and animal husbandry has become the national development strategy. Tibet is in the stage of accelerating the modernization and urbanization of the plateau. It is facing many serious problems such as shortage of resources and environmental destruction. In order to further seize the important development opportunities and promote the early construction of a well-off society in an all-round way, the development and construction of agriculture and animal husbandry in Tibet should follow the principle of Based on the benchmark of "agricultural construction and development", this paper focuses on the "optimum utilization mode of agricultural and animal husbandry resources in mountainous ecotone" as the research object. According to the principles of "reduction, reuse and resource utilization", scientific development and construction will achieve maximum economic benefits and sustainable development of agriculture and animal husbandry with the least waste and waste discharge, and make rational and effective use of Tibet's unique rich natural resources. Landscape resources and human resources will turn waste from agriculture and animal husbandry and agricultural and pastoral areas into treasures to realize the recycling and reuse of agriculture and animal husbandry, promote the healthy development of agriculture and animal husbandry, realize the unity of economic, environmental and social benefits, and build a resource-saving and environment-friendly society. This paper briefly expounds the optimum utilization mode of agricultural and animal husbandry resources in mountainous ecotone from four main aspects. Among them, the modern high-efficiency utilization mode of new energy mainly focuses on the optimum utilization of advanced equipment and devices to obtain abundant natural renewable resources, and the green fusion development mode of "chain melt" is mainly based on the cultivation and production of agriculture and animal husbandry, animal husbandry and agricultural products. Deep processing is a horizontal and vertical integration mode, with the development of agricultural and animal husbandry industry market and economic development as the main body; the efficient utilization mode of multi-level agricultural and animal husbandry resources is the full combination and rational utilization of the waste resources generated by planting, animal husbandry and production and living in agricultural areas as the main body; the leisure and sightseeing tourism mode of mountain agriculture and animal husbandry is a comprehensive regional human, landscape, tourism and entertainment. The dominant characteristic resources such as music and catering are the main body of comprehensive development. The theory, model and practice of circular economy development in agriculture and animal husbandry is a pioneering and innovative research work. This article, supported by the open subject of the Key Laboratory of Renewable Energy Development and Utilization in Rural Areas of the Ministry of Agriculture and Rural Areas, expounds the basic framework of the optimum utilization model system of agricultural and animal husbandry resources in mountainous ecotone through visits and investigations in some areas of Tibet. Because my professional and level limitations only provide my own personal views on the development of circular agriculture and animal husbandry, I hope to provide reference ideas and successful experience for the future development and construction of circular agriculture and animal husbandry economy in Tibet.

Acknowledgement
Subsidized Project: Open Fund of Key Laboratory of Rural Renewable Energy Development and Utilization of Ministry of Agriculture and Rural Areas, Research on Circular Development Model of Agriculture and Animal Husbandry in Tibet Based on the Concept of Ecological Development (Project No. 2017015)

Reference

[1] Tang H J 2008 Development model and strategic countermeasures of circular agriculture and animal husbandry in China. *China Agric. Sci. Tech. Report*, **08**(01): 12-18
[2] Zhou Z F 2008 Study on the development model of circulating agriculture. *Study Agric. Moderniz.*, **09**(01): 19-25
[3] County Economic Observer of this magazine. Practice Model of Three Characteristics of Circulating Agriculture. Leadership Decision-making Information, 2009, **12** (11): 15-21
[4] Zhou Y, Yin C B, Qiu J J, et al. 2008 Classification of development patterns of circular agriculture in China. *Chin. J. Eco-Agric.*, **03** (06): 54-62
[5] Yin C B, Zhou Y and Liu L H, et al 2013 Theory and practice of circular agriculture development in China. *Chin. J. Ecolog. Agricul.*, **10** (01): 26-32
[6] Zhang X M and Jiang Z D 2009 Comparison between eco-agriculture and circulating agriculture. *Study Agric. Mechaniz.*, **03**(06): 55-61

[7] Xu Q 2008 Sustainable development of circulating agriculture in China. *Modern Agric. Sci. Tech.*, **05**(20): 33-38
[8] Chen H B, Lu J D, Zhao L Y etc 2007 Origin and development of circular agriculture. *Agric. Resour. Regional. China*, **11** (06): 32-36
[9] Gao W S, Chen Y Y, Liang L, et al 2007 On the basic principles and technical system of sustainable agriculture development. *Res. Agric. Moderniz.*, **07** (06): 46-53
[10] Zhang H Y 2017 A Brief analysis of low carbon agricultural economy and circulating agricultural economy. *Knowledge Econ.*, **10**(17): 18-21

Characterization of protease activity from hepatopancreas of blue crab

W Makkapan[1, 2] and P Narkthewan[1]

[1]Food and Agricultural Biotechnology Program, Department of General Science and Liberal Arts, King Mongkut's Institute of Technology Ladkrabang Prince of Chumphon Campus, Pathiu, Chumphon, 86160, Thailand

E-mail: eang_ang@hotmail.com

Abstract. Proteolytic enzymes play an important role in determining the quality of blue crab during postmortem storage. Activity of endogenous proteases is involved in the texture softening and autolysis of blue crab, which limits the customer acceptance and marketing price. This research aimed to characterize the protease activity of crude enzyme extract from the hepatopancreas of blue crab. The optimum activity of crude protease extract was found at pH 7.0 and 50°C. The crude protease enzyme was highly stable over a wide pH range of 4.0-11.0 and showed high stability at temperatures below 40°C. In addition, the protease activity continuously decreased with an increasing concentration of NaCl (0-15% w/v). Therefore, an understanding of the endogenous proteases in the blue crab could be used to develop appropriate storage methods during its distribution process.

1. Introduction

Local distribution of fish and shellfish, especially marine animals, generally involves being stored in ice. Therefore, a major problem occurring during local distribution of fish and shellfish is a rapid deterioration due to autolytic degradation and microbial spoilage. Autolytic degradation is related with the activity of endogenous proteolytic enzymes [1-3]. Proteases are a group of enzymes responsible for protein hydrolysis. The endogenous proteases in fish and shellfish can be classified into four groups: cysteine proteases, aspartic proteases, serine proteases and metalloproteases, which are associated with a loss of textural quality during postmortem storage [3]. The proteolytic enzymes from the hepatopancreas played a probable role in the muscle softening of prawn during postmortem handling and storage [4], and proteolytic activity was suggested as a factor influencing textural quality of salmon meat during iced storage [5]. As a result, understanding the role of proteases during postmortem storage can be valuable for developing techniques to safely preserve the quality of marine food products. Blue crab is a popular local marine food in Pathiu, a district in Chumphon province, and it is generally distributed using iced storage. Postmortem quality loss tends to decrease the taste and texture of the blue crab, which leads to reduction in economic value. However, there is little information associated with the activity of proteases in blue crab. This study aimed to characterize the proteolytic activity from the hepatopancreas of blue crab in order to develop better preservation methods during local distribution.

2. Materials and methods

Content from this work may be used under the terms of the Creative Commons Attribution 3.0 licence. Any further distribution of this work must maintain attribution to the author(s) and the title of the work, journal citation and DOI.
Published under licence by IOP Publishing Ltd

2.1. Preparation of crude enzyme

Blue crabs were obtained from port of Pathiu district, Chumphon province, Thailand. Their weight ranged between 140 and 160 g. The hepatopancreas was isolated and homogenized in 50 mM Tris-HCl (pH 8.0) at a ratio of 1:3 w/v. After centrifugation for 30 min at 10,000xg at 4°C, the collected supernatant was defined as "crude enzyme extract" and used for protease activity analysis.

2.2. Characterization of proteolytic activity in crude extract from hepatopancreas

2.2.1. pH profile and stability. The optimum pH of the crude protease was investigated by the modified method from [6,7] using 1% (w/v) casein as a substrate. The following buffer system was used for different pH conditions: 100 mM Citrate buffer (pH 4.0-6.0), 100 mM Tris–HCl (pH 7.0-9.0), 100 mM Glycine–NaOH (10.0-11.0) [8]. The crude extract (1 ml) was mixed with the 1% casein (1 ml) and incubated for 15 min at 37°C. The enzymatic reaction was ceased by adding 20% (w/v) trichloroacetic acid (2 ml) and the mixture was kept on ice for 1 h. After centrifugation for 10 min at 10,000xg at 4°C, the collected supernatant was determined the absorbance at 280 nm. One unit of protease activity was defined as releasing 1 μmol of tyrosine/min. The relative activity was computed and reported compared to condition showing the highest value (100% activity).

The pH stability of crude enzyme activity was analyzed by keeping the crude enzyme at different pH values (4.0-11.0) using several buffers as previously described at 4°C for 1 h. The residual activity was assayed under standard conditions.

2.2.2. Thermal profile and stability. The activity of crude protease was determined over the various temperatures (0, 4, 30, 37, 40, 50, 60, and 70°C). The protease activity was carried out as previously described at the optimum pH.

The thermal stability of crude protease was analyzed by keeping the crude enzyme at various temperatures (0, 4, 30, 40, 50, 60 and 70°C) for 1 h. Aliquots were taken at 15 min intervals to assay the residual activity under standard conditions.

2.2.3. Effect of NaCl. The influence of NaCl on the protease activity was measured by adding the NaCl in the reaction mixture to gain the final concentration of 0-20% (w/v). The protease activity was analyzed under standard conditions.

3. Results and discussion

3.1. pH profiles and stability

The influence of pH on the protease activity of the crude enzyme from hepatopancreas was analyzed over a pH range of 4.0-11.0 as represented in figure 1(a). The crude protease exhibited the highest activity at pH 7.0. The crude protease was also active between pH 8.0-10.0, showing more than 80% activity compared to the highest activity observed at pH 7.0. The proteolytic activity consistently decreased when pH was below pH 7.0. Optimum pH value of crude enzyme from the hepatopancres of the blue crab was in accordance with that of the crude extract from *Macrobrachium rosenbergii* hepatopancreas [4], which had an optimal pH at 7.0. The proteolytic enzyme from the midgut gland of *Farfantepenaeus paulensis* [9] and viscera of *Portunus segnis* [10] displayed the maximum activity at pH 8.0. Many proteases from fish, such as *Oreochromis niloticus* [7,11], *Stolephorus indicus* [1] and *Raja clavata* [12], showed a high activity within pH ranges of 7.0-9.5. The decline in the proteolytic activity at a pH outside of the optimal pH might be involved in the change of charge distribution and conformation of both the enzyme and substrate [8,13].

The pH stability of the crude protease was represented in figure 1(b). Crude protease from the hepatopancreas was extremely stable over a wide pH range between 4.0-11.0, retaining over 90% of its initial activity after 60 min incubation at 4°C. The pH stability of proteases from fish viscera, such as *R. clavata* [12,14], *Zosterisessor ophiocephalus* [14], *and Scorpaena scrofa* [14,15], was extremely stable over a wide broad pH range between pH 5.0-12.0.

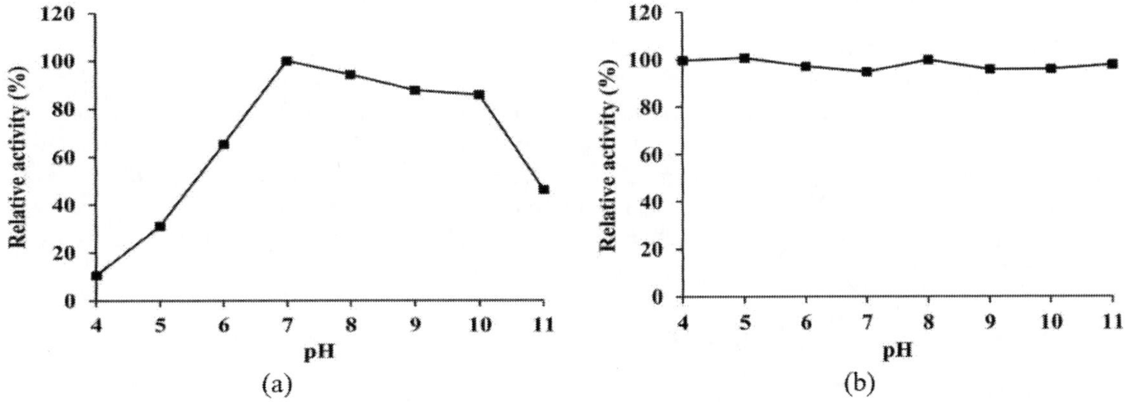

Figure 1. (a) pH profiles and (b) stability of crude protease from blue crab hepatopancreas.

3.2. Thermal profiles and stability

The protease activity of the crude enzyme from the hepatopancreas measured at different temperatures at the optimum pH was shown in figure 2(a). The crude protease demonstrated the highest proteolytic activity at 50°C. The crude enzyme showed activity above 80% at 37, 40 and 60°C. The relative activity at 70°C was approximately 62% of that at 50°C. The protease activity sharply declined more than 80% at low temperatures (0 and 4°C). Similar results on the optimum temperature of protease activity were observed in *R. clavata* [12, 14] and *Z. ophiocephalus* [14]; the enzyme extracts had the highest activity at 50°C. The optimum temperature of the crude protease from the hepatopancreas of blue crab was lower than the value reported from *M. rosenbergii* [4], *P. segnis* [10], *S. scrofa* [14, 15], and *Salaria basilisca* [16], which had optimum temperatures between 55-60°C.

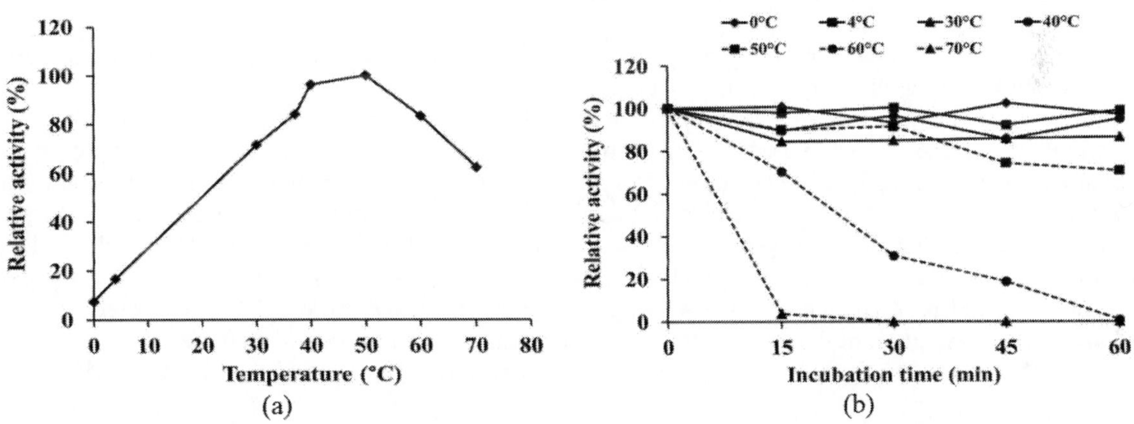

Figure 2. (a) Thermal profiles and (b) stability of crude protease from blue crab hepatopancreas.

Thermal stability profile of the crude protease was represented in figure 2(b). The crude protease extract was extremely stable below 40°C, retaining over 80% of its original activity after 60 min incubation. The crude protease extract was quite stable at 50°C with its activity retained about 70% of its original activity after 60 min incubation. A continuous decrease of protease activity was found at 60°C since incubation for 15 min until no protease activity remained after incubation for 60 min. The proteolytic activity was almost absolutely inactivated at 70°C after incubation for 15 min. The thermal stability of the crude protease extract from the blue crab hepatopancreas showed a similar result with the protease from viscera of *P. segnis* [10], *Z. ophiocephalus* [14] and *Sardinella aurita* [17].

Moreover, the crude protease from hepatopancreas showed a higher stability than the visceral protease of *R. clavata* [12,14] and *S. scrofa* [14].

3.3. Effect of NaCl concentration

The influence of NaCl on the activity of the crude protease from the hepatopancreas was represented in figure 3. The activity of the crude protease continuously decreased with an increment of NaCl concentration up to 15%. Thereafter, the enzyme activity was constant with a further increment of NaCl concentration at 20%. In accordance with several reports on fish, such as *Pangasianodon gigas* [8], *Katsuwonus pelamis* [13] and *S. aurita* [17], the reduction of protease activity was observed with an increasing NaCl concentration. The decline in activity might be due to an increase in ionic strength, which is described by the salting out effect [18,19].

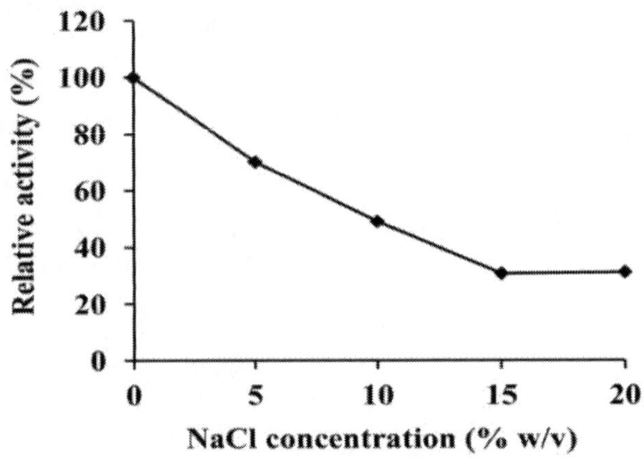

Figure 3. Influence of NaCl concentrations on the activity of crude protease from hepatopancreas.

According to [20], the preservation methods used in fish and shellfish can be divided into three groups; physical (e.g. chilling, freezing, ozonation), chemical (e.g. curing with salt or preservatives), and bio-preservation methods. Nevertheless, the use of ice is a popular method for storage and local distribution because it is simple and cheap. Our results revealed that the protease from the blue crab hepatopancreas was also active at pH of ice (pH 6.5-8.5) and highly stable at chilling temperature (0 and 4°C). Moreover, the concentration of salt affected the activity of the blue crab protease. Therefore, an understanding of protease characteristics is valuable for controlling blue crab quality during storage and distribution. To inhibit protease activity during blue crab preservation, the storage condition should be regulated at pH lower than 6.0, chilled between 0 and 4°C, and stored in salt. However, one limitation of the study is that the spoilage and shelf life of blue crab are associated with several other factors, so further research is needed to help improve storage and distribution methods.

4. Conclusions

The crude protease extract from the blue crab hepatopancreas was highly active at pH 7.0 and a temperature of 50°C. The crude protease was extremely stable at a pH range of 4.0-11.0 and at temperatures below 40°C. These results are a beneficial data for developing preservation methods during storage and local distribution of the blue crab.

Acknowledgments

This work was supported by King Mongkut's Institute of Technology Ladkrabang Prince of Chumphon Campus, Chumphon, Thailand. The authors would like to thank Ms. Phornthip Kriangkraison for assisting this research.

References

[1] Siringan P, Raksakulthai N and Yongsawatdigul J 2006 *Food Chem.* **98** 678-84
[2] Sriket C, Benjakul S, Visessanguan W and Kishimura H 2011 *Food Chem.* **124** 29-35
[3] Sriket C 2014 *Int. Food Res. J.* **21** 433-45
[4] Sriket C, Benjakul S and Visessanguan W 2010 *J. Sci. Food Agric.* **91** 52-9
[5] Hultmann L and Rustad T 2004 *Food Chem.* **87** 31-41
[6] Mehrotra S, Pandey P K, Gaur R and Darmwal N S 1999 *Bioresour. Technol.* **67** 201-3
[7] Chaijaroen T and Thongruang C 2016 *Int. Food Res. J.* **23** 1432-8
[8] Vannabun A, Ketnawa S, Phongthai S, Benjakul S and Rawdkuen S 2014 *Food Biosci.* **6** 9-16
[9] Buarque D S, Castro P F, Santos F M S, Lemos D, Junior L B C and Bezerra R S 2009 *Aquac. Res.* **40** 861-70
[10] Hamdi M, Hammami A, Hajji S, Jridi M, Nasri M and Nasri R 2017 *Int. J. Biol. Macromol.* **101** 455-63
[11] Bezerra S R, Lins E J F, Alencar R B, Paiva P M G, Chaves M E C, Coelho L C B B and Carvalho Jr L B 2005 *Process Biochem.* **40** 1829-34
[12] Lassoued S, Hajji S, Mhamdi M, Jridi A, Bayoudh A and Barkia M 2015 *Int. J. Biol. Macromol.* **80** 668-75
[13] Klomklao S, Kishimura H, Nonami Y and Benjakul S 2009 *Food Chem.* **115** 155-62
[14] Nasri R, Younes I, Lassoued I, Ghorbel S, Ghorbel-Bellaaj O and Nasri M 2011 *J. Amino Acids* **2011** 913616 http://dx.doi.org/10.4061/2011/913616
[15] Younes I, Nasri R, Bkhairia I, Jellouli K and Nasri M 2015 *Food Res. Int.* **94** 453-62
[16] Ktari N, Khaled H B, Younes I, Bkhairia I, Mhamdi S, Hamza I and Nasri M 2014 *J. Food Sci. Technol.* **51** 3094-103
[17] Khaled H B, Ghorbel-Bellaaj O, Hmidet N, Jellouli K, El-Hadj Ali N, Ghorbel S and Nasri M 2011 *Food Chem.* **128** 847-53
[18] Klomklao S, Benjakul S and Visessanguan W 2004 *J. Food Biochem.* **28** 355-72
[19] Klomklao S, Benjakul S, Kishimura H and Chaijan M 2011 *Food Chem.* **129** 739-46
[20] Mahmud A, Abraha B, Samuel M, Mohammedidris H, Abraham W and Mahmud E 2018 *MOJ Food Process Technol.* **6** 303-10

IOP Conf. Series: Earth and Environmental Science **346** (2019) 012033 doi:10.1088/1755-1315/346/1/012033

Study on influence and compensation for soil compactness on volumetric water content measurement

H Tian[1,2,3], C Gao[1,2,3], Z T Gao[4] and Y D Zhao[1,2,3,5]

[1]School of Technology, Beijing Forestry University, Beijing, China
[2]Beijing Laboratory of Urban and Rural Ecological Environment, Beijing Municipal Education Commission, Beijing, China
[3]Key Lab of State Forestry Administration for Forestry Equipment and Automation, Beijing, China
[4]Institute of Disaster Prevention Science and Technology, Hebei, China

E-mail: yandongzh@bjfu.edu.cn

Abstract. Volumetric water content measurements based on soil dielectric properties are affected by soil compaction. Today's commercial products produce substantial measurement error as they do not account for the influence of soil compaction. This paper presents a portable soil volumetric moisture content sensor based on the principle of standing wave rate and simultaneous soil compactness measurement. The coupling relationship between soil compactness and soil volumetric moisture content can be easily observed by converting them into polar coordinates. A modified soil volumetric moisture content model based on the soil compactness is also established to support the operation of the proposed sensor.

1. Introduction

Soil moisture content is an important physical factor affecting plant growth and crop yield, and is thus of crucial significance to agricultural and forestry production professionals [1,2]. The changes in soil water content form an important basis for evaluating plant health and guiding irrigation. Soil volumetric water content is typically measured by drying methods, dielectric methods, or neutron meter methods [3-7]; drying methods are most common among them as they are convenient, accurate, and simply operated [8-10]. Dielectric methods, however, have become increasingly popular as they facilitate rapid, efficient, and continuous measurement of soil volume moisture on either portable or fixed equipment [11-13]. The effects of soil compaction on the measured data are rarely considered in plant health assessment or irrigation guidance applications [14-18]. Soil compaction, an important factor affecting soil volumetric water content, is meaningful only when comparing soil volumetric water content measurements under a unified compactness level.

The cone index is widely used to characterize soil compactness. Ayers [19], Vaz [20,21], and Busscher [22] have studied the relationship between soil moisture content, bulk density, and cone index in the laboratory environment and established semi-empirical models accordingly. Lin *et al* [23] designed a device for the simultaneous measurement of soil cone index and volume water content and to assess the performance of the semi-empirical model. Soil volume water content is closely related to soil cone index; that is to say, soil volume water content is closely coupled with soil compactness. Previously proposed models may not properly account for soil bulk density, however, and come with oversize census monitoring workload in real-world production scenarios. The bulk density needs to be

Content from this work may be used under the terms of the Creative Commons Attribution 3.0 licence. Any further distribution of this work must maintain attribution to the author(s) and the title of the work, journal citation and DOI.
Published under licence by IOP Publishing Ltd

calculated by test sampling, so rapid measurement is not possible.

In this study, we designed a synchronous measuring device for soil compaction and volumetric water content. We defined the coupling relationship between soil water content and compactness and built a semi-empirical model based on compaction compensation soil moisture content accordingly, then tested the accuracy and stability of the corrected measurement results. Converting the measured soil volumetric water content to the soil volumetric moisture content at the same compactness level gives the soil volumetric water content data measured in an actual agroforestry production environment strong comparability, and provides a highly accurate basis for agroforestry production and ecological monitoring.

2. Materials and methods

2.1. Soil volumetric water content measurement method

There is a monotonous relationship between the soil volumetric content and the soil dielectric constant, so the former can be measured indirectly by measuring the latter. In this study, we recorded the dielectric constant according to the standing wave principle to measure the soil volumetric water content (figure 1). We used a 100 MHz sinusoidal signal as the oscillating signal source to generate the excitation with a transmission line characteristic impedance of 50 Ω. When the electromagnetic wave propagates along the transmission line, the impedance around the ring probe does not match the impedance of the transmission line and a standing wave forms on the transmission line [24-26]. The standing wave can be detected at both ends of the transmission line and the small signal can be amplified via differential signal amplifier to determine the output signal as follows:

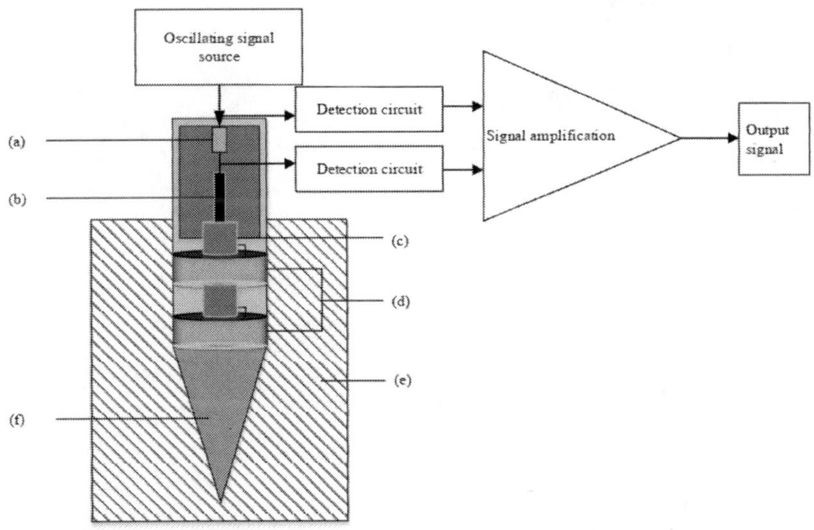

(a) Transmission line characteristic impedance, (b) coaxial transmission line cable, (c) metal ring fixed column, (d) ring probe, (e) soil to be tested, (f) stainless steel cone.

Figure 1. Soil volumetric moisture content detection based on standing wave principle.

$$\Delta U = 2A\sigma = 2A\frac{Z_P - Z_L}{Z_P + Z_L} \tag{1}$$

where ΔU is output signal voltage difference (V), A is electromagnetic wave amplitude (V), σ is the transmission line reflection index, Z_P is the probe impedance (Ω), and Z_L is the transmission line impedance (Ω). Z_L is dependent on the physical dimensions and dielectric constant of the insulating

IOP Conf. Series: Earth and Environmental Science **346** (2019) 012033 doi:10.1088/1755-1315/346/1/012033

material, which is 50 Ω here. The transmission line reflection index is a fixed value. Formula (1) indicates that the changes in impedance around the ring probe cause the output signal voltage disparity to change. The impedance around the ring probe is determined by the soil dielectric around the constant, probe size, and oscillating signal source frequency. The soil volumetric water content can be obtained by measuring the output signal voltage difference once the probe size and frequency are determined.

2.2. Soil compactness measurement based on cone index

Cone index (CI) is often used to describe soil compactness as an indicator of the soil's resistance to conical penetration at uniform velocity. When a standard cone penetrates the soil at a uniform speed, the average pressure per unit floor area is the CI. ASAE specifies respectively the specification of two kinds of cones with diameters of 12.83 mm and 20 mm and a conical head cone angle is 30°. CI is expressed as follows:

$$CI = N / S \tag{2}$$

where N is the cone resistance to uniform penetration into soil and S is the conical head bottom area. In the process of uniform penetration, the resistance of cone penetration into the soil is equal to the pressure applied on the cone meter, so CI can be calculated by measuring the pressure applied on the cone meter and the area of the cone bottom.

2.3. Compactness synchronization measuring device

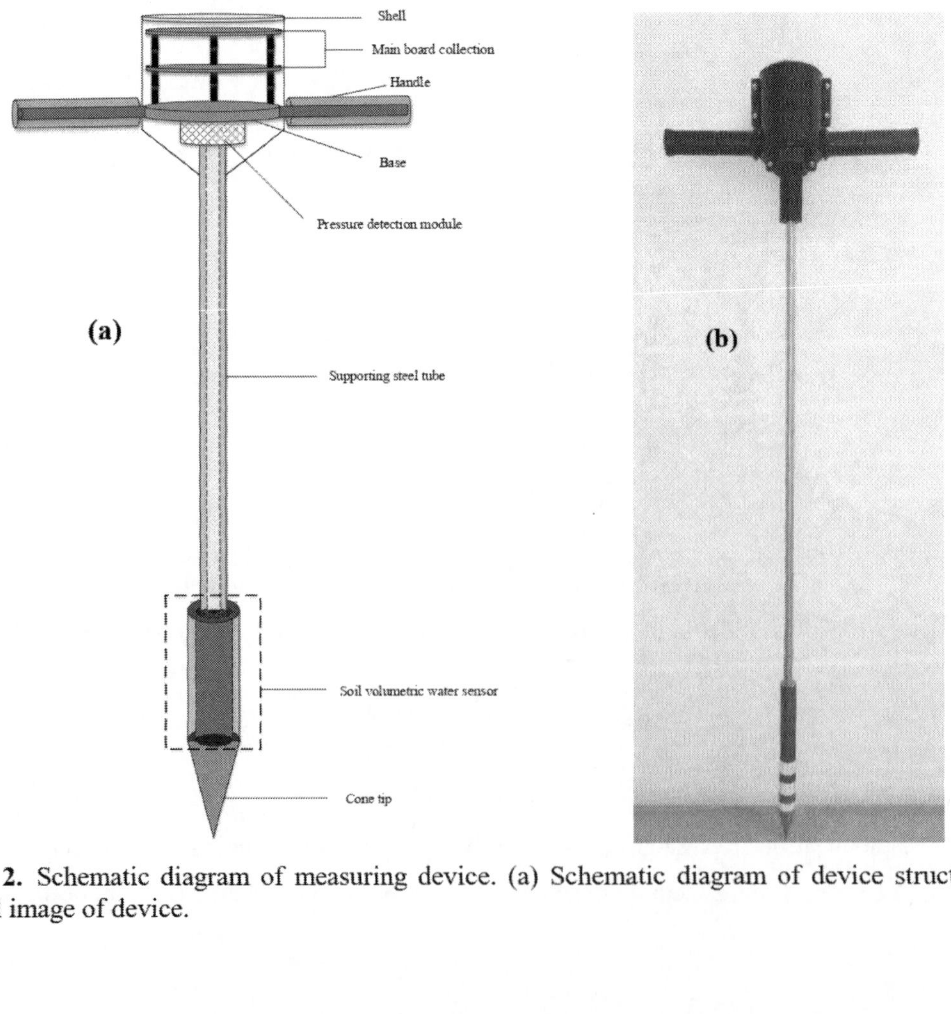

Figure 2. Schematic diagram of measuring device. (a) Schematic diagram of device structure, (b) physical image of device.

A diagram of the proposed device is shown in figure 2. A measuring probe is connected to a moisture measuring circuit to collect soil volumetric water content data. The measuring probe is comprised of two stainless steel rings with outer diameter of 20 mm and inner diameter of 18 mm. The rings are installed on a PVC support column (diameter 18 mm) and a conical head (conical angle is 30°, diameter is 20 mm) is installed at the lower end of the support column. The measuring probe is installed at the lower end of the supporting steel pipe (outer diameter is 10 mm, inner diameter is 5 mm). The output signal of the volumetric water content measuring circuit board is transmitted to the acquisition motherboard through a connecting cable inside the steel pipe. The upper part of the supporting steel pipe is equipped with a pressure detection module (JLBT-M2, Bengbu Sensor System Engineering Company, China; range 0-7000 N, output voltage 0-2500 mV, accuracy 2.8 N/mV) and the base. The penetration pressure is measured by the pressure module and CI is calculated according to the data it provides.

2.4. Experimental sample

Soil samples were collected from the Gongqing Forest Farm (116.73°E, 40.11°N) in Shunyi District, Beijing, from Zhenyuan Country in Qingyang City (107.03°E, 35.54°N), Gansu Province, and from the Bejing Forestry University Sanqingyuan Experimental Base (116.34°E, 40.00°N). Samples were collected randomly from three areas each with a depth of 0-40 cm. A total of 30 kg of soil was taken back to the laboratory. The soil samples were dried in a drying oven (105°C, 48h) and passed through a 40 mesh screen to remove stones, roots, and impurities. The samples include clay loam soil (11% sand, 71% silt, 18% clay), loess soil (15% sand, 65% silt, 20% clay), and sand soil (85% sand, 10% silt, 5% clay).

2.5. Experimental procedure

2.5.1. Soil volumetric moisture sensor calibration. Each soil sample was weighed 5 kg with precision electrons. They were recorded as clay loam soil, loess soil and sand soil. A certain amount of water was added to each sample soil and stirred for 10 min until the water and soil were thoroughly mixed and the soil moisture content was uniform. Each type of soil was then equally divided into eight samples; the first sample was placed into a polyvinyl chloride (PVC) calibration barrel (40 cm diameter, 25 cm height) and compacted with nylon rods (50 mm diameter, 50 cm length) freely falling from the surface of the sample about 3 cm to compact it after adding different amount of water, with slight rotation from the periphery to the middle to ensure that the soil in calibration barrel was completely integrated. These steps were performed on the remaining seven samples, and an experimental sample was configured in the barrel. Then the calibration barrel was sealed for 48 hours to ensure the balanced movement of water in the barrel. The sensor was then inserted into the barrel for measurement at an insertion depth of 20 cm.

Each soil sample was measured in 10 replications. The maximum and minimum voltage values were removed, then the remaining eight values were averaged to obtain the current soil sample measurement result. The volumetric moisture content of the soil sample in the calibration barrel was measured simultaneously via drying method. Soil samples of varying volumetric moisture content can be configured by adding different masses of water to the soil samples. Each soil sample in the experiment was configured with eight different volumetric moisture content characteristics. We recorded the volumetric moisture content and voltage value measured by the sensor, then tested the linear fit of the values.

2.5.2. Soil volumetric moisture content measurement under different compactness. Each soil sample was weighed to 10 kg with precision electrons and recorded as clay loam soil, loess soil, and sand soil. According to the soil moisture degree in table 1, 338.0 g, 829.0 g, 1356.0 g, 1800.0 g, and 2212.0 g of water were added to the soil samples to ensure various levels of dryness (or "moisture content"). As described in Section 2.5.1, samples of varying compactness were configured by controlling the

IOP Conf. Series: Earth and Environmental Science **346** (2019) 012033 doi:10.1088/1755-1315/346/1/012033

number of nylon rod compaction cycles (1, 2, 4, 6, 8, 10, 12, 14, 16, and 20). Ten samples were configured for each soil moisture degree and 150 samples were configured for each of the three different soil types. The sensor was uniformly penetrated into the sample to 20 cm and the corresponding soil compactness and soil volumetric moisture content were recorded.

Table 1. Soil moisture classification.

Soil moisture classes	Gravimetric water content /%
Higher soil water	>20
Suitable soil water	15-20
Mild dry	12-15
Moderate dry	5-12
Severe dry	<5

3. Results and discussion

3.1. Soil volumetric moisture content sensor calibration

The output voltage of sensor and soil volumetric moisture content were linearly fitted as shown in figure 3. The determination coefficients of the linear fitting curves of three soil samples are 0.97, 0.99, and 0.96, which indicates that the output voltage of the sensor has a good linear relationship with the soil volumetric moisture content of the tested sample under the same degree of compactness. In other words, the soil volumetric moisture content can be calculated by the sensor output voltage.

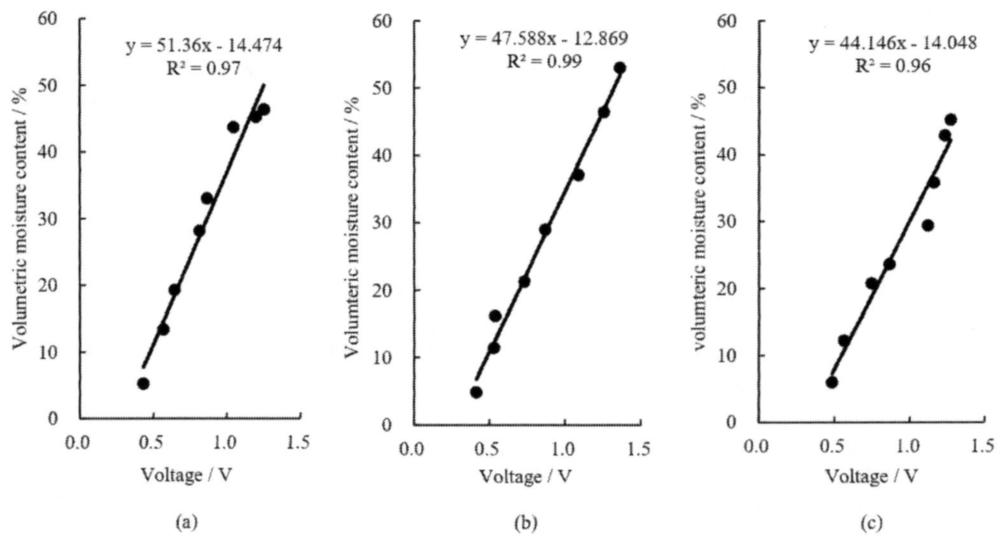

Figure 3. Fitting curve between output voltage values and soil moisture content. (a) Clay loam soil, (b) loess soil, (c) sand soil.

3.2. Changes in soil volumetric moisture content for different compactness

The changes in soil volumetric moisture content corresponding to different compactness (CI) for the three types of soil samples are shown in figure 4. We found that the change rate of soil volumetric moisture content caused by soil compaction differs under different soil moisture degrees. Soil volumetric moisture content does not change with soil compaction for "severe dry" soil, but increases slowly with soil compactness when the soil is characterized by "moderate drought" and increases rapidly with soil compactness for "mild drought" and "suitable" soil. When soil moisture is very high, the soil volumetric moisture content and soil compactness remain stable [27,28]. These change trends also appeared to differ for different soil types. The soil volumetric moisture content of Sample 1 increases with soil compactness, while both qualities remain basically constant in Samples 2 and 3 as soil moisture increases.

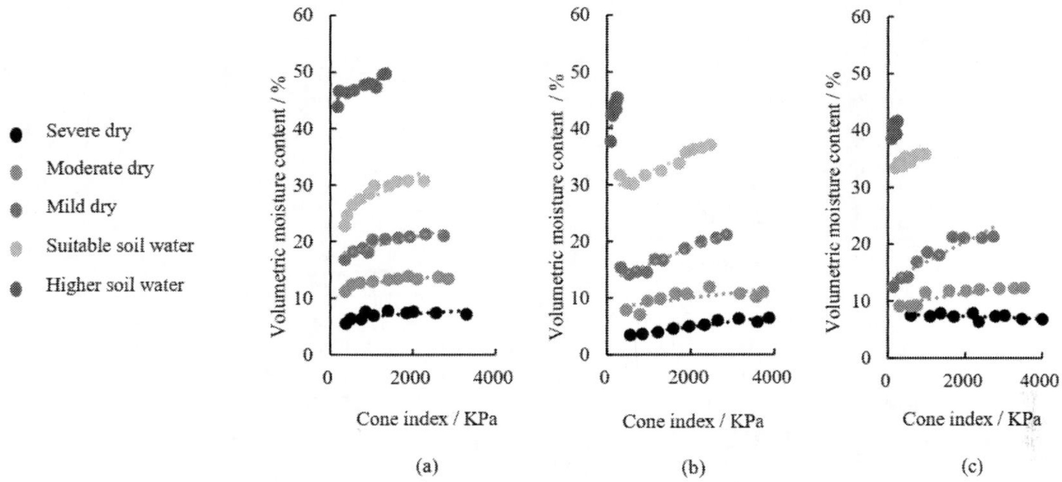

Figure 4. Changes in volumetric water content with changes in compactness. (a) Clay loam soil, (b) loess soil (c) sand soil.

3.3. Relationshiop between volumetric moisture content and compactness in polar coordinates

Figure 4 shows that the slope of the curve increases as soil moisture increases, but it is difficult to observe the specific coupling relationship between with soil volumetric moisture content and soil compactness in the planar coordinate system. We converted the soil volumetric moisture content and soil compactness from the planar coordinate system data to the polar coordinate system as follows:

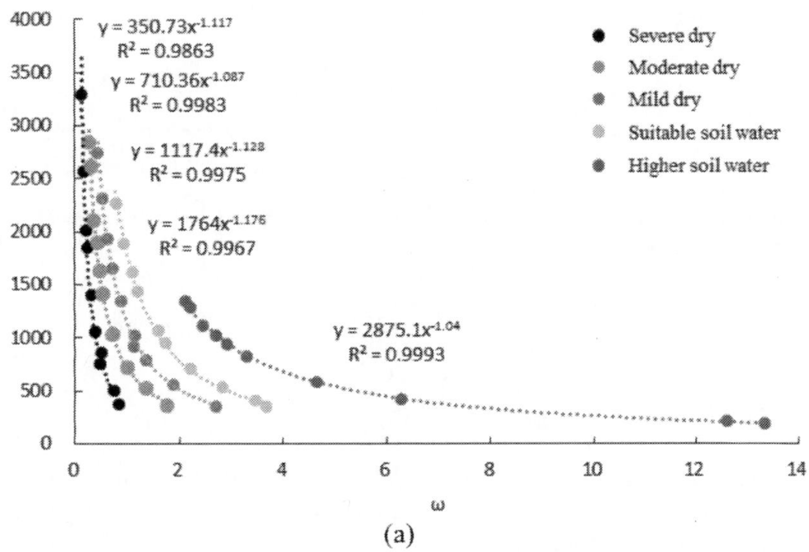

(a)

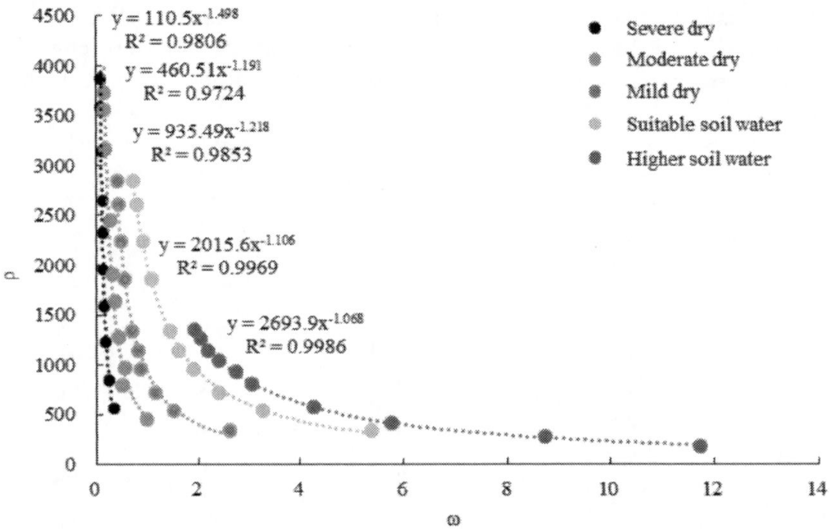

(b)

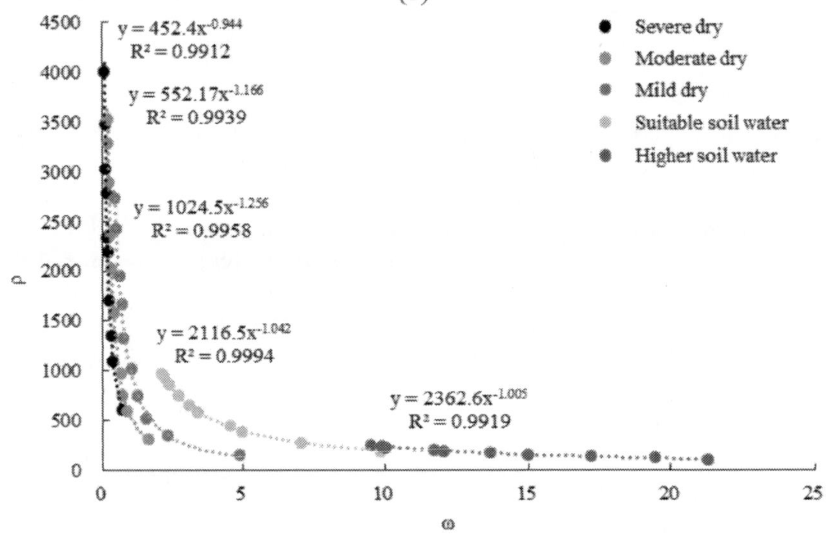

(c)

Figure 5. Changes in polar radius with polar angle and ρ is polar radius, ω is polar angle. (a) Clay loam soil, (b) loess soil, (c) sand soil.

$$\rho = \sqrt{\theta^2 + CI^2} \tag{3}$$

$$\omega = \arctan(\frac{\theta}{CI}) \tag{4}$$

Where ρ is polar radius, ω is polar angle, θ is the soil volumetric moisture content and CI is soil compactness. As shown in figure 5, the fitting determination coefficients are all greater than 0.98; the polar angle has a good linear relationship with the polar radius in the polar coordinate system.

3.4. Soil volumetric moisture content modified model based on soil compactness

The polar radius and polar angle satisfy a power function relationship as per the fitting results shown in figure 5. The relationship can be expressed as follows:

$$\rho = a \bullet \omega^{b} \tag{5}$$

where a and b are constants. The difference in soil types and moisture degrees makes the power function curve coefficients obtained by fitting the power exponential function are also different, so a and b need to be calibrated for different soil types and moisture degrees. Plugging equations (3) and (4) into equation (5) yields the following:

$$\theta^2 + CI^2 = a^2 \bullet (\arctan(\frac{\theta}{CI}))^{2b} \tag{6}$$

and further simplification produces equation (7):

$$(\frac{\theta}{CI})^2 + 1 = (\frac{a}{CI})^2 \bullet (\arctan(\frac{\theta}{CI}))^{2b} \tag{7}$$

We propose the following modified soil volumetric moisture content model based on soil compactness:

$$(\frac{\theta}{CI})^2 + 1 - (\frac{a}{CI})^2 \bullet (\arctan(\frac{\theta}{CI}))^{2b} = 0 \tag{8}$$

According to different soil compactness, substituting into equation (8) and solving the soil volumetric moisture content produces the corresponding volumetric moisture content under the given compaction condition. Converting the measured soil volumetric water content to the soil volumetric moisture content at the same compactness level gives soil volumetric water content data measured in an actual agroforestry production environment high comparability.

4. Conclusions

In this study, we designed a bimetallic ring probe and pressure sensor for simultaneous soil moisture and compactness measurement. The linear fitting determination coefficient determined in our calibration experiment exceeds 0.96, which indicates that the proposed device – which works based on the standing wave principle – accurately measures soil volumetric moisture content. We observed changes in soil volumetric moisture content with changes in soil compactness and built a coordinate transformation to make the coupling relationship between them easier to observe, and established a modified soil volumetric moisture content model accordingly based on soil compactness.

Acknowledgment

This work is supported by the Fundamental Research Funds for the Central Universities (Grant No. 2019YC15), National Key Research and Development Program of China (Grant No. 2017YFD0600901) and Special Fund for Beijing Common Construction Project, Design of Frozen Soil Sensing Sensor Based on Dielectric Method (Grant No. QN2018317).

References

[1] Dean T J, Bell J P and Baty A J B 1987 Soil moisture measurement by an improved capacitance technique: Part I sensor design and performance *J. Hydrol. (Amsterdam, Neth.)* **93** 67-78
[2] Gaskin G J and Miller J D 1996 Measurement of soil water content using a simplified impedance measuring technique *J. Agric. Eng. Res.* **63** 153-9
[3] Bell J P, Dean T J and Hodnett M G 1987 Soil moisture measurement by an improved capacitance technique, part II. Field techniques, evaluation and calibration *J. Hydrol. (Amsterdam, Neth.)* **93** 0-90
[4] Pellet C, Christin H, Antoine M *et al* 2016 Soil moisture data for the validation of permafrost models using direct and indirect measurement approaches at three alpine sites *Front. Earth Sci.* **3** 1-21

[5] Patil S, Ramgir N, Mukherji S *et al* 2017 PVA modified ZnO nanowire based micro sensors platform for relative humidity and soil moisture measurement *Sens. Actuators, B* **253** 1071-1078
[6] Aravind P, Gurav M, Mehta A *et al* 2016 A wireless multi-sensor system for soil moisture measurement *Sensors. IEEE* **2015** 1-4
[7] Gao Z, Hao T, Zhao Y *et al* 2017 Linear scale measurement method for soil profile moisture *Trans. Chin. Soc. Agric. Mach* **48** 257-64
[8] Boada M, Lazaro A, Villarino R *et al* 2018 Battery-less soil moisture measurement system based on a NFC device with energy harvesting capability *IEEE Sens. J*, **18** 5541- 5549.
[9] Lekshmi S U S, Singh D N, Tarantino A *et al* 2018 Evaluation of the performance of TDR and capacitance techniques for soil moisture measurement *Geotech. Test. J* **41** 292-306
[10] Rahimi-Ajdadi F, Abbaspour-Gilandeh Y, Mollazade K *et al* 2018 Development of a novel machine vision procedure for rapid and non-contact measurement of soil moisture content *Measurement* **121** 179-89
[11] Cao D F, Shi B, Wei G Q *et al* 2018 An improved distributed sensing method for monitoring soil moisture profile using heated carbon fibers *Measurement* **123** 175-84
[12] Cao D F, Shi B, Zhu H H *et al* 2018 A soil moisture estimation method using actively heated fiber Bragg grating sensors *Eng. Geol.* **242** 142-9
[13] Matula S, Bã Å K and Legese W L 2016 Laboratory performance of five selected soil moisture sensors applying factory and own calibration equations for two soil media of different bulk density and salinity levels *Sensors* **16** 1912
[14] Pramuhadi G and Sembiring E N 2011 Soil - metal adhesion due to the variation of soil compactness and the soil moisture content *J. Keteknikan Pertanian* **15** 57-71
[15] Niu J Z, Zhou H and Shi D K 2006 Effect of moisture content, degree of bale compactness, and urea preservation on baled alfalfa hay quality *Acta Agrestia Sin.* **14** 34-8
[16] Cui F, Wu Z Y, Wang L *et al* 2005 Application of the ground penetrating radar ARMA power spectrum estimation method to detect moisture content and compactness values in sandy loam *J. Geol. Appl. Geophys.* **120** 26-35
[17] Bronick C J and Lal R 2005 Soil structure and management: A review *J. Geoderma* **124** 3-22
[18] Luo X and Perumpral J V 1995 Study on soil strength with microwave reflection loss *Trans. Chin. Soc. Agric. Mach* **11** 46-51
[19] Ayers P 1982 Moisture and density effect on cone index *Trans. ASAE* **25** 1169-72
[20] Vaz C M P and Hopmans J W 2001 Simultaneous measurement of soil penetration resistance and water content with a combined penetrometer-TDR moisture probe *Soil Sci. Soc. Am. J.* **65** 4-12
[21] Ohu J O 1988 Cone index prediction of compacted soils *Trans. ASAE* **31** 0306-10
[22] Busscher W J 1990 Adjustment of flat-tipped penetrometer resistance data to a common water content *Trans. ASAE* **33** 519-24
[23] Lin J, Sun Y and Lammers P S 2014 Evaluating model-based relationship of cone index, soil water content and bulk density using dual-sensor penetrometer data *Soil Tillage Res.* **138** 9-16
[24] Yi L T, You K Y, Dimon M N *et al* 2016 A modified microstrip ring resonator sensor with lumped element modelling for soil moisture and dielectric predictions measurement *Measurement* **94** 119-25
[25] Goutal N, Keller T, Défossez P *et al* 2013 Soil compaction due to heavy forest traffic: measurements and simulations using an analytical soil compaction model *Ann. For. Sci.* **70** 545-56
[26] Zegelin S J, White I and Jenkins D R 1989 Improved field probes for soil water content and electrical conductivity measurement using time domain reflectometry *Water Resour. Res.* **25** 2367-76
[27] Zhao Y and Wang Y 2002 Analysis of sensitivity of soil moisture measurement based on standing-wave ratio *Trans. Chin. Soc. Agric. Mach.* **18** 5-8

[28] Mossadeghi-Björklund M, Arvidsson J, Keller T *et al* 2016 Effects of subsoil compaction on hydraulic properties and preferential flow in a Swedish clay soil *Soil Tillage Res.* **156** 91-8

Information technology promotes the development of agriculture in China: the mechanism, bottleneck problems and solutions

R Feng[1]

E-commerce Department, Guangdong University of Foreign Studies, Guangzhou, China

E-mail: fengran@oamail.gdufs.edu.cn

Abstract: With the purpose of maximizing productive efficiency by means of leveraging information technology in agriculture, this study creates agriculture information system which constitutes factors from supply-side based on the theories of economics. Then, Gray Correlation Analysis has been used to analyze agro- industrialization. The wide application of the Internet and financial support in agriculture exert significant positive effect on the development of agriculture. At the same time, less agriculture weather stations, developing agriculture futures have no obviously positive effect on it. Low level education of the rural labour has the negative effect on it. Therefore, suggestions like increasing the investment on rural education, giving more support on land conservation, integrating the data from different relevant fields to develop agriculture futures, developing meteorological monitoring technology are given based on the experience of developed countries.

1. Introduction

Traditional agriculture in China is mainly promoted through increasing productive factors. Specifically speaking, the factors are: high investment in agriculture resources, high emission of pollutants, relatively high yield of agriculture products, and low usage of energy. Such "extensive" business model cannot keep a long-term and sustainable growth since solid agriculture resources in China are in short. And the model does not match the thought that China's economic transformation should be high-level, clustering, informatized, green and low-carbon. Therefore, it is urgent for China's agriculture development to shift from being "extensive" to "intensive".

Among all the driving factors of "intensive" agriculture, information technology is a key one that cannot be neglected. It is not only the main tool to realize agriculture modernization, but also a

Content from this work may be used under the terms of the Creative Commons Attribution 3.0 licence. Any further distribution of this work must maintain attribution to the author(s) and the title of the work, journal citation and DOI.
Published under licence by IOP Publishing Ltd

significant way to stimulate national GDP (Gross Domestic Product, GDP). According to survey data, the GDP of developed countries has increased by 5%-9% because of information technology, and the figure is as high as 15%-25% in developing countries [1]. Therefore, to enhance the application efficiency of information technology and agriculture informatization is of great significance for facilitating agriculture modernization [2]. However, nowadays, China is suffering from many adverse factors of agriculture informatization, for example, the education in rural areas is "of low quality" and the agriculture information seems to be "isolated", etc [3]. How to solve these problems become the key for developing agriculture and rural new kinetic energy. Therefore, in this study, the current condition and existed problems will be combed through and analyzed. Also, on the basis that information technology and agriculture modernization mechanism has been well illustrated, quantitative analysis will be performed to locate adverse factors on agriculture informatization. Then, the experience in agriculture of developed countries will be referred to before coming up with corresponding measures for improvement to accelerate the process of agriculture modernization.

2. Current status and problems of agriculture informatization in China

In the area of agriculture production, there is a relatively large gap between the Internet of Things in China and that in the United States. However, since 2010, when the National Development and Reform Commission launched the industrial plan of the Internet of Things covering fine husbandry, until 2015, when the Ministry of agriculture promoted 116 items of the Internet of Things model to save water, fertilizer and labor, which can "save costs and improve performance". Within a short period of over ten years, during the research and development stage of the technology of Internet of Things, China has completed the task of 3S (Include Remote sensing, Geography information systems and Global positioning systems) integrated technology and application software and developed in the direction of manufacturing production equipment and agriculture machines or tools with intelligent control on a large scale. As for the implementation, in the past, the information was collected manually and the mechanical operation was regular, but now it is semi-automated, and will be fully-automated. During the promotion stage, it has been shifting from large-scale farms and high-tech agriculture pilot zone which are less affected by natural conditions and of high-level industrialization degree to capable rural and farming households [4]. However, when the technology of Internet of Things is developing fast, the agriculture yield capacity becomes relatively low. According to UN's Food and agriculture Organization (FAO), the amount of fertilizer use and average crops (grain) production index in 43 countries show that the fertilizer amount in China is 364.4 (ton/1000 hectares). It is far more than the figure of Australia, 50.9, and that of the United States, 131.9. China's figure also surpasses the average use amount of major countries, which is 284.96 tons. On contrast, the per capita yield of crops (grain) in China is only 948.51 (kilo/hectare) in 2016. It is only about 1/15 of that of Australia, 13925.05, and about ¼ of the average level at 3765.88. Therefore, from the perspective of yield, information technology hasn't effectively stimulated agriculture production efficiency.

In terms of agriculture operation, as the increase of agriculture cooperative and large agriculture-related enterprises, information integration system has extended from pure production plan and management to various areas including supply and sales, financial management and personnel management. An integrated agriculture ERP (Enterprise Resource Planning, ERP) system covering various agriculture areas has been formed. For government organs of agriculture, agriculture information system can not only be used to collect the information about price, environment, quality control and farming. These data will be released to the public through databases like "agriculture Economic Information Calendar", "agriculture Macro Data Enquiry", and so on. These have also been applied to daily administrative management by all levels of agriculture departments in the form of "electronic administration". In this way, the efficiency of the government will be enhanced. For example, in 2017, the website of the Ministry of agriculture has undertaken as many as 49 online administrative business items and the daily capacity of case handling has reached about 250 [5]. Despite of the fact that China's e-government has already reached the above-average level [6], but there is a distinct imbalance in the management of agriculture information. Such imbalance is not only manifested geographically, but also in each part of supply chains. To be specific, the e-government level in eastern China, southern China, and northern China is relatively high, but that in the southwest and northwest is relatively low. There is an obvious disadvantage in future information management of high-end agriculture products. Currently, there are 192 categories of future trading around the globe, but in China, there is only 22 categories future trading of agriculture products.

In the area of agriculture multi-industrial fusion, in 2015, the Ministry of agriculture started the e-commerce pilot projects for agriculture products in 10 provinces and cities including Beijing, Guangdong, Hainan, and Hebei. In 2016, the cold chain network for agriculture products of the leading enterprise "Yiguo Fresh Products" has covered 27 provinces and 310 large and medium cities in China. The new industrial chain model of "the Internet+ the New Three Rural Issues" which makes the delivery of agriculture products possible within the day or order has been achieved. In 2017, the number of electronic platforms for China's agriculture products has reached 585. The average yearly increase rate of online sales volume has reached 52.58% (during the period from 2013 to 2017) [7]. In the meantime, the leisure tourism in rural areas has seen prominent increase. Supported by the "No.1 Document" issued by the Central Government and the "Poverty-Relief Plan by Rural Tourism" issued by National Tourism Administration, there has been 2.1 billion person time of tourists visiting rural areas in 2016, with the business revenue increasing 570 billion yuan, accounting for about 14.47% of total domestic tourism income [8]. However, according to "China's Unicorn Enterprise Development Report" released by the Ministry of Science and Technology in 2016, among 30 "Unicorn" e-commerce enterprises in China, only 2 serve for agriculture products. Another example is that in 2017, the number of model county for e-commerce entering the rural areas has reached 36, with 1946 e-commerce service stations built. However, the 800 million yuan of e-commerce volume only accounted for 0.46% [9] of the total value of agriculture output which was 175.25 billion yuan in Guangxi Province hat year. From the above analysis, despite of the fact that information technology and the Internet has been applied into agriculture production, operation and multi-industrial integration, it still shows distinct geographical imbalance. That is to say, successful cases are mostly in the southeastern and coastal areas or the key agriculture pilot industries in China; moreover, it does not show significantly positive impact on agriculture production efficiency or multi-industrial integration.

Obviously, information technology hasn't given play for enhancing China's agriculture production, operation and service. Then, what indeed are the problems? For agriculture operation, there are many procedures, which are independent from each other but also overlap. As a result, to locate the problems becomes difficult. But if looking at the issue from economic perspective, it can be found that information technology is affected by several factors like the current level of domestic information technology, how IT has penetrated agriculture area, the labor force, tools, subjects who carry, spread and use the information, as well as social system. Therefore, from the supply-side perspective, the factors affecting agriculture informatization will be pinpointed.

3. Mechanism of how information technology enhances agriculture modernization

Agriculture modernization mainly refers to the process of enhancing farmers' quality through modern equipment, management approaches, service system and knowledge. These are mainly manifested in the procedures of agriculture production, operation and service. In the modernization drive of agriculture, information technology plays an important role, improving the efficiency of agriculture production, operation and service and facilitating the multi-industrial structural integration of agriculture (details in Figure 1 below) [10].

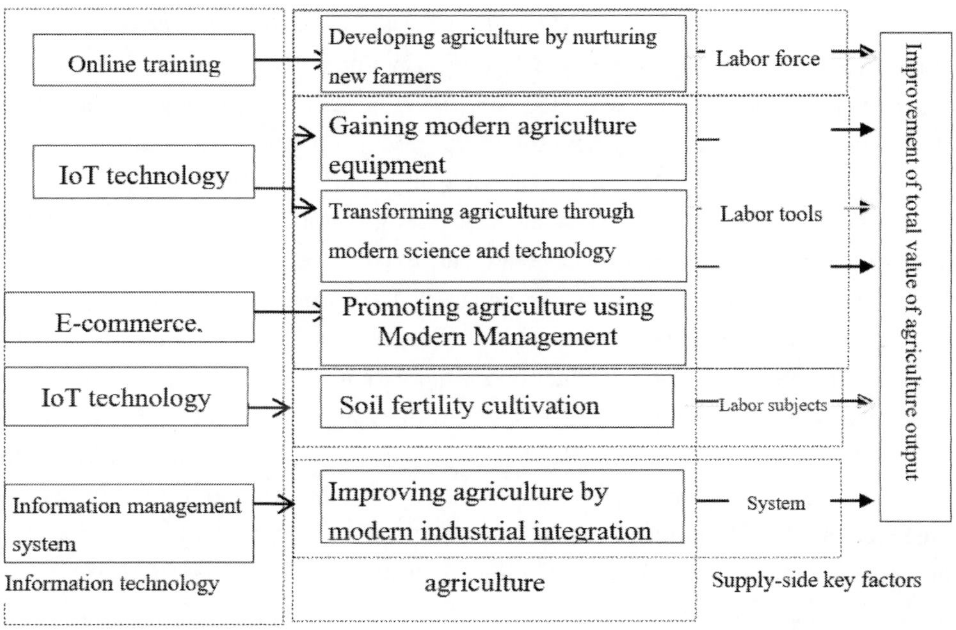

Figure 1. Relation scheme of information technology and agriculture modernization.

3.1. Function of information technology in enhancing labor force and labor subject potency

The Internet is the infrastructure carrying and spreading information and the important channel and physical basis for arranging the Internet of Things and cultivating rural new labor force. The features of the Internet make it possible for information transmission to break the limitations of time and region. Thus, Metcalfe's rule can be well functioned. The agriculture information and corresponding overflow value can show an increase of N^2 online. Therefore, as the advancement of projects of "Broadband China" and "Broadband Village Plus", 201 million users (by the end of 2017) have been involved in rural Internet, showing a penetration rate of 34.0% [11]. Particularly, the balanced development of regional Internet, i.e. 62.7% in the eastern costal area, 43.0% of central area, and 45.8% of western area, will all enhance the educational level of farmers. Then the means of labor will help improve the efficiency of labor subjects [12], and there will be overflow effect of voluntary innovation. In the meantime, the Internet of Things will be effectively leveraged to monitor the soil condition.

3.2. Function of information technology in the enhancing efficiency of labor tools

To enhance the efficiency of labor tools include the in-depth integration of first and second industry, as well as to improve the efficiency of labor tools in a limited sense and to expand the function of labor tools of broad sense through the integration of first and tertiary industry.

To enhance the efficiency of labor tools of limited sense through information technology refers to the process that advanced biological technology, information technology and industrial equipment are put into application in agriculture production to improve agriculture production efficiency and output by better farming, water conservancy, mechanization and informatization. A typical example is the technology of the Internet of Things, which centers on information technology and composed of sensors, network transmission and software analysis. It can monitor the indicators of farming and cultivation like water content [13-14], temperature [15], nutrient [16], salination degree [17], and so on in a real-time manner. It will adjust the land fertility, save much labor costs, and improve crop output. Moreover, it is demonstrated that when the application rate of the technology of the Internet of Things is 80% in the farms of Untied States, the grain production has realized an output of 7.34 tons/hectare, which is far higher than the average global level of 3.85 tons/hectare [18].

To improve the potency of labor tools of broad sense through information technology refers to the process of gaining as much as agriculture development, extending the industrial chain, and fully leveraging the features and functions of the tertiary industry. The purpose is to increase sales channels of agriculture products, improve corresponding added value, reduce agriculture production risks and stabilize price in the market [19]. For example, under the external effect of network, the marginal effect will show progressive increase, which is different from traditional economy [20]. To be specific, the larger the number of users in the network, the lower the marginal effect of each user, and the stronger the potency will be. Therefore, in the rural areas, the e-commerce platform can be used to sell agriculture products and develop tourism, as well as online trading of forward futures of agriculture products. Such high-end financial tools of agriculture industrial chain can be used to protect and guarantee the benefits of farmers.

3.3. Function of information technology in enhancing systematic potency

Reasonable agriculture system can help to achieve large scale operation of agriculture lands. The costs before, during and after the agriculture production, trading and operation can be lowered. At present, as the proposal of the concept of "three separate rights" of lands, the issue that restricts the scale expansion of China's agriculture has been shifted from land property dispute to agriculture management. In the meantime, since agriculture operation is easily affected by market supply and demand, price fluctuation and other external factors, its management requires the collection and fast handling of various kinds of information. According to the bounded rationality theory of Herbert Simon, complicated and uncertain external environment will affect the rationality of human beings. In order to observe the matters more objectively, human will collect more information, which will increase the rational costs [21]. The new institutional economics holds the opinion that social activities will generate trading costs between enterprises and between the enterprise and government. Therefore, the Internet is not limited by time and space and can spread as fast as light. The automatic approaches of information management system can help people to collect information in a more convenient way, enhance their cognitive ability, improve the management efficiency of agriculture matters, lower the imbalance of trading information, and reduce trading costs, so that various trading risks can be avoided [22].

4. Quantitative analysis on factors affecting China's agriculture informatization drive

In this study, the agriculture informatization development index system has been established based on the agriculture modernization index system proposed by Literature Information Center of Chinese Academy of agriculture Sciences from the perspective of four key factors of supply-side in economics, i.e. labor force, labor subjects, means of labor and labor mechanism. In the meantime, quantitative analysis has been performed on the relations among these factors, with the purpose of locating adverse factors on the full functioning of information technology in the agriculture area.

4.1. Establishment of agriculture informatization development index system

Since China's agriculture informatization statistics is not complete, the factors of authority, system and collectability of the data are taken into consideration for the agriculture informatization quantitative index system designed in this study (details specified in Table 1 below). The index system includes: 1. The Internet infrastructure construction status, which reflects the development of information of technology; among which, the index of the total number of national IP4 and IP6 addresses can reflect the network construction status; 2. The Internet popularization rate, which reflects the penetration degree of IT in agriculture areas; among which, the penetration rate can be reflected by the number of national IP addresses and the population with access to the Internet; 3. The rural labor forces carrying information and imparting knowledge; among which, the number of graduates from technical training school and high school and rural labor force can reflect the information conversion capacity; 4. Labor tools of integrating and applying information technology; among which, the electricity amount in rural areas and rural mechanical total power can reflect the popularization degree of modern labor tools; the information socialization degree, agriculture product future trading volume and the number of agriculture meteorology station can reflect how modern operation can facilitate agriculture development; 5. Means of labor (land) which accept informatization; among which, the ratio of total grain output against total fertilizer amount can reflect the fundamental abilities of China's lands; 6. System and industrial structure environment which supports agriculture development; among which, the portion of agriculture costs of total fiscal costs and the portion of the first industry in total GDP can reflect how supportive the government is for the new form of agriculture development and the degree of multi-industrial integration.

Table 1. Index system to evaluate agriculture informatization.

Overall Index	Main Index		Group Index
Degree of agriculture informatization	Popularization of information technology		Number of IP address in the country
			Population with access to the Internet
	Labor force	Developing agriculture by nurturing new farmers	Number of graduates from technical training school
			Number of graduates from middles school in rural area
	Means of labor	Equip agriculture with modern materials	Used electricity of rural areas
			Total mechanical power in rural areas
		Facilitate agriculture by modern operation form	Index of information socialization
			Futures trading volume of agriculture products (bean pulp, corn)
			Number of agriculture meterology stations
	Labor subjects	Transforming agriculture through modern science and technology	Unit output of agriculture products (grain)/fertilizer amount
	System	Government supports the development of new agriculture	Proportion of agriculture costs against total fiscal costs
		agriculture enhanced by modern industrial integration	Trading volume of agriculture products*
agriculture informatization benefits		Increase of agriculture GDP	Total production value of agriculture, forestry, animal husbandry and fishery

Note: The agriculture modernization index system proposed by Literature Information Center of Chinese Academy of agriculture Sciences has been referred to; *The e-commerce trade volumes were only recorded in 2016 and 2017. Therefore, it is unable to measure the integration degree of first and tertiary industry facilitated by informatization.

4.2. Effect of analyzing informatizing factors

4.2.1. Data description. The data includes the gross value of agriculture production (the gross value of agriculture, forestry, animal husbandry and fishery production in real terms) as the explained variable, number of IP addresses, online population of rural areas, graduates from technical training school, number of secondary school graduates, electricity for rural use, total machinery power, information socialization index and agriculture product futures trading volume (bean pulp, corn), the ratio of grain production to chemical fertilizer application, the number of national agriculture meteorological observatories and the proportion of fiscal agriculture expenditure to total fiscal expenditure as the explanatory variables. The above indicators are selected from 2005 to 2015 (see Table 2 for details), as the statistics of online population in rural areas started in 2005 and the data of 2017 China agriculture Statistical Yearbook is up to 2015.

Table 2. Agriculture Informatization.

Year	AgricultureGDP(100 million yuan)	Number of IP Address (unit: 10,000)	Online Population in Rural Areas (unit: 10,000)	Number of secondary school graduate (unit: 10,000)	Graduates from technical training school (unit: 10,000)	Electricity for Rural Use (myriawatt)	Total Machinery Power (myriawatt)	NIQ	Corn futures trading volume * (unit: 100 million yuan)	Grain Yield/Fertilizer	Agriculture Meteorological Stations **	Proportion of agriculture Expenditure***
2005	23537.	7439	1931.4	75.1	4793.2	4375.7	68549.	-	-	0.0895	836	7.2
2006	24418.	9802	2284.9	79.7	4520.6	4895.8	72636	-	1.35	9.1521	839	7.9
2007	25732.	13527	2311	115.7	4670.3	5509.9	76878.	-	1.19	8.9338	808	6.8
2008	26809.	18127	7460	75.9	4358.2	5713.2	82190.	-	1.12	9.1329	720	7.2
2009	27954.	23245	10681	51.6	4130.7	6104.4	87496.	-	0.33	8.9105	721	8.7
2010	40530.	27764	12500	59.8	3813.1	6632.3	92780.	-	0.72	8.9248	721	9.0
2011	42253.	33045	13600	59.5	3794.7	7139.6	97734.	0.32	0.54	9.1054	721	9.1
2012	44174	33054	15600	56.4	3563.2	7508.5	102559	0.36	0.76	9.2372	721	9.5
2013	45923.	33033	17700	47.1	3416.0	8549.5	103906	0.39	0.27	9.3488	721	-
2014	47849.	33201	17800	56.8	3207.7	8884.4	108057	0.42	0.19	9.2964	721	-
2015	49786.	33654	19500	-	-	9026.9	111728	0.43	0.84	9.5022	723	-

Data source: "China's agriculture Statistical Yearbook" and "China's Information Statistical Yearbook" during the period of 2006 to 2017 [23-24]. Note: *Limited by the data, the corn futures trading volume only include the annual trading volume in Dalian Future Exchange; ** The amount of agriculture meteorological observation stations includes the number of agro-meteorological observatories plus the number of ecological and agriculture meteorological stations; ***The proportion of agriculture expenditure refers to the proportion of agriculture fiscal expenditure to total fiscal expenditure; ****Reciprocal of proportion of agriculture GDP in the total GDP.

4.2.2. Model design. To further verify changes in the relationship between agriculture informatization indicators from the quantitative perspective and avoid endogenous problems encountered in the use of multi-consumption model, this study adopts the Grey System Theory and T's Correlation Degree Analysis suitable for analyzing "small sample" data to get the correlation degree. Specifically, the relationship between gross farm production and total machinery power are exemplified to explain the solution process of T's Correlation Degree.

Set x_1 as the reference sequence that represents the agriculture GDP, x_j as the comparison sequence (j=2, 2...12) that represents the remaining eleven variables (listed in the table 2) such as the Number of IP Address, and values of the reference sequence and the comparison sequence between

$[t_1, t_n]$ are $x_1 = \{x_1(t_1), x_1(t_2), \cdots, x_1(t_n)\}$ and $x_j = \{x_j(t_1), x_j(t_2), \cdots, x_j(t_n)\}$ respectively, where the number of periods is n which is from 2005 and 2015. According to the improvement of Grey Correlation Analysis by Tang Wuxiang [25], the basic calculation steps of T's correlation degree analysis are as follows:

① Standardize to make the values of each sequence comparable:

$$D_1 = \frac{1}{n-1}\sum_{j=2}^{12}|x_1(t_n) - x_1(t_{n-1})|$$

$$D_j = \frac{1}{n-1}\sum_{j=2}^{12}|x_j(t_n) - x_j(t_{n-1})|$$

② Get the standardization sequence:

$$y_1 = \{x_1(t_n)/D_1, n = 1, 2, \cdots, 11\}$$

$$y_j = \{x_j(t_n)/D_j, n = 1, 2, \cdots, 11, j = 2, 3, ..., 12\}$$

③ Get the Incremental sequence:

$$\Delta y_1 = \{\Delta y_1(t_n) = y_1(t_n) - y_1(t_{n-1}), n = 1, 2, \cdots, 11\}$$

$$\Delta y_j = \{\Delta y_j(t_n) = y_j(t_n) - y_j(t_{n-1}), n = 1, 2, \cdots, 11; j = 2, 3, ..., 12\}$$

④ Calculate correlation coefficients of each period:

$$\mathrm{sgn}(\Delta y_1(t_n) \cdot \Delta y_j(t_n)) \cdot \frac{\min(|\Delta y_1(t_n)|, |\Delta y_j(t_n)|}{\max(|\Delta y_1(t_n)|, |\Delta y_j(t_n)|}$$

$$\xi(t_n) = 0 \qquad (当\Delta y_1(t_n) \cdot \Delta y_j(t_n) = 0时)$$

⑤ Calculation of correlation:

$$r(x_1, x_j) = \frac{1}{n-1}\sum_{j=2}^{12}\Delta t_n \cdot \xi(t_n)$$

Table 3. Correlation of agriculture Informatization Index.

Variables	Correlation	Variables	Correlations
GDP* & NIQ	0.71	GDP &Soybean futures trading volume	0.06
GDP & Machinery Power	0.57	GDP &Grain Yield/Fertilizer	0.04
GDP &Online Population in Rural Areas	0.38	GDP &Number of secondary school graduate	0.03
GDP &Proportion of agriculture Expenditure	0.28	GDP &agriculture Meteorological Stations	-0.01
GDP &Number of IP Address	0.19	GDP &Graduates from technical training school	-0.43
GDP &Corn futures trading volume	0.07		

Note: *GDP represents agriculture GDP.

4.2.3. Result. The degree of correlation between the explained and explanatory variables is calculated to explore the underlying reasons for restriction of the development of agriculture informatization profoundly (see Table 3). According to the final results, there is a positive correlation between the total agriculture production and the social informatization index, rural electricity consumption, total mechanical power, the number of online population, per unit output of agriculture products, the proportion of agriculture fiscal expenditure to total fiscal expenditure and the number of IP addresses. Specifically, the effects of increase of online population in rural areas on the increase of agriculture GDP are significantly higher than that of nationwide network promotion; Its relationships with the number of secondary school graduates, the number of agriculture meteorological stations, the volume of agriculture futures trading and the ratio of grain yield to the amount of chemical fertilizer utilized are not significant; and there a negative correlation between it with the number of graduates of agriculture technology training.

4.3. Conclusion and explanation

Obviously, the increased financial support in the agriculture sector of China, rural labor mechanization presented in the combination of agriculture and modern industry and the rather advanced means of labor for agriculture development resulted from the informatization and networking of key rural construction promoted the increase in the total value of agriculture production effectively. In the meanwhile, the correlation between the number of agriculture meteorological stations, the turnover of major agriculture products (corn, soybean meal) futures and the proportion of grain yield and fertilizer application and the development of agriculture is not significant, indicating that the informationized, marketized and high-tech construction of high-end sections of the agriculture industrial chain should be strengthened in combination of agriculture with means such as high-tech meteorological observation and financial futures trading to further promote the development of agriculture modernization. In addition, according to the research, the significant negative relationship between the number of graduates in secondary school and agriculture technology trainees reduced over the past ten years, especially for the agriculture technology trainees. The conclusion is in line with the idea of numerous scholars that the lower education of rural residents will hinder the deep integration of agriculture and informatization [26]. Strengthening training and delivery of agriculture technicians in rural areas should be emphasized for the promotion of agriculture modernization development.

5. Strategies for promoting the development of agriculture informatization of China

According to analysis, the effects of network popularization in rural areas on agriculture development are significantly higher than that of nationwide network popularization, so the government should continue to increase the construction of network facilities in rural areas, with the improvement of rural labor education, continuous cultivation of fundamental productivity and the application of high-end agriculture financial instruments and high-tech agriculture monitoring methods emphasized on this basis.

5.1. Enhancing the quality of rural agriculture labors significantly

The improvement of peasant education will be restricted by the scarcity of educational resources, thereby affecting their acceptation and application of information technology, as well as the creation of spontaneous innovation environment [27]. For example, according to the survey conducted by China Internet Information Center, among the reasons leading to the low network penetration in rural areas, in addition to the population of too young or old age, the proportion of not being computer and network literate is up to 60.0%, that of lacking Internet devices such as computer is only 9.4%. At present, the level of rural labor education in China is lower with small proportion of education resource input in rural areas. According to the data of *2016 Educational Statistics Yearbook of China*, the number of graduates in secondary school and agriculture technology trainees has been falling from year to year at an average rate of 15.30% and 5.28% respectively. The falling range is far more than the average natural decline of rural population, which is 0.31% [28]. In the meanwhile, although China's share of education expenditure in GDP of 4.26% in 2015 exceeded the target of 4% set up in the National Medium- and Long-Term Education Reform and Development Plan (2010-2020) formulated in 2010, it is still lower than the world average of 4.67%, especially agriculture developed countries such as 6.0% of Israel, 5.6% of the United States and 5.1% of Germany. As the total national education expenditure is lower than the global average, the education resources utilized by junior high school of rural compulsory education are only half of that of city, and the education resources utilized by senior high school of rural non-compulsory education are less than one tenth of city [29].

In order to improve the spillover effect of information technology and solve the contradiction between the demand for agriculture informatization talents and the lack of rural education, the Chinese government should increase the financial investment in education, increase the quality of agriculture labor force in multiple channels and approaches and enhance the soft power and hard power of "labor" in the supply side to achieve zooming of the efficiency of all agriculture factors. Specifically, all-free vocational training of agriculture of Britain, Germany and Sweden, delivery of agriculture talents in the form of pedagogical education of France and all-round talent training for agriculture production, development, maintenance, sales and tourism of the United States worth learning from; at the moment, the Chinese government, in the context of land "separation of the three powers" reform, can loosen the admittance criterion for agriculture business entities, grant corporate entities with land management qualification and deliver highly-educated employees to rural areas for agriculture production through enterprises to improve agriculture production efficiency.

5.2. Adjusting the input structure of maintenance fund, increase cultivation of fundamental productivity of arable land

The increase in crop yields in China is largely dependent on the input of large amounts of fertilizer. According to the agriculture data in Table 1, the ratio of cereal production to fertilizer use between 2005 and 2015 continued to be about 9, indicating that grain yields not only increased with fertilizer application, but remained essentially the same weight. Since the productivity of arable land consists of fundamental productivity and fertilizing amount, soil with high fundamental productivity has higher potential for high yield, and soil with low fundamental productivity can only achieve high yield at higher fertilization level [24,30]. As the increase of grain production is almost proportional to the increase in fertilizer input in a linear regression manner in China, it is obvious that the increase of China's grain production is mainly resulted from the use of fertilizer, rather than the improvement of fundamental productivity of arable land.

China can start from strengthening the cultivation of fundamental productivity of land to achieve a good combination of land and information technology and increase the output of crops. By far, agriculture developed countries have invested a large amount of funds to protect the agriculture ecological environment. Taking the United States for instance, it spends a lot of money each year to encourage farmers to maintain their land. For example, according to the 2014-2018 agriculture budget published in the 2014 Farm Bill, USA will invest $28.165 billion in land conservation, accounting for 5.8% of total agriculture expenditure. As China has established the awareness of protecting the ecological environment, according to the data of 2016 China agriculture Statistical Yearbook, 2.005 billion yuan was invested in agriculture resources and ecological protection, accounting for 6.69% [31] of the total investment in agriculture construction of 29.974 billion yuan in 2015. The proportion is basically the same with the financial investment in agro-ecological protection of USA. However, 2 billion yuan of the 2.005 billion yuan of resource protection funds was used for returning grazing land to grassland, and 5 million yuan was used to manage the prominent environmental problems, with the funds used for pasture and arable land conservation accounting for 0.06% and 0.00009% of the gross value of livestock and agriculture production that year respectively, indicating that the proportion of funds invested in ecological protection is not proper. In addition, the funds for agriculture science and technology innovation capacity in 2015 were 470 million yuan, accounting for only 1.57% of the investment in agriculture construction. Although it is close to 2% of USA's R&D in total agriculture expenditure, due to the completely different agriculture modernization progress, China, which is in the early stage of agriculture modernization, should continue to increase financial investment in agriculture science and technology innovation to support the development and application of information technology in land productivity improvement.

5.3. Making up the weakness in informatization of high-end sections of agriculture

Agriculture commodity futures trading is one of the important symbols of modernization and marketization of agriculture development. Since agriculture is vulnerable to natural disasters and market price fluctuations, hedging in the futures market is rather useful for developed countries to transfer planting risks, guarantee farmers' income and stabilize farming. Among the numerous agriculture developed countries, USA has an over 80% proportion of farmers who hedges with futures. In addition to the rich variety of futures and active trading, USA has established an improved legal system with laws [32] such as the Futures Management Law, Speculation Restricting Act and Commodity Futures Modernisation Act. In particular, the US Department of agriculture has provided farmers and agriculture operators with comprehensive data support for getting involved in futures trading by supplying detailed information such as the total yield, acre yield, market price, market sales, warehouse inventory and demands of USA and other countries via technical means such as satellite high-angle shot and picture analysis, with the stability of US agriculture and the enthusiasm of farmers promoted effectively.

The agriculture futures trading history of China is short and the information of auxiliary transactions is incomplete at present. China only established the first agriculture futures exchange in 1993 and the government proposed the "insurance + futures" policy pilot in the No.1 Central Document in 2016, resulting to problems such as insufficient financial support in the agriculture futures market, high premium amount paid by farmers; only 22 agriculture commodity futures, few variety of active trading and small market scale; short supply of rural financial services and limited risk management awareness of farmers, etc., farms haven't fully utilized futures to stabilize the forward price of agriculture products.

China should pay much attention to the informatization construction needs of agriculture financial derivatives to make up the weakness in informatization of high-end sections of agriculture. Firstly, China shall encourage farmers to get access to the relevant information of agriculture commodity futures, guide farmers to get involved in the agriculture futures market in the form of rural cooperatives, make agriculture marketed and form the price formation mechanism for agriculture; secondly, China shall set up a modern information disclosure system for agriculture products, deepen the organic connection between the spot market and the futures market of agriculture products and provide information-based channels for predicting the price of agriculture products. It will establish unified and authoritative information dissemination channels and platforms to the greatest extent to integrate and analyze cross-border agriculture-related data of authorities such as meteorological departments, agriculture departments, environmental protection departments, industrial and commercial departments and foreign trade departments and information such as increase and decrease of planting area and changes in climatic conditions. Moreover, the system will provide channels for agribusiness and large farms to acquire real-time data and reference trend analysis reports via the Internet and database, so that reasonable agriculture product prices and production expectations can form. Thirdly, China shall improve the legal system of agriculture futures trading, guarantee the authority of agriculture products futures trading commitments and arouse the enthusiasm of farmers and agriculture commodity operators to participate in futures trading.

5.4. Strengthening supply of high-tech agriculture supervision information

The level of informatization and automation of agro-ecological climate monitoring plays a key role in all aspects of agriculture development. Prior to agriculture production, agro-meteorological information can help farmers to identify the best planting time; during agriculture production, real-time information on agro-meteorology can ensure that agriculture products meet expected standards and avoid losses resulted from climate. For example, since rain is unfavorable for cherries at maturity, orchards in USA will prepare the drone in time after getting informed of rainstorm and lift it off when the rain starts to accelerate the evaporation of water on surface of cherries with the falling airflow generated by the wing during the flight, so as to avoid cherry rot caused by excessive watering[33]. After agriculture production, the timely supply of eco-information can help farmers to get informed of the fundamental productivity of arable land and identify the cycle of agriculture recuperation. For example, with methods for measuring the bearing capacity and the natural force of land such as economic model [34], trend analysis [35-36] abandoned, Geography Information System (GIS), which was originally used for urban construction planning [37], is applied to agriculture in the United States. In other words, data of various types monitored by the remote sensor is input to decision equations with different weights with results displayed by graphical techniques, achieving measurement, monitoring and prediction of land pollution, fundamental productivity improvement, forests and biodiversity [38-39], so as to assist agriculture production and income increasing by virtue of monitoring and improving land fundamental productivity.

China should pay much attention to the positive effects of supplying high-tech agriculture monitoring information on the development of agriculture informatization. Firstly, since meteorological observatory is the basis for timely mastering the agro-meteorological situation, integrated weather monitoring station should be increased in key agriculture production areas such as the northwest and northeast to change the imbalance of meteorological information supply between the East and the West. Secondly, agro-meteorological observation projects and means should be appropriately added, for example, to observe weather changes in combination of meteorological stations, satellites, radars and meteorological balloons; multiple items such as climate, ice and snow, hydrology, earth surface, paleoclimate, remote sensing and solar activity can be included in observation to guarantee all aspects of agriculture production; thirdly, active promotion of disastrous meteorological information should be enhanced to minimize the damage caused by catastrophic natural disasters to agriculture; finally, knowledge of meteorological science should be promoted via multiple channels, so that farmers can be aware of the importance of disaster prevention and mitigation.

Reference

[1] Rachel F F and ShayerahIlias A and Wayne M M 2017 Digital Trade and U.S. Trade Policy *Congressional Research Service* **6**(6) p 3-5

[2] Han H B and Zhang L 2015 Research on the function of agriculture information to the economic growth *Chin. Rural Econ.* **8** p 11-21

[3] Wang R J 2013 Bottleneck of agriculture Informatization development in China and the thinking of coping strategies *Bull. Chin. Acad. Sci.* **3** p 337-347

[4] State Statistical Bureau Rural social and economic investigation division 2016 *China Rural Statistics Yearbook* China Statistics Press

[5] Website of agriculture Ministry 2017 Work programme of promoting "Internet agriculture government service, http://jiuban.moa.gov.cn/zwllm/tzgg/tz/201701/t20170109_5426799.htm

[6] Website of UN Survey Report on E-government in 2016, http://www.chinacio.org/art/2016/8/2/art_1035_4882.html

[7] Chinese E-commerce Research Center 2018 2018th Chinese E-commerce Development Report of agriculture Product ,http://b2b.toocle.com/detail--6441580.html

[8] Statistics Bureau of the peoples' Republic of China 2017 *China Statistical Yearbook* China Statistics Press

[9] The Guangxi Zhuang Autonomous Region Statistics Bureau 2018 The development of Rural Economy in Guangxi Province http://www.gxtj.gov.cn/tjsj/jdfx/qq/201802/t20180211_143746.html

[10] Deng X 2013 Modern agriculture and agriculture Development *J. Huazhong Agricul. Univ. (Social Science)* **12** p 1-4

[11] China Internet Network Information Center 2017 The 40th Chinese Network Development Report, http://www.cnnic.cn/hlwfzyj/hlwxzbg/hlwtjbg

[12] Yu W and Zhang P 2015 Study on the coordinated development of provincial rural education and agriculture modernization of China's Provinces *J. South China Agricul. Univ. (Social*

Science Edition) **1** p 16-24

[13] Zhang X F 2016 Intelligent data transmission and field monitoring applied based on the internet of things *PHD Thesis of Donghua University* **9**

[14] Song Y and Huang L S 2017 Based on the virtual prototype of a kind of watermelon collecting de-seeding combine research *J. Agricul. Mechan. Res.* **7** p 189-191

[15] Cui N N, Liu B and Li S J 2017 IOT Test simulation research agriculture temperature and humidity data in the database *Comp. Simul.* **2** p 338-341

[16] Shi Z G, Liu Q C, Bai M J, Shi Y and Zhang S D 2017 Water and fertilization integrated intelligent irrigation system design and benefit analysis based on the Internet of Things *J. Water Resour. Water Engin.* **7** p 221-227

[17] Fu T F, Chen G Q, Xu X Y, Liu W Q and Su Q 2017 Design of a real-time monitoring and management system for soil salinization based on the internet of things *J. Ocean Tech.* **12** p 71-77

[18] Andrew Meola.Why IoT, big data & smart farming are the future of agriculture 2016, http://www.businessinsider.com/internet-of-things-smart-agriculture-2016-10

[19] Du Y 2018 Small-scale production and agriculture modernization 2018 *Chin. Rural Econ.* **12** p 2-6

[20] Feng R 2017 Competition constraint, operation paradigm and oligopoly governance of network platform *Reform* **5** p 106-113

[21] Lu X X 2007 *Neoinstitutional Economics* Beijing University Press, China

[22] Lu Y Y and Lu W 2017 Scale operation of the rural land based on the agriculture information construction *Scient. Technol. Managem. Land Resour.* **2** p 48-56.

[23] National Bureau of Statistics 2016 *China's agriculture Statistical Yearbook* China Statistics Press

[24] National Bureau of Statistics *China's Information Statistical Yearbook* during the period of 2006 to 2017 Publishing House of Electronics Industry, China

[25] Tang W X 1995 The calculated methods of the grey relational analysis *Appl. Statist. Manag.* **1** p 34-39

[26] Wang R J 2018 The importance of agriculture sensor and intelligent detection technology *China Rural Sci. Technol.* **1** p 32-36

[27] Song Y and Huang L S 2017 Monitoring and fertilization system system of agriculture soil nutrients based on internet of thing *J. Agricul. Mechan. Res.* **7** p 189-191

[28] National Bureau of Statistics *China agriculture Statistical Yearbook* from during the period of 2008 to 2016 China Statistics Press, China

[29] National Bureau of Statistics 2016 *Educational Statistics Yearbook of China of 2016* China Statistics Press, China

[30] Xia S Y 1998 The relation between crop yield and soil basic ferility with flue-cured tobacco *Shanghai Agricul. Sci. Technol.* **1** p 6-8

[31] Lei L G, Yuan H M Agriculture Input 2016 *China agriculture Statistical Yearbook* China Statistics Press, China

[32] Yang M 2016 Development process comparison and enlightenments of agriculture commodity

futures market in China *World Agricul.* **7** p 165-169

[33] Zhou J F, Lav R K, Troy P, Matthew D W, Zhang Q and David G 2016 Efficacy of unmanned helicopter in rainwater removal from cherry canopies *Comp. Electron. Agricul.* **124** p 161-167

[34] Veldkamp A and Fresco L O 1996 CLUE-CR: An integrated multi-scale model to simulate land use change scenarios in costa rica *Ecolog. Modelling* **91**(3) p 231-248

[35] Verburg P H, Koning G De, Kok K, Veldkamp A and Bouma J 1999 A spatial explicit allocation procedure for modelling the pattern of land use change based upon *Actual Land Use Ecolog. Modelling* **116**(1) p 45-61

[36] Verburg P H, Soepboer W, Veldkamp A, Limpiada R and Espaldon V 2002 Modeling the spatial dynamics of regional land use: the CLUE-S mode *Environ. Manag.* **30**(3) p 391-405

[37] Church R L 2002 Geographical information systems and location science *Comput. Operations Res.* **29**(6) p 541-562

[38] Martine N 2012 GIS and multi-criteria decision analysis for land use resource planning *J. Geograp. Inform. Syst.* **4** p 341-348

[39] GerleePuntsag. Land suitability analysis for urban and agriculture land using gis: case study in HVITA to HVITA, ICELAND *Land Restoration Training Programme* United Nations University

Preliminary study on development of city animal husbandry industrialization

S H Zhang[1,2]

[1]Animal Husbandry and Veterinary Station of Gongchang Town, Dingxi city, Gansu Province,748100, China

E-mail: 564846411@qq.com

Abstract. As an important part of agriculture, animal husbandry is not only the sun-rising industry, but also the advantage industry to increase farmers' income. Today, the industrialization of animal husbandry as a product of the socialist market economic development, has become key means for a sustained and stable development of animal husbandry. Dingxi City has a long history of the development of animal husbandry and livestock natural resources which provide a solid foundation, plus there are good transport conditions. Advantages of natural resources, location and other aspects of animal husbandry make Dingxi as dominant industries. Using of the existing advantages to achieve industrialization and intensification. In this paper, the development of the rural animal husbandry industrialization is preliminarily explored in the case of Dingxi City. Combined with the current situation of the animal husbandry industrialization in Dingxi City, the paper puts forward some problems. In view of the mentioned problems, this paper focuses on the industrialization of rural animal husbandry, and puts forward the following suggestions: cultivating and strengthening leading enterprises to play their leading role in animal husbandry industry chain; enhance the construction of animal husbandry professional cooperation organizations; continue to promote the training of talents and constantly strengthen the construction of socialized service system, and so on.

1. General situation of animal husbandry development in China

Animal husbandry is an indispensable industry in human society with a long history, since it was separated from agriculture, the relationship between animal husbandry and people's daily life has become more and more prominent. It offers a range of animal products that are highly nourished foods, because they contain higher proteins than vegetable foods. The living standard of people in a country or region can be measured according to the adequacy of animal food and animal protein in dietary structure. China's animal husbandry has made remarkable achievements since the steady development of reform and opening up. The national output of meat, eggs and milk increased from 65.15 million tons, 19.88 million tons and 13.37 million tons in 2015 to 87.067 million tons, 28.939 million tons and 38.412 million tons in 2017 respectively [1]. The increasing animal products, to a large extent, enrich the vegetable basket of residents, which basically meet the people's demand for animal products. However, with the reform of the market economy system, coupled with the change of people's diet concept and the improvement of living standards, the traditional animal husbandry can no longer meet the needs of people for the daily gain and growth of livestock products.

The transformation of livestock and animal husbandry industry into an important part of agricultural industrialization is an important means to promote the sustainable development of animal husbandry and to increase the income of farmers and herdsmen. The industrialization management of animal husbandry combines all links of production, processing and sales, and forms a mechanism of sharing

Content from this work may be used under the terms of the Creative Commons Attribution 3.0 licence. Any further distribution of this work must maintain attribution to the author(s) and the title of the work, journal citation and DOI.
Published under licence by IOP Publishing Ltd

interests and sharing risks through the complete industrial chain. According to the market demand, the leading enterprises carry out the processing of animal products, and the market information is transmitted to farmers in time, so that farmers can adjust the production structure of planting and animal husbandry in time, which guide the farmers to reduce blind production and solve the problem of disconnection between production and market; Closely connecting a series of processes from production to sales, effectively linking the production and market of small and scattered farmers and herdsmen, enabling the vast number of farmers to enter the large market, arousing the enthusiasm of farmers and herdsmen in production, and improving production efficiency, then meet people's demand for animal products [2].

2. Concept of animal husbandry industrialization

Industrialization refers to the specialized service and quality management which form the serialization and brand management of a certain industry under the condition of market economy, guided by market demand and aimed at realizing economic benefits. Animal husbandry industrialization is an important part of agricultural industrialization [3]. It refers to the various links in the process of economic reproduction of animal husbandry, that is, prenatal, middle and post-natal production as a complete industrial system. Its basic connotation is to take the market as the guide, take the economic benefit as the center, take the backbone enterprise as the "leader", take the thousands of families as the foundation, take the cooperative system and other intermediary organizations as the link and to carry out one-stop production and management system in the leading industry of animal husbandry in a region. It is characterized by close combination of production, supply and marketing, pastoral industry and commerce, animal husbandry science and education. Its essence is to form a complete industrialization pattern of market-intermediary organization-farmer by connecting the enterprises with the market and taking the animal products as the leading role according to the demand of the market so as to strengthen the effective connection of the production, supply and marketing processes. In order to promote the development of high-yield, high-quality and high-efficiency animal husbandry, and to promote the modernization of animal husbandry, it is necessary to carry out the operation mechanism of sharing risks and sharing interests, and to make animal husbandry enter the benign track of self-development and self-regulation.

3. Present situation and advantages of animal husbandry industrialization in Dingxi City

In recent years, in the face of the new normal situation of domestic economy from high speed growth to medium speed growth, Dingxi City has constantly improved its development ideas, deepened reform and people's livelihood construction, and made great efforts to adjust and optimize the industrial structure, so as to realize the steady development of the city's economy.

According to the statistical bulletin data of Dingxi City in 2017, The city's gross domestic product (GDP) was 106 billion yuan, an increase of 16.1 percent over the previous year. Among them, the output value of the primary industry was 24.11106 billion yuan, an increase of 0.97 percent over the previous year, the output value of the secondary industry was 51.76095 billion yuan, an increase of 14.65 percent over the previous year, and the output value of the tertiary industry was 30.21513 billion yuan, an increase of 22.29 percent over the previous year. The added value of the three industries is 25.897 billion yuan, 52.576 billion yuan and 27.535 billion yuan respectively. The three industries boosted economic growth by 1.3, 12.3 and 2.4 percentage points respectively. The three industrial structures are 24.4 m, 49.6 and 26.0. The per capita gross domestic product (GDP) was 66031 yuan, an increase of 15.9 percent over the same period last year.

From **Table 1** data show, the GDP of Dingxi City has increased year by year. Among the three industries, the output value of the primary industry and the third industry increased steadily, while the output value of the secondary industry increased at a higher rate and contributed the largest. Compared with various regions of Gansu, the added value of the secondary industry in Dingxi City was also in the forefront. The three industries show the industrial structure of "two, three, one", and the proportion of the second industry is the largest.

IOP Conf. Series: Earth and Environmental Science **346** (2019) 012035 doi:10.1088/1755-1315/346/1/012035

Table 1. GDP and output value of three industries in Dingxi City in 2017 (10,000 Yuan).

Year	2015	2016	2017	2017
Total output value	7029369	8185631	9373116	10608714
Primary industry	1903414	2167200	2387831	2411106
Proportion (%)	27.1	26.5	25.5	22.7
Secondary industry	3311095	3887740	4514582	5176095
Proportion (%)	47.1	47.5	48.2	48.8
Tertiary industry	1814859	2130691	2470703	3021513
Proportion (%)	25.8	26.0	26.3	28.5

Source: Dingxi Statistical Yearbook (2017)

3.1. General situation of animal husbandry development in Dingxi City

Since the founding of the people's Republic of China, the animal husbandry in Dingxi City has gone through several different stages of development [4]. The first stage was 1949-1957. The Party's policies related to animal husbandry liberated and developed the productive forces in rural areas. In the early 1960s, due to the adjustment of rural economic policies and systems, animal husbandry has been unprecedented development. The second stage is the period of the "Cultural Revolution" before the reform and opening up. In order to meet the grain demand, the farmers' grain self-sufficiency led to the destruction of grass and grain planting, and finally seriously hindered the development of animal husbandry production. The third stage is after the reform and opening up. Because the restriction of keeping livestock has been lifted, the purchase of live livestock according to the fixed base, the return of livestock prices and the joint production contract responsibility system have greatly stimulated the enthusiasm of farmers and herdsmen in production. At the same time, with the implementation of the separate household management system of grassland, the establishment and improvement of animal husbandry science and technology promotion service institutions [5]. The establishment of grassland supervision organization and veterinary health supervision organization and the strengthening of animal husbandry legal system have greatly promoted the vigorous development of animal husbandry in Dingxi City. By 2017, the livestock production in Dingxi City had reached 10.0967 million, the total number of livestock storage bars had reached 4.2275 million by the end of the year, and the total meat output in the city had reached 212900 tons, ranking first in the province. The output value of animal husbandry in the city was 3.721 billion yuan, accounting for 44.8% of the total output value of agriculture, forestry, animal husbandry and fishery. In recent years, Dingxi City has taken the sustained income of farmers and herdsmen as the core, strengthened the construction of bases, market development, processing and transformation, scientific and technological support, and made great efforts to build bases such as grain, livestock products, seed production, vegetables, and so on. Through a variety of preferential policies, the local government effectively mobilizes the enthusiasm of farmers. According to the statistical yearbook data of Dingxi City, in 2017, animal husbandry accounted for 17.03% of the per capita GDP of farmers and herdsmen. The added value of agriculture, forestry, animal husbandry and fishery in Dingxi City was 18.79791 billion yuan, of which the added value of animal husbandry was 10.54158 billion yuan, accounting for 56.08% of the total added value.

While strengthening the construction of animal husbandry transformation project in Dingxi City, we should focus on farming in agricultural areas, work hard to increase the quantity and improve the quality, and take the further income of farmers and herdsmen as the ultimate goal to promote the development of beef and sheep industry. The production capacity of animal husbandry in the whole city has also been significantly improved. In 2017, the output value of animal husbandry in Dingxi City accounted for 27.4% of Gansu Province. At the end of the year, there were 436.61 head in livestock storage, an increase of 0.7% over the previous year, and 944.5 000 livestock were released at the end of the year, an increase of 6.8%. The total output of meat in the city was 525100 tons, up 4.5 percent, milk output was 858300 tons, down 1.2 percent, and egg production was 85200 tons, up 2.5 percent. The proportion of the number of columns, meat, milk and eggs were 9.7%, 22.87%, 35.31%, 58.18% and 27.92%, respectively. Dingxi City has become an important animal production base in the province.

3.1.1. Animal husbandry in agricultural areas in Dingxi. The aquaculture industry in agricultural areas is mainly raised in captivity. It is characterized by:

- Food consumption is the main type of aquaculture. The main livestock species are pigs, poultry, livestock and goats, which consume more grain. The feed sources are agricultural products, feed grain, straw, wild grass, wild vegetables and so on.
- the dual-purpose aquaculture industry is more developed, such as cattle raising which is used for milk or labor, or for meat or labor, and so on.
- It is mainly raised in hostel. Except for short-term stubble grazing after harvest, artificial feeding was carried out in livestock houses for the rest of the time. Feed cost accounts for a relatively high proportion, generally accounting for more than 65% of livestock costs.
- If it can fully realize the combination of agriculture and animal husbandry, the management is meticulous, the production level is high, then the aquaculture industry in agricultural area is the key to realize the industrialization of animal husbandry.

As one of the development bases of animal husbandry in Gansu agricultural areas, Dingxi City has steadily improved the level of standardized animal husbandry breeding in recent years, and the scale of livestock cultivation is also expanding, and the number of livestock stocks is in the forefront of the province. The number of beef and sheep farms in the city has increased from 2 to 9, and 100 meat sheep breeding farms have been increased, and 5 standardized demonstration farms for beef and sheep breeding have been added. The income of meat cattle and meat sheep accounted for nearly 50% of the per capita income of animal husbandry.

From **Table 2** data, it can be seen that in 2017, the number of cattle, sheep and pigs accounted for 11.8%, 13.1% and 62.3% of the total livestock, respectively. As shown in table 2, the number of cattle in Dingxi City at the end of 2017 was 517300, accounting for 11.84% of the total number of livestock in the county at the end of the year and 13.48% of the total number of cattle in the county at the end of the year. Among the counties and cities, Longxi County and Lintao County account for a large proportion of the total number of livestock. There were 130000 cattle in Longxi County, accounting for 25.1% of the total number of cattle in the county, and 103600 cattle in Lintao County, accounting for 20.0% of the total number of cattle in the county. 206200 pigs in Longxi County, accounting for 36.1% of the total number of pigs in the county, accounted for the largest proportion compared with other counties and cities. There were 732700 sheep in Lintao County, accounting for 26.7% of the total number of sheep in the county. Compared with other counties and cities, the proportion of sheep in Longxi County was the largest.

Table 2. The number of livestock in Dingxi City (include 6 counties) at the end of 2017 (10,000 heads).

	Total	cattle	horses	donkeys	pigs	goats	sheep
Dingxi city	436.86	51.73	5.03	0.66	57.14	45.52	274.71
Longxi	86.04	13.00	0.90	0.01	20.62	11.31	39.79
Tongwei	29.32	4.70	0.41	0.04	3.85	6.87	13.31
Weiyuan	55.13	9.55	1.21	0.03	4.22	7.87	31.96
Zhangxian	56.64	4.89	0.38	0.06	9.36	1.47	40.22
Lintao	100.19	10.36	0.73	8.60	6.90	73.27	
Minxian	58.42	7.22	0.64	0.30	8.69	5.11	36.35

Source: Dingxi Statistical Yearbook (2017)

3.1.2. Animal product processing industry. (1) Meat industry. Dingxi City has great advantages in meat industry, and its meat production is the first compared with other cities of Gansu province. In recent years, due to the continuous introduction and variety improvement of new varieties in the city, the wide application of animal husbandry science and technology and the improvement of breeding level, the quality of livestock and poultry meat has also been improved. A large number of leading enterprises of livestock products have made outstanding contributions to the introduction of improved breeds of livestock and poultry. In recent years, leading enterprises of animal husbandry have created local fine livestock and poultry brands. The total number of improved cattle in the city was 138500, the rate of improved breeds was 58%, the rate of improved sheep was 544000, the rate of improved breeds of sheep was 64%, and the rate of improved breeds of pigs and poultry was above 90%.

According to **Table 3** data of Dingxi City Statistical Communique at the end of 2017, the number of new beef sheep, beef cattle, live pigs and poultry was 602000, 60300, 2.42 million and 6850 respectively. The output of meat, milk and egg is 541500 tons, 870000 tons and 85000 tons respectively. The total number of livestock at the end of the year was 4.3685 million. Among them, 517300 cattle, 50300 horses, 6600 donkeys, 571400 pigs, 455200 goats and 2.7471 million sheep.

Table 3. Meat Production in Dingxi City in 2017(t).

	Beef	sheep	mutton	pork	poultry	horse meat
Dingxi city	79542	92691	10111	178389	124252	4585
Longxi	20513	14856	3359	43510	20032	1118
Tongwei	5220	9950	2250	16442	21683	577
Weiyuan	17176	12820	895	21143	30425	1270
Zhangxian	7488	15699	245	13125	19508	660
Lintao	15533	19690	1382	74080	17758	180
Minxian	10408	11855	1042	7779	13694	307

Source: Dingxi Statistical Yearbook (2017)

Judging from the distribution of all kinds of meat products in various counties, the total meat production of Lintao County, Weiyuan County and Longxi County was 317254 tons, accounting for 64.49% of the total meat production of the county, of which the meat yield of Lintao County accounted for 26.17% of the total meat output of the county. Longxi County had the highest beef yield, accounting for 25.79% of the total county, Lintao County had the highest mutton yield, accounting for 21.24% of the total mutton output, and Lintao County had the highest pork yield, accounting for 41.53% of the total pork output. Weiyuan County has the highest poultry meat yield, accounting for 24.49% of the total poultry meat output in Weiyuan County. And goat meat and horse meat production at county level is relatively small.

(2) Dairy industry. Dairy products have become an ideal food for people because of their easy absorption and high nutrition. Not only developed countries but also developing countries strongly advocate milk consumption [6]. Since Dingxi City has taken animal husbandry as the leading factor to adjust the agricultural structure, Dingxi City has continuously increased the development of dairy industry, and a number of dairy farms and dairy enterprises have emerged in the city. Tianliyuan Science and Technology Co., Ltd. has become a breeding demonstration base with teaching and scientific research, and has been awarded the title of "Top 200 Dairy cattle breeding demonstration Farm" by the Ministry of Agriculture of China. Every year, more than 300 breeding experts are trained in Dingxi City, which gives full play to the radiation driving effect of demonstration cattle farm. With the rapid development of dairy industry in Dingxi City, dairy processing enterprises provide more and more products. Shengou Luoji Dairy Group Co., Ltd., Gansu Tianliyuan Dairy Co., Ltd., a series of dairy enterprises of pure fresh milk, chocolate milk, peanut walnut milk, original yoghurt, strawberry, corn, wheat flavor yoghurt, Cereal sour milk, full-fat milk powder and so on are favored in the whole province, occupying the Gansu market steadily. In 2017, Dingxi dairy production was 126390 tons, including 112093 tons of liquid milk and 6573 tons of milk powder. Dingxi City produced 876832 tons of milk, of which the output of milk was 872302 tons, accounting for 59.13 percent of the total milk output in Gansu Province. The total milk output of Weiyuan County, Lintao County and Longxi County was 597187 tons, accounting for 68.11% of the total milk output of the county. Among them, the milk output of Weiyuan County accounted for 28.80% of the whole county, which was the highest in the whole

county. The output value of milk products in 2017 was 3.40032 billion yuan, accounting for 18.50% of the total animal husbandry output value of the city. The total output value of milk products in Weiyuan County, Lintao County and Longxi County is 2.01394 billion yuan, accounting for 59.23% of the total output value of dairy products in Weiyuan County, among which the output value of milk products in Weiyuan County accounts for 22.51% of the total output value of dairy products in the whole city. It shows that Dingxi dairy industry is mainly concentrated in these three counties, while Weiyuan County has the highest milk commodity rate, which has obvious advantages in the whole city milk industry. By 2017, 108 dairy farms have been built in the city, including 30 cattle farms, 66 standardized dairy farms, 84 milking farms and 870000 tons milk production. There are 194500 cows in the barn and there are 85 milk stations, 243 breeding stations and 157 medicine baths; The government department do their best to implement the people's livelihood project, comprehensively promote the full coverage of the student milk drinking program, and all 126366 primary and secondary school students in the nine-year compulsory education stage in the city enjoy the student milk drinking program. Dingxi dairy processing industry is developing rapidly, and many brands are loved by the masses. However, due to the lack of processing enterprises to play a leading role, it can be said that the basis for the production and processing of other animal products in Dingxi City is still relatively weak. Under the increasing trend of people's demand for animal products, The existing production level cannot fully meet the needs of people. Therefore, it is necessary to cultivate the "leader", improve the quality, expand the scale, focus on improving the added value of the product, build the brand of animal products and enhance the competitiveness of the market.

4. Advantages of animal husbandry industrialization in Dingxi City

- Variety advantage. Dingxi City is rich in livestock breeding fine variety resources. The main breeds of cattle are Holstein, Simmental, brown cattle and so on [7]. Holstein cattle have the characteristics of strong climate domestication ability and cold tolerance. Simmental cattle are characterized by fast growth rate, weight gain of more than 1.0 kg per day, and cold tolerance in winter and heat resistance in summer. Simmental cattle are suitable for grazing all year round. Sheep owners have famous breeds such as Suffolk. After the Suffolk sheep and the local sheep are hybridized, the meat quality of the Suffolk sheep exceeded that of the local sheep. Pig breeds are mainly Kexia, Duroc, Landrace, Gansu black pigs and so on. The poultry farmed are mainly laying hens and broilers. By 2017, there were 138500 improved cattle in the city, the rate of improved cattle was 58%, the rate of improved sheep was 544000, the rate of improved breeds of sheep had reached 64%, and the rate of improved breeds of pigs and poultry was more than 90%.
- Location advantage. The geographical location of Dingxi is very important. It is not only an important town on the ancient "Silk Road", but also a necessary place for the New Eurasian Continental Bridge. It is also the east gate of Lanzhou. It is known as "Gansu throat, Lanzhou Portal," only 98 kilometers from Lanzhou, adjacent to Guanzhong-Tianshui Economic Zone and Lanbai Metropolitan Economic Circle. With the construction of the city and the acceleration of the construction of the transportation system, Dingxi will be fully integrated into the Lanbai urban economic circle. Longhai Railway, Baolan double Line, 312, 310, 212, 316 National Highway and Shiliu Expressway, Lanlin Expressway, Pingding Expressway, Tianding Expressway all pass by Dingxi city. The construction of Lanyu Railway and Baolan passenger special Line have passed through Dingxi, which has become an important transportation hub in western China.Convenient transportation conditions broaden the sales channels for animal products in Dingxi City.
- Production advantages. By 2017, there are 16 livestock products processing enterprises, 10 meat sheep slaughtering and processing enterprises, 6 pig slaughtering and processing enterprises and 4 broiler processing enterprises in Dingxi City. The number of animal husbandry farmers and herdsmen breeding cooperatives established in the city has reached 196. The 512 large-scale farms are established, of which 12 are standardized demonstration farms at the national level.

These production infrastructures have laid a solid foundation for the realization of standardized breeding, specialized production and industrial management in Dingxi City.

- Policy support advantages. In order to increase the income of farmers and herdsmen and improve the lives of farmers and herdsmen, the government has issued a series of policies, such as animal husbandry subsidy, animal epidemic prevention subsidy, grassland ecological protection subsidy reward policy and so on. The implementation of these policies is helpful to improve the rate of improved livestock breeding, to provide a strong guarantee for the health and safety of livestock, to strengthen the ecological protection of grassland, and to improve the level of dairy production, quality and safety. Based on the existing animal husbandry resources and development advantages, animal husbandry in Dingxi City has developed to a certain extent, gradually forming large-scale cultivation, specialized production, and the income of farmers and herdsmen has also increased obviously. In 2017, more than 600 yuan of the average newly-added income of farmers and herdsmen was from animal husbandry. At the same time, the development of animal husbandry has created a large number of employment opportunities, absorbed unemployed people in agricultural areas, and played a positive role in promoting the whole social economy. However, from the overall situation of local animal husbandry, its overall industrialization level still has room for further improvement, and there is a big gap with the areas with good animal husbandry development in China. Therefore, the author believes that it is necessary to find out the problems that need to be further strengthened and improved in the process of animal husbandry industrialization. This is of great decisive significance to the realization of the ultimate goal of this paper.

5. Problems of animal husbandry industrialization in Dingxi city

5.1. Leading enterprise's leading role needs to be strengthened
In the process of animal husbandry industrialization in Dingxi city, there are some influential enterprises engaged in dairy products processing, forage processing and meat products processing, but at present the market demand is increasing constantly, and the quality requirements of animal products are also continuously improved, the leading role needs to be strengthened. If "Leading enterprises" can not play the role of driving farmers to increase income, it will become the primary factor restricting the development of animal husbandry industrialization.

5.2. The function of specialty cooperative economic organization of animal husbandry is weak
Since the implementation of the Cooperative Law in 2007 in China, a number of professional cooperatives have been set up one after another in Dingxi city, and the number of cooperatives has also been increasing. There have 278 city-wide farming professional cooperatives. Although new cooperatives are constantly springing up throughout the city and according to the overall situation of all kinds of cooperatives, there is a low degree of organization, less large-scale farming, low level of technology, a lack of talent, and a small scale of production and with the market connection is not close, the organization function is weak and so on.

5.3. The socialized service system is not perfect
Social service for animal husbandry requires the provision of a series of services before, during and after birth, specifically, financial services, skills training services, improved breed services, marketing services. Livestock farmers cannot find a real platform, path and so on. Production can only be arranged according to past experience and lagged market demand information, resulting in a long-term passive relationship between production and marketing. Thus, production, supply, sales are disjointed and did not form an integrated service. The construction of technicians cannot keep pace with the development of animal husbandry, but large-scale, standardized farming requires livestock technicians to have high-end diagnosis and treatment techniques. At present, high-level technicians are still very few, which cannot meet the needs of farmers, and the form of service is passive and single, which cannot adapt to

large-scale, standardized farming needs. The effect of animal disease prevention and control of technicians has not formed an interest relationship with the farmers, and the vast number of farmers have misgivings about livestock diseases. The quality of scientific and technological service personnel is on the low side, the circulation service system is not perfect, the traditional service institutions do not meet the needs of commodity economy. These not only seriously affect the overall benefits and large-scale benefits of service, but also restrict the continuous development of animal husbandry industrialization.

5.4. The capital investment is insufficient, and the scientific and technological content of animal husbandry is low

The fund is the "catalyst" to improve the development level of animal husbandry science and technology. In recent years, with the continuous development of animal husbandry in Dingxi city, the local government has also increased the support for further development of local animal husbandry, but from the overall situation of animal husbandry development, the existing capital investment is still insufficient. At present, the local animal husbandry lacks sufficient capital input, thus the infrastructure is weak; the farmers' investment ability is limited, but the loan is difficult; and it is difficult to attract social financing. However, the industrialization of animal husbandry must be guided by technological progress, requiring science and technology to run through every link from forage cultivation, production, processing, livestock farming, processing of livestock products to transportation, storage and marketing. Without scientific and technological support, animal husbandry industrialization cannot achieve the expected goal, and then cannot achieve the transformation of agricultural growth mode. With the continuous improvement of social and technical level, although Dingxi city has begun to use some science and technology in animal husbandry management methods and livestock farming, animal product processing and other aspects, the lack of capital investment leads to the low level of animal husbandry science and technology content, compared with the areas where animal husbandry industrialization has been better developed. These are mainly reflected in the following aspects: the large-scale, intensive farming of animal husbandry is not high, the cultivation level of the farmers is low; the degree of good breeding of livestock and poultry is not high; the meat industry system is not perfect; the situation of animal epidemic prevention and control is not optimistic; the construction of science and technology support system is still lagging behind, the level of technical equipment is low, and the coverage of the popularization and application of new technologies is low. It is necessary to introduce advanced technology and advanced equipment at home and abroad to improve the level of science and technology.

Whether it is the development of large-scale breeding of livestock or poultry, or the construction of infrastructure facilities, the construction of the production and supply capacity of forage, the construction of the improved breeding system, the construction of the system for the prevention and control of animal epidemics, the construction of animal product quality safety system and control capacity, the construction of animal product processing capacity, the construction of socialized service system, and so on, all need the support of funds very much. Only through the sufficient capital input can the above-mentioned construction be realized. These constructions are indispensable and important means to promote the industrialization of animal husbandry.

5.5. The environmental problems caused by animal husbandry industrialization

In the process of animal husbandry industrialization, animal husbandry has become the most active growth point and main pillar industry in rural economy. With the rapid development of large-scale livestock and poultry breeding industry, the scale of livestock farms is getting larger and larger, the degree of intensification and mechanization is getting higher and higher, due to the imperfect environmental laws and regulations and the shortage of funds, The vast majority of farms did not consider a series of problems, such as the treatment of livestock and poultry faeces, resulting in the accumulation of faeces in many farms and the arbitrary discharge of sewage, which seriously polluted the surrounding environment. It also directly affects the health and epidemic prevention of the farm

IOP Conf. Series: Earth and Environmental Science **346** (2019) 012035 doi:10.1088/1755-1315/346/1/012035

itself, and reduces the quality of animal products. Animal husbandry is also the source of air pollution gas emissions, which will produce Greenhouse Effect [8]. The continuous growth of animal husbandry production will aggravate the pressure on the environment and natural resources. The environmental problems caused by animal husbandry are becoming more and more serious harm to the living environment of human beings.

6. Countermeasures and suggestions on industrialization of animal husbandry in Dingxi city

6.1. Vigorously cultivate and support leading enterprises

Leading enterprises as the bearers of animal husbandry industrialization, their competitiveness level represents the competitiveness of the whole region. The strength of the leading enterprise also determines the level of the industrial operation of animal husbandry in a region. Therefore, it is necessary to cultivate and strengthen the leading enterprises, break the limitations of small-scale production, rely on the leading enterprises to effectively protect the interests of farmers, and achieve animal husbandry industrialization.

First, government should give full play to the support and guidance of the leading enterprises, vigorously develop the leading enterprises with great potential and competitive power, and give necessary investment. What's more, government should also implement preferential policies, rectify market order by legal means, encourage leading enterprises to introduce and adopt advanced technology and equipment, and vigorously carry out quality certification in order to improve their market competitiveness; encourage small and medium-sized enterprises to carry out joint, merger, equity and other business methods for integration, in order to form leading groups; guide enterprises to develop new products, encourage large and small enterprises to cooperate in production and promote each other.

Second, government should actively coordinate the relationship of interests between farmers and leading enterprises, and improve the mechanism of interest connection; enhance the sense of compliance between the two sides, form a stable purchase-sale relationship, and achieve a win-win situation for farmers and enterprises; pay attention to the construction of production base and solve the problems of incoordination and synchronization between leading enterprises and raw materials base; in the construction of production base, government should make great efforts to promote the standardization of products, perfect the mechanism of benefit connection between enterprises and farmers, and expand the coverage of animal husbandry products by order; encourage leading enterprises to establish risk funds, minimum acquisition, protection prices, return of interests and other systems; allowed farmers to take shares in land contractual management rights and land use rights as well as products, technologies and resources in order to form an interest community with leading enterprises.

Third, government should rely on leading enterprises to create animal products brand. The competition of leading enterprises is the competition of brand in essence. Brand is an asset that can guarantee the long-term stable production and operation of enterprises and absorb new capital for a long time. The government should actively cultivate, develop and strengthen the brand, and take the brand as the starting point to occupy the market. Therefore, government should speed up the development of large-scale, standardize livestock and poultry breeding industry, focus on improving the quality of products, especially cultivate, support the production of beef and mutton, increase the supply of beef, mutton and poultry milk and eggs, create green brands and speed up the process of animal husbandry industrialization.

6.2. Strengthening the construction of specialized cooperatives

In order to promote the development of animal husbandry industrialization, it is very urgent to cultivate and strengthen their own professional cooperatives. By participating in professional cooperatives, farmers can not only properly deal with the relationship with enterprises, but also improve the negotiating status of farmers. Although the professional cooperative economic organizations run by farmers have achieved great performance, there is still a large deviation compared with the general requirements of farmers in Dingxi city. In addition, the establishment of cooperative economic

IOP Conf. Series: Earth and Environmental Science **346** (2019) 012035 doi:10.1088/1755-1315/346/1/012035

organizations is an important policy of the country, so Dingxi city government should attach great importance to the construction of cooperatives, which should be supported by the government and credit concessions. In order to achieve the further development of professional cooperatives, the government should do the following:

First, strengthen publicity efforts to deepen farmers' correct understanding of the function of professional cooperatives. It is necessary to make the farmers understand the cooperative law gradually and strengthen the propaganda to the cooperative. Government should organize cadres and relevant staff at all levels to enter the breeding areas to popularize knowledge on the management of cooperatives, promote technical work, adopt one-on-one exchanges to implement the laws and regulations relating to cooperatives, make farmers understand the organizational function of cooperatives and recognize the importance of cooperatives.

Second, expand cooperatives and increase the amount of funds invested. At present, one of the problems of cooperatives is that they are small in scale and weak in competitiveness. Most cooperatives have the phenomenon of small scale and decentralization, few personnel, single products and a lack of influential leading cooperatives that can play a leading role. Cooperatives can be driven by leading enterprises to expand their size. In order to avoid individual breeders from standing in a weak position, cooperatives should act as a link between enterprises and farmers, conduct production guidance for farmers and negotiate with enterprises on behalf of farmers so as to ensure the return of the profits for farmers, achieve equal benefits and share the risks. Enterprises promote cooperatives, cooperatives guide farmers, they have their own advantages, complement each other, promote each other, thus form a virtuous circle, gradually realize the scale of cooperatives, and enhance their competitiveness. In the area of funds, the government should give full support at the same time to clarify the preferential tax policies, and make timely adjustments in accordance with the development of cooperatives, focus on solving the problem of credit difficulties, and optimize the lending policy. Banking institutions and rural credit cooperatives are urged to give priority support to cooperatives with strong motivation and standard operation according to their operating conditions.

Third, pay close attention to the work of education and training. The relevant government departments should adopt appropriate methods for local farmers to impart new knowledge and technology of modern animal husbandry management among cooperative members. The government should select the leaders who have the ability to organize the cooperatives, carry out targeted training and education, and focus on enhancing their abilities in entrepreneurial management, organizational management, risk control, cultural construction, and so on; encourage the leaders of the breeding cooperatives go out to visit, regular exchanges and other ways to expand their experience, increase their knowledge, and then promote the upgrading of the technical level of other members by leaders, thus build a team of farmers who are diligent, know how to cooperate with each other, good at organizing and operating, care for cooperative interests and willing to contribute to the professional cooperative.

6.3. Establish and perfect the socialized service system

A sound socialized service system is the key to promote the coordinated development of the animal husbandry industrialization. The construction of socialized service system for the development of animal husbandry industrialization, starting from the level of development of regional productive forces. Government should establish diversified agricultural socialized service system with multiple communication channels, such as the establishment of a variety of economic components and forms of service adapted to the needs of the development of animal husbandry industrialization; the government and leading enterprises will vigorously foster and support professional cooperative economic organizations, farmers' associations, farmers' brokers and other intermediary organizations, vigorously develop market circulation, and strengthen the construction of market networks.

First, government should strengthen the construction of the market sales and service system, foster intermediary cooperation organizations for the circulation of livestock products, set up specialized cooperative organizations with various functions, such as production, processing, marketing, storage, transportation, technology, and so on, so that they can become bridges between the market and farmers.

In order to solve the contradiction between small-scale operation and large-scale market, government provide farmers with systematic services in all aspects of prenatal, labor and post-natal period.

Second, government and leading enterprises should strengthen the construction of science and technology extension service system, carry out technical services such as improved breed, feed, breeding, epidemic disease prevention and control and set up a system of animal husbandry technical services that integrates epidemic prevention, diagnosis and treatment, scientific and technological training. Through training technicians and farmers, establishing demonstration bases for scientific research, inviting experts and scholars to go deep into the first-line production guidance and impart professional knowledge and skills, farmers can master breeding skills with high level of quality; They should also make full use of all kinds of science, technology and educational resources, carry out post training to animal husbandry practitioners through further study, impart advanced science and technology in modern animal husbandry production, and through various forms of publicity and promotion of scientific research results, increase the construction of animal epidemic prevention personnel, animal disease diagnosis and treatment personnel and breeding technician.

Third, the government, the economic organization, the individual farmers and the multi-party joint effort to set up the investment and financing service system, increase the input to the animal husbandry through various channels, and gradually form the investment and financing service system with the government investment as the guide, combined with the financial loan, the farmers and the enterprise investment.

Fourth, government should strengthen the construction of information consulting service system to ensure the interests of farmers. By providing the market information in a timely manner, the production, processing and circulation of the livestock products are guided, so that the farmers are prevented from being blindly produced; encourage farmers and professional households to work together to establish professional associations for self-service. In fact, as long as the market information can be accurately grasped, farmers will soon be able to put into production; support leading enterprises to establish social services in breeding bases and perfect the information network service to solve the problem of small-scale production and market disconnection, and improve the information level of animal husbandry.

6.4. Increase the input of funds and support to improve the science and technology content of animal husbandry

Financial aspects: Governments at all levels and relevant units should actively support scientific and technological innovation, introduce and use advanced technologies in forage silage, scientific breeding, breeding of improved variety of livestock, prevention and treatment of epidemic diseases, and promote the overall technical level of animal husbandry; Actively strive for the development of industrialization, science and technology to promote agriculture, scientific research and promotion, poverty alleviation, agricultural comprehensive development, small town construction, scientific and technological training, grassland ecological compensation mechanism and other projects. Through relying on the scientific bundling of project funds, make the project construction funds play a role, gradually improve the scientific and technological content of animal husbandry, and inject vitality into the development of animal husbandry; Farmers are the main investors in animal husbandry. While actively guiding the vast number of farmers to invest in animal husbandry, local finance should also increase their financial support to farmers, and financing institutions should give active credit support and encourage them to master advanced technology in order to improve economic efficiency.

Policy aspects: First, government vigorously support large-scale farming farms and breeding bases, give full play to the role of farmers, bases and leading demonstration leading role, so that project funds become a booster to improve the content of science and technology and improve the use of policies and encourage enterprises and farmers to build large-scale farms and intensive feeding areas in wasteland. Second, by adjusting the structure of the planting industry, expanding the planting area of forage crops and forage, making financial subsidies to purchase forage processing machinery, encouraging growers to plant, harvesting, storing and processing mechanically, and raising the utilization rate of straws, to realize mechanization, science and technology; Encourage the establishment of cooperative economic organization such as forage grass planting, processing, marketing to promote the production and specialization; Strengthen the supervision of feed safety in order to build a safe, high-quality and efficient feed production network. Third, government should further optimize the environment for attracting foreign investment, increase the intensity of attracting investment, and guide enterprises within and outside the region to implement a strong association through mergers and acquisitions, ownership, holding, and construction, so as to form large-scale animal husbandry enterprise groups. Government will increase the injection of funds into the process of animal husbandry cultivation, breeding, processing and circulation, use new technologies and mechanical equipment to upgrade the level of science and technology in leading enterprises, and promote the industrialization of new technologies in animal husbandry and the technological progress of the industry.

6.5. Thoughts on the countermeasures of controlling environmental pollution in the process of animal husbandry Industrialization

First, strengthening the leadership of ecological protection of animal husbandry. It is necessary to attach great importance to the pollution caused by the livestock and poultry industry to the environment, draw lessons from the pollution before prevention and control existing in the development of animal husbandry, and bring the ecological protection of animal husbandry into the main contents of environmental management. Raise the awareness of animal product safety, organize animal husbandry production in accordance with international practice, strengthen the safety management and quality certification of livestock and their products. It is also necessary to strictly examine and approve the environmental protection of farm construction and strengthen the environmental monitoring of livestock pollution.

Second, establishing artificial ecological animal farm. It is necessary to establish a moderately large-scale farm according to the local land area and make full use of the land reduction method to melt the fecal waste, which is the most feasible and economical method of fecal pollution treatment at present. Organize the production of original and new livestock farms according to the principles of ecology and ecological economics, make full use of local natural resources, make use of the interdependence between animals, plants and microorganisms, and modern technology, Carry out waste-free and pollution-free production. Through the promotion of various ecological breeding models, such as the combination of agriculture and animal husbandry, the combination of fruit and animal husbandry, and the combination of pastoral and vegetable husbandry, we should strengthen the harmless treatment and comprehensive utilization of livestock faeces, and improve the overall scientific and technological level, environmental protection level and sustainable development level of animal husbandry to realize the virtuous circle within animal husbandry. Reasonable planning of breeding areas should consider the convenience of transportation, but it should not be too close to the highway, and should be built as far away from the city, industrial and mining areas and dense population as possible in order to solve the problem of good water use. Restricted areas should be designated in water sources, scenic tourist areas, residential areas, schools, hospitals and other public places.

Third, carry out the research and popularization of pollution control technology. Poor treatment of emissions is pollutants, good treatment is resources, and the utilization of emissions from aquaculture is a new subject [9]. It is necessary to organize scientific research, teaching, production, promotion and other aspects of technical research, in-depth development of breeding layout, farm construction, feed production, feeding methods, manure treatment, machinery and equipment, agriculture and animal husbandry and other key links of technical research. We should speed up the popularization and utilization of pollution control technology, and apply advanced and practical technologies such as rain pollution separation, dry and dilute separation technology, low nitrogen feed production and use, low cost sewage treatment and resource utilization to production practice. We should provide strong technical support for livestock pollution control by extensive exchange and cooperation of pollution control technology, strengthening the introduction and innovation of advanced technology and equipment. Gradually realize the industrialization and standardization of animal husbandry pollution control in engineering design, construction, maintenance, Management and other links.

7. Conclusion

First, animal husbandry continues to grow in Dingxi City, but there is still a gap compared with other areas. Judging from the overall situation of animal husbandry industrialization, leading enterprises have promoted the pace of industrialization, the scientific and technological content has been continuously improved, the level of improved breeds of livestock and poultry varieties has been significantly improved, and the ability of animal epidemic situation monitoring and prevention and control has been continuously enhanced. The socialized service system has been established initially, the construction of technical staff has been strengthened, and the mechanism of linking the industrial interests between enterprises and farmers has been gradually formed. However, the overall scale and development level have a great gap compared with some areas of the rapid development and with similar natural conditions and economic structure of the domestic animal husbandry. This paper holds that this condition is mainly caused by the slow development of the forage industry, the weak traction of the leading enterprises, the shortage of funds and the imperfection of the socialized service system, the harm to the environment and etc.

Secondly, the integration of animal husbandry industry refers to the process of animal husbandry economic reproduction, namely, prenatal, productive and postnatal integration into a complete industrial system. Its basic characteristics are scientific structure, industrialization of production, socialization of service and commercialization of products. According to the requirement of animal husbandry industrialization, it is necessary to concentrate population, improve quality, increase investment, build base and strengthen service. Under the condition of the increasing demand of the market, and the quality requirements of the animal products, the development of the animal husbandry in Dingxi city is not strong, the organization of the farmer is not high, and the scientific and technological level of the animal husbandry is still to be improved. The organization of the cooperative economic organization of the animal husbandry is low and the large-scale cultivation is small due to the shortage of funds, the lack of the talents, the connection between the farmers and the market are not tight, and the social service system is not sound.

Finally, the development of forage industry should be regarded as the key in the process of animal husbandry industrialization, focusing on adjusting the planting structure, strengthening the construction of artificial forage base, expanding the planting area of forage, improving the supply capacity of forage to meet the forage demand of farmers. Government should cultivate and support the leading enterprises which are the main body of animal husbandry industrialization, promote its forage, livestock products processing capacity, improve its competitiveness, so that the leading enterprises fully play its role [7]; strengthen the construction of cooperatives and strengthen the work-sharing cooperation among the farmers and carry out specialized cultivation; increase the investment of funds, improve the whole level of science and technology of animal husbandry, strengthen the technical training, and continuously perfect the socialized service system. The above-mentioned are effective ways to promote the industrialization development of animal husbandry in Dingxi City.

References

[1] Han J W and Ding S L 2009 *Animal Husbandry Economic Management* 2nd edi. (Beijing: Agricultural Publishing House)
[2] National Animal Husbandry Economic Research Association 1998 Study on industrialization of animal husbandry *Proceedings of the eighth National Animal Husbandry Economic Research Association* (Beijing: China Agricultural Science and Technology Publishing House)
[3] Yue L 2013 *Industrial Economics* 3rd edi. (Dalian, China: Northeast University of Finance and Economics Press)
[4] Deng R, Yan X J and Sun B C S 2010 The development trend of animal husbandry viewed from the evolution of agricultural internal industrial structure China Animal Husbandry 2010, 9
[5] Li W Q and Sun Z Q 2006 *Innovative Management Model of Animal Husbandry* (Beijing: China Social Publishing House)
[6] An J Z and An B 2018 Thoughts on constructing the system of sci-tech innovation in grass animal

husbandry—taking Dingxi city of Gansu Province as an example *China Ani. Industry* **2018** 28-31
[7] The Government of Dingxi City 2018 The famous hometown of Chinese bacon available at: www.Dingxi.gov.cn/
[8] Yan C L 2007 Analysis on the influence of animal husbandry industrialization on grassland ecological environment *Northern Econ.* **2007** 121-135
[9] Zhang J M 2016 Current situation and countermeasures of environmental pollution in animal husbandry in China *China Agri. Sci. Tech. Guide* **5** 71-74

Removal of cyfluthrin by fine bubble technology in oranges (*Citrus reticulata* Blanco)

J T Vuthijumnonk[1, 2] and W Shimbhanao[1]

[1]College of Integrated Science and Technology, Rajamangala University of Technology Lanna, Chiang Mai, Thailand

E-mail: vjanyawat@hotmail.com

Abstract. Cyfluthrin, a synthetic pyrethroid insecticide, is used widely for orange cultivation. Prolong exposure to cyfluthrin causes several human health issues such as kidney inflammation, weight loss and diarrhoea. This study aimed to investigate the fate of cyfluthrin residue in oranges after treatment with fine bubble produced with air or ozone. Micro-bubbles produced from different types of gas (air, oxygen and ozone) were used to treat oranges contaminated with cyfluthrin. Cyfluthrin residue was tested using GPO-TM kit. The pesticide residue was not detected after micro-bubble treatments. An air micro-bubble technology was chosen for further analysis by HPLC. HPLC analysis confirmed that more than 90% of contaminated cyfluthrin was removed by air micro-bubble treatment. Moreover, the insecticide residue was not detected in washing water.

1. Introduction

Oranges (*Citrus reticulata* Blanco) are widely consumed as fresh fruits or ready-to-drink juice. The total area under orange cultivation in Thailand is around 35,600 acres with production of 135,898 tons and productivity of 4.32 tons per acre [1]. To maintain quantity and quality of the produce, a lot of pesticide, mainly cyfluthrin, a synthesis pyrethroid pesticide is used.

Cyfluthrin, a synthesis pyrethroid pesticide, has a similar chemical structure as dichlorodiphenyltrichloroethane (DDT) (figure 1).

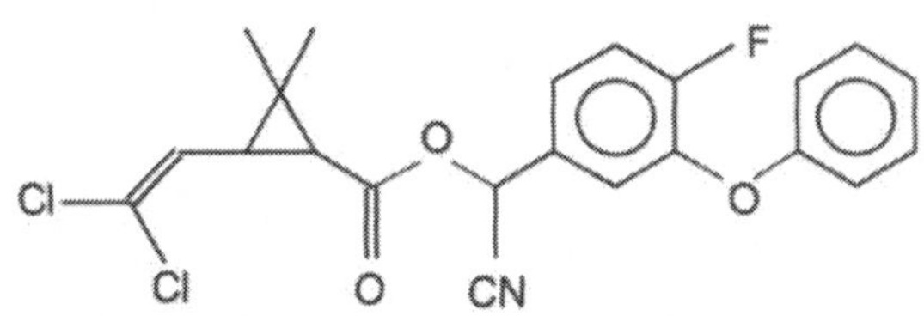

Figure 1. Cyfluthrin (Cyano(4-fluoro-3-phenoxy-phenyl)methyl3-(2,2-dichloroethenyl)-2,2-dimethylcyclopropanecarboxylate).

The mode of action of cyfluthrin is an inhibition of calcium transporting enzyme which causes accumulation of acetylcholine at synapse and continuous muscle contraction [2]. The use of cyfluthrin in orange cultivation said to cause several illness [3] and also toxicity of this pesiticide has been investigated on several laboratory animals. Prolong exposure to the pesticide in a very low level (150

Content from this work may be used under the terms of the Creative Commons Attribution 3.0 licence. Any further distribution of this work must maintain attribution to the author(s) and the title of the work, journal citation and DOI.
Published under licence by IOP Publishing Ltd

ppb) caused weight loss, vomiting, diarrhoea and kidney inflammation [2]. According to EU pesticide database, Maximum Residue Levels (MRLs) of cyfluthrin in oranges is 0.02 mg/kg [4]

Fine bubble technology is a novel technology in Thailand. Fine bubbles can be generated in in liquid medium using different types of generator. Sizes of fine bubble range from one micrometer to one hundred millimeters. Fine bubbles produced from different type of gas show different properties however, shrinking of fine bubbles generate free radicals [5]. Micro-bubble technology exhibited pesticide removal efficacy in several studies [6,7].

GPO TM-Kit (Petty patent no. 7554), is a test kit produced by Government Pharmaceutical Organization (GPO), Thailand. It is used for a rapid detection of insecticide using Thin Layer Chromatography (TLC) coupled with UV (254 nm) reaction. This kit was used for screening the optimal condition in insecticide reduction. The most suitable condition was further confirmed using High Performance Liquid Chromatography (HPLC).

Keeping in view of the above facts, an effort has been made to reduce cyfluthrin residue in oranges by using micro-bubble technology.

2. Materials and methods

2.1. Materials

Oranges were purchased from a local market, Chiang Mai. The fruits were kept at 4°C and used for the experiment within 24 hours after purchase. Cyfluthrin (2.5% w/v) was purchase from Global-crops, Thailand. GPO TM-kits was purchased from the Northern Government Pharmaceutical Organization, Thailand.

2.2. Preparation of cyfluthrin-contaminated oranges

Insecticide was prepared according to the commercial label. Orange fruits were immersed in an insecticide solution for 15 min and then dried at room temperature for 30 min.

2.3. Treatment of cyfluthrin-contaminated oranges using fine bubble technology

After pesticide treatment, the fruits were separated to make five groups *viz*., control group (C: no further treatment), a positive control group (ROW: treated with Reverse Osmosis-water), an AMB group (treated with air micro-bubble), an OMB group (treated with oxygen micro-bubble) and an OzMB group (treated with ozone micro-bubble). Micro-bubble was generated by an in-house decompression-type micro-bubble generator (Model: RMUTL-KVM-01) with water flow rate of 1.7 L/min, pressure of 0.25 - 0.4 mPa, gas flow rate of 0.1 L/min. AMB was produced by adding five liters of Reverse Osmosis-water (RO-water) in a cylindrical vessel of the generator. OMB and OzMB was conjured in the same manner with addition of oxygen or ozone, respectively. Orange fruits (200 g) were immersed in the vessel containing bubbling micro-bubble for 15 minutes.

2.4. Determination of cyfluthrin contamination using GPO-TM kits and HPLC

Five groups of experimental products were tested for cyfluthrin contamination using GPO-TM kit according to the commercial label.

Briefly, the sample (5 g) was finely chopped and place in 25 mL container then the extracting solution (5 mL) was added. No charcoal powder used in the extraction process. The container was closed, shook for one minute and left at an ambient temperature for five minutes. The clear solution was taken out and dried at 48°C. The extracting solution (20 uL) was added to the dried sample and used for TLC analysis. The TLC sheet was put in a mobile phase until the mobile phase reach a solvent front line (8.5 cm). The sheet was dried at room temperature and sprayed with GPO-TM4 solution. When the sheet dried, it was put under UV light (254 nm) for 3 minutes. The sheet presented with brown spot reveals that the tested sample is contaminated with organochlorine or pyrethroid insecticides.

The most suitable condition was selected and further analysed using HPLC. Orange skin was included for the experiment.

2.5. Determination of cyfluthrin contamination using HPLC

Orange fruit samples (peel included) were cut into small pieces and homogenized by means of a kitchen blender. Pesticide extraction method was carried out according to a previous study with slight modification [8]. The blended fruit (100 g) sample was mixed with anhydrous sodium sulphate (50 g) and extracted with ethyl acetate (200 ml) in conical flask. The content was allowed to settle down for half an hour and the ethyl acetate extract was then filtered through a Buchner-funnel fitted with a filter paper covered by (20 g) of anhydrous sodium sulfate. After filtration, the extract was evaporated to dryness and re-dissolved in acetonitrile (5 mL) Then, the volume was reduced to 0.5 mL using stream of liquid nitrogen. The extract was then transferred to a graduated test tube and the final volume was adjusted at exactly 1 mL using acetonitrile. Solutions were then centrifuged and filtered.

The extracts from all experimental groups were analyzed by a HPLC with UV/Visible detector (Agilent Technologies) for identification and quantification of pesticides. Separation was performed on reversed phase C-18 column. The mobile phase: linear gradient of solvent A deionzied water: solvent B was acetonitrile (70:30, v/v), flow rate was 0.3 mL min^{-1}. Injection volume was 25 μL. The wavelength of the UV/visible detector was fixed at 254 nm for the pesticides. The wavelength was set at 230 nm.

The pesticide residue was identified by comparing its retention time with respect to technical grade reference standard. The concentration of pesticide residues in fruits washed with micro-bubbles were compared to the unwashed (control) to determine the percentage of pesticide residue reduction.

2.6. Data analysis

All data were analyzed by analysis of variance (ANOVA) and the differences between fine bubble treatments were analyzed by a least significant difference (LSD) comparison. The level of significant difference was indicated with the following: $*P \leq 0.05$ for all comparisons.

3. Results and discussion

After cyfluthrin and micro-bubble treatments, oranges from five experimental groups had normal appearance. Determination of cyfluthrin contamination using GPO-TM kits revealed that oranges from C and ROW groups were contaminated with pesticide residue while oranges treated with micro-bubble regardless of gas type showed no pesticide contamination. Therefore, air micro-bubble treatment was selected for further study because it has less complication during preparation process and did not required additional equipment. Determination of cyfluthrin residue using HPLC showed that cyfluthrin and its derivatives were detected. The retention time of cyfluthrin I, II, III and IV were 18.60, 18.86, 19.06 and 19.18, respectively (figure 2). The cyfluthrin level in the control group was 0.52 mg/kg. Treatment with ROW reduced the level of cyfluthrin to 0.48 mg/kg. Oranges treated with air micro-bubble was found with 0.02 mg/kg of cyfluthrin. Interestingly, no cyfluthrin residue was detected in air micro-bubble water used for treatment.

In the present study, approximately 90% of cyfluthrin residue was successfully removed by micro-bubbles. The level of cyfluthrin residue in the sample treated with micro-bubbles met EU MRLs. Like this study, micro-bubbles have been studied for their insecticide removal effects. Organophosphate pesticide, fenitrothion, was reduced to 32% and 52% after treated with ozone micro-bubble in lettuce and cherry tomato, respectively [9]. Also carbamate insecticide contamination in green persimmon leaves was removed after treated with ozone micro-bubbles [7]. Moreover, micro-bubbles conjugated with ultrasonic radiation, an advance oxidative process, reduced more than 70% of pesticide contamination in tangerine [10]. According to the research reports, studies on the efficacy of cyflutrin removal of micro-bubble technology have not yet occurred.

Cyfluthrin is a persistent pesticide. In California orange production, its abolition rate was almost 100% after 19 days of application [11]. Therefore, it is important to discover a rapid and simple method to eliminate the pesticide residue from agricultural products. Tap water was effective for removing significant amount of cyfluthrin in olive however, extension of washing time was not positively correlated with cyfluthrin removal ability [12]. Cyfluthrin degradation through oxidation results in 4-fluro-3-phenoxybenzoic acid (FPB acid) as a major by-product [13].

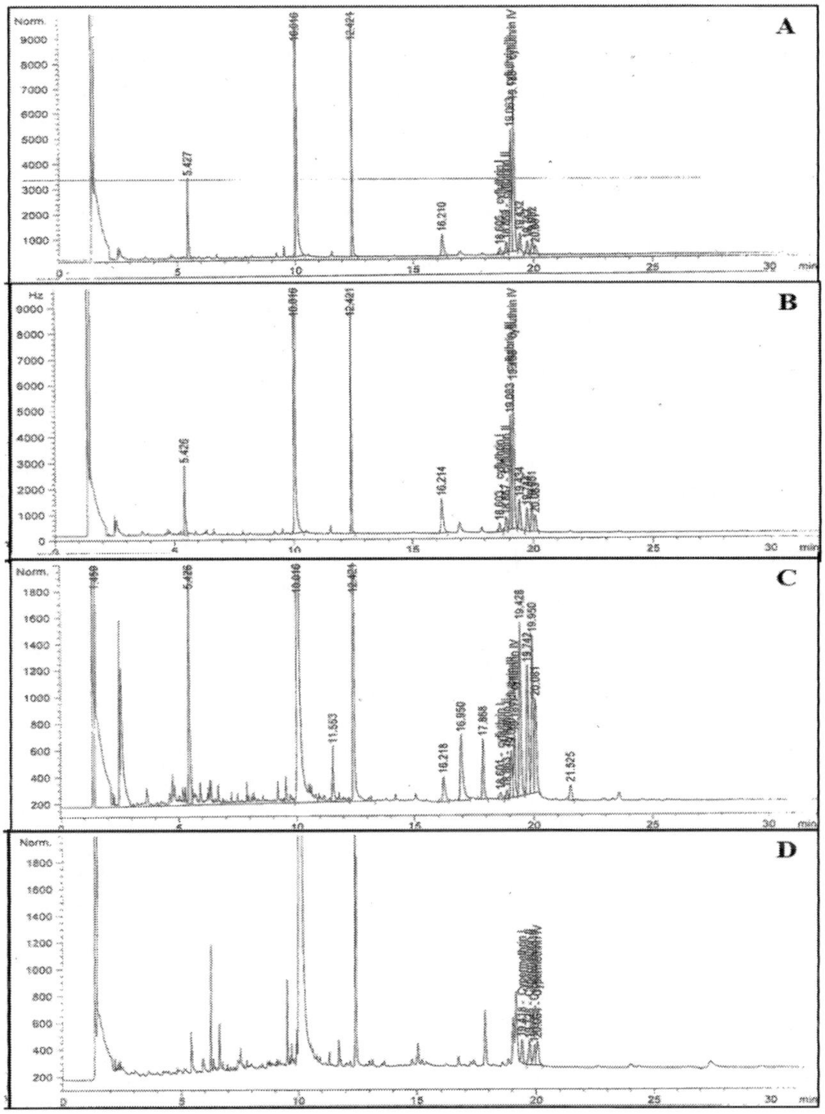

Figure 2. HPLC chromatograms of cyfluthrin and its derivatives in A: Cyfluthrin-contaminated oranges without further treatment, B: Cyfluthrin-contaminated oranges treated with reverse-osmosis water, C: Cyfluthrin-contaminated oranges treated with air micro-bubble water and D: Air micro-bubble water.

Even though, the pesticide-removal mechanism of micro-bubble technolgy is still unclear anm required further investigation, a possible mechanism has been proposed. Free radical generation is one of the most discuss property of micro-bubbles. Free radicals are generated after the bubbles disintegrated under non-thermal condition by an augmentation of the ion concentration around the shrinking gas–water interface. Under acidic condition, the generated radical species could break down several organic compounds [14]. Also, Hydroxyl radicals were also hypothesized for the efficacy to reduce pyrethroid contamination [15].

4. Conclusion

Micro-bubble reduced the level of cyfluthrin contamination by 90% and met the EU MRLs. Cyfluthrin decontamination effect could attributed to hydroxyl radical generated during micro-bubble collapsing. This is the first study reported cyfluthrin reduction effect of micro-bubble technology. Howsoever, oxidative property of micro-bubbles, and insecticide degradation mechanism of free radical species required further investigation.

Acknowledgments

This work has been granted by National Research Council of Thailand and supported by Rajamangala University of Technology, Lanna.

References

[1] Maisukklang K 2017 Department of Agricultural Extension, Thailand available at: http://www.agriinfo.doae.go.th/year60/plant/rortor/fruit2/orange.pdf
[2] Cox C 2014 *J. Pest. Reform* **14** 28-34
[3] Spencer J and O'Malley M 2006 *Rev Environ Contam Toxicol* **186** 64-5
[4] EU pesticide database available at: http://ec.europa.eu/food/plant/pesticides/eu-pesticides-database/public/?event=pesticide.residue.selection&language=EN
[5] Agarwal A, Jern Ng W and Liu Y 2011 *Chemosphere* **84** 1175-80
[6] Ikeura H, Kobayashi F and Tamaki M 2011 *J. Hazard Mater.* **186** 956-9
[7] Ikeura H, Hamasaki S and Tamaki M 2013 *Food Chem.* **138** 366-71
[8] Lamees M A-L and Iyad A-M 2016 *Int. J. Sci. Nat.* **7** 433-7
[9] Ikeura H, Hamasaki S and Tamaki M 2013 *J. Food Sci.* **78** T350-5
[10] Whangchai K, Uthaibutra J, Nuanaon N and Aoyagi H 2017 *Int. Food Res. J.* **24** 1135-9
[11] Khan I and Morse J G 1997 *J. Agric. Entomol.* **14** 409-20
[12] Kiriş S and Velioglu Y S 2016 *Food Addit. Contam. Part A* **33** 128-36
[13] Casjens H 2002 California Department of Pesticide Regulation, Sacramento, California available at: https://www.ca.gov/
[14] Takahashi M, Chiba M and Li P 2007 *J. Phys. Chem. B* **111** 1343-7
[15] Tran N, Trinh T, Hoang N M and Ngo T M 2014 *APCBEE Procedia.* **8** 151-6

Simple, rapid and cost-effective DNA extraction method for high quality DNA suitable for PCR based downstream application in mungbean [*Vigna radiata* (L.) Wilczek]

P Pookhamsak[1]**, P Pornbungkerd**[2] **and P A Tantasawat**[1,3]

[1]School of Crop Production Technology, Suranaree University of Technology, Muang District, Nakhon Ratchasima 30000, Thailand
[2]The Center for Scientific and Technology Equipment, Suranaree University of Technology, Muang District, Nakhon Ratchasima 30000, Thailand

E-mail: piyada@sut.ac.th

Abstract. Mungbean is a widely cultivated pulse crop. It is a source of high-quality protein for human consumption. Several new mungbean varieties have been developed employing molecular breeding tools. The quality and quantity of extracted DNA are very important for PCR amplification, particularly for the inter-simple sequence repeat-anchored resistance gene analog (ISSR-RGA) marker. In addition, reducing the time required for DNA extraction is essential in cases when large number of genotypes are analysed. Thus, the aim of this study was to compare eight combinations of two homogenization methods (manual grinding and Bullet blender® homogenizer) and four modified DNA extraction protocols. The effectiveness of DNA extraction for PCR amplification was evaluated using 3 PCR based markers, simple sequence repeat (SSR), ISSR and ISSR-RGA, on 6% polyacrylamide gel. Using homogenizer with modified protocol 2 (T6) resulted in a high DNA concentration (1032.60 ng/µl) and an A260/A280 ratio of 1.80, indicating high DNA purity. The PCR amplification of the resulting DNA with three types of molecular markers showed clear DNA bands, suggesting that DNA quality was appropriate for various molecular studies. In addition, homogenizers allowed processing of large number of samples in one go with minimal cost. These results suggest that this simple, rapid and cost-effective DNA extraction method is useful for marker-assisted breeding.

1. Introduction

Mungbean or green gram (*Vigna radiata* (L.) Wilczek) is an important economic crop in South and Southeast Asia. Currently, it is cultivated in more than 6 million hectares worldwide [1]. Almost 90% of world's mungbean production comes from Asia—India, China, Pakistan and Thailand are among the most important producers [2]. It is grown mainly for its edible seeds, which are fermented, cooked, roasted, sprouted, or milled [3]. However, mungbean production has reduced due to poor harvest index, lack of genetic variability, abiotic and biotic stresses [4]. Therefore, plant breeders need to develop new varieties to increase yields. Several molecular techniques are currently used in mungbean breeding to facilitate genotyping and selection. Molecular markers are useful in evolutionary, phylogenic, taxonomical, ecological and genetic studies of plants [5]. Mungbean genetic diversity has been evaluated using several DNA markers to identify polymorphisms among varieties such as SSR, ISSR and amplified fragment length polymorphism (AFLP) [6,7]. Moreover, marker-assisted selection

Content from this work may be used under the terms of the Creative Commons Attribution 3.0 licence. Any further distribution of this work must maintain attribution to the author(s) and the title of the work, journal citation and DOI.
Published under licence by IOP Publishing Ltd

(MAS) has been used to improve disease resistance using several molecular markers including ISSR and ISSR-RGA markers, which improve breeding program speed and accuracy [8-11]. Quality of DNA generally influences target DNA amplification therefore identifying a suitable extraction method for plant genomic DNA is a pre-requisite for molecular plant breeding, especially in mungbean plants with prominent oxidation problems [12].

Although many plant genomic DNA extraction protocols have been developed previously [13-20], they are often specific to tissues and plant species. Some protocols also have high costs, complex procedures, consume excessive time, or result in poor quality and the DNA insufficient for various polymerase chain reaction (PCR) based downstream applications. Recently, genomic DNA extraction protocols of legume crops including mungbean have been developed to generate high quality DNA and were suitable for amplification with diverse DNA markers such as RAPD, RFLP, SSR and ISSR [12,21-24]. However, these procedures still require long manual grinding using a mortar and pestle. This method is not suitable for screening large numbers of progenies in breeding populations. Moreover, some protocols resulted in unclear DNA bands especially with amplification of ISSR-RGA markers. In addition, a high throughput DNA extraction method has never been reported in mungbean. Thus, the present study aimed to find the optimal method of mungbean DNA extraction by comparing eight combinations of two homogenization methods (manual grinding and Bullet blender® homogenizer) and four modified DNA extraction protocols.

2. Materials and methods

2.1. *Plant materials*

Genomic DNA was extracted from young seedlings [15-20 days after sowing (DAS)] of the Chainat 72 (CN72) variety grown in the Suranaree University of Technology Farm greenhouse, Nakhon Ratchasima, Thailand. Harvested fresh emerging leaves were immediately taken to the laboratory and kept in cool conditions (4°C).

2.2. *DNA extraction buffers*

The reagents and chemicals used in DNA extraction buffers were partly modified from four published protocols (table 1).

Table 1. Composition of DNA extraction buffers from four published protocols.

Protocols	Extraction buffer
(1) Modified from Lodhi *et al* [25]	3% (w/v) cetyltrimethylammonium bromide (CTAB), 1.4 M NaCl, 0.1 M Tris-HCL (pH 8), 0.02 M ethylenediaminetetraacetic acid (EDTA; pH 8), 2% (w/v) polyvinylpyrrolidone (PVP), 0.2% (v/v) β-mercaptoethanol
(2) Modified from Dharajiya *et al* [24]	2.5% (w/v) CTAB, 1.4 M NaCl, 0.1 M Tris-HCL (pH 8), 0.02 M EDTA (pH 8), 2% (w/v) PVP, 0.2% (v/v) β-mercaptoethanol
(3) Modified from Agbagwa *et al* [23]	3% (w/v) CTAB, 4 M NaCl, 0.1 M Tris-HCL (pH 8), 0.02 M EDTA (pH 8), 2% (v/v) β-mercaptoethanol
(4) Modified from Inglis *et al* [26]	Sorbitol wash buffer: 0.35 M sorbitol, 0.1 M Tris-HCL (pH 8), 0.05 M EDTA (pH 8), 1% (w/v) PVP, 1% (v/v) β-mercaptoethanol CTAB buffer: 3% (w/v) CTAB, 4 M NaCl, 0.1 M Tris-HCL (pH 8), 0.02 M EDTA (pH 8), 1% (w/v) PVP, 1% (v/v) β-mercaptoethanol

2.3. *Genomic DNA extraction methods*

Two homogenization methods consisting of manual grinding (with a pestle and mortar) and homogenizer (Bullet blender®, Next Advance, Inc., NY, USA) were combined with four modified genomic DNA extraction protocols. Extraction buffers used were shown in table 1. In Bullet blender® homogenizer method, 3.2 mm stainless steel beads were used with the extraction buffer without liquid nitrogen in all the protocols. While manual grinding was done in the presence of liquid nitrogen (protocol 1 and 4) or without liquid nitrogen (protocol 2 and 3). The experiment was conducted following a 2 x 4 factorial completely randomized design (CRD) with 8 treatments and 5 replications. The procedure for each genomic DNA extraction protocol was as follows:

Protocol 1: Leaf samples (50 mg) were ground to fine powder with liquid nitrogen using mortar and pestle. A CTAB extraction buffer (600) µl was added into the ground tissue and vortexed for 30 s to obtain a homogeneous solution and the mixture was then incubated at 65°C for 30 min. Chloroform:isoamyl alcohol (24:1; 600 µl) was added to the mixture and centrifuged at 13,000 rpm for 20 min at 25°C. The supernatant was then transferred to a new micro-centrifuge tube. This step was repeated once more. The DNA pellets were precipitated with 200 µl of 5 M NaCl and 600 µl of chilled absolute isopropanol and kept at -20°C for 60 min. The samples were centrifuged at 13,000 rpm for 15 min at 25°C. The DNA pellets were washed with 500 µl of chilled 70 % and absolute ethanol, and centrifuged at 13,000 rpm for 5 min at 25°C. The supernatant was then discarded and the pellets were air dried. After air drying, the pellets were re-suspended in 100 µl sterile water. Finally, the RNA was removed by adding 40 µl of 1 mg/ml RNase and incubating at 37°C for 30 min, and the DNA was finally stored at -20°C.

Protocol 2: Leaf samples (50 mg) were ground with 1 ml CTAB extraction buffer without liquid nitrogen using a mortar and pestle. The mixture was then incubated at 65°C for 45 min and allowed to return to room temperature for 5 to 10 min. Chloroform:isoamyl alcohol (24:1; 600 µl) was added to the mixture and centrifuged at 11,000 rpm for 10 min at 25°C. The supernatant was transferred to a new microcentrifuge tube, and 5 µl of 30 mg/ml RNase were added and incubated at 37°C for 20 min. The extraction step with 24:1 chloroform:isoamyl alcohol was repeated once more. The DNA pellets were precipitated with 150 µl of 3 M sodium acetate and 600 µl of chilled absolute isopropanol and kept at -20°C for 60 min. The samples were centrifuged at 10,000 rpm for 15 min at 4°C. The DNA pellets were washed with 500 µl of chilled 70% and absolute ethanol, and centrifuged at 7,000 rpm for 5 min at 4°C. The supernatant was then discarded and the pellets were air dried. After air drying, the pellets were resuspended in 100 µl TE buffer. The resulting DNA was stored at -20°C.

Protocol 3: Leaf samples (50 mg) were ground with 1 ml CTAB extraction buffer without liquid nitrogen using a mortar and pestle. The mixture was incubated at 65°C for 30 min and allowed to return to room temperature for 5 to 10 min. Chloroform:isoamyl alcohol (24:1; 600 µl) was added to the mixture and centrifuged at 10,000 rpm for 15 min at 22°C. The supernatant was transferred to a new microcentrifuge tube. This step was repeated once more. The DNA pellets were precipitated with 200 µl of 5M NaCl and 600 µl of chilled absolute isopropanol and kept at -20°C for 60 min. The samples were centrifuged at 6,000 rpm for 8 min at 4°C. The DNA pellets were washed with 500 µl of chilled 70% and absolute ethanol, and centrifuged at 7,000 rpm for 5 min at 4°C. The supernatant was then discarded and the pellets were air dried. After air drying, the pellets were resuspended in 100 µl TE buffer. Finally, the RNA was removed by adding 40 µl of 1 mg/ml RNase and incubating at 37°C for 30 min, and the DNA was stored at -20°C.

Protocol 4: Leaf samples (50 mg) were ground to a fine powder with liquid nitrogen using a mortar and pestle. Sorbitol wash buffer (1 ml) was added into ground tissue and vortexed for 30 s to obtain a homogeneous solution. The mixture was centrifuged at 5,000 rpm for 5 min at 25°C. The supernatant was then transferred to a new microcentrifuge tube, and CTAB extraction buffer (600 µl) was added. The mixture was incubated at 65°C for 60 min and allowed to return to room temperature for 5 to 10 min. Chloroform:isoamyl alcohol (24:1; 600 µl) was added to the mixture and centrifuged at 5,100 rpm for 10 min at 25°C. The supernatant was transferred to a new microcentrifuge tube. This step was repeated once. The DNA pellets were precipitated with 200 µl of 5 M NaCl and 600 µl of chilled absolute isopropanol and kept at -20°C for 60 min. The samples were centrifuged at 13,270 rpm for 10 min at 25°C. The DNA pellets were washed with 500 µl of chilled 70% and absolute ethanol, and centrifuged at 13,270 rpm for 5 min at 25°C. The supernatant was then discarded and the pellets were air dried. After air drying, the pellets were resuspended in 100 µl sterile water. Finally, RNA was removed by adding 40 µl of 10 mg/ml RNase and incubating at 37°C for 30 min, and the DNA was stored at -20°C.

The eight treatment combinations for genomic DNA extraction methods were as follows:

- T1 = Manual grinding and genomic DNA extraction protocol 1
- T2 = Manual grinding and genomic DNA extraction protocol 2
- T3 = Manual grinding and genomic DNA extraction protocol 3
- T4 = Manual grinding and genomic DNA extraction protocol 4
- T5 = Bullet blender® homogenizer and genomic DNA extraction protocol 1
- T6 = Bullet blender® homogenizer and genomic DNA extraction protocol 2
- T7 = Bullet blender® homogenizer and genomic DNA extraction protocol 3
- T8 = Bullet blender® homogenizer and genomic DNA extraction protocol 4

2.4. DNA quantification and purity estimation
The concentration and purity (Optical Density (OD) at 260 and 280 nm) of genomic DNA from mungbean were measured with a ND-1000 spectrophotometer (NanoDrop Technologies, Inc., Wilmington, DE, USA) and adjusted to a final concentration of 50 ng/μl for PCR amplification.

2.5. DNA quality evaluation by PCR amplification
The quality of DNA was evaluated by PCR amplification using SSR, ISSR and ISSR-RGA markers with 2 primers/ primer pairs for each marker type. The SSR markers were amplified with primer VR393 (Forward: 5'-TGGCACTTTCCATAACGAATAC-3', Reverse: 5'-ATCAGCCAAAAGCTC AGA AAAC-3') and CEDG084 (Forward: 5'-ATCAACTGAGGAGCATCATCGA-3', Reverse: 5'-CAA CATTTCAACCTTGGGACAG-3'). The ISSR makers were amplified with primer ISSR 885 (5'-BHB $(GA)_7$-3'; B = C,G,T; H = A,C,T) and ISSR 816 (5'-$(CA)_8$T3'). The ISSR-RGA makers were amplified with primers ISSR 827 (5'-$(AC)_8$G-3') and RLKfor (5'-GAYGTNAARCCI GARAA-3') and primers ISSR 842 (5'-$(GA)_8$YG-3') and P-Loop (5'-$(GGI)_2$GTIGGIAAIACIAC-3'). The composition of PCR mixtures are described in table 2. SSR markers were amplified in a T100TM Thermal Cycler (Bio-Rad Laboratories, Inc., California, USA) as follows: initial denaturation at 94°C for 2 min; 35 cycles of denaturing at 94°C for 30 s, annealing at 50-65°C for 30 s, extension at 72°C for 1 min; and final extension at 72°C for 10 min. The amplification of ISSR and ISSR-RGA markers was performed in a T100TM Thermal Cycler, programmed as follows: initial denaturation at 95°C for 5 min; 35 cycles of denaturing at 95°C for 1 min, annealing at 50°C for 1 min, extension at 72°C for 1 min; and final extension at 72°C for 10 min. The PCR products were separated on 6% denaturing polyacrylamide gel at 200 V for 50 and 70 min for SSR, ISSR and ISSR-RGA, respectively. DNA bands were detected with silver nitrate staining [27]. These markers are linked to powdery mildew or Cercospora leaf spot resistance genes in mungbean and are useful for marker-assisted selection for disease resistance [28,29].

Table 2. PCR mixture composition of SSR, ISSR and ISSR-RGA markers.

Markers	PCR mixture						
	DNA (ng)	Primers (μM)	*Taq* DNA polymerase (unit)	Buffer	$MgCl_2$ (mM)	dNTP (μM)	Total volume (μl)
SSR	150	0.5 Forward and Reverse primers	1	1X Buffer[a]	2	200	20
ISSR	150	0.4	1	1X Buffer	3.5	250	20
ISSR-RGA	150	0.4 ISSR and 1 RGA primer	1	1X Buffer	3.5	250	20

[a]1X Buffer consists of 50 mM KCl, 10 mM Tris-HCl (pH 9.1) and 0.01% (v/v) Triton™ X-100.

3. Results and discussion
When the effect of homogenization methods on DNA concentrations was evaluated, DNA concentrations were significantly variable ($F = 398.91$, $df = 1$, $P < 0.01$) between the manual grinding and Bullet blender® homogenizer methods. Overall, the homogenizer gave approximately 2.1-fold

IOP Conf. Series: Earth and Environmental Science **346** (2019) 012037 doi:10.1088/1755-1315/346/1/012037

higher DNA concentrations (932.1 ng/µl) than manual grinding (447 ng/µl). Similarly, genomic DNA extraction protocols also significantly affected DNA concentrations ($F = 83.36$, $df = 3$, $P < 0.01$). On average, protocol 1 gave the highest DNA concentration (850.91 ng/µl), followed by protocol 2 (707.89 ng/µl). Significantly lower DNA concentrations were obtained from protocol 3 (619.07 ng/µl) and protocol 4 (580.32 ng/µl). The interaction between homogenization methods and genomic DNA extraction protocols was significant ($F = 83.36$, $df = 3$, $P < 0.01$) on DNA concentration. Thus, the combination of both factors should be considered simultaneously to obtain high total DNA yield in mungbean.

The quantity and estimated quality of genomic DNA extracted by different combinations of homogenization methods and genomic DNA extraction protocols are shown in table 3. From the eight treatments, T7, T6 and T1 gave significantly higher DNA concentrations than other treatments (1069.26, 1032.58 and 920.12 ng/µl, respectively). The concentrations of CTAB and β-mercapto-ethanol commonly used in DNA extraction protocols of T1 and T6 were 2.5-3% and 0.2%, respectively, which have been shown to maximize total DNA yield in cereal crops, while, higher concentrations of both caused a reduction on DNA yield [30]. Similar results were observed in our experiment, however, we found that a high DNA yield can be achieved when using high concentrations of β-mercaptoethanol (2%) in the T7 using the homogenizer. Note that T3 which used similar DNA extraction protocol as T7 but with manual grinding gave a significantly lower DNA yield, substantiating the effects of homogenization methods and the interaction between homogenization methods and genomic DNA extraction protocols.

Table 3. Concentrations and OD (A_{260}/A_{280}) of mungbean genomic DNA isolated from different genomic DNA extraction methods and the time required for extraction.

Genomic DNA extraction methods (Treatments)	DNA concentrations (ng/µl)	OD (A_{260}/A_{280})	Sample grinding time (min/sample no.)	Total time for 24 samples (min)
(1) Manual, Protocol 1	920.12 ± 37.12abc[a]	1.90	3/1	260
(2) Manual, Protocol 2	383.20 ± 57.40d	1.73	3/1	265
(3) Manual, Protocol 3	168.88 ± 11.29e	1.83	3/1	198
(4) Manual, Protocol 4	315.76 ± 5.03d	1.43	3/1	250
(5) Homogenizer, Protocol 1	781.70 ± 12.25c	1.90	3/24	191
(6) Homogenizer, Protocol 2	1032.58 ± 17.86ab	1.80	3/24	196
(7) Homogenizer, Protocol 3	1069.26 ± 44.32a	1.95	3/24	129
(8) Homogenizer, Protocol 4	844.88 ± 46.70bc	1.67	3/24	181

[a]Results are presented as mean ± SE, n =5. Means identified with the same letters are not significantly different ($P > 0.05$).

Both T6 and T7 gave high DNA quantity; Nevertheless, the highest purity of DNA estimated from the A_{260}/A_{280} ratio was observed in T6. The A_{260}/A_{280} ratio of T6 was 1.8, suggesting the best quality of extracted genomic DNA for molecular studies—it likely had a very low contamination level from other compounds such as phenol, proteins and polysaccharides [24]. On the other hand, the ratio of nearly 2 in T7 may indicate RNA contamination because RNA can greatly absorb UV light at 260 nm [31]. By contrast, the aromatic amino acids in proteins can absorb UV light at 280 nm, resulting in a lower A_{260}/A_{280} ratio as shown in T4 and T8. Both types of contaminants may be found in the DNA solution of some treatments, affecting DNA quality. In addition, the presence of guanidine may lead to high absorbance at 260 nm. If the A_{260} is used to calculate DNA concentration, it may be overestimated [32]. Although PVP has been suggested to play a major role on extracted DNA purity by removing polyphenolic contamination, preventing polyphenol oxidation and/or browning of DNA [33], the omission of PVP in T3 and T7 did not result in browning. And T7 also gave the highest DNA concentration in our experiment, possibly due to high concentration of β-mercaptoethanol that prevent the oxidation problem.

Moreover, our study also focused on reducing the extraction time with a Bullet blender® homogenizer. When comparison was made between manual grinding using mortar and pestle (T1-T4) and Bullet blender® homogenizer (T5-T8), the homogenizer did not require liquid nitrogen and can homogenize much faster (3 min per 24 samples), while manual grinding required 3 min per sample and needed liquid nitrogen in T1 and T4. The combination of using the Bullet blender® homogenizer with protocol 2 (T6) gave both high yields and purity of DNA, and required only 196 min/ 24 samples, indicating that it was the most efficient DNA extraction method from this experiment.

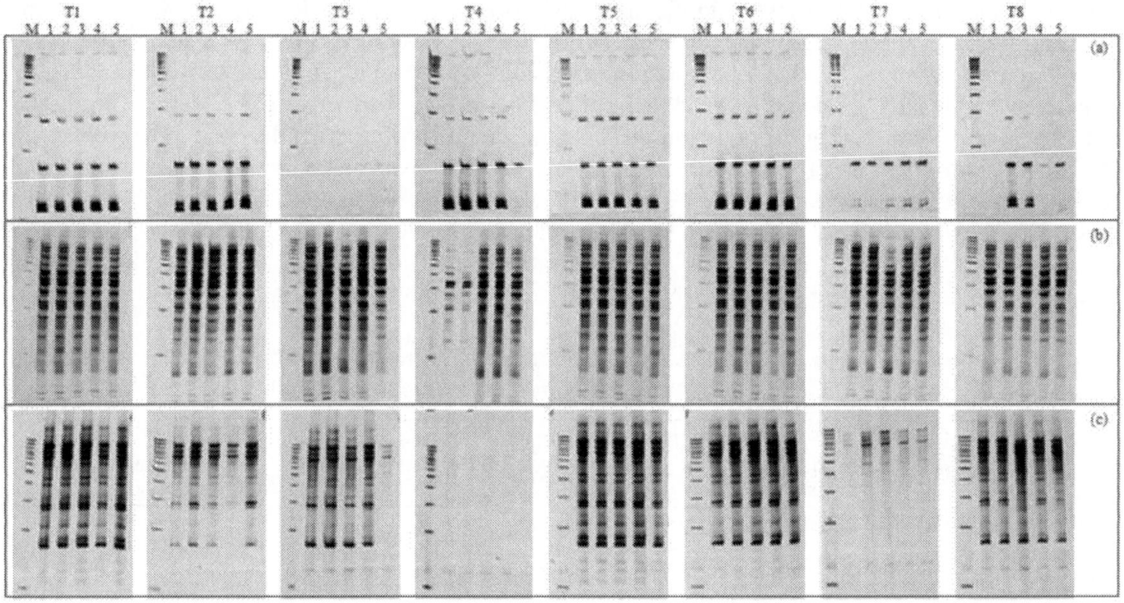

Figure 1. PCR products of genomic DNA extracted by different extraction methods, amplified using SSR, ISSR and ISSR-RGA primer/ primer pairs, and separated on 6% polyacrylamide gel. (A, B and C), using SSR (CEDG084), ISSR (885) and ISSR-RGA (827 and RLKfor) primer/ primer pairs, respectively. T1-T4 is the combinations of manual grinding with 4 modified DNA extraction protocols. T5- T8 is the combination of Bullet blender® homogenizer with 4 modified DNA extraction protocols.

The quality of genomic DNA isolated from various protocols was further confirmed by amplifying them with various molecular markers. Downstream PCR-based amplification results using 3 types of markers are presented in figure 1. The genomic DNA was amplified using SSR, ISSR and ISSR-RGA primers. The T1, T5 and T6 showed clear DNA bands with all molecular makers. These results indicated that the genomic DNA isolated using these modified methods is of decent quality and purity and can be applied with diverse molecular markers. However, the remaining treatments showed no or unclear DNA bands when amplified with same SSR, ISSR and/or ISSR-RGA primers/ primer pairs; This phenomenon might be due to the contamination of plant DNA with polyphenolics, polysaccharides, RNA, and/or proteins, which inhibited DNA amplification in PCR reactions [12,34-36]. This is most obvious with T4 which appeared to have amplification problems in all marker types. Note that the DNA extracted by this method also had the lowest A_{260}/A_{280} ratio of 1.43, implying contamination by proteins. When the three methods were compared based on all criterions, T5 was found to give lower DNA yield than T1 and T6. In addition, T1 required much longer extraction time (260 min/ 24 samples) than T6 (196 min/ 24 samples) and needed liquid nitrogen for grinding the samples, which is costly. Therefore, T6 was found to be the optimal genomic DNA extraction method for mungbean, it was quick, and yielded both high DNA concentrations and high purity (A_{260}/A_{280} ratio) suitable for diverse PCR applications. It was also found cost-effective because liquid nitrogen was unnecessary for this method.

4. Conclusion

The genomic DNA extraction method that combines the Bullet blender® homogenizer with modified DNA extraction protocol 2 (T6) is optimal for genomic DNA extraction of mungbean as it yielded abundant and pure genomic DNA. At the same time, this procedure was also simple, rapid and cost-effective as well as suitable for application of diverse molecular markers. Therefore, it can be used efficiently to carry out large scale marker analysis and genotyping, reducing time required and cost in molecular breeding.

Acknowledgment

We thank Dr. Colin Thomas Strine for editing the manuscript. This work was supported by the Office of the Higher Education Commission under NRU project of Thailand and grants from Suranaree University of Technology and Agricultural Research Development Agency (Public Organization), Thailand.

References

[1] World Vegetable Center 2016 Establishing the international mungbean improvement network [on-line] Available: https://avrdc.org/intl-mungbean-network/

[2] Lambrides C J and Godwin I D 2007 Mungbean *Genome Mapping and Molecular Breeding in Plants-Pulses, Sugar and Tuber Crops* ed C Kole (Heidelberg: Springer) pp 69-90

[3] Malik B A 1994 Grain Legumes *Crop Production* ed S Nazir *et al* (Islamabad: National Book Foundation) pp 277-8

[4] Tantasawat P, Trongchuen J, Prajongjai T, Thongpae T, Petkhum C, Seehalak W and Machikowa T 2010 Variety identification and genetic relationships of mungbean and blackgram in Thailand based on morphological characters and ISSR analysis *Afr. J. Biotechnol.* **9** 4452-64

[5] Agarwal M, Shrivastava N and Padh H 2008 Advances in molecular marker techniques and their applications in plant sciences *Plant Cell Rep.* **27** 6173

[6] Singh R, Adriaan W, Heusden V and Ram C Y 2013 A comparative genetic diversity analysis in mungbean (*Vigna radiata* L.) using inter-simple sequence repeat (ISSR) and amplified fragment length polymorphism (AFLP) *Afr. J. Biotechnol.* **12** 6574-82

[7] Kaur G, Joshil A and Jainl D 2018 SSR-marker assisted evaluation of genetic diversity in mungbean (*Vigna radiata* (L.) Wilcezk) genotypes *Braz. Arch. Biol. Technol.* **61** e180613

[8] Maiti S, Basak J, Kundagrami S, Kundu A and Pal A 2010 Molecular marker-assisted genotyping of mungbean yellow mosaic india virus resistant germplasms of mungbean and Urdbean *Mol. Biotechnol.* **47** 95-104

[9] Kabi M, Das T R, Baisakh B and Swain D 2017 Resistant gene rnalogous marker assisted selection of yellow mosaic virus resistant genotypes in Greengram (*Vigna radiata*) *Int. J. Curr. Microbiol. App. Sci.* **6** 3247-52

[10] Poolsawat O, Kativat C, Arsakit K and Tantasawat P A 2017 Identification of quantitative trait loci associated with powdery mildew resistance in mungbean using ISSR and ISSR-RGA markers *Mol. Breed.* **37** 150-61

[11] Chathiranrat N, Nitisit S, Chaiyapan C, Wansuriwong N, Papan P and Tantasawat P A 2018 Selection of mungbean resistant to powdery mildew in BC_1F_1 progenies based on ISSR and ISSR-RGA markers *Proc. of Advances in Science Engineering and Technology* (Macau, Chaina) pp 73-7

[12] Ambawat S, Kumar R, Singh S and Yadav R 2017 An easy, quick and cost effective method of

high quality DNA extraction from mungbean [*Vigna radiata* (L.) Wilczek] without liquid nitrogen *Int. J. Curr. Microbiol. App. Sci.* **6** 2695-703

[13] Saghai-Maroof M A, Soliman K M, Jorgensen R A and Allard R W 1984 Ribosomal DNAsepacer-length polymorphism in barley: mendelian inheritance, chromosomal location and population dynamics *Proc. Nat.l Acad. Sci.* **81** 8014-9

[14] Doyle J J and Doyle J L 1990 Isolation of plant DNA from fresh tissue *Focus* **12** 13-5

[15] Scott K D and Playford J 1996 DNA extraction technique for PCR in rain forest plant species *Bio. Techniques* **20** 974-9

[16] Haymes K M 1996 Mini-prep method suitable for a plant breeding program *Plant Mol. Biol. Rep.* **14** 280-4

[17] Pirttilä M A, Hirsikorpi M, Kämäräinen T, Jaakola L and Hohtola A 2001 DNA isolation methods for medicinal and aromatic plants *Plant Mol. Biol. Rep.* **19** 273

[18] Drábková L, Kirschner J and Vlcek C 2002 Comparison of seven DNA extraction and amplification protocols in historical herbarium specimens of Juncaceae *Plant Mol. Biol. Rep.* **20** 161-75

[19] Shepherd M, Cross M, Stokoe R L and Scott L J 2002 High throughput DNA extraction from forest trees *Plant Mol. Biol. Rep.* **20** 425

[20] Mogg J and Bond J M 2003 A cheap, reliable and rapid method of extracting highquality DNA from plants *Mol. Ecol. Notes.* **3** 666-8

[21] Chakraborti D, Sarkar A, Gupta S and Das S 2006 Small and large scale genomic DNA isolation protocol for chickpea (*Cicer arietinum* L.), suitable for molecular marker and transgenic analyses *Afr. J. Biotechnol.* **5** 585-9

[22] Krasova-Wade T and Neyra M 2007 Optimization of DNA isolation from legume nodules *Lett. Appl. Microbiol.* **45** 95-9

[23] Agbagwa I O, Datta S, Patil P G, Singh P and Nadarajan N 2012 A protocol for high-quality genomic DNA extraction from legumes *Genet. Mol. Res.* **4** 4632-9

[24] Dharajiya D T, Khadia S M, Pagi N K, Khatrani T J, Jasani S V, Khunt A D and Ravindrababu Y 2017 Modified method of high quality genomic DNA extraction from mungbean [*Vigna radiata* (L.) Wilczek] suitable for PCR based amplification. *Indian J. Sci. Technol.* **10** 1-7

[25] Lodhi M A, Ye G N, Weeden N F and Reisch B I 1994 A simple and efficient method for DNA extraction from grapevine cultivars and Vitis species *Plant Mol. Biol. Rep.* **12** 6-13

[26] Inglis P W, Pappas Md C R, Resende L V and Grattapaglia D 2018 Fast and inexpensive protocols for consistent extraction of high quality DNA and RNA from challenging plant and fungal samples for highthroughput SNP genotyping and sequencing applications *PLoS One* **13** e0206085

[27] Sambrook J and Russell D W 2001 *Molecular Cloning: A Laboratory Manual* (New York: Cold Spring Harbor Laboratory Press)

[28] Arsakit K, Papan P, Tharapreuksapong A and Tantasawat P A 2017 Simple sequence repeat markers associated with cercospora leaf spot and powdery mildew resistance in mungbean (*Vigna radiata* L. Wilczek) *Proc. Forum-Agriculture, Biology and Life Science* (Kyoto, Japan) pp 27-9

[29] Poolsawat O, Kativat C, Arsakit K and Tantasawat P A 2017 Identification of quantitative trait loci associated with powdery mildew resistance in mungbean using ISSR and ISSR-RGA markers *Mol. Breed.* **37** 150-61

[30] Mace E S, Buhariwalla H K and Crouch J H 2003 A high-throughput DNA extraction protocol for tropical molecular breeding programs *Plant Mol. Biol. Rep.* **21** 459a-h

[31] Gallagher S 2001 Quantitation of nucleic acids with absorption spectroscopy *Curr. Protoc. Protein Sci.* **13** A.4K.1-3

[32] Promega Corporation 2019 How do I determine the concentration, yield and purity of a DNA sample [on-line] Available: https://worldwide.promega.com/resources/pubhub/enotes/how-do-i-determine-the-concentration-yield-and-purity-of-a-dna-sample/

IOP Conf. Series: Earth and Environmental Science **346** (2019) 012037 doi:10.1088/1755-1315/346/1/012037

[33] Jadhav K P, Ranjani R V and Senthil N 2015 Chemistry of plant genomic DNA extraction protocol *BioInfo*. **12(3A)** 543-8

[34] Murray M G and Thompson W F 1980 Rapid isolation of high molecular weight plant DNA *Nucleic Acids Res.* **8** 4321-5

[35] Shioda M and Murakami-Murofushi K 1987 Selective inhibition of DNA polymerase alpha by a polysaccharide purified from slime of *Physarum polycephalum Biochem. Bioph. Res. Co.* **146** 61-6

[36] Richards E 1988 Preparation of genomic DNA from plant tissue *Current Protocols in Molecular Biology* ed F M Ausubel *et al* (New York: Greene Publishing Associates and Wiley-Interscience)

Development status and suggestions for e-commerce platform of traditional Chinese medicine in Henan Province

L N Zhao[1,2], Q R Liu[2] and Y S Wang[2,3]

[1]Zhengzhou Railway Vocational and Technical College, Zhengzhou City, Henan Province, China
[2]Henan University of Chinese Medicine, Zhengzhou City, Henan Province, China

E-mail: 895027475@qq.com

Abstract. With the further implementation of the "Internet plus" and "healthy China 2030" national strategy, the development environment of the Chinese herbal medicine industry has improved. In order to accurately understand the development situation of Henan Province's e-commerce platform of traditional Chinese medicine(TCM) and promote the development of the industry of TCM, this paper investigated the planting situation of TCM, the basic situation of Chinese medicinal farmers, the demand for technical services and the use of e-commerce platform of TCM in the main planting areas of Henan Province by means of on-site visits, questionnaires and expert interviews. The results showed that the development of E-commerce of TCM in Henan Province was still in the extensive stage, and the development of E-commerce was lagged behind.Therefore, this paper put forward that Henan Province's drug administration departments should actively promulgate supporting policies to establish a standardized and modern distribution system of TCM seamlessly joined by both producers and sellers, set up public trading platform and modern logistics distribution system of TCM E-commerce, and adopt the policy of assisting agriculture by science and technology to promote the communication between farmers and technical experts.

1. Introduction

In china, Henan Province is the hometown of Zhang Zhongjing. With convenient transportation, it has a long history of traditional Chinese medicine(TCM) culture. The variety, output and planting area of TCM were in the forefront of the country. So, it has a Yuzhou specialized market of TCM, who has been approved as one of 17 specialized markets of TCM in China. However, some reports [1-3] suggest that there were many problems in the specialized market of TCM, such as low entry threshold, uneven quality of personnel, loose management, poor hygiene, disorderly competition in the market and unclear supervision subjects, which directly affect the quality of TCM. Under the technical background of "Internet+" and big data, new business forms such as e-commerce have been rising constantly, which has impacted the specialized market of TCM. Therefore, the author had investigated the current situation of the transaction of TCM market and the development of electronic commerce platform of TCM in Henan Province, combed the key problems that need to be solved in the development, and put forward some suggestions to provide decision-making basis for the development of electronic commerce of TCM in Henan Province.

2. Methods

- Documentation method

Content from this work may be used under the terms of the Creative Commons Attribution 3.0 licence. Any further distribution of this work must maintain attribution to the author(s) and the title of the work, journal citation and DOI.
Published under licence by IOP Publishing Ltd

By means of literature search, the current situation, existing problems, development trend, information level and e-commerce development of traditional Chinese medicine industry in China and Henan Province were retrieved.

- Expert interview method

By means of special interview, we consulted the experts of Henan University of Traditional Chinese Medicine, Henan Science and Technology Department, Nanyang Science and Technology Bureau and Luoyang Kangxin Traditional Chinese Medicine Decoction Co., Ltd. to understand the situation of electronic commerce of TCM in Henan Province and listen to opinions and suggestions.

- Questionnaire survey

According to the feedback from expert interviews, the questionnaire was designed by using the questionnaire Star software to conduct an online survey on the development and using of e-commerce in Henan Province.

- Field visits

Field visits were conducted in towns and cities of Henan Province to survey the planting area, varieties, yields and sales of TCM of recent years in various regions, so as to give guidance to the planting situation of TCM.

- Mathematical statistics

The collected data was statistically processed.

3. Contents and results

- Basic information of the respondents

From the age distribution of the respondents, the majority of them were under 40 years old, accounting for 89.95% of the total number. These people were skilled in using smartphones to buy online and accepted new things quickly. From the cultural level of the respondents, most of them have college education or higher. From the native place of the respondents, 87.23% of them came from Zhumadian, Jiaozuo, Shangqiu, Puyang and Nanyang. Most of these areas were the main planting areas of TCM. Farmers were more concerned about the development information of traditional Chinese medicine industry and wanted to know more about the development of electronic commerce of TCM.

- Internet usage of the respondents

The results showed that the respondents mainly used the Internet to watch videos, chat, watch news and information, shop online, listen to music and other activities, among which online shopping ranked 4th, indicating that online shopping was the main use of the Internet for the respondents. (see table 1).

Table 1. Internet usage of the respondents.

Options	Proportion
See the video	60.05%
Listen to the music	35.05%
Play games	14.4%
Get news	48.64%
Learn financial management knowledge	4.08%
Chatting	50.54%
Online shopping	45.11%
Open online shop	1.09%
Others	10.87%

- Current Situation of Planting Area, Variety, Yield, Sale and Planting Technology of TCM

The survey found that most of the areas planted TCM, mostly sporadically, while the large-scale planting was less, accounting for only 24.27%. The main varieties planted were *dioscoreae rhizoma, lonicerae japonicae flos, chrysanthemi flos, paeoniae radix alba* and more than 20 other TCM, which were suitable for planting in Henan Province (see figure 1). The sale mode of TCM mainly relied on the farmers themselves to contact the purchaser or the purchaser to purchase directly from the farmers' home. Only 7.77% of the sales through the Internet, so the sales mode was relatively blocked, and it was easy to follow the trend of planting and unsalable phenomenon (see table 2). The planting management were extensive and backward, with only 7.77% of them equipped with technicians to guide planting, and 2.91% of them employing technical experts to guide planting.

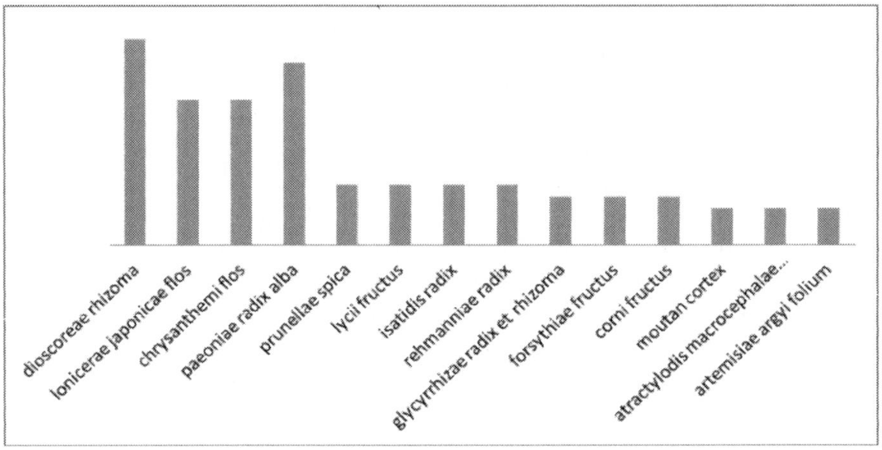

Figure 1. Planting varieties of TCM.

Table 2. The sale mode of TCM.

Options	Proportion
The purchaser to purchase directly from the farmers'home	40.78%
The farmers themselves to contact the purchaser	44.66%
Poor sales when production is large	17.48%
Sales through the Internet	7.77%
Not clear	35.92%

- Respondents' understanding and participation of e-commerce platform of TCM

The survey found that only 18.75% of the respondents had purchased or sold TCM online. Among them, 81.16% chose Taobao and other popular e-commerce platforms to buy, while only 17.39% purchased chose Chinese herbal medicines websites such as Drug Bo Mall and China Pharmaceutical Material Network , which also had the trading function of TCM. This showed that on the one hand, TCM as medicines, people generally believed that it was difficult to guarantee the quality of online purchases, on the other hand, it also showed that people have a low degree of understanding and participation in the e-commerce platform of TCM (see table 3). However, people were not disgusted with the promotion and application of a special e-commerce platform for traditional Chinese medicines. 62.77% of the respondents would actively use it when all conditions were ripe. 32.61% of respondents think that the use of e-commerce platform for traditional Chinese medicines would bring convenience. (see table 4).

Table 3. The situation of purchased or sold TCM online.

Options	Proportion
Taobao and other popular e-commerce platforms	81.16%
Chinese herbal medicines websites such as Drug Bo Mall and China Pharmaceutical Material Network	17.39%
Not clear	8.7%

Table 4. Promotion and application of TCM e-commerce platform.

Options	Proportion
Convenient and Healthy	32.61%
Too risky and unreliable in transportation	4.62%
Actively use when all conditions were ripe	62.77%

- Expectations and Current Difficulties of TCM E-commerce Platform

The survey results showed that the respondents generally hoped that the e-commerce platform of TCM could provide quality inspection, traceability information, logistics transportation, planting technology and price guidance services, 90.22% of them believed that providing quality inspection services was particularly important. The main reason was that the current market of TCM was mixed and inconsistent, and the quality was not uniform. How to build a mass accredited quality inspection platform should be the main problem to be solved in the electronic commerce platform of TCM (see table 5). There aslo have many other problems, such as the backward concept of farmers' e-commerce, the backward construction of rural network infrastructure, the backward logistics service system in rural areas, the difficulty in establishing traceability system of TCM (see table 6).

Table 5. Expectations of TCM e-commerce platform.

Options	Proportion
Traceability information	70.11%
Quality inspection	90.22%
Logistics transportation	55.16%
Planting technology	53.53%
Price guidance	54.62%
Supply and demand information management	49.18%
Others	8.7%

Table 6. Current Difficulties of TCM e-commerce platform.

Options	Proportion
The backward concept of farmers'e-commerce	57.61%
The backward construction of rural network infrastructure	44.02%
The backward logistics service system in rural areas	47.28%
The quality of TCM is difficult to guarantee	69.02%
The difficulty in establishing traceability system	56.52%
Backward planting technology of traditional Chinese medicine and so on	9.51%

- Initiative of Chinese Medicine Professional and Technical Personnel to Participate in E-Commerce Platform of TCM

The survey results showed that 57.34% of Chinese medicine professionals and technicians were willing to join the expert consultation database of e-commerce platform and provide technical services for farmers through the platform.

4. Suggestions

The competent pharmaceutical authorities of our province should actively introduce support policies to promote the rapid development of the Chinese herbal medicine industry, encourage large-scale cultivation of Chinese herbal medicines, promote the business model of "company + farmers + bases", rely on enterprises, and rationally plan and plant varieties. Establish a Chinese herbal medicine planting base and implement order production. On the one hand, it can form a stable industrial chain, promote the increase of the scale and trade scale of the Chinese herbal medicine industry in our province, and help to cultivate leading enterprises; on the other hand, the implementation of order production can reduce the planting risks of farmers, ensure the income of farmers, and mobilize The enthusiasm of the farmers has led to the cultivation of Chinese herbal medicines in our province.

Actively carry out the construction of Chinese medicine e-commerce trading platform and modern logistics distribution system, and explore the establishment of a standardized and modernized Chinese medicine material circulation system that seamlessly connects the production and marketing parties, so that the drug farmers can conduct information inquiry through the platform at any time, and establish an online Chinese herbal medicine trading center. Logistics centers, etc., make Chinese herbal medicine online shopping prosper in Henan Province. The e-commerce platform can provide technical services such as quality inspection of medicinal materials, traceability information of TCM, logistics and transportation, planting and processing technology and price guidance. It is especially important to provide quality inspection services.

Adopting an excellent talent policy, encouraging Chinese herbal medicine technicians to move to the grassroots level and publicizing and promoting the cultivation techniques of commonly used Chinese herbal medicines. The Chinese medicine e-commerce platform should widely absorb the technical experts of Henan Agricultural University, Henan University of Traditional Chinese Medicine, Henan Province and the Municipal Science and Technology Bureau, establish a communication platform between pharmaceutical farmers and technical experts, and adopt a one-to-one approach to communicate.

Acknowledgments

I express my gratitude to all those who helped me during the writing of this thesis. The work described in this review supported by the Outstanding Youth Project of Scientific and Technological Innovation in Henan Province (No. 184100510017); Henan Province "Three Districts" Talent Support Program Science and Technology Personnel Special Plan 2017 (No. 107); Henan Science and Technology Development Plan 2019 (No. 192102310443).

References

[1] Wei J H, Tu P F, Li G *et al* 2015 Analysis of the status quo and development trend of Chinese traditional medicine agriculture in China *Mod. Chin. Med.* **17** 94-8
[2] Cao H L and Wang W Q 2015 Current status and development suggestions of Chinese medicinal planting industry *Chin. Mod. Med.* **17** 753-5
[3] Gao B 2013 Status quo and objectives of traditional Chinese medicine agriculture development in Henan Province *Henan Agricult.* **1** 49-50

Two-step chemical mutagen treatment to convert Fuhua 12, a normal oleic Spanish peanut cultivar, into its high oleic version

S T Yu[1#], C T Wang[2,3,4#], H Wang[1#], G Q Yu[1], P X Shi[1], L Ren[1], H B Yu[1] and Z B Du[3]

[1]Institution of Sandy Land Management and Utilization of Liaoning, Fuxin 123000, China

[2]Shandong Peanut Research Institute, Qingdao 266100, China

[3]Luhua Group Company, Laiyang 265200, China

[#]All these three authors contributed equally to this work.

E-mail: chinapeanut@126.com

Abstract. High oleic peanut food products have proved to have better keeping quality and several health benefits. Most of the high oleic acid peanut cultivars thus far released were bred through hybridization. Breeding through mutagenesis directly provides a possibility of keeping the main agronomic characteristics from the parental cultivars/lines unchanged in the new cultivars, which is of significance to their use in commercial production. The aim of the present study was to convert Fuhua 12, a normal oleic Spanish peanut cultivar, into a high oleic version through EMS (ethyl methane sulfonate) mutagenesis. The effect of the first round of EMS treatment of Fuhua 12 on oleic acid content was limited. However, after the second round of EMS treatment, we were able to identity a high oleic plant through near infrared spectroscopy (NIRS) and further studies by gas chromatography and sequence comparison of the mutated type and wild type *FAD2A/FAD2B* genes. The mutant had the same mutations in *FAD2A* and *FAD2B* as UF435. A G448A substitution in *FAD2A* and an A insertion (441_442insA) in *FAD2B* together contribute to the high oleic phenotype of the mutant. In this study, Fuhua 12 was successfully converted into a high oleic version without changing most of the agronomic

Content from this work may be used under the terms of the Creative Commons Attribution 3.0 licence. Any further distribution of this work must maintain attribution to the author(s) and the title of the work, journal citation and DOI.
Published under licence by IOP Publishing Ltd

characters, demonstrating the utility of a two-step chemical mutagen treatment protocol in inducing useful mutations in the cultivated peanut. The novel method may expedite genetics and breeding studies in the cultivated peanut, and is of reference to other polyploid plant species.

1. Introduction

Peanut (*Arachis hypogaea* L.) is cultivated as an important oilseed crop globally. Fatty acid composition is of relevance to its quality as seed oil content of peanut may be as high as 50% or even more. Oleic (C18:1) and linoleic (C18:2) acids together account for about 80% of the total fatty acids in peanut [1-2]. High oleic peanut generally has more than 72% oleic acid and less than 7.7% linoleic acid [3]. In contrast, normal oleic peanut contains less oleic acid and more linoleic acid. Linoleic acid is prone to oxidization, whereas oleic acid is more stable. With increased oxidative stability, high oleic peanut has an extended shelf life [3-4]. As shown by previous studies, consumers benefit from intake of high oleic peanut food, resulting in lowered blood low-density cholesterol levels and ease of control of body weight and blood glucose [3-5]. In addition, alkyl oleate esters are useful as lubricant, surfactant, emulsifying agent, emollient, fuels additive and biodiesel [6]. High oleic peanut has been well recognized by leading food and oil processors and health-conscious consumers. Therefore, high oleic acid trait has become a must-have in peanut varietal releases [3].

Of the high oleic peanut cultivars thus far released worldwide for commercial production, most were developed through hybridization. C458 and M2-225, two high oleic peanut cultivars released in the USA, however, were directly selected from EMS and DES (diethyl sulfate) mutagenized populations, respectively [3].

Hybridization may cause a large scale of genetic recombination. To overcome the uncertainty frequently encountered in hybrid populations, backcrossing is often used to incorporate desirable trait(s) into a cultivar with wide use in food processing. Mutagenesis is another way to the same goal worthwhile trying. Breeding high oleic peanut cultivars through mutagenesis directly provides a possibility of keeping the main agronomic characteristics from the parental cultivars/lines unchanged in the new cultivars, which is of significance to their use in commercial production. Choice of appropriate parents is therefore of vital importance to the success of this breeding method, however, as the parents used should still be viable in productivity and adaptability even several years afterwards in a world full of competitions.

Widely cultivated in Liaoning province, Fuhua 12 is a Spanish-type normal oleic peanut cultivar bred for export at Sandy Land Amelioration and Utilization Research Institute of Liaoning, China [7]. The aim of the present study was to obtain its high oleic acid version through EMS mutagenesis.

2. Materials and methods

2.1. Materials

Fuhua 12, a high-yielding Spanish-type peanut cultivar with wide adaptability [7], was used. Maturing within 125-128 days, it is suitable for cultivation in northeast provinces of China (125°-135°E, 38°-56°N) [7], a major peanut production region with low risk of aflatoxin contamination possibly due to its cool climate.

2.2. Methods

2.2.1. Chemical mutagen treatment and identification of mutants with elevated oleic acid content. 1,000 Fuhua 12 seeds were soaked in 0.4% EMS (Solarbio, Beijing). Only the M_3 single plants with morphology like the parent were chosen. M_4 seeds from the selected single M_3 plants were then analyzed for oleic acid content in bulk seed samples using near infrared spectroscopy (NIRS) on a NIR machine (Matrix I, Bruker Optics) as described earlier[3]. Of these, seeds with elevated oleic acid were mixed, and 200 well-filled seeds were randomly selected for further EMS (0.4%) treatment. Seeds from the M_3 single plants of the second round of EMS treatment were screened for high oleic acid using NIRS [3]. Bulk seed samples from single plants were first used in search of high oleic plants; once identified, the high oleic single plant(s) were then analyzed for oleic and linoleic acid in individual single seeds [3]. High oleic acid phenotype was further confirmed by gas chromatography on an Agilent 7820A system. Injection volume was 2 μl. A silica capillary column (DB-wax, 30m×0.25mm×0.25μm) was programmed for an initial temperature of 210°C (for 9 min), then increasing at a heating rate of 20°C/min, to a final temperature of 230 °C (held for 5 min). A split ratio of 15:1 was employed. Temperature of detector was set at 300°C. Gas flow rates were 40ml/min for hydrogen, 400ml/min for air, and 66.5ml/min for helium (carrier gas), respectively.

2.2.2. Cloning and sequence analysis of FAD2A/FAD2B

2.2.2.1. Peanut DNA extraction

Genomic DNA was extracted from peanut cotyledonary slices distal to embryo ends, without affecting seed germinability, following the procedure described by Yu *et al.* [8].

PCR amplification of FAD2A/FAD2B from peanut genome

PCR amplification was performed on a Veriti^{TM} 96-Well PCR machine using peanut genomic DNA. Gene-specific PCR primer pairs, aF19/R1 and bF19/R1[8], were utilized to amplify the FAD2A and FAD2B from the cultivated peanut genome, respectively (Table 1). The reactions were performed in a 50μl volume containing 2μl of DNA, 25μl of Gflex buffer, 1μl of each primer, 1μl of TKS Gflex polymerase (Tiangen, Beijing), and 20μl of sterile double distilled water. The thermal cycling profile was as follows: 94 °C for 6 min; 30 cycles consisted of 94 °C for 30 s, 53°C for 1 min, 72 °C for 2 min; and a final extension of 72 °C for 4 min. The PCR product was resolved on a 1% agarose gel.

Table 1. Sequence information of PCR primers used in this study[8].

Primer name	Sequence (5' to 3')
aF19	GATTACTGATTATTGACTT
bF19	CAGAACCATTAGCTTTG
R1	CTCTGACTATGCATCAG

2.2.2.2. Ligation, E. coli transformation and identification of PCR positive colonies

PCR products of expected size from the genomic DNA amplifications were ligated into the pEASY-Blunt Zero vector (Tiangen, Beijing), and the ligation products were then used to transform the Trans-T1 Phage Resistant Chemically Competent Cell (Tiangen, Beijing) using the heat shock method following the manufacturer's instructions. Putative transformants were identified through blue white screen, and the right recombinants were confirmed by PCR. The reaction (20μl) consists *E. coli*, 1μl of each primer, 10μl of 2×Bench Top Tap MasterMix (Biomiga, San Diego) and 8μl of sterile double distilled water. Primer pairs and PCR program were the same as aforementioned for peanut genomic DNA.

2.2.2.3. DNA Sequencing and sequence analysis

PCR positive colonies were sent to Shanghai Sunny Biotechnology Co., Ltd (Shanghai) for sequencing the inserts on ABI 3730XL. Bio-XM 2.6 and DNAMAN were used to analyze the DNA sequences.

2.2.3. Characterization of the high oleic mutant and Fuhua 12 for main agronomic traits. At harvest, 10 plants of high oleic mutant and 10 plants of Fuhua 12 were sampled randomly, and plant height, branch length, no. of pods per plant, 100-pod weight and 100-seed mass were measured and recorded following the methods of Yu et al [9].

3. Results and discussion

3.1. Identification of a high oleic mutant plant of Fuhua 12

The first round EMS treatment of Fuhua 12 had some effects on oleic acid content. Even though no high oleic plants were found, single plants with elevated oleic acid, but still lower than 75%, were identified in M_3 generation. These were named 12M. Selected seeds from 12M were subjected to the second round EMS treatment. A high oleic peanut plant of the M_3 generation of the second round EMS treatment, WC-103 (Figure 1), was initially identified as with over 75% oleic acid by NIRS, and its high oleic phenotype was later confirmed by GC analysis of its 44 single seeds (Table 2). Seeds of WC-103 proved to contain 76.91%~83.94% oleic acid in their total fatty acids. For example, oleic acid in WC12 was as high as 80.7%, in contrast to only 35.9% in Fuhua 12 (Figure 2, Table 2)

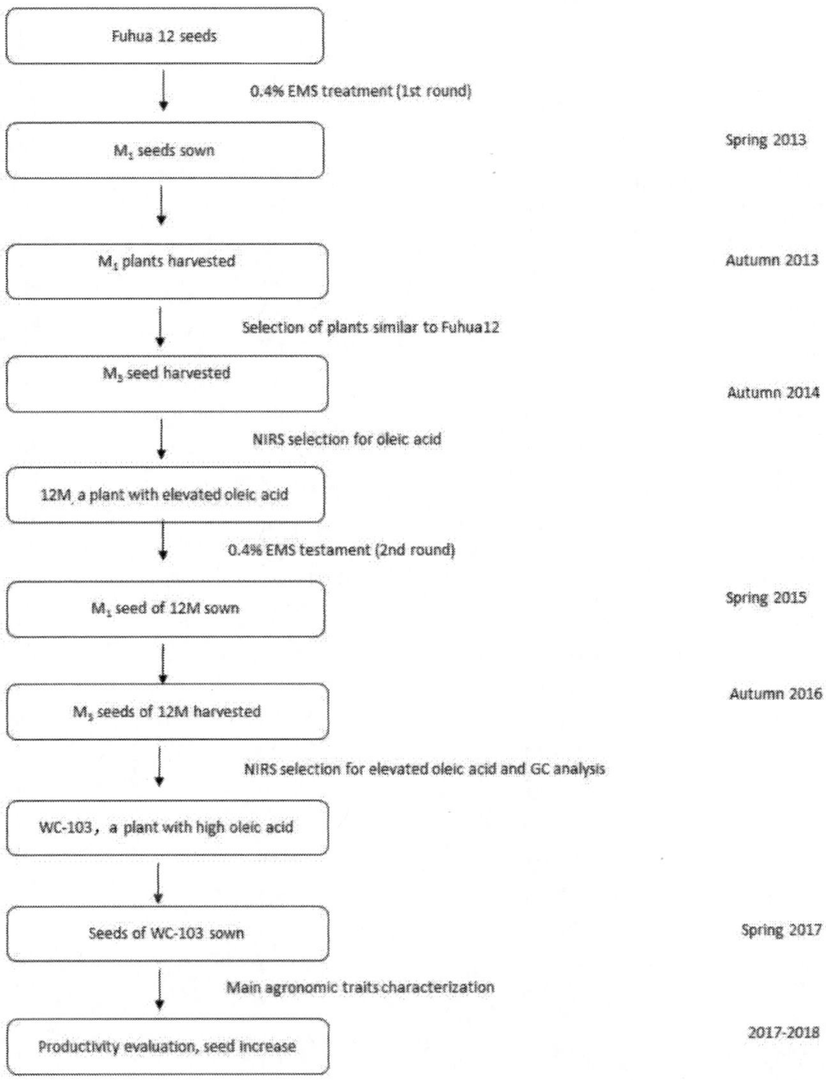

Figure 1. Breeding of "WC-103" with 2 rounds of EMS treatment and NIRS-aided selection.

Table 2. Oleic, linoleic acid in seeds of the high oleic peanut mutant plant, WC-103, revealed by GC.

Seed serial No.	Oleic acid (O) (%)	linoleic acid (L) (%)	O/L
WC1	83.94	2.40	34.98
WC2	82.31	2.99	27.53
WC3	82.86	3.03	27.35
WC4	82.66	2.56	32.29
WC5	83.05	2.83	29.35
WC6	79.21	4.81	16.47
WC7	78.95	2.92	27.04

IOP Conf. Series: Earth and Environmental Science **346** (2019) 012039 doi:10.1088/1755-1315/346/1/012039

WC8	77.23	5.44	14.20
WC9	80.18	4.13	19.41
WC10	81.75	3.08	26.54
WC11	79.53	4.76	16.71
WC12	80.70	3.47	23.26
WC13	82.34	2.89	28.49
WC14	82.71	2.83	29.23
WC15	83.00	3.02	27.48
WC16	83.30	2.91	28.63
WC17	83.06	3.21	25.88
WC18	82.53	2.68	30.79
WC19	81.42	3.33	24.45
WC20	81.39	3.41	23.87
WC21	83.83	2.41	34.78
WC22	82.61	2.74	30.15
WC23	77.06	7.19	10.72
WC24	83.20	2.38	34.96
WC25	82.74	3.09	26.78
WC26	81.97	3.38	24.25
WC27	79.74	4.33	18.42
WC28	81.12	3.31	24.51
WC29	82.93	2.77	29.94
WC30	80.34	3.79	21.20
WC31	82.67	2.90	28.51
WC32	79.36	4.52	17.56
WC33	81.01	3.98	20.35
WC34	79.68	4.54	17.55
WC35	83.10	2.76	30.11
WC36	83.17	2.74	30.35
WC37	79.86	4.32	18.49
WC38	82.52	3.20	25.79
WC39	77.49	5.44	14.24
WC40	76.91	5.58	13.78
WC41	77.16	6.15	12.55
WC42	78.94	4.61	17.12
WC43	81.02	4.05	20.00
WC44	81.55	3.92	20.80

Notes: a Oleic acid and linoleic acid reported as percentage of total fatty acids.

b Peanuts with an O/L ratio no less than 9 can be classified as high oleic.

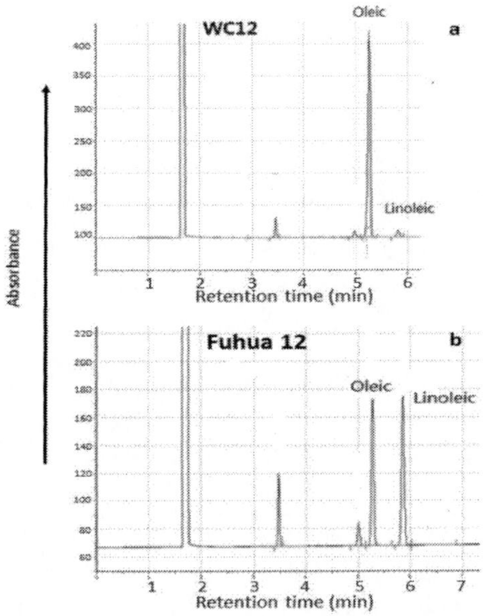

Figure 2. GC Chromatograms showing increased oleic acid and reduced linoleic acid in the high oleic mutant WC12 (a) as compared with the wild type Fuhua12 (b).

3.2. Comparison of FAD2A/FAD2B sequences of WC12, a high oleic mutant and Fuhua 12

The *FAD2A* and *FAD2B* from the mutated seeds (WC1, WC 2, WC3, WC4) and their parental cultivar (Fuhua 12) were cloned and sequenced. Multiple sequence alignment showed that the differences were the same as previously reported in UF435. A G448A substitution in *FAD2A* and an A insertion (441_442insA) in *FAD2B* were found in all the mutated seeds assayed, but absent in the wild type parent, Fuhua 12 (Figure 3).

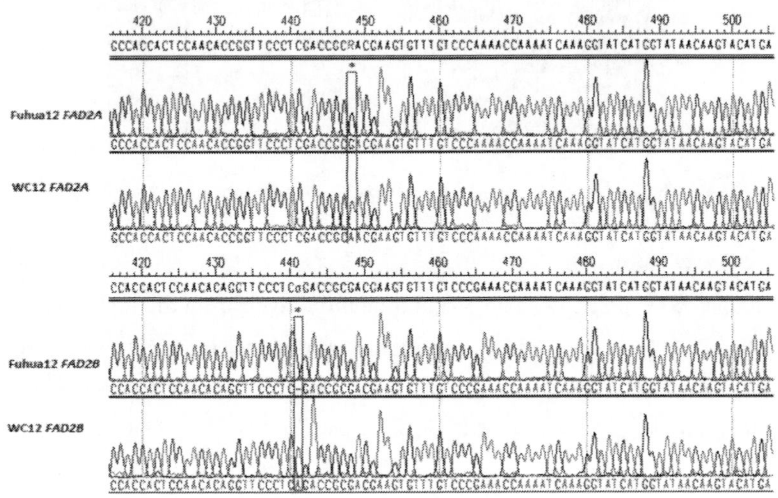

Figure 3. Comparison of *FAD2A* and *FAD2B* partial DNA sequences from normal oleic Fuhua 12 and its high oleic mutant WC12, showing a G448A substitution in *FAD2A* and an A insertion (441_442insA) in *FAD2B* in WC12.

3.3. Main agronomic characters of WC-103 and Fuhua 12

As shown in Table 3, WC-103 was quite similar to Fuhua 12 in the main agronomic characters investigated. Of the 6 traits listed, only 1 (branch length) differed at 0.05 level. Indeed, the plant morphology of the mutant line and its parental cultivar was nearly identical (Figure 4).

Table 3. Main agronomic characters of the high oleic mutant (WC-103) and its parent, Fuhua 12.

Material	Plant height (cm)	Branch length (cm)	No. of branches per plant	No. of pods per plant	Pod weight per plant(g)	100-pod weight (g)	100-seed mass (g)
WC-103	39.2±1.3	43.8±1.9*	6.8±0.8	19.4±2.1	15.9±1.6	206.6±3.2	74.3±1.8
Fuhua 12	38.6±2.1	41.2±2.3	6.5±1.5	18.2±2.5	14.6±2.1	201.1±5.4	75.1±1.3

Note: Figure with an asterisk (*) indicated significant difference at 0.05 probability level according to LSD (Least Significant Difference) test.

Figure 4. Plant morphology of WC-103 (left) and its parental cultivar, Fuhua 12 (right).

4. Discussion and conclusions

In this study, we obtained a high oleic acid peanut mutant derived from the high-yielding, normal oleic food use peanut cultivar, Fuhua 12. This new version was almost identical to the original version in plant morphology and the main agronomic characters investigated, except for oleic and linoleic acid and branch length. The productivity and adaptability of the new line still need further evaluation.

Previously reported peanut induced mutants mostly, if not all, resulted from a single treatment with chemical/physical mutagen(s), and composite treatment was seldom studied. The present study, however, demonstrated, for the first time as far as we know, the utility of a two-step chemical mutagen treatment protocol in inducing useful mutations in the cultivated peanut. As a tetraploid species with two sub-genomes and duplications in its genome [10], the cultivated peanut sometimes was considered not amenable to induced mutagenesis. A large number of mutations may be difficult to be identified due to gene redundancy. *FAD2A* from the A sub-genome and *FAD2B* from the B sub-genome are genes conditioning oleic acid content in the cultivated peanut [3,10]. Expression of the high oleic phenotype requires both genes to be dysfunctional [3,10]. In this study, though not confirmed, it is likely that the first round of EMS treatment mutated one of the *FAD2A/FAD2B* gene, and the second round mutated the other. Actually, we obtained more mutations in the mutagenized populations after the second round of EMS treatment as compared with those in one round of EMS treatment (data not shown). The two-step mutagenesis may expedite genetics and breeding studies in the cultivated peanut, and is of reference to other polyploid plant species. The present study suggests that, to get better results, more than one round of mutagen treatment or composite treatment may be necessary for a crop species with sub-genomes, if the outcome of a single round of mutagen treatment is not satisfactory.

Acknowledgments
Authors wishing to acknowledge assistance or encouragement from Shandong Peanut Research Institute and financial support from Foundation items: China Agricultural Research System (CARS-13); The Central Government Guiding Local Financial Special Project (2017108006).

References
[1] Ahmed E M and Young C T 1982. Composition, quality, and flavor of peanuts. *Peanut Sci. Tech.* **25**:65-88
[2] Norden A J, Borget D W and Knauft D A 1987. Variability in oil quality among peanut genotypes in the Florida breeding program. *Peanut Sci.* **4**:7–11.
[3] Wang C T and Zhu L G 2017. High oleic acid peanut. Shanghai Science and Technology Press, Shanghai China. p. 28-35.
[4] Braddock J C, Sims C A and O'keefe S F 1995. Flavor and oxidative stability of roasted high oleic acid peanuts. *J. Food Sci.* **60**: 489-93.
[5] Moreira Alves R D, Boroni Moreira A P, Macedo V S, Bressan J, Alfenas B C G, Mattes R and Costa M B 2014 High-oleic peanuts: New perspective to attenuate glucose homeostasis disruption and inflammation related obesity. *Obesity*, **22**(9): 1981-8.
[6] Whelan L 2014. Oleic acid: Production, uses and potential health effects. Nova Science Publishers. New York, USA 83-98.
[7] Yu H B 2007. Breeding and cultivation of Fuhua 9, Fuhua 12 and Fuhua 13 peanut cultivars. *Miscellaneous Food Crops*. **27**:189-90.
[8] Yu S T, Wang C T and Wang X Z 2010. Simple method to prepare DNA templates from a slice

of peanut cotyledonary tissue Polymerase Chain Reaction. *Electr. J. Biotech.* **13**:1-4.

[9] Yu S L 2008. Peanut varieties in China and their genealogy. Shanghai Science & Technology Press, Shanghai, China. 792-7.

[10] Chu Y, Holbrook C C and Ozias-Akins P 2009. Two alleles of *ahFAD2B* control the high oleic acid trait in cultivated peanut. *Crop Sci.* **49**:2029–36.

IOP Conf. Series: Earth and Environmental Science **346** (2019) 012040 doi:10.1088/1755-1315/346/1/012040

Cellulase activity of *Bacillus velezensis* isolated from soil in a dairy farm

P Narkthewan[1, 2] and W Makkapan[1]

[1]Food and Agricultural Biotechnology Program, Department of General Science and Liberal Arts, King Mongkut's Institute of Technology Ladkrabang, Prince of Chumphon Campus, Pathiu, Chumphon, 86160, Thailand

E-mail: kpapatch@hotmail.com

Abstract. The aim of the present study was to isolate and screen cellulase producing bacteria from soils including characterizing the cellulase activity of the identified bacterial strain. Screening of the cellulase producing bacteria was performed on a carboxymethylcellulose (CMC) agar (pH 7.0) plate at 37°C for 3 days. Bacterial isolate, namely CC1-1, showed the highest cellulase activity on the CMC agar plate. Subsequently, it was identified as *Bacillus velezensis* based on 16S rDNA gene sequence analysis. Cellulase production was carried out under submerged fermentation. The maximum CMCase and FPase activity of 0.030 U/ml and 0.047 U/ml were obtained after 20 h and 32 h, respectively. The optimal pH and temperature of CMCase activity were 6.0 and 60°C, respectively. The cellulase activities were active in a broad pH and temperature range. The CMCase activity was stable at pH 6.0-7.0, 37°C for 120 min and the stability of CMCase was revealed at 50-60°C for 90 min with over 80% remaining activity. However, the optimum pH and temperature of FPase activity were 7.0 and 50°C, respectively. The FPase activity was stable at pH 6.0, 37°C for 90 min and its activity was stable at 60°C for 60 min with over 80% remaining activity. Due to its particular properties, cellulase-producing *B. velezensis* could be a potential candidate for the composting process.

1. Introduction

Cellulose is the most abundant carbohydrate in the plant cell wall. It is a linear polymer of D-glucose residues linked with β-1, 4 glucosidic bonds [1]. Cellulases have been applied in various industries, such as the paper and pulp industry, textile industry, detergent industry, fruit juice extraction, food processing, and animal feed additives as well as in bioethanol production [2]. In nature, complete cellulose degradation requires the synergy of three major types of cellulases: endoglucanase, exoglucanase and β-glucosidase. Endoglucanases randomly cleave the β-1,4 bonds at internal amorphous points of the cellulose polysaccharide chain, generating various lengths of oligosaccharides. Exoglucanase is essential for the cutting of the non-reducing or reducing end of a cellulose chain and liberating either cellobiose or glucose as main products. β-glucosidase catalyzes the hydrolysis of cellobiose to glucose [3].

A number of microorganisms are able to hydrolyse cellulose such as bacteria, actinomycetes and fungi. Most cellulase production has been explored in fungi [4-8]. Compared to fungi, bacteria have considerable benefits for industrial utilization, i.e. easy cultivation, high growth rate and adaptability to substantial genetic manipulations [9]. Extracellular cellulase production from bacteria such as *Cellulomonas* sp. YJ5 [10], *Brevibacillus* sp. DUSELG12, *Geobacillus* sp. DUSELR7, *Paenibacillus*

Content from this work may be used under the terms of the Creative Commons Attribution 3.0 licence. Any further distribution of this work must maintain attribution to the author(s) and the title of the work, journal citation and DOI.
Published under licence by IOP Publishing Ltd

terrae ME27-1 [11], *Bacillus* sp., *Pseudomonas* sp., *Serratia* sp. [12], *B. brevis* [13], *B. subtilis* [14-16], *B. thuringiensis* [14, 16], *B. mojavensis* [15] and *B. pumilus* [16] etc. Cellulose degrading bacteria have been surveyed to obtain more proficient cellulases from various sources such as soils, decomposed plant materials, feces of ruminants, hot springs, and composts [17]. Researchers have been looking for microorganisms with higher cellulase activity. Therefore, the objectives of this study were to isolate and screen the cellulase producing bacteria from soils in the agricultural area and animal farms in a variety of places in Thailand. Moreover, the most active strain from preliminary screening was identified and its cellulase properties were studied. The selected cellulase producing strain has potential benefits for the composting process.

2. Material and methods

2.1. Collection of soil samples
Twenty soil samples were obtained and collected from several sites in Chumphon, Prachuap Khiri Khan and Phetchaburi province, Thailand. The sample was randomly taken to a depth of 10 cm from the area of the lime, banana, pineapple, durian and oil palm tree field including the soil samples from the goat and dairy farms.

2.2. Isolation and screening of cellulolytic bacteria
One gram of sample was added to 9 ml of 0.85% NaCl and serially diluted and the spread plate technique was done using nutrient agar (NA). The culture was grown on 37°C for 24 h and streak plate technique was used for the isolation to obtain the pure colonies. The bacterial isolate was further point inoculated on a carboxymethylcellulose agar (CMC agar) containing (g/L): peptone 10; K_2HPO_4 2; $MgSO_4.7H_2O$ 0.3; $(NH_4)_2SO_4$ 2.5; CMC 10 and agar 15 and cultivated at 37°C for 3 days. A 1% Congo red solution was poured onto the agar plate and left for 15 min at room temperature. It was then destained with 1 M NaCl solution for 15 min. A clear zone was shown around bacterial colony displaying cellulose degradation [18]. The maximum ratio of the clear zone diameter to colony diameter was investigated so as to choose the best cellulase producing strain.

2.3. Molecular identification of bacterial isolate
The selected cellulolytic bacterial isolate was sent to Apical Scientific Sdn Bhd., Malaysia for sequence analysis. Molecular identification of the bacterial isolate was investigated from the nucleotide sequence of the 16S ribosomal DNA (rDNA). Sequence homology was analyzed using the nucleotide BLAST tool (https://blast.ncbi.nlm.nih.gov/Blast.cgi). Phylogenetic construction was performed using the neighbor-joining method using MAGA 7.0 with 1,000 replicates of bootstrap values.

2.4. Cellulase production
The bacterial isolate was inoculated in 150 ml CMC broth containing (g/L): CMC 10; $FeSO_4.7H_2O$ 0.02; $MgSO_4.7H_2O$ 0.2; KNO_3 0.75; K_2HPO_4 0.5; $CaCl_2$ 0.04; yeast extract 2; and D-glucose 1. The culture was then cultivated at 37°C, 150 rpm for 24 h for inoculum preparation. The inoculum (5%) was transferred into the same medium in a 5 L fermenter. The bacterium was grown at 37°C, airflow rate of 1.0 vvm, stirring rate of 200 rpm, for 72 h and the sample was withdrawn every 4 h. The crude enzyme was harvested by centrifugation at 14,000×g, 4°C for 10 min. The culture filtrate was analyzed for cellulase activity assay.

2.5. Cellulase assay
Endocellulase (CMCase) activity and filter paper activity (FPase) were analyzed using the modified method proposed by Wood and Bhat [19]. The CMCase activity was investigated by mixing the enzyme solution (0.5 ml) with 1.0 (w/v) CMC in 0.05 M phosphate buffer (pH 7.0) 0.5 ml and incubated at 37°C for 30 min. The reducing sugar liberated was assayed using the dinitrosalicylic (DNS) method [20] with glucose as the sugar standard. The total cellulase activity or FPase activity assay in the culture filtrate

was the same as the CMCase assay method while for the substrate a 1x6 cm strip of Whatman No.1 filter paper immersed in 1.0 ml of 0.05 M phosphate buffer (pH 7.0) was used and incubated at 37°C for 30 min. The reducing sugar was assayed using the DNS method. One unit of cellulase (CMCase or FPase) was the quantity of enzymes which produced 1 μmole of glucose or reducing sugar per minute under the specified conditions. All experiments of cellulase activity assay were performed in three replications and the results were reported as mean±SD.

2.6. Characterization of crude cellulase activity

2.6.1. pH profile and stability. The various buffers such as 50 mM of glycine-HCl buffer (pH 3.0), citrate buffer (pH 4.0-5.0), phosphate buffer (pH 6.0-7.0), Tris-HCl (pH 8.0) and glycine-NaOH (pH 9.0-10.0) were used to investigate the influence of pH on the activity and stability of CMCase and FPase. Crude enzyme was mixed with 1% CMC or a 1x6 cm strip of Whatman No.1 filter paper prepared in each buffer solution. Determination of CMCase and FPase activities was done after the reaction assay was kept at 37°C for 30 min. The highest activity was defined as 100% relative activity and the others were calculated by comparison with the highest activity.

pH stability was performed by keeping crude enzyme in different buffers as above described at 37°C for 3 h. The sample was taken every 30 min for determination of the residual CMCase and FPase activities according to the standard assay method.

2.6.2. Temperature profile and stability. Crude enzyme was mixed with 1% CMC or a 1x6 cm strip of Whatman No.1 filter paper prepared in a suitable buffer solution. The CMCase and FPase activities were assayed after the reaction mixture was kept in the temperature range of 30-90°C for 30 min. The highest activity was defined as 100% relative activity and others were calculated by comparison with the highest activity.

Regarding the thermostability measurement, the crude enzyme was incubated in a variety of temperatures as described above for 3 h. The sample was taken every 30 min for determination of the residual CMCase and FPase activities according to the standard assay method.

3. Results and discussion

3.1. Isolation and screening of cellulase producing bacteria

Thirty-six bacterial isolates were obtained from 20 soil samples. After selection of the cellulase producing strains on the CMC agar plates, the results showed that three bacterial isolates (CC1-1, CC2-1 and GF-4) produced extracellular cellulase for the hydrolysis of CMC. The highest ratio of clear zone size to colony size was found in CC1-1 isolated from soil taken from the dairy farm in Phetchaburi province.

3.2. Identification of bacterial isolate by using molecular technique

The selected bacterial isolate, CC1-1, was sequenced approximately 1,500 bp of 16S rDNA gene. The nucleotide sequence was aligned and compared with others from GenBank using the nucleotide BLAST tool. CC1-1 was demonstrated to have 99% similarity with *Bacillus* sp. Phylogenetic analysis showed that CC1-1 was identified as *Bacillus velezensis* as shown in figure 1. Several cellulose degrading bacteria have been found in different ecological niches. In addition, *Bacillus* was reported to be the dominant cellulose degrading bacteria in samples collected from soils such as *B. brevis* [13], *B. subtilis* [14-16], *B. cereus* [14-15], *B. thuringiensis* [14, 16], *B. mojavensis* [15] *B. pumilus* [16].

3.3. Production of cellulase under submerged fermentation

Cellulase production from *B. velezensis* was carried out in CMC broth at 37°C for 72 h. The highest CMCase and FPase activities of 0.030 U/ml and 0.047 U/ml were obtained after 20 and 32 h, respectively. In the previous reports, several bacterial cellulases activities were similar to this study.

CMCase and FPase activities were 0.079 U/ml and 0.011 U/ml, respectively in *B. pumilus* EB3 after 24 h fermentation [21]. Subsequently, Rastogi et al. [22] reported that CMCase activity of 0.020 and 0.058 U/ml was received from *Brevibacillus* sp. DUSELG12 and *Geobacillus* sp. DUSELR7 after 10 and 7 days of cultivation, respectively. However, the higher CMCase activities were found in *B. cereus* B7 (0.440 IU/ml/min), *B. cereus* B37 (0.410 IU/ml/min), *B. subtilius* B20 (0.357 IU/ml/min), and *B. thuringiensis* B49 (0.334 IU/ml/min) [14].

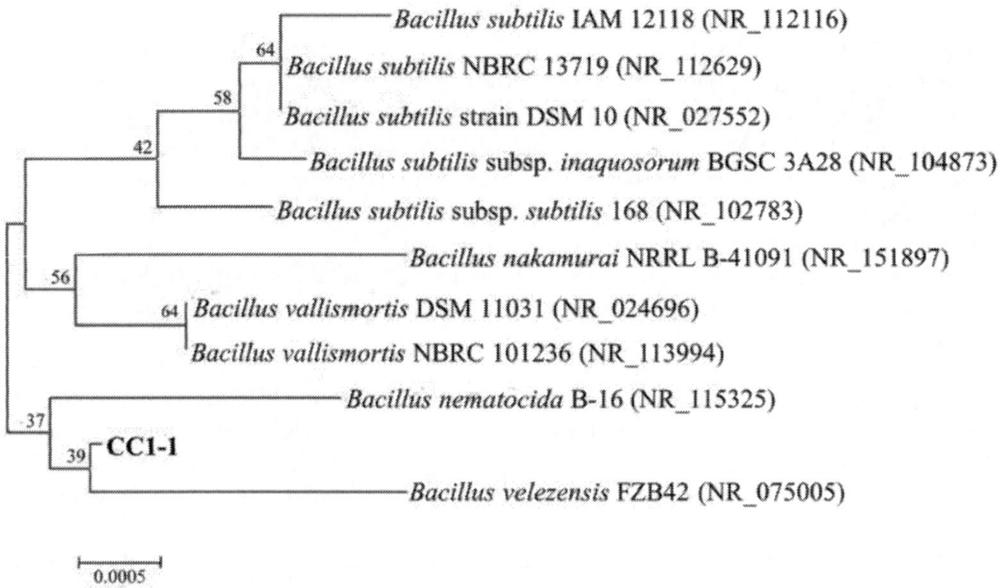

Figure 1. Phylogenetic analysis of bacterial isolate, CC1-1 based on 16S rDNA gene sequences was created by the neighbor-joining method. Numbers represent bootstrap values and the accession numbers of the strains are given in brackets.

3.4. Characterization of crude cellulase activity

3.4.1. pH profile of CMCase and FPase activity. The effect of pH value on cellulase activity was studied in a wide pH range of 3.0-10.0 as shown in figure 2. The highest CMCase activity was achieved at pH 6.0. Remaining activities of 98%, 80% and 72% was found at pH 5.0, 4.0 and 3.0, respectively. At pH 8.0-9.0, the enzyme activities remained at 64% with the remaining activity of 24% at pH 10.0. The optimum pH of CMCase from *B. velezensis* in this study was similar to *B. licheniformis* 2D55 [23]. Several research works revealed that the optimum pH of CMCase from bacteria was 5.0-5.5. The CMCase activity from the composite microbial system (FH3) showed the highest value at pH 5.0 [24]. The maximum CMCase activities of *Geobacillus* sp. DUSELR7 and *Brevibacillus* sp. DUSELG12 were pH 5.0 and pH 5.5, respectively [22]. Moreover, the optimum pH of CMCase activity from *P. terrae* ME27-1 was 5.5 [11]. However, the optimum purified CMCase from *Cellulomonas* sp. YJ5 was pH 7.0 [10]. The highest FPase activity of *B. velezensis* was at pH 7.0 and its activity decreased in both the acidic and basic pH ranges. In contrast, the optimum pH of FPase activity from the composite microbial system (FH3) was pH 4.5 [24] and that from *B. licheniformis* 2D55 was pH 6.0 [23].

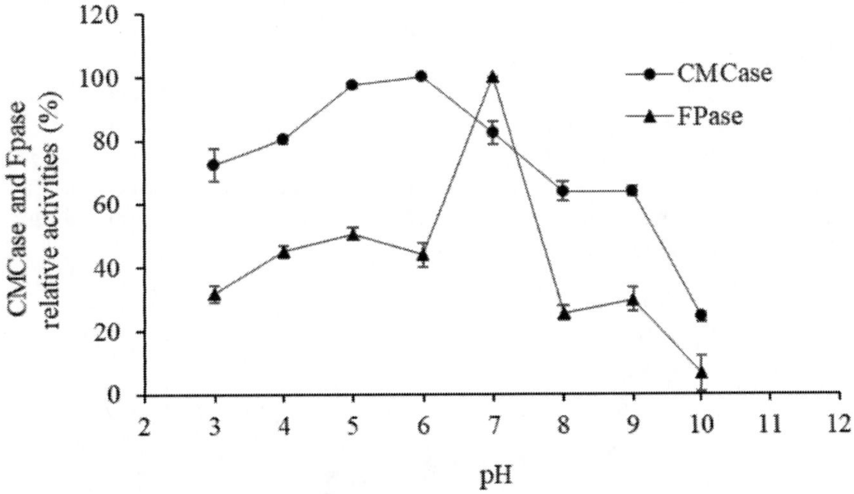

Figure 2. pH profile of CMCase and FPase activities from *B. velezensis* at 37°C for 30 min.

3.4.2. Temperature profile of CMCase and FPase activity. The optimum temperature of cellulase was studied ranging from 30-90°C as shown in figure 3. The optimal temperature of CMCase activity was 60°C. The remaining CMCase activity was even 95% at 70°C, but its activity rapidly decreased at 80-90°C. The maximum temperature of the purified CMCase activities from *B. pumilus* EB3 [21] and *Cellulomonas* sp. YJ5 [10] were similar to this study. However, those of CMCase activities from *P. terrae* ME27-1, *Brevibacillus* sp. DUSELG12 and *Geobacillus* sp. DUSELR7 were 50°C, 55°C and 75°C, respectively [11, 22]. The optimum temperature of FPase activity of *B. velezensis* was 50°C and its activity sharply decreased after 50°C, whereas that of *B. licheniformis* 2D55 was 80°C [23].

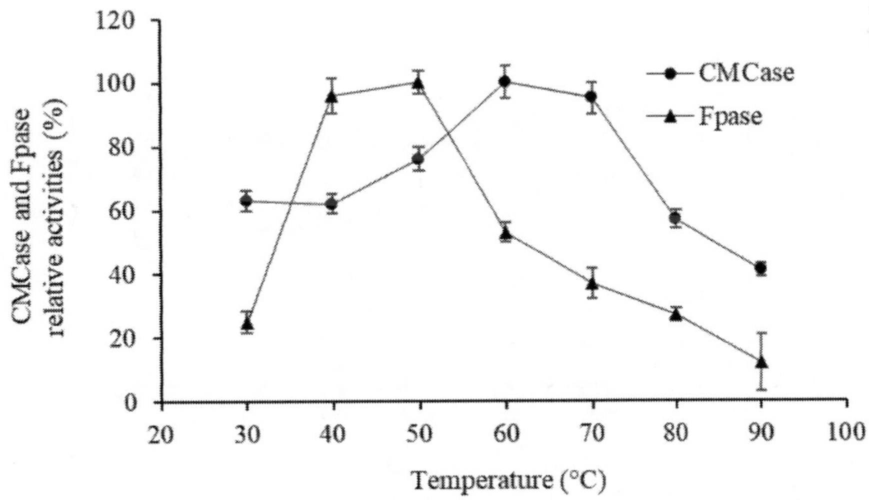

Figure 3. Temperature profile of CMCase and FPase activities from *B. velezensis* at optimum pH for 30 min.

3.4.3. Effect of pH on CMCase and FPase stability. Incubation of cellulase at various times in buffer solutions with pH values between 3.0-8.0 at 37°C. The cellulase activities were active in a wide pH range. The CMCase stability was at pH 6.0-7.0 for 120 min and the activity was maintained at over 80% as shown in figure 4. The residual activity of CMCase was almost 70% under pH 5.0 and its activity maintained 50% at pH 8.0 after 120 min. However, CMCase from *Cellulomonas* sp. YJ5 was stable at pH 7.5-10.5 [10]. The CMCase activity from *P. terrae* ME27-1 displayed stability in a wide pH range of 5.0-9.5 [11]. The FPase stability of all pH ranges gradually decreased after a longer incubation period. The remaining activity of approximate 66-70% was found in the pH range of 5.0-7.0 for 180 min as shown in figure 5. At pH 3.0 and pH 8.0, the residual FPase activity was approximately 50% after 120 min. On the contrary, the FPase activities of *Bacillus* strain MAM-29 and *Bacillus* strain MAM-38 were the best at high pH values but low at low acidic pH [25].

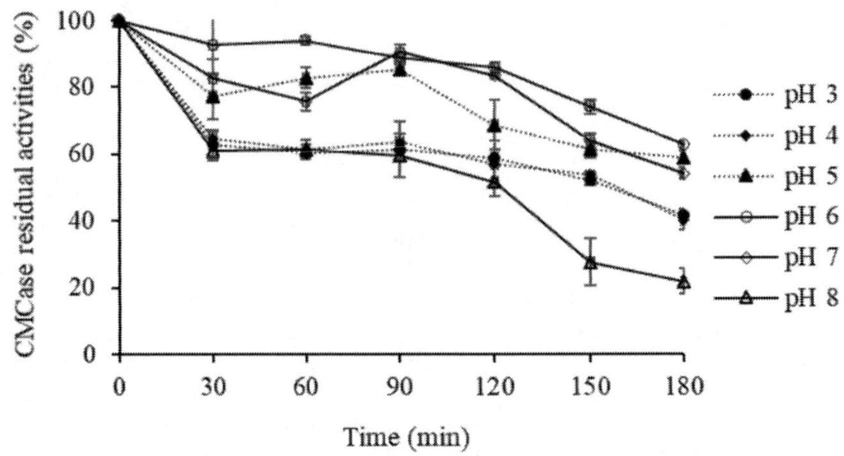

Figure 4. Effect of pH on CMCase stability after incubation in various pH buffer solutions at 37°C for 180 min.

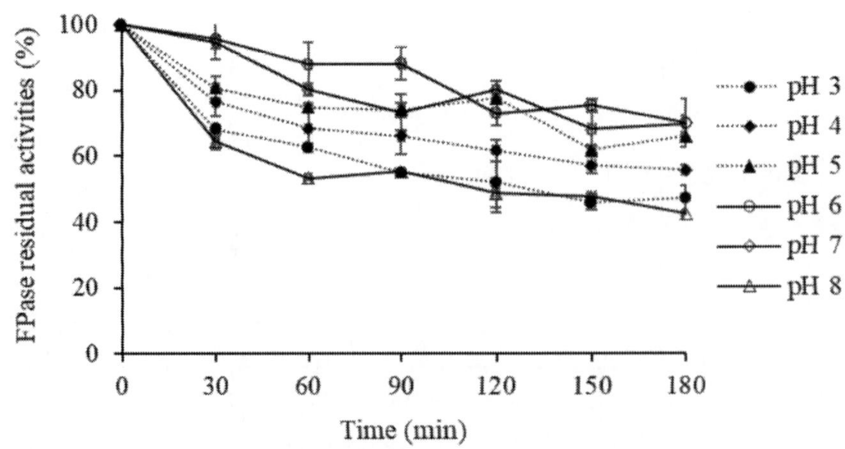

Figure 5. Effect of pH on FPase stability after incubation in the various pH buffer solutions at 37°C for 180 min.

3.4.4. Effect of temperature on CMCase and FPase stability. The thermostable enzymes are a very important parameter for industrial applications. The cellulase activities were active in the broad range of temperature. The CMCase activities were stable at 50-60°C for 90 min and their activities decreased with prolonged incubation as shown in figure 6. The remaining CMCase activity was approximately 50% at 70°C for 90 min whereas the activities were unstable at higher temperatures (80°C and 90°C). The remaining FPase activity at approximately 80% was found at 60°C for 60 min as shown in figure 7. On the other hand, FPase activity decreased with the increasing temperature (70-90°C). Regarding the temperature profile of CMCase and FPase, the optimum temperatures of CMCase and FPase were 60°C and 50°C, respectively. Therefore, the remaining CMCase and FPase activities after incubation at 30-40°C were lower than those at 50-60°C for 180 min. The activity of cellulases of *B. velezensis* was more thermostable than previous studies. The CMCase and FPase activities from the composite microbial system (FH3) were stable at 30-45°C for 3 h, and then decreased sharply above 50°C [24]. The purified CMCase activities from *Cellulomonas* sp. YJ5 [10] and *P. terrae* ME27-1 [11] were stable at temperatures below 50°C. Similar results with the present study were observed from *Brevibacillus* sp. DUSELG12, CMCase activity was stable at 50°C above 1 h [22].

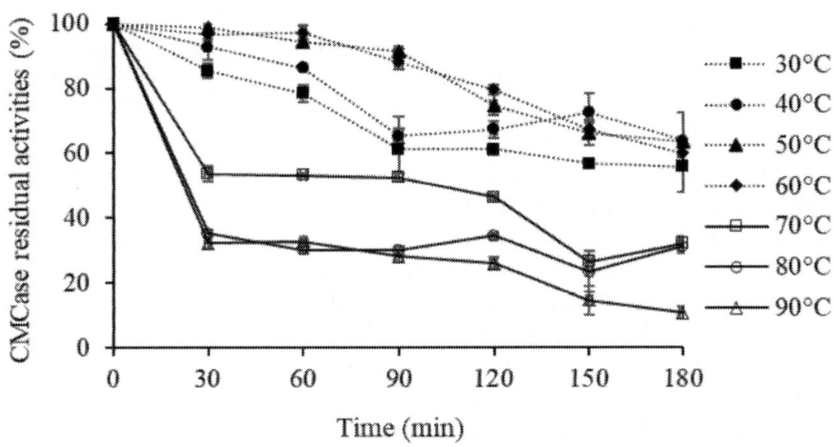

Figure 6. Effect of temperature on CMCase stability after incubation at different temperatures for 180 min.

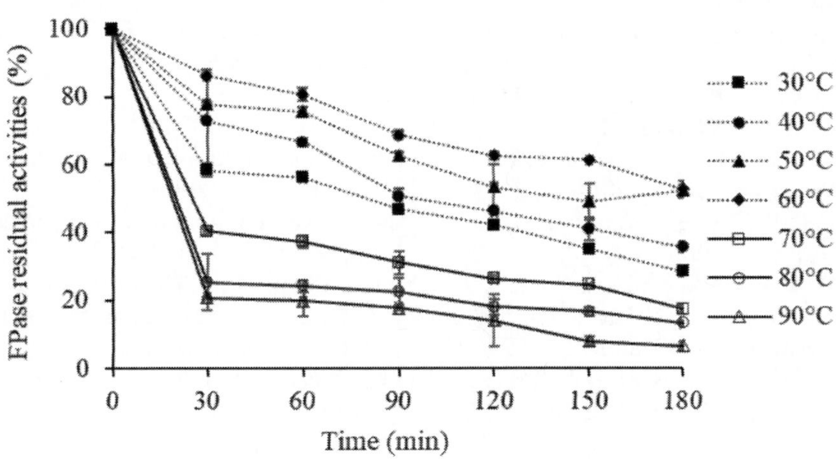

Figure 7. Effect of temperature on FPase stability after incubation at different temperatures for 180 min.

4. Conclusions

Bacterial isolate, CC1-1, was isolated from soil collected from the area of a dairy farm in Phetchaburi province, Thailand. It showed the maximum hydrolysis capacity of CMC by cellulase. Subsequently, it was identified as *B. velezensis*. The properties of cellulases were investigated after cellulase production under submerged fermentation process. The optimum pHs of CMCase and FPase activities were pH 6.0 and 7.0, respectively. The optimum temperatures of these activities were 60°C and 50°C, respectively. Because of the ability to retain good activity in a broad pH and temperature range, the strain has tended to be used in the composting process. Further studies are recommended on *B. velezensis* and strain improvement for increasing cellulose hydrolysis in agricultural wastes which reduces composting time.

Acknowledgments

The authors greatly appreciate Biotechnology Laboratory, King Mongkut's Institute of Technology Ladkrabang Prince of Chumphon Campus for providing the laboratory facilities and Miss Khwanruthai Chowbankoh for assisting the research.

References

[1] Lynd L R, Weimer P J, Zyl W H and Pretorius I S 2002 *Microbiol. Mol. Biol. R.* **66** 506-77
[2] Kuhad R C, Gupta R and Singh A 2011 *Enzyme Res* **Article ID 280696** 1-10 doi:10.4061/2011/280696
[3] Teeri T T 1997 *Trends Biotechnol.* **15** 160-7
[4] Umikalsom M S, Ariff A B, Shamsuddin Z H, Tong, C C, Hassan M A and Karim M I A 2013 *Appl. Microbiol. Biotechnol.* **47** 590-5
[5] Bansal N, Tewari R, Soni R and Soni K S 2012 *Waste Manag.* **32** 1341-6
[6] Maeda R N, Barcelos C A, Anna Santa A L M and Pereira N Jr 2013 *J. Biotechnol.* **163** 38-44
[7] Kim D M, Cho E J, Kim J W, Lee Y W and Chung H J 2013 *Afr. J. Biotechnol.* **13** 145-55
[8] Narra M, Dixit G, Divecha J, Madamwar D and Shah A R 2014 *Bioresour. Technol.* **121** 355-61
[9] Amore A, Pepe O, Ventorino V, Birolo L, Giangrande C and Faraco V 2013 *FEMS Microbiol. Lett.* **339** 93-101
[10] Yin L J, Huang P S and Lin H H 2010 *J. Agric. Food Chem.* **58** 9833-7
[11] Liang Y L, Zhang Z, Wu M, Wu Y and Feng J X 2014 *BioMed Res. Int.* **Article ID 512497** 1-13 http://dx.doi.org/10.1155/2014/512497
[12] Khatiwada P, Ahmed J, Sohag M H, Islam K and Azad A K 2016 *J. Bioprocess. Biotech.* **6** 1-5
[13] Fatema K, Manchur M A 2015 *Int. J. Res. Agric. For.* **2** 14-22
[14] Patagundi B I, Shivasharan C T and Kaliwal B B 2014 *Int. J. Curr. Microbiol. App. Sci* **3** 59-69
[15] Reddy K V, Vijayalashmi T, Ranjit P and Raju M N 2017 *Braz. Arch. Biol. Technol.* **60** 1-6
[16] Rawway M, Ali S G and Badawy A S 2018 *J. Eco. Heal. Env.* **6** 15-24
[17] Doi R H 2008 *Ann. N.Y. Acad. Sci.* **1125** 267–79
[18] Teather R M and Wood P J 1982 *Appl. Environ. Microbiol.* **43** 777-80
[19] Wood T M and Bhat K M 1998 Methods in Enzymology Cellulose and Hemicellulose vol 160, ed W A Wood and J A Kellogg (New York: Academic Press) pp 87-112
[20] Miller G L 1959 Use of dinitrosalicylic acid reagent for determination of reducing sugar. *Anal Chem.* **31** 426-28
[21] Ariffin H, Abdullah N, Umi Kalsom M S, Shira, Y and Hassan M A 2006 *Int. J. Eng. Technol.* **3** 47-53
[22] Rastogi G, Muppidi G L, Gurram R N, Adhikari A, Bischoff K M, Hughes S R, Apel W A, Bang S S, Dixon D J and Sani R K 2009 *J. Ind. Microbiol. Biotechnol.* **36** 585-98
[23] Kazeem M O, Shah U K M, Baharuddin A S and Rahman N A A 2016 *BioResources* **11** 5404-23
[24] Zhang Y M, Lu X B, Dan H B and Sun Y K 2009 *Braz. Arch. Biol. Technol.* **52** 35-43
[25] Abo-State M A M, Fadel M, Abdellah E M and Ghaly M F 2013 *Am Eurasian J. Agric. Environ. Sci.* **13** 1568-75

Streptomyces spp. culture filtrates reduced T4 infectivity to *E. coli*

M J Lee[1], B L Liu[1], H R Kuo[1], C J Chen[2], K M Chang[3] and C A Chang[1, 4]

[1]Department of Applied Chemistry, Chaoyang University of Technology, Taichung, Taiwan
[2]Neural Regeneration Laboratory, Department of Neurosurgery, Neurological Institute, Taipei Veterans General Hospital, Taipei, Taiwan
[3]Department of Research, Koo Foundation Sun Yat-Sen Cancer Center, Taipei, Taiwan

E-mail: cachang@cyut.edu.tw

Abstract. A *Streptomyces* spp. culture was isolated and identified. Its filtrate was found to be able to destroy ORSV and CYmMV virus coat protein, and could be used to reduce the transmission of plant virus disease in vivo, in our previous study. We also tested and found that it can destroy the coat proteins of 13 other of plant viruses, suggesting that the mechanism of action is not specific to a few plant viruses. Aim: To test if this culture filtrate can destroy or affect the infectivity of non-plant viruses. Materials and methods: *Streptomyces* spp. were cultured in soybean-based culture medium for 14 days, and the culture was collected and filtered. T4 phage infection of *E. coli* was used as a model. *E. coli* was cultured on 0.75% TSA agar plates. The T4 phage was incubated with various concentrations of *Streptomyces* spp culture filtrate for 30 minutes, before adding to the *E. coli* lawn. Culture filtrates of two strains, the C5-6 and the SML-1, were used. T4 phage incubated with 300 ppm virusbom (a known anti-viral agent) was used as a positive control. The formation of T4 lysis colony was calculated for the plaque-forming unit (PFU). Results: The dilution of 1 in 8 of the culture filtrates reduced the number of phage colony on *E. coli* lawn. The infectivity was significantly reduced when T4 was incubated with the culture filtrate at 1/8 dilution when compared to the non-treated groups. The virusbom treated PFU was significantly reduced, in comparison with C5-6 treated group (P=0.013), and with SML-1 treated group (P=0.028) as per ANOVA test followed by Tukey post-hoc comparison. Summary: These data demonstrated that culture filtrates of *Streptomyces* spp. reduced T4 to *E. coli* infectivity and suggested that the anti-viral compounds in the filtrate is not specific to plant viruses. The application of the culture filtrate and its content might have broader applications.

1. Introduction

Antiviral agents are chemicals that inhibit the spread of virus, by blocking replication of the genome, preventing entry to host cells, or reducing viral protein synthesis or viral assembly. Although there are small synthetic molecules or biotech engineered molecules available for popular, specific viral disease such as HIV/AIDS or HPV/cervical cancer, amongst many others [1-3], broad range antiviral agents are still in demands. They were not used for specific virus but rather a general anti-viral, environmental cleansing reagent that would prevent the spreading of the virus. Traditionally chemicals that denature proteins are often used for this purpose. For example, for prevention of influenza

Content from this work may be used under the terms of the Creative Commons Attribution 3.0 licence. Any further distribution of this work must maintain attribution to the author(s) and the title of the work, journal citation and DOI.
Published under licence by IOP Publishing Ltd

IOP Conf. Series: Earth and Environmental Science **346** (2019) 012041 doi:10.1088/1755-1315/346/1/012041

spreading, the hands were washed by detergent such as soap, or and the floor was mopped by diluted bleach. However, in certain area of application there is still needs for different chemicals with variety of working mechanism. For example, in a laboratory working bench, the sterilizing reagent of hands and gloves is 70% alcohol, which might not be ideal for technician with sensitive skin. Further, methods that apply heat, bleach, or alkaline solution are detrimental to the environment and were impossible to use on plants or human skin. A broad range antiviral agent could be used in cases like this.

It has been discovered that a sterilizing material of *Streptomyces* spp. culture filtrate was able to destroy *Odontoglossum ringspot virus* (ORSV) and *Cymbidium mosaic virus* (CYmMV) virus coat protein, and potentially act as an anti-viral agent [4,5]. Previous tests found that the concentrated filtrate could destroy the infectivity of both viruses to the orchid within one minute of incubation in culture filtrates [4]. Therefore, it is possible to us for immediate disinfection of equipment, bench surface, or even technician's skin.

On a broader scale, some of the plant derived anti-viral reagents were reported to target more than just plant viruses [6]. Viral proteases, themselves are protein in nature, are target for protease [6-8]. In this sense, proteases are good candidate for anti-viral agents against specific viruses. Furthermore, bacteria of the *Streptomyces* genus were known to secrete various enzymes against biological big molecules [9-11]. We are interested to know whether our culture filtrate could target viruses other than plant viruses. In our case, the active ingredient was possibly a protein, as it was sensitive to treatment temperature higher than 70°C [12]. Proteases has been used to act against viral spreading or replication, as many of the enzymes that were vital to viral survival are proteins [7]. For example, the addition of serine protease inhibitors improves the treatment outcome of patients infected with hepatitis C virus genotype [13]. The activity of our product was stable for up to 6-months' storage at 4°C.

We have previously found that the virus eradication of our culture filtrate was not restricted to ORSV and CymMV, as other 13 viruses were tested and found effective [4,12]. In this project, we aim to test if the application of this anti-viral agent could be extended to viruses other than plant viruses.

2. Materials and methods

2.1. Preparation of the test agent

Streptomyces spp. was cultured in a soybean based medium for 14 days, and the medium collected, filtered through 0.22 μm membrane (Amicon). *Escherichia coli* DH5-α was regrown from a commercial kit of competent cells and T4 bacteriophage (BCRC no. 70041) was purchased from Bioresource Collection and Research Center (BCRC) Taiwan. Difco™ Luria-Bertani (LB), Bacto™ trypsic soy broth (TSB) were purchased as powders from Becton, Dickinson and Company (DB) (LB broth, 244620, TSB, 211825). American bacteriological agar was purchased from Laboratorios CONDA (Madrid, Spain).

2.2. Treatment groups

Control treatment was prepared by using the bacterial culture medium incubated at 37°C for 14 days, without inoculation with bacteria. The C5-6 strain was isolated by Chen and Chang [14] and SML-1 by Chien and Chang, 2018. Cultured medium was collected and filtered at 0.22 μm and used. T4 phage incubated with virusbom was used as a positive control for anti-viral effect.

2.3. Zymography

The zymography was performed by a method modified from Michaud [15]. Samples were separated, using 10% SDS-gel with 0.1% gelatin (Invitrogen, Carlsbad, CA). After separation, the gel was renatured by washing in 2.5% Triton X-100 and incubated in substrate buffer (50 mMTris-HCl, pH 7.5, 10 mM $CaCl_2$, and 0.01% Sodium Azide) for 48 hr at 37°C to enable the possible enzyme to cleave the gelatin. Each gel was then stained with Coomassie Blue for 1 hr and destained in a solution of 30% methanol and 10% acetic acid. Negative staining indicated the locations of active protease bands.

2.4. Plaque forming unit (PFU)

This is performed by an adoption to Adams [16] and Andersons *et al* [17]. Briefly, LB agar in Petri dishes were streaked with DH5-α strain of *E. coli* and incubated overnight at 37°C. A 50mL aliquots of sterile LB broth was inoculated with a single colony and incubated overnight with shaking, until the optical density reached an OD600 >1. The petri-dish was lined with 10 ml of 1.5% TSA. The overlaying agar was prepared by mixing 3 mL of overnight culture of DH5-α *E. coli* and 50mL of 0.75% warm TSA. 5mL mixture was used for 1 dish. The T4 phage was incubated with treatment material for 30 minutes, followed by serial dilution by 1% peptone. 20 μl of the diluted T4 phage was spread evenly on the surface of 0.75% TSA by glass beads and incubated at 37°C overnight. Petri dishes, exhibiting between 20 to 200 well isolated plaques, were used to calculate the phage titers and for subsequent statistical analyses. The PFU was calculated as: pfu/m=number of colony * dilution * (1ml/phage volume used)

2.5. Treatment groups

T4 phage was passed through 0.22 μm filter, and 10 μl of T4 phage and 10 μl of the treatment liquid were mixed and incubated for 30 minutes, before 10 μl of mixture was spread on DH5-α *E. coli*-containing top agar layer. The treatments were: 1/8 of culture filtrates of SML-1 strain of the *Streptomyces* spp., 1/8 of culture filtrate of C5-6 strain of the *Streptomyces* spp., 1/8 of 300 ppm virusbom (commercial product line from Money Marketing Communication Ltd), or LB broth only.

3. Results

3.1. Biochemical characteristics

The anti-ORSV or anti-CYmMV activities were lost when incubated at temperature higher than 70°C, and these activities were sustained when kept at 4°C for 6 months. This suggested that at least one of the active ingredients in the culture filtrate was protein. We went on to test whether the active ingredient was a protease. In the zymography experiments, the culture filtrate was concentrated by DEAE sepharose column, and was separated on the 10% SDS-gel with 0.1% gelatin. The gel was renatured for 48 hours for the possible enzyme to activate. The commassie blue staining revealed a negative staining band, indicating the presence of protease in the culture filtrate (figure 1).

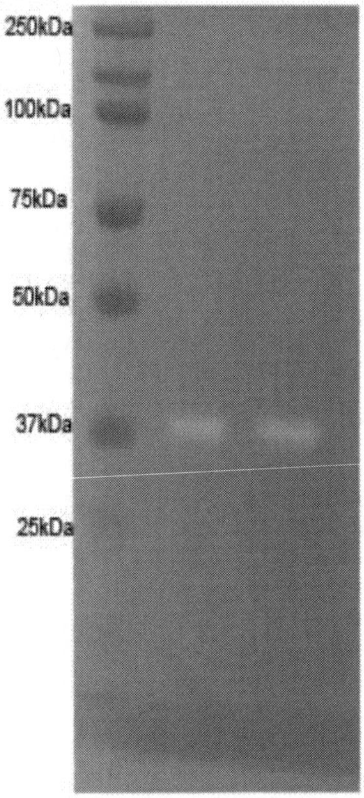

Figure 1. A Commassie Blue staining for a SDS-gel containing 1% gelatin which has been digested by the proteinase activity within the culture filtrate of C5-6 strain. The proteinase activity was located to roughly 37 kDa area of the gel, suggesting there may be a proteinase of this size in the culture filtrate. It is not known whether this is the same activity that affected the *E. coli* infectivity.

3.2. Bacteria infectivity

Three treatments (SML1, C5-6, virusbom) were diluted 1/2, 1/4 and 1/8 and tested for *E. coli* infectivity. The results were similar among different dilutions, suggesting that further dilution may be needed. In this paper we only listed the results for the 1/8 dilutions. The phage was mixed with treatment agents for 30 minutes before it was serial diluted 10^2 to 10^{10} for plating. In the 4 tests, the phage colonies were about 20 for the phage diluted at 10^6, and 100-200 colonies for the phage diluted at 10^5 as shown in figure 2.

The T4 phage was incubated with treatment material (see following) for 30 minutes, followed by serial dilution by 1% peptone. 20uL of the diluted T4 phage was spread evenly on the surface of 0.75% TSA by glass beads and incubated at 37°C overnight. Plates that has 10^5 diluted phage mixture typically yield about 200 colonies and were chosen to calculate phage forming unit (PFU).

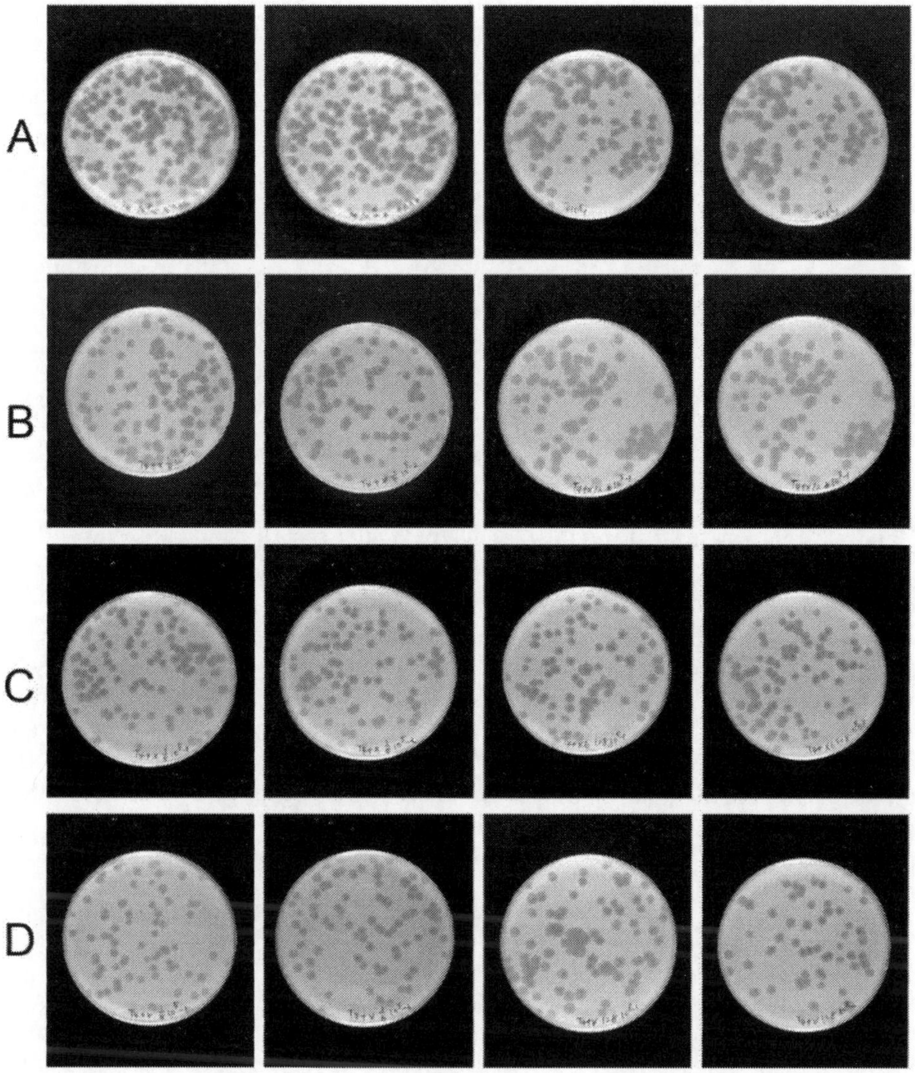

Figure 2. Row A: T4 phage without treatment. Row B: T4 phage treated with 1/8 SML-1 strain filtrate. Row C: T4 phage treated with 1/8 C5-6 strain filtrate. D: T4 phage treated with 1/8 300ppm virusbom. N=4.

3.3. PFU

We calculated the PFU from those of the 10^5 dilutions of phage. The PFU was listed in the table 1 and figure 3. The standard deviation (SD) was between 0.05-0.29, which was typical of the experiments and was acceptable (table 1).

Table 1. PFU for T4 with different treatments.

treatments	test 1	test2	test3	test4	mean	SD
T4	3.18	3.06	2.66	2.60	2.88	0.29
T4+SML1	2.24	2.16	1.76	1.74	1.98	0.26
T4+C5-6	2.08	2.1	1.8	1.84	1.96	0.16
T4+virusbom	1.52	1.58	1.66	1.58	1.59	0.05

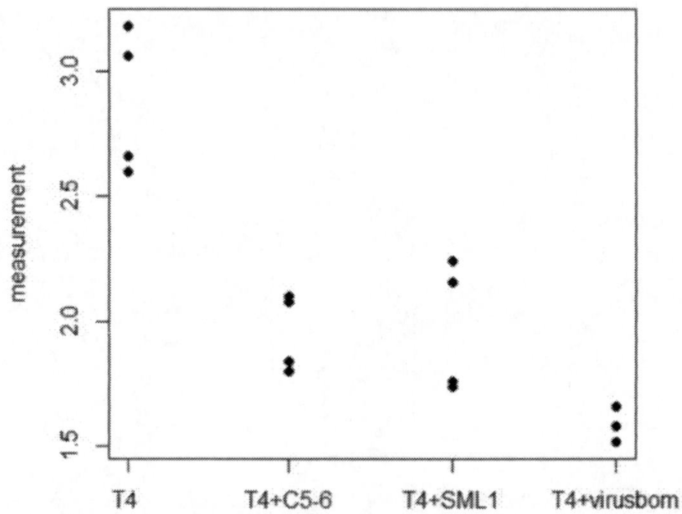

Figure 3. Distribution of PFU values for T4 with different treatments . N=4.

3.4. *Statistics*

ANOVA and Tukey HSD test were used to assess whether measurements were different between treatments and where the source of major variation (see supplementary file). Multiple one-tailed two-sample t tests were conducted to assess our major concerns: whether T4+virusboom, T4+C5-6, or T4+SML1 were better than T4, and whether T4+virusbom is better than T4+C5-6 or T4+SML1. The analyses were computed in R v3.5.

The results of 5 one-tailed t-tests indicated that the measurements of T4+C5-6, T4+SML1, and T4+virusbom are significantly less than the measurements of T4 (*, p value=0.006, 0.006 and 0.006), and the measurements of T4+C5-6 and T4+SML1 were significantly higher than the measurements of T4+virusbom (#, p value=0.013, 0.028) (figure 4).

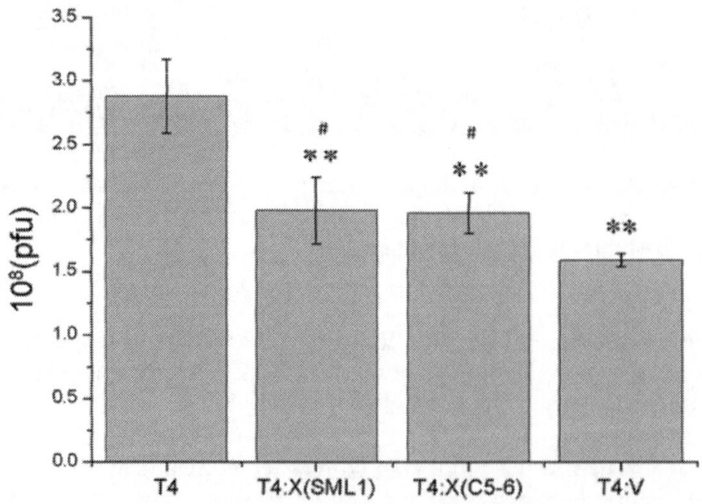

Figure 4. Comparison of the treated with non-treated. SMLA-1 treated and C5-6 treated are significantly lower than T4 none-treated (*). Those treated with virusbom showed significantly smaller PFU than T4 treated with C5-6, as well as with SML1 treated (#) . N=4, ANOVA and Tukey HSD followed by one tail test.

IOP Conf. Series: Earth and Environmental Science **346** (2019) 012041 doi:10.1088/1755-1315/346/1/012041

4. Discussion

4.1. Biochemical characteristics
A gelatinase activity was demonstrated by the C5-6 culture filtrate. The proteinase activity was located to roughly 37 kDa area of the gel, suggesting a proteinase of this size was present in the culture filtrate. Although the activity that eradicated the plant virus was lost when the concentrated culture filtrate was heated over 70°C, this only suggested that part of the activity could be contributed to a protein, and our present data do not support a hypothesis that this gelatinase activity is the major activity that reduced the *E. coli* infectivity. Further study is underway to search for the identify for the possible active molecule.

4.2. Information for further safety evaluation
The culture filtrate itself was a mixture of rich nutrient aimed for providing the growth of bacteria, and part of the activity was attributed to a protein. This conclusion provided information to the safety evaluation for further development for this product. One of the possible usage of this product would be application on the technician's skin. We have done a separate project on the cytotoxicity and genotoxicity effects of the filtrate using cell culture model (manuscript submitted).

4.3. Extension of target to T4 phage merit further development for human disease related virus
Our data demonstrated that the culture filtrate from SMLA-1 treated and C5-6 treatment reduced the T4 to *E. coli* infectivity. Please note that these were only the culture filtrate without concentration or isolation. The way that un-concentrated filtrate could work on infectivity of a phage suggest that the concentration of the active ingredient is high enough in the culture filtrate to merit further development to make an anti-viral product. We have previously demonstrated that the targets of the concentrated culture filtrate extended to more than 13 plant viruses. In this project the extension of targets to the T4 phage suggested a broader application on viruses related to human disease could be explored.

4.4. Comparison to virusbom
Virusbom, a patented anti-viral product sold in Taiwan was used as a positive control. This chemicals collapsed viral envelop, change the structure of the bacteria and reduced bacteria growth [18]. The structure of the virusbom was not published, but it was reported in news to be a synthetic enzyme. 300-1000 ppm was the concentration reported to kill NDM positive Klebsiela Pneumonae [18]. In our tests, 300 ppm of virusbom, when diluted 1 in 8, was able to reduce the infectivity of T4 phage to the *E. coli*, and the effect was better than that of C5-6 and SML-1 filtrate. The C5-6 and SMLA-1 was not concentrated and the final concentration of usage was not determined. Hence this piece of information does not provide evidence to support the superiority of virusbom over our future product. It would be interesting to carry out side-by-side comparison when we further develop this into a more mature form of product.

5. Summary
These data demonstrated that the culture filtrates from both strains of the culture reduced the T4 to *E. coli* infectivity, and suggested that the anti-viral compounds in the filtrates was not specific to plant viruses. The application of the culture filtrate and its content might have broader application than we previously known.

References
[1] Puhl A C *et al* 2019 New targets for HIV drug discovery *Drug Discov. Today,* **24** (5) 1139-1147.
[2] McIntosh J, Sturpe D A and Khanna N 2008 Human papillomavirus vaccine and cervical cancer prevention: Practice and policy implications for pharmacists *J. Am. Pharm. Assoc.* **48** e1-16

[3] Barbas C F 3rd and Burton D R 1996 Selection and evolution of high-affinity human anti-viral antibodies *Trends Biotechnol.* **14** 230-4
[4] Chang C A, Chen H R and Chiu C H 2010 Disinfection of odontoglossum ringspot virus and cymbidium mosaic virus from tools used during orchid cultivation (International Society for Horticultural Science (ISHS), Leuven, Belgium)
[5] Chang C A, Purcifull D E and Hiebert E 1988 Purification, characterization, and immunological analysis of nuclear inclusions induced by bean yellow mosaic and clover yellow vein potyviruses *Phytopathology* **78** 1266-75
[6] Hegde V R *et al* 2003 Two antiviral compounds from the plant Stylogne cauliflora as inhibitors of HCV NS3 protease *Bioorg. Med. Chem. Lett.* **13** 2925-8
[7] Hellen C U and Wimmer E 1992 Viral proteases as targets for chemotherapeutic intervention *Curr. Opin. Biotechnol.* **3** 643-9
[8] Patick A K and Potts K E 1998 Protease inhibitors as antiviral agents *Clin. Microbiol. Rev.* **11** 614-27
[9] Hui Z *et al* 2004 Alkaline serine protease produced by Streptomyces sp. degrades PrP(Sc) *Biochem. Biophys. Res. Commun.* **321** 45-50
[10] Jarerat A, Tokiwa Y and Tanaka H 2006 Production of poly(L-lactide)-degrading enzyme by Amycolatopsis orientalis for biological recycling of poly(L-lactide) *Appl. Microbiol. Biotechnol.* **72** 726-31
[11] Manna A, Giri P and Paul A 1999 Degradation of poly(3-hydroxybutyrate) by soil streptomycetes *World J. Microbiol. Biotechnol.* **15** 705
[12] Chiu C H 2006 Screening and application of actinomycete secreting potease that can degrade capsid protein and abolish infectivities of odontoglossum ringspot virus and cymbidium mosaic virus (Taichung, Taiwan: Institute of Biochemical Sciences and Tehcnology, Chaoyang University of Technology)
[13] Razonable R R 2011 Antiviral drugs for viruses other than human immunodeficiency virus *Mayo. Clin. Proc.* **86** 1009-26
[14] Chen H R 2009 Production and application of an orchid virus degrading protease secreting by an Actinomycetes, in Graduate institute of biochemical science and technology (Taiwan: Chaoyang University of Technology) p 91
[15] Michaud D 1998 Gel electrophoresis of proteolytic enzymes *Analytica Chimica Acta.* **372** 173-85
[16] Adams M H 1959 Enumeration of bacteriophage particles *Bacteriophages* (London: Interscience Publishers, Ltd) pp 27-34
[17] Anderson B *et al* 2011 Enumeration of bacteriophage particles: Comparative analysis of the traditional plaque assay and real-time QPCR- and nanosight-based assays *Bacteriophage* **1** 86-93
[18] Wang S-S 2011 Method for killing NDM-1 Klebsiella pneumoniae (P. National Intellectual Property Administration, China)

Comparison of 2-acetyl-1-pyrroline production between *Aspergillus awamori* and *Aspergillus oryzae*

T Wongsadee[1], S Vatanyoopaisarn[1,4], B Thumthanaruk[1], C Puttanlek[2], D Uttapap[3] and V Rungsardthong[1]

[1]Department of Agro-Industrial, Food, and Environmental Technology, King Mongkut's University of Technology North Bangkok, Bangkok, Thailand
[2]Department of Biotechnology, Silpakorn University, Nakhon Pathom, Thailand
[3]Division of Biochemical Technology, School of Bioresources and Technology, King Mongkut's University of Technology Thonburi, Bangkok, Thailand

E-mail: savitri.v@sci.kmutnb.ac.th

Abstract. 2-Acetyl-1-pyrroline (ACPY) is a key flavor compound in fragrant rice and widely exploited in food flavoring. It is produced by various microorganisms. This study focused on ACPY production by two fungi (*Aspergillus awamori* and *A. oryzae*). The volatile compounds derived from mold cultivation in synthetic medium 18 were identified by Gas chromatograph-mass spectrometer (GC-MS). Seven volatile substances were detected in the liquid culture of *A. awamori*, i.e. ACPY, one ketone (1-hydroxy-2-propanone), two acids (acetic acid and 4-hydroxybutanoic acid), two alcohols (2,3-butanediol and 2,5-dimethyl-4-hydroxy-3(2H)-furanone) and one saponin glycoside (2,3-dihydro-3,5-dihydroxy-6-methyl-4H-pyran-4-one). Further, 12 volatile compounds were detected in *A. oryzae* : ACPY, four fatty acids (tetradecanoic acid, *n*-hexadecanoic acid, octadecanoic acid and oleic acid), one alcohol (1-butanol), two benzenes (ethylbenzene and benzene, 1,3-bis(1,1-dimethylethyl)), three alkanes (pentadecane, heptadecane and 5-methoxy-2,2,6-trimethyl-1-(3-methyl-buta-1,3-dienyl)-7-oxa-bicyclo[4.1.0] heptane) and one sesquiterpenes (1,6-dimethyl-4-(1-methylethyl)-naphthalene). Both *A. awamori* and *A. oryzae* produced the highest amount of ACPY in the stationary phase when cultivated for 72 and 80 h, with the product yield of 0.914 and 1.323 mg/L, respectively. The supplementation with spermidine and spermine resulted in a more than fivefold increase in ACPY production by *A. awamori*. However, the production of ACPY by *A. oryzae* was lower when supplemented with spermine or spermidine than without spermine or spermidine. This indicated that the intermediates involved in ACPY production were different between these two species of fungi.

1. Introduction

The food industry has increased the demand for flavoring compounds due to the aroma of food affects consumer acceptance. The volatile components contributing to the aroma of food affect consumer acceptance. Rice flavor is one of the most favorite flavors for consumers. 2-Acetyl-1-pyrroline (C_6H_9NO) or ACPY is the main volatile compound that contributes to the rice fragrance and is also a key flavor compound in many other foods, such as pandan leaf (*Pandanus amaryllifolius come*), popcorn and bread [1]. It is a heterocyclic compound. Its odor threshold in water and air has been measured to be 0.1 ng/L and 0.02 ng/L, respectively [2]. The extraction of ACPY from plant sources is complicated and requires various instruments [3]. Several methods are used for the production of ACPY such as

Content from this work may be used under the terms of the Creative Commons Attribution 3.0 licence. Any further distribution of this work must maintain attribution to the author(s) and the title of the work, journal citation and DOI.
Published under licence by IOP Publishing Ltd

chemical synthesis and microbial biotechnology. The synthesis of ACPY requires organic chemicals. Its synthesis is difficult because of the unstable nature of this compound, which degrades very rapidly [4]. Microorganisms are used to carry out fermentation to improve the taste, and texture of products that can be considered natural. ACPY is produced by various bacteria such as *Bacillus cereus* isolated from cocoa fermentation boxes (about 0.03-0.075 mg/kg) on plate count agar [5], *B. cereus* ATCC 27522 and *B. cereus* ATCC 14737 (about 26.2 and 6.0 μg/kg, respectively) on plate count agar [6], *B. cereus* ATCC 10702 (85 μg/kg) [7] and *Lactobacillus hilgardii* DSM 20176 in off-flavor characterization of wine (0.005 mg/L) [8]. Also, several fungal strains have been found to produce ACPY. *Aspergillus awamori* and *Acremonium nigricans* produced ACPY in synthetic liquid medium 18 (Syn 18) at about 2.8 mg/L and 1.11 mg/L, respectively. The yield of ACPY was about 1,000-fold higher than the aroma of jasmine rice [9]. A previous study exploring the effects of nitrogen sources (putrescine, glutamic acid, proline, and ornithine) on ACPY production by *A. awamori* showed that putrescine was the best nitrogen source for ACPY production. Further, the supplementation with spermidine and spermine resulted in a significant increase in ACPY production. The results indicated that putrescine, spermidine and spermine were the major intermediates of ACPY production [10]. However, the information on the optimal cultivation time for the highest ACPY production by *A. awamori* and the reports on the profile of volatile compounds in the liquid culture of fungi are lacking. The purpose of the present study was to identify the volatile compounds in the culture medium of *A. awamori* and *A. oryzae* and compare the effect of the major nitrogen substrates (putrescine, spermidine and spermine) on ACPY production by these two fungal species.

2. Materials and methods

2.1. Microorganisms and inoculum preparation

Strains of *A. awamori* TISTR 3123 and *A. oryzae* TISTR 3256 (collected from Thailand Institute of Scientific and Technological Research) were studied. The two fungi were inoculated into potato dextrose agar slant and incubated at 28°C for 7 days. Fungal spores were suspended in 10 mL of 0.1% Tween 80. Then the 5 mL of spore suspension (2 x 10^9spores/L) was added into 200 mL of potato dextrose broth and incubated in a shake flask (200 rpm) at 28°C for 3 days. The mycelia in the liquid medium were mixed using a sterile blender and used as a starter culture. Further, 10 mL of the starter culture was inoculated into 90 ml of synthetic medium 18 (Syn 18); the composition of Syn 18 medium is displayed in table 1 and incubated with shaking (120 rpm) at 30°C for 24 h. Spermine ($C_{10}H_{26}N_4$) or spermidine ($C_7H_{19}N_3$) was incorporated into some flasks (20 mg/L each). The Syn 18 medium without supplement was used as control. All flasks were further incubated with shaking (120 rpm) at 30°C [10]. The samples were collected and filtrated through Whatman No.1 filter paper. The amount of ACPY in the filtrate was measured using GC-MS and GC-FID.

Table 1. The composition of the synthetic medium 18 (Syn 18) for fungal cultivation.

Composition	Concentration
Glucose	15.0 g/l
Putrescine	2.0 g/l
$MgSO_4{\cdot}7H_2O$	0.5 g/l
NaCl	0.1 g/l
$CaCl_2{\cdot}2H_2O$	0.1 g/l
Phosphate buffer, pH 7.2	20 mM
Vitamin solution	1.0 ml
- Biotin 50 mg/l	
- Ca-pantothenate 400 mg/l	
- Inositol 2000 mg/l	
- Pyridoxine·HCl 400 mg/l	
- Thiamine·HCl 500 mg/l	

Mineral solution	1.0 ml
- $MnSO_4 \cdot H_2O$ 100 mg/l	
- $ZnSO_4 \cdot 7H_2O$ 300 mg/l	
- $CuSO_4 \cdot 5H_2O$ 100 mg/l	
- $FeCl_3 \cdot 6H_2O$ 250 mg/l	
- Ammonium molybdate 50 mg/l	
- $H_3BO_3$300 mg/l	

The 10 ml of liquid culture (pH~2) was adjusted to pH 8.0 with 0.1 N NaOH and then 100 µl of 2,4,6-Trimethylpyridine (Internal standard) and 2 ml of diethyl ether were added and mixed. The amounts of ACPY and 2,4,6-Trimethylpyridine in the ether layer were analyzed by GC-MS and GC-FID.

2.2. Identification of volatile compounds from liquid culture by GC-MS

The extract was injected into an HP-INNOWAX part no. 19091N-133 (length 30 m, inner diameter 0.25 mm, and film thickness 0.25 µm) and an Agilent 7000C Triple Quadrupole GC/MS (Agilent Technologies Inc., USA). The injection was conducted in a split ratio of 25:1. The injector temperatures were set at 180°C. Helium gas was used as the GC carrier gas. The column temperature was held initially at 40°C for 3 min, and then the temperature was increased at a rate of 10°C/min to 120°C and was further increased to 250°C at a rate of 25°C /min. Finally, the column temperature was maintained isothermally at 250°C for 10 min. The ion source temperature was set at 250°C, the GC-MS transfer line was set to 230°C and the ionization energy was 70 eV. The compounds were identified mainly by comparing their mass spectra with the mass spectral data of the NIST2011 library [11].

2.3. Quantification of ACPY from the liquid culture using GC-FID

Gas chromatography – flame ionization detector (GC2010; Shimadzu, Japan) analysis of ACPY was carried out. The injection was conducted in the splitless mode for 5 min at 155°C. The GC was equipped with a D-WAX capillary column (length 30 mm, inner diameter 0.32 mm and film thickness 0.50 µm). The oven temperature was held initially at 40°C for 1 min and then programmed from 40°C to 120°C at a rate of 9°C/min and 120°C to 230°C at a rate of 5°C/min. Helium was used as the carrier gas, at a flow rate of 1.1 mL/min [12]. The concentration of ACPY in the liquid culture (mg/L) was calculated as follows [9]:

ACPY in liquid culture (mg/L) = (Area of ACPY / Area of collidine) x (1,000/10) x (added internal standard (μg) /1,000) x RRF

where RRF is the relative recovery between 125 ppm collidine and ACPY = 1.3.

2.4. Biomass determination

The mycelium of each treatment was harvested on a filter paper (Whatman No.1), washed three times with distilled water, and dried for 24 h at 80°C. The dry weight of mycelium was subjected to the following analysis.

Cell dry weight (g/L*)* = [(weight of filter paper + dried residue) - weight of filter paper] x (1,000/100)

3. Results and discussion

3.1. Profile of volatile components in the liquid culture of fungi

The ability of *A. awamori* and *A. oryzae* to produce volatile components from the Syn 18 medium was analyzed and is displayed in tables 2 and 3, respectively. As shown in table 2, the profile of volatile compounds produced by *A. awamori* revealed eight compounds: 2-acetyl-1-pyrroline, one ketone (1-hydroxy-2-propanone), two acids (acetic acid and 4-hydroxybutanoic acid), two alcohols (2,3-butanediol and 2,5-dimethyl-4-hydroxy-3(2H)-furanone) and one saponin glycoside (2,3-dihydro-3,5-

dihydroxy-6-methyl-4H-pyran-4-one). Six out of seven compounds reflected the flavor of food ingredients.

Table 2. Volatile components in the liquid culture of *Aspergillus awamori*.

Compound	Odor description	Retention time	% Area
1. Ketones	Hazelnuts [A]		
1-Hydroxy-2-propanone	Fragrant rice [B]	9.379	3.50
2-Acetyl-1-pyrroline		9.503	1.40
2. Acids	Vinegar [C]		
Acetic acid	Red wine [D]	11.252	33.57
4-Hydroxybutanoic acid		12.952	1.40
3. Alcohols	Fruity [E]		
2,3-Butanediol	Buckwheat [F]	12.408	27.27
2,5-Dimethyl-4-hydroxy-3(2H)-furanone		15.104	2.10
4. Saponin glycoside	Unknown		
2,3-Dihydro-3,5-dihydroxy-6-methyl-4H-pyran-4-one		16.143	12.59
*2,4,6-Trimethylpyridine (internal standard)	-	11.479	18.18

[A] Nicolotti *et al* [13]
[B] Paule and Power [1]
[C] Ubeda *et al* [14]
[D] Elliott and Burgess [15]
[E] Pho-am *et al* [16]
[F] Janes *et al* [17]

Table 3. Volatile components in the liquid culture of *Aspergillus oryzae*.

Compound	Odor description	Retention time	%Area
1. Ketones			1.48
2-Acetyl-1-pyrroline	Fragrant rice [A]	9.503	
2. Fatty acids			
Tetradecanoic acid	Unknown	17.889	2.22
n-Hexadecanoic acid	Unknown	19.406	31.86
Octadecanoic acid	Unknown	21.741	16.70
Oleic acid	Unknown	22.221	17.15
3. Alcohols			0.52
1-Butanol	Alcohol-like, wine [B]	5.405	
4. Benzenes			
Ethylbenzene	Benzene	6.002	3.03
Benzene, 1,3-bis(1,1-dimethylethyl)	Benzene	10.795	1.40
5. Alkanes			
Pentadecane	Alkane	7.996	0.81
Heptadecane	Alkane	10.916	1.18
5-Methoxy-2,2,6-trimethyl-1-(3-methyl-buta-1, 3-dienyl)-7-oxa-bicyclo[4.1.0]heptane	Unknown	15.350	1.03
6. Sesquiterpenes			
1,6-Dimethyl-4-(1-methylethyl)- naphthalene	Unknown	19.997	6.95
*2,4,6-Trimethylpyridine (internal standard)	-	11.395	15.67

[A] Paule and Power [1]
[B] Pho-am *et al* [16]

Further, 12 volatile compounds were detected in the fermentation liquid produced by *A. oryzae* : 2-acetyl-1-pyrroline, four fatty acids (tetradecanoic acid, *n*-hexadecanoic acid, octadecanoic acid and oleic acid), one alcohol (1-butanol), two benzenes (ethylbenzene and benzene, 1,3-bis(1,1-dimethylethyl)), three alkanes (pentadecane, heptadecane and 5-methoxy-2,2,6-trimethyl-1-(3-methyl-buta-1,3-dienyl)-7-oxa-bicyclo[4.1.0] heptane) and one sesquiterpenes (1,6-dimethyl-4-(1-methyl ethyl)-naphthalene). In a previous study 2,3-butanediol and 1-butanol were reported in coconut water and KDML 105 jasmine rice solution, respectively [16].

ACPY was particularly detected by GC-MS which confirmed the fragrant rice aroma scent by the human perception from the cultured liquid by both fungi. The presence of ACPY assured as the major aroma compound.

3.2. Comparison of fungal growth and production of ACPY in Syn 18 medium

The comparison between the growth (dry weight of the mycelia) of the two fungi during fermentation and the amount of their ACPY are shown in figure 1. The cell dry weight of both fungi continuously increased during the first 66 h. Thereafter from 66 to 90 h the cell dry weight did not increase further as the fungi reached the stationary phase. The ACPY began to appear after 48 h and gradually rose to the maximum level after 72 and 80 h of cultivation. *A. awamori* produced the highest ACPY amount after 72 h (0.914 mg/L) when the cell dry weight was 6.23 g/L. In *A. oryzae,* the highest production of ACPY was obtained with a yield of 1.323 mg/L and 6.58 g/L of cell dry weight after 80 h of cultivation. The results indicated that the highest amount of ACPY from both fungi was obtained in the stationary phase and thus ACPY was a secondary metabolite.

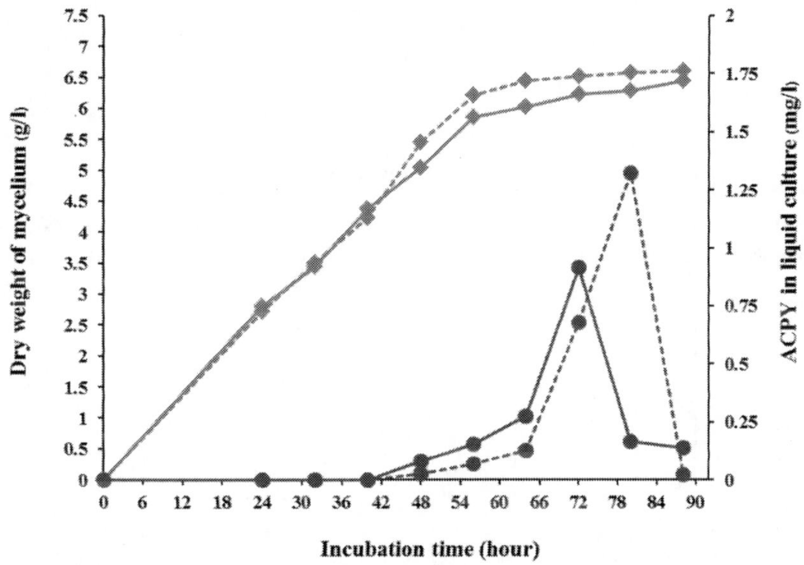

Figure 1. Comparison of growth and production of ACPY in the fermentation liquid produced by the two fungi; *(—, —) A. awamori, (- - -, - - -) A. oryzae;*◆ dry weight of mycelium; ● ACPY in liquid culture.

3.3. Effect of polyamine supplement on ACPY production by the two fungi

Nitrogen sources were found to be essential for the production of volatile compounds. In the Syn 18 medium, putrescine ($C_4H_{12}N_2$) was a nitrogen source proved to enhance the production of ACPY in *Acremonium nigricans* [18]. Spermine ($C_{10}H_{26}N_4$) and spermidine ($C_7H_{19}N_3$) are simple aliphatic primary amines and essential constituents of eukaryotic cells [19]. These two polyamines can be converted into putrescine in plants [20]. Therefore, to see whether this conversion existed in fungi, the effect of the addition of either spermine or spermidine during the mid-log phase of growth was investigated. At each time interval, ACPY in the liquid culture of both fungi were quantified using GC-FID. Figure 2 shows the effect of polyamines on the production of ACPY by *A. awamori*. It was clear that the addition of spermidine or spermine led to at least fivefold higher ACPY production compared with that in the medium with no supplement. The highest amount of ACPY obtained was up to 5.854 mg/L after 68 h, in the medium with spermine. In contrast, the result was opposite in *A. oryzae* (figure 3); the supplementation with spermine or spermidine did not improve the production of ACPY. The amount of ACPY was lower than that in the presence of such polyamines. Thus, for *A. oryzae* the highest amount of ACPY was merely 1.323 mg/L in the Syn 18 medium alone after 80 h of fermentation.

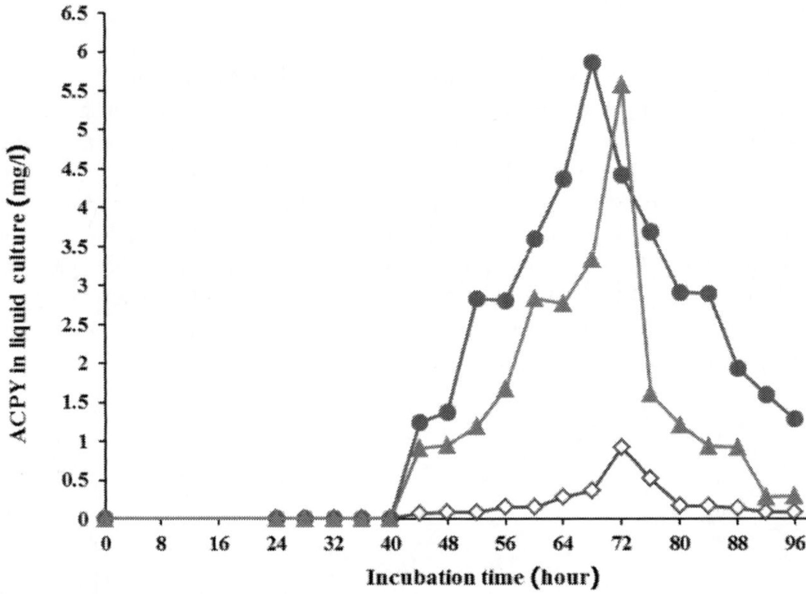

Figure 2. Effect of spermidine and spermine on ACPY production by *A. awamori*; ▲ spermidine; ● spermine; ◇ without spermine/spermidine

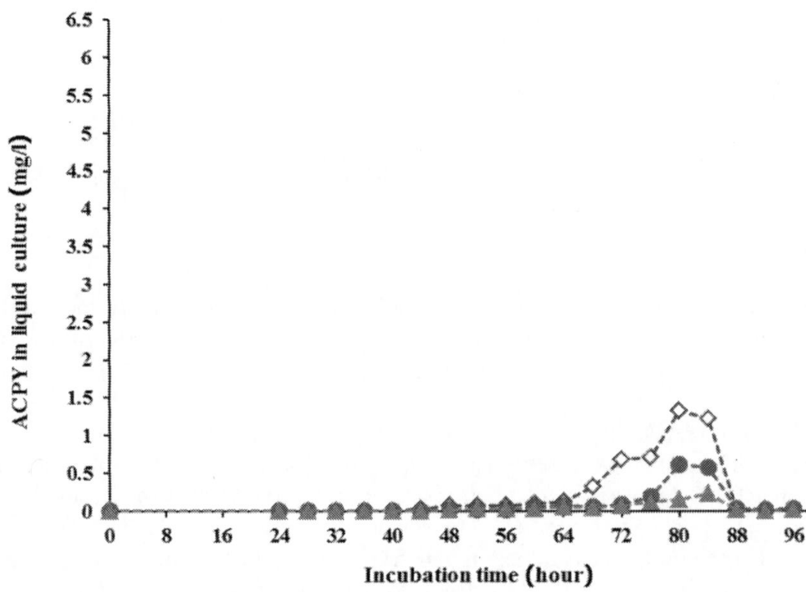

Figure 3. Effect of spermidine and spermine on ACPY production by *A. oryzae*; ▲ spermidine; ● spermine; ◇ without spermine/spermidine

The results confirmed that spermine and spermidine were major intermediates of the ACPY production by *A. awamori*. Through the conversion of spermine into spermidine and spermidine into putrescine, at each step, aminopropyl residues were removed by polyamine oxidase from spermine and spermidine, both of which were acetylated by spermidine/spermine N^1-acetyltransferase [20,21]. This result was consistent with previous findings, which explained the pathway for ACPY production by *L. hilgardii* DSM 20176, and the interaction of intermediates from two disparate pathways: 1) N-heterocyclic or 1-pyrroline derived from the catabolism of L-ornithine; and 2) acetyl side-chain accumulated from the heterolactic fermentation [6,8]. Previous studies have reported that putrescine was the decarboxylation product of ornithine; thus 1-pyrroline could be formed through putrescine [22,23].

However, the conversion of polyamine into putrescine did not occur in *A. oryzae*. Further, the volatile components, such as four fatty acids, three alkanes, two bezenes, and one sesquiterpenes were only found in liquid culture of *A. oryzae* (as shown in table 3) but not in the liquid culture of *A. awamori*, thereby suggesting that *A. oryzae* might form these volatile components from spermine and spermidine. Therefore, the differences in the pathway of ACPY production or enzymes involved need to be elucidated in the future.

4. Conclusion

The highest production of ACPY by *A. awamori* and *A. oryzae* was 0.914 mg/L (72 h of cultivation) and 1.323 mg/L (80 h of cultivation) in Syn 18 medium, respectively. The amount of ACPY reached its maximum value in the stationary growth phase of both fungi. The supplementation with spermidine and spermine was shown to increase ACPY production by *A. awamori*. While the effect was reverse in *A. oryzae*. Spermidine and spermine were the major intermediates of ACPY production by only *A. awamori*.

Acknowledgment

This research was financially supported by the National Research Council of Thailand (grant number KMUTNB-61-GOV-B-21) and the authors wish to acknowledge King Mongkut's University of Technology North Bangkok, Bangkok, Thailand and Bansomdejchaopraya Rajabhat University, Thailand for laboratory support.

References

[1] Paule C M and Power J J 1989 *J Food Sci Technol* **54** 343-5
[2] Schieberle P 1991 *J Agric Food Chem* **39** 1141-4
[3] Yahya F, Lu T, Santos R C D, Fryer P J and Bakalis S 2010 *J Supercrit Fluids* **55** 200-7
[4] Wei X, Handoko D D, Pather L, Methven L and Elmore J S 2017 *Food Chem* **232** 531-44
[5] Romanczyk L J Jr., McClelland C J, Post L S and Aitken W M 1995 *J Agric Food Chem* **43** 469-75
[6] Adams A and Kimpe D N 2006 *Food Chem* **101** 1230-8
[7] Deshmukh Y, Khare P and Patra D D 2014 *Biotechnol Prog* **30** 1356-63
[8] Costello P J and Henschke P A 2002 *J Agric Food Chem* **50** 7079-87
[9] Rungsardthong V and Noomhoom A 2005 *Flavour Fragr J* **20** 710-4
[10] Sillapapakdee J, Rungsardthong V and Prathumpai W 2014 Pathway of 2-Acetyl-1- Pyrroline formation by *Aspergillus awamori* TISTR 3193 (King Mongkut's University of Technology North Bangkok, Bangkok, Thailand)
[11] Yoshihashi T, Huong N T T and Inatomi H 2002 *J Agric Food Chem* **50** 2001-4
[12] Grimm C C, Champagne E T, Lloyd S W, Easson M, Condon B and McClung A 2011 *Cereal Chem* **88** 271-7
[13] Nicolotti L, Cordero C, Bicchi C, Rubiolo P, Sgorbini B and Liberto E 2013 *Food Chem.* **138** 1723-33
[14] Ubeda C, Callejón R M, Hidalgo C, Torija M J, Mas A, Troncoso A M and Morales M L 2011 *Food Res. Int.* **44** 259-68

[15] Elliott S and Burgess V 2005 *Forensic Sci. Int.* **151** 289-92
[16] Pho-am S, Kerdchoechuen O and Laohakunjit N 2015 *Agricultural Sci J* **46** 393-6
[17] Janes D, Kantar D, Kreft S and Prosen H 2008 *Food Chem* **112** 120-4
[18] Rungsardthong V 1995 Production of 2-acetyl-1-pyrroline, A Major Component of Aromatic Rice Flavours, by *Acremonium nigricans* (Asian Institute of Technology, Bangkok, Thailand)
[19] Ahmed N 1987 *Biochem Educ* **15** 106-10
[20] Alcázar R, Altabella T, Marco F, Bortolotti C, Reymond M, Koncz C, Carrasco P and Tiburcio A F 2010 *Planta* **231** 1237-49
[21] Jeon K W 2002 *International Review of Cytology* (Academic Press, U.S.A.)
[22] Fothergill J C and Guest J R 1976 *J Gen Microbiol* **99** 139-55
[23] Jacoby W B and Fredericks J 1959 *J Biol Chem* **234** 2145-50

IOP Conf. Series: Earth and Environmental Science **346** (2019) 012043 doi:10.1088/1755-1315/346/1/012043

Product development of dried noodle from wheat flour and riceberry rice flour by extrusion

P Chokchaithanawiwat[1], V Rungsardthong[1], B Thumthanaruk[1], C Puttanlek[2], D Uttapap[3], S Boonraksa[4] and J Wonngsa[5, 6]

[1]Department of Agro-Industrial, Food and Environmental Technology, Faculty of Applied Science, Food and Agro-Industry Research Center, King Mongkut's University of Technology North Bangkok, Bangkok, Thailand
[2]Deparment of Biotechnology, Faculty of Engineering and Industrial Technology, Silapakorn University, Nakhon Pathom, Thailand
[3]Division of Biochemical Technology, School of Bioresources and Technology, King Mongkut's University of Technology Thonburi, Bangkok, Thailand
[4]Bioplus Innovate Company Limited, Bang Phli, Samut Prakan, Thailand
[5]Department of Agricultural Engineering for Industry, Faculty of Industrial Technology and Management, King Mongkut's University of Technology North Bangkok Prachinburi Campus, Prachinburi, Thailand

E-mail: jittimon.w@fitm.kmutnb.ac.th

Abstract. Riceberry rice (*Oryza sativa* L.), a cross-breed of Khao Hom Nin, a local non-glutinous purple rice and Khao Hom Mali 105 contains antioxidants and other significant bioactive constituents in its bran fraction. This research aimed to produce noodle from riceberry rice and wheat flour by direct extrusion. In this study, the effects of barrel temperature, riceberry flour ratio, and food additives (sodium carbonate, guar gum and xanthan gum) on appearance and cooking qualities of extruded riceberry noodles were examined. Riceberry flour at 5%, 10%, 15% and 20% (by weight) was mixed with wheat flour via a single screw extruder with 0.8 mm diameter die, 30% feed moisture content (wet weight basis), screw speed of 40 rpm and the third barrel temperatures at 90:100:100 °C, respectively. Experiments were conducted to find appropriate food additives by using mixture design of 3 components (sodium carbonate, guar gum and xanthan gum) and barrel temperature controlled between 90-110 °C. The result showed that riceberry flour substitution could be used up to 15% with acceptable properties. The addition of guar gum and xanthan gum at 0.8% and 2.0%, respectively, could be used leading to rice noodle with cooking time 3.0 min and cooking loss 3.63%, comparable to the product made by conventional method (as the control formula).

1. Introduction

Riceberry, dark purple grain; (*Oryza sativa* L.), is a cross-bred strain developed from Hom Nin Rice variety with high antioxidant, and the well-known fragrant rice Khao Dawk Mali 105 developed by the Rice Research Center, Kasetsart University (Kamphaeng Saen Campus), Nakhon Pathom, Thailand [1]. Riceberry rice is a aromatic new rice variety that has black/purple in color, light and fluffy in texture when cooked. In addition, it is rich in antioxidants which support immunity, has anti-ageing properties and has high fiber and bran oil to aid digestion [2]. Nowadays, people are getting more and more

Content from this work may be used under the terms of the Creative Commons Attribution 3.0 licence. Any further distribution of this work must maintain attribution to the author(s) and the title of the work, journal citation and DOI.
Published under licence by IOP Publishing Ltd

interested in the properties of antioxidants that promote good health. Naturally extracted antioxidants have become popular as they can prevent cell damage from free radical reactions and reduce the risk of various chronic diseases like cardiovascular disease, arthrosclerosis, cancer and diabetes [3].

Asian people consume noodles as a staple food since ancient time. It is convenient, easy to cook and delicious product. A noodle strand was made from rice flour that was kneaded with water and salt which resulted in dough that can be pressed to produce a thin sheet before being streamed and then sliced into thin strands [4].

The production of noodles with traditional methods has many complex long production processes. Therefore, the risk of microbial contamination exists with high waste and large production space requirement. Extrusion has been used to produce noodle strands due to its multi-function ability including, in successive order, mixing, kneading, cooking, and forming the noodles. The process can be quickly executed by a single machine that runs continuously [5]. Moreover, extrusion offers significant economic advantages accruing from decreased energy and less floor space requirement and product quality can be improved through the better process uniformity and control [5]. Presently, food additives like emulsifiers (glycerol monostearate, soya lecithin and sodium stearoyl lactylate), gums (xanthan gum, guar gum, acacia gum and carrageenan) and dough strengthener (sodium carbonate) are used to improve the cooking properties and texture of the extruded rice-flour-based products [6-7]. Therefore, the objective of this study was to develop noodle with substitution of riceberry flour to wheat flour by using direct extrusion. In the present study ratio of riceberry rice flour, extrusion temperature and use of food additives on the noodle properties were investigated.

2. Material and methods

2.1. *Materials*

Riceberry rice (*Oryza sativa* L.) was obtained from Bioplus Innovate Co., Ltd (Thailand), and sodium carbonate, guar gum and xanthan gum were supplied from M Good Ingredient Co., Ltd (Thailand). Wheat flour, Kite brand was purchased from MC food Co., Ltd., Thailand.

2.2. *Chemical analysis*

The chemical compositions of riceberry flour and wheat flour were analyzed according to AOAC method [8].

2.3. *Preparation of noodles by direct extrusion*

The preparation of flour blend was performed by adding water at 30%wb to wheat flour (85%) and riceberry flour (15%). The mixture was mixed slowly at low speed (level 2) for 15-20 min in a mixer (Kenwood) before feeding into a single screw extruder (Barbender -19/20 Barrel Bore (D) 19.1 mm, Barrel Length (L) 20D, screw compression ratio 4:1) with the die diameter at 0.8 mm. Screw speed was operated at 40 rpm and barrel temperatures were maintained at 90 °C (first zone), 100 °C (second zone) while the effect of the third barrel temperature on the properties of extruded noodle was investigated at 90, 100 and 110°C.

2.4. *Riceberry flour ratio and use of food additives*

Riceberry flour, instead of wheat flour, has been used to produce noodles by extrusion process at 5, 10, 15 and 20% (by weight). The extruded samples were placed in the dryer with 12% moisture content for 25°C 24 h at room temperature. The noodle quality was improved by using food additives such as sodium carbonate, guar gum and xanthan gum by the mixture design experiment.

Wheat and riceberry noodles prepared by traditional method served as control. Wheat flour was mixed with riceberry flour at 20% (by weight) by using a Kenwood mixer and kneaded in to a dough, and incubated at room temperature for 30 min. The sheet was cut into 2.0 mm width by using a noodle slitter and steamed. To achieve dried product, the streamed noodles was left in a hot air oven for 4 hours at the drying temperature of 50°C.

2.5. Cooking qualities

Cooking qualities of the cooked noodles including cooking time and cooking loss were evaluated following Wandee et al, [9]. Cooking time of noodle was evaluated by observing the time of disappearance of the core of the noodle strand during boiling (every 30 s) by squeezing the noodles strand between two glass slides. Quantification of cooking loss was achieved by measuring residue left in water that was used in the cooking process. The experiment involved boiling of 10 g noodle in 300 ml of distilled water in a 500 ml beaker. Consequently, the cooking water was transferred into an aluminium container and placed in an oven at 105 °C. After the water was completely evaporated, the remaining material was weighted and calculated against the original quantity of the material. Mean value of the experiments was assessed to evaluate the cooking loss.

2.6. Noodle color analysis

Hunter Lab, Color Quest 45/0, (Australia) with a D65 illuminant using the CIE L*a*b* system was used to determine the colour of dried noodle strands. The obtained values signified the colours which were classified as lightness, redness and yellowness, respectively. The assessments were conducted 3 times for each sample.

2.7. Statistical analysis

Statistical Package for the Social Sciences version 22.0 software (SPSS Inc., Chicago, IL, USA) was used to evaluate statistical information. The results revealed mean average values of the triplicate experiments with ± the standard deviation (S.D.). However, the results did not apply to the textural assessment. Variance was determined by one-way analysis where differences between the mean values were assessed by Duncan's multiple range tests at a significant level of $p < 0.05$.

3. Result and discussions

3.1. Proximate compositions

Chemical compositions of riceberry flour and wheat flour are shown in Table 1. It was found that ash and lipid content of riceberry flour were higher than that of wheat flour while protein of riceberry flour was higher than that of wheat flour. [10] reported that protein, lipid, crude fiber and ash contents of riceberry flour were 10.94, 0.91, 2.73 and 1.13 g/100 g, respectively while Fari *et al.* (2011) [11] and Han *et al.* (2011) [12] found 6.84-11.18% of protein content in different variety of rice. On the other hand, wheat was found to have higher protein content at 11.65-14.93% [13-14]. Quantity of protein in rice may differ, depending on the variety of rice [12]. As pasting properties formed a strong network of protein and starch molecules, flour with higher protein content exhibits more binding sites for the starch which consequently resulted in higher water holding capacity [15-17].

Table 1. Chemical compositions of riceberry flour and wheat flour

Chemical compositions (%)	Riceberry flour	Wheat flour
Moisture content (ns)	9.67±0.39	9.24±3.03
Ash	1.36 ± 0.02^{a}	0.50 ± 0.01^{b}
Protein	8.33 ± 0.04^{b}	10.92 ± 0.05^{a}
Lipid	3.79 ± 0.03^{a}	1.81 ± 0.02^{b}
Carbohydrate (ns)	76.85±0.04	77.53±0.03

Values are mean ± SD of triplicate samples (n=3)
Different superscript letters in the same column indicate significant difference ($P \leq 0.05$)
ns = no significant difference ($P < 0.05$).

IOP Conf. Series: Earth and Environmental Science **346** (2019) 012043 doi:10.1088/1755-1315/346/1/012043

3.2. Effect of third barrel zone temperature

In this study, the use of third barrel zone temperature between 90-100°C significantly affected the characteristic of the extruded noodles as shown in Table 2. The surface of noodle strands extruded at 90°C gave wave-like noodle strands, while the temperature of 110°C, the noodle strand's surface was slightly inflated and porous due to the high temperature. However, noodles from all third zone temperature yielded the products with the same cooking time, 4.00 min which might come from same level of gelatinization of the flour mix. Consequently, the barrel temperature of zone 1: 2: 3 was controlled at 90, 100 and 110°C, respectively for following experiment.

Table 2. Effects of third zone temperature on cooking time and appearance of extruded noodle.

Barrel Temperature (°C)			Appearance	Cooking Time (min)
Zone 1	Zone 2	Zone 3		
90	100	90	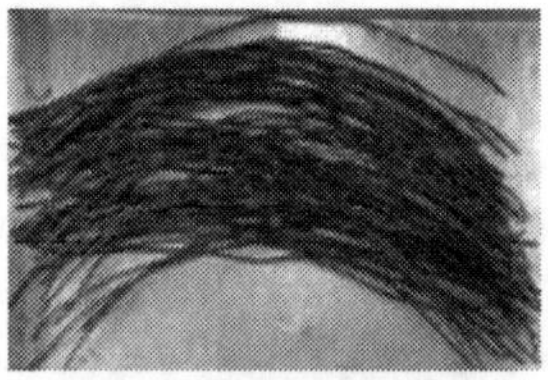	4.00±0.01
90	100	100		4.00±0.01
90	100	110	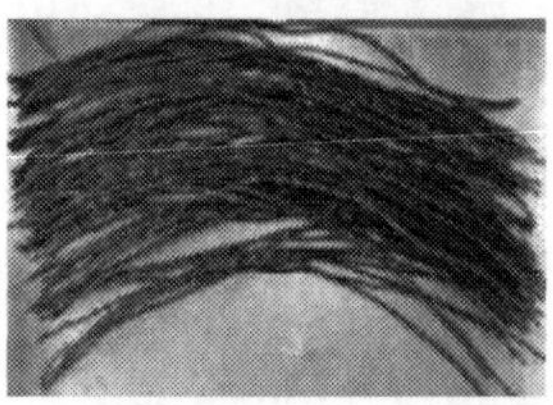	4.00±0.01

Values are mean ± SD of triplicate samples (n=3).

3.3. Riceberry flour ratio

Appearance of dried noodles and cooking time of cooked noodles from the extrusion of wheat flour with riceberry flour substitution at 5, 10,15 and 20% (by weight) are shown in Table 3.

Color appearance is one of the characters that influence consumer's preferences. Color, L*, a*, b* of riceberry-made noodles are displayed in Table 3. Riceberry noodle produced by traditional method was used as the control and be compared with the riceberry flour to wheat flour at 5, 10, 15 and 20% (by weight). L* and b* value of the noodles were reduced while a* value increased with the increase of riceberry flour. Higher content of riceberry resulted in more reddish color due to anthocyanin pigment in the riceberry flour [18]. Considering the appearance and color acceptance, the substitution of riceberry flour at 15% (by weight) was selected for further study.

Cooking time of noodles made from wheat flour had longer cooking time than noodles that use partial rice flour instead of wheat flour. The differences in cooking quality were attributed primarily to the gluten content. Gluten fraction dilution was caused by higher riceberry flour content, thus, required lower heat for the cooking process. As a result, dough properties and the cooking quality were affected by higher riceberry flour content [19].

Table 3. Color characteristics of dried noodles from riceberry rice.

Sample	L*	a*	b*	Appearance
5% RF	25.74±0.15^{b}	1.89±0.01^{d}	7.55±0.03^{b}	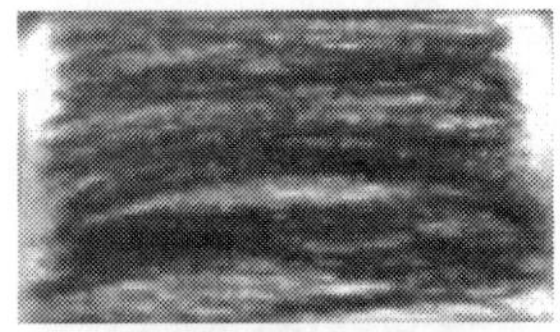
10% RF	21.60±0.09^{c}	2.78±0.02^{b}	4.17±0.01^{c}	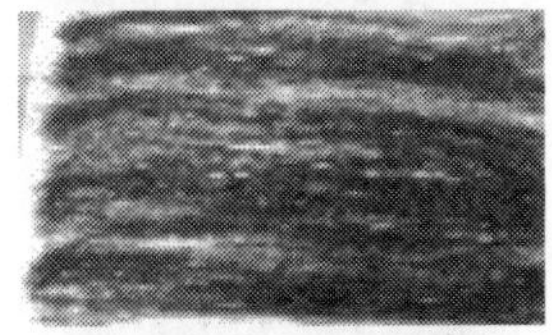
15% RF	17.27±0.08^{d}	2.95±0.02^{a}	1.90±0.05^{d}	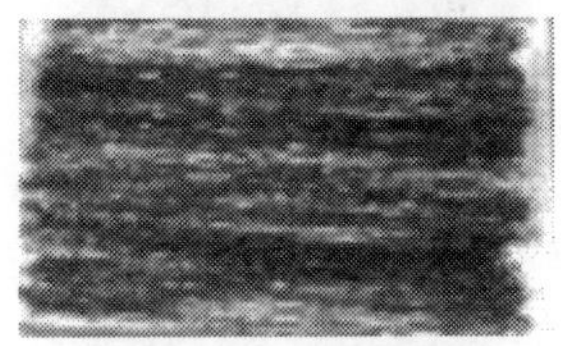
20% RF	16.38±0.08^{e}	2.16±0.13^{c}	0.51±0.16^{e}	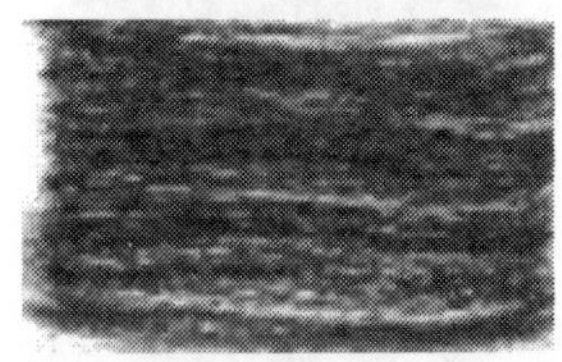
20% RF (control)	30.87±0.16^{a}	2.74±0.11^{b}	9.47±0.20^{a}	

Values are mean ± SD of triplicate samples (n=3)
Different superscript letters in the same column indicate significant difference (P≤0.05)

3.4. Use of food additives on noodle properties

The effects of food additive (sodium carbonate, guar gum and xanthan gum) on the cooking time and cooking loss of extruded noodles explored by mixture design are shown in Table 4. It was found that the noodle with riceberry flour at each formula gave color and appearance slightly different due to different amounts of food additives used in the experiments. Cooking times of noodle ranged from 3.00 to 4.00 min.

Cooking loss is the amount of dry matter in the cooking water at each noodle. An increase in the cooking loss with noodles containing food additives in formula 1, 5-9 (Table 5) may be due to interaction of the food additives at a certain amount, resulted to the property starch gel network and more solid leaching out from the noodles into the cooking water. Formular 2 exhibited lowest cooking time and

IOP Conf. Series: Earth and Environmental Science **346** (2019) 012043 doi:10.1088/1755-1315/346/1/012043

cooking loss and had better texture properties than other the formula, determined by preliminary sensory testing. Kaur et, al. (2015) [20] reported that the noodles prepared with guar gum and xanthan gum showed significantly lower cooking loss than that prepared without the guar gum and xanthan gum.

Table 4. Appearance and cooking time of extruded noodles prepared from wheat flour substituted with riceberry flour at different percentage.

Samples	Appearance	Cooking Time (min)
0% RF	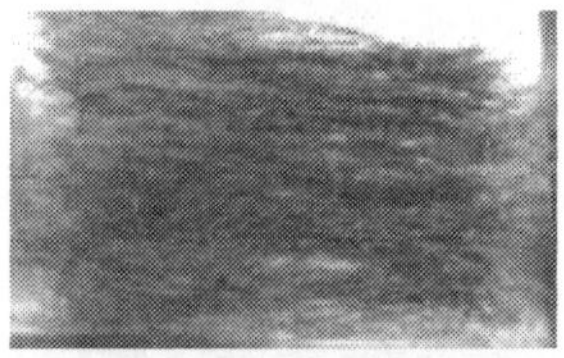	4.50±0.01
5% RF	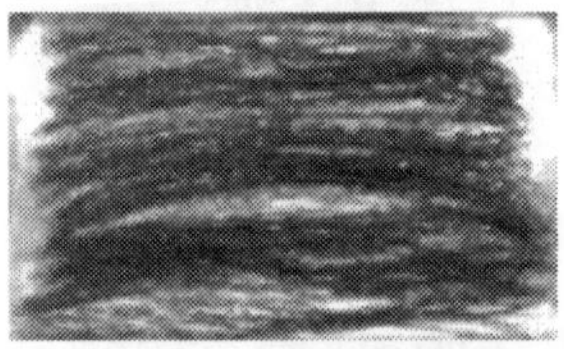	4.00±0.01
10% RF	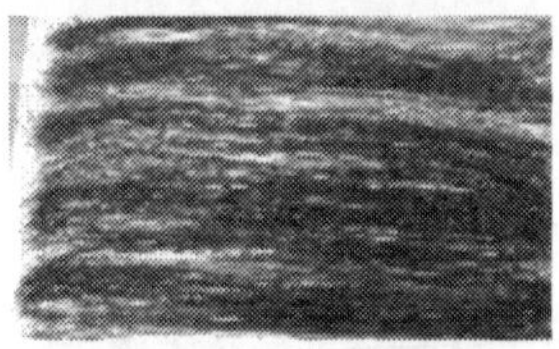	4.00±0.01
15% RF	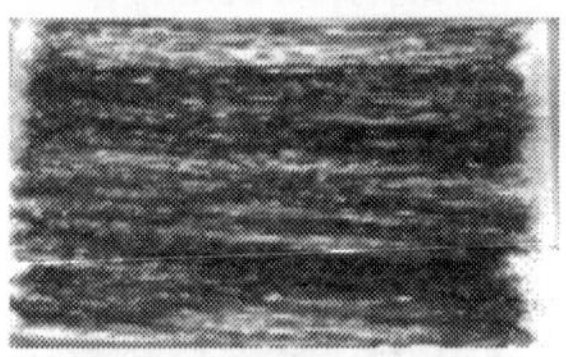	4.00±0.01
20% RF	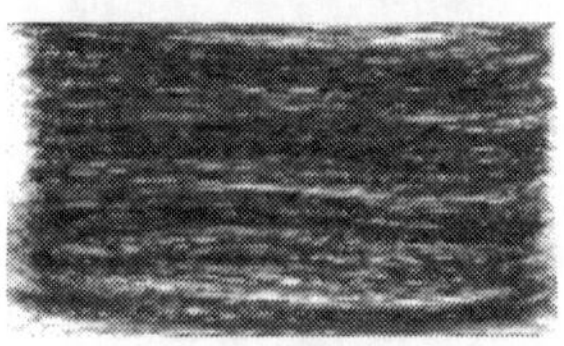	4.00±0.01
20% RF (control)		3.50±0.01

Values are mean ± SD of samples (n=3), RF = riceberry flour

Table 5. Cooking time, cooking loss and cooking yield of noodles from wheat flour and riceberry flour with various food additives.

Formula	Food Additive (g)			Cooking time (min)	Cooking loss (%)
	Na_2CO_3	Guar gum	Xanthan gum		
1	-	1	-	3.50±0.01	45.27±0.06[c]
2	-	0.8	2	3.00±0.00	3.63±0.02[f]
3	0.3	0.7	-	4.00±0.00	6.40±0.03[ef]
4	0.3	0.5	2	3.50±0.01	6.72±0.08[e]
5	0.15	0.75	1	4.00±0.00	76.08±0.01[b]
6	0.075	0.875	0.5	3.50±0.01	74.82±0.03[b]
7	0.075	0.775	1.5	3.00±0.00	80.35±0.01[a]
8	0.225	0.725	0.5	4.00±0.00	27.79±0.03[d]
9	0.225	0.625	1.5	3.50±0.01	26.77±0.07[d]

Values are mean ± SD of triplicate samples (n=3) Different superscript letters in a column indicate significant difference. ($P < 0.05$)

4. Conclusion

This research explored the possibility of using up to 20 % of riceberry flour to produce noodles. Extrusion process at barrel temperatures of zone 1: zone 2: zone 3 at 90:100:100°C was used in the process. According to the results, riceberry flour at 15% (by weight basis) produced acceptable color and cooking quality compared to the product from traditional method. Guar gum and xanthan gum at 0.8% and 2.0% respectively were additives used for rice noodles that resulted in the cooking time of 3.0 min and the cooking loss of 3.63%.

Acknowledgment
This research was supported by Office of the Higher Education Commission and Bioplus Innovate Co., Ltd, under the Talent Mobility Program.

References

[1] Rice Science Center of Kasetsart University 2016 Riceberry rice Retrieved from http://dna.kps.ku.ac.th/index.php/articles-ricersc-rgdu-knowledge/29-2015-03-27-02-0415/53-riceberry.
[2] Leardkamolkarn V, Thongthep W, Suttiarporn P, Kongkachuichai R, Wongpornchai S and Wanavijitr A 2011 Chemopreventive properties of the bran extracted from a newly developed Thai rice: The Riceberry. *Food Chem* **125** 978-985.
[3] Pannangrong W, Wattanathorn J, Muchimapura S, Tiamkao S and Tong-Un T 2011 Purple riceberry is neuroprotective and enhances cognition in a rat model of alzheimer's disease. *J. Med. Food* **14**(7) 688-694.
[4] Ahmed I, Qazi I M, Li Z and Ullah J 2016 Rice noodles: Materials, processing and quality evaluation. Proceedings of the Pakistan Academy of Sciences: B. Life and Environmental Sciences **53**(3) 215-238.
[5] Miller R C 1988 Continuous cooking of breakfast cereal. *Cereal Foods World* **33** 284-291.

[6] Wongsa J, Uttapap D, Lamsal B P, Puttanlek C and Rungsardthong V 2017 Effect of extrusion conditions, monoglyceride and gum arabic addition on physical and cooking properties of extruded instant rice. *KMUTNB Int J. Appl Sci Technol* **10**(1) 23-30.

[7] Wang J P, An H Z, Jin Z Y, Xie Z J, Zhuang H N and Kim J M 2011 Emulsifiers and thickeners on extrusion-cooked instant rice product. *J. Food Sci Technol* **50**(4) 655-566

[8] AOAC 2005 Official Method of Analysis, Association of Official Agricultural Chemist. 12th Ed. Washington, D.C., USA.

[9] Wandee Y, Uttapap D, Puncha-arnon S, Puttanlek C, Rungsardthong V and Wetprasit N 2015 Quality assessment of noodles made from blends of rice flour and canna starch. *Food Chem* **179** 85-93.

[10] Jamal S, Qazi I M and Ahmed I 2016 Comparative studies on flour proximate compositions and functional properties of selected Pakistani rice varieties. *Pakistan Acad Sci* **53**(1) 47-56.

[11] Fari M J M, Rajapaksa D and Ranaweera K K D S 2011 Quality characteristics of noodles made from selected varieties of Sri Lankan rice with different physicochemical characteristics. *J. Natl Sci Found Sri* **39** 53-60.

[12] Han H M, Cho J H and Koh B K 2011 Processing properties of Korean rice varieties in relation to rice noodle quality. *Food Sci Biotechnol* **20** 1277-1282.

[13] Ngozi A A 2014 Effect of whole wheat flour on the quality of wheat-baked bread. *J. Glo. Fd Sci. Technolo.* **2(3)** 127-135.

[14] Han L, Zhou Y, Tatsumi E, Shen Q, Cheng Y and Li L 2013 Thermo-mechanical properties of dough and quality of noodles made from wheat flour supplemented with different grades of tartary buckwheat flour. *Food Bioprocess Tech* **6** 1953-1962.

[15] Rosniyana A and Hazila K K 2013 Nutritional properties and organoleptic acceptability of traditional cakes made from mr 220 rice flour. *J. Trop Agric Fd Sci* **41**(1) 41-52.

[16] Chung H J, Cho A and Lim S T 2012 Effect of heat-moisture treatment for utilization of germinated brown rice in wheat noodle. *J. Food Sci Technol* **47** 342-347.

[17] Martinez O M, Ayerdi S S, Acevedo E A, Goñiand I, Pérez L A B 2009 Un-ripe banana flour as an ingredient to increase the undigestible carbohydrates of pasta. *Food Chem* **113**(1) 121-126.

[18] Yodmanee S, Karrila T T and Pakdeechanuan P 2011 Chemical and antioxidant properties of pigmented rice grown in southern Thailand. *Int. Food Res J* **18**(3) 901-906.

[19] Li P H, Huang C C, Yang M Y and Wang C C R 2012 Textural and sensory properties of salted noodles containing purple yam flour. *Food Res Int.* **47** 223-238.

[20] Kaur A, Shevkani K, Singh N, Sharma P and Kaur S 2015 Effect of guar gum and xanthan gum on pasting and noodle-making properties of potato, corn and mung bean starches. *J. Food Sci Technol* **52**(12) 8113-8121.

On the loss regularity of water and fertilizer in sandy soil with abundance of underground water but lack of cultivated layer

Y Luo[1], H F Yue[1], S H Li[1], H Yi[2], J H Zhao[1], T N Wuyun[1], F D Li[1], N Zhang[3], H Zhao[1,4] and G P Zhu[1,4]

[1a]Non-timber Forestry Research and Development Center, Chinese Academy of Forestry, [b]China Paulownia Research and Development Center, and [c]Key Laboratory of Non-timber Forest Germplasm Enhancement & Utilization of State Forestry Administration, Zhengzhou 450003, P. R. China
[2]Liangyuan County Forestry Bureau, Shangqiu, 476000 Henan, P. R. China
[3]Henan Provincial Institute of Science of Technology Information, Zhengzhou 450003, P. R. China
Y Luo and H F Yue are joint first author.

E-mail: Zdh707@163.com (H Zhao); Zhugaopu@163.com (GP Zhu)

Abstract. To explore the loss regularity of water and fertilizer in the ancient course of the Yellow River in cultivation of *Chaenomeles sinensis* and *Prunus armeniaca* × *sibirica*, three water management measures were employed: mulching with plastic films (Wmp), mulching with cornstalk (Wmc) and conventional management (Wck). Meanwhile, the effects of irrigation times, namely slow-release N, P and K fertilizer (NPK SRFs), compound N, P and K fertilizer (NPK CFs), urea fertilizer (Urea Fs), and unfertilized (Fck) were studied. Results showed that the soil water content was effectively increased using Wmp and Wmc, Wmp effectively extended the water retention time and soil temperature stability. After three irrigations, NPK SRFs, NPK CFs and Urea Fs were completely lost. The compound fertilizer had a significantly higher nutrient maintenance capacity than urea. Furthermore, the application of urea had to be balanced to avoid an excessive rise in soil pH. The growth of *C. sinensis* and *P. armeniaca* × *sibirica* was significantly better than traditional model. It is suggested to use crop straw to maintain water levels and improve soil organic matter, and replace urea by applying a compound fertilizer in the early and late growing season in sandy soil.

1. Introduction

The ancient course of the Yellow River stretches for hundreds of kilometers from western to eastern China, and its breadth exceeds 10 kilometers on both sides of the river. Its route passes through Inner Mongolia, Shanxi, Henan and Shandong provinces, and it has a wide range, covers a large area and has considerable influence [1]. The total area is about 903 thousand hm^2 and contains about 316 thousand hm^2 of sand area and 45 thousand hm^2 of sediment area. The soil types are mainly riverbank sand, middle-low yield saline-alkali land and low-lying waterlogged land [1,2]. The soil in this area is sandy and barren, causing water and fertilizer leakage. There are very few tree species in the area, with a low forest coverage and poor biodiversity, and it therefore has a poor environmental carrying capacity and is easily affected by wind erosion. In turn, this leads to a dry topsoil and sandstorms formations in winter and spring. The ancient course of the Yellow River has been listed as a key

Content from this work may be used under the terms of the Creative Commons Attribution 3.0 licence. Any further distribution of this work must maintain attribution to the author(s) and the title of the work, journal citation and DOI.
Published under licence by IOP Publishing Ltd

desertification area in China [3-5]. However, in this region the land is soft, broad and flat which makes it convenient for the use of mechanization, and the underground water level is high. Therefore, it will be an important future source of arable land in China with a scarce reserve of soil.

The moisture content in the soil is the primary factor limiting plant growth and development [6]. The inability of ancient course of the Yellow River to retain water and fertilizer leads to insufficient moisture and nutrient contents in the farming layer of the soil, and this has been the main factor restricting agricultural production in this area. A series of important achievements have been made globally in the field of water and fertilizer research. For instance, the quality and yield of the sand cucumber [7] and potato [8] was effectively improved through appropriate integrated measures of irrigation and fertilization. The water retention of sandy soil was also improved by adding other substances [9]. Current research on the ancient course of the Yellow River sand has focused on the ecological benefits of wetlands [10], soil biological characteristics and quality [4,5], controlling wind erosion by planting vegetation [5] etc. In the field of water and fertilizer research, Chen *et al* [2] studied the effects of different irrigation methods on the *Robinia pseudoacacia* seeding, Hou *et al* [11] studied the dynamic effects of orchard grass on the soil fertility of a pear garden. In these studies, the scholars focused on how to improve the effective water content of sandy soil and the effects of single (water or fertilizer) factors on trees in the soil. However, there has been no relevant research on the regularity of water and fertilizer loss in the ancient course of the Yellow River or analogous areas. As a result, it is difficult to devise a precise management strategy for agricultural crops in this area. Combined with the current frequent application of fertilizers and indiscriminate irrigation measures of the soil and underground water pollution [12], secondary salinization [13] has become an urgent agricultural ecological and environmental problems to be solved in this area.

Our previous research [1,14,15] has shown that unlike the site conditions in desert areas, due to the sandy area of the ancient course near the Yellow River, long-term river erosion and multiple road diversions lead to a lack of water in the cultivated layer (0~30 cm) and an abundance of underground water (250~300 cm). It can be reasonably speculated that the trees in this area can grow and develop by forming a large root system after entering the adult stage, using natural precipitation in normal years and deep groundwater in other years, without the need of a large economic investment in the construction of an irrigation system. The critical period for the survival of the trees is the period between one to three years on this sandy land.

In this study, we used a sandy region with middle to low yield saline-alkali land, which is the second largest region of the Yellow River, as the research site to uncover the loss regularity of water and fertilizers. Through the analysis of these regularities, the water and fertilizer management measures of papaya and kernel-apricot were optimized. This work can provide a reference for the other species of fruit trees in the region.

2. Materials and methods

2.1. Location and materials

- Experimental site

The experimental site is in the town of Xiguo, Mengzhou city, Henan province, which is a long-term sample plot of the Non-timber Forestry Research and Development Center, Chinese Academy of Forestry (112°42'51" E, 34°51'35" N). This region has a warm temperate climate and is located in the continental monsoon climate. It experiences four distinct seasons. The annual average duration of sunshine and the average temperature are 2160.4 hours and 14.2°C, respectively. The average annual rainfall is 586.9 millimeters with an average annual evaporation of 1599.0 mm. The frost-free period is 209 days. The soil type is the sandy land of the Yellow River and the vertical profiles can be divided into seven layers (figure 1). Within 25 cm of the soil surface, the total nitrogen content of the soil contains is 33 mg kg^{-1}, the hydrolyzed nitrogen content is 48.79 mg kg^{-1}, the available phosphorus content is 4.80 mg kg^{-1}, the available potassium content is 156.62 mg kg^{-1}, the total phosphorus content is 0.05%, the total potassium content is 0.10%, and the organic matter content is 0.41%. The soil pH value is 8.67, and the underground water level is 250~300 cm. The period from June to August, in this region is the fast-growing period for most non-timber forests such as papaya and apricot. This period exhibits high rainfall and evaporation, alternating occurrence of drought and flood, and obvious soil and fertilizer leakage, which are quite detrimental to crop production.

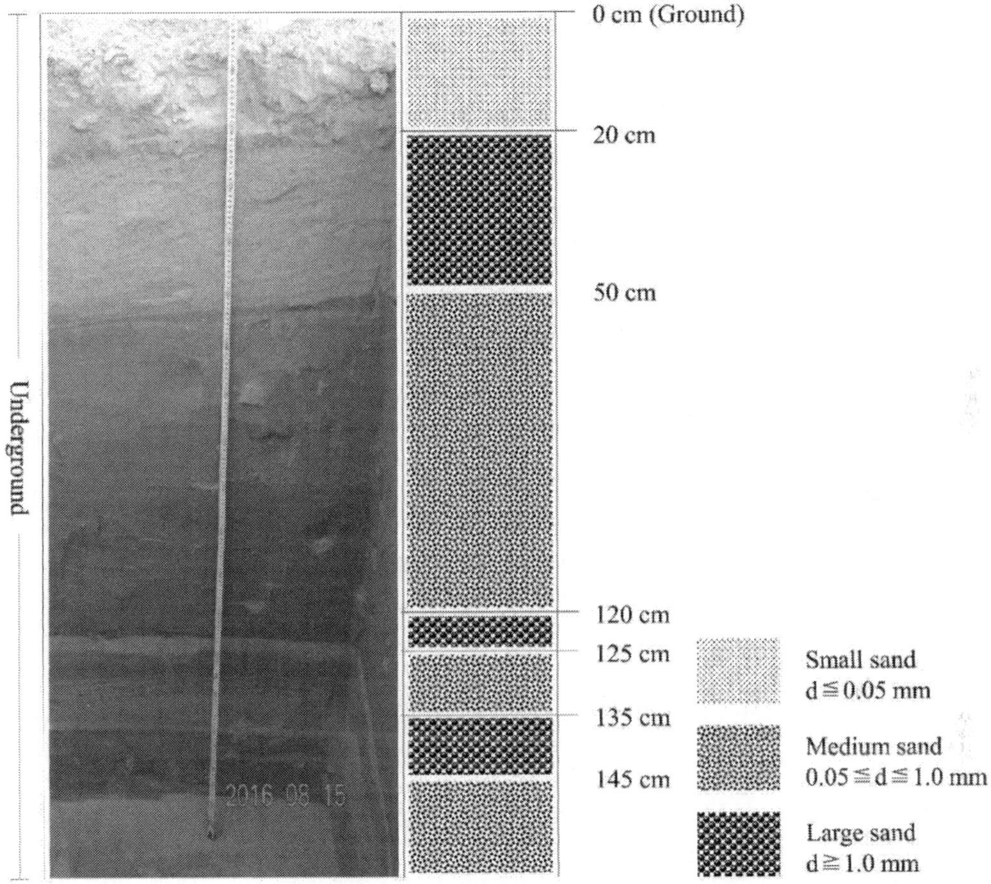

Figure 1. Soil profile in the test area.

- Plant materials

The planting materials were papaya (*Chaenomeles sinensis)* and kernel-apricot (*Prunus armeniaca* × *sibirica*). The variety of papaya is *C. sinensis* cv Jinshuai （Jinshuai） which has a large golden fruit with middle drought resistance, approved by the Non-timber Forestry Research and Development Center, Chinese Academy of Forestry in 2015. The kernel-apricot variety is *P. armeniaca* × *sibirica* cv Zhongren No. 1 (Zhongren No. 1). It is characterized by a sweet kernel, with a strong drought resistance, and high yield and quality, and was approved by the Non-timber Forestry Research and Development Center, Chinese Academy of Forestry in 2011.

2.2. Methods

- Experimental design of the soil water conservation at different irrigation treatments

Three soil irrigation treatments were designed. Treatment Wck (Control check) is a conventional management method, which simulated normal farmland water management measures, with a construction of a furrow at 1 m wide, 25 ~ 30 cm deep, as the daily management of the irrigation and nutrition supply. Treatment Wmp (Mulching with plastic-films) used a black polyethylene film (grass control) with a thickness of 0.06 mm and a width of 70 cm to lay the soil film along the central line of the nutrient belt, the other conditions were the same as WcK. In the Wmc treatment (Mulching with cornstalk), the maize straw was crushed into 5.0 cm long fragments and was covered with soil in the nutrient zone, the straw layer thickness was 8.0~10.0 cm and was covered with a layer 5.0~8.0 cm thick of soil. The three treatments were fully irrigated to the soil saturated state, and the ventilation windows around the plastic greenhouse were opened after the treatment to keep similar conditions to the field climate. The detection of soil volume water was set layer 5 cm and 20 cm depth, the temperature of the soil at 20cm depth and the air humidity was placed 50 cm above the surface which was measured every 5 days. The measurement time was from June to August in 2014 and 2015, with sunny and windless conditions at 17:00 PM, taking 30 days as the test period. Each treatment was randomly sampled using the 5-point sampling method and repeated 3 times. The total experimental area was about 190 m^2.

- Experimental design of the soil fertilizer loss at different chemical fertilizers

Four types of fertilizer were set in the plastic greenhouse for the simulation of fertilization. Treatment NPK SRFs is a slow-release N, P and K fertilizer. The fertilizer nutrient content was 50% or more, and the ratio of N: P: K was 30:10:10. It is commercially available, from Hebei. Treatment NPK CFs is a compound of N, P and K fertilizer. The fertilizer nutrient content was greater than or equals to 45%, the ratio of N: P: K was 15:15:15, it is commercially available and produced in Yunnan province. Treatment Urea Fs is a urea fertilizer. The fertilizer nitrogen nutrient content was 46% and it is a commercially available also, produced in Yunnan; Treatment Fck, the control check, did not have any chemical fertilizer applied.

The soil sample were dug according to the cultivation density of 3.0 m × 4.0 m, with a depth and diameter of 20-25 cm. 0.50 kg of the several types of chemical fertilizers were applied to each site. After fertilizing, the soil was fully irrigated to the saturated state. Samples were collected at a depth of 30 cm in the soil at 17:00 PM on the second day after each irrigation, and soil samples were collected randomly using 5-point sampling method for each treatment, with 3 repetitions. The test area had an area of 600 m^2.

- Preparation of test area and materials

In March 2015 (20 days before the planting of the papaya and kernel apricot) 75 $t \cdot hm^{-3}$ rotting cow dung was applied as the base fertilizer in the land. The papaya and kernel apricots were planted with a density of 3.0 m × 4.0 m, and the cultivation of the tree was conducted according to conventional field management measures. From May 2016, the papaya and apricot kernel with uniform growth were selected as samples, and the management measures were optimized according to the simulated experiment results in this study.

2.3. Data acquisition and analysis

- Determination of soil moisture, soil temperature and air humidity

The soil moisture speed measurement system with soil temperature and air humidity sensors (TZS-3X, Zhejiang Top Instrument Co., Ltd, Hangzhou, China) was adopted.

- Soil nutrient content determination

The soil samples were dried, ground into powder and sifted in preparation of the experiment. The determination of hydrolysis nitrogen (using the alkali hydrolysis diffusion method), available phosphorus (using the sodium bicarbonate extraction - molybdenum anticolorimetry method), available potassium (using the ammonium acetate extraction - flame photometry), and soil pH value (using the glass electrode method) were measurement at the Institute of Soil, Fertlizer and Water-saving Agriculture, Gansu Academy of Agriculture Sciences.

- Measurement of biomass indexes

In the October 2015 and 2016, the growth of new shoots and crown width were measured by a meter scale, and the diameter of the trunk of the trees at 50 cm above the ground were measured by a circumference scale. Six trees were randomly selected for each treatment and measured in four directions of the crown.

IOP Conf. Series: Earth and Environmental Science **346** (2019) 012044 doi:10.1088/1755-1315/346/1/012044

- Data processing

The Microsoft excel 2016 software was used for data entry and construction. The DPS (Data Processing System) 7.05 software [16] was used for data processing. The data analyzed using one-way analysis of variance and Duncan's new complex range method was used to analyze the difference between mean values. The significance level was set as 0.05 and 0.01. The Student's t test was used to compare the two groups.

3. Results

3.1. Effects of three treatments on water retention and agricultural microclimate

- Effects of the three treatments on water retention

Irrigation is the most important method to improve the soil water content. This study revealed that different treatment methods led to different abilities to maintain soil water (figure 2A). The soil moisture content at 5 cm underground in treatment Wmp and Wck showed a trend of continuous decrease. While treatment Wmc showed first a decrease, then increase and finally a drop wave form trend (figure 2A). After 15 days of irrigating the Wck treatment the soil moisture content at 5 cm almost approached the drought threshold (4%) and 20 days after it reached to the extent of severe drought (3%). Treatment Wmp can adequately maintain the effective moisture content at a 5 cm depth in the soil, the soil moisture remained at around 8.0% after 15 days of irrigation, and then rapidly declined to the level of severe drought at 4.0% by the 20th day. Wmc effectively extended the soil moisture retention time at 5 cm, the soil moisture content remained at around 10.0% after 15 days. The soil moisture content first increased and then decreased after 20th days. In general, Treatment Wmc had the strongest water-retaining capacity at a 5 cm depth of soil for more than 30 days, followed by Wmp for about 15 days, and Wck was the least effective at about 10 days.

All three treatments showed a trend of continuous decrease in soil moisture content at 20 cm underground, but the rate of decline was different (figure 2B). Treatment Wck showed two obvious peaks for the declining trend of soil moisture content at 20 cm: The water content dropped from 20% to below 10% from the 5th day to the 10th day; The second rapid decline peak occurred from the 20th day to the 30th day, and soil moisture content decreased from 9.0% to about 3.5%. The decreasing trend of water content at the 20 cm soil depth that was treated using treatment Wmp and Wmc was gentle, but the two-time nodes at the 5th day and 25th day after irrigation showed an extremely significant difference ($p < 0.01$) (figure 2B). On the 5th day after irrigation, the soil moisture content at 20 cm treated by Wmc reached 17.0%, significantly higher than the treatment Wmp level of 15% (figure 2B), and the difference between the two treatments was not significant from the 10th day to the 20th day. On the 25th day and 30th day after irrigation, the Wmc soil moisture content was 10.0% and 9.0% respectively, extremely significant difference higher than treatment Wmp ($p < 0.01$). It indicates that Wmc had extremely significant higher soil moisture retention capacity at 20 cm among the three treatments, followed by Wmp, and Wck which was the weakest.

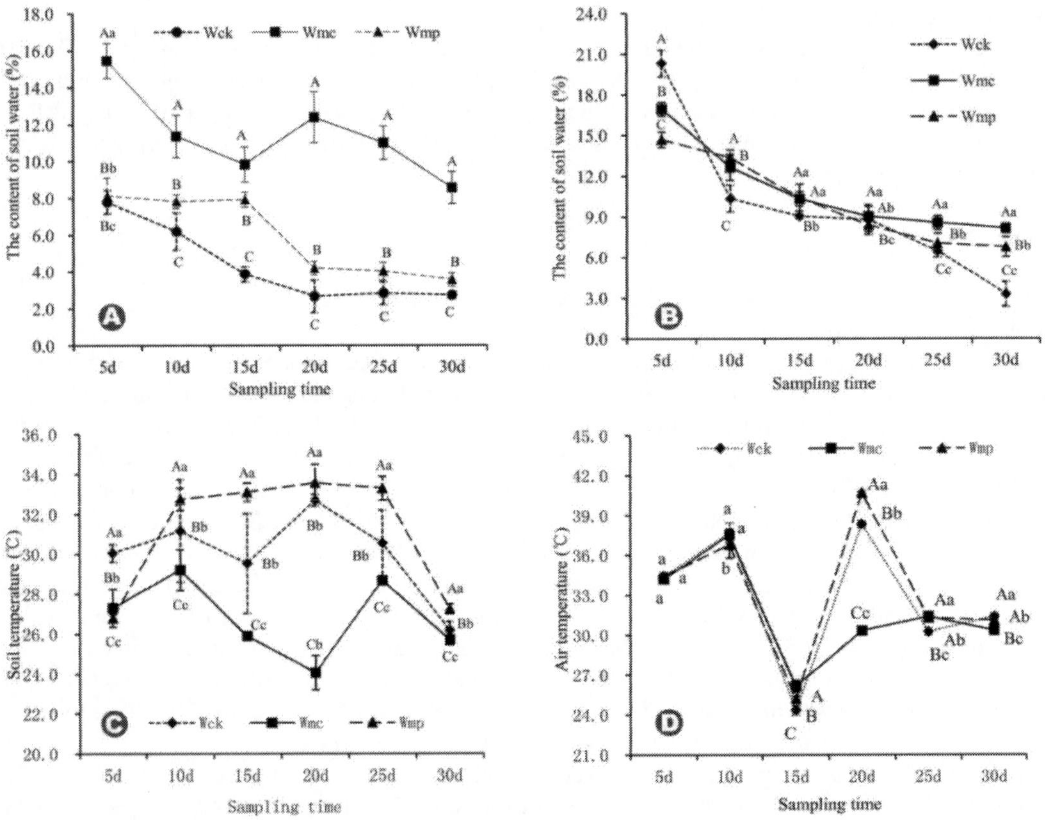

Figure 2. The changes of soil water and temperature at different water treatments. A. The changes of soil water underground 5 centimeters; B. The changes of soil water underground 20 centimeters; C. The changes of soil temperature underground 5 centimeters; D. The changes of air temperature aboveground 0.5 meters. For each column, mean values with different letters are significantly different, lowercase letters show significant difference ($p < 0.05$), capital show highly significant ($p < 0.01$). The same as below.

- Effects of the three treatments on temperature in the soil and air

A suitable soil temperature has a significant positive effect on plant root development, and water has a large specific heat capacity and a strong temperature maintenance ability. In this study, the three irrigation strategies significantly affected the change of soil temperature within 20 cm (figure 2C). On the 15th day and 25th day after irrigation, the effects of three treatments at the 20 cm soil depth temperature were extremely significantly different ($p < 0.01$). The Wmp treatment had the highest temperature (33.1°C, 33.3°C), followed by Wck (29.5°C, 30.6°C) and Wmc which had the lowest temperature (25.9°C, 25.9°C). On the 20th day, Wmp had a peak at 20 cm soil depth with a temperature of 33.6°C, followed by Wck (32.7°C) while Wmc (24.1°C) was significantly ($p < 0.01$) lower than the other two. Overall, treatment Wmc had the least fluctuation in temperature and especially improving the 20 cm soil temperature.

Air temperature directly affects the change of soil temperature. The near surface air temperature fluctuations in Wmp and Wck both showed a "high to low to high to low" bimodal curve, the temperature change range was 40.7°C (25th day) to 24.4°C (15th day). The difference between the two treatments on the first 10 days was not significant, and the difference on the following 20 days was significant ($p < 0.05$) (figure 2D). The Wmc treatment had a stronger ability to maintain air temperature (figure 2D). The changes of near surface air temperature showed a gentle trend with "high to low to high" single curve which ranged between 36.8°C (25th day) and 26.2°C (15th day) (figure

2D). It indicated that straw mulching can significantly reduce the air temperature variation within the range of 0.5 m.

- Effects of rapid nutrient leaching on soil pH

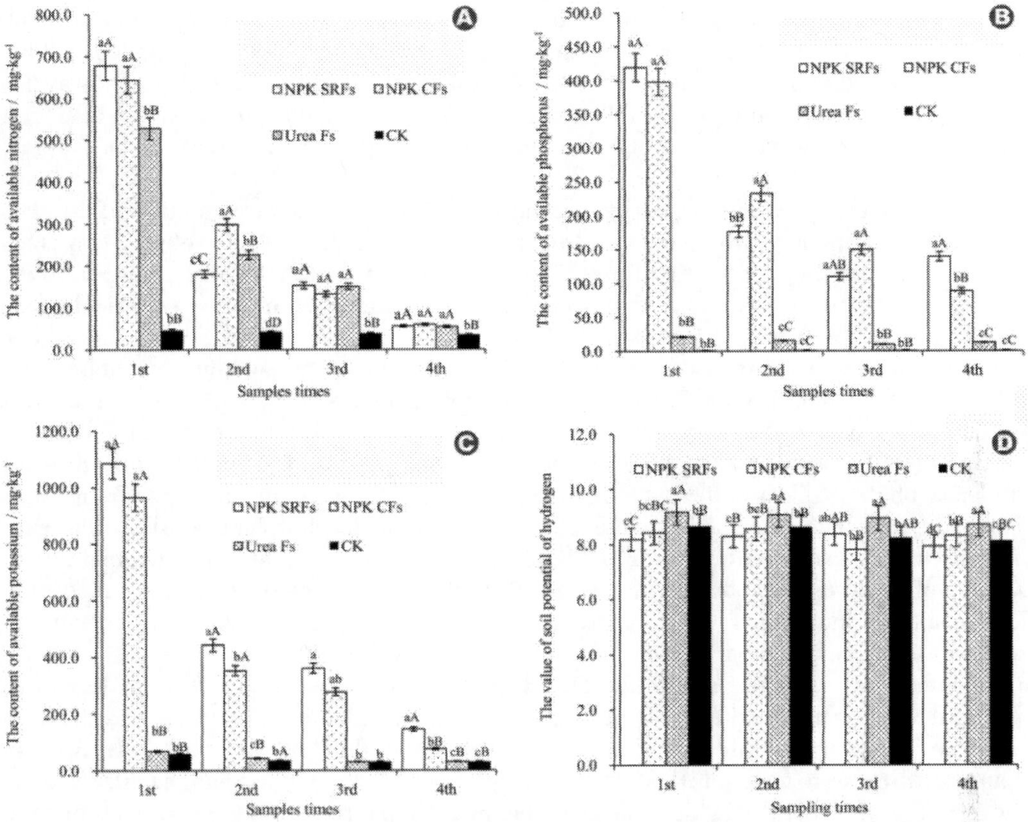

Figure 3. The content changes of hydrolyzable nitrogen, phosphate, potash and pH value at different sampling times. A. The loss rule of available nitrogen; B. The loss rule of available phosphrous; C. The loss rule of available potassium. D. The changes of soil pH.

The growth rate of trees is determined by the content of available nutrients during the fast-growth stage. In our study, the available states of N, P and K showed a rapid downward trend with the increase of irrigation, but there were certain differences in different nutrient types (figures 3A-3C). The content of available nitrogen in each treatment after the first irrigation was maintained in the range of 528.1 to 678.5 mg·kg^{-1}, which was more than 11 times the control (figure 3A). The nitrogen content of urea was the highest of the three fertilizers, but the level of nitrogen content after the first irrigation was extremely significant difference ($p < 0.01$) lower than that of the compound fertilizer. It indicated that the available nitrogen maintenance ability of the compound fertilizer was significant ($p < 0.05$) better than the single nitrogen element. After the second irrigation, the content of available nitrogen in each treatment was maintained in the range of 180.3 to 298.7 mg·kg^{-1}, which was more than 4 times the Fck treatment, and the rate of available nitrogen decrease was different in each fertilizer type. The available nitrogen in NPK SRFs decreased by about 74.0%, Urea was down 57.0%, and the NPK CFs descent rate was the slowest, at about 53.6%. Whereas Fck treatment only dropped by about 8.5 % and the difference between treatments was extremely significant difference ($p < 0.01$). The results showed that the fertilizer applied after the second watering still had a higher fertilizer efficiency and among them NPK SRFs had the strongest release capacity. After the third irrigation, the rate of available nitrogen declines between different treatments all slowed down significantly. The content of available nitrogen in each treatment was maintained in the range of 130 to 150 mg·kg^{-1}, more than 3 times the Fck treatment, and the difference between the treatments was not significant but still extremely significantly ($p < 0.01$) higher than the control. It showed that after three events of irrigation, the levels of available nitrogen and nutrients in each treatment were consistent but still maintained a higher level of fertility than the control group. After the fourth irrigation, the content of available nitrogen in each treatment was maintained in the range of 53.9 to 59.4 mg·kg^{-1}. Although it was significantly higher than the Fck treatment (34.8 mg·kg^{-1}), the overall fertility level was the same as that of the foundation fertility level of 48.8 mg·kg^{-1} which was in a lacking state and could not meet the needs of the trees for available nitrogen in the fast-growth period and needed to be supplemented (figure 3A).

In terms of available phosphorus, except urea in itself contains only trace amounts of the phosphate element and is at a low level, the other two treatments showed a decline rule roughly consistent with that of available nitrogen after the first three events of irrigation, and the difference appeared after the fourth irrigation (figure 3B). After the first irrigation, the nutrient level of the two composite fertilizers was maintained at 398.2 to 419.3·mg·kg^{-1} and the difference was not significant. After the second irrigation, the nutrient level rapidly decreased to 117.1 to 232.9 mg·kg^{-1}, and the content between NPK SRFs was extremely significant difference ($p < 0.01$) lower than NPK CFs. After the third irrigation, the available phosphate in the two combined fertilizers was maintained in the range of 110.4 to 150.2 mg·kg^{-1} and the difference between them was not significant ($p < 0.05$). After the fourth irrigation, the available phosphate fertilizer in the two combined fertilizers still maintained a higher range of 88.3 to 139.7 mg·kg^{-1}, in which NPK SRFs contained a higher available phosphate fertilizer that extremely significantly higher than that of NPK CFs ($p < 0.01$). On the one hand, this indicated that the mobility of available phosphorus in sandy soil was weak, and four applications of irrigation can still maintain abundant available phosphorus. On the other hand, it revealed that NPK SRFs had a better sustained release effect on the phosphate fertilizer.

In the case of available potassium, the difference observed between the previous three irrigation events and the potassium fertilizer was the same as that of available phosphorus, and the difference also increased after the fourth irrigation (figure 3C). After the first irrigation, the two types of compound fertilizer were maintained at the level of 905.2 to 1084.9 mg·kg^{-1} and the difference was not significant. After the second irrigation, the available potassium fertilizer of the two types of compound fertilizer maintained an elevated level of 353.2 to 442.8 mg·kg^{-1}, but NPK SRFs treatment was significantly ($p < 0.05$) higher than NPK CFs. After the third irrigation, the rate of rapid decline of available potassium of the two combined fertilizers was significantly reduced and it still maintained at an elevated level of 276.2 to 360.9 mg kg^{-1} and there was no significant difference. However, this is different for rapidly available phosphorus. The content of rapidly available phosphorus in NPK CFs treatment was significantly lower than that of treatment NPK SRFs after the fourth irrigation, but remained in the range of medium and superior fertility between 74.9 and 144.5mg·kg^{-1}. This indicated that treatment NPK SRFs had a strong capability of available phosphorus retention.

Different fertilizer types led to the fluctuation of soil pH value but were less affected by irrigation times (figure 3D). Applying Urea Fs for topdressing increased the soil pH in the ancient course of the Yellow River, while treatment NPK SRFs decreased the soil pH (figure 3D). After the first irrigation, the application of Urea Fs significantly increased the soil pH value from 8.67 to 9.17, while the combined fertilizer reduced the soil pH value among which NPK SRFs had the most significant effect which decreased from 8.67 to 8.19. After the second irrigation, the effect of the different treatments on soil pH value showed a similar trend to the first irrigation, and there was no significant difference between their changes. After the third irrigation, treatment NPK CFs caused a significant decrease in soil pH value (7.81), but there was no significant difference compared to Fck (8.23), and no significant change in soil pH value between the other two treatments compared to the second irrigation. After the fourth irrigation, although all treatments showed a downward trend compared with the first irrigation, they were kept in the alkaline range from 7.9 to 8.7. There was a remarkable fact that urea caused the significant increase of soil pH in treatment Fs compared with Fck but there was no significant difference ($p < 0.05$) between Fs and Fck in the other two treatments.

- Improved method of field management based on the loss regularity of water and fertilizer in papaya and apricot

The appropriate soil nutrient and moisture content played a key role in obtaining a higher economic income. Nitrogen fertilizer could be significant in promoting plant growth and development, especially for nutrient growth [13,17,18], meanwhile, the elements phosphorus and potassium are important guarantees in promoting the balanced development of above and underground sections of plants [15]. A suitable degree of drought can significantly improve plant root development [1], quality and yield [17,19]. The balance of soil water and nutrient content determines the growth and development of plants and is especially important in the juvenile stage of plants [18]. Currently, the juvenile stage management of papaya and apricot kernels had been affected by traditional concepts. The heavy application of nitrogen fertilizer and light application of compound fertilizer together with frequent irrigation resulted in serious nutrient loss (table 1) and caused the low efficiency of fertilizer utilization. It was not conducive to plant growth and had destroyed the soil ecological balance. In view of these problems, based on the research results of the water and fertilizer loss rule of sand in this paper and the existing management methods of papaya and apricot kernel in the field, we have developed new management methods of papaya and apricot in its juvenile stage (table 1).

Table 1. Methods of traditional and improved management between *C. sinensis* and *P. armeniaca* × *sibirica*.

Plant types	Traditional management (CK)	Improved management (Treatments)
'Jinshuai'	TM-CS_{CK}: From May to September, irrigated every 20 days. In combination with watering, urea was applied once every month from May to August. In September, NPK CFs fertilizer was applied at 100 g each time. Fertilize five times in total. Appropriately delay watering if there was rainfall during the period.	IM-CS: From May to September, cover the soil surface with straw and irrigate it every 30 days. In combination with watering, NPK CFs fertilizer was applied at 150 g for each tree in the early May. In middle of July, applied urea 150g for each tree. Suspended fertilization in August and 150g of NPK CFs compound was applied to each plant from late September to early October. Fertilize three times in total. Appropriately delay watering if there was rainfall during the period.
'Zhongren No.1'	TM-PS_{CK}: From May to September, irrigated every 30 days. In combination with watering, urea was applied once every month from May to August. NPK CFs fertilizer was applied at 100 g each time in the middle of September. Fertilize four times in total. Appropriately delay watering if there was rainfall during the period.	IM-PS: From May to September, cover the soil surface with straw and irrigate it every 40 days. In combination with watering, NPK CFs fertilizer was applied at 150 g for each tree in the early May. In middle of July, applied urea 150g for each tree. Suspended fertilization in August and 150g of NPK CFs compound was applied to each plant from late September to early October. Fertilize three times in total. Appropriately delay watering if there was rainfall during the period.

When exploring the regularity of water and fertilizer loss in the ancient course of the Yellow River, we reduced the irrigation frequency 1 to 2 times and increased the fertilizer amount by 50 g each time. It was proven by two consecutive years of measurement that the length, ground diameter and crown width of papaya increased by 86.1%, 42.2% and 50.0% respectively compared with the control and kernel-apricot increased by 37.5%, 54.5% and 72.7% respectively. The growth rates were extremely significantly ($p < 0.01$) higher than the traditional cultivation method (table 2). Our improved water and fertilizer management reduced the erosion of fertilizer by irrigation, reduced labor intensity and economic input (labor use decreases by 1 or 2 person-times, 80 RMB$\cdot d^{-1}$ per person, average price of fertilizer is 1,500 RMB$\cdot t^{-1}$) simultaneously. Therefore, their comprehensive benefits were better than traditional method.

Table 2. Effect of improved management for biomass between *C. sinensis* and *P. armeniaca* × *sibirica*.

Plant types	Treatments	Length of shoots / cm	Range of variation / %	Ground diameter / cm	Range of variation / %	Crown breadth / m	Range of variation / %
'Jinshuai'	IM-CS	80.2±5.3A	↑86.1±11.9	6.4±1.9A	↑42.2±5.5	1.8±0.2A	↑50.0±6.3
	TM-CS_{CK}	43.1±2.5B		4.5±0.7B		1.2±0.5B	
'Zhongren No. 1'	IM-PS	63.1±3.2A	↑37.5±3.8	8.5±0.9A	↑54.5±4.7	3.8±0.5A	↑72.7±9.1
	TM-PS_{CK}	45.9±2.1B		5.5±0.5B		2.2±0.5B	

4. Discussions

In the long-term of evolution, plants have formed the ability to adjust their behavior systematically according to environmental changes as well as, initiating different physiological and biochemical stress responses to guarantee its growth and development [20]. Soil moisture, nutrients [21] and the pH value are direct impact factors on plant growth and development of the most important series of soil environmental factors. Suitable water and fertilizer conditions can not only satisfy the growth and development of plants [18] and obtain higher economic benefits [22] but could also reduce the load on the ecological environment. Especially for sandy soil, a reasonable decrease in irrigation times can maintain the same crop yield meanwhile reducing water leakage and nutrient leaching significantly, Its economic benefit is optimal.

Crop straws, such as corn straws and wheat straws, show the rejection of water due to their smooth surface or attached lipids. However, with the accelerated process of biological fermentation, hydrophobic substances are degraded and the ability of absorbing water is gradually improved, making it an ideal natural water-retaining material. Wu *et al.* 's study found that after 6 months of being buried (0.5 m underground), the soil water retention capacity of straw could be significantly improved [23]. The difference between this study and Wu *et al.* was that we covered the top of the plants' roots with corn stalks. Because the sand soil surface exposure can be more than 45°C after in summer, but the optimum temperature for an apricot root system to develop is 15°C to 25°C [24]. Maize straw can isolate heat and prevent rapid evaporation of underground water, simultaneously, the heat of biological fermentation can be used in the morning and evening to form a higher temperature difference with the atmosphere, which promotes condensation of moisture in the air. As a result, the soil moisture content remained around 10.0% after 15 days of irrigation, and the soil moisture content increased at a 5 cm depth after 20 days due to the dew gathered, 30 days after the irrigation, the soil moisture content remained around 8%. Although the water retention performance of corn stalk is weaker than that of a water retention agent such as sewage sludge, biochar etc. [23], but as microbial activity accelerates, its advantages of water retention and low cost will expand. At the same time, the application of maize straw significantly improves the physical structure and exhibits characteristics of increasing the soil capacity of sandy soil [25] which is worth promoting and applying in production. Non-timber forestry such as papaya and apricot have a strong main root system. During the period of 1 year after transplantation, the above-ground section developed slowly, and the main roots of the underground section grew rapidly and vertically, reaching a depth more than 2 m, but its lateral roots were mainly distributed in the 20 to 30 cm tillage layer, which was the main layer that absorbed water and fertilizer. Crops such as cucumber [7], potato [8,21], etc. growing on sand are usually inferior in developing main roots, and their roots are relatively shallow in the soil concentration distribution so that they must rely on external water and nutrients to meet their own needs and complete the life cycle. But papaya and apricots can use strong root absorption and natural rainfall to complete their own growth and development as soon as they surpass their juvenile stage. A high biomass accumulation can be achieved without the investment on constructing a large integrated water and fertilizer infrastructure. It has high practical value for the ancient course of the Yellow River, which has poor infrastructure and a high groundwater level.

5. Conclusions

In the ancient course of the Yellow River, the treatment including maize straw with a thickness of 5 cm and covered with soil thickness 8 to 10 cm on the soil surface layer can not only efficiently extend the effective water retention time of soil at a depth of 5 cm and 20 cm but also maintain a constant soil temperature. The compound fertilizer applied in this area had a higher nutrient retention capacity than urea. The application of available nitrogenous fertilizer, phosphorus or potassium need be supplemented after 3 irrigation events. It should be noted that the nitrogen fertilizer should be applied as evenly as possible to avoid the high local soil pH cause the decay of the root system. For papaya, kernel-apricot and other fruit trees it is suggested to use a crop straw cover on the ground to increase soil effective water content and duration. According to the regularity of the tree's needs, using a compound fertilizer to replace urea in the early and late growing season of their juvenile stage, the economic benefits can be maximized.

Acknowledgments

This work was supported by "the Fundamental Research Funds for the Central Non-profit Research Institution of CAF (CAFYBB2017MB027)".

References

[1] Xu M S, Zhao H, Zhou X X, Wuyun T N, Li F D and Zhu G P 2016 Responses of photosynthetic physiology and biomass accumulation of sweet kernel apricot (*Prunus armeniaca × sibirica*) seedling to soil drought stress in the ancient course of the middle Yellow River *Taiwan J. Forest. Sci.* **31** 271-84

[2] Tian Y, Li Y G and Jiang H 2010 On the regional ecological planning of the ancient yellow river course in Shangqiu City *J. Landsc. Res.* **2** 35-9

[3] Chen J P, Lan Z P, Yang S J, Zhang X P, Sun S W and Wu P J 2015 Water use efficiency and plantation forest under different irrigation methods in old course of the Yellow River area *Acta Ecol. Sin.* **35** 252-256

[4] Zhu X Y, Hu Y C and Lu J 2014 Evaluation on soil biological characteristics and quality of the wetlands in ancient bed of the Yellow River of east He'nan province *Res. Soil Water Conserv.* **21** 27-32

[5] Liu J S, Wang R Q, Dai J L, Zhang Y L and Wang Q 2008 Soil environmental background concentrations in old course of the Yellow River in Shandong province *Environ. Sci.* **29** 1699-705

[6] Yu X N, Huang Y M, Li E G, Li X Y and Guo W H 2018 Effects of rainfall and vegetation to soil water input and output processes in the Mu Us sandy land, northwest China *Catena* **161** 96-103

[7] Yang J H, Yan F, Sun L L, Zhou X T, Hu X H and Zou Z R 2014 Effects of different water-fertilizer combinations on yield and quality of greenhouse cucumber cultivated in sand *J. Northeast Agricul. Forest. Univ.* **42** 111-8

[8] Feng Z W, Kang Y H, Wan S Q and Liu S P 2017 Effect of drip fertigation levels on potato growth and the water and fertilizer efficiency on sandy soil in Inner Mongolia *Agricul. Res. Arid Area* **35** 242-9

[9] Machekposhti M F, Pla C and Valdes-Abellan J 2018 Improving hydraulic properties of sandy soil with lime powder addition *Eur. Geosci. Union* **20** 12195

[10] Li H, Dong Z, Li Y Q and Cui G S 2007 Research on effect of wind erosion control after planting grasses in wind-sand region of the Yellow River old riverway *Chinese J. Grassl.* **29** 61-5

[11] Hou Q C, Mei S W and Tang L L 2009 Analysis of soil fertility varability in pear orchards in the old course of the Yellow River area *Jiangsu Agricul. Sci.* **12** 329-30

[12] Liu R, Kang Y, Zhang C, Pei L, Wan S Q, Jiang S F, Liu S P, Ren Z Y and Yang Y 2014 Chemical fertilizer pollution control using drip fertigation for conservation of water quality

in Danjiangkou Reservoir *Nutr. Cycl. Agroeco.* **98** 295-307

[13] Li S, Tian Y H, Wu K, Ye Y F, Yu J P, Zhang J Q, Liu Q, Hu M Y, Li H, Tong Y P, Harberd N P and Fu X D 2018 Modulating plant growth–metabolism coordination for sustainable agriculture *Nature* **560** 595-600

[14] Xu M S, Zhu G P, Fu G Q, Zhou X X, Zhao H, Wuyun T N and Li F D 2017 Effects of N, P and K deficiency on growth and nutrient uptake of *Prunus armeniaca* seedlings *J. Northeast Agr. Forest. Univ.* **45** 81-90

[15] Xu M S 2016 The physiological role of N, P, K in sweet kernel apricot leaf during transition from juvenile stage to adult (Beijing: Chinese Academy of Foresty)

[16] Tang Q Y and Zhang C X 2013 Data Processing System (DPS) software with experimental design, statistical analysis and data mining developed for use in entomological research *Insect Sci.* **20** 254-60

[17] Zhou Z J, Plauborg F, Liu F L, Kristian K and Andersen M N 2018 Yield and crop growth of table potato affected by different split-N fertigation regimes in sandy soil *Eur. J. Agricul.* **92** 41-50

[18] Yin C, Palmroth S, Pang X, Tang B, Liu Q and Oren R 2018 Differential responses of *Picea asperata* and *Betula albosinensis* to nitrogen supply imposed by water availability *Tree Physiol.* **38** 1694-1705

[19] Cantore V, Wassar F, Yamaç S S, Sellami M H, Albrizio R, Stellacci A M and Todorovic M 2014 Yield and water use efficiency of early potato grown under different irrigation regimes *Int. J. Plant Prod.* **8** 409-28

[20] Xu W Z, Zhu L X and Zhao G 2006 Application of plant ecological adaptability in hydroponics *Subtrop. Plant Sci.* **35** 28-32

[21] Badra M A, El-Tohamy W A and Zaghloul A M 2012 Yield and water use efficiency of potato grown under different irrigation and nitrogen levels in an arid region *Agricul. Water Manag.* **110** 9-15

[22] Zhang F, Yue G, Jiao W and Wenhui H U 2017 Effects of water and fertilizer supply on growth, water and nutrient use efficiencies of potato in sandy soil of Yulin area *Chinese Soc. Agricul. Machinery* **48** 270-8

[23] Wu C F, Wen Q, Han Y L, Wang Q, Jiang Y and Li P 2017 Effect of water retaining agent and strawapplications on water and fertilizer status and wheat yield in sandy Chao soil *Chinese J. Soil Sci.* **48** 692-700

[24] Li F D, Wuyun T N and Zhu G P 2018 *Cultivation Operative Technology of Kernel-Apricot* (Beijing: Chinese Forest Press)

[25] Głąb T, Żabiński A, Sadowska U, Gondek K, Kopeć M, Mierzwa–Hersztek M and Tabor S 2018 Effects of co-composted maize, sewage sludge, and biochar mixtures on hydrological and physical qualities of sandy soil *Geoderma* **31** 27-35

Production of vinegar from organic broken rice noodles

J Jaikang[1], V Rungsardthong[1], B Thumthanaruk[1], C Puttanlek[2], D Uttapap[3], V Plathanaporn[4] and S Vatanyoopaisarn[1, 5]

[1]Deparment of Agro- Industrial, Food, and Environmental Technology, Faculty of Applied Science, Food and Agro-Research Center, King Mongkut' s University of Technology North Bangkok, Bangkok, Thailand
[2]Deparment of Biotechnology, Faculty of Engineering and Industrial Technology, Silapakorn University, Nakhon Pathom, Thailand
[3]Division of Biochemical Technology, School of Bioresources and Technology, King Mongkut's University of Technology Thonburi, Bangkok, Thailand
[4]Foodle Noodle Co., Ltd, 98/38 Moo 2, Bueng Nam Rak, Thanyaburi, Pathum Thani 12110, Thailand

E-mail: savitri.v@sci.kmutnb.ac.th

Abstract. Organic broken rice noodles (OBRN) are generally considered as waste material from noodle processing factories and are of low value. The purpose of this research study was to increase the value of OBRN by converting it to fermented vinegar. In the first step, the hydrolysis methods of OBRN were studied. The conversion of starch into reducing sugar using acid hydrolysis, mold inoculation, and rice cake starter (loog-pang) were compared. For acid hydrolysis, 0.6 M sulfuric or hydrochloric acid at 80, 100, and 121 °C was used. For biological hydrolysis, steamed OBRN was soaked in water at a ratio of 1:2 for 1 h, then mixed with 0-0.4 % (w/w) koji or ricecake starter and left to ferment. The latter methods yielded the highest glucose of 167.66 and 178.94 g/L, respectively, when 0.2 % of starters were inoculated for 3 days. In a second step, OBRN wine was made from a mixture of hydrolyzed OBRN liquid and 20 °Brix of pineapple juice at a 1:1 ratio and subsequently fermented for 7 days. The alcohol produced from OBRN showed the highest concentration of 10.05 % (v/v) at 7 days of incubation. Lastly, wine was prepared at various alcohol concentrations of 4-6 % (v/v) and at a pH of 5.5 for further vinegar fermentation using *Acetobacter pasteurianus* TISTR 102, by shaking at 150 rpm for 7 days. The result showed that the appropriate alcohol concentration for producing vinegar was 4 % (v/v) and the OBRN vinegar exhibited clear, light yellow color, and had total acidity content of 3.52 % (v/v) and a pH of 3.28 after 4 days of incubation.

1. Introduction

Vinegar is produced from raw materials containing sugar or starch by submerged culture. The production of vinegar typically involves two steps of fermentation. First, the sugars in the raw material are converted to alcohol using yeast, such as *Saccharomyces cerevisiae*, *Saccharomyces fragillis*, and *Kluyveromyces fragillis,* etc... The resulting alcohol is further oxidized to acetic acid by acetic acid bacteria such as *Acetobacter pasteurianus* and *Gluconobacter oxydans,* etc... in a second step [1]. Vinegar is used in food preparation including as a preservative for vegetables, fish, and meat, as well as in the preparation of mayonnaise, salad dressings, mustard, and other food condiments [2-3]. In addition, vinegar has been proven to have medicinal advantages by virtue of its physiological effects,

Content from this work may be used under the terms of the Creative Commons Attribution 3.0 licence. Any further distribution of this work must maintain attribution to the author(s) and the title of the work, journal citation and DOI.
Published under licence by IOP Publishing Ltd

such as regulating blood glucose, blood pressure, aiding digestion, stimulating the appetite, promoting calcium absorption, and promoting recovery from exhaustion [4-6]. According to Thai Industrial Standards, vinegar is defined as a sour solution that must contain not less than 4 g of acetic acid per 100 cm^3 and be produced from alcohol and successive acetic fermentation from sugary and starchy substrates.

In general, raw materials used for the production of fermented vinegar are fruits such as grapes, apples, plums, coconuts, and pineapples, etc... Carbohydrate ingredients including rice malt starch and various other agricultural materials are also used [7]. The conversion of starch into sugar can also be carried out by hydrolysis of food processing by-products such as broken noodles, with acid or mold or using rice-cake starter (loog-pang).

Thailand is one of the world leaders for food-product export; consequently, a large quantity of by-products from food processing plants, such as fruit and vegetables, rice, fat, and oil, are available [8]. These by-products cause a major disposal problem for the industry. Rice noodles are one of the main export products from Thailand with a value of about 555 million Thai baht in 2018 [9]. Organic food products including organic rice noodles are gaining popularity due to consumers' health consciousness. The broken noodles of such products are considered as a by-product which results from the rolling and cutting process. Most of these broken rice noodles are often used as animal feed, fertilizers, or treated as waste, all of which have low value [10]. Therefore, if the waste or by-product from the rice processing plant can be used as a raw material for the production of other value-added products, including fermented vinegar, it would generate a higher value product.

Rice-cake starter is a traditional starter used to produce alcoholic beverages or indigenous food products in many Asian countries. These traditional starters have various names, such as marcha or murcha in India, ragi in Indonesia, bubod in the Philippines, Chinese yeast in Taiwan, nuruk in Korea [11], banh men in Vietnam, koji in Japan, ragi tapai in Malaysia, and loog-pang in Thailand [12]. Microorganisms in loog-pang include mold, bacteria, and yeast. Most studies found *Amylomyces rouxii* and *Saccharomycopsis fibuligera* are common mold and yeast, respectively, in loog-pang [13]. This research study aims at employing the by-product from the organic rice noodle processing as a substrate for vinegar fermentation. The suitable methods of hydrolysis and alcohol concentration were investigated, as well as some of the properties of the final product.

2. Material and methods

2.1. Materials

Organic broken rice noodles were obtained from Foodle Noodle Co., Ltd., Thailand. The rice- cake starter was purchased from a market in Nan Province, Thailand. The mold, *Amylomyces rouxii* TISTR 3182 used for the koji starter [14], and *Acetobacter pasteurianus* TISTR 102, used for acetic acid fermentation, were purchased from the Culture Collection Center, Thailand Institute of Scientific and Technological Research, Thailand.

2.2. Starch hydrolysis

Acid hydrolysis was performed using HCl or H_2SO_4 at 0.6 M. OBRN in acid solutions was submitted to a thermal treatment at different temperatures (80, 100, and 121 °C) for 15, 30, and 45 min. The samples were prepared in 250 ml flasks, using 10 g of OBRN and 90 ml of the acid solutions. The hydrolysis of OBRN using rice-cake starter and koji was started by soaking 100 g of OBRN in water at a ratio of 1:2 for 1 h, and steaming for 15 minutes, followed by mixing with various ratios of rice-cake starter and koji at 0, 0.1, 0.2, 0.3, and 0.4 % (w/w), and finally incubated at 30 °C for 3 days. Reducing sugar in the fermentation supernatant after centrifugation was determined by the dinitrosalicylic acid method (DNS method) [15].

IOP Conf. Series: Earth and Environmental Science **346** (2019) 012045 doi:10.1088/1755-1315/346/1/012045

2.3. Alcohol fermentation

Alcohol fermentation was carried out by adding 20 °Brix of pineapple juice into the liquid obtained from starch hydrolysis, using rice-cake starter at a ratio of 1:1 and incubated at 30 °C for 7 days. The samples were collected every day for physical and chemical analysis as stated in 2.5.

2.4. Acetic acid fermentation

The alcohol-containing liquid was adjusted to alcohol concentrations of 4, 5 and 6 % (v/v), with pH of 5.5, and further inoculated with 5 % *of Acetobacter pasteurianus* TISTR 102, and then shaking was carried out at 150 rpm, at 30 °C for 7 days. The samples were collected every day and subjected to analysis, as follows.

2.5. Physical and chemical analysis

Reducing sugar in the fermentation supernatant after centrifugation was determined by the DNS [15]. The pH was measured according to AOAC 2000 [16] by using a pH meter. Total acidity content in the vinegar was analyzed by titration with 0.1 M NaOH until a pH end-point of 8.4, and the acidity was calculated as acetic acid in vinegar according to AOAC 2000 [16]. The alcohol content was determined by an alcohol vapor sensor from Vernier Software & Technology, USA. The color of the vinegar was measured as L*, a* and b* using a Hunter lab.

3. Result and discussions

3.1. Starch hydrolysis between chemical and biological treatment

Figure 1. shows the reducing sugar concentration obtained from OBRN hydrolysis using HCl and H_2SO_4. The optimum conditions were 0.6 M HCl, with a reaction time of 30 min at 121 °C, which resulted in 118.80 g/L (92.64 % of theoretical yield) of reducing sugar. Acid hydrolysis was applied to digest various organic polysaccharides, in order to release the sugar molecules. Woiciechowki *et al.* (2002) [17] reported that when using 1 % hydrochloric acid to digest cassava bagasse, it was almost completely converted to glucose under optimum conditions (100 °C, for 30 min). Such conditions represented 94.5 % of recovery of reducing sugar from the starch available in cassava bagasse. Hot acid pre-treatment was also successful in Napier grass straw, where 0.7 % HCl was applied at 105 °C for 1 h [18].

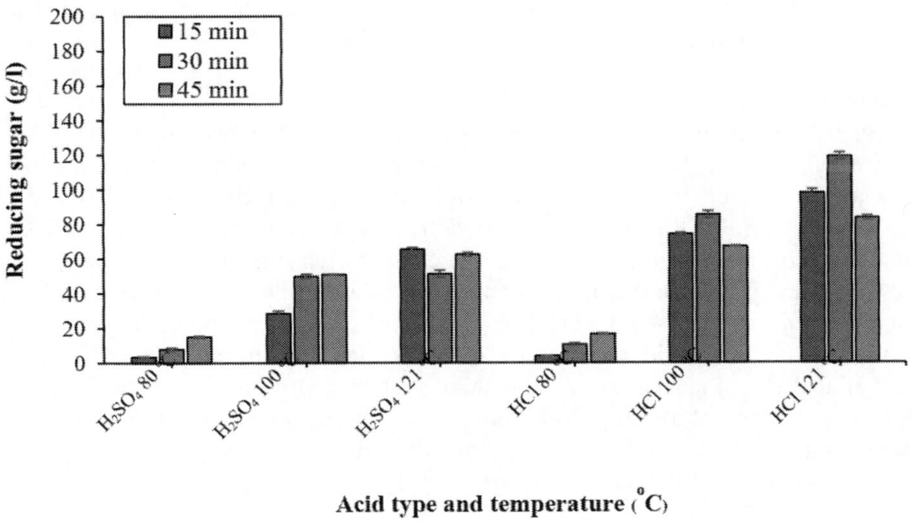

Figure 1. Reducing sugar obtained from the hydrolysis of OBRN by sulfuric acid and hydrochloric acid at 80, 100 and 121 °C for 15, 30 and 45 minutes.

Reducing sugar obtained from the hydrolysis of OBRN by rice-cake starter and koji at different percentages of inoculum are shown in Figure 2. The optimum concentration of rice-cake starter or koji was 0.2 % (w/w) with the highest amount of reducing sugar obtained at 178.95 ±0.05 g/L (45.71% of theoretical yield) and 167.67 ±0.05 g/L (42.83 % of theoretical yield), respectively, at 3 days of cultivation. Molds in rice-cake starter were usually *Rhizopus* sp., *Mucor* sp., *Amylomyces* sp. and *Aspergillus* sp., and mold of koji can produce amylases and these enzymes hydrolyze starch into sugars [19-21].

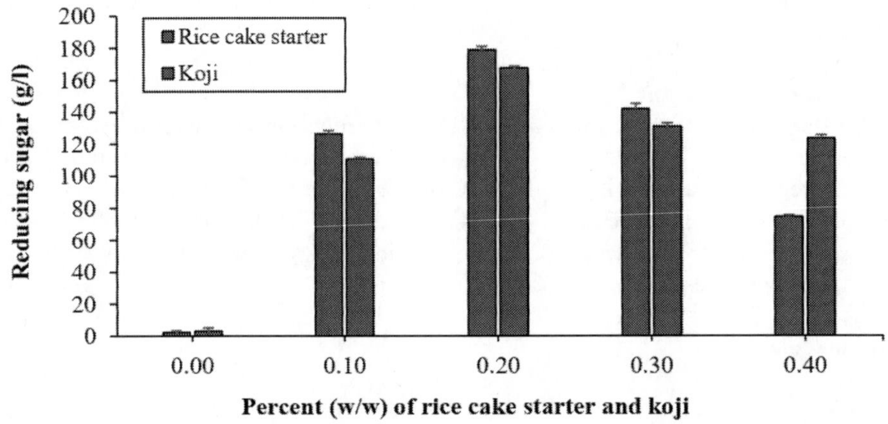

Figure 2. Reducing sugar obtained from the hydrolysis of OBRN using rice cake starter or at various concentration.

The study showed that the process yields of recovered reducing sugar from OBRN hydrolysis by acid, rice-cake starter, and koji were 92.64 %, 45.71 %, and 42.83 %, respectively. Although starch hydrolysis by acid was the most efficient method concerning the yield based on the reducing sugar recovered from the OBRN, high process temperature during hydrolysis caused toxic compounds which might produce derivatives of furfur (2-furaldehyde and 5-hydroxymethyl-2-furaldehyde) [17]. Therefore, starch hydrolysis using rice cake starter was used for further study as it would be safe and easy to perform by the noodle processing plant.

3.2. Fermentation to alcohol

At the beginning of fermentation, the sugar concentration of samples was adjusted to 20 °Brix. The results of total soluble solid, alcohol content, pH, total acidity, and reducing sugar during alcoholic fermentation are presented in Table 1. The sugar level gradually declined inversely to the incrementation of alcohol. At the end of fermentation, 7 days, the samples contained 10.05±0.05 % (v/v) alcohol. It has been shown that yeasts in rice cake starter such as *Saccharomyces cerevisiae* mainly consumed sugar in the alcohol fermentation [22]. The titratable acidity of samples during the fermentation increased slightly from 0.32 to 1.08 %, and pH changed slightly. This increase in acidity provided optimal growth conditions to initiate acetification. The alteration in pH attributed to the accumulation of acetic acid and other organic acids such as propionic and tartaric acid, which are important for the development of flavor and aroma [23]. Adebayo-Oyetoro *et al.* (2017) [24] studied the production of vinegar from mango and found that the pH increased slightly from 3.98 to 4.02 and the total acidity increased from 0.02 to 4.02 % after 30 days of fermentation.

Table 1. Some chemical properties of organic broken rice noodles fermented by rice cake starter for 7 days.

Time (Days)	Total soluble solid (°Brix)	Alcohol (%)	pH	Total acidity (%)
0	21.33±0.15[b]	4.43±0.01[e]	3.54±0.02[cd]	0.32±0.03[f]
1	21.63±0.12[a]	3.84±0.01[g]	3.54±0.02[cd]	0.62±0.03[e]
2	20.93±0.12[c]	3.66±0.01[h]	3.52±0.02[de]	0.88±0.03[d]
3	18.27±0.12[d]	4.27±0.01[f]	3.51±0.01[e]	0.88±0.03[d]
4	15.13±0.12[e]	6.08±0.01[d]	3.55±0.01[c]	0.94±0.03[cd]
5	10.17±0.06[f]	7.50±0.01[c]	3.60±0.01[b]	1.00±0.03[bc]
6	9.17±0.06[g]	8.04±0.00[b]	3.59±0.02[b]	1.04±0.03[ab]
7	8.11±0.01[h]	10.05±0.05[a]	3.66±0.02[a]	1.08±0.03[a]

Data represent the average of triplicate analysis, [a-h] different letters in the same column are significantly different ($p<0.05$)

3.3. Acetic acid fermentation

The effects of different alcohol concentrations on acetic acid production from OBRN through vinegar fermentation with *Acetobacter pasteurianus* TISTR 102 are shown in Figure 3. The total titratable acidity calculated as acetic acid in all samples increased with fermentation time. After 2-4 days of fermentation, the acidity remained constant. The higher the alcohol concentration, the lower the acid concentration obtained. The initial alcohol content at 4 % (v/v) generated the highest percentage of acidity at 3.52± 0.02 %, compared to those of 5 and 6 %, which resulted in 2.92±0.03 and 2.02±0.03 % (v/v), respectively. Ory *et al.* (2002) [25] studied the influence of ethanol concentration in vinegar fermentation and found that the optimal concentration for vinegar fermentation was 3.55-4.7 % (v/v). It has been proven that when ethanol concentration is over 4 % (v/v), it is very unlikely that any growth of bacteria will be seen [26]. In addition, the lag phase increases proportionally with the ethanol concentration. Therefore, the concentration of ethanol was adjusted to 4 % (v/v) for further study.

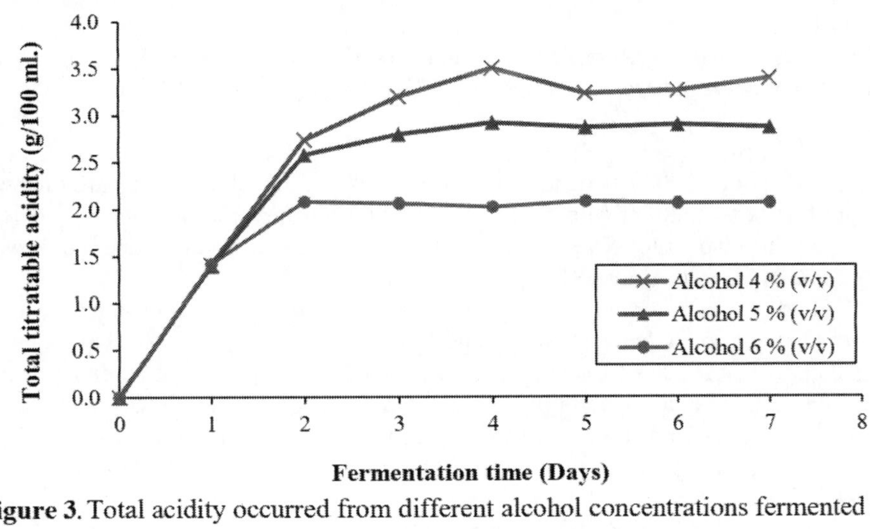

Figure 3. Total acidity occurred from different alcohol concentrations fermented from organic broken rice noodles.

3.4. Physical and chemical analysis

Table 2 shows some properties of the final product of fermented vinegar. The vinegar produced had a total soluble solid of 7.20±0.02 °Brix, pH of 3.28±0.02 and 3.52±0.02 % total acidity (acetic acid). This fell below the 4 % acetic acid level allowed in the vinegar according to Thai industrial standards, as well as AOAC [16]. This could be attributed to an interruption of the air supply during the oxidation process, as acetic acid bacteria are reported to be very susceptible to air interruption during oxidative processes [27]. Therefore, in order to increase the acetic acid concentration, it is suggested that the oxygen concentration should be increased by aeration during acetous fermentation in the OBRN vinegar. The alcohol content continued to decrease with time from 4.00 % to about 0.24 % at day 4, and the vinegar was clear and has a light-yellow color (39.89±0.83 of *L).

Table 2. Some properties of the vinegar obtained from the fermentation of OBRN.

Parameters	
Total soluble solids (°Brix)	7.20±0.02
Titratable acidity (%)	3.52±0.02
pH	3.28±0.02
Alcohol content (%V/V)	0.28±0.04
L*	39.89±0.83
a*	1.65±0.24
b*	0.28±0.06
h	9.72±0.98

*Values are means ±SD of triplicate determinations.

4. Conclusion

This research study provides evidence showing that the use of organic broken rice noodles in alcohol fermentation by rice-cake starter (loog-pang) and acetic acid, using *Acetobacter pasteurianus* TISTR 102, is possible. The vinegar produced from organic broken rice noodles also has acceptable properties and quality. It is also implied that broken rice noodles which are conventionally regarded as waste in noodle processing can be converted into a value-added commodity, and also facilitate environmental safety.

Acknowledgment

We would like to thank Thailand Research Fund and Foodle Noodle Co., Ltd. for their kind support under the Research and Researchers for Industries (RRi) program (MSD59I0101).

References

[1] Gullo M and Giudici P 2008 Acetic acid bacteria in traditional Balsamic vinegar: phenotypic traits relevant for starter cultures selection. *Int. J. Food Microbiol* **125** 46-53.
[2] Turker I 1963 Sirke Teknolojisi ve Teknikte Laktik Asit Fermantasyonları. *Schoolbook of Faculty of Agriculture* Ankara Univ 181.
[3] Tan S C 2005 Vinegar fermentation [*Master of Science thesis*]. Louisiana State University *Department of Food Science* Baton Rouge 101.
[4] Fushimi T and Sato Y 2005 Effect of acetic acid feeding on the circadian changes in glycogen and metabolites of glucose and lipid in liver and skeletal muscle of rats. *Br. J. Nutr* **94** 714-9.
[5] Kondo S, Tayama K, Tsukamoto Y, Ikeda K and Yamori Y 2001 Antihypertensive effects of acetic acid and vinegar on spontaneously hypertensive rats. *Biosci. Biotechnol. Biochem* **65** 2690-4.
[6] Xu Q, Tao W and Ao Z 2007 Antioxidant activity of vinegar melanoidins. *Food Chem* **102** 841-

9.
[7] Adams M R 1985 Microbiology of fermented foods. *New York Institute of Food Technologists* 1-45.
[8] Korkerd S, Wanlapa S, Puttanlek C, Uttapap D and Rungsardthong V 2016 Expansion and functional properties of extruded snacks enriched with nutrition sources from food processing by-products. *J. Food Sci. Technol* **53** (1) 561–570.
[9] Kasikorn Research Center 2011 Thai rice innovations. from http://www.kasikornresearch.com.
[10] Moonsarn Y and Moonsarn P 2011 Improvement of noodle waste by yeast for use as animal feed. *J. Sci.* **2** 1905-3193.
[11] Tsuyoshi N, Fudou R, Yamanaka S, Kozaki M, Tamang N, Thapa S and Tamang J P 2005 Identification of yeast strains isolated from marcha in Sikkim: a microbial starter for amylolytic fermentation. *Int. J. Food Microbiol.* **99** 135-146.
[12] Limtong S, Sintara S, Suwanarit P and Lotong N 2005 Species diversity of molds in Thai traditional fermentation starters (Loog-Pang). *KU J. Soc Sci.* **39** 511-518.
[13] Limtong S, Sintara S, Suwanarit P and Lotong N 2002 Yeast diversity in Thai traditional alcoholic starter. *KU J. Soc Sci.* **36** 149-158.
[14] Chou C C and Rwan J H 1995 Mycelial propagation and enzyme production in koji prepared with *Aspergillus oryzae* on various rice extrudates and steamed rice. *J. Biosci. Bioeng.* **79** 509-512.
[15] Miller G L 1959 Use of dinitrosalicylic acid reagent for determination of reducing sugars. *Anal Chem* **31** 426-429.
[16] AOAC International 2000 Official Method of Analysis, Association of Official Agricultural Chemist. 17th Ed. AOAC International. Gaithersburg: MD.
[17] Woiciechowski A L, Nitsche S, Pandey A and Saccol C R 2002 Acid and enzymatic hydrolysis to recover reducing sugars from cassava bagasse: an economic study. *Braz. arch. biol. technol* **45** 393-400.
[18] Amnuaycheewa P, Rodiahwati W, Sanvarinda P, Cheenkachorn K, Tawai A and Sriariyanun M 2017 Effect of organic acid pretreatment on napier grass (*Pennisetum purpureum*) straw biomass conversion. *KMUTNB Int. J. Appl Sci. Technol* **10** 107-117.
[19] Dung N T P, Rombouts F M and Nout M J R 2006 Functionality of selected strains of moulds and yeasts from vietnamese rice wine starters. *Int. J. Food Microbiol* **23** 331-340.
[20] Hernández-Orte P, Ibarz M J, Cacho J and Ferreira V 2006 Addition of amino acids to grape juice of the Merlot variety: Effect on amino acid uptake and aroma generation during alcoholic fermentation. *Food Chem* **98** 300-10.
[21] Negi S and Benerjee R 2006 Optimization of amylase and protease production from Aspergillus awamori in single bioreactor through EVOP factorial design technique. *Food Sci. Biotechnol* **44** 257-61.
[22] Taillandier P, Portugal F R, Fuster A and Strehaiano P 2007 Effect of ammonium concentration on alcoholic fermentation kinetics by wine yeasts for high sugar content. *Food Microb* **24** 95-100.
[23] Byarugaba-Bazirake G, Byarugaba W, Tumuslime M and Kimono D A 2014 The technology of producing banana wine vinegar from starch of banana peels. *Afr. J. Food Sci. Technol* **5** 1-5.
[24] Oyetoro A O, Denubi E, Ogundipe O O, Bankole B O and Olalekan Adeyeye S A 2017 Production and quality evaluation of vinegar from mango. *Cogent Food Agric.* **3** 1278193.
[25] De Ory I, Romero L E and Cantero D 2002 Optimum starting-up protocol of a pilot plant scale acetifier for vinegar production. *J. Food Eng.* **52** 31-7.
[26] Drysdale G S and Fleet G H 1988 Acetic acid bacteria in winemaking: a review. *Am. J. Enol. Vitic.* **39** (2) 143-154.
[27] Gonz´alez A, Hierro N, Poblet M, Mas A and Guillamon J M 2005 Application of molecular methods to demonstrate species and strain evolution of acetic acid bacteria population during wine production. *Int. J. Food Microbiol.* **102** 295-304.

IOP Conf. Series: Earth and Environmental Science **346** (2019) 012046 doi:10.1088/1755-1315/346/1/012046

Effect of freshness and salt on quality of white shrimp

N Kundacha[1], V Rungsardthong[1], W Kiatponglarp[2], S Vatanyoopaisarn[1], C Puttanlek[3], D Uttapap[4], C Hirunarj[5] and B Thumthanaruk[1, 6]

[1]Deparment of Agro - Industrial, Food, and Environmental Technology, Faculty of Applied Science, Food and Agro-Industry Research Center, King Mongkut's University of Technology North Bangkok, Bangkok, Thailand
[2]Synchrotron Light Research Institute, 111 Moo 6, University Avenue, Muang, Nakhonratchasima 30000, Thailand
[3]Deparment of Biotechnology, Faculty of Engineering and Industrial Technology, Silapakorn University, Nakhon Pathom, Thailand
[4]Division of Biochemical Technology, School of Bioresources and Technology, King Mongkut's University of Technology Thonburi, Bangkok, Thailand
[5]Marine Gold Products Ltd., 57/37 Moo 4, Ekkachai Road, Khok Kham, Mueang Samut Sakhon, Samut Sakhon 74000, Thailand

E-mail: benjawan.t@sci.kmutnb.ac.th

Abstract. The concentration of salt (NaCl) in white shrimp, which is one of the main export products of Thailand, is very important in quality control. In this research study, Na and Cl distribution in white shrimp was investigated using the X-ray fluorescence (XRF) technique to examine salt absorption and distribution within samples at two freshness levels, freshness levels 3 and 4. White shrimp of both freshness levels were soaked in a salt solution at 9 % (w/v) for 90 min and boiled for 110-130 sec, in accordance with work instructions from a factory. Additional samples were cooked after soaking, and still other samples of raw shrimp (not soaked, not cooked) were examined by the XRF technique and the results obtained were compared with quantitative salt analysis by the titration method. The concentration of Na and Cl can be matched to a color scale to represent the level of Na and Cl absorption. The results show that white shrimp with freshness level 3 absorbed a higher concentration of salt solution (2.49±0.11 %) than the samples with freshness level 4 (2.20±0.14 %). The results were similar to those resulting from the titration method, although the samples at freshness level 3 indicated higher salt concentration. However, white shrimp with freshness level 4 also indicated a higher TVB-N value than that of freshness level 3, but still meets the Thai standard for frozen shrimps and prawns.

1. Introduction

Frozen white shrimp is one of the main export products of Thailand to many countries, thanks to its quality and variety. White shrimp (*Litopenaeus vannamei*), or vannamei, is smaller than black tiger shrimp but is a strong breed and therefore popular for cultivation in many other countries, including the USA, Mexico, Colombia and Peru [1]. White shrimp has 8 segments in the body, including 1 segment for the head, 6 segments for the body and 1 segment for the tail.

The production of frozen white shrimp consists of several steps, including raw materials receiving, cleaning, sizing, trimming, soaking, freezing, and packing, as well as cooking in the case that the product is the cooked type [2]. One of the important steps in the production process, which increases productivity

Content from this work may be used under the terms of the Creative Commons Attribution 3.0 licence. Any further distribution of this work must maintain attribution to the author(s) and the title of the work, journal citation and DOI.
Published under licence by IOP Publishing Ltd

and improves product quality, is soaking in a salt solution. Generally, salt and phosphate compounds are used to increase water holding capacity and improve product quality after thawing and cooking of the shrimp [3].

Salt or sodium chloride (NaCl) contains sodium ions and chloride ions that are bonded with ionic bonds. Salt is very important to the food industry, as it is inexpensive and can be used for various purposes, including cooking and preservation. The purity of salt used in industry is 99.9 % and its moisture content does not exceed 0.15 % [4]. Salt helps animal muscles absorb more water due to the increased strength of the ionic bond in the muscle protein [5]. Lopkulkiaert *et al.* (2009) [6] studied the effects of sodium bicarbonate containing traces of citric acid in combination with sodium chloride on yield and freezing loss of white shrimp. Fresh shrimp soaked in sodium bicarbonate containing traces of citric acid at 4 g and sodium chloride at 3 g per 100 mL leads to an increase of the yield. The opposite result was found, in that the freezing loss decreased 6.83-10.28 % and 6.41-12.4 % for uncooked and cooked products, respectively. The study of Wachirasiri *et al.* (2016) [7] reported that shrimps treated with arginine at 1 g per 100 mL with NaCl 3 g per 100 mL had increased cooking yields from 75.02 to 97.59 g per 100 g of fresh shrimp. And Wachirasiri *et al.* (2016) [8] studied changes in processing yield and physical properties of frozen white shrimp *(Penaeus* vannamei) treated with lysine and sodium bicarbonate. It was found that combinations of lysine, $NaHCO_3$, and NaCl increased both the thawing yield and cooking yield of shrimp, comparable to the use of sodium tri-polyphosphate (STPP), by providing more negative charge resulting in increased electrostatic repulsion between the myofibrillar proteins. Salt concentration in frozen shrimp, both cooked and uncooked products, is very important since it affects the product quality and taste, as well as the heath concerns for salt concentration. The control and monitoring of salt concentration in shrimp after soaking in salt solution is required in the frozen shrimp industry. The penetration of salt into shrimp depends on many factors such as soaking time and the size of the shrimp.

X-ray fluorescence technique (XRF) is used to study the composition of the elements in a sample, based on the difference of the energy layers of each element. The advantage of this technique is the ease of sample preparation without product destruction [9], and XRF research can be applied in various fields of study. The study of the composition of cigarette smoke and ash by Wavelength Dispersive X-Ray Fluorescence Spectrometer (WDXRFS) showed that heavy elements found in ash did not contaminate in the smoke [10]. The study of skin permeability of titanium oxide and zinc oxide nanoparticles, which are added to sunscreen, showed that two nanoparticles gathered in the stratum corneum with no permeation into the epidermis and dermis [11]. Pengthamkeerati *et al.* (2018) [12] reported on the content of silica (95.6 %) extracted from biomass fly ash by using alkaline hydrothermal treatment and analyzed by the XRF technique. XRF was applied in many research projects and yielded reliable results for the component analysis [13]. The objective of the present research project was to study the effect of salt and freshness on the quality of white shrimp. The concentration of salt in the shrimp meat after soaking and cooking was determined by titration. The absorption and distribution of the sodium and chloride in the shrimp meats were investigated using the XRF technique.

2. Material and Methods

2.1. Materials and quality determination

White shrimp was obtained from Marine Gold Products Ltd. (Samut Sakhon, Thailand). Generally, the shrimp freshness could be divided into 5 levels, following the transportation or logistic period, as follows: level 1 is "just harvested" shrimp, level 2 corresponds to shrimp during transportation to the processing plant, levels 3 and 4 concern shrimp in the processing plant, and level 5 denotes shrimp with deteriorated quality that can still be used for the processing. Consequently, the shrimp arriving at the plant indicated their freshness mostly as level 3. The shrimp of freshness level 3 and 4 was evaluated by the specialists in the factory with a weight of 16 g per pc, as shown in figure 1. The samples were cleaned with cool water, washed with chlorine water, de-headed and de-veined with the skin on, and soaked in a salt solution at 8-12 % (w/v) for 60-120 min. For the cooking, the samples were further boiled for 110-130 sec by heated circulators. Finally, all samples were peeled and frozen.

The quality of shrimp in terms of freshness is judged by its appearance; head color, adhesion of head and body, eyes, body and skin, meat texture, and the odor of the samples according to the factory standard. Moisture content of the samples was determined in accordance with AOAC (2005) [14]. The pH level of the samples was determined by weighing 5 g of shrimp meat in a beaker with distilled water 45 mL, homogenized and measured by a pH meter. Total volatile base nitrogen (TVB-N) of the samples was measured in accordance with Siang and Kim (1992) [15] using 4 g of shrimp meat in a beaker with 4 % trichloroacetic acid (TCA), homogenized and filtered through a Whatman paper filter no 41 to obtain clear extract. Subsequently, 2 mL of the extract and 2 mL of K_2CO_3 were added by pipette into the outer ring of a Conway's dish, while 2 mL of 1% boric acid with indicator (Bromocresol green + Methyl red) was added by pipette into the inner ring. The dish was then placed in an incubator at 37 °C for 60 min and titrated with 0.02 N HCl to the endpoint indicated by the change of the color of the indicator from green to pink.

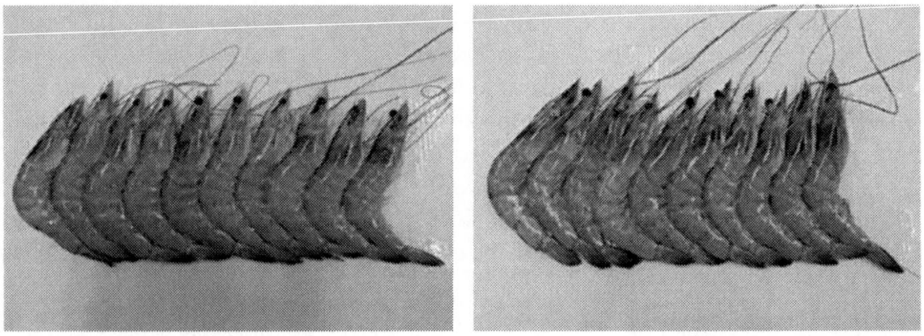

(a) (b)

Figure 1. White shrimp with different freshness levels; (a) freshness level 3 and (b) freshness level 4.

2.2. Na and Cl distribution by XRF technique

Four shrimp samples, comprising raw white shrimp (a); white shrimp with freshness level 3 after salt solution soaking (b); white shrimp with freshness level 4 after salt solution soaking (c); and freshness level 3 after salt solution soaking and cooking (d), were peeled and frozen. Frozen shrimp meat from the third abdominal segment were cut into slices of 50 µm-thick in cross section with the use of a cryomicrotome (CM1950 Cryotome from Leica, Singapore) and mounted on aluminium sample cups. Samples were air-dried for 3 days prior to the XRF analysis. Samples used for these studies were obtained without fixative media as shown in figure 2. For the XRF measurement, a wavelength dispersive XRF (ZSX Primus IV XRF, Rigaku, Japan) was used. The XRF uses a Rh-anode X-ray tube operated at 4 kW. Primary X-ray filter Ni400 excited the sample at two energy levels, 1.04 and 2.62 keV, and for analysis an area diaphragm of 0.5 mm was used. A gas flow proportional counter (F-PC) was used as a detector. For analysis of the shrimp specimens, the samples were measured at an area scan of 450 x 450 µm with a 50 µm step size in the x- and y-directions. Under field conditions, only one replicate was analyzed.

2.3. % salt determination by titration method

Four samples as described above were peeled, ground, and measured for the salt concentration. The whole shrimp was homogenized and 2 g of the homogenized sample was placed in a beaker, 2 % HNO_3 20 mL was added and stirred with a magnetic bar for 5 min, then titration was performed with 0.1 N $AgNO_3$ using an automatic titrator.

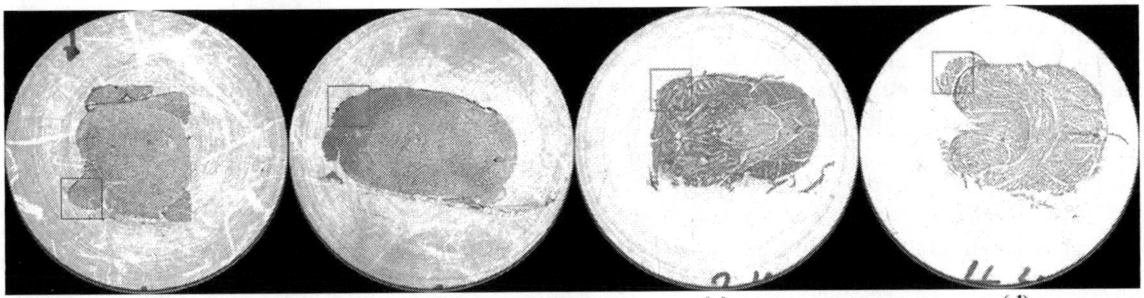

Figure 2. White shrimp meat in different conditions after cutting with a cryomicrotome; (a): raw white shrimp materials, (b): white shrimp with freshness level 3 after soaking in salt solution, (c): white shrimp with freshness level 4 after soaking in salt solution, and (d): white shrimp with freshness level 3 after soaking in salt solution and cooking.

2.4. Statistical analysis
Analysis of variance was performed to analyze the statistical significance using SPSS 23, and the mean value comparison was carried out by Duncan's multiple range test.

3. Results and discussion

3.1. Raw material quality
The freshness of white shrimp can be divided into 5 levels during the whole storage time starting with catching from the pond (the best quality, level 1) or natural sources to a level that is not acceptable for processing (level 5). However, the level of shrimp freshness used in the factory can be divided into 2 levels (corresponding to level 3 and 4 on the basis of each freshness level), by observing appearance such as color and adhesion of head and body (Table 1). The appearance and odor were different due to different storage times before processing, as summarized in Table 1. Shrimp are marine animals that display quality deterioration easily, and that constantly changes with time by the enzymatic reaction and/or bacteria [16].

Table 1. Appearance and odor of raw white shrimp at both freshness levels.

Appearance and odor	Freshness level 3	Freshness level 4
Head color	Blue-gray cheeks. Some pieces begin to be slightly black and begins to be yellow in the middle of the head	Orange and black head color Yellow in the middle of the head
Adhesion of head and body	Head and body begin to separate from each other	Head and body are separated
Eyes	Opaque eyes	Increased opaque eyes
Body and skin	Slightly glossy skin	Slightly glossy skin
Meat	Opaque meat color	Increased opaque meat color
Odor	Very slightly fishy smell odor	Slightly fishy smell odor

The moisture content of shrimp at freshness levels 3 and 4 were 78.60±0.01 and 80.06±0.01 %, respectively (*p<0.05*), as shown in Table 2. Sriket *et.al* [17] reported the chemical composition of white shrimp with moisture content of 77.2 %, which might be different due to several factors such as the method of raw material handling after death of the marine animals, soaking with ice and cold water, and storage time.

The pH of white shrimp with freshness levels 3 and 4 were 7.05±0.04 and 7.74±0.06, respectively (*p<0.05*). A slightly basic pH of the samples might be caused by autolysis by protease that generates amines and ammonia formation [3].

The TVB-N values of shrimp samples were 10.11±1.35 and 15.55±1.35 mg of N per 100 g of samples from freshness levels 3 and 4, respectively. The TVB-N values increased with storage time for marine products, due to the autolysis of the animals after death [3]. However, TVB-N values were less than 20 mg of N or 30 mg of N per 100 g of samples, which is still considered as fresh shrimp [18] and acceptable according to the Thai standard for frozen shrimps and fresh prawns [19].

Table 2. Some properties of shrimp at different freshness.

Parameters	Freshness level 3	Freshness level 4
Moisture content (%)	78.60±0.01[b]	80.06±0.01[a]
pH	7.05±0.04[b]	7.74±0.06[a]
TVB-N (mg of N per 100 g of samples)	10.11±1.35[b]	15.55±1.35[a]

*Means of shrimp at different freshness (n=3).

[a, b] mean with different letter in same row are significantly different (*p<0.05*).

3.2. Na and Cl distribution by XRF technique

Figure 2. shows sample preparation for XRF analysis of four shrimp samples. The samples are of four different conditions; raw white shrimp material, freshness level 3 after salt solution soaking, freshness level 4 after salt solution soaking, and freshness level 3 after salt solution soaking and cooking. The samples were analyzed for Na and Cl distribution using µ-XRF. The intensity of the sodium and salt concentration shown in the XRF mapping was normalized for all shrimp samples, as shown in figure 3. The results of element distributions were in accordance with the quantitative salt analysis performed by automatic titrator as shown in Table 3. The color scale, blue and dark blue, was observed in the fresh raw shrimp (3a). The shrimp with better quality, freshness level 3 after soaking (3b) in the salt solution, absorbed more sodium chloride than the sample from lower qualities (the samples with freshness level 4, 3c). However, the salt concentration in the soaked and cooked shrimps decreased sharply, as can be seen from the titration method and exhibited clearly by the µ-XRF (3d). Shrimp at both freshness levels lost their salt content after cooking and cooling. Protein denaturation was apparent from the shrimp meat appearance. The intensity of the Na and Cl was high near the de-veining area, resulting from the penetration and distribution of Na and Cl through the shrimp meat. In addition, these Na and Cl hot spots were rarely observed in the cooked shrimp samples, a result of loss of salt content during the cooking and cooling process. Na co-localization with Cl was also observed in µ-XRF maps, inferring that NaCl salt was obtained, which arises from undissolved NaCl or recrystallization of NaCl during drying. Such co-localization of Na with Cl might be attributed to NaCl crystals, and additional research will be needed to resolve and characterize the implications for co-localization. A study by Chantarat *et al.* (2005) [20] reported a decrease of myosin in fresh fish with increased storage time. Myosin can hold the water in a sample. Therefore, more red area was found in the sample with freshness level 3 after soaking (3b), while that of freshness level 4 showed less red area.

IOP Conf. Series: Earth and Environmental Science **346** (2019) 012046 doi:10.1088/1755-1315/346/1/012046

Table 3. Salinity of white shrimp samples with different freshness levels (by titration method).

White shrimp samples	Salinity (%)	
	Freshness level 3	Freshness level 4
Raw shrimp^{ns}	0.25 ± 0.02^{C}	0.28 ± 0.01^{C}
After soaking^{ns}	2.49 ± 0.11^{A}	2.20 ± 0.14^{A}
After cooking^{ns}	1.50 ± 0.07^{B}	1.54 ± 0.03^{B}

*Mean of white shrimp samples salinity (n=3). $^{A, B, C}$ and $^{a, b, c}$ mean with different letter in the same column and row are significantly different (*p<0.05*), respectively. ^{ns} mean in same row is not significantly different (*p<0.05*).

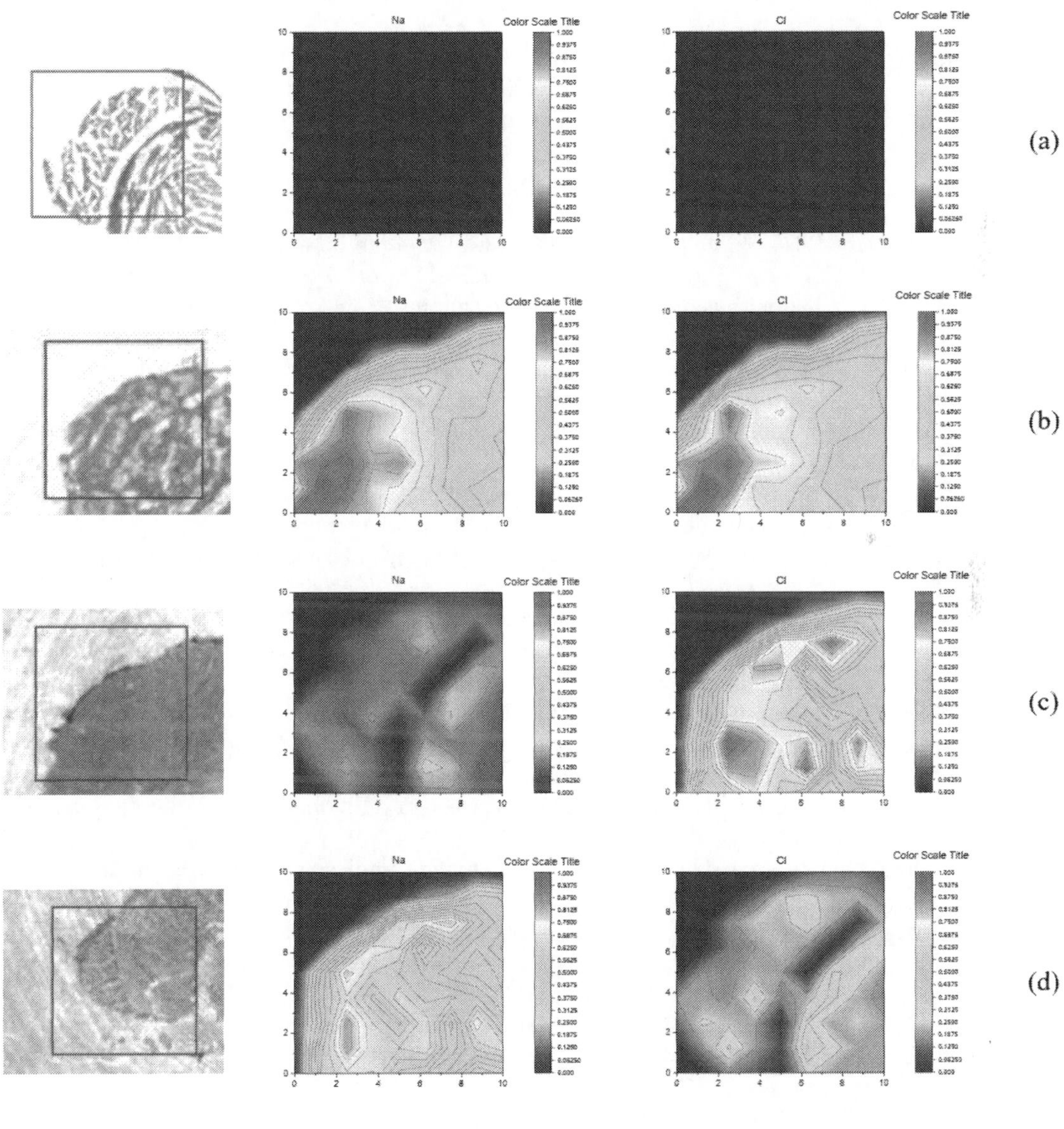

Figure 3. Concentration of Na and Cl by the X-ray fluorescence (XRF) technique on (a) raw white shrimp material, (b) white shrimp with freshness level 3 after salt solution soaking, (c) white shrimp with freshness level 4 after salt solution soaking, and (d) white shrimp with freshness level 3 after salt solution soaking and cooking. In the color scale, red represents a high concentration of the element while

blue represents a low concentration of the element. The red box indicates the sample area for the XRF analysis, 450x450 μm.

4. Conclusion

In conclusion, the X-ray fluorescence (XRF) technique can be used to monitor the distribution of salt solution within the shrimp meat. It can give an accurate result and is very close to the results obtained by the titration method. The result of saline distribution analysis in the samples shows that differences of freshness level affect absorption and distribution of salt solution.

The freshness of white shrimp affected the absorption of salt and water holding capacity. The white shrimp at freshness level 3 showed a larger area of salt distribution, as investigated by XRF, and higher salt content by titration.

Acknowledgment

We would like to thank the Thailand Research Fund and Marine Gold Products Ltd. for their kind support under the Research and Researchers for Industries (RRi) program (MSD60I0128).

References

[1] Wanichpongpan P 2012 White Shrimp (*Litopenaeus vannamei*). *Marine Animals* **158** 87-90.
[2] Wongjinda N 2015 Frozen Shrimp Production Process. *Depart. Fish*. Bangkok Thailand.
[3] Benjakul S 2011 Chemistry and quality of marine animals. Bangkok Odeon Store Thailand.
[4] Pattavarakorn D 2013 Industrial Process Chemistry. Chiangmai Chiangmai University Thailand.
[5] Wongwiwat P, Yanpakdee S and Wattanachant S 2007 Effect of mixed spices in lemon glass marinade cuisine on changes in chemical physical and microbiological quality of ready-to-cook Thai indigenous chicken meat during chilled storage. Songklanakarin, *J. Food Sci. Technol.* **29** (6) 1619-1632.
[6] Lopkulkiaert W, Prapatsornwattana K and Rungsardthong V 2009 Effects of sodium bicarbonate traces of citric acid in combination with sodium chloride on yield and some properties of white shrimp (*Penaeus vannamei*) frozen by shelf freezing, air-blast and cryogenic freezing. *J. Food Sci. Technol.* **42** 768-776.
[7] Wachirasiri K, Wanlapa S, Uttapap D and Rungsardthong V 2016 Use of amino acids as a phosphate alternative and their effects on quality of frozen white shrimps (*Penaeus vannamei*). *J. Food Sci. Technol.* **69** 303-311.
[8] Wachirasiri K, Wanlapa S, Uttapap D, Puttanlek C and Rungsardthong V 2016 Changing in processing yield and physical properties of frozen white shrimp (*Penaeus vannamei*) treated with lysine and sodium bicarbonate. *Int. J. Food Sci. Technol* **52** (3).
[9] Synchrotron Light Research Institute 2017 Micro-X-ray Fluorescence. Nakhon Ratchasima Thailand.
[10] Erbim E 2006 Quantitative elemental analysis of smoke samples using wavelength dispersive x-ray fluorescence spectrometer. Bangkok Srinakharinwirot University.
[11] Kiatponglarp W, Jearanaikoon N. and Wongsangta N 2018 Study of skin permeability of titanium dioxide and zinc oxide nanoparticles by micro-synchrotron based x-ray fuorescence imaging technique. Synchrotron Light Research Institute Nakhon Ratchasima Thailand.
[12] Pengthamkeerati P, Satapanajaru T, Sananwai N, Boonrite A and Welutung P 2018 Extracting silica from biomass fly ash by using alkaline hydrothermal treatment and silica precipitation by using organic acids. *J. KMUTNB.* **28** (1).
[13] Lambongang J Y and Suwanpinij P 2018 Materials characterization techniques for the analyses of components of port fuel injectors. *KMUTNB Int. J. Appl. Sci. Technol.*
[14] AOAC 2005 Determination of Moisture, Ash, Protein and Fat. Official method of analysis of the association of analytical chemists. 18th Edition. AOAC, Washington DC.
[15] Siang N C and Kim L L 1992 Determination of trimethylamine oxide, trimethylamine and total

volatile basic nitrogen by Conway's microdiffusion method. Southeast Asia Fisheries Development Center. B3.1-B3.6.

[16] Pedraja R R 1970 Change of composition of shrimp and other marine animals during processing. *Food Technol.* **24** (12) 37-42.

[17] Sriket P, Benjakul S, Visessanguan W and Kijroongrojana K 2007 Comparative studies on chemical composition and thermal properties of black tiger shrimp (*Penaeus monodon*) and white shrimp (*Penaeus vannawei*) meats. *Food Chem.* **103** 1199-1207.

[18] Mendes R 2006 Guidebook on melanosis inhibitors and processing technology of crustaceans. In *Iniap/Ipimar. Project QLK1-CT-2002-*71517 (New approaches to the crustaceans prevention of melanosis and quality indices) **41**.

[19] TIS 1986 Thai industrial standard for quick-frozen shrimps or prawns. Bangkok Thailand TIS 115-2529

[20] Chantarat P, Kijroongrojana K and Vittayanont M 2005 Effect of ice storage on muscle protein properties and qualities of emulsion fish sausage from bigeye snapper (*Priacanthus tayenus*) and lizardfish (*Saurida undosquamis*). Songklanakarin, *Food Sci. Technol.* **27** (1) 123-138.

Effect of culture conditions and medium compositions on kojic acid production by *Aspergillus oryzae* ATCC 10124

A Promsang[1], V Rungsardthong[1], B Thumthanaruk[1], C Puttanlek[2], D Uttapap[3], T Foophow[4], V Phalathanaporn[5] and S Vatanyoopaisarn[1, 6]

[1]Department of Agro-Industrial, Food and Environmental Technology, Faculty of Applied Science, Food and Agro-Industry Research Center, King Mongkut's University of Technology North Bangkok, Bangkok, Thailand
[2]Department of Biotechnology, Faculty of Engineering and Industrial Technology, Silapakorn University, Nakhon Pathom, Thailand
[3]Division of Biochemical Technology, School of Bioresources and Technology, King Mongkut's University of Technology Thonburi, Bangkok, Thailand
[4]Deparment of Food Processing Technology, School of Culinary Arts, Suan Dusit University, Bangkok, Thailand
[5]Foodle Noodle Co., Ltd., 98/38 Moo 2, Bueng Nam Rak, Thanyaburi, Pathum Thani 12110, Thailand

E-mail: savitri.v@sci.kmutnb.ac.th

Abstract. Kojic acid is a secondary metabolite produced by some strains of *Aspergillus* spp and has been exploited commercially in food and cosmetic products. The objective of this work was to optimize the cultivation conditions of *Aspergillus oryzae* ATCC 10124 prior to triggering the fermentation process using organic broken rice noodles (OBRN) as a carbon source. Trials with various carbon and nitrogen sources show that 10% glucose and 0.05% yeast extract with ammonium sulfate was favorable for A. *oryzae* to produce kojic acid (1.58 g/L at day 4 of cultivation). Initial pH and agitation rate significantly affected the kojic acid formation. The maximum quantities of kojic acid were obtained when the pH of the medium was at 2.5, and shaking was at 200 rpm, at 1.60 g/L and 1.65 g/L, respectively. When the carbon source was changed to OBRN, the substrate was saccharified by alpha-amylase and glucoamylase, and 100% of the hydrolysate was used to replace glucose for the kojic acid fermentation. Kojic acid content increased when using OBRN, at 1.52 g/L, and was comparable to that yielded from the fermentation with glucose (1.58 g/L).

1. Introduction

Kojic acid was first isolated by Satio (1907) [1] from the mycelium of *Aspergillus oryzae* on steamed rice. The chemical structure of kojic acid was identified as 5-hydroxy-2-(hydroxymethyl)pyran-4-one [2]. Kojic acid is a secondary metabolite produced by different types of fungi utilizing various sub-

Content from this work may be used under the terms of the Creative Commons Attribution 3.0 licence. Any further distribution of this work must maintain attribution to the author(s) and the title of the work, journal citation and DOI.
Published under licence by IOP Publishing Ltd

strates during aerobic fermentation. It has many potential industrial applications in the food, agriculture, cosmetics, and pharmaceutical industries [3].For instance, kojic acid can be used in preserving agricultural products, particularly vegetables and fruits, during storage by inhibiting the action of polyphenol oxidase (PPO). It can suppress hyperpigmentation in human skin and restrains the formation of melanin in human skin by inhibiting the tyrosinase enzyme that is responsible for skin pigmentation [3].

Kojic acid was produced from various carbon sources as the secondary metabolite synthesized via phosphorylated intermediates, as reported by Kahn *et al.* (1997) [4] and medium composition is one of the crucial factors for high kojic acid production. Several monosaccharides have been investigated; however, glucose was reported to generate the highest yield of kojic acid in A. *oryzae* (24.2 g/L) [5] and A. *flavus* (39.9 g/L) [6], when 100 g/L of glucose was supplied. The type of nitrogen sources influenced both fungus growth and kojic acid production. Organic nitrogen sources were preferable to inorganic nitrogen sources, as was shown in the work of Rosfrizan and Ariff (2000) [6] in which 5 g/L yeast extract resulted in the maximum kojic acid concentration (39.9 g/L).

A previous study showed that the optimal pH for kojic acid production by a different strain of *Aspergillus oryzae* ranged between 3 to 5. Lui *et al.* (2016) [7] reported the highest kojic acid production in the medium with pH 3 (35.27 g/L) by *Aspergillus oryzae* BCRC 30010.Optimization of oxygen provision could enhance the effectiveness of glucose catabolism by the aerobic microorganism in the early stage of fermentation. Consequently, the intermediates obtained during the growth phase are essential for the biosynthesis of kojic acid. Examples of aeration and agitation on kojic acid production were reported by Ariff *et al.* (1996) [8]. They stated that in A. *flavus*, the production of kojic acid decreased significantly when the dissolved oxygen tension levels during the initial phase of growth were lower than 80% [8].

There are many advantages to using microorganisms for the production of natural products; for example, microbial products can be produced by large-scale fermentation; the fermentation process is not subject to seasonal variation or climatic factors; and new cultivation techniques and microbial strain selection will help to improve the productivity. Production costs can be minimized by using cheap starting materials or industrial by-products. Several fungi, such as *Aspergillus awamori*, *Acremonium nigricans*, and *Aspergillus oryzae*, were used for 1-(3,4-dihydro-2H-pyrrol-5-yl) ethanone (2-acetyl-1-pyrroline) formation (a major flavour compound in aromatic rice) [9]. In addition, *Aspergillus oryzae* has been reported as being of low pathogenic potential, and does not produce aflatoxins or any other carcinogenic metabolites; therefore, it is an excellent vehicle for the safe production of low-risk products. Spores of *Aspergillus oryzae* NRRL 484 (ATCC 10124) were entrapped in calcium alginate gel beads for the production of kojic acid [10]. The immobilized cells in flask cultures were accumulated with a concentration as high as 83 g/L. Liu *et al.* (2016) [7] reported that high kojic acid production (83.47 g/L) could be obtained in a PCS-immobilized bioreactor by *Aspergillus oryzae* BCRC 30010. However, all kojic acid formation in all reports mentioned above was measured by spectrophotometry, which depends on the color reaction that may be subject to interference from other ingredients in the medium.

Organic broken rice noodles (OBRN) are a by-product of organic rice noodle processing. The cost of organic rice is high compared to normal rice used in noodle production, but this by-product is sold cheaply as animal feed ingredients. The OBRN are good sources of carbon, as they consist of 86.65% carbohydrate, which can be used for kojic acid production. Enzymatic hydrolysis could be applied to digest polysaccharide into fermentable sugars, as well as offering the advantages of higher conversion, minimal by-product formation, low energy requirements, and mild operating conditions, compared to other chemical conversions [11]. Consequently, the objective of this research was to study the effect of the culture conditions and medium compositions, including carbon sources and nitrogen sources, on the formation of kojic acid. The potential use of organic broken rice noodles as the carbon source for fermentation was also investigated.

2. Material and methods

2.1. Microorganism and spore preparation

Aspergillus oryzae ATCC 10124 was purchased from American Culture Collection Center, USA. The liquid spores were prepared by cultivation on PDA for 4-7 days and harvested from PDA agar plates with 1% (v/v) of tween solution. The spore suspension was counted using a counting chamber to adjust spore suspension to around 10^6 spores/mL. Kojic acid was obtained from Alfa Aesar (USA) and all chemicals used were of analytical grade.

2.2. Medium and culture conditions for kojic acid production

A base medium with the following composition was used: 100 g/L of each carbon source, 0.5 g/L of each nitrogen source, 0.5 g/L KH_2PO_4, 0.5 g/L K_2HPO_4, 0.5 g/L $MgSO_4.7H_2O$, and 1 ml/L of $MnSO_4.H_2O$, $FeSO_4.7H_2O$, and $ZnSO_4.7H_2O$ [10]. To investigate the effect of the different carbon sources (glucose, sucrose, and fructose) and nitrogen sources (yeast extract, peptone, ammonium sulfate, yeast extract with ammonium sulfate, and peptone with ammonium sulfate), different sugars and nitrogen sources were substituted. The effect of medium pH was investigated at pH levels of 2, 2.5, and 3. All submerged fermentations were carried out with 50 ml of medium in a 250 ml Erlenmeyer flask, and incubated at 30 °C and shaken at 150, 200 and 250 rpm on a shaking incubator (LabTech, Korea). OBRN were hydrolyzed using alpha-amylase and glucoamylase prior to being used as a carbon source for kojic acid production.

2.3. Analytical method

Three ml of mold suspension was taken every day from the culture media described above. The sample was filtered using Whatman paper No. 1 while the residues were dried in a hot-air oven at 80 °C overnight in order to measure the cell dry weight. The residual sugar in the filtrate was analyzed using the modified dinitrosalicylic acid (DNS) method [12]. The pH of the culture suspension was measured using a pH meter. The filtrate was again filtered through a 0.45 micron paper filter and degassed before injection to high-performance liquid chromatography (HPLC) to analyse the kojic acid content in the culture broth with a C18 column, 4.6x150 mm (Luma 5 um), and a UV detector at 270 nm. The mobile phase was carried out with a mixture of 1% acetic acid (solvent A, in MilliQ water) and acetonitrile (solvent B). The elution was performed from 0 to 10 min using isocratic 5% solvent B, 10.1-14 min, isocratic 60% solvent B, and a re-equilibration period of 6 min with 5% solvent B used between individual runs. A standard curve for kojic acid was prepared for the calculation. In addition, kojic acid was measured in parallel by the colorimetric method [13]. One ml of cultured medium was added to 4 ml of 1% ferric chloride, and absorbance was read at 540 nm by a spectrophotometer.

2.4. Statistical analysis

All experiments were performed using one-way analysis of variance, and mean comparison was tested using Duncan's multiple range test for the differences between mean at $p<0.05$.

IOP Conf. Series: Earth and Environmental Science **346** (2019) 012047 doi:10.1088/1755-1315/346/1/012047

3. Results and discussion

3.1. Effect of carbon and nitrogen sources on kojic acid production

Figure 1. shows the HPLC chromatogram of the standard kojic acid, which indicates a retention time of 4.247 min. For the kojic fermentation, significant differences in kojic acid formation with different carbon and nitrogen sources are shown in Table 1. It was clear that when glucose was a carbon source, the kojic acid concentration was significantly higher than with the other C-sources. The amount of kojic acid was enhanced when yeast extract and $(NH_4)2SO_4$ were added in equal amounts (1.58 g/L at day 4). Glucose can be directly converted to kojic acid through multi-step enzyme reactions [6]. Fructose and sucrose in this work contributed to lowering the kojic acid concentration significantly. This was in agreement with the work of Kitada *et al.* (1967) [5] and Rofarizan and Ariff [6]. Sucrose is a disaccharide, which must be hydrolyzed to glucose and fructose by activitating an invertase enzyme secreted by the fungus used in the fermentation. The C_6 precursor derived from glucose, which was in the form of pyranose, was more favourable to conversion directly to kojic acid than the furanose form of fructose [5, 6].

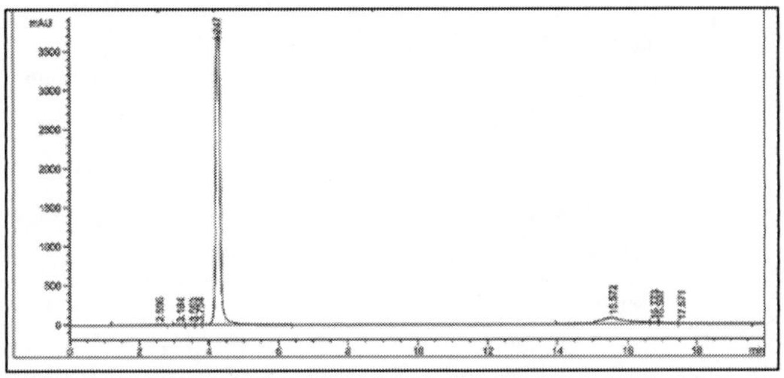

Figure 1. HPLC Chromatogram of 5-hydroxy-2-(hydroxymethyl)pyran-4-one (Kojic acid).

The medium with fructose as the carbon source resulted in a very low kojic acid concentration (0.02 g/L). The selection of nitrogen sources for the fermentation medium is important for kojic acid fermentation. Yeast extract and peptone contain vitamins, minerals, and oligoelements for kojic acid production [14-15]. On the other hand, Kwak and Rhee (1992) [10] reported that conidia of Aspergillus oryzae NRRL 484 immobilized in Ca-alginate beads gave a high level of kojic acid, up to 83 g/L (measured by a spectrophotometer), when the fungus was cultivated using 0.5 g/L yeast extract and 0.75 g/L (NH4)2SO4) as nitrogen sources. In the work of Liu *et al.* (2016) [7] *Aspergillus oryzae* BCRC 30010 was immobilized in a plastic composite support (PCS) bioreactor with a nitrogen-deficient medium for kojic acid production. The concentration of immobilized culture for kojic acid production was 83.47 g/L after three cycles of cultivation [7]. Several studies have reported high content of kojic acid production analyzed by spectrophotometry and a few reports covered the determination of kojic acid using HPLC [16].

The measurement of kojic acid in our experiment using HPLC was 1.5 g/L, while it was calculated at 30.19 g/L when the same sample was analyzed by the spectrophotometry method. Though spectrophotometry is widely accepted as a useful tool in quantifying many organic substances, when being used with multi-components, particularly the culture media, higher absorbance may occur. Hence, we preferred to express the concentration of kojic acid derived from HPLC measurement rather than that shown by the spectrometer.

Table 1 Kojic acid production influenced by different carbon and nitrogen sources.

Carbon sources (100 g/L)	Nitrogen sources (0.5 g/L)	Kojic acid content (g/L)
Glucose	Yeast extract	1.13±3.54^{c}
	Peptone	0.92±5.66^{e}
	$(NH_4)_2SO_4$	1.14±14.14^{b}
	Yeast extract+$(NH_4)_2SO_4$	1.58±1.07^{a}
	Peptone+$(NH_4)_2SO_4$	1.03±7.07^{d}
Sucrose	Yeast extract	0.21±8.49^{f}
	Peptone	0.10±1.41^{j}
	$(NH_4)_2SO_4$	0.15±4.24^{hi}
	Yeast extract+$(NH_4)_2SO_4$	0.16±2.12^{g}
	Peptone+$(NH_4)_2SO_4$	0.14±5.67^{i}
Fructose	Yeast extract	0.02±0.02^{m}
	Peptone	0.06±0.06^{k}
	$(NH_4)_2SO_4$	0.02±0.02^{m}
	Yeast extract+$(NH_4)_2SO_4$	0.03±0.03^{m}
	Peptone+$(NH_4)_2SO_4$	0.05±0.02^{l}

Values are mean ± SD of triplicate samples
$^{a-m}$ = the different letters within a column for each attribute constituent are significantly different ($p<0.05$).

The quantity of kojic acid increased gradually from the first days to a maximum value (1.58 g/L), along with a decrease in the reducing sugar concentration, while pH did not change much during the fermentation (Figure 2). The fungus growth was highest at day 2 (5.95 g/L). The kojic acid fermentation was categorized as a non-growth associated route, where the fermentation comprised both a growth phase and a production phase [15]. However, some kojic acid accumulated in the culture may have been consumed in order to produce other organic acids, after all supplies of glucose have been utilized [17]. Hence, the occurrence of kojic acid reduction appears [18].

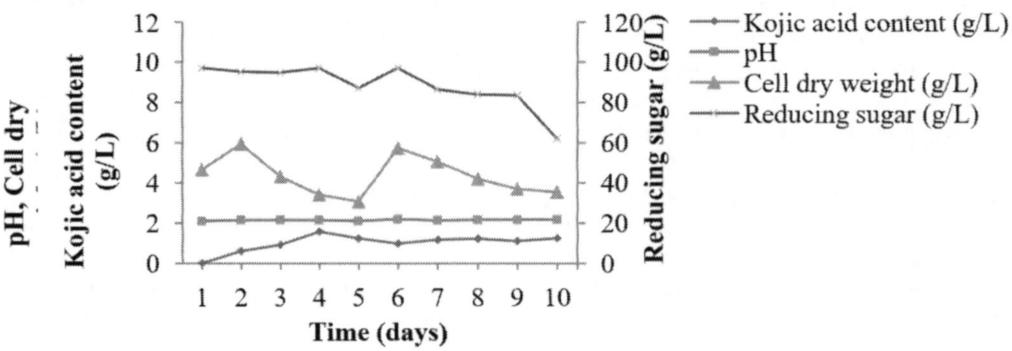

Figure 2. Time course of kojic acid production on pH, cell dry weight, kojic acid content, and reducing sugar from the fermentation medium with glucose, yeast extract, and ammonium sulfate and cultivated at 30 °C, 200 rpm.

3.2. Effect of pH and the agitation rate on kojic acid production

Generally in fungi, secondary metabolites are produced in the late log and stationary phases, in which the culture condition has already been acidified by several primary metabolites. In addition, the quantity of kojic acid increased while glucose concentration decreased linearly with incubation time. The results of kojic acid production at different pH levels and agitation rates are shown in Figure 3. The highest amount of kojic acid appeared when the mold had been grown in a pH 2.5 medium (1.62 g/L) at day 4 of the cultivation. The results were in accordance with the study by Rosfarizan *et al.* (2000) [6] who reported that the optimal pH for the kojic acid producing enzyme was around pH 3.5, during which the pH stress may affect the production of secondary metabolism in both *Aspergillus oryzae* and *Aspergillus flavus* [19]. In addition, the secondary metabolites are formed in the late log and stationary phases, during which the fermentation has already been acidified by primary metabolites [20].

The oxygen requirement is one of the approaches to maximize kojic acid production. The results in Figure 2 (d-f) indicate that the oxygen requirement at 200 rpm significantly yielded higher kojic acid quantities (1.65 g/L at day 4) than those at 150 and 250 rpm. However, there were no significant differences in the mycelia dry weight. The secondary metabolites and enzymes produced by various fungal species are subject to the culture conditions, temperature and pH of the medium, composition of the substrate, types of inducers, and inoculum used during the experiments [21]. This was in agreement with this work in order to find the optimum conditions required for A. *oryzae* to produce kojic acids.

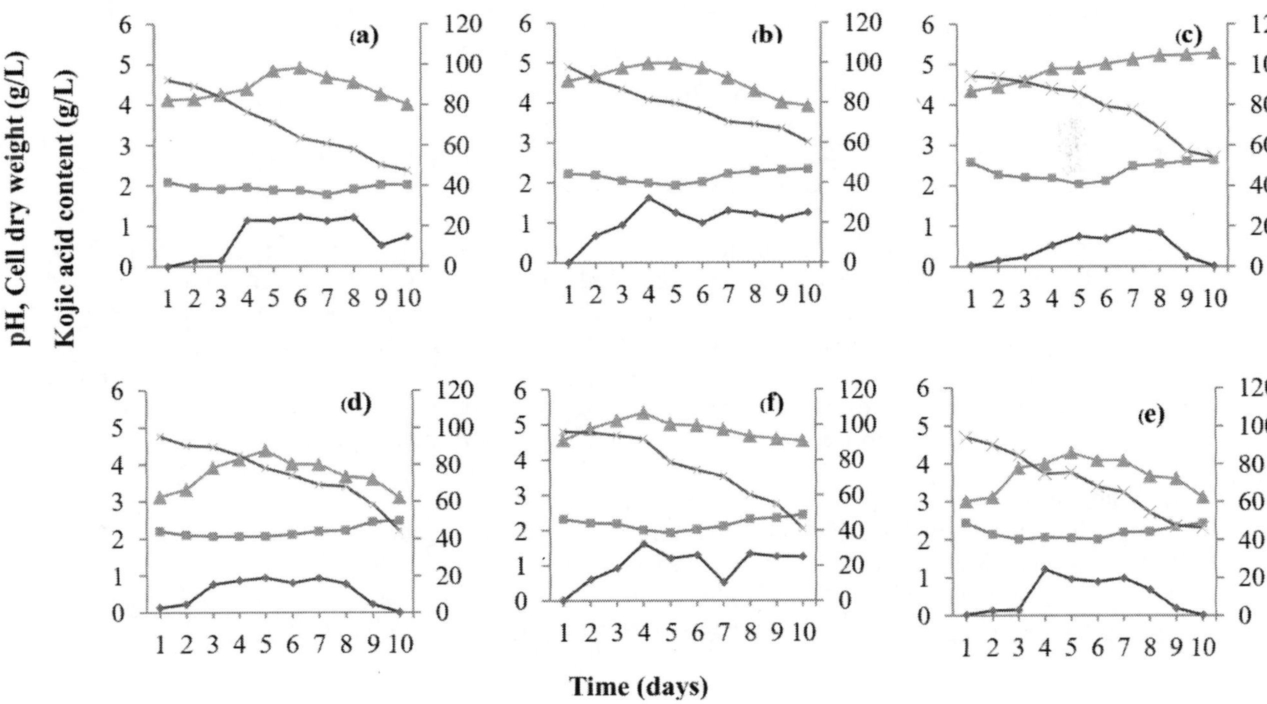

Figure 3. Time course of kojic acid production on pH, cell dry weight, kojic acid concentration and reducing sugar (a: pH 2, b: pH2.5, and c: pH 3) and agitation rate (d: 150 rpm, e:200 rpm, and f:250 rpm) cultivated at 30 °C.

3.3. Use of OBRN as a carbon source in kojic acid fermentation
In order to investigate the potential use of OBRN as a carbon source (instead of glucose), OBRN were hydrolysed in the optimum conditions obtained from earlier experiments. OBRN were hydrolyzed using 1 % alpha-amylase at pH 6.5 in a water bath controlled at 85°C, and shaken at 100 rpm for 1 h. Further reaction was carried out with 0.5% glucoamylase at pH 4.5, in a water bath at 65°C, and shaken at 100 rpm for 21 h. The kojic acid produced, as well as the reducing sugar consumption and growth of A.*oryzae* ATCC 10124 for up to 8 days of incubation, are displayed in Figure 4.

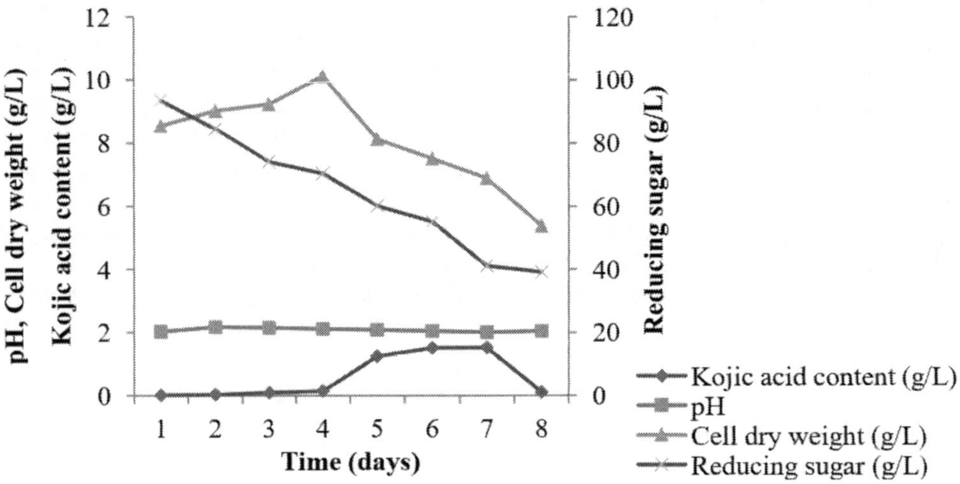

Figure 4. Time course of kojic acid production on pH, cell dry weight, kojic acid content, and reducing sugar, in fermentation using OBRN cultivated at 30 °C, 200 rpm.

The fungus growth was rapid during the initial stage of the fermentation and decreased from day 4, while the highest amount of kojic acid produced was 1.52 g/L at day 7 of the cultivation. Kojic acid was the secondary metabolite found after the rapid growth of the fungi. Rostarizan *et al.* (1998) [22] reported kojic acid production at 23.5 g/L (measured by spectrophotometer) when 100 g/L of sago starch was used from *Aspergillus flavus* strain S33-2. The production of kojic acid in our experiment was not as high, since we express the measurement of kojic acid by HPLC. However, when the same sample was read by a spectrophotometer, it was much higher, at about 30 g/L.

4. Conclusion
Kojic acid fermentation by *Aspergillus oryzae* ATCC 10124 was optimized when cultured in 100 g/L of glucose and 0.5 g/L of yeast extract with ammonium sulfate at an initial pH of 2.5 and incubated at 30 ₒC and 200 rpm shaking. The highest kojic acid concentration obtained was 1.61 g/L. When the organic broken rice noodles were hydrolyzed and used as the carbon source, 1.52 g/L of kojic acid was

obtained at day 7 of the fermentation. The results show the possibility of using organic broken rice noodles as a carbon substrate for kojic acid production.

Acknowledgment
This research was sponsored by the Thailand Research Fund (TRF) and Foodle Noodle Co., Ltd, under the Research and Researchers for Industries (RRi) Program (MSD59I0102).

References

[1] Saito K *1907* Über die Säurebinding von *Aspergillus oryzae. The Botanical Magazine Tokyo* **21** *7–11.*
[2] Yabuta T *1924* The constitution of kojic acid, a δ-pyrone derivative formed by *Aspergillus flavus* from carbohydrates. *Chem. Soci.* ***125*** *575-587.*
[3] Saeedi M, Eslamifar M and Khezri K *2019* Kojic acid applications in cosmetic and pharmaceutical preparations. *Biomed. Pharmacother.* ***110*** *582-93.*
[4] Kahn V, Lindner P and Zakin V *1997* Effect of kojic acid on the oxidation of N-acetyldopamine by mushroom tyrosinase. *J. Agric. Food Chem.* ***45 (11)*** *4460-65.*
[5] Kitada M, Ueyama H, and Fukimbara T *1967* Studies on kojic acid fermentation (I) cultural condition in submerged culture. *Ferment. Technol.* **45** 1101-07.
[6] Rofarizan M and Ariff AB *2000* Kinetics of kojic acid fermentation by *Aspergillus flavus* using different types and concentrations of carbon and nitrogen sources. *J. Indus. Microbiol. Biotech.* ***25*** *20-4.*
[7] Lui J M, Yu T C, Lin S P, Hsu R J, Hsu K D and Cheng K C *2016* Evaluation of kojic acid production in a repeated-batch PCS biofilm reactor. *J. Biotechnol.* ***218*** *41-8.*
[8] Ariff A B, Salleh M S, Ghani B, Hassan M A, Rusul G and Karim MIA *1996* Aeration and yeast extract requirements for kojic acid production by *Aspergillus flavus. Enzyme Microb. Technol.* ***19*** *545-50.*
[9] Rungsardthong V and Noomhorm A *2005* Production of *2*-acetyl-*1*-pyrroline by microbial cultures. *Flavour Fragr. J.* ***20*** *710-14.*
[10] Kwak M and Rhee J S *1992* Cultivation characteristics of immobilized *Aspergillus oryzae* for kojic acid production. *Biotechnol. Bioeng.* ***39(8)*** *903-6.*
[11] Amnuaycheewa P, Rodiahwati W, Sanvarinda P, Cheenkachorn K, Tawai A and Sriariyanun M *2017* Effect of organic acid pretreatment on Napier grass (*Pennisetum purpureum*) straw biomass conversion. *KMUTNB Int. J. App. Sci. Technol.* ***10(2)*** *107-17.*
[12] Miller G L *1959.* Use of dinitrosalicylic acid reagent for determination of reducing sugar. *Anal. Chem.* ***31(3)*** *426-28.*
[13] Bentley R *1957* Preparation and analysis of kojic acid. *Methods Enzymol.* ***24*** *238-41.*
[14] Lin M T, Mahajan J R, Dianese J C and Takatsu A *1976* High production of kojic acid crystals by *Aspergillus parasiticus* UNBF A*12* in liquid medium. *Appl. Environ. Microbiol.* ***32*** *298-9.*
[15] Chaudhary J, Pathak A N and Lakhawat S *2014* Production technology and applications of kojic acid. *Annu. Res. Rev. Biol.* ***4(21)*** *3165-96.*
[16] Masse M O, Duvallet V, Borremans M and Goeyens L *2001* Internat. J. Cosmetic Sci. ***23*** *219-232.*
[17] Clevström, G and Ljunggren H *1985* Aflatoxin formation and the dual phenomenon in *Aspergillus flavus* Link. *Mycopathol.* ***92*** *129-39.*
[18] Durga Devi K B, Vijayalakshmi P, Shilpa V and Kumar BV *2015* Optimization of cultural

parameters for cost effective production of kojic acid by fungal species isolated from soil. *MRJI.* ***7(5)*** *255-68.*

[19] Rosfarizan M, Ariff A B, Hassan M A and Karim M I A *2000* Influence of kojic acid on kojic acid fermentation by *Aspergillus flavus*. *Pakistan J. Biol. Sci.* ***3(6)*** *977-82.*

[20] Shwab E K., Keller N P *2008* Regulation of secondary metabolite production in filamentous ascomycetes. Mycological Res ***112(2)*** *225-230.*

[21] Isanapong J, Krailoekpaiboon T and Noiniyom W *2017* Utilization of organic wastes for Laccase production by *Pleeurotus ostreatus*. *KMUTNB Int. J. Appl. Sci. Technol.* ***10(2)*** *239-44.*

[22] Rosfarizan M, Ariff A B, Hassan M A and Karim M I A *1998* Kojic acid production by *Aspergillus flavus* using gelatinized and hydrolyzed sago starch as carbon sources. *Folia Microbio. (Praha)* **43(5)** *459-64.*

IOP Conf. Series: Earth and Environmental Science **346** (2019) 012048 doi:10.1088/1755-1315/346/1/012048

Effect of egg white on physicochemical properties of mixed fish sausage

T Klankklin[1], P Banjongsinsiri[2], S Vatanyoopaisarn[1], V Rangsardthong[1] and B Thumthanaruk[1, 3]

[1]Department of Agro-Industrial Food and Environmental, Faculty of Applied Science, Food and Agro-Industry Research Center, King Mongkut's University of Technology North Bangkok, Bangkok, Thailand
[2]Expert Center of Innovative Health Food, Thailand Institute of Scientific and Technological Research. 35 Technopolis, Tambon Khlong Ha, Amphoe Khlong Luang, Pathum Thani 12120

E-mail: benjawan.t@sci.kmutnb.ac.th

Abstract. Fish sausage can be produced using a mixture of multiple species of fish, but such products are not generally well-received by customers due to their fishy flavor and soft texture. The objective of this research was to develop fish sausage using three types of fish (salmon, spotted knife fish and seven-striped carp). In addition, the effect of the addition of egg white on the chemical and physical properties of the fish sausage was also studied. Results showed that the formula FS7, which contains salmon, spotted knife fish, and seven-striped carp at a ratio of 0.33:0.33:0.33% (w/w), had a moisture content of 68.19%, emulsion stability of 97.15%, water holding capacity of 86.37%, and cooking loss of 4.63%. Formula FS7 was, therefore, chosen for improvement of its texture by adding fresh egg white. The effect of egg white at all levels had an impact on the color of the mixed fish sausage. Results show that increasing egg white from 1% to 3% in the mixed fish sausage formula FS7 slightly increased both L* and h°. However, increasing egg white did not affect the hardness of the mixed fish sausage. The addition of egg white at all levels increased the emulsion stability, but not the water holding capacity. Based on the results of this experiment, the use of 1% egg white helps to increase the lightness of mixed fish sausage but decreases the percentage of the water holding capacity.

1. Introduction

Salmon (*Salmo salar*), a marine fish from the family of Salmonidae, has high levels of eicosapentaenoic (EPA) and docosahexaenoic acid (DHA), which are known for their anti-thrombogenic and anti-inflammatory properties [1-2]. Spotted knife fish (*Notopterus chitala*) and seven-striped carp (*Probar busjullieni*), which are freshwater fish in the family of Notopteridae and Cyprinidae, are used in Thai cooking recipes

Content from this work may be used under the terms of the Creative Commons Attribution 3.0 licence. Any further distribution of this work must maintain attribution to the author(s) and the title of the work, journal citation and DOI.
Published under licence by IOP Publishing Ltd

such as Thai steamed fish curry in banana leaves, deep fried fish cakes, fish balls, and fish sausage [3]. In general, commercial sausages are prepared using chicken, pork, or beef emulsified with fat compounds from various sources. During mixing and blending, bundles of protein fibers, myofibrils, and filaments are broken down and the size of fatty tissues is gradually reduced. Myofibrillar proteins become soluble and migrate to the fat globule surface, concentrate, and form a protein matrix at the fat/water interface when salt is added [4]. Apart from the use of theses main ingredients, phosphates are added to increase water holding capacity (WHC) of animal proteins during processing and storage. Nitrites are used as a preservative for inhibiting the growth of *Clostridium botulinum.* Flavors and flavor enhancers such as salt, monosodium glutamate (MSG) [5], guanosine monophosphate (GMP) [6] or 5'-ribonucleotide salts [3] are also used to increase palatability and consumer acceptance. Juntachote (2018) reported that low-fat chicken sausages containing 1.5 and 2% carrageenan enhanced the quality characteristics of low-fat chicken sausages [7]. However, the consumption of these chicken, pork, or beef sausages, which are high in saturate fat and cholesterol, could lead to high blood cholesterol, hypertension, heart disease, or other health problems. Fish sausage is an alternative choice for good health that has low saturate fat and cholesterol, but high EPA or DHA if marine fish like salmon is used [2]. The drawback of fish sausage products is the soft texture of fish. The objective of this research was to develop fish sausage by varying the ratio of the three types of fish (salmon, spotted knife fish, and seven-striped carp). The addition of fresh egg white on the chemical and physical properties of the mixed-fish sausage was also investigated.

2. Material and methods

2.1. Preparation of fish sausage

The fillets of salmon (*Salmo salar*), spotted knife fish (*Notopterus chitala*), and seven-striped carp (*Probar busjullieni*), along with back-fat pork, soy sauce, salt, fish sauce, and sugar were purchased from a retail store in Nonthaburi, Thailand. Sodium tripolyphosphate and sodium nitrite were purchased from Sigma-Aldrich (USA). The fish samples were stored on ice for 1 hour during transportation to the laboratory. As described by Pakoawnoy (2014) [8], the fish sausage was prepared with some modification. The control formula of the fish sausage consisted of each type of fish fillets (salmon, spotted knife fish or seven-striped carp) (71.22%), back-fat pork (17.81%), ice (7.33%), soy sauce (1.07%), salt (0.85%), fish sauce (0.71%), sugar (0.71%), and sodium tripolyphosphate (0.29%) w/w. Sodium nitrite was also added at a concentration of 125 ppm. The preparation of the sausages started with cleaning of the fish fillets (salmon, spotted knife fish and seven-striped carp) with tap water and drained for 5 minutes, then cut into small pieces and blended in a mixer at high speed for 2 min. Then all other ingredients were added and mixed for another 10 minutes. The temperature was maintained at 12±3°C throughout the preparation. The resulting batter was then stuffed into a collagen casing (25 mm in diameter), baked at 70±2°C for 60 minutes, boiled at 80±2°C for 60 min, and cooled in cold water (15°C). All fish sausage samples were packed individually in polyethylene bags and kept frozen at -18°C.

2.2. Development of mixed-fish sausage (FS) formulations

The ratio of salmon (X1), spotted knife fish (X2) and seven-striped carp (X3) for producing mixed- fish sausage in this experiment was varied as shown in the mixture design chart in Table 1. Ten mixed-fish formulations of fish sausage were produced as described in 2.1 above. In addition, three single-fish formulations were prepared in FS as controls (FS1- FS3).

IOP Conf. Series: Earth and Environmental Science **346** (2019) 012048 doi:10.1088/1755-1315/346/1/012048

Table 1. The formula of fish sausage having different ratios of fish fillet.

Sample	Ratio of flesh fish (%)			Quantity (g/100 g)	
	X1	X2	X3	X1	X2
FS 1	1	0	0	71.22	0
FS 2	0	1	0	0	71.22
FS 3	0	0	1	0	0
FS 4	0.5	0.5	0	35.61	35.61
FS 5	0.5	0	0.5	35.61	0
*FS 6	0	0.5	0.5	0	35.61
FS 7	0.33	0.33	0.33	23.74	23.74
FS 8	0.66	0.16	0.16	47.48	11.87
FS 9	0.16	0.66	0.16	11.87	47.48
FS 10	0.16	0.16	0.66	11.87	11.87

*FS6 is a control.

2.3. *Effect of egg white addition on the chemical and physical properties of mixed-fish sausage*

Chicken eggs were purchased from a retail store in Nonthaburi, Thailand. The eggs were cleaned with tap water and separated into egg yolk and egg white. Only the fresh egg white was used in this experiment. The FS (fish sausage) formula having the highest results in terms of emulsion stability and water holding capacity (as described in 2.2 above) was selected for further study of the effect of addition of egg white to the sausage. Three concentrations (1%, 2%, and 3% (w/w)) of fresh egg white were added and labeled as EW1, EW2 and EW3, as shown in Table 2.

Table 2. Quantity of all ingredients used in mixed-fish sausage with egg white.

Ingredient	Quantity (g/100 g)			
	Control	EW1	EW2	EW3
Egg white	0	1	2	3
Salmon	23.74	23.50	23.27	23.03
Spotted knife fish	23.74	23.50	23.27	23.03
Seven stripped carp	23.74	23.50	23.27	23.03
Pork back-fat	17.81	17.63	17.45	17.28
Ice	7.33	7.26	7.18	7.11
Fermented soy sauce	1.07	1.06	1.05	1.04
Salt	0.85	0.84	0.83	0.82
Fish sauce	0.71	0.70	0.69	0.68
Sugar,	0.71	0.70	0.69	0.68
Sodium Tripolyphosphate	0.29	0.29	0.28	0.28
Sodium nitrite	0.00125	0.00125	0.00125	0.00125

2.4. *Physicochemical determination*

The color of fish sausage was determined using the CIE system and reported as L* value (lightness), a* value (redness/greenness), and b* value (yellowness/blueness), using a Hunter Lab colorimeter [9]. Texture profile analysis (TPA) was performed using a TA-XT2 texture analyzer (TA.XT2 Stable Micro System, UK), as described by Zakaria (2018) [10] with some modification. Sausage samples were cut into a cylindrical shape (20 mm height × 25 mm diameter) in order to fit on the platform. A compression probe was used (P/6). Calibration of the load cell was also performed. The TPA textural parameters were as follows: pre-test speed 1 mm/s; test speed of 1 mm/s; and post test speed of 1 mm/s. The settings for probe height calibration were as follows: 30 mm distance speed; 8 mm/s return speed; and 5 g contact force. The samples were then placed on the center of the platform and the analysis was run with 75% of strain. The texture profile results in terms of hardness, springiness, cohesiveness, gumminess and chewiness were monitored. The emulsion stability (ES) was determined by centrifugation of samples obtained after the emulsification step at 4,000 rpm for 3 minutes at 25 °C and was calculated as described by Huang *et. al.* (2001) [10]. Cooking loss was measured according to the method of Gnanasambanbam and Zayas (1992) [11], as follows: 10 g of fish sausage was placed in a centrifuge tube and heated at 80°C for 30 min; after cooling to room temperature, the weight of the drained sample was measured; finally, the percentage of cooking loss was calculated from the weight difference between the unheated and heated sample. Water holding capacity (WHC) of samples was measured using the method of Motohiro (1981) [12], as follows: 10 g of fish sausage sample was placed in a centrifuge tube and centrifuged at 10,500 rpm for 15 minutes at 15 °C; the WHC was then expressed as percentage differences of the weight calculated from the sample before and after centrifugation.

2.5. Statistical analyses

All results were expressed as mean ± standard deviation. Results were analyzed by one-way ANOVA. Mean values were tested with Duncan's test for paired comparison, with a significance level of $\alpha = 0.05$, using the SPSS statistics software version 22.

3. Result and discussions

3.1. Quality of mixed-fish sausages

The fish sausage can be produced by mixing together different fish species (salmon, spotted knife fish, and seven-striped carp). Color results of all mixed fish sausages had the value of L* (68.42-74.66), a* (1.64-8.70), b* (12.97-18.82), h° (63.14-83.59), and chroma (13.64-20.94) as shown in Table 3.

Table 3. Effect of different fish types and ratio in mixed-fish sausages on color.

Samples	Lightness (L*)	Redness (a*)	Yellowness (b*)	Hue angle (h°)	Chroma (c*)
FS 1	74.66 ± 0.76^{a}	8.70 ± 0.54^{a}	18.62 ± 0.50^{a}	65.91 ± 0.34^{de}	20.94 ± 0.41^{a}
FS 2	69.05 ± 0.62^{e}	6.13 ± 0.41^{c}	12.81 ± 0.68^{e}	65.37 ± 0.81^{e}	14.69 ± 0.40^{f}
FS 3	66.93 ± 0.38^{f}	1.64 ± 0.15^{f}	15.37 ± 0.38^{c}	83.59 ± 0.36^{a}	15.71 ± 0.14^{e}
FS 4	72.86 ± 0.69^{bc}	8.27 ± 0.19^{a}	16.45 ± 0.14^{b}	63.14 ± 0.14^{f}	18.54 ± 0.09^{b}
FS 5	72.80 ± 0.69^{bc}	6.69 ± 0.20^{b}	16.35 ± 0.15^{b}	66.59 ± 0.15^{d}	17.78 ± 0.08^{c}
FS 6	71.68 ± 0.78^{d}	4.20 ± 0.09^{e}	12.97 ± 0.69^{e}	72.25 ± 0.11^{b}	13.64 ± 0.30^{g}
FS 7	72.05 ± 0.85^{cd}	5.28 ± 0.14^{d}	14.60 ± 0.26^{d}	70.44 ± 0.02^{c}	15.47 ± 0.07^{e}

FS 8	73.20±0.80^{b}	6.84±0.86^{b}	14.13±0.93^{d}	63.63±0.10^{f}	16.43±0.14^{d}
FS 9	68.42±0.86^{e}	5.89±0.62^{c}	14.30±0.52^{d}	66.29±0.79^{d}	15.60±0.12^{e}
FS 10	69.40±0.84^{e}	4.72±0.35^{de}	14.89±0.46^{cd}	72.01±0.39^{b}	15.68±0.15^{e}

*FS6 is a control.

*Values with the same superscript (a-g) within columns are not significantly different (p>0.05).

The results in terms of hardness, springiness, cohesiveness, gumminess, and chewiness of all mixed-fish sausages were in the range of 105.11-200.06 g, 0.93-0.98 sec, 0.47-0.59, 45.8-109.73 g, and 42.97-106.12 g•sec, respectively. The highest value of all texture characteristics was found in the FS7 formulation, as shown in Table 4.

Table 4. Effect of different fish types and ratio in mixed-fish sausages on texture analysis.

Samples	Texture profile analysis (TPA)			
	Hardness (g)	Springiness (sec)	Cohesiveness	Gumminess (g)
FS 1	163.39±0.40^{d}	0.95±0.00^{c}	0.53±0.01^{d}	95.77±0.14^{d}
FS 2	179.29±0.80^{c}	0.98±0.01^{ab}	0.55±0.01^{bcd}	102.39±0.98^{c}
FS 3	128.90±0.56^{h}	0.99±0.00^{a}	0.54±0.02^{d}	54.94±0.27^{h}
FS 4	196.69±0.59^{b}	0.97±0.01^{bc}	0.55±0.01^{cd}	103.56±0.56^{b}
FS 5	108.99±0.69^{i}	0.97±0.01^{abc}	0.49±0.01^{e}	51.68±0.25^{i}
FS 6	146.01±0.60^{f}	0.98±0.01^{ab}	0.58±0.01^{ab}	78.30±0.85^{f}
FS 7	200.06±0.69^{a}	0.93±0.03^{d}	0.59±0.03^{a}	109.73±0.56^{a}
FS 8	137.72±0.86^{g}	0.97±0.00^{abc}	0.48±0.01^{e}	69.01±0.24^{g}
FS 9	155.78±0.77^{e}	0.98±0.00^{ab}	0.57±0.01^{abc}	88.56±0.86^{e}
FS 10	105.11±0.47^{j}	0.97±0.02^{abc}	0.47±0.01^{e}	45.84±0.61^{j}

*FS6 is a control.

*Values with the same superscript (a-i) within columns are not significantly different (p>0.05).

The emulsion stability (ES) of all FSs did not show any significant differences (p<0.05) (Table 5), with values of ES in the range of 96.26-97.57%. The water holding capacity determines the protein quality that can bind or retain inherent water in the structure. The formulation FS 7 (salmon, spotted knife fish and seven-striped carp in the ratio of 0.33, 0.33, 0.33% (w/w)) showed the highest water holding capacity at 86.37±0.38%. After cooking, the quality of sausage emulsion was determined by the cooking loss. The eating quality is related to cooking loss; in other words, it should be reduced by the high percentage of cooking loss, thereby reducing product satisfaction. Results showed the values of cooking loss of all samples were significantly different ($p \leq 0.05$). However, the formula FS7 had the lowest percentage of cooking loss. According to the results of WHC and cooking loss compared to all of the developed formulation, the FS7 formulation was chosen for improving the texture of mixed-fish sausage, as shown in Table 5.

3.2. Effect of egg white addition on the chemical and physical properties of mixed-fish sausage

Egg white is generally used as a binder in food products. The addition of egg white changes the physico-chemical properties of mixed-fish sausage. The moisture content of egg white was 87.72 ± 0.22%. Due to this high-water content in the fresh egg white, increasing egg white from 1% to 3% in the mixed-fish sausage

FS7 formulation slightly increased both L* and h°, as shown in Table 6, but did not affect the hardness values of the mixed-fish sausage. However, the texture of EW1 showed the highest values for hardness (265.92 g), springiness (0.95 sec), cohesiveness (0.59), gumminess (95.60 g), and chewiness (98.47 g•sec), as shown in Table 7.

Table 5. Effect of different fish types and ratio in mixed-fish sausages on moisture contents, emulsion stability, water holding capacity, cooking loss, and pH value.

Samples	Moisture (%)	Emulsion stability (%)^{ns}	Water holding capacity (%)	Cooking loss (%)	pH Value
Control	74.19±0.53^{a}	97.04±0.55	77.39±0.87^{d}	9.62±0.11^{e}	6.86±0.00^{b}
FS 1	58.32±0.74^{f}	96.81±0.92	79.74±0.34^{c}	8.26±0.79^{cd}	6.54±0.01^{h}
FS 2	71.48±0.87^{b}	97.42±0.99	82.34±0.92^{b}	7.11±0.74^{bc}	6.74±0.01^{g}
FS 3	67.03±0.73^{d}	97.38±0.64	78.79±0.40^{cd}	8.78±0.54^{de}	6.82±0.01^{cd}
FS 4	67.51±0.62^{d}	96.30±0.95	81.95±0.36^{b}	6.85±0.65^{b}	6.81±0.00^{e}
FS 5	63.78±0.40^{e}	96.26±0.67	82.02±0.41^{b}	4.92±0.64^{a}	6.83±0.00^{c}
FS 6	74.19±0.53^{a}	97.04±0.55	77.39±0.87^{d}	9.62±0.11^{e}	6.86±0.00^{b}
FS 7	68.19±0.67^{d}	97.15±0.60	86.37±0.38^{ab}	4.63±0.37^{a}	6.82±0.00^{d}
FS 8	67.14±0.86^{d}	97.51±0.27	78.19±0.09^{cd}	9.29±0.57^{de}	6.79±0.00^{f}
FS 9	67.75±0.52^{d}	96.56±0.68	83.40±0.40^{b}	8.31±0.74^{cd}	6.78±0.01^{f}
FS 10	69.53±0.86^{c}	96.50±0.66	77.14±0.97^{d}	9.00±0.59^{de}	6.87±0.00^{a}

*FS6 is a control.
*Values with the same superscript (a-f) within columns are not significantly different (p>0.05).

Table 6. Effect of egg white in mixed-fish sausages on color.

Samples	Lightness (L*)	Redness (a*)	Yellowness (b*)	Hue angle (h°)	Chroma (c*)
Control	67.79±0.59^{d}	5.87±0.31^{d}	14.71±0.46^{d}	68.26±0.60^{a}	15.84±0.53^{d}
EW 1	68.60±0.79^{c}	10.97±0.20^{a}	17.38±0.39^{a}	57.70±0.43^{b}	20.56±0.41^{a}
EW 2	70.81±0.26^{b}	10.36±0.48^{b}	16.35±0.36^{b}	57.67±0.70^{b}	19.36±0.56^{b}
EW 3	71.97±0.38^{a}	9.71±0.37^{c}	16.20±0.48^{c}	58.96±0.46^{a}	18.73±0.37^{c}

*Values with the same superscript (a-d) within columns are not significantly different (p>0.05).

Table 7. Effect of egg white ratio in mixed fish sausages on texture profile.

Samples	Texture profile analysis (TPA)			
	Hardness (g)	Springiness (sec)	Cohesiveness	Gumminess (g)
Control	201.02±0.55^{b}	0.93±0.09^{a}	0.44±0.05^{b}	89.59±0.93^{c}
EW 1	265.92±0.52^{a}	0.95±0.02^{a}	0.59±0.05^{a}	95.60±0.94^{a}
EW 2	265.49±0.93^{a}	0.93±0.02^{a}	0.54±0.04^{a}	93.06±0.81^{b}
EW 3	265.40±0.71^{a}	0.90±0.03^{a}	0.54±0.02^{a}	93.34±0.85^{b}

*Values with the same superscript (a-d) within columns are not significantly different (p>0.05).

The addition of egg white at all levels increased the emulsion stability. Results of WHC of mixed-fish sausage gradually decreased with increased amounts of added of egg white. Increased egg white amounts changed the percentage of cooking loss. When the ovalbumin protein and fish protein were heated, sulfur containing amino acid formed a disulfide bond and conjugated with the hydrophilic group of fish protein, turning into a three-dimensional structure. Thus, this phenomenon helps to stabilize the protein structure [13]. During heating, gelation of egg white and mixed-fish emulsion occurred, resulting in adhesion between the egg white protein and the fish protein. As a result, the gelation of the sausage became brittle and easily breakable. The higher quantity of egg white reduced the appearance and softened the texture of fish sausage, due to the fresh egg white having a higher moisture content than fish fillet. Based on the results in terms of hardness, emulsion stability, cooking loss, and water holding capacity, the EW1 formulation, with 1% egg white in mixed-fish sausage, showed the proper characteristics of fish sausage as shown in Table 8.

Table 8. Effect of egg white in mixed fish sausages on physicochemical properties.

Samples	Moisture (%)	Emulsion stability (%)	Water holding capacity (%)	Cooking loss (%)	pH Value
Control	67.81±0.39^{c}	99.79±0.05^{b}	84.30±0.48^{a}	1.96±0.19^{a}	6.89±0.00^{b}
EW 1	68.55±0.39^{b}	99.90±0.04^{a}	81.97±0.86^{b}	1.42±0.45^{b}	6.87±0.00^{d}
EW 2	68.47±0.69^{b}	99.85±0.05^{ab}	81.10±0.81^{bc}	1.49±0.26^{b}	6.89±0.00^{c}
EW 3	69.37±0.20^{a}	99.83±0.07^{ab}	80.53±0.12^{c}	1.52±0.24^{b}	6.91±0.00^{a}

*Values with the same superscript (a-d) within columns are not significantly different ($p>0.05$).

4. Conclusions

The mixed-fish sausages of formulation FS7, with salmon, spotted knife fish and seven striped carp at the ratio of 0.33:0.33:0.33% (w/w), had the highest water holding capacity (86.37%), but lowest cooking loss (4.63%). Egg white was chosen to improve the texture property of FS7 mixed-fish sausage. The addition of egg white at 1% to FS7 fish sausage increased the moisture content but reduced the water holding capacity and cooking loss. Therefore, in this experiment, the maximum usage of egg white in mixed-fish sausage was not greater than 1%. Dried egg white could be further studied for its effect on the texture properties of mixed-fish sausage.

Acknowledgement

The research was funded by Thailand institute of scientific and technology research (*(6116301010* and graduate college, King Mongkut's University of Technology North Bangkok. The authors would like to thank Mr. Peter Blumenfeld for proof reading.

References

[1] Pigott G M and Tuck B E 1990 Seafood: effect of technology in nutrition. *Marcel Dekker Inc* NewYork 362.
[2] Connor W E 2000 Importance of *n*-3 fatty acid in health and disease. *Am J Clin Nutr* **71** 171-172.
[3] Hutakovit W, Soiraya B, Suthibut C, Suthibut N, Sakulyunyongsuk N, Bunyasawat J and Sotarayam

T 2008 Development of value added food product from freshwater fish. Faculty of Home Economics Technology Rajamangala University of Technology Phra Nakhon 126.

[4] Intarasirisawat R, Benjakul B and Visessanguan B 2014 Effects of skipjack roe protrin hydrolysate on properties and oxidative stability of fish emulsion sausage. *Lebensmittel-Wissenschaft und-Technologie-Food Sci Tech* **58** 280-286.

[5] Chumreang S, Sangjindavong M and Pluksawanich K 2014 Uses of potassium bromate and egg white for increasing gel forming ability of surimi. *Kasetsart J* **33 (1)** 102-110.

[6] Namatsila Y 2006 Development of Chinese style sausage from salmon. Master Thesis Prince of Songkla University Thailand 135.

[7] Juntachote T 2018 Influence of carrageenan Addition on Physico-chemical and sensory properties of low fat chicken sausages. *The Journal of KMUTNB* **28(3)** 605-616.

[8] Pakoawnoy N and Gawborism S 2014 Utilization of fish trim meat for material cost reduction in tilapia sausage processing. *Khon Kaen AGR J* **42(1)** 38-43.

[9] Zakaria N and Sarbon N 2018 Physicochemical properties and oxidative stability of fish emulsion sausage as influenced by snakehead (*Channa striata*) protein hydrolysate. *Food Sci Tech* **94** 13-19.

[10] Huang S Tsai Y C and Chen C 2011 Effects of wheat fiber oat fiber and inulin on sensory and physicochemical properties of Chinese-style sausages *Asian Australasian J. Ani Sci* **24** 875-880.

[11] Gnanasambandam R and Zayas J F 1992 Functionality of wheat germ protein in comminuted meat products as compared with corn germ and soy protein. *J Food Sci* **57** 829-833.

[12] Motohiro T 1981 General aspect of processing marine food Science of Processing Marine Food Products Vol ll Kanagawa International Fisheries Training Center Japan International Cooperation Agency Japan.

[13] Krasaechol N 2017 Effect of calcium chloride and egg white powder on quality of Marinated Ground-Fish from Indian Mackerel (*Rastrelliger kanagurta*) *Burapha Sci* **22(3)** 569-580.

Anethum graveolens as a possible modulator of testicular steroidogenesis

E Tvrda[1,4], T Brenkus[1], M Duracka[1], R Kirchner[2] and J Arvay[3]

[1]Department of Animal Physiology, Faculty of Biotechnology and Food Sciences, Slovak University of Agriculture in Nitra, Tr. A. Hlinku 2, 94976 Nitra, Slovakia
[2]University Vivarium, Botanical Garden, Slovak University of Agriculture in Nitra, Tr. A. Hlinku 2, 94976 Nitra, Slovakia
[3]Department of Chemistry, Faculty of Biotechnology and Food Sciences, Slovak University of Agriculture in Nitra, Tr. A. Hlinku 2, 94976 Nitra, Slovakia

E-mail: evina.tvrda@gmail.com

Abstract. The aim of this study was to evaluate the possible *in vitro* effects of *Anethum graveolens* L. (dill) extract on the steroidogenesis in testicular tissue. Specifically, we focused to determine the dominant chemical components of the dill extract followed by a subsequent assessment of its *in vitro* effects on the production of cholesterol, dehydroepiandrosterone, androstenedione and testosterone by rat testicular fragments. Dill extract was subjected to high performance liquid chromatography (HPLC) which determined cyranosid, carvone, chlorogenic acid, p-hydroxybenzoic acid and salicylic acid among the predominant biomolecules. For the *in vitro* experiments, testicular fragments from 12 sexually mature Wistar rats were incubated in the absence (control) or presence of 10, 100 or 1000 μg/mL extract for 24 h. Cholesterol levels were quantified via photometry and the hormones were evaluated by Enzyme-Linked Immunosorbent Assay (ELISA). Our data showed that the release of cholesterol and androstenedione (but not dehydroepiandrosterone and testosterone) by the testicular fragments obtained from seminiferous tubules was significantly affected by the dill extract in a dose-dependent manner. Exposure to the extract led to a significant decrease of both cholesterol as well as androstenedione ($P<0.01$ in case of 10 μg/mL; $P<0.001$ with respect to 100 and 1000 ug/mL extract). Our results suggest that the bioactive molecules present in *Anethum graveolens* L. could have dose-dependent impact on the secretion of selected male reproductive hormones, playing a role in the regulation of testicular steroidogenesis. Nevertheless, the exact effect of individual biomolecules on the endocrine testicular function has yet to be investigated.

1. Introduction

The role of medicinal herbs in the management of health issues has been known for centuries, and plant-based therapy has contributed significantly to the creation of modern drugs and nutraceuticals. Moreover, more than 60 % of the world's population relies on bioactive ingredients from herbs used for medical purposes [1]. The World Health Organization is a strong advocate in favour of the use of medicinal herbs and encourages scientists to emphasize on a rational use of a vast array of plants for the development of new pharmaceutical remedies [2]. A variety of plant extracts are nowadays used in the prevention or treatment of male subfertility, including hypoactive sexual desire disorder, hormonal imbalance, erectile and ejaculation dysfunction as well as sub-standard semen parameters [3].

Multiple plants have been traditionally used to counteract male infertility, based on their enhancing

Content from this work may be used under the terms of the Creative Commons Attribution 3.0 licence. Any further distribution of this work must maintain attribution to the author(s) and the title of the work, journal citation and DOI.
Published under licence by IOP Publishing Ltd

effects on the testicular steroidogenesis and spermatogenesis. Rose mallow (*Hibiscus*) and Malabar Spinach (*Basella alba*) have been shown to stimulate the testosterone secretion in adult males and are currently recommended as a complementary therapy to counteract sexual fatigue [4]. Extracts obtained from Peruvian ginseng (*Lepidium meyenii*) have been shown to activate different stages of spermatogenesis in healthy rats, as well as to alleviate testicular stress caused by heavy metal intoxications [5]. In the meantime, pomegranate (*Punica granatum*) juice administered to rats has exhibited the ability to increase sperm count and motion characteristics, density of spermatogonia and spermatocytes, as well as the volume of seminal vesicles and male reproductive cells [6]. These examples indicate emphasize on the need to further investigate the potential modulatory effects plant extracts may exhibit on the testicular structure and function.

Anethum graveolens L. (dill) has been known as an ethnomedicinal plant since ancient times and it is still a popular herb widely used as a spice and as a staple for the production of essential oils. Dill is an aromatic herb found in the *Apiaceae* family. The most prominent volatile biomolecules found in dill include carvone (dill seeds), α-phellandrene, limonene, dill ether and myristicin (dill herb) [7, 8].

Dill is used to relieve colic pain and flatulence in young children. The essential oil relieves intestinal spasms and griping, improves appetite and aids digestion. Dill triggers milk production in lactating females, and it also helps to treat urinary complaints, piles and mental disorders [8, 9].

A number of *in vivo* studies have suggested that dill extracts have the ability to stimulate the spermatogenic as well as steroidogenic capacity in rodents [10, 11]. *In vitro* studies on primary cell cultures and tissues are nevertheless a useful strategy to understand the behaviour of potentially beneficial or harmful biomolecules because of their consistent functional characteristic and increased sensitivity [12, 13]. Taking into account the essential roles of sex steroids in the growth, differentiation and function of the male reproductive system, it is important to investigate the impact of plant extracts on male steroidogenesis. We are unconscious of any reports that have investigated the direct *in vitro* effect of the dill extract on the endocrine processes involved in reproductive functions. As such, our aim was to assess the impact of selected concentrations of the dill extract on the secretion of cholesterol, testosterone, dehydroepiandrosterone (DHEA) and androstenedione by rat testicular fragments. At the same time, we determined the predominant bioactive compounds present in the dill extract and associated their occurrence with the potential to modulate testicular steroidogenesis.

2. Materials and methods

2.1. Plant collection and preparation of the extract

Dill was harvested at the Botanical Garden, Slovak University of Agriculture (Nitra, Slovakia) in the summer of 2016. Fresh plant material was dried and crushed, weighed and immersed in 96% ethyl alcohol (Centralchem, Bratislava, Slovakia). The maceration took two weeks at 20-22°C in the absence of light. The resulting extract was placed into a rotary evaporator (Cole-Parmer, Stone, UK) and processed under reduced pressure at 40°C (vacuum pump, KNF Neuberger, Freiburg, Germany) in order to dispose of any remaining ethyl alcohol. The obtained plant material was resuspended in DMSO (dimethyl sulfoxide; Sigma-Aldrich, St. Louis, Missouri, USA). The final concentration of the stock solution was set to 1000 mg/mL [14].

For the HPLC assessment, fresh plant material was freeze dried and pulverized. Twenty-five mL of 80% methanol (suitable for HPLC; Sigma-Aldrich, St. Louis, Missouri, USA) were added to 1 g of the plant material. The resulting mixture was agitated using a shaker (250 RPM; 20-22°C; 8 h), filtered and stored at 5°C.

2.2. HPLC-DAD analysis

Specific HPLC standards, methanol (suitable for HPLC), acetonitrile (suitable for HPLC) and phosphoric acid (ACS grade) were provided by Sigma-Aldrich (St. Louis, Missouri, USA). Double deionized water was obtained using a purification system (Simplicity 185; Millipore SAS, France). Each standard (0.5 mg) was dissolved in 10 mL methanol. The methanolic standards and plant extracts were filtered through a filter paper (no. 390, Munktell & Filtrak, Niederschlag, Germany) followed by the Q-Max syringe filter (0.22 mm diameter; Frisenette, Knebel, Denmark).

HPLC assessment of the dill extract was performed with the 1260 Infinity II LC System (Agilent Technologies, Santa Clara, California, USA). The analysis was executed using a Purosphere reverse phase C18 column (4 mm x 250 mm x 5 mm; Merck, Kenilworth, New Jersey, USA). The mobile phase was composed of acetonitrile and 0.1% phosphoric acid in double deionized water. The starting flow rate was set to 1 mL/min with 10 mL of the injection volume. The column temperature was set to 30°C while the standards and samples were in the sample manager at 4 °C. Data collection and analysis was carried out using the Agilent OpenLab ChemStation software for LC 3D Systems [15]. The extract was analysed in triplicate.

2.3. Collection of testicular fragments and in vitro culture
Male Wistar rats (n=12, 120 days old) obtained from the Institute of Experimental Pharmacology (Dobrá Voda, Slovakia) were kept in plastic cages at 24±1 °C and 12 h light/12 h dark regime. The animals were provided with a standard laboratory feed and water *ad libitum*. Institutional and national regulations for the care and use of laboratory animals were followed, and all procedures were approved by the Slovak State Veterinary and Food Institute and Ethics Committee (no. 3398/11-221/3).

The rats were anesthetized by intraperitoneal injection with chloral hydrate, the testes were removed, deposited into a sterile Petri dish with fresh, sterile-filtered Dulbecco's phosphate buffered saline (PBS; Sigma-Aldrich, Missouri, USA) and dissected into 4 approximately equal fractions.

Seminal tubules were isolated from the interstitial tissue and mechanically dissociated with the help of tweezers and a blade knife. In order to confirm the steroidogenic activity within the heterogenous cell suspension collected from the testicular fragments, the 3β-hydroxysteroid dehydrogenase (3β-HSD) staining was performed following the protocol introduced by Klinefelter et al. [12] and modified by Jambor et al. [13]. Sample aliquots were incubated with a mixture consisting of nitrotetrazolium blue chloride (Sigma-Aldrich, Missouri, USA), dehydroisoandrosterone-3sulfate (Sigma-Aldrich, Missouri, USA) and nicotinamide adenine dinucleotide (Sigma-Aldrich, Missouri, USA) at 37°C for 2 h. Subsequently the cell suspensions were observed under a light microscope (Leica, Wetzlar, Germany) for the occurrence of blue formazan crystals, indicating the activity of 3β-HSD.

The collected fragments were washed twice in PBS and placed into 1 ml DMEM (Dulbecco's Modified Eagle Medium, Sigma Aldrich, Missouri, USA) enriched with 10% foetal bovine serum (Sigma Aldrich, Missouri, USA), 1% penicillin-streptomycin solution (Sigma Aldrich, Missouri, USA) and without (control) or with the *Anethum* extract supplemented at various concentrations (10, 100, 1000 µg/mL) for 24 h. The *in vitro* culture was carried out in 24-well plates at 37 °C and 5% CO_2. Four independent experiments were carried out on 4 individual days using separate pools of testes collected from 12 rats.

2.4. Assessment of cholesterol and steroid hormones
Following incubation, the culture medium was collected and centrifuged (300 x g, 4 °C, 10 min). The supernatant was subjected to the quantification of cholesterol and sex hormones. Cholesterol levels were assessed using the DiaSys commercial kit (Diagnostic Systems, Holzheim, Germany) and the Rx Monza photometer (Randox Laboratories, Crumlin, UK). Testosterone, DHEA, and androstenedione were quantified using commercial ELISA (enzyme-linked immunosorbent assay) kits (Dialab, Wiener Neudorf, Austria). Concentrations of steroid hormones were measured at 450 nm using the Glomax Multi+ plate spectrophotometer (Promega, Wisconsin, USA). The analysis was performed in duplicates.

2.5. Statistical analysis
GraphPad Prism 3.02 (GraphPad Software, San Diego, California, USA) was used for the statistical data evaluation. One-way analysis of variance (ANOVA) and the Dunnett's test were selected for advanced statistical calculations. Bartlett's test was used to test the assumption that the obtained data came from populations with equal variances, while the F test assessed the homogeneity of the variance. The results

are expressed as the mean±standard deviation (SD). The level of significance was set at ***P<0.001; **P< 0.01 and *P< 0.05.

3. Results and discussion

Concentrations of the specific biomolecules identified by HPLC are displayed in table 1. The predominant biomolecule found in the extract was cyranosid (618.15±9.94 mg/kg). With respect to the phenolic acids; chlorogenic, neochlorogenic, p-hydroxybenzoic, salicylic acid and vanillic acid were identified, amongst which chlorogenic acid was the most abundant (533.36±8.89 mg/kg).

Table 1. Major biomolecules determined [mg/kg] in the dill extract.

Cynarosid	618.15±9.94
Carvone	45.60±4.09
Limonene	17.51±2.66
Myristicin	11.35±1.22
Chlorogenic acid	533.36±8.89
Neochlorogenic acid	107.67±1.98
p-hydroxybenzoic acid	12.04±2.27
Salicylic acid	17.27±1.89
Vanillic acid	8.77±2.01

(Mean±SD)

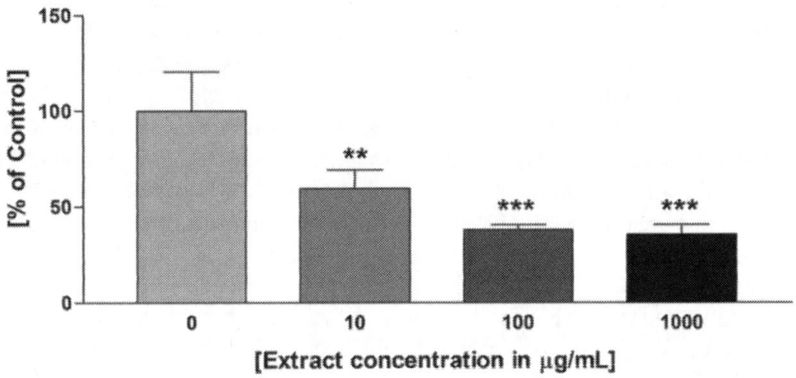

Figure 1. The in vitro impact of the dill extract on the cholesterol levels in cells collected from rat seminiferous tubules. Each bar represents mean (±SD) values for cholesterol expressed as % of the control (0 µg/mL extract) and experimental groups (exposed to 10, 100 and 1000 µg/mL extract). ***P< 0.001; **P< 0.01.

As a progenitor of steroid molecules, cholesterol plays an essential role in the synthesis of sex hormones. As revealed in figure 1 the cholesterol levels were noticeably different between the control (0.22±0.11 µg/mL) and experimental groups (0.14±0.05 µg/mL in case of 10 µg/mL extract; 0.08±0.02 µg/mL with respect to 100 µg/mL extract; 0.07±0.03 µg/mL in relation to 1000 µg/mL extract). We detected a significant decrease of the cholesterol quantity, depending directly on the dose of extract used (P<0.01 with respect to 10 µg/mL; P<0.001 in case of 100 and 1000 µg/mL extract).

Similarly to cholesterol, androstenedione production of testicular fragments exposed to the dill extract (figure 2) was affected significantly (P<0.01 in terms of 10 µg/mL; P<0.001 in case of 100 and 1000 µg/mL extract). The levels of androstenedione decreased depending on the concentration of the extract used in the experiment (18.29±2.26 ng/mL in case of 10 µg/mL extract; 15.25±1.99 µg/mL with respect to 100 µg/mL extract; 5.61±0.67 µg/mL in relation to 1000 µg/mL extract) in comparison with the control (23.08±3.01 ng/mL).

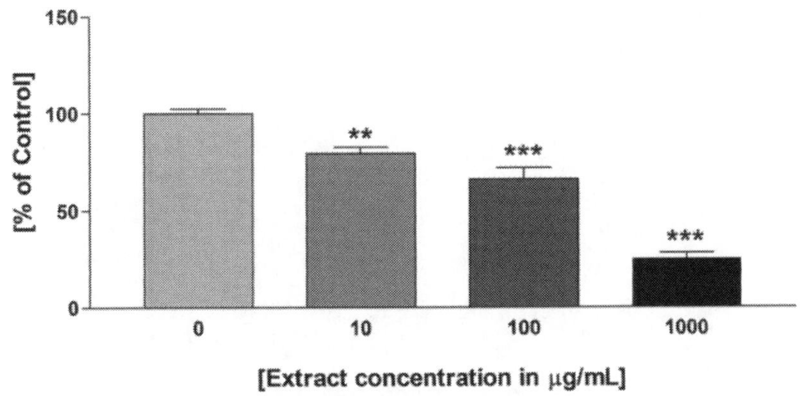

Figure 2. The in vitro impact of the dill extract on the androstenedione production in cells collected from rat seminiferous tubules. Each bar represents mean (±SD) values for cholesterol expressed as % of the control (0 μg/mL extract) and experimental groups (exposed to 10, 100 and 1000 μg/mL extract). ***$P< 0.001$; **$P< 0.01$.

As shown in figure 3, dehydroepiandrosterone concentrations were not significantly different between the control and experimental groups. A small decrease was observed following exposure to 100–1000 μg/mL (9.07±0.84 – 8.55±0.91 μg/mL) of the extract, however without significant differences ($P>0.05$) in comparison with the control (9.57±0.74 μg/mL). The analysis showed that the *Anethum graveolens* extract did not affect the concentration of dehydroepiandrosterone in the culture medium.

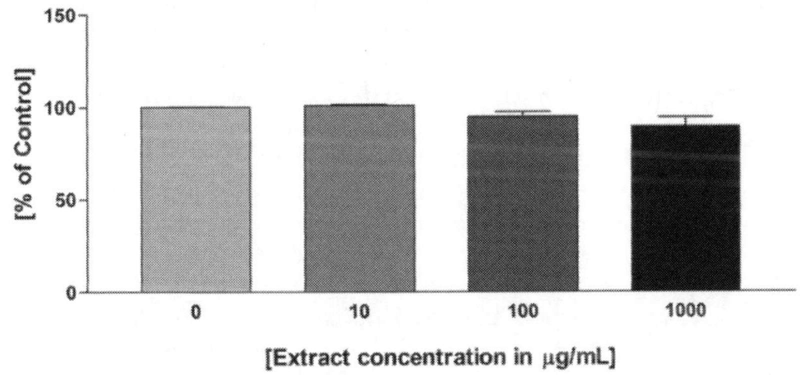

Figure 3. The in vitro impact of the dill extract on the dehydroepiandrosterone levels in cells collected from rat seminiferous tubules. Each bar represents mean (±SD) values for cholesterol expressed as % of the control (0 μg/mL extract) and experimental groups (exposed to 10, 100 and 1000 μg/mL extract).

Similarly, testosterone concentrations were not significantly affected by the exposure to the *Anethum* extract (figure 4; $P>0.05$). A minor decrease of testosterone production was recorded in all experimental groups (14.87±2.09 ng/mL in case of 10 μg/mL extract; 15.18±2.50 ng/mL with respect to 100 μg/mL extract; 14.22±2.91 ng/mL in relation to 1000 μg/mL extract), however without significant differences ($P>0.05$) in comparison with the control (15.82±1.92 ng/mL). The analysis showed that the dill extract did not have an impact on the quantity of testosterone in the culture medium.

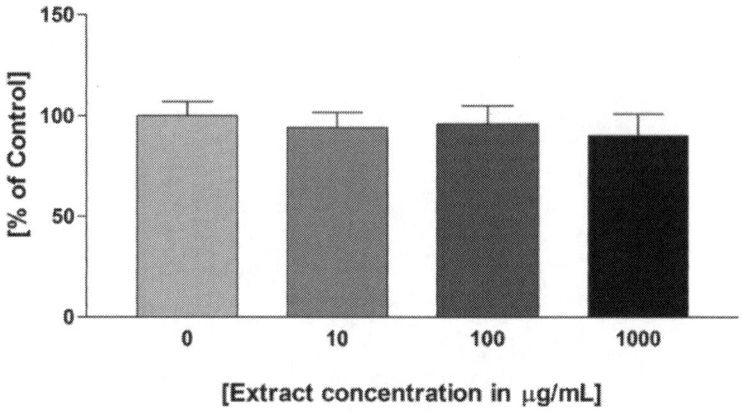

Figure 4. The in vitro impact of the dill extract on the testosterone levels in cells collected from rat seminiferous tubules. Each bar represents mean (±SD) values for cholesterol expressed as % of the control (0 μg/mL extract) and experimental groups (exposed to 10, 100 and 1000 μg/mL extract).

Outcomes from scientific research focused on the use of plants and their extracts traditionally used in etnotherapy may be appropriately interpreted following a hierarchy of significance. Data obtained from properly designed *in vivo* studies primarily contribute to the overall understanding of the beneficial or adverse effects plant extracts have on the living organism. Nevertheless, we chose an in vitro approach, which may provide invaluable information and serve as a reasonable predictor of possible clinical outcomes. Such in vitro strategy has the ability to clarify the contradictory and often controversial knowledge gathered so far on the impact of plant extracts on the living organism [16]. In this case, we focused to assess the chemical composition of the *Anethum graveolens* extract and subsequently to investigate its effects on the steroidogenic activity of rat testicular tissue.

Preliminary data are available on the biochemical description and possible biological effects of dill and its extracts or essential oils. A complex assessment of the biomolecules present in 8 dill cultivars was provided by Al Ahl and Omer [17]. Similarly to our results, limonene (9.60–18.0%) and carvone (81.35–89.98%) were the predominant biomolecules detected in European cultivars. Interestingly, and correspondingly to our data, neither dillapiole nor piperitone were detected in essential oils prepared form European cultivars, whereas abundant amounts of these biomolecules were found in African dill cultivars. As such, we may agree that the exact biocomposition of medicinal plants and their products depends heavily on the cultivar, region, climate conditions as well as the age of the plant during collection [17, 18]. Szopa and Eikert [19] furthermore state that *in vitro* cultures of *Anethum graveolens* could become a suitable model to understand the accumulation of a variety of phenolic acids, as well as a possible source of p-hydroxybenzoic acid and o-hydroxybenzoic (salicylic) acid for biotechnological use. The occurrence of both acids was confirmed by our HPLC analysis as well.

To our knowledge, no report exists on the analysis of the production of reproductive hormones by testicular tissue in vitro following exposure to dill extract. A number of *in vivo* studies have however assessed the effects of aqueous *Anethum* extract on the male reproductive system in rats. Contrary to our observations, neither Iamsaard et al. [11] nor Monsefi et al. [10] detected no negative or harmful effects of the exposure to the dill extract on the body growth, sperm concentration, motility or morphology, and as such, the extract did not exhibit any negative effect on the epididymal sperm maturation. Based on both studies, we may speculate that the spermatogenic function of testicular tissue may not be affected by the presence of the dill extract or its biomolecules. Furthermore, morphometrical evaluations revealed that the diameter of seminiferous tubules and density of spermatogenic cells were not different between the control and experimental groups, indicating no significant damage to the structural integrity of male reproductive organs potentially caused by the *Anethum* extract. Surprisingly, both studies also observed normal testosterone levels in the groups receiving treatment in comparison with the control, suggesting that the complex hypothalamic-pituitary-gonadal axis in males was not disrupted by the dill extract. Nevertheless, our *in vitro* study shows that the *Anethum* extract dos exhibit significant modulatory effects on the endocrine activity by decreasing the levels of cholesterol as well as androstendione by rat testicular fragments.

Assuming that the dill extract does not exhibit a substantial effect on the response of male gonads towards the luteinizing hormone (LH) and follicle-stimulating hormone (FSH), the decrease of cholesterol and androstenedione detected in our study may be linked to the actual testosterone biosynthesis within testicular Leydig cells. Testosterone as a steroid hormone, is derived from cholesterol, with two subsequent pathways that may be followed. While the Δ-4 pathway is based on the conversion of androstenedione, the most important intermediate of the Δ-5 pathway is DHEA [20].

In our study, a significant dose-dependent decrease of cholesterol was recorded. Logically, its decrease should trigger a subsequent decline of all observed hormones, as recorded in a number of previous reports [12, 21-23], however that was not the case in our experiments. We may speculate that the decreasing concentration of cholesterol did not affect the Δ-5 pathway, being reflected on non-significant changes in the production of DHEA. This may subsequently become the primary source for further conversion to testosterone as previously suggested by Miller [24]. On the other hand, the decreasing trend of cholesterol was well-mirrored by a significant decrease of the most important intermediate of the Δ-4 pathway, androstenedione. As such, we may hypothesize that the bioactive compounds present in the dill extract have led to a shift in the balance of the testicular steroidogenesis towards the Δ-5 pathway via DHEA-S, leading to lower levels of cholesterol available for androstenedione synthesis. A similar observation was reported by Zimmermann et al. [25]. Cholesterol is considered to be the building block of all the steroid hormones, and it is well known that low cholesterol levels contribute to poor hormone production [26]. In our case, the decreasing levels of cholesterol might have had an impact on the dynamics of male steroidogenesis, however the end product (testosterone) was not significantly affected. We may assume that although lower, the levels of cholesterol were still high enough to provide substrate for subsequent reactions of steroid hormone biotransformation [25, 27]. At the same time, based on the fact, that two pathways exist to produce testosterone, we may hypothesize that the Δ-5 pathway was less affected by the biomolecules present in *Anethum* than the Δ-4 pathway, and thus compensated for the imbalance caused by the administration of dill extract. Similar changes in the balance of testosterone biosynthesis were reported by Jambor et al. [13]. Nevertheless, to verify our hypothesis, it would be necessary to quantify the intermediates of both pathways, i.e. progesterone and androstenediol. Furthermore, although a non-significant, but a decreasing trend was observed in case of both DHEA and testosterone, which may become more pronounced with the increasing time of *in vitro* culture. As such, it may be important to prolong the culture of testicular fragments in the presence of the dill extract in order to re-evaluate its effect on the steroidogenesis from a long-term point of view.

The interplay of bioactive molecules present in the dill extract is another point to be addressed in further studies. While it has been shown that cynarosid, carvone, limonene and myristicin present in the dill extract may exhibit beneficial effects on the male reproductive system *in vivo* [28-30], it is important to assess their synergy or antagonism in a complex mixture such as a plant extract. Moreover, the exact *in vitro* effect of individual biomolecules on the endocrine testicular function has yet to be investigated. Finally, future *in vitro* studies are highly anticipated in order to assess the impact of medicinal plants and their properties in specialized cells and tissues in order to predict their behavior in a complex organism.

4. Conclusion

Our preliminary results offer new evidence on the biological impact of the *Anethum graveolens* extract on male reproductive tissues. The application of medicinal plants with a wide array of beneficial properties into practical andrology is unquestionably necessary. Our data evidently cannot predict an indisputable *in vivo* or *in vitro* conclusion as a direct impact of dill extract on fertility in males has to be addressed further. To transform our observations into practice, reports on the possible toxicity, pharmacokinetic behaviour and bioavailability of dill and its products in relation to the functional activity of the male reproductive system are more than necessary.

Acknowledgment
This study was supported by the European Community Project no. 26220220180: Building Research Centre "AgroBioTech", by the VEGA Project no. 1/0039/16, and by the APVV Grant no. APVV-15-0544.

References

[1] Kumar S, Paul S, Walia Y K, Kumar A and Singhal P 2015 Therapeutic potential of medicinal plants: A review *J. Biol. Chem. Chron.* **1** 46-54
[2] WHO 2002 *WHO Traditional Medicine Strategy 2002–2005* J D Quick, J Sawyer, P Thorpe, D Whitney and X Zhang edis. (Geneva: World Health Organization)
[3] Kamal R, Gupta R S and Lohiya N K 2003 Plants for male fertility regulation *Phytother. Res.* **17** 579-90
[4] Moundipa F P, Kamtchouing P, Koueta N, Tantchou J, Foyang N P and Mbiapo F T 1999 Effects of aqueous extracts of *Hibiscus macranthus* and *Basella alba* in mature rat testis function *J. Ethnopharmacol.* **65** 133-9
[5] Rubio J, Riqueros M I, Gasco M, Yucra S, Miranda S and Gonzales G F 2006 *Lepidium meyenii* (Maca) reversed the lead acetate induced-damage on reproductive function in male rats *Food Chem. Toxicol.* **44** 1114-22
[6] Lin W H, Tsai M T, Chen Y S, Hou R C, Hung H F, Li C H, Wang H K, Lai M N and Jeng K C 2007 Improvement of sperm production in subfertile boars by *Cordyceps militaris* supplement *Am. J. Chin. Med.* **35** 631-41
[7] Blank I and Grosch W 1991 Evaluation of potent odorants in dill seed and dill herb (*Anethum graveolens* L.) by aroma extract dilution analysis *J. Food Sci.* **56** 63-7
[8] Jana S and Shekhawat G S 2010 *Anethum graveolens*: An Indian traditional medicinal herb and spice *Pharmacogn. Rev.* **4** 179-84
[9] Duke J A 2001 *Handbook of Medicinal Herbs* (London: CRC Press)
[10] Monsefi M, Zahmati M, Masoudi M and Javidnia K 2011 Effects of *Anethum graveolens* L. on fertility in male rats *Eur. J. Contracept. Reprod. Health Care* **16** 488-97
[11] Iamsaard S, Prabsattroo T, Sukhorum W, Muchimapura S, Srisaard P, Uabundit N, Thukhammee W and Wattanathorn J 2013 *Anethum graveolens* Linn. (dill) extract enhances the mounting frequency and level of testicular tyrosine protein phosphorylation in rats *J. Zhejiang Univ. Sci B.* **14** 247-52
[12] Klinefelter R G, Hall F P and Ewing L L 1987 Effect of luteinizing hormone deprivation *in situ* on steroidogenesis of rat Leydig cells purified by a multistep procedure *Biol. Reprod.* **36** 769e783
[13] Jambor T, Greifova H, Kovacik A, Kovacikova E, Tvrda E, Forgacs Z, Massanyi P and Lukac N 2018 Parallel effect of 4-octylphenol and cyclic adenosine monophosphate (cAMP) alters steroidogenesis, cell viability and ROS production in mice Leydig cells *Chemosphere* **199** 747-54
[14] Tvrdá E, Michalko J, Matušíková I and Lukáč N 2016 *In vitro* effects of the *Chlamydomonas reinhardtii* extract on bovine spermatozoa *JMBFS* **6** 972-5
[15] Luksic L, Árvay J, Vollmannová A, Tóth T, Skrabanja V, Trcek J, Germ M and Kreft I 2016 Hydrothermal treatment of Tartary buckwheat grain hinders the transformation of rutin to quercetin *J. Cer. Sci.* **72** 131-4
[16] Brandt K and Seal C 2006 Advantages and pitfalls of different types of studies for investigations of the impact of food on health Paper presented at Joint Organic Congress (Odense, Denmark: University of Newcastle upon Tyne, School of Agriculture)
[17] Al Ahl H S and Omer E 2016 Essential oil content and chemical composition of eight dill

(*Anethum graveolens* L) cultivars cultivated under Egyptian conditions *Int. J. Pham. Pharm. Sci.* **8** 227-31

[18] Aminzadeh M, Amiri F, Ashor Abadi E, Mahdevi K and Fadai S 2010 Factors affecting on essential chemical composition of *Thymus kotschyanus* in Iran *World Appl. Sci. J.* **8** 847-56

[19] Szopa A and Eikert H 2015 *Anethum graveolens* L. *in vitro* cultures – a potential source of bioactive metabolies, phenolic acids and furanocoumarins *Acta Biol. Cracoviens Botanica* **57** 29-37

[20] Winters S J and Clark B J 2003 Testosterne synthesis, transport, and metabolism *Androgens in Health and Disease Contemporary Endocrinology* C J Bagatell and W J Bremner eds. (Totowa, NJ: Humana Press) pp 3-22

[21] Hou J W, Collins D C and Schleicher R L 1990 Sources of cholesterol for testosterone biosynthesis in murine Leydig cells *Endocrinology* **127** 2047-55

[22] Sriraman V, Sairam M R and Rao A J 2003 Evaluation of relative roles of LH and FSH in regulation of differentiation of Leydig cells using an ethane 1,2-dimethylsulfonate-treated adult rat model *J. Endocrinol.* **176** 151-61

[23] Gupta R S, Kachhawa J B S, Khushalani V, Tanwar K and Joshi Y C 2006 Effect of *Cressa cretica* methanol extract on testicular function of albino rats *Pharm. Biol.* **44** 382-8

[24] Miller W L 2002 Androgen biosynthesis from cholesterol to DHEA *Mol. Cell Endocrinol.* **198** 7-14

[25] Zimmerman Y, Foidart J M, Pintiaux A, Minon J M, Fauser B C, Cobey K and Coelingh Bennink H J 2015 Restoring testosterone levels by adding dehydroepiandrosterone to a drospirenone containing combined oral contraceptive: I. Endocrine effects *Contraception* **91** 127-33

[26] Kelly D M and Jones T H 2013 Testosterone: a metabolic hormone in health and disease *J. Endocrinol.* **217** R25-45

[27] Zimmerman Y, Eijkemans M J, Coelingh Bennink H J, Blankenstein M A and Fauser B C 2014 The effect of combined oral contraception on testosterone levels in healthy women: A systematic review and meta-analysis *Hum. Reprod. Update* **20** 76-105

[28] Tajuddin, Ahmad S, Latif A, Qasmi I A and Amin K M 2005 An experimental study of sexual function improving effect of *Myristica fragrans Houtt.* (nutmeg) *BMC Complement. Altern. Med.* **5** 16

[29] Wilson M J, Lindgren B R and Sinha A A 2008 The effect of dietary supplementation with limonene or myo-inositol on the induction of neoplasia and matrix metalloproteinase and plasminogen activator activities in accessory sex organs of male Lobund-Wistar rats *Exp. Mol. Pathol.* **85** 83-9

[30] Wang X, Tao R, Yang J, Miao L, Wang Y, Munyangaju J E, Wichai N, Wang H, Zhu Y, Liu E, Chang Y and Gao X 2017 Compounds from *Cynomorium songaricum* with estrogenic and androgenic activities suppress the oestrogen/androgen-induced BPH process *Evid. Based Complement. Alternat. Med.* **2017** 6438013

IOP Conf. Series: Earth and Environmental Science **346** (2019) 012050 doi:10.1088/1755-1315/346/1/012050

Effect of barrel temperatures and starch type on some properties of extruded glass noodles

C Thapnak[1], V Rungsardthong[1], S Vatanyoopaisarn[1], C Vimolchalao[2], C Puttanlek[3], D Uttapap[4], J Wongsa[5] and B Thumthanaruk[1, 6]

[1]Deparment of Agro-Industrial, Food, and Environmental Technology, Faculty of Applied Science, Food and Agro-Industry Research Center, King Mongkut's University of Technology North Bangkok, Bangkok, Thailand
[2]Thai Vermicelli Industry Co., Ltd.775, 777, 779, Song Wat Rd, Khwaeng Chakkrawat, Khet Samphanthawong, Bangkok 10100, Thailand
[3]Deparment of Biotechnology Technology, Faculty of Engineering and Industrial Technology, Silapakorn University, Nakhon Pathom, Thailand
[4]Division of Biochemical Technology, School of Bioresources and Technology, King Mongkut's University of Technology Thonburi, Bangkok, Thailand
[5]Department of Agricultural Engineering for Industry, Faculty of Industrial Technology and Management, King Mongkut's University of Technology North Bangkok Prachinburi Campus, Prachinburi, Thailand

E-mail: benjawan.t@sci.kmutnb.ac.th

Abstract. Traditional production of glass noodles is a long and complicated processes mainly uses mung bean starch which is costly. This research aimed to study the extrusion effect of barrel temperature of zone 3 on size, cooking time and stability time (time until the noodles tear in boiling water) compared to the control, the commercial glass noodle produced by the traditional processes. The effects of using starch blends from mung bean starch (MBS) and pea starch (PS), potato starch (PTS), cassava starch (CS) and modified cassava starch (MCS) at the ratio of 80:20, 70:30, 60:40 and 0:100 (w/w) on the size and cooking properties of the glass noodles were investigated and compared to the control. Results showed that the increased barrel temperature of zone 3 of the extruder increased the size of the dried and cooked glass noodles. The starch blends of MBS and PTS glass noodle resulted to the highest cooking loss (7.94-25.09%). The cooking weights (%) of glass noodle were lower with the decrease of MBS followed with the increase of cooking losses (%) in most cases. However, the starch blends of MBS and MCS showed the lowest values of cooking loss at all ratios used (5.30-5.83%). The extruded glass noodles showed similar appearance with lower cooking time (2 min) and stability time (8 min) compared to the control (3 min and 20 min, respectively). Based on the results of this study, it can be concluded that the starch blends of MBS and MCS glass noodle yields cooking properties better than other types of starch blends for the extruded glass noodles.

1. Introduction

Glass noodles, also known as vermicelli, cellophane noodles, crystal noodles, is a popular type of noodles generally consumed in East and Southeast Asian countries, as well as in western countries. Glass noodles have generally made from mung bean (*Vigna radiate*) starch [1-2]. The most important characteristics of glass noodles are clear, transparent and glossy. For cooked or boiled glass noodles

Content from this work may be used under the terms of the Creative Commons Attribution 3.0 licence. Any further distribution of this work must maintain attribution to the author(s) and the title of the work, journal citation and DOI.
Published under licence by IOP Publishing Ltd

with a short cooking time, the product should have bland taste, firm and elastic texture and minimal surface stickiness and minimal cooking loss [1]. The traditional production of glass noodles is a long, non-continuous processes, requires high energy and time-consuming. In addition, the traditional process generates a lot of waste water and large volume of solid loss [3]. The traditional method for the glass noodle production starts from the mixing of uncooked mung bean starch (85%) and cooked mung bean starch (15%) in a single speed pinhead mixer for 3 min. Then the dough will be extruded through dies having a diameter of about 1.60 mm by a cylindrical forming machine (without heating). Extruded glass noodles are immersed in boiling water for 10 s, cooled in cooling water (4°C) for 3 to 5 min and stored at -10°C for 24 h. Finally, glass noodles will be dried on the polyethylene sheet in a hot air drier at 40 °C and transferred to a covered airtight container [4].

Extrusion is the continuous process combining mixing, cooking and forming in the extruder under high temperature and shorter time (HTST). The advantages are high production capacity, versatility, but minimal use of energy, less water and time requirement and low production cost per unit [3][5][6]. Extrusion process has long been produced either commercial snack products or noodles. There are many reports on extruded food products such as pea starch noodle [2], instant rice [5], high nutritional snack [7], pea andlentil starch blend noodle [8] and paste-like product from pea starch [9]. Several types of extruders in the market generally divided based on the number of screw into a single-screw and a twin-screw extruder. Twin screw extruder offers better mixing, quality control and processing flexibility [10] but single screw extruders are relatively inexpensive and easy to maintain. Typically, a single screw barrel can be divided into three processing zones: feeding zone, kneading zone and the final cooking zone, while the temperature of each zone will influence the properties of the raw material inside the barrel [11]. The major parameters which influence the property of the extrudate are process variables including temperature, screw speed, and initial moisture content of the feed, while the composition of the feed is also very important [5][6][12].

Mung bean starch (MBS) is considered as the best raw material for high quality glass noodles because of its high amylose content, restricted swelling during gelatinization, and high shear resistance of its paste [1][13][14]. Amylose content is one of the most important factor affecting the starch gel strength due to its prompt association, retrogradation and interaction with lipid to form helical complex, as well as the complex with amylopectin to give strong gel networks [6]. However, the cost of MBS is quite high compared to other starches such as pea starch, potato starch, cassava starch, and modified starch. In this study, a more cost competitive starch type will be chosen and used as a starch blend to produce the extruded glass with comparable properties to the noodle produced by the long traditional processes. Besides MBS, pea starch noodles produced by a twin-screw extruderhas a bright color similar to that of the commercial mung bean starch noodles prepared by traditional method [2]. Cooking time and surface stickiness of cooked pea starch noodles were within the ranges exhibited by commercial mung bean starch noodles. Upon comparing with commercial products, cooked pea starch noodles had a bland taste and firmer texture [2].

Therefore, this study aimed to explore the use of mung bean starch and other alternative ingredients to produce glass noodles by direct extrusion process. Physical properties, cooking quality, texture analysis as well as rehydration time and noodle stability (time until the noodles tear in boiling water) of the extruded noodles were also evaluated compared to the commercial glass noodles produced by traditional long process.

2. Material and methods

2.1. Materials and chemical composition

Mung bean starch (MBS), pea starch (PS), potato starch (PTS), cassava starch (CS) and cross-linked modified cassava starch (MCS) were kindly supplied from Thai Vermicelli Industry Co., Ltd. (Bangkok, Thailand). Glass noodles produced by the traditional process used as the control. Chemical analysis of the moisture, ash, protein, and lipid contents of each starch was carried out following AACC 2012 [15].

2.2. Pasting properties

Pasting properties of MBS, PS, PTS, CS, and MCS were determined according to the method described by Charutigon *et al.* (2008) [5] using a rapid viscosity analyzer (RVA-4SA, Newport Scientific, Narrabeen Australia). Each sample (3.0 g, dry matter) was prepared into a RVA canister and adjusted to 28 g by adding distilled water. Peak time (PT), peak viscosity (PV), holding strength (HS), final viscosity (FV), breakdown (BD), setback (SB), and pasting temperature (PTT) were determined.

2.3. Effects of barrel temperature on properties of extruded MBS glass noodle

Direct extrusion for the glass noodle production was carried out by using a single screw extruder (Brabender 19/20, DGE 330, PL 2200, Germany) with the following conditions: barrel bore (D), 19.1 mm, barrel length (L) = 20D, 4:1 screw compression ratio, and the screw speed was operated at 40 rpm. The barrel consists of three independent temperature zones, zone 1 (entering zone), zone 2 (kneading zone) and at the barrel end, indicated as zone 3. According to our experience and preliminary study in extrusion of the noodle from various starch and flour, the screw speed and initial moisture content of the feed as well as the temperature for the zone 1 and zone 2 were fixed in this study. In this study, the effect of the third zone temperature and starch blend on the quality of the extruded glass noodle were investigated. Barrel temperatures were maintained at 50 °C (zone 1), 70 °C (zone 2). The effect of the third zone of barrel temperature at 70, 80, 90, 100, 110, 120, and 130 °C was investigated. The initial moisture content of MBS was adjusted to 35% (wwb). The hydrated starch was extruded through a circular die of 0.50 mm round opening and cut into 30 cm length as they exited the extruder die. Then, the noodles were dried at 80 °C for 10 min. The dried noodles were kept in sealed polyethylene bags until analysis. The diameter of glass noodles was measured by a Vernier caliper. The stability time and cooking time was performed in accordance with AACC 2012 [15]. The stability time is the boiling time used until glass noodles were broken. The optimum cooking time is the time required for the disappearance of opaque of central core of the noodles when gently squeezed the cooked noodles between two glass plates.

2.4. Effects of type of starch blends and ratio of starch blends on properties of extruded glass noodle

Five starch types were used for the formulation to produce the glass noodle by the extrusion, MBS, PTS, MBS, CS, and MCS. The glass noodles were also extruded by using 100% MBS. Four types of starch blends (MBS: PTS, MBS: PS, MBS: CS and MBS: MCS) at different ratios at 80:20, 70:30, 60:40 and 100:0 (w/w) were compared and extruded using the prior condition described in 2.3. The barrel temperature of the third zone at 120 °C was chosen to run this experiment. The analysis of size, cooking time, cooking weight and stability time were determined. The noodles (25 g) were cooked in a beakerfilled with 300 ml water. Then, the noodles were rinsed in cold water (25 °C ± 1 °C) and drained using a sieve for 15 min before being weighed. The remain solid contents in the cooking water were dried at 105 °C overnight. The cooking weight was the weight of cooked sample compared to the weight of dry sample before cooking while cooking loss was evaluated by determining the amount of solids lost into the cooking water. The extruded products also compared with the commercial glass noodle which made by the traditional long processes

2.5. Statistical analysis

All experiments were carried out for triplication and the data was expressed as means±standard deviation. The results were statistically analyzed by Duncan's and SPSS 21.0 with significant difference at $p<0.05$.

3. Result and discussions

3.1. Chemical properties of each type of starch

The moisture, protein, and ash content of MBS, PTS, PS, CS, and MCS ranged from 9.96-15.38%, 0.18-0.39%, and 0.04-0.24%, respectively as shown in Table 1. The moisture content of PTS was slightly higher than those of other starches. MBS contained highest amylose content followed with pea starch, potato starch, modified cassava starch, and cassava starch, respectively. MBS is suitable for the production of glass noodle as it contains high amount of amylose content. No lipid content of all starch samples was detected.

Table 1. Chemical compositions of each starch used for the extrusion of glass noodles.

Starch	Moisture (%)	Protein (%)	Ash (%)	Amylose (%)
Mung bean starch (MBS)	11.14±0.05[b]	0.38±0.01[a]	0.22±0.09[a]	50.78±0.07[a]
Potato starch (PTS)	15.38±0.24[c]	0.28±0.01[b]	0.18±0.07[a]	37.09±0.04[c]
Pea starch (PS)	9.97±0.43[a]	0.39±0.00[a]	0.24±0.04[a]	40.76±0.10[b]
Cassava starch (CS)	9.96±0.22[b]	0.18±0.00[c]	0.20±0.08[a]	33.94±0.04[d]
Modified cassava starch (MCS)	10.93±0.11[c]	0.19±0.01[c]	0.04±0.03[b]	35.62±0.10[e]

Values are mean ± SD of triplicate samples (n=3)
ns = means within a column for each attribute constituent with the same letter are not significantly different (p<0.05)

The inorganic compounds found in the ash depend on the origin of starch. Cereal starches were reported to contain less inorganic composition (potassium, sodium, calcium and magnesium) but the starch from tuber crop was associated with potassium and phosphorus [16]. Negative charge of phosphorus on surface starch along with other negative charges affected the swelling property of the starch gel [7].

3.2. Pasting properties of each type of starch
Starch gelatinization is a process of breaking down intermolecular bonds of starch molecules in the presence of water and heat which allows hydrogen bonding sites (the hydroxyl hydrogen and oxygen) to engage with more water molecules. Penetration of water thus increased the randomness in the starch granule structure, and caused swelling. Eventually, amylose molecules leakinto the surrounding water led to disintegrations of granule structure.The starch granules dissolved irreversibly in water act as a plasticizer [17]. The pasting viscosity (PV) indicates the functional property of starch and varies by type of starch. The PV of potato starch (PTS) and cross-linked modified cassava starch (MCS) were 834.90 and 829.33 RVU and those of mung bean starch (MBS), cassava starch (CS) and pea starch (PS) were 629.19, 434.07 and 364.36 RVU, respectively (Figure 1).

The holding strength (HS) of MCS, PS, MBS, PTS and CS were 596.83, 278.53, 260.39, 159.61 and 139.17 RVU, respectively. Breakdown (BD), the difference between PV and HS, indicates an ability of starch molecule resist to temperature and agitation of PS, MCS, CS, MBS, and PS were 85.83, 232.50, 294.68, 368.80, and 675.30 RVU, respectively. Setback (SB), the difference between FV and HS, indicatesthe re-association between starch molecules during cooling commonly involving retrogradation, or re-ordering of the starch molecules resulting in texture property of various starch gel. The highest setback values (347.74 RVU) of PS indicates highest retrogradation, while PTS, MCS, CS, and MBS showed the lower retrogradation (183.22, 147.80, 140.83, and 139.22 RVU, respectively). The setback value is related to amylose content of starch. Cereal starch (for example MBS) having higher amylose content dictates higher setback value than tuber crop starch (PTS and CS) [18].

Pasting temperature (PTT) is the temperature at which initial swelling of starch granules suspended in water resulting in gelatinization of starch molecule (the breakdown of bonds within starch molecules). The pasting temperature of MBS, PS, MCS, CS, and PTS were 74.76, 72.73, 68.66, 64.28, and 64.28 °C, respectively. Swelling and soluble capacity of starch from tuber crop (CS, PTS) is higher than those of cereal starch due to its lower gelatinization temperature [19]. In addition, phosphate group of PTS could affect high swelling due to electromotive force formation.

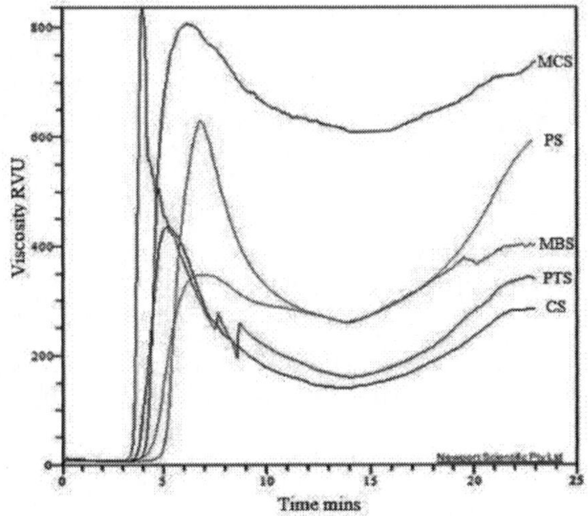

Figure 1. Pasting properties of different starch type.
Note: MBS (mung bean starch), PS (pea starch), PTS (potato starch), CS (cassava starch), and MCS (modified cassava starch)

Table 2. Temperature effects of third zone barrel of extruder on some properties of extruded glass noodles.

Barrel temperature (°C) of zone 3	Size (mm)		Cooking time(min)	Stability time (min)
	Dried product	Cooked product		
70	0.57±0.01^{e}	0.69±0.01^{g}	2.00±0.01^{b}	5.00±0.01^{e}
80	0.62±0.03^{ed}	0.72±0.01^{f}	2.00±0.01^{b}	5.00±0.01^{e}
90	0.63±0.05^{ed}	0.75±0.01^{f}	2.00±0.01^{b}	6.00±0.01^{d}
100	0.68±0.05^{d}	0.82±0.05^{e}	1.00±0.01^{c}	7.00±0.01^{c}
110	0.77±0.10^{c}	0.98±0.02^{d}	1.00±0.01^{c}	8.00±0.01^{b}
120	1.05±0.05^{b}	1.19±0.02^{c}	1.00±0.01^{c}	8.00±0.01^{b}
130	1.12±0.03^{a}	1.25±0.03^{b}	1.00±0.01^{c}	7.00±0.01^{c}
Commercial product (produced by traditional processes)	1.03±0.03^{b}	1.46±0.07^{a}	3.00±0.01^{a}	20.00±0.01^{a}

Values are mean ± SD of triplicate samples (n=3)
ns = means within a column for each attribute constituent with the same letter are not significantly different (p>0.05)

3.3. Effects of barrel temperature of zone 3 on quality of extruded MBS glass noodles

Results showed that the increased barrel temperature of the extruder significantly affected the size of dried and cooked extruded noodles as well the cooking time and stability timeas shown in Table 2. The increased size of extruded glass noodle could be due to a sudden change in the pressure at the die exit resulting in the expansion of the extruded products [20-21]. The noodle surface was smooth when extruded at the temperature of no greater than 120°C. If the barrel temperature of third zone increased from 90 to 100°C or higher, the stability time increased except for the product extruded using the temperature of 120 and 130 °C but the cooking time decreased from 2.00 to 1.00 min. The short cooking time would benefit to the product for grouping into an instant product. However, the cooking time of commercial product was 3 min and had long stability time of 20 min. The increased temperature as well as the geometry of the extruder including screw design, shallow groove and slope could increase shear forces and mixing led to the gelatinization of the starch gel. When samples areto

IOP Conf. Series: Earth and Environmental Science **346** (2019) 012050 doi:10.1088/1755-1315/346/1/012050

extrude under different temperature and pressure conditions of the extruder, starch will gelatinize and melt, and water will evaporate, thereby increased product expansion with stable swelling at atmospheric pressure [6]. According to the size similarity of extruded glass noodle and control, the extrusion condition of the barrel temperature of zone 3 at 120 °C was chosen for further study.

Table 3. Properties of glass noodles extruded with the barrel temperature of third zone at 120 °C by using different starch blends.

Starch blends sample	Size (mm)		Stabilitytime (min)	Cooking weight (%)	Cooking loss (%)
	Dried product	Cooked product			
MBS: PTS					
100: 0	1.05 ± 0.05^{cde}	1.19 ± 0.08^{ijk}	8.00^{b}	357.15 ± 1.07^{b}	6.59 ± 0.94^{h}
80: 20	0.74 ± 0.03^{k}	1.17 ± 0.08^{jk}	5.00^{e}	348.45 ± 1.04^{c}	7.94 ± 0.65^{g}
70: 30	0.83 ± 0.07^{ij}	1.11 ± 0.04^{kl}	4.00^{f}	340.86 ± 1.01^{d}	13.96 ± 0.43^{c}
60: 40	0.89 ± 0.06^{hi}	1.28 ± 0.06^{hi}	4.00^{f}	317.65 ± 1.13^{e}	19.20 ± 0.36^{b}
0: 100	1.04 ± 0.02^{edf}	1.25 ± 0.03^{hij}	3.00^{g}	256.43 ± 1.34^{i}	25.09 ± 0.16^{a}
MBS: PS					
80: 20	1.32 ± 0.04^{a}	1.70 ± 0.05^{bc}	7.00^{c}	299.67 ± 1.29^{f}	6.41 ± 0.46^{hi}
70: 30	1.11 ± 0.06^{cde}	1.64 ± 0.06^{cd}	6.00^{d}	289.62 ± 1.19^{h}	7.73 ± 0.67^{g}
60: 40	1.32 ± 0.03^{ae}	1.89 ± 0.05^{a}	5.00^{e}	272.59 ± 1.28^{j}	7.89 ± 0.50^{g}
0: 100	1.23 ± 0.04^{b}	1.78 ± 0.03^{b}	5.00^{e}	133.36 ± 1.20^{p}	10.40 ± 0.56^{e}
MBS: CS					
80: 20	0.93 ± 0.03^{gh}	1.58 ± 0.06^{de}	3.00^{g}	294.05 ± 0.99^{j}	7.61 ± 0.75^{g}
70: 30	0.97 ± 0.04^{fg}	1.08 ± 0.03^{l}	3.00^{g}	258.28 ± 1.11^{k}	9.04 ± 0.29^{f}
60: 40	1.04 ± 0.04^{def}	1.32 ± 0.03^{h}	3.00^{g}	245.31 ± 1.14^{m}	11.82 ± 0.33^{d}
0: 100	0.98 ± 0.04^{efg}	1.42 ± 0.03^{g}	3.00^{g}	187.66 ± 1.30n	13.54 ± 0.59^{c}
MBS: MCS					
80: 20	0.78 ± 0.03^{jk}	1.04 ± 0.08^{l}	8.00^{b}	289.39 ± 1.21^{h}	5.83 ± 0.74^{hij}
70: 30	1.12 ± 0.04^{c}	1.51 ± 0.03^{ef}	8.00^{b}	277.15 ± 1.03^{k}	5.70 ± 0.23^{hij}
60: 40	1.29 ± 0.02^{ab}	1.45 ± 0.03^{fg}	8.00^{b}	252.27 ± 1.10^{i}	5.53 ± 0.24^{ij}
0: 100	1.03 ± 0.03^{f}	1.42 ± 0.05^{g}	8.00^{b}	146.59 ± 0.87^{o}	5.30 ± 0.37^{j}
Commercial product	1.03 ± 0.03^{def}	1.46 ± 0.07^{fg}	20.00 ± 0.00^{a}	452.31 ± 1.14^{a}	2.66 ± 0.02^{k}

Note: MBS (mung bean starch), PS (pea starch), PTS (potato starch), CS (cassava starch), and MCS (modified cassava starch)
Values are mean ± SD of triplicate samples (n=3)
ns = means within a column for each attribute constituent with the same letter are not significantly different ($p>0.05$)

3.4. Effects of type of starch blends and ratio of starch blends on properties of extruded glass noodle
The raw material factor used starch blends (type and ratio) was investigated. Four types of starch blends (MBS:PTS, MBS:PS, MBS:CS and MBS:MCS) and four different ratios of starch blends (MBS:PTS, MBS:PS, MBS:CS and MBS:MCS) varying at 80:20, 70:30, 60:40 and 100:0 (w/w) were extruded at the barrel temperature of 50, 70 and 120 °C for the first, second and third zone. Results showed that the size of extruded glass noodles were slightly different depend on the type and ratio of starch blends. However, when boiling the dried sample, the size of boiled glass noodle of all starch blends used increased. The use of 100% MBS yielded glass noodle with cooking weight at 357.15%

indicated highest water absorbing capacity, followed with the use of PTS, CS, MCS and PS which showed cooking weight at 256.43, 187.66, 146.59 and 133.36%, respectively. Increasing MBS content resulted in gradual rises of amylose content, retrogradation degree, storage modulus (G') and loss modulus (G″) while the solubility, swelling power and pasting viscosities decreased [3]. Each starch has different granulesize, water absorption ability. The pasting profile affects the cooking properties of the noodles [3]. Starch granuleis fabricated fromsemi-crystalline network which amylose and amylopectin molecule are arranged in crystallite structure and amorphous or gel phase. Short amylopectin chain arrangements are formed as double helices or crystallite structure while granule amorphous starch consisted of amylose and long chains of amylopectin [8] [22].

Results showed the lowest stability time of 3 min of glass noodles found in the extruded starch blends of MBS:CS at all ratios while the highest stability time of 8 min was from the extruded starch blends of MBS: MCS. The chemical property of each starch type indicates the product functionality. Cross-linking starch is normally lower the swelling capacity and water absorption of the starch granules, thereby decreasing cooking weight but maintaining the strength of gel network of the noodle and noodle stability [5][22]. Results revealed that, at the same ratio of starch blends used, the starch blends of MBS and PTS glass noodle yielded the highest cooking weight (%). The lowest cooking loss (%) was obtained by the use of starch blends of MBS: MCS (5.30-5.83%) at all ratios. However, if lower ratio of MBS was used, the cooking losses (%) were higher in most cases (Table 3).

Cooking loss was mainly due to solubilization of loosely bound gelatinized starch on the noodle surface. The degree of cooking loss generally depends on the degree of starch gelatinization and the strength of gel network-like structure of the noodles. The surface gelatinization of traditional extruded noodle was critical for hot-water stability of the noodles [9]. The cooking loss of glass noodles from traditional method was 1.66%. For the commercial perspective, consumers can accept the cooking loss of noodles no greater than 10% [20].

4. Conclusion

Glass noodles were successfully prepared using mung bean starch and other starch blends by using direct extrusion process. Processing variables including the barrel temperature of zone 3 had significant effects on physicochemical, cooking quality, and textural characteristics of theglass noodles. The optimum barrel temperature of zone 3 used for the extrusion was 120 °C. The extruded glass noodles showed similar appearance with lower cooking time compared to the control. The highest cooking weight was obtained at 357.15% from MBS while lowest cooking loss was obtained 5.30% from MCS with cooking time 2 min. For all starch blends used, the blend of MBS and MCS yielded the lowest % cooking loss (5.30-5.83%) and longest stability time (8 min) at all ratios of starch blends.

Acknowledgement

The authors would like to thank Thailand Research Fund (TRF) under Research and Researchers for Industries (RRi) program (MSD59I0103) and Thai Vermicelli Industry Co., Ltd. for their financial support.

References

[1] Kasemsuwan T, Bailey T and Jane J 1998 Preparation of clear noodles with mixtures of tapioca and high-amylose starches1*Carbohydr. Polym.* **36**(4) 301-2
[2] Wang N, Maximiuk L and Toews R 2012 Pea starch noodles: Effect of processing variables on characteristics and optimisation of twin-screw extrusion process *Food Chem.* **133**(3) 742-53
[3] Wu F, Meng Y, Yang N, Tao H and Xu X 2015 Effects of mung bean starch on quality of rice noodles made by direct dry flour extrusion *LWT-Food Sci. Technol.* **63**(2)1199-205
[4] Li Z, Liu W, Shen Q, Zheng W and Tan B 2008 Properties and qualities of vermicelli made from sour liquid processing and centrifugation starch. *J Food Engin.* **86**(2), 162-6.
[5] Wongsa J, Uttapap D, Lamsal B P, Puttanlek C and Rungsardthong V 2017 Effect of extrusion

conditions, monoglyceride and gum arabic addition on physical and cooking properties of extruded instant rice *KMUTNB Int J Appl Sci Technol* **10**(1) 23-30

[6] Charutigon C, Jitpupakdree J, Namsree P and Rungsardthong V 2008 Effects of processing conditions and the use of modified starch and monoglyceride on some properties of extruded rice vermicelli *LWT-Food Sci Technol* **41**(4) 642-51

[7] Galliard T and Bowler P 1987 Morphology and composition of starch In T Galliard (Ed), Starch properties and potential Chichester: Wiley pp. 57-78

[8] Wang N, Warkentin T D, Vandenberg B and Bing D J 2014 Physicochemical properties of starches from various pea and lentil varieties, and characteristics of their noodles prepared by high temperature extrusion *Food Res Int* **55** 119-27

[9] Wang N, Bhirud P R, Sosulski F W and Tyler R T 1999 Pasta-like product from pea flour by twin-screw extrusion *J Food Sci* **64**(4) 671-7

[10] Lusas E and Riaz M 1994 An introduction to extruders and extrusion principles. *Extrusion communique* 7(4) 9-34.

[11] Mercier C, Linko P and Harper J M 1989 Extrusion cooking St Paul MN American Association of Cereal Chemists.

[12] Sereewat P, Suthipinittham C, Sumathaluk S, Puttanlek C, Uttapap D and Rungsardthong V 2015 Cooking properties and sensory acceptability of spaghetti made from rice flour and defatted soy flour, *LWT - Food Sci Technol* **60** 1061-7.

[13] Li C Y and Chang S M 1981 Characterization of red bean *(Phaseolusradiatus*var .Aurea*)* starch and its noodles quality *J Food Sci* **46**(1)78-81

[14] Muhammad K, Kusnandar F, Hashim D M and Rahman R A 1999 Application of native and phosphorylated tapioca starches in potato starch noodles *Inter. J. Food Sci. Technol.* **34**(3) 275-80.

[15] AACC 2012 Approved methods of the AACC international *(*11thed*)* Methods 08-16.01,44-17.01. 76-13.01 and 76-21.01. The association :St .Paul, M.

[16] Lineback DR 1984 The starch granule organization and properties *Bakers Dige* .hicago

[17] Sanders J PM 1996 Starch manufacturing in the world *Advanced Post Academic Course on Tapioca Starch Technology* 22-6

[18] Swinkels, J M 1985 Sources of Starch, Its Chemistry and Physics. In: Van Beynum, G.M. and Roel, J.A., Eds., Starch Conversion Technology, Marcel Dekker Inc., New York, 15-46.

[19] Tan H Z, Li Z G and Tan B 2009 Starch noodles: History, classification, materials, processing, structure, nutrition, quality evaluating and improving *Food Res Int* **42**(5-6) 51-576

[20] Waramboi J G, Gidley M J and Sopade P A 2014 Influence of extrusion on expansion, functional and digestibility properties of whole sweet potato flour *LWT-Food Sci Technol* **59** (2) 1136-45

[21] Korkerd S, Wanlapa S, Puttanlek C, Uttapap D and Rungsardthong V 2016 Expansion and functional properties of extruded snacks enriched with nutrition sources from food processing by-products *J Food Sci Technol* **53(1)** 561-70

[22] Rickard J E, Asaoka M and Blanshard J M V 1991 The physico-chemical properties of cassava starch *Review Trop Sci* **2** 189-207

IOP Conf. Series: Earth and Environmental Science **346** (2019) 012051 doi:10.1088/1755-1315/346/1/012051

The use of Akwa-Biot-Norm biogenic feed additive in the cultivation of Lena sturgeon (*Acipenser baeri, Brandt*)

V G Semenov[1], F P Petryankin, N I Kosyaev, D A Nikitin, G M Toboev, A P Nikitina and G P Tikhonova

Federal State Budgetary Educational Institution of Higher Education "Chuvash State Agricultural Academy", Cheboksary, K. Marx street, 29, 428003, Russia

E-mail: semenov_v.g@list.ru

Abstract. Growing fish in industrial conditions is accompanied by the pressure of various environmental factors on it. Any changes in the environment affect the physiological state and general resistance of the fish, resulting in reduced growth rates, resistance to adverse factors of different nature, and increased mortality. In such conditions, the issue of finding the means and methods of enhancing the protective functions of objects of breeding and cultivation is particularly relevant. Currently, the drugs that have pronounced biological activity are used to improve the functional state of the fish, as well as to increase the value of the artificial feed. In the conditions of the fish farm of LLC YUTAS of the Chuvash Republic, extensive production studies were conducted to assess the effectiveness of the use of nutrient feed additive Akwa-Biot-Norm to the Lena sturgeon, developed by scientists of the Chuvash State Agricultural Academy. Research results indicate that the use of nutrient feed additives in the technology of growing Lena sturgeon (*Acipenser baeri, Brandt*) has a positive effect on some fish-breeding indicators, such as growth rates and survival rates for fish, etc. Thus, the use of biogenic feed additives contributed to a significant increase in the overall weight gain of fish by 5.06% and safety by 2.6% for the period of the experiment (240 days). The results of hematological and immunological studies show a general improvement in the physiological state of the growing objects. The use of biogenic feed supplements had a positive effect on the number of erythrocytes and leukocytes, hemoglobin content, the ratio of lymphocytes and neutrophils and the concentration of total protein in blood serum. In fish, there was an increase in the level of hemoglobin in the blood, as well as an increase in the total protein content in the blood serum by 10.9% to the control indicator, which is a favorable prerequisite for optimizing metabolic processes and guaranteeing high nonspecific resistance. The application of nutrient feed additives has improved the resistance of fish to diseases associated with stress, such as saprolegnia. The conducted studies confirm the efficiency of application of nutrient feed additives Akwa-Biot-Norm for growth stimulation and disease prevention of fish due to the activation of nonspecific resistance of the organism in terms of the pressing environmental and technological stress factors.

1. Introduction

It is possible to meet the population's demand for marketable fish products, including sturgeon, in modern conditions of a decrease in commercial fish production, due to aquaculture, which is a reliable source of increasing fish food production and serves as a guarantor of Russian food security. Currently, industrial fish farming, based on intensive methods of cultivation in cages and swimming pools with direct and closed water supply, is gaining more and more development. At the same time, the technology

Content from this work may be used under the terms of the Creative Commons Attribution 3.0 licence. Any further distribution of this work must maintain attribution to the author(s) and the title of the work, journal citation and DOI.
Published under licence by IOP Publishing Ltd

of sturgeon cultivation with a high density of planting in warm-water farms is used, which allows growing fish all year round and get maximum indicators on fish productivity and its quality [1- 6].

However, in conditions of artificial reproduction of fish under high biomass loads per unit volume, artificial feeding of farmed objects, water pollution with organic substances, various technical operations, etc. are constant stress factors. Their continuous influence contributes to reducing the overall resistance of the organism of fish, resulting in reduced growth rates, resistance to the action of adverse factors of different nature and, as a result, preservation decreases [4, 7, 8, 9].

In such conditions, the issue of finding the means and methods of enhancing the protective functions of objects of breeding and cultivation is particularly relevant. Currently, the introduction of antibacterial, probiotic and other biologically active drugs into the rations for fish is practiced [10, 11].

The determination of the physiological and immunological state of the fish organism provides valuable information that allows evaluating the effectiveness of the use of biogenic preparations for enhancing nonspecific resistance [12-14].

The purpose of this work is to study the physiological and immunological state of the body of fish against the background of the application of a biogenic feed additive based on a polysaccharide complex of yeast cells.

2. Material and methods

Research work was carried out in a specialized full-system fish farm LLC YUTAS, Cheboksary, Russian Federation, in which sturgeon fish are grown. In the farm for feeding fish using extruded feed Efiko Sigma, manufactured by BioMar, Denmark. Scientists of the Chuvash State Agricultural Academy have developed a technology for introducing the developed biogenic feed additive into the feed to improve the biological usefulness of animal feed, ensuring that its initial biological activity is preserved, and eliminating the loss of the drug as a result of washing out with water.

The objects of study were individuals of the Siberian sturgeon of the Lena population (Acipenser baerii Brant, 1869). Two groups of fish were formed for the experiment, 500 copies each, experimental and control, according to the principle of groups of analogs, taking into account the clinical and physiological state and body weight. The average live weight of fish in the control group was 245.1 ± 3.59 g, in the experimental group 243.6 ± 3.48 g. The biogenic feed additive Akwa-Biot-Norm, developed by scientists of the Chuvash State Agricultural Academy, Russian Federation, was additionally introduced into the diet of fish from the experimental group, in two courses lasting 5 days with a break of 2 days. At the beginning of the experiment and every decade of the experiment, control weights of the fish were carried out with a threefold random sample of 10 specimens to determine the growth dynamics. In addition, before the experiment and upon completion, blood samples were taken from 10 samples from each group for hematological, biochemical and immunological studies. Morbidity and safety of fish were recorded in the groups.

The biogenic feed additive Akwa-Biot-Norm is a suspension of 50 mass parts of a 1% gelatin solution, 1 mass part of the concentrate of the purified Saccharomyces cerevisiae polysaccharide complex and 3 mass parts of levomizol.

Research work was performed using the following methods:

1) **hematological** - determined the number of erythrocytes, erythrocyte sedimentation rate, hemoglobin concentration, leukocyte count, additional leukogram differentiation was performed microscopically, making blood smears and staining using Romanovsky method, Giemsa modification, blood for analysis was taken in vivo from the tail artery, immediately after extraction fish out of water;
2) **biochemical** - the level of total protein in the blood serum was studied using an IRF-454B-2M refractometer (A.M. Akhmedov), the protein spectrum using a turbidimetric method (S.A. Karpyuk);
3) **immunobiological** - phagocytic activity of leukocytes was determined using a daily agar culture of Staph. aureus, strain 0-55 (V.S. Gostev), lysozyme activity of blood plasma using daily agar culture of M. lysodeiticus, strain ML-43-29-1 (V.G. Dorofeychuk), bactericidal activity of serum

using daily agar E.coli cultures (O.V. Smirnova et al.);

4) **zootechnical** - determined the live weight and average daily gain of the Lena sturgeon by periodic weighing, exterior measurements;

The digital material of the experiments was processed by the method of variation statistics on the reliability of the difference between the compared indicators (P <0.05-0.01) using the Microsoft Office Excel 2007 software package.

Diagnostic studies and data processing were performed under the Department of Parasitology, epizootiology and veterinary-sanitary examination of FBOU HE the Chuvash State Agricultural Academy, BU CR "Chuvash Republican Veterinary Laboratory" of the state veterinary service of the CR.

3. Research results

The results of monitoring the safety and increase in live weight of the Lena sturgeon are shown in Table 1.

Table 1. Growth and safety of Lena sturgeon.

Indicator		Group	
		control n=500	experimental n=500
Average fish weight, g	At the beginning of the experiment	245.1±3.59	243.6±3.48
	Upon completion of the experiment	930.8±5.98	964.2±6.39**
Live weight gain, g		685.7	720.6
Safety, %		94.4	97.0
Ichtyomass, kg	At the beginning of the experiment	122.55	121.80
	Upon completion of the experiment	439.33	467.68

** P<0.01

From the data obtained it can be seen that upon the completion of the experiment, the sturgeon of the control group had an average weight of 930.8 g, experienced - 964.2 g, which is 33.4 g more, or - by 3.65%. The total weight gain in the control group was 685.7 g, the experimental group - 720.6 g, which is more by 34.9 g. High preservation of fish had a positive effect on the ichthyomass of the experimental group, which amounted to 467.68 kg, against 439.33 kg of the control group, which is 28.35 kg more. These results indicate a positive effect of feed supplementation on fish growth and survival.

At the beginning of the experiment, the number of erythrocytes in the blood of fish in the control group was 0.70×10^{12} / l, which is within the physiological norm. Upon the completion of the experiment, this indicator did not change significantly and had a value of 0.79×10^{12} / l. In the blood of the fish of the experimental group, this indicator significantly increased by 26.4% and amounted to 0.86×10^{12} / l.

The hemoglobin concentration in the blood of the fish of the experimental group at the beginning of the experiment had a value of 33.3 ± 1.07 g / l, the control - 32.73 ± 1.21 g / l (Table 2). Upon the completion of the experiment, this indicator increased, the hemoglobin content in the blood of fish in the control group was 75.7 ± 2.23 g / l, and in the experimental one - 83.2 ± 1.91 g / l, which is 9.9% statistically significantly more benchmark.

An analysis of the dynamics of hemoglobin concentration indicates that, against the background of the use of biogenic feed additive Akwa-Biot-Norm in the technology of feeding Lena sturgeon, its increase is observed. When comparing this indicator with the number of erythrocytes, it becomes obvious that an increase in hemoglobin concentration occurs both due to an increase in the number of erythrocytes and due to an increase in their functional activity, which indicates activation of hemopoiesis, an increase in the intensity of redox processes and metabolism in general.

Table 2. Morphological and biochemical parameters of the blood of Lensky sturgeon.

Indicator	At the beginning of the experiment		Upon completion of the experiment	
	group		group	
	Control	experimental	Control	experimental
Red blood cells, 10^{12} / l	0.70±0.01	0.68±0.04	0.79±0.02	0.86±0.01*
Hemoglobin, g/ l	32.73±1.21	33.3±1.07	75.7±2.23	83.2±1.91*
Leukocytes, $\times 10^9$ / l	24.3±0.53	24.9±0.69	34.10±2.10	33.8±2.01
Neutrophils, %	13.2±0.39	13.3±0.99	16.6±1.17	16.9±1.54
Eosinophils, %	9.8±0.27	9.9±0.31	5.75±0.34	5.04±0.55
Basophils, %	0.3±0.01	0.3±0.03	0.4±0.02	0.4±0.01
Lymphocytes, %	74.7±1.53	74.6±1.23	74.65±1.65	75.36±1.58
Monocytes, %	1.91±0.04	1.83±0.03	2.6±0.67	2.3±0.55
Total protein, g/ l	19.08±0.57	19.41±0.78	25.4±1.63	28.18±1.77

* $P<0.05$

When determining the dynamics of the content of leukocytes in the blood of fish against the background of the use of a biogenic feed additive, leukocytosis was established, which was within the physiological norm in the blood of the fish of the experimental group. At the beginning of the experiment, the number of leukocytes in the blood of the Lena sturgeon of the experimental group was 24.9 ± 0.69 x 10^9 / l, and the control group - 24.3 ± 0.53 x 10^9 / l. Upon the completion of the experiment, this indicator did not have significant differences from the initial and control indicators in the blood of fish of the experimental group.

The percentage of eosinophils in the blood of the Lena sturgeon of the experimental group at the end of the experiment was lower than the initial and control indicator. It should be noted that at the same time it was within the physiological norm.

The absolute majority were lymphocytes in the leukocyte formula of sturgeon blood. At the beginning of the experiment, this component in the blood of the fish in the control group was 74.7%, and that of the experimental group was 74.6%. Upon the completion of the experiment, the number of lymphocytes in the control group was lower than that of the experimental group and was equal to 74.65%. In the sturgeon of the experimental group, this indicator increased to 75.36% by the end of the experiment.

At all periods of observation, the relative amount of monocytes in the blood of fish did not have statistically significant differences between the groups and was within the physiological norm.

Analysis of the leukocyte picture shows that when using the biogenic feed additive Akwa-Biot-Norm in fish, an increase in the absolute number of leukocytes and a decrease in the relative number of eosinophils is observed. There is an increase in lymphocytes in the blood of fish of the experimental group. The decrease in the relative amount of eosinophils can be explained by an increase in the proportion of lymphocytes.

The tests indicate leukocytosis within the physiological norm, with pronounced lymphocytosis, having an absolute majority in the leukocyte picture. These data confirm the high degree of activity of the cellular component of the immune system.

At the beginning of the experiment in the blood serum of Lensky sturgeon of the experimental group, the amount of total protein was 19.41 g/l. At the end of the experiment, this figure increased and amounted to 28.18 g / l or 10.9% more than in the control, but statistically unreliable.

An increase in the protein concentration in the blood serum of the Lena sturgeon of the experimental group correlates with an increase in ichthyomass, which indicates the activation of synthetic processes in the body of fish and, as a consequence, an increase in their growth.

IOP Conf. Series: Earth and Environmental Science **346** (2019) 012051 doi:10.1088/1755-1315/346/1/012051

The results of the study of the immunological properties of the blood of Lensky sturgeon against the background of the use of biogenic feed additives based on the polysaccharide complex of yeast cells are presented in Table 3.

Table 3. Immunological properties of the blood of Lena sturgeon.

Indicator	At the beginning of the experiment		Upon completion of the experiment	
	group		group	
	Control	experimental	Control	experimental
Bactericidal activity of blood serum,%	13.7±0.56	13.8±0.84	14.3±1.02	16.1±0.98*
Phagocytic activity of neutrophils,%	18.33±0.18	18.48±0.16	19.3±0.27	20.92±0.47*
Lysozyme, μg / l	7.18±0.16	7.2±0.18	9.4±0.29	10.12±0.37

* $P<0.05$

The bactericidal activity of the blood serum of the control and experimental fish at the beginning of the experiment did not differ significantly and was 13.7 ± 0.56% and 13.8 ± 0.84%, respectively. Upon the completion of the experiment, this indicator in the fish of the control group increased by 4.38% of the initial value but had no statistically significant differences. The bactericidal activity of the blood serum of the experimental fish at the end of the experiment was significantly higher than in the control and was 16.1 ± 0.98%. This indicator was higher than the original by 16.67%, and the control by 12.58%.

From the table 3, it follows that the phagocytic activity of blood neutrophils in the control group increased, but the value of this indicator is not statistically significant. The results of the experiment indicate that the phagocytic activity of the blood neutrophils in the fish of the experimental group increased. At the beginning of the experiment, this indicator was 18.48 ± 0.16%, and at the end of 20.92 ± 0.47%, which was significantly higher than the control by 8.39%.

Consequently, the biogenic feed additive Akwa-Biot-Norm activated cellular factors of nonspecific protection of the body of fish, with a more pronounced effect of the latter.

The concentration of the proteolytic enzyme at the beginning of the experiment in the blood serum of the control and experimental fish groups was 7.18±0.16 μg/l and 7.2±0.18 μg / l, respectively, within the physiological norm. Upon the completion of the experiment, lysozyme concentration in the control and experimental groups were also within physiological norms. In the blood of fish of the experimental group, the indicated indicator of nonspecific resistance increased by 40.55% and amounted to 10.7 ± 0.12 μg / l, but not statistically significant.

The efficiency of freshwater aquaculture depends on many factors, such as the sanitary and hygienic regime of the aquatic environment, the quality of animal feed, compliance with technical processes, etc. These requirements are not always met in a production environment. Fish are constantly exposed to various stress factors associated with the use of low-quality feed, violation of the temperature, oxygen regime, etc. Favorable conditions are created for the development of an infectious process caused by the association of microorganisms, while not necessarily pathogenic, and often even a complex of saprophytic bacteria that contaminate the fish's body as a result of a decrease in its resistance.

One of these diseases is saprolegniosis. During the work, we noted fish diseases with the following clinical signs: on the skin and fins, the presence of white thin filaments perpendicularly extending from the surface of the fish body. After a few days, a white bloom of white color consisting of interlaced hyphae appears on these places. Microscopic examination of scrapings from the skin showed clearly visible hyphae of the fungus. Based on these signs, a preliminary diagnosis was made - saprolegniosis. Selected pathological material for identification of the fungus in the laboratory.

The percentage of lesion of the Lena sturgeon in the control and experimental groups was established with careful examination of the fish. The results of the examination showed that the level of fish affection by saprolegnosis in the control group was 33.68%, and in the experimental group - 15.46%. Fishes of both groups are subjected to treatment-and-prophylactic treatment.

In the laboratory, in the study of pathological material, after obtaining a clean culture and growing on the environment of Chapek, the causative agent of saprolegniosis is established.

Other infectious and invasive diseases, as well as non-infectious etiology, were not revealed. The safety of the fish in the experimental group for the entire experimental period was 97.0% against 94.4% in the control group.

The results of the experiment showed that the use of the nutrient feed additive Akwa-Biot-Norm to the Lena sturgeon increases their resistance to diseases associated with stress, such as saprolegniosis.

Thus, the use of the Akwa-Biot-Norm biogenic feed additive to the Lena sturgeon with two courses of 5 days duration with a break of 2 days has a positive effect on their body, activating nonspecific resistance factors and causing protection of fish from the negative effects of various environmental factors.

4. Conclusion

Against the background of the application of the biogenic feed additive Akwa-Biot-Norm based on the Saccharomyces cerevisiae polysaccharide complex of yeast cells in Lena sturgeon, an increase in live weight of fish is observed by 5.09%.

A significant increase in hemoglobin concentration by 9.9% was found due to an increase in the number of red blood cells and an increase in their functional activity.

A significant increase in the absolute number of leukocytes by 35.7% was noted, with a decrease in the relative amount of eosinophils. The relative number of lymphocytes, although not statistically significant, increased by 1.02%. Thus, leukocytosis within physiological norms, with pronounced lymphocytosis, with an absolute majority in the leukocyte formula and a slight decrease in the proportion of eosinophils, while maintaining their high numbers indicates a high degree of development of the cellular link of the immune system.

The amount of total protein in the blood serum of the Lensky sturgeon of the experimental group relative to the control indicator was higher by 10.9 %, which indirectly indicates the activation of synthetic processes in the body of sturgeons and, as a consequence, an increase in the intensity of fish growth.

The study of the immunological properties of the blood of Lensky sturgeon against the background of the use of biogenic feed additive Akwa-Biot-Norm revealed a significant increase in the value of such indicators of nonspecific resistance of the fish organism as bactericidal activity of blood serum by 16.6%, phagocytic activity of neutrophils by 13.2%, as well as an increase in the concentration of the proteolytic enzyme lysozyme by 40.5 %. These indicators are higher than the corresponding indicators of the control group. These changes of the immunological properties of blood Lensky sturgeon indicate activation of cellular and humoral components of nonspecific resistance of the organism of the fish, and confirm effectiveness of application of nutrient feed additives Akwa-Biot-Norm based on polysaccharide complex of yeast cells Saccharomyces cerevisiae to enhance nonspecific resistance of the organism of sturgeon fish in conditions of environmental and technological pressures.

References

[1] Vasilieva L M 2015 *Technologies of food and processing industry of the agro-industrial complex - healthy food products* **1** pp 18-23

[2] Kurovskaya L, Lysenko V and Neborachek S 2015 Fisheries Science of Ukraine **1** **(31)** pp 108-19

[3] Kurovskaya L Ya 2013 Parasitology in a changing world: V Congress of the Parasitological Society of the Russian Academy of Sciences: All-Russia. conf. with Intern. participation: theses of reports (Novosibirsk) p 101

[4] Matishov G G, Matishov D G, Ponamareva E N et al. 2006 Experience of growing sturgeon in a closed water supply system for farms (Rostov-on-Don) p 72

IOP Conf. Series: Earth and Environmental Science **346** (2019) 012051 doi:10.1088/1755-1315/346/1/012051

[5] Saurabh Shailesh and Sahoo P K 2008 *Aquacult. Res.* **39** pp 223-39
[6] Firth M A, Sheven P E and Hodgins D S 2005 *Anim. Health. Res. Rev.* **6(2)** pp 143-58
[7] Pronina G I 2014 Proceedings of the Orenburg State Agrarian University (Orenburg) **3** pp 180-2
[8] Popyk I M and Vischur O I 2013 *Anim. Biol.* **15(3)** pp 105-110
[9] Subramanian S, MacKinnon S L and Ross N W 2007 *Comp. Biochem.Physiol.* **148** pp 256-63
[10] Grosesku Yu N, Bakhareva A A and Shulga E A 2011 *Fish Farming Fisher.* **4** pp 49-52
[11] Shulga E A 2009 Probiotics in the feeding of sturgeon by commercial farming: Thesis of the candidate of biological sciences (Astrakhan) 120
[12] Lapirova T B and Flerova E A 2013 Bulletin of Astrakhan State Technical University. Series: Fisheries (Astrakhan) **1** pp 140-5
[13] Gavrilin K V, Mikryakov D V, Silkina N I and Suvorova T A 2010 Veterinary Kuban (Krasnodar) **6** pp 14-6
[14] Semenov V G, Mudarisov R M and Vasilyev V A 2014 "European Conference on Innovations in Technical and Natural Sciences". Proceedings of the 1st International scientific conference (Vienna, Austria) pp 176-81

IOP Conf. Series: Earth and Environmental Science **346** (2019) 012052 doi:10.1088/1755-1315/346/1/012052

The possibility of using immunogenetic criteria for the characterization of the breed, to predict the productive qualities and results of the selection of pigs of the tsivilsky breed

N V Evdokimov[1, 3], A Y Lavrentyev[1], V A Alekseev[1], V S Sherne[1], E Y Nemtseva[1], N S Petrov[1], N V Danilova[1], A A Novikov[2] and G M Toboev[2]

[1]Federal State Budgetary Educational Institution of Higher Education "Chuvash State Agricultural Academy", Cheboksary, K. Marx street, 29, 428003, Russia
[2]All-Russian Research Institute of Pedigree Work, Forest Glades, Moscow Region, Russia

E-mail: evdonikvit@mail.ru

Abstract. The study of the occurrence of certain alleles and genotypes in certain loci of blood groups of large white and Tsivilsky pig breeds was carried out. It has been established that there are differences in the frequency of occurrence of both certain alleles and genotypes of pig blood groups in the studied livestock of pigs. So, in pigs of the Tsivilsky breed, such genotypes as Eabg / aeg, Eaef / bdf, Ebdf / deg are quite common, more often than in pigs of the large white breed. It is also established that there are animals with the genotypes Eabg / bdg, Eabg / bdf only among pigs of the Tsivilsky breed, which are absent in pigs of the large white breed. Analysis of the boars in the context of their linear affiliation showed that the aeq / aeq genotype is inherent only in the clad line boars, the aeq / bdq genotype is most prevalent in Satur line boars. We have analyzed the results of selection of boars with certain genotypes of blood groups with uterus - carriers of certain genes of blood groups. As a result, the most successful combinations of boars with a uterus have been established. Another important economic indicator studied in the context of carrier pigs of certain genotypes of blood groups was the study of fattening and meat qualities of pigs. The results indicate that the G-system of blood groups is the most informative when studying the fattening qualities of pigs. So, pigs - heterozygotes for this system, the highest rates of average daily gains for fattening (851 grams), costing 1 kg gain of 3.3 k.un. And spending only 178 days on achieving a body weight of 100 kg. Conducted researches have allowed concluding that chromagraphemia factors can be successfully used in characterizing breeds of pigs, the definition of linear facilities boars and for the selection of parental pairs of pigs, taking into account the possibility of forecasting the future productivity of the offspring of pigs.

1. Introduction

Any breed bred in our country has its own biological characteristics, including productive qualities, physique, type of constitution and exterior, color and weight factors [1-3]. Recently, many scientists have recommended to use their interior features, including genotypes of blood group systems, types of polymorphic proteins when describing the features of rocks [4,5]. Publication on pigs of many cultivated breeds is sufficient in the available literature, but data on local, few breeds are scarce or

Content from this work may be used under the terms of the Creative Commons Attribution 3.0 licence. Any further distribution of this work must maintain attribution to the author(s) and the title of the work, journal citation and DOI.
Published under licence by IOP Publishing Ltd

practically absent. As a result of numerous researches, it is established that enrichment of traditional methods of selection by the latest achievements of modern genetics, biochemistry, physiology, mathematics and other sciences considerably modernizes practice and facilitates selection work [6, 7] at selection of more highly productive, having the best adaptive qualities of animals [8].

The purpose of this work is to study the comparative blood group factors of pigs of large white and civil breeds, the characteristics of pigs of civil breed and different lines of this breed on immunogenetic parameters (the frequency of occurrence of certain genotypes of blood groups in the population) the relationship of the main indicators of productivity of pigs of civil breed (dam and boars) with the carrier of certain genotypes of blood group loci, the dependence of productive qualities of the parent pair on the results of selection of genotypes of blood groups, as well as comparison of changes in the frequency of genes of blood groups of civil pigs breed occurred over the past 18 years studying.

2. Material and methods

The objects of the study were pigs of large white and tsivil breeds of Morgaushsky (large white) farms and the gene pool of FSUE Kolos of Tsivilsky districts in the amount of 549 animals. Testing of pigs by blood groups was carried out in the laboratory of pig genetics of the All-Russian Research Institute of Pedigree Breeding with generally accepted methods using standard antigens, the frequency of occurrence of certain alleles and genotypes, and the level of homozygosity were calculated using the methods of L. Zhivotovsky [9], N. Sukhova [10]. To determine the difference in the indicators, we compared the productivity of pigs with certain genotypes among themselves. The reliability of the difference in the indicators was established using the Student table. Biometric processing of the obtained results was carried out according to Plokhinsky.

3. Research results

For a long time, we have been studying the blood types of planned breeds of pigs of the Chuvash Republic (large white and Tsivilsky breeds). Our studies allow us to conclude that these pigs, unlike the pigs that participated in their hatching, have many distinctive features in the A, E, H, K, L blood system loci.

According to the E-system of blood groups, genotypes are found that are characteristic only for Tsivilsky breed pigs. It should be noted that there are animals with the genotypes Eabg / bdg, Eabg / bdf, only among the pigs of the Tsivilsky breed which are generally absent in pigs of large white breed. The Eabg / aeg, Eaef / bdf, Ebdf / deg genotypes found in breed pigs, are much less common in large whites, and it should be emphasized that this is not in all the populations studied. The homozygous genotype Ebdf / bdf attracts attention, which has a frequency of occurrence equal to 0.090 in Tsivilsky breed pigs. This genotype was detected in only one of the 362 studied pigs of a large white breed (Table 1).

Table 1. The frequency of occurrence of genotypes in Tsivilsky breed pigs in comparison with some populations of the large white breed.

Locus	Genotype	Tsivilsky n=187	Large white n=362
A	a /.	0.620	0.370***
	o /.	0.208	0.406***
	. /.	0.171	0.224
B	a / a	0925	0.881***
	a / b	0.0	0.119**
D	a / b	0.086	0.094
	b / b	0.914	0.906
E	abg / aeg	0.011	0.003

	abg / bdg	0.011	-
	abg / bdf	0.011	-
	edg / edg	-	0.005
	adg / adf	0.005	0.008
	adg / bdg	0.005	0.050***
	aeg / bdf	0.123	0.061***
	aeg / def	0.096	0.126
	aeg / deg	0.011	0.111
	aef / bdg	0.011	0.003
	aef / def	0.011	-
	aef / deg	-	0.003
	bdg / bdg	-	-
	bdg / bdf	-	0.039
	bdg / deg	0.021	0.003
	bdg / def	0.021	0.133***
	bdi / bdf	-	0.008
	bdf / def	0.090	0.003***
	bdf / deg	0.037	0.014
	bdf / deg	0.353	0.092***
	beg / beg	0.027	0.133***
	def /def	-	0.008
	def /deg	0.160	0.179
F	a / o	0.177	0.022***
	b / o	0.822	0.978*
G	a / a	0.118	0.209
	a / b	0.577	0.412
	b / b	0.304	0.378
H	a /.	0.104	0.080
	a / b	-	0.016
	b /.	0.016	0.258**
	. /.	0.064	0.040
K	ad /.	0.070	0.197**
	ad / bf	0.630	0.494
	bf /.	0.300	0.42
	. /.	-	0.066
L	adhi / adhi	0.066	0.006
	adhi / bdfi	0.045	0.024
	adh / bcgi	0.098	0.053
	agi / bdfi	-	0.128
	agi / bcgi	0.054	0.025
	bcgi / bcgi	0.503	0.472
	bdfi / bdfi	0.059	0.067
	bdfi / bcgi	0.174	0.225
	bcgi / bcgi	-	-
Tf	AA	0.059	0.075
	AB	0.546	0.495
	BB	0.395	0.430
Pa	AA	0.487	0.216
	AB	0.434	0.521
	BB	0.079	0.263

In addition, we found that the pigs of the Tsivilsky breed are characterized by the complete absence of homozygous genotypes at the E locus, such as bdg / bdg, def / def, deg / deg and heterozygous for the H locus - a / b.

The comparison revealed a statistically significant difference between the animals of the studied breeds in the distribution of genotypes in the loci A, L, and blood group systems.

By distributing the pig population of the gene pool in terms of their linear affiliation and calculating the frequency of occurrence of certain genotypes of the E-system of blood groups in boar lines, we were able to identify certain markers inherent in pigs of certain lines (Table 2).

Table 2. Genotypical characteristics of Tsivilsky breed boar lines by E-blood group system.

Genotype	The occurrence of genotypes, %							
	Robust fellow p=17	Satur p=18	Salagay p=11	Clan n=90	Treasure p=66	Sozar p=12	Levan p=31	Moulded p=7
aeg/aeg	-	-	-	-	1.52	-	-	2.49
aeg/ вdg	11.10	15.0	-	3.33	4.55	-	-	3.19
aeg/ bdg	-	-	18.18	2.22	1.52	58.33	6.45	2.96
aeg/ edf	5.23	10.32	-	-	-	-	-	0.78
bdg/ bdg	22.89	15.88	-	36.67	28.79	25.0	19.35	21.4
вdg / edg	22.89	27.73	9.09	22.22	16.67	16.67	16.13	27.70
вdg / edf	5.23	-	9.09	21.44	7.58	-	3.23	2.33
edg / edg	5.23	-	9.09	11.11	19.70	-	32.26	9.33
edg / edf	11.10	10.32	9.09	1.11	3.03	-	-	3.66
bdf / edf	-	-	-	3.33	-	-	0.5	13.0
вdg / aeg	11.10	15.88	9.01	-	4.55	-	-	13.9

Thus, we carried out the studies on 8 lines of pigs of the Tsivilsky breed available as of January 1, 2019. We were able to identify that the aeq / aeq genotype is inherent only in Treasure line boars, the aeq / вdq genotype is most prevalent in Satur line boars. Taking into account the fact that high-yielding animals are of some value for science, which steadfastly transmit these qualities to the offspring and these qualities of sows depend not only on the mother's genotype, but also on the immunological compatibility of the parent pair by antigens of blood groups, we set the task: to study the possibility of the influence of individual compatibility of boars and dams on separate systems of blood groups (G, H, K, L systems) on the indicators of production (Table 3).

Analysis of the obtained data allows us to conclude that the selection of animals based on the genotype for locus G blood groups is the best for such indicators as the number of live pigs at birth, the number of pigs at the age of 21 days was the combination of boars with the genotype G a / в with uterus G a / в, with a result of 11.3 ± 1.31 and 9.00 ± 0.51 heads, respectively, combinations of boars with genotypes G a / в with uterus with genotypes G в / в differ in the mass of the nest.

When analyzing various combinations of H systems, we managed to find out that the best indicators were for combinations of boars with H - / - genotypes with uterus with H a / - genotypes, at which 12.43 ± 0.50 piglets were born at birth, at the age of 21 days 10.12 ± 0.33 piglets with nest weight 44.00 ± 2.60 kg.

When considering different combinations of boars and dams by the K - blood group system, it was found that the best indicators were obtained when the boars combined with the K a / в genotype and the K в / в uteri. It should be noted that according to this system, the lowest indices are obtained when boars are combined with K / i genotype with uterus with K a / a genotype.

According to the L-system, the best indicators were obtained from a combination of boars with genotypes L a / в with uterus genotype L a / в, with the result: multiple births - 12.45 ± 0.68 piglets; at the age of 21 days - 10.3 ± 0.41 piglets with a mass of 46.9 ± 3.2 kg. All this makes it possible to recommend t pig breeders specialists when drawing up a plan for fixing boars to the uterus to take into

IOP Conf. Series: Earth and Environmental Science **346** (2019) 012052 doi:10.1088/1755-1315/346/1/012052

account all these factors of compatibility of couples in order to increase multiplicity and safety. Another important indicator of productivity is the fattening qualities of pigs. These qualities were studied depending on the genotypes of blood groups by the A, B, D, G, F, M systems and separately from the carriage of genotypes by the E - system.

Table 3. Indicators of productivity of pigs, depending on the combination of boars and dams, taking into account G.H.K.L genotypes of blood group systems.

Genotype				Productivity indicators		
				Number of living piglets at birth	At the age of 21 days	
♂	n	♀	n		Number of pigs	The mass of a nest
G ав	6	ав	9	11.30±1.31	9.00±0.51	41.30±1.50
		вв	4	11.08±0.43	8.77±0.64	42.90±2.20
вв	5	ав	7	11.10±0.80	9.01±0.48	41.80±2.80
		вв	13	11.70±0.41	9.43±0.31	42.30±1.30
H a/-	3	a/-	9	11.13±9.61	9.37±0.83	43.0±1.90
	6	-/-	17	11.40±0.41	9.41±0.45	43.80±2.00
-/-		a/-	12	12.43±0.50	10.12±0.33	44.00±2.60
		-/-	41	11.37±1.21	9.30±9.28	41.20±1.90
K ав	4	ав	9	9.80±0.67	9.40±0.43	46.70±2.30
		aa	4	11.70±1.37	8.32±0.65	41.03±3.50
		вв	13	13.18±0.75	9.78±0.36	42.8±2.900
вв	6	ав	14	11.20±0.49	10.09±0.43	44.00±1.90
		aa	7	10.90±1.07	8.87±0.89	36.40±2.10
		вв	23	11.10±0.81	9.02±0.37	38.20±1.70
L ав	7	ав	13	12.45±0.68	10.3±0.41	46.90±3.20
		вв	19	11.23±0.47	9.37±0.31	41.80±2.10
вв	9	вв	23	11.00±0.80	8.90±0.26	41.0±1.70

Studies have shown that the best indicators for such an indicator as the age attaining a live weight of 100 kg (180.0 ± 1.2 vs. 188.0 days) were pigs with genotype A - / -, they are much smaller (0.48 feed units) spent feed per unit weight gain, and 100 g more than their average daily gain (table. 4).

The difference in performance is significant in the first and second cases ($P < 0.05$).

According to B-pig blood group system. homozygotes reach a weight of 100 kg for 188.0 ± 0.9 days. at a cost of 3.60 feed units. adding 699.0 ± 6.0 grams per day in the group. which is lower than that of heterozygotes for 5 days. 0.21 feed unit and 78 grams of average daily gain.

Pigs with в/в genotype of the blood group system reach the age of 100 kg 4 days earlier. significantly exceed the cost of feed (by 0.13 feed units) and the average daily growth (by 59 grams) over pigs with D a/в genotype.

Heterozygous pigs for the F system reach the age of 100 kg for 183.0 ± 1.1 days. at a cost of 1 kg of weight gain of 3.71 ± 0.1 fodder units of feed. with an average daily gain of 821.0 ± 8.2 g. And the difference is: by age when reaching 100 kg 7 days. at a feed cost of 0.07 feed units and an average daily gain of 180 grams.

When analyzing these indicators. depending on the genotypes of the G-system. it was revealed that heterozygous pigs had the best performance. reaching a body weight of 100 kg at the age of 178.0 ± 1.1 days. spending 1 kg of gain of 3.30 kg of fodder units with an average daily gain of 851.0 ± 8.2 g. The difference in the indications G a / в and G a / a. G a / в and G в / в by age reaches 100 kg is significant at $P < 0.001$ and $P < 0.05$. respectively. at the cost of feed G a / в and G a / a - at $P < 0.05$.

When studying the fattening qualities of pigs from the H genotypes of the system. we found that this index varies depending on the factors studied.

Table 4. The fattening quality of pigs, depending on the genotypes for A, B, D, F, G, H, M - blood group systems.

Genotype	Number of heads	Age of achievement of live weight of 100 kg, days.	Feed costs per unit of production, K. units	Average daily gain, g
A cp/-	15	188.0±1.1	3.81±0.1	701.0±6.3
-/-	17	180.0±1.2	3.33±0.2	801.0±5.8
B a/a	24	188.0±0.9	3.60±0.1	699.0±6.0
a/в	18	193.0±0.6	3.90±0.1	621.0±3.9
Д a/в	21	191.0±1.5	3.7 8±0.2	710.0±4.3
в/в	13	187.0±1.2	3.65±0.1	651.0±4.7
F a/в	17	183.0±1.1	3.71±0.1	821.0±8.2
в/в	15	190.0±0.7	3.78±0.2	640.0±3.3
G a/a	12	192.0±0.9	3.80±0.1	623.0±4.7
a/в	18	178.0±1.1+++	3.30±0.1+++	851.0±4.3+++
в/в	11	191.0±0.8	3.77±0.2	637.0±2.7
H a/-	19	193.0±1.2	3.81±0.2	626.0±5.9
a/в	17	181.0±1.9	3.71±0.1	810.0±4.0
в/-	21	195.0±0.7	3.95±0.2	618.0±5.1
в/в	13	188.0±1.2	3.88±0.2	701.0±7.3
-/-	9	191.0±0.5	3.76±0.2	636.0±6.9
M a/-	18	194.0±1.2	3.81±0.1	620.9±4.1
d/-	19	181.0±1.9	3.72±0.2	801.0±4.7
-/-	27	190.0±1.4	3.91±0.3	650.0±6.1

In the next sample. namely. depending on the genotypes of the H - blood group system. it was established that the age at which body weight reaches 100 kg varies from 181.0 ± 1.9 days in pigs with the genotype H a /в to 195.0 ± 0. 7 days with genotype H в / -; feed costs from 3.71 ± 0.1 kg with genotype H a / - up to 3.95 ± 0.2 kg in pigs with genotype H a / -; average daily gain from 626.0 ± 5.9 g in pigs with genotype H a / - to 810.0 ± 4.0 g with genotype H a / в. Differences in feed consumption per unit of production and average daily growth. respectively. are significant at P <0.05 and P <0.001.

When studying the fattening quality of pigs. depending on the genotype of the M-type blood group system. it was revealed that pigs with the M a / - genotype reached weight of 100 kg 13 days earlier than pigs with the M d / - genotype and 9 days earlier than with M - / - (the difference in the first variant is significant at P <0.001). Feed costs were lowest in pigs with the M d / - genotype. they also had the highest average daily gain (801.0±4.7). The difference is also reliable in terms of feed costs and average daily gain between pigs with the genotype M a / - and M - / - at P <0.01 and P <0.001.

When studying the relationship of genotypes E of the blood group system for fattening and slaughter qualities. it was found that pigs with the edg / edg genotype have the best performance. having reached a live weight of 100 kg after 186.4 ± 6.4 days (Table 5).

This indicator varied depending on the carriage of certain genotypes of the studied system from 186.4 ± 6.4 days (with the edg / edg genotype) to 205.3 ± 3.3 (in pigs with the bdf / edf genotype). The difference is significant at P <0.001. The smallest amount of feed per unit of production was spent by pigs with the bdg / bdg genotype (3.73 ± 0.1). and the highest - 4.05 ± 0.1 feed units - by pigs with the bdf / edf genotype (the difference is significant at P < 0.001).

The average daily gain of pigs. depending on the genotypes of the E-system of blood groups. varies from 673.0 ± 13.20 grams to 844.4 ± 11.8 g.

Table 5. The fattening qualities of Tsivilsky breed pigs with different genotypes of E - blood group systems.

Genotypes E systems	Number of animals. heads	The age of reaching 100 kg. days	Feed costs. feed	Average daily gain. g
edg / edg	5	186.4±6.4	3.75±0.2	768.6±6.9
bdg/edg	14	199.0±4.5	3.90±0.1	742.6±4.4
bdg/bdg	17	190.6±3.9	3.73±0.1	844.4±11.8
acd/bdg	5	194.2±6.6	3.76±0.2	760.4±8.2
acd/cdg	5	195.8±4.4	3.95±0.1	713.4±7.2
edf/edf	2	193.0±2.5	3.90±0.3	673.0±13.2
bdf/edg	3	205.3±3.3	4.05±0.1	682.6±9.6

When analyzing the meat qualities of pigs. depending on the carriage of certain genotypes of the E-system of blood groups. a significant difference ($P < 0.001$) was obtained from the thickness of the fat of pigs with the genotypes edg / edg and aeg / bdg (Table 6).

Table 6. Meat qualities of pigs with different genotypes of E - blood group systems.

Genotypes of E system	Number of heads	Carcass length. cm	Backfat thickness. mm	Eye of loin area. $cm2^2$	Ham weight. kg
edg/edg	5	97.0±0.89	27.80±0.97	31.97±0.91	9.56±0.42
bdg/edg	14	97.86±0.82	27.64±0.85	33.64±0.39	10.43±0.32
bdg/bdg	17	97.11±0.59	25.53±1.06	31.76±0.76	10.18±0.25
aeg/bdg	5	98.20±0.66	24.60±2.35	32.40±0.87	10.68±0.15
aeg/edg	5	97.33±1.85	27.33±1.30	33.0±0.51	10.23±0.15
bdf/edf	2	97.50±1.50	26.50±0.05	33.5±0.09	10.05±0.85
bdf/edg	3	99.0±2.0	30.0±0.57	32.0±0.57	10.00±0.87

Pigs with the bdg / edg genotype had the largest area of the "eye of loin. and the smallest - with bdg / bdg (the difference is significant at ($P < 0.01$).

Depending on the factors studied. no significant difference was obtained for the length of the carcass and the mass of the back half, but it should be noted that pigs with the aeg / bdg genotype had the greatest length of the carcass, and their ham was better made.

The next stage of work was the study of sperm production of boars-manufacturers from the carriage of certain genotypes of the E-system of blood groups for which all the population of boars was divided into samples depending on the E genotype of the blood group system (Table 7)

Table 7. Indicators of sperm production of boars with different genotypes of the E-system of blood groups.

Genotype of E-system	Number. heads	Ejaculate volume. ml	Number of spermatozoids. mlrd.	Conc. sperm. in ejaculate. mln in 1 ml	Activity of spermatoz. point
edg / edg	12	244.0±12.30	51.0±4.3	225.5±2.9	7.42±0.06
bdg/edg	14	271.0±13.48	49.0±3.6	218.0±4.3	7.51±0.03
bdg/bdg	18	240.0±16.13	59.0±4.7	211.0±3.7	7.30±0.05
spermatozo id	6	259.0±11.20	52.0±3.1	198.0±3.5	7.50±0.07
aeg/edg	5	248.0±17.10	44.0±4.2	203.0±4.3	7.60±0.06
edf/edf	3	234.0±12.40	47.0±3.2	195.0±7.6	7.35±0.08
bdf/edg	4	268.0±7.90	50.1±2.9	198.0±7.2	7.45±0.07

The analysis revealed that by the volume of ejaculate boars with genotype bdg / edg stand out for the better. bdf / edg from which ejaculate volume was obtained in the amount of 271.0 ± 12.4 and 268.0 ± 7.90 respectively. This indicator varies from 234.0±12.4 to 271.0 ±13.48 ml (with edg/edf genotype) and bdg/edg respectively). The difference in indicators is significant at $p < 0.05$ (Table 7).

The largest number of spermatozoids was found in the ejaculate of boars - producers with the genotype bdg/ bdg with the number of 59.0±4.7 billion spermt, and boars with the genotype aeg/edg – the lowest number (44.0±4.2 billion) with a significant difference at ($P < 0.001$).

The concentration of spermatoz in the ejaculate changes upwards from 195.0 ± 7.6 million in 1 ml in boars with the edf / edf genotype to 225.5 ± 2.9 million in 1 ml in boars with the edg / edg genotype (the difference is significant ($P < 0.001$).

At the ballpoint assessment of spermatoz activity it was found that spermatoz obtained from boars with aeg / edg genotype was estimated at 7.6 ± 0.06 (the highest score). and boar spermtaz with bdg / bdg genotype at 7.30 ± 0.05 score (lowest score). The difference in indicators of boars with aeg / edg and bdg / bdg; bdg / edg and bdg / bdg; aeg / bdg and bdg / bdg genotypes is significant at $P < 0.001$.

The presence of definitely significant differences in the indices of boars with different genotypes of the E-blood group system suggests that there is a relationship between the studied parameters and the sperm productivity of boars, but it should be emphasized that more confident conclusions require additional studies that we have yet to conduct.

It is known that the genetic structure of the herd may change under the influence of breeding aimed at increasing the productivity of animals. We have been tracing the genetic structure of the population for 18 years. The results of the analysis showed: the structure of the Tsivilsky breed pig remains at about the same level as 18 years ago. In the dynamics there are insignificant deviations in the concentration of alleles which confirms the fact of the connection of blood groups with the reproductive qualities of animals and their involvement in natural selection. When analyzing the dynamics of the genetic structure by genotypes of blood groups the following should be noted: polymorphic the most informative locus E shows a variety of genotypes the frequency of occurrence of which varies from 0.100 bdf / def to 0.500 bdg / deg. The frequency of the heterozygous genotype E bdg / deg remains the highest compared with other genotypes of this locus since 2000 and remains approximately at the level of 0.500. The presence of animals with genotypes E bdg / bdg. Ebdg / def, Edeg / deg, Ede / def remains constant, but significantly with a lower frequency.

Among diallel loci the D system is different, where an increase in the concentration of the Da / b genotype is observed with a frequency of 0.220. This requires further research on this locus According to our research and literature data the Da allele is rarely found in domestic pigs (3.71). The ratio of the main genotypes and their frequency of occurrence in the population with small fluctuations in the dynamics as well as for alleles are preserved. In total 32 basic genotypes were identified in this population of the Tsivilsky breed pig, theoretically there are much more, it depends on the number of reagents used to determine blood groups.

Another indicator such as the level of herd homozygosity compared with previous years has now slightly declined (0.7%). which is noted as a positive sign in the characteristics of this herd of pigs. An analysis of the genetic and genealogical structures of the herd (farm) of pigs of the FGUP Kolos shows that a consolidated gene pool of the Tsivilsky breed has been formed as a result of long-term breeding according to the closed population method. It is necessary to eliminate the existing shortcomings in the organizational structure of this population, such as number, placement and some others, since the expediency of preserving pigs of a given breed is confirmed by their distinctive features according to their economic and biological characteristics.

4. Conclusion

As a result of this work, we came to the conclusion that the blood groups of the studied populations of pigs have a number of features. Also, features were identified by the frequency of occurrence of certain genotypes in boar lines, plus genotypes inherent in pigs of certain lines that can be used as a marker of pigs belonging to a certain line were identified. It has been established that when taking into account the genes of blood groups when compiling parental pairs, it allows increasing the multiplicity of sows and the mass of the nest. It can also be stated that the fattening qualities and indicators of sperm production of boars depend on the carriage of certain genotypes of blood groups, which we recommend to use when selecting young animals for breeding purposes (preference is given to young animals that have desirable genes with high productivity).

References

[1] Glazko V I and Sozinov A A 1993 Genetics of animal and plant isoenzymes. Kiev: Harvest 1993. - 526 p.
[2] Serdyuk G N 2000 Immunogenetic markers. and their use to improve the efficiency of pig breeding: authors abstract of Dr. of biol. Sciences. SPb .; Pushkin; VNIIRGZH. 2000. - 58 p.
[3] Serdyuk G N 2002 Immunogenetic pigs: Theory and practice - St. Petersburg. 2002. - 399 p.
[4] Gorelov I G and Ermolaev V 2000 Population-genetic studies of the Tsivilsky breed pig. *Pig Breeding*. - 2000. - No. 3.
[5] Evdokimov N V, Novikov A A and Zavada A N 2017 Tsivilsky breed of pigs: immuno - and cytogenetics. Cheboksary New Time. 260 p.
[6] Evdokimov N V and Petrov N S 2017 Gene pool and productive qualities of Tsivilsky breed pigs. Beau Bassin .: LAP Lambert Academic. 374 p.
[7] Tikhonov V N 1991 Immunogenetics and biochemical polymorphism of domestic and wild pigs - Novosibirsk. 1991. - 303 p.
[8] Mashurov A M 1980 Genetic markers in animal breeding. Science. 1980. - 318 p.
[9] Zhivotovsky L A and Mashurov A M 1974 Guidelines for the statistical analysis of immunogenetic data for use in animal breeding.
[10] Sukhova N O, Burlak Z K and Dmitriev G L 1981 The use of immunogenetic analysis in pig breeding -Novosibirsk: The All-Union Academy of Agricultural Sciences. 1981. - 58 p.

IOP Conf. Series: Earth and Environmental Science **346** (2019) 012053 doi:10.1088/1755-1315/346/1/012053

The efficiency of the use of "Biostrong 510" additive in the technology of broiler chicken meat production

A Y Lavrentyev[1], A I Nikolaeva, N V Evdokimov, V G Semenov, V S Sherne, E Y Nemtseva, N V Danilova, N S Petrov and G M Toboyev

Federal State Budgetary Educational Institution of Higher Education "Chuvash State Agricultural Academy", K. Marx Street, Cheboksary, 29, 428003, Russia

E-mail: lavrentev65@list.ru

Abstract. Growing chickens for meat is the main link in the technological chain of broiler production. Broilers – the final hybrid obtained as a result of interbreeding of different chicken breeds lines (meat parent forms), tested for compatibility. Because of this, the bird has an intense growth rate, high taste of meat. Consumer demand for broiler meat has been increasing steadily in recent decades. Biostrong 510 (BIOSTRONG 510) herbal feed supplement to improve palatability and increase consumption or feed for farm poultry. The mechanism of action is based on the joint action of several plant substances, in which there are active substances, which, when combined, affect a certain category of animals. Four groups of broiler chickens (1 control and 3 experimental) of the Cobb 500 cross at the daily age of 50 heads each were formed to conduct the experiment by the method of analogues. The experiments were carried out with the daily up to 40 days of age. Expenses of feeds in the experimental groups was lower than in the control group by 0.07, 0.13, 0.12 kg, respectively. The purpose of this work was to establish the feasibility and effectiveness of enrichment of animal feed with vegetable feed additive "Biostrong 510" used in the production of meat broiler chickens. Assessing the results of scientific and economic experience, we can safely say that in all the experimental groups in which "Biostrong 510" was used as an additive, the gain in the live weight of broiler chickens was increased compared to the control group. In total, 4020 g of compound feed was spent for the period of an experiment for each experimental group of broiler chickens. Due to the improved taste of feed, the palatability in the experimental groups was higher than in the control group by 0.3 %; 0.9 %; 0.8 %. The cost of feed in the experimental groups was lower than in the control group. For example, in the control – 1, 87 kg per 1 kg of body weight gain. In experimental – below 0.07; 0.13; 0,12 kg, respectively. As a result of the experiment, it was found that the viability of experimental broiler chickens was almost 100%. The natural loss of livestock was only due to technological injuries. The obtained data on the productivity of broiler chickens indicate a favorable effect of plant feed additive on the live weight of broiler chickens. Based on the results of the research, we recommend to introduce a plant feed additive based on essential oils and plant substances "Biostrong 510" in the composition of complete feed in the amount of 150 g/t to increase live weight, growth intensity, preservation, meat productivity, reduce feed costs per unit of broiler chickens of the cross "Cobb 500".

1. Introduction

The successful development of the broiler industry is determined by achievements in breeding, use of specialized feed and compliance with the recommended conditions of farming. Poultry farms and

Content from this work may be used under the terms of the Creative Commons Attribution 3.0 licence. Any further distribution of this work must maintain attribution to the author(s) and the title of the work, journal citation and DOI.
Published under licence by IOP Publishing Ltd

organizations-members produce more than 85% of eggs and poultry meat of the total volume produced by agricultural enterprises of the Russian Federation [1-8].

The emphasis on the development of poultry has to be done not only because of the fact that it is the most "precocious" industry, but also because it gives the greatest return per unit of spent feed, labor and other resources in conditions of limited grain resources. The cost of feed and labor is 2-3 times lower in poultry than in pig and cattle breeding. Dietary poultry products are significantly cheaper than other products containing animal protein. Thus, from the point of view of the population and state interests in terms of solving the food problem, the development of poultry farming should be carried out on a priority basis [9-17].

The nutritional value of broiler meat as a food product, the availability of technological equipment, the use of highly productive meat crosses is the basis for increasing the production of broiler meat [18-26].

The share of poultry meat in the total production of livestock and poultry in all categories of farms is more than 20 %. Poultry products account for 30-35% in the large industrial centers of Russia in the structure of the population nutrition.

Providing the population of Russia with poultry products of its own production determines the food independence of the country, which depends on the development of the national agricultural sector. In this case, an important role is played by the possibility of increasing the poultry productivity with the lowest production costs. The world experience of this branch successful introduction of proves a need to solve a feed problem first of all. Only with a complex feeding of a bird, the genetic potential of productivity is realized [27-30].

Today there is a necessity for a correct and scientifically based clinical and morphological assessment of agricultural, wild, exotic birds organisms metabolism with the search and use of drugs in industrial poultry, correcting metabolism, safety, productivity of livestock and improving the quality of poultry products [31-33] in modern socio-economic conditions.

To this end, the use of "Biostrong 510" as a vegetable feed additive to increase productivity and reduce the cost of feed per unit of production is relevant in industrial poultry farming.

The purpose of this work was to establish the feasibility and effectiveness of enrichment of animal feed with vegetable feed additive "Biostrong 510" used in the production of meat broiler chickens.

2. Material and methods

The scientific-economic experiment was held to examine the efficiency of adding herbal feed additive "Biostrong 510" in compound feeds for broiler chickens under the conditions of the poultry farm LLC "Ptitsefabrika Akashevskaya".

Four groups of broiler chickens (1 control and 3 experimental groups) were formed to conduct the experiment by the method of analogues of the "COBB 500" cross at the daily age of 50 heads each. The experiments were carried out with birds aged 1 day up to 40 days of age. The experiment was carried out according to the scheme presented in table 1.

Table 1. Scheme of the experiment.

Group	Livestock chickens, heads.	Main diet, feed	Quantity of "Biostrong 510", % of dry matter of compound feed
Control	50	PK + of "Biostrong 510"	-
I Experimental	50	PK + of "Biostrong 510"	0.01
II Experimental	50	PK + of "Biostrong 510"	0.015
III Experimental	50	PK + of "Biostrong 510"	0.02

The peculiarity of feeding was that a different amount of vegetable feed additive "Biostrong 510" was introduced in the feed of experimental groups broiler chickens.

Complete feed was used in feeding of experimental broiler chickens: PK 5-0 (pre-starter, 1 to 7 days), 5-1 PK (starter, from 8 to 15 days), 5-2 PK (grower/growth, from 16 to 22 days), PK 6-1 (finisher 1, from 23 to 34 days), PK 6-2 (finisher 2, Art. 35 days.).

Feeding of poultry was carried out with dry balanced feed with nutritional parameters corresponding to the recommended norms of the cross "COBB 500". Feeding front was 2.5 cm per 1 head (deviations up to ±5% are allowed). The quality of water for drinking met the requirements of GOST R 51232-98. The choice of water supply sources was carried out in accordance with SanPiN 2.1.4.559-96. Water hardness was within the permissible limits (no more than 30°). The drinking front was not less than 1 cm per head (deviations up to ± 5% are allowed). Access to water was constant.

During the scientific and economic experience, experimental broiler chickens were placed on a deep bedding (in specially fenced sections) when kept on the floor.

The poultry houses had a mixed ventilation system. The air flow was carried out both through the roof, where roof fans are installed, and through the supply valves. Exhaust fans are installed in the wall openings. There were plastic ventilation shafts on the roof. Lighting was provided by smooth adjustment, localized in the area of feeders.

Materials of researches are processed by method of variation statistics on the personal computer with use of the software of Microsoft Office of Excel 2007.

Bistrong 510 (BIOSTRONG 510), herbal feed Supplement, is to improve palatability and increase consumption or feed for farm poultry. The mechanism of action is based on the joint action of several plant substances, in which there are active substances, which, when combined, affect a certain category of animals.

The content of the main active substances with the possibility of their detection in premixes and compound feeds by analytical methods both in quantity and quality is guaranteed.

The composition of the feed additive "Biostrong 510" includes essential oils, herbal plants and flavorings. Carrier and auxiliary substances are wheat bran, limestone, silica, and starch.

The feed additive "Biostrong 510" does not contain genetically modified products.

The composition of this additive: essential oils (microencapsulated), sharp substances, bitter substances, saponins.

"Biostrong 510" contains: anise oil-5.33%; thyme oil – 3.33%; gentian powder – 20%; chili paprika powder – 10%; killaya extract – 23.34%; wheat bran – 21.5%; limestone – 10%; silicon dioxide – 5%; starch – 1.5%.

The maximum permissible deviations in the content of ingredients did not exceed 15%.

Feeding of broiler chickens was organized in accordance with the norms of feeding for the cross "Cobb 500". When growing the experimental broiler chickens, the calculation of food consumption was carried out: per head per day and per 1 kg of live weight gain. During the experiment, the broilers were placed on deep bedding. The relative humidity, the temperature in the poultry house, the light supply in the room were automatically controlled by the "Francom" company microclimate computer system/

4020 g of feed for each study group of broiler chickens was spent during the scientific and economic experiment.

For the entire period of farming, the palatability of the feed in experimental groups exceeded the eaten one in the control group. In the experimental groups, the feed palatability was higher, since the inclusion of "Biostrong 510" increased the palatability of feed, due to its composition.

3. Research result

For 1 kg of broiler live weight gain, 1.87 kg of feed was consumed in the control group. The feed costs of the experimental groups were lower than that of the control group by 0.07; 0.13; 0.12 kg.

Safety was determined for 40 days in order to determine the effect of feeding in the rations of experimental broiler chickens on the body's stability of the experimental chickens of various doses of the herbal feed additive "Biostrong 510".

The survival rate of the 1st experimental group was 98 %, the 2nd experimental group - 100%, the 3rd experimental group – 100%.

The effectiveness of the use of plant feed additive in the experiment was established by taking into account changes in the live weight of birds during the accounting period (figure 1).

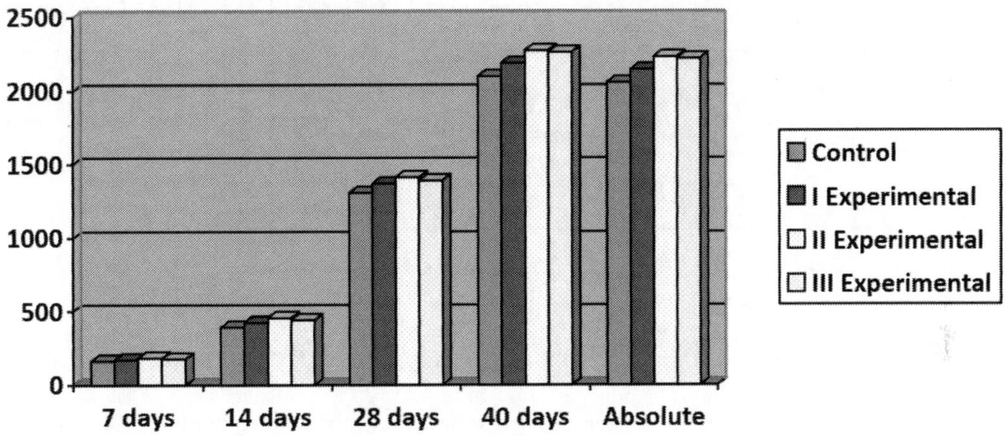

Figure 1. Dynamics of live weight gain of broiler chickens, g (P<0.05; P<0.01).

In the course of our research, it was found that the intensity of experimental broiler growth chickens was high, as evidenced by the average daily growth of live weight. The experimental groups exceeded the control group in terms of the average daily growth (figure 2).

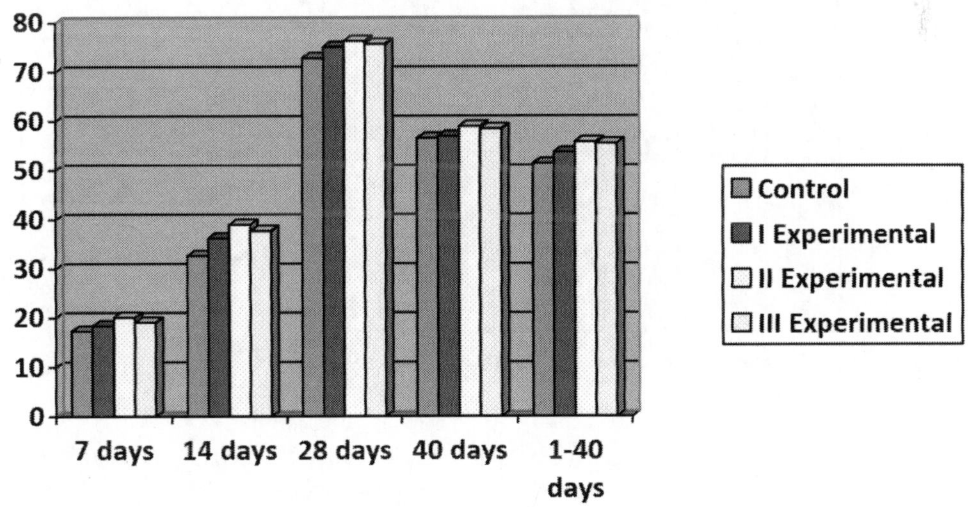

Figure 2. Dynamics of average daily growth, g (P<0.05; P<0.01).

The absolute increase was calculated to study the indicator of live weight gain completely. The absolute gain in live weight in broiler chickens of all experimental groups was greater than in the control group (figure 3).

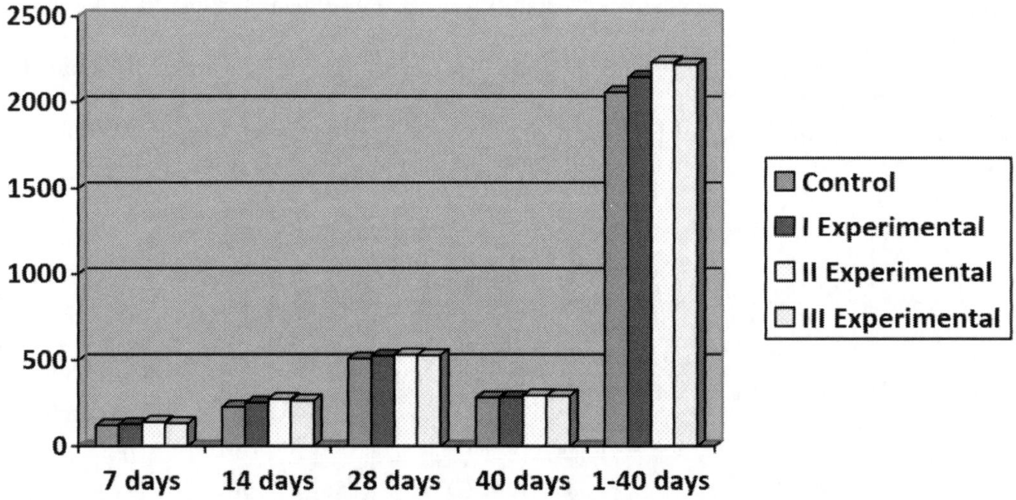

Figure 3. Dynamics of average daily gain, g (P<0.05; P<0.01).

The addition of a plant-based feed additive to the main diet had a positive effect on the absolute gain in live weight and allowed to increase it in the experimental groups.

One of the birds' development signs is growth. Growth rates reflect its intensity for a single segment. The greatest relative gain was obtained from the chickens of the second and third experimental groups, receiving vegetable feed additive "Biostrong 510" in their diets in the amount of 0.015 and 0.02% per 1 ton of feed, respectively 192.76 and 192.71% (figure 4).

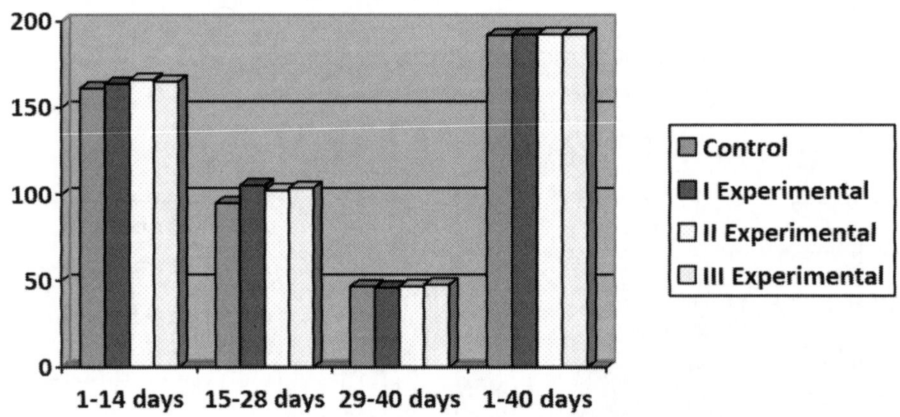

Figure 4. Dynamics of the relative growth rate of broiler chickens, % (P<0.05; P<0.01).

In order to establish the effect of feeding vegetable feed additives on meat productivity by the method of T. M. Polivanova (1987) in the slaughter shop of the poultry farm at the end of the experiment, control slaughter of broiler chickens was carried out.

Broiler chickens of experimental groups 1,2 and 3 had superiority over the control in pre-slaughter live weight, which was higher by 4.3; 8.3; 7.7 %, respectively (As shown in Table 2). According to the results obtained in the half-eviscerating, it can be seen that the experimental groups of broilers by weight exceeded the control group by 4.5; 8.9; 8.2 %. The mass of the eviscerated carcass in the experimental groups prevailed over the control group. The mass of eviscerated carcass in the first experimental group was higher than in the control group by 119.5 g , in the second experimental group – by 181.7 g , in the third experimental group – by 169.8 g .

Table 2. Control slaughter of experimental broiler chickens (on average by groups).

Group	Pre-slaughter live weight of 1 head, g	Half-eviscerated carcass		Eviscerated carcass	
		weight, g	%	weight, g	%
Control	2037.6±24.4	1626.0±36.2	79.8	1323.2±19.8	63.6
1st experimental	2150.4±20.7**	1690.2±26.1	78.6	1423.6±18.6*	66.2
2nd experimental	2206.7±21.2**	1771.9±29.2	80.3	1504.9±17.7**	68.2
3rd experimental	2195.6±26.0**	1758.7±31.1	80.1	1493.0±24.4**	68.0

Note: * P<0.05; * * P<0.01

Thus, the introduction of the vegetable feed additive "Biostrong 510" into the basic ration of broiler chickens from the experimental groups contributes to an increase in the pre-slaughter body weight, the weight of the half-gutted carcass, the mass of the gutted carcass and the slaughter yield, compared to the control. Moreover, the best indicators were established in the second experimental group.

For the second experimental group of chickens, the boning of the carcass thoracic part showed that muscle tissue occupied 368.7 g (24.5 %) of the total weight and exceeded 75 g (25.5 %) of the control group. The third experimental group also exceeded this figure by 67.6 g (23%) and amounted to 361.3 g (figure 5).

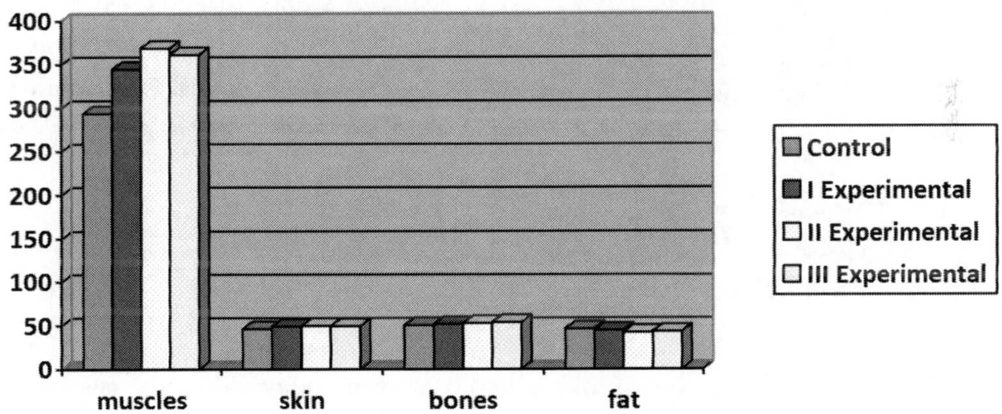

Figure 5. Morphological composition of the breast of experimental broiler chickens, g (P<0.05; P<0.01).

During the boning of the carcass femoral part it was revealed that with almost identical outputs of muscle tissue, skin and bone, the advantage in absolute terms was on the experimental groups broilers side.

With equal bone mass, they were superior to the young birds of the control group in terms of muscle tissue in the first experimental group by 21 g (12.5 %), skin by 1.8 g (5.0 %); in the second experimental group by 30.6 g (18.2%), skin by 3.4 g (9.5%); in the third experimental group by 26 g (15.5%), skin by 1.6 g (4.5%) (figure 6).

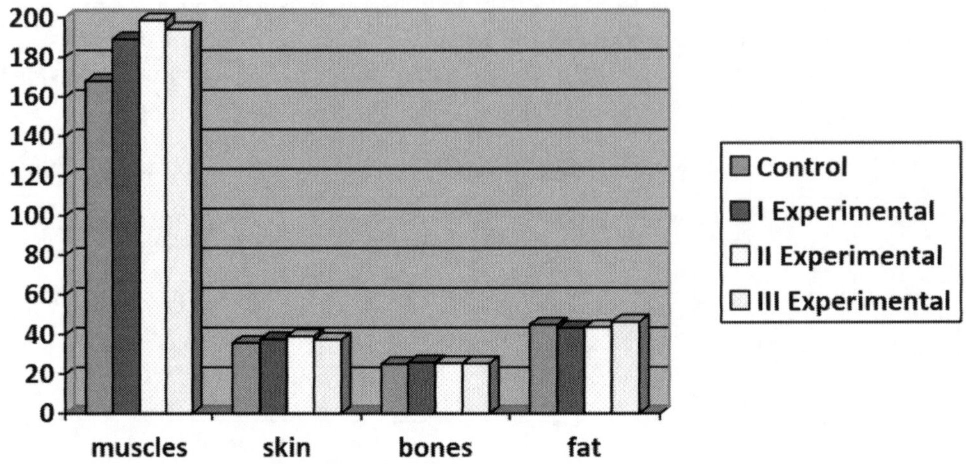

Figure 6. Morphological composition of the breast of experimental broiler chickens, g (P<0.05; P<0.01).

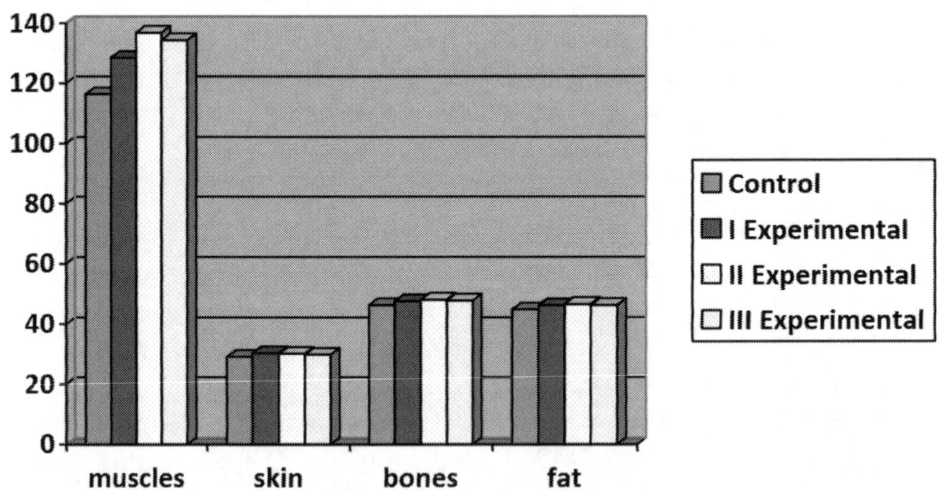

Figure 7. Morphological composition of the lower leg of experimental broiler chickens (P<0.05; P<0.01).

Undoubtedly, the number of experimental groups had a number of advantages in comparison with the control group, in particular, in the study of the lower leg morphological composition, and comparing the meat qualities of poultry (As shown in Table 3). The output of muscle tissue in the first experimental group was 128.4 g, which is more than the control group by 12 g (10.3%). In the second experimental group, the output of muscle tissue also prevailed over the control group and amounted to 136.9 g, which is more by 20.5 g (17.6%). In the third experimental group, this figure was higher than in the control group by 18 g (15.5%). Skin yield in the first experimental group was greater than in the control by 1.2 g, in the second group-by 1 g, in the third-by 0.7 g. The same situation was with the yield of bones and fat, the results of the experimental groups prevailed over the control group. Figure 7 shows a diagram of the experimental broiler chicken lower leg morphological composition.

Table 3. Economic Indicators of Experiment (rbl).

Indicator	Group			
	Control	1-st Experimental	2-nd Experimental	3-rd Experimental
Average number heads.	50	50	50	50
Survival rate, %	98	98	100	100
Live weight when putting on the experiment, g	42.1±0.04	42.2±0.02	41.9±0.03	42.0±0.03
Live weight at the end of the experiment, g	2100.6±23.6	2190.4±28.8	2275.0±29.6	2263.5±28.7
The absolute gain, g	2058.5±23.6	2148.2±28.8	2233.1±29.6	2221.5±28.7
Additional growth was obtained, g	-	89.7	174.6	163.0
%	100	104.3	108.5	107.9
The selling price of 1 kg, rbl.	96.73	96.73	96.73	96.73
The cost of gain of 1 head, RUB	116.61	122.87	128.54	128.56
Revenue from the sale of 1 head, RUB	203.1	211.83	220.0	218.89
Profit, RUB	86.51	88.96	91.46	90.33
Additional revenue, RUB	-	2.45	4.95	3.82
Realized for 1 RUB add. costs, RUB	-	1.77	2.43	1.40

Additional products (1 ruble spent) for the purchase of plant feed additives were received for 1.77 rubles in the first experimental group, for 2.43 rubles in the second experimental group and 1.40 rubles in the third experimental group.

Thus, the plant feed additive based on essential oils and plant substances "Biostrong 510" contains the main active substances with the possibility of their detection in feed by analytical methods, both in quantity and quality. Contains active ingredients exclusively of plant origin that, unlike synthetic, have a more intense effect on the body of the animal. Active plant substances that are part of the plant feed additive based on essential oils and plant substances "Biostrong 510", serve to improve the digestibility of nutrients. The mechanism of action is based on sinergizm of several herbal substances which are active substances, in mutual combination, affect the growth and development of broiler chickens.

It was possible to improve the taste of the feed and increase secretion in the digestive tract with the help of "Biostrong 510" during the experiment. Due to this, the intestinal microflora is stabilized, the accumulation of toxins decreased, inflammatory processes were regulated. The immune system began to work more actively. Assessing the results of scientific and economic experience, we can safely say that in all the experimental groups in which "Biostrong 510" was used as an additive, the gain in the live weight of broiler chickens was increased compared to the control group. The highest values of gain were observed in the second experimental group and amounted to 2275 g, which is more than the gain of the control group by 8.3 %.

In total, 4020 g of compound feed was spent for the period of an experiment for each experimental group of broiler chickens. Due to the improved taste of feed, the palatability in the experimental groups was higher than in the control group by 0.3 %; 0.9 %; 0.8 %. The cost of feed in the experimental groups was lower than in the control group. For example, in the control – 1, 87 kg per 1 kg of body weight gain. In experimental – below 0.07; 0.13; 0,12 kg, respectively.

As a result of the experiment, it was found that the viability of experimental broiler chickens was almost 100%. The natural loss of livestock was only due to technological injuries.

The obtained data on the productivity of broiler chickens indicate a favorable effect of plant feed additive on the live weight of broiler chickens.

The health status of broiler chickens was controlled by the determination of blood parameters: the content of red blood cells, white blood cells, hemoglobin, total protein, total calcium, phosphorus, cholesterol, and glucose. The studies conducted by us did not reveal the effect of plant feed additive on hematological parameters of broiler chickens in the experimental groups and were generally within the physiological norm, but there was a tendency to increase the total protein.

4. Summary

Based on the results of the research, we recommend to introduce a plant feed additive based on essential oils and plant substances "Biostrong 510" in the composition of complete feed in the amount of 150 g/t to increase live weight, growth intensity, preservation, meat productivity, reduce feed costs per unit of broiler chickens of the cross "Cobb 500".

References

[1] Davleev A D 2012 Prospects and problems of the poultry sector of Russia and the customs of Union countries in the world market *Poult. Poultr. Prod.* - Rzhavki, **5**. p. 15-9.
[2] Egorov I L 2013 Drugs Koreton and Biokoreton-Forte in compound feeds for broiler chickens *Poultry*. - M., **1**. p. 23-7.
[3] Egorov I A 2015 Modern approaches to feeding highly productive birds *Poult. Poultr. Prod.* - Rzhavki, **3**. p. 27-9.
[4] Fisinin V I 2014 Manual optimization of feed recipes for poultry. Sergiev Posad: VNITIP.
[5] Ivanov N G 2018 The immunity of birds on a background of application of probiotic "Biosporin" In the collection: the development of agricultural science as an essential condition for the effective functioning of the agro-industrial complex of the country: materials of the all-Russian scientific and practical conference dedicated to the 70th anniversary of the honored worker of the higher school of the Chuvash Republic and the Russian Federation, doctor of veterinary Sciences, Professor Nikolai Kirillov. Cheboksary, Chuvash State Agricultural Academy. p. 138-142.
[6] Lavrentyev A Y and Ivanova E Y 2016 Feed with domestic enzyme preparations for laying hens *Agricultural science*. - M., **1**. p. 20-1.
[7] Tyurin V G, Semenov V G, Ivanov N G and Ivanova E E 2017 Correction of nonspecific resistance and specific immunogenesis of the organism in the implementation of the bird biopotential In the collection: Modern problems of veterinary pathology and biotechnology in the agro-industrial complex materials of the International scientific and practical conference dedicated to the 95th anniversary of RUE " Institute of Experimental veterinary named after S. N. Vyshelessky." Institute of Experimental Veterinary Medicine. S. N. Vyshelessky. p. 390-4.
[8] Zemlyanaya Z E and Radkevich V S 2011 The development of the poultry industry in the Russian Federation in 2010 and growth prospects *Poult. Poultr. Prod.*- Rzhavki, **1**. p. 13-6.
[9] Alekseev V A and Nemtseva E Y 2017 Increasing the productivity of broiler chickens when used in their diets zeolite-containing drug "Permait" *Vestnik of Ulyanovsk State Agricultural Academy*. - Ulyanovsk, **3** (39). p.105-9.
[10] Annex: *Development and Feeding of Cross Cobb broilers 500*. - 2012.
[11] Antipov A A, Fisinin V I and Egorov V I 2010 The effectiveness of probiotic based on BacillusSubtilis and BacillusLicheniformis in the cultivation of broiler chickens *Zootechnics*. No. **4**. p. 10-2.
[12] Bobyleva G A 2014 The state of the poultry complex in Russia and the prospects for its development *Poult. Poult. Prod.* **6**. p. 18-23.
[13] Buyarov V C, Kleymenova I S and Shalimova O A 2012 Status and prospects of poultry meat *Vestnik Orel GAU*. **1**. p. 49-61.
[14] Fisinin V I 2015 Modern approaches to feeding highly productive birds *Poult. Poultr. Prod.* Rzhavki, **3**. p. 120-7
[15] Ivanova E Y and Lavrentyev A Y 2015 The effect of enzyme preparations on egg production and egg weight of laying hens *Bulletin of the Ulyanovsk State Agricultural Academy*. -

IOP Conf. Series: Earth and Environmental Science **346** (2019) 012053 doi:10.1088/1755-1315/346/1/012053

Ulyanovsk, **1** (29). p. 94-7.

[16] Sobol N 2016 Successful poultry farms choose "Cobb" *Poultry*. M., **5**. p. 12-5.

[17] Volkova E A and Senko A Y 2010 Effect of probiotic and vitamin preparations on meat productivity and quality of poultry meat *Poultry, and Poultry Products*. - Rzhavki, **3**. p. 33-5.

[18] Egorova A V 2007 The increasing method of broilers productivity. Farming of broilers separated by sex *Animal Husbandry of Russia*. - M., **3**. p. 15-9.

[19] Filonenko V I 2006 Productivity and meat quality of broilers of "Cobb 500" cross depending on the density of planting and age of slaughter *Sat. science. labour. VNITIP*. - Sergiev Posad, **81**. p. 39-49.

[20] Fisinin V I 2010 Innovations in the poultry industry in Russia *Bulletin of the Russian Academy of Agricultural Sciences*. **1**. p. 9 – 12.

[21] Lavrentyev A Y 2007 Improving the technology of growing young farm animals using feed additives and biologically active substances: abstract of the thesis for the degree of doctor of agricultural Sciences The Chuvash State Agricultural Academy. – Cheboksary. p. 24

[22] Lavrentyev A Y 2018 Egg laying hens when used in animal feed enzyme preparations Scientific Bulletin of the Ural. Uralsk, **3**. - p. 86-9.

[23] Lavrentyev A Y and Nikolaeva A I 2018 Vegetable fodder additive for broiler chicken *Feed*. **10**. p. 80-1.

[24] Likhachev I V 2015 "Cobb" confirms its advantage *Poultry*. M., **6**. p. 6-9.

[25] Motovilov K Ya and Ivanova O V 2011 Effect of feed additives on the growth and survival rate of broiler chickens *Feeding of farm animals and feed production*. M., **5**. p. 36-43.

[26] Savateeva E A, Petryankin F P and Ivanov A 2012 Evaluation of the quality of poultry meat using probiotic *Coredon Scientific notes of Kazan state Academy of veterinary medicine. N. E. Bauman*. Kazan, **210**. p. 186-9.

[27] Kochish I I, Myasnikova O V and Galkin V A 2018 The increase in output of broiler chickens by monitoring the "output window" and adjust the time of incubation *Veterinary, zootechnics, and biotechnology*. - M., **3**. p. 43-7.

[28] Kuznetsova E Y 2004 The use of zeolite-containing additives in the technology of growing broiler chickens Author's abstract, Candidate of Agricultural Sciences. – Cheboksary. p. 23.

[29] Lavrentyev A Y and Sherne V S 2016 The effect of enzyme preparations on the productivity of gooslings Compound Feed. - M., **7-8**. p. 78-9.

[30] Lavrentyev A Y 2016 Domestic enzyme preparations in mixed feeds of laying hens In the collection: Fundamental and applied aspects of feeding farm animals and feed technology Materials of the conference dedicated to the 120th anniversary of M. F. Tomme Dubrovitsy. p. 134-9.

[31] Petryankin F P, Lavrentyev A Y and Sherne V S 2017 Use of biologically active substances of natural origin in *poultry* In the collection: Rational nature management and socio-economic development of rural areas as the basis for the effective functioning of the agro-industrial complex of the region Materials of the all-Russian scientific and practical conference with international participation, dedicated to the 80th anniversary of the honored worker of agriculture of the Russian Federation, honorary citizen of the Chuvash Republic Aidak Arkady Pavlovich. Chuvash State Agricultural Academy. Cheboksary. p. 309-14.

[32] Nikolaeva A I, Lavrentyev A Y and Sherne V S 2018 Vegetable feed additive in feed broilers *Poultry*. **11-12**. p. 43-4

[33] Semenov V G, Lyagina E E and Tyurin V G 2018 Productive performance of laying hens of parental flock of broilers on the background of immune In the collection: The development of agricultural science as an essential condition for the effective functioning of the agro-industrial complex of the country materials of the all-Russian scientific and practical conference dedicated to the 70th anniversary of the honored worker of the higher school of the Chuvash Republic and the Russian Federation, doctor of veterinary Sciences, Professor Nikolai Kirillov. p. 188-93.

IOP Conf. Series: Earth and Environmental Science **346** (2019) 012054 doi:10.1088/1755-1315/346/1/012054

Formation of elements of the harvesting structure of spring durum wheat in agroecological conditions of the Chuvash Republic

A G Lozhkin[1, 2], L G Shashkarov[1], L V Eliseeva[1] and A N Alexandrova[1]

[1]Federal State Budgetary Educational Institution of Higher Education "Chuvash State Agricultural Academy", Cheboksary, K. Marx street, 29, 428003, Russia

E-mail: lozhkin_tmvl@mail.ru

Abstract. The article presents for the first time data on the productivity of varieties of spring durum wheat of different morphotypes under the conditions of the northern regions of the Chuvash Republic, in order to select the most promising of them and determine the possibilities of selective and technological adaptation in the region. The experiment involved 5 varieties of spring durum wheat: 1) Bezenchuk 200 (control), 2) Bezenchuk Niva, 3) Bezenchuk 205, 4) Bezenchuk 209 and5) Luch 25. The results of biometric and structural analysis of spring durum wheat plants on average for 2015-2017 found that the height of plants by varieties on average for three years ranged from 74.6 to 104.8 cm. Indicators of the best productive bushiness were noted for the variety Bezenchuk Niva and Bezenchuk 209. Weather conditions in 2015 and 2016 provided more favorable conditions for the formation of elements of the structure of the crop, in contrast to 2017, characterized by an abundance of precipitation and lower temperatures during the growing season. It was found that the number and weight of seeds in the head depending on the variety varied in proportion to the length of the head. The best indicators of the crop structure were found in the Bezenchuk 209 and Bezenchuk Niva varieties, where the number of seeds in the main spike was 21.8 - 22.9 and the mass of seeds in the spike- 1.0-1.1 g. The mass of 1000 seeds amounted to options from 42.32 to 53.3 grams. The Bezenchuk Niva and Luch 25 varieties significantly exceeded the control of this indicator by 8.9 and 10.8%, respectively. Accounting for biological yields revealed the advantage of the Bezenchuk Niva and Bezenchuk 209 varieties.

1. Introduction

The annual deficit of durum wheat in the world market is currently estimated at 1.5 million tons. The shortage, especially of quality grain, will increase in the near future. This is due to the fact that there are almost no regions in the world where the expansion of durum wheat (the fastest way to increase gross production) is environmentally and economically feasible, with the exception of Russia and Kazakhstan [1]. At the same time, the need for durum wheat grain and products from it increases. The growing welfare of the population in the world creates an increased demand for quality food, including pasta made from durum [2]. The consumption of pasta in China and other Southeast Asian countries is increasing rapidly. The Arab population is growing, which has a significant migration potential, which leads to the spread and popularization of traditional dishes of the Middle East – bulgur and couscous, which are prepared from durum wheat. The so-called "flatbread" baked from durum wheat flour is popular in Italy and in a number of regions of the Middle East and North Africa, which also expands the market for durum wheat producers [3].

Content from this work may be used under the terms of the Creative Commons Attribution 3.0 licence. Any further distribution of this work must maintain attribution to the author(s) and the title of the work, journal citation and DOI.
Published under licence by IOP Publishing Ltd

The average annual production of durum wheat in Russia for the last 3-4 years ranges from 500-600 thousand tons, about 800 thousand tons were produced in 2017. Export amounts to 100-150 thousand tons [4]. At the same time, grain imports from Kazakhstan in 2016 exceeded 250 thousand tons [5]. Import of pasta (mainly from Italy) annually exceeds 100 thousand tons. The demand of the Russian market for high-quality pasta (from durum) is estimated at 750-800 thousand tons, in high-quality wheat cereals (from durum and spelt) in 100 thousand tons, which is equivalent to 1.5 million tons of durum wheat. Taking into account the prospects for the development of export potential and import substitution, the volume of production of durum wheat in Russia should be at least 2.0-2.5 million tons annually, which is 3 times more than the record harvest in 2017 [6]. In this regard, along with solving the problems of increasing the production of durum wheat in the traditional regions of its cultivation, it is advisable to assess the possibility of expanding the area of this culture in the Northern regions of the Volga and Ural regions with a more favorable hydrothermal regime of the growing season. Each agricultural region has certain environmental resources for the formation of the appropriate size and quality of the harvest of a particular cultivated crop [7, 8]. The Chuvash Republic is one of these regions. Cultivation of durum wheat in a new region can be successful only on the basis of an adequate varietal type [9].

The purpose of this work is to study the productivity of varieties of spring durum wheat of different morphotypes in the Northern regions of Chuvashia and the selection of the most promising of them and to determine the possibilities of breeding and technological adaptation of spring durum wheat in the region.

2. Material and methods

Experiments on the various testing of spring durum wheat are carried out at the Student's Training and Production Center of the Chuvash State Agricultural Academy since 2015 on a medium loamy dark gray forest soil. The topsoil of the test plot has a soil reaction close to neutral, the humus content is low, mobile phosphorus is high, and exchangeable potassium is high. Microplot trials were laid according to the scheme: 1) Bezenchuk 200 (control), 2) Bezenchuk 205, 3) Bezenchuk 209, 4) Bezenchuk Niva, 5) Luch 25. The first 4 varieties are bred in the Samara NIISH, and Luch 25 - the All-Russia Scientific-Research Institute of Agriculture of the South-East. Seeds of all varieties were of the elite category. The varieties Bezenchuk 200, Bezenchuk 205 and Bezenchuk 209 belong to the variety Gordeiform, Bezenchuk Niva and Ray 25 - leukurum. For the duration of the period of "sprouting-earing" they are arranged in order of increasing period as follows: Bezenchuk 205, Luch 25, Bezenchuk 200, Bezenchuk Niva, Bezenchuk 209. Bezenchuk 209 variety carries a strong gene reduction of plant height RhtB1b and belongs to the stunted morphotype. All other varieties do not have strong genes that control plant height, its value depends on the duration of vegetation and environmental conditions during the growth of internodes and their number.

Bookmark and design of experiments, observations, and records during the growing season, harvesting and accounting of the crop were carried out according to the method of state variety testing. Mathematical processing of data was carried out according to Dospekhov. Agricultural machinery of cultivation was generally accepted for the Chuvash Republic: autumn plowing to a depth of 25-27 cm, spring harrowing and presowing cultivation. Predecessor - potatoes. Sowing was carried out in May at a soil temperature at a seed embedment depth of 8-10°C, with a norm of 5 million pieces of germinating seeds per 1 ha.

Weather conditions in 2015 were generally favorable for the growth and development of spring durum wheat plants. The second decade of May was characterized by moderately warm weather with frequent precipitation. Their sum exceeded two perennial norms. The average monthly temperature was 3.8 degrees above the long-term. In June, due to elevated temperatures, there was a lack of precipitation. Spike formation took place in conditions of insufficient moisture supply. In July, the weather was kept cool for this season with frequent and heavy rains. Wax ripeness of cereals was noted in mid-August. In general, during the period of the active growing season (May-August), the average temperature exceeded the multi-year indicator by 1.8°C and amounted to 18.2°C, the amount of precipitation was 249 mm or 115% of the multi-year norm.

Sowing of spring durum wheat in 2016 was performed in late April-early May. This year, the growth and development of field crops came under conditions of lack of moisture against the background of high temperature during the entire growing season. The least amount of precipitation was observed in the first two months of the growing season: in May, 17% and in June, 45% of the multiyear monthly norm. In July and August were - 88% - 80% of the perennial amounts, respectively. The reserves of productive moisture in the soil were remained low level. It had been raining for more than 22 days in September and 9 days were marked with precipitations more 51 mm. The total precipitation for the month amounted to 134 mm, 290% of the multi-year norm. In this regard, the harvesting work in 2016 was completed only in October [10]. In general, during the period of the active growing season of plants (May-August), the average temperature of the air was 19.6 ° C, exceeding the long-term one by 3.2°C. 134.3 mm of precipitation fell, 54% of the long-year norm.

In 2017, sowing was carried out in the second decade of May. The growth and development of durum wheat plants at the beginning of the growing season (May-June) took place under conditions of excess moisture against the background of low temperatures. The second half of the growing season (July-August) in temperature and precipitation differed little from the average multiyear norms. The period of ripening of the crop (the third decade of August) coincided with heavy precipitation, accompanied by strong winds, which caused lodging of the crops. In general, during the growing season of durum wheat (May-August), the average daily temperature (15.8°C) was lower than the average multiyear values by 0.6°C. These circumstances and the relatively late date of sowing led to a prolongation of the growing season and late harvesting periods. In both years, in terms of other spring grain crops (spring soft wheat, barley), it was quite competitive in terms of grain yield and gross income per unit of cultivated area.

3. Research results

Analysis of the structure of the crop was carried out on trial sheaves taken before harvesting. The results of biometric analysis of spring durum wheat plants (Table 1) it was noted that the height of the plant varieties ranged from 74.6 to 104.8 cm. Most tall varieties are reliably marked in the variety Bezenchuk 205 and low-growing – Bezenchuk 209. Bezenchuk 209 variety carried a strong gene reduction of plant height RhtB1b and belonged to the stunted morphotype. However, a great height was not always advantage and low-growing plants were usually more resistant to lodging. In all likelihood, the lodging problem was solved without short stem forms due to directional selection for the thickness and strength of the straw, if it was only about stem lodging. Also, root lodging was extremely widespread in spring wheat with its was weak knot roots. Here, short-stalk played a positive role, especially when lodging caused by soil spreading after rains, such as in 2017, when the strength of the connection between roots and soil showed sharply weakens and short. Light stem showed undoubted advantages in plant resistance.

Table 1. Biometric indicators of varieties of spring durum wheat on average for the years 2015 - 2017.

Varieties	A number of plants before harvesting(pcs / sq. m)	Plant height (cm)	Bushiness		Number of internodes(pcs)
			General	Productive	
Bezenchuk 200	349	76.6	2.0	1.9	3.8
Bezenchuk 205	314	104.8	2.1	2.1	3.8
Bezenchuk 209	343	74.6	2.4	2.3	3.5
Bezenchuk Niva	357	93.2	2.4	2.3	3.5
Luch 25	370	94.4	2.0	1.7	3.6
NSR 05	8.9	6.5	0.3	0.3	0.4

Also, in our studies, we determined the number of internodes of durum wheat stem. However, many researchers assign a large role in the stalk strength not so much to the number of internodes as to the thickness of the internodes and their anatomical structure [11]. In our studies, the number of internodes by varieties on average over two years varied in the range of 3.5- 3.8, however, the difference in options could not be considered reliable, since changes were within the margin of field experience.

The number of shoots that form wheat plants was commonly called business, and the its process was called tillering energy.

However, not all shoots produce spikelets, therefore they distinguished between total bushiness (the number of shoots per plant) and productive bushiness (the number of spikelets on the plant), which was an important factor shaping the productivity of plants. In the experiment under study, the indicators of total and productive bushiness varied within 2.0 - 2.4 and 1.7 - 2.3, respectively. The results obtained reliably indicated that the best results of total and productive bushiness were formed by plants of the Bezenchuk Niva varieties, 2.4 - 2.3, respectively, and Bezenchuk 209, 2.4-2.3. Good tillering energy allowed these varieties, respectively, to form a larger number of productive stems, by grade Bezenchuk 209 - 788 pcs/m^2 and Bezenchuk Niva - 821 pcs/m^2. 663 productive stalks per 1 m^2 were formed in the control variant. The weakest tillering was observed in the Luch 25 variety, where the index of productive bushiness was 1.7 and the number of productive stems was 629 pcs / m^2.

Analysis of the structure of the crop is an important method for assessing the development of cultivated plants; it allowed establishing patterns of yield formation and tracing its dependended on the variety of environmental factors [12, 13]. The results of the analysis of the structure of the crop on average for the studied year were given in Table 2. It followed from them that the number and weight of seeds in a head varied in proportion to the length of a head, depending on the variety. The length of the head of durum wheat varied from 3.85 to 5.1 cm. Also in our studies, we determined the productivity of the main spike, counting the number of grains and their mass. The best indicators of the structure of the crop compared with the control were reliably observed in the Bezenchuk Niva variety, where the head length was 5.1 cm, respectively, the number of seeds per head was 22.9, and the weight of seeds per head was 1.1 g. The smallest indices were recorded for the Bezenchuk 205 variety, a head length was 3.4 cm, the number of seeds per head was 16.8 pieces, and the seed weight was 0.7 g.

Table 2. Yield structure and yield of spring durum wheat varieties on average for the years 2015 - 2017.

Varieties	Head length (cm)	The number of grains in the main spike(pcs)	The number of grains in the main spike(pcs)	Weight of 1000 seeds(g)	Yield(t / ha)
Bezenchuk 200	4.6	20.2	0.9	48.1	2.63
Bezenchuk 205	3.4	16.8	0.7	42.3	1.96
Bezenchuk 209	4.4	21.8	1.0	46.7	2.96
Bezenchuk Niva	5.1	22.9	1.1	52.4	3.41
Luch 25	5.0	18.6	1.0	53.3	2.98
NSR 05	0.7	1.4	0.2	3.5	4.7

The formation of grains in the spike was affected by many factors, such as, for example, temperature. A puny grain was formed at low temperatures, lack of moisture or when it was in excess, and the mass of 1000 seeds decreased accordingly [14]. The mass of 1000 seeds was determined to characterize the fullness and size of seeds. Large, heavy seeds, having a larger supply of nutrients, produce powerful seedlings in field conditions during germination, which later develop well and provide a higher yield. In the studied the year 2015, the weather was more favorable for growing durum wheat than the conditions of 2016 and 2017, characterized by drought in 2016 and an abundance of precipitation at low temperatures during the growing season of 2017. For example, this indicator for varieties in the conditions of 2015 varied from 45 to 60 grams, in 2016 from 41 to 53 grams and in 2017 from 38 to 46 grams, respectively. As a result, durum wheat plants formed a more complete grain with the best mass index of 1000 seeds in the agro-climatic conditions of 2015. The mass of 1000 seeds amounted to options for an average of three years from 42.3 to 53.3 grams. The Bezenchuk Niva and Luch 25 varieties reliably exceeded the control and the rest options on this indicator. Bezenchuk 205 was inferior to all other varieties in terms of the mass of 1000 seeds, which averaged 42.3 grams over the three a typical agroclimatic years.

Yield is an important criterion for assessing the cultivation of crops. For all the years studied, we managed to get a full-fledged harvest of spring durum wheat, which was almost as good as the traditional spring wheat in the Republic. The yield data of spring durum wheat for years of research was presented in Figure 1. Durum wheat varieties reacted differently to the agro-ecological conditions of atypical years of research. For example, year 2017 was the most favorable for yield, for the yield of Bezenchuk 200 variety (control) was maximum compared to the other varietiesand amounted to 3.56 t / ha. Year2017 conditions were the most favorable for shaping the yield of the Bezenchuk 205 and Luch 25 varieties. In 2015, the yields of Bezenchuk 209 and Bezenchuk Niva were 3.1 and 3.6 t / ha, respectively.

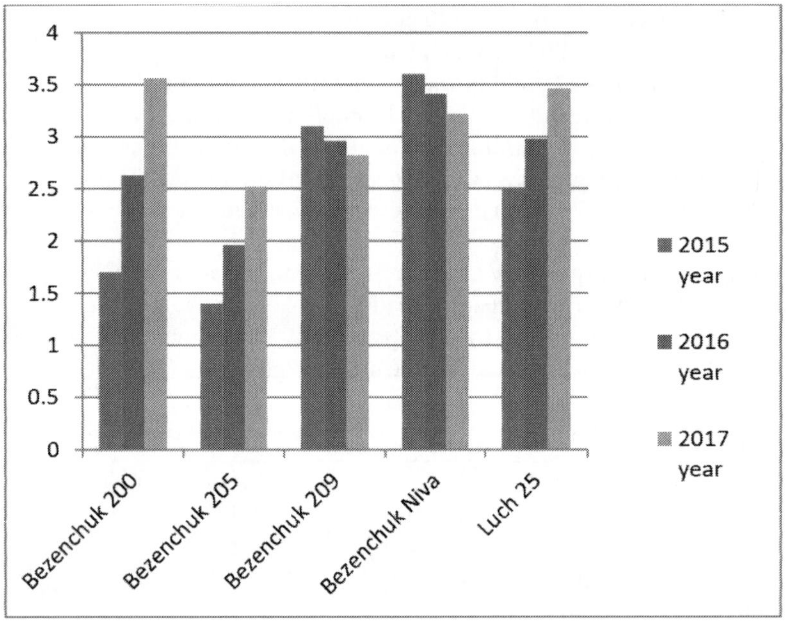

Figure 1. The yield of varieties of spring durum wheat in 2015-2017.

An analysis of the average yield data over the years of research revealed that the highest biological yield was reliably formed by the Bezenchuk Niva variety - 3.4 t / ha. The lowest yield (1.9 t / ha) was formed by the Bezenchuk 205 variety compared with the control.

4. Conclusion

Thus, according to the results of three-year studies, it could be concluded that the climatic conditions of the Chuvash Republic as a whole meet the biological requirements of durum wheat. All the varieties studied during 3 years formed a full grain harvest. The value of the yield, its structure identified for the production and further research (cultivation technology, selection) Bezenchuk Niva variety. Prospects for the commercial use of new varieties of durum wheat in general in the Chuvash Republic are estimated at 50-75 thousand hectares, which will ensure the needs of local pasta and cereal enterprises and export grain to other regions.

References

[1] Malchikov P N, Vyushkov A A and Myasnikova M G Formation of models of varieties of durum

wheat for the Middle Volga region of Russia *Monograph* (Samara: SamNC, Russian Academy of Sciences) p 112

[2] Lozhkin A G and Malchikov P N 2018 The productivity of varieties of spring durum wheat in the Chuvash Republic *Agricul. Scient. J.* **12** 31

[3] Lozhkin A G 2018 Prospects for the cultivation of varieties of spring durum wheat in the conditions of the forest-steppe zone of the Chuvash Republic 2018 *Bullet. Bashkir State Agrarian Univer.* **2** 40

[4] Shashkarov L G and Malov N P 2018 Plant density, field germination and survival of spring wheat plants depending on the variety *Bullet. Kazan State Agrarian Univer.* **3** 65

[5] Pushkarenko N N and Ivanshikov Y V 2018 Problems of technical service in agriculture and possible solutions *Prospects for the development of technical service in the agro-industrial complex. Materials of the all-Russian scientific-practical conference. Chuvash State Agricultural Academy* (Cheboksary) p 228

[6] Lozhkin A G, Malchikov P N and Myasnikova M G 2018 Spring durum wheat in the conditions of the forest-steppe zone of the Chuvash Republic *Grain Econ. Russia* **4** 59

[7] Ilyin A N and Ilyina T A 2017 Air regime of the arable layer of gray forest soil with different technologies of its processing in the conditions of the Chuvash Republic *Agroecological and organizational-economic aspects of the creation and effective functioning of environmentally stable territories materials of the All-Russian scientific-practical conference* (Cheboksary) p 76

[8] Makushev A E and Abrosimova M S 2018 State regulation of the activities of agricultural enterprises *The development of agrarian science as the most important condition for the effective functioning of the agro-industrial complex of the country materials of the All-Russian Scientific and Practical Conference dedicated to the 70th anniversary of the Honored Worker of the Higher School of the Chuvash Republic and the Russian Federation, Doctor of Veterinary Sciences, Professor Kirillov Nikolay Kirillovich* (Cheboksary) p 491

[9] Malchikov P N and Myasnikova M G 2015 Varieties of spring durum wheat for the Middle Volga and Ural regions of the Russian Federation 2015 *Achievem. Sci. Tech. Agricul.* **10** 58

[10] Ilyina T A, Ilyin A N and Vasiliev O A 2017 Influence of processing technologies on the reserves of moisture in the gray forest soil in Chuvash Republic *Bullet. Kazan State Agrarian Univer.* **4** 8

[11] Shashkarov L G and Lebedeva Z G 2016 Formation of planting density and crop structure of spring wheat depending on the variety and presowing treatment of seeds *Bullet. Kazan state agrarian Univer.* **1** 30

[12] Vasiliev O A, Ilyina T A, Zaitseva N N 2017 The effectiveness of biofertilizers in the cultivation of spring wheat *Bullet. Chuvash State Agricul. Acad.* **3** 5.

[13] Eliseeva L V, Kayukova O V and Eliseev I P 2018 Influence of seed size on the productivity of grain legumes *Innovative technologies in the field and ornamental plant growing Collection of articles on the materials of the II all-Russian (national) scientific-practical conference* p 51

[14] Kuznetsov N N, Pushkarenko N N, Medvedev V I, Zaitsev P V, Vasiliev A O, Andreev R V 2018 Model of the functioning of the technological process of post-harvest grain processing in the reception and preliminary cleaning of the heap of grain *Bullet. Kazan State Agrarian Univer.* **4** 114

IOP Conf. Series: Earth and Environmental Science **346** (2019) 012055 doi:10.1088/1755-1315/346/1/012055

Utility of ion-exchange substrates to grow potato seedlings under pot conditions

S V Filippova[1], L V Eliseeva[1, 2], G A Mefodev[1], A A Makushev[1] and L G Shashkarov[1]

[1]FSBEI HEI "Chuvash State Agricultural Academy ([1]FSBEI HEI CSAA), Cheboksary, Russia

E-mail: ludmilaval@yandex.ru

Abstract. Present study reports the cultivation of potato seedlings on ionite substrates in pot culture. The possibility of using these substrates to preserve the qualitative characteristics of the potato genotype was studied. The research was conducted on hybrid seed potatoes produced in VNIIKH named after A. G. Lorch. During the growing season, the intensity of seedlings development in terms of stem height (every 2 weeks) was taken into account. At the end of the growing season, the number of leaves was analyzed, and after harvesting, the yield and its contributing components were taken into account. Dispersion analysis was performed to study the quantitative assessment of phenotypic factors. In peat soil, depending on their origin and year of cultivation, the percentage of seedling death varied from 26.0 to 41.6%. On ionite substrates, the survival rate of seedlings significantly increased - only 4.0-14.4% died when grown in cuvettes. After picking in pots, the full survival of seedlings was observed both in the control and in the experimental variants. The best of the experimental options for growing seedlings in pot culture was the substrate IS-2E, prepared on the basis of IS-2V and clinoptilolite in a volume ratio of 3:1. Growth and survival rate of seedlings on ionite substrates significantly increased by 19.3-30.2% over control and thus reduces percentage of seedlings elimination. Growing seedlings is one of the first and most important stages in the general scheme of potato breeding [1,2]. The pot culture of seedlings in greenhouses, which provides separate cultivation of genotypes, high planting density and the ability to eliminate the seasonality of this process, has acquired great importance [3]. However, the mixtures used as a substrate for growing plants do not provide optimal conditions for the growth and development of seedlings and the preservation of all genotypes [4,5]. Identifying optimal substrates and methods for obtaining hybrid seedlings of potatoes and improving the efficiency of selecting high-yielding genotypes at an early stage of selection is considered as an important researchable issue to increase the effectiveness of the entire selection process [6,7]. All this leads to strengthen the research in this direction, which has not yet developed and reported [8].

1. Purpose of research

The purpose of the research was to study the possibility of using ionite substrates for growing potato seedlings in pot culture.

2. Materials and methods

To see the impact of substrates and methods of growing potato seedlings, hybrid seeds were procured from the department of breeding and genetics of the All-Union Scientific Research Institute of the Microbiology named after A. G. Lorch. They were:

Content from this work may be used under the terms of the Creative Commons Attribution 3.0 licence. Any further distribution of this work must maintain attribution to the author(s) and the title of the work, journal citation and DOI.
Published under licence by IOP Publishing Ltd

1. 374-55 x Aksenovsky (middle x early)
2. Nevsky x 733-65 (medium early x early)
3. Aksenovsky x Sotka (early x middle late)
4. 889-99 x Smena (medium late x medium)
5. 1169-2 x Zarevo (medium late x medium late)

Potato seeds were first germinated on petri dishes under diffused light at room temperature for 2 weeks on filter paper (control) and on ion-exchange substrates (experimental variants).

The following substrates were used in the studies:
1. Peat and soil - control (filter paper in experiments on the germination of potato seeds);
2. Ionite substrate IS-2V based on ion exchangers KU-2 and EDE-10P with pH - 5.6;
3. Ionite substrate IS-2S based on ion exchangers KU-2 and EDE-10P with pH - 5.9;
4. Ionite substrate IS-2E, which is a mixture of IS-2V and clinoptilolite (modified zeolite) in a volume ratio of 3: 1.

Peat and soil used were characterized by the following indicators: The pH of the salt extract was 5.2, the content of mobile phosphorus was 22.2 mg per 100 g of soil, potassium 30.0 mg per 100 g of soil, and magnesium 23.6 mg per 1 kg of soil.

In all cases, seed germination was taken into account. The seedlings were transplanted into plastic cuvettes filled with ion-exchange substrates (experimental variants) or into cuvettes filled with peat soil (the control variant). In addition, in the following experiment, seeds were sown directly into cuvettes on the ion-exchange substrate IS-2E and on peat soil to account for greenhouse germination of seeds. The cuvettes with seedlings were placed in film-gauze houses.

After 30–40 days, transplantation of seedlings having 3-4 true leaves was done in plastic pots having a diameter of 12 cm and height of 10 cm. The survival rate of seedlings was taken into account when they were grown in cuvettes.

Part of the seedlings grown in cuvettes on ion exchangers, i.e. seedlings swoop in pots with peat ground (option IT). The pots were installed in plastic-gauze houses. During the period of research, seedlings were grown in spring and summer. Care of seedlings was carried out according to the generally accepted methodology of VNIIKH named after A.G. Lorch. Harvesting was carried out in the first week of September.

During the growing season, the intensity of seedlings development by stem height (every 2 weeks) was taken into account. At the end of the growing season, the number of leaves was recorded, and after harvesting, the yield and other attributes were taken into account. Quantitative assessment of factors in phenotypic manifestation of signs was carried out by the method of dispersion analysis.

3. Results

We have studied seeds of 5 hybrid populations with respect to germination on three ionite substrates compared with control (filter paper). Results as obtained from petri dish experiment indicated that the germination of seeds of different origin on all ionite substrates in comparison with the control did not give obvious advantages (Table 1). Perusal of Table 1 indicate that increase in seed germination on ionite substrates in comparison to control was insignificant (0.8-4.0%).

Table 1. Germination of hybrid seeds of different origin of the used substrates.

Origin	Option	Seeds sown, PCs.	Germination, %
	2015		
374-55 x Aksenovsky (middle x early)	Control	310	81,6±2,2
	IS-2V	310	80,0±2,3
	IS-2S	125	80,8±3,5
	IS-2E	160	81,2±3,1

Aksenovsky x Sotka (early x middle late)	Control	310	93,6±1,4
	IS-2V	310	94,2±1,3
	IS-2S	125	93,6±2,2
	IS-2E	160	93,1±2,0
1169-2 x Zarevo (medium late x medium late)	Control	310	97,0±1,0
	IS-2V	310	97,3±0,9
	IS-2S	125	93,6±2,2
	IS-2E	160	95,0±1,7
2016			
374-55 x Aksenovsky (middle x early)	Control	150	80,7±3,2
	IS-2V	120	81,6±3,5
	IS-2S	50	82,0±5,4
	IS-2E	60	81,7±5,0
1169-2 x Zarevo (medium late x medium late)	Control	150	92,0±2,2
	IS-2V	120	94,2±2,1
	IS-2S	50	96,0±2,8
	IS-2E	60	95,0±2,8
2017			
Nevsky x 733-65 (medium early x early)	Control	50	74,0±6,2
	IS-2E	50	78,0±5,9
889-99 x Smena (medium late x medium)	Control	50	100.0
	IS-2E	50	100.0
HCP05		2.9-12.4	

Sowing of seeds in cuvettes with substrates without pre-germination led to significant decrease in germination (Table 2). However, compared with the control (torfogrunt), substrate IS-2E, this method of growing of seedlings exhibited an increase in seed germination by 12.3-17.0 % depending on their origin.

Table 2. Greenhouse germination of seeds of different origin depending on the growing substrate.

Origin	Option	Seeds sown, PCs.	Germination, %
2016			
374-55 x Aksenovsky	Control	100	54,0±5,0
	IS-2E	100	71,0±4,5
1169-2 x Zarevo	Control	100	68,0±4,7
	IS-2E	100	83,0±3,8
2017			
Nevsky x 733-65	Control	60	56,7±6,4
	IS-2E	60	70,0±5,9
889-99 x Smena	Control	60	78,3±5,3
	IS-2E	60	90,0±3,9
HCP05		11,7-13,2	

The influence of the substrate was manifested in the subsequent period and significantly affected the survival of seedlings when seedlings were grown in cuvettes with substrates (Table 3). On peat soil, depending on their origin and year of cultivation, the percentage of seedling death varied from 26.0 to 41.6%. On ionite substrates, the survival rate of seedlings significantly increased and only 4.0-14.4% seedlings died when grown in cuvettes. After placing them in pots complete survival of seedlings was observed both in the control and in the experimental variants.

Table 3. The effect of the substrate on the survival of seedlings of different origin in pot culture.

Origin	Option	Seeds sown, PCs.	Germination, %
	2015		
374-55 x Aksenovsky	Control	253	70,0±2,9
	IS-2V	248	97,1±1,1
	IS-2S	101	94,6±2,2
	IS-2E	130	95,4±1,8
Aksenovsky x Sotka	Control	290	61,4±2,9
	IS-2V	292	86,3±2,0
	IS-2S	117	85,6±2,4
	IS-2E	149	86,7±2,8
1169-2 x Zarevo	Control	301	58,4±2,8
	IS-2V	302	87,7±1,9
	IS-2S	117	93,2±2,3
	IS-2E	152	95,4±1,7
	2016		
374-55 x Aksenovsky	Control	121	63,6±4,3
	IS-2V	98	93,8±2,4
	IS-2S	41	92,7±4,1
	IS-2E	49	91,8±3,9
1169-2 x Zarevo	Control	138	68,1±4,0
	IS-2V	113	94,7±2,1
	IS-2S	48	91,7±4,0
	IS-2E	57	96,5±2,4
	2017		
Nevsky x 733-65	Control	37	73,0±7,3
	IS-2E	39	92,3±4,3
889-99 x Smena	Control	50	74,0±6,2
	IS-2E	50	96,0±2,8
HCP05		6,0-16,7	

Results indicated that when seedlings were grown on peat soil it was desired to sow the seeds without prior germinating which had increased the survival rate of seedlings by 2.7-4.3%, whereas on ionite substrates, prior germination of seeds increased the survival rate of seedlings by 2.0-9.7%.

The research results showed that all varieties tested on ion exchangers proved to be favorable for the growth and development of seedlings (Table 4). This was due to the fact that under normal conditions root contributes to increase the supply of minerals to the above-ground part and thus increased the intensity of photosynthesis [9].

In the course of research, it was also found that seedlings on such substrates had much more developed above ground part compared to seedlings on soil culture. The magnitude of the increase of stem height reached 5.0-51.1 cm and number of the leaves to 2.6-11.9 /bush.

Table 4. The height of seedlings (cm) of different combinations at the end of the growing season depending on the substrate of cultivation.

Origin	Option	X±Sx	V
	2015		
374-55 x Aksenovsky	Control	26,8±0,6	25.4
	IS-2V	70,1±1,3	10.1
	IS-2S	70,0±3,8	23.4
	IS-2E	75,6±4,3	22.8
	IT	29,3±1,6	24.6
Aksenovsky x Sotka	Control	32,2±0,9	27.8

	IS-2V	71,3±2,4	16.8
	IS-2S	65,5±2,8	18.3
	IS-2E	83,3±2,8	12.8
	IT	34,5±2,0	25.2
	Control	39,2±1,1	29.8
	IS-2V	61,4±2,4	17.1
1169-2 x Zarevo	IS-2S	58,4±3,7	25.3
	IS-2E	69,1±5,0	27.3
	IT	37,4±2,4	28.3
HCP05		5.7	
	2016		
	Control	28,6±1,4	38.9
	IS-2V	49,1±4,1	37.0
374-55 x Aksenovsky	IS-2S	50,5±1,1	9.5
	IS-2E	65,5±2,3	15.9
	IT	33,0±2,7	36.1
	Control	34,5±1,1	23.7
	IS-2V	43,8±2,2	22.6
1169-2 x Zarevo	IS-2S	39,5±1,5	16.9
	IS-2E	50,0±2,2	19.9
	IT	32,1±2,3	32.1
HCP05		11.3	
	2017		
	Control	53,4±2,2	18.4
Nevsky x 733-65	IS-2E	74,5±2,8	16.5
	IT	55,0±2,3	18.9
	Control	56,5±3,6	28.1
889-99 x Smena	IS-2E	76,2±3,9	22.6
	IT	54,8±3,7	29.7
HCP05		19.4	

Seedlings of variant IS-2E compared with those of other variants, differed in development (stem height and number of leaves) and had surpassed them in both of these indicators (Table 5).

The variability of weakly measurable traits, including the stem height and the average number of leaves per seedling, allowed us to estimate the beneficial effect of ion-exchange substrates on the development of seedlings. The coefficient of variation in stem height at the end of crop cycle in control varied from 18.4 to 38.9%, and the number of leaves from 12.5 to 19.0% depending on the origin and year of cultivation of seedlings.

Table 5. Number of leaves per seedling at the end of crop cycle depending on the growing substrate.

Origin	Option	X±Sx	V
	2015		
	Control	12,1±0,1	12.5
	IS-2V	23,8±0,2	5.5
374-55 x Aksenovsky	IS-2S	22,6±0,2	8.8
	IS-2E	22,7±0,6	10.1
	IT	12,9±0,4	12.4
	Control	13,6±0,2	14.8
	IS-2V	22,9±0,3	6.5
Aksenovsky x Sotka	IS-2S	22,7±0,7	13.2
	IS-2E	25,0±0,7	10.4
	IT	13,2±0,5	15.2
1169-2 x Zarevo	Control	14,3±0,2	18.1
	IS-2V	21,8±0,4	7.3

	IS-2S	20,8±0,7	12.9
	IS-2E	23,4±0,6	10.3
	IT	14,0±0,5	16.4
HCP05		5.2	
	2016		
374-55 x Aksenovsky	Control	12,5±0,4	15.0
	IS-2V	18,1±0,6	13.8
	IS-2S	19,3±0,6	13.0
	IS-2E	22,5±0,6	12.4
	IT	15,1±0,5	15.9
1169-2 x Zarevo	Control	15,7±0,7	19.0
	IS-2V	18,3±0,7	17.5
	IS-2S	18,2±0,6	13.9
	IS-2E	19,2±0,5	10.9
	IT	12,1±0,6	23.1
HCP05		2.6	
	2017		
Nevsky x 733-65	Control	17,5±0,5	13.7
	IS-2E	25,6±0,6	10.5
	IT	18,3±0,6	13.7
889-99 x Smena	Control	17,9±0,7	16.8
	IS-2E	25,4±0,7	12.2
	IT	17,8±0,6	15.2
HCP05		12.6	

The seedlings grown on the ion-exchange substrates IS-2V and IS-2E were characterized not only by a powerful above-ground part but also had a higher yield compared to the controls (Table 6). In most cases, the version of IS-2S also exceeded control.

Based on the data obtained, the best substrate for crop accumulation was a mixture of IS-2V with clinoptilolite.

Study also indicated that when seedlings were grown on ion-substrates, not only the average yield increased but also the breadth of the seedlings compared to control [10]. This increase might be due to a more complete manifestation of the quality characteristics of the genotype on ionite substrates.

On ionite substrates number of tubers was more to those grown on control cndition. Due to the increase in the proportion of small tubers in the fractional composition in 2015, the average mass of the tuber in the variants of the ionite substrate in all populations was slightly lower than that observed in the control. At the same time, in the number of medium and large sized tubers obtained from seedlings grown on the variant IS-2V were not inferior to the control, and in the variant IS-2E, the size of tubers was significantly higher. The average weight of the tuber seedlings of the ionite substrate variants in the remaining years of cultivation was significantly higher than the values of the control variants.

Table 6. Characteristics of seedling populations of different origin according to the yield and its components, depending on the substrate of cultivation.

Option	Number of seedlings, PCs.	Yield, g/Bush		The number of tubers per 1 bush, pcs.	Average tuber weight, g
		X±Sx	V	X±Sx	X±Sx
1	2	3	4	5	6
		2015			
		374-55 x Aksenovsky			
Control	150	13,7±0,4	38.7	3,4±0,1	5,5±0,2
IS-2V	32	18,8±2,0	60.6	16,3±2,1	1,3±0,2
IS-2S	19	28,2±6,1	94.2	18,0±2,3	1,7±0,3

IS-2E	16	61,9±6,2	40.0	20,5±2,6	3,6±0,7
IT	20	17,9±3,9	97.2	6,9±0,8	3,0±0,4
		Aksenovsky x Sotka			
Control	110	16,4±0,7	42.6	3,9±0,2	5,0±0,2
IS-2V	25	25,6±4,0	77.7	14,8±2,6	2,0±0,4
IS-2S	18	10,6±1,8	72.6	17,7±4,0	0,6±0,1
IS-2E	15	20,1±4,2	80.0	11,6±2,6	2,8±0,6
IT	20	20,3±3,4	74.9	7,4±0,9	3,4±0,3
		1169-2 x Zarevo			
Control	100	21,0±1,6	73.8	2,9±0,2	8,3±0,7
IS-2V	20	22,0±4,5	91.8	8,2±1,1	3,1±0,5
IS-2S	16	24,5±4,5	73.1	11,9±2,6	2,8±0,6
IS-2E	14	36,9±7,8	78.0	5,9±0,4	6,3±1,3
IT	20	17,6±3,3	83.0	2,9±0,4	7,3±0,3
HCP05		8.9		8.3	2.1
		2016			
		374-55 x Aksenovsky			
Control	60	8,5±0,5	45.9	4,4±0,3	2,0±0,1
IS-2V	20	16,2±1,7	46.9	6,6±0,6	2,5±0,2
IS-2S	20	13,1±1,5	50.4	5,4±0,6	2,7±0,2
IS-2E	20	16,8±1,8	47.6	6,0±0,5	3,0±0,3
IT	20	9,3±1,0	47.3	6,5±0,5	2,3±0,2
		1169-2 x Zarevo			
Control	60	4,0±0,2	48.9	3,2±0,2	1,3±0,1
IS-2V	20	11,1±1,5	60.2	4,8±0,5	2,2±0,2
IS-2S	20	10,3±1,6	70.4	3,9±0,4	2,8±0,3
IS-2E	20	14,1±1,7	52.1	4,2±0,5	3,2±0,3
IT	20	6,3±0,8	58.7	3,4±0,5	2,1±0,3
HCP05		3.7		0.6	2.2
		2017			
		Nevsky x 733-65			
Control	20	16,5±2,8	75.2	3,0±0,4	5,3±0,7
IS-2E	20	44,8±8,0	79.7	3,2±0,4	15,3±2,4
IT	20	12,4±2,5	89.7	4,0±0,4	5,0±1,0
		889=99 x Smena			
Control	20	10,7±1,3	55.1	2,6±0,3	4,0±0,5
IS-2E	20	47,0±7,6	71.7	3,0±0,4	18,5±2,2
IT	20	9,4±1,5	69.6	2,9±0,2	3,7±0,5
HCP05		15.7		10.6	1.8

Analysis of factors affecting the manifestation of traits a significant contribution of the type of composition of the environment was observed in the overall variability of traits. The height of the stem increased by 46.5-82.7% and the number of leaves by 49.5-81.4% depending on the what type of substrate seedlings were grown. Yield and its contributing components also had a significant impact on the origin of seedlings and its interaction with the substrate of cultivation.

4. Conclusion

Growing seedlings on ionite substrates significantly increased the survival rate of seedlings and reduced the percentage of elimination when compared to those seedlings grown on peat culture and this difference was 19.3-30.2%.

Seedlings on ionite substrates showed more developed aboveground in comparison to those grown on peat, so it eventually led to increased productivity. The best of the experimental options for growing seedlings in pot culture was IS-2E substrate prepared on the basis of IS-2V and clinoptilolite in a volume ratio of 3:1.

When seedlings were grown on ionite substrates, the variability among hybrid populations about the yield was more due to more favorable conditions leading to more realization of the potential of the genotype. As a result, the assessment of populations and selection of hybrids with respect to yield became possible when they were grown from seeds.

References

[1] Accafino P and Malagamba P 1983 Growing potata from TPS Gurrent Agronomie *Knowledge And Future Prospects* **6** pp 174-201
[2] Akeley R V and Stevenson F J 1943 Yleld, specific gravity and starch content in a potato program *Amer. Pot. J.* **20** pp 203-17
[3] Moller K H 1956 Samlingeaneucht in gewachshaus zur zuchtung fruhreifer kartoffeln *Zuchter.. Bd.* **26** pp 243-8
[4] Perry B A V, Aceley R V and Cook W B 1960 A potato breeding techigue used for Texas *Amer. Pot. J.* **37** pp 306-7
[5] Schick R and Hopfe A 1962 Die Zuchtung der Kartofell (Die Kartofell) Berlin.: Veb. Dtsch. Iandwirtsch pp 1476-8
[6] Moller K H and Pfeffer C H 1965 Methoden, selection und ernteverfahren in der kartoffeleuchtung *Zuchter Bd.* **35** pp 118-23
[7] Preffer C H and Griess H 1976 Eine neue technologie den samlingsanzucht von kartoffelin: arch. zuchtunetorsch Berlin Bd.**6-H** pp 233-46
[8] Ludwig H and Kahnts G 1957 Zur vereinheitlichtuhg der methodik bei der augenstecklibsprufung von pflanskartoffeln z. landw. *Ver. u.Unters.wesen Bd.* **3** 59-65
[9] Wrazildo W 1959 Zur anzucht von augenatecklingspflanzen fur die beurteilung des sesundheitswertes von kartoffelpflanzgut in zies – kultur Z. Londw. Vers.u. Unters. wesen. Bd. **5** pp 14-21
[10] Griess H 1976 Sum Finsats kunetlicher subetrate im Gefassversuch mit Kartoffeln Arch. Zuchtungeforcech Berlin,. - Bd. **6** pp 77-90

Ion-exchange substrates as the basis for growing seedlings of potato in tube culture

S V Filippova[1], L V Eliseeva[1, 2], G A Mefodev[1], A A Makushev[1] and L G Shashkarov[1]

[1]FSBEI HEI "Chuvash State Agricultural Academy ([1]FSBEI HEI CSAA), Cheboksary, Russia

E-mail: ludmilaval@yandex.ru

Abstract. Studies had been carried out to study the possibility of using ionite substrates for growing potato seedlings in tube culture, to analyze the variability of hybrid potato populations by morphobiological and economically valuable characteristics depending on the substrate, to analyze the effectiveness of the selection of high-yielding forms at the early stage of the breeding process. The experiment included 7 different modifications of the nutrient medium composition. Seed germination, seedling survival rate, tuber formation rate, percentage of tuber-forming seedlings, number of tubers per seedling, size, and shape of tubers were taken into account. Quantitative assessment of factors in phenotypic manifestation of signs was carried out by the method of dispersion analysis. Sterile microtubers were stored at 2-4 °C in the dark before planting. As a result of the research, a different tuber-forming ability was found in seedlings of populations of different variants. The tube culture of seedlings ensures full survival of seedlings. For the growth and development of seedlings in a tube culture, the best of those studied was the nutrient medium based on the ion-exchange substrate IS-2V, containing 1% sucrose and 500 mg / l CCC. Two-stage cultivation of seedlings (the initial increase in the above-ground mass, then the induction of tuberization) had a positive effect on the pace and quality of tuberization. 30–40 days after induction, tuberization was observed in 90–100% of seedlings, regardless of their origin, and the size of microtubers was on average 6.7–8.7 mm.

1. Introduction

In the implementation of the food program, great importance is attached to the application of scientific achievements [1]. In this regard, the role of research in the direction of accelerating and improving the efficiency of the breeding process to create highly productive forms of plants resistant to extreme environmental conditions, pests, pathogens, as well as suitable for mechanized cultivation [2,3].

For the development of seedlings, the most favorable is the environment containing the minimum amount of sucrose in the composition of which there are no growth-stimulating substances [4]. This indicates that sprouts and micro-seedlings do not need stimulants, since the seed itself is a nutrient base [5].

Some studies indicate that the main factors of tuber formation in the cultivation of plants *in vitro* culture are photoperiod, temperature, sucrose concentration in the medium [6]. The experiments of scientists indicate that the shortened photoperiod under *in vitro* conditions had a decisive effect on the process of tuber formation – shortening of the photoperiod caused an increase in the number of tubers [7].

Content from this work may be used under the terms of the Creative Commons Attribution 3.0 licence. Any further distribution of this work must maintain attribution to the author(s) and the title of the work, journal citation and DOI.
Published under licence by IOP Publishing Ltd

It is believed that tubers can be obtained in an environment designed for plant growth [8]. However, most researchers argue that for mass tuber formation it is necessary to use specially developed for this purpose nutrient media [9].

A number of researchers note that in the process of tuberous material reproduction obtained from test tubes microtubers, retains the advantages of the main features necessary for the successful conduct of the breeding process [10].

2. Purpose of research

The main purpose of research on this topic was to study the possibility of using ionite substrates for growing seedlings in a test tube culture.

3. Materials and methods of research

The studies used hybrid seeds obtained in the department of breeding and genetics VNIIKH named after A. G. Lorch. The experiment included the following combinations:

1. 374-55 x Aksenovsky (middle x early)
2. Nevsky x 733-65 (medium early x early)
3. Aksenovsky x Sotka (early x middle late)
4. 889-99 x Smena (medium late x medium)
5. 1169-2 x Zarevo (medium late x medium late)

Seedlings were grown in glass jars of 50 ml in the period from October to April. Research on the cultivation of seedlings in a test-tube culture was carried out according to the guidelines for Trofimzu L. Sterilization was carried out by immersing the seeds for 1-2 s in alcohol and 7 minutes in a 0.1% diacide solution. The experiment included the following options:

1. Options AS-1 and AS-2. Seedlings were cultivated on an agar medium with a mineral base according to Murasige and Skoog in the modification of the Scientific Research Institute of Chemical Achievements named after A.G. Lorch.
2. Option P-IAS. Seedlings cultivated an agarized agar medium based on the ionic substrate IS-2V;
3. Options OP-IS and 1P-IS. Seedlings were cultivated on the medium based on the ionite substrate IS-2V;
4. Options 2P-IS and 3P-IS. Two-stage cultivation of seedlings on the medium, based on the ion-exchange substratum IS-2V;

The components of nutrient media are as follows:

1. AS-1 – vitamins by White, kinetin – 0.4 mg/l, adenine – 4 mg/l, sucrose – 80 g/l, agar – 7 g/l.
2. AS-2 – vitamins by Staba, casein hydrolysate – 40 mg/l, kinetin – 0.5 mg/l, alenin – 5 mg/l, sucrose – 80 g/l, agar – 7 g/l.
3. OP-IS – vitamins by White, kinetin – 0.4 mg/l, adenine – 4 mg/l, sucrose – 80 g/l.
4. P-IAS – vitamins by Staba, casein hydrolysate – 40 mg/l, kinetin – 0.5 mg/l, adenine – 5 mg/l, sucrose – 80 g/l, agar – 7 g/l.
5. 1P-IS – vitamins by Staba, casein hydrolysate – 40 mg/l, kinetin – 0.5 mg/l, adenine – 5 mg/l, sucrose – 80 g/l.
6. 2P-IS – vitamins by Staba, casein hydrolysate – 40 mg/l, gibberellinic acid – 0, 5 mg/l, heteroauxin – 1.5 mg/l, kinetin – 0.5 mg/l, adenine – 5 mg/l, sucrose – 20 g/l.
7. 3P-IS – chloroindole – 500 mg/l, sucrose 10 g/l.

In variants 2P-IS after 50 – 100 days and in 3P-IS after 50 days of cultivation of seedlings for inducing tuber formation, the concentration of sucrose in the medium was brought to 8%, and in addition, in 3P-is the concentration of kinetin – up to 0.5 mg/l, the concentration of adenine – up to 4 mg/l.

Variants AS-1 and OP-IS were grown in room lighting, the rest– at a photoperiod of 16 hours with illumination of 6000-8000 Lux at room temperature. After inducing tuber formation options 2P-IS and 3P-IS in 2017 for 3 days were grown in the dark, then – at room conditions.

IOP Conf. Series: Earth and Environmental Science **346** (2019) 012056 doi:10.1088/1755-1315/346/1/012056

Seed germination, seedling survival rate, tuber formation rate, percentage of tuber-forming seedlings, number of tubers per seedling, size, and shape of tubers were taken into account. Quantitative assessment of factors in phenotypic manifestation of signs was carried out by the method of dispersion analysis. Microtubers were stored at 2-4 °C sterile in the dark before planting.

To study the differences between the seedlings of forming and non-forming tubers in the test tube culture in yield, 10 seedlings of these two groups were cloned by microferencing in 2017, receiving 2 plants from each seedling.

Test-tube plants of seedlings and microtubers were planted in a film-gauze house in torfogrunt with a feeding area of 30x30 cm. In 2017, microtubers were planted in the open ground with a power area of 70x30 cm.

Tuberous reproductions of pot culture seedlings (first and second), as well as 2 tuberous reproductions of test tube culture were grown in the field. Nutrition area for 1 tuberous reproduction is 70x60 cm, for 2 reproductions – 70x30 cm. Field-planting was carried out in the first half of May. Hybrids of 1 tuberous reproduction were represented by one tuber, hybrids of 2 reproductions – by three tubers (selected by economic and valuable characteristics – by ten tubers). Among the studied hybrids, there were plots with parental forms and standard varieties (10 plants each). The following varieties were used as standards: Nevsky - mid-early, Zavorovsky - mid-season, Ramensky - mid-late.

4. Research results.

The germination of seeds according to the variants was taken into account 4 weeks after sowing. This was due to the fact that the germination seeds invariants containing sucrose in a concentration of 8% was somewhat delayed. In 2 weeks after sowing seeds in variants 2P-IS and 3P-IS, the nutrient media of which contain 2% and 1% sucrose, respectively, there was friendly germination, while in other variants – in 3-4 weeks (table 1).

Table 1. Germination of seeds of different origin in tube culture on different media.

Origin	Option	Seeds sown, PCs.	Germination, %
		2016	
374-55 x	AS-1	150	82,3±3,1
Aksenovsky	OP-IS	150	82,3±3,1
Aksenovsky x	AS-1	150	90,0±2,4
Sotka	OP-IS	150	91,6±2,3
1169-2 x Zarevo	AS-1	150	96,7±1,5
	OP-IS	150	98,7±0,9
		2017	
374-55 x Aksenovsky	AS-2	25	88,0±6,5
	1P-IS	30	86,7±6,2
	2P-IS	22	81,8±8,2
	P-IAS	20	90,0±6,7
Aksenovsky x Sotka	AS-2	25	84,0±7,3
	1P-IS	40	95,0±3,4
	2P-IS	27	96,3±3,6
	P-IAS	17	100.0
1169-2 x Zarevo	AS-2	25	76,0±3,5
	1P-IS	38	97,4±2,6
	2P-IS	40	100.0
	P-IAS	25	100.0
		2018	
Nevsky x 733-65	2P-IS	20	95,0±4,9
889-99 x Smena	2P-IS	20	100.0
	3P-IS	100	99,0±1,0
HCP05			3,4 – 21,1

When comparing the variants of the test tube culture, it was found that the agar medium based on M-S (variants AC-1 and AC-2) did not have the same effect on the germination of seeds depending on their origin. So, in the combinations Aksenovskaya x Sotka and 1169-2 x Zarevo in these embodiments, the seed germination was below 0.7-24.0 %, compared to the options based on the is-2 and 2 374-55 x Aksenovskaya was on their level.

With artificial lighting of 6000-8000 lux, full survival of seedlings in all variants as noted. In low-light indoor conditions, the percentage of seedling deaths in the AS-1 and OP-IS variants varied from 6.9 to 13.3% and from 51.9 to 53.3%, respectively, depending on the combination.

The test-tube culture based on the ion-exchange substrate IS-2V made it possible to increase the germination of seeds, as compared to the agar medium based on M-S, and to achieve the maximum survival rate of seedlings. There were two main ways to use test tube seedlings:

1. Transplanting them at the age of 30 days in pots with peat or ionite substrate for further cultivation.
2. The induction of tuber formation in tube culture.

More promising was to obtain microtubers in a test tube culture.

In this work, we evaluated 5 hybrid combinations of seedlings by tuber-forming ability in a test-tube culture on the ion-exchange substrate IS-2V.

As a result of the research, a different tuber-forming ability was found in seedlings of populations of different variants. This was primarily due to the light mode of growing seedlings and the composition of nutrient media.

Table 2. Characteristics of seedling populations by the ability of tuber formation in the test tube culture, depending on the substrate of cultivation.

Origin	Option	Total number of seedlings, pcs.	The number of seedlings with tubers, pcs.	% of tuberization
		2016		
374-55 x Aksenovsky	AS-1	112	0	0
	OP-IS	58	12	20,7±5,3
Aksenovsky x Sotka	AS-1	117	0	0
	OP-IS	63	9	14,3±4,4
1169-2 x Zarevo	AS-1	135	32	23,6±3,7
	OP-IS	72	12	16,7±4,4
		2017		
374-55 x Aksenovsky	AS-2	22	14	63,6±10,3
	1P-IS	26	2	7,7±5,2
	2P-IS	18	17	94,4±5,4
	P-IAS	18	8	44,4±11,7
Aksenovsky x Sotka	AS-2	21	13	61,9±10,6
	1P-IS	38	6	15,8±5,9
	2P-IS	26	26	100,0±0
	P-IAS	17	10	58,8±11,9
1169-2 x Zarevo	AS-2	19	9	47,4±11,5
	1P-IS	37	10	27,0±7,3
	2P-IS	40	36	90,0±4,7
	P-IAS	25	17	60,0±9,3
		2018		
Nevsky x 733-65	2P-IS	19	19	100±0
889-99 x Smena	2P-IS	20	20	100±0
	3P-IS	99	99	100±0
HCP05				19,5-30,9

When grown in 2015 in low light conditions, the seedlings were strongly drawn out and had an underdeveloped leaf apparatus. At the same time, on an agar medium (variant AS-1) in populations 374-55 x Aksenovsky and Aksenovsky x Sotka, tuberization was not observed in seedlings. Only in the population of 1169-2 x Zarevo in 32 seedlings out of 135, the formation of micro-tubers was observed, which amounted to 23.6% (Table 2).

Under the same conditions on a nutrient medium based on an ionic substrate (OP-IS variant), tuber formation was observed in all populations. Its percentage depending on the origin of seedlings varied from 14.3 to 20.7%.

Insufficient lighting, thus, lead to tuber formation only in a small number of seedlings in the environment of OP-IS in all, and on AS-1 only in individual populations, that is, only single genotypes were capable of tuber formation in a test tube culture when growing seedlings under room lighting. Therefore, artificial lighting was used for the cultivation of seedlings in the future.

Cultivation with a 16-hour photoperiod with illumination of 6000-8000 lux allows obtaining seedlings with a well-developed leaf apparatus. In all variants in 2017, regardless of origin, the formation of tubers was noted. At the same time, the nutrient medium of 2P-IS was more favorable. Already on the 90th day after sowing, the seeds in the population of 374-65 x Aksenovsky 5.6% and Aksenovsky x Sotka 15.6% of seedlings had tubers. After inducing tuber formation (after 100 days of cultivation) 94.4% of seedlings in the population of 374-55 x Aksenovsky, 90.0% of seedlings 1169-2 x Zarevo and all seedlings Aksenovsky x Sotka formed tubers. In other variants, the percentage of tuber formation is much lower.

Table 3. Characteristics of the populations of seedlings on the number, shape and the size of the tubers in the test tube culture.

Origin	Option	A number of tubers per 1 seedling, pcs.	Tuber shape, score	The size of the tubers, mm
		2016		
374-55 x Aksenovsky	OP-IS	1,0±0	3,0±0,3	5,6±0,7
Aksenovsky x Sotka	OP-IS	1,0±0	4,8±0,3	4,5±0,4
1169-2 x Zarevo	AS-1	2,0±0,2	4,2±0,2	5,4±0,3
	OP-IS	1,2±0,1	4,0±0,3	4,0±0,4
		2017		
374-55 x Aksenovsky	AS-2	2,1±0,4	3,4±0,4	5,8±0,8
	1P-IS	1,5±0,7	3,5±0,3	3,5±0,7
	2P-IS	1,2±0,1	3,2±0,2	8,7±0,4
	P-IAS	1,6±0,2	2,8±0,5	6,5±0,8
Aksenovsky x Sotka	AS-2	1,5±0,3	3,9±0,3	5,0±0,6
	1P-IS	1,6±0,2	4,5±0,3	4,3±0,5
	2P-IS	1,5±0,1	3,9±0,1	7,9±0,3
	P-IAS	1,5±0,3	4,2±0,3	4,7±0,5
1169-2 x Zarevo	AS-2	1,4±0,3	3,8±0,4	5,4±0,7
	1P-IS	1,2±0,1	4,6±0,2	4,1±0,4
	2P-IS	1,5±0,1	3,9±0,1	7,6±0,2
	P-IAS	1,7±0,2	3,9±0,3	5,7±0,6
		2018		
Nevsky x 733-65	2P-IS	1,2±0,1	3,7±0,2	7,3±0,6
889-99 x Smena	2P-IS	1,4±0,2	3,9±0,3	6,7±0,5
	3P-IS	1,3±0,1	4,0±0,1	8,1±0,3
HCP_{05}		2.3	3.5	2.3

The agar medium, both ordinary and based on IS-2V (variants AS-1, AS-2, and P-IAS), compared with the others, delays the rate of tuberization by an average of 3 weeks. Therefore, when growing seedlings in in tube culture in order to obtain microtubes, it is advisable to use non-agarized media, which make it possible to obtain microtubers in earlier periods.

In addition to the ability of the seedling population to form tubers in different variants of the test tube culture, the number, size, and shape of the tubers were analyzed (table 3).

By the number of tubers per plant between the variants of all populations, no significant differences were identified. In all variants, on average, 1 seedling accounts for 1-2,1 tuber. Also, the options do not differ in the form of tubers. Tubers of more rounded shape have seedlings of populations Aksenovsky x Sotka and 1169-2 x Zarevo.

Seedlings of variant 2P-IS of all three populations as a whole were characterized by larger tubers than seedlings of other variants (the differences are significant).

Analysis of the correlation between the size and shape of tubers shows that in the agar medium AS-2 in all populations this relationship is highly negative (the correlation coefficient ranged from -0.63 to -0.83). In seedlings on 2P-IS, there was a weakening of this connection. All this suggests that in tube culture when growing seedlings on the ion-exchange substrate IS-2V, it was possible to more accurately estimate populations by the shape of the tubers, since the shape of the tubers, in this case, was less dependent on their size.

When studying the degree of influence of the studied factors on the manifestation of signs, it was found that the variability of variations in the number of tubers by 41.2% due to the interaction of the cultivation substrate and the origin of seedlings. The size of tubers by 35.6% depended on the cultivation substrate and by 29.3% on the origin of the seedlings. The proportion of uncontrollable factors in the phenotypic manifestation of these characters was also large (32.6 - 40.6%).

In 2018, when growing seedlings of Nevsky x 733-65 and 889-99 x populations, tuber formation was induced after 50 days of seedling cultivation. The formation of tubers was observed already 70 days after sowing the seeds, and on the 90th day, all the seedlings of both populations had tubers, the average size of which was 6.7-7.3 mm. In this case, there was also a weak negative relationship between the size and shape of the tubers (from -0.24 to -0.31).

Analysis of the data obtained showed that for the growth and development of seedlings in a test-tube culture, the best medium among those studied in 3P-IS. This medium did not contain vitamins and growth regulators.

An insignificant dependence of tuber-forming ability on the group of maturity of parental forms was revealed. Its nature depends heavily on the variant of growing seedlings. So, on the usual agarized medium AS-2, the tuber-forming ability decreased in the following order: 374-55 x Aksenovsky (medium early x early) - 63.6%, Aksenovsky x Sotka (early x middle late) - 61.9%, 1169-2 x Zarevo (medium late x medium late) - 47.4%. In options 1P-IS and P-IAS the reverse picture was observed.

In variant 2P-IS, all seedlings of the Aksenovsky x Sotka populations, Nevsky x 733-65 and 889-99 x Aksenovsky tuberization percentage was 94.4%, and in 1169-2 x Zarevo - 90.0%. Consequently, the tuber-forming ability in variants of the ionite substrate largely depends not on the group of maturity of the parental forms, but on the composition of the nutrient medium and the conditions for growing seedlings in a test-tube culture.

Germination of microtubers obtained in test tube culture was of great importance for the preservation of all genotypes of seedling populations. As the data obtained show, the germination of tubers in 1 tuber reproduction, both when grown in closed and open ground, was quite high - from 73.7 to 100% (Table 4). 100% germination in closed ground was marked in option 2P-IS. Also, all the tubers rose in variants 1P-IS and P-IAS of the Aksenovsky x Sotka population and in the 1P-IS variant of the 374-55 x Aksenovsky population. The AS-1 and AS-2 variants in terms of germination of tubers were inferior not only to the 2P-IS, but also to the other variants of the ion-exchange substrate.

When growing 1 tuber reproduction in open ground, the germination of tubers was slightly lower compared to the closed ground. So, for the population of Nevsky x 733-65, it was 73.3%, for 889-99 x Smena - 88.1%. Apparently, this was due to the fact that under conditions of moisture deficiency, which differed during the growing season during the years of research, a more reliable supply of water to plants was ensured in greenhouses.

Table 4. Germination of tubers of test tube culture in 1 tuberous reproduction.

Option	Total planted, pcs.	Came up, pcs.	Germination of tubers, %
	2017		
	374-55 x Aksenovsky		
OP-IS	12	10	83.3
1P-IS	2	2	100.0
2P-IS	17	17	100.0
P-IAS	8	7	87.5
AS-2	14	12	85.7
	Aksenovsky x Sotka		
OP-IS	9	7	77.8
1P-IS	6	6	100.0
2P-IS	26	26	100.0
P-IAS	10	10	100.0
AS-2	13	11	84.6
	1169-2 x Zarevo		
AS-1	32	28	87.5
OP-IS	12	11	91.7
1P-IS	10	9	90.0
2P-IS	36	36	100.0
P-IAS	17	16	94.1
AS-2	9	8	88.9
	2018		
	Nevsky x 733-65		
2P-IS	19	14	73.7
	889-99 x Smena		
P-IS[a]	143	126	88.1

[a] – tubers of all variants are combined

Thus, the cultivation of seedlings on an ionite substrate in a test tube culture in comparison with the agar medium based on M-S allows increasing the germination of tubers in 1 tuber reproduction by 11.9 – 15.4%. In order to preserve all genotypes of the populations of test tube culture seedlings, 1 tuberous reproduction should be grown in closed soil conditions, or in the open ground, provided that the optimal soil moisture was reliably ensured.

5. Conclusion

The tube culture of seedlings ensures full survival of seedlings. For the growth and development of seedlings in a tube culture, the best of those studied was the nutrient medium based on the ion-exchange substrate IS-2V, containing 1% sucrose and 500 mg / l CCC. Two-stage cultivation of seedlings (the initial increase in the above-ground mass, then the induction of tuberization) has a positive effect on the pace and quality of tuberization. 30–40 days after induction, tuberization was observed in 90–100% of seedlings, regardless of their origin, and the size of microtubers was on average 6.7–8.7 mm.

Growing microtubers from a test-tube culture under greenhouse conditions allows preserving all genotypes in a population in 1 tuber reproduction.

References

[1] Madec P. 1958 Le role die tubercule – mere dausi evolution des germs de la pomme de terre *Ann. Inet.Nat.Rech. Agron.* **8** pp 5-30
[2] Gregory L E 1956 Some factore for tuborisation in the potato plants *Amor.J.Pot.* **43** pp 281-8

[3] Chapman H W 1958 Fuberisation in the potato plant *Husiol. Plantarum.* **2** pp 215 – 24
[4] Okazawa Y 1973 Physiological aspect of tuberisation in potato plant *YARQ* **7** pp 11-4
[5] Oslund C R and Id P H 1972 Metabolism of nucleic acid in potato stolon tips of the onset of tuberization inducer by long nights and low temperatures *Potato Ros* **221** pp 279-80
[6] Blanc A 1979 Belation entre lavance de la tuberisation de gerimes de pomme deterre cultives in vitro sous l effect D un traitement termique et la modification de l activite de L invertase *C. B. Acad. Sc. Ser. D.* **289** pp 279-82
[7] Mix G 1982 In vitro preservation of potato material *Plant Genet. Res.- Rome* **51** pp 6-8
[8] Husseu G and Stacey R I 1984 Factors affecting the formation of in vitro tubers of potato *Ann.Bot.* **53** pp 565-78
[9]. Menzel C M 1980 Tuberization in potato at high temperature respontes to gibberelin and growth inhibitors *Ann. Bot.* **46** pp 259-265
[10] Tovar P, Estrada R, Schide-Rentschler L and Dodds J H 1985 Induction and use in vitro potato tubers *CIP Circular.* **13** pp 1-5

Methods for the effective treatment of animal burns

O I Efimova[1], A I Dimitrieva, O P Nesterova, A V Aldyakov, A V Obukhova and T N Ivanova

Federal State Budgetary Educational Institution of Higher Education, Chuvash State Agricultural Academy, K. Marx St. 29, Cheboksary, 428003, Russia

E-mail: inna.efimova@mail.ru

Abstract. Burns (combustio) - tissue damage arising from thermal, chemical, electrical or radiation exposure. Burn disease – a complex violation of the organs and systems, developing due to extensive burns. The cause of burn disease is the loss of all types of the skin functions, loss of plasma, the collapse of red blood cells, as well as metabolic disorders. The probability of development, severity, and prognosis in this pathology are determined by the age of the patient, the General condition of his body and some other factors, but the leading role is played by the area of the lesion. 500 thousand animals are burned according to statistics in Russia for the year, and as a rule, mortality among burned animals is more than 11%. Treatment includes antibiotic therapy, infusion, and detoxification therapy, correction of all organs and systems. Treatment of burns of the first and second degree is most often limited to the local use of drugs that reduce pain sensitivity, promote faster tissue regeneration and help prevent infection, which leads to lengthening of healing processes and other complications. In the case of third and fourth-degree burns, regular thorough cleaning of the affected areas is also necessary, sometimes surgical intervention is required. Today, burns are widespread, so the search for effective treatments remains relevant. In this paper, we describe the etiology, pathogenesis of burns of small pets, the definition of the most effective and cost-effective method of treatment. The task of the research was also to determine the degree of burns and diagnosis of burns.

1. Introduction

Burn - damage to body tissues caused by the action of high temperature or the action of certain chemicals (alkalis, acids, salts of heavy metals).

The sudden formation of extensive burns causes the release of a large number of toxins and elements of decaying cells into blood. The blood levels of prostaglandins, serotonin, histamine, sodium, potassium and proteolytic enzymes increase dramatically. This leads to increased capillary permeability. Plasma comes out of the vascular bed, accumulates in the tissues, as a result, the exchange of circulating blood is significantly reduced. In response, the body emits hormones into bloodstream that cause vasoconstriction – noradrenaline, adrenaline, and catecholamines.

The mechanism of blood circulation centralization starts. Peripheral parts of the body, and then the internal organs begin to suffer from a lack of blood? which leads to the development of hypovolemic shock. Along with this, there is a thickening of blood and disorders of water-salt metabolism. All of the above leads to malfunctions of various organs. Oligoanuria develops. Subsequently, pathological changes are exacerbated by the depletion of the immune and endocrine systems, as well as the toxic effect of tissue breakdown products on the internal organs. Degenerative changes occur in heart and

Content from this work may be used under the terms of the Creative Commons Attribution 3.0 licence. Any further distribution of this work must maintain attribution to the author(s) and the title of the work, journal citation and DOI.
Published under licence by IOP Publishing Ltd

liver, ulcers are formed in the gastrointestinal tract, possible intestinal paresis, embolism, and thrombosis of mesenteric vessels, pneumonia is detected in the lungs.

In the first 24-72 hours after the burn, the number of red blood cells increases to 10-15 million in 1 mm^3, leukocytes – up to 20-30 thousand, the hemoglobin content increases to 120-130%. Metabolic disorders are characterized by dehydration, acidosis, low chloride content in the blood and a disorder of oxidative processes. Thickening of blood with extensive burns is explained by a significant loss of the liquid part of blood (plasma) and enhanced regeneration of red blood cells, resulting from irritation of the bone marrow protein breakdown products. In the future, the condensed blood gradually begins to change in the direction of liquefaction, and 1-2 weeks after the burn, anemia develops, which is associated with the intoxication of the body and a large loss of protein through the burnt surface.

Reflex processes in the pathogenesis of burns are crucial ones. Excessive thermal stimulus, acting on the receptor apparatus of the skin and other tissues, at the time of burn injury causes a huge flow of pain impulses to the central nervous system, the functional change of which determines the nature of the clinical picture of burn disease. It should be noted that the flow of pain impulses into the central nervous system does not stop with the elimination of the high temperature action on the tissue, it is supported by squeezing the inflammatory-swollen tissues of peripheral nerves and their endings, irritation with their products of tissue decay and toxins of microbes.

The burn is not only local tissue damage but also a serious general disease of the body (burn disease). By localization they are divided into burns of the skin; eye burns; inhalation injuries and burns of the respiratory tract.

The severity of the burn, the prognosis and the choice of therapeutic measures depend not only on the depth but also on the area of the burn surfaces. When calculating the area of burns, the «rule of nines» is used.

In accordance with the «rule of nines»: the area of the neck and head is 9% of the entire body surface; chest – 9%; abdomen – 9%; back surface of the trunk – 18%; one upper limb – 9%; one thigh – 9%; one shin along with the foot – 9%; external genitals and perineum – 1%.

There are 4 degrees of burn: 1 - redness of the skin, 2 - blistering, 3 - necrosis of the entire thickness of the skin, 4 - charring of tissues.

The severity of the burn is determined by the size of the area and depth of tissue damage. The larger the area and the deeper the tissue damage, the more severe the burn injury. With a burn that occupies 10% or more of the body surface, there are significant changes in the composition of blood, it is noticeably thickened, metabolism is disturbed. Metabolic disorders are characterized by dehydration, acidosis, low chloride content in the blood and a disorder of oxidative processes

First aid for burns is to halt the impact of external factors and the treatment of the wound.

Burns can develop due to the following reasons: thermal effects; chemical effects; electrical effects; radiation exposure.

A thermal burn is caused by the action of high temperature in fires, ignition of the gasoline, as well as from the impact on the tissue of boiling water and hot steam

The fire is a red-hot gaseous medium consisting largely of partially ionized particles, in which chemical interaction and physicochemical transformations of the particles of fuel, oxidizer, impurity particles occur, accompanied by «glow» and heat release. The area of the burn is relatively large, the depth is mainly of the 2nd degree. The primary treatment of the wound represents the complexity of the charred wool residues removal and can subsequently serve as foci of infection. The organs of vision and upper respiratory tract may be affected [1, 2].

Boiling water - water kept to the boiling point, that is, reached the boiling point (100 degrees). The area of the burn is relatively small but relatively deep, mainly that of 2-3 degrees.

Steam. When exposed to steam, a skin-deep tissue lesion and affection of the upper respiratory tract happen in most cases.

Hot items. When skin contacts with hot objects, a clear boundary of the object remains in the place of exposure. These burns are deep and are characterized by the second or fourth degree of damage.

Additional injuries may occur when you remove the object that caused the injury. There is a detachment of the affected skin layers.

The degree of skin damage during thermal exposure depends on the following factors: the temperature of influence (the higher the temperature, the stronger the lesion); the duration of exposure to the skin (the longer the contact time, the heavier the degree of burn); thermal conductivity (the higher it is, the stronger the degree of damage); skin condition and health of the sufferer. Animals often have thermal burns that occur in fires in livestock buildings, railway cars, vehicles, forests, peat bogs, etc. Burns in different animals is manifested differently, this factor depends on the structure of the skin. The most sensitive to burn injury are sheep and cattle, less - horses and pigs. Young animals endure burns worse than adult animals. The reaction to the burn is much stronger, and the course of the disease is more severe in those who have undergone any disease, had injuries with significant loss of blood or are in the last stage of pregnancy

Burns in the head, groin, genitals, udders, and limbs are the most dangerous and severe. So, in case of head burns and inhalation of hot air, suffocation (asphyxia) occurs due to edema of the mucous membrane of the larynx and trachea or subsequently complications occur in the form of purulent necrotic inflammation of the trachea, bronchi or bronchopneumonia. Lesions in the groin and genitals are accompanied by shock, a disorder of urination, and in the udder – deep cracks, suppuration, mastitis, general blood infection (sepsis) [3, 4].

The burnt areas of the extremities are polluted, suppressed, in the subsequent development of contraction, the impossibility of extension in the joints (contracture), partial or complete loss of function. Extensive and deep burns of chest are complicated by purulent inflammation of the pleura and lungs (pleurisy and pneumonia).

Chemical burns occur as a result of the action on the tissue of strong acids, alkalis, salts of heavy metals, quicklime, phosphorus. In animals there are chemical burns of the skin of the mucous membrane of the mouth, esophagus, and stomach; this happens due to careless storage and use of chemicals and fertilizers. When exposed to strong acids and salts of heavy metals, coagulation necrosis of skin, mucous membranes, and underlying tissues occur. A thick fabric scab, that quite quickly forms on the place of impact of such chemicals, prevents further penetration of them to the tissue depth. Alkali and other similar chemical compounds, dissolving proteins and saponifying fats cause tissue necrosis colliquation. As a result, alkali burns are deep; necrotic tissue at the site of the burn turns into a soft scab usually white. When healing chemical, especially alkaline, burns, usually formed powerful deep scars. Chemical burns are sluggish, without a florid inflammatory reaction; cleansing from dead tissues is slower than in case of thermal burns; changes in the general condition of a body are absent or weakly expressed, arise as a result of contact with the skin of aggressive chemicals (for example, acids, alkalis).

The degree of damage depends on its concentration and duration of contact with the skin. The influence of acids on the surface of the skin causes superficial lesions. A burn crust, which prevents further penetration of acids into the skin, forms after exposure to the affected area in a short time. Skin is deeply penetrated due to the influence of caustic alkali on the surface. Salts of some heavy metals (e.g. silver nitrate, zinc chloride) cause surface burns in most cases.

Electrical burns occur in contact with a conductive material. Trauma is associated with technical electric current or lightning coming through a body. The electric current spreads through the tissues with high electrical conductivity through the blood, cerebrospinal fluid, muscles, to a lesser extent - through the skin, bones or adipose tissue.

There is always a current mark (point of entry and exit) *in case of electric shock on the body of the victim.* Burns of this type are characterized by a small area of the lesion, but they are deep. They can be caused by non-insulated wiring, therefore rodents and rabbits are more likely to suffer from such injuries.

Burns due to radiation exposure can be caused by ultraviolet radiation, ionizing radiation, infrared radiation.

Ultraviolet skin lesions mainly occur in summer period. Burns, in this case, are shallow, but characterized by a large area of damage. When exposed to ultraviolet light, surface burns of the first or second degree often occur. Ionizing radiation causes damage not only to the skin but also to nearby organs and tissues. Burns in such case are characterized by a shallow form of the lesion. Burns are usually shallow, but their treatment is difficult due to the damaging effects of radiation on the underlying organs and tissues. Vascular fragility and bleeding increase but ability to regenerate reduce. Infrared radiation can cause damage to the eyes, mainly the retina and cornea, as well as the skin. The degree of damage, in this case, will depend on the intensity of the radiation, as well as on the duration of exposure.

Depending on the strength and duration of the cause of the burn, as well as local and general changes in the body, there are four degrees of burn. Uncomplicated burns with an area of up to 5% of the body surface do not greatly affect the overall condition of a body. Burns of the second, third and fourth degree, exceeding 5% of the skin area, proceed as a serious disease, and the second and third degree, affecting more than 10% of the skin area, are accompanied by nervous reactions (shock) and can result in the death of animals.

A first-degree burn: the affected area of light skin is red or pink (hyperemia), slightly compacted, sensitive to external irritations; slight swelling (swelling) of skin, the hair is tousled; in case of thermal burns, not completely burnt hair is visible; pain is expressed slightly. The animal sometimes licks the injured part of the body or rubs it on objects.

A second-degree burn: The skin is compact, sometimes sedentary; touching with fingers and prick with a needle is painful; spilled swelling of the skin and subcutaneous tissue. The greatest development is reached in 24-48 h after the heat injury. Extensive edema falls on the breastbone, lower part of the chest, abdomen, and limbs. Serous exudate in the form of yellowish-pink droplets sweats out on the borders of the burn, which, drying up, form loose crusts; on the tender areas of the skin, bubbles with pinkish transparent content and necrosis of the surface layers of the skin can form; the wool in the affected area is scorched, in places charred hair awns are visible. The surface layer of cells exfoliates in pigs in the first minutes after the injury from the burned surface, hanging in the form of dirty rags. The skin in these places has a wet, contaminated surface. The hurt animal worries, falls to the ground, rubs with a damaged surface on objects. These characteristics are especially strong in dogs and horses.

Third degree burn: it is characterized by necrosis of skin and deep-lying tissues. Skin is compact, immobile, insensitive to pain stimuli (touching, feeling the prick of the needle) when the patient is in the move, it gets pleated – «goffer skin». Pressing your finger in the burn area is painful. After 15-40 minutes after the burn, subcutaneous tissue in the area of the lesion and in the circumference swells, and the swelling falls on the underlying parts of the body by 3-4 day. Dead skin areas begin to reject, cracks appear on the 7th-15th day from which serious, then purulent drainage is pouring out. As a result of healing, large scars are formed. Long-term exposure to concentrated acids leads to the formation of a dense crust – scab of dead tissue. Concentrated alkalis melt tissues often to the bone and, unlike acids, cause very deep burn lesions. General changes in third-degree burns are more significant than in second-degree burns. The stage of excitation of burn shock can last for 3-6 hours. The animal lies down, jumps to feet, makes finical movements, staggers. The animal has increased sweating, muscles are trembling, it has frequent urination, defecation. Dogs and pigs are aggressive at this time. They have increased body temperature, increased heart rate, and breathing, blood pressure briefly increases. Then the stage of oppression comes: all reflexes are weakened, the animal refuses to eat, it is thirsty, movements are uncoordinated, blood pressure is decreased. The oppression develops in sheep and cattle in the fastest way.

After 6-12 hours after the burn organism is getting intoxicated with products of protein decomposition, the activity of all systems and organs are impaired. These signs may occur with extensive burns of the second and third degree.

Fourth degree burn: occur from prolonged exposure to high temperature and leads to the charring of a body part or organ. The skin and deep-lying tissues are wrinkled, dry, very compact, completely insensitive to external stimuli crust. Deep cracks are formed in some places, from which bloody fat-like liquid often flows.

Significant in width and depth burns, starting from the 4-8 day, can be complicated by general blood infection and intoxication (toxemia) due to the development of pyogenic infection. The condition of the hurt animal deteriorates. Periods of elevated temperature alternate with periods of normal temperature (intermittent fever), cardiac activity is weakened, there are deep metabolic disorders of all kinds. This condition often ends with the death of the hurt animal.

When determining the severity of the burned condition, factors such as depth, breadth (area of injury) and localization of the burn, as well as the age, type and initial state of the animal are taken into account.

The size of the burning surface is not difficult to identify, but to determine the depth of the lesion, particularly with burns II and III degree, is possible only on the 5-8-th day, when there will be a delimitation of the necrotic tissue and start of their rejection. Therefore, when determining the degree of burn in the first hours, the following indicators are taking into account.

In case of the I-st degree burns, wool is ablaze not on the entire length, skin is slightly compacted and becomes very sensitive to the outside stimuli. Severe hyperemia develops on depigmented areas and there is a slight swelling of the skin.

In case of the II-nd degree burns, skin becomes compact, sometimes inactive, with a needle and palpation, pain is expressed. At the edge of the burn, there are drops of serous fluid. Bubbles may form in areas with delicate skin. Edema develops in the area of burn in the first hours.

In case of the III-rd degree burns skin becomes very dense, motionless and insensitive to external stimuli (palpation, needle angle). When moving, the skin gets pleated -»corrugations». Large swelling is developing in the area of burn and its circumference.

In case of the IV-th degree burns, charring of the affected area and even the organ occurs. Skin and underlying tissues become very dense, immobile and completely insensitive to external stimuli. In some affected areas, deep cracks are formed, of which there is often a bloody fat-like effusion.

Based on the fact that the treatment of burns requires a lot of time and a lot of work of veterinary workers, as well as certain material costs, which with extensive burns do not always pay off, we believe that the following animals should get treatment: animals which got burns of I and partially II degree; animals that got deep burns on the area up to 5% of the body surface and in some cases (highly productive and breeding cattle) animals with deep burns on the area up to 10% of the body surface can be treated.

Treating animals with more extensive and deep burns is not economically feasible. Such animals should, if possible, be butchered, and their meat must be used for other purposes. In determining the severity of injuries and deciding the treatment of the animal it is necessary to consider the localization of the burn.

Treatment of animal burns is one of the most difficult and less developed problems in veterinary surgery.

Determining the essence of burns, many authors have been considering this type of injury as a local process up to the present time. Thus, Fabritius Gilden in his book, published in 1607, wrote the following: «Burn is a violation of the integrity of the epidermis, skin, and muscle meat, veins, arteries, nerves caused by the power of fire.» I. G. Rufanov says that a burn is tissue damage caused by the action of high temperature (thermal burns) or the action of chemicals (chemical burns).

First aid for burns is very important. It includes prevention of shock; fight against dehydration, blood thickening, and intoxication; primary treatment of the burned area; prevention of infection.

In order to prevent shock, novocaine was administered intravenously in a dose of 1 ml of 0.25% or 0.5% solution per 1 kg of animal weight, or a paranephral blockade was carried out. In addition to anti-shock action, novocaine eliminates increased capillary permeability and thereby reduced blood plasma loss. To prevent dehydration, blood thickening and intoxication the following is recommended: abundant introduction of water into the body (through the mouth, with enemas, subcutaneously), transfusion of large doses of blood (3-5 liters for cattle and horses), intravenous isotonic sodium chloride solution (1-2 liters for injection for cattle and horses). As an alkalizing agent for acidosis, a 5% solution of sodium bicarbonate (400-600 ml for horse and cattle) is used. To combat burn intoxication, in addition, it is recommended to use serum from animals that have previously suffered a burn.

The nature of the primary treatment of the burnt surface depends on what method of treatment is chosen. There are open, closed and mixed methods of treatment. In veterinary practice, the most common was an open method of treatment in combination with the use of tanning, fixing and coagulating substances (2-3% solution of potassium permanganate, 5% alcohol solution of tannin, 1-2% solutions of methylene blue or diamond green, 5% or 10% solution of silver nitrate, 5% or 10% alcohol solution of iodine). When processing (lubrication) with these solutions, the burnt surface is covered with a solid crust, which prevents the loss of plasma and protects the damaged tissues from infection.

With a closed method of treating burns, bandages with antiseptic agents are used: Vishnevsky emulsion (tar – 1 part of benzocaine and the xeroform 3 parts castor oil – 100 parts), sintomitsinovoy and streptocide emulsions. In recent years, in the treatment of burns includes covering of a burnt surface with fibrin and plastic films that protect damaged tissues from the impact of the external environment and infection.

The prevention and control of microflora in case of burns should be started as soon as possible. A common antiseptic therapy should be applied with this purpose in conjunction with the local application of the above-mentioned means.

The basic principles of burns subsequent treatment are as follows:

1) treatment of infected burn wounds is similar to that of conventional purulent wounds. In this case, use the same tools and methods as in the treatment of wounds. It should be noted that antibiotics are effective only during the first decade. By the end of this period, their appointment is impractical, since the microflora of the burned surface already becomes insensitive to antibiotics;

2) in the treatment of burns methods of epidermization (ultraviolet radiation, tissue therapy, etc.) stimulation are of particular importance. The transplantation of free flaps is the only technique for saving the life of an animal in case of the extensive burn;

3) in relation to the high protein loss through the wound surface and the development of post-burn anemia, it is important to provide sick animals with full and fortified food in a timely manner, to perform a blood transfusion, apply the blood plasma hydrolysate and other substitutes.

At the time of the burn, it is necessary to stop the exposure to high temperature as soon as possible. The burnt surface is carefully cleaned of burnt wool, hay particles, sawdust, chips, earth and washed with water. To reduce pain and limit inflammation, use cold or lotions with alcohol, cologne, 1-3% solution of potassium permanganate. For pigs, in the first minutes after the burn the upper layer of skin cells is exfoliated, there is a very strong soreness, so the burnt surface is not mechanically cleaned, but only irrigated with the above antiseptics.

As for chemical burns, the surface is poured first with a large amount of water for 30 minutes, then acids are neutralized with 1-2% alkaline solutions (soap, soda water) or sprinkle with chalk and magnesia when burns with strong acids. In case of a burn with alkalis, the burned place is poured with 1-2% solution of acetic or citric acid. The burn caused by quicklime is treated by milk. Burns with phosphorus, where the chemical effect is combined with thermal, are abundantly washed with water and lotion of 5% copper sulfate is applied to them. Pieces of phosphorus are removed from the skin.

A doctor or paramedic renders urgent complex medical care to the patient performing local and general treatment of organism. Good results in burns of any localization are obtained from intravenous administration of 0.5% solution of novocaine (0.2 g of novocaine per 100 kg of animal weight). Neuroleptic and ganglioplegic drugs are successfully used: aminazine (subcutaneously large animals 0.5-1 mg, small 2.5 mg per 1 kg of their weight), dimedrol (subcutaneously large animals 0.3-0.5 mg, small 0.02-0.03 mg per 1 kg of weight). It is better to use 2% solution of promedol (large animals 0.1 g, small 0.2 mg per 1 kg of weight) subcutaneously for cattle and horses. Post-burn excitation is more quickly removed as a result of the combined use of Novocain with aminazine, dimedrol, promedol [5].

In subsequent days novocaine anesthesia can be combined with sleeping pills (sodium bromide 0,02-0,05 g per 1 kg of animal weight 2 times a day). Polyglucin (1.5-2 l for large animals and 100-500 ml for small animals), etc is injected intravenously as antishock fluids.

It is advisable to treat animals only with burns of the first and partly second degree, but not with burns of the third and fourth degree, affecting 5-10% of the body area. Such animals are used on meat in the first 1-2 days after burn and only on medical conclusion.

In the modern world people in their home conditions often manage a wide variety of pets, whose main function is animal therapy. But often animals are not adapted to the conditions offered by a man. Animals are exposed to various hazards, including being exposed to burns.

The aim of the research was to develop an effective and cost-effective method of treating burns of small rodents, managed in the home conditions.

2. Material and methods

Experimental work was carried out in the veterinary clinic «4 paws» Novocheboksarsk City, Chuvash Republic and at the Department of Morphology, Obstetrics and Therapy of the Federal State Budgetary Educational Institution of Higher Education, the Chuvash State Agricultural Academy.
The task of our research was:

Conducting experiments to determine the effective method of treatment of burns of small animals;

Control over the effectiveness of the method according to the following indicators:

- duration of treatment;
- economic costs;
- the outcome of the disease.

Djungarian hamsters of the same age and sex were selected as the material for research. Animals were kept in the same conditions (cell content, nutrition, and exercise), they were the same in weight, sex, age. Physiology of hamsters was determined in the normal range: body temperature 37.5-38.5'C respiration rate 33-127 movements per minute, pulse 280-412 beats per minute. Deviations in development and in health were not observed. All hamsters were 2 months old, male. They were kept in three cages in groups. The filler in the cages is sawdust. Hamsters had free access to water and food «Little One». The water in the drinking bowl was changed every two days, food - every day. There was a wheel in each cage where hamsters actively run.

Research methods. The animals were divided into three groups of 3 hamsters in each group. Wool sized 2*2 cm was shaved on the dorsal part of the hamster's body (dorsal portion). Burn of the exposed skin surface was carried out by chemical means (peroxide solution with a 5% alcohol solution of iodine). Gauze swab moistened in a burn solution was used to treat the prepared skin of the animal's back. The resulting solution was applied to the shaved area of the hamsters' body by gauze swab. First, they wiped, then they pressed the swab to the skin for a few seconds. Immediately after applying the solution to the skin changes in its color were not noticeable. After several hours the characteristic features of first-degree burn (swelling skin) was clearly visible in animals.

Animals were taken under observation and we assigned the treatment of each experimental group in different ways.

The first method of treatment was to treat the edges of the affected area with wild camomile, the burn wound itself was treated with sea buckthorn oil. Such treatment was repeated in the morning and in the evening for five days.

Wild camomile preparations have anti-inflammatory, antispasmodic, moderate antimicrobial, diaphoretic, carminative, soothing, analgesic properties, reduce the fermentation process, increase the secretion of digestive glands, accelerate the regeneration of the epithelium in ulcers.

Sea buckthorn oil has an anti-inflammatory effect, accelerates wound healing and acts as an analgesic, increases immunity, has a healing effect on vision, improves skin condition, rejuvenates, restores the hormonal balance, reduces cholesterol and blood sugar, improves blood circulation and strengthens the walls of blood vessels, is a laxative.

The second method of treatment was the use of novocainic and 3% tetracycline ointment. In the morning the wound was treated with novocaine ointment, and in the evening with tetracycline ointment, for five days.

Novocainic 5% ointment has a local anesthetic and anti-arrhythmic effect. It blocks sensitive nerve fibers and endings, causing their anesthesia, has a wide breadth of therapeutic properties. When absorbed and when directly injected into the bloodstream, novocaine has a general effect on the body, reduces the formation of acetylcholine and reduces the excitability of peripheral cholinoceptive systems, has a blocking effect on the autonomic ganglia, reduces spasms of smooth muscles, reduces the excitability of muscles and excitability of the motor areas of the cerebral cortex, but causes excitement in toxic doses, then paralysis of the central nervous system [6].

Tetracycline ointment is a bacteriostatic antibiotic that stops the reproduction of pathogenic microorganisms with the suppression of their vital activity at the protein level. The drug has a broad spectrum of action. It perfectly copes with staphylococcal, streptococcal, gonorrheal, chlamydial, salmonella, and other bacterial infections. Tetracycline antibiotic is used in the treatment of a wide range of diseases, but in the form of ointment, its use is limited. The drug is prescribed for the treatment of two types of diseases: inflammation in the eye area and infectious skin pathologies [7, 8].

As for the the third method of treatment, procaine block with 0.25% solution of novocaine (diffuse blockade) was applied – once a day and the burn wound were treated with Levomekol ointment. The ointment was applied twice a day. Treatment also was continued for five days.

A short procaine block is a local anesthetic with moderate anesthetic activity and a large breadth of therapeutic action especially in the acuity. The essence of the block lies in the fact that the solution of novocaine is injected around the lesion and under its base, on the border of healthy and diseased tissues. It is necessary to avoid strong pressure and the application of unnecessarily large volumes of solution at the time of the injection of the solution, since this may cause even greater squeezing of blood vessels and tissues. Novocain solution reduces the excitability of the motor areas of the cerebral cortex, myocardium and peripheral cholinergic system, ganglioplegic action, including antispasmodic on smooth muscles. Used to treat wounds, ulcers, fistula, myositis, papillomatosis (intradermally or intravenously).

Levomekol ointment is used as an antimicrobial, anti-inflammatory and regenerating agent. Levomekol ointment is used externally for the treatment of a number of pathologies: sores and skin ulcers, infected with pathogenic microflora; surgical wounds; traumatic injuries of the skin; burns second and third degree; trophic ulcers, furuncles, carbuncles, abscesses, pimples, calluses, sores, eczema and dry oozing, necrosis, frostbite, diaper rash, prickly heat and other.

A first-degree burn was performed in our experiment and we treated it in three different ways. The first method of treatment consisted in treating the edges with wild camomile extract, the wound itself was treated with sea buckthorn oil. This treatment was repeated in the morning and evening for five days. The second method of treatment was the use of novocainic and 3% tetracycline ointment. The burn wound was treated with novocainic ointment in the morning, and with tetracycline ointment in the evening, for five days as well. In the third treatment method, procaine block with a 0.25% solution of novocaine (short procaine block) was applied once a day and the burn wound were treated with levomekol ointment. Ointment was applied twice a day. The treatment lasted for the same period of time.

3. Research result

We observed a painful response to the application of therapeutic drugs at the beginning of treatment of animals of the first group. The body temperature was slightly increased from the physiological norm, from 39.0 to 39.30 ^{0}C, the pulse and respiration rates were also slightly, but higher than the norm. On the second day of treatment, the affected area had redness, painful on palpation. Hamsters behaved inactive, most of the time they were in the sleep phase, the appetite was absent. On the fourth day of treatment, the animals of this group had almost no redness in the burn area, the pain persisted, and the appetite was weak. One of the hamsters was trying to walk on the gym wheel. Other rodents led an inactive lifestyle most of the day. The seventh day of treatment was characterized by the recovery of experimental animals: temperature, pulse and respiration were normal. The appetite was good, the fur on the side of the burn wound was growing, the hamsters were active.

In the second method of treating rodents, the first three days did not differ much from the previous group, but the appetite and activity in this group of experimental animals came one day earlier, the burn wound was not so clearly expressed, the wound itself had all the signs of regenerative healing.

The animals getting the third method of treatment, namely the use of procaine block and ointment levomekol, on the third day of the treatment procedures had a less reddened area of the burn wound, although there was still pain, but the hamsters actively ran in the wheel, actively ate food and drunk water. The skin on the affected area also had a denser character than in animals of the compared groups.

The results of the studies made it possible to conclude about the effectiveness of the treatment of burn wounds by the third method of treatment when we used procaine block and Levomekol ointment. So, the pulse, breathing, and temperature on the second day of treatment were normalized using this method. We noticed improvement in the burn wound on the third day of treatment. There was no swelling or pain response. Animals showed less concern, had a good appetite, actively played and ran in the wheel.

With regard to the economic component of treatment, the clinical examination, consultation, preparation of the animal for the procedures (shaving, sanitary and hygienic processing), the use of gauze tampons, stationary maintenance of sick animals, the cost of consumables (gloves) had the same cost for all methods of treating animals. Drugs used in the treatment had a distinctive cost. For example, with the first method of treatment, that is, with the use of wild camomile decoction and sea buckthorn oil, the economic costs amounted to 175 rubles. In the second method - the use of novocaine and tetracycline ointment amounted to 148 rubles, and the third treatment option - the use of the short procaine block and levomekol ointment - 85 rubles.

Economic costs of the first method was 2025 rubles, the second one - 2008 rubles and the third one - 1960 rubles. The third method of treatment, where the amount was 1960 rubles, is the most cost-effective option [9, 10].

As a result of our studies to determine the most effective method of treatment of burns of the first degree of small rodents, it was found that the use of medicinal ointment «Levomekol», two times per day, and the short procaine block allowed to achieve recovery in 100% of rodents within five days with the disappearance of clinical signs of burn. Hair grew on the skin of all the hamsters a month after the application of a chemical burn.

As prevention of this disease, we recommend keeping rodents in the room where all electrical wires are isolated, not available for gnawing. Animals are protected from the effects of hot sources, open fire and chemicals of various applications (household, agricultural and others).

References

[1] Danilevskaya N V, Deltsov A A, Tuhvatova R F, Bessarabova E V and Parasyuk L P 2018 Workshop on General and Private Pharmacology, Moscow, p.94

[2] Maltseva B M 2004 The use of wound coatings in burns [experiments on white rats] *Veterinary. Abstract J* - Moscow, p. 485

[3] Pulniashenko P R, Kozjy R S and Fedorov V N 2006 Infusion therapy animals *Veterin. Sci. Kuban*, **2** pp. 5-9

[4] Stepanov A A and Yermolaev V A 2016 Modern methods of treatment of burns in veterinary medicine The International student's scientific bulletin. No. **4-3**; URL: http://eduherald.ru/ru/article/view?id=16261

[5] Levomekol; URL:http://medside.ru/levomekol

[6] Danilevskaya N V, Deltsov A A and Parasyuk L P 2016 Fundamentals of pharmaceutical technology. Part 2. Industrial production. Manual - Moscow, p. 124

[7] Schelgachev V V 2006 Technology of preparation of the drug Biogel B, intended for the

treatment of skin lesions of animals Dissertation. - Moscow, p. 18

[8] Pharmaceutical camomile: medicinal properties, instruction for application, contraindications; URL:http://zdravotvet.ru/romashka-aptechnaya-lechebnye-svojstva-instrukciya-po-primeneniyu-protivopokazaniya/

[9] Efimova I O, Dimitrieva A I and Mochalova E Y 2018, Methods of treatment of chemical burns Proceedings of the all-Russian scientific and practical conference dedicated to the 70th anniversary of the honored worker of the higher school of the Chuvash Republic and the Russian Federation, doctor of veterinary Sciences, Professor Nikolai Kirillov. «The development of agricultural science as an essential condition for the effective functioning of the agro-industrial complex of the country.» Cheboksary 2018. pp. 104-109

[10] Efimova I O and Mochalova E Y 2017 Treatment of burns All-Russian student scientific-practical conference with participation of pupils of 10-11 128 forms «Student science – the first step and academic science» March 22-23, 2017. Cheboksary – p. 185

Activization of adaptogenesis and realization of meat qualities of Aberdeen-Angus bulls by biological products

I I Kochish[1], V G Tyurin[2], A F Kuznetsov[3], V G Semenov[4, 5], A Y Lavrentiev[4], A V Yakimov[4], N I Kosyaev[4] and I VTsarevsky[4]

[1]Federal State Budgetary Educational Institution of Higher Education "Moscow State Academy of Veterinary Medicine and Biotechnology named after K.I. Scryabin", Moscow, Akademika Skryabina St., 23, 109472, Russia
[2]All-Russian Scientific Research Institute of Veterinary Sanitation, Hygiene and Ecology - a branch of the Federal State Budgetary Scientific Institution "Federal Research Center - All-Russian Scientific Research Institute of Experimental Veterinary Medicine named after K.I. Scryabin and Y. R. Kovalenko of the Russian Academy of Sciences", Zvenigorodskoe highroad, 5, 123022, Moscow, Russia
[3]Federal State Budgetary Educational Institution of Higher Education "St. Petersburg State Academy of Veterinary Medicine", St. Petersburg, Chernigovskaya St., 5 196084, Russia
[4]Federal State Budgetary Educational Institution of Higher Education "Chuvash State Agricultural Academy", K. Marx street, 29, 428003, Cheboksary, Russia

E-mail: semenov_v.g@list.ru

Abstract. The feasibility of using Prevention-N-E, a newly developed and tested biological product, in comparison with the previously tested PS-6 preparation in the adaptive technology of keeping specialized beef cattle of imported breeding, had been scientifically substantiated and experimentally proved. Under the conditions of production tests, biopreparations PS-6 and Prevention-N-E stimulated the production of red blood cells and increased the hemoglobin concentration in the blood of bulls, that is, improved hematopoiesis; they caused physiological eosinophilia, moderate neutrophilopenia with a shift of the neutrophilic nucleus to the right and lymphocytosis; increased protein metabolism, mainly due to the synthesis of albumin and γ-globulin fractions; intensified nonspecific resistance of an organism. Against the background of the use of drugs in animals of the 1st and 2nd experimental groups, the incidence of the digestive and respiratory organs decreased by 1.4 and 2.3 times, the recovery time reduced by 3.36 and 4.88 days. accordingly, compared with the control ($P <0.05$). By the end of the growing period, the animals of the 1st and 2nd experimental groups exceeded in live weight control subjects by 6.6 and 9.2 kg, rearing by 10.4 and 14.8 kg, and fattening by 14.2 and 22.2 kg respectively ($P <0.05$-0.01). In the cuts of the carcasses of bulls of the 1st and 2nd experimental groups, there was a bigger amount of the highest grade flesh: in spin-breast - by 1.0 and 1.0 kg, in the lumbar - by 0.2 and 0.4 kg, and in the hip - by 2.3 and 4.2 kg ($P <0.05$-0.001), rather than in the control.

1. Introduction

The development strategy of beef cattle in the Russian Federation is aimed at increasing the share of domestic meat production in shaping meat resources in accordance with scientifically based consumption standards, increasing the competitiveness and investment attractiveness of the sub-

Content from this work may be used under the terms of the Creative Commons Attribution 3.0 licence. Any further distribution of this work must maintain attribution to the author(s) and the title of the work, journal citation and DOI.
Published under licence by IOP Publishing Ltd

industry, and provides for solving the most important socio-economic task of providing the population with biologically valuable and good quality beef [1].

World experience shows that meeting the demand for beef in sufficient quantity is impossible without developed specialized beef cattle breeding, whose share in the total cattle population in Europe and North America ranges from 40 to 85%. In Russia, at present, beef production is 90% based on the sale of livestock of dairy and combined breeds [2-7].

The same methods of cattle breeding are used practically in all countries of the world, in all climatic zones. However, when transporting animals from continent to continent, from one country to another, even if the countries are close in climatic conditions, time and efforts of specialists are needed for the animals to adapt [8, 9].

To enhance the adaptation of imported specialized beef cattle to the natural temperature regime of the habitat and to realize the biological resource potential of the organism, the veterinary market offers a wide range of pharmacological agents, but many of them have chemical origin, the bioavailability of which is low [10-12]. In 1991, the research and development center of P.E. Ignatov developed a biogenic drug Dostimum, representing a 0.5% aqueous suspension of a purified yeast cell polysaccharide complex (glucan). The closest prototype for the problem to be solved is the immobilization of polysaccharides of yeast cells in 10-14% saline-injected solution of polyvinylpyrrolidone (RU 2137480 C1 A61 K 31/79, 35/72, 20.09.99) or in an agar gel (TU 10.07.236-91 on the production of the drug we obtain), which increases the activity of immunocompetent cells. The disadvantages of drugs include the absence of antibacterial activity of animal pathogens and insufficient stimulating activity against some parts of the immune system [13, 1].

Therefore, the development and introduction of complex, polytropic and cost-effective pharmacological tools based on natural raw materials to enhance the protective and adaptive functions of the body of beef cattle imported selection to adaptive technology of growing, additional growing and fattening and, as a result, realizing the biological potential of the organism, is an important issue of the modern veterinary science and practice [10, 14, 15].

The purpose of this work was the activation of adaptogenesis and the realization of the bioresource potential of specialized Aberdeen-Angus breed beef cattle with biological preparations PS-6 and Prevention-N-E.

2. Material and methods

The experimental part of the research work was carried out in the breeding reproducer LLC Agrofirma Myaskom, Lyskovsky district of the Nizhny Novgorod region, specializing in breeding Aberdeen-Angus breed beef cattle (2500 heads). In accordance with the research plan of the Chuvash State Agricultural Academy, processing of materials was carried out in the Lyskovsky interdistrict veterinary laboratory of the state-financed medical institution SFI "State Department of Lyskovsky District of the Nizhny Novgorod Region" and in the laboratory of the Department of Morphology, Obstetrics and Therapy of FSBEI HE Chuvash State Agricultural Academy from 2014 to 2017.

The objects of research were purebred bulls Aberdeen-Angus breed, imported from the United States (Virginia). In the scientific and production experience, three groups of bulls were formed on the principle of groups-analogues of 15 animals in each group, taking into account the physiological state, age, live weight. Animals of all groups in the rearing period up to 210 days old were kept on suction with mother cows in pens in the open air, and later in the rearing period up to 360 days old and in the fattening period up to 540 days old (duration of experiments) - in open areas under shelters, that is, by adaptive technology.

Hygienic conditions of keeping animals corresponded to natural temperature and humidity conditions of atmospheric air in all seasons of the year.

Diets for bulls during the periods of growing up to 210 days in pens, growing up to 360 days and fattening up to 540 days in open areas under canopies on adaptive technology provided the needs of the body in energy and nutrients, mineral elements and vitamins according to the norms of feeding. On average, the diet of suckling bulls up to 7 months of age consisted of 5.3 kg of alfalfa hay and 1.5 kg of feed. The animals were fed twice a day: in the morning 60 % of normal total giving of feed and in the evening – 40% of the feed. The structure of the diet for animals in the period from 7 to 12 months included: roughage – 13 %, juicy food – 46.5 %, concentrated – 40.5 %. In the period from 13 months to the end of fattening the diet of bulls consisted of fodder mixture, in which the share of roughage was 10.5 %, juicy – 39.0 % and concentrated – 50.5 %. Feed was given in amount of 4.0-4.5 kg per 1 head per day.

In order to intensify the adaptogenesis of beef cattle imported to the climatic conditions of the Nizhny Novgorod region and the most complete realization of the bioresources potential of the bull-calves organism under the conditions of the natural temperature of the habitat, environmentally safe complex biological products developed by scientists from FSBEI HE Chuvash GSHA, PS-6 and Prevention-N-E were used (V.G. Semenov, F. P. Petryankin, D.A. Nikitin et al.).

PS-6 - a complex biological product, an aqueous suspension containing a polysaccharide complex of yeast cells immobilized in an agar gel with the addition of a benzimidazole derivative - (-) 2,3,5,6-tetrahydro-6-phenylimidazo [2,1-b] -thiazole hydrochloride and bactericidal drug from the group of aminoglycosides - kanamycin monosulfate - 0-3-Amino-3-deoxy-alpha-D-glucopyranosyl- (1"6) -0- [6-amino-6-deoxy-alpha-D -glucopirinazyl- [1"4)] - 2-deoxy-D-streptamine (in the form of sulfate).

Prevention-N-E - a complex preparation, an aqueous suspension containing a saccharomyces cerevisiae polysaccharide complex immobilized in an agar gel with the addition of the benzimidazole derivative - (-) 2,3,5,6-tetrahydro-6-phenylimidazo [2,1-b] -thiazole hydrochloride and macrolide bactericidal drug - erythromycin - [3R- (3R *, 4S *, 5S *, 6R *, 7R *, 9R *, 11R *, 12R *, 13S *, 14R *)] - 4 [(2,6-Dideoxy-3-O-methyl-3-0-methyl-alpha-L-ribohexopyranosyl)oxy]-14-ethyl-7,12,13-trihydroxy-3,5,7,9,11,13-hexamethyl-6-[[3,4,6-trideoxy-3-(dimethylamino)-beta-D xylohexopyranosyl) oxy] oxycyclotetradec n-2,10-dione.

Animals of the 1st experimental group were injected intramuscularly with a biological preparation PS-6 at a dose of 3 ml for 2-3 and 7-9 days of life, the 2nd experimental group - Prevention-N-E at the indicated dose and at the same time, the control groups - biological preparations were not administered.

Growth, morbidity and preservation, clinical and physiological status, morphological and biochemical profile of the blood, as well as non-specific resistance of the body of bulls were studied for 1-, 210-, 240-, 360-, 450- and 540th days according to generally accepted veterinary medicine methods. After controlling slaughter, bulls were assessed for slaughter quality, and veterinary and sanitary examination of beef was carried out.

Research work was carried out using the following methods:

1) **clinical and physiological** – body temperature, heart rate and respiration were measured by conventional veterinary methods;

2) **hematological** - number of erythrocytes, concentration of hemoglobin, total number of leukocytes and types on the PCE 90 Vet automatic veterinary hematology analyzer were determined;

3) **biochemical** - level of total protein in blood serum was investigated - using an IRF-22 refractometer according to A.M. Akhmedov, protein spectrum - by turbidimetric method according to S.A. Karpyuk;

4) **immunobiological** - phagocytic activity of leukocytes was determined according to V.S. Gostev using the daily agar culture of St.aureus, lysozyme activity of blood plasma according to V.G. Dorofeychuk using daily agar culture M. lysodeiticus, bactericidal activity of blood serum according to O. V. Smirnova et al. using daily agar culture of E. coli, as well as the amount of antibodies in the serum of photoelectrocolorimeter FEK-56M by D. A. Mac-Ewan et.al.;

5) **zootechnical** - live weight and average daily gain of animals by monthly weighing were determined. Meat productivity and slaughter qualities were assessed by the results of the control slaughter of bulls (5 animals from each group) at the age of 540 days according to the standard technique. At the same time, pre-slaughter live weight, mass of hot and cooled carcass and internal raw fat, relative yield of carcass, fat, slaughter mass and slaughter yield were taken into account. After cooling for 24 hours, the right halves of the bulls carcasses were weighed and boned to determine the absolute and relative content of the flesh, fat, tendons and bones, flesh output by grades, in terms of carcass;

the meat index is the ratio between the flesh mass of the carcass and the bone mass, that is, the amount of flesh per 1 kg of bone;

6) **veterinary and sanitary examination** of meat by organoleptic (appearance, color, texture, smell, condition of fat and tendons, degree of bleeding), cooking sample (smell, transparency, taste of broth) and biochemical parameters (pH, amino-ammonia nitrogen, formal acid reaction, reaction to peroxidase and with copper sulfate) in accordance with the "Rules of veterinary examination of slaughter animals and veterinary-sanitary examination of meat and meat products" (M., 1988);

7) **spectrometric** - the content of heavy metals (lead, cadmium, arsenic, copper, zinc, mercury) was determined in muscle tissue using atomic absorption spectrophotometry;

8) **economic** - determined the effectiveness of activating the adaptogenesis of imported bulls with PS-6 and Prevention-N-E biological preparations according to the method of I.N. Nikitin et al.

Therapeutic and prophylactic efficacy of biological products was determined by monthly registration of the diseased (taking into account the diagnosis and duration of the disease) and culled animals according to the veterinary statistical reporting with the subsequent calculation of the Mellenberg coefficient using the formula

$$MC = \frac{sick\_animals\_number \times duration\_of\_disease}{number\_of\_animals\_in\_group \times experiment\_duration} \times 100$$

The digital material of the experiments was processed by the method of variation statistics on the reliability of the difference between the compared indicators (P <0.05-0.001) using the Microsoft Office Excel software package.

3. Research results

It was found that the double intramuscular injection of PS-6 and Prevention-N-E on 2nd...3rd and 7th...9th day of life in a dose of 3 ml did not affect the clinical and physiological state of the body.

During the growing period, both in the control and in the experimental groups, cases of bull diseases were identified, which are presented in the Table 1.

Table 1. Incidence of bulls.

Indicator	Group of animals		
	Control	1 experimental	2 experimental
Number of animals in groups	15	15	15
Got sick	7	5	3
Recovered	7	5	3
Duration of disease, day	7,23±1,26	3,87±1,14*	2,35±0,65*
Morbidity, %	46.6	33.3	20.0
Mellenberg coefficient	1.61	0.61	0.22

It follows from the presented tabular data that the animals of the experimental groups decreased in the incidence of the digestive and respiratory organs by 1.4 and 2.3 times, the recovery time was reduced by 3.36 and 4.88 days. and the Mellenberg coefficient decreased by 2.6 and 7.3 times, respectively, compared with the control (P <0.05).

The selective mobilization of the morphological and biochemical profiles of blood, cellular and humoral factors of nonspecific resistance of the bulls body in the conditions of adaptive content technology in open areas has been revealed. Proven biological products have a wide range of bioeffect:

- stimulated the production of red blood cells and increased the concentration of hemoglobin in the blood of bulls, that is, improved hematopoiesis, but did not affect the production of white blood cells (table 2);

Table 2. Hematological parameters of bulls.

Group of animals	Age, days	Erythrocytes, $\times 10^{12}$/l	Hemoglobin, g/l	Leukocytes, $\times 10^{9}$/l
Control	1	8,18±0,24	117±2,54	9,38±0,29
	210	7,24±0,29	106±2,30	7,27±0,38
	240	7,45±0,21	107±2,44	6,93±0,31
	360	7,61±0,26	110±2,22	7,04±0,37
	450	7,68±0,24	108±2,67	6,80±0,24
	540	7,43±0,23	109±2,51	6,66±0,32
1 experimental	1	8,26±0,17	119±2,62	9,22±0,24
	210	8,14±0,26*	118±2,35**	7,59±0,32
	240	8,33±0,22*	119±2,63*	7,41±0,34
	360	8,47±0,24*	121±2,48*	7,51±0,39
	450	8,57±0,20*	123±2,69**	7,32±0,24
	540	8,24±0,19*	120±2,35*	7,02±0,28
2 experimental	1	8,10±0,21	115±2,56	9,18±0,25
	210	7,97±0,26	116±2,36*	7,47±0,33
	240	8,17±0,26	118±2,40*	7,33±0,29
	360	8,30±0,19	119±2,11*	7,41±0,38
	450	8,46±0,22*	120±2,40*	7,16±0,32
	540	8,15±0,19*	118±2,57*	7,04±0,25

* P≤0,05, ** P≤0,01.

- caused physiological eosinophilia, moderate neutrophilopenia with a shift of the neutrophilic nucleus to the right and lymphocytosis (table 3);

Table 3. Leukocyte formula of bovine blood.

Group of animals	Age, days	Group and type of leukocytes						
		granulocytes, %					agranulocytes, %	
		Basophils	Eosino Phils	Neutrophils: Young	Neutrophils: Nuclear Stick	Neutrophils: Segmento-Nuclear	Lymphocytes	Monocytes
Control	1	0,2±0,20	0,4±0,24	2,0±0,32	22,4±1,03	37,2±1,93	35,6±1,60	2,2±0,58
	210	0,8±0,37	1,4±0,24	0,6±0,24	13,4±1,12	25,6±1,25	52,6±1,66	5,6±0,51
	240	0,8±0,37	1,2±0,49	1,2±0,37	15,0±1,67	27,8±1,16	49,4±1,54	4,6±0,40
	360	1,2±0,37	1,2±0,20	0,8±0,20	14,0±1,58	24,4±1,17	53,6±1,63	4,8±0,49
	450	1,6±0,51	1,6±0,24	0,6±0,40	9,8±1,59	24,6±1,08	56,4±1,57	5,4±0,51
	540	2,0±0,45	2,0±0,32	0,0±0,00	5,6±1,12	24,6±1,33	59,8±1,24	6,0±0,55
1 experimental	1	0,0±0,00	0,2±0,20	1,4±0,51	22,2±0,86	38,2±1,98	36,4±1,96	1,6±0,75
	210	0,2±0,20	2,0±0,32	0,2±0,20	7,4±1,57	27,0±1,22	60,0±1,30**	3,2±0,37
	240	0,4±0,24	1,8±0,37	0,4±0,40	7,6±1,47	29,4±1,21	57,4±1,63**	3,0±0,55
	360	0,6±0,40	2,2±0,37*	0,2±0,20	6,8±1,56	27,8±0,86*	59,0±1,30*	3,4±0,51
	450	0,8±0,37	2,8±0,37*	0,2±0,20	2,4±1,25	28,2±1,11*	61,2±1,16*	4,4±0,81
	540	0,8±0,37	3,4±0,40*	0,0±0,00	1,4±0,51	28,4±0,40*	61,8±1,02	4,2±0,37
2 experimental	1	0,2±0,20	0,4±0,24	1,6±0,24	23,2±0,80	36,6±2,06	36,0±1,76	2,0±0,32
	210	0,4±0,40	2,2±0,37	0,4±0,24	6,6±1,60	27,4±1,12	59,4±1,03**	3,6±0,75
	240	0,2±0,20	2,4±0,24	0,6±0,24	8,0±1,48	30,4±1,17	55,6±1,57*	2,8±0,58
	360	0,4±0,40	2,80±0,49*	0,2±0,20	4,6±1,69	27,6±1,03	60,8±1,53*	3,6±0,40
	450	0,6±0,24	3,2±0,49*	0,0±0,00	2,6±0,75	27,8±0,80*	61,0±0,84*	4,8±0,37
	540	1,0±0,32	3,6±0,51*	0,0±0,00	1,8±0,58	28,4±0,68*	60,0±0,84	5,2±0,37

- increased protein metabolism, mainly due to the synthesis of albumin and γ-globulin fractions (table 4);

Table 4. Dynamics of total protein and protein fractions in bovine serum.

Group of animals	Age, day	Total protein, g/l	Protein fractions, g/l			
			albumins	α-globulins	β-globulins	γ-globulins
Control	1	62,8±1,30	26,5±0,81	13,5±1,22	13,2±0,65	9,6±0,87
	210	61,0±1,11	27,0±0,96	6,9±0,73	6,6±0,91	20,4±0,75
	240	61,4±1,15	28,1±0,97	6,9±0,63	7,1±0,89	19,3±0,71
	360	61,8±1,19	28,7±0,98	6,1±0,86	6,5±0,94	20,5±0,62
	450	61,7±1,17	29,7±1,03	5,5±1,14	5,1±0,91	21,3±0,64
	540	62,2±1,24	30,0±0,91	5,6±1,08	5,6±1,21	21,0±0,71
1 experimental	1	62,4±1,33	26,0±0,94	13,4±1,20	13,6±0,99	9,4±0,71
	210	65,8±1,12*	30,8±0,90*	5,3±0,82	5,4±0,61	24,2±0,71**
	240	65,5±1,30*	31,6±0,99*	5,2±0,36	5,2±0,63	23,5±0,65**
	360	66,3±1,27*	32,1±0,93*	5,4±0,91	5,1±0,75	23,8±0,89*
	450	66,5±1,25*	32,9±0,88*	4,7±0,86	4,5±0,70	24,3±0,90*
	540	66,7±1,10*	33,2±0,82*	5,5±0,92	4,7±0,42	23,3±0,60*
2 experimental	1	63,2±1,43	27,4±0,87	12,7±0,82	13,1±0,95	10,0±0,73
	210	64,7±1,16	31,4±0,95*	5,2±1,34	4,4±1,19	23,7±0,70*
	240	65,0±1,01*	32,6±1,01*	4,7±1,24	4,6±1,23	23,1±0,76**
	360	65,6±1,16	33,0±1,02*	4,9±1,35	5,1±1,47	22,6±0,64*
	450	65,9±1,12*	33,6±1,03*	3,9±1,27	3,7±1,14	24,7±0,59**
	540	66,2±1,04*	33,6±0,77*	4,6±0,94	4,2±0,85	23,9±0,55*

- activated cellular and humoral factors of nonspecific resistance of the organism.

The dynamics of the bulls body nonspecific resistance cellular and humoral links main indicators is shown in Figure 1 – 4.

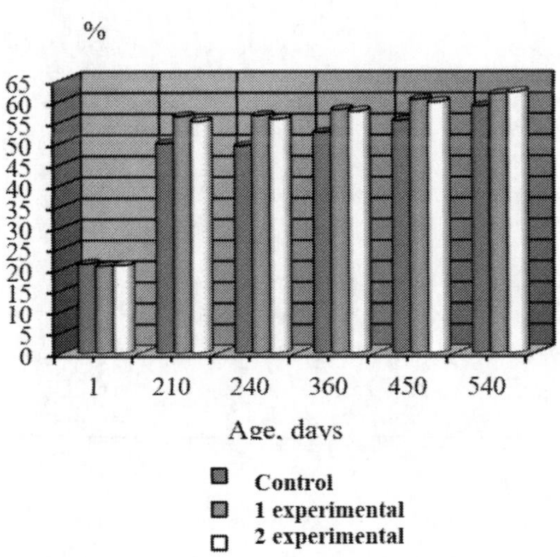

Figure 1. Dynamics of phagocytic activity

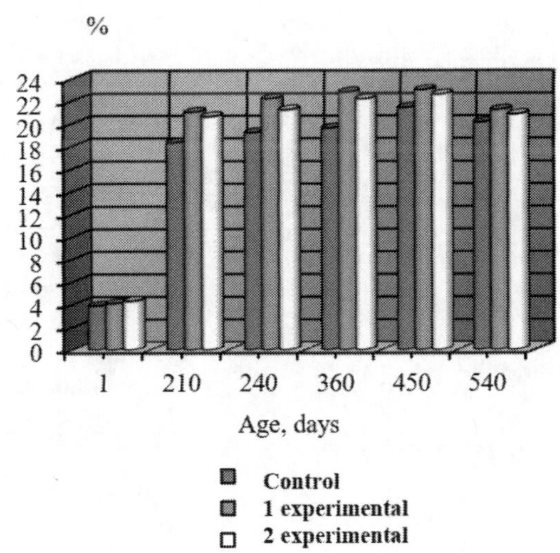

Figure 2. Dynamics of lysozyme activity.

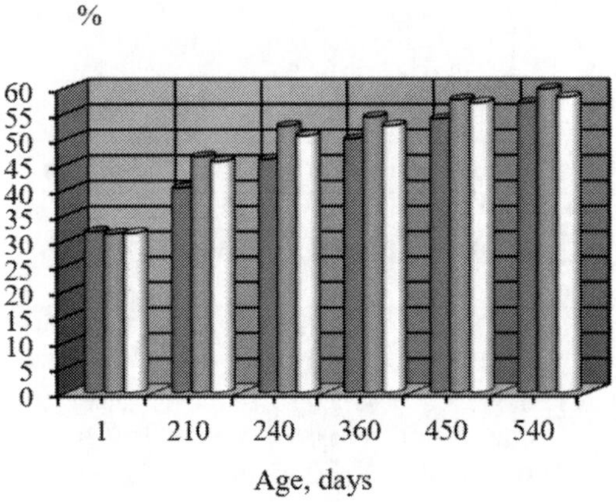

Figure 3. Dynamics of bactericidal activity.

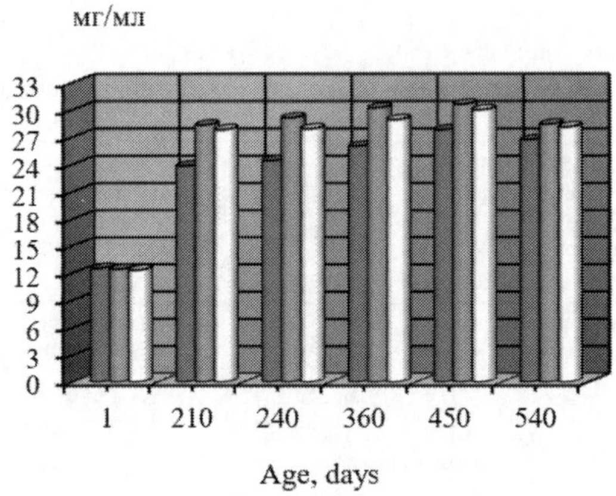

Figure 4. Dynamics of immunoglobulinsin blood serum.

The dynamics of live weight, average daily gain and growth rate of experimental groups bulls are presented in table 5.

Table 5. Dynamics of growth and development of bulls.

Group of animals	Age, days	Live weight, kg	Average daily gain, g	Coefficient of growth
Control	1	27,8±0,58	–	–
	210	202,8±2,06	972±13,83	7.23
	240	230,6±2,32	927±16,33	8.29
	360	335,6±2,62	875±14,56	12.07
	450	421,0±2,95	949±21,26	15.14
	540	497,2±3,37	847±18,16	17.88
1 experimental	1	28,0±0,63	–	–
	210	209,4±2,11	1008±13,45	7.48
	240	238,4±2,14*	967±18,26	8.51
	360	346,0±2,88*	897±16,77*	12.36
	450	433,8±3,12*	976±22,88	15.49
	540	511,4±3,44*	862±16,67	18.26
2 experimental	1	27,4±0,75	–	–
	210	212,0±2,61*	1026±13,56**	7.73
	240	242,6±3,06*	1020±20,00*	8.85
	360	350,4±3,37**	898±10,00**	12.79
	450	440,6±3,61**	1002±27,76	16.08
	540	519,4±3,87**	876±21,49	18.96

* P≤0,05, ** P≤0,01.

It was established that by the end of the growing period, the 210-day bulls of the 1st and 2nd experimental groups exceeded the live weight of control ones by 6.6 and 9.2 kg, respectively, of growing by 10.4 and 14, 8 kg and fattening - by 14.2 and 22.2 kg (P <0.05-0.01).

A similar pattern occurred in the nature of changes in average daily growth and growth rate (the ratio of body weight in certain age periods to that at birth) of the compared groups animals. Consequently, intramuscular administration of biological products to bulls stimulates their growth, due to increased protein metabolism (mainly albumin fraction of protein – plastic material) on the background of activation of cellular and humoral factors of nonspecific resistance of the body.

The research results of the slaughter qualities of calves are shown in the Table 6.

Table 6. Indicators of control slaughter of bulls.

Indicator	Group of animals		
	Control	1 experimental	2 experimental
Live weight for the removal of fattening, kg	497,2±3,37	511,4±3,44*	519,4±3,87**
Pre-slaughter live weight, kg	483,4±3,56	498,8±3,95*	505,4±4,13**
The steam mass carcass, kg	269,8±1,93	279,4±2,16*	289,4±2,38***
Carcass output, %	55.8	56.0	57.3
Weight of internal fat, kg	6,5±0,25	7,1±0,33	7,0±0,25
Internal fat yield, %	1.34	1.42	1.38
Slaughter weight, kg	287,1±2,06	301,2±2,60**	308,8±2,66***
Slaughter yield, %	59.4	60.4	61.1

* P≤0,05, ** P≤0,01, *** P≤0,001.

The results of the control slaughter showed that the bulls of the 1st (498.8 ± 3.95 kg) and the 2nd (505.4 ± 4.13 kg) experimental groups were superior to the ones of the control group (483.4 ± 3.56 kg) by pre-slaughter body weight by 15.4 kg or by 3.2% (P <0.05) and by 22.0 kg, i.e. by 4.5% (P <0.01). It was established that the mass of the steam carcass of bulls grown on the background of intramuscular injection of PS-6, with further rearing and fattening in open areas under sheds, exceeded the same figures in the control by 9.6 kg or 3.5% (P <0.05), and with the use of a biological product Prevention-N-E - by 19.6 kg, i.e. by 7.3% (P <0.001). At the same time, the bulls of the 1st and 2nd experimental groups outperformed their ones in control by 1.0 and 1.7%.

The morphological composition of bull carcasses is presented in Table 7.

Table 7. Morphological composition of carcasses of bulls.

Indicator	Group of animals		
	Control	1 experimental	2 experimental
Chilled carcass weight, kg	260,2±2,27	269,8±2,35*	278,6±3,23**
The mass of the flesh, kg	206,8±2,35	214,8±2,33*	222,4±3,11**
The yield of flesh, %	79.48	79.61	79.82
Including weight of fat by weight of flesh, kg	15,4±0,58	16,3±0,31	16,1±0,29
Fat yield, %	5.92	6.04	5.78
Tendon weight, kg	8,9±0,17	9,2±0,12	9,3±0,17
The yield of tendons, %	3.42	3.41	3.34
Bone weight, kg	44,5±0,75	45,8±0,66	46,9±0,74
Bone yield, %	17.10	16.97	16.83
Flesh yield per 100 kg of pre-slaughter live weight	42,78±0,12	43,06±0,24	44,04±0,29**
Fleshing index	4,65±0,15	4,69±0,11	4,74±0,08

* P≤0,05, ** P≤0,01.

The process of carcasses cooling inevitably leads to a loss of mass (drying), due to evaporation of moisture from 3.5% to 3.8%. Therefore, the mass of cooled carcasses of bulls of the control, 1st and 2nd experimental groups turned out to be lower, respectively, by 9.6 kg, 9.6 and 10.8 kg, rather than paired carcasses.

As a result of the carcasses deboning, it was established that, by the weight of the flesh of carcasses, the bulls of the 1st and 2nd experimental groups exceeded the ones of the control group by 8.0 and 15.6 kg (P <0.05-0.01), and the fat yield by 0.9 and 0.7 kg (P> 0.05). The yield of tendons in the section of the experimental groups of bulls revealed no definite pattern, and it varied in an insignificant range - from 8.9 ± 0.17 to 9.3 ± 0.17 kg. The absolute yield of bones from animal carcasses of the 1st and 2nd experimental groups was higher by 1.3 and 2.4 kg (P>0.05), respectively, than in the control group. However, bone yield, expressed as a percentage relative to the mass of the carcass, in the bulls of the experimental groups, on the contrary, was lower by 0.13 and 0.27 %, respectively. The flesh yield per 100 kg of pre-slaughter weight of bulls in the 1st experimental group was 43.06±0.24 kg, i.e. it was more by 0.28 kg (P>0.05), and in the 2nd experimental group – 44.04±0.29 kg, i.e. it was more by 1.26 kg (P<0.01) than in the control group – 42.78±0.12 kg. The meat index (the ratio between the mass of carcass flesh and bone mass, that is, the amount of flesh per 1 kg of bones) of the bulls of the 2 experimental group was 4.74, which is more than in the control and 1st experimental groups by 0.09 and 0.05.

Thus, we can conclude that the bull carcasses of all groups were distinguished by the good development of muscle and fat tissues. With an increase in the weight of the carcasses of experimental animals, the specific gravity of the flesh increased, while the bones, on the contrary, decreased. This indicates the well developed meat qualities of animals of all groups. The greatest amount of flesh differed in bull calves of the experimental groups, especially the 2nd one.

When assessing the meat productivity of the animal, not only the ratio of the carcass tissue is imporant, but also the ratio of the anatomical parts from which different varieties of meat.

Weight and output of cuts from carcasses of bulls are shown in Table 8.

It was established that bull carcasses of the 1st and 2nd experimental groups had an advantage in terms of the most valuable cuts mass: spin-breast - by 8.0 and 11.5 kg, lumbar - by 1.8 and 3.2 kg, hip - by 3.2 and 7.4 kg (P <0.01-0.001), respectively, than in the control. The yield of these cuts from the carcasses of bulls of the 1st and 2nd experimental groups was higher by 1.9 and 2.2%, by 0.2 and 0.3%, by 0.2 and 0.8%, respectively, than in control.

Table 8. Weight and yield of cuts from carcasses of bulls.

Indicator	Group of animals		
	Control	1 experimental	2 experimental
Carcass weight, kg	260,2±2,27	269,8±2,35*	278,6±3,23**
including the cut:			
cervical, kg	29,1±0,15	27,2±0,23	26,5±0,24
%	11.2	10.1	9.5
humeroscapular, kg	48,7±0,21	47,2±0,12	47,6±0,22
%	18.7	17.5	17.1
spinal chest, kg	74,9±0,72	82,9±0,54***	86,4±0,62***
%	28.8	30.7	31.0
lumbar, kg	33,3±0,40	35,1±0,37*	36,5±0,60**
%	12.8	13.0	13.1
hip, kg	74,2±0,59	77,4±0,62**	81,6±0,71***
%	28.5	28.7	29.3

In order to determine the varietal composition of carcasses, the classification was used, according to which the beef was divided into 4 grades: the highest, I, II and fat one. The highest grade included pieces of beef without visible tendon and fat particles, first grade - pieces containing no more than 6% of connective tissue, second grade - beef containing up to 20% fat and connective tissue, fatty beef includes pieces of meat with subcutaneous fat and marbling.

The grade of flesh of carcasses of experimental bulls is presented in Table 9.

Table 9. Grade of carcass bull meat.

Indicator	Group of animals		
	Control	1 experimental	2 experimental
The flesh mass, kg	206,8±2,35	214,8±2,33*	222,4±3,11**
The mass of the highest grade flesh,	49,0±0,77	52,2±0,63*	54,7±0,65***
kg	23.7	24.3	24.6
The yield of the highest grade flesh, %	107,3±1,40	112,1±1,17*	117,4±1,53**
The mass of the first grade flesh, kg	51.9	52.2	52.8
The yield of the first grade flesh, %	50,5±0,53	50,5±0,59	50,3±0,60
The mass of the second grade flesh, kg	24.4	23.5	22.6
Second grade flesh yield, %			

* $P \leq 0,05$, ** $P \leq 0,01$, *** $P \leq 0,001$.

It follows from the table that the highest content of the highest grade flesh was characterized by bulls carcasses of the 1st (52.2 ± 0.63 kg) and 2nd (54.7 ± 0.65 kg) experimental groups, respectively, by 3.2 and 5, 7 kg compared with the control (49.0 ± 0.77 kg; P <0.05-0.001). The yield of high grade flesh was higher in animals of the experimental groups by 0.6 and 0.9%.

The results of trimming cuts are given in table 10

It follows from the table that the cervical cut along the seventh vertebra inclusive mainly consists of the flesh of the first and second grades. It was established that the bulls of the 1st and 2nd experimental groups were lower than the control ones in mass of first-grade flesh by 0.7 and 0.8 kg, of the second grade - by 0.7 and 0.9 kg, respectively (P> 0.05).

As a result of the varietal separation of the shoulder-cut cuts of bull-calves carcasses, it was found that the intergroup differences were insignificant (P> 0.05). The highest mass of top grade flesh was in the shoulder cuts of the bulls of the 2nd experimental group and amounted to 5.5 ± 0.12 kg, i.e. it turned out to be higher than that in animals of the control and 1st experimental groups by 0.1 and 0.4 kg, respectively.

The highest content of the highest grade flesh was characterized by spin-breast cuts of bull-calves of the 1st (9.5 ± 0.17 kg) and 2nd (9.5 ± 0.14 kg) experimental groups, which is 1.0 kg higher than in control (8.5 ± 0.21 kg; P <0.01). At the same time, in the experimental groups, the yield of the highest grade flesh was more by 0.1 and 0.3%. The first-grade flesh content in spin-breast cuts of bull-calves of the 1st and 2nd experimental groups was 3.2 and 2.5 kg more, the flesh yield was 0.3 and 0.1% compared to the control (P <0.001).

The highest-grade flesh was the highest in lumbar cuts of bull-calf carcasses of the 2nd experimental group - 4.7 ± 0.19 kg, which is 0.4 and 0.2 kg more, respectively, than in the control and 1st experimental groups. At the same time, the highest grade flesh exceeded that of the 1st experimental group bulls of the control and 2nd experimental groups ones, respectively, by 0.3 and 0.1%. In terms the first-grade flesh content, bulls of the 1st and 2nd experimental groups were superior to their ones in the control group by 0.8 (P> 0.05) and 1.6 kg (P <0.05). The yield of the first grade flesh in the lumbar cuts of the experimental groups bulls was almost the same and amounted to 65.4%, 65.4% and 65.3% in the control, 1st and 2nd experimental groups, respectively.

The amount of the highest grade flesh in the hip bullhead cuts of the 1st (33.1 ± 0.49 kg) and 2nd (35.0 ± 0.53 kg) experimental groups was 2.3 and 4.2 kg more (P <0.01-0.001) than in the control (30.8 ± 0.38 kg). At the same time, the output of the highest grade flesh in the control group was 49.9%, in the 1st test group - 50.6% and in the 2nd test group - 48.7%. In terms of flesh content of the first grade, bulls of the 1st (27.5 ± 0.46 kg) and 2nd (32.0 ± 0.58 kg) experimental groups also exceeded their ones in control (25.0 ± 0.29 kg) by 2.5 and 7.0 kg (P <0.01-0.001). And the yield of first grade flesh was higher in the experimental groups by 1.5 and 4.0%, rather than in the control.

Table 10. Grade flesh of bulls carcass cuts.

Indicator	Group of animals		
	Control	1 experimental	2 experimental
	Collar butt		
The flesh mass, kg	24,6±0,37	23,2±0,42	22,9±0,31
The mass of the first grade flesh, kg	14,7±0,29	14,0±0,36	13,9±0,34
The mass of the second grade flesh, kg	9,9±0,22	9,2±0,25	9,0±0,35
	Scapulohumeral butt		
The flesh mass, kg	36,4±0,38	34,6±0,29	35,8±0,29
The mass of the highest grade flesh, kg	5,4±0,11	5,1±0,05	5,5±0,12
The mass of the first grade flesh, kg	22,8±0,34	21,8±0,33	22,6±0,42
The mass of the second grade flesh, kg	8,2±0,16	7,7±0,09	7,7±0,12
	Spinal butt		
The flesh mass, kg	55,6±1,03	62,0±0,80**	60,9±0,46**
The mass of the highest grade flesh, kg	8,5±0,21	9,5±0,17**	9,5±0,14**
The mass of the first grade flesh, kg	26,2±0,22	29,4±0,23***	28,7±0,27***
The mass of the second grade flesh, kg	20,9±0,20	23,1±0,48**	22,7±0,28***
	Lumbar butt		
The flesh mass, kg	28,5±0,29	29,6±0,38	30,9±0,40**
The mass of the highest grade flesh, kg	4,3±0,12	4,5±0,17	4,7±0,19
The mass of the first grade flesh, kg	18,6±0,37	19,4±0,26	20,2±0,39*
The mass of the second grade flesh, kg	5,6±0,26	5,7±0,24	6,0±0,23
	Hip butt		
The flesh mass, kg	61,7±0,52	65,4±0,70***	71,9±0,92***
The mass of the highest grade flesh, kg	30,8±0,38	33,1±0,49**	35,0±0,53***
The mass of the first grade flesh, kg	25,0±0,29	27,5±0,46**	32,0±0,58***
The mass of the second grade flesh, kg	5,9±0,15	4,8±0,18	4,9±0,20

Thus, the highest content of the highest grade flesh was characterized by carcasses and cuts, especially lumbar and hip, bulls of the 1st and 2nd experimental groups grown on the background of intramuscular injection of biologics in conditions of adaptation to the natural temperature and humidity regime of atmospheric air.

The results of assessing the quality of beef with organoleptic, biochemical and spectrometric indicators are given in Table 11.

In terms of organoleptic, biochemical and spectrometric indicators, beef complied with the requirements of the Technical Regulations of the Customs Union "On Food Safety" of the TP TC 021/2011 TP and Technical Regulations of the Customs Union "On the Safety of Meat and Meat Products" TP TC 034/2013 TR.

Table 11. Evaluation of the quality of beef.

Indicator	Group of animals		
	Control	1 experimental	2 experimental
Organoleptic:			
the appearance and color of the surface	samples of meat are covered with a dried crust of pale pink color		
muscles on the cut	slightly damp, do not leave a wet spot on the filter paper; the color is light red		
consistency	the meat is dense, elastic on the cut; formed by pressing the finger hole is quickly aligned		
smell	specific to fresh beef		
surface fat	it has a yellowish color; the consistency is solid, when pressed crumbles		
the condition of the tendons	elastic, dense, the surface joints smooth, brilliant		
transparency and flavor of the broth	transparent, fragrant, on the surface of the broth large drops of fat		
Biochemical:			
pH (5,6 – 6,2)	5,78±0,03	5,76±0,02	5,74±0,02
amino-ammonia nitrogen, mg (in 10 ml of extract from fresh meat no more than 1,26)	1,21±0,01	1,19±0,01	1,20±0,01
shaped reaction		negative	
reaction to the peroxidase		positive	
reaction with copper sulfate		negative	
Spectrometric, mg/kg			
lead (not more than 0.5)	0,06±0,01	0,07±0,01	0,05±0,01
cadmium (not more than 0.05)	not detected	similar to control	similar to control
arsenic (not more than 0,1)	not detected	similar to control	similar to control
copper (not more than 5,0)	0,74±0,03	0,77±0,04	0,76±0,02
zinc (not more than 70)	25,4±0,24	26,0±0,16	26,4±0,21
mercury (not more than 0.03)	not detected	similar to control	similar to control

4. Conclusions

The results of studies on the use of biological products to enhance the protective and adaptive functions of the body of bulls to the conditions of adaptive technologies for growing, additional growing and fattening, and realizing the biological potential of the organism indicated that under the influence of PS-6 and Prevention-N-E, the adaptive plasticity of the body to low temperatures of the environment, hematopoiesis, cellular and humoral factors of nonspecific resistance were activated, diseases of the respiratory and digestive organs were reduced, and growth accelerated, and meat productivity also increased. It should be noted that Prevention-N-E had a more pronounced stimulating effect on nonspecific resistance of the body, shows preventive effectiveness in these diseases of non-infectious etiology, improves the fattening and slaughter qualities of bulls.

The selective mobilization of the morphological and biochemical profiles of blood, cellular and humoral factors of bulls body nonspecific resistance in the context of adaptive content technology in open areas aimed at the adaptation of the organism was revealed.

It had been established that a double intramuscular injection of biopreparations PS-6 and Prevention-N-E to bulls on the 2nd ... 3rd and 7th ... 9th day of life at a dose of 3 ml did not affect the clinical and physiological state of the body, but increased the effectiveness of preventive measures. In animals, the incidence of the organs of the digestive and respiratory organs decreased by 1.4 and 2.3 times, the recovery time was reduced by 3.36 and 4.88 days, and the Mellenberg coefficient decreased by 2.6 and 7.3 times, respectively, compared with the control ($P < 0.05$).

Against the background of an increase in non-specific resistance of the body of bulls under the influence of biopreparations, the activation of their growth and development has been established. By the end of the nursing period, 210-day bulls of the 1st and 2nd experimental groups outnumbered, by live weight, the control ones by 6.6 and 9.2 kg, and the rearing (360 days) by 10.4 and 14.8 kg and fattening (540 days) - by 14.2 and 22.2 kg ($P < 0.05$-0.01).

It was found that bull calves of the 1st and 2nd experimental groups had an advantage in terms of the mass of the most valuable cuts: spin-breast - by 8.0 and 11.5 kg, lumbar - by 1.8 and 3.2 kg, hip - by 3.2 and 7.4 kg ($P < 0.01$-0.001), respectively, than in the control.

The highest content of premium meat was characterized by bulls carcasses of the 1st (52.2 ± 0.63 kg) and 2nd (54.7 ± 0.65 kg) experimental groups, respectively, by 3.2 and 5.7 kg compared to control (49.0 ± 0.77 kg; $P < 0.05$-0.001).

References

[1] Semenov V G, Kosyaev N I, Lavrentyev A Yu, Larionov G A, Evdokimov N V, Toboyev G M and Nikitin D A 2018 *Inter. J. Engin. Tech.* **7** pp 648–55
[2] Amerkhanov H A, Kochetkov A and Sharkaev V 2008 *Dairy and beef cattle* (Moscow) **1** pp 2–4
[3] Babushkin V A, Shemetyuk S A, Avdalyan Ya V, Zizyukov I V and Shchegolkov N F 2018 *Bulletin of Michurinsky State Agrarian University* (Michurinsk) **1** pp 62-4
[4] Japaridze T G 2008 *Chief zootechnician* (Moscow) **8** pp 39–41
[5] Dunin I M, Shichkin G I and Kochetkov A A 2014 *Dairy and Beef Cattle* (Moscow) **1** pp 2–5
[6] Dyuldina A V 2016 *Dairy and Beef Cattle* (Balashikha) **8** pp 31–3
[7] Legoshin G P 2003 *Zootechnics* (Moscow) **3** pp 24–6
[8] Gizatullin R S and Sedykh T A 2016 *Monograph* (Saarbrücken) **119**
[9] Goldobin M I, Grigoriev A G and Azatov R M 1994 *Husbandry* (Moscow) **11** pp 26–7
[10] Nikitin D A and Semenov V G 2013 *Russian J. "Problems of veterinary sanitation, hygiene and ecology"* (Moscow) **1(9)** pp 59–63
[11] Topuria G M, Topuria L Yu, Donnik I M and Shkuratova I A 2017 *Agrarian Bullet. Urals* (Ekaterinburg) **9 (163)** pp 67–70
[12] Shukanov A A 2000 *Veterinary* (Moscow) **10** pp 48–52
[13] Vasilyev V A, Semenov V G and Mudarisov R M 2014 *European Conference on Innovations in Technical and Natural Sciences. Proceedings of the 1st International scientific conference «East West» Association for Advanced Studies and Higher Education* (GmbH. Vienna, Austria) pp 176–81
[14] Gorlov I F, Slozhenkina M I, Zakurdaeva A A, Randelin A V, Mosolova D A and Miroshnik A S 2017 *Recommendations* (Volgograd) p 19
[15] Yakimov A V, Mudarisov F J and Salakhov V V 2016 *Bullet. Ulyanovsk State Agricul. Acad.* (Ulyanovsk) **3 (35)** pp 165–9

IOP Conf. Series: Earth and Environmental Science **346** (2019) 012059 doi:10.1088/1755-1315/346/1/012059

Case analysis of artificial snowfall enhancement by aircraft in Winter Olympic Games Area

W X Yang[1,2,4], Z X Hu[3], X B Dong[1,2], Y Yang[1,2] and L W Zhao[1,2]

[1]Hebei Artificial Weather Office, Shijiazhuang 050021, China
[2]Hebei Key Laboratory of Meteorology and Ecological Environment, Shijiazhuang 050021, China
[3]Institute of Atmospheric Physics, Chinese Academy of Sciences, Beijing 100082, China

E-mail: ywx418@163.com

Abstract. It was dry in Hebei Province in winter of 2017, with less snow. But on November 25, 2017, snowfall occurred in Zhangjiakou and Chengde areas of Hebei Province by the influence of low trough-cold front weather process. Hebei Artificial Weather Office used cloud physics model products issued by National Artificial Weather Office for timely and accurate early warning, and successfully implemented artificial snowfall enhancement operation by an aircraft. Aircraft was King of the Air 350, and carried series of cloud physics and environment detection instruments produced by SPEC Company. In this paper, the case was analyzed, and technical conditions for artificial snowfall enhancement were summarized from water content and vertical velocity in clouds by combining several rainfall processes from winter of 2017 to spring of 2018.

1. Introduction

In the 1930s, Bergeron-Findeison proposed Begeron-Fendeison cold cloud precipitation theory, and artificial rainfall enhancement test on this basis was carried out in the world. Since entered into the 21st century, improvement of various detection methods promoted overall development of cloud physics in the world. In 2001, Hu *et al* proposed that attention should be paid to the role of supercooled water vapor. They pointed out that artificial ice crystal made supercooled cloud water transform into rainfall via Begeron process and a part of the supersaturated water vapor on the ice surface converted into precipitation [1]. With the improvement of assimilation technology of observation data in recent years, observation and research on mechanism of artificial rainfall enhancement developed rapidly [2-7], but there were fewer studies on snowfall process. Since formation mechanism of rainfall in snowfall process is unique, it is necessary to explore and study snowfall enhancement technology by aircraft.

2. Weather system

It was low trough-cold front weather process, and high-altitude and ground weather situation was shown as figure 1. It was clear that weather system mainly affected north Hebei Province.

Content from this work may be used under the terms of the Creative Commons Attribution 3.0 licence. Any further distribution of this work must maintain attribution to the author(s) and the title of the work, journal citation and DOI.
Published under licence by IOP Publishing Ltd

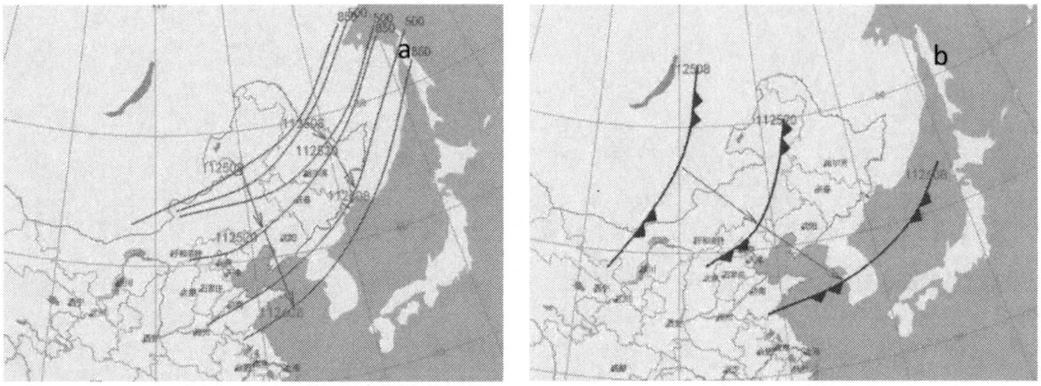

Figure 1. High-altitude and ground weather situation from 08:00 on November 25, 2017 to 08:00 on 26th.

3. Methodology used: Forecast situation of numerical model

At 17:00 on November 24, 2015, forecast result of horizontal potential zone in the process by CPEFS mode of National Artificial Weather Office was shown as figure 2. Whole weather process moved from northwest to southeast, and water condensate cloud band affecting Hebei Province contained multiple large-value centers, and the maximum could reach 3 mm. Main influence period was 08:00-22:00 on May 25, 2018, and it was consistent with movement situation of weather system.

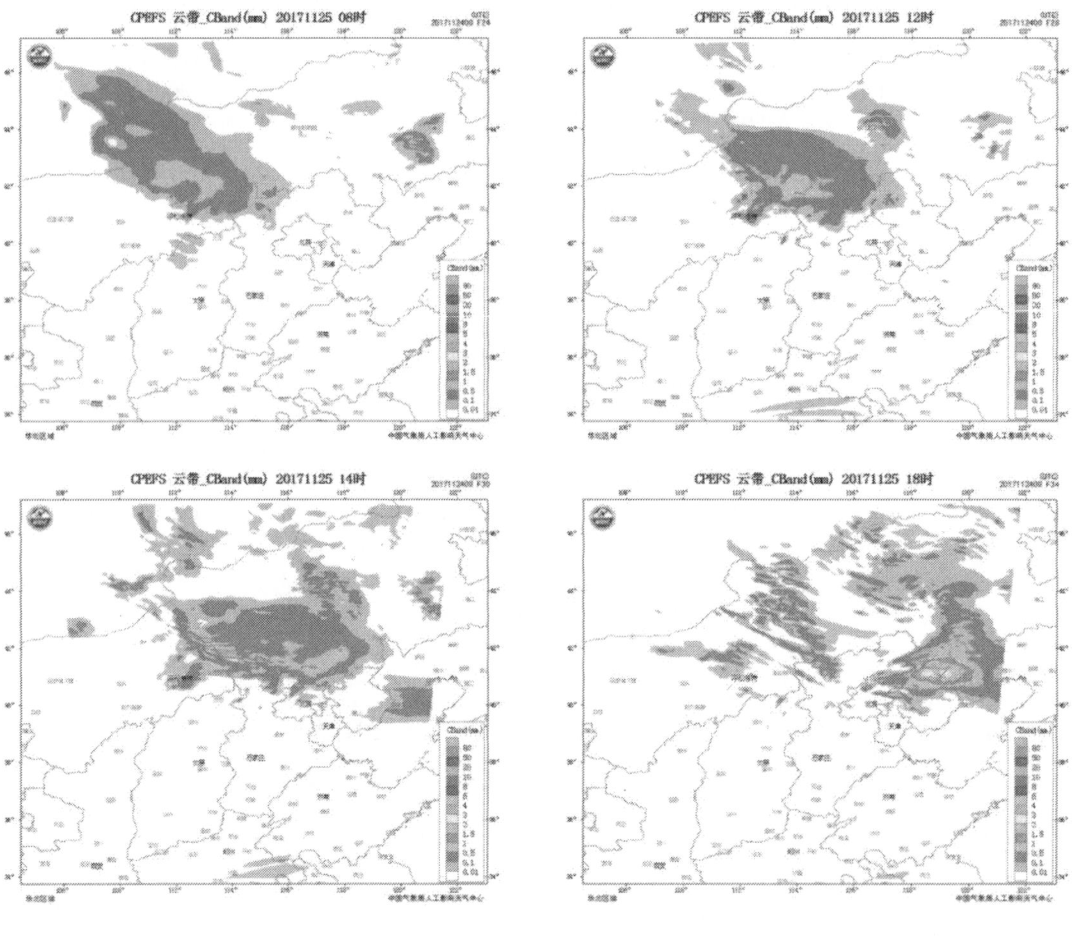

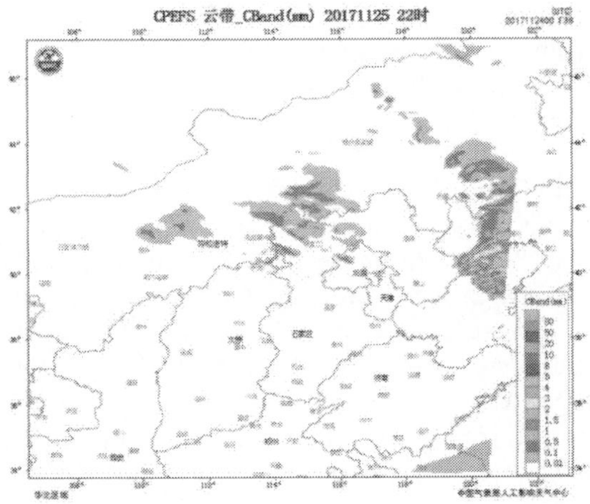

Figure 2. Horizontal movement situation of stratiform cloud belt during 08:00-22:00 on November 25, 2017 (unit: mm, 云带: cloud band).

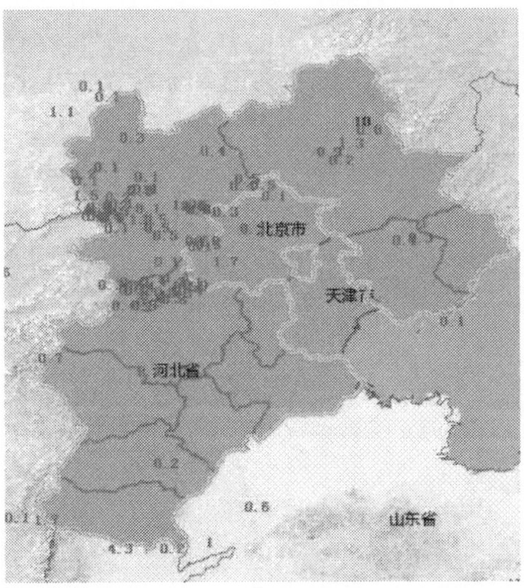

(a)

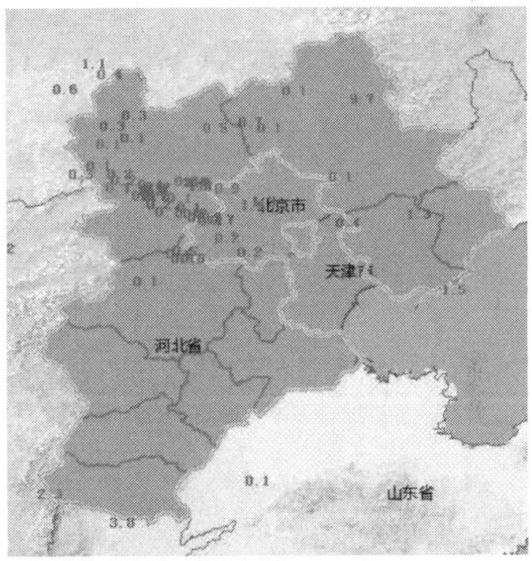

(b)

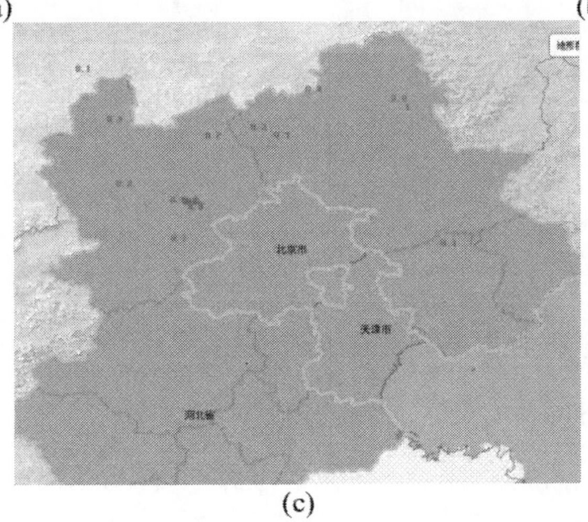

(c)

Figure 3. Observation data of surface precipitation at national and regional stations in the weather process. (a) 08:00-14:00 on 25th; (b) 14:00-20:00 on 25th and (c) 20:00-24:00 on 25th.

Figure 3 showed observation data of surface precipitation at national and regional stations in the weather process. Compared with figure 2, rainfall range and period forecasted by CPEFS model were consistent with actual situation. Seen from figure 4, the content of supercooled water was very rich in the process, and supercooled water in heavy precipitation center of clouds appeared at 750-500 hPa (-20- -35°C), and concentration of natural ice crystal corresponding to supercooled water was higher (≥1/L). The maximum content of supercooled water reached 0.05 g/kg, and horizontal range of large-value zone was obviously larger than other snowfall cases, with better artificial snowfall enhancement potential. Therefore, two snowfall enhancement operations by aircraft on 25th were designed, and a sortie of snowfall enhancement was carried out in the afternoon, and it obtained more complete information.

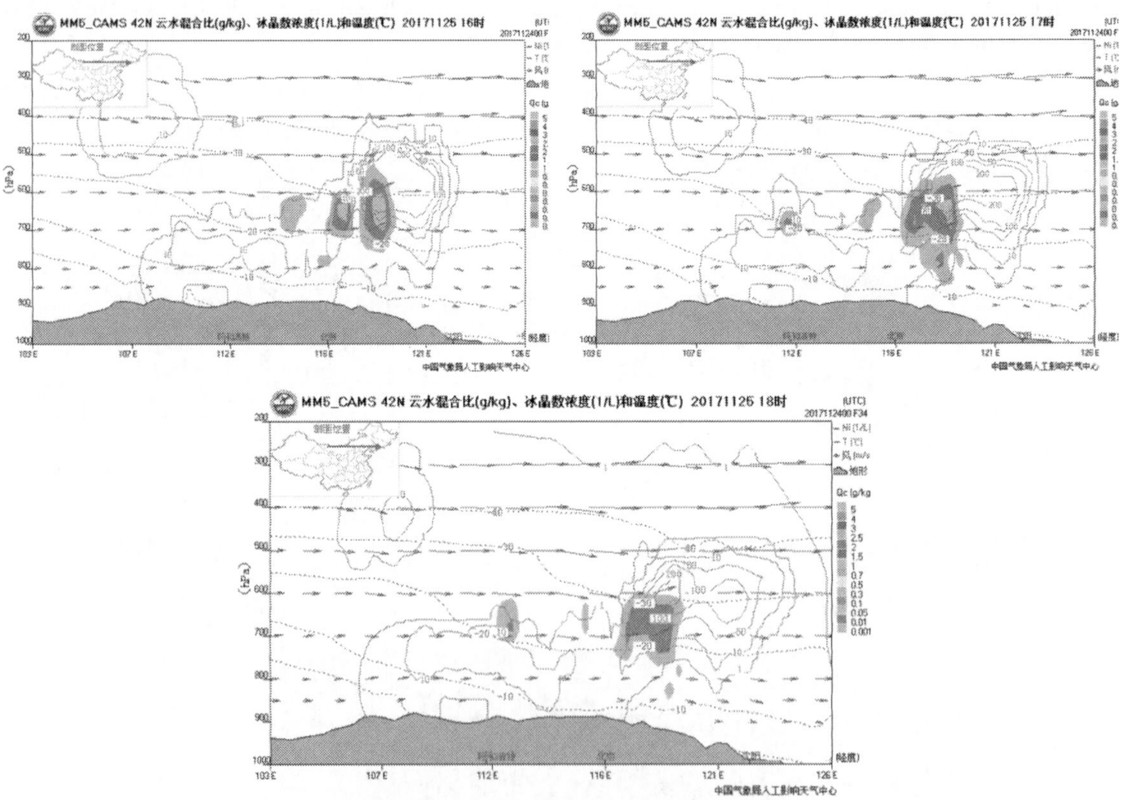

Note: Colouring shadow showed cloud water; red contour showed ice crystal; purple contour showed isotherm.

Figure 4. Vertical distribution of supercooled water at large-value center of moisture content of cloud band forecasted by MM5 model at 16:00-18:00 on November 25, 2017, (云水混合比：cloud water mixing ratio;冰晶数浓度：number concentration of ice crystal;温度:temperature)

4. Detection situation of aircraft

Hebei Artificial Weather Office implemented artificial snowfall enhancement by King of the Air 350 in northwest area of Hebei Province during 16:19:24-18:54:21, and the aircraft carried series of cloud physics detection instruments produced by SPEC Company. Measurement range of cloud physics

IOP Conf. Series: Earth and Environmental Science **346** (2019) 012059 doi:10.1088/1755-1315/346/1/012059

detection instrument was shown as Table 1, and horizontal and vertical routes of aircraft were shown as figure 5. At 16:37, aircraft entered into Laiyuan County of Baoding City. At this time, it was snowing on the ground, and aircraft started to fly horizontally and performed catalytic operation. At 17:00, aircraft entered into Yuxian of Zhangjiakou City. During 17:19-17:43, aircraft carried out vertical detection and spiral descent at 3 800-5 000 m of Chongli, and then flied horizontally and sowed artificial ice crystal in clouds. At 18:03, aircraft returned back to the south of Xuanhua District of Zhangjiakou City. At 18:29, aircraft started to descend after flied horizontally to Quyang of Baoding City, and returned back to Shijiazhuang. Therefore, main period of aircraft in cloud area was during 16:27-18:29. Seen from figure 5(b), the flight mainly consisted of two level flights and one oblique landing process. There was a little vertical detection in Chongli, and relative humidity of cloud area was mainly 70%-90%, while temperature was between -35 and 0°C. It was basically consistent with model forecast result of figure 4.

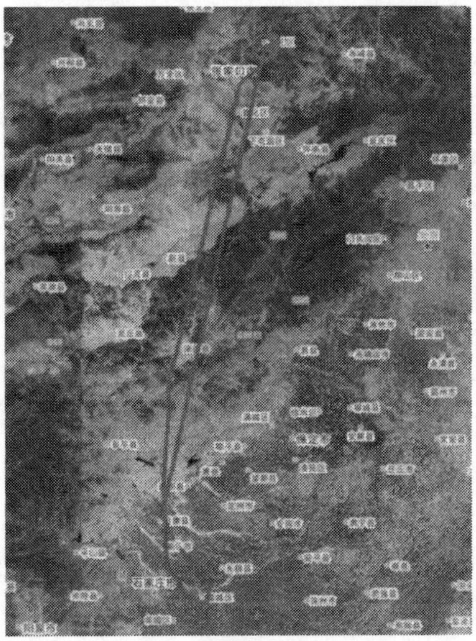

(a)

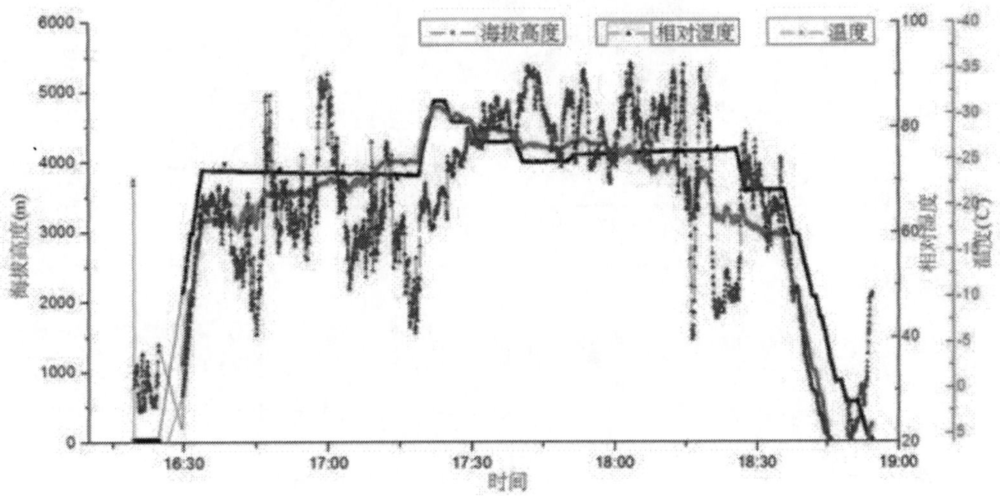

(b)

Figure 5. Horizontal (a) and vertical tracks (b) of aircraft detection during 16:19:24-18:54:21 (海拔高度: altitude; 相对湿度: reletive humidity; 温度: temperature)

Table 1. Cloud physics detection instruments produced by SPEC Company and their measurement ranges.

Name of probe	Measurement range and interval
FCDP	2-50 μm
2D-S	10 μm*128=1.28 mm
HVPS	150 μm*128=19.2 mm
CPI	2.4 μm*1024*1024
CCP	2-50 μm
	15 μm*62=930 μm
PCASP	0.1-3.0 μm
AIRMS	Environmental factors

5. Data analysis at level flight stage

Compared with PMS series instruments, cloud physics detection series instruments produced by SPEC Company are improved greatly, especially increasing detection of vertical airflow in clouds. At present, detection of vertical airflow only could be applied in level flight stage. When the aircraft has the speeds of ascending and descending, the detection values will be seriously distorted. The case had two stages of longer level flight data. As shown in figure 6, both two level flight stages occurred in snowfall region of Zhangjiakou. It was clear that content of supercooled water and large-value zone of relative humidity showed better positive correlation with vertical ascending velocity. The maximum vertical ascending velocity was 6 m/s, and corresponding peak of supercooled water content was 0.6 mg/m^3. In the process, the maximum peak of supercooled water was 0.85 mg/m^3, with rich content.

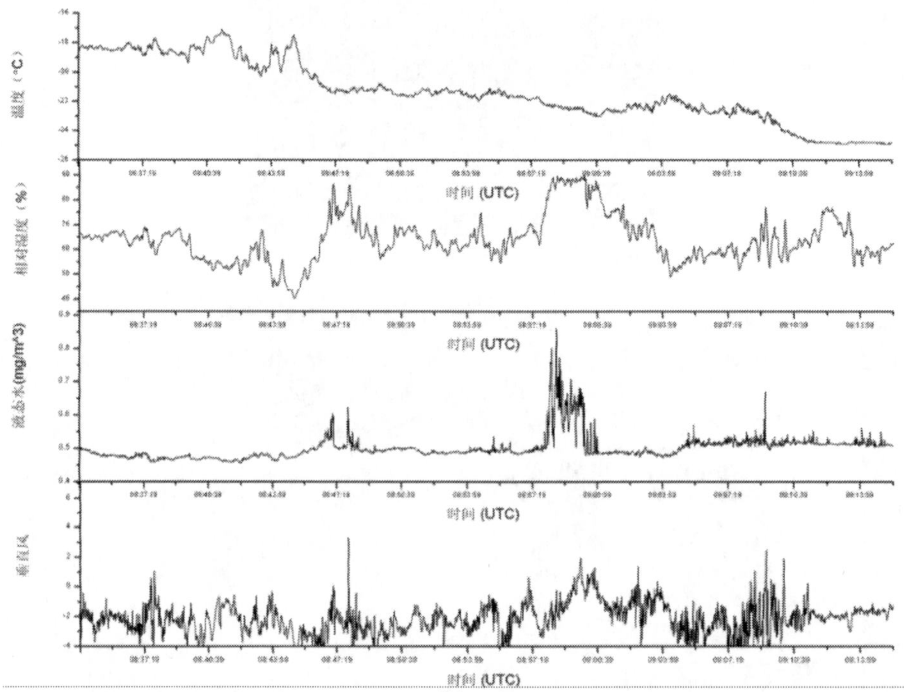

(a)

IOP Conf. Series: Earth and Environmental Science **346** (2019) 012059 doi:10.1088/1755-1315/346/1/012059

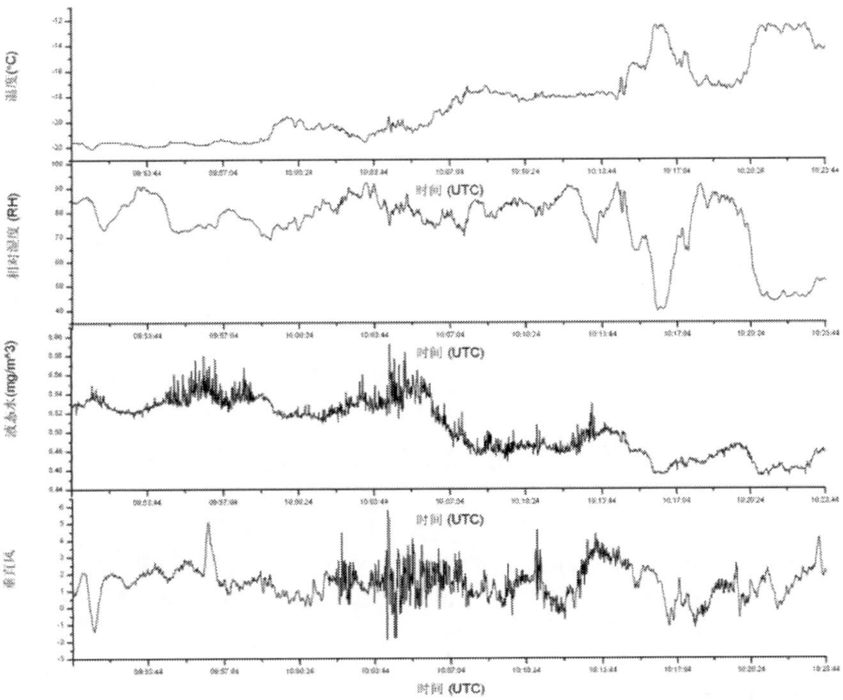

(b)

Figure 6. Analysis on level flight data on November 25, 2017. (a) 16:33:58-17:18:58 and (b) 17:50:23-10:25:27,(温度:temperature; 相对湿度:reletive humidity;液态水:liquid water; 垂直风:vertical wind)

6. Analysis on vertical structure at descent stage

After aircraft flied horizontally to Quyang of Baoding City, it started to descend and returned back to Shijiazhuang at 18:29. Vertical structure at descent stage was shown as figure 7. Peak areas of relative humidity and supercooled water content existed at the height of 3 600 m, and there was a section of inversion zone of -10--15°C. After aircraft flied to Shijiazhuang, temperature started to decline, while relative humidity gradually rose. It needed considering that aircraft has left snowfall region after leaving Zhangjiakou City, and clouds did not have snowfall conditions.

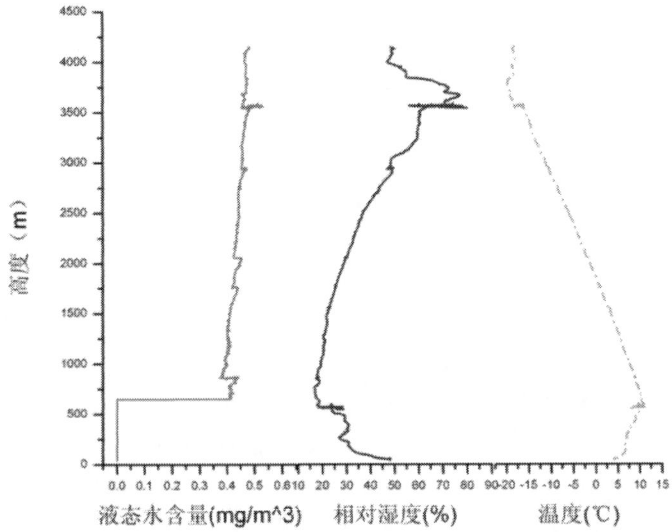

Figure 7. Vertical structure probed by aircraft during 10:25:27-10:54:21 on November 25, 2017(高度: altitude; 液态水: liquid water; 相对湿度: reletive humidity; 温度: temperature)

7. Conclusions and discussion

Compared with artificial rainfall enhancement operation, there were fewer studies on artificial snowfall enhancement. The artificial snowfall enhancement by aircraft was successful, and it was scientific and reasonable from model forecast product, formulation of operation plan and acquisition of aircraft data. A significant feature of the case was that content of supercooled water was higher. Compared with other snowfall enhancement cases in Hebei Province in winter of 2017 and spring of 2018, content of supercooled water was higher, and its area was larger. Since ice and snow crystals are rich in snowfall process, content of supercooled water in clouds is a direct and reliable index for identifying artificial snowfall enhancement potential. When content of supercooled water in clouds is lower and close to zero, it needs considering contribution of supercooled water vapor, declining sowing quantity of artificial ice crystal, and selecting large-value zone of vertical updraft for sowing.

Acknowledgment

Supported by Hebei Science and Technology Plan Project (14227003D, 17227001D); the National Natural Science Foundation of China (41475121).

References

[1] Hu Z J 2001 Discussion on mechanisms, conditions and methods of precipitation enhancement in stratiform clouds *Quarterly J. Appl. Meteorol.* **12(S)** 10-3 (in Chinese)

[2] Yang W X, Niu S J, Wei J G *et al* 2005 Airborne observation for microphysical structure of precipitation system of stratiform cloud in Hebei *Plateau Meteor.* **24** 84-90 (in Chinese)

[3] Zhang D G, Guo X L, Gong D L *et al* 2011 The observational results of the clouds microphysical structure based on the data obtained by 23 sorties between 1989 and 2008 in Shandong Province *Acta Meteorol. Sin.* **69** 195-207 (in Chinese)

[4] Hou T J, Hu Z X and Lei H C 2011 A study of the structure and microphysical processes of a precipitating stratiform cloud in Jilin *Acta Meteorol. Sin.* **69** 508-20 (in Chinese)

[5] Hong Y C 2012 Research progress of stratiform cloud structure and precipitation mechanism and discussion on artificial precipitation problems *Climatic Environ. Res.* **17** 937-50 (in Chinese)

[6] Yang W X, Zhou Y Q, Sun J *et al* 2014 Observational studies of distribution characteristics of supercooled cloud water during a westerly trough process *Acta Meteorol. Sin.* **69** 583-95 (in

Chinese)
[7] Sun J, Yang W X and Zhou Y Q 2015 Numerical simulations of cloud structure and seed ability of a precipitating stratiform in Hebei *Plateau Meteor.* **34** 1699-710 (in Chinese)

Analyzing the possible sources of heavy metals in farmland soil around Weihe coal-fired plant based on multivariate statistical analysis

X L Zhang[1,3], X Ding[2] and X W Lu[2,3]

[1]School of Fine Arts, Shaanxi Normal University, Xi'an 710119, China
[2]School of Geography and Tourism, Shaanxi Normal University, Xi'an 710119, China

E-mail: zhangxiaolan@snnu.edu.cn (X L Zhang); luxinwei@snnu.edu.cn (X W Lu)

Abstract. Contents of heavy metals Cu, Pb, Zn, As, Ni, Mn, V, Co and Cr in farmland soil around Weihe coal-fired plant were surveyed by applying X-ray fluorescence spectrometry to estimate the impact of coal-fired industrial activity on the surrounding soil environment. The possible sources of heavy metals analyzed in the soil were discriminated by utilizing multivariate statistical analysis methods, including principal component analysis, correlation analysis and cluster analysis, and the contribution of each source was apportioned applying principal component analysis subsequently multiple linear regression method. The results revealed that the content ranges of Cu, Pb, Zn, As, Ni, Mn, V, Co, and Cr were 18.1 to 45.9, 20.1 to 64.5, 38.7 to 69.7, 6.0 to 17.9, 23.9 to 39.1, 218.6 to 681.6, 64.5 to 104.2, 18.9 to 61.0, and 94.7 to 128.1 mg/kg with the average of 23.3, 28.3, 48.7, 9.4, 31.2, 564.7, 78.3, 29.1, and 110.7 mg/kg, respectively. Compared with the element background value of Shaanxi soil, the mean contents of Cu, Cr, Co, Pb, Ni and V were larger than their background values, while the mean contents of As, Zn and Mn were less than or close to their background values. Three main sources were discriminated for heavy metals investigated in the surrounding farmland soil of Weihe coal-fired plant. Zn, Cu, Pb, Ni and V derived from traffic and natural source. Co and Cr primarily derived from coal-fired plant discharge. Mn and As primarily came from natural source. The contributions of traffic and nature mixing source, natural source and coal-fired source to heavy metals in farmland soil around Weihe thermal power plant were 21.7%, 49.9% and 28.4%, respectively.

1. Introduction

Heavy metals, a kind of common and toxic pollutant, widely exist in environmental media, such as soil, water, sediment and atmospheric particulates [1-3]. Heavy metal contamination of soil is very serious in China owing to the rapid development, particularly in agricultural and industrial areas [4]. In terms of the latest Chinese national survey of soil pollution bulletin, the total over-limit ratio of pollutants in national soil is 16.1% and the over-limit ratio of investigation sites in farmland soil is 19.4%. Heavy metals were found the main pollutant of over-limit pollutants in Chinese arable soil. In China, there is about 10,000,000 tons' loss of crop output every year due to farmland soil contaminated by heavy metal [5]. Heavy metals are indestructible, toxic, persistent, cumulative and abundant in environment [2]. Heavy metals accumulated in farmland soil could enter into human body by food chains and result in serious health risks to human beings [1]. Therefore, pollution degree, environmental hazard and source of heavy metal in farmland soil have drawn increasing attention.

Content from this work may be used under the terms of the Creative Commons Attribution 3.0 licence. Any further distribution of this work must maintain attribution to the author(s) and the title of the work, journal citation and DOI.
Published under licence by IOP Publishing Ltd

IOP Conf. Series: Earth and Environmental Science **346** (2019) 012060 doi:10.1088/1755-1315/346/1/012060

The contents of heavy metals in soil including farmland/arable soil and urban soil were mainly impacted by soil parent material and human activities [1]. Excessive accumulating of heavy metals in surface soil primarily comes from man-made sources such as refining and mining of ores, metal processing and smelting, equipment manufacturing, chemical production, fertilizer, pesticides, wastewater irrigation, fossil fuel combustion, and vehicular exhaust, etc. [1,4]. In China, coal accounts for 70% of the energy structure and is mainly used for power generation [3]. Heavy metals contained in coal release into the surrounding soil of coal-fired plant with the coal combustion for power generation. The effect of a coal-fired plant on its surrounding soil environment is related to the used coal quality, plant capacity, stack height, local meteorological conditions and the pollution control technology of the coal-fired plant. Owing to the diversity of the aforementioned factors in each coal-fired plant, it is essential to research the pollution characteristics and the possible sources of heavy metals in surrounding farmland soil of different coal-fired plant for crops safety and environmental protection.

Four objects of this research are: (1) to investigate the content characteristics of heavy metals in farmland soil around Weihe coal-fired plant; (2) to analyze the relationship among the investigated heavy metals; (3) to differentiate the possible sources of heavy metals investigated in the farmland soil around the plant; and (4) to quantify the contribution of every source. The results of this study would offer a scientific basis for environmental protection and pollution control of farmland soil around coal-fired plant.

2. Materials and methods

2.1. Sample and analytical methods

Soil samples were gathered from the farmland around Weihe coal-fired plant (WCFP) (108°54'21"E, 34°26'19"N), which is located in the northwest 20 km away from Xi'an downtown. The WCFP with a 210 m-high chimney and 1.2×10^6 kWh annual energy production was built in 1991. The annual coal consumption of WCFP is approximately 145×10^4 t and the consumed coal is mainly from Weibei coal mines of Shaanxi province. WCFP has installed an electric precipitator system in 2015 to reduce particle emission. The prevailing wind direction in the investigation area is southwest in spring and summer, and is northeast in autumn and winter. In the study area, the zonal soil is cinnamon soil and the primary crops are vegetables and wheat. The west of WCFP is the airport expressway, the northeast is the approach track and the southeast is coal railway line.

Forty one soil sampling sites were distributed around WCFP at the distances of 100 m, 500 m and 1000 m. At each sampling site, about 1 kg surface soil (0-20 cm depth) sample was collected by using a stainless steel shove from the corners of a 4 m^2 (2 m × 2 m) sampling plot and packed in a polyethylene bag. All collected soil samples were air-dried naturally in the laboratory for two weeks, and then sifted through a 1 mm nylon mesh to eliminate small stones and other debris. About 100 g of every sieved soil sample (< 1 mm) was gotten by quartering method, and then milled to less than 0.075 mm for heavy metals measuring.

The contents of Cu, Pb, Zn, As, Ni, Mn, V, Co, and Cr in soil samples were determined by using X-ray fluorescence spectrometry (PW2403 apparatus, PANalytical) [3]. Standard reference materials (GSS-1, GSD-12) and 10% duplicate samples were applied for quality control in the experiment [6]. The measured values of all investigated heavy metals content in standard reference materials are in 95-103% the official values. The relative errors of all heavy metals measured in the duplicate samples are in 3-8%.

2.2. Statistical analysis methods

Content characteristics of heavy metals investigated in farmland soil around WCFP were described using maximum (Max), minimum (Min), arithmetic mean (AM), standard deviation (SD), and coefficient of variation (CV). Principal component analysis, correlation analysis and cluster analysis, widely used in literatures [6-9], were applied to judge the interrelationships of the heavy metals and to discriminate their potential sources in the investigated soil. Principal component analysis subsequently multiple linear regression method was used to quantify the contribution of each source.

Correlation analysis can determine the correlation relationship between two elements. Principal component analysis can reveal the relationships among the observed variables by reducing the number of correlated complex variables and extracting latent factors (principal components). Principal component analysis depends on an eigenvector decomposition of correlation matrix or covariance matrix. The initial data matrix is first decomposed into factor loading matrix, factor score matrix and residual matrix. Factor loading matrix is then rotated to attain the rotated component loading matrix and factor score is obtained by principal component analysis. In the study, principal component analysis was conducted by the method of varimax rotation with Kaiser Normalization of variables [10]. Cluster analysis can classify a group of observations into two or more incompatible clusters on basis of a combination of internal variable. It is often used to test the principal component analysis results [10]. Multiple linear regression analysis was performed after principal component analysis with the total heavy metals contents as the dependent variable and the factor scores of principal component analysis for heavy metal content as independent variables [9]. The detail method of principal component analysis subsequently multiple linear regression can be found in the literatures [9,11]. SPSS 19.0 software was used in the study to carry out all statistical analyses.

3. Results and discussion

3.1. Heavy metal contents in farmland soil

The descriptive statistics of Cu, Pb, Zn, As, Ni, Mn, V, Co and Cr contents in farmland soil around Weihe coal-fired plant (WCFP), and the element background values of Shaanxi soil [12], are displayed in table 1. The AMs of Cu, Pb, Zn, As, Ni, Mn, Cr, Co and V contents in the investigated soil samples were 23.3, 28.3, 48.7, 9.4, 31.2, 564.7, 110.7, 29.1 and 78.3 mg/kg, respectively. Compared to the element background value of Shaanxi soil, farmland soil around WCFP had elevated contents of Cu, Pb, Co, Cr, V and Ni. Their maximum contents were 45.9, 64.5, 61.0, 128.1, 104.2 and 39.1 mg/kg, which were 2.1, 3.0, 5.8, 2.1, 1.6 and 1.4-folds the element background values of Shaanxi soil, respectively. Coefficient of variation (CV) can reflect the disparity degree of observed variable. The CV values of Co, Pb and Cu content in the investigated soil samples are greater than other heavy metals, indicating that the varieties of these 3 heavy metals contents in the soil samples are bigger than other heavy metals.

Table 1. The contents of heavy metals in surrounding farmland soil of WCFP and reference value (mg/kg).

Heavy metal	Cu	Pb	Zn	As	Ni	Mn	Cr	Co	V
Min	18.1	20.1	38.7	6.0	23.9	218.6	94.7	18.9	64.5
Max	45.9	64.5	69.7	17.9	39.1	681.6	128.1	61.0	104.2
AM	23.3	28.3	48.7	9.4	31.2	564.7	110.7	29.1	78.3
SD	4.8	6.2	6.2	1.2	3.1	68.3	7.7	7.9	7.1
CV(%)	20.4	22.1	12.7	12.8	10.0	12.1	6.9	27.0	9.1
Reference value	21.4	21.4	69.4	11.1	28.8	557.0	62.5	10.6	66.9

Compared with the contents of heavy metals in the surrounding soil of Baqiao coal-fired plant [3], another coal-fired plant situated in the northeast of Xi'an city, the contents of Zn, Cu and Pb in the surrounding soil of Baqiao coal-fired plant are remarkably higher, and the contents of Cr and Co in the surrounding soil of Baqiao coal-fired plant are lower, while the contents of other heavy metals in the surrounding soil of two power plants are proximity. The content diversities of heavy metals in the surrounding soil of two power plants maybe related with their sources. Baqiao power plant is closer to Xi'an city than WCFP. There are many roads around Baqiao power plant and where traffic congestion is frequent. Therefore, traffic emission plans a great influence on Zn, Cu and Pb contents in the surrounding soil of Baqiao power plant.

3.2. Heavy metal sources in farmland soil

Table 2. Rotated component matrix for heavy metal contents of farmland soil around WCFP.

Element	Principal component			Communality
	1	2	3	
As	0.448	0.711	0.407	0.872
Co	0.111	-0.285	0.699	0.807
Cr	-0.042	0.004	0.935	0.876
Pb	0.738	-0.339	-0.043	0.945
Zn	0.823	-0.147	0.155	0.725
Cu	0.876	-0.364	0.041	0.901
Mn	-0.076	0.894	-0.015	0.804
Ni	0.779	0.346	-0.185	0.761
V	0.936	0.043	0.005	0.878
Eigenvalue	3.556	2.660	1.354	
Variance contribution rate (%)	39.508	29.557	15.041	
Cumulative variance contribution rate (%)	39.508	69.065	84.106	

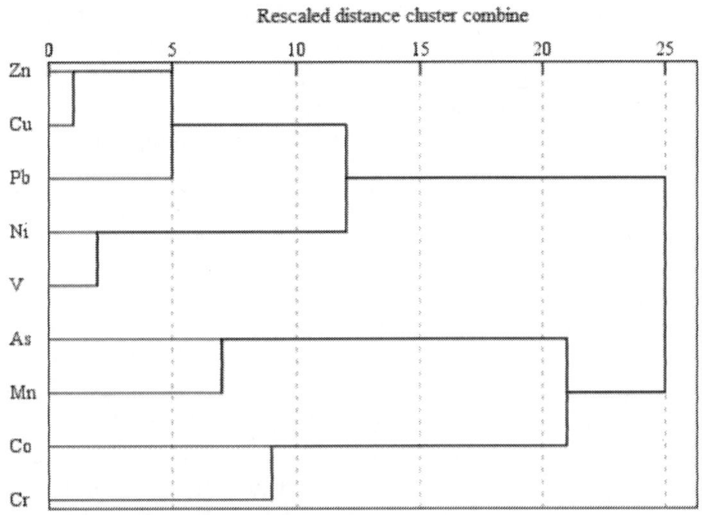

Figure 1. Cluster dendrogram of heavy metals determined in the farmland soil around WCFP.

The principal component analysis results of Cu, Pb, Zn, As, Ni, Mn, V, Co and Cr contents in farmland soil around WCFP are shown in table 2. The results proved that three principal components were extracted, which account for 84.106% of the total variance. Principal component 1, explaining 39.508% of the total variance, mainly loads on V (0.936), Cu (0.876), Zn (0.823), Ni (0.779) and Pb (0.738), and moderately loads on As (0.448). Cu, Pb, Zn, V and Ni are remarkably positive correlation at $P<0.01$. Figure 1 shows that Zn, Cu, Pb, Ni and V are classified in one cluster. The results of principal component analysis, correlation analysis and cluster analysis disclose that Cu, Pb, Zn, V and Ni may have a similar source in the farmland soil. The contents of Zn in the investigated samples are less than its background value in Shaanxi soil except for in one sample approaches to the background value and the contents of Pb in the investigated soil samples are bigger than its background value in Shaanxi soil except for in one sample its content is close to the background value. The contents of Cu, Ni and V respectively in 78%, 75% and 54% soil samples approach to their corresponding background values in Shaanxi soil. The higher value sites of Zn, Cu, Pb, V and Ni content are same, which are situated near the airport expressway and the approach track. Many researches have proven that traffic source can discharge these metals into surface environment [6,13,14]. Therefore, according to the content characteristics of these five metals in the investigated soil samples and the results of multivariate statistical analysis, we consider that Zn, Cu, Pb, V and Ni in farmland soil around WCFP maybe primarily came from nature and traffic mixing sources.

Principal component 2, mainly loaded by As and Mn, accounts for 29.557% of the total variance (table 2). As and Mn are classified together in cluster analysis (figure 1). They are remarkably positively correlation at $P<0.001$. These results demonstrate that As and Mn may have a similar source in farmland soil around WCFP. The contents of As and Mn in all investigated soil samples are less than or close to the background values in Shaanxi soil, which suggest that As and Mn in farmland soil around WCFP mainly derived from natural source.

Principal component 3 is loaded by Cr and Co, explaining 15.041% of the total variance (table 2). Cr and Co are remarkably positive correlation at $P<0.01$, and they belong to a cluster in cluster analysis (figure 1). The contents of Cr and Co in all investigated soil samples are respectively 1.5-2.0 times and 1.8-5.8 times their background values in local soil. These results show that Cr and Co contents in farmland soil around WCFP were impacted by local human activities. The coal used in WCFP came from Weibei coal fields and the contents of Co and Cr in Weibei coal are 5 to 39 mg/kg and 4 to 96 mg/kg, respectively [15]. Cr and Co in coal mainly exist in sulphate and clay minerals [16], which are incombustible components of coal. In the combustion process of coal, incombustible components come to fly ash and slag. In this investigation, the higher Co and Cr content samples distribute in the southwest of WCFP (downwind direction). Therefore, it is can be inferred that Co and Cr in farmland soil around WCFP mainly derived from coal-fired power generation discharge.

On the base of the sources identification of heavy meals in farmland soil around WCFP, the contribution of each source was apportioned using principal component analysis subsequently multiple linear regression method and the results indicated that the percentage contribution is 21.7% for traffic and nature mixing source, 49.9% for natural source and 28.4% for coal-fired source.

4. Conclusions

The contents of Co, Cr, Pb, V, Cu and Ni in farmland soil around Weihe coal-fired plant were greater than their background values in Shaanxi soil, while the contents of Zn, As and Mn in the farmland soil were less than or close to their background values in Shaanxi soil. Heavy metals investigated in farmland soil around Weihe coal-fired plant presented different variation characteristics. The variations of Pb, Cu and Co content are greater than other heavy metals in the farmland soil. Zn, Cu, Pb, Ni and V had the mixing source of traffic emission and nature. As and Mn originated from natural source. Cr and Co mainly originated from coal-fired plant discharge. The total amount of heavy metals investigated in the farmland soil around Weihe coal-fired plant mainly originated from natural source (contribution 49.9%). The percentage contribution was 28.4% for coal-fired power plant discharge and 21.7% for traffic and nature mixing source. This research indicates that the farmland soil around Weihe coal-fired plant was impacted by local traffic and coal-fired plant. Therefore, it is essential in the future that the continuous monitoring of heavy metal in the farmland soil and local crops.

Acknowledgments

The work was supported by the Fundamental Research Funds for the Central Universities of China (No. GK201601009) and the National Natural Science Foundation of China (No. 41271510). We would like to thank Long Pang, Shigang Chao, Ni Zhao, Xue Xu and Mengmeng Zhang for their help with during field sampling and experiments.

References

[1] Khalid S, Shahid M, Niazi N K, Murtaza B, Bibi I and Dumat C 2017 A comparison of technologies for remediation of heavy metal contaminated soils *J. Geochem. Expl.* **182** 247-68

[2] Wang L, Yang D, Li Z T, Fu Y H, Liu X M, Brookes P C and Xu J M 2019 A comprehensive

mitigation strategy for heavy metal contamination of farmland around mining areas-screening of low accumulated cultivars, soil remediation and risk assessment *Environ. Pollut.* **245** 820-8

[3] Lu X W, Liu W, Zhao C F and Chen C C 2013 Environmental assessment of heavy metal and natural radioactivity in soil around a coal-fired power plant in China *J. Radioanal. Nucl. Chem.* **295** 1845-54

[4] Yang Q Q, Li Z Y, Lu X N, Duan Q N, Huang L and Bi J 2018 A review of soil heavy metal pollution from industrial and agricultural regions in China: Pollution and risk assessment *Sci. Total Environ.* **642** 690-700

[5] Wu G, Kang H B, Zhang X Y, Shao H B, Chu L Y and Ruan C J 2010 A critical review on the bio-removal of hazardous heavy metals from contaminated soils: Issues, progress, eco-environmental concerns and opportunities *J. Hazard. Mater.* **174** 1-8

[6] Chen X D, Lu X W and Yang G 2012 Sources identification of heavy metals in urban topsoil from inside the Xi'an Second Ringroad, NW China using multivariate statistical methods *Catena* **98** 73-8

[7] Wang P C, Li Z G, Liu J L, Bi X Y, Ning Y Q, Yang S C and Yang X J 2019 Apportionment of sources of heavy metals to agricultural soils using isotopes fingerprints and multivariate statistical analyses *Environ. Pollut.* **249** 208-16

[8] Strbova K, Ruzickova J and Raclavska H 2019 Application of multivariate statistical analysis using organic compounds: source identification at a local scale (Napajedla, Czechia) *J. Environ. Manage.* **238** 434-41

[9] Pan H Y, Lu X W and Lei K 2017 A comprehensive of heavy metals in urban road dust of Xi'an, China: Contamination, source apportionment and spatial distribution *Sci. Total Environ.* **609** 1361-9

[10] Lu X W, Wang L J, Li L Y and Lei K 2010 Multivariate statistical analysis of heavy metal in street dust of Baoji, NW China *J. Hazard. Mater.* **173** 744-9

[11] Nasir M F M, Samsudin M S, Mohamad I, Awaluddin M R A, Mansor M A, Juahir H and Ramli N 2011 River water quality modelling using combined principal component analysis (PCA) and multiple linear regressions (MLR): A case study at Klang River, Malaysia *World Appl. Sci. J.* **14** 73-82

[12] CNEMC (China National Environmental Monitoring Center) 1990 *The Background Values of Elements in Chinese Soils* (Beijing: Environmental Science Press of China) (In Chinese)

[13] Chen H, Lu X W, Li L Y, Gao T N and Chang Y Y 2014 Metal contamination in campus dust of Xi'an, China: A study based on multivariate statistics and spatial distribution *Sci. Total Environ.* **484** 27-35

[14] Han X F, Shi D Q and Lu X W 2018 Concentration and source of trace metals in street dust from an industrial city in semi-arid area of China *J. Environ. Sci. Manage.* **2** 90-9

[15] Lu X W, Luo K L, Wang L Z and Wang W Y 2003 The content of trace elements of coals in Weibei area, Shaanxi province *J. Jinlin Univ. (Earth Sci. Edition).* **33** 178-82 (In Chinese)

[16] Yin L Q and Guan X Y 2005 The study of pollution of heavy metals from coal-fired power plant and its control *Elect. Power Tech. Environ. Prot.* **21** 31-4 (In Chinese)

Effect of different aging treatments on the vigor of high-oleic acid peanut seeds

H Wang [1,3], S T Yu[1,3,4], C T Wang [2,3], G Q Yu [1], X Y Cui [1], S L You[1], Z Y Gao[1], P X Shi [1], H B Yu[1] and L Ren[1]

[1]Institution of Sandy Land Management and Utilization of Liaoning, Fuxin 123000, China

[2]Shandong Peanut Research Institute, Qingdao 266100, China

[3]Shenyang Agriculture University, Shenyang 110866, China

Email: shutaoyu@163.com

Abstract. Accelerated aging is known to reduce seed viability and vigor in many crop species. This study was undertaken to evaluate the effect of accelerated aging on germination potential, germination percentage, electric conductivity, acid value and peroxide value. Compared with common peanut, high oleic acid peanut were more beneficial to health and can significantly improve the storability of seed. Eight high oleic acid (range of oleic acid content: 74.95%-80.6%) and two common varieties were used in this experiment. Six treatments were contained, natural aging of kernels, natural ageing of pods, accelerated aging was achieved by incubating seed at 40°C and 100% relative humidity in a closed chamber for 4, 6, and 8 days. The results showed that: the changes of germination potential and germination percentage of natural aging were not significant for high oleic acid peanuts, smaller than that of the common varieties. No matter artificial aging or natural aging treatment, compared with non-aging treatment, electric conductivity gradually increased, but high oleic acid varieties remained at low levels. The acid value of each variety increased gradually with the aging time, comparison between aging treatments, the difference between high oleic acid varieties was not significant, not the same with the common varieties. The peroxide value of 8 high oleic acid varieties was zero under each aging treatment, but the peroxide value of two common varieties increased with the aging time. The variation rules of all indicators of artificial aging for 4 days and 6 days were the same with

Content from this work may be used under the terms of the Creative Commons Attribution 3.0 licence. Any further distribution of this work must maintain attribution to the author(s) and the title of the work, journal citation and DOI.
Published under licence by IOP Publishing Ltd

the two natural aging treatments. So artificial aging for 4 days and 6 days can be used for high oleic acid aging test, replace natural aging, which not only shortens the test time but also ensuring the best effect.

1. Introduction

Seed aging relates to the phenomenon that seed vigor naturally declines during storage [1]. Seed vigor reflects the ability of the potential germination, emergence and storage of seeds, it reaches its highest level at physiological maturity, then decline. The seed deteriorating and aging, seed quality will be affected, including seed germination, seedling growth and development, so attention has been put on seed aging research.

The four seasons of Liaoning province in the northeast peanut production area are distinct, peanut cultivation system is one harvest a year, oleic acid content of currently extended peanut varieties is less than 45%, so higher demand for safe storage of peanut seeds is needed by peanut seed producers and operators. Previous studies on the aging of common peanuts were more frequent, while there were fewer studies on high oleic acid peanuts. Screening peanut germplasm which is resistant to accelerated aging with relative germination percentage [2], it was found that the aging resistance of peanut was positively correlated with oleic acid content in some treatments, but negatively correlated with linoleic acid content, but no study on high oleic acid peanut.

High temperature and high humidity aging are the most commonly used artificial accelerated aging method, which is widely used in research on storability of various seeds [3-4]. In this study, 8 high oleic acid peanut varieties with different oleic acid content and 2 common varieties were used. By measuring the vigor and quality of the seeds, inquiring the influence of artificial aging and natural aging on the vigor and quality of seeds. So, providing a basis for the safe storage of high oleic acid peanuts and technical support for the promotion of high oleic acid peanuts.

2. Materials and methods

2.1. Materials

Seeds of peanut were obtained from Institution of Sandy Land Management and Utilization of Liaoning. Ten varieties including 8 high oleic acid peanuts with different oleic acid content and a certain gradient and two common varieties Fuhua20 and Fuhua24.The content of oleic acid and linoleic acid were detected by gas chromatography(Agilent 7820), O/L is the content of oleic acid divide the content of linoleic acid. 100-seed mass was weight of 100 randomly selected seeds. The quality data were shown in table 1:

IOP Conf. Series: Earth and Environmental Science **346** (2019) 012061 doi:10.1088/1755-1315/346/1/012061

Table 1. Quality characteristics.

	L95	L88	L92	S52	L107	L14	L34	L30	Fuhua24	Fuhua20
Oleic acid (%)	80.6	78.94	78.96	78.24	76.2	76.5	74.73	74.95	42.9	38.2
O/L	20.30	20.29	20.19	17.01	15.74	14.83	11.92	10.89	1.30	0.91
100-seed mass(g)	71.19	70.26	69.94	63.18	69.01	82.73	72.86	73.41	85.49	98.70

2.2. Treatments

Five treatments were contained, natural aging of kernels, natural ageing of pods, artificial aging of kernels for 4 days, artificial aging of kernels for 6 days, artificial aging of kernels for 8 days, and no aging for kernels as a control.

The method of natural aging: after harvested, the pods were dried naturally until the water content was lower than 10%, then were preserved including kernels and pods, the total storage time was 20 months (October 2015 to June 2017).

The method of artificial aging and the control: the kernels used for artificial aging was the same with the control. According to local production habits, after drying naturally (the water content was lower than 10%) the pods were sent to storage for 8 months (October 2016 to June 2017), it was the control. Then the kernels were put under high temperature and high humidity conditions (T(temperature) = 40°C, RH (relative humidity) = 100%) for four, six and eight days separately and then dried for 3-4 days naturally.

2.3. Measuring methods

Germination percentage and germination potential: germination percentage (%) = Germination number on the 7th day/ total number of experimental seeds×100%; germination potential (%) = number of seed germination on 4th day / total number of experimental seeds × 100% [5].

Electric conductivity (EC) and relative conductivity (RC): three replicates per variety, 20 seeds per variety, after weighed (W), washed with deionized water, soaked in 30ml distilled water for 24h, then electric conductivity been measured (d1). Then boiled for ten minutes and measured conductivity again (d2), use the same amount of distilled water as a control (CK). They were calculated using the following formulas:

$EC = (d1\text{-}d2)/W$

$RC = (d1\text{-}CK)/ (d2\text{-}CK) \times 100\%$

Acid value and peroxide value were measured based on the analytical method of GBT5009.37-2003 edible vegetable oil hygiene standard.

2.4. Data analysis

Statistics and correlation analysis were performed using Excel software and DPS software.

3. Results

3.1. Analysis of variance

It can be seen from Table 2 that the effects of different aging treatments 、 different varieties and their interactions on peanut germination potential, germination percentage, electrical conductivity, relative conductivity and acid value were significant.

Table 2. Two-factor variance of different aging treatments and different varieties on peanut seed vigor.

Variation source	df	Germination potential	Germination percentage	Electric conductivity	Relative conductivity	Acid value
		MS	MS	MS	MS	MS
Treatments	5	1.03**	0.76**	121393.64**	1.74**	0.06**
Varieties	9	0.13**	0.08**	10557.07**	0.09**	0.06**
Treatments x Varieties	45	0.002**	0.01**	2282.39**	0.02**	0.02**

Notes:**significance level at 0.01

3.2. Changes of germination potential and germination percentage

As shown in Figure 1, under different aging treatments, germination potential and germination percentage of high oleic peanut varieties was higher than that of the common varieties, and the variation trend of each variety were consistent. The control > natural aging of pods> natural aging of kernels> artificial aging of kernels for 4 days > artificial aging of kernels for 6 days >artificial aging of kernels for 8 days.

Compared in varieties, germination potential of S52 and L107 was the highest in all aging treatments except for L30 whose was the highest in artificial aging for 8 days. The lowest germination potential under each treatment was common varieties Fuhua20 and Fuhua24, which were significantly lower than high oleic acid variety.

The results of natural aging of pods were similar with the natural aging of kernels. The germination percentage gradually decreased with the aging time increased. Except for L14, trend of artificial aging was similar with natural aging. Germination percentage decreased with the decrease of oleic acid content.

Compared with the control, germination percentage of varieties with an O/L higher than 10 under all aging treatments were all above 50%, and the smallest was 51%. While the germination percentage of two common varieties Fuhua20 and Fuhua24 significantly reduced, after artificial aging for 6 days and 8 days. And the germination percentage decreased to 35.2% and 41.6% respectively of artificial aging for 8 days.

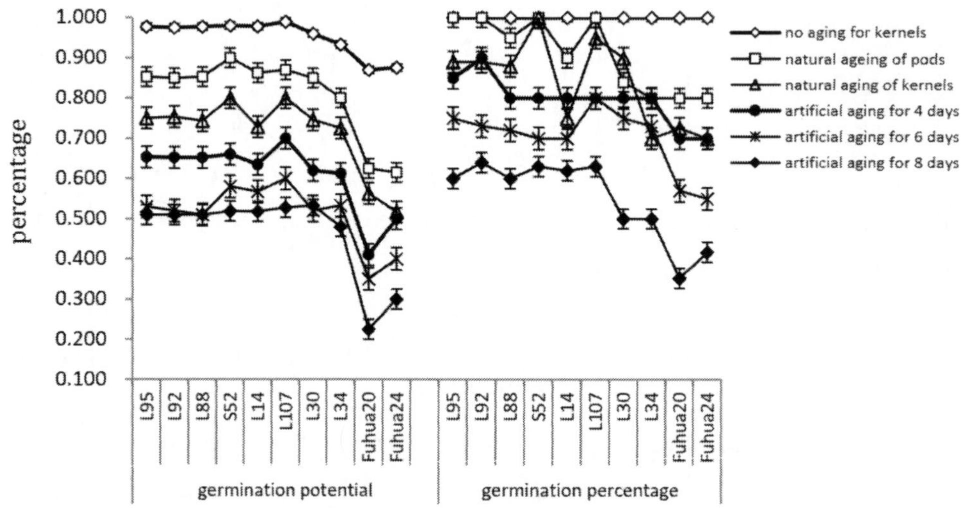

Figure 1. Changes of germination potential and germination percentage of various varieties under different aging treatments.

3.3. Comparison of electric conductivity under different aging treatments

As shown in Figure 2, under different aging treatments, electric conductivity of high oleic acid peanut varieties was generally lower than that of the common varieties under the influence of saturated fatty acids, and it was lower than that of artificial aging, and the trend was the same: The control < natural aging of pods < natural aging of kernels < artificial aging of kernels for 4 days < artificial aging of kernels for 6 days < artificial aging of kernels for 8 days.

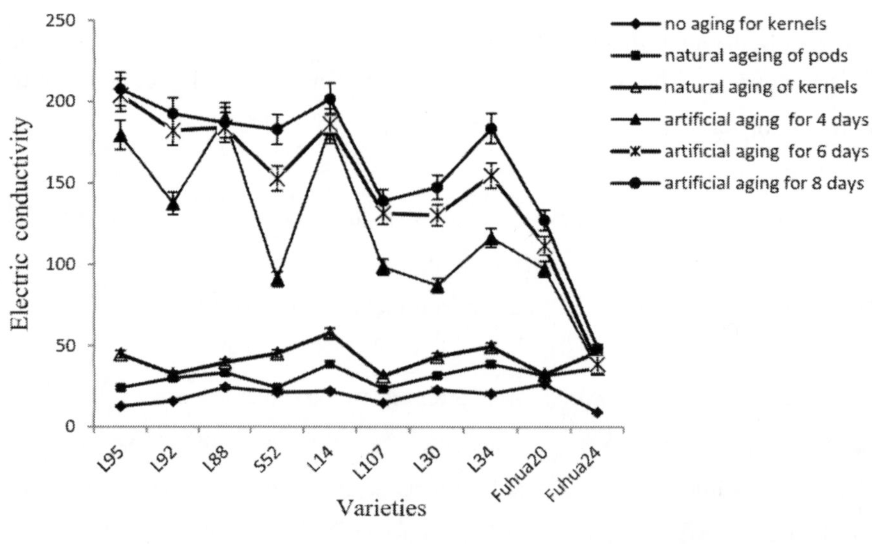

Figure 2. Comparison of electric conductivity under different aging treatments.

It is worth noting that under all aging treatments, electric conductivity of Fuhua24 was always kept at the minimum, probably due to the high crude fat content of it (detected by the Ministry of Agriculture Oil and Products Quality Supervision, Inspection and Testing Center (Wuhan, China) up to 56.08%).

3.4. Comparison of acid value and peroxide value of different varieties

As shown in Figure 3, under all aging treatments, peroxide value of common varieties was higher than high oleic acid varieties. Compared between the treatments, The control < natural aging of pods < natural aging of kernels < artificial aging of kernels for 4 days < artificial aging of kernels for 6 days < artificial aging of kernels for 8 days. With the aging time prolonged, the acid value became higher.

Compared between varieties, common varieties Fuhua20 and Fuhua24 was the highest in all treatments. The acid value of L92 and L95 whose O/L was more than 20 was the lowest in each aging treatments.

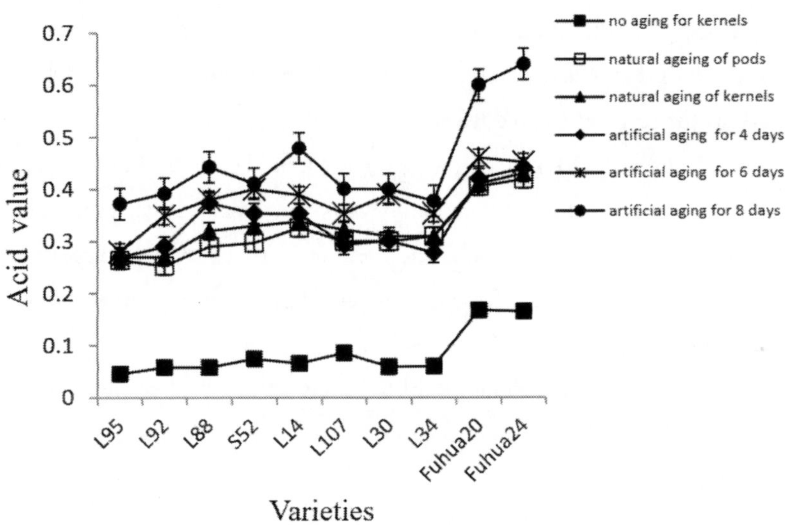

Figure 3. Comparison of acid value under different aging treatments.

In this study, the peroxide value had been tested through the redox titration method. The result showed that peroxide value of all high oleic acid varieties was zero under all aging treatments, whereas the common varieties Fuhua20 and Fuhua24 were all oxidized.

4. Discussion

There are few studies on the research of artificial aging and natural aging of high oleic acid home and abroad. [6] conducted a natural aging study on high oleic acid peanut, and studied on germination percentage, electric conductivity, field emergence and yield, which confirmed that high oleic acid peanut has good storability. This study combined natural aging and artificial aging of different days, the same results with Wang. Shen Yi found that, germination percentage of the common peanut was less than 50% after artificial aging for 2 days [2]. This study found the germination percentage of high oleic acid peanut would be reduced to 60% after 8 days of artificial aging. It can be seen that high oleic acid peanut has less damage to the cell membrane after aging, which better maintains the integrity of the seeds and make the seed vigor still at a high level. Significantly higher than the common peanut. This phenomenon is due in part to aging-induced lipid perosidation, which lias the potential to damage membranes of the seed tissues [1].

The germination potential, germination rate, vigor index and germination index of seeds are reliable indicators reflecting seed vigor [7]. By comparing natural aging with artificial aging, two treatment methods could reflect the vigor of the seed, and with the prolongation of artificial aging time, the storability of the seed is significantly different. Therefore, compared with the long-term natural aging method, artificial aging method can make up for the shortcomings, which can reflect the seed vigor characteristics well, also the same with other research results [8]. The variation rules of various indicators of artificial aging for 4 days and 6 days and the two natural aging treatments were consistent. So artificial aging for 4 days and 6 days can be used for high oleic acid aging test, replace natural aging, which not only shortens the test time also ensuring the best effect.

Eight high oleic acid lines selected have a gradient in oleic acid content. From the result, germination percentage, germination potential, electric conductivity and relative conductivity under different aging treatments were positively correlated with oleic acid content, and negatively correlated with acid value. However, there was also one strain with different changes-S52, whose weight was just 63.18g, and also was resistant to cold tolerance.

In addition, comparing between natural aging of the pods and natural aging of the kernels, the pods have better storability than kernels, because of the pods provide a relative safe space for the kernels. So, keeping shells when storing.

Acknowledgments

Authors wishing to acknowledge assistance or encouragement from Shandong Peanut Research Institute and financial support from Foundation items: China Agricultural Research System (CARS-13); The Central Government Guiding Local Financial Special Project (2017108006).

References

[1] Sung J M and Jeng T L 1994. Lipid peroxidation and peroxide-scavenging enzymes associated with accelerated aging of peanut seed. *Physiol. Plantarum*. **91**: 51-5.

[2] Shen Y, Liu Y H and Chen Z D 2013. Identification and estimation of aging resistant varieties in peanut (Arachis hypogaea L.). *Chin. Agricul. Sci. Bull.* **29**(18):67-71..

[3] Xie H, Chen X Z, Qi J, Liu K, Yang L, Wang J L and Nan Z J 2006. Study on seed vigor of the artificial aged seeds in soybean. *J. Beijing Agricul. College.* **21**(3):15-7.

[4] Liu M J, Wang T G, Chen S L, Wang C H and Zhao X L 2008. Physioloycial and seed vigour

changes of maize seeds during artificial aging course. *J. Nuclear Agricul. Sci.* **22**(4):510-3.

[5] Gao H W, Man Q, Pan J, Lei Y Q, Yu H B, Wu M, Zhang W H and Liu L K 2015. Differences in properties of seed vigor vetween artificially and naturally aged soybean seeds. *Seed*, **34**(1):14-9.

[6] Wang C T, Zhang Q Y, Tang Y Y, Wang X Z, Wu Q, Sun Q X, Li J, Liu F and Wang H 2016. Effects of natural ageing on yield-related traits of high-oleic peanut. *Shandong Agricul. Sci.* **48**(1):44~6.

[7] Yan Q H 2001. *Seed Sci.* Beijing: China Agriculture Press.110- 6; 425-31.

[8] Gao H W, Man Q, Wu M and Lei Y Q 2014. Screening of soybean germplasm with good seed longevity. *Soybean Sci.* **33**(1):6-12.

IOP Conf. Series: Earth and Environmental Science **346** (2019) 012062 doi:10.1088/1755-1315/346/1/012062

Quality evaluation of bakery products enriched with spinach

A N Sapozhnikov[1], L N Rozhdestvenskaya and A V Kopylova

Department of Technology and Organization of Food Industries, Faculty of Business, Novosibirsk State Technical University, 20 Prospekt K. Marksa, Novosibirsk, 630073, Russian Federation

E-mail: a.sapozhnikov@corp.nstu.ru

Abstract. Garden spinach (*Spinacia oleracca*) is a well-known leaf vegetable in the world. It contains biologically active substances and thus can be used for enrichment of different kinds of food production, including bakery products. However, spinach is not yet deeply studied from this point of view. Recent studies show that spinach is advisable for introducing into the dough in state of infrared-dried powder. The paper describes the results of quality evaluation of developed bakery products enriched with spinach powder by sensory, physico-chemical characteristics and nutritional value. Three control samples (roll № 1, curd and lingonberry patties № 2 and № 3) were prepared using high-grade and second-grade wheat flour in 1:3 ratio. All experimental samples (№ 1, № 2, № 3) were enriched with infrared-dried spinach powder in the amount of 4% from flour weight. All bakery products samples were prepared according to standard technology and passed quality evaluation. The sensory characteristics of experimental samples didn't change significantly, though some of values decreased by 0.1 points or increased by 0.1…0.3 points. The dry matters content values either decreased by 1.45…3.38% or increased by 0.36%, the acidity and ash content values increased by 0.2…0.6 deg and 0.16…0.74% accordingly. All samples were evaluated for estimated content of proteins, fats and carbohydrates and their relative approximation to ratio of 1:1:4. The ratio in control and experimental samples № 1 and № 2 are close to the recommended ratio value. In control and experimental samples № 3 the ratio value is not kept but can be improved in different ways. Thus, the results of study show, that the enrichment of bakery products with spinach can save or improve their sensory characteristics, increase organic acids and minerals content. The enrichment also doesn't have negative effect on products' nutritional value. The ratios of proteins, fats and carbohydrates allow to use bakery products enriched with spinach in daily diet of adults and children of preschool and school age.

1. Introduction

Bakery products are the essential part of Russian population daily diet. They are produced almost everywhere in the country at enterprises of different capacity.

One of the most important directions in Russian bread baking industry is expansion of its production range, which can be achieved by developing and introducing of functional and specialized varieties of bakery products enriched by vitamins, minerals and dietary fibers. This direction meets the purposes of the "Bread is Health" Concept, which full name is "The Concept of Ensuring the Sanitary and Epidemiological Well-Being of the Population Through the Development of Functional and Specialized Bread Baking in the Russian Federation until 2020" [1].

Content from this work may be used under the terms of the Creative Commons Attribution 3.0 licence. Any further distribution of this work must maintain attribution to the author(s) and the title of the work, journal citation and DOI.
Published under licence by IOP Publishing Ltd

According to the direction, a lot of studies concerned with the development of bakery products with additives, which are obtained from local plant raw materials, have been done in Russia and other countries.

At the same time, some kinds of plant raw materials, which can be successfully applied in bakery products formulations and serve as functional ingredients, are studied insufficiently. One of these raw materials is garden spinach (*Spinacia oleracca*). It is known, that spinach in its fresh and processed states is a source of bioflavonoids, vitamins, minerals and antioxidants. Thus, these beneficial features allow to use it in dietary nutrition, as well as in the prevention and treatment of a wide range of diseases (anemia, diabetes, helminth infections, obesity, inflammations, nervous diseases, etc.) [2–4].

It should be noted, that spinach has specific taste and scent. Its introduction into bakery products formulations in natural state gives the products moist and fibrous texture, thereby their sensory characteristics are lowered. This indicates the feasibility of enrichment of bakery products with spinach in a powder state, preliminary reducing its moisture content through various types of drying: convective [5], freeze-drying [4], by infrared rays [6], by contact method from puree state on a roller dryer, followed by grinding in a ball mill [7]. From mentioned drying methods, infrared drying is considered as optimal, since it has simple hardware design and saves useful substances in products for a rather high degree.

To date, several studies that have shown the positive effect of spinach in natural and processed states on bakery products (including national cuisines products) quality have been conducted [5, 8, 9].

According to the information above, the objective of our study was the development of formulations of bakery products enriched with infrared-dried spinach powder. All obtained bakery product samples passed quality evaluation on their sensory, physico-chemical characteristics and nutritional value.

2. Materials and methods

2.1. Materials

The main raw materials for bakery products samples preparation were high-grade flour and second grade flour, which were obtained on "Avangard Flour Mill Plant" in Novosibirsk, Russian Federation. The other raw materials (spinach, curds, lingonberry, eggs, sugar, salt, margarine, yeast, milk, vanillin, sunflower oil) were purchased from local retail network in Novosibirsk.

2.2. Bakery products samples preparing

The control samples of bakery products were based on formulations from "The Digest of Dishes and Culinary Products Formulations for Children Nutrition in Preschool Organizations" (2005). According to our previous experimental studies, two kinds of flour were used in formulations in 1:3 ratio, i.e. 25% of high-grade flour and 75% of second-grade flour. This ratio allows to use second-grade flour with higher content of dietary fibers, B-group vitamins and other useful substances in bakery products formulations in larger amount. Therefore, high-grade flour can be used in lower amount. At the same time, it's enough to obtain production with high sensory characteristics, such as regular shape, well-developed porosity, smooth crust surface, etc.

The dough preparing was conducted by straight method (without using leavened dough), which allows to reduce preparing time of bakery products without significant change of their sensory characteristics. The kneading process was carried out manually. The dough fermentation was held out at 38…40 °C and moisture content of 70…75% in proofing cabinet of Sveba Dahlen S200 rotary oven (Sweden). From the fermented dough the products stocks were formed and proofed during 15…20 min in the same proofing cabinet.

For the experimental samples, the infrared-dried spinach powder was introduced into the dough in the amount of 4% from flour weight. This amount was obtained by our previously carried out theoretical and practical studies and was considered by us as optimal.

The primary processed spinach was dried in the electrical infrared compact dryer (Russian patent № 2265169) at 50…60 °C during 2.5…3 h in pulse mode. After that, the dried spinach was cooled at 20…25 °C and mechanically grinded into fine powder.

IOP Conf. Series: Earth and Environmental Science **346** (2019) 012062 doi:10.1088/1755-1315/346/1/012062

Thus, the following control and experimental bakery products samples were obtained:

- wheat flour roll – control sample № 1;
- curd patty – control sample № 2;
- lingonberry patty – control sample № 3;
- wheat flour roll with spinach (4% from flour weight) – experimental sample № 1;
- curd patty with spinach (4% from flour weight) – experimental sample № 2;
- lingonberry patty with spinach (4% from flour weight) – experimental sample № 3.

All product samples were baked in Sveba Dahlen S200 rotary oven at 180…190 °C during 20…30 min. After that, they were cooled at 20…25 °C during 1…2 h and evaluated for their sensory, physico-chemical characteristics and nutritional value.

2.3. Sensory characteristics

Sensory evaluation of obtained bakery product samples was carried out according to Russian national standard GOST 31986-2012 "Public catering service. Method of sensory evaluation of catering products" for the following characteristics:

- appearance (in whole products and on the cut);
- scent;
- taste;
- color;
- consistency.

Every indicator was evaluated on the 5-point scale (5 indicates the best and 1 implies the worst) by 4 semi-trained panelists. As a result of the evaluation, the average points for each characteristic were obtained.

2.4. Physico-chemical characteristics

2.4.1. Dry matters content. This characteristic was evaluated according to GOST 21094-75 "Bread and bakery products. Method for the determination of moisture". The drying of prepared and weighted samples to constant weight at 130±2 °C was carried out in the ShS-80-01 SPU drying cabinet (Russia). The samples before and after drying were compared by their weights and dry matters content D (%) was determined using the following equation:

$$D = 100\left(1 - \frac{m_1 - m_2}{m}\right) \quad (1)$$

where m_1 (g) is weight of cup with sample before drying, m_2 (g) is weight of cup with sample after drying, m (g) is weight of sample before drying. In this study, m is 5 g.

2.4.2. Acidity. This characteristic was evaluated according to GOST 5670-96 "Bread, rolls and buns. Methods for determination of acidity" by titration of extracts from samples with sodium hydroxide solution with a concentration of 0.1 mol/dm^3. Based on received data, the acidity value X (deg) was determined using the following equation:

$$X = 2V \cdot K \quad (2)$$

where V is volume (cm^3) of acid solution of of 0.1 mol/dm^3 concentration used for titration, K is correction ratio of used sodium hydroxide solution adduction to solution of exact molar concentration of 0.1 mol/dm^3. In this study, K is 1.

2.4.3. Ash content. This characteristic was evaluated according to GOST 27494-87 "Flour and bran. Methods for determination of ash content". The sample weights were ignited in crucibles at the hotplate at 400…500 °C with following incinerating at the PM-10M 732 muffle furnace (Russia) at 350…400 °C during 1.5…2 days to constant weight. After incineration, the ash content (A, %) was determined using the following equation:

$$A = m_1 \cdot \frac{100}{m_0} \cdot \frac{100}{100-H} \quad (3)$$

where m_0 (g) is weight of sample before incineration, m_1 (g) is weight of sample after incineration, H (%) is humidity of sample before drying. In this study, m_1 is 5 g. The humidity of samples was obtained from their dry matters content according to equation (1) and using following equation:

$$H = 100 - D \quad (4)$$

2.5. Nutritional value

The nutritional values of bakery products samples were evaluated by calculations of proteins (g), fats (g), carbohydrates (g) and calorie (kcal) content according to reference data from "Chemical Compositions of Russian Food Products Manual" (2002).

After calculations, all samples were evaluated for relative approximation of proteins, fats and carbohydrates ratio to value of 1:1:4.

3. Results and discussion

3.1. Sensory characteristics

The results of bakery products samples evaluation for appearance, color and consistency are shown on Figure 1.

It was noted, that all samples held their shapes after baking and had uniform porosity structure on the cut. The high quality of control and experimental samples showed, that the replacement of high-grade flour with flour of second grade in amount of 75% didn't have negative effect on technological process parameters, as well as the appearance and color of crust and crumb.

The enrichment of experimental bakery products samples with infrared-dried spinach powder showed, that there weren't any changes of parameters in dough kneading and fermentation, stocks proofing and baking processes. The crumb of all experimental samples had nice green color, while patties stuffing (curds, lingonberry) were not colored.

The Figure 2 shows results of bakery product samples sensory evaluation for each characteristic. Since our study evaluated the effects of infrared-dried spinach powder on various types of bakery products, it was advisable to compare each experimental sample with its corresponding control sample.

According to the Figure, spinach powder increased the consistency value of wheat flour roll (control and experimental samples № 1) by 0.2 points (from 4.8 to 5.0 points), though the scent and taste were decreased by 0.1 point for each characteristic (from 4.9 to 4.8 points and 4.8 to 4.7 points accordingly), which can be explained by specific spinach flavor. The appearance and color remained at the highest value (5.0 points).

Figure 1. Results of sensory evaluation of bakery products samples.

The sensory characteristics increase was more specific for curd patties (control and experimental samples № 2). Though, the scent value of experimental sample № 2 in comparison with the corresponding control sample was decreased by 0.1 points (from 4.9 to 4.8 points) because of specific spinach flavor. The color of experimental sample increased by 0.2 points (from 4.9 to 4.7 points), which can be explained by specific combination of colored crumb and curd stuffing.

At the same time, the experimental sample № 2 in comparison with all other control and experimental samples had got high taste value (4.9 points), i.e. there were not expressed specific spinach flavor, which can be explained by fat component in formulation. It is known, that spinach contains a large amount of fat-soluble substances. Therefore, adding fat components to bakery product formulations will not only increase the digestibility of nutritional substances contained in spinach, but also can improve sensory characteristics of products [10].

The consistency and appearance values of the samples remained at the same highest values of 5.0 and 4.9 points accordingly.

IOP Conf. Series: Earth and Environmental Science **346** (2019) 012062 doi:10.1088/1755-1315/346/1/012062

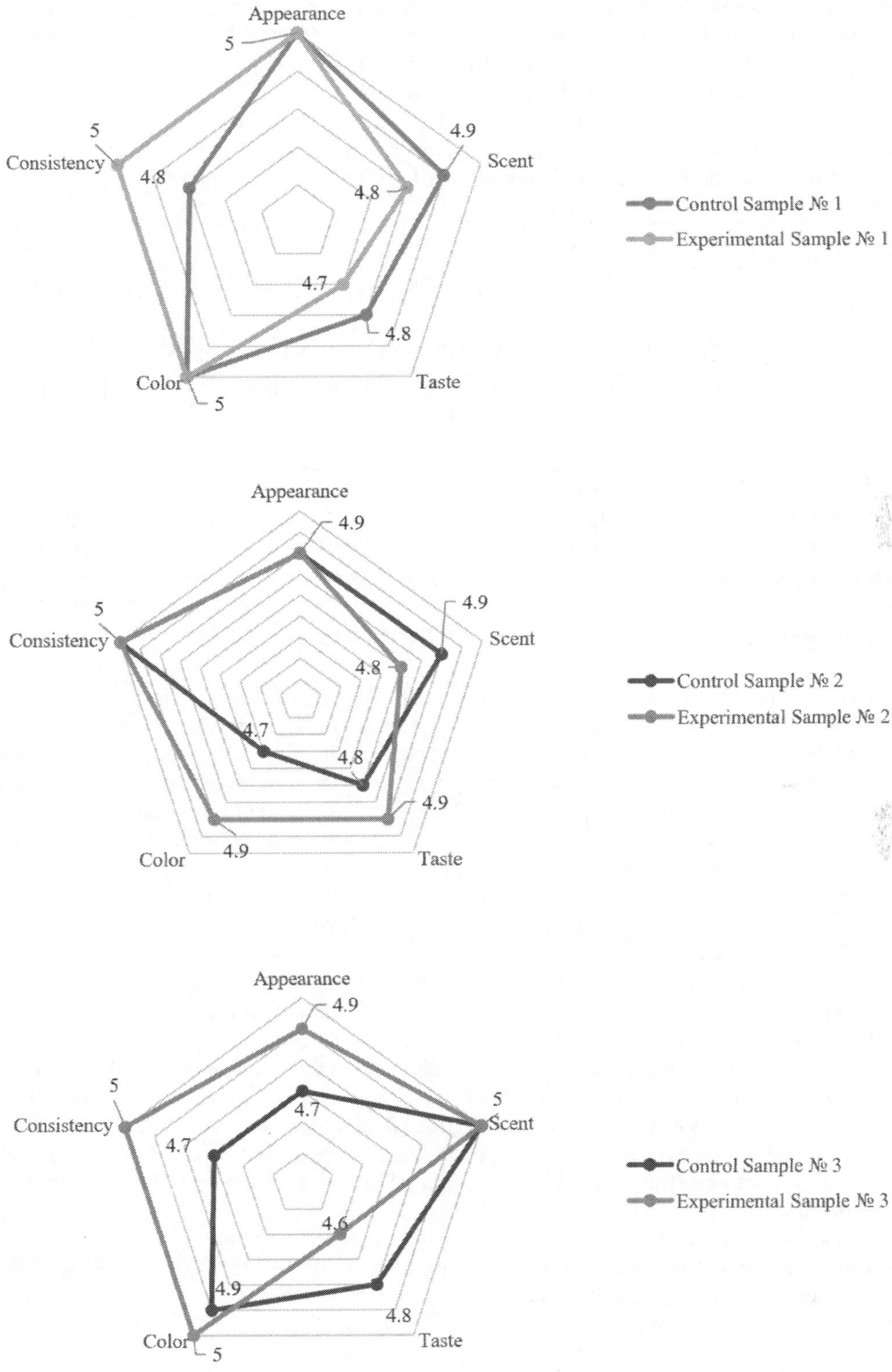

Figure 2. Sensory characteristics of bakery products.

The enrichment of lingonberry patties (control and experimental samples № 3) with infrared-dried spinach powder increased the product appearance by 0.2 points (from 4.7 to 4.9 points), consistency – by 0.3 points (from 4.7 to 5.0 points), color – by 0.1 points (from 4.9 to 5.0 points). The decrease of taste value by 0.2 points (from 4.8 to 4.6 points) can be explained by bitter aftertaste flavor of lingonberry, which strengths in spinach presence. Herewith, the scent value remains at the highest point (5.0 points) without any changes, because specific spinach scent is not felt in the presence of lingonberry. Therefore, using of natural flavoring additives can mask the undesirable aftertaste and improve the taste of bakery products in general.

3.2. Physico-chemical characteristics
In the Table 1 the results of physico-chemical characteristics of bakery products evaluation are shown.

Table 1. Physico-chemical characteristics of bakery products.

Samples	Dry matters content, %	Acidity, deg	Ash content, %
Control sample № 1	20.16±1.21	2.1±0.14	0.97±0.01
Experimental sample № 1	20.52±0.44	2.7±0.14	1.71±0.02
Control sample № 2	38.36±2.4	2.4±0.07	0.75±0.01
Experimental sample № 2	34.98±4.45	2.6±0.28	0.91±0.02
Control sample № 3	31.02±0.68	0.9±0.14	0.63±0.02
Experimental sample № 3	29.57±2.09	1.4±0.28	0.97±0.01

According to the obtained results, the enrichment of experimental sample № 1 with infrared-dried spinach powder practically didn't change dry matters content in ready product, though it increased by 0.36% (from 20.16 to 20.52%). It can be explained through similar humidity content both in wheat flour and spinach powder, and through low content of spinach powder in the products. Herewith, the acidity increased by the average of 0.6 deg (from 2.1 to 2.7 deg), and ash content increased by 0.7% (from 0.97 to 1.71%). It can be explained, that infrared-dried spinach powder is a concentrated product. It contains organic acids and minerals in larger amount than spinach in its natural form. The infrared radiation also preserves most of nutritional substances in spinach. Infrared rays remove the largest portion of water from drying materials at low temperatures without destruction of other substances, so their amounts in dried products are increased [6].

The dry matters content in experimental sample № 2 decreased on average by 3.38% in relation to the corresponding control sample (from 38.36 to 34.98%). The acidity increased on average by 0.2 deg (from 2.4 to 2.6 deg), and ash content – by 0.16% (from 0.75 to 0.91%).

The dry matters content in experimental sample № 3 on average decreased by 1.45% in relation to the corresponding control sample (from 31.02 to 29.57%). The acidity increased on average by 0.5 deg (from 0.9 to 1.4 deg), and ash content – by 0.34% (from 0.63 to 0.97%).

In this case, the dry matters content decrease can be explained by the peculiarities of patty dough formulation and water absorption by the spinach powder in combination with its interaction with the fatty component [10].

The increasing of bakery products samples acidity values indicates the presence of organic acids remained in the spinach powder after infrared drying. Herewith, the acidity increase corresponds to finished products taste improvement, except control and experimental samples № 3, which can be explained by high acid content in lingonberry. The increasing of ash content in experimental samples points to the enrichment of products with minerals contained in spinach powder.

Thus, the results show that infrared-dried spinach powder can enrich bakery products with organic acids and minerals.

3.3. Nutritional value

The Table 2 shows the results of estimating and analyzing of bakery products nutritional value.

Table 2. Analysis of nutritional value of control and experimental samples of bakery products.

Samples	**Proteins, g**	**Fats, g**	**Carbohydrates, g**	**Calorie content, kcal**	**Protein/Fats/Carbohydrates ratio**
Control sample № 1	6.0	6.5	22.3	144	1:1.1:3.7
Experimental sample № 1	6.0	6.2	20.9	138	1:1:3.5
Control sample № 2	12.1	9.8	36.8	262	1.2:1:3.8
Experimental sample № 2	12.8	10.0	38.1	271	1.3:1:3.8
Control sample № 3	5.9	9.1	48.7	285	0.6:1:5.4
Experimental sample № 3	6.0	9.2	49.0	289	0.7:1:5.3

The figures from the Table show, that using spinach powder in bakery products formulations practically doesn't have effect on products' nutritional value, because it is introduced into dough in a low amount. The proteins, fats and carbohydrates ratio in control and experimental samples № 1 shows, that it is closest to the recommended ratio of 1:1:4. The ratio in control and experimental samples № 2 shows that it is also close to the same value. In control and experimental samples № 3 the ratio is not kept, because these samples have a lower proportion of proteins and a higher proportion of carbohydrates. Thus, the formulation of lingonberry patties can be recommended for the improvement. The lingonberry patties may be also consumed with products, which are both rich with proteins and have low carbohydrates content (for example, kefir and similar dairy products).

It should be noted that experimental samples № 1 and № 2 with proteins, fats and carbohydrates ratio closest to recommended value have maximum demand for the organization of intermediate food and snacks for children of preschool and school age and can be embedded in their weekly diets.

4. Conclusion

The conducted theoretical and practical studies show that infrared-dried spinach powder is a perspective ingredient for bakery products enrichment. The obtained results allow to use combination of high-grade and second-grade wheat flour in ratio 1:3 and 4% spinach powder from flour weight in bakery products formulations. These changes in classical formulations of bakery products don't have effect on technological process parameters.

The sensory characteristics of experimental samples didn't change significantly, though some of values decreased by 0.1 points and increased by 0.1…0.3 points. The lowest characteristic is taste of experimental sample № 3 (4.6 points), but the same sample has the highest consistency value (5.0 points). So experimental samples № 1 and № 2 have the best sensory characteristics, and sensory characteristics of experimental sample № 3 must be improved through using of natural flavoring additives, which won't have negative effect on its nutritional value. The highest value (5.0 points) have appearance and color of control and experimental samples № 1, consistency of control and experimental samples № 2 and consistency of experimental sample № 3.

The dry matters content values in the samples either decreased by 1.45…3.38% or increased by 0.36%, the acidity and ash content values increased by 0.2…0.6 deg and 0.16…0.74% accordingly. This indicates the presence of high amount value of organic acids and minerals, which are remained in the spinach powder after infrared drying.

All samples were evaluated for estimated content of proteins, fats and carbohydrates and their relative approximation to ratio of 1:1:4. The ratio in control and experimental samples № 1 and № 2 are close to the recommended value. In control and experimental samples № 3 the ratio is not kept but can be improved by consuming with products with high protein and low carbohydrate content. Thus, the developed bakery products enriched with spinach can be recommended for children of preschool and school age, as well as for adults.

However, a more detailed study of bakery products chemical composition will make possible more accurately reveal their nutritional value, including functional properties.

References

[1] Bogomolova I P and Belimova E A 2014 Directions and mechanisms of state regulation of the production of functional bakery products *Proceedings of the Voronezh State University of Engineering Technologies* **2** 177–83

[2] Jaime L, Vázquez E, Fornari T, Carmen López-Hazas M, *et. al.* 2015 Extraction of functional ingredients from spinach (*Spinacia oleracea L.*) using liquid solvent and supercritical CO_2 extraction *J. Sci. Food Agricul.* **95(4)** 722–9

[3] Jiraungkoorskul W 2016 Review of neuro-nutrition used as anti-alzheimer plant, spinach, spinacia oleracea *Pharmacognosy Review* **10(20)** 105–8

[4] Neascu M, Vaughan N, Raikos V, Multari S, *et. al.* 2015 Phytochemical profile of commercially available food plant powders: their potential role in healthier food reformulations *Food Chem.* **179** 159–69

[5] Galla N R, Pamidighantam P R, Karakala B, Gurusiddaiah M R and Akula S 2017 Nutritional, textural and sensory quality of biscuits supplemented with spinach *Intern. J. Gastron. Food Sci.* **7** 20–6

[6] Sarimiseli A and Yuceer M 2015 Investigation of infrared drying behavior of spinach leaves using ANN methodology and dried product quality *Chem. Process Engi.* **36(4)** 425–36

[7] Rodionova L Ya, Sokol N V, Shubina L N and Olhovatov E A 2017 Technology and application of powdered food additives from vegetable raw materials *Scient. J. Kuban State Agrarian Univ.* **131(07)** 1389–404

[8] Khan M A, Mahesh C, Semwal A D and Sharma G K 2015 Effect of spinach powder on physico-chemical, rheological, nutritional and sensory characteristics of chapati premixes *J. Food Sci. Technol.* **52(4)** 2359–65

[9] López-Nicolás R, Frontela-Saseta C, González-Abellán R and Barado-Piqueras R 2014 Folate fortification of white and whole-grain bread by adding Swiss chard and spinach Acceptability by consumers *LWT – Food Sci. Technol.* **59(1)** 263–9

[10] Xi Y, Xiaojuan L, McClements D J, Yong C and Hang X 2018 Enhancement of phytochemical bioaccessibility from plant-based foods using excipient emulsions: impact of lipid type on carotenoid solubilization from spinach *Food Funct.* **9(8)** 4352–65

IOP Conf. Series: Earth and Environmental Science **346** (2019) 012063 doi:10.1088/1755-1315/346/1/012063

The association of SNP rs42688595 in LOC514211 gene with Indonesian-Holstein cow's reproductive traits

A P Rahayu[1,2,4], T Hartatik[3], A Purnomoadi[1] and E Kurnianto[1]

[1]Faculty of Animal and Agricultural Sciences, Diponegoro University, Semarang, Indonesia
[2]Permanent Address: Agriculture, Fisheries and Food Center of Semarang Regency, Ungaran, Indonesia
[3]Faculty of Animal Sciences, Universitas Gadjah Mada, Yogyakarta, Indonesia

E-mail: amaliapujirahayu.kabsmg@gmail.com

Abstract. The aim of this study was to study the association between SNP (single nucleotide polymorphism) rs42688595 in LOC514211 gene with the reproductive traits on Indonesian Holstein dairy cows. Blood samples and reproduction records from 89 cows were used. The DNA analysis was carried out using the PCR-RFLP method. Cows with AC genotypes showed the lowest days open (152.06 ± 20.54 days) and the lowest calving interval (427.06 ± 20.26 days) ($P<0.05$) compared to those of CC genotypes (236.46 ± 23.34 and 514.95 ± 22.86 days, respectively) and AA genotypes (270.14 ± 26.27 and 552.06 ± 26.29 days, respectively). The studied SNP was not associated with the age at first calving and service per conception ($P>0.05$). The LOC514211 rs42688595 shows a potency to be used as a genetic marker for a selection program to produce cows with better reproductive performance. Selection could be done by increasing the genetic frequency of heterozygous animals (AC).

1. Introduction

Improving the genetic quality of livestock through selection is one of the efforts to increase the productivity of dairy cows. In addition to selecting the characteristics of milk production, the selection is also needed in improving the genetic quality of reproductive traits. This needs to be considered because the reproductive traits also have economic value [1].

Molecular biology techniques can be used to access genomic information based on sequences of bovine genomes with differences in nucleotide sequences [2]. The difference in one single nucleotide is called SNP (single nucleotide polymorphism). SNPs can be used as selection markers to carry out genomic selection. Genomic selection is a revolution in dairy cow breeding that makes it possible to estimate breeding values based on the genotype of the SNP [3].

Junjing [4] and Anggraeni *et al* [5] found that SNP rs42688595 in the LOC514211 gene could be a potential marker for the characteristics of production and reproduction of dairy cattle in the Chinese Holstein cattle population. No other study has reported the effect of this gene on the reproductive traits of dairy cows. The research on SNPs on these genes are still limited. Meanwhile, to be defined as genetic markers, in-depth research needs to be done, so this research is important to be performed. Based on various studies of other genes [2,6], the different results of the genetic marker were known in various breeds. The purpose of this study was to examine the effect of LOC514211 rs42688595 SNP on Indonesian-Holstein cow population.

Content from this work may be used under the terms of the Creative Commons Attribution 3.0 licence. Any further distribution of this work must maintain attribution to the author(s) and the title of the work, journal citation and DOI.
Published under licence by IOP Publishing Ltd

2. Materials and methods

2.1. Materials
The materials used in this study were the first lactation reproduction record of 89 Indonesian-Holstein cows at Baturraden Dairy Breeding and Forage Center (BBPTUHPT Baturraden). The materials used for DNA analysis were cow blood samples, DNA extraction kits, PCR kits, forward primers (F) 5'-acggtgtttgggttcctg-3' and reverse (R) 5'-ctgtcttgccctgtttcg-3', *TaqI* restriction enzymes (10 U/μL), agarose gel, reagents for electrophoresis, and 50 bp DNA markers. The DNA analysis was conducted in the Laboratory of Animal Breeding and Genetics, Universitas Gadjah Mada, Yogyakarta, Indonesia.

2.2. Methods
The method used was PCR-RFLP (Polymerase Chain Reaction-Restriction Fragment Length Polymorphism) based on Anggraeni *et al* [5] (modified). A total of 1-3 ml of blood samples were taken through the tail vein. Genomic DNA is extracted from the blood using a kit according to the manufacturing protocol. PCR mix was made by mixing 2 μl DNA templates, 12.5 μl PCR kits, 0.5 μl forward primers (10 pmol/μl), 0.5 μl reverse primer (10 pmol/μl) and 9.5 μl ddH_2O, so the total volume was 25 μl. The mixture was put into a thermal cycler machine (Peqlab, Germany) in accordance with the program as follows: Predenaturation of 94°C for 10' followed by 36 cycles consisting of denaturation 94°C, 30"; annealing 58.8°C, 45"; extensions 72°C, 45"; and final extension 72°C, 25'. 4 μL of the PCR product was cut by adding 0.3 μL of the *TaqI* restriction enzyme; 1.5 μL *TaqI* Buffer and 9.2 μL ddH2O, then incubated 65°C for 3 hours. The RFLP results were then evaluated by 3% agarose gel electrophoresis using 50 V voltage for an hour. DNA bands were observed in UV transilluminator and compared with DNA markers.

2.3. Statistical analysis
Association analysis of the SNPs with traits was performed using the general linear model procedure in SPSS 20 program.

3. Results and discussions

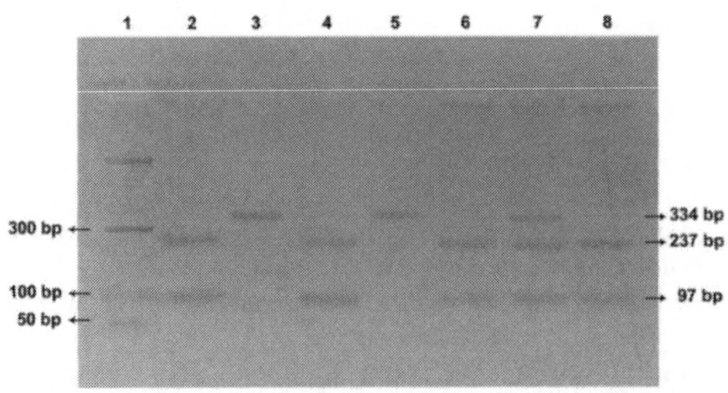

Figure 1. PCR-RFLP result. (1) 50 bp DNA ladder; (2,4,6,8): CC genotype; (3,5): AA genotype; (7): AC genotype.

Gene amplification was based on nucleotide sequences from GenBank no. access SNP rs42688595. The PCR product produced was 352 bp. PCR products were cut with restriction enzymes resulting in a different pattern of band lengths that indicate variation in one locus (figure 1). Mutations (SNP) occur in the +501 nucleotide sequence or the 238th order from the forward primer. Three genotypes were found, i.e. AA, AC, and CC. In the CC genotype, the C allele (normal allele) does not change so that the restriction enzyme can recognize the cutting site and cut it into two bands (97 bp and 237 bp). In the AA genotype, the C allele mutates into A allele so that the restriction enzyme cannot recognize the cutting site and only form one band (334 bp). This point mutation is occurred because of transversal substitution from pyrimidine to purine bases cause changes in codons from CGA (arginine) to AGA (arginine). A change in a base pair that causes a change in genetic code but does not result in amino acid changes (synonymous) is categorized as a silent mutation [7]. The AC genotype contains A allele and C allele, so it formed 3 bands: 97 bp, 237 bp, and 334 bp.

The association of LOC514211 rs42688595 with reproductive traits are presented in table 1. Heterozygote animals (AC genotype) showed the lowest days open (DO) (152.06±20.54 days) compared to homozygous AA genotype (270.14±26.27 days) and CC genotype (236.46±23.34 days) (P=0.024). AC genotype also showed the lowest calving interval (CI) (427.06±20.26) compared to AA (552.06±26.29) and CC (514.95±22.86) (P=0.014). On the other hand, genotypes were not associated with age at first calving (AFC) and service per conception (S/C) (P>0.05).

Table 1. Association of LOC514211 rs42688595 with reproductive traits.

Parameter	Genotype ($\bar{Y}$±SEM)			P
	AA (n=36)	AC (n=16)	CC (n=37)	
AFC	845.72±28.07	818.69±26.21	816.14±15.83	0.599
DO	270.14±26.27^b	152.06±20.54^a	236.46±23.34^b	0.024
S/C	2.89±0.30	1.94±0.27	2.86±0.33	0.176
CI	552.06±26.29^b	427.06±20.26^a	514.95±22.86^b	0.014

AFC= age of first calving (days), DO= days open (days), S/C= service per conception, CI= calving interval (days). Different superscripts on same row show significant differences (P<0.05).

In the previous study [5], A allele was called T, so the TT and TC genotypes in that study were the same as AA and AC genotypes in this study. In contrast to this study, the genotypes in the previous study were not associated with DO and CI, both at parity 1, 2, and 3, but associated with days to first service (DFS). The AC genotype showed the lowest DFS compared to the other two genotypes at the second parity. This is in line with this current study that heterozygous genotypes show the best reproductive performance. If the DFS is low, the DO and CI can also be lowered, assuming if the S/C is the same value. The similarity of the two studies is that genotypes are not associated with the AFC.

Genotype does not affect the S/C, possibility because the S/C is more influenced by environmental factors. This is indicated by the generally small S/C heritability value. S/C heritability values from various studies including 0.001 [8]; 0.02 [9]; and 0.03 [10]. Low S/C heritability value means that genetic additives that affect S/C are small so that improvements through direct selection will be slow [11]. The intended environmental factors are feed (quality and quantity), the number of cows, reproductive disorders, and reproductive management including frozen semen handling, heat detection, skills, the optimum timing of insemination, etc [12,13].

The LOC514211 gene is located on chromosome 13 in the *Bos taurus*. Although its role has not been known with certainty but based on the GWAS (genome-wide association) study by Junjing [4], SNPs on this gene have a good effect on the production traits without worsening their reproductive performance. GWAS is an observational study of a genome-wide set of genetic variants to see if any variant is associated with a trait [14]. From various studies [15-17], there was an unfavorable genetic correlation between milk production and fertility, so improvements in production traits will generally have a negative effect on reproductive traits, and *vice versa*. However, with GWAS technology, where many traits can be analyzed at once, candidate genes that have beneficial effects on production and reproduction, can be chosen, including in this LOC514211 gene. Mutations that occurred in the SNP position were synonymous. However, Hunt *et al* [18] stated that synonymous mutations can still influence gene expression by affecting the enzymatic structure.

4. Conclusion

The LOC514211 rs42688595 is associated with Indonesian-Holstein cow's reproductive traits. It shows a potency to be used as a genetic marker for breeding selection program based on reproductive traits. Selection could be done by increasing the genetic frequency of heterozygous animals (AC) to reduce days open and calving interval. Exploration of SNP in other positions of this gene is a prospective study opportunity for cattle selection.

Acknowledgments
This study was financed by the Indonesia Endowment Fund for Education through *Doctoral Programs Scholarship* in the year 2016, contract number 20160222035511.

References

[1] Rahayu A P, Kurnianto E and Johari S 2015 *J. Indonesian Trop. Anim. Agric.* **40** 79-86
[2] Bartoň L, Bureš D, Kott T and Řehák D 2016 *Meat. Sci.* **114** 18-23
[3] De Mello F, Kern E L and Bertoli C D 2014 *Adv. Dairy Res.* **2** 1-2
[4] Junjing W 2012 Genome-wide association study of reproduction and milk production traits in dairy cows and water buffalo (China: Huazhong Agricultural University)
[5] Anggraeni A S, Kurnianto E, Johari S, Sutopo and Shujun Z 2015 *J. Indonesian Trop. Anim. Agric.* **40** 191-8
[6] Matsumoto H, Inada S, Kobayashi E, Abe T, Hasebe H, Sasazaki S, Oyama K and Mannen H 2012 *Livest. Sci.* **144** 281-4
[7] Widayanti R, Solihin D D, Sajuthi D and Perwitasari R R D 2006 *J. Vet. Sci.* **24** 126-31
[8] Prahanisa H, Sumadi and Adiarto 2011 *Buletin Peternakan* **35** 1-10
[9] Biffani S, Canavesi R and Samore A B 2005 *Stočarstvo.* **59** 145-53
[10] Cassell B 2009 *Using Heritability for Genetic Improvement* (Virginia: Virginia Tech)
[11] Kealey C G, MacNeil M D, Tess M W, Geary T W and Bellows R A 2006 *J. Anim. Sci.* **84** 283-90
[12] Caraviello D Z 2004 *Reprod. Genet.* **611** 1-8
[13] Cassandro M 2014 *Acta Agr. Kapos.* **18** 11-23
[14] Bertolini F, Schiavo G, Galimberti G, Bovo S, D'Andrea M, Gallo M, Buttazzoni L, Rothschild M F and Fontanesi L 2018 *Animal.* **12** 1777-82
[15] Chagunda M G G, Bruns E W, Wollny C B A and King H M 2004 *Livest. Res. Rural Dev.* **16** 47-54
[16] González-Recio O., Alenda R, Chang Y M, Weigel K A and Gianola D 2006 *J. Dairy Sci.* **89** 4438-44
[17] Gunawan A, Sari R, Parwoto Y and Uddin M J 2011 *J. Indonesian Trop. Anim. Agric.* **36** 83-90
[18] Hunt R, Sauna Z E, Ambudkar S V, Gottesman M M and Kimchi-Sarfaty C 2009 *Methods Mol. Biol.* **578** 23-39

Development of sauce production technology based on cultivated oyster mushrooms (*Pleurotus ostreatus* (Jacq.) P. Kumm)

A A Dril[1], L N Rozhdestvenskaya and A N Sapozhnikov

Department of Technology and Organization of Food Industries, Faculty of Business, Novosibirsk State Technical University, 20 Prospekt K. Marksa, Novosibirsk, 630073, Russian Federation

E-mail: dril@corp.nstu.ru

Abstract. Cultivated oyster mushrooms (*Pleurotus ostreatus* (Jacq.) P. Kumm) are common for many countries of the world mainly because of their high yield. Their growing doesn't require special conditions. The mushrooms contain useful nutritional substances and have high sensory characteristics, so they can be suitable for developing of new types of culinary production. In the paper, the technology of sauce production using paste-like semi-finished products from oyster mushrooms is described. While obtaining the semi-finished products, it is advisable to use chopping of raw mushrooms before their heat treatment in combi oven by 100 °C during 30 min. The hot and cold sauce production samples based on the semi-finished products have got mainly high points for their sensory evaluation. The estimated nutritional value of sauces shows that they appear to be low-calorie products containing vitamins and mineral substances. Thus, the developed sauce production based on cultivated oyster mushrooms can be recommended for further researches and practical implementation at public catering enterprises.

1. Introduction

Oyster mushroom (*Pleurotus ostreatus* (Jacq.) P. Kumm) is one of the most common cultivated edible mushrooms in the world. It belongs to *Pleurotus* species, which represent a well-defined group of *Basidiomyceteous* fungi of the *Agaricales* order and *Tricholomataceae* family. The mushrooms are characterized by the production of fruiting bodies with an eccentric stalk and a wide cap shaped like an oyster shell, with the widest portion of the cap being away from the stalk. They grow over a wide range of temperatures and can colonize wide spectrum of unfermented, natural, lignin-cellulosic wastes. Because of fast mycelium growth rate, they colonize substrates rapidly; the yield of fruiting bodies is also high. The bifactorial inheritance, observed in many of the species, suggests the likelihood of a high degree of genetic variability, and, hence, considerable breeding potential. Oyster mushrooms can be grown on a commercial scale, without the need for composting and artificial conditioning of the ambient temperature [1]. The further processing of cultivated mushrooms can be carried out by their pickling, salting, drying, obtaining of fried and frozen semi-finished products, etc. [2]. They can also be sold in retail in fresh state.

At the same time, among existing technologies of oyster mushrooms processing there is no technology, which can be used at public catering enterprises for preparing culinary production. This technology proposes decreasing the amount of technological operations during preparation. Thus, the process can be faster and more effective.

Content from this work may be used under the terms of the Creative Commons Attribution 3.0 licence. Any further distribution of this work must maintain attribution to the author(s) and the title of the work, journal citation and DOI.
Published under licence by IOP Publishing Ltd

Puree soups and sauces from cultivated edible mushrooms are common dishes in many countries. However, the heat treatment and grinding of mushrooms can take a lot of time and effort. For the optimization of these operations we suggest sauce production technology based on paste-like semi-finished products from oyster mushrooms, which is described in the paper.

2. Theoretical researches

The study of cultivated oyster mushrooms chemical composition is carried out widely enough. The information on their chemical composition and nutritional value varies significantly. According to the information from different authors, oyster mushrooms contain proteins ranging from 15 to 34%, lipids from 1.5 to 8.7%, carbohydrates from 19.5 to 52%, ash from 5 to 11.4% and moisture from 85 to 90% [1, 3–8].

It is well known, that oyster mushrooms have a complete set of essential amino acids such as tryptophan, cystine, aspartic acid, lysine, alanine [3, 5], as well as vitamins B_1 (0.35–0.92 mg/100g), B_2 (1.17–1.74 mg/100g), B_3(1.72–2.75 mg/100g), B_5 (6.85–8.0 mg/100g), C (3.27–3.65 mg/100g), D (3.8–4.22 mg/100g), etc. [4].

Macro- and microelements are presented by nitrogen (4.5 mg/100g), phosphorus (0.81 mg/100g), potassium (1.3 mg/100g), calcium (32.25 mg/100g), magnesium (10.62 mg/100g), iron (46.8 mg/100g), copper (3.5 mg/100g), manganese (2.43 mg/100g) and zinc (12.88 mg/100g) [6].

Fatty acids in oyster mushrooms are mainly presented by monounsaturated oleic acid (C18:1) and polyunsaturated linoleic acid (C18:2n–6c). These fatty acids provide normal tissue growth and prevent the cholesterol level increase [4].

The high content of pure protein in oyster mushrooms fruiting bodies helps to prevent and treat hepatitis, stomach ulcers, reduces the amount of cholesterol in the blood, helps to normalize blood pressure during both hyper- and hypotension, has antitumor effect and boosts the body's immune resistance. The bactericidal characteristics of oyster mushrooms give them anti-radiation and radio-protective properties. Isolated extracts obtained from oyster mushrooms are used to prevent cardiovascular diseases [4, 9–11]. Carbohydrates contained in mushrooms are used to produce several substances: melanin, beta-glucans, that are used as anti-cancer biologically active supplements and drugs [12].

The taste and smell of mushrooms are composed from volatile metabolic products compounds of the group of aliphatic aldehydes, ketones, alcohols (among which octane derivatives are most often identified) and the group of nitrogen-containing compounds, including simple amines, amides, amino acids, derivatives of glutamic acid, etc. These substances create specific taste and smell of heat-treated oyster mushrooms, which can be attractive for consumers [13].

All these beneficial features of oyster mushrooms allow to increase their cultivation in many countries using various substrates [4–6, 14–17]. The world leader in cultivated edible mushroom production (mainly two-sided champignons, oyster mushrooms and shiitake) is China, where there are more than 30 billion kg of edible mushrooms (almost 87% of global production) are produced every year [14]. At the same time, the edible mushroom cultivation is increasing in Russia mainly because of food embargo on products imported from USA, Europe and other countries. Thus, oyster mushrooms can be available for Russian consumers both in fresh and processed forms because of their high yield and low cost [15–17], and it is necessary to develop various approaches to their integrated processing technology.

Processed oyster mushrooms can be quite interesting for public catering enterprises. One of the main tasks for public catering is to gain the increase through improving consumer demands. This can be reached by using semi-finished products obtained from oyster mushrooms in culinary production formulations. These semi-finished products allow to reduce preparation time of culinary production and expand its range. They may have different forms (dried, salted, pickled, etc.) and can be used in soups, main course dishes, pastry and bakery production [2]. Among that, the processed oyster mushrooms can be used in a paste-like form in sauce production formulations, which we have developed in our researches.

3. Experimental researches

Mushroom sauces are suitable for many kinds of food production: such as meat, fish and vegetable dishes, as well as pasta or cereals. They are used both in hot and cold state.

It is known that traditional mushroom sauces formulations are based on mushroom broth or extract. In the developed formulations, paste-like semi-finished products from oyster mushrooms were used as base ingredients. The oyster mushroom fruiting bodies parts were ground and heat-treated. Further, the starch-containing thickening agents (potato puree, rice puree or heat-treated wheat flour) were added and the mixtures were homogenized. Herewith, the quality of used thickening agents can be evaluated only after the sauce is ready. The evaluation was carried out by sensory method during acceptable shelf life and realization of sauce production for 3 hours. The main criterium for choosing of thickening agent was sauce consistency (stability of sauce emulsion).

The formulations and technology of developed sauces are based on semi-finished products from oyster mushrooms with further addition of ingredients used in existing sauce formulations. For obtaining of hot sauces the semi-finished products were mixed with milk, cream, butter, species, mushroom salts, etc. and heated up. For obtaining of cold sauces the semi-finished products were mixed with dairy products, vegetable oil, green vegetables, beets, etc.

The general model of obtaining semi-finished products from oyster mushrooms is shown on Figure 1.

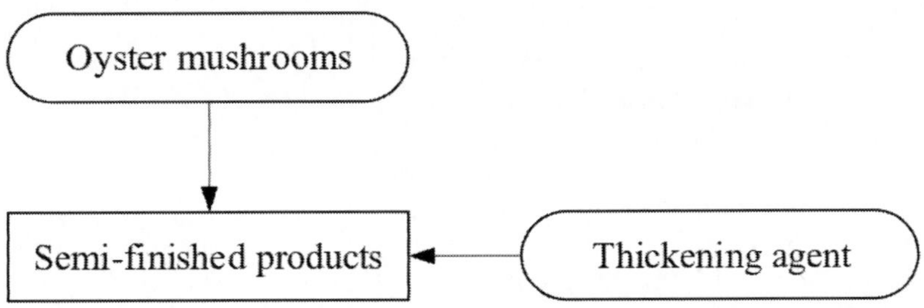

Figure 1. Obtaining of semi-finished products from oyster mushrooms general model.

In the first step of obtaining semi-finished products the caps were cut from druses. It is recommended to use oyster mushroom fruiting bodies without a cap, since:

- such processing methods as pickling, frying, stuffing preparing, etc. are more applicable for caps; they can be sold fresh too;
- sensory characteristics of paste-like semi-finished products from caps have low values (dark color, liquid consistency, strong smell, etc.);
- oyster mushroom stalks contain more vitamins, macro- and micro elements and amino acids;
- losses during heat treatment are greater for caps than for stalks, because dry substances content in caps (including cellulose) is 1.4–1.5 times less [7].

The flow chart of semi-finished products from obtaining oyster mushrooms based on formulation models and technological properties of mushrooms was developed (Figure 2).

The technological properties of oyster mushrooms substantially depend on proteins, fibers and water content [13]. This causes a different behavior of mushroom parts during their heat treatment. It is possible to predict large losses during heat treatment and its shorter duration for caps than for stalks. This also indicates that caps and stalks of oyster mushrooms should not be used simultaneously for the same methods of heat treatment – such as steam cooking in a combo oven. This method was chosen because most food substances are preserved in the product, and they are not extracted into the broth.

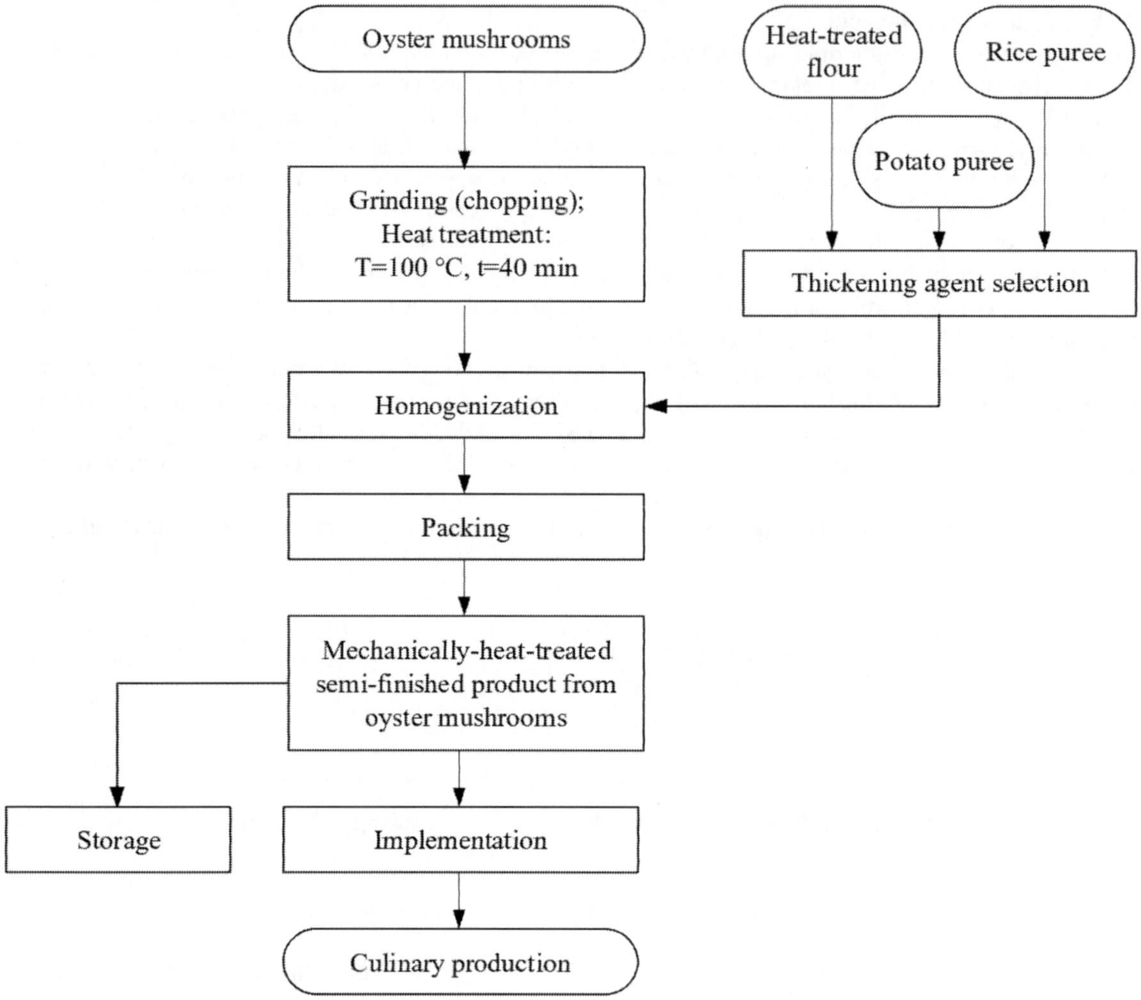

Figure 2. Obtaining of semi-finished products from oyster mushrooms flow chart.

The obtaining of paste-like semi-finished products from oyster mushrooms was carried out in two following ways:

Technology № 1. Stages:

- Heat treatment of raw oyster mushrooms;
- Grinding of heat-treated mushrooms;
- Homogenization with thickening agent.

Technology № 2. Stages:

- Chopping raw oyster mushrooms;
- Heat treatment;
- Homogenization with thickening agent.

The weight losses, which were evaluated visually, were less while using technology № 2. This can be explained due to increasing of the heated surface area of mushrooms and decreasing of heat treatment duration. Thus, the technology № 2 is preferable, because it saves more nutritional substances contained in mushrooms together with the chosen heat treatment method.

While the technology of obtaining semi-finished product from oyster mushrooms was developed, their technological properties were considered through the different heat treatment duration for caps and stalks. The weight losses of the oyster mushroom fruiting bodies parts after heat treatment are given in the Table 1. The dynamics of the change in the weight of oyster mushrooms parts depending on heat treatment duration and temperature are shown in the Figure 3.

Table 1. Changes of the oyster mushroom parts weights after heat treatment (HT).

Mushroom part	HT by 100 °C, 30 min			HT by 100 °C, 60 min			HT by 130 °C, 60 min		
	Weight before HT, g	Weight after HT, g	Losses, %	Weight before HT, g	Weight after HT, g	Losses, %	Weight before HT, g	Weight after HT, g	Losses, %
Caps	3010±5.0	2220±7.0	26.24± 1.06	278± 4.0	186± 2.0	33.09± 0.89	295± 2.0	146± 3.0	50.51± 0.39
Stalks	2490±8.0	2290±6.0	8.03± 0.37	261± 7.0	177± 4.0	32.18± 0.72	258± 3.0	131± 3.0	49.22± 0.98

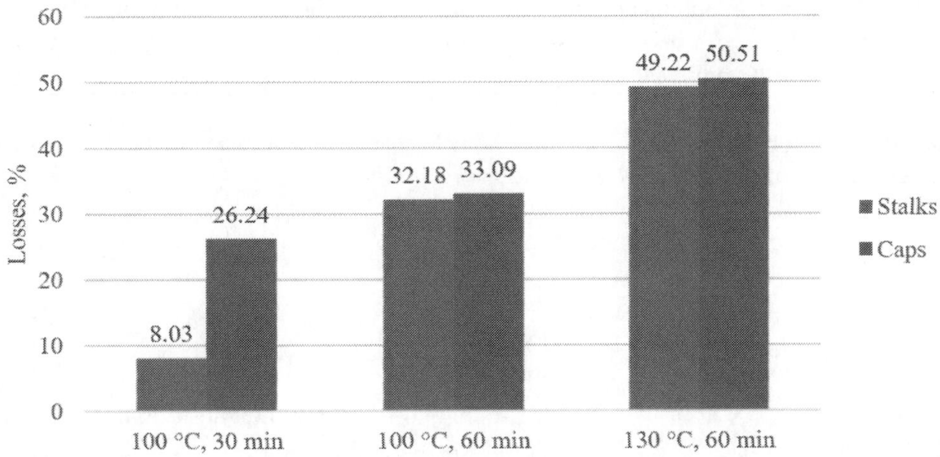

Figure 3. The magnitude of weight losses of caps and stalks of oyster mushrooms after heat treatment.

The results from Table and Figure show that the highest losses were for both caps (50.51%) and stalks (49.22%) heat-treated by 130 °C during 60 min. A better result was for both caps (33.09%) and stalks (32.18%) heat-treated by 100 °C during 60 min. Thus, the temperature lowering caused decrease for weight losses of heat-treated mushrooms. The least weight losses were found out for stalks (8.03%) heat-treated by 100 °C during 30 min. They are less approximately on 18%, than for caps (26.24%). Thus, 30 min for heat treatment duration of stalks is enough to obtain the product for further grinding and homogenization together with the thickening agent.

The obtained semi-finished products were the base ingredients for 6 developed sauce production formulations. The basis for determining the optimal ratio of the original sauce ingredients in formulations was sensory characteristics of the sauce production samples.

For imparting the aroma of forest mushrooms to developed production, white mushrooms and chanterelles were introduced into the samples in the form of mushroom salts (fine-ground dried white mushrooms or chanterelles).

For sensory evaluation, the sauce production samples were divided into two groups – hot and cold sauces. It was carried out by 5 semi-trained panelists. As a result, the total scores for each sensory characteristic (appearance, consistency, smell, taste and color) were obtained.

In the first group there were white mushroom sauce (sample № 1), chanterelle sauce (sample № 2), cheese sauce (sample № 3); in the second group (cold sauces) were tartar (sample № 4) and pesto sauces with olive oil (sample № 5) and sour cream with beets puree (sample № 6).

The Table 2 presents the developed requirements for sensory evaluation of new sauce production.

Table 2. Requirements for sensory evaluation of sauce production.

Quality indicators	**Characteristics**
Appearance and consistency	Homogeneous, viscous, non-stratified wiped mass, without extraneous inclusions, without films on the surface, with evenly distributed ingredients.
Smell	Pleasant, pronounced, specific for incoming ingredients, without extraneous smells.
Taste	Pleasant, pronounced, specific for incoming ingredients, slightly salted, moderately sour (for sauces with dairy products), without bitterness or extraneous unspecific taste.
Color	Pronounced, uniform, from beige to light gray, red purple (for sauces with beets), appropriate to the color of the incoming ingredients and additives.

The results of sensory evaluation of samples № 1, № 2 and № 3 are presented in Figure 4, the results of sensory evaluation of samples № 4, № 5 and № 6 – in Figure 5.

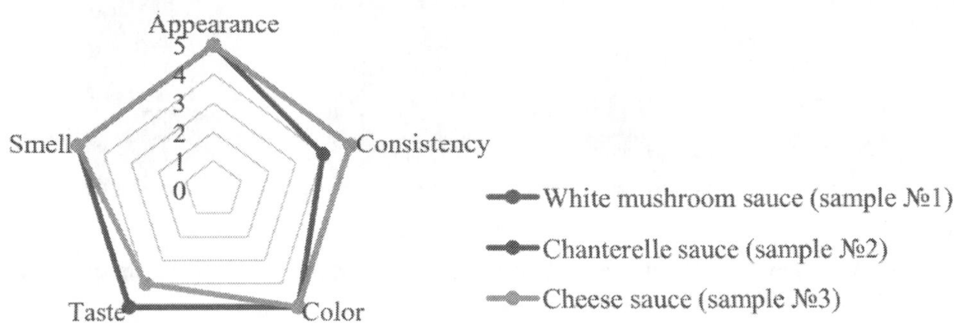

Figure 4. Results of sensory evaluations for the first group of sauces.

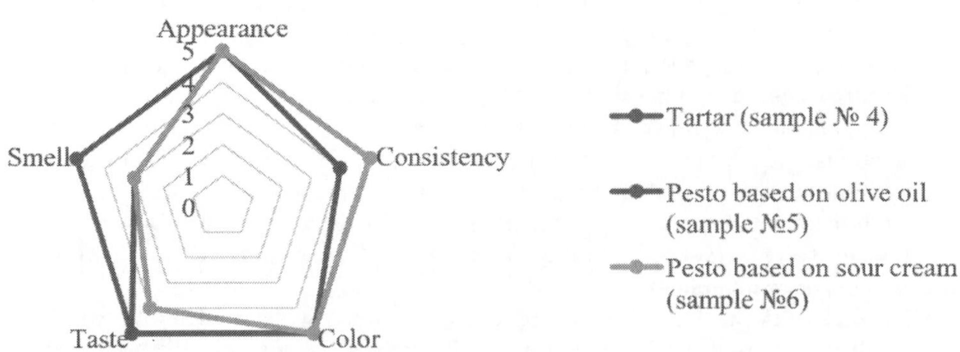

Figure 5. Results of sensory evaluations for the second group of sauces.

All samples from the first group received high points for sensory evaluation. The best one was the sample № 1. The samples № 2 and № 3, according to panelists' opinion, are characterized by a less distinguishable taste.

In the second group, each sample received lower points for sensory characteristics compared to the first group, due to their specific smell and taste of dill and parsley.

The results of the evaluation show that sensory characteristics of developed sauces are high. Therefore, the formulations of the samples from the second group may be improved through correcting the ratios of the used sauce ingredients.

The Table 3 shows the nutritional value of developed mushroom sauces, which was calculated based on theoretical data from "Chemical Compositions of Russian Food Products Manual" (2002).

Table 3. Calculation of nutritional value of sauces (in 100 g of production).

Sauces samples	Proteins, g	Fats, g	Carbohydrates, g	Calorie, kcal	Ash, g	K, mg
Sample № 1	5.4	4.0	13.5	111	1.72	1.85
Sample № 2	6.6	4.4	15.3	127	1.86	1.79
Sample № 3	7.7	7.6	13.1	152	1.0	0.43
Sample № 4	4.3	3.1	7.2	74	1.49	0.12
Sample № 5	4.5	4.9	7.6	92	1.66	0.6
Sample № 6	4.8	2.6	8.0	75	1.48	0.12
Sauces samples	**Carotene, mg**	**B_1, mg**	**B_2, mg**	**PP, mg**	**B_5, mg**	
Sample № 1	1.75	390.2	297.89	1376.12	41.41	
Sample № 2	1.69	422.74	276.74	1403.39	41.41	
Sample № 3	0.03	0.11	0.29	4.04	0.86	
Sample № 4	4.36	186.27	399.3	890.77	178.48	
Sample № 5	0.04	0.14	0.39	5.09	1.29	
Sample № 6	5.01	193.48	402.54	880.5	192.65	

As shown in the Table, all sauce samples are appeared to be low-calorie products. The highest protein, fats and carbohydrates content has sample № 3 because of cheese ingredient, the lowest substance content have samples № 4 and № 6. All samples also contain different amounts of such substances as potassium, carotene vitamins B_1, B_2, PP, B_5. The obtained results of calculation correspond to results of other authors [1, 3–8]. For their further confirmation, the practical researches of sauce production nutritional value must be carried out.

4. Conclusion

The results of carried out researches have shown that paste-like semi-finished products from cultivated oyster mushrooms are perspective ingredients for new low-calorie sauce production with high sensory characteristics.

The researches have also shown that the most useful parts of mushrooms for paste-like semi-finished products and sauce production are mushrooms stalks. The production samples from them had high sensory characteristics and food substances content because of less weight loss in stalks during heat treatment.

Thus, the developed sauce production line can expand the mushroom sauces range and improve sensory characteristics of various culinary production, which will be carried out in further researches.

References

[1] Rajarathnam S, Bano Z and Miles P G 1987 *Pleurotus* mushrooms. Part I A. morphology, life cycle, taxonomy, breeding, and cultivation *C R C Critical Reviews Food Sci. Nutri.* **26(2)** 157–

223

[2] Thakur M P 2018 Advances in post-harvest technology and value additions of edible mushrooms *Indian Phytopathol.* **71(3)** 303–15

[3] Velichko N A and Berickashvili Z N 2008 The chemical compostition of *Pleurotus Ostreatus (Fr) Kumm* fruiting bodies *Bulletin of KrasGAU* **4** 274–8

[4] Johathan S G, Okon C B, Oyelakin A O and Oluranti O O 2012 Nutritional values of oyster mushroom (*Pleurotus ostreatus*) (Jacq. Fr.) Kumm. cultivated on different agricultural wastes *Nature Sci.* **10(9)** 186–91

[5] Deepalakshmi K and Mirunalini S 2014 Pleurotus ostreatus: an oyster mushroom with nutritional and medicinal properties *J Biochem Tech* **5(2)** 718–26

[6] Ahmed M, Abdullah N and Nuruddin M M 2016 Yield and nutritional composition of oyster mushrooms: an alternative nutritional source for rural people *Sains Malaysiana* **45(11)** 1609–15

[7] Oluwafemi G, Seidu K T and Fagbemi T N 2016 Chemical composition, functional properties and protein fractionation of edible oyster mushroom (*Pleurotus ostreatus*) *Annals. Food Sci. Techol.* **17(1)** 218–23

[8] Tolera K D and Abera S 2017 Nutritional quality of Oyster Mushroom (*Pleurotus Ostreatus*) as affected by osmotic pretreatments and drying methods *Food Sci Nutr* **5(5)** 989–96

[9] Corrêa R C G, Brugnari T, Bracht A, Peralta R M *et. al.* 2016 Biotechnological, nutritional and therapeutic uses of *Pleurotus* spp. (Oyster mushroom) related with its chemical composition: A review on the past decade findings *Trends Food Sci. Technol.* **50** 103–17

[10] Shah S, Ghosh D, Mallick S K, Sarangi I *et. al.* 2007 Immunomodulatory and antitumor activities of water-soluble proteoglycan isolated from the fruiting bodies of culinary-medicinal oyster mushroom *Pleurotus ostreatus* (Jacq.: Fr.) P. Kumm. (*Agaricomycetideae*) *Internat. J. Medic. Mushrooms* **9(2)** 123–8

[11] Llaurado G, Morris H J, Tamayo V, Lebeque Y, Beltran Y *et. al.* 2015 Haematopoiesis radioprotection in Balb/c mice by an aqueous mycelium extract from the Basidiomycete Pleurotus ostreatus mushroom *Nat Prod Res* **29(16)** 1557–61

[12] Golak-Siwulska I, Kałużewicz A, Spiżewski T, Siwulski M *et. al.* 2018 Bioactive compounds and medicinal properties of Oyster mushrooms (*Pleurotus* sp.) *Folia Hort* **30(2)** 191–201

[13] Misharina T A, Muhutdinova S M, Zharikova G G, Terenina M B and Krikunova N I 2009 The composition of volatile components of cepe (*Boletus edulis*) and oyster mushrooms (*Pleurotus ostreatus*) *Applied Biochemistry and Microbiology* **45(2)** 187–93

[14] Kapahi M 2018 Recent advances in cultivation of edible mushrooms *Water Resources in Central Asia: International Context*, ed S S Zhiltsov, I S Zonn, A G Kostianoy and A V Semenov (Springer International Publishing) pp. 275–86

[15] Mandeel Q A, Al-Laith A A and Mohammed S A 2005 Cultivation of oyster mushrooms (*Pleurotus* spp.) on various lignocellulosic wastes *World J. Microbiol. Biotechnol.* **21(4)** 601–7

[16] Pant D, Gangi Reddy U and Adholeya A 2006 Cultivation of oyster mushrooms on wheat straw and bagasse substrate amended with distillery effluent *World J. Microbiol. Biotechnol.* **22(3)** 267–75

[17] Girmay Z, Gorems W, Birhanu G and Zewdie S 2016 Growth and yield performance of *Pleurotus ostreatus* (Jacq. Fr.) Kumm (oyster mushroom) on different substrates *AMB Expr* **6** 87

IOP Conf. Series: Earth and Environmental Science **346** (2019) 012065 doi:10.1088/1755-1315/346/1/012065

Food security in Java Island, Indonesia: Panel data of ordinary least square approach

I M Y Prasada[1,3] and Masyhuri[2]

[1]Agricultural Economics, Faculty of Agriculture, Universitas Gadjah Mada, Yogyakarta, Indonesia
[2]Faculty of Agriculture, Universitas Gadjah Mada, Yogyakarta, Indonesia

E-mail: imade.yogap@gmail.com

Abstract. Food security over the past few years has become a major problem faced by developing countries, including Indonesia, and needs to be resolved properly. Therefore, this study aims to determine the factors that influence the food security of the population in Java Island. The data used in this study were secondary data taken from the Central Bureau of Statistics (BPS) in 2008-2017, namely data on real prices of cooking oil, real sugar prices, real beef prices, real price of rice, food and non-food expenditures, real per capita income, and area of agricultural land. The data was analyzed using a fixed effect regression panel data model. The results of the study show that the factors influencing the share of food expenditure in Java Island were the variables of real prices of rice and real per capita income.

1. Introduction

Food is a very basic human need because it influences human existence and survival, both in terms of quantity and quality. The Law of the Republic of Indonesia Number 18 of 2012 defines food as everything derived from biological sources of agricultural, plantation, forestry, fisheries, water and water products, both processed and unprocessed which are intended as food or beverages for human consumption, including ingredients of additional food, food raw materials, and other materials used in the process of preparing, processing and making food or drinks [1]. Food needs that can be fulfilled properly will have consequences for maintaining the level of food security of the population, especially the staple food of the population, i.e. rice [2].

Food security of population is a problem that is often faced by developing countries including Indonesia. In Indonesia, food security is a problem that needs to be resolved properly since the high level of food security will reflect the better quality of life of the population. The concept of food security keeps developing from time to time. In Indonesia, it was initiated for the first time through the Law of the Republic of Indonesia Number 7 of 1996 on Food stating that food security is a condition of fulfilling household food which is reflected in the availability of adequate food, both in quantity and quality, safe, equitable and affordable [3]. In this concept, food security only considers aspects of food availability for the population in Indonesia. However, in 2012, the law was revised into the Law of the Republic of Indonesia Number 18 of 2012. It defines food security as a condition of fulfilling food for the country up to the individual level, which is reflected in the availability of adequate food, both in number and quality, safe, diverse, nutritious, equitable and affordable and does not conflict with religion, beliefs and culture of society, to be able to live a healthy, active and productive life in a sustainable manner [1]. Based on the Law of the Republic of Indonesia Number 18 of 2012, it can be

Content from this work may be used under the terms of the Creative Commons Attribution 3.0 licence. Any further distribution of this work must maintain attribution to the author(s) and the title of the work, journal citation and DOI.
Published under licence by IOP Publishing Ltd

concluded that the focus of measuring the current level of food security is not only on aspects of food supply, but also aspects of food access for the population, and aspects of utility or absorption of food, both at the regional and level household and individual [4,5].

The level of food security of a region can be determined using various methods. One of those methods is calculating the share of food expenditure (SFE) [6]. It is the proportion of food expenditure to total household expenditure, and it has a negative relationship with household income which has been empirically proven that the higher the household income, the lower the share of food expenditure.

Studies related to the level of food security of the population in a region have been carried out. With primary data and OLS (Ordinary Least Square) method, a research in Manokwari Regency measures household food security with a share of food expenditure. The results show that the factors affecting household food security in Manokwari Regency are income, cooking oil prices and kerosene prices; while variables of the number of family members, farmer education, rice prices, sugar prices, vegetable prices, fish prices and egg prices do not affect the share of food expenditure [7]. Furthermore, research carried out in Iran using the binary logistic method concluded that the average per capita income has a negative significant effect [8]. Research in Nepal and Bangladesh concluded that the area of land tenure is a factor that has a positive effect on the level of household food security [9].

Java Island is one of the islands of rice production centers in Indonesia [10]. Based on data from the Central Bureau of Statistics of the Republic of Indonesia, rice production averaged 33,511,839 tons per year in Java, equivalent to rice of 19,316,491 tons per year in the period 2004-2015. Rice production in Java Island accounts for 53.63% of national rice production, with an average rice consumption per capita population reaching 98.03 kg / capita / year. Therefore, it can be seen that the island of Java plays a very important role for the availability of food for the population in Indonesia. Although Java Island is a food production center with high food availability, based on studies conducted by BKP in 2017, it shows that there are still three provinces included in the category of food insecurity, namely East Java, West Java and Banten [11]. Therefore, this study aims to determine the factors that influence the level of food security of the population in Java Island.

2. Materials and methods

The data used in this study are secondary data which includes data on real prices of beef (BP), sugar (SP), cooking oil (COP), rice (RP), real per capita income (PCI), and area of agricultural land (AL). Data sources were obtained from the Central Bureau of Statistics (BPS) and the method used in this research was panel data regression analysis.

Panel data regression analysis is an analysis that combines two data, namely cross section data and time series which use observational data on one or more variables in one unit continuously over a period of time. This regression analysis is known to have several advantages, such as control heterogeneity of individuals with cross section data assumed to be homogeneous (no other influences entered); for example, in this case time, and in time series data, data will change at every interval period of time. This regression is also known to provide more informative data and can minimize the bias that occurs because it categorizes units into larger groups [12].

The time series data used is the data from 2008-2017 with cross section data retrieved from three provinces in Java, namely West Java, East Java, and Banten. The three provinces were selected by purposive sampling method considering that the three provinces were rice production centers, but at the same time included in the category of food insecurity regions [11]. Data were analyzed using Eviews 9.0 software. The panel data regression equation to find out the factors that influence the food security of the population in Java can be seen in the equation as follows:

$$Ln\ (SFE) = \alpha_0 + \alpha_1 Ln(BP) + \alpha_2 Ln(SP) + \alpha_3 Ln(COP) + \alpha_4 Ln(RP) + \alpha_5\ Ln(PCI) + \alpha_6 Ln(AL) + u_i$$

Sign of expected estimate $\alpha_5, \alpha_6 < 0$; $\alpha_1, \alpha_2, \alpha_3, \alpha_4 > 0$

Where: SFE= Share of food expenditure (%), BP= Real prices of beef (Rp/Kg), SP= Real prices of sugar (Rp/Kg), COP= Real prices of cooking oil (Rp/Kg), RP= Real prices of rice (Rp/Kg), PCI= Real per capita income (Rp/capita), AL= Area of agricultural land (Ha), u_i= Error term

Estimates with panel data regression models are carried out with three approaches, including Common Effect Model (CEM), Fixed Effect Model (FEM), and random effect model (REM). CEM is the simplest approach assuming the intercepts of each variable are the same, and the coefficients on all-time series and cross section units. FEM approach is assumed that there is no time specific effect and only focuses on the specific individual effects where the intercepts of each individual are different, but the intercept for time series units is constant. Meanwhile, REM approach involves correlation between error terms due to changes time and research unit. The selection of the best panel data regression estimation model to be used among these models is based on the chow test, hausman test, and Breush Pagan Lagrange Multiplier test [13]. In this study, the random effect model cannot be done because the number of cross sections is smaller than the number of independent variables. Therefore, choosing the best model is done to choose between the common effect or fixed effect models using Chow Test [14]:

2.1. Chow test

The testing of the common effect model is better than the fixed effect model determined by the Chow test; so, the hypothesis is: H_0: $\alpha_1 = \alpha_2 = \ldots.. = \alpha_N$ (*common effect*); H_1: there is at least one different α_i i: 1,2,…,N. Test statistics is:

$$F = \frac{(RSSS\text{-}URSSS)/(N\text{-}1)}{URSS/(NT\text{-}N\text{-}K)}$$

Where: RSSS: *sum square residual* OLS model, URSS: sum square residual fixed model, N: number of cross section units, T: number of time units, K: the number of parameters to be estimated.

H_0 is rejected if F count> F (N-1, NT-N-K; α) which means that intercept for all unit cross sections is not the same, then the regression model equation is a fixed effect model.

3. Result and discussion

This study uses secondary data, i.e. food security of population in Java Island from 2008 to 2017. Before being analyzed, the model used in this study has fulfilled all classical assumption tests. Therefore, a further test can be done to select the regression model using the best panel data through two estimation approaches, including the common effect and fixed effects models. The random effect model cannot be done because it does not fulfill the requirement for the random effect model, i.e. the number of cross sections is greater than the number of dependent variables in the model. The results of the analysis to determine the best model, namely between the common effect model and the fixed effect, can be done using the Chow Test. The results of the analysis can be seen in table 1 as follows.

Table 1. Chow test.

Effects Test	Statistics	d.f	Prob.
Cross-section F	6.151	(2.22)	0.008
Cross-section chi-square	13.325	2	0.001

Source: Secondary Data Analysis, 2019

According to table 1, it can be seen that the probability of the chi-square cross-section is 0.001 <α (0.05); thus, H_0 is rejected. In addition, it can be concluded that the fixed effect model is better than the common effect model. Therefore, established the best model between both, further analysis can be carried out in the form of multiple linear regression to determine the factors that influence food security of the population on Java. The regression results can be seen in table 2.

According to table 2, it can be seen that the Adjusted R^2 value is 0.520, indicating that 52.00% of the variation in food security of population variables in Java can be explained by the independent variables contained in the regression model. Meanwhile, the probability of an F test of 0.002 <α (0.01) can be interpreted that the variable real price of beef, real price of sugar, real price of cooking oil, real price of rice, real per capita income, and agricultural land area altogether affect food security variable significantly.

Table 2. Determinants of food security (share of food expenditures) population on Java Island.

Variables	Coefficient	Standard Error	t-statistics	Sig.
C	15.575*	7.948	1.959	0.063
BP	0.049	0.094	0.522	0.607
SP	0.001	0.056	0.017	0.986
RP	0.286**	0.111	2.578	0.018
COP	0.174	0.162	1.070	0.297
AL	-0.469	0.570	-0.822	0.420
PCI	-0.678**	0.262	-2.588	0.017
Adjusted R^2	0.520			
F statistics	4.934			
F. Sig	0.002			

Source: Secondary Data Analysis, 2019
Notes:** refers to significant at the error rate 5% (α=0.05)
* refers to significant at the error rate 10% (α=0.10)

The constant value that has a probability of 0.063 <α (0.10) indicates that if the independent variable contained in the model is considered constant or zero, the effect of other variables outside the model on the food security variable is 15.575. If other variables outside the model increase by 1%, food security will increase by 15.575%. In addition, as observed in table 2, it is known that the variables that have a significant effect on food security include rice prices and real per capita income of the population.

The variable of real price of rice shows a positive influence on the share of food expenditure. This positive influence shows that the increasing real price of rice will cause an increase in the share of food expenditure. The regression coefficient of real price of rice is 0.286, indicating that for every 1% increase in the real price of rice, the share of food expenditure will increase by 0.286%. The higher share of food expenditure indicates a lower level of food security. This can occur due to increasing food prices will cause food to be difficult to reach for the population as well as will lead to a decrease in food security. High food prices have an impact on price stability which will later affect food security and nutrition [15]. Food prices are important matters that need to be considered. Since even though food is available in large quantities, if the food cannot be bought by the community, the food needed to meet the living needs of the population is difficult to fulfill properly due to expensive food prices. Therefore, to maintain food security of the population, the stability of food prices needs to be considered, so that food prices become affordable, and the population can access food well and food security can be improved.

According to table 2, it can be seen that the regression coefficient of real per capita income is -0.678. Based on the results of regression analysis, it can be seen that the real per capita income variable has individually a significant effect on the share of food expenditure. It means that for every 1% increase in income, the share of food expenditure will decrease by 0.678%. On the other words, the higher the income of the population, the lower the share of food expenditure. It might occur since income is inversely proportional to the share of food expenditure. The declining share of food expenditure shows that food security is increasing. Then, increasing the income of the population will lead to an increase in the ability of households to buy various consumption needs with better quality. This result is consistent with research conducted in Nepal, Malawi, Tanzania, Uganda and Madagascar. The study shows that food security of population will be closely related to calories from food consumed by community, so that lower income will cause someone to spend their income more for food consumption than for non-food consumption. This situation will reverse when the population experiences an increase in income, so that there is residual income that can be spent on non-food consumption [16].

4. Conclusion
The food security of the population in Java is influenced by the variables of real price of rice and real per capita income of the population. The variable of real price of rice has a positive influence on the variable of share of food expenditure, while the income variable of the real per capita income of the population has a negative influence on the variable of share of food expenditure. The higher share of food expenditure will indicate a lower level of food security. Efforts to improve food security can be carried out by stabilizing affordable food prices for the population and increasing the per capita income of the population in Java Island.

Acknowledgments
Thank you, the authors convey to the Education Fund Management Institute (LPDP) for their assistance, both financially and non-financially so that the authors can publish the results of research related to the food security of the population in Java Island, Indonesia.

References
[1] UU RI No. 18 2012 Law No. 18 of 2012 Law of the Republic of Indonesia pp 1-83
[2] Negoro W J and Hadayani M M 2014 Analysis of the policy of rice stock availability (Case study on rice warehousing, Perum Bulog Kota Palu) *E-J. Agrotekbis* **2** 62-8
[3] UU RI No. 7 1996 Law No. 7 of 1996 Law of the Republic of Indonesia pp 1-56
[4] Nugroho C P and Mutisari R 2015 Analysis of Probolinggo City food security indicators: Spatial approach *Agrise* **XV** 166-81
[5] Triwindiyanti Q A F, Tertius E P and Mahmudiono T 2018 Differences and influence of food security indicators on the proportion of LBW in coastal areas of Java Island (Bangkalan District and Tulungagung Regency) *Amerta Nutr* **2018** 37-43
[6] Rahmi R D, Suratiyah K and Mulyo J H 2013 Food security of farmer households in Ponjong District, Gunungkidul Regency *Agro Ekonomi* **24** 190-200
[7] Sianipar J E, Hartono S and Hutapea R T P 2012 Analysis of farm household food security in Manokwari Regency *SEPA* **8** 51-182
[8] Rahim S, Saeed D, Rasool G A and Saeed G 2011 Factors influencing household food security status *Food Nutr. Sci.* **2** 31-4
[9] Faridi R and Wadood S N 2010 An econometric assessment of household food security in Bangladesh *Bangladesh Dev. Stud.* **XXXIII** 97-111
[10] BPS 2016 *Survey of Agricultural Production of Rice and Secondary Crops in Indonesia* A Achmad & J Lumban Eds. 1st ed. (Jakarta: Badan Pusat Statistik Indonesia)
[11] BKP 2017 *Food Security Statistics* 2016 (Jakarta: Badan Ketahanan Pangan Kementerian Pertanian)
[12] Ratnasari N P A M, Kencana I P E N and Gandhiadi G K 2014 The application of panel data regression with the fixed effect model approach (Case study: PT PLN Gianyar) *E-Journal Matematika* **3** 1-7
[13] Melliana A and Zain I 2013 Statistical analysis of factors that influence the Human Development Index in the district / city of East Java Province using panel regression *Journal Sains Dan Seni POMITS* **2** D237-42
[14] Astuti W I, Ratnasari V and Wibowo W 2017 Analysis of factors that influence the level of open unemployment in East Java Province uses panel data regression *Journal Sains Dan Seni ITS* **6** 150-6
[15] Obayelu A E 2010 Global food price increases and nutritional status of Nigerians: The determinants, coping strategies, policy responses and implications *ARPN J. Agri. Biol. Sci.* **5** 67-80

[16] Hjelm L, Mathiassen A and Wadhwa A 2016 Measuring poverty for food security analysis: Consumption-versus asset-based approaches *Food Nutr. Bull.* **37** 275-89

IOP Conf. Series: Earth and Environmental Science **346** (2019) 012066 doi:10.1088/1755-1315/346/1/012066

Application of the technology of encapsulation of rowanberry extract to develop a functional cottage cheese dessert

I V Matseychik[1], S M Korpacheva[1], E G Martynova[1, 3] and I O Lomovsky[2]

[1]Department of Technology and Organization of Food Industries, Faculty of Business, Novosibirsk State Technical University, 20 Prospekt K. Marksa, Novosibirsk, 630073, Russian Federation

[2]Institute of Solid-State Chemistry and Mechanochemistry of the Siberian Branch of the Russian Academy of Sciences, 18 Kutateladze str., Novosibirsk, 630128, Russian Federation

E-mail: liz.martynova@mail.ru

Abstract. The possibility of using the technology of encapsulation of rowanberry extract to remove the natural bitter taste of raw material is discussed in this paper. Fruits of the rowan native to Novosibirsk city and the Altai region (Russia) were subjected to infrared (IR) drying (the rowanberries were collected in September 2017 and 2018). A plant extract produced using the IR-dried rowanberry powder was utilized for encapsulation. Two encapsulation technologies, spray drying and freeze drying, were discussed and compared. Polysaccharides (konjak gum and guar gum) were used as encapsulating agents. It was found that encapsulation could significantly improve properties of the finished product. The developed encapsulated powder can be used to produce value-added functional cottage cheese dessert. The functional properties of the samples before and after encapsulation were verified using physicochemical methods. The resulting powder had increased contents of vitamin C, β-carotene, flavonoids, and antioxidants.

1. Introduction

Development of the technologies and recipes of functional food products is a very promising new trend in modern food industry. Addition of food products containing functional ingredients to the food consumption pattern is one of the key objectives of developing novel added-value food products that can improve the nutritional status of the population. The content of functional ingredients in these food products is $\geq$ 15% per serving. These products are known as functional foods.

Today, there is a steady trend towards using natural plant-based raw materials and their derivatives in functional food technology [1]. It has been suggested that fruits of the rowan (*Sorbus Aucuparia* L.), which is widespread throughout Siberia, can be used for developing a functional ingredient. Rowanberries can be utilized as an important source of vitamin P and carotene, as well as macro- and microelements [2]. However, red rowan berries are bitter, due to the high content of phenolic compounds, therefore their use as functional food ingredients is limited [3].

The polyphenolic complex contained in the fruits of mountain ash, helps to restore the suppressed activity of the immune system, providing a stimulating effect on the formation of antibodies and the bactericidal activity of immunocompetent cells. This property researchers substantiate the presence in the fruits of a complex of phenolic compounds, primarily proanthocyanidins, which have antioxidant, immunotropic, anti-inflammatory and capillary strengthening effect [4].

Content from this work may be used under the terms of the Creative Commons Attribution 3.0 licence. Any further distribution of this work must maintain attribution to the author(s) and the title of the work, journal citation and DOI.
Published under licence by IOP Publishing Ltd

IOP Conf. Series: Earth and Environmental Science **346** (2019) 012066 doi:10.1088/1755-1315/346/1/012066

The objective of this study was to use the technology of encapsulation of rowanberry powder produced by infrared (IR) drying in order to improve its sensory properties and to utilize the resulting powder for production of a functional cottage-cheese dessert.

Cottage cheese is a valuable raw material in terms of eliminating the existing nutrient deficiency in the human diet. The advantage of encapsulation is that when using this technology on bitter, but useful, powder, you can add more useful substance to the recipe of curd dessert.

2. Experimental

2.1. Characteristics of the raw materials used

Rowanberries and their derivatives contain a broad range of biologically active substances, such as bioflavonoids (239.1±0.3 mg/100g), anthocyanins (105.4±0.4 mg/100g), vitamin C (377.4±0.3 mg/100g), β-carotene (15.0±0.2 mg/100g), etc. Furthermore, rowanberries exhibit a high antioxidant capacity (AOC = 4.2±0.5 mg quercetin/g) [5].

Rowanberries reach their peak ripeness in September or October. At this point, the content of biologically active substances, vitamins, and minerals in them is the highest. However, most of these substances are destroyed during storage so that only non-nutritive ballast components are left. Biological activity of the plant-based supplement can be preserved by using infrared drying, which is especially relevant for Siberia [6].

In human nutrition, cottage cheese is of great importance as a source of proteins (14–17%), fats (9–18%), carbohydrates (lactose occupies an important place – 1.3–2.8%), mineral substances (1%) and vitamins (A, B_1, B_2). Cottage cheese is rich in calcium, phosphorus, iron, magnesium – the substances necessary for the growth and proper development of the young organism. Its salts of Ca and P are in an optimal ratio. Protein and fat cottage cheese are easily digested. The water content in cottage cheese is 65–80%.

The encapsulation technology (entrapment of one material in another one giving rise to microparticles) was used to neutralize the bitter flavor of rowanberry powder (Figure 1) [7].

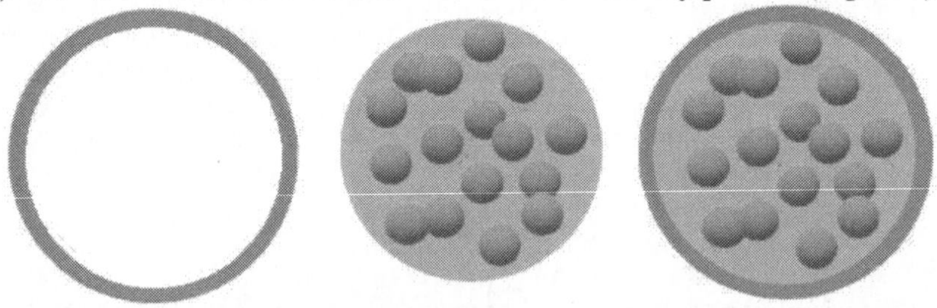

Figure 1. Types of encapsulation: a – encapsulation; b – incorporation of a substance in another substance; c – a combination of these two technologies.

In collaboration with the researchers from the Institute of Solid-State Chemistry and Mechanochemistry of the Siberian Branch of the Russian Academy of Sciences, we employed two technologies: spray drying and freeze drying.

2.2. Preparation of rowanberries for the study

Rowanberries harvested in the forest stands in the city of Novosibirsk and Ust-Muny rural locality (the Altai region, Russia) were subjected to infrared drying (SEDONA Express SD-6780). Drying was carried out at T ~ 55°C; ≥ 90% of biologically active substances was preserved.

The residual moisture content was measured using a WPS 50SX moisture analyzer. It was ≤ 5.7% [8].

The product was finely disintegrated on a DESI-11 disintegrator at 14,000 rpm under liquid nitrogen cooling. Intense mechanical treatment of IR-dried rowanberries gave rise to a brick-red fine-grained powder with the average grain size of 70 μm. The resulting product was used as the reference sample; its chemical composition was determined. Table 1 summarizes the results of the analyses.

Table 1. Chemical composition of IR-dried rowanberry powder.

Parameters	Actual value
Moisture content, g /100 g	5.7±0.01
Crude protein, g / 100 g	5.5±0.01
Crude fat, g / 100 g	4.5±0.01
Crude fiber, g / 100 g	5.7±0.03
Crude ash, g / 100 g	2.8±0.04
Sugar, g / 100 g	25.4±0.21
Calcium, g / 100 g	0.3±0.01
Vitamin C, mg / 100 g	312.9±0.29
β-carotene, mg / 100 g	12.5±0.15
Flavonoids, mg / 100 g	224.2±0.02
Iron, mg / 1000 g	102.0±0.18
Zinc, mg / 1000 g	10.5±0.09
AOC, mg quercetin / 100 g	5.6 ± 0.31

The content of water-soluble substances in the reference powder sample was determined for further experiments. Functional ingredients were extracted from the raw materials into a water phase on an ultrasonic bath at a 1:25 solid-to-liquid ratio (1 g of solid and 250 g (ml) of bidistilled water). Characteristics of an ultrasonic bath: model PS-40, volume 10 L, frequency 40 KHz, power 240W, manufacturer Omsk, Russia. Processing time – 1 hour. The extraction yield of water-soluble substances was 76% of the weight of the dry reference sample [9].

2.3. Preparation of encapsulation shells (polysaccharides)

Konjac gum was used as an encapsulating agent in spray drying encapsulation. This substance was chosen because it is stable at elevated temperatures that can be reached during spray drying. Guar gum was utilized for freeze drying encapsulation. Guar gum was chosen in this case because this substance retains its properties when the temperature is decreased to –30°C.

These two polysaccharides were prepared using the same scheme: the gum was added to water until it swelled. The as-prepared solutions of guar and konjac gums were mixed with rowanberry extract:

- *for Spray drying encapsulation* – 4 g of konjac gum + 250 g (ml) of bidistilled water and 5.3 g of rowanberry powder + 250 g (ml) of bidistilled water. Next, combine the two aqueous solutions and get 500 g of the working solution.
- *for Freeze drying* – 5,0 g of guar gum + 250 g (ml) of bidistilled water and 13,1 g of rowanberry powder + 250 g (ml) of bidistilled water. Next, combine the two aqueous solutions and get 500 g of the working solution for encapsulation.

2.4. Spray drying encapsulation

The working solution was spray-dried on a Buchi B290 spray dryer (inlet air temperature, 130 °C; outlet temperature, 70 °C) (Figure 2). After spraying, the resulting particles had a spherical shape; the average particle size was 10 μm [10].

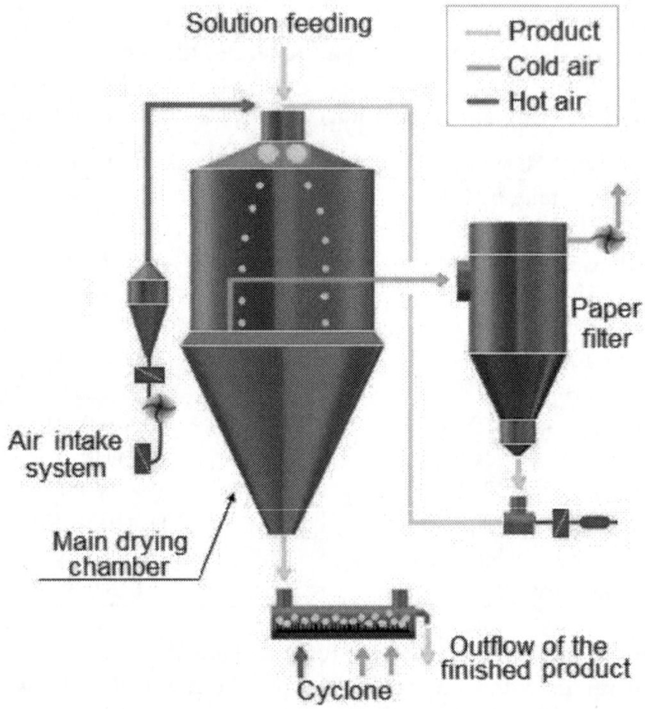

Figure 2. The operating principle of a spray dryer.

Drying gave rise to a cream-colored powder with neutral flavour. However, its yield was < 15% of the anticipated value, so another method, freeze-drying encapsulation, was also tested.

2.5. Freeze drying encapsulation

Freeze drying (lyophilization) was conducted using an Inei-4 setup (chamber pressure, 3 Pa; coolant temperature, – 40°C; sample temperature ≤ – 5°C) (Figure 3). The working solution consisted of an aqueous solution of guar gum and the rowanberry extract [11].

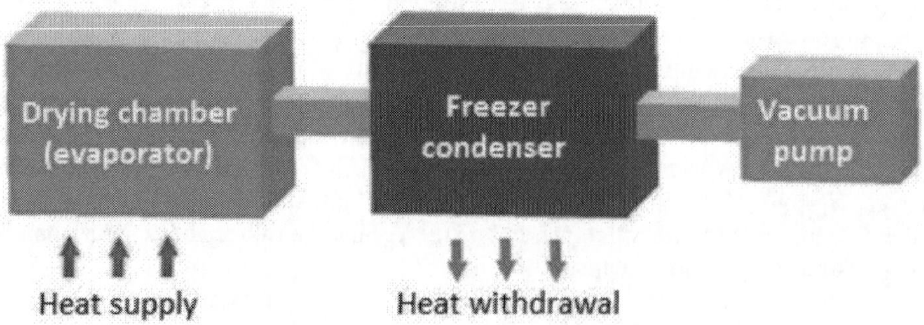

Figure 3. Flow diagram of a freeze dryer.

A cream-colored powder with a neutral taste was obtained with the ~ 100% yield of the anticipated value.

3. Functional properties of the encapsulated powder and the dessert

3.1. Antioxidant capacity (AOC)

A series of studies was conducted to verify that the powders have functional properties. The antioxidant capacity was measured using a Cvet-Yauza 01-AA liquid chromatography system (Table 2) [12].

Table 2. Antioxidant capacity of rowanberry powders.

Sample	AOC, µg quercetin / g	
	2017	**2018**
Unprocessed rowanberries	0,56 ± 0.02	0,72 ± 0.01
Rowanberries after disintegration	0,46 ± 0.02	0,87 ± 0.02
Encapsulated powder	0,26* ± 0.01	0,64* ± 0.01

* with allowance for 2: 1 dilution, the AOC remains virtually unchanged (admissible error, 8%).

The AOC was elevated because the cell walls of the berries were ruptured during disintegration. When AOC was measured for the disintegrated powder, more substances exhibiting antioxidant properties were extracted.

Furthermore, a comparative analysis of the content of antioxidants in rowanberries stored throughout a year was carried out. The antioxidant capacity declined after storage. However, the loss of antioxidant capacity was less significant for the encapsulated powder. A plausible reason is that the outer shell of the encapsulated powder, which is made of polysaccharide, prevents oxidation of the encapsulated ingredient.

3.2. Flavonoids

The most characteristic reaction (with iron (III) chloride) was chosen for qualitative detection of flavonoids; complexes were analyzed spectrophotometrically [13].

A flavonoid forming a complex whose absorption peak lies closest to the absorption peak of the complex formed between the test sample and aluminum chloride was used as a reference sample (Figure 4).

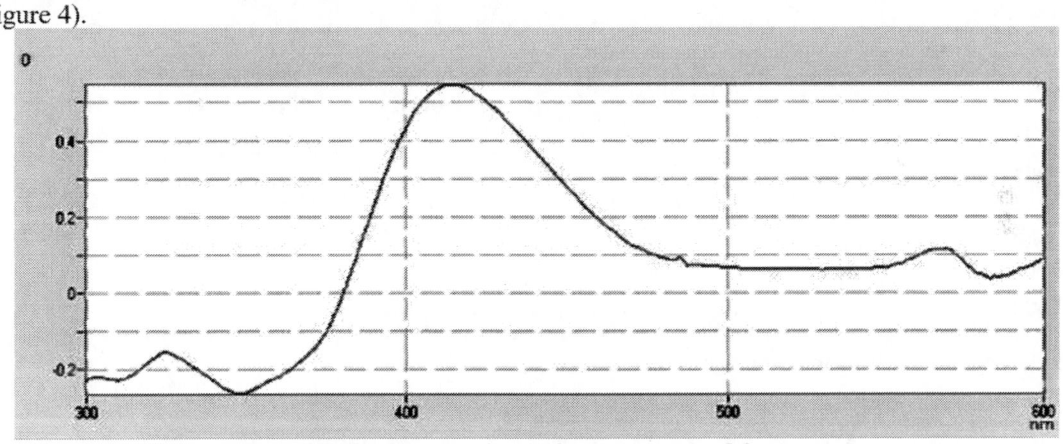

Figure 4. An absorption spectrum of the extract.

The calibration curve was plotted using a solution of the reference sample in 60% ethanol (Table 3, Figure 5).

Table 3. The data used to plot the calibration curve.

Absorbance (D), * 10^{-3}	Concentration (C), mg/ml * 10^{-4}
2.1	0.8
3.7	1.6
5.4	2.4
6.3	3.2
8.6	4.0
1.1	4.8

The line of best fit (functional dependency, Figure 5) was estimated using Microsoft Excel:

- linear equation: $y = 21.577x + 0.0178$;
- the coefficient of determination is a figure showing how well the line of best fit represents the actual data; R_2=0.9819.

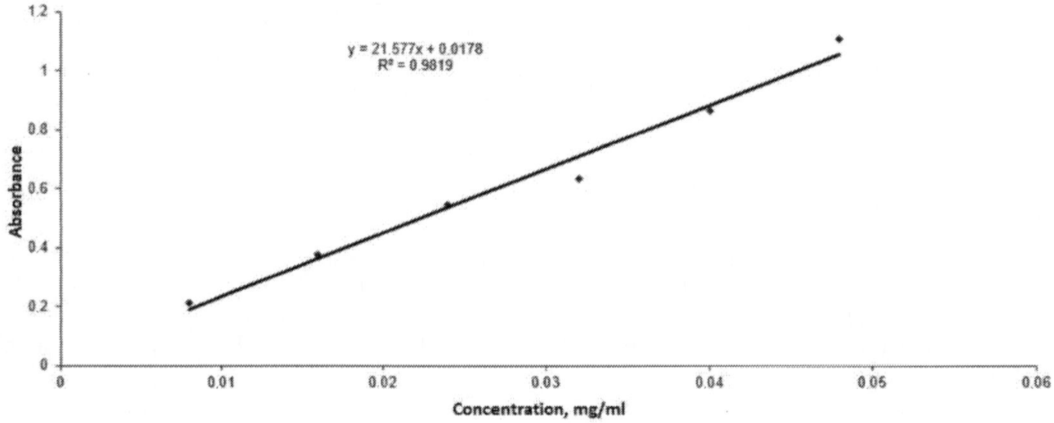

Figure 5. Absorbance (*D*) as a function of rutin concentration (*C*) in the solution analyzed spectrophotometrically.

The results are summarized in Table 4.

Table 4. The calculated contents of flavonoids in the rowanberry extract.

	Absorbance (*D*)	Concentration (*C*), mg/ml * 10^{-3}	Rowanberry sample weight, g	Weight of flavonoids, mg	Content of flavonoids, mg/g
IR-dried rowanberry powder	0.21±0.03	9,00±0.01	4.12±0.1	9.00±0.3	4.24±0.02
Encapsulated powder	0.14±0.02	5,80±0.01	3.11±0.1	7.25±0.3	2.33*±0.02

* with allowance for dilution of the solution containing rowanberry and the polysaccharide (2: 1), the total content of flavonoids remained unchanged.

3.3. Determination of vitamin C and β-carotene in powders

3.3.1. Vitamin C. The method is based on the extraction of vitamin C solution of hydrochloric acid. After extraction, titration with a solution of sodium 2,6-dichlorophenolindophenolate is carried out until a light pink color is obtained. Processed results are presented in Table 5.

Table 5. Vitamin C content in powders.

Sample	Vitamin C, mg/g
IR-dried rowanberry powder	3.62±0.1
Encapsulated powder	2.51±0.1

3.3.2. β-carotene. The method is based spectrophotometric determination of the mass concentration of carotenoids. The proportions of individual carotenoids (of the total content of carotenoids) are determined by spectrophotometric measurement in fractions obtained during chromatographic separation of the extract. Processed results are presented in Table 6.

Table 6. β-carotene content in powders.

Sample	β-carotene, mg/g
IR-dried rowanberry powder	0.17±0.1
Encapsulated powder	0.13±0.1

With allowance for dilution of the solution containing rowanberry and the polysaccharide (2: 1), the total content of β-carotene and vitamin C remained unchanged.

3.4. Sensory evaluation of the developed rowanberry powders
The descriptor-profile method was used for sensory evaluation of the developed functional ingredients. It demonstrated that the encapsulated powder had better sensory properties (Figure 6) [14].

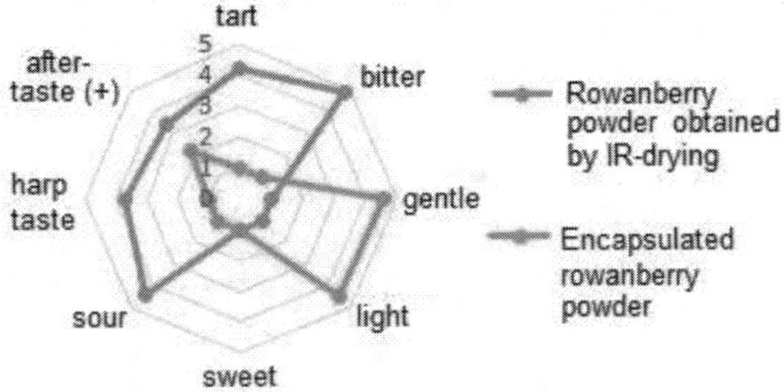

Figure 6. Flavor profile.

The technology of encapsulating plant-based rowanberry extract gives rise to a powder having no bitter, pungent, or astringent taste, unlike the IR-dried rowanberry powder.

3.5. Mathematical modeling of the formulations of cottage cheese desserts
Taking into account the enhanced properties of the samples under study, a series of functional cottage cheese desserts containing IR-dried rowanberry powder (2%) and encapsulated rowanberry powder (9%) was developed at the Technology and Organization of Food Industries Department of the Novosibirsk State Technical University. The optimal ratio between the main components was found using mathematical modeling by solving linear programming problems using MatLab software (Table 7) [15].

The table shows the content of flavonoids, vitamin C, β-carotene in each ingredient, which is included in the recipe of cottage cheese dessert.

Table 7. Mathematical modelling.

Ingredients in the formulation	Range of variation	Vitamin C, mg/100g	β-carotene, mg/100g	Flavonoids, mg/100g	Index, X_i	Energy value, kcal
Fat-free cottage cheese	15…20	0.5	0	0	X_1	76
Cream	25…30	0	0	0	X_2	322
Milk	10…18	1.32	0.01	0	X_3	54
Granulated sugar	3…6	0	0	0	X_4	399
Water	6…26	0	0	0	X_5	0
Encapsulated rowanberry powder	6.5…14	251.31	13.09	233.07	X_6	264
Gelatin	1…5	0	0	0	X_7	0

X_1–X_7 stand for the unknown specific weights of each type of raw material within the product. One needs to determine the unknown values at which:

- $F(x) = \min \{76 \cdot X_1 + 322 \cdot X_2 + 54 \cdot X_3 + 399 \cdot X_4 + 0 \cdot X_5 + 264 \cdot X_6 + 0 \cdot X_7\}$ provided that the following conditions are met:
  - Vitamin C content is ≥ 13.5 mg (15% of the recommended daily intake): $0.5 \cdot X_1 + 0 \cdot X_2 + 1.3 \cdot X_3 + 0 \cdot X_4 + 0 \cdot X_5 + 152.12 \cdot X_6 + 0 \cdot X_7 \geq 10.5$;
  - β-carotene content is ≥ 0.75 mg (15% of the recommended daily intake): $0 \cdot X_1 + 0 \cdot X_2 + 0.01 \cdot X_3 + 0 \cdot X_4 + 0 \cdot X_5 + 6.86 \cdot X_6 + 0.3 \cdot X_7 \geq 0.9$;
  - Total flavonoid content is ≥ 7.5 mg (15% of the recommended daily intake): $0 \cdot X_1 + 0 \cdot X_2 + 0 \cdot X_3 + 0 \cdot X_4 + 0 \cdot X_5 + 145.38 \cdot X_6 + 0 \cdot X_7 \geq 7.5$;

Per product unit: $X_1 + X_2 + X_3 + X_4 + X_5 + X_6 + X_7 = 1$.

This problem is solved as follows: $X_1 = 20.0$ g; $X_2 = 25.12$ g; $X_3 = 14.93$ g; $X_4 = 7.84$ g; $X_5 = 19.06$ g; $\mathbf{X_6 = 8.97}$ **g**; $X_7 = 4.92$ g.

Mathematical modeling of the formulation using linear equations showed that 100 g of the cottage cheese dessert needs to contain at least 8.97 g of the encapsulated rowanberry powder to make the finished product functional.

3.6. Mathematical modeling of the formulations of cottage cheese desserts

Functional cottage cheese desserts containing IR-dried rowanberry powder and encapsulated rowanberry powder were developed at the Division of Technology and Organization of Food Production Facilities of the Novosibirsk State Technical University according to the data obtained by mathematical modeling of the dessert formulation. Four desserts have been developed: the ones containing IR-dried rowanberry powder at concentrations of 2% and 7% and the ones containing encapsulated rowanberry powder at concentrations of 9% (the rowanberry: polysaccharide ratio in the powder being 2:1) and 14% (1:1 ratio).

The formulation of the dessert containing encapsulated powder (2:1) at concentration of 9% is shown in Table 8.

Table 8. Mathematical modelling.

Raw materials/food products	Gross weight	Net weight	Weight of the finished product
Fat-free cottage cheese, g	20.2	20.0	–
Non-dairy creamer, g	25.0	25.0	–
Milk (2.5%), g	15.0	15.0	–
Sugar, g	8.0	8.0	–
Gelatin, g	4.0	4.0	–
Encapsulated powder, g	9.0	9.0	–
Water, g	19.0	19.0	–
Yield	–	–	**100.0 g**

Physicochemical analysis and sensory evaluation of the finished product were carried out to verify that the developed dessert has functional properties.

4. Results and Discussion

4.1. Sensory evaluation of the developed desserts

The sensory and physicochemical properties of the finished product samples were analyzed (Figure 7 and Table 9) [16].

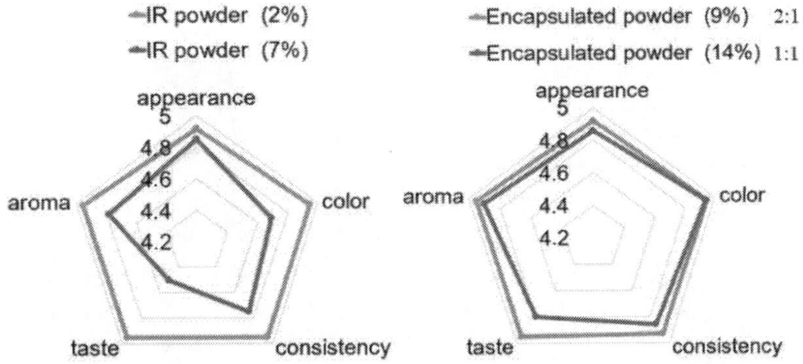

Figure 7. Sensory evaluation of cottage cheese desserts: (a) desserts containing IR-dried rowanberry powder and (b) desserts containing encapsulated rowanberry powder.

The cottage cheese dessert containing encapsulated rowanberry powder (b) has better sensory properties than those of the dessert containing IR-dried rowanberry powder (a). In turn, sensory evaluation of the desserts containing encapsulated rowanberry powder only demonstrated that the cottage cheese dessert with 9% powder content was given the highest score.

Table 9. Physicochemical evaluation of the quality of cottage cheese desserts.

Parameters	Control sample	Dessert containing the IR-dried rowanberry powder (2%)	Dessert containing the encapsulated powder (9%)
Dry solids content, g/100g	69.89 ± 0.04	68.36 ± 0.04	68.32 ± 0.04
Acidity, °T	54.00 ± 0.5	72.00 ± 0.5	62.00 ± 0.5
Ash content, g/100g	0.04 ± 0.02	0.16 ± 0.02	0.21 ± 0.3
Vitamin C, mg/100g	0.54 ± 0.02	6.17 ± 0.15	21.86 ± 0.2
β-carotene, mg/100g	–	0.32 ± 0.02	1.01±0.05
AOC, mg/100g	0.06 ± 0.01	2.05 ± 0.08	5.30 ± 0.2
Flavonoids, mg/100g	–	4.48 ± 0.12	20.30 ± 0.2

The recommended daily intake of biologically active substances is as follows: vitamin C ~70 mg; β-carotene ~ 5 mg; flavonoids ~50 mg; antioxidants (quercetin) ~30 mg [17].

4.2. Functional properties of desserts

The physicochemical studies have verified that the desserts had functional. The results are shown in Figs. 8–11 (the line shows the amount above which the product is considered functional: ≥ 15% of the recommended daily intake).

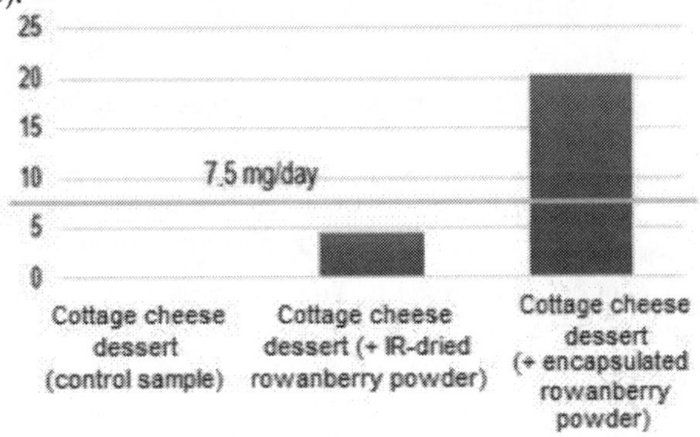

Figure 8. Total flavonoid content in cottage cheese desserts.

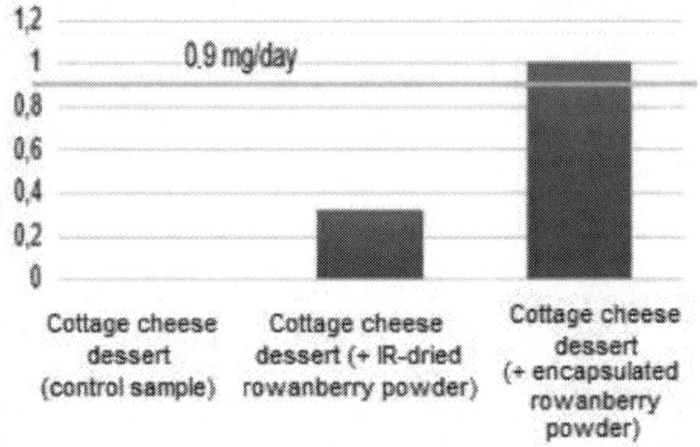

Figure 9. β-carotene content in cottage cheese desserts.

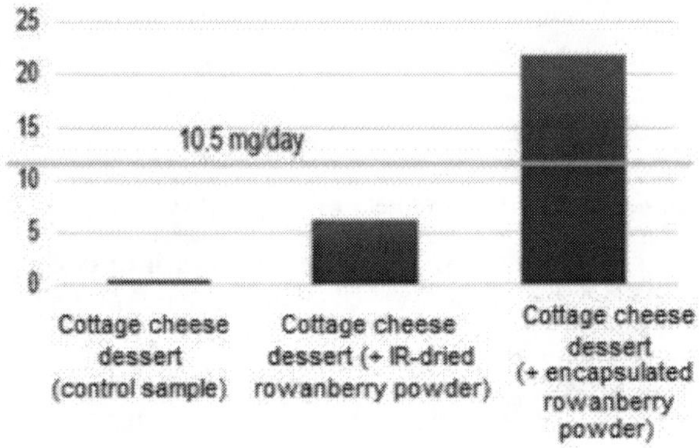

Figure 10. Vitamin C content in cottage cheese desserts.

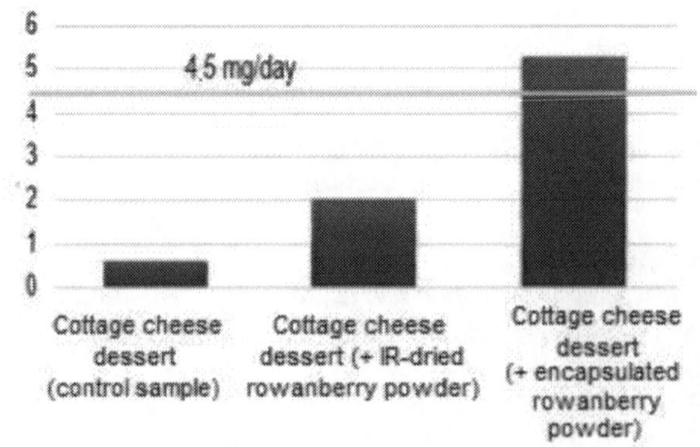

Figure 11. Antioxidant capacity of cottage cheese desserts.

Because of the bitter taste of IR-dried rowanberry powder, the maximum content of this ingredient in the dessert is 2%.

Such a low content does not make the food product functional. Hence, the IR-dried rowanberry powder cannot be used as a functional ingredient. The encapsulation stage has increased the amount of functional component that can be added to the cottage cheese dessert sevenfold, thus ensuring the required amount of biologically active substances in the finished food product. The sample containing the encapsulated rowanberry powder satisfies more than 15% of the daily requirements for antioxidants, vitamin C, β-carotene, and flavonoids for humans.

IOP Conf. Series: Earth and Environmental Science **346** (2019) 012066 doi:10.1088/1755-1315/346/1/012066

5. Conclusions

A technology for producing encapsulated powders of IR-dried rowanberries by extracting functional ingredients from raw material has been developed. This method was used for the first time to produce functional food product from rowanberries. The encapsulation technology neutralizes the taste of the finished food product by masking the natural bitterness of the added raw material. The food product with improved sensory properties has been obtained, while its nutritional value and the total content of biologically active substances has not been reduced.

References

[1] Maceichik I V, Karpacheva S M, Tkach A N and Suvorova E A 2018 Use of complex natural additives for the development of new bakery recipes *Izvestiya Vuzov. Prikladnaya Khimiya i Biotekhnologiya* **8(4)** 158–65

[2] Maceichik I V, Lomovskiy I O and Martynova E G 2018 Using the technology of encapsulation of the extract of mountain ash ordinary to create a functional curd product *Eurasian Scientific Association* **10(44)** 66–9

[3] Sõukand R and Kalle R 2016 Use of wild food plants *Changes in the Use of Wild Food Plants in Estonia* 29–136

[4] Isaykina N V, Kolomiets N E, Abramets N Y and Bondarchuk R A 2017 Study of phenolic compounds in the extracts of berries of *Sorbus aucuparia Chem. Plant Raw Materials* **(3)** 131–9

[5] Hasbal G, Yilmaz-Ozden T and Can A 2015 Antioxidant and antiacetylcholinesterase activities of Sorbus torminalis (L.) Crantz (wild service tree) fruits *J. Food Drug Anal.* **23**(1) 57–62

[6] Adak N, Heybeli N and Ertekin C 2017 Infrared drying of strawberry *Food Chem.* **219** 109–16

[7] Đorđević V, Paraskevopoulou A, Mantzouridou F, Lalou S *et. al.* 2015 Encapsulation technologies for food industry *Emerging Tradit. Technol. Safe, Healthy Quality Food* **1** 329–82

[8] Rigon R T and Zapata Noreña C P 2016 Microencapsulation by spray-drying of bioactive compounds extracted from blackberry (*rubus fruticosus*) *J Food Sci Technol.* **53(3)** 1515–24

[9] Nielsen S S 2017 Moisture content determination *Food Analysis Laboratory Manual. Food Science Text Series* (Springer, Cham) pp. 105–15

[10] Fang Z and Bhandari B 2016 Spray drying of bioactives *Engin. Foods Bioactives Stability Delivery* (Springer, New York) 261–284

[11] Yamashita C, Chung M, Santos C, Mayer C R M *et. al.* 2017 Microencapsulation of an anthocyanin-rich blackberry (*Rubus* spp.) by-product extract by freeze-drying *LWT – Food Sci. Technol.* **84** 256–62

[12] Deineka L A, Makarevich S L, Deineka V I and Chulkov A N 2015 HPLC of anthocyans with an amperometric detector: Evaluation of the antioxidant activity *J. Anal. Chem.* **70(8)** 989–94

[13] da Silva L A, Pezzini B R and Soares L 2015 Spectrophotometric determination of the total flavonoid content in *Ocimum basilicum* L. (Lamiaceae*)* leaves *Pharmacogn Mag* 2015 **11(41)** 96–101

[14] L de Cock H 2019 Sensory evaluation, an important tool for understanding food and consumers *Encyclopedia Food Security Sustainabil.* **1** 546–9

[15] Tarczyńska-Łuniewska M 2018 The procedure of building stable fundamental database with the use of Matlab software *Procedia Computer Sci.* **126** 2163–72

[16] Kumalasari R, Ekafitri R and Indrianti N 2017 Quality assessment of physical and organoleptic instant corn rice on scale-up process *IOP Conf. Series: Earth Environm. Sci.* **101** 012025

[17] Maceichik I V, Lomovskiy I O and Martynova E G 2018 The use of encapsulation technology for improvement of functional properties of curd desserts *Fundam. Bases Mechanochem. Technol.* (Novosibirsk: IPC NSU) p 150

IOP Conf. Series: Earth and Environmental Science **346** (2019) 012067 doi:10.1088/1755-1315/346/1/012067

The competitiveness of indonesian cloves commodity in international market

A P Pratama[1,3] and D H Darwanto[2]

[1]Agricultural Economics, Faculty of Agriculture, Universitas Gadjah Mada, Yogyakarta, Indonesia
[2]Faculty of Agriculture, Universitas Gadjah Mada, Yogyakarta, Indonesia

Email: adnanputrapratama07@gmail.com

Abstract. Clove is one of the plantation commodities that contributes to the Indonesian economy. Indonesia, with a large potential of land area, is expected to be able to increase domestic clove production to provide domestic and international demand, increasingly. The purpose of this study is to analyze the competitiveness of Indonesian clove commodities in the international market. Using the Revealed Comparative Adventage (RCA) indicator, the Trade Specialization Index and the Acceleration Ratio (AR). The results of the analysis show that Indonesian clove commodities have a comparative advantage, and play a role as net clove exporters and are able to compete and reduce the international market share.

1. Introduction

Clove (*Syzigium aromaticum*. L) is one of the strategic commodities whose plantations are managed by many people in Indonesia and contribute to the Indonesian economy. The clove plant itself comes from the Maluku region, which is one of the spice-producing regions and clove itself, which is one of the important commodities in the economy in Indonesia. However, the fact behind the superiority and the high selling price, this plantation commodity from Maluku is still unable to become part of Indonesia's main commodities. One of the causes of declining clove production is climate or weather. The development of clove plantations to date has not yet experienced a recovery like the conditions of its heyday. This can be seen from the area which only reached around 500,000 ha from the largest 700,000 ha in the early 1990s, with production still fluctuating around 60,000 to 100,000 tons each year [1]

The main products of cloves are clove flowers or cloves. Most cloves are used in Indonesia as a raw material for the clove cigarette industry. Historically, the clove cigarettes industry very much needed cloves which developed in Java from 1927. Clove cigarette factories were built in Kudus, Kediri, Blitar, Tulungagung and Mojokerto. Clove needs are increasing rapidly, so Indonesia sometimes has to import [2].

Clove plants (*Syzygium aromaticum*. L) are also one of the producers of essential oils. Clove oil as a commodity that has great potential in Indonesia. Clove plants can produce three types of essential oils, including clove oil, clove stalk oil and clove leaf oil. The oil content in clove flowers ranges from 17-18%, on the stalk or clove handle around 5%, while on the leaves around 2-3%. There are at least 70 types of essential oils which have been traded on the international market and 40 of them can be produced in Indonesia, but only a small proportion of essential oils have been cultivated in Indonesia. The oil is obtained by distillation [3].

Content from this work may be used under the terms of the Creative Commons Attribution 3.0 licence. Any further distribution of this work must maintain attribution to the author(s) and the title of the work, journal citation and DOI.
Published under licence by IOP Publishing Ltd

IOP Conf. Series: Earth and Environmental Science **346** (2019) 012067 doi:10.1088/1755-1315/346/1/012067

Based on data from the Ministry of Agriculture, the trend of the development of production and the area of Indonesian clove land in recent years has tended to fluctuate. In 2011, the lowest of Indonesia clove domestic production was only 72.246 tons. Although fluctuating, production tends to increase every year, but in the period of 2015 to 2017 the increase in land area has not yet resulted in a significant increase in production, expected production to rise turns out to stagnate and even decline in 2017 [4]

Price uncertainty is a problem that is often faced by most farmers in Indonesia, including clove farmers, erratic selling prices lead to farmers' reluctance to maintain their crops so that crops are vulnerable to pests and diseases such as Clove Wood Bacteria (CPC), Clove Leaf Smallpox (CDC), Deciduous Clove Leaf (GDC) and clove stem borers. In 1995 national clove production reached 90,007 tons fell to 52,903 tons at the time of the small harvest in 1999 and only reached 79,009 tons at the time of the big harvest in 2002. Other problems faced were in terms of supporting infrastructure, almost all of the infrastructure in the clove production centers area was inadequate, therefore the farming costs become high and the selling price becomes less competitive. For example, the clove production center in Maluku islands and Toli-toli which only have one alternative transportation, namely water transportation. This condition resulted in supporting institutions tending to pressure farmers, such as marketing institutions that tended to be monopsony, financial institutions dominated by ijon systems which tended to harm farmers. Although there has been a Clove Farmers Association (APCI), farmers' access to information technology and markets has not gone well. On the other hand, the need for clove for industry rose to an average of 92,133 tons / year. The supply shortage is a challenge for farmers and entrepreneurs to be able to fulfill it. The balance of supply to demand can be done through intensification, rehabilitation and rejuvenation of plants, supported by a reasonable purchase price by the industry [5]

Along with the development of international market needs, clove demand from Indonesia continues to experience a significant increase. This is indicated by the average number of exports from various major world clove producing countries in the last ten years Indonesia ranks first, showing the highest difference from other competing countries such as Madagascar and Tanzania [6].

During 2002 to 2016 Indonesia has exported cloves continuously to 21 countries. The main markets country that dominate Indonesia's clove exports are Saudi Arabia, Vietnam, Malaysia, Pakistan, the United Arab Emirates, the Netherlands, Thailand, Germany, Egypt and Australia. During the fifteen years the volume of Indonesian clove exports in the market fluctuated; however, the average export grew [7]. Therefore, this research will try to describe to the reader the development of Indonesian clove commodity exports in the International Market and the competitiveness of Indonesian clove commodities in the international market.

2. Materials and methods

2.1. Types and data sources

This research on clove competitiveness provides the latest information regarding the strength of the competitiveness of Indonesian clove commodities in the international market since 1994 - 2017. The data used is in the form of quantitative secondary data which is translated into time series data. Sources of data were obtained from various related institutions including Directorate General of Plantation, Ministry of Agriculture, Food and Agriculture Organization (FAO), and United Nation of Commodity Trade Database (UNComtrade) using HS code 0907.

2.2. Data analysis

The data analysis method used in this study is adjusted to the research objectives by using indicators of mathematical approaches to measures of commodity competitiveness in the international market. The indicators used are Revealed Comparative Adventage (RCA), Trade Specialist Index (ISP) and Acceleration Ratio (AR).

2.2.1. Revealed Comparative Adventage (RCA). RCA is one method that can be used to see the market share of a commodity in a country by looking at the comparative advantage and competitiveness of these commodities on the world market. In this analysis, the development of the RCA index of Indonesian clove commodities will be calculated from year to year. Mathematically RCA is formulated as follows:

$$RCA = \frac{Xijt/Xjt}{Wit/Wt}$$

Information:
Xijt: Value of export of commodity i by country j in year t
Xt: Value of total exports of country j in year t
Wit: Value of world commodity exports i in year t
Wt: Value of total world exports in year t

The results of the RCA calculation indicate the occurrence of two possibilities or criteria to describe the competitiveness of a commodity, that is if the RCA value is more than one (RCA> 1) then the commodity in that country has a comparative advantage and high competitiveness in the world market and the higher RCA showing stronger competitiveness. If the RCA value is less than 1 (RCA <1), the commodity in that country has a comparative advantage that is low or still below the world average and is considered weak competitiveness.

2.2.2. Trade Spialization Index (ISP). ISP is used to identify the position or stage of development of a product or commodity so that it can be known the tendency of a country to act as an exporter or importer. Mathematically the ISP can be formulated as follows:

$$ISP = \frac{Xi-Mi}{Xi+Mi}$$

Information:
Xi: Value of Export of Commodities i
Mi: Import Value of Commodities i

The ISP will identify the growth rate of a product in trading into the following five stages based on figure 1 below [8]:

1. Introduction phase, if the ISP value is between -1 and -0.50 (Net Importer)
2. Import substitution stage, if the ISP value is between -0.50 to 0.00 (Importer)
3. Growth stage, if the ISP value is between 0.01 and 0.80 (Export Expansion)
4. Maturity stage, if the ISP value is between 0.81 to 1.00 (Net Exporter)
5. The stage of re-importing, if the ISP value returns decreases from 1.00 to 0.00

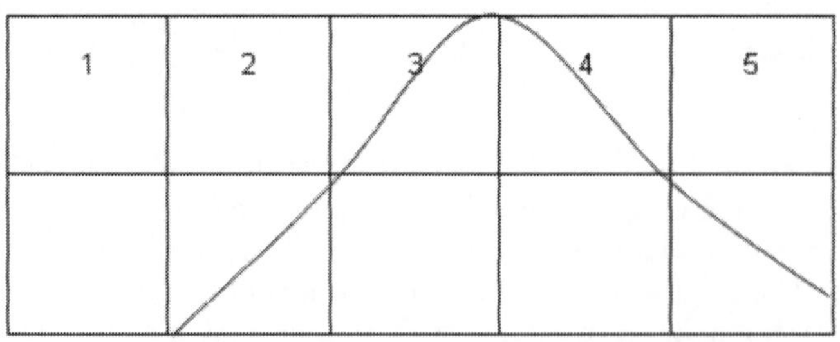

Figure 1. ISP curve according to product cycle theory

2.2.3. Acceleration Ratio (AR). The Acceleration Ratio (AR) method is used to see a comparison between the acceleration of a country's export growth towards the acceleration of world import growth. Mathematically formulated as follows:

$$AR = \frac{Trend\ Xij+100}{Trend\ Mi+100}$$

Information
Xij: Value of commodity exports i in country j

Mi: World import value for commodities i

AR calculation results can be stated if AR approaches or more than one (AR> 1) then the commodity from that country can capture the market, whereas if AR is less than zero (AR <0) or close to -1 means that there is a market share of the supplier earlier could not capture market share

3. Result and discussion

3.1. Development of land area, production and productivity of indonesian clove commodities and competitor countries

The pattern of clove plantations in Indonesia is still dominated by the pattern of People's Plantation (PR) whose percentage reaches 98% of the total, the remaining 1% each comes from the Large Private Plantation (PBS) and the State Large Plantation (PBN). Indonesian Clove Plantations are spread in several main areas including Central Sulawesi, South Sulawesi, North Sulawesi, Southeast Sulawesi, Maluku, Central Java, East Java, West Java and other Provinces [9]. Indonesia has a much wider clove plantation area compared to other countries which are the main countries producing world cloves such as Madagascar and Tanzania. The potential of this high land area should be a great opportunity for Indonesia to optimize national clove production, which at present Indonesia is a producer country, as well as the largest clove consumer in the world. The following figure describes the development of the clove plantation area of main producer countries based on the data source from Food and Agricultural Organization from 1994 - 2017.

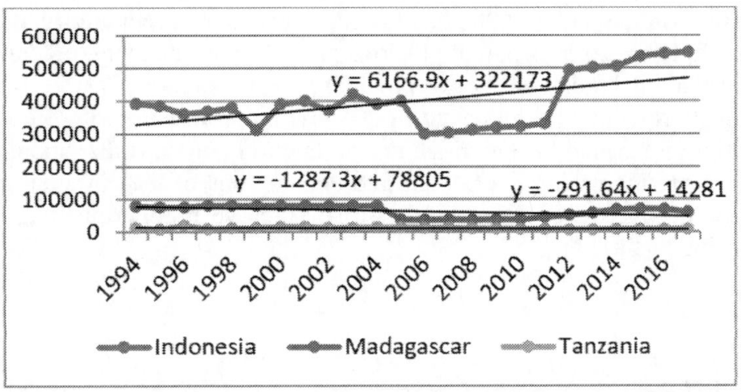

Figure 2. Developments of the clove areas in Indonesia, Madagascar and Tanzania in 1994-2017.

Based on figure 2, it can be seen that the increase in the area of Indonesian clove plantations from year to year tends to increase where in 1994 the area was only 392,000 ha and by 2017 it had reached 548,091 ha. There was a significant decrease in the area in the period 2000 and 2006, this occurred because farmers abandoned their clove plants and even cut them down, resulting in a drastic decrease in the area of clove. Madagascar itself has also experienced a decline in clove plantation areas from 2005 to 2010, while Tanzania itself tends to experience a decline in the area of clove plantations from the period 2008 to the present.

This neglect was caused by a drop in world clove prices, causing farmers to abandon their clove plants and even cut them down and replace them with more profitable crops, even though the government had established a CPC (Marketing and Clove Marketing Agency) to regulate national clove prices but this has not yet had a positive impact on improving domestic clove prices. The increase in the area of clove plantations is expected to be a factor driving the increase in national clove production.

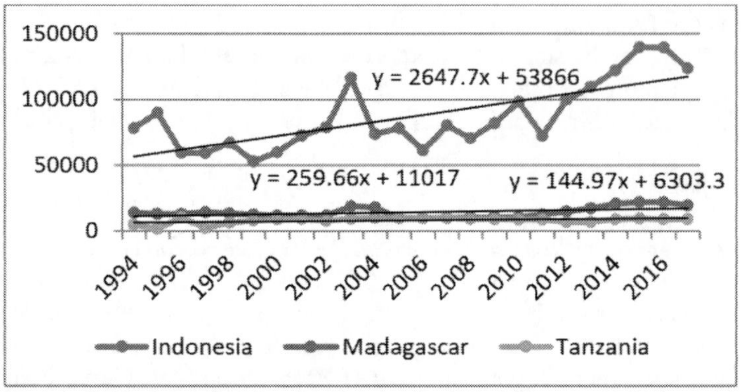

Figure 3. Developments of clove production in Indonesia, Madagascar and Tanzania in 1994-2017.

Based on figure 3, the development of Indonesian clove production in the period 1994-2017 based of the data source from Food and Agricultural Organization from 1994 – 2017 was generally fluctuating and tended to increase, where in 1994 the production of cloves was only 78,379 tons then until 2017 it had reached 123,773 tons. While from competing countries so far, the highest production produced by Madagascar amounted to 22,000 tons in 2016 while Tanzania itself experienced instability in production, this can be seen from the decline in production in the last 3 years where production only reached 8000-9000 tons per year. Clove production in Indonesia is currently estimated at around 120,000-130,000 tons per year, resulting from around 330,000 ha of produce. Thus the productivity of clove plants ranges from 400-450kg / ha. Judging from its potential, clove plants can produce up to 1,000 kg / ha so that the national productivity achieved today has not reached 50% of its potential [10].

The highest production so far was achieved in 2015 with a production of 139,641 tons. Therefore, more intensive handling is needed to maximize the production potential that should be achieved. The industrial needs which are increasing every year and become one of the cornerstones of the country's revenue sources require consistency from the increase in clove production itself so that it remains available on an on going basis [11].

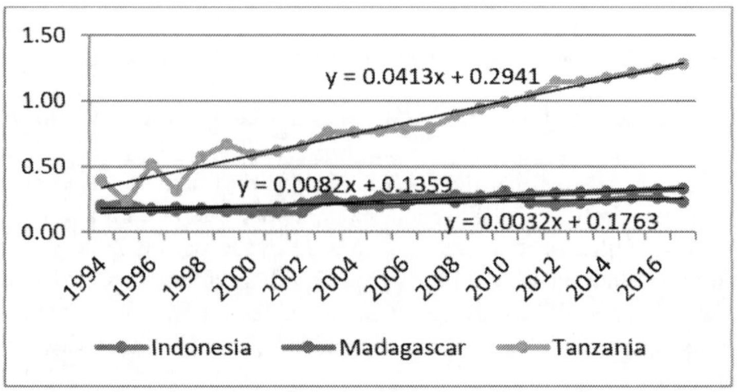

Figure 4. Developments of productivity of cloves in Indonesia, Madagascar and Tanzania in 1994-2017.

Based on figure 4, it can be seen that in terms of productivity based of the data source from Food and Agricultural Organization from 1994 – 2017, Indonesia still lags behind the two competing countries. Tanzania looks far superior to Indonesia and Madagascar in terms of productivity despite the area and production of cloves Tanzania is not as big as Indonesia but they are superior in terms of productivity where the average annual yield is 1-1.2 tons / ha / year while Indonesia can only produce an average of 0.2-0.3 tons / ha / year. The low productivity of Indonesian cloves is partly due to the low intensity of care carried out by farmers and the lack of rejuvenation measures to replace damaged or dead plants.

Until now, efforts to increase clove production and productivity are still being encouraged in an effort to meet the increasing needs, especially in the industrial sector. Most of Indonesia's clove production is used to meet the raw material needs of clove cigarette manufacturers and when viewed from its contribution to the country's revenues is quite large, this commodity occupies a quite important position in the national economy.

3.2. Development of clove import exports in Indonesia

Indonesia plays a very strategic role in international trade. This can be seen from the consistency of Indonesia as a producer which in addition to fulfilling the demand of the domestic industry as well as the world's largest clove exporter in order to meet the international market demand. The fluctuations in clove prices, especially in the export market, affect farmers' motivation in maintaining and fertilizing crops [12]. The chart below shows the development of Indonesia's clove import exports in the period 2000-2017 based of the data source from International Trade Statistic Database (Processed).

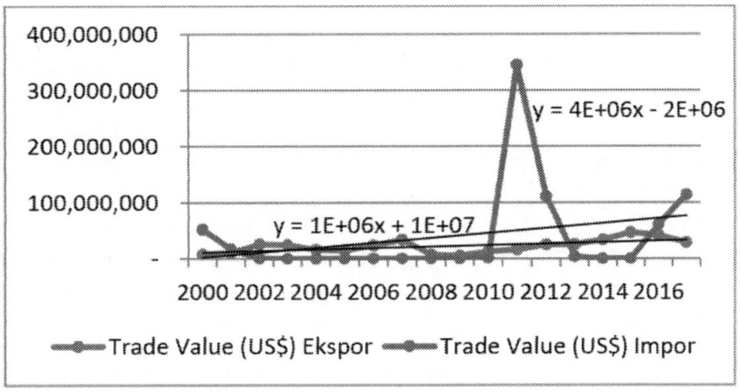

Figure 5. Developments of value of indonesian clove exports and imports 2000-2017.

Based on figure 5, the development of the value of Indonesian clove exports in the period of 2000-2017 was fluctuated but tended to increase. During this period the export value of Indonesia's highest occurred in 2015 reaching US $ 46,483,663 while from the Indonesian import side it had not imported cloves from outside at all in 2008 and 2014 but even worse things have happened in Indonesia where very high imports of clove occurred in 2011 to reach US $ 345,150,592, this was due at that time the high demand for domestic industries which at the same time decreased the production of clove in the country due to climate and weather anomalies at that time.

3.3. Revealed Comparative Adventage (RCA) analysis

The chart below shows the development trend of Indonesia's clove RCA index during the period 1994-2017 based of the data source from International Trade Statistic Database (Processed). Based on figure 6, it can be seen that the development of the RCA index of Indonesian clove commodities fluctuated and tended to rise even though the drastic decline had occurred during the 1998-1999 period, 2003-2004, 2007-2009, and 2015-2017, but the average clove RCA index Indonesia is 9.53 which means that Indonesian clove commodities have competitiveness from comparative advantage [13]. This figure can still be improved considering the potential of the Indonesian state in terms of land area and efforts to increase efficiency in the cultivation process. This position made Indonesia one of the determinants of the price of clove commodity exports because Indonesia was one of the largest clove producers and exporters on the international market. In addition, the world market began to realize about the multipurpose function of cloves, so that clove consumption continued to increase. Importing countries also tend to choose to buy high quality organic cloves, where Indonesian cloves are cloves which are recognized as having the best quality standards in the international market [7].

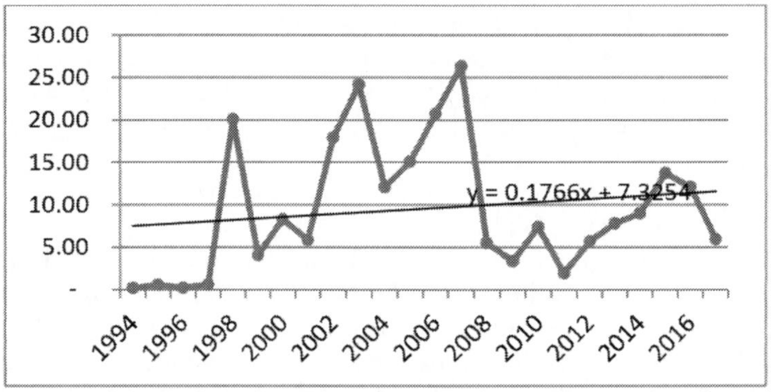

Figure 6. Developments of the Indonesian Clove RCA Index on International Markets in 1994-2017.

3.4. Trade Specialization Index (ISP)

Based on figure 7, the analysis of the Trade Specialization Index of Indonesian Clove Commodities based of the data source from International Trade Statistic Database (Processed) shows that Indonesia's position in world clove trade is likely to be an exporting country. This is indicated by the ISP value in the period 1994-1998 which was positively ranging from 0.84 - 0.99 the period 2002-2010 and 2013-2015 ranged from 0.77 - 1 [13].However, Indonesia also imported several times from other countries, this can be seen from the ISP values that were negatively marked in the period 1999-2001, 2011-2012 and 2016-2017. The fluctuations in ISP values were caused by fluctuations in domestic clove production, which led to the rise and fall of the value of Indonesian clove exports and imports, but overall Indonesian clove commodities had enormous potential to capture world market share.

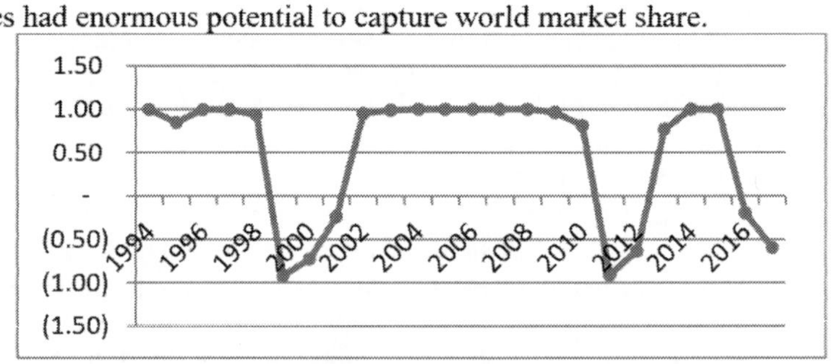

Figure 7. Developments of Indonesian Clove Trade Specialization Index (ISP) on International Markets in 1994-2017.

3.5. Acceleration Ratio (AR)

Analysis using Acceleration Ratio (AR) describes the ability of a country to capture market share. If the AR value is smaller than 1 (<1) then this indicates the acceleration of the growth of total exports of a commodity in a country is lower than the total imports of these commodities in the world, so it can be said that a country is difficult to compete and capture world market share.

Table 1. Value of Acceleration Ratio (AR), 1994-2017.

Country	*Acceleration Ratio* (AR)
Indonesia	1.0011
Madagascar	1.0002
Tanzania	1.0036

Based on table 1, the analysis of AR values from Indonesia, Madagascar and Tanzania based of the data source from International Trade Statistic Database (Processed) shows a positive value of more than 1 (> 1). Indonesia as the highest producer showed positive AR results of 1.0011 (> 1), which means that Indonesia was able to capture the foreign clove market followed by its two competing countries, Madgascar and Tanzania [13]. This is because the growth of clove exports from the three main producing countries is greater than the growth of world clove imports so that these three countries have a very large role in the availability of cloves in the world.

4. Conclusion

Cloves are Indonesian native plant that was originally an export commodity has now turned into a commodity that must be imported because of the rapid development of the industry, the problems faced by producers, in this case farmers as the main producers need full attention from the government as policy makers to protect farmers from various risks. Given the enormous potential of the clove plantation area which is still not able to be optimally optimized by it, strategic steps are needed to anticipate it. Things that can be done in the form of an intensified, rehabilitation and rejuvenation program of clove plants. Development of clove agribusiness systems in an effort to increase profits and competitiveness through strengthening small and medium enterprises (SME) as downstream industries. With the opportunity to develop downstream industries for food, pharmaceuticals and vegetable pesticides, including exports, the private sector is expected to be able to invest in clove agribusiness that includes upstream agribusiness, for example in seed breeding, on-farm large plantation establishment (PBS) and rejuvenation replanting as well as downstream agribusiness in the oil refining industry, food and pharmaceutical industry and processing of clove vegetable pesticides. Equally important is institutional strengthening at the farm level and access to capital and financial institutions and pricing policies.

Acknowledgments

Thank you, the authors convey to the Education Fund Management Institute (LPDP) for their assistance, both financially and non-financially so that the authors can publish the results of research related to the competitiveness of Indonesian cloves commodity in international market.

References

[1] Ndiba T A, Wullur M and Tumade P 2016 Evaluation of supply chain performance of clove commodities (study in lalumpe village, minahasa regency). J. *EMBA*, **4**(1), 153–164.
[2] Semangun H 2014. History of cloves in clove books: history, cultivation and industry. indesso and master of biology. Universitas Kristen Satya Wacana. Salatiga
[3] Azhari A 2009. Prototype of steam distillation tower for distilling clove leaf essential oil. Poli Teknik Negeri Sriwijaya.
[4] Agricultural Statistics. 2018. Agricultural Statistics. Data Center and Agricultural Information System. Ministry of Agriculture of the Republic of Indonesia
[5] Agricultural Research and Development Agency. 2007. Propek and Direction of Clove Agribusiness Development Second Edition. Ministry of Agriculture, Jakarta. http://www.deptan.go.id
[6] Food and Agriculture Organization of United Nation (FAO). 2019. http://faostat.fao.org
[7] Nurhayati E, Hartoyo S and Mulatsih S 2018. Analysis of Indonesian clove export development. *J. Econ. Develop. Pol.*, **7**(1), 21–42.
[8] Department of Trade. 2008. Trade Specialization Index (ISP). http://www.kemendag.go.id
[9] Data Center and Agricultural Information System. 2014. Outlook for Clove Commodities. Ministry of Agriculture. http://www.pusdatin.setjen.pertanian.go.id/
[10] Directorate General of Plantation. 2017. Indonesian Plantation Statistics. Ministry of Agriculture. Jakarta. http://ditjenbun.pertanian.go.id
[11] Wahyudi A 2016. Stabilization strategy of national clove market performance. *J. Perspektif* **15**(1), 73–86 DOI: http://dx.doi.org/10.21082/psp.v15n1.2016.73-85

[12] Kingu J 2014. Cloves Export response to trade liberalization in tanzania: a cointegration analysis. *J. Econom. Sustain. Develop.* **5**(1): 99–120.
[13] UNComtrade. 2019. *International Trade Statistic Database*. https://comtrade.un.org/

IOP Conf. Series: Earth and Environmental Science **346** (2019) 012068 doi:10.1088/1755-1315/346/1/012068

Optimization conditions of adventitious buds regeneration from petiole explants of *Echinacea purpurea*

Y Liu[1], M F Gao[1], G S Rao[1], Y B Xue[1,3] and J P Chen[2,3]

[1]Department of Biotechnology, Faculty of Agricultural Science, Guangdong Ocean University, P. R. China

[2]Department of Food Science and Engineering, College of Food Science and Technology, Guangdong Ocean University, P. R. China

E-mail: yingbinxue@yeah.net (Y B Xue); cjp516555989@126.com (J P Chen)

Abstract. *Echinacea purpurea* as a perennial herb has high ornamental and medicinal value. While the tissue culture technologies of *E. purpurea* are contradictory. In order to improve the culture conditions of the adventitious buds regeneration of *E. purpurea*, petioles from the aseptic seedlings were used as the explants. Different concentrations of sodium nitrophenolate (SN), copper sulfate ($CuSO_4$), amine fresh ester (DA-6), sodium nitroprusside (SNP) and ascorbic acid (Vc) were added into MS medium containing cytokinin and auxin, respectively, to investigate their effects on inducing regeneration of adventitious buds. The results showed that the best induction effects of adventitious buds were obtained under 1 mg/L $CuSO_4$, 1 mg/L DA-6, 0.2 mg/L SN, 2 mg/L SNP and 20 mg/L Vc conditions. Our results can provide references for the induction of adventitious buds regeneration and rapid propagation of *E. purpurea*.

1. Introduction

Echinacea purpurea is a perennial herb of Echinacea in Compositae [1], which has highornamental and medicinal value [2]. Although there are many reports on the tissue culture of *E. purpurea*, many of them are contradictory, and the regeneration efficiency of adventitious bud of numerous culture technology is still very low [3], so it is necessary to optimize the regeneration system of culture and improve induction efficiency of adventitious bud regeneration in *E. purpurea*. It has been reported that during the *Cotinus coggygria* tissue culture, adding SN and GA3 will improve the growth, and shorten the culture cycle [4]. The Cu^{2+} has been known as an key cofactor of some enzymes which play important roles in plant tissue culture [5]. On the other hand, SNP, nitric oxide donor, has been shown helpful in improving plant regeneration in chrysanthemums [6]. To date, most studies on the culture of *E. purpurea in vitro* are compared and optimized for the concentrations of the hormone components, but few studies concerning with the factors such as SN, DA-6, $CuSO_4$, Vc and SNP on the culture effect of *E. purpurea* have been reported. This study can provide references for the induction of adventitious buds regeneration and rapid propagation of *E. purpurea*.

2. Methods

2.1. Plant sources

In this study, the purple coneflower seeds were provided by professor Yue-sheng Yang from College

Content from this work may be used under the terms of the Creative Commons Attribution 3.0 licence. Any further distribution of this work must maintain attribution to the author(s) and the title of the work, journal citation and DOI.
Published under licence by IOP Publishing Ltd

of Life Sciences, South China Agricultural University.

2.2. Preparation of explants

The purple coneflower seeds were sterilized by 75% ethanol solution for 1 min, and 2% sodium hypochlorite (NaClO) for 20 min, respectively, finally, using sterile distilled water rinsed 5 times. Aseptic seedlings could be gained by inoculating sterile seeds on MS (Murashige and Skoog) medium [7] after 45 days of culture. Petioles were isolated from the 45-day-old sterile seedlings, and cut them into 5 cm pieces as petiole explants.

2.3. Preparation of mediums and condition of culture

The same conditions of culture were suitable for all experiments. The medium pH was controlled from 5.8 to 6.0 by 1 mol/L NaOH before the mediums were autoclaved (LX-C75, Hefei, China) for 20 min at 1.4 $kg \cdot cm^{-2}$. The mediums were placed at 25 ± 2°C with 12 h photoperiod of cool white fluorescent tubes with a 60-80 $\mu mol \cdot m^{-2} \cdot s^{-1}$ intensity.

2.4. Adventitious buds regeneration

In order to induce new adventitious buds, the explants of petiole were inoculated on different concentrations of SN (0, 0.02, 0.2 and 2 mg/L), DA-6 (0, 0.01, 0.1 and 1 mg/L), $CuSO_4$ (0, 0.1, 1 and 10 mg/L), Vc (0, 20, 40 and 80 mg/L) and SNP (0, 2, 4 and 8 mg/L) were added into MS medium supplemented with 0.5 mg/L 6-BA and 0.01 mg/L NAA for inducing adventitious buds regeneration respectively for 30 days of culture.

2.5. Data analysis

Adventitious buds regeneration rate (%) = (number of adventitious buds differentiated/number of petiole explants inoculated) ×100%.

Average number of buds (number of buds) = total number of adventitious buds/number of petiole explants inoculated.

The test data was analyzed by statistical analysis method. All the tables are made by Microsoft Excel.

3. Results

3.1. Results of $CuSO_4$ on regeneration of adventitious buds from petiole explants

As shown in table 1, the regeneration efficiency of adventitious buds was affected by concentration of $CuSO_4$ distinctly. The biggest regeneration percentage of adventitious buds (50%) was acquired, when 1 mg/L of $CuSO_4$ was added into mediums on the induction of adventitious buds (table 1). Moreover, although the maximum number of buds was obtained at 0.1 mg/L, the buds were relatively small and poor quality.

Table 1. Effects of different concentrations of $CuSO_4$ on adventitious buds regeneration of petiole explants in *E. purpurea*.

Concentration (mg/L)	Adventitious buds regeneration rate (%)	Average number of buds
0	15	0.30
0.1	30	0.85
1	50	0.45
10	25	0.50

3.2. Influences of DA-6 on regeneration of adventitious buds from petiole explants

From the results summed up in table 2, regeneration of adventitious buds might be promoted by adding DA-6 in mediums. While the regeneration efficiency would be raised with adding high concentrations (0-1 mg/L) of DA-6 in the induction mediums (table 2). When 1 mg/L of DA-6 was applied into mediums on the induction of adventitious buds, the bset regeneration percentage of adventitious buds (35%) and the maximum number of buds (0.9) were gained (table 2).

Table 2. Results of different concentrations of DA-6 on adventitious buds regeneration of petiole explants in *E. purpurea*.

DA-6 concentration (mg/L)	Adventitious buds regeneration rate (%)	Average number of buds
0	15	0.30
0.01	15	0.30
0.1	20	0.55
1	35	0.90

3.3. Effects of sodium nitrophenolate (SN) on regeneration of adventitious buds
As shown in table 3, the concentrations of SN influenced adventitious buds regeneration visibly. From the results summarized in table 3 and figure 1(A-D), it's explicit clear that when the concentration of SN was 0.2 mg/L, the best results of regeneration of adventitious buds (35%) and the biggest buds number (0.8) were obtained respectively (table 3 and figure 1C).

Table 3. Effects of different concentrations of SN on adventitious buds regeneration of petiole explants in *E. purpurea*.

Concentration of SN (mg/L)	Adventitious buds regeneration rate (%)	Average number of buds
0	15	0.30
0.02	25	0.45
0.2	35	0.80
2	33.3	0.73

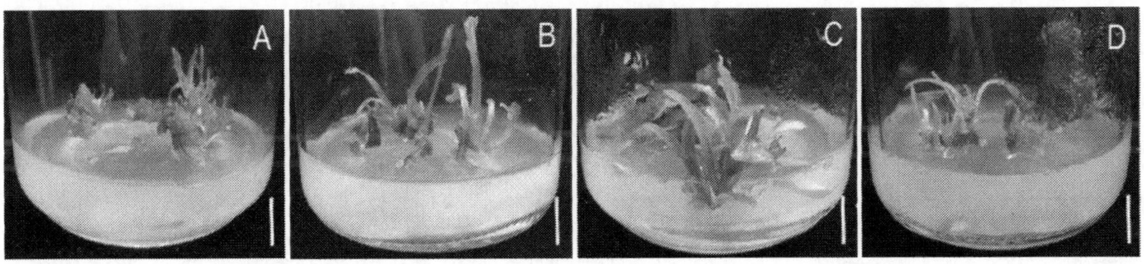

Figure 1. Effect of SN on adventitious buds regeneration. A: 0 mg/L; B: 0.02 mg/L; C: 0.2 mg/L; D: 2.0 mg/L. (Bars=1 cm).

3.4. Effects of SNP on adventitious buds induction from petiole explants of E. purpurea
The responses of adventitious buds induction were significantly influenced by the concentrations of SNP in mediums (table 4). The addition of 2 mg/L SNP made the highest adventitious buds regeneration rate and the most buds (1.35) (table 4). While, if SNP concentration was more than 2 mg/L, the adventitious buds regeneration rate was dramatically decreased (table 4).

Table 4. Effects of different concentrations of SNP on adventitious buds regeneration of petiole explants in *E. purpurea*.

Concentration of SNP (mg/L)	Adventitious buds regeneration rate (%)	Average number of buds
0	15	0.30
2	65	1.35
4	30	0.40
8	0	0.00

3.5. Effects of vitamin C (Vc) on adventitious buds induction by petiole explants of E. purpurea

The results shown that Vc affected the induction of adventitious buds from petiole explants of *E. purpurea*, significantly (table 5). When 20 mg/L Vc was added in the mediums, the highest adventitious buds induction percentage (35%) and the maximum number of buds (1.45) were achieved (table 5).

Table 5. Results of different concentrations of Vc on adventitious bud regeneration of petiole explants in *E. purpurea*.

Concentration of Vc (mg/L)	Adventitious buds regeneration rate (%)	Average number of buds
0	15	0.30
20	35	1.45
40	30	0.70
80	15	0.25

4. Discussion

Cu^{2+} is one of the trace elements essential for plant growth, but it can be toxic to plants when excess. Previous studies shown that the regeneration efficiency of adventitious buds can be promoted by the addition of a small amount of copper sulfate into mediums [3]. In the present study, when the concentration of copper ions exceeded 1 mg/L, the number of regenerated adventitious buds decreased observably. Furthermore, the regeneration of adventitious buds was facilitated at low concentrations of copper ions added into mediums.

DA-6 is a new type of plant growth regulator in plants. It is suitable for many plants and their entire growth period, and widely used in vegetable production [8,9]. However, DA-6 is rarely used in plant tissue culture. Sodium nitrophenolate (SN), a potent cell activator, could significantly increase cell viability [10]. In plant tissue culture, ascorbic acid (Vc) is commonly used as an antioxidant to prevent browning, but it can also promote the growth of buds [11]. It's found that the regeneration rate of adventitious buds could be increased at the right amount addition of DA-6, SN and Vc in mediums.

Studies have reported that SNP could significantly induce the differentiation of adventitious buds in tea (*Camellia sinensis*) [12]. Our results found that the addition of an appropriate amount of SNP in the medium containing cytokinins significantly induced the regeneration of adventitious buds, and the inducing effects were more prominent than the effects of using only cytokinins. SNPs as exogenous NO donors could signally facilitate adventitious buds regeneration, probably because cytokinins could rapidly stimulate the production of NO [13].

This study aims to provide a method for the regeneration and rapid propagation of adventitious buds in *E. purpurea*. It is hoped that more researchers will optimize the regeneration system of *E. purpurea*, which will make the largest development and utilization of the ornamental and medicinal value of *E. purpurea*.

Acknowledgments

This work were supported by the Chinese Postdoctoral Science Foundation (2018M633059), Natural Science Foundation of Guangdong Province (2018A030310057), Program for Nanhai Youth Scholar Project of Guangdong Ocean University, Program for Scientific Research Start-up Funds of Guangdong Ocean University (R17023), the Project of Science and Technology of Zhanjiang City (2016B01004), the Project for Innovation and Strong School of Department of Education of Guangdong Province (2016KQNCX067), Education Teaching Reform Project of Guangdong Ocean University (524210393), Teaching Project of Innovation and Strong School Engineering of Guangdong Ocean University (524210441), National Natural Science Foundation of China (No. 21602034), Natural Science Foundation of Guangdong Province (No. 2016A030310332), Scientific Research Foundation of Guangdong Ocean University (No. R17034), Innovation Strengthens University Project of Guangdong Ocean University.

References

[1] Chen R, Yang Y S and Wu H 2012 Study on the yield of *Echinacea* and the dynamic changes of cichoric acid in different harvesting stages *Chinese Tradit. Herb. Drugs* **43** 1186-90
[2] Hao T J 2007 Seedling cultivation and cultivation techniques of *Echinacea Rural Sci. Technol.* **2007** 61
[3] Zhao F C, Nilanthi D and Yang Y S *et al* 2006 Anther culture and haploid plant regeneration in purple coneflower (*Echinacea purpurea* L.) *Plant Cell Tiss. Org.* **86** 55-62
[4] Yang Y, Jia H and Pan Y 2017 Effects of combination sodium nitraphenolate + GA3 on *Cotinus coggygria* tissue culture plants *J. South Agr.* **48** 1653-8
[5] Sparks C A, Doherty A and Jones H 2014 Genetic transformation of wheat via *Agrobacterium*-mediated DNA delivery *Cereal Genomics: Methods and Protocols. Series Methods in Molecular Biology* R J Henry and A Furtado eds. vol 1099 (New York: Springer) pp 235-50
[6] Muthukrishnan A, Aung H N, Su M J, Trinh N A, Thanda A and Chang K K 2017 Sodium nitroprusside stimulates growth and shoot regeneration in Chrysanthemum *Hortic. Environ. Biotech.* **58** 78-84
[7] Murashige T and Skoog F 1962 A revised medium for rapid growth and bioassays with tobacco tissue cultures *Physiol. Plantarum* **15** 473-9
[8] Yang X Y, Li S H and Mei Y C *et al* 2017 Effects of DA-6 on the growth and physiological characteristics of hydroponic lettuce *J. Agro-Environ. Sci.* **36** 32-8
[9] Sun X H, Li C L and Chen J Q *et al* 2017 Effects of different amine esters (DA-6) concentration and application methods on spinach growth *North Hortic.* **2017** 122-8
[10] Liu X Y 2007 Novel plant growth regulator sodium nitrophenolate and its application *Pestic. Mark. Inform.* **2007** 35-44
[11] Gao J S, Zhang W F and Ou Y H *et al* 2003 Study on the factors affecting the rapid propagation of Xinli No. 7 *Xinjiang Agr. Sci.* **2003** 42-5
[12] Han X J, Yang H Q, Yu S Z *et al* 2008 Adventitious bud regeneration and NO effect in leaves of Pingyi sweet tea *J. Hortic.* **35** 419-22
[13] Tun N N, Holk A and Scherer G F 2001 Rapid increase of NO release in plant cell cultures induced by cytokinin *Febs. Lett.* **509** 174-6

Evaluation of water spinach (*Ipomoea aquatica*) as forage substitution on *in vitro* gas production, digestibility, and kinetic fermentation

H Hasanah[1,2,4], J Achmadi[2], E Pangestu[2] and A Agus[3]

[1]Department of Biology, Faculty of Mathematic and Natural Science, Yogyakarta State University, Yogyakarta, Indonesia
[2]Faculty of Agriculture and Animal Science, Universitas Diponegoro, Semarang, Indonesia
[3]Faculty of Animal Science, Universitas Gadjah Mada, Yogyakarta, Indonesia

E-mail: hasanahhimmatul07@gmail.com

Abstract. The aim of this study was to evaluate water spinach as forage substitution on *in vitro* gas production, digestibility and kinetics fermentation. The experiment was arranged in a completely randomized design with four treatments and five replications. The Treatments were: T1= control of *Pennisetum purpureum* cv Mott grass + 0% of water spinach, T2= + 20% of water spinach, T3: 40% of water spinach and T4: 60% of water spinach. Results showed that substitution of 20% Water Spinach as forage increased gas production after 48 h of incubation. There is no difference in the potential fraction degraded by the addition of a water spinach at the level of 20% compared to the control, although there is a decrease in the addition of 40 and 60% of the water spinach. Substitution of water spinach was not significant on kinetics fermentation of readily soluble fraction (a) and rate constant per hour of gas production (c). The substitution of 20% water spinach increased dry matter digestibility reaching 10% compared to control. The water spinach substitution increased organic matter digestibility ($P<0.05$). It is concluded that 20% of water spinach had the highest gas production and nutrient digestibility and thus, it can be used as forage alternative for ruminant.

1. Introduction

Ruminant animals have the ability to utilize forage into meat and milk. The quality of forage will affect consumption, nutrient utilization, and quality of product. The challenge of tropical countries, such as Indonesia is the low quality of forage due to high fiber content which leads to the inefficiency of nutrient utilization causing a decrease in the productivity of ruminants [1-4]. Therefore, forage alternative with high nutrient content is needed that it can be utilized by ruminants.

Water spinach is forage as source of fiber which is widely grown in Indonesia. The area planted with water spinach reached 47,805 ha with production reaching 276,976 ton or 5.79 ton per ha [5]. According to its nutrient content, water spinach is potential to be forage. The results of Khamparn and Preston [6], Opene et al. [7], Umar et al. [8] showed that water spinach contained crude protein ranging from 5.18 to 24.6%, fiber 13-17.67%, dry matter 10.3% and organic matter 87.6%.

The high quality of nutrient content makes water spinach forage alternative that can be used to supply ruminant requirements. However, the effects of water spinach substitution in the diet is needed to be carried out to evaluate water spinach as forage substitution on *in vitro* gas production, digestibility and

Content from this work may be used under the terms of the Creative Commons Attribution 3.0 licence. Any further distribution of this work must maintain attribution to the author(s) and the title of the work, journal citation and DOI.
Published under licence by IOP Publishing Ltd

kinetics fermentation. The aim of this study was to evaluate water spinach as forage substitution on *in vitro* gas production, feed nutrient digestibility and kinetics fermentation

2. Materials and method

2.1 Experimental site

The study was carried out at the Laboratory of Nutritional Biochemistry, Department of Animal Nutrition and Feed Science, Faculty of Animal Scicence, Universitas Gadjah Mada, Yogyakarta, Indonesia.

2.2 Experiment design and basal diet

Analysis of nutrient content carried out on feed ingredients includes dry matter (DM), organic matter (OM), crude fiber (CF), crude protein (CP), ether extract (EE), hemicellulose, cellulose and lignin by AOAC [9]. The nutrient content of feedstuffs used in this study are shown in table 1.

Table 1. Nutrient content of feedstuffs used in the study (% DM)*.

Feedstuffs (%)	*Pennisetum purpureum* cv Mott grass	Concentrate	Water spinach
DM	92.93	88.87	89.42
OM	83.43	89.77	84.60
CF	31.49	17.59	21.62
CP	6.70	12.60	10.65
EE	1.35	2.66	1.86
Ash	16.57	10.23	15.4
Hemicellulose	21.18	28.48	27.24
Cellulose	30.70	25.35	26.88
Lignin	12.67	11.47	12.22

DM =Dry Matter, OM =Organic Matter, CF =Crude Fiber, CP =Crude Protein, EE =Extract Ether
*Analysis in Laboratory of Nutritional Biochemistry, Faculty of Animal Science

The study consists of four treatments, T1: control or 0% water spinach, T2: with 20% water spinach, T3: 40% water spinach, and T4: 60% water spinach. This experiment was designed in completely randomized design with 5 replicates per treatment. The formulation and nutrient composition of diets are shown in table 2.

Table 2. Nutrient composition of diet substituted by adding water spinach.

Feedstuffs	T1	T2	T3	T4
Formulation (%)				
Pennisetum purpureum cv Mott grass	60	40	20	0
Water Spinach	0	20	40	60
Concentrate	40	40	40	40
Total	**100**	**100**	**100**	**100**
Nutrient composition (%)				
DM	90.74	90.39	90.13	88.08
OM	86.27	87.91	89.94	89.23
CP	13.72	14.36	15.06	15.04
EE	2.14	2.02	1.87	2.18
CF	21.43	22.03	21.29	20.97
Ash	13.73	12.09	10.06	10.77
Hemicellulose	19.77	22.86	24.04	26.02
Cellulose	18.60	23.83	28.15	31.23
Lignin	12.48	10.55	11.10	11.77

T1: control; T2:+water spinach 20%; T3: + water spinach 40%; T4: + water spinach 60%
DM =Dry Matter, OM =Organic Matter, CP =Crude Fiber, EE = Extract Ether, CF = Crude Fiber

2.3 *In vitro fermentation*

In vitro method used the [10] method for gas production. Rumen fluid was collected from fistullated cows fed with forage and concentrate (60:40) for one week before collecting it.

2.4 *Sampling and analysis*

After 48 h of fermentation process, the gas production was calculated using Neway Program (Excel), a program by Chen [11]. The calculation of gas production gained the equation $Y= a+b(1-e^{-ct})$, where Y = gas volume produced at time 't', a = soluble fraction, b = fraction of potential degraded or potential gas production, c = degradation rate of fraction b [11,12]. The residue of *in vitro* fermentation was used to determine nutrient degradability of dry matter (DMD) and organic matter (OMD) based on AOAC [9].

2.5 *Statistical analysis*

The data of gas production, fermentation parameter, and feed nutrient digestibility were statistically measured using a completely randomized design (*One Way ANOVA*). Then, to determine the difference of means, *Duncan's Multiple Range Test* (DMRT) was applied [13].

3. Results and discussions

The results of the effect of water spinach substitution as fiber source on feed nutrient digestibility and kinetics fermentation of feed are shown in table 3. This study indicates that 20% of water spinach substitution increased DMD up to 10% compared to the control. The other levels of substitution at 40%, 60% of water spinach did not significantly have any effect. Increasing of OMD occurred in feed substituted by water spinach (P<0.01) which ranged from 3 to 4% (T3 and T4) compared to the control. The substitution of 20% water spinach did not show difference in OMD compared control, T3 and T4.

Table 3. Digestibility and feed degradation rate with substitution of water spinach as fiber sources.

Parameter	T1	T2	T3	T4	SEM	P value
DMD (%)	57.97b	64.03a	60.15b	57.80b	0.74	<0.01
DOM (%)	81.42a	83.79ab	86.39bc	87.53c	0.37	<0.01
a (%)	-0.77	-0.88	-1.11	-0.38	0.15	0.45
b (%)	81.10a	84.06a	76.67b	76.78b	0.87	<0.01
c (%)	0.046	0.047	0.049	0.052	0.00	0.06

a,b,c Different superscripts in the same column show significant difference among treatments
T1: control; T2:+water spinach 20%; T3: + water spinach 40%; T4: + water spinach 60%
a: soluble fraction; b: degraded potential fraction c: rate constant per hour of gas production.
DMD = dry matter digesbility, OMD = organic matter digesbility

The study conducted by Aderao et al. [14] stated that the substitution of forages by the leaves produced higher DMD than hay (79,5-82,0%). The enhancement of DM and OM digestibility were possible because the content of water spinach lignin was lower than that of forage in control treatment. The result study of Jančík et al. [15] and Jagadeesh et al. [16] represent that the distinction of forages cell wall content would affect the DMD and OMD in rumen. Higher cell wall content lowers the feed digestibility. Chumpawadee and Pimpa [17] mentioned that the high lignin content in fiber sources led to the difficulty of microorganism to attach with low fermentation rates, low digestibility and limited feed consumption.

The substitution of water spinach as fiber sources influenced fraction b (degraded potential fraction), although it did not affect fraction a (soluble fraction in feed) and fraction c (gas production rate). The substitution level of 20% water spinach did not show the difference in fraction b, while the level of substitution 40% and 60% was lower in fraction b. Gado et al. [18] reported that forage with difference fiber content has variation in fraction b. Tan et al. [19] claimed that kinetics parameters of feedstuff degradation were affected by the type of carbohydrate in forages and ratio of structural carbohydrate and non-structural carbohydrate.

The data of gas production can be seen in table 4. The highest gas production ($P<0.01$) was produced by diet with 20% water spinach substitution at 48 h. On the other side, the substitution level of 40% and 60% water spinach resulted in lower gas production than control. The highest increase of gas production from feed degradation occurred at 12 – 24 h (figure 1). The difference of gas production in each treatment started at 2 h.

Table 4. *In vitro* gas production of feed with the water spinach substitution as fiber sources.

Treatment	h 0	h 1	h 2	h 4	h 6
T1	29.60±0.89	33.30±0.76	35.80±0/91^{ab}	40.30±0.67^{b}	44.40±0.82^{b}
T2	29.50±0.50	33.62±0.65	36.50±0.86^{b}	41.38±0.82^{c}	45.63±0.96^{c}
T3	29.50±0.35	33.13±0.54	35.38±0.41^{a}	38.87±0.41^{a}	42.37±0.41^{a}
T4	29.80±0.45	33.80±0.57	36.30±0.48^{ab}	40.00±0.71^{b}	43.70±0.90^{b}
Treatment	**h 8**	**h 12**	**h 24**	**h 36**	**h 48**
T1	49.10±0.89^{c}	58.70±1.20^{b}	81.50±1.11^{b}	89.80±0.84^{b}	92.90±2.60^{b}
T2	50.75±0.75^{d}	61.13±1.67^{c}	84.67±1.08^{c}	91.83±1.14^{c}	96.33±0.82^{c}
T3	46.25±0.25^{a}	56.37±0.65^{b}	78.50±1.11^{a}	84.63±0.65^{a}	87.00±1.87^{a}
T4	48.00±0.70^{b}	60.00±0.70^{bc}	79.50±1.50^{a}	87.71±3.48^{a}	88.40±1.67^{a}

^{a,b,c} Different superscripts in the same column show significant difference among treatments
T1: control; T2:+water spinach 20%; T3: + water spinach 40%; T4: + water spinach 60%

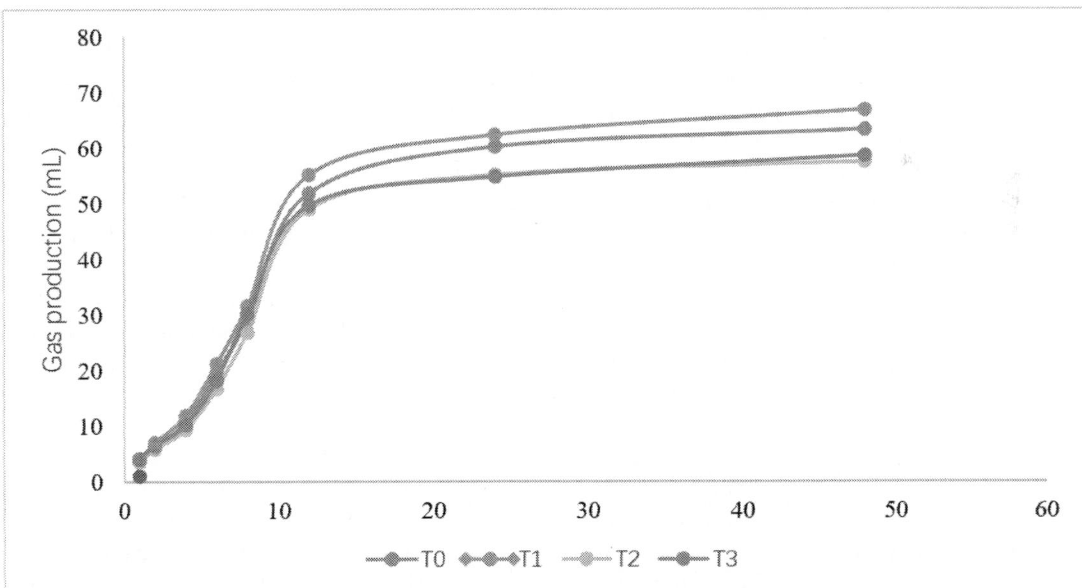

Figure 1. Gas production with water spinach substitution as fiber sources.

In this study, there was positive correlation between DMD and gas production produced in 20% water spinach substitution. The substitution of 20% water spinach resulted in highest DMD and gas production. The research had similarity the research by Sommart et al. [20], reporting that the increase of feed digestibility would affect the total increase of gas production. Soto et al. [21] stated that gas production in diet with vegetable waste substitutions showed an improvement (172-188 ml/g) for 24 h of incubation. Gas production was parameter used to estimate digestibility value and fermentation products [14,22].

4. Conclusion

The utilization of 20% water spinach as fiber sources can increase dry matter and organic matter digestibility and total gas production with no difference for rumen fermentation kinetics. Thus, it is concluded that the level of 20% water spinach substitution is potential to be fiber sources to substitute forage.

References

[1] Agus A and Widi T S M 2018 Current situation and prospect of beef cattle production in Indonesia — A review *Asian-Australas J Anim Sci.* **31** 1-18
[2] Harper K J and McNeill D M 2015 The role iNDF in the regulation of feed intake and the importance of its assessment in subtropical ruminant systems (the role of iNDF in the regulation of forage intake) Agriculture 5 778-90
[3] Warny E L, Fariani A, Ichinohe T, Abdulrazak S A and Fujihara T 2004 Comparative rumen degradability of some legume forages between wet and dry season in West Sumatra, Indonesia *Asian-Aust. J. Anim. Sci.* **17** 1107-11
[4] Nasrullah, Niimi M, Akashi R and Kawamura O 2003 Nuritive evalution of forage plants grown in South Sulawesi, Indonesia *Asian-Aust. J. Anim. Sci.* **16** 693-701
[5] BPS-Statistics Indonesia 2017 *Statistics of Seasonal Vegetable and Fruit Plants* (Jakarta)
[6] Khamparn P and Preston T R 2006 Effect of a supplement of fresh water spinach (*Ipomoea aquatica*) on feed intake and digestibility in goats fed a basal diet of cassava foliage *Livest. Res. Rural Dev.* **18** http://www.cipav.org.co/lrrd/lrrd18/3/kham18035.htmL.
[7] Opene A O, Ogunka-Nnoka C U and Uwakwe A A 2018 Comparative study on the nutrient composition and *in-vitro* antioxidant properties of leaves and stems of *Ipomoea involucrata Int. J. Agric. Innov. Res.* **7** 272-9
[8] Umar K J, Hassan L G, Dangoggo S M and Ladan M J 2007 Nutritional composition of water spinach (*Ipomoea aquatica* Forsk.) leaves *J. Appl. Sci.* **7** 803-9
[9] AOAC 2005 *Official Method of Analysis of the Association of Official Analitycal Chemist* W Harwitz edi. 18th ed. (Maryland, United Stated of America: AOAC International)
[10] Menke H H and Steingass H 1988 Estimation of the energetic feed value obtained from chemical analysis and *in vitro* gas production using rumen fluid *J. Anim. Res. Dev.* **28** 7-55
[11] Chen X 1994 Neway program *International Feed Resources Unit* (Backburn, Aberdeen: Rowett Research Institute)
[12] Kustantinah A, Hartadi H, Suhartanto B, Fathoni Hadi R and Melano R 2012 The effect of supplement and forage offering to goats on nutrients intake and digestibility *Proc. The 15th AAAP Animal Science Congress* (Thailand)
[13] Gomez K A and Gomez A A 1978 Statistical procedures for agricultural research with emphasis on rice *Biometrics* **34** 1-294
[14] Aderao G N, Sahoo A, Bhatt R S, Kumawat P K and Soni L 2018 *In vitro* rumen fermentation kinetics, metabolite production, methane and substrate degradability of polyphenol rich plant leaves and their component complete feed blocks *J. Anim. Sci. Tech.* **60** 1-9
[15] Jančík F, Koukolová V, Kubelková P and Čermák B 2009 Effects of grass species on ruminal degradability of silages and prediction of dry matter effective degradability *Czech J. Anim. Sci.* 54 315-23
[16] Jagadeesh C H, Reddy Y R, Nagalakshmi D, Mahender M, Kumari N N, Sridhar K and Devi K B S 2017 Effect of stage of harvest on the yield, chemical composition, in vitro and in sacco digestibility of hybrid napier (Pennisetum purpureum) variety APB N Indian *J. Anim. Res.* 51 116-20
[17] Chumpawadee S and Pimpa O 2008 Effect of non forage high fibrous feedstuffs as fiber sources in total mixed ration on gas production characteristics and in vitro fermentation *Pak. J. Nutr.* 7 459-64

[18] Gado H M, Elghandour M M Y, Cipriano M, Odongo N E and Salem A Z M 2016 Rumen degradation and nutritive utilization of wheat straw, corn stalks and sugarcane bagasse ensiled with multienzymes *J. Appl. Anim. Res.* 45 485-9

[19] Tan Z-L, Lu D-X, Hu M, Niu W-Y, Han C-Y, Ren X-P, Na R and Lin S-L 2002 Effect of dietary structural to non-structural carbohydrate ratio on rumen degradability and digestibility of fiber fractions of wheat straw in sheep *Asian-Aust. J. Anim. Sci.* 15 1591-8

[20] Sommart K, Parker D S, Rowlinson P and Wanapat M 2000 Fermentation characteristics and microbial protein sintesis in an in vitro system using cassava, rice straw and dried ruzi grass as substrates *Asian-Aust.J.Anim.Sci.* 13 1084-13

[21] Soto E C, Yanez D R R, Khelil H and Molina A E 2011 An in vitro evaluation of the potential use of greenhouse wastes to replace barley in goats' diets Challenging Strategies to Promote the Sheep and Goat Sector in the Current Global Context M J Ranilla, M D Carro, H Ben Salem and P Moran-Fehr eds. (Zaragoza: CIHEAM / CSIC/Universidad de Leon/FAO) pp 285-91

[22] Kamalak A, Canbolat O, Gurbuz Y and Ozay O 2005 Comparison of *in vitro* gas production technique with *in situ* nylon bag technique to estimate dry matter degradation *Czech J. Anim. Sci.* **50** 60-7

Jack bean as tempe ingredients: The safety study and fate of protein against gastrointestinal enzymes

E Puspitojati[1,2], R Indrati[2,3], M N Cahyanto[2] and Y Marsono[2]

[1]Magelang Agricultural Extension College, Jl. Kusumanegara, 55167, Umbulharjo, Yogyakarta, Indonesia
[2]Department of Food and Agricultural Product Technology, Faculty of Agricultural Technology, Universitas Gadjah Mada, Jl. Flora, Bulaksumur, 55281 Yogyakarta, Indonesia

E-mail: endahpuspitojati@gmail.com

Abstract. This study evaluated the change of hydrogen cyanide (HCN) content and concanavalin A (Con A) activity during the processing of jack bean tempe. In addition, the fate of protein of cooked tempe following in vitro gastrointestinal digestion was also investigated. The jack bean tempe was prepared by the step of the soaking-boiling-soaking-boiling-fermentation process. The mature tempe was harvested on 48 h of fermentation and followed by the cooking process for 10 min (T10) and 20 min (T20). The results showed the soaking, soaking-boiling, soaking-boiling-soaking, and soaking-boiling-soaking-boiling were able to decrease the HCN content of jack bean as much as 3.98; 85.88; 92.88 and 97.95%, respectively. It did not significantly change during fermentation with content for mature tempe was as low as 0.71 ppm, which is categorized as a safe level according to FAO. The haemagglutination assay of Con A showed the protein extracted from raw jack bean provided red blood cell clotting, while it was not found in the protein extracted from boiled jack bean and jack bean tempe. In the in vitro digestion model, both T10 and T20 samples showed that there was no significant change in peptides content before and after hydrolysis using pepsin ($p<0.05$). The action of gastrointestinal enzymes improved the released of the bioactive peptides which have the ability to inhibit angiotensin –I– converting enzyme (ACE). It can be concluded that the jack bean was the safe protein source for consumption as tempe. Furthermore, the protein of cooked tempe was highly hydrolyzed following in vitro gastrointestinal digestion produced bioactive peptides with high ACE inhibitory activity.

1. Introduction

Jack bean (*Canavalia ensiformis*) is one of under-exploited legume that can grow in various ecological conditions. The utilization of jack bean as a food source is still very limited due to anti-nutritional substances such as concanavalin A (Con A), canavanine, canatoxine, polyamine, protease inhibitors, flatulence compounds, cyanogenic glycosides, tannins, saponins, urease, L-Dopa [1,2]. Most of these substances give rise to dangerous biological responses, while some are widely applied in pharmacologically active nutrition and agents [3]. Con A is a single-cell protein which binds to the non-reducing terminal α-D-mannosyl and α-D-glucosyl groups of sugars, glycoproteins, and glycolipids. It can agglutinate red blood cells by interacting with immunoglobulin glycol-peptides [4,5]. HCN is a toxic component which causes death if consumption reaches a dose of 0.5-3.5 mg

Content from this work may be used under the terms of the Creative Commons Attribution 3.0 licence. Any further distribution of this work must maintain attribution to the author(s) and the title of the work, journal citation and DOI.
Published under licence by IOP Publishing Ltd

HCN/kg weight. It could be reduced by soaking, heat treatment, and fermentation during food processing [6].

Jack bean is a promising source of vegetable protein due to its high protein content. In addition, the essential amino acids (isoleucine, leucine, and tyrosine) were higher than legumes in general [2]. One of the uses of jack bean in Indonesia is in the production of tempe, although it is still restricted in a certain region [7,8]. Most people are still hesitant to consume it because they are still worried about the toxicity Even though, it has been reported that jack bean protein during tempe fermentation produced bioactive peptides which found to have ACE inhibitory activity [9].

The bioactive peptides are able to undergo physiological changes during the digestive tract which determines its biological activity. The peptides that are hydrolyzed by digestive enzymes can lose their biological activity or even increase activity due to the formation of active fractions of new peptides [10]. The gastrointestinal enzymes such as pepsin, chymotrypsin, and peptidase produce peptides in various length that define their biological activities including ACE inhibitor peptides. The fate of protein and stability of ACE inhibitory peptides during *in vitro* gastrointestinal digestion have been widely reported in pea seeds, whey protein, peanut, and grape skin [11-14].

There is little information about the fate of jack bean tempe protein during human digestion. The information about the degradation of the toxic component of the jack bean during tempe processing was also very limited. The objective of this study was to evaluate the safety of tempe made from jack bean in the context of HCN content and Con A activity. Furthermore, in this study, tempe samples were cooked at various time according to the way people consumed tempe. The fate of protein of cooked tempe following in vitro gastrointestinal digestion was also investigated in terms of the production of the bioactive peptides.

2. Materials and methods

2.1. Materials

Jack bean seeds (*Canavalia ensiformis*) and Raprima commercial inoculum were obtained from the local market. The human erythrocytes were provided by the Institute of Health Science, Yogyakarta, Indonesia. Angiotensin-I-converting enzyme from rabbit lung, HHL (Hippuryl-L-Histidyl-L-Leucine) from Sigma–Aldrich Company, O- phtaldialdehyde (OPA), KCN from Merck and Pepsin from porcine gastric mucosa (P7012), and Porcine pancreatin (P7545) were from Sigma–Aldrich Company. All other chemicals were of analytical grade.

2.2. Preparation of jack bean tempe

The making of jack bean tempe was adopted from the method of Puspitojati *et al* [9]. The inoculated seeds were wrapped by banana leaves and fermented at room temperatures (30°C) for 48 h. The matured tempe was cooked for 10 (T10) and 20 minutes (T20). The boiled jack bean was used as a control. Experimental samples were lyophilized for further analysis.

2.3. Protein extraction

Protein extraction was performed using the modified method of Rusdah and Puspitojati *et al* [9,15]. The protein extraction was conducted on the raw and boiled jack bean; the raw and cooked tempe. The protein was concentrated to 1 mg/mL (calculated by Lowry methods) and used for haemagglutination assay.

2.4. Simulated in vitro digestion

Enzymatic hydrolysis of jack bean tempe was conducted by the method of Mineskus *et al* and Sun *et al* [16,17] with slight modification. The sample was firstly hydrolyzed with pepsin (2000 U/mL) at 37°C for 120 min and followed by pancreatin (100 U/mL Trypsin) for 120 min. The sample was incubated in boiling water for 10 min for stopping the enzymatic reaction. The cool sample was centrifuged at 8000 *xg*, 4°C for 15 min (5424 R, Eppendorf, Germany). The supernatant was used for further analysis.

2.5. *Haemagglutination assay*

Haemagglutination activity of Con-A was detected using human erythrocytes in the presence of anti-coagulant according to the method of Marndi [4]. The protein sample was serially diluted in a microtiter plate that was further mixed with 50 µl of the human erythrocytes. The positive control contained 50 µl of raw jack bean protein extract and 100 µl of blood, while negative control contained 50 µl of PBS solution and 50 µl of blood. Hemagglutination titers were examined after incubation for 60 s at room temperature.

2.6. *Determination of HCN content*

The HCN content was determined by the method of William [18]. The sample was pounded and mixed with the distilled water, then incubated for 4 h. 1 mL of filtrate was added by 1 mL of KOH 2%, followed by 5 ml of alkaline picrate. The solution was dipped in boiling water for 15 s. The absorbance was measured at 510 nm using a Spectrophotometer UV-VIS (Dynamica Scientific, Halo SB 10, UK) with potassium cyanide (KCN) used as a standard.

2.7. *Assay of the peptide content and degree of hydrolysis*

Determination of peptide content was performed by the method of Church [19], while the degree of hydrolysis was performed by the method of Lin *et al* [20]. Tryptone was used as a standard curve [21].

2.8. *Determination of ACE-inhibitory activity*

The ACE-inhibitory activity of jack bean tempe was calculated by Cushman and Cheung method with slight modifications [22].

2.9. *Statistical analysis*

The data were analyzed by analysis of one-way variance (ANOVA) and the average difference between treatments was analyzed by Duncan Multiple Range Tests (DMRT). The data from ACE inhibition were evaluated using an independent T-test. The statistical analysis was conducted using SPSS IBM 23 while the significant differences were determined with a 95% confidence interval.

3. Results and discussion

3.1. *HCN content during jack bean tempe processing*

The HCN content decreased significantly during the tempe pre-fermentation processes ($p<0.05$). It has been described in table 1. The pre-fermentation process was the stages before fermentation, included the first soaking, the first boiling, the second soaking, and the second boiling. The soaking of jack bean seeds for 24 h followed by boiling for 30 min resulted in the highest decrease of HCN content, with the value from 57.56 to 8.46 ppm. The addition of wood ash was conducted in this first boiling. It has been known that ash can be used as an absorbent [6]. Rice husk ash could be used as an adsorbent to remove cyanide, good scavenger, and low cost. The high reduction of HCN content using wood ash and rice husk ash has been carried out in the previous studies on gadung (*Discorea hispida*) [23] and rubber seeds (*Hevea brasiliensis*) [6]. The effect of rice husk ash on the degradation of HCN as a function of contact time, pH, adsorbent concentration and temperature [6]. In addition, the boiling process itself was able to reduce HCN content because of its properties. HCN is a volatile compound and evaporates rapidly in the temperature over 28°C, it also dissolves readily in water. HCN content did not change significantly during tempe fermentation ($p>0.05$).

Standard threshold levels of HCN produced by cyanogenic glycogen in plants (tubers, nuts, and seeds) is 50 ppm. The Codex Alimentarius Commission has developed and published standards for a product suitable for direct human consumption (edible cassava flour) with the total HCN content must not exceed 10 mg/kg (10 ppm) [24]. The HCN content of mature jack bean tempe in this study was 0.71 ppm, therefore, the final product is safe for consumption.

Table 1. HCN content during tempe processing.

Treatment	HCN content (ppm)	HCN reduction (%)
Initial	59.95±2.67[e]	
S	57.56±0.69[d]	3.98±1.15
SB	8.46±0.19[c]	85.88±0.32
SBS	4.70±0.08[b]	92.16±0.14
SBSB	1.23±0.03[a]	97.95±0.06
SBSBF24	1.08±0.03[a]	98.19±0.05
SBSBF48	0.71±0.06[a]	98.86±0.11

Data are the mean ± SD of n=3,
Different letters in the same column indicate significant statistical differences ($p < 0.05$).
Initial is raw jack bean; S is jack bean seeds soaked for 24 h; SB is S followed by boiling for 30 min; SBS is SB followed by soaking for 48 h; SBSB is SBS followed by boiling for 30 min; SBSBF24 is SBSB followed by fermentation for 24 h or tempe 24 h; and SBSBF48 is SBSB followed by fermentation for 48 h or tempe 48 h.

3.2. Haemagglutination activity during jack bean tempe processing

The agglutination activity of the jack bean protein was tested with the human red blood cells. Figure 1 shows the protein extracted from raw jack bean had blood agglutination activity (line 1). The previous study investigated that all the *Canavalia* spp. were found to have high hemagglutination activity against cattle and human red blood cells [2]. That's why we used the protein extracted from raw jack bean as a positive control in this study. It was noted that red jack bean (*Canavalia gladiata*) had more ability to agglutinate human blood than other *Canavalia* spp.

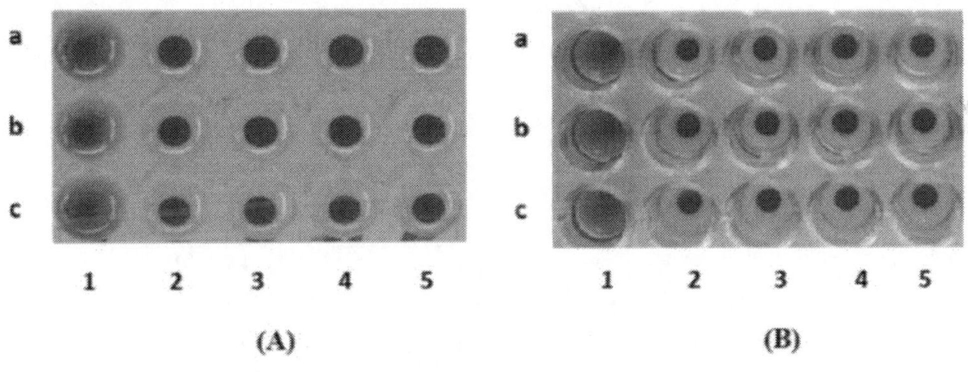

Figure 1. Agglutination activity of protein observed in microtiter plate from the bottom (A) and the top (B); Lane 1: positive control; lane 2: negative control; lane 3: boiled jack bean; lane 4-5: jack bean tempe 24 and 48 h; lane a-c: protein content 50, 25, and 10 µg.

It was found in the previous study that cracking the seeds into 2-7 pieces helped to eliminate Con A. In addition, cooking for 2 h or pressure-cooking for 45 min was effective in the elimination of hemagglutination activity in whole seeds [5]. The jack bean seed was sliced into 4-6 pieces during tempe processing in this study. The combination of cracking and cooking twice for each 30 min did not show any red blood cell clotting, it also found in tempe 24 h and 48 h. Jack bean tempe could be categorized as safe for consumption in terms of Con A content.

3.3. The peptide content and degree of hydrolysis following in vitro digestion

The fate of protein from tempe cooked for 10 and 20 min following gastrointestinal digestion was investigated. The peptides content in both samples increased during digestion simulation (figure 2). However, hydrolysis using pepsin did not show any significant increase. Pepsin breaks down large protein molecule into small peptides and converts almost all structural protein into the component with soluble small molecules [25]. Furthermore, it has a certain preference to cleave peptide bonds that contain the amine group of aromatic amino acids [26]. Tempe is a fermented product that has naturally undergone a process of protein hydrolysis during fermentation. In the previous study, it was reported that tempe fermented for 48 h only had less than 22% protein with a molecular weight more than 14 kD, the rest was a relatively small molecular weight peptide (unpublished data). Table 2 shows the degree of hydrolysis of jack bean tempe following in vitro digestion model. The level of protein hydrolysis using pepsin was very low in both samples. T20 had lower hydrolysis degree than T10. The time of cooking affected the type of peptides produced, which resulted in a difference in pepsin action.

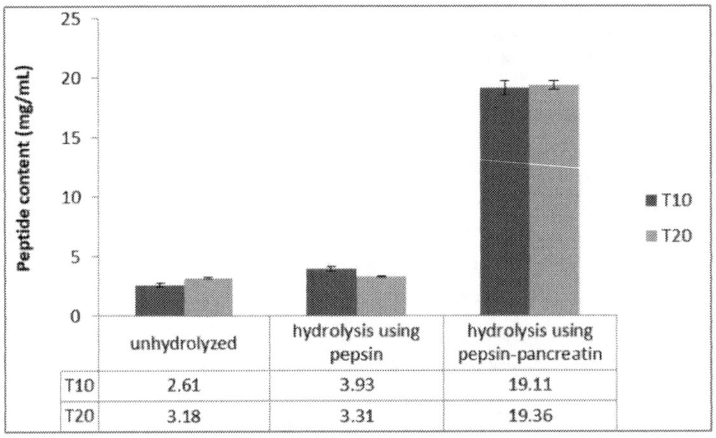

Figure 2. The peptide content of jack bean tempe following in vitro digestion model.

Table 2. The degree of hydrolysis of jack bean tempe following in vitro digestion model.

Sample	Degree of hydrolysis	
	Pepsin	Pepsin-pancreatin
T10	4.34±0.61	54.44±2.01
T20	0.45±0.20	54.44±1.18

Data are the mean ± SD of n=3
T10 is tempe cooked for 10min; T20 is tempe cooked for 20 min

Furthermore, the peptide content and hydrolysis degree were observed to have increased sharply in both samples after the hydrolysis using pancreatin. Pancreatin is mixtures enzymes that have wide specificity of hydrolytic action. It has been reported to have a greater hydrolytic action in lysine or arginine, tryptophan or tyrosine, leucine, alanine, and phenylalanine [27]. The wide specificity of pancreatin caused the high degree of protein hydrolysis, therefore, resulting in more small size peptides. The increase of peptides content was almost 5-fold after tempe protein was hydrolyzed using pancreatin (figure 2).

3.4. ACE inhibitory activity
The ACE inhibitory of peptides increased following in vitro gastrointestinal digestion (figure 3). The protein degradation during gastrointestinal digestion may release new peptides with stronger ACE inhibition. The protein hydrolysis using pepsin-pancreatin resulted in more small-size peptides than using pepsin. The peptide size is associated with the nutritional value of hydrolyzed protein [28].

ACE-inhibitory peptides were generally short chain peptides with 3-20 amino acids [10]. The protein hydrolysis using pepsin-pancreatin produced more small-size peptides which influenced the increase in ACE inhibitory activity. The activity of ACE inhibition was significantly different from both T10 and T20 in each stage of hydrolysis ($p<0.05$). The strongest of ACE inhibition activity was found to T10 after sequential hydrolysis using pepsin-pancreatin, with the value as high as 86.85%.

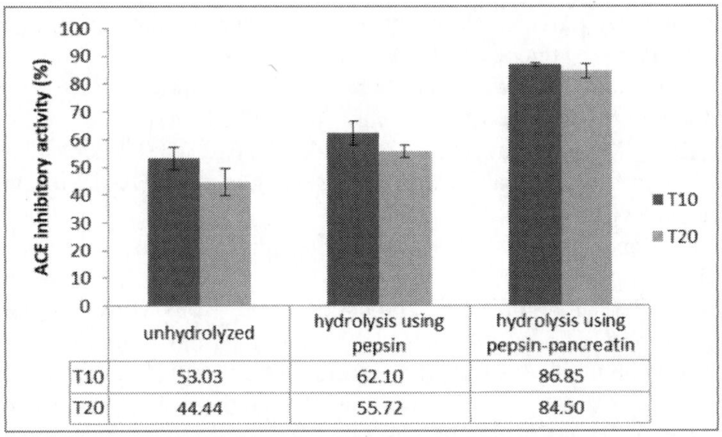

Figure 3. ACE inhibitory activity of jack bean tempe following in vitro digestion model.

4. Conclusion

The decrease of HCN occurred during the processing of jack bean tempe due to the treatment such as soaking and boiling. The resulting tempe had a very low HCN content and did not show agglutination activity when reacting with red blood cells. It can be concluded that the jack bean protein was safe to consume in the form of tempe. Furthermore, the protein of cooked tempe was highly hydrolyzed by pepsin-pancreatin following in vitro gastrointestinal digestion. The increase of bioactive peptides with ACE inhibitory activity was found during digestion. The highest inhibition was obtained in cooked tempe for 10 min with a value of 86.85 %.

Acknowledgments

The authors acknowledge the financial support of the Doctoral Program from the Indonesia Endowment Fund for Education (Lembaga Pengelola Dana Pendidikan), the Ministry of Finance. The authors also appreciate the Ministry of Research, Technology and Higher Education for supporting this project through Hibah Tim Pascasarjana 2018 on behalf of Dr. Ir. Retno Indrati, M.Sc.

References

[1] Widaningrum, Sukasih E and Purwani E Y 2015 Introductory study on processing of fermented jack bean *J. Penelit Pasca Panen Pertan* **12** 129-36
[2] Sridhar K R and Seena S 2006 Nutritional and antinutritional significance of four unconventional legumes of the genus *Canavalia* - A comparative study *Food Chem.* **99** 267-88
[3] Cheeke P R 1989 *Toxicants of Plant Origin* (CRC Press) Boca Raton Florida United States
[4] Marndi R 2012 Isolation and characterization of concanavalin a from the seeds of *Canavalia ensiformis* (Departement of Life Science, National Institute of Technology, India) p 44
[5] Udedibie A B and Carlini C 1998 Crack and cook: A simple and quick process for elimination of concanavalin A (Con A) from *Canavalia* seeds *Anim. Feed Sci. Technol.* **74** 179-84
[6] Fortuna D, Rahimsyah A and Puspitasri Y 2015 Degradation of acid cyanide poison in rubber seed (*Hevea brasiliensis*) after treatment with rice husk ash *Int. J. Adv. Sci. Eng. Inf. Technol.* **5** 291-3

[7] Suciati A 2012 Pengaruh lama perendaman dan fermentasi terhadap kandungan HCN pada tempe kacang koro (Canavalia ensiformis L) (Food Sci. Technol. Dep. Universitas, Hasanudin) p 83
[8] Bintari S H and Nugraheni K 2017 The Potential of tempeh as a chemopreventive and chemotherapeutic agent targeting breast cancer cells *Pakistan J. Nutr.* **16** 743-9
[9] Puspitojati E, Cahyanto M N, Marsono Y and Indrati R 2019 Production of angiotensin-I-converting enzyme (ACE) inhibitory peptides during the fermentation of jack bean (*Canavalia ensiformis*) tempe *Pakistan J. Nutr.* **18** 464-70
[10] Escudero E, Mora L and Toldrá F 2014 Stability of ACE inhibitory ham peptides against heat treatment and in vitro digestion *Food Chem.* **161** 305-11
[11] Jakubczyk A, Karaś M, Baraniak B and Pietrzak M 2013 The impact of fermentation and in vitro digestion on formation angiotensin converting enzyme (ACE) inhibitory peptides from pea proteins *Food Chem.* **141** 3774-80
[12] Luo Q, Boom R M and Janssen A E M 2015 Digestion of protein and protein gels in simulated gastric environment *Food Sci. Technol.* **63** 161-8
[13] Quist E E, Phillips R D and Saalia F K 2009 Angiotensin converting enzyme inhibitory activity of proteolytic digests of peanut (*Arachis hypogaea* L.) flour *Food Sci. Technol.* **42** 694-9
[14] Fernández K and Labra J 2013 Simulated digestion of proanthocyanidins in grape skin and seed extracts and the effects of digestion on the angiotensin I-converting enzyme (ACE) inhibitory activity *Food Chem.* **139** 196-202
[15] Rusdah 2016 Antioxidative of tempe peptides from Indonesia (Department Food Sci. Technol. Bogor. Agric. Univ.) p 50
[16] Minekus M *et al* 2014 A standardised static *in vitro* digestion method suitable for food – an international consensus *Food Funct.* **5** 1113-24
[17] Sun Y *et al* 2016 An investigation into the gastrointestinal stability of exenatide in the presence of pure enzymes, everted intestinal rings and intestinal homogenates *Biol. Pharm. Bull.* **39** 42-8
[18] Williams H J and Edwards T G 1980 Estimation of cyanide with alkaline picrate *J. Sci. Food Agric.* **31** 15-22
[19] Church F C, Swaisgood H E, Porter D H and Cantignani G L 1983 Spectrophotometric assay using o-phthaldialdehyde for determination of proteolysis in milk and isolated milk proteins *J. Dairy Sci.* **66** 1219-27
[20] Lin H and Alashi A M 2017 Antihypertensive properties of tilapia (*Oreochromis* spp.) frame and skin enzymatic protein hydrolysates *Food Nutr. Res.* **61** 1-11
[21] Wulandari R D, Rahayu E, Marsono Y ad Utami T 2017 Aktivitas antioksidan dan angiotensin-I converting enzyme inhibitor oleh yogurt dengan ekstrak daun ficus glomerata Roxb *Agritech* **37** 246-55
[22] Cushman. D W H S C 1971 Spectrophotometric assay and properties of the angiotensin converting enzymes from rabit lung *Biochem. Pharmachol.* **20** 1637-48
[23] Alma'arif A L, Wijaya A and Murwono R P D 2012 Penghilangan racun asam sianida (HCN) dalam umbi gadung dengan menggunakan bahan penyerap abu *J. Teknol. Kim. Dan. Ind.* **1** 14-20
[24] Codex 2018 Discussion paper on maximu level for hydrocyanic acis and mycotoxin contamination in cassava and cassava-based products *Joint FAO/WHO Food Standards Programme* 12th Session Utrecht (The Netherlands) 12 - 16 March 2018
[25] Lowe J S and Anderson P G 2015 *Stevens & Lowe's Human Histology* 4th Edi Elsevier Ltd. New York United States
[26] Ahn J, Cao M-J and Engen J R 2013 Assesing the reproducibility and specifity of pepsin and other aspartic protease *Biochim. Biophy. Acta* **1834** 1222-9
[27] Evnin L B, Vasquezt J R and Craik C S 1990 Substrate specificity of trypsin investigated by using a genetic selection *Biochemistry* **87** 6659-63

[28] Silvestre M P C, Morais H A, Silva V D M and Silva M R 2013 Degree of hydrolysis and peptide profile of whey proteins using pancreatin *Nutrire* **38** 278-90

IOP Conf. Series: Earth and Environmental Science **346** (2019) 012071 doi:10.1088/1755-1315/346/1/012071

Light interception and yield of some maize varieties grown in a double-row pattern under different urea applications

I K D Jaya[1,2], S Sudirman[1] and I W Sudika[1]

[1]Faculty of Agriculture University of Mataram, Jl. Majapahit 62 Mataram 83125, Lombok, Nusa Tenggara Barat, Indonesia

E-mail: ikdjaya@unram.ac.id

Abstract. To increase light interception in maize by improving the population density using a double-row pattern planting often causes difficulty in applying fertilizers. This study aimed to assess the light interception and yield of some maize hybrid varieties treated with and without a third application of urea fertilizer. The varieties tested were NK7328, BISI18, BISI2, and P35. The first application of urea was performed at planting, the second at 35 days after planting (DAP), and the third at 56 DAP. The experiment was conducted in an arid area with a sandy loam soil in the North Lombok region. Spacing for the double-row pattern was 35 × 20 cm^2 within a double row and 70 cm between double rows, resulting in approximately 98,000 plants/ha. The results showed that light interception was significantly affected by the variety of maize but not by the third-time application of urea, being the highest (96.7%) in NK7328 and the lowest (92.0%) in BISI18. The yield was significantly affected by both the variety and the third application of urea. The highest yielding variety was P35 (9.7 ton/ha), followed by NK 7328 (9.5 ton/ha), BISI18 (8.8 ton/ha), and BISI2 (8.2 ton/ha). The third-time application of urea in a high maize population density improved the yield by 21%.

1. Introduction

A recent increase in air temperature resulting from climate change affects the production of food crops, including maize. Maize that grows well at an optimum temperature range of 30°C and 36°C can experience heat stress that affects its productivity by a delay in silking if the temperature exceeds the optimum [1]. The other effects of climate change, such as erratic rainfalls, also affect crop growth and yield. In arid and dryland areas, dry spells that occur during a rainy season reduce maize yield severely, especially when the dry spell happens during silking or the grain-filling stage [2]. However, climate change that triggers an increase in CO_2 concentration in the atmosphere can benefit maize crop as long as the other factors, such as water, nutrition, and temperature, are not limited. CO_2 is one of the growth factors that play an important role in the photosynthesis process, together with water and sunlight. An abundant amount of CO_2 and sunlight should be managed to increase the production of a crop, especially maize.

Maize is one of the agricultural products that have a high potential to increase country foreign exchange. Demand on maize keeps rising, and Indonesia, as one of the maize-producing countries in the world, has started exporting maize since 2018. To continue to improve the country's foreign exchange and the welfare of farmers through maize production, in addition to expanding the planting area, the technology of maize cultivation, especially the one that is adaptive to climate change, also needs to be continuously updated. The existence of modern hybrid varieties of maize with very high yield potentials must be used to the maximum extent by correct planting technology. As previously

Content from this work may be used under the terms of the Creative Commons Attribution 3.0 licence. Any further distribution of this work must maintain attribution to the author(s) and the title of the work, journal citation and DOI.
Published under licence by IOP Publishing Ltd

reported, modern hybrid varieties of maize have upright leaves due to their narrow leaf angle [3], and hence can be planted in a high population density. A high population of maize plants can increase yield per unit of the land area and also optimize the use of sunlight and CO_2, which are abundant. The population of maize plants is said to be optimum for maximum productivity if the plant canopy during the grain-filling phase can capture 95% of sunlight [4].

The population of maize plants is increased by a double- or twin-row pattern of planting [5]. In a double row, the spacing used is 35 cm between rows and 20 cm in a row, while the distance between double rows is 70 cm, yielding 98,000–100,000 plants/ha. Meanwhile, in conventional planting patterns, such as a single row, the spacing used is 70 cm between rows and 20 cm in a row, producing around 64,000–70,000 plants/ha. In terms of the role of maize leaves in capturing sunlight and absorbing CO_2 from the atmosphere, the double-row planting pattern is adaptive to the current climate change conditions because the plants can convert more CO_2 and sunlight into biomass and crop yields. However, which modern hybrid varieties of maize available in Indonesian markets have the best performance in double-row planting patterns is still not known.

The yield of maize is determined by the ability of plants to convert CO_2 and sunlight into carbohydrates (in the presence of sufficient water) and also the availability of nutrients. One of the important nutrients for maize plants is nitrogen, sourced from urea. Nitrogen has been reported to improve yields for maize crops, especially in a high crop population [6]. Sufficient nitrogen in the leaves during the grain-filling stage plays an important role in determining the maize yield [7]. In practice, most maize farmers are reluctant to apply urea for the third time to provide sufficient nitrogen during grain filling, especially in a high maize population density. The reason is mainly the technical difficulties encountered while going through the thick maize canopy. This study aimed to examine light interception and yield performance of some modern hybrid varieties of maize (production of PT BISI, Du Pont, and Syngenta), which were planted in a double-row pattern and treated with and without the third application of urea in an arid area with sandy loam soil.

2. Materials and method

A field trial was carried out in an arid area of Gumantar village, Kayangan subdistrict, North Lombok (8,253654 S, 116,285695 E), Indonesia, from May to October 2018. Soil properties at the experimental site were as follows: pH 6.6, C-organic 0.6%, N-total (Kjeldhal) 0.06%, available P (Spectro) 8.4 ppm, and exchangeable K 0.49 meq%. The soil type was Entisol with sandy loam texture. Two treatments were tested: (1) modern maize hybrid variety and (2) application of urea for the third time. The modern variety of maize consisted of four types, NK7328 (product of Syngenta), BISI18 and BISI2 (products of PT BISI), and P35 (product of Du Pont). The third application of urea comprised two levels: without the third-time application and with the third-time application of urea. All treatments were arranged factorially using a randomized, block design with three replications.

The 70 × 35 × 20 cm double-row planting pattern was used, which meant that the distance between two double rows was 70 cm, the distance within the double row was 35 cm, and the spacing in a double row was 20 cm. The plot size of each treatment was 455 × 200 cm^2, and with such a plot size, three double rows of maize were planted in each plot with one single row in each edge of the plot. The planting orientation was east–west.

The plants were fertilized with basic urea fertilizer and NPK (nitrogen, phosphor, potassium) Phonska (15-15-15) at doses of 150 and 190 kg/ha, respectively. Furthermore, the first supplementary fertilizer was applied when the plants were at 35 days after planting (DAP) with 200 and 190 kg/ha doses of urea and Phonska, respectively. A 150 kg/ha dose of urea was applied for the third time at tasseling (56 DAP). Weekly irrigation water was obtained from a nearby small river. The maintenance of the crops, such as mechanical weeding, was done before the first supplementary fertilizer application. A systemic fungicide (Tebukonazol + Trifloksistrobin) was applied at 70 DAP to prevent the spread of an unidentified fungus that caused discoloration of the lower leaves.

The variables measured were plant height, number of leaves, and interception of sunlight by the plant canopy (using AccuPAR/LAI Ceptometer LP-80, METER Group, Inc. USA) at tasseling. The crop yield and its components, such as cob length, cob weight, number of rows of seeds in a cob, weight of 1000 seeds, and weight of sundried seeds per cob and per plot, were measured after the cobs were sundried for 2 days without husk and shank. Analysis of variance for all collected data was run on Minitab 15, and correlation analyses were performed in Microsoft Excel.

3. Results and discussion

No rain occurred during the period of the experiment, and therefore water for the plants was obtained from the weekly irrigation water. The plants received sufficient water during the early vegetative stage, but after 6 weeks, the plants started to experience slight wilt in the top leaves on every sixth day after irrigation. The maximum temperature recorded during the experimental period was 37°C with a minimum of 29°C. All the plants also experienced a minor unidentified fungus disease attack starting from 70 DAP, which caused discolouring of lower leaves and could also affect yield and yield components.

Table 1. Effects of maize variety and the third application of urea on some parameters measured at tasseling and harvest.

	Measured variables				
Treatments	Plant height (cm)	Leaves number	Leaf angle (°)	Intercepted light (%)	Stover weight (ton/ha)
Variety					
NK7328	233.2	16.9	31.5 a	96.0 a	5.1
BISI18	236.0	15.9	32.0 a	95.1 a	4.6
BISI2	241.0	16.3	23.6 b	91.8 b	4.8
P35	235.4	15.6	30.7 a	96.9 a	4.6
Fertilizer					
With the third application of urea	238.5	16.3	29.8	95.1	5.0
Without the third application of urea	234.6	16.1	29.1	94.8	4.6

Mean values within a column in the same treatment followed by different letters are significantly different at P <0.05 according to Tukey's range test.

All the tested modern maize hybrid varieties performed similarly in terms of plant height and leaf number at tasseling (table 1). This is a common result because modern maize hybrid varieties have small genetic variations, and their performance is affected mainly by the growing environment. However, in terms of leaf angle (the third leaf down from the flag leaf), the narrowest angle (23.6°) was recorded in BISI2 variety, which was much lower than that of the other three types as presented in table 1. This showed that BISI2 had the most erect or upright leaves, and therefore its population density had the highest possibility to be improved. With the current double-row planting pattern, the achieved population density was 98,000 plants/ha. The plant height, leaf number, leaf angle, and intercepted light by canopy were not affected by the third application of urea. This was obvious because urea was applied at the end of the vegetative stage (tasseling) and the target of the fertilizer application was the yield, not the growth parameters.

The third application of urea did not have a significant effect on light interception, but the canopy of the three maize varieties, P35, NK7328, and BISI18, intercepted ≥95% of sunlight, being the highest in P35. The other two varieties intercepted less but with no significant difference (table 1). The lowest light interception was recorded in BISI2, which was significantly different from that in the other varieties. A positive and significant correlation ($r^2 = 0.67$) was found between the leaf angle and light interception. At tasseling up to the grain-filling stage, the optimum sunlight interception for an optimum yield was suggested as 95% [8]. Maize hybrid varieties grown in a single-row pattern with a population density of fewer than 84,000 plants/ha have been reported to intercept less than 86% of sunlight at tasseling [9]. However, light interception can be improved by improving the population density [10], and the double-row pattern is an effective way to achieve this [6]. The canopy of P35, NK7328, and BISI18 grown at a population density of 98,000 plants/ha intercepted sunlight more than 95% at the tasseling stage, indicating that the population densities of these three varieties reached their optimum when planting a double-row pattern.

Plant stover (including leaves, leaf sheaths, stalks, and tassels) after harvest was not affected by either third application of urea or maize variety (table 1). The results showed that the four modern maize hybrid varieties tested in this study performed the same in terms of plant biomass after excluding their kernels (grains, cobs, and husks) as they had a very similar height and the number of leaves. The plant stover weight at harvest, which ranged from 4.6 to 5.1 ton/ha in this study, was within the range of data reported in an earlier study on eight varieties of modern maize at different planting dates [11]. The stover weight of those treated and not treated with the third application of urea was same because of the lowering rate of photosynthesis after tasseling and use of nitrogen to mainly provide more carbohydrate to the newly developed strong sinks, the ears [12].

Table 2. Effect of maize variety and the third application of urea on some cob parameters measured at shelling dry condition.

Treatments	Measured variables				
	Cob weight (g)	Cob length (cm)	Cob diameter (cm)	Grain rows per cob	Percentage of grain to cob (%)
Variety					
NK7328	158.3 a	13.9 a	4.2 a	16.3 a	81.6
BISI18	155.2 a	13.6 ab	4.2 a	15.9 ab	81.6
BISI2	134.4 b	11.9 b	3.7 b	13.2 b	82.1
P35	161.9 a	13.6 ab	4.3 a	16.3 a	83.4
Fertilizer					
With the third application of urea	163.1 a	13.8	4.2 a	15.7	82.5
Without the third application of urea	141.8 b	12.7	3.9 b	15.2	81.8

Mean values within a column in the same treatment followed by different letters are significantly different at $P<0.05$ according to Tukey's range test.

Significant differences were found in cob parameters (without husk and shank and measured at shelling dry condition), such as cob weight, cob length, cob diameter, and number of grain rows in a cob, but not the percentage of grains to cob. Overall, BISI2 produced the smallest cob size, which was significantly lower than that of the other three varieties, but with no significant difference (table 2). In terms of number of seed rows in a cob, the two BISI varieties produced a very similar number of seed rows. Plant biomass, especially after tasseling, has been reported to affect the cob parameters, but the sink–source ratio, especially the sink strength, determines the cob parameters the most [13]. In this study, the stover weight was not significantly different for the four varieties (measured at harvest), but the lowest cob parameter was measured in BISI2. It was possible that the sink strength of BISI2 was not as good as that of the other three varieties when grown under arid and non-optimum irrigation water conditions. The third application of urea affected the cob weight and cob diameter but not the cob length, number of grain rows in a cob, and percentage of grains to cob (table 2). The cob weight and cob diameter are parameters that determine the maize yield and are very much affected by the nitrogen status, especially during the grain-filling stage [14].

The grain weight per cob was affected by both variety and third application of urea. The P35 variety showed the highest grain weight but with no significant difference compared with those of NK7328 and BISI18, while the lowest was recorded in BISI2 (table 3). The BISI2 variety was supposed to produce two cobs per plant under favorable conditions. In this study, not all BISI2 plants produced two cobs per plant, and for those that produced two, only one had a length of more than 8 cm or cob diameter of more than 3.0 cm. Arid and non-optimum water irrigation conditions, especially after the end of the vegetative stage, might contribute to the non-optimum performance of the BISI2 variety. Slight water stress, especially during the critical stage of tasseling or cob formation, caused severe loss of maize yield [15]. In some cases, mild water stress could generate 13%–63% of maize yield [16]. High carbohydrate allocation was needed to produce more cobs during cob initiation [17], and this condition might not be achieved by the BISI2 variety at the experimental site with a limited source of water. However, the other three varieties performed better under such conditions compared with BISI2. This result can help in selecting the variety to be grown under arid conditions with a limited amount of irrigation water.

Table 3. Effect of maize variety and the third application of urea on yield and yield components.

Treatments	Measured variables			
	Grains weight per cob (g)[a]	1000 grains weight (g)	Grains weight per plot (kg)	Grains weight per ha (ton)
Variety				
NK7328	129.2 a	276.3 a	8.8 a	9.5 ab
BISI18	125.8 a	270.0 a	7.9 ab	8.8 ab
BISI2	105.2 b	255.0 b	7.4 b	8.2 a
P35	132.7 a	269.3 a	8.8 a	9.7 b
Fertilizer				
With the third application of urea	135.9 a	279.5 b	9.0 b	9.8 b
Without the third application of urea	110.5 b	256.1 a	7.5 a	8.2 a

Mean values within a column in the same treatment followed by different letters are significantly different at $P < 0.05$ according to Tukey's range test.

Correlation analysis results showed that the grain weight per cob had significant and medium correlations with leaf angle ($r^2 = 0.57$), light interception by canopy ($r^2 = 0.50$), and number of grain rows per cob ($r^2 = 0.67$), and high correlations with cob length ($r^2 = 0.76$), cob weight ($r^2 = 0.97$), and cob diameter ($r^2 = 0.87$). Light interception by plants' canopy had a better correlation with grain weight per cob compared with the leaf angle. An earlier study showed that the grain yield of maize increased with the increase in light interception by the canopy of maize grown in a narrow row [4]. A correlation value of 0.70 was reported for the relationship between grain yield and light interception, which was higher than the results reported in this study. Location differences that lead to sun angle differences may contribute to the different values of the relationship. The positive and significant correlation among grain weight per cob, cob length, and number of grain rows per cob has also been reported earlier [18], with slightly higher correlation values in this study.

The smaller cob size in BISI2 resulted in lower grain yield per plot compared with the other three varieties, but with no significant difference compared with BISI18. The positive and medium correlation values between grain yield per plot and cob diameter ($r^2 = 0.72$) as well as with cob length ($r^2 = 0.60$) contributed to this result. Less light interception by canopy at the grain-filling stage might also contribute to the less yield in BISI2 compared with the other three varieties. Grain yield per plot had a positive but nonsignificant correlation ($r^2 = 0.40$) with light interception by plant canopy. The third application of urea increased the grain weight per cob by 18.7% and the grain weight per plot by 21% compared with those resulting from crops that did not receive a third application of urea. This result was in line with the earlier finding that nitrogen affected the yield of modern maize hybrid varieties [13]. Modern maize that experienced nitrogen deficiency, especially during the grain-filling stage, yielded less compared with varieties having sufficient nitrogen. Furthermore, sufficient nitrogen during the grain-filling stage was able to maintain photosynthesis capacity of the maize leaves [19]. The weight of 1000 seeds treated with a third application of urea, which was around 9% higher compared with those that did not receive a third application of urea, contributed to the higher yield.

4. Conclusion and suggestions

Maize varieties grown in a double-row pattern with a leaf angle of less than 30°, such as in BISI2, intercepted less than 95% of sunlight at tasseling. Less light interception coupled with a smaller cob size in BISI2 resulted in a lower yield compared with those in P35, NK7328, and BISI18 varieties. Yield of maize grown in an arid area with limited irrigation water improved by 21% when urea was applied for the third time at a rate of 150 kg/ha. These results indicated that BISI2, as a modern hybrid variety, performed less optimally in an arid area with less optimal irrigation water. To improve the yield of maize in a high population density, urea should be applied for the third time at tasseling. Further studies are needed to estimate the economic benefits of applying urea for the third time and improving BISI2 population density so that the canopy can intercept at least 95% of sunlight at tasseling.

Acknowledgment

The authors thank the University of Mataram for funding the study under contract number 1457.N/UN18.L1/PP/2018 and Applied Research and Innovation Systems in Agriculture (ARISA) for other supports.

References

[1] Cicchino M, Ediera J I R and Otegui M E 2010. *Crop Sci.* **50** 1431-7.
[2] Jaya I K D and Sudirman R 2017 *Int. J. Biosci. Biotechnol.* **5** 1-11.
[3] Pugano E and Maddonni G A 2004 *Field Crop Res.* **101** 306-20.
[4] Andrade F H, Calviño P, Cirilo A and Barbieri P 2002. *Agron. J.* **94** 975-80.
[5] Shakarami G and Rafiee M 2009 *Am-Eurasian J. Agr. Envir. Sci.* **5** 69-73.
[6] Dahmardeh M 2011 *Am. J. Plant Physiol.* **6** 44-9.
[7] Paponov I A, Sambo P, Presteri T, Geiger H H and Engels C 2005. *Plant Soil* **272** 111-23.
[8] Puntel L A 2012 Field characterization of maize photosynthesis response to light and leaf area index under different nitrogen level: A modelling approach (Iowa State University, Ames, IA, USA).
[9] Jaya I K D, Sudirman and Jayaputra 2015 *Agroteksos* **25** 144-50.
[10] Jeschke M R 2017 *Managing Corn for Greater Yield* (Johnston: DuPont Pioneer) **24** p 6
[11] Daynard T B and Hunter R B 1975 *Can. J. Plant Sci.* **55** 77-84.
[12] Ding L, Wang K J, Miang G, Biswas D K, Xu H, Li L F and Li Y H 2005. *Ann. Bot.* **96** 925-30.
[13] Borràs L and Otegui M E 2001 *Crop Sci.* **41** 1816-22.
[14] Uhart S A and Andrade F H 1995 *Crop Sci.* **35** 1384-89.
[15] Çakir R 2004 *Field Crop Res.* **89** 1-16.
[16] Earl H J and Davis R F 2003 *Agron. J.* **95** 688-96.
[17] Lee E A and Tollenaar M 2007 *Crop Sci.* **47** S202-15.
[18] Silva A R, Souza Jr C L, Aguiar A M and de Souza A P 2004 *Maydica* **49** 65-71.
[19] Tollenaar M and Lee E A 2006 *Maydica* **51** 399-408.

IOP Conf. Series: Earth and Environmental Science **346** (2019) 012072 doi:10.1088/1755-1315/346/1/012072

Experimental analyses on shearing properties of hop stem

J Yang[1,3] and X K Li[2]

[1]College of Electron-mechanics and Automobile Engineering, Tianshui Normal University, Tianshui, 741001, China
[2]Institute of Agricultural Machinery, Gansu Institute of Machinery Research Co., Ltd., Lanzhou, 730030, China

E-mail: lwg_yj@163.com

Abstract. In order to solve the problems of low cleaning rate and high breakage rate of the existing hop harvesting mechanical cutting devices, single factor and multi-factor orthogonal tests were carried out on hop stems of different varieties, such as Qingdao big flower, Marco Polo and Fragrant flower, by using the universal testing machine. The tests were focused on the influence of the factors, including moisture content, shearing mode and shearing velocity on shearing force. The results show that the influence of the above three factors are all significant, among which the first two are more significant. The shearing force will increase with the decrease of moisture content and it will decrease with the increase of shearing velocity. When the shearing velocity is kept between 100 mm/min and 200 mm/min, it will be the most advantageous to the cutting, and the slip cutting is the most labour-saving mode. The tests show that the cutting effect will be the best during the course of harvesting when the hop stems are fresh. The shearing velocity arrives at 100mm/min and the slip cutting mode is adopted. The results can provide a theoretical basis for improving the cutting devices of the existing hop harvesting machinery.

1. Introduction

Hops are climbing plants of the mulberry family [1]. Hops are oval and puffy in shape, which contain formic acid. When formic acid is added to beer, beer has aromatic bitterness and special fragrance, which is an important raw material for beer production. Hop cultivation in China is mainly distributed in Gansu and Xinjiang provinces. The warm temperate drought climate in Hexi Corridor of Gansu Province has promoted the fine cultivation of hops. At the same time, beer production enterprises have adopted the industrialization mechanism of "company + base + peasant household", which has made the production of hops increase continuously [2]. Therefore, Gansu Province has become the largest province of hop cultivation in China. The shape of hops is shown in Figure 1.

The main hop varieties in Gansu Province are Qingdao big flower, Marco Polo and Fragrant Flower. Due to the high content of formic acid, Qingdao big flower accounts for more than 80% of the total planting area in Hexi Corridor [3]. Field tests have showed that the mechanized harvesting performance of Qingdao big flower is poor, and it is not suitable for mechanized harvesting. Although hop harvesting machinery has been developed successfully for a long time, the difficulty of mechanized harvesting is different due to the different varieties of hop planting, which makes the existing machinery have the defects of low cleaning rate and high breakage rate. As a result, the imperfection of mechanized harvesting technology has become a bottleneck restricting the "13th Five-Year Plan for Agricultural Development of Gansu Province".

Content from this work may be used under the terms of the Creative Commons Attribution 3.0 licence. Any further distribution of this work must maintain attribution to the author(s) and the title of the work, journal citation and DOI.
Published under licence by IOP Publishing Ltd

IOP Conf. Series: Earth and Environmental Science **346** (2019) 012072 doi:10.1088/1755-1315/346/1/012072

Figure 1. The shape of hops.

In order to improve the defects of hop harvesting machinery, it is necessary to determine the physic-mechanical properties hop stems such as shearing resistance strength. There are many studies on the mechanical properties of crop stems at home and abroad, and the main purpose is aimed at the mechanical properties of cash crops such as wheat, forage, corn, ramie, sugarcane and so on. Guo [4, 5] studied the relationship between the tensile strength of wheat stems and their chemical components, which provided a theoretical basis for improving lodging resistance of wheat. Su [6] studied the tensile and compressive properties of the xylem of ramie stem, which provided a reference for mechanized harvesting of ramie. Zhao [7-9] carried out the tests of tensile, bending and shearing of legume forage stems. The results of the tests provided basic information for selecting reasonable matching power and cutting system structure parameters. Liu [10, 11] studied the damage tests of sugarcane stems under torsion, compression and tension loads, which had an important influence on the cutting mechanism of sugarcane. Yu and Chen [12, 13] conducted the tensile test of maize straw to provide technical support for mechanized harvesting of maize. Zhou [14] obtained the elastic modulus, Poisson's ratio, bending and shear modulus of industrial hemp stem through tensile, bending and compression tests. Luo [15] obtained the optimal combination of cutting devices through the reciprocating cutting experiment of eulaliopsis binata stems. Song [16] studied the cutting performance of cotton stems through average speed, cutting inclination and cutting speed ratio. Li [17] measured the process of shearing stress of cotton stem changing with moisture content on a universal testing machine, and determined the optimum harvesting time of cotton stem. The purpose of the studies is also mostly focused on lodging resistance and mechanism. As one of the characteristic pillar industries in Gansu province, the research on the mechanical properties of hop stems is still blank. To improve the existing hop harvesting machinery, it is necessary to research the shearing mechanical properties of hops stems. In view of this, this paper takes hops stems of different varieties (Qingdao big flower, Marco Polo, Fragrant flower) as the research subjects, and chooses different shearing mode under different moisture content conditions. The optimum cutting parameters are determined by the orthogonal test, which provides a basis for improving the existing cutting device.

2. Test materials and methods

2.1. Test materials

The experimental materials were picked in the hop base of Yumen Agricultural Reclamation Construction Company in Gansu Province in mid-August 2018. In the experiment, the stems of Marco Polo, Qingdao big flower and Fragrant flower which were mature, intact and undamaged were selected. The diameter and wall thickness of the samples were measured by vernier caliper. Then, the stems were classified according to the diameter. The stems were made 150 mm length as specimens to complete the shearing test.

2.2. Test instrument

ETM105D, a universal testing machine manufactured by Shenzhen Wance company, was used for shearing test. During the test, the force, displacement, deformation, loading velocity and test curve can be displayed dynamically in real time by using the universal testing machine. The maximum test loading of this machine was 100KN, the relative error of force was less than 0.18%, the relative error of displacement and velocity were less than 0.19%, and the accuracy level was 0.5. During the test, the stems were fixed on the self-made fixture, and the self-made cylindrical fixture support was matched with the cylindrical hollow base of the universal testing machine. The cutting tool was in the form of a light knife with a thickness of 2 mm. The shearing modes were divided into transverse cutting, oblique cutting and slip cutting. In transverse cutting, the blade was kept vertical to the stems in the horizontal plane. In oblique cutting, the direction of the blade was changed. The cutter blade and the stems should have an angle in the horizontal plane. The test chose 45 degrees angle in oblique cutting. In sliding cutting, the angle between the fixture and the horizontal plane were changed to make it 30 degrees angle with the horizontal plane [18].

2.3. Test methods

2.3.1. Single factor test. The shearing force of hop stems were affected by many factors, such as shearing mode, shearing velocity, moisture content, holding conditions, blade form and so on. According to the working conditions of the existing hop harvesting machinery, the shearing mode, shearing velocity and moisture content were selected as the main factors affecting the shearing force of stems. The test conditions were shown in table 1.

Table 1. Factors and levels of single-factor test.

Test level	Factors		
	Shearing mode A	Shearing velocity B	Moisture content C
1	transverse cutting	50 mm/min	fresh samples
2	oblique cutting	100 mm/min	semi-dry samples
3	slip cutting	150 mm/min	dry samples
4		200 mm/min	
5		250 mm/min	

In the single-factor test of shearing force by velocity, 120 samples of different varieties (Qingdao big flower, Fragrant flower and Marco Polo) with similar diameters were selected as the test subjects. Under different shearing velocity in the range of 50 to 250mm/min, the shearing test was conducted by transverse mode. The test index was maximum shearing force. The test was repeated for eight times at each location, and the data were recorded and the average value was obtained.

In the single factor test of shearing mode and moisture content on shearing force, 135 samples with the same conditions were selected. Stems quality were measured by balance, and then the moisture content of the stems was determined by 105°C drying method. Among them, the moisture content of fresh samples remained between 70% and 80%, the moisture content of semi-dry samples was between 40% and 60% after one day in the ventilated and dried laboratory, and the moisture content of dry samples was between 20% and 30% after two days in the ventilated and dried laboratory. The test was repeated for five times at each location. Data were recorded and their average values were taken.

2.3.2. Multivariate factor test. In order to verify the influence degree and optimum level of each factor on shearing force, the orthogonal test of shearing velocity, shearing mode and moisture content were carried out.135 samples were selected as the same as the single factor test, and the test conditions were the same as the single factor test conditions.

3. Results and analysis

3.1. Single factor test

3.1.1. Effect of shearing velocity on shearing force. In the velocity range of 50 to 250 mm/min, the shearing force tests of Qingdao big flower, Marco Polo and Fragrant flower were carried out. Test data were shown in table 2.

Table 2. Experimental data of shearing force of hop stems at different velocities.

Varieties	Maximum Shearing force at different shearing velocities (N)				
	50 (mm/min)	100 (mm/min)	150 (mm/min)	200 (mm/min)	250 (mm/mi n)
Qingdao big flower	258.01	237.22	224.24	200.61	246.14
Marco Polo	215.05	203.30	254.53	197.22	229.40
Fragrant flower	209.14	197.03	192.63	159.55	167.98

[a] Using fresh samples and transverse cutting.

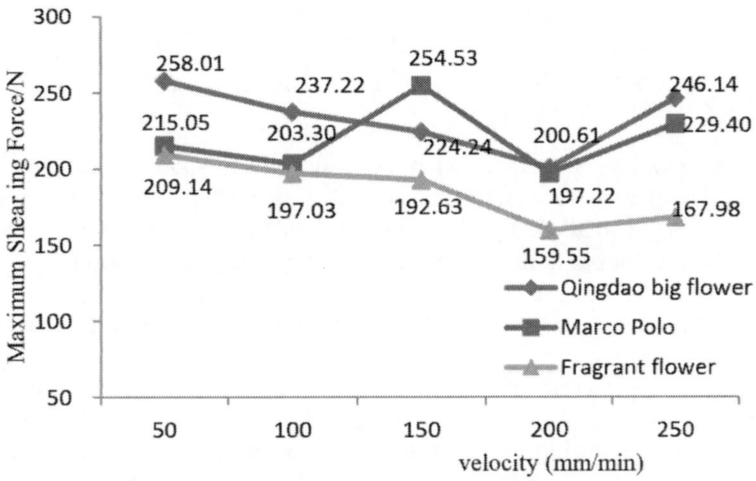

Figure 2. Shearing force test at different velocities.

The results show that the shearing force of Qingdao big flower and Fragrant flower decreases with the increase of velocity in the range of 50 to 200 mm/min. The shearing force of Qingdao big flower changes 57.40N. The shearing force of fragrant flowers changes 49.59N. When the velocity is 200mm/min, the shearing force reaches the minimum, Qingdao big flower is 200.61N and Fragrant flower is 159. 55N.When the velocity increases gradually from 200 mm/min to 250 mm/min, the shearing force increases. The shearing force of Qingdao big flower increases by 45.53N and that of Fragrant flower by 8.43N.The reasons for this phenomenon are as follows: with the increase of shearing velocity, the proportion of extrusion stage before cutting decreases gradually, that is, the distance of cutter extruding the stem decreases, which leads to the decrease of the shearing force of stems. When the shearing velocity exceeds a certain value and continues to increase, the change of the distance of cutter extruding the stem is no longer significant [16]. The shearing force of Marco Polo varies smoothly with the increase of velocity and changes 57.31N. When the velocity is 150 mm/min, the maximum shearing force of Marco Polo is 254.53N, which decreases with the increase of velocity.

From the above single factor test data, it can be seen that too high or too low velocity will increase the shearing force, and the velocity in the middle section is the most favourable for shearing. Therefore, the optimum shearing velocity range of three varieties is from 100 mm/min to 200 mm/min.

3.1.2. Effect of shearing mode and moisture content on shearing force. According to the influence of shearing velocity on shearing force, the optimum shearing velocity range is 100 mm/min to 200 mm/min. Therefore, the shearing test of hop stems of different varieties with different shearing modes and different moisture content was carried out under the conditions of 200 mm/min and diameter 6~7 mm. The experimental data were shown in table 3. The results show that the shearing force of three varieties increases with the decrease of moisture content, that is, fresh samples need the least shearing force, followed by Semi-dry samples, and dry samples need the largest shearing force.

Table 3. Experimental data of shearing force in different shearing mode and moisture content.

Varieties		Maximum Shearing Force (N)		
		transverse cutting	oblique cutting	slip cutting
Qingdao big flower	fresh samples	175.48	220.10	161.96
	semi-dry samples	296.20	321.68	208.32
	dry samples	326.49	456.04	228.42
	fresh samples	207.08	210.11	177.14
Marco Polo	semi-dry samples	255.37	270.33	217.60
	dry samples	398.54	475.42	226.30
	fresh samples	169.18	189.82	144.29
Fragrant flower	semi-dry samples	321.62	208.80	159.59
	dry samples	360.86	215.80	189.02

[a] Shearing velocity is 200mm/min.

As far as shearing mode is concerned, whether fresh, semi-dry or dry samples, oblique cutting of Qingdao big flower and Marco Polo needs the largest shearing force, followed by transverse cutting, and slip cutting requires the smallest shearing force. When fresh fragrant flowers are cut, the shearing force required by oblique cutting is the largest, followed by transverse cutting, and the shearing force needed by slip cutting is the smallest. The shearing force of semi-dried and dry samples of fragrant flowers is the greatest in transverse cutting, followed by oblique cutting, and the most labour-saving in slip cutting.

According to the single factor data, with the decrease of moisture content, the shearing force increases no matter what kind of shearing mode is. Therefore, in the same shearing mode, fresh samples are the most labour-saving. Three kinds of shearing modes are used to cut hop stems under three different varieties and different moisture content respectively. The shearing force needed for slipping fresh samples is the smallest, which is the best shearing mode. Compared with three different varieties of samples, under the same shearing mode, Marco Polo is the most laborious, followed by Qingdao big flower, and fragrant flower is the most labour-saving.

3.2. *Multivariate test*

Three factors including shearing mode, shearing velocity and moisture content were selected for orthogonal test. Compared with single factor test, the shearing force of hop stems of three varieties decreased gradually with the increase of velocity in the range of 50 mm/min to 250 mm/min. When the velocity exceeded 200 mm/min, the shearing force would increase. Therefore, the levels of selected velocity were 100 mm/min, 150 mm/min and 200 mm/min.

In the test, the moisture content of fresh sample was 70%~80%, the moisture content of semi-dry sample was 40%~60%, the moisture content of dry sample was 20%~30%, and the shearing mode was transverse cutting, oblique cutting and slip cutting. The L_9 (3^4) orthogonal table [19-21] was selected. The test index was the maximum shearing force. The level of test factors was shown in table 4. The following tests were repeated five times, and the test data were shown in table 5.

Table 4. Multivariate orthogonal table.

Number	Shearing mode A	Shearing velocity B (mm/min)	Moisture content C (%)
1	1(transverse cutting)	1(100)	1(fresh samples)

2	1(transverse cutting)	2(150)	2(semi-dry samples)
3	1(transverse cutting)	3(200)	3(dry samples)
4	2(oblique cutting)	1(100)	2(semi-dry samples)
5	2(oblique cutting)	2(150)	3(dry samples)
6	2(oblique cutting)	3(200)	1(fresh samples)
7	3(slip cutting)	1(100)	3(dry samples)
8	3(slip cutting)	2(150)	1(fresh samples)
9	3(slip cutting)	3(200)	2(semi-dry)

[a] Unit: Newton

Table 5. Test data.

Test varieties	Test times	Test number 1	2	3	4	5	6	7	8	9
	1	230.57	307.24	312.92	219.88	389.11	241.63	222.32	154.14	198.80
Qingdao	2	230.39	314.33	303.70	242.05	265.39	231.42	204.60	140.58	206.62
Big flower	3	216.94	324.84	377.01	251.52	387.46	240.65	194.95	129.34	219.69
	4	204.18	274.19	355.57	274.56	260.63	215.84	193.67	192.93	222.99
	5	247.13	238.94	326.79	254.15	243.40	219.08	190.67	170.88	200.88
Marco Polo	1	181.08	182.18	458.27	292.76	367.48	197.82	282.87	186.58	237.59
	2	174.85	206.80	408.17	194.46	319.28	202.83	229.90	171.86	220.06
	3	219.39	226.11	416.36	221.04	372.37	191.04	283.54	161.10	185.60
	4	187.19	218.35	374.20	259.47	340.42	180.17	283.96	165.63	179.13
	5	207.96	205.46	425.15	246.76	333.76	172.16	293.37	205.28	199.41
Fragrant flower	1	199.11	265.51	446.23	281.95	351.23	131.78	210.29	129.82	160.62
	2	201.06	331.80	410.98	273.76	399.92	166.24	243.64	144.43	181.27
	3	202.83	359.97	321.35	215.36	375.30	182.24	228.98	139.17	137.89
	4	255.25	356.67	311.52	296.24	278.34	186.21	248.04	157.07	140.21
	5	213.16	338.22	354.96	368.76	295.94	162.14	226.05	124.57	177.97

The results of analysis of variance by table 6 show that for Qingdao big flower, shearing mode A and moisture content C have a highly significant effect on shearing force ($P < 0.001$), and shearing velocity ($P < 0.05$) B is more significant. Among them, shearing mode has the greatest influence on shearing force, and slip cutting is the most labour-saving. It shows that the appropriate shearing mode is an important way to improve the shearing quality and reduce the breakage rate of Qingdao big flower. The effect of moisture content is extremely significant. With the decrease of moisture content, the flexibility of stems decreases, brittleness increases, and shearing force increases in turn. Therefore, fresh samples' shearing force is the smallest. Through the above analysis, the order of the factors affecting the shearing force of Qingdao big flower is: shearing mode, moisture content and shearing velocity. The optimum combination is slipping cutting, fresh samples, at a speed of 100 mm/min.

For Marco Polo, the effect of moisture content C and shearing mode A on the maximum shearing force is the most significant ($P < 0.001$), and the effect of shearing velocity B on the shearing force is more significant ($P < 0.05$). Moisture content has the greatest influence on Marco Polo, with the decrease of moisture content, the shearing force increases in turn; Secondly, slip cutting is the most labour-saving, then oblique cutting, transverse cutting is the most laborious. Shearing velocity also has a significant impact on the shearing force, with the increase of velocity, the shearing force increases. Therefore, when choosing the working velocity range of blade, the vibration and production efficiency of the unit should be considered comprehensively. For Marco Polo, the order of influence of each factor on shearing force is moisture content, shearing mode and shearing velocity. The optimal combination scheme is slipping cutting, fresh samples at a speed of 100 mm/min.

For Fragrant flower, shearing mode A and moisture content C have the most significant effect on shearing force ($P < 0.001$), and shearing velocity ($P < 0.05$) B is more significant. Moisture content has the most significant effect. Fresh samples need the least shearing force, followed by semi-dry samples, and dry samples need the largest shearing force. The effect of shearing mode is also very significant.

IOP Conf. Series: Earth and Environmental Science **346** (2019) 012072 doi:10.1088/1755-1315/346/1/012072

Sliding cutting is the most labour-saving, followed by oblique cutting, and transverse cutting needs the largest shearing force. Therefore, the order of the factors affecting the shearing force of fragrant flowers is moisture content, shearing mode and shearing velocity. The best combination scheme of shearing force is slipping cutting, fresh samples at a speed of 100 mm/min.

Table 6. Variance analysis.

Varieties	Factors	Sum of squares of deviations	Freedoms	Mean square deviations deviation	F	P
Qingdao big flower	A	73888	2	36944	37.83	$9.11e^{-10***}$
	B	9453	2	4726	4.84	0.0134^{*}
	C	45524	2	22762	23.31	$2.48e^{-7***}$
Marco Polo	A	23469	2	11734	10.18	2.89 $e\text{-}^{4***}$
	B	8861	2	4431	3.84	0.0302^{*}
	C	212632	2	106316	92.19	2.64 e^{-15***}
Fragrant flower	A	128336	2	64168	46.75	1.39 e^{-10***}
	B	11476	2	5738	3.91	0.0285^{*}
	C	150581	2	75290	51.33	1.59 e^{-11***}

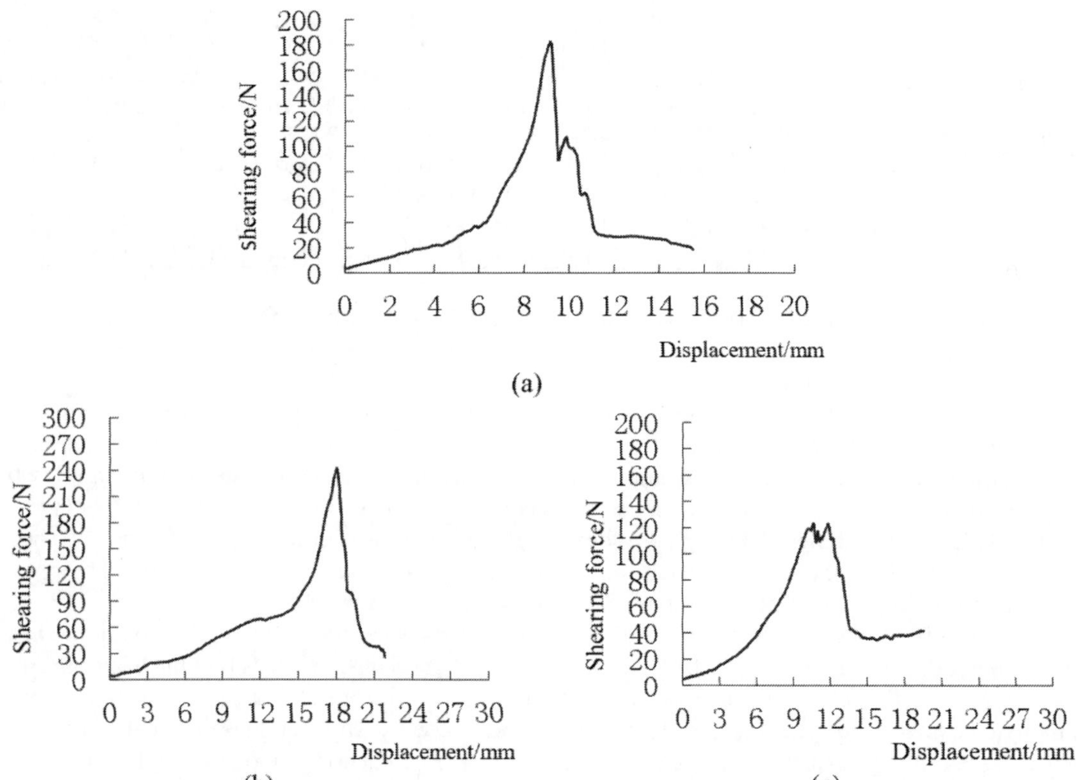

Figure 3. Force and displacement curve of different varieties hop stems.

Based on the above experimental analysis, it is concluded that moisture content and shearing mode have rather significant effects on the shearing force of different varieties, and shearing velocity has also relatively significant effects on the shearing force. Based on the analysis of the above three kinds of stems, the optimal combination of shearing force is that fresh samples and slip cutting are adopted, and the velocity of 100mm/min is the most labour-saving. When slip cutting velocity is 100 mm/min, the shearing force-displacement curves of the three varieties are shown in figure 3. The curve (a) is force and displacement curve of Qingdao big flower and the curve (b) is force and displacement curve of Marco Polo. The curve (c) is force and displacement curve of Fragrant flower. With the increase of displacement, the shearing force increases continuously. When it reaches the maximum value, stems are cut off and the shearing force drops rapidly to 0. It can be seen from the curve that when the shearing force reaches the maximum value, the displacement and the shearing force produced by Marco Polo are the largest among the three varieties, the displacement is 18 mm and the maximum shearing force is 242.12N. When Qingdao big flower is cut off, the displacement is 9mm and the maximum shearing force is 182.49N. When fragrant flower is cut, the shearing force increases with the increase of cutter stroke. When it reaches a certain peak, it keeps changing around the peak until it is cut off, and the maximum shear force is 123. 23N. This phenomenon can be attributed to the differences of three different varieties´ lignin, cellulose content and micro-structure which lead to the difference of shearing force. Figure 4 shows the cross-section microstructures of three kinds of stems. Figure 4(a) is the microstructure map of Qingdao big flower (magnify 180 times). Figures 4(b) and 4(c)are the microstructure map of Marco Polo (magnify 190 times) and the microstructure map of Fragrant flower (magnify 190 times) respective.

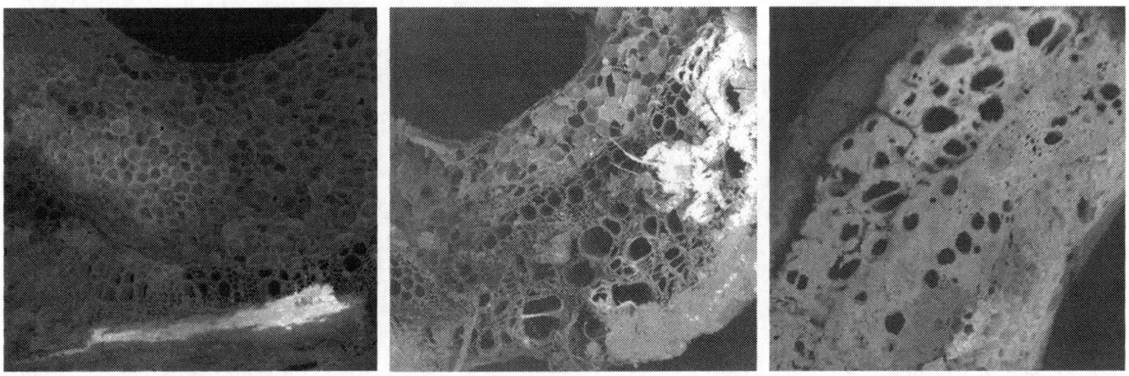

(a) (b) (c)

Figure 4. SEM images of transverse in hop stems.

4. Conclusions and discussions

The paper selects the structural parameters and material parameters related to shear to carry out simulation tests. The results show that:

- The effect of moisture content on shearing force is extremely significant. With the decrease of moisture content in stems, the flexibility decreases and brittleness increases, the shearing force increases gradually. Therefore, in hop harvesting, the shorter the time for hop feeding material to be separated from their parents and the sunlight, and the lower the temperature for doing this, the better it will be. That is to say, hops should usually be harvested in the morning and evening with soft materials, high moisture content, low shearing force and low breakage rate.
- Tests are carried out on three different varieties of hop stems. The effect of shearing mode on shearing force is extremely significant. Sliding cutting mode is the most labour-saving, followed by transverse cutting mode, and oblique cutting mode requires the largest shearing force. Therefore, sliding cutting is an important way to improve the cutting quality of hop harvesting machinery.
- Combining the single factor test and the multi-factor test, the best shearing mode is fresh sample, slip cutting with the speed of 100 mm/min.
- No matter what shearing mode, shearing velocity and moisture content are adopted, fragrant flowers' shearing force is the smallest, followed by Qingdao big flower and Marco Polo. This is related to the micro-structure and tissue arrangement of different varieties of stems.

On the basis of single factor and multi-factor tests, the optimal cutting mode of hop stems was obtained by the orthogonal test. There are many factors affecting hop stem cutting. In addition to the three factors mentioned in the paper, there are also factors such as cutter type and cutting position and so on. Because of the complexity of external factors in the process of stem shearing, so a dynamic loading process instead of the quasi-static one in the test is more rational. Moreover, the research group is discussing further. The conclusion of the tests provides technical parameters for the difficult problem of mechanized harvesting of Qingdao big flower and the reasonable cutting mode of hop harvesting machinery to reduce energy consumption.

Acknowledgments
The work is supported by research grants from scientific research projects of colleges and universities in Gansu Province (2018B-46) and Research projects for Young and Middle-aged Teachers in Tianshui Normal University (YB-2017-07). The author is thankful to the anonymous referees for the constructive comments in enhancing the content.

References

[1] Wang C X 1998 Analysis of sustainable development potential of hops in Gansu Province*Gansu Agric. Sci. Tech.* **07**4-5
[2] Ran S B 2013 Current situation, existing problems and development countermeasure of hop industry in Gansu Province *J. Zhejiang Agric. Sci.* **11** 1532-5
[3] Yang J 2016Study on hops production situation and mechanized harvesting technology *Agric. Machinery* **11** 115-17+19
[4] Wang F E, Huang G B, Guo W J *et al* 2009 Mechanical properties and micro-structure of wheat stems *Transact. CSAE* **40** 92-5
[5] Guo W J, Huang G B, Wang F E *et al* 2010 Mechanicalproperties andmicro-structure of wheatroots *Transac. CSAE* **41** 92-5
[6] Su G B, Liu J Y, Wang S C *et al* 2007 Study on mechanical properties of xylem of ramie stalk *Transac. CSAE***38**62-5
[7] Zhao C H, Zhang F W and Cao Z Z 2009 Experiment on stalk mechanical properties of legume forage and grasses *Transactions CSAE* **25** 122-5
[8] Zhao C H, Han Z S,Wang F E *et al* 2010 Experiment on biomechanical properties of bottom stems offorage in harvesting period *Transac. CSAE* **41** 65-9
[9] Zhao C H, Zhang F W, Han Z S *et al* 2010 Stems biomechanical properties experiments of creeping tangled forageinharvesting period *Transac. CSAE* **41** 85-89+100
[10] Liu Q T, Qv Y G, Qing S L *et al* 2007 Cutting force test of sugarcane stalk *Transac. CSAE* **07** 90-4
[11] Liu Q T, QvY G, Qing S L *et al* 2007Study progress on mechanics properties of crop stalks *Transac. CSAE* **07** 172-6
[12] Yu Y, Lin Y, Mao M *et al* 2012 Experimental study on tensile properties of corn straw *Transa. CSAE* **28** 70-6
[13] Chen Z G,Wang D F, Li L Q *et al* 2012 Experiment on tensile and shearing characteristics of rind of corn stalk *Transac. CSAE* **28** 59-65
[14] Zhou Y, Li X W, Shen C *etal* 2016 Experimental analysis on mechanical model of industrial hemp stalk *Transac. CSAE* **32** 22-9
[15] Luo H F, Tang C Z, Zou D S *et al* 2012 Experiment on reciprocating cutting of eulaliopsisbinatastem *Transac. CSAE* **28** 13-7
[16] Song Z H, Song H L, GengA J *et al* 2015 Experiment on cutting characteristics of cotton stalk withdouble supports *Transac. CSAE* **31** 37-45
[17] Li X Q, Wang F E, Guo W J *et al* 2013 Influencing factor analysis of cabbage root cutting force based onorthogonal *Transact. CSAE* **29** 42-8
[18] Li Y D, Du X J, Song Z H *et al* 2011 Test of shear mechanical properties of cotton stalks *Transac.*

CSAE **27** 124-8

[19] Chatopadhyay P S and Pandey K P 1999 Mechanical properties of sorghum stalk in relation to quasi-static deformation *J. Agric. Engng. Res.* **73** 199-206

[20] Hirai Y, Inove E and Mori K 2004 Application of quasi-static stalk bending analysis to the dynamicresponse of rice and wheat stalks gathered by acombine harvester reel *Biosys. Engin.* **88** 281-94

[21] Chen K 2005 *Test Design and Analysis* Beijing: Tsinghua University Press Beijing China

Amelioration of genetic diversity and its assessment in *Brassica napus – carinata* introgression lines

I Dhaliwal[1,4], A Mason[2,3], S Banga[1], S Bharti[1] and S S Banga[1]

[1]Department of Plant Breeding and Genetics, Punjab Agricultural University, Ludhiana- 1004, India
[2]School of Agriculture and Food Sciences, University of Queensland, Brisbane, QLD 4072, Australia
[3]Department of Plant Breeding, Research Centre for Biosystems, Land Use and Nutrition, Justus Liebig University, Heinrich-Buff-Ring 26-32, 35392, Giessen, Germany

E-mail: dhaliwalinderpreet@gmail.com

Abstract. One of the breeding goals in *Brassica napus* has been to facilitate introgression of novel traits from wild or closely related species through inter- and intra-specific crosses. The present study is based on the evaluation of *B. napus – carinata* introgression lines (ILs) for introgressed morphological and genetic diversity in comparison to the parental *B. napus* lines. A set of 81 *Brassica napus – carinata* ILs, developed previously following a back cross strategy, was assessed for introgressed genetic diversity by comparing them with the 15 *B. napus* recipient parents under two environments (E1: timely sown and E2: late sown) for phenotypic expression of 13 morpho-physiological traits. Trait averages as well as trait variations were generally higher for ILs, indicating introgression of variability in the desired direction. In comparison to the parents, the ILs were observed to be early to flower, early to mature, higher yielding, had more siliquae on the main axis and the siliquae were longer. ANOVA revealed highly significant differences amongst genotypes for 10 traits of which 4 traits, *viz.,* days to 50% flowering, days to 100% flowering, plant height and pod length were highly significant in both the environments. SNP-based chromosome-wise gene diversity, as estimated using software SELECTIONTOOL, indicated high estimates for individual chromosomes in ILs as compared to the parental lines. The phenotypic variability estimated for majority of the traits in ILs is being utilized for developing new high yielding rapeseed varieties.

1. Introduction

The allopolyploid *Brassica napus* L. (AACC, 2n= 38) is considered to have originated ~7500 years ago [1] by way of natural hybridizations betwixt *Brassica rapa* (AA, 2n = 20) and *Brassica oleracea* (CC, 2n= 18) [2,3]. However, these dual impediments of polyploidy and domestication, coupled with intensive breeding, especially for canola quality, have tenably culminated in a very narrow genetic base of the crop [4,5]. This is apparent by the absence of exploitable variation as well as declining response to selection for several traits of breeding value and agronomic interest. Furthermore, *B. napus* does not have any known wild forms. Therefore, one of the breeding goals is to enhance variability through inter- and intra-specific hybridizations with progenitor or related species.

In advanced breeding lines, morphological, phenological and agronomical traits have customarily been used as a benchmark for the detection of directly introgressed novel variation. However, new

Content from this work may be used under the terms of the Creative Commons Attribution 3.0 licence. Any further distribution of this work must maintain attribution to the author(s) and the title of the work, journal citation and DOI.
Published under licence by IOP Publishing Ltd

genetic variations are also produced in early generations after inter- and intra-specific hybridizations [6,7] which too are of equal interest. There is ample documentation to prove that inter-specific hybridization (including allopolyploidy) leads to genome rearrangements [8,9], elimination of parental DNA sequences and gene conversions [10], transposon activation/transposon-induced insertional mutagenesis [11,12] and epigenetic changes [13-17], which in turn translate to phenotypic changes [18,19].

The B-genome containing *Brassica* species (*B. nigra, B. carinata* and *B. juncea*) are a treasure trove of valuable agronomic traits and of these, the allotetraploid *B. carinata* (BBCC; 2n = 34) is important as a potential source for introgression of desirable variability. The present investigation was thus planned to assess the *B. napus - carinata* introgression lines (ILs) for introgressed morphological and genetic diversity in comparison to the parental *B. napus* lines.

2. Materials and methods

2.1. Plant material

The germplasm, developed previously [20] following a back cross strategy (Figure 1), comprised of 81 *Brassica napus – carinata* ILs and 15 distinct cultivars of *B. napus* that were initially hybridized with *B. carinata cv.* PC5 (Appendix Table A). The study was conducted under two environments [timely-sown {first week of October (E1)} and late-sown {last week of November (E2)}] during the winter season of 2012-2013. Since varied field measurements necessitated a different layout plan for each environment, therefore the experiment was carried out in an alpha lattice design with two replications. Thus, E1 had 12 blocks while E2 had 16 blocks in each replication. Each genotype, randomized for each replication using software CropStat v.7.2, was sown in a 2.5 m x 1.8 m plot having 4 rows, 45 cm apart with plant-to-plant spacing of 15 cm. The crop was cultivated following recommended cultural practices for irrigated conditions.

2.2. Phenotyping for quantitative traits

The data were recorded for 13 agro-morphological traits, namely: days to 50% flowering (DTF50%), days to 100% flowering (DTF100%), number of leaves before bolting (LN), plant height in cm (PH), raceme height in cm (RH), number of primary branches per plant (PB), number of secondary branches per plant (SB), number of pods on main axis (PMA), pod length on main axis in cm (PL), seeds per pod (S/P), seed yield in g (SY), biological yield in kg (BY) and harvest index (HI). Data on DTF50% (when ≈50% of the plants have initiated flowering), DTF100% (when ≥ 90% of the plants have initiated flowering), SY and BY were recorded on plot basis. For remaining characters, the data were recorded from five random competitive plants from the two middle rows.

2.3. DNA extraction and quantification

Following standard procedure [21], DNA was isolated from 3-4 juvenile leaves from nearly a month-old seedling of each genotype that were frozen immediately after collection till further processing. Using liquid nitrogen, the leaves were homogenized to a fine powder and pre-warmed (65°C) modified CTAB extraction buffer (900 µl) with high concentrations of NaCl and Poly-Vinyl Pyrrolidone (PVP) for removal of polysaccharides [22] and polyphenols [23], respectively, was added. After incubating the sample at 65°C for 45 min with intermittent gentle shaking, chloroform: isoamyl alcohol (1:2) mixture (800 µl) was added and the sample then shaken thoroughly to ensure complete phase emulsification. Following centrifugation (12000 rpm for 20 min), the supernatant was decanted and subsequent addition of ice-cold ethanol (600 µl) lead to DNA precipitation, which was looped out and transferred to a fresh tube. The pellet was washed twice (70% ethanol) to remove residual CTAB and air-dried (37°C) overnight. Dissolved DNA (≈100 µl 1x TE buffer) was treated with RNAse (incubation at 37°C for 1 hr) and its quality and quantity assessed using "Eppendorf BiophotometerPlus".

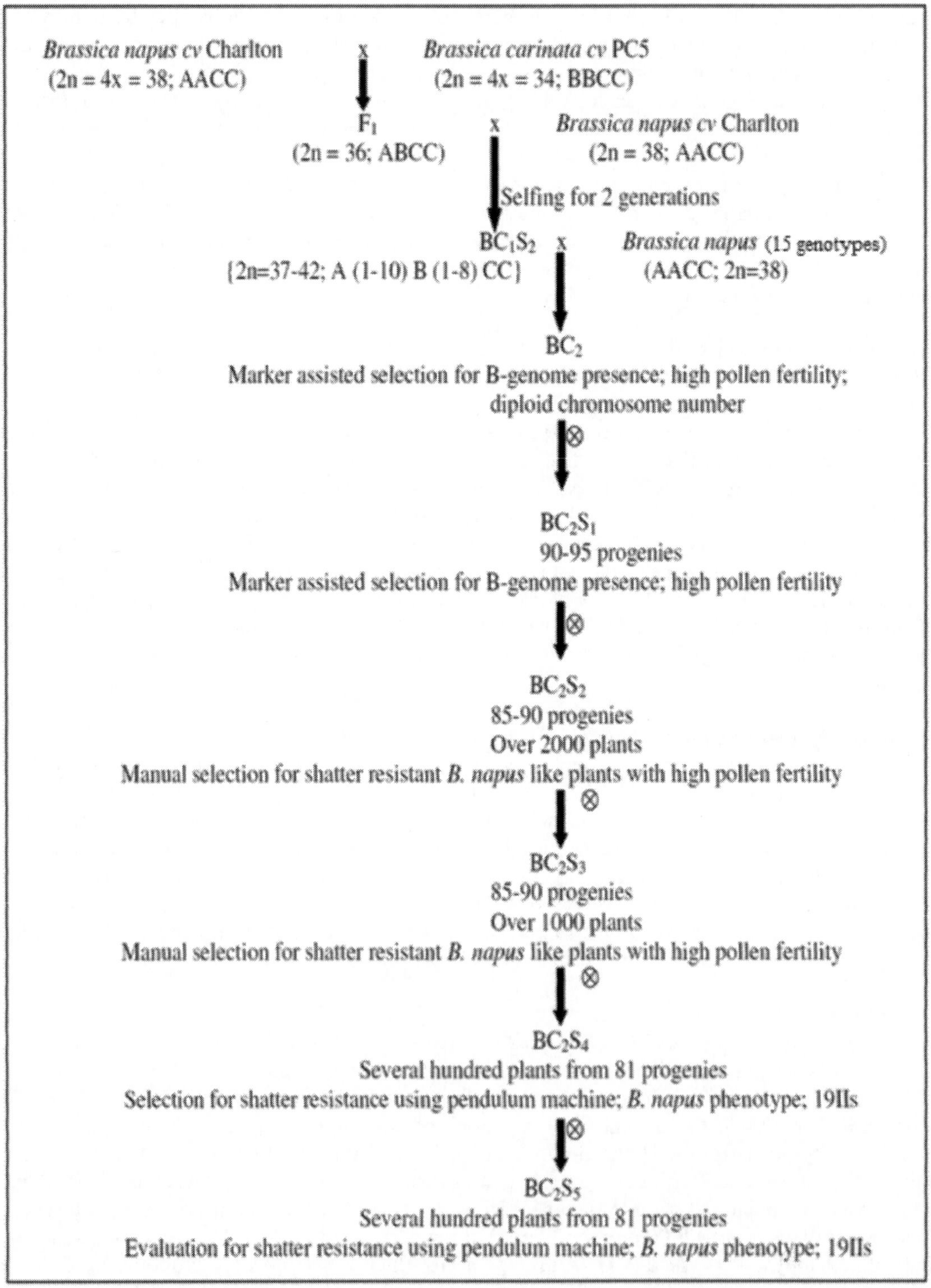

Figure 1. Development of *Brassica napus-carinata* introgression lines (ILs). Hybridizing diverse *B. napus* genotypes to *B. carinata* cv. PC5; backcrossing F_1s to respective *B. napus* parent; selfing BCs for 2 generations; again, backcrossing BC_1S_2 to respective *B. napus* parent; selfing BC_2 and practising selection in each subsequent generation.

IOP Conf. Series: Earth and Environmental Science **346** (2019) 012073 doi:10.1088/1755-1315/346/1/012073

2.4. *SNP genotyping*

SNP genotyping requires high-quality DNA; of the 81 ILs and 15 parents, only 69 ILs and 5 parents (marked with an asterisk (*) in Appendix Table A) yielded such high-quality DNA, which was lyophilized and shipped to The CILR Laboratory, University of Queensland, Australia for genotyping. A 60k SNP array (Illumina Infinium), carrying 52,157 SNPs for all the A- and C-genome chromosomes of *B. napus,* was used to assess the quantum of gene diversity in the ILs. All protocols were run as per the manufacturer's specifications. An Illumina HiScanSQ scanned the chips and data were visualized using Genome Studio V2011.1 (Illumina, Inc., San Diego, CA, USA).

2.5. *Statistical analysis*

The phenotypic data were analyzed with SAS v.9.3 (SAS Institute Inc., Cary, NC). The presence of statistically significant differences among the genotypes for each trait was checked with the GLM procedure. Contrast analysis was performed to check for significant variation in the ILs as compared to the parents by conducting all simulations in R version 3.0.0 [24]. Software package SELECTIONTOOL [25] was utilized for estimating the quantum of gene diversity introgressed into the lines.

3. Results

3.1. *Phenotypic variations of quantitative traits in introgression lines*

The phenotypic data with respect to the thirteen traits recorded under two environments, *viz.* E1 (timely-sown) and E2 (late-sown) are depicted in Figure 2. The data show wide variation for genotype means over the two environments. Averages as well as trait variations were generally higher for environment E1 as compared to environment E2. Comparison of the means of parental lines and introgression lines indicated introgression of variability in the desired direction. As compared to the parental lines, introgression lines were observed to be early to flower (109.8 *vs.* 117.1 days in E1 and 103.6 *vs.* 111.5 days in E2); were taller (183.3 cm *vs.* 162.8 cm in E1 and 174.9 cm *vs.* 161.1 cm in E2); had more siliquae on the main axis (68.6 *vs.* 59.3 in E1 and 70.6 *vs.* 61.5 in E2); exhibited longer pod length in E1 (7.0 cm *vs.* 6.8 cm) but were at par with the parents in E2 (6.9 cm *vs.* 6.9 cm); yielded higher in E1 (1.3 kg/plot *vs.* 0.9 kg/plot) but were almost at par in E2 (0.7 kg/plot *vs.* 0.8 kg/plot). The seed size trait could not be evaluated since the E1 seed was accidently mixed. In such a scenario, the test weight (1000 seed weight) of parents and ILs used for sowing the trial was evaluated. It was found that the seed size of the parents ranged from 2.00 gm to 3.68 gm, giving a mean weight of 3.04 gm (Figure 2). The seed size of the ILs ranged from 3.11 gm to 5.06 gm, giving a mean weight of 3.76 gm. This trait has not been included in the analysis.

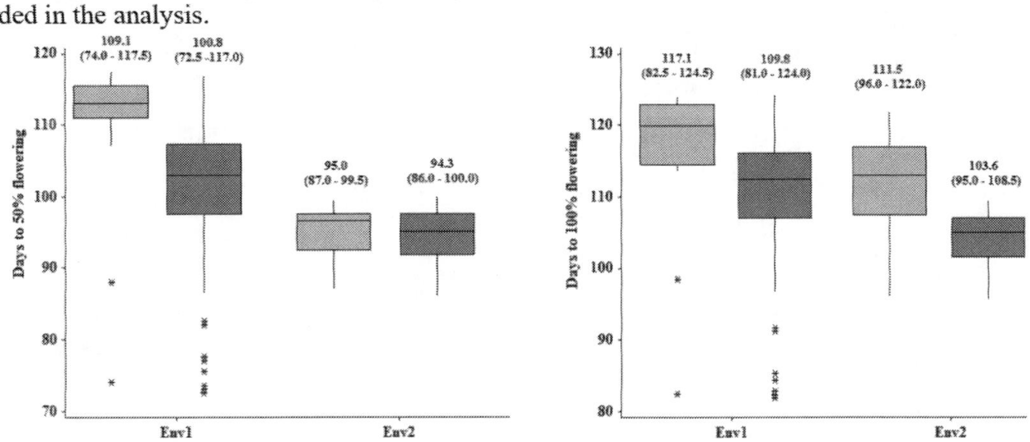

IOP Conf. Series: Earth and Environmental Science **346** (2019) 012073 doi:10.1088/1755-1315/346/1/012073

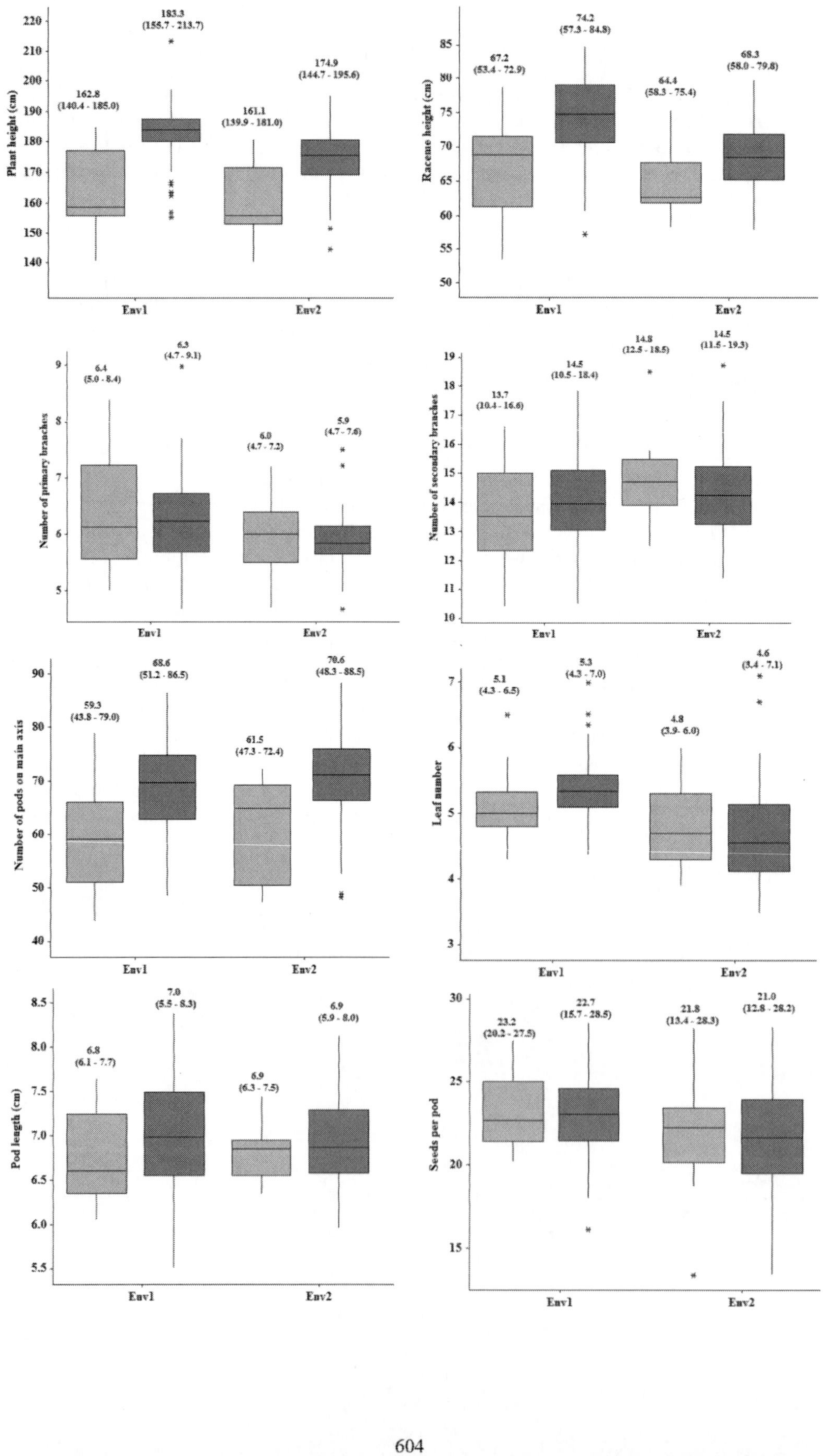

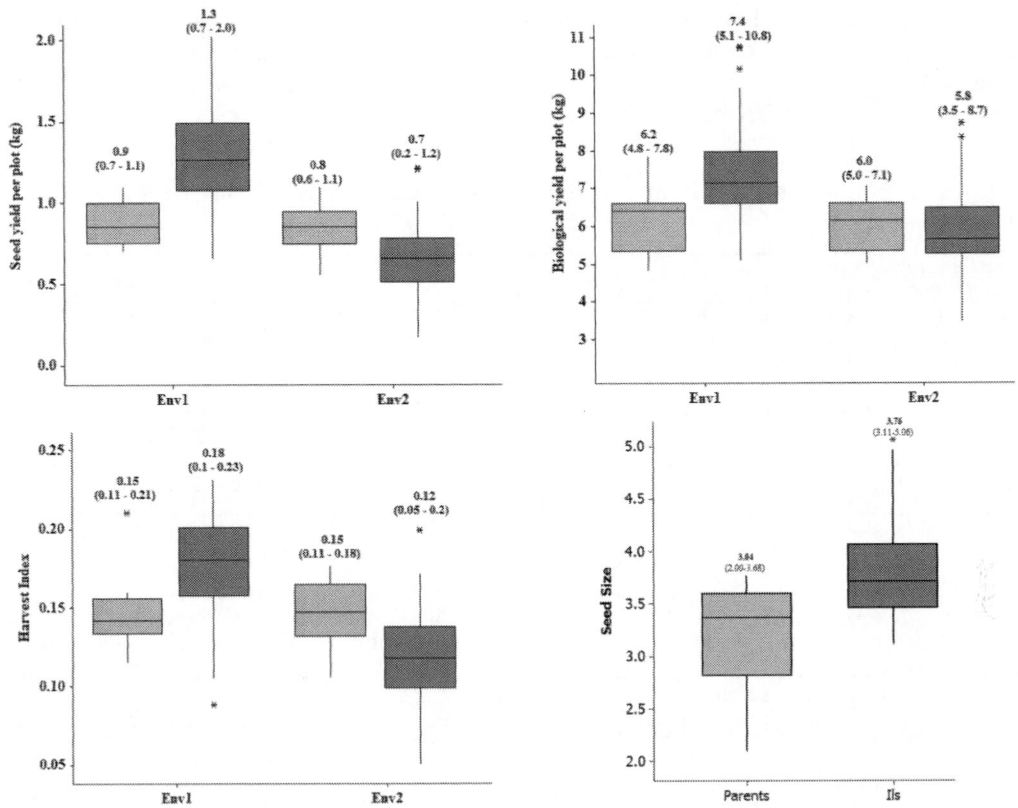

Figure 2. Box plots showing phenotypic variation for indicated traits in the parental and introgression lines of *B. napus*. (*Light grey boxes represent parental genotypes; dark grey boxes represent introgression lines; figures above each box are averages; range values are given in parenthesis*).

3.2. Analysis of variance

The Analysis of variance (ANOVA) for the phenotypic trait data of the 96 test genotypes grown under two environments is given in Table 1. Highly significant differences (α=0.01) were found amongst genotypes with respect to 4 quantitative traits; *viz.* DTF50%, DTF100%, plant height and pod length in both the environments. Differences amongst the genotypes were also highly significant (α = 0.01) for seed yield and harvest index under E1 and for number of pods on main axis under E2. The genotypes were at par with each other for the rest of the traits in either of the environment; exceptions being the number of primary branches per plant under E1 and seed yield per plot under E2, both of which traits were significant at α = 0.05. Pooled analysis of variance suggested significant differences across the two environments. Genotype x environment effects were highly significant (α=0.01) for DTF50%, DTF100%, pod length and seed yield.

3.3. Genetic diversity in B. napus - carinata introgression lines

In a population under Hardy-Weinberg equilibrium, the proportion of heterozygous genotypes expected for a given locus is characterized by gene diversity or expected heterozygosity [26]. Accordingly, an unbiased estimator of gene diversity was devised for random samples of unrelated, non-inbred individuals [27]. However, when inbred lines are included, this unbiased estimator of gene diversity exhibits a downward bias [28,24]. Therefore, in a sample of diploid individuals, the expected value of gene diversity was derived to account for the effects of inbreeding [28,29], thereby producing an unbiased estimator that makes use of the mean inbreeding coefficient across sampled individuals. This

IOP Conf. Series: Earth and Environmental Science **346** (2019) 012073	doi:10.1088/1755-1315/346/1/012073

inbreeding coefficient of an individual for a randomly chosen locus is defined as the probability that the two alleles of the individual are inherited identically by descent from a common ancestor. This estimator was extended to account for the bias produced in samples containing close relatives [24], which is implemented in the SELECTIONTOOL software.

Table 1. Analysis of variance for 13 morphological traits studied in the 81 introgression lines and 15 parents of *B. napus*.

Environment E1 (timely-sown)														
Source	df	DTF50%	DTF100%	LN	PH	RH	PB	SB	PMA	PL	S/P	SY	BY	HI
Rep	1	56.333	39.422	0.594	1957.77**	23.840	3.783	113.62**	2506.64*	0.003	47.084	0.119	9.707	0.0026
Blk (rep)	22	29.826	24.458	0.590	139.260	31.490	0.964	5.455	135.730	0.205	10.427	0.103	3.749	0.0007
Trt	95	219.95**	186.24**	0.447	263.74**	82.53*	1.11*	4.906	142.53*	0.477**	11.619	0.15**	1.971	0.002**
Error	73	25.623	28.507	0.276	68.580	38.598	0.608	4.784	69.300	0.095	14.553	0.046	1.714	0.0007
h^2		0.8	0.7	0.2	0.6	0.4	0.3	0.0	0.4	0.7	0.0	0.5	0.1	0.5
Environment E2 (late-sown)														
Source	df	DTF50%	DTF100%	LN	PH	RH	PB	SB	PMA	PL	S/P	SY	BY	HI
Rep	1	10.55	57.42	2.30	5.92	81.64	0.04	25.23	21.00	0.02	1.47	0.12	35.54**	0.0047
Blk (rep)	30	4.57	7.39	0.81	52.85	28.26	0.42	6.95	72.60	0.36	9.73	0.09	3.27	0.0009
Trt	95	24.56**	47.72**	0.97	212.08**	49.99	0.41	3.90	145.76**	0.29**	20.14	0.049*	1.54	0.0012
Error	65	3.23	5.97	0.83	85.04	35.06	0.27	5.47	56.85	0.15	10.38	0.03	1.41	0.0007
h^2		0.8	0.8	0.1	0.4	0.2	0.2	0.0	0.4	0.3	0.3	0.2	0.0	0.3
Pooled														
Source	df	DTF50%	DTF100%	LN	PH	RH	PB	SB	PMA	PL	S/P	SY	BY	HI
Env	1	4565.08**	2821.91**	26.51**	2785.66**	2094.56**	10.14**	20.83	352.83	0.12	190.19**	15.46**	75.06**	0.17**
Rep	1	57.82	96.00	0.28	874.23	96.86	2.30	15.88	1034.38	0.002	32.60	0.23	41.19	0.01
Blk	15	18.08	18.10	0.76	110.65	16.38	0.37	9.13	99.65	0.30	11.94	0.23	7.70	0.001
Trt	95	181.81**	184.03**	0.90	469.80**	101.78**	1.09**	5.34	252.45**	0.74	17.67	0.12**	2.19	0.002**
Env x Trt	95	72.84**	60.46**	0.62	58.96	32.69	0.68	5.54	67.25	0.22**	14.26	0.096**	1.62	0.001
Error	176	14.84	16.81	0.58	83.03	36.36	0.52	5.79	79.10	0.16	11.90	0.04	1.63	0.001
h^2		0.6	0.7	0.3	0.9	0.7	0.4	0.0	0.7	0.7	0.2	0.2	0.3	0.4

*: Significant at $\alpha = 0.05$;　　**: Significant at $\alpha = 0.01$

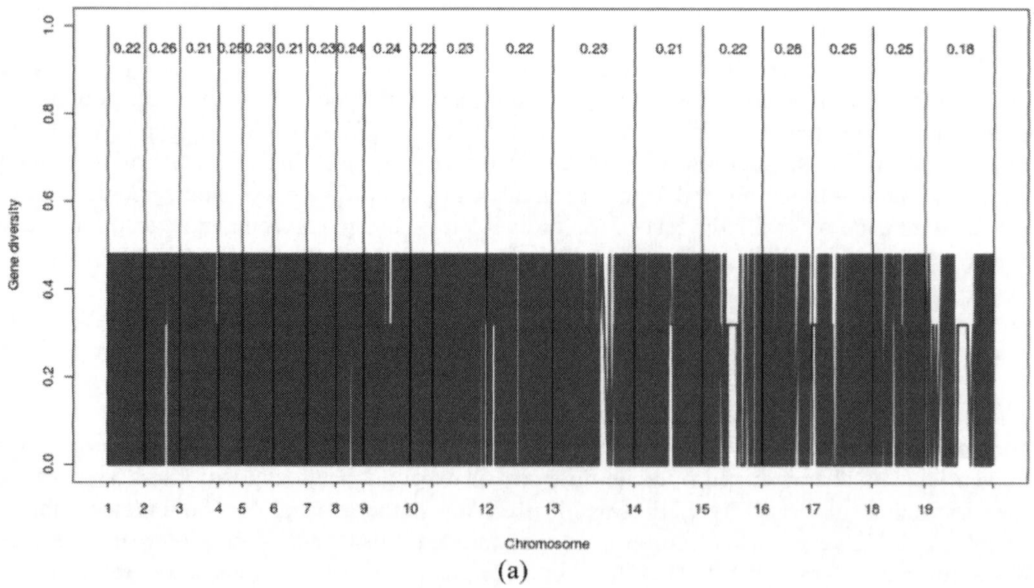

(a)

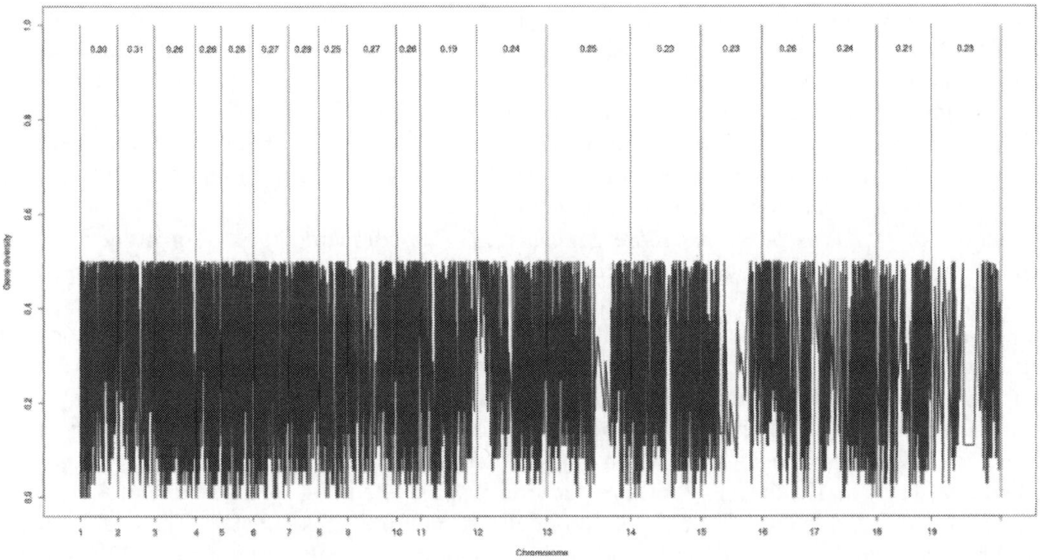

(b)

Figure 3. Enhanced genetic diversity of the introgression lines as compared to the parental lines, highlighted with the use of SELECTIONTOOL software. (A-genome chromosomes are represented by numbers 1-10 while C-genome chromosomes are represented by numbers 11-19). (a) Parental lines and (b) Introgression lines.

Figure 3 (obtained by using SELECTIONTOOL software) depicts the estimates of chromosome-wise gene diversity for parents as well as for ILs. The estimates for individual chromosomes were high in ILs as compared to the diversity present in individual chromosomes of the parental *B. napus* lines. This was especially so for A-genome chromosomes of the ILs; for the C-genome chromosomes, enhanced gene diversity was recorded only for chromosome 19 (C9). Increased genetic diversity could partly be attributed to B-genome translocations/introgressions replacing chunks of chromosomes of the recipient *B. napus* as also observed with SSR and GISH analyses [30]. During the course of the present investigation, we have observed that most of the ILs carried B-genome segments from B. carinata [30], but this diversity of B-genome introgressions could not be accounted for by A-/C-genome SNPs (Unpublished), since the B-genome SNPs were unavailable at that point of time.

4. Discussion

The existence of variability is imperative for any breeding program. A number of studies have shown that *B. napus* possesses a narrow genetic base [31-36,15], which has been further restricted through aggressive plant breeding efforts towards canola quality [37]. This is apparent by the absence of exploitable variation as well as declining response to selection for several traits of breeding value and agronomic interest.

The present set of 81 *B. napus- carinata* introgression lines was evaluated for introgressed variability. Comparison of phenotypic means indicated introgression of desirable variability for DTF50% (less), days to maturity (less), number of siliquae on the main shoot (more), siliqua length (more) and seed yield (more). Significant differences were found amongst genotypes with respect to DTF50% and DTF100%. Ostensibly, early flowering and early maturing genotypes exhibit high performance due to lack of heat stress at grain filling stage. Early flowering *B. napus* genotypes are of immense relevance since they can escape terminal heat stress and also adapt to avoid pest pressure [38].

Pooled analysis suggested significant differences across the two environments thereby indicating the differential response of genotypes to environment. The lower and higher range for majority of the traits was found in the introgression lines rather than the *B. napus* parents. This is indicative of the transgressive segregants that were obtained during the course of development of ILs. A collective appraisal of the data in Table 1 and Figure 2 strongly suggests that three agro-morphological traits, viz. pod length, seeds per pod and seed yield can profitably be used for selection breeding to bring about quantitative improvements in rapeseed. Higher heritability of a trait is helpful in breeding new varieties with desirable agronomic traits. The phenotypic variability estimated for a majority of traits showed that these introgression lines can indeed be utilized for developing new high yielding rapeseed varieties, especially those possessing high pod shatter tolerance.

A-genome of *B. napus* possesses more gene diversity as compared to its C-genome is very well known. A major objective of developing introgression lines of *B. napus* was to enhance gene diversity by introgressing novel variation from *B. carinata*. It appears that the enhanced gene diversity of A-genome chromosomes in the introgression lines developed in the present study possibly accrued from introgressions from B-/C-genome of *B. carinata*. Pairing and genetic exchange between A- and C-genome chromosomes of Brassica species has also been earlier reported [39].

That the alien gene introgressions can potentially increase productivity was first observed in oats [40]. Then followed the collegial association of QTL mapping and alien gene introgression, and the AB-QTL strategy [41]. The ILs described in this study ferry substituted B-genome segment(s) on the A-/C-genome of *B. napus*, which may bear crucial productivity and/or defense related QTLs, and can thus be utilized for its genetic and functional genomic analysis. Currently, the genes/QTLs controlling productivity and pod shattering in *B. napus* are being mapped using these ILs.

Acknowledgements
S.S.B endorses the monetary grant and salary remuneration from the ICAR, New Delhi and GRDC, Australia funding for the project "Expanding the Brassica germplasm base through collaboration with China and India." I.D. recognizes salary support from PAU, Ludhiana during the course of her Ph.D. A.M. acknowledges salary support from ARCD Early Career Researcher Award (DE120100668), Australia and Deutsche Forschungsgemeinschaft Emmy Noether award, Germany.

References

[1] Chalhoub B *et al* 2014 Early allopolyploid evolution in the post-Neolithic *Brassica napus* oilseed genome *Science* **345** 950-3
[2] Kimber D and McGregor D I 1995 *Brassica* Oilseeds: Production and Utilization (Wallingford, UK: CAB)
[3] Prakash S *et al* 2009 *Brassica* and its close allies: cytogenetics and evolution *Plant Breed Rev* **31** 21-187
[4] Hasan M *et al* 2006 Analysis of genetic diversity in the *Brassica napus* L. gene pool using SSR markers *Genet Resour Crop Ev* **53** 793-802
[5] Gyawali S *et al* 2013 Genetic diversity and population structure in a world collection of *Brassica napus* accessions with emphasis on South Korea, Japan and Pakistan *Crop Sci* **53** 1537-45
[6] Comai L 2000 Genetic and epigenetic interactions in allopolyploid plants *Plant Mol Biol* **43** 387-99
[7] Hegarty M J and Hiscock S J 2005 Hybrid speciation in plants: New insights from molecular studies *New Phytol* **165** 411-23
[8] Song K M *et al* 1995 Rapid genome change in synthetic polyploids of *Brassica* and its implications for polyploid evolution *PNAS (USA)* **92** 7719-23
[9] Pontes O *et al* 2004 Chromosomal locus rearrangements are a rapid response to formation of the allotetraploid *Arabidopsis suecica* genome *PNAS (USA)* **101** 18240-5
[10] Han F P *et al* 2005 Rapid and repeatable elimination of a parental genome-specific DNA repeat (pGcIR-1a) in newly synthesized wheat allopolyploids *Genetics* **170** 1239-45
[11] Kashkush K *et al* 2003 Transcriptional activation of retrotransposons alters the expression of adjacent genes in wheat *Nat Genet* **33** 102-6
[12] Wang H Y *et al* 2010 Transpositional reactivation of two LTR retrotransposons in rice-*Zizania*

recombinant inbred lines (RILs) *Hereditas* **147** 264-77
[13] Chen Z J 2007 Genetic and epigenetic mechanisms for gene expression and phenotypic variation in plant polyploids *Annu Rev Plant Biol* **58** 377-406
[14] Ha M *et al* 2009 Small RNAs serve as a genetic buffer against genomic shock in *Arabidopsis* interspecific hybrids and allopolyploids *PNAS (USA)* **106** 17835-40
[15] Pignatta D and Comai L 2009 Parental squabbles and genome expression: Lessons from the polyploids *J Biol* **8** 43
[16] Soltis P S and Soltis D E 2009 The role of hybridization in plant speciation *Annu Rev Plant Biol* **60** 561-88
[17] Parisod C *et al* 2010 Impact of transposable elements on the organization and function of allopolyploid genomes *New Phytol* **186** 37-45
[18] Pires J C *et al* 2004 Flowering time divergence and genomic rearrangements in resynthesized *Brassica* polyploids (*Brassicaceae*) *Biol J Linn Soc* **82** 675-88
[19] Gaeta R T *et al* 2007 Genomic changes in resynthesized *Brassica napus* and their effect on gene expression and phenotype *Plant Cell* **19** 3403-17
[20] Banga S *et al* 2011 Transfer of resistance to seed shattering from *Brassica carinata* to *B. napus* *Proc 13th Internat Rapeseed Cong* (Prague, Czech Republic) pp 863-6
[21] Doyle J J and Doyle J L 1990 Isolation of plant DNA from fresh tissue *Focus* **12** 13-15
[22] Lodhi M A *et al* 1994 A simple and efficient method for DNA extraction from grapevine cultivars and *Vitis* species *Plant Mol Biol Rep* **12** 6-13
[23] Maliyakal E J 1992 An efficient method for isolation of RNA and DNA from plants containing polyphenolics *Nuc Acids Res* **20** 23-81
[24] R Core Team 2013 R: A Language and Environment for Statistical Computing. R Foundation for Statistical Computing, Vienna, Austria available at http://www.R-project.org/
[25] DeGiorgio M and Rosenberg N A 2009 An unbiased estimator of gene diversity in samples containing related individuals *Mol Biol Evol* **26** 501-12
[26] Nei M 1973 Analysis of gene diversity in subdivided populations *PNAS (USA)* **70** 3321-3
[27] Nei M and Roychoudhury A K 1974 Sampling variances of heterozygosity and genetic distance *Genetics* **76** 379-90
[28] Weir B S 1989 Sampling properties of gene diversity *Plant Population Genetics, Vol 16, Breeding and Genetic Resources* A H D Brown *et al* eds (Sunderland, MA, Sinauer) pp 23-42
[29] Weir B 1996 Genetic Data Analysis Vol II, Sinauer Associates
[30] Dhaliwal I *et al* 2016 Cytogenetic and molecular characterization of B-genome introgression lines of *Brassica napus* L *G3: Genes Genom Genet* doi:10.1534/g3.116.036442
[31] Diers BW *et al* 1996 Relationship between heterosis and genetic distance based on restriction fragment length polymorphism markers in oilseed rape (*Brassica napus* L.) *Crop Sci* **36** 79-83
[32] Lombard V *et al* 2000 Genetic relationships and fingerprinting of rapeseed cultivars using AFLP: Consequences for varietal registration *Crop Sci* **40** 1417-25
[33] Cartea M E *et al* 2005 Relationships among *Brassica napus* (L.) germplasm from Spain and Great Britain as determined by RAPD markers *Genet Resour Crop Evol* **52** 655-62
[34] Zhou W J *et al* 2006 Genetic survey of Chinese and Swedish oilseed rape (*Brassica napus* L.) by simple sequence repeats (SSRs) *Genet Resour Crop Evol* **53** 443-7
[35] Chen S *et al* 2008 Divergent patterns of allelic diversity from similar origins: the case of oilseed rape (*Brassica napus* L.) in China and Australia *Genome* **51** 1-10
[36] Li Y *et al* 2012 QTL analysis of yield-related traits and their association with functional markers in *Brassica napus* L *Aust J Agric Res* **58** 759-66
[37] Østergaard L *et al* 2007 Dehiscence *Plant cell separation and adhesion* Roberts JA and Gonzalez-Carranza ZH eds (Blackwell Publishing, Victoria) pp 137–63
[38] Aien A 2006 Study on yield potential of advanced rapeseed varieties in Jiroft area *PajouhheshSazandegi* **77** 119-24

[39] Leflon M *et al* 2006 Pairing and recombination at meiosis of *Brassica rapa* (AA) x *Brassica napus* (AACC) hybrids *Theor Appl Genet* **113** 1467-80
[40] Frey KJ *et al* 1983 Increasing cereal yields with genes from wild and weedy species *Proc 15th Internat Genet Cong* M S Swaminathan ed (Oxford and IBH Publishing Company, New Delhi, India) pp 51-68
[41] Tanksley SD *et al* 1996 Advanced backcross QTL analysis in a cross between an elite processing line of tomato and its wild relative *L. pimpinellifolium Theor Appl Genet* **92** 213-24

IOP Conf. Series: Earth and Environmental Science **346** (2019) 012073 doi:10.1088/1755-1315/346/1/012073

Appendix

Table A. List of *B. napus– carinata* Introgression Lines (ILs) and *B. napus* parental lines used in the study; the lines marked with * were used for molecular analysis.

Code	Entry
* IL 1	CHARLTON/NCA/DWARF/CARINATA HARD
IL 2	CHARLTON/NCA-3/SH9
* IL 3	CHARLTON/NCA-3/SH8
* IL 4	CHARLTON/NCA-3/SH8+
* IL 5	CHARLTON/NCA-3/SH7+
* IL 6	CHARLTON-MONTY/NCA-2/DT/SH 10
* IL 7	CHARLTON-MONTY/NCA-2/SH10
IL 8	MYSTIC/NCD-2/SH 10
* IL 9	MYSTIC/SH9+
* IL 10	MYSTIC/NCA-4/SH8+
* IL 11	MYSTIC/NCA-4/SH8
* IL 12	MYSTIC/NCD-2/SH7
* IL 13	RAINBOW/NCA /LONG PODS/SH10
* IL 14	RAINBOW/NCA/SH10
IL 15	RAINBOW/NCD-4/SH9+
* IL 16	RAINBOW/NCA/SH9+
* IL 17	RAINBOW/NCD-4/SH9+
* IL 18	RAINBOW/NCD-4/SH 9
* IL 19	RAINBOW/NCA-3/SH8
* IL 20	RQ-001/NCA-2/SH10
* IL 21	RQ-001/NCA-2-2/SH10
* IL 22	RQ-001/NCA/TRILOCULAR
* IL 23	RQ-001/NCA-2/SH9+
* IL 24	RQ-001/NCA 2-2/SH9+
IL 25	RQ-001/NCC-2-2/SH9
* IL 26	RQ-001/NCA-2-2/SH8+
* IL 27	RQ-001/NCA-2-2/SH8
* IL 28	RQ-011/NCB/SH10
IL 29	RQ-011/NCB/SH10
* IL 30	RQ-011/NCB/SH10
* IL 31	RQ-011/NCA/SH10+
* IL 32	RQ-011/NCB/SH8
* IL 33	RQ-011/NCB/SH8+
* IL 34	RQ-011/NCA/SH9+
IL 35	RQ-011/NCB/SH9
IL 36	RR-001/NCA-4/Alt DT/SH10
* IL 37	RR-001/NCA-4/SH10
* IL 38	RR-001/NCA-4/SH10
* IL 39	RR-001/NCB/SH9+
* IL 40	RR-001/NCB-4/SH9+
* IL 41	RR-001/NCA/HEAVY BEARING/SH8
* IL 42	RR-002/NCA/SH9+
* IL 43	RR-002/NCA/Alt/SH9
* IL 44	RR-002/NCA/SH8+
* IL 45	RR-005/NCA-1/SMALL PODS/CARINATA HARD
* IL 46	RR-005/NCA/CARINATA HARD
* IL 47	RR-005/NCA-1/SH10
* IL 48	RR-005/NCA-1/SH10
* IL 49	RR-005/NCA-1/SH10

IOP Conf. Series: Earth and Environmental Science **346** (2019) 012073 doi:10.1088/1755-1315/346/1/012073

* IL 50	RR-005/NCA/SH10
* IL 51	RR-005/NCA/SH10
* IL 52	RR-005/NCA/SH9+
* IL 53	RR-005/NCA-1/SH9+
* IL 54	RR-005/NCA/SH9+
IL 55	RR-005/NCA/SH9
* IL 56	RR-005/NCA/SH8
* IL 57	RR-005/NCA-1/SH7
IL 58	RR-009/NCA/SH9+
* IL 59	RR-009/NCB/SH9+
IL 60	RR-009/NCA-2/SH9+
* IL 61	RR-009/NCA-1/SH9+
* IL 62	RR-009/NCA/SH9+
* IL 63	RR-009/NCA-1/SH 9+
* IL 64	RR-009/NCA/SH9+
* IL 65	RR-009/NCB/SH9
* IL 66	RR-009/NCA/SH7
* IL 67	RR-009/NCA/SH8+
* IL 68	RR-013/NCA/SH10
* IL 69	RR-013/NCA-3-3/SH8
* IL 70	RR-013/NCA-3-3/SH7
* IL 71	SKIPTON/NCC-4-1/SH10
* IL 72	SKIPTON/NCC-4-1/SH9+
* IL 73	SKIPTON/NCA/SH9+
* IL 74	SKIPTON/NCC/SH9
* IL 75	SKIPTON/NCC-4-1/DWARF/SH8
* IL 76	SKIPTON/NCC-4-1/INTENSE PODDING/SH8
IL 77	SURPASS 400/NCA/SH9+
* IL 78	SURPASS 400/NCB/BOLD PODS/SH8
* IL 79	SURPASS 400/NCB-3/SH9
* IL 80	TRIGOLD x GSC-6/NCA-2-1/RADISH PODS/SH8
IL 81	TRIGOLD x GSC-6/NCA-2-1/DWARF/SH 8
Parent 82	CHARLTON
Parent 83	MONTY
* Parent 84	MYSTIC
* Parent 85	RAINBOW
Parent 86	RQ-001
* Parent 87	RQ-011
Parent 88	RR-001
Parent 89	RR-002
* Parent 90	RR-005
Parent 91	RR-009
Parent 92	RR-013
Parent 93	SKIPTON
* Parent 94	SURPASS 400
Parent 95	TRIGOLD
Parent 96	GSC-6 (*napus* parent, used as check)
Parent 97	PC 5 (*carinata* parent, used for trait introgression)

Study of pre-treatment and frying condition for seasoned deep fried shredded bamboo shoots *(Thyrsostachys siamensis)*

S Rajchasom[1, 2], J T Vuthijumnonk[1], P Pianpayoungpong[1], P Kantawong[1] and P Thingakhuea[1]

[1]College of Integrated Science and Technology, Rajamangala University of Technology Lanna, Chiang Mai, Thailand

E-mail: sureewan@rmutl.ac.th

Abstract. The objective of this research is to develop a new product of bamboo shoots, which is a seasoned deep fried shredded bamboo shoot. The study was divided into three parts which were the assessment of pre-treatment process for seasoned deep fried bamboo shoot prior to frying, the frying conditions and the consumer acceptability of the seasoning and flavour of the final product. Three pre-treatment processes assessed were freezing, vacuum drying and freeze drying. Three frying temperature levels were 160, 170, and 180°C for 3 and 5 seconds. Two seasoning flavours, corn cheese and paprika, were chosen in this study for a consumer acceptance evaluation. The results showed that the appropriate pre-treatment and frying conditions were the vacuum drying at 60°C for 3 hours and frying at 180°C for 3 second. This condition provided the good characteristics of inflated light golden yellow of the product with a low moisture content of 3.40±1.08% wet basis (w.b.), crispness value of 7.58 N and colour value in L^* a^* b^* term were 51.57±0.91, 6.44±0.19 and 21.96±1.09, respectively. The consumer acceptability of the product was evaluated by sensory test using 9-points hedonic scale, it was found that the overall acceptability on corn cheese and paprika flavours of the product were moderate to much with the score of 7.08±1.44 and 7.18±1.25, respectively. It can be concluded that the bamboo shoots processed in this way have the potential to be developed into a new product.

1. Introduction

Bamboo shoots are one of the favorite spring season vegetables in the East Asian region, particularly in China, Taiwan, Japan and other South-East Asian countries. In the northern part of Thailand, there are many regions planting bamboo for the wooden craft industry. In the raining season, a new shoot arises from underneath the root system which is then gathered, and eaten as a vegetable [1].

Bamboo shoots are a very low-calorie commodity. 100 grams of fresh cane has only 27 calories. Bamboo shoot contain moderate levels of soluble and non-soluble (NSP- non-starch carbohydrates) dietary fiber. 100 g of fresh shoots provides 2.2 grams of fiber. Dietary fiber helps maintain good bowel function, and can decrease bad (LDL) cholesterol levels by binding to it in the intestines. Studies suggest that a high-fiber diet can help cut down risk of colon-rectal cancer by protecting digestive organs from the toxic compounds in the food. Regular consumption of dietary fiber showed various health benefits such as prebiotic activity, lower risk of coronary heart disease, hypertension, obesity and some gastrointestinal diseases [2].

Bamboo shoots are also rich in the B-complex group of vitamins such as thiamin, riboflavin, niacin, vitamin B-6 (pyridoxine), and pantothenic acid, these are essential for optimum cellular enzymatic and metabolic functions. Bamboo shoots contain important minerals, especially manganese and copper. It

Content from this work may be used under the terms of the Creative Commons Attribution 3.0 licence. Any further distribution of this work must maintain attribution to the author(s) and the title of the work, journal citation and DOI.
Published under licence by IOP Publishing Ltd

has small amounts of some essential minerals and electrolytes such as calcium, iron, and phosphorus. Manganese is utilized by the human body as a co-factor for the antioxidant enzyme, superoxide dismutase. Copper is employed in the production of red blood cells. Iron is essential for cellular respiration and red blood cell formation [3].

In addition, bamboo shoots contain excellent levels of potassium. 100 grams of fresh shoot offers 533 mg or 11% of daily required levels of potassium. Potassium is an important component of cell and body fluids that helps controlling heart rate and blood pressure by countering effects of sodium [4].

There bamboo shoots germinate in the raining season, which causes an oversupply of bamboo shoots in the community in this season but a lack of supply throughout the rest of the year. To process the bamboo shoots, the villagers normally apply an easy and simple method to preserve the bamboo shoots such as making a pickled bamboo shoot or boiling the bamboo shoot and keeping it in a vacuum-packed plastic bag. This can extend the shelf life of bamboo shoot but the price of these products is very low. The bamboo shoot is an abundant and costless plant crops in Thailand. Therefore, this study aims to determine if a new product could be produced from the bamboo shoot which would increase its value. The seasoned deep-fried shredded bamboo shoot was designated because the main characteristic of bamboo shoot is that it has a lot of fiber content. The pre-treatment processes for frying bamboo shoots was applied in order to promote a novel product with added-value with subsequent acceptance by local consumers.

2. Materials and methods

2.1. Materials

Bamboo shoots *(Thyrsostachys siamensis)* were provided from the villagers in Mae-On district, Chiangmai province, Thailand. The bamboo shoots were preliminary prepared by boiling at 100°C for 10 minutes to eliminate cyanide, which is a toxic in bamboo shoots. It was reported in the literature that boiling process can reduce cyanide for 91% [5]. The initial weight of the sample was controlled to an average of 92.52±13.9 grams per one bamboo shoot. The moisture content of the sample was preliminary analysed as 80.06±1.29 g/100 g bamboo shoots. After that the boiled bamboo shoots were cut into small pieces with a dimension of 0.3 x 7 x 0.3 cm. They were then prepared according to the different pre-treatment processed outlined below.

2.2. A study of pre-treatment process for preparing the bamboo shoots before frying

A Completely Randomized Design (CRD) was used to design the experimental plan. Three pre-treatment conditions were freezing at -18°C for 24 hours, vacuum drying at 60°C for 3 hours and freeze drying at -30°C for 3 hours [6,7]. 800 grams of bamboo shoots were prepared for each of the three pre-treatment conditions. After the process, the moisture content of the bamboo shoots was determined using moisture analyzer (MB45, Ohaus). Then, all bamboo shoots were fried at 180°C for 3 seconds. The moisture content and texture of the fried bamboo shoots were measured.

2.3. A study of frying condition of the shredded bamboo shoots

After the most optimal pre-treatment condition was established, next, the most suitable frying condition for the shredded bamboo shoots was also investigated. To design the experiment, 3×2 Factorial Design in CRD was used. There were three levels of frying temperature and two levels of frying time, which were 160 170 and 180°C for 3 and 5 seconds [7]. Therefore, there were six conditions for this study. After frying, the chemical and physical properties (moisture content, texture and colour) of the products were measured, all analyses were performed in triplicate.

2.4. Physical and chemical characteristics

2.4.1. Moisture content. The moisture content was measured using moisture analyzer (MB45, Ohaus).

IOP Conf. Series: Earth and Environmental Science **346** (2019) 012074 doi:10.1088/1755-1315/346/1/012074

2.4.2. Texture. The texture of the fried sample was measured using Texture Analyzer CT3 in a compression test mode. A spherical probe with diameter of 13 mm was applied. A speed of compression was of 0.05 mm/min [8]. The base size was of 38 mm width and 14 mm thickness.

2.4.3. Colour. Colour is a key factor indicating an attractive appearance of the sample. In this study the colour of the sample was measured using a colorimeter (MINOLTA, CR-400) and expressed in term of L^* a^* and b^* indicating the lightness, redness and yellowness, respectively.

2.5. A study of consumer acceptability in seasoning flavours

To investigate consumer acceptability of the seasoned deep-fried shredded bamboo shoots, two flavor seasoning were chosen, which were corn cheese and paprika. A sensory evaluation was conducted using a 9-point hedonic scale with 50 untrained panelists ranging from 18 to 65 years of age. To prepare the sample, the deep-fried shredded bamboo shoot sample was mixed with the seasoning powder in a ratio of 2% w/w and served to panelists seated separately in booths in order to allow unbiased evaluation of the sensory attributes. Samples of seasoned deep-fried shredded bamboo shoot were evaluated in terms of product specific colour, taste, flavour, texture (crispness and dryness) and overall acceptability on a 9-point hedonic scale. The acceptability attributes were valued from the lowest score 1 (dislike extremely) and the highest score 9 (like extremely).

2.6. Statistical analysis

Experimental data was statistically analyzed using EXCEL (version 2016) software program. One-way analysis of variance (ANOVA) was performed with pre-treatment and frying condition as factors and $P < 0.05$ was taken as indicator of statistical differences between the means. The differences among mean values of all data were also tested using EXCEL software program using Turkey's multiple comparison test method at 95% confidence intervals.

3. Results and discussion

3.1. A study of pre-treatment process for preparing the bamboo shoots before frying

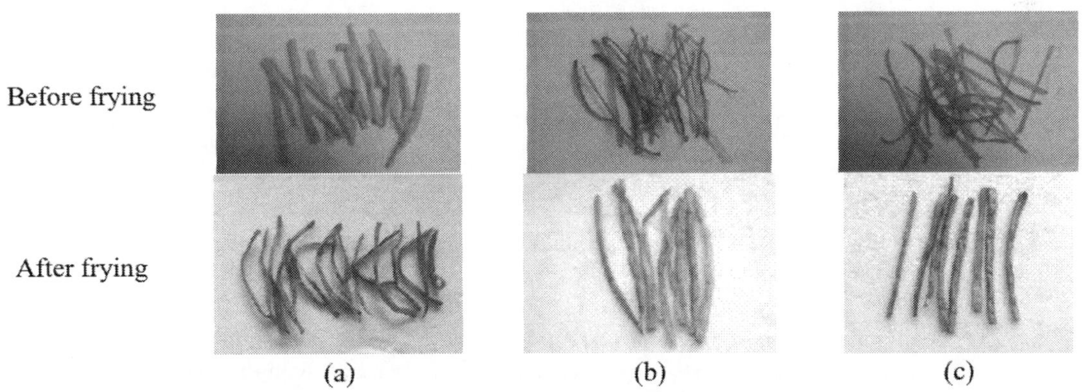

Figure 1. The characteristics of the shredded bamboo shoots after three pre-treatment conditions. (a) Freezing at -18°C, (b) Vacuum drying at 60°C and (c) Freeze drying at -30°C.

The appearance of the shredded bamboo shoots after pre-treatment are illustrated in figure 1. It can be seen that the appearance of the shredded bamboo shoot under freezing was moist and light yellow colour which was different from the other two samples. The sample under vacuum drying and freeze drying were withered and darker in appearance. This can be explained by the fact that the freezing process did not cause any changes to these characteristics or obvious mass loss of the sample because the freezing process changes the water in the sample into ice but does not remove water from the sample. In the other hand, vacuum and freeze drying are water removal process by evaporation and sublimation, respectively. The process of water removal contributed to the contraction of cell plant matrix that was previously occupied by the water [9]. The samples were dried but had a small change in colour. After pre-treatment, the samples were fried at 180°C for 3 seconds to preliminary observe the characteristic after frying of the sample and the fried samples were also shown in figure 1. The result showed that the sample from the freezing process had uneven colour and developed a curve shape. However, the other two sample from vacuum drying and freeze drying were straight and golden yellow colour. It can be assumed that pre-treatment had an effect on the final characteristics of the sample after frying, the vacuum drying and freeze drying provided a better appearance of the sample.

To further determine the influence of the pre-treatment on the physical and chemical properties of the samples, the moisture content and texture were measured. The moisture content of the samples from all three pre-treatment processes before and after frying were compared with the control sample without any pre-treatment and shown in figure 2. The initial moisture content of the sample from a control, freezing, vacuum drying and freeze-drying processes before frying were 80.06±1.29, 79.50±0.66, 6.57±0.77 and 6.50±1.08%w.b. respectively. The moisture content of two samples from vacuum frying and freeze drying were much lower than that of the control and the sample from the freezing process because drying process removed the water from the sample. After frying, the moisture content of all samples decreased significantly to 7.50±0.39, 7.44±0.42, 2.88±0.13 and 2.88±0.34%w.b., respectively. This was because the water inside the sample evaporated due to the high temperature of frying process [10]. The lowest moisture contents were found in the sample from vacuum drying and freeze drying because their moisture content before frying were lower as a result of the pre-treatment process.

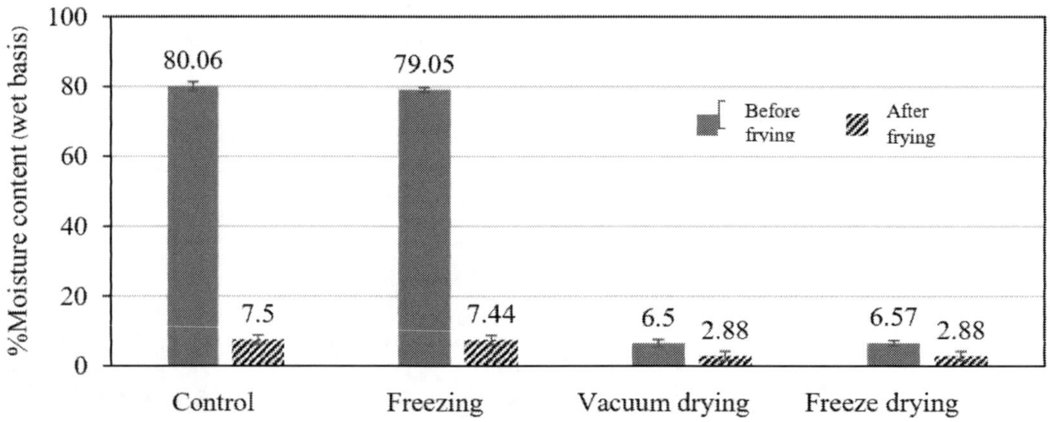

Figure 2. The moisture content of the shredded bamboo shoots from all three pre-treatment processes before and after frying.

A compression test was done to give an indication of the crispness of the fried sample and the results are shown in figure 3. From the force vs displacement profile, it can be seen that the fried sample from vacuum drying and freeze-drying processes showed a fluctuating pattern, which is typical of a brittle and crispy sample. However, there was no pattern for the control and the sample from the freezing process and no breakage of the shoots was observed during compression. The texture profile showed a smooth curve. It means that these two samples were not crispy according to a high initial moisture content and no crust occurring on their surface during frying. High moisture content absorbed oil during frying resulting a soft texture [11,12].

From the result, it can be concluded that vacuum drying and freeze drying were able to use in order to prepare the sample in the pre-treatment process before frying because these two methods offered the required characteristics of the fried shredded bamboo shoot. However, freeze drying is a costly method and take considerably longer than the other processes. Therefore, the suitable pre-treatment method for

IOP Conf. Series: Earth and Environmental Science **346** (2019) 012074 doi:10.1088/1755-1315/346/1/012074

the fried shredded bamboo shoot was vacuum drying at 60°C for 3 hours.

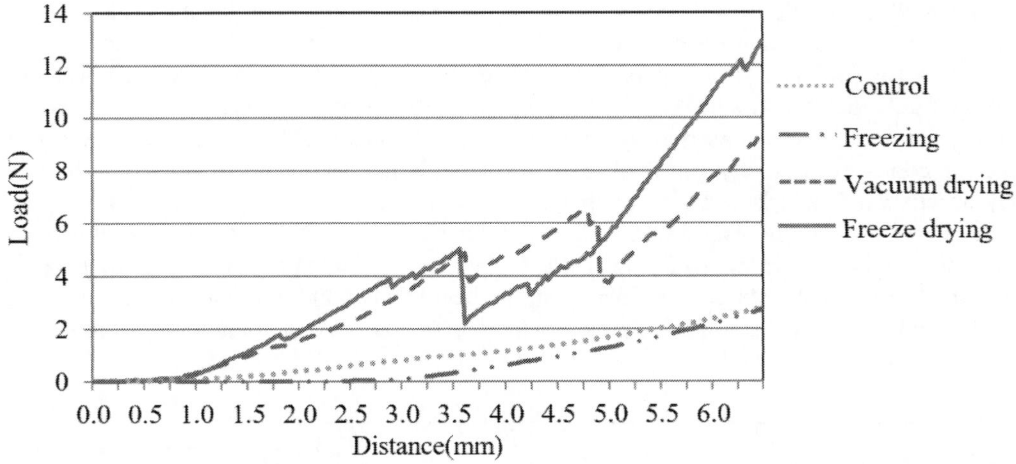

Figure 3. The force displacement profile from the compression test of fried shredded bamboo shoots from all three pre-treatment processes.

3.2. A study of frying condition of the shredded bamboo shoots

The dried shredded bamboo shoot from the pre-treatment process in the previous step was further studied to investigate the frying condition. Six frying conditions were applied, which were 160, 170 and 180°C for 3 and 5 seconds. The physical and optical properties of the final fried shredded bamboo shoots were measured namely colour moisture content and texture. The moisture content and colour values in terms of *L** (lightness) *a** (redness) and *b** (yellowness) are shown in table 1. It was found that the moisture content of the fried samples decreased with increasing frying temperature and time because water rapidly evaporated with higher amount at higher temperature and longer time. The lowest value was found in the highest frying temperature with the longest time. The same results were found in many studies about frying e.g. mushroom [7,13], potato chips [14] and banana [15]. The final moisture content of samples fried at 160, 170 and 180°C for 3 were 4.40±0.88, 4.28±1.25, 3.96±1.10 and 5 seconds were 3.65±0.69, 3.40±1.08 and 1.77±0.40%w.b., respectively. These moisture content values were below the critical moisture content (10% w.b.) for microbrial growth in a dried food product [16]. The lightness (*L**) and yellowness (*b**) of the fried sample decreased with increasing time and temperature, but the redness (*a**) decreased. This was seen as the sample became darker because the non-enzymatic browning reaction took place on the sample surface [17-22]. The higher the temperature and longer the time, the higher non-enzymatic browning reaction the more browning occurred [23-25].

Table 1. The moisture content and colour value of fried shredded bamboo shoots.

Frying Conditions	3 seconds				5 seconds			
	MC (%w.b.)	*L**	*a**	*b**	MC (%w.b.)	*L**	*a**	*b**
160°C	4.40±0.15^{a1}	61.71±1.48^{a}	4.29±0.05^{b}	29.38±0.11^{a}	3.65±0.57^{a}	61.70±1.16^{a}	3.12±0.28^{c}	25.63±0.58^{b}
170°C	4.28±0.18^{a}	60.72±1.35^{a}	5.28±0.34^{a}	25.30±0.99^{b}	3.40±2.43^{a}	51.58±1.35^{b}	4.22±0.31^{b}	21.96±0.77^{c}
180°C	3.96±0.28^{a}	51.57±0.91^{b}	6.44±0.19^{a}	21.96±1.09^{c}	1.77±0.32^{b}	50.15±0.91^{b}	5.28±0.22^{b}	16.83±1.02^{d}

[1]Mean ± standard deviation (n = 3). Means within a column bearing the same superscript letter were not significantly different ($p > 0.05$).

However, from those previous two properties, the moisture content and colour, it cannot be concluded that which frying condition was the best. Therefore, the texture of the samples was also analyzed. The texture profile was shown in figure 4. From the profile, it was found that all samples showed the fluctuation pattern presenting all samples were crispy and brittle. It was also found that the texture profiles of the samples frying at 170°C for 5 seconds, 180°C for 3 and 5 seconds were higher and had more fluctuation than that of other three frying conditions, which were lower temperature and shorter time. The first compression force was found as 3.07, 4.68 and 5.71 N for frying at 160, 170 and 180°C for 3 seconds, respectively and 3.44, 5.16 and 5.37 N for frying at 160, 170 and 180°C for 5 seconds, respectively. The highere frying temperature and time, the higher crispness was obtained. The higest crispness of the sample was found in the sample fried at 180°C for 3 seconds. However, the crispness of the sample dropped when frying at 180°C for a longer time of 5 second since too long frying time resulted in more oil absorbtion [8]. Therefore, the best frying condition was 180°C for 3 seconds because this condition provided a good characteristics in the moisture content, colour and texture of sample.

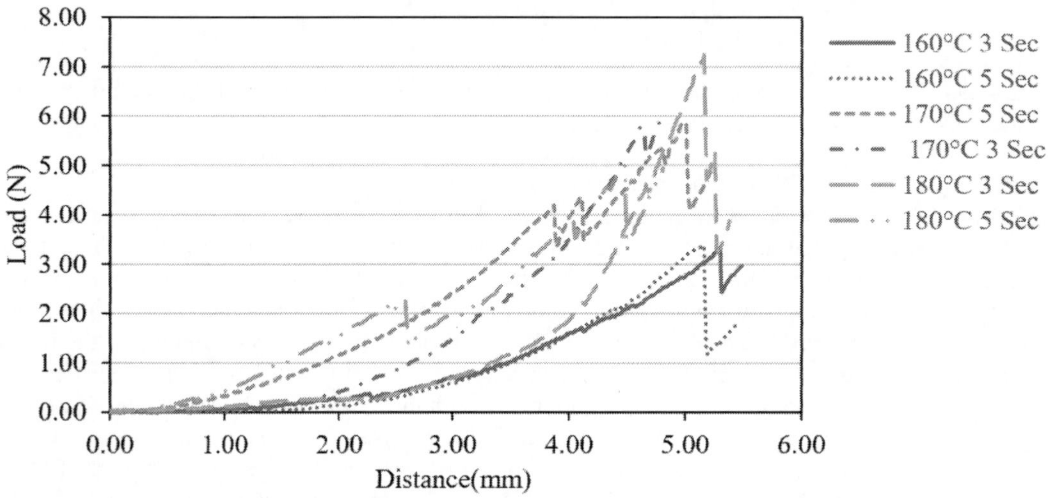

Figure 4. The force displacement profiles from the compression test of fried shredded bamboo shoots at different frying temperature and time.

3.3. A study of consumer acceptability in seasoning flavours

A sensory evaluation was conducted on the seasoned deep fried shredded bamboo shoots to investigate the acceptability of the consumer on the developed product. The sensory evaluation showed a high score for all sensory attributes, the scores were between 94.5±1.44 and 7.24±1.44 as shown in figure 5. The consumer acceptability was like moderately to like very much and showed similar scores in every sensory attributes for these two flavours. However, the customers liked paprika flavour more than corn cheese flavour in some attributes e.g. taste, flavour and overall acceptance. This may be because paprika flavour delivered more taste than corn cheese flavour. The sensory evaluation in deep fried shredded bamboo shoot corn cheese flavour showed the highest score in overall acceptability (7.08±1.44) followed by colour (6.82±1.34), flavour (6.74±1.68), taste (6.68±1.52), dryness (6.54±1.83), crispness (6.24±2.50) and odour (5.94±2.05). For the paprika flavour, the highest score was obtained in flavour (7.24±1.44) followed by overall acceptability (7.18±1.25), taste (6.92±1.55), dryness (6.34±1.88), colour (6.26±1.55), crispness (6.22±1.89) and odour (5.94±1.80). It can be seen that the customers like less the odour and crispness of the product because the bamboo shoot has an unpleasant smell. Thus this problem should be solved in an additional study. From the result, it can be concluded that corn cheese and paprika seasoning were possible to use inorder to produce the seasonned deep fried bamboo shoots as a commercial product.

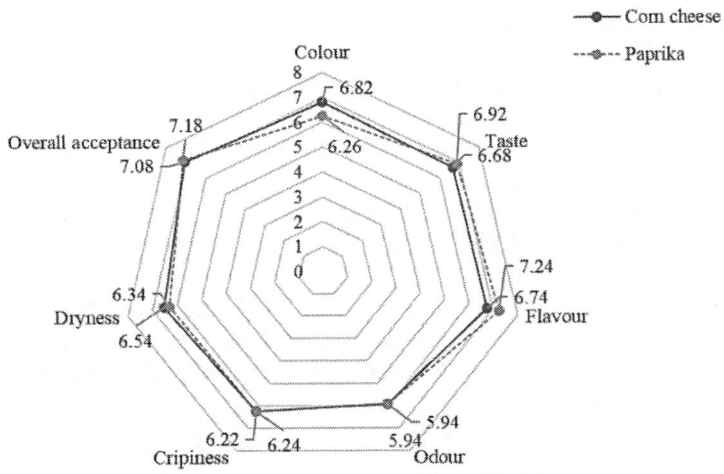

Figure 5. Spider web diagram of the sensory evaluation of the seasoned deep-fried shredded bamboo shoots in two flavours.

4. Conclusion

From the result, it can be concluded that the pre-treatment process had influence on the final characteristics of the fried shredded bamboo shoot e.g. texture and color. In this case, the appropriate pre-treatment condition was vacuum drying at 60°C for 3 hours since vacuum drying reduced the moisture content of the sample without any adverse effect to the appearance e.g. colour. The low moisture content of the sample resulted in good characteristics for colour and texture of the product after frying. The suitable frying condition was 180°C for 3 seconds. The sensory evaluation on seasoned fried shredded bamboo shoot in two flavours (corn cheese and paprika) showed that the consumer accepted both these flavours with overall moderate approval.

References

[1] Debangana C, Jatindra K S and Sharma G D 2012 Value addition to bamboo shoots: A review *J. Food Sci. Technol.* **49** 407-14

[2] Satya S, Poonam S, Lalit M B and Sudhakar P 2011 Bamboo shoot: A potential source of food security *Med. J. Nutrition Metab.* **5** 1-10

[3] Nirmala C, Madho S B and Sheena H 2011 Nutritional properties of bamboo shoots: Potential and prospects for utilization as a health *Compr. Rev. Food Sci. F.* **10** 153-68

[4] Nongdam P and Leimapokpam T 2014 The nutritional facts of bamboo shoots and their usage as important traditional foods of northeast India *International Scholarly Research Notices* **1** 1-17

[5] Suvanee T *et al* 2011 Risk assessment of cyanide in bamboo shoots *Bureau of Quality and Safety of Food* **53** 67-79

[6] Wongchanapai P and Chinnasarn S 2011 Effect of frying condition and pre-drying treatment on oil uptake and quality of deep-fried restructured taro strips *Agricultural Sci. J* **44** 441-4

[7] Tunyaporn S and Nattiga S 2016 Effect of pre-frying treatments on moisture content, effective moisture diffusivity and oil uptake content of deep-fat fried Shitake mushrooms *(Lentinusedodes) J. Food Technol, Siam University* **1** 57-66

[8] Sarayuth S 2008 Effects of temperature and reuse of frying oil on oil uptake, texture and color of fried shrimp chips *Thesis Report* (Chiang Mai University Thailand)

[9] Somchart S 1997 *Drying of Seed and Food* 7th ed. (King Mongkut's University of Technology Thonburi Bangkok)

[10] Vilai R 2014 *Food Processing Technology* 6th ed. (King Mongkut's University North Bangkok)

[11] Rice P and Gamble M H 1988 The effect of slice thickness on potato crisp yields and composition *J. Food Eng.* **8** 31-46

[12] Mariska M 2003 Mechanism and reduction of fat uptake in deep-fried foods *Trends Food Sci. Technol.* **14** 364-73

[13] Chanya W, Nanthana S and Chularat H 2101 Effects of boiling temperature and frying time on quality of flavor-shredded straw mushroom *(Volvarilla volvacea) Proceedings of 50th Kasetsart University Annual Conference: Agro-Industry*, Kasetsart University Bangkok, Bangkok, Thailand, pp 216-23

[14] Patcharee L, Pornthip V and Akanith P *2006* Research and development of process technology for durian french fries *Full Report of Crop Processing Research and Development Project* p 150

[15] Taksina I, Ratchada T, Waraporn P and Danupol K 2017 Effect of frying conditions on physical and chemical properties of deep fried ripe banana *(Musa sapientum L. cv. Namwa)* crisps *The National and International Graduate Research Conference,* Khon Kaen University, Thailand pp 558-69

[16] Fellows P J 2009 *Food Processing Technology: Principles and Practice.* 3rd ed. (Cambridge, U.K.: Elsevier Science & Technology) vol 1 44-7

[17] Maillard L C and Gautier M A 1912 The reaction of amino acids with sugars: Mechanisms of melanoid formation *CR Seances Academic Science III.* **154** 66-8

[18] Labuza T P, Reineccius G A, Monnier V M, O'Brien J and Baynes J W 1998 *Maillard Reactions in Chemistry, Food and Health* 1st ed. (Cambridge U.K.: Woodhead Publishing) pp 126-200

[19] Lopez A, Pique M T, Boatella J, Parcerisa J, Romero A, Ferrá A and Garci J 1997 Influence of drying conditions on the hazelnut quality. III. Browning *Dry Technol* **15** 989-1002

[20] Lee J W, Oh S H, Kim J H, Byun E H, Kim R, Baek M and Byun M W 2006 The effect of irradiation temperature on the non-enzymatic browning reaction in cooked rice *Radiat Phys. Chem.* **76** 886-92

[21] Ajandouz E H, Desseaux V, Tazi S and Puigserver 2008 A Effects of temperature and pH on the kinetics of caramelisation, protein cross-linking and Maillard reactions in aqueous model systems *Food Chem.* **107** 1244-52

[22] deMan J M 1999 *Principles of Food Chemistry* 3rd ed. (Gaithersburg, MD: Aspen Publishers) vol **1** 23-30

[23] Ames J M 1998 Applications of the Maillard reaction in the food industry *Food Chem* **62** 431-9

[24] Fayle S E and Gerrard J A 2002 *The Maillard Reaction* (Cambridge, U.K.: Royal Society of Chemistry) vol **1** 1-8

[25] Ames J M and Benjamin C 2003 Nonenzymatic browning *Encyclopedia of Food Sciences and Nutrition* B Caballero, L Trugo and P Finglas eds. (London, UK: Academic Press) pp 665-72

Impacts of supplementation of probiotics on the prevalence of grape marc derived polyphenols in colonic digesta using *in vitro* digestion model

M J Li[1], Y T Loo[1], L Cheng[2], K Howell[1] and P Z Zhang[1, 3]

[1]Faculty of Veterinary and Agricultural Sciences, The University of Melbourne, Building 184, Parkville Campus, Victoria 3010, Australia.
[2]Faculty of Veterinary and Agricultural Sciences, The University of Melbourne, Dookie Campus, Victoria 3647, Australia.

E-mail: pangzhen.zhang@unimelb.edu.au

Abstract. Grape marc is a by-product of winemaking that contains numerous polyphenols, the bioavailability of polyphenol is largely dependent on their metabolism by probiotics or gut microbiome. The two-way phenolic-microbiota interaction has been postulated as the microorganisms and specific polyphenols could affect each other and exert beneficial effects on the host health. However, the oral taken probiotics showed a low survival rate in the gastrointestinal tract, while using encapsulation technique can provide them with a protective microenvironment. The objective of this study was to investigate the bioaccessibility of polyphenols originated from whole grape marc, consisting of seeds, skins and stems, combined with encapsulated oral taken probiotics, using *in vitro* digestion model and analytical chemistry techniques. 23 phenolic compounds were evaluated in this study, which included phenolic acids, anthoxanthins and stilbenes. This study showed that the encapsulated probiotics were protected by microcapsules until being liberated in the large intestine, and the metabolism of some polyphenols derived from selected grape marc samples was promoted by probiotics. These results indicate the potential of combining grape marc polyphenols with probiotics to develop a new health product. Furthermore, this provides the wine industry with an opportunity to make a profit by using by-products.

1. Introduction

10 million tons of grape marc, which consisted of seeds, skins and stems, are produced annually worldwide by the wine industry after the fermentation of grapes [1]. Conventionally, they were used as compost and animal feed, or ingredients to produce a liquor called 'Pomace Brandy' [2,3]. Previous studies have found that part of the phenolic compounds, with several health-promoting potentials, such as antioxidant, anti-inflammatory, anti-carcinogenic and neuroprotective ability, is retained in the grape marc [4-7]. However, these benefits can only be achieved if the polyphenols are digested and absorbed by human. Direct absorption of dietary polyphenols by the human body in the small intestine is only 5 to 10 percent of total consumption, while the remaining 90 to 95 percent of unabsorbed polyphenols are delivered to the large intestine and potentially utilized by the gut microbiota [8]. Studies suggested that the poor bioavailability of dietary polyphenols might be due to its polymerized structure where dimers and polymers were difficult to be absorbed in the small intestine. Besides, glycosylation of polyphenols also compromised their bioavailability for the absorption in small intestine [9]. Research found that the

Content from this work may be used under the terms of the Creative Commons Attribution 3.0 licence. Any further distribution of this work must maintain attribution to the author(s) and the title of the work, journal citation and DOI.
Published under licence by IOP Publishing Ltd

bioaccessibility of the polyphenols in the human body could be improved through their reaction with microorganisms, including gut microbiota and probiotics [10]. Gut microbiota could metabolise polyphenols by cleaving the glycosidic linkages in the phenolic compounds and breaking down the heterocyclic backbone in the phenolic structure, while the oral taken probiotics could enhance these effects if they can reach the gastrointestinal tract and modulate the gut microflora [11-13]. *Lactobacillus plantarum* is implemented as the model probiotics in this study because it is easy to produce, has a human origin, and has relatively higher tolerance to the low pH environment and bile salts exposure as compared to the other commonly used probiotics [14-15]. However, their oral delivery shows a low efficiency due to the low survival rate of free bacterial cells during digestion [16]. To protect the probiotics from being damaged, encapsulation techniques have been developed. Encapsulation consisted of the entrapment of target substance with a carrier agent, which could provide target substance with a micro-environment and therefore enhance its stability [17-18]. With the use of encapsulation, probiotics will not be released until they reach their desired destination [10].

In this study, alginate was chosen as the encapsulation material, and extrusion as the encapsulation technique. Alginate is a type of carbohydrate polymers extracted from algae [19]. Alginate contains mannuronic and guluronic acid; with the presence of calcium ions, it can lead to complex formation of compact and strong gels [20]. The effects of alginate have been demonstrated by some previous studies, for example, the non-encapsulated *Bifidobacterium longum* BIOMA 5920 was not able to survive in simulated gastric fluid, whereas the alginate encapsulated *B. longum* survived with a four log cycles reduction after two hours of gastric fluid treatment [21]. Similar results have been reported by Annan et al., as the survival rate of the free bacteria showed two log units lower çompared with alginate encapsulated *Bifidobacterium adolescentis* 15703T [22]. For the extrusion technique, it has the characteristics of simple operation, cost-effective equipment, and no deleterious materials. This gentle operation process costs minimal damage to the cells with high cell viability achieved, under both aerobic and anaerobic conditions. However, the main drawback of this technique is that the slow formation of microcapsules, which makes it unsuitable to be adopted in large scale production [23]. Besides, the formation of high-diameter particles (>1mm) using extrusion method limits its applications in food industry, where mouthfeel is a crucial factor [24].

The interactions between phenolic compounds and probiotics, particularly the probiotic effect on the release of polyphenols derived from grape marc, is important factor for the polyphenols to exert beneficial health effects. Besides, the understanding of the metabolism and bioavailability of grape marc polyphenols requires further in-depth investigation. There are insufficient studies up to date to illustrate these issues. Therefore, in order to explore these research gaps, this study aims to investigate the bioaccessibility of polyphenols originated from grape marc combined with oral taken probiotic L. plantarum, using *in vitro* digestion model and analytical chemistry techniques. In details, the main objectives focus on the aspects as the following: i) to select one grape marc sample from red and white grape respectively collected from different wineries, by determining their total phenolic content and antioxidant activity; ii) to encapsulate probiotics using extrusion technique and investigate their effect; iii) to investigate the effects of probiotics on the release of polyphenols in the gastrointestinal tract using *in vitro* digestion model; iv) to determine the phenolic compounds of grape marc and the digesta of the combination of grape marc and probiotics with HPLC technique.

2. Materials and methods

2.1. Materials and chemicals

The following reagents and chemicals were purchased from Sigma-Aldrich (Castle Hill, NSW, Australia): MRS broth, calcium chloride, sodium alginate, sodium chloride, methanol, hydrogen chloride, 2.0N FCR, sodium carbonate solution, gallic acid, glacial acetic acid, sodium hydroxide, 2,4,6-Tris(2-pyridyl)-s-triazine, iron (III) chloride anhydrous, iron (II) sulfate heptahydrate, 2,2-Diphenyl-1-picrylhydrazyl, trolox, potassium chloride, monopotassium phosphate, sodium bicarbonate, magnesium chloride, ammonium carbonate, salivary α-amylase, porcine pepsin, pancreatin, sodium phosphate dibasic, sodium phosphate monobasic, bile salts, peptone, yeast extract, casein, L-Cysteine HCl, Tween-80, guar, soluble starch, ,tryptone, pectin, mucin, sodium bicarbonate, magnesium sulfate, dipotassium phosphate, acetic acid 50% solution, 5-(hydroxymethyl)furfural, protocatechuic acid, p-hydroxybenzoic acid, caftaric acid, caffeic acid, p-coumaric acid, syringic acid, ferulic acid, trans-sinapic acid, procyanidin B1, procyanidin B2, resveratrol, epicatechin, catechin, epicatechin gallate, polydatin, quercetin, quercetin 3-O-glucoside, quercetin 3-O-galactoside, quercetin 3-O-glucuronide, quercetin 3-O-rhamnoside, kaempferol and kaempferol 3-O-glucoside.

LAMVAB plates were purchased from media preparation unit of the department of microbiology & immunology, The University of Melbourne (VIC, Australia). All water used was prepared by Milli-Q Gradient water purification system (Millipore Australia Pty. Ltd., North Ryde, NSW, Australia). HPLC grade Acetonitrile was obtained from Merck Pty. Ltd. (Bayswater, VIC, Australia).

Eight grape marc samples, which consisted of seeds, skins and stems, were collected from 3 vineyards located in Victoria, Australia. 6 different red grape marc samples were collected from Mount Langi Ghiran winery in Grampians wine region. Middle Old Blocks marc (OM) and West Old Blocks marc (OW) were collected in 2016 vintage, the OM grapes were collected by hand picking while OW grapes were collected by machine. The grape marc of cabernet sauvignon (CS), shiraz marc from Contour Block (TR), shiraz marc from South Block (SS) and shiraz marc from the House Block (HB) were collected in 2018 from the same winery. White grape marc used were from Chardonnay grape collected in 2018 vintage from the Bowen Ridge Winery (BR) and Leura Park winery (LP) in Geelong wine region, Victoria, Australia.

2.2. Grape marc selection

The harvested grape marc samples were freeze-dried, blended into powder, extracted by methanol, centrifuged and filtered before further analyses [25]. One red and one white grape marc sample with the highest antioxidant and scavenging capacity were identified from the red and white grape marc collection for the further *in vitro* digestion experiment.

2.2.1. Determination of total phenolic content (TPC). The spectrophotometric detection of chromogens was the basis of TPC's quantification. When the functional hydroxyl groups of phenols reacted with the reagent containing phosphomolybdic/ phosphotungstic acid complexes, the chromogens formed [26]. The absorbance of the reacted mixture was measured against the blank sample at the wavelength of 765nm using a Multiskan™ GO Microplate Spectrophotometer (Thermo Fisher Scientific Oy, Vantaa, Finland). The TPC of samples were expressed as gallic acid equivalents (GAE) [mg GAE/g sample dry mass (DM)]. All the determinations were triplicated.

2.2.2. Determination of ferric reducing antioxidant power (FRAP). The ability of antioxidants as reductants to reduce the colourless ferric tripyridyltriazine complex to blue ferrous-(2,4,6-tripylridyl-s-triazine)2 in the oxidation-reduction reaction was the basis of FRAP test. The absorbance of the reacted mixture was measured at the wavelength of 593nm [27]. The FRAP value of samples was expressed as Fe2+ equivalents (mM Fe2+ eq./g sample DM). All the determinations were triplicated.

2.2.3. Determination of DPPH radical scavenging capacity. The mechanism of donation of hydrogen to the DPPH radical by antioxidants, which acted as radical scavengers was the basis of DPPH assay. The absorbance of the reacted mixture was measured at the wavelength of 515nm [28-29]. The DPPH radical scavenging capacity of samples was expressed as Trolox equivalents (mg Trolox eq./g sample DM). All the determinations were triplicated.

2.3. In vitro digestion

2.3.1. Experimental design. Group A was the treatment group with faeces added in the in vitro digestion, which comprised of 2 types of grape marc powder, namely the white grape marc powder (WC) and red

IOP Conf. Series: Earth and Environmental Science **346** (2019) 012075 doi:10.1088/1755-1315/346/1/012075

grape marc powder (RC); and 4 treatment groups with the incorporation of probiotics, namely the white grape marc powder plus probiotics powder (WP), red grape marc powder plus probiotics powder (RP), white grape marc powder plus encapsulated probiotics (WPE), and red grape marc powder plus encapsulated probiotics (RPE). Group B was the treatment control group with no faeces added in the in vitro digestion, while other conditions were similar with group A, and named as WCn, RCn, WPn, RPn, WPEn, RPEn.

2.3.2. Probiotic preparation. The probiotics were sub-cultured and incubated in MRS-broth. After centrifugation, wash, resuspension and freeze-dry, the probiotics were processed into powder [30-31].

2.3.3. Encapsulation. 0.25 g freeze-dried probiotic powder was dissolved into 600 μl sterile saline solution, and then suspended in 800 μl 2% sodium alginate solution, which acted as hydrocolloid solution. 7% calcium chloride solution which acted as hardening solution was prepared and sterilised prior to the extrusion process. For each treatment, 25 mL of 7% calcium chloride solution was used to harden the droplets, and it was used only once for each treatment to ensure the same effect among treatments. Via a syringe, the viscous bacteria suspension was passed through an 18-gauge needle and allowed to free-fall onto a culture dish with 25 mL calcium chloride solution to generate droplets. The droplets were allowed to gel for 1 minute to form the encapsulated microcapsules. The hardening solution together with the microcapsules was then filtered by a sterilised cell tissue strainer and the microcapsules were transferred into the 50mL centrifuge tubes and ready for in vitro digestion [12,16].

2.3.4. In vitro digestion. The in vitro digestion experiment was carried out following two previous studies [32-33], with modifications to better fit the research objectives. There were 4 stages in this experiment, namely oral, gastric, small intestinal and large intestinal phases. The pig faeces, which collected freshly from 12-weeks old pigs in the animal house at the Dookie campus of the University of Melbourne, were used to simulate the microbial environment of the large intestine.

2.3.5. Microbial count. Drop plate method based on a published protocol was adopted for the quantification of lactobacilli, with modifications to better fit the research objectives [34].

2.3.6. Characterisation of polyphenols by HPLC analysis. The HPLC analysis of polyphenols was performed by the Waters 2690 Alliance HPLC separation module (Waters, Rydalmere, NSW, Australia) collocated with Waters 2998 Photodiode Array (PDA) detector, following a published protocol. The external standard method and standard addition HPLC calibration method were applied to identify and quantify the phenolic compounds in the sample extracts.

The preparation of samples for high-performance liquid chromatography (HPLC) analysis was conducted in three methods depending on the sample. For the grape marc samples, 5 g freeze-dried grape marc powder together with 50 mL 1% HCl v/v methanol were mixed and extracted for 24 h at 20 °C by the shaking incubator at 150 rpm, then the mixture was centrifuged at 6500 rpm for 10 minutes at 20 °C, and the supernatant was collected. The supernatants were filtered by 30mm × 0.45μm nylon syringe filter (Thermo Fisher Scientific, VIC, Australia) prior to being evaporated by a Hei-Vap rotary evaporator and vacuum pump (Heidolph Instruments GmbH & CO. KG, Schwabach, Germany). The residue was redissolved in 10 mL of HPLC grade methanol, then filtered by 30mm × 0.45μm nylon syringe filter and transferred to amber vials [35]. For the small intestinal digesta supernatant, they were 2-fold diluted by HPLC grade methanol, then filtered by 30mm × 0.45μm nylon syringe filter and transferred to amber vials. For the colonic fermentation digesta supernatant, 1mL of each sample were mixed thoroughly with 3mL of acetone: methanol (1:1, v/v) by a vortex for 30 s, and were then centrifuged at 5400g for 5 minutes at 20 °C. the resulted supernatant was collected in a 10mL-glass bottle and dried under nitrogen gas flow at room temperature. The residue was redissolved in 1.5 mL of HPLC grade methanol, then filtered by 30mm × 0.45μm nylon syringe filter and transferred to amber vials [36].

Two systems were performed separately to identify the phenolic acids (System I) and anthoxanthins and stilbenes (System II) in the sample extracts. For both systems, the mobile phases were the same, namely acetic acid in Milli-Q water (2:98, v/v) as solvent A, and 0.5% (v/v) acetic acid in water and acetonitrile (50:50, v/v) as solvent B. All the mobile phases were degassed using an ultrasonic degasser. In System I, the total run time was 60 minutes, all samples were injected with a volume of 10 μL, and the synchronous detection was conducted for hydroxybenzoic acids at 280nm and hydroxycinnamic acids at 320nm. In the elution process, the flow rate was kept constant at 1.0 mL/min, and a gradient program was adopted as, in the first 10 minutes, the proportion of solvent A decreased from 90 to 85%, followed by an isocratic period at 85% for 3 minutes. After that, the proportion of solvent A decreased from 85% to 75% for 7 minutes, and then further declined from 75% to 45% for 30 min. In the following 1 minute, its proportion decreased from 45 to 0% and kept at 0% for the next 5 minutes for column washing purpose. Finally, the proportion of solvent A was increased from 0% to 90% in last 4 minutes for equilibration. In System II, the total run time was 72 minutes, all samples were injected with a volume of 10 μL, and the synchronous detection was conducted for flavanols at 280nm, stilbenes at 320nm, and flavonols at 370nm. In the elution process, the flow rate was kept constant at 1.0 mL/min, and a gradient program was adopted as, from 90 to 76% solvent A (20 min), from 76 to 70% solvent A (20 min), from 70 to 45% solvent A (20 min), from 45 to 0% solvent A (5 min), 0% solvent A isocratic period (5 min) for column wash, and finally, from 0 to 90% solvent A in the last 2 minutes for equilibration [25].

2.3.7. Statistical analysis. Statistical analysis was conducted using Minitab 17 software (Minitab, Inc., NSW Australia). The significance of differences among the generated data was performed by doing the one-way analysis of variance (ANOVA). Results were demonstrated as mean ± standard deviation and noted with different letters when significant differences were found in the mean at the confidence level of $p < 0.05$.

3. Results and discussion

3.1. Total phenolic content and antioxidant activities of grape marc

The analyses of TPC, FRAP, DPPH (Table 1) showed that, among the red grape marc samples, sample CS showed a higher TPC and FRAP values compared to the other five samples, with the TPC of 30.9 ± 0.70 mg GAE/g sample DM, and FRAP value of 385 ± 17.6 mM Fe 2+ eq/g sample DM (DM, dry mass of the freeze-dried grape marc sample). Its DPPH radical scavenging capacity showed no difference with SS, TR and HB, but had the highest value of 27.4 ± 0.44 mg Trolox eq/g sample DM. In comparison with the reported value from the study by Rockenbach and his team, sample CS exhibited lower TPC and DPPH radical scavenging capacity than the Carbernet Sauvignon grape pomace with a TPC of 74.7 ± 2.22 mg GAE/g sample DM and 505 ± 4.62 μmol Trolox eq/g sample DM, respectively. Besides, TPC and DPPH activity of the grape pomaces from other varieties such as Merlot, Bordeaux, and Isabel were also higher than the sample CS [37]. However, as compared to another study, DPPH activity of sample CS of 109 ± 1.75 μmol Trolox eq/g sample DM (equivalent to 27.4 ± 0.44 mg Trolox eq/g sample DM) is higher than the grape pomaces from varieties such as Sangiovese (81.4 ± 1.14 μmol Trolox eq/g sample DM), Primitivo (77.9 ± 0.91 μmol Trolox eq/g sample DM), and Negro Amaro (26.9 ± 0.32 μmol Trolox eq/g sample DM) [38].

Between white grape marc samples, the sample BR had a higher TPC, FRAP and DPPH values compared with sample LP. Consequently, sample CS and sample BR were selected as the model grape marc in the further experiment.

Table 1. Total phenolic content and antioxidant activities of red and white grape marc samples.

Sample Name		TPC (mg GAE/g sample DM)	FRAP (mM Fe2+ eq/g sample DM)	DPPH (mg trolox eq/g sample DM)
Red Grape Marc	CS	30.9 ± 0.70 [a]	385 ± 17.6 [a]	27.4 ± 0.44 [a]
	SS	23.3 ± 1.67 [b]	260 ± 17.6 [b]	27.3 ± 0.17 [a]
	TR	22.0 ± 0.04 [b]	256 ± 18.4 [b]	27.1 ± 0.23 [a]
	HB	19.9 ± 1.39 [c]	210 ± 15.8 [c]	26.3 ± 0.56 [a]
	OM	13.2 ± 0.89 [d]	122 ± 8.54 [d]	17.8 ± 3.46 [b]
	OW	10.7 ± 0.48 [e]	100 ± 13.7 [d]	16.7 ± 3.25 [b]
White Grape Marc	BR	17.8 ± 0.52 [a]	161 ± 3.52 [a]	27.1 ± 0.16 [a]
	LP	14.1 ± 1.81 [b]	123 ± 18.2 [b]	24.6 ± 2.11 [b]

Different letters within each row indicate statistically significant differences between treatments by one-way ANOVA ($P < 0.05$).

3.2. Microbial content of digesta

Table 2 showed the microbial counts of small and large intestinal digesta of different treatments, with the data being subtracted against the microbial counts of the blank group. According to the small intestinal digesta results, treatments with un-encapsulated probiotics showed a significantly higher value than those with encapsulated probiotics, which indicated that encapsulation could protect probiotics from releasing in the small intestine. Some previous research showed that encapsulated probiotics can better survive in simulated gastric and intestinal environment with bile salt added compared with free cells [39,40]. It is postulated that encapsulation technique can improve the low pH tolerance of probiotics passing through the *in vitro* gastric digestion which is acidic. Encapsulation also provides a protective barrier for the probiotics during the *in vitro* small intestinal digestion where the presence of bile salt that can cause damage to the probiotics [41]. Regarding large intestinal digesta results, treatments with encapsulated probiotics showed a significantly higher value than those without encapsulation, which indicated that encapsulation could effectively deliver probiotics to the large intestine, and the probiotics were released after colonic fermentation.

Table 2. Microbial counts of small and large intestinal digesta.

	Small intestinal digesta (CFU/g)	Large intestinal digesta (CFU/g)
RM	0.00 [b]	$1.29\times10^7 \pm 7.27\times10^6$ [b]
RP	$2.50\times10^7 \pm 6.56\times10^6$ [a]	$9.80\times10^7 \pm 4.91\times10^7$ [b]
RPE	$4.23\times10^4 \pm 3.28\times10^4$ [b]	$1.60\times10^9 \pm 1.04\times10^9$ [a]
RMn	0.00 [b]	$6.00\times10^5 \pm 3.61\times10^5$ [b]
RPn	$2.30\times10^7 \pm 2.65\times10^6$ [a]	$7.67\times10^7 \pm 5.51\times10^7$ [b]
RPEn	$4.03\times10^4 \pm 4.08\times10^3$ [b]	$6.83\times10^8 \pm 4.54\times10^8$ [a]
WM	0.00 [b]	$3.53\times10^8 \pm 1.17\times10^8$ [a]
WP	$6.73\times10^7 \pm 3.06\times10^6$ [a]	$9.50\times10^8 \pm 5.22\times10^8$ [a]
WPE	$7.47\times10^5 \pm 1.12\times10^5$ [b]	$6.07\times10^9 \pm 6.87\times10^9$ [a]
WMn	0.00 [b]	$7.00\times10^7 \pm 4.36\times10^7$ [b]
WPn	$6.97\times10^7 \pm 5.51\times10^6$ [a]	$4.70\times10^8 \pm 4.59\times10^8$ [b]

WPEn	$7.03\times10^5 \pm 6.51\times10^4$ [b]	$1.60\times10^9 \pm 7.00\times10^8$ [a]

Different letters within each row indicate statistically significant differences between treatments by one-way ANOVA ($P < 0.05$).

3.3. Polyphenol changes after in vitro digestion

Table 3 showed the 10 identified polyphenols out of 23 standard compounds, other polyphenols that were not identified from the grape marc samples and the supernatant of intestinal digested samples are: gallic acid, 5-(hydroxymethyl) furfural, caftaric acid, caffeic acid, sinapic acid, polydatin, quercetin 3-O-galactoside, quercetin 3-O-glucuronide, quercetin 3-O-glucoside, kaempferol 3-O-glucoside, quercetin 3-O-rhamnoside, epicatechin gallate and quercetin. Due to the detection limit for the methodology and equipment used, most of the target compounds were not identified. In the following relevant projects and further research, LC-MS analysis is recommended as an efficient alternative to evaluate the digestion of phenolic compounds.

3.3.1. Small Intestinal digesta. According to the results obtained in the HPLC analysis of small intestinal digesta, three major changes were observed. Firstly, the polyphenols that found in the grape marc did not exist in the digesta. This phenomenon was hypothesized as the parental polyphenols found in the grape marc were bio-transformed by the probiotics used into other phenolic metabolites after digestion, where those metabolites were not included in the standard compounds in this experiment. For example, L. plantarum IFPL935 was shown to metabolize the flavan-3-ol enriched grape seed extract with the action of galloyl-esterase, decarboxylase and benzyl alcohol dehydrogenase that led to the formation of gallic acid, pyrogallol and catechol, respectively [42]. Secondly, the metabolites of polyphenols, such as protocatechuic acid was not found in the grape marc sample, but in the digesta. This is in agreement with some previous research, which showed that protocatechuic acid was a metabolite of catechins, protocatechuoyl ester or cyanidin-glucosides 1–3. As protocatechuoyl ester can be degraded by the esterase of L. plantarum to protocatechuic acid, the protocatechuic acid detected in small intestine digesta might be due to the degradation of protocatechuoyl ester [43-45]. In addition, the third observation was that the polyphenols' concentrations increased after digestion, this may be due to they were metabolites originated from other polyphenols. To support this, p-coumaric acid is reported to be attached to the cell wall covalently, and it is necessary to break down the ester linkages between polymers to release p-coumaric acid from the cell wall. Feruloyl esterases are reported to be the important enzymes in this reaction [46,47].

Comparing the effect of probiotics in the treatments, concentrations of polyphenols demonstrated a higher value among treatments without probiotics added than those with probiotics added. For example, by comparing the concentration of p-coumaric acid in RM, RP and RPE, it was found that p-coumaric acid in RM was significantly higher. According to a previous study, in controlled fermentation, the presence of lactobacillus could lead to the conversion of simple phenolic and the depolymerization of high-molecular-weight phenolic compounds [48]. Therefore, RP, which had the highest lactobacillus amount, had the lowest concentration of p-coumaric acid, which indicated that probiotics could promote the conversion of polyphenols into other forms that were easier to be absorbed.

Comparing the treatments with and without encapsulated probiotics, the concentrations of most of the polyphenols among treatments with encapsulation technique applied was lower than those without encapsulation. Therefore, the probiotic mainly plays a role in the treatments without encapsulation in the small intestine.

3.3.2. Colonic fermentation digesta. According to the results obtained in the HPLC analysis of the colonic fermentation digesta, syringic acid, and protocatechuic acid in red grape marc were detected in the large intestine. These polyphenols might be the metabolites originated from other phenolic compounds. A previous study showed that cyanidin-3-glucoside could undergo a significant degradation during digestion [49]. One of its major phenolic acid degradation products is protocatechuic acid.

Although cyanidin-3-glucoside was not analysed in this experiment, they have been found in grapes as an anthocyanin pigment [50]. Therefore, it is assumed that protocatechuic acid and ferulic acid were derived from cyanidin-3-glucoside in grape.

Comparing the treatments with and without probiotics, in majority treatments without probiotics added showed a significantly lower value than those with probiotics added. For example, the concentration of syringic acid among RMn, RPn and RPEn were 4.59 ± 0.01, 5.07 ± 1.02 and 7.59 ± 1.93 ug/g sample DM respectively. This might because of the release of encapsulated probiotics in the large intestine and therefore increased the production of syringic acid, as syringic acid is a product of microbial metabolism of anthocyanins and other polyphenols [51].

3.3.3. Small Intestinal digesta VS Colonic fermentation digesta. Comparing the difference of polyphenols' concentrations between small intestinal digesta and colonic fermentation digesta, most of the polyphenols existed in colonic fermentation digesta exhibited increased concentrations. Take protocatechuic acid as an example, it showed an increase in RP and RPE in the large intestine, while it was not identified in RM after colonic fermentation. This phenomenon could be explained by the metabolism of protocatechuic acid from catechins, protocatechuoyl ester or cyanidin-glucosides by the action of lactobacillus. According to the results shown in Table 3, the concentration of protocatechuic acid increased in the large intestine, while catechin existed in grape marc samples but could not be identified in digesta. Therefore, catechin is suggested to be metabolized under the effect of L. plantarum with gut microbiota to produce protocatechuic acid.

Table 3. Polyphenols found in the grape marc, small and large intestinal digesta.

Polyphenols	**Red grape (ug/g sample DM)**							**White grape (ug/g sample DM)**						
	Red grape marc	**RM**	**RP**	**RPE**	**RMn**	**RPn**	**RPEn**	**White grape marc**	**WM**	**WP**	**WPE**	**WMn**	**WPn**	**WPEn**
								Small intestinal digesta						
***p*-hydroxybenzoic acid**	13.64 ± 0.53	—	—	—				25.48 ± 0.77	—	—	—			
Syringic acid	5.32 ± 0.29	53.99 ± 2.06 ^{a}	49.01 ± 7.92 ^{a}	45.58 ± 0.64 ^{a}				15.38 ± 0.18	13.82 ± 1.33 ^{A}	12.98 ± 0.67 ^{AB}	11.39 ± 1.03 ^{B}			
Ferulic acid	—	—	—	—				0.14 ± 0.02	—	—	—			
Protocatechuic acid	—	23.10 ± 0.11 ^{a}	16.65 ± 3.65 ^{ab}	19.23 ± 3.58 ^{ab}				—	—	—	—			
***p*-coumaric acid**	0.49 ± 0.01	4.92 ± 0.34 ^{a}	3.93 ± 0.52 ^{b}	4.30 ± 0.40 ^{ab}				—	—	—	—			

Procyanidin B2	14.78 ± 8.00	—	—	—				8.79 ± 1.98	—	—	—			
Catechin	29.85 ± 2.46	—	—	—				49.64 ± 0.07	—	—	—			
Epicatechin	18.80 ± 1.71	—	—	—				69.79 ± 0.91	—	—	—			
Resveratrol	1.10 ± 0.11	—	—	—				1.09 ± 0.16	—	—	—			
Kaempferol	2.10 ± 0.06	—	—	—				0.34 ± 0.00	—	—	—			
					Colonic fermentation digesta									
syringic acid	5.32 ± 0.29	14.02 ± 1.72 a	22.30 ± 5.12 a	—	4.59 ± 0.01 b	5.07 ± 1.02 b	7.59 ± 1.93 a	7.59 ± 1.93 a	—	—	—	—	—	—
protocatechuic acid	—	—	21.78 ± 6.88 a	26.03 ± 6.07 a	—	—	—	—	—	—	—	—	—	—

4. Conclusion

According to the microbial count results and the physical situation of the digested samples, this study demonstrated that extrusion technique using sodium alginate and chloride solution could encapsulate freeze-dried *L. plantarum* powder in small microcapsules and assisted in the delivery through digestive tract until their release in the large intestine. The HPLC analysis showed that 10 out of 23 common polyphenols compounds were abundant in the grape marc samples and the supernatant of intestinal digested samples. With the existence of *L. plantarum*, some complex phenolic compounds were metabolised to the simpler structure, and this effect was enhanced among the treatments with encapsulation technique applied during the colonic fermentation stage. Last but not least, the winery by-product grape marc showed a pharmaceutical potential due to its content of anti-oxidative phenolic compounds, which could benefit the winery industry in regard to waste utilisation.

References

[1] Schieber A 2017 Side streams of plant food processing as a source of valuable compounds: Selected examples *Annual Review Food Sci. Technol.* **8** 97-112

[2] Laufenberg G, Kunz B and Nystroem M 2003 Transformation of vegetable waste into value added products::(A) the upgrading concept;(B) practical implementations *Bioresour. Technol.* **87**(2) 167-98

[3] Torres J L, Varela B, García M T, Carilla J, Matito C, Centelles J J, Cascante M, Sort X and Bobet R 2002 Valorization of grape (Vitis vinifera) byproducts. Antioxidant and biological properties of polyphenolic fractions differing in procyanidin composition and flavonol content *J. Agricultur. Food Chem.* **50**(26) 7548-55

[4] Fontana A R, Antoniolli A and Bottini Rn 2013 Grape pomace as a sustainable source of bioactive

compounds: extraction, characterization, and biotechnological applications of phenolics *J. Agricultur. Food Chem* **61**(38) 8987-9003
[5] Zhu F, Du B, Zheng L and Li J 2015 Advance on the bioactivity and potential applications of dietary fibre from grape pomace *Food Chem.* **186** 207-12
[6] Jennings A, Welch A A, Fairweather-Tait S J, Kay C, Minihane A M, Chowienczyk P, Jiang B, Cecelja M, Spector T and Macgregor A 2012 Higher anthocyanin intake is associated with lower arterial stiffness and central blood pressure in women *Americ. J. Clinic. Nutrit.* **96**(4)781-8
[7] Zamora-Ros R, Agudo A, Luján-Barroso L, Romieu I, Ferrari P, Knaze V, Bueno-de-Mesquita H B, Leenders M, Travis R C and Navarro C 2012 Dietary flavonoid and lignan intake and gastric adenocarcinoma risk in the European Prospective Investigation into Cancer and Nutrition (EPIC) study *Americ. J. Clinic. Nutrit.* **96**(6) 1398-408
[8] Manach C, Williamson G, Morand C, Scalbert A and Rémésy C 2005 Bioavailability and bioefficacy of polyphenols in humans. I. Review of 97 bioavailability studies *Americ. J. Clinic. Nutrit.* **81**(1) 230S-42S
[9] Bohn T 2014 Dietary factors affecting polyphenol bioavailability. *Nutrit. Reviews* **72**(7) 429-452
[10] Allan-Wojtas P, Hansen L T and Paulson A T 2008 Microstructural studies of probiotic bacteria-loaded alginate microcapsules using standard electron microscopy techniques and anhydrous fixation *LWT-Food Sci. Technol.* **41**(1) 101-8
[11] Aura A M, Martin-Lopez P, O'Leary K A, Williamson G, Oksman-Caldentey K M, Poutanen K and Santos-Buelga C 2005 *In vitro* metabolism of anthocyanins by human gut microflora *European J. Nutrit.* **44**(3) 133-42
[12] Burgain J, Gaiani C, Linder M and Scher J 2011 Encapsulation of probiotic living cells: From laboratory scale to industrial applications *J. Food Engin.* **104**(4) 467-83
[13] Shori A B 2017 Microencapsulation improved probiotics survival during gastric transit *HAYATI J. Biosci.* **24**(1) 1-5
[14] De Vries M C, Vaughan E E, Kleerebezem M and de Vos W M 2006 Lactobacillus plantarum-survival, functional and potential probiotic properties in the human intestinal tract *Internat. Dairy J.* **16**(9) 1018-28
[15] Reid G, Kim S O and Köhler G A 2006 Selecting, testing and understanding probiotic microorganisms *FEMS Immunol. Medic. Microbiol.* **46**(2) 149-57
[16] Prakash S and Jones M L 2005 Artificial cell therapy: New strategies for the therapeutic delivery of live bacteria *BioMed Res. Internat.* **2005**(1) 44-56
[17] Kechagia M, Basoulis D, Konstantopoulou S, Dimitriadi D, Gyftopoulou K, Skarmoutsou N and Fakiri E M 2013 Health benefits of probiotics: a review *ISRN Nutrit.* **2013**
[18] Chávarri M, Marañón I and Villarán M C 2012 Encapsulation technology to protect probiotic bacteria *Probiotics InTech*
[19] Prakash S and Jones M L 2005 Artificial cell therapy: New strategies for the therapeutic delivery of live bacteria *BioMed Res. Internat.* **2005**(1) 44-56
[20] Simpson N E, Stabler C L, Simpson C P, Sambanis A and Constantinidis I 2004 The role of the CaCl2–guluronic acid interaction on alginate encapsulated βTC3 cells *Biomaterials* **25**(13) 2603-10
[21] Sultana K, Godward G, Reynolds N, Arumugaswamy R, Peiris P and Kailasapathy K 2000 Encapsulation of probiotic bacteria with alginate–starch and evaluation of survival in simulated gastrointestinal conditions and in yoghurt *Internat. J. Food Microbiol.* **62**(1-2) 47-55
[22] Annan N, Borza A and Hansen L T 2008 Encapsulation in alginate-coated gelatin microspheres improves survival of the probiotic Bifidobacterium adolescentis 15703T during exposure to simulated gastro-intestinal conditions *Food Res. Internat.* **41**(2) 184-93
[23] Krasaekoopt W, Bhandari B and Deeth H 2003 Evaluation of encapsulation techniques of probiotics for yoghurt *Internat. Dairy J.* **13**(1) 3-13

[24] Leong J Y, Lam W H, Ho K W, Voo W P, Lee M F X, Lim H P and Chan E S 2016 Advances in fabricating spherical alginate hydrogels with controlled particle designs by ionotropic gelation as encapsulation systems *Particuology* **24** 44-60

[25] Kammerer D, Claus A, Carle R and Schieber A 2004 Polyphenol screening of pomace from red and white grape varieties (Vitis vinifera L.) by HPLC-DAD-MS/MS *J. Agricult. Food Chem.* **52**(14) 4360-7

[26] Singleton V L and Rossi J A 1965 Colorimetry of total phenolics with phosphomolybdic-phosphotungstic acid reagents *Americ. J. Enol. Viticult.* **16**(3) 144-58

[27] Benzie I F and Strain J J 1996 The ferric reducing ability of plasma (FRAP) as a measure of "antioxidant power": the FRAP assay *Anal Biochem* **239**(1) 70-6

[28] Bondet V, BrandWilliams W and Berset C 1997 Kinetics and mechanisms of antioxidant activity using the DPPH free radical method *Food Sci. Technol.-Lebensmittel-Wissenschaft Technologie* **30**(7) 772

[29] Wu G, Johnson S K, Bornman J F, Bennett S J and Fang Z 2017 Changes in whole grain polyphenols and antioxidant activity of six sorghum genotypes under different irrigation treatments *Food Chem.* **214** 199-207

[30] Jalali M, Abedi D, Varshosaz J, Najjarzadeh M, Mirlohi M and Tavakoli N 2012 Stability evaluation of freeze-dried Lactobacillus paracasei subsp. tolerance and Lactobacillus delbrueckii subsp. bulgaricus in oral capsules *Res. Pharmaceut. Sci.* **7**(1) 31

[31] Saarela M, Virkajärvi I, Alakomi H L, Mattila-Sandholm T, Vaari A, Suomalainen T and Mättö J 2005 Influence of fermentation time, cryoprotectant and neutralization of cell concentrate on freeze-drying survival, storage stability, and acid and bile exposure of Bifidobacterium animalis ssp. lactis cells produced without milk-based ingredients *J. Applied Microbiol.* **99**(6) 1330-9

[32] Minekus M, Alminger M, Alvito P, Ballance S, Bohn T, Bourlieu C, Carriere F, Boutrou R, Corredig M and Dupont D 2014 A standardised static *in vitro* digestion method suitable for food–an international consensus *Food Funct.* **5**(6) 1113-24

[33] Sirisena S, Ajlouni S and Ng K 2018 Simulated gastrointestinal digestion and *in vitro* colonic fermentation of date (Phoenix dactylifera L.) seed polyphenols *Internat. J. Food Sci. Technol.* **53**(2) 412-22

[34] Herigstad B, Hamilton M and Heersink J 2001 How to optimize the drop plate method for enumerating bacteria *J. Microbiol. Methods* **44**(2) 121-9

[35] Kammerer D, Claus A, Carle R and Schieber A 2004 Polyphenol screening of pomace from red and white grape varieties (Vitis vinifera L.) by HPLC-DAD-MS/MS *J. Agricult. Food Chem.* **52**(14) 4360-7

[36] Rocchetti G, Chiodelli G, Giuberti G and Lucini L 2018 Bioaccessibility of phenolic compounds following *in vitro* large intestine fermentation of nuts for human consumption *Food Chem.* **245** 633-40

[37] Rockenbach I I, Rodrigues E, Gonzaga L V, Caliari V, Genovese M I, Gonçalves A E d S S and Fett R 2011 Phenolic compounds content and antioxidant activity in pomace from selected red grapes (Vitis vinifera L. and Vitis labrusca L.) widely produced in Brazil *Food Chem.* **127**(1) 174-179

[38] Rockenbach I I, Gonzaga L V, Rizelio V M, Gonçalves A E d S S, Genovese M I and Fett R 2011 Phenolic compounds and antioxidant activity of seed and skin extracts of red grape (Vitis vinifera and Vitis labrusca) pomace from Brazilian winemaking *Food Res. Internat.* **44**(4) 897-901

[39] Krasaekoopt, Bhandari and Deeth 2004 The influence of coating materials on some properties of alginate beads and survivability of microencapsulated probiotic bacteria *Internat. Dairy J.* **14**(8) 737-43

[40] Valero-Cases E and Frutos M J 2015 Effect of different types of encapsulation on the survival of Lactobacillus plantarum during storage with inulin and *in vitro* digestion *LWT-Food Sci.*

Technol. **64**(2) 824-8

[41] Shi L E, Li Z H, Li D T, Xu, Chen H Y, Zhang Z L and Tang Z X 2013 Encapsulation of probiotic Lactobacillus bulgaricus in alginate–milk microspheres and evaluation of the survival in simulated gastrointestinal conditions *J. Food Engin.* **117**(1) 99-104

[42] Tabasco R, Sánchez-Patán F, Monagas M, Bartolomé B, Moreno-Arribas M V, Peláez C and Requena T 2011 Effect of grape polyphenols on lactic acid bacteria and bifidobacteria growth: resistance and metabolism *Food Microbiol.* **28**(7) 1345-1352

[43] Ordoudi S A and Tsimidou M Z 2006 Crocin Bleaching Assay (CBA) in Structure– radical scavenging activity studies of selected phenolic compounds *J. Agricultur. Food Chemi.* **54**(25) 9347-56

[44] Pietta P G, Simonetti P, Gardana C, Brusamolino A, Morazzoni P and Bombardelli E 1998 Catechin metabolites after intake of green tea infusions *Biofactors* **8**(1,2) 111-8

[45] Vitaglione P, Donnarumma G, Napolitano A, Galvano F, Gallo A, Scalfi L and Fogliano V 2007 Protocatechuic acid is the major human metabolite of cyanidin-glucosides *J. Nutrit.* **137**(9) 2043-8

[46] Esteban-Torres M, Reverón I, Mancheño J M, de las Rivas B and Muñoz R 2013 Characterization of the first feruloyl esterase from Lactobacillus plantarum *Appl. Environm. Microbiol.* **79**(17) 5130-6

[47] Muñoz R, de las Rivas B, de Felipe F L, Reverón I, Santamaría L, Esteban-Torres M, Curiel J, Rodríguez H and Landete J 2016 Biotransformation of phenolics by Lactobacillus plantarum in fermented foods *Fermented foods in health and disease prevention, Elsevier* 63-83

[48] Othman N B, Roblain D, Chammen N, Thonart P and Hamdi M 2009 Antioxidant phenolic compounds loss during the fermentation of Chétoui olives *Food Chem.* **116**(3) 662-9

[49] Kay C D, Kroon P A and Cassidy A 2009 The bioactivity of dietary anthocyanins is likely to be mediated by their degradation products *Molecular Nutrit. Food Resear.* **53**(S1) S92-S101

[50] Kim H B 2003 Quantification of cyanidin-3-glucoside (C3G) in mulberry fruits and grapes *Korean J. Sericultur. Sci.* **0440-2332**

[51] Bogert M T and Ehrlich J 1919 The synthesis of certain substituted pyrogallol ethers, including a new acetophenetide derived from the ethyl ether of syringic acid *J. American Chem. Soci.* **41**(5) 798-810

Experimental study on mechanical properties of reed stalk

K P Tian[1], C Shen[1,2,3], B Zhang[1,2], X W Li[1], J C Huang[1] and Q M Chen[1]

[1]Nanjing Research Institute for Agricultural Mechanization, Ministry of Agriculture and Rural Affairs, Nanjing 210014, China
[2]School of Mechanical Engineering, Southeast University, Nanjing 211189, China

E-mail: shencheng1989@cau.edu.cn

Abstract. Reed is a kind of important natural resources. The inherent mechanical properties of reed stalk are the basis for development of its production equipment. As the research on the basic mechanical properties of reed stalk is still in blank both at home and abroad, this paper makes the compression, tensile, and bending experiments respectively on the bottom reed stalk in the harvest time with an electronic universal testing machine, obtains the corresponding stress-strain curves under the compression, tensile, and bending loads, and confirms that the mean compression strength is 2.69 MPa, the mean compression elasticity modulus is 75.52 MPa, the mean tensile strength is 16.45 MPa, the tensile elasticity modulus is 433.19 MPa, the mean bending strength is 2.15 MPa, and the mean fracture deflection is 5.54 mm. The experimental results may provide a reference for the development and parameter optimization of the reed production equipment.

1. Introduction

Reed is a perennial gramineous plant with strong adaptability and resistance, as an important kind of natural resources [1,2] that spreads around the world. Due to the tough material and high fiber content, it is not only a high-quality paper making raw material but also a superior environmental protection material widely used in the paper making industry, the construction industry, and other industries [3,4]. In China where the forest resources are relatively rare, reed has been preferred to wood and plays a vital role in the national economy. It is hailed as the "Second Forest" [5].

As reed is tall and its bottom stalk is hard, its harvest, mostly dependent on the traditional manual harvest due to poor machinery, is a major challenge in the production process all the time. Compared to manual harvest which, high in labor intensity and cost but low in efficiency, goes against the reed industrialization development, it is significant to enforce mechanized harvest which can raise labor productivity, lower production cost, guarantee the material supply of the paper making industry, and improve the utilization ratio of such a resource.

By conducting an experimental study on the mechanical properties of reed stalk and acquiring the parameters such as compression strength, tensile strength, bending strength, and elasticity modulus, we may provide not only a theoretical reference and basic technical parameters for the selection of the cutting power, cutter type, and cutting ways of the reed harvester and the subsequent procedures such as packaging and smashing but also a basis for the structural design of the reed production equipment. But through literature search, it is found that the research at home and abroad on the reed production equipment simply focuses on the structural design of the major components of reed harvester [1,6-9], and no relevant research on the mechanical properties of reed stalk has been reported. Therefore, in order to provide a reference for development and parameter optimization of the reed production

Content from this work may be used under the terms of the Creative Commons Attribution 3.0 licence. Any further distribution of this work must maintain attribution to the author(s) and the title of the work, journal citation and DOI.
Published under licence by IOP Publishing Ltd

equipment, this paper intends to carry out an experimental study on the mechanical properties of reed stalk and gain the parameters of the bottom reed stalk in the harvest time.

2. Experimental materials, equipment, and methods

2.1. Experimental materials

2.1.1. Material collection. We collected the experimental materials in the wild reed lands located at High and New Tech Development Zone, Wuxi on November 3, 2017. See figure 1 for the reeds. They were all 3~5 m high and 12~16 mm wide in bottom diameter. We selected those straight from top to bottom and about 14mm wide in bottom diameter without pests and damages randomly as the experimental materials. We cut off the reeds at the bottom, making sure their remaining height was 200 mm, sealed them in plastic bags, and finished the experiments within 72 hours.

Figure 1. Reeds used for experimental materials.

2.1.2. Specimen preparation. As this paper intends to study the compression, tensile and bending properties of reed stalk in the harvest time, we prepared the compression, tensile, and bending specimens respectively.

See figure 2(a) for the compression specimen. D_1, k, and h were the external diameter, wall thickness, and height of the specimen. D_1 and k were the actual sizes, while h was set at 30 mm (±2 mm) uniformly.

We prepared the tensile specimens by regarding the preparing methods of the specimens of bamboo wood for tensile property as the standards. See figure 2(b) for the tensile specimen. Hereinto: l_1, n, q, r, and t were the length, active length, width of the effective segment, transition radius, and thickness of the specimen. We set l_1 at 70 mm, n at 18 mm, q at 2 mm, and r at 3 mm, while t was the actual thickness of the specimen.

See figure 2(c) for the bending specimen. D_2, c, and l_2 were the external diameter, wall thickness, and the length of the specimen. D_2 and c were the actual sizes, while l_2 was set at 180 mm (±15 mm) uniformly.

The finished experimental specimens were shown in figure 3.

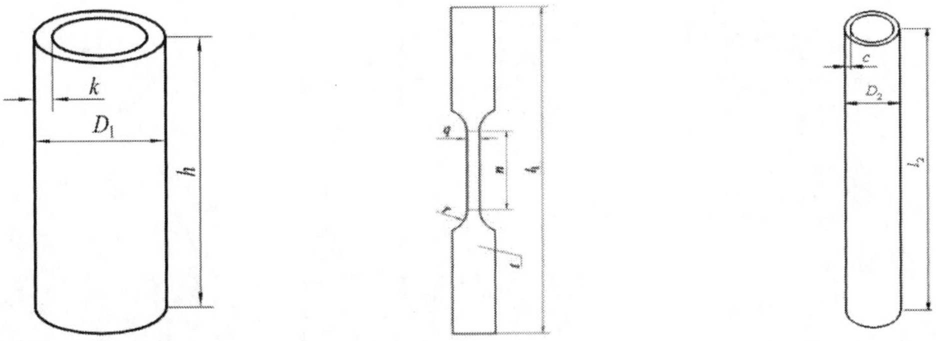

(a) (b) (c)

Figure 2. Structures of experimental specimens. (a) Compression specimen, (b) tensile specimen and (c) bending specimen.

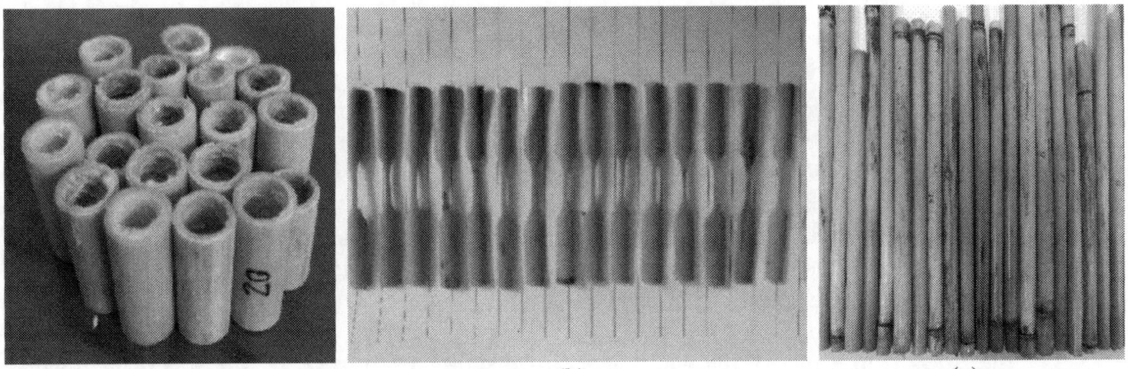

(a) (b) (c)

Figure 3. Preparations of experimental specimens. (a) Compression specimens, (b) tensile specimens and (c) bending specimens.

2.2. Experimental equipment

We conducted the experiments basically with the WDW-10 electronic universal testing machine (with the fixtures suitable for the compression, tensile, and bending experiments) manufactured by Jinan Chuanbai Instrument & Equipment Co., Ltd. The machine has 5 KN strength, level-1 precision, 5% force accuracy, and 0.1% displacement accuracy. We also used other devices in the experiments, such as moisture content meters, art knives, and vernier calipers.

2.3. Experimental methods

We chose the experimental methods by reference to that of the physical and mechanical properties of bamboo wood [10].

During the experiments, we firstly designed an experimental scheme with the machine, measured the size parameters of the specimens with vernier calipers, recorded them into the system, and then fixed the specimens respectively with the compression, tensile, and 3-point bending fixtures. Setting the pre-tightening force for system start-up at <1 N, the loading speed at 50 mm/min, We reset the pre-tightening force and pushed the start button, after which the system software of the machine automatically made a record of the experimental data, drew the stress-strain curves, and eliminated the invalid experiments to make sure the number of the valid experiments of each kind was 20.

See figures 4-6 for the initial clamping positions of the specimens in the compression, tensile, and bending tests.

Figure 4. Compression experiment.

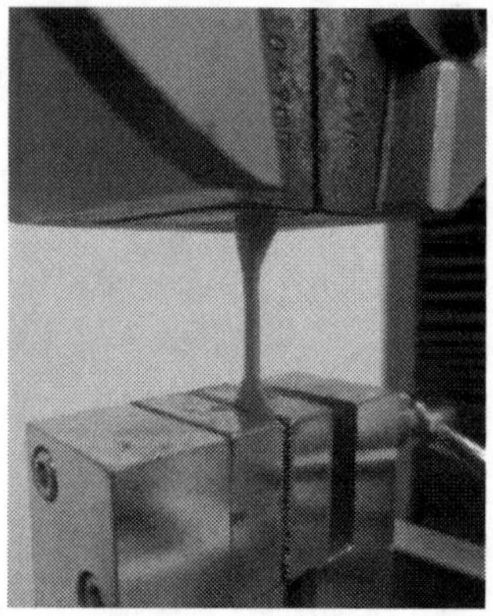

Figure 5. Tensile experiment.

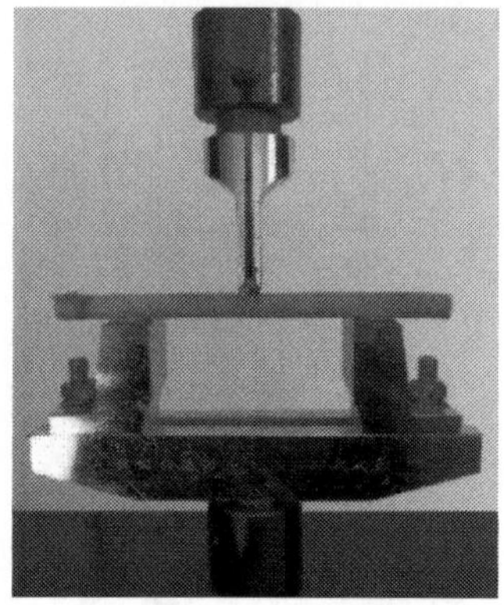

Figure 6. Bending experiment.

3. Results and analysis

3.1. Results of compression experiment and analysis

See figure 7 for the stress-strain curve of reed stalk in the compression experiment. See table 1 for the data of the compression experiment.

We can see from figure 7 that the compression stress-strain curve of reed stalk can be divided into two stages: straight climb of the stress and fluctuant decline of the stress. In the initial phase of the compression process when the strain rate maintains within 6%, the stress goes up straightly as the strain increases, and the reed stalks show the property of elasticity deformation, which conforms to Hooke's law. At the same time, the internal organizational structures of the reed stalks are compressed evenly, making the loose texture firm.

When the stress is maximized, the walls of the reed stalks begin to crack, the specimens begin to sink, and then the stress goes down sharply with fluctuation. After the stress reaches the strength set by the machine, the experiment stops automatically.

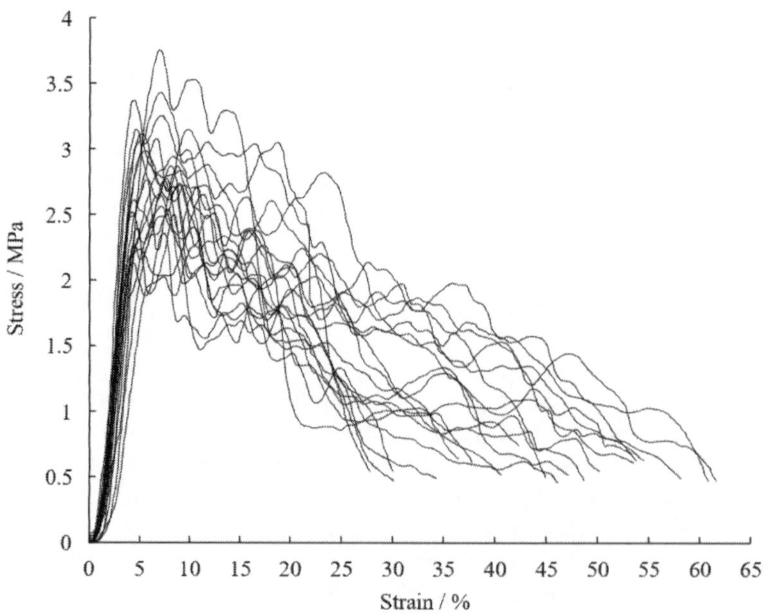

Figure 7. Compression stress-strain curve.

Table 1. Data of compression experiment.

No.	External Diameter /mm	Wall Thickness /mm	Height /mm	Max Compression Force /N	Compression Strength /MPa	Elasticity Modulus /MPa
1	11.90	1.50	29.05	135.28	2.76	83.95
2	12.32	1.74	31.70	146.68	2.54	86.99
3	13.83	2.20	31.29	190.78	2.37	68.43
4	11.98	1.32	44.21	122.54	2.77	67.16
5	13.58	2.07	29.51	176.22	2.35	77.57
6	12.58	1.47	30.94	161.56	3.15	85.27
7	12.39	2.01	29.63	194.72	2.97	84.36
8	13.70	2.12	30.00	174.20	2.26	62.60
9	12.73	1.76	29.30	197.28	3.25	99.01
10	13.38	2.14	30.89	185.96	2.46	69.29
11	12.74	1.85	28.14	194.68	3.08	85.85
12	13.28	1.96	28.22	211.84	3.04	63.37
13	12.66	1.92	29.92	177.08	2.73	44.73
23	11.90	1.50	29.05	135.28	2.76	83.95
15	12.15	1.50	30.68	120.44	2.40	65.81
16	12.26	1.66	30.10	152.44	2.76	97.17
17	12.86	1.91	31.04	146.30	2.23	76.18
18	12.07	1.62	28.44	141.04	2.65	71.10
19	12.31	1.43	30.27	139.44	2.85	81.23
20	12.35	1.43	30.94	122.38	2.49	56.32
Mean	12.65	1.76	30.67	161.31	2.69	75.52
Standard Deviation	0.61	0.28	3.35	28.69	0.30	13.60

We can see from the data above that the mean external diameter is 12.65 mm, the mean wall thickness is 1.76 mm, the mean height is 30.67 mm, the mean max compression force is 161.31 N, the means max compression strength is 2.69 MPa, and the mean elasticity modulus is 75.52 MPa.

3.2. Results of tensile experiment and analysis

See figure 8 for the stress-strain curve of reed stalk in the tensile experiment. See table 2 for the data of the tensile experiment.

Table 2. Data of tensile experiment.

No.	Width /mm	Thickness /mm	Parallel Length /mm	Tensile Force /N	Tensile Strength /MPa	Elasticity Modulus /MPa
1	2.25	1.51	15.97	45.50	13.39	297.14
2	1.98	1.34	16.01	44.60	16.81	391.47
3	2.66	1.43	22.45	41.90	11.02	343.63
4	2.10	1.35	20.4	36.70	12.95	541.87
5	2.80	1.32	17.54	50.70	13.72	279.25
6	2.54	1.56	15.84	44.62	11.26	576.73
7	1.76	1.61	19.43	57.38	20.25	442.83
8	2.10	1.57	17.00	60.98	18.50	424.33
9	2.37	1.50	18.06	43.88	12.34	592.93
10	1.82	1.24	17.87	48.26	21.38	284.69
11	2.67	1.78	17.30	79.04	16.63	471.71
12	2.45	1.45	18.08	62.38	17.56	435.92
13	1.71	1.58	14.07	46.80	17.32	440.59
14	1.96	1.08	15.00	56.12	26.51	433.56
15	2.38	1.62	20.02	48.18	12.50	545.59

16	2.07	1.19	18.47	59.06	23.98	471.56
17	2.39	1.71	19.36	48.50	11.87	567.76
18	2.54	1.53	20.87	44.58	11.47	509.63
19	1.97	1.26	17.31	42.30	17.04	299.34
20	1.72	0.88	15.33	30.01	19.38	313.27
Mean	2.21	1.43	17.82	49.87	16.45	433.19
Standard Deviation	0.34	0.20	2.15	10.14	4.51	103.63

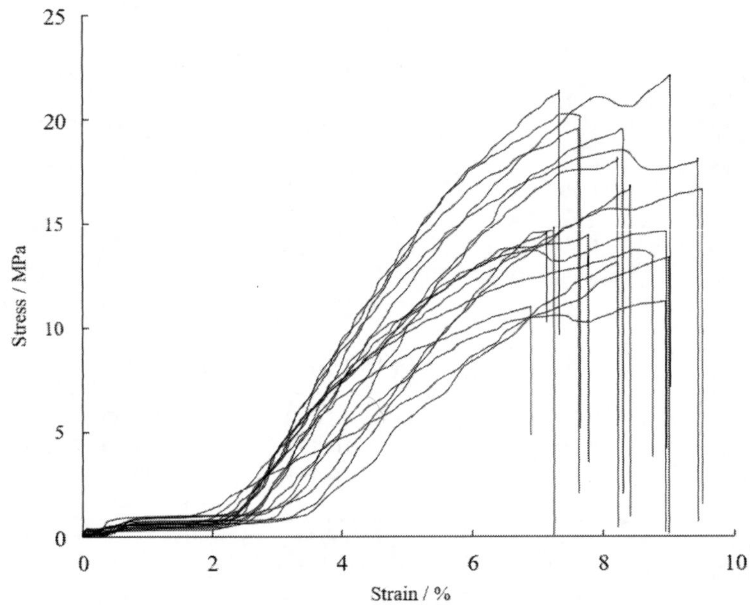

Figure 8. Tensile stress-strain curve.

We can see from figure 8 that the tensile stress-strain curve of reed stalk can be divided into three stages: constant stress, steady climb of the stress, and sudden drop of the stress. In the first state where the strain rate maintains 0~2%, the stress remains unchanged as the strain increases. After analyzing the experimental process, we can see that it is caused by the small slips of the chunk of the machine. In the second stage, the stress rises like a semi-parabola as the strain increases, and the organizational structures of the specimens show plastic deformation under the tensile load. In the third stage where the stress is maximized, the specimens are snapped, and the stress falls to zero suddenly.

We can see from the data above that the mean tensile force is 49.87 MPa, the mean tensile strength is 16.45 MPa, and the mean elasticity modulus is 433.19 MPa.

3.3. *Results of bending experiment and analysis*

See figure 9 for the stress-strain curve of reed stalk in the three-point bending experiment. See table 3 for the data of the bending experiment.

We can see from figure 9 that in the initial state, the stress-strain curve is nearly linear. After analyzing the experimental process, we can see that the reed stalks show elastic deformation under the bending load but no bending failure.

When the yield stress occurs (i.e. the strain rate is around 7%), the stress goes down quickly as the strain increases, and the reed stalks are snapped under the bending load. After that, the stress goes down quickly as the strain increases and tends to be constant when the strain is about 20%.

We can see from the data above that the mean max force is 17.92 N, the mean bending strength is 2.15 MPa, and the mean fracture deflection is 433.19 MPa.

IOP Conf. Series: Earth and Environmental Science **346** (2019) 012076 doi:10.1088/1755-1315/346/1/012076

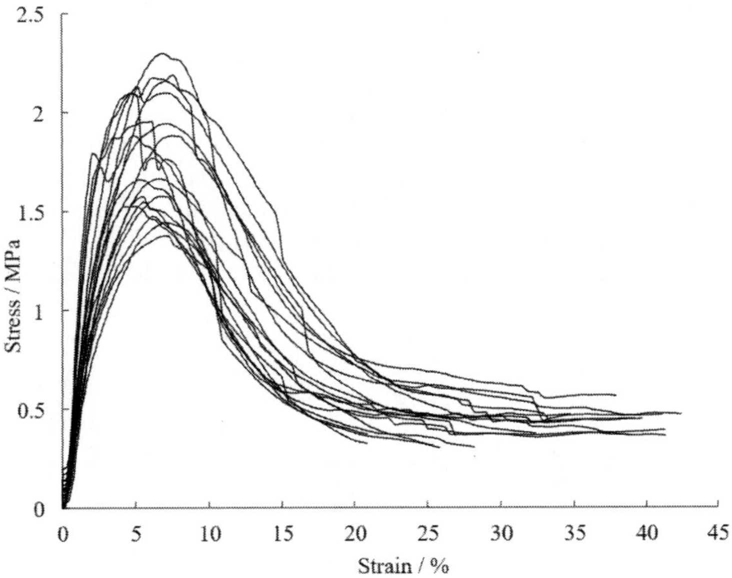

Figure 9. Bending stress-strain curve.

Table 3. Data of bending experiment.

No.	External Diameter /mm	Wall Thickness /mm	Span /mm	Max Force /N	Bending Strength /MPa	Fracture Deflection /mm
1	12.48	1.64	90	16.54	2.37	5.12
2	12.17	1.94	90	20.40	2.62	5.50
3	14.37	1.8	90	24.00	2.70	6.25
4	14.15	1.54	90	17.38	2.28	6.10
5	12.41	1.61	90	15.94	2.34	4.43
6	12.08	1.57	90	12.38	1.91	4.80
7	13.03	1.21	90	11.00	1.96	5.67
8	12.74	2.09	90	22.72	2.60	5.64
9	13.39	1.86	90	16.16	1.92	5.09
10	13.60	2.04	90	14.38	1.55	6.40
11	12.93	1.72	90	19.66	2.60	4.45
12	14.69	2.04	90	19.66	1.94	6.81
13	13.47	1.69	90	16.44	2.10	4.93
14	13.81	2.11	90	13.70	1.41	6.41
15	15.77	1.76	90	25.42	2.63	6.70
16	14.86	2.14	90	20.30	1.90	6.47
17	13.91	2.25	90	16.46	1.60	6.28
18	15.32	2.06	90	22.24	2.07	4.65
19	12.39	2.13	90	11.8	1.37	4.53
20	11.83	1.69	90	21.90	3.26	4.17
Mean	13.47	1.84	90	17.92	2.15	5.54
Standard Deviation	1.13	0.27	0	4.16	0.49	0.88

4. Conclusion and discussion

Through the compression, tensile, and bending experiments of the bottom reed stalks in the harvest time, we acquire the corresponding stress-strain curves, and through the deformations of the reed stalks in the experiments, we know the failure form and law of the reed stalks and identify compression strength, compression elasticity modulus, tensile strength, tensile elasticity modulus, bending strength, fracture deflection, and other parameters. The experimental results may provide a reference for development and parameter optimization of the reed production equipment.

This paper conducts the experiments only on the samples in a certain region. As China is vast in territory, the forms and mechanical properties of the reeds in one place and of one type differ largely from that of the reeds in other places and of other types. As this paper does not take the variation into consideration, the experimental results are not so highly representative. In the future studies, it will pay more attention to the variation and establish a database of the mechanical properties of the reeds in different places and of different types, with a view to providing a more comprehensive reference for the reed production equipment.

Acknowledgments
We greatly appreciate the careful and precise reviews by the anonymous reviewers and editors. This work was supported by This research was financially supported by the National Key R & D Program of China (2016YFD0701404) and the Agricultural Science and Technology Innovation Program of Chinese Academy of Agricultural Sciences (ASTIP, CAAS).

References
[1] Tian L, Yuan M Q, Ma Q Y *et al* 2011 Adaptability design of southern higher stalk bulrush harvester *J. Cent. South U. Forestry Tech..* **31(11)** 153-6
[2] Chen M J, Guo W, Qu H L *et al* 2013 Current situation and development proposals of reed harvesting equipment in China *J. Chinese Agric. Mech.* **34(4)** 29~31,41
[3] Ma W L, Cao Z P and Zang Y P 2014 Experimental study on reed used to produce dissolved pulp *Aritf. Fiber* **49(1)** 2-7
[4] Li G J, Ceustermans B F *et al* 2002 The possibilities of Chinese reed fibres as an environmentally sound organic substrate *Acta Agric. Zhejiangensis* **14(2)** 27-34
[5] Zhao Y Q, He T H, Xia G J *et al* 2015 Growth and photosynthetic characteristics of phragmite saustralis under different water control conditions *J. Northwest Forestry U.* **30(1)** 69-74
[6] Su X M and Ji T H 2017 Designing and development of reed harvester in wetland *China Water Power Electrification* **2017(4)** 65-7
[7] Jin J 2014 Design and study of key components of harvesting ship of bulrush (Baoding, Hebei, China: Agriculture University of Hebei)
[8] Ma Q Y, Tian L, Wu X D *et al* 2010 Research on structure of cutting platform of the 4WS2000 high bar bulrush harvester *J. Cent. South U. Forestry Tech.* **33(10)** 138-42
[9] Shu C X, Liao Q X, Min G Z *et al* 2002 Development of 4WG 1.5 high-stalk reed harvester *Transactions Chinese Soc. Agric. Mach.* **33(5)** 129-30
[10] GB/T 15780-1995 *Test Method for Physical and Mechanical Properties of Bamboo*

IOP Conf. Series: Earth and Environmental Science **346** (2019) 012077 doi:10.1088/1755-1315/346/1/012077

Analysis of spatial-temporal evolution characteristics of abandoned cropland in Yunnan Province based on multi-temporal MODIS global land cover product

W J Xiao[1], J S Wang[1,2,3] and Y C Liu[1]

[1]School of Information Science and Technology, Yunnan Normal University, Kunming, 650500, China

[2]The Engineering Research Center of GIS Technology in Western China, Kunming, 650500, China

E-mail: wjerson@foxmail.com

Abstract. With the outflow of rural labour forces, the abandonment of cropland is widespread, which seriously affects food security. Obtaining the spatial and temporal evolution characteristics of abandoned cropland is particularly important for formulating cultivated land protection policies and the land transfer measures. However, Yunnan province lacks the data sets of space-time distribution of abandoned land. In order to enrich the data of abandoned cropland in Yunnan province and grasp its spatial-temporal evolution characteristics to manage abandoned cropland more conveniently. Based on MODIS global land cover products and by applying the method of change detection, we have already obtained the dataset of the abandoned land in Yunnan province. Also, analysed the evolution characteristics with gravitational center computation. The result shows that the average abandoned cropland is 1284 km^2 every year, among which 2003 witnessed the least abandoned cropland, which was 582.94 km^2 and 2011 witnessed the most abandoned land, which was 1961.21 km^2. It is found that the changes fluctuate obviously. The abandoned cropland is mainly distributed in the eastern plateau area of the study area. And the gravity of abandoned land shifted from Northwest Kunming to the border of Kunming city, Chuxiong and Yuxi city. But it located in Yuxi most years.

1. Introduction

Driven by the development of urbanization and industrialization, a large number of rural labor forces has moved to cities, resulting in the shortage of rural labor forces and the abandonment of cultivated land. The abandoned land directly leads to the decrease of actual cultivated land area and grain yield, which has an important impact on food security. Therefore, mastering the characteristics of spatial and temporal distribution of abandoned land can provide decision-making basis for cultivated land protection and food security. The abandonment of cultivated land is common all over the world, especially in developed countries. China began to experience the abandonment of land in the 1980s. Many domestic and international scholars carried on a great deal of research to this issue.

The phenomenon is the most common in developed countries [1]. The movement of a large number of rural labour force to the city has become the direct cause of the abandoned arable land [2]. Europe is the main research area of idle farmland [1,3]. Such as Lesiv *et al* [4] used the Soviet Union Countries as the study area , blended a variety of land use data based on Bayesian network， For example, the data of abandoned cultivated land in Europe [5], global cultivated land data with 30m resolution [6], MODIS

Content from this work may be used under the terms of the Creative Commons Attribution 3.0 licence. Any further distribution of this work must maintain attribution to the author(s) and the title of the work, journal citation and DOI.
Published under licence by IOP Publishing Ltd

Collection 5 global land cover with 500m resolution [7] and so on. For that, the 10 arc-second resolution cultivated land products and abandoned cultivated land products in 2010 were obtained. More often, MODIS time series data are used for the extraction and analysis of abandoned cultivated land [5,8]. However, it is limited to the extraction of wasteland information in a short time range, and it is impossible to obtain long-term characteristics of abandoned land.

Most studies have been carried out to analyze the driving force of cultivated land abandonment [9,10], but the few studied for the spatial and temporal distribution of abandoned farmland. And these are mainly limited to small areas (counties). For example, Shi *et al* [11], used the existing land use map, data on conversion of farmland to forests and land for forestry etc., obtained the abandoned farmland In result after overlaying the cultivated land layers and removing the map spots of returning farmland to forest and forest engineering in typical counties of Chongqing city based on ArcGIS software. In addition, Yang *et al* [12] used the methods of inter-annual change detection and intra-annual change detection to extract abandoned arable land in Linger County, Hohhot, and Inner Mongolia. By comparison, the change detection method is more reliable than the former layer superposition method and avoids the problem of layer superposition mismatch. Compared with visual interpretation [13-15] assisted interpretation [16,17] unassisted interpretation [15] and other methods based on remote sensing image classification, this method avoids the problem of ground class confusion in the classification process. And it's the workload can be reduced greatly compared with the sampling survey method [5,18,19]. In general, the change detection method is more suitable for the information extraction of long time series of abandoned farmland, and the results are reliable without a lot of work.

Yunnan province is located in the southwest of China, mainly in plateau and mountainous areas, with fragmented cropland and single income source of rural households. It is a province with more migrant workers in China, and its abandonment phenomenon is more common. At present, there is no study on the spatial-temporal distribution of abandoned land in Yunnan province. Therefore, based on the multi-temporal MODIS (Moderate-resolution Imaging Spectroradiometer) land cover products from 2001 to 2016 to analyze the spatial and temporal evolution characteristics of abandoned land in Yunnan province by using change monitoring and gravity center analysis method.

2. Data sources

The data used in this study is the global land cover type data of MODIS with medium resolution (MODIS Collection 6 global land cover) in 2001-2016, which is the longest land cover product in the world at present. It includes 17 international geosphere-biosphere program (IGBP) land cover classes. The data details are shown (see table 1). This comes from the Google Earth Engine platform (https://earthengine.google.com).

Table 1. Basic information of remote sensing data of MODIS global land cover types.

	MODIS global land cover types
spatial resolution	500 m
time range	2001-2016
temporal resolution	1year
17 land cover classes	Needle leaf Forests(2 classes), Broadleaf Forests(2 classes), Mixed Forests(1 class), shrub lands(2 classes), Savannas(2 classes), Grasslands(1 class), Wetlands(1 class), Cropland(2 classes), The Impervious(Urban and Built-up Lands: 1 class), Snow and Ice(1 class), Bare Land(1 class),Water Bodies(1 class)

3. Methods

In this paper, the abandoned land refers to the cultivated land which has not been planted for one year. The transfer direction of cultivated land is mainly woodland, grassland and bare land, hereinafter referred to as transferred land. Therefore, the basic idea of this paper is to extract the cultivated land each year first, and then judge the cultivated land abandoned situation of the next year by using the transferred land type in adjacent years. In MODIS Collection 6 global land cover, the transferred land corresponds to shrub lands, savannas, grasslands and bare land.

The technique route for the study is shown as below (figure 1).

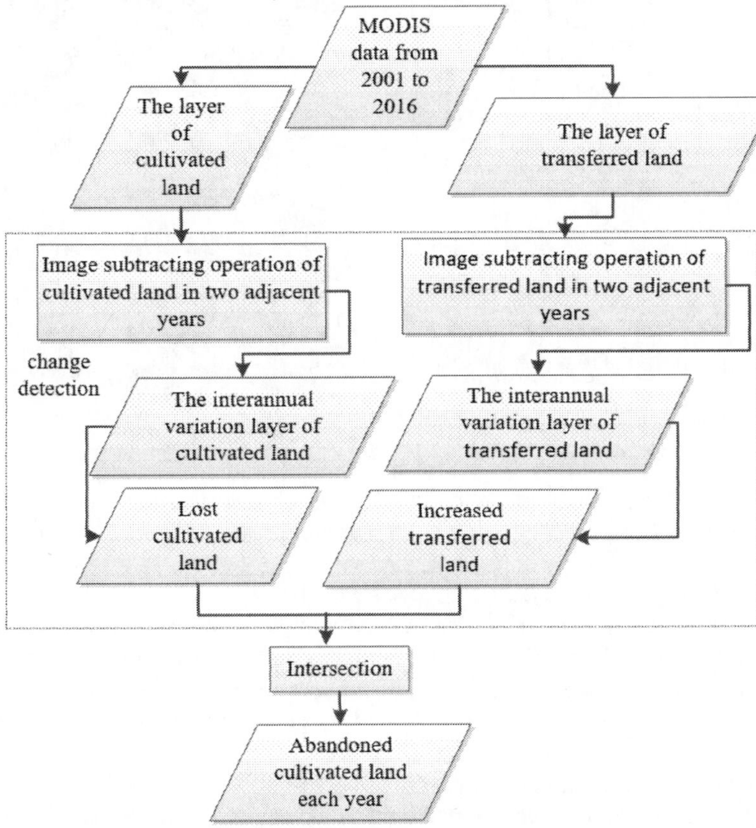

Figure 1. Technique route.

3.1. Extraction of the range of cultivated land and transferred land

MODIS Collection 6 image dataset is composed of 17 refined land cover classes. It can be seen from step 2 that the cultivated land type includes 2 classes. shrublands include 2 classes: closed shrublands and open shrublands. Similarly, depending on density, grasslands also include two classes. In this study, the 17 refined land types were remapped into 3 categories, the 2 classes cultivated land were classified into the first category, the 6 classes that shrublands, savannas, grasslands and bare land were classified into the second category, and the remaining 9 classes land types were classified into the third category. After the remapping and classification of MODIS dataset, the first and second categories were masked respectively, and the cultivated land mask (figure 2) and transferred land (the shrublands, savannas, grasslands and bare) mask (figure 3) were obtained. It can be seen from figure 2 that the cultivated land in the study area is mainly distributed in Qujing city, Zhaotong city, Hani-Yi Autonomous Prefecture of Honghe and Wenshan city in East Yunnan. A small part of the distribution in Dali Bai Autonomous Prefecture, Dehong Autonomous Prefecture and the city of Baoshan and Lincang in West Yunnan. During the study period, the cultivated land in the study area shows a decreasing trend on the whole. But between 2001 and 2004, the cultivated land increased slightly. As can be seen from figure 3, the study area is covered by shrublands, savannas, grasslands and bare land in a large scale and distributed around cultivated land. During the study period from 2001 to 2016, the total area of shrublands, savannas, grasslands and bare land in the study area decreased year by year, which was roughly the same as the trend of cultivated land.

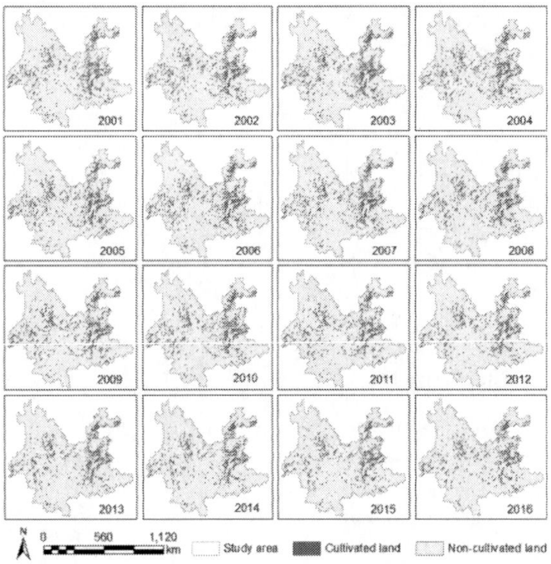

Figure 2. Distribution of cultivated land in Yunnan province from 2001 to 2016.

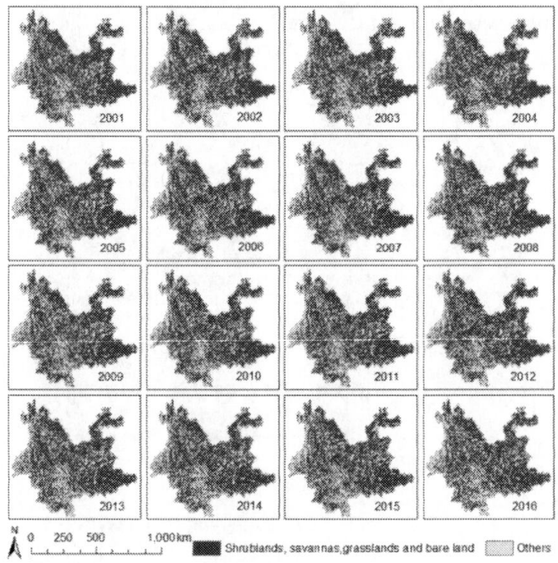

Figure 3. Distribution of transferred land in Yunnan province from 2001 to 2016.

3.2. Extraction of abandoned land based on change detection

Changes in one type of land cover are accompanied by changes in other types. For example, when the first type of land (cultivated land) is transformed into the second type of land (transferred land), the area of the first type of land will decrease, while that of the second type of land will increase in the same region. Therefore, the abandoned land can be obtained by detecting the changes of these two types of land types within the same time interval.

This method is change detection. By the way, the part of cultivated land that flowing to shrublands, savannas, grasslands and bare land as abandoned land. Making use of the two mask images obtained in step 2 (cultivated land and transferred land), the steps for utilizing this method are as shown below:

Step 1: First of all, the interannual change result between the previous and the after were obtained by subtracting calculation. The calculation formula is as in equation (1):

$$Change_{(class,T+1)} = Class_{T+1} - Class_{T} \tag{1}$$

For the formula, $\mathrm{Change}_{(\mathrm{class},\,T+1)}$ is the interannual change result of the masked image in T+1 (time range from T to T+1 year). Class_{T} and Class_{T+1} are the masked images in T and T + 1 respectively. Taking the change in 2016 as an example, the results of interannual variation of two types of masked images of are as follows. The change of cultivated land is shown (see figure 4), while transferred land is shown (figure 5). By comparing the lost cultivated land (dark green part) in figure 4(c) with the newly increase shrublands, savannas, grasslands and bare land (black part) in figure 5(c), it can be concluded that almost all the decreasing cultivated land in Yunnan province in 2016 was converted into shrublands, savannas, grasslands and bare land.

Step 2: Interannual variation results of the two types of masked images obtained in step 1 are calculated to obtain the overlapping part of the images as the abandoned land as in equation (2).

$$AL_{T+1} = Change_{-(class1,T+1)} \cap Change_{+(class2,T+1)} \quad (2)$$

In the formula, AL_{T+1} represents abandoned land in T+1. $Change_{-(class1,T+1)}$ is the loss of the first category of land in T+1. $Change_{+(class2,T+1)}$ is the second category of new parts in T+1. An example of this step acquiring abandoned cultivated land is shown below (figure 6).

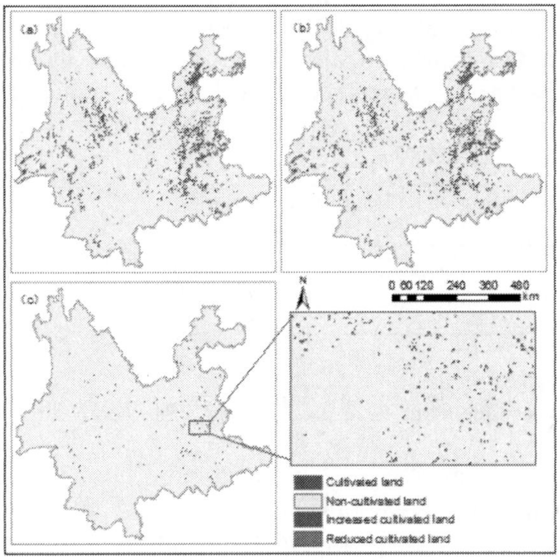

Figure 4. Sample results of change detection step 1: (a) Cultivated land distribution in 2015; (b) Cultivated land distribution in 2016; (c) Interannual change of cultivated land in 2016.

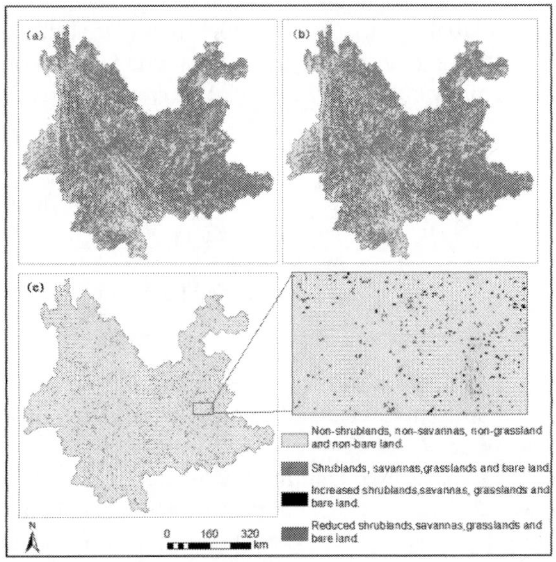

Figure 5. Sample results of change detection step 1: (a) Distribution of transferred land in 2015; (b) Distribution of transferred land in 2016; (c) Interannual change of transferred land in 2016.

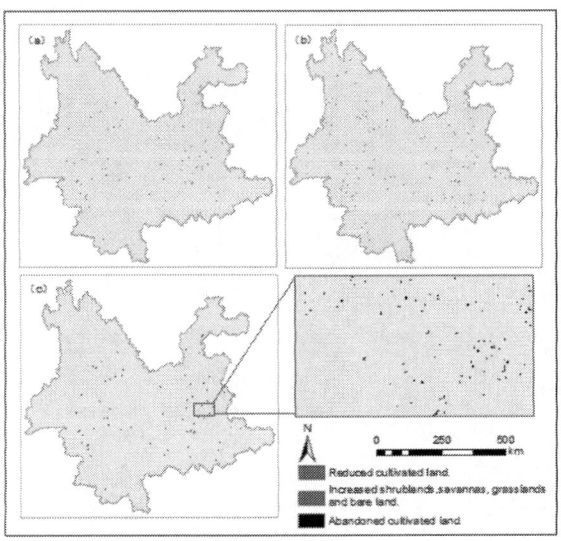

Figure 6. Sample results of change detection step 2: (a) Distribution of lost cultivated land in 2016; (b) Distribution of increased transferred land in 2016; (c) Distribution of abandoned land in 2016.

4. Results and analysis

4.1. Validation of the results

Based on the abandoned land data in 2016, the accuracy of the results was verified by visual interpretation with Google Earth. A total of 300 sample points of abandoned land and 100 sample points of non-abandoned were obtained. The confusion matrix obtained by comparing the sample points with the extracted results are shown (table 2). The overall accuracy is 78.5% with confusion matrix. It shows that the distribution of abandoned lands in Yunnan province can be basically reflected by the research results. The reason why the accuracy is not high is that the spatial resolution of the data used is 500 m, while many cultivated lands in Yunnan province are scattered with small plots and easy to be confused.

Table 2. Confusion matrix details.

Actual category / Extracted category	Abandoned land	Non- abandoned land	Total
Abandoned land	124	2	126
Non- abandoned land	46	51	97
total	170	53	223

4.2. Extracted results of abandoned land

The distribution results of abandoned land in Yunnan province from 2002 to 2016 were obtained through change detection, as shown below (figure 7).

The abandoned land mainly located in the mountainous and hilly areas of East Yunnan. Such as the west of Qujing city, the northwest of Hani-Yi Autonomous Prefecture of Honghe, the east of Kunming city, the west of Wenshan Zhuang and Miao Autonomous Prefecture and the southwest of Zhaotong city.

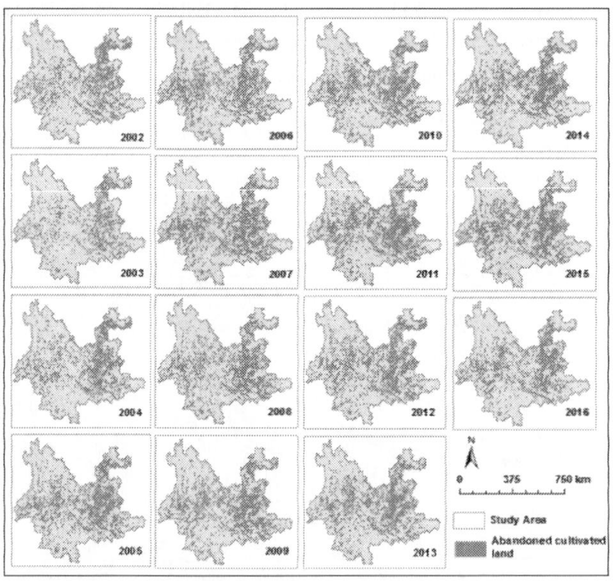

Figure 7. The distribution of abandoned land in Yunnan province from 2002 to 2016.

4.3. Analysis of area changes of abandoned land

- Analysis of variation characteristics of abandoned land area in Yunnan province from 2002 to 2016 are show below (figure 8). As we can be seen from figure 8, the abandoned cultivated land changed a lot in the three years of 2003, 2011 and 2016, while the fluctuation was less in other years. This is because since the implementation of "regulations on returning farmland to forests" in 2003, part of the project has occupied the abandoned land and the abandoned land decreased eventually. After 2004, the intensity of returning farmland to forests weakened, and the development of industrialization and urbanization increased the degree of abandonment. By 2011, that was at highest. After 2011, the implementation of the "land reclamation regulations", abandoned arable land has decreased. And after 2015, a series of measures, such as the use of idle land and the measures to supplement the quantity of cultivated land, improve the quality of cultivated land, and achieve the balance between the occupation and compensation of cultivated land, have reduced the degree of abandoned land.

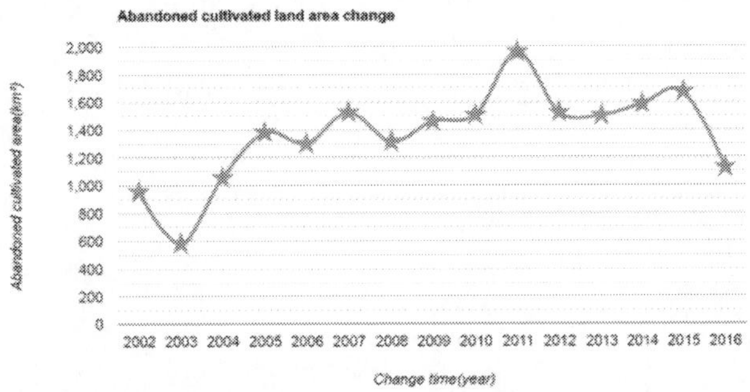

Figure 8. The variation of abandoned land area in Yunnan province from 2002 to 2016.

- Analysis on the relationship between the area of reduced cultivated land and abandoned land (see table 3). As we can be seen from table 3, the area of lost cultivated land is the same as that of abandoned land basically. It shows that during the change of cultivated land, almost all the lost parts are turned into shrublands, savannas, grasslands and bare land. The part of the reasons for the change is that cultivated land spread over flat areas. That is easy to cultivate and irrigate, and is used commonly, with little chance of fallow. The more likely change is to be for construction land, which is not the extent of fallow land. Compared with the cultivated land at higher elevations, it is less likely to be occupied by construction and more likely to become shrublands, savannas, grasslands and bare land.

Table 3. The area of reduced cultivated land and abandoned land.

Area (km^2) / Year	Reduced cultivated land	Increased abandoned land
2002	957.38	956.25
2003	584.99	582.94
2004	1057.90	1056.54
2005	1382.29	1380.97
2006	1305.12	1300.59
2007	1532.53	1527.12
2008	1316.80	1312.49
2009	1467.24	1460.44
2010	1516.08	1506.79
2011	1967.56	1961.21
2012	1530.43	1524.16
2013	1507.77	1503.90
2014	1588.95	1587.72
2015	1676.23	1673.50
2016	1128.93	1127.80

- Statistics of the change of abandoned area in the cities and states of Yunnan province. In order

IOP Conf. Series: Earth and Environmental Science **346** (2019) 012077 doi:10.1088/1755-1315/346/1/012077

to find the areas with serious abandoned land in the study area, the variation of abandoned land in various cities and states in the study area was analyzed (figure 9). As we can be seen from figure 9, the abandoned areas in Kunming, Qujing and Hani-Yi Autonomous Prefecture of Honghe are abandoned seriously. Among them, the change of the area of abandoned land in Qujing is highly consistent with that in the whole study area. Qujing is the second largest city in Yunnan province, and the development of urbanization and characteristic agriculture is in the forefront of the province. It is difficult to maintain the balance of between cultivated land occupation and compensation in some mountainous areas.

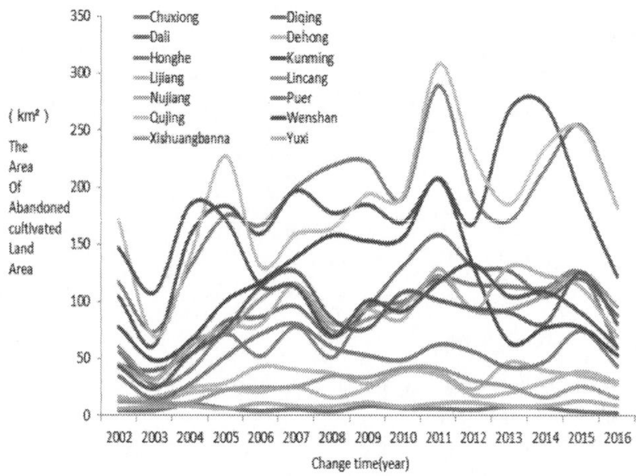

Figure 9. The change of abandoned land area in the cities and states of Yunnan Province.

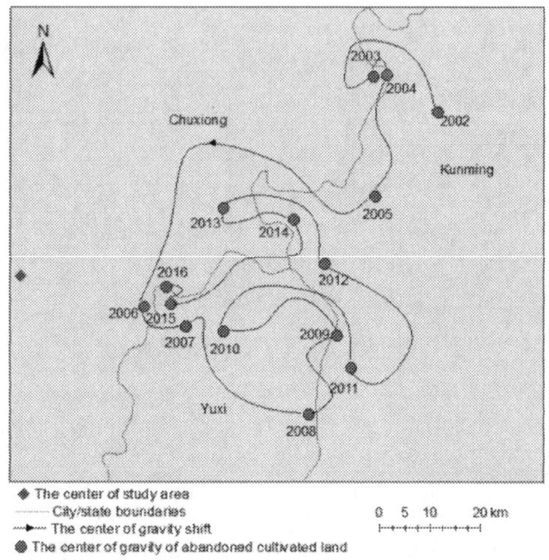

Figure 10. The gravity centers curve of abandoned land from 2002 to 2016.

4.4. Analysis of center of gravity changes

Analyzing the Center of gravity changes is a necessary work to further demonstrate spatial agglomeration characteristics [20,21] and migration characteristics [22] of abandoned land. The Mean Center in Spatial Statistics tools of ArcGIS software was used to calculate the Center of gravity of the changes of abandoned cultivated land from 2002 to 2016, that is, the changed curve of the Center of gravity of abandoned cultivated land was obtained (figure 10). As we can be seen from figure 10, the distribution of abandoned land is uneven, so the center of gravity of abandoned land does not overlap with the center of the study area. The center of abandoned land remained in the eastern part of the study area, changing back and forth among Kunming, Chuxiong and Yuxi. Among them, from 2004 to 2006, the center of gravity of abandoned land shifted to the southwest, from the border of Fumin County in Kunming city to Shuangbai County in Chuxiong. Then it moves east.

According to the analysis in figure 10 of step 4.3, it can be concluded that in the western or the western part of the study area, such as Baoshan, Dali Bai Autonomous Prefecture, Dehong Autonomous Prefecture and Lincang city, the abandoned cultivated land increased from 2004 to 2006. In Yuxi in Central Yunnan and Hani-Yi Autonomous Prefecture of Honghe in Southeast Yunnan, abandoned land is also growing. Therefore, the center of gravity of abandoned land shifted to the southwest, which is the result of the change of abandoned farmland in these two parts. After that, the growth in the east became apparent. As we can be seen from figure 10, till 2008, there is an obvious dividing line between the eastern and western parts of the abandoned land. In the west, the area of abandoned land is less than 86 km^2, while in the east, it is more than 168 km^2. The center of gravity shifted eastward.

5. Discussion

Based on the land cover products of MODIS from 2001 to 2016, this study uses the change detection method to study the spatial and temporal changes of abandoned lands in Yunnan province, and draws the following conclusions:

- From 2002 to 2016, the abandoned land in Yunnan province fluctuated significantly, with an average uncultivated area of 1284.07 km^2 per year. In 2003, the least arable land was left uncultivated, and in 2011, the most.
- The abandoned lands in Yunnan province are mainly distributed in East Yunnan. Among them, Kunming, Qujing and Honghe have a high degree of abandoned land every year.
- The center of gravity of abandoned land changes back and forth among Kunming, Chuxiong and Yuxi. From 2002 to 2006, the center of gravity was abandoned from east to west. From 2006 to 2008, the center of gravity was abandoned from west to east. However, due to the coarse spatial resolution of the data used in this paper, the accuracy of the extraction results is insufficient. In the future work, more accurate spatial and temporal analysis of abandoned land needs to be carried out based on high-resolution remote sensing data and products.

Acknowledgments
This work was supported by Scientific Research Foundation for Doctoral and Research and Innovation Fund for the Graduate of Yunnan Normal University (No. ysdyjs2019157).

References

[1] Li S F and Li X B 2016 Progress and prospect on farmland abandonment *Acta Geogr. Sin.* **71** 370-89 (In Chinese)
[2] Shi T C and Li X B 2013 Farmland abandonment in Europe and its enlightenment to China *Geogr. Geo-Inf. Sci.* **29** 101-3 (In Chinese)
[3] Ustaoglu E and Collier M 2018 Farmland abandonment in Europe: An overview of drivers, consequences, and assessment of the sustainability implications *Environ. Rev.* **26** 1-21
[4] Lesiv M, Schepaschenko D, Moltchanova E *et al* 2018 Spatial distribution of arable and abandoned land across former Soviet Union Countries *Sci. Data* **5** 180056
[5] Estel S, Kuemmerle T, Alcántara C *et al* 2015 Mapping farmland abandonment and recultivation across Europe using MODIS NDVI time series *Remote Sens.* **163** 312-25
[6] Xiong J, Thenkabail P, Tilton J C *et al* 2017 Nominal 30-m cropland extent map of continental Africa by integrating pixel-based and object-based algorithms using sentinel-2 and landsat-8 data on google earth engine *Remote Sens.* **9** 1065
[7] Friedl M A, Sulla-Menashe D, Tan B *et al* 2010 MODIS Collection 5 global land cover: Algorithm refinements and characterization of new datasets *Remote Sens. Environ.* **114** 168-82
[8] Alcantara C, Kuemmerle T, Baumann M *et al* 2013 Mapping the extent of abandoned farmland in Central and Eastern Europe using MODIS time series satellite data *Environ. Res. Lett.* **8** 35035
[9] Xiao T 2016 Study on the influence of rural labor force transfer on abandoned land in Jiangyang District (Sichuan: SAU) (In Chinese)
[10] Wu Y Y 2012 Reasons and solutions for abandoned land in Xiushan county *South Chin. Agr.* **6** 54-7 (In Chinese)
[11] Shi T C and Xu X H 2016 Extraction and validation of abandoned farmland parcel in typical counties of Chongqing *Trans. Chin. Soc. Agric.* **32** 261-7 (In Chinese)
[12] Yang T, Guo X D, Yue D P *et al* 2019Information extraction and driving factor assessment of farmland abandonment based on joint change detection *Trans. Chin. Soc. Agric. Mach.* **2019** 201-8 (In Chinese)
[13] Zhang B R, Hou Z H, Duan P *et al* 2018 Abandoned land extraction methods for different resolution images *Geoma. Spati. Inf. Tech.* **41** 176-179 (In Chinese)
[14] Feng M Q, Pan H Y, Zhu F *et al* 2018 Spatial pattern of different abandoned farmland and its affecting Southwest China *J. Agr. Sci.* **31** 1260-6 (In Chinese)
[15] Wang Y Q, Qin F Y, Yin S *et al* 2019 Land cover information extraction of dalinuoer wetland

nature reserve: Based on GF-1 satellite data and object-oriented classification *Chin. Agr. Sci. Bull.* **35** 137-41 (In Chinese)

[16] Ma S J, Pei Z Y, Wang F *et al* 2019 Application on remote sensing survey of abandoned farmlands in winter along the Huaihe River based on GF-1 image *Trans. Chin. Soc. Argic.* **35** 227-33 (In Chinese)

[17] Li J J, Chen G J, Li Z J *et al* 2018 Spatial distribution pattern of abandoned land based on GIS: A case study of Qinglong Township in Quxian County of Sichuan Province *Guizhou Agr. Sci.* **46** 129-32 (In Chinese)

[18] Alcantara C, Kuemmerle T, Prishchepov A V *et al* 2012 Mapping abandoned agriculture with multi-temporal MODIS satellite data *Remote Sens. Environ.* **124** 334-47

[19] Li S F, Li X B, Xin L J *et al* 2017 Extent and distribution of cropland abandonment in Chinese mountainous areas *Resour. Sci.* **39** 1801-11 (In Chinese)

[20] Ge M L and Feng Z M 2009 Population distribution of China based on GIS: Classification of population densities and curve of population gravity centers *Acta Geogr. Sin.* **64** 202-10 (In Chinese)

[21] Guan X L, Fang C L and Lu S S 2010 Analysis of satial dstribution and gravity centers curve dynamic cultivated Lnd changes in China *J. Nat. Resour.* **25** 1997-2006 (In Chinese)

[22] Liu B T, Song H P, Guo B *et al* 2012 Study on annual variation of rainfall erosivity in southwest China using gravity center model *Trans. Chin. Soc. Agric.* **28** 113-20

Streptomyces spp. culture filtrates deconstruct *Odontoglossum ringspot virus* (ORSV) and *Cymbidium mosaic virus* (CymMV) virus coat

M J Lee[1], B L Liu[1], C J Chen[2], K M Chang[3], S D Huang[4] and C A Chang[1, 5]

[1]Department of Applied Chemistry, Chaoyang University of Technology, Taichung, Taiwan
[2]Neural Regeneration Laboratory, Department of Neurosurgery, Neurological Institute, Taipei Veterans General Hospital, Taipei, Taiwan
[3]Department of Research, Koo Foundation Sun Yat-Sen Cancer Center, Taipei, Taiwan
[4]Department of Applied Mathematics, National Tunghua University, Hualien, Taiwan

E-mail:cachang@cyut.edu.tw

Abstract. *Odontoglossum ringspot virus* (ORSV) and *Cymbidium mosaic virus* (CymMV) are very resistant viruses and survive without cells in soil or bench environment for up to 10 years. Two strategies may prevent the infection of these virus to the plants. Firstly, to rid the cultivation environment and agriculture tools of the virus particles, and secondly, to apply the anti-viral agent on the plant to prevent the infection of the viruses. These requires development of new type of anti-viral agents. Previously *Streptomyces* spp was isolated, and culture filtrate was demonstrated to be anti-viral. Two strains, SML-1 and CA5-NO6 were isolated. In this paper we presented tests to compare the effects of these two strains and other anti-viral agents, on ORSV and CymMV. The culture filtrate of *Streptomyces* spp. CA5-NO6 strain, as well as SML-1 strain, was tested for their ability to destroy coat protein of ORSV. Morphology of ORSV or CymMV was examined by atomic force microscopy (AFM) after treatment of the following: (1) The culture filtrate of CA5-NO6 strain (2) Trypsin, a known proteinase, (3) virusbom, an proprietary product sold in Taiwan claimed to work as an enzyme. Our results demonstrated that CA5-NO6 strain destroyed the coat protein slightly better than SML-1 strain. CA5-NO6 act on the ORSV virus in a different way to that of trypsin or virusbom, while different buffers result in different morphology of the same virus under AFM. Our data suggested the *Streptomyces* spp. culture filtrates deconstruct ORSV and CymMV virus coat by a mechanism not similar to those of trypsin or virubom, and both strains works, with CA5-NO6 slightly better than SML-1 strain. This provide evidence to support the further development for commercialization of the culture filtrates.

1. Introduction

Among the thousands of viruses that infect plants, many viruses could survive in adverse environment and keep the ability to propagate. The viruses in *tobamovirus* genus are typical of this type of resistant viruses. Viruses in this genus include the *Tobacco mosaic virus* (*TMV*) that infect tobacco and many economical crops, *Tomato mosaic virus* (*ToMV*), *Pepper mild mottle virus* (*PeMMoV*) and *Cucumber green mottle mosaic virus* (*CGMMV*), which do great damage to tomato, pepper and melons. In this

Content from this work may be used under the terms of the Creative Commons Attribution 3.0 licence. Any further distribution of this work must maintain attribution to the author(s) and the title of the work, journal citation and DOI.
Published under licence by IOP Publishing Ltd

genus are also *Odontoglossum ringspot virus (ORSV)* and *Cymbidium mosaic virus (CymMV)*, which infect orchids worldwide and did damage to the horticulture industry.

ORSV could survive *in vitro* for more than 10 years and is still capable of infecting the orchid. These viruses usually propagate via insect. However, insect bites are not necessary for the disease spreading. The viruses could also spread via residual tissue or secretion of the infected plants that left the plant cell. The viruses exist in the carrier, be it solid or liquid, and contaminate the tools used to manipulate the plant, the hands that handle the plants, the soil that plants were cultivated in, or even in the irrigation water.

For viruses as resistant as these, two strategies could be applied: firstly, to rid the cultivation environment and agriculture tools of the virus particles. This reduce the virus count in the environment and reduce chances of orchid being infected. Another strategy is to apply the anti-viral agent on the plant to prevent the infection of the viruses. To achieve these goals, the present anti-viral agent that were used in the tissue culture room and orchid farm were not ideal, especially for the latter.

To rid the tissue culture room or the orchid farm from the virus, the wall and floor could be mopped with diluted bleach, which denature the virus coat and destroy the infectivity of viruses. The equipment that come in contact with the cultivar could be sterilized by autoclave or 70% ethanol. However, some of these were detrimental to the environment (bleach), some expensive and is health hazard and cannot be widely applied in environment (the alcohol), and some were time consuming and cannot be used for large areas (the autoclave). All of them cannot be applied on technician's hands, and may cause allergic reaction to human. Further, they could not be applied directly on plants for preventing the infection. Therefore, a new type of anti-viral agents is needed.

In recent years many anti-viral agents were discovered from natural sources [1, 2]. Many plants extracts were demonstrated to inhibit the virus entry into the cells [3]. The glycoprotein from Celosia cristata is inhibitory to the infection of tobacco mosaic virus (TMV) to the tomato leaves [4, 5]. There are similar phytoprotein reported for *Dianthus caryophyllus, Mirabilis jalapa, Clerodendrum inerme,* and *Bougainvillea spectabilis* [1, 6, 7]. These are proof of concept that naturally derived anti-viral agents existed from various sources.

Orchids are vulnerable to several diseases caused by mechanically transmissible and highly *in vitro* stable plant viruses such as ORSV or CymMV. These viruses can be very stable in soil or desk tops for up to years without infecting live cells, and sanitation steps during preparation of tissue culture and cultivation of orchid is not effective in ridding these viruses. We have previously identified culture of *Streptomyces* spp that was capable of degrading orchid virus capsid proteins and inhibiting virus infectivity [8-10]. ORSV infectivity could be largely reduced by simple incubation with culture supernatant containing the microbial origin compound. In this project we report the isolation of different culture strains, and study the mechanism of virus degradation with atomic force microscopy.

2. Material and methods

2.1. Streptomyces spp

CA5-NO6 and SML1 were used throughout the study. To reveal antiviral activities *Streptomyces* strains were grown on soybean milk agar. The composition of the basic medium per liter was as follows: 5 g of soybean milk, 5 g of glucose, and 0.4 g of $CaCO_3$. Growth was carried out at 30°C for 7 or 25 days in 250-mL Erlenmeyer flasks placed in a model G-25 Superohm rotary shaker (150 rpm) containing 100 mL of culture medium and 5 mL of inoculum.

2.2. In-directed ELISA assay of antiviral activity

The method is based on general procedure of in-direct ELISA assay, with modification [11]. The ORSV infected orchid tissue was grinded, filtered, and loaded into a 96-well pre-coated ELISA plates (100 μL in each well), and standing at 37°C for 2 hrs. The wells were then rinsed 3 times with 1X phosphate buffered saline with Tween-20 (PBST) solution (allowed to stand for 3 mins between each rinses) and then patted dry. The culture filtrate sample was incorporated into the well, and the reaction was carried

IOP Conf. Series: Earth and Environmental Science **346** (2019) 012078 doi:10.1088/1755-1315/346/1/012078

out at 37°C for 30 mins. Subsequently, a 100 μL of 1,000X diluted ORSV antibody (primary antibody) was incorporated into each well. After two hours of reaction at 37°C (or overnight at 4°C), the well was rinsed 3 times with 1X PBST solution. An alkali phosphatase substrate kit was then added for coloring and the reaction was allowed to stand for another hour. The absorption value at a wavelength of 405 nm was measured by an ELISA reader. The lower the absorption value, the better the ability to degrade the virus capsid protein. The tissue fluid from ORSV infected orchid tissue and that from the non-infected orchid, were used as controls.

2.3. SDS-PAGE electrophoresis

For SDS-PAGE, the sample (50 μL) was dissolved in sample buffer (0.125 M Tris-HCl buffer, pH 6.8, containing 20% glycerol, 3 μM bromophenol blue, and 1% 2-mercaptoethanol) and heated at 95°C for 5 mins. The electrophoresis was conducted by the Laemmli method with a 10 μL sample on the 10% polyacrylamide gel using a vertical slab electrophoresis apparatus (Mini-Protein III, Bio-Rad Laboratories, Inc., Hercules, CA). After the electrophoresis (~2 h), the gel was stained with Coomassie brilliant blue (50 mL glacial acetic acid, Coomassie brilliant blue 1.5 g, 250 mL distilled water, 250 mL methanol) for 5 mins and then destained with destaining buffer (100 mL methanol, 200 mL glacial acetic acid, 1700 mL distilled water).

2.4. Sample preparation and atomic force microscope (AFM)

The still unknown microbial origin compound in the bacterial culture filtrate was referred to as 'X' in this project. To tested how the plant virus morphology was affected by X, the plant virus ORSV and CymMV were prepared by filtering grounded infected plant leaves with 0.22 μM filter (Millipore). Fourteen-day liquid culture of bacteria grown from single colony was used. Virusbom, which was demonstrated effective in collapsing bacteria and viruses [12], was used as positive control at 150 ppm. Bacterial culture from both strains was used at 2 times dilution (0.5X). Culture medium without bacteria was used as negative control. Test samples were mixed with plant virus diluent for various incubation time (0, 5, 90 minutes), and mounted on mica for observation by AFM (CSPM 5500 being nano-instruments, LTD). For the same virus another model of AFM was used (Veeco Dimension 3100). Tapping mode was used, and 3 μM x 3 μM area in channels including surface, amplitude, phase was recorded.

2.5. Treatment groups

For study of CymMV, the treatment groups are as follows: (1) A: CymMV only (2): CymMV+X, 5 minutes (3): CymMV+X, 90 minutes (4) CymMV+ trypsin, 90 minutes (5): CymMV+virusbom, 90 minutes. Note that virusbom 150 ppm, a known anti-viral agent was used as a control. Trypsin (0.125%) was used as proteinase control. For the study of ORSV, the treatment groups are as follows: (1) A: ORSV only (2): ORSV+X, 30 minutes.

2.6. Statistics

Non-parametric methods for hypothesis testing were used at significance level 0.05. P-values of multiple comparisons were adjusted according to Hochberg's method. All hypothesis testings were run on R v3.5.

3. Results

3.1. The ORSV virus particle was destroyed by incubation of culture filtrate of Streptomyces spp in 30 minutes: ELISA

Indirect ELISA methods using antibody against ORSV coat protein demonstrated that the immunoreactivity demonstrated by OD 405 was significantly reduced, although not completely, after both filtrate treatments (table 1).

IOP Conf. Series: Earth and Environmental Science **346** (2019) 012078 doi:10.1088/1755-1315/346/1/012078

Table 1. Indirect ELISA of ORSV protein

							Ave	stdev
V	2.969	2.938	3.276	3.080	3.015	3.104	3.064	0.12173
SML 1	0.625	0.678	0.642	0.647	0.623	0.624	0.640	0.02129
CA5-NO6	0.453	0.425	0.410	0.408	0.423	0.427	0.424	0.01614
H	0.173	0.153	0.042	0.044	0.100	0.107	0.103	0.05408

Table 1 OD reads for non-treated (V), SML-1 strain (SML1), CA5-NO6 strain (CA5-NO6), and healthy, non-infected tissue fluid (H). Each column represent the OD read for one experiments, and there are 5 separate experiments were done. (N=5).

3.2. The ORSV virus particle was destroyed by incubation of culture filtrate of Streptomyces spp in 30 minutes: statistics

One-way ANOVA (Kruskal-Wallis rank sum test) was conducted to see whether measurements were different with different treatments (V, SML 1, CA5-NO6, and H). The result (p-value=0.0001) indicated that the measurement-levels were significantly different with different treatments. Next, multiple one-tailed comparisons (Wilcoxon rank-sum test) were proceeded to check whether measurements of CA5-NO6 or SML 1 are lower than those of V, and whether measurements of CA5-NO6 are lower than those of SML 1. The p-values of all one-tailed tests were 0.0011 (both raw and adjusted). Therefore, we concluded that measurement-level of CA5-NO6 or SML 1 were significantly lower than those of V, and measurement-level of CA5-NO6 are significantly lower than that of V. It was found that, firstly, both the filtrate of SML-1 and CA5-NO6 strain reduced the immunoreactivity of the ORSV coat protein, compared to the non-treated one (* in figure 1). Secondly, the effect of CA5-NO6 was significantly better than that of SML1 (# in figure 1).

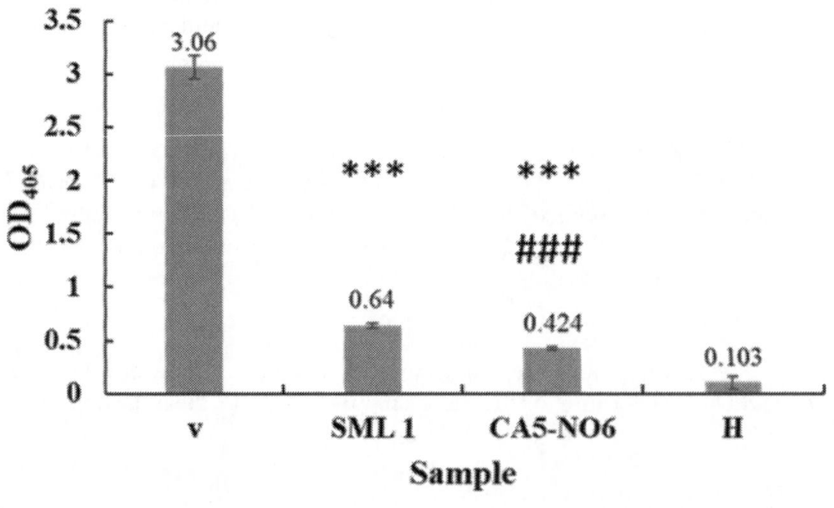

Figure 1. Analysis of ORSV coat protein degradation activity of *Streptomyces* (SML 1, 25 days, CA5-NO6, 7 days) through ELISA assay. non-treated (V), SML-1 strain (SML1), CA5-NO6 strain (CA5-NO6), and healthy, non-infected tissue fluid (H). Both the SML1 and CA5-NO6 were compared to the non-treated group (V) and demonstrated difference (*** for P<0.001). CA5-NO6 was significantly different to SML-1 (### for p <0.001), and the clearance was better in CA5-NO6, since the OD (reactivity) was further reduced).

3.3. CA5-NO6 culture filtrate was rich in protein of lower molecular weight

After precipitation and re-suspension, the x containing solution was loaded onto SDS PAGE and stained with commassie blue. The solution was rich in molecules of small molecular weight. Of note, there is several bigger bands around 50, 27, and 20 KD. The identify of these protein bands were not known (Figure 2).

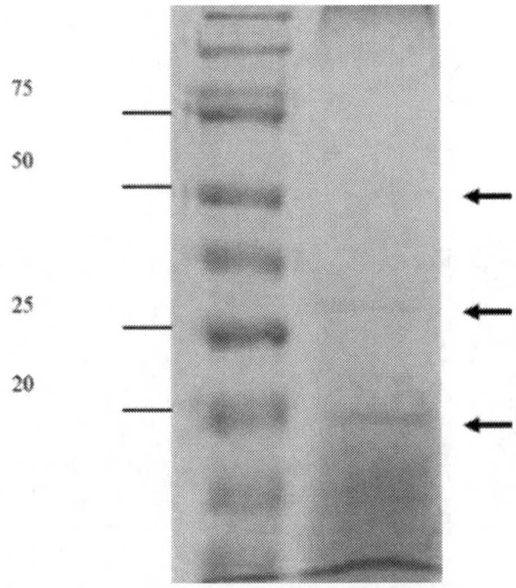

Figure 2. SDS PAGE stained with commassie blue for the precipitated molecules in the CA5-NO6 filtrate.

3.4. ORSV morphology was affected when treated with bacterial culture containing X (CA5-NO6)

Following this, we used bacterial culture filtrate from CA5-NO6 to treat the ORSV, and record the morphology using AFM. For non-treated ORSV, the virus morphology was intact with obvious rod shape [13], while after X treatment the virus structure become flat and occupy more area per virus, and the virus structure become a diffuse clouds of unknown substance (figure 3).

Figure 3. A: ORSV receive treatment of negative control (bacterial culture without bacteria). B: ORSV received treatment of X containing bacterial culture (0.5x). The estimated length of the virus structure was therefore 500 – 1000 nM, and width about 10-20 nM. This corresponds to the known ORSV size of 500 nM x 15 nM. Left panel: phase, middle panel: amplitude, right panel: plane. Left panel demonstrated the texture of the material, while the middle panel demonstrated the thickness or height along the Z axis

3.5. CymMV morphology was affected when treated with CA5-NO6 strain bacterial culture: comparison to trypsin and virusbom

The estimated length of the virus structure was 1000-2000 nM, and width about 10-20 nM. This corresponds to the known CymMV size. The fibrous structure of the virus also confirmed the identity of the virus (figure 4, A). CymMV treated with X for 5 minutes already resulted in the destruction of the virus (figure 4, B) to a degree similar to those treated for 90 minutes (figure 4, C). The structure of destructed virus was not like that in trypsin treated ones (figure 4, D), which seemingly lack the virus protein structure and only threads of supposedly nucleic acids were observed. The CA5-NO6 treated virus seem has fragmented structures remained (figure 4, arrows in B, C), unlike that treated with virusbom that has flattened structure with textured surface (figure 4, E).

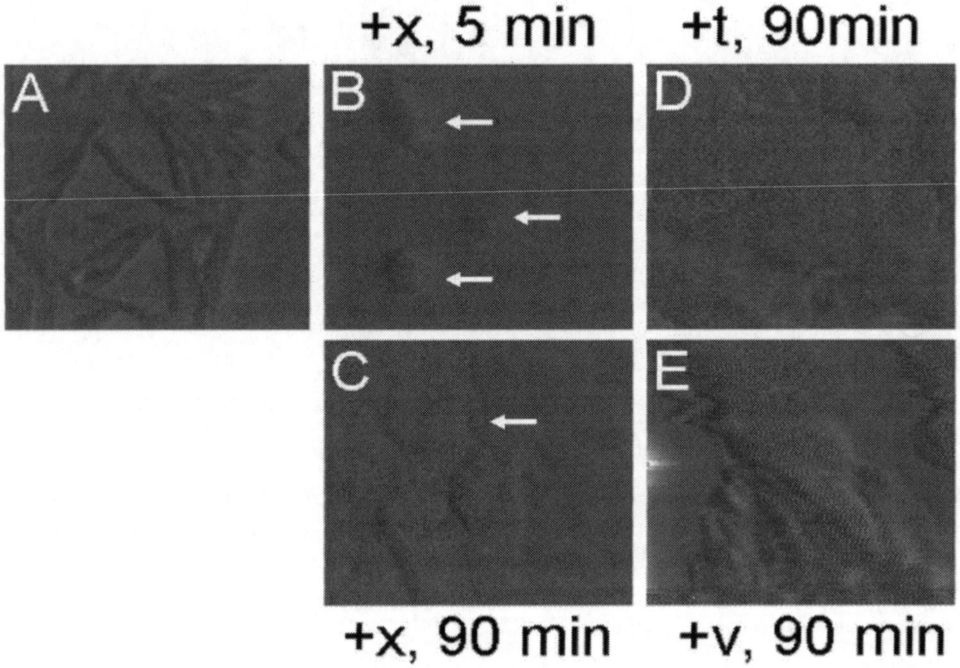

Figure 4. A: CymMV without treatment. B: CymMV treated with CA5-NO6 (0.5x) for 5 minutes. C: CymMV treatmented with CA5-NO6 (0.5x) for 90 minutes. D: CymMV treated with trypsin (0.125%) for 90 minutes. E: CymMV treated with virusbom (150ppm) for 90 minutes. The estimated length of the virus structure was 1000-2000 nM, and width about 10-20 nM. This corresponds to the known CymMV size. The fibrous structure of the virus also confirmed the identity of the virus (A).

3.6. ORSV morphology was affected when treated with bacterial culture containing X (CA5-NO6): observation using other model of AFM

Following this, we used bacterial culture filtrate from CA5-NO6 to treat the ORSV, and record the morphology using another model of AFM (Veeco Dimension 3100). For non-treated ORSV, the virus morphology was intact with obvious rod shape, while after X treatment the virus structure become flat and occupy more area per virus, and the virus structure become a diffuse clouds of unknown substance. The following in a panel of AFM result we obtained for the ORSV. The top panel (figure 5, A, B, and C) demonstrated the virus without filtrate treatment, and the bottom panel (figure 5, D, E, F) demonstrated the virus after CA5-NO6 treatment. Three different channels were noted. The result was more or less the same as those demonstrated in figure 3.

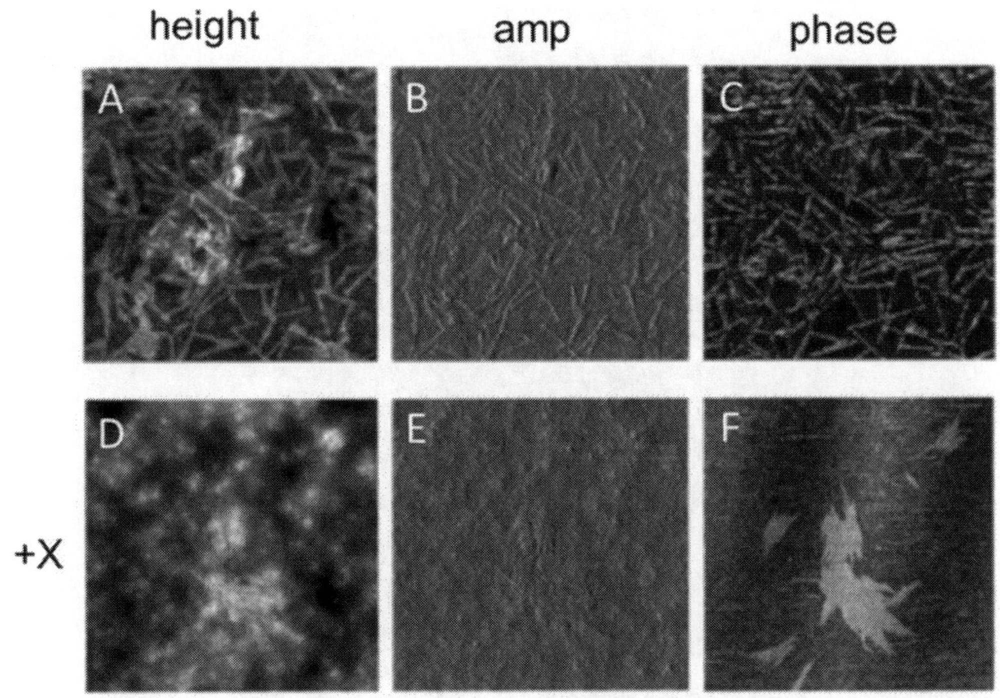

Figure 5. A, B, C: ORSV without treatment. D, E, F: ORSV treated with CA5-NO6 (0.5x) for 5 minutes. The estimated length of the virus structure was therefore 500-1000 nM, and width about 10-20 nM. This corresponds to the known ORSV size of 500 nM x 15 nM · The fibrous structure of the virus also confirm the identity of the virus (A, B, C).

3.7. Preparation of buffer for the AFM affect the morphology: ORSV as an example

After the filtrate treatment, the viruses were diluted 10000 times before applying to the mica and allowed to air dry. During our preparation of viruses for the AFM observation, we found that the solution viruses were diluted in influenced the morphology of these fibrous viruses. Although these pictures were obtained in different model of AFM (CSPM 5500 being nano-instruments vs Veeco Dimension 3100), these difference aside, for those diluted in pure water, the morphology of the ORSV was wigglier, the virus was flat and seemingly squashed to the mica, and the virus occupied more dimension in the x-y axis of the planes. For those diluted using the bacterial culture medium, which contain various nutrients as well as salts, the ORSV were more needle like, straight and rigid in their appearance, and occupy less area in the x-y axis of the plane. In the phase channel, which give the information of the elastic texture of the virus (figure 5, F), salt crystals demonstrated by the white color was apparent.

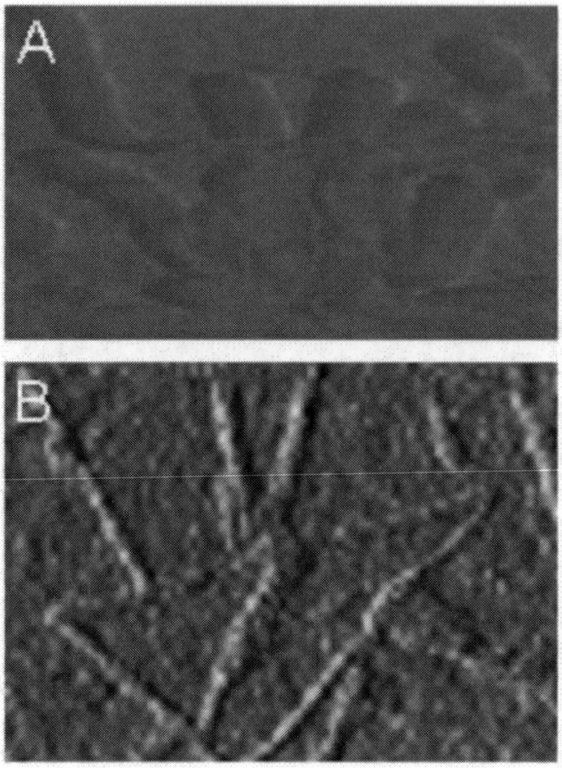

Figure 6. ORSV with different preparation taken in different AFM. A: ORSV diluted in water, under CSPM 5500 being nano-instruments. B: ORSV diluted in bacterial culture medium, under Veeco Dimension 3100.

4. Discussion

Our results demonstrated that culture filtrates of both SML-1 and CA5-NO6 strains are capable of deconstructing the ORSV after 30-minute incubation. This is done using the isolated virus from the infected plants, therefore it can only be used as an indication that application of these culture filtrate on the plants might render productive effect against virus infection. The coast-effect of applying the culture filtrate on a whole plant constantly of course needs further evaluation. The result actually simulated better to the situation where the bench top, or the technicians' hands, or the instruments that manipulate the plant and for tissue culture, was contaminated. In these situation the virus did not infect the plant or cell, and was in a free form, albeit capable of infection later on. Our data suggested that this product may be used in disinfecting these surface areas quite effectively.

The fact that 5 minutes is enough for the virus to change shape to an extent that is visible under the atomic microscope is significant, for sometimes subtle changes in several bonds or atoms could affect the function of a molecule, not to mention an alteration so big that is physically observed by a microscope. This support the further development of these filtrate into commercial product. Although CA5-NO6 is slightly better than SML-1 strain, the culture condition for these two strains that yield maximal effect or active ingredient was not the same. Further research on the culture condition is needed for further development of the product.

When the effects of the CA5-NO6 culture filtrates was compared to trypsin, a peptidase, and virusbom, a synthetic compound that collapsed the virus capsid, the morphology of the resultant viruses were very different among the 3 different treatment. After 30- minute incubation the CA5-NO6 resulted in fragmented viruses, trypsin results in collapse of coat protein and seemingly exposed nucleic acid, while virubom result in collapse of coat proteins. This suggested that our culture filtrate act on the virus using a different mechanism as peptidase or a capsid collapser. This is important, as it suggested that the product developed from our discovery will not be very similar to the existing product, and has the possibility to synergistically work with other product to produce even better result.

Acknowledgments

Authors wishing to acknowledge assistance Dr Chen in Research Center for sustainable Energy and Nanotechnology, NCHU, Taiwan,

References

[1] Karthikeyan G K, Gandhi S, Doraisamy R, Rabindran R and Rabindran T G 2009 Evaluation of antiviral principles for the induction of systemic resistance in blackgram (Vigna mungo) against Urdbean Leaf Crinkle Virus. *Archi. Phytopathol Plant Protect.* **42**(12): p. 1172-1186.

[2] Patick A K and Potts K E 1998 Protease inhibitors as antiviral agents. *Clin. Microbiol. Rev.*, **11**(4): p. 614-27.

[3] Verma H N and Varsha Baranwal B K 1995 Endogenous antiviral inhibitors from plants, their physical and biological properties. Antiviral proteins in Higher Plants., ed. D B D. Chessin M, Zipf A. CRC press Boce Raton Florida.

[4] Baranwal V K and Verna a H N 1992 Localized resistance against virus infection by leaf extract of Celosia cristata. *Plant Pathol.* **41**(5): p. 633-638.

[5] Balasubrahmanyam A, Baranwal V K, Lodha M L, Varma A and Kapoor H C 2000 Purification and properties of growth stage-dependent antiviral proteins from the leaves of Celosia cristata. *Plant Sci.* **154**(1): p. 13-21.

[6] Baranwal V K and Verma H N 2000 Antiviral phytoproteins as biocontrol agents for efficient management of plant virus diseases., in biocontrol potential and its exploitation in sustainable agriculture, M.K.G. Upadhyay R.K., Chamola B.P. Editor. Springer: Boston, MA.

[7] Balasaraswathi R, Sadasivam S, Ward M and Walker J M 1998 An antiviral protein from Bougainvillea spectabilis roots; purification and characterisation. Phytochemistry, **47**(8): p. 1561-5.

[8] Chang C A, Chen H R and Chiu C H 2010 Disinfection of odontoglossum ringspot virus and cymbidium mosaic virus from tools used during orchid cultivation. International Society for Horticultural Science (ISHS), Leuven, Belgium.

[9] Chen H R 2009 Production and application of an orchid virus degrading protease secreting by an Actinomycetes 2009 Master Theses, Graduate institute of biochemical science and technology. Chaoyang University of Technology Taiwan. p. 91.

[10] Chiu C H 2006 Screening and application of actinomycete secreting protease that can degrade capsid protein and abolish infectivities of odontoglossum ringspot virus and cymbidium mosaic virus. Master thesis, institute of biochemical sciences and Technology. Chaoyang University of Technology: Taichung, Taiwan.

[11] Chang C A, Purcifull D E, and Hiebert E 1988 Purification, characterization, and immunological analysis of nuclear inclusions induced by bean yellow mosaic and clover yellow vein potyviruses. Phytopathology **78**: p.1266-1275.

[12] Wang S S 2011 Method for killing NDM-1 Klebsiella pneumoniae, P. National intellectual property administration, Editor. China

[13] Jensen, D D and Gold A H 1951 A virus ring spot of odontoglossum orchid: symptoms, transmission and electron microscopy. Phytopathology **41**: p. 648–653

Effect of urban green biomass (pruning materials) compost on the growth of komatsuna (*Brassica rapa L. var. perviridis LH Bailey*) in different soils

E X Liu[1, 2] and T Takahashi[1]

[1]Graduate School of Horticulture, Chiba University, Japan

E-mail: lexgxs@gmail.com

Abstract. Mineralization of organic matter in soil is affected by the total carbon (C) and nitrogen (N) contents and microbial activity. Therefore, we hypothesized that the application of compost made from pruning materials (PMs) would have different effects on plant growth depending on the properties of the soil. We grew komatsuna (*Brassica rapa L. var. perviridis LH Bailey*) plants for 1 month in different soils and found that there was a significant positive correlation between plant growth and the total C, N contents and microbial activity of the soil ($p < 0.05$). However, although PMs compost could be used as conditioners to increase the C and N contents and microbial activity of infertile soils, it was difficult to promote plant growth during the early stages of PMs compost application in these soils. Together, our results indicate that PMs compost is appropriate compost for sustaining or increasing the total C and N contents and microbial activity of the soil and increasing the yield of plants growing in fertile soil.

1. Introduction

The rapid development of urban green spaces has seen a consistent increase in the quantity of plant waste being produced [1], the disposal of which has become a major problem that affects both the environment and sustainable development [2]. Pruning materials (PMs) have traditionally been disposed of through burning. However, sustainable approaches that allow the PMs to be recycled, such as composting or mulching, can allow them to be fully used and effectively reduce CO_2 emissions [3-4].

The application of compost made from PMs compost to soil has been shown to effectively increase the soil porosity and water holding capacity [5]. PMs compost also have nearly the same capacity as conventional compost made from dung, leaves, and bark to improve soil chemical properties, such as fertility, as indicated by the cation exchange capacity (CEC), base cation saturation, and total carbon (C) and nitrogen (N) contents [6]. Furthermore, because PMs compost have a high content of cellulose and hemicellulose mixed with large amount of lignin, they have a high content of mineralizable C compared with conventional compost; thus, its application may have a priming effect on the organic matter mineralization that is present in the soil [7].

The priming of organic matter mineralization can have both positive and negative effects on soil properties [8]. For example, although it can improve the chemical, physical, and biological properties of the soil, it can also inhibit plant growth [9]. Thus, mixing PMs compost into the soil can effectively increase the soil organic matter content but a high content of mineralizable C can inhibit plant growth due to N immobilization [10], which results from the rapid increase in microbial activity [11]. Therefore, it is necessary to determine the growth of plants in soil to which PMs compost have been applied.

Content from this work may be used under the terms of the Creative Commons Attribution 3.0 licence. Any further distribution of this work must maintain attribution to the author(s) and the title of the work, journal citation and DOI.
Published under licence by IOP Publishing Ltd

The C and N cycling dynamics in compost-supplemented soils can be affected by the compost feedstock, processing conditions, and time [11]. Furthermore, the mineralization of organic matter in the soil is affected by N availability and microbial activity [12-13], and the available N content is positively correlated with the total C and N contents of the soil [14]. Therefore, we hypothesized that the application of PMs compost would have different effects on plant growth depending on the total C and N contents and microbial activity of the soil.

The aim of this study was to evaluate the effect of PMs compost on plant growth in soils with different total C and N contents and microbial activities. In addition, we investigated whether PMs compost affect plant growth during the early stages of application. For this, we applied PMs compost to komatsuna (*Brassica rapa L. var. perviridis LH Bailey*) growing in different soils and used the dry weight of individual plants as an indicator of plant growth.

2. Materials and methods

2.1. Soil and compost

To investigate how the application of PMs compost affect plant growth in different soils, a planting experiment was conducted using forest soil, subsoil, and sand (Table 1). The forest soil (0–10 cm depth) was collected from evergreen broad-leaved forest growing on the Matsudo Campus of Chiba University (Matsudo, Chiba Prefecture, Japan) and the subsoil (>30 cm depth) was collected from an experimental field on the Matsudo Campus of Chiba University. The sand was purchased. Different types of soils were produced by mixing varying proportions of different soils (Table 2).

A (volume ratio of forest soil is 100%), B (volume ratio of forest soil : subsoil is 1:1), C (volume ratio of forest soil : subsoil is 1:2), D (volume subsoil is 100%), E (volume ratio of subsoil : sand is 1:1), F (volume ratio of subsoil : sand is 1:2) and G (volume ratio sand is 100%), with the resulting soil types A to G representing high to low total C and N contents and microbial activities as determined by fluorescein diacetate (FDA) hydrolysis [15] (Table 3).

The compost was made from PMs obtained from Agora Landscape Architecture Corporation (Tokyo, Japan). The PMs was consisted of leaves and branches from different species in urban green space. The composting process was divided into two stages. In the first stage, 30% volume of PMs compost as microorganism source was mixed with fresh wood chips (particle size < 25 mm) made from the PMs and was kept for 30 days, airing the mixture once per week. In the second stage, the wood chips mixed Table 1. Properties of the different soils and PMs compost used in this experiment with PMs compost were kept for a further 100 days for maturation. [16] and [2] reported that PMs compost were composted sufficiently after 120 days. Therefore, the compost used in this experiment, which had been composted for 130 days, should have been sufficiently composted.

Table 1. Properties of the different soils and PMs compost used in this experiment.

Soil/compost	Water content (kg/kg)	Dry density (kg/L)	Fresh density (kg/L)	Total carbon content (g/kg)	Total nitrogen content (g/kg)	C/N ratio
Forest soil	0.28	0.57	0.79	110.4	8.50	12.98
Subsoil	0.25	0.74	0.98	37.0	2.10	17.61
Sand	0.09	1.21	1.33	2.6	0.20	13.00
PMs compost	0.44	0.17	0.3	396.0	16.30	24.29

PMs: Pruning materials.

Table 2. Composition of the different soil types use.

Soil	Experimental materials	Volume ratio
A	Forest soil	1
B	Forest soil:subsoil	1:1
C	Forest soil:subsoil	1:2
D	Subsoil	1
E	Subsoil:sand	1:1
F	Subsoil:sand	1:2
G	Sand	1

Table 3. Total carbon (C) and nitrogen (N) contents and microbial activities of the different soils used in experiment.

Soil	Total C (g/kg)	Total N (g/kg)	C/N ratio	FDA[†]
A	110.4 ± 9.01a	8.5 ± 0.73a	12.99	1.40 ± 0.044a
B	74.6 ± 6.09b	5.6 ± 0.49b	13.32	1.20 ± 0.04b
C	65.4 ± 5.34c	5 ± 0.43c	13.08	0.85 ± 0.028c
D	37 ± 3.02d	3.1 ± 0.24d	11.94	0.66 ± 0.022d
E	12 ± 0.98e	0.9 ± 0.08e	13.33	0.29 ± 0.01e
F	5.1 ± 0.42f	0.6 ± 0.03f	8.50	0.21 ± 0.007f
G	2.6 ± 0.21g	0.2 ± 0.01g	13.00	0.19 ± 0.009f

Different lower-case letters within a column indicate significant differences among soil types ($p < 0.05$).
Fluorescein diacetate (FDA) hydrolysis was used to indicate microbial activity. The values are absorbance values of the sample liquid at 490 nm [15].

Rectangular pots (volume: 2.2 L; length: 25 cm; width: 11 cm; depth: 8 cm) were filled with soils and PMs compost as treatments or only soils as a control. Each of the soil types were mixed with different ratios of PMs compost, as shown in Table 4.

2.2. Measurement of plant growth

Komatsuna was used as the test plant in this experiment. A total of 12 komatsuna seeds (Sakata Seed Corporation) were planted in each pot on August 20, 2016 and were watered once per day for 1 month. The plants were then harvested on September 20, 2016. After harvesting, the roots and leaves of each individual plant were washed with water, placed in a paper bag, and dried in an oven, to obtain the dry weight of the plant.

2.3. Total C and N contents, C/N ratio, and microbial activities in the soils and PMs compost

The total C and N contents and C/N ratio of the soils and compost were analyzed using a measured using a CN corder（YANACO Analytical Industry Co., Ltd.MT-700). Microbial activity was measured using the FDA hydrolysis method [7]. The FDA value used in this experiment was based on the amount of FDA hydrolyzed per 1 g dry weight of each sample.

2.4. Statistical analysis

To examine the difference in the parameters among the composts studied, one-way analysis of variance (ANOVA; multiple comparisons) was performed, followed by the least significant difference test. The results were considered significantly different at $p < 0.05$. All the values were presented as the mean. SPSS 19.0 program was used to perform all statistical analyses.

IOP Conf. Series: Earth and Environmental Science **346** (2019) 012079 doi:10.1088/1755-1315/346/1/012079

Table 4. Details of the treatments used.

Soil†	Volume of PM added to soil (%)	Treatment	Volume ratio
A	0	Forest soil	100%
	7.5	Forest soil: PM	70%:30%
	15	Forest soil: PM	85%:15%
	30	Forest soil: PM	92.5%:7.5%
B	0	Forest soil: subsoil	50%:50%
	7.5	Forest soil: subsoil: PM	35%:35%:30%
	15	Forest soil: subsoil: PM	42.5%:42.5%:15%
	30	Forest soil: subsoil: PM	46.25%:46.25%:7.5%
C	0	Forest soil: subsoil	33.3%:66.7%
	7.5	Forest soil: subsoil: PM	23.3%:46.7%:30%
	15	Forest soil: subsoil: PM	28.33:56.67:15%
	30	Forest soil: subsoil: PM	30.83:61.67:7.5%
D	0	Subsoil	100%
	7.5	Subsoil: PM	70%:30%
	15	Subsoil: PM	85%:15%
	30	Subsoil: PM	92.5%:7.5%
E	0	Subsoil: sand: PM	50%:50%
	7.5	Subsoil: sand: PM	35%:35%:30%
	15	Subsoil: sand: PM	42.5%:42.5%:15%
	30	Subsoil: sand: PM	46.25%:46.25%:7.5%
F	0	Subsoil: sand	33.3%:66.7%
	7.5	Subsoil: sand: PM	23.3%:46.7%:30%
	15	Subsoil: sand: PM	28.33:56.67:15%
	30	Subsoil: sand: PM	30.83:61.67:7.5%
G	0	Sand	100%
	7.5	Sand: PM	70%:30%
	15	Sand: PM	85%:15%
	30	Sand: PM	92.5%:7.5%

† A: forest soil alone; B: 1:1 ratio of forest soil: subsoil; C: 1:2 ratio of forest soil: subsoil; D: subsoil alone; E: 1:1 ratio of subsoil: sand; F: 1:2 ratio of subsoil: sand; G: sand alone. PM: compost made from pruning materials.

3. Results

The total C and N contents and microbial activities of the different treatments are shown in Table 5. In soils A, B, and C, which had the highest total C and N contents, the application of 7.5% PMs compost had no significant effect on the total C and N contents ($p < 0.05$), and in soils A and B, the application of 15% PMs compost also had no significant effect on the total C and N contents ($p < 0.05$). By contrast, the application of 7.5% and 15% PMs compost significantly increased the total C and N contents of soils D to G ($p < 0.05$), and the application of 30% PMs compost significantly increased the total C and N contents of all seven soil types ($p < 0.05$). The application of PMs compost significantly increased soil microbial activity in all seven soil types ($p < 0.05$), with larger amounts of PMs compost having greater effects.

Growth of the komatsuna plants, as indicated by the dry weights of individual plants, is shown in Table 6. In soils A and B, the application of 7.5% and 15% PMs compost significantly promoted plant growth ($p < 0.05$), while in soil C, only 7.5% PMs compost had a significant effect ($p < 0.05$), with 30% PMs compost inhibiting plant growth. In soils D to G, all levels of PMs compost significantly inhibited plant growth ($p < 0.05$).

IOP Conf. Series: Earth and Environmental Science **346** (2019) 012079 doi:10.1088/1755-1315/346/1/012079

Table 5. Total carbon (C) and nitrogen (N) contents, C/N ratios, and microbial activities in different soils after the addition of compost.

Soil†	Volume of PM added to soil (%)	Total C (g/kg)	Total N (g/kg)	C/N ratio	FDA‡
A	0	110.4 ± 9.01a	8.5 ± 0.73a	12.99	1.20 ± 0.044a
	7.5	116.1 ± 9.63a	8.7 ± 0.78a	13.34	1.59 ± 0.056b
	15	124.4 ± 10.66ab	9.0± 0.87a	13.82	1.58 ± 0.055b
	30	142.2 ± 11.72b	9.6 ± 0.95b	14.81	1.70 ± 0.069c
B	0	76.0 ± 6.09a	5.7 ± 0.49a	13.33	1.04 ± 0.040a
	7.5	83.6 ± 6.34a	6.0 ± 0.51a	13.93	1.32 ± 0.044b
	15	92.7 ± 6.75a	6.3 ± 0.55a	14.71	1.35 ± 0.053c
	30	112.4 ± 8.69b	7.2 ± 0.70b	15.61	1.60 ± 0.055d
C	0	64.8 ± 5.34a	4.3 ± 0.43a	15.07	0.69 ± 0.028a
	7.5	68.9 ± 5.32a	4.7 ± 0.43a	14.66	0.96 ± 0.036b
	15	83.1 ± 6.27b	5.4 ± 0.51b	15.39	1.08 ± 0.044c
	30	96.5 ± 7.06c	5.8 ± 0.57b	16.64	1.14 ± 0.048d
D	0	43.3 ± 3.02a	3.1 ± 0.24a	13.97	0.42 ± 0.022a
	7.5	49.4 ± 4.03b	3.4 ± 0.32a	14.53	0.86 ± 0.038b
	15	61.4 ± 5.36c	3.9 ± 0.43b	15.74	1.03 ± 0.035c
	30	83.9 ± 7.29d	4.9 ± 0.59b	17.12	1.06 ± 0.046c
E	0	18.0 ± 0.98a	1.3 ± 0.08a	13.85	0.29 ± 0.01a
	7.5	24.4 ± 1.01a	1.6 ± 0.08b	15.25	0.38 ± 0.012b
	15	33.7 ± 1.45b	2.0 ± 0.11b	16.85	0.47 ± 0.018c
	30	52.4 ± 3.18c	2.9 ± 0.25c	18.07	0.58 ± 0.022d
F	0	11.9 ± 0.42a	0.8 ± 0.03a	14.88	0.20 ± 0.007a
	7.5	18.8 ± 0.72b	1.1 ± 0.05b	17.09	0.32 ± 0.011b
	15	26.4 ± 1.09c	1.5 ± 0.08b	17.60	0.38 ± 0.018b
	30	44.8 ± 2.39d	2.4 ± 0.19c	18.67	0.46 ± 0.015c
G	0	2.6 ± 0.21a	0.2 ± 0.01a	13.00	0.19 ± 0.009a
	7.5	7.9 ± 0.61b	0.4 ± 0.04b	19.75	0.24 ± 0.007b
	15	15.4 ± 0.82c	0.8 ± 0.06b	19.25	0.27 ± 0.009bc
	30	32.1 ± 1.64d	1.6 ± 0.13c	20.06	0.32 ± 0.012c

† A: forest soil alone; B: 1:1 ratio of forest soil: subsoil; C: 1:2 ratio of forest soil: subsoil; D: subsoil alone; E: 1:1 ratio of subsoil: sand; F: 1:2 ratio of subsoil: sand; G: sand alone.
‡ Fluorescein diacetate (FDA) hydrolysis was used to indicate microbial activity. The values are absorbances of the sample liquid at 490 nm [15].
Different lower-case letters in the same column indicate significant differences among treatments ($p < 0.05$).

Table 6. The mean (± SD) dry weight of the plants in each treatment group.

Soil†	**0%‡**	**7.50%‡**	**15%‡**	**30%‡**
A	0.83 ± 0.03	1.08 ± 0.04*	1.10 ± 0.04*	0.92 ± 0.06
B	0.75 ± 0.04	0.89 ± 0.02*	0.92 ± 0.04*	0.72 ± 0.05
C	0.64 ± 0.03	0.74 ± 0.02*	0.65 ± 0.04	0.48 ± 0.01*
D	0.56 ± 0.05	0.49 ± 0.03*	0.42 ± 0.02*	0.27 ± 0.03*
E	0.51 ± 0.05	0.45 ± 0.01*	0.36 ± 0.05*	0.24 ± 0.1*
F	0.39 ± 0.05	0.33 ± 0.02*	0.29 ± 0.01*	0.21 ± 0.01*
G	0.31 ± 0.01	0.25 ± 0.01*	0.21 ± 0.01*	0.15 ± 0.04*

† A: forest soil alone; B: 1:1 ratio of forest soil: subsoil; C: 1:2 ratio of forest soil: subsoil; D: subsoil alone; E: 1:1 ratio of subsoil: sand; F: 1:2 ratio of subsoil: sand; G: sand alone.

IOP Conf. Series: Earth and Environmental Science **346** (2019) 012079 doi:10.1088/1755-1315/346/1/012079

‡ Percentage of compost made from pruning materials (PM) added to the soil.
* Significantly different from the corresponding control group ($p \leq 0.05$).

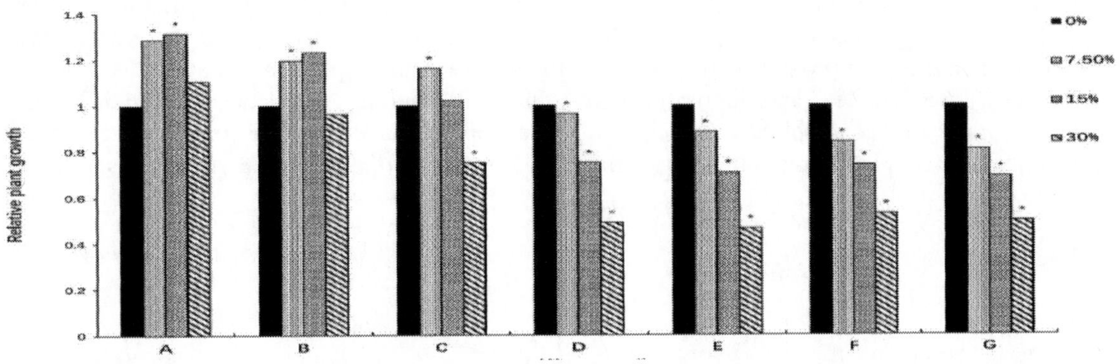

Figure 1. Relative growth of komatsuna (*Brassica rapa* var. *perviridis*) plants in each treatment group compared with its corresponding control group. Four concentrations of compost made from pruning materials (PM) were applied to seven different soil types: (A) forest soil alone, (B) 1:1 ratio of forest soil: subsoil, (C) 1:2 ratio of forest soil: subsoil, (D) subsoil alone, (E) 1:1 ratio of subsoil: sand, (F) 1:2 ratio of subsoil: sand, and (G) sand alone. Relative plant growth was calculated as the ratio of the dry weight of an individual plant in a particular treatment group compared with the corresponding control group. * Significant difference compared with the control group ($p \leq 0.05$).

The ratio of the dry weight of a particular treatment group to the dry weight of the corresponding control group was used as an indicator of the effect of PM application on plant growth, whereby a ratio of >1 indicates plant growth promotion and a ratio of <1 indicates plant growth inhibition (Figure 1).

4. Discussion

The application of 7.5% PMs compost significantly promoted plant growth in soils A–C, and 15% PMs compost also significantly promoted plant growth in soils A and B (Figure 1). However, plant growth was significantly inhibited by the application of 7.5% and 15% PMs compost in soil types D–G and 30% PMs compost in all seven soil types. Compost contains a large number of microorganisms [17] and PMs compost contain a large amount of mineralizable carbon, which increases soil microbial activity. Consequently, the application of PMs compost increased microbial activity in the soil (Table 5), which resulted in the consumption of inorganic N in the soil [7] and N starvation.

The relationship between relative plant growth and the total N content of the soil is shown in Fig. 2. This demonstrates that when the same amount of PMs compost was applied to the soil, there was a positive correlation between the total N content of the soil and relative plant growth, indicating that the N concentration regulated plant growth in this experiment. Plants and microorganisms compete for N in the soil, so the consumption of available N by microorganisms decreases the amount that is available for plant growth [18]. Consequently, the application of PMs compost to soils that had a low total N content caused a sharp increase in microbial activity and contributed to N starvation [19], which significantly inhibited plant growth (Figure 1). As more amounts of PMs compost are applied to the soil, more microorganisms and mineralizable carbon are added, resulting in higher levels of plant growth inhibition.

The relationship between soil microbial activity and relative plant growth is shown in Figure 3. There was a positive correlation between the original soil microbial activity and relative plant growth. The application of PMs compost to soil that already had a relatively high microbial activity (soils A–C) had a greater effect on soil microbial activity because more microorganisms were present (Table 5). There is a positive relationship between net N mineralization and N uptake by plants [18] and the application of PMs compost enhances nutrient release [20]. Consequently, in soils A–C, the high original microorganism content in the soil contributed to the shortened process of organic matter mineralization and the reduced time until the net mineralized N was released from the applied PMs compost to the soil, thus increasing plant N uptake and enhancing nutrient release. Therefore, the application of 7.5% and 15% PMs compost increased net N mineralization and provided the necessary nutrients for plant growth, resulting in the immediate promotion of plant growth. However, the application of 30% PMs compost added excess mineralizable C and microorganisms to the soil, resulting in the consumption of N being higher than the amount supplied. Consequently, the application of 30% PMs compost did not significantly promote plant growth in soils A and B, and appeared to inhibit plant growth in soil C.

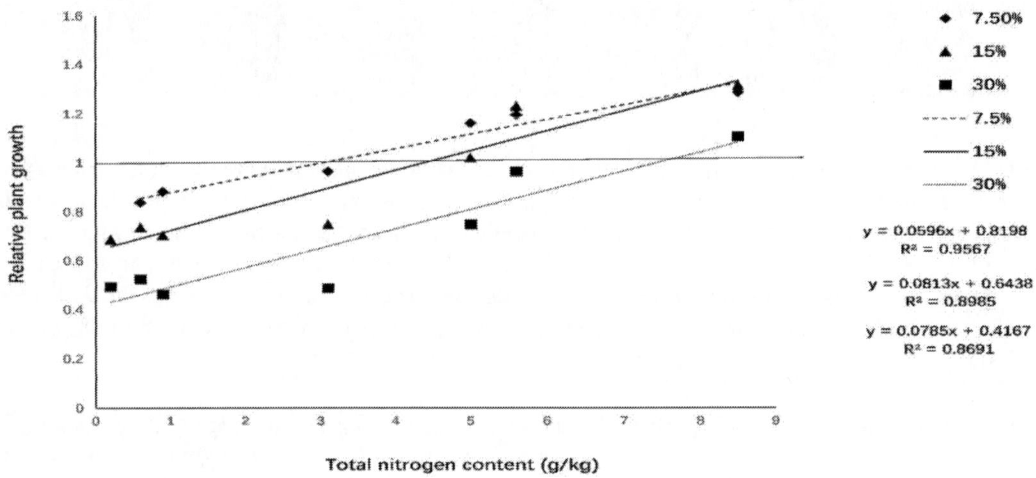

Figure 2. Relationship between the total nitrogen content and relative plant growth. Relative plant growth was calculated as the ratio of the dry weight of an individual plant in a particular treatment group to the corresponding control group.

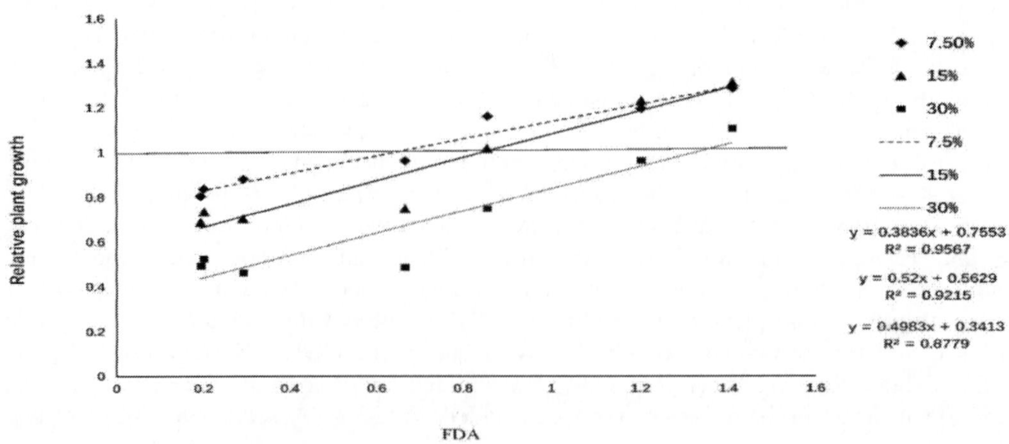

Figure 3. Relationship between microbial activity and relative plant growth. The fluorescein diacetate (FDA) value is shown as the amount of FDA hydrolyzed per 1 g dry weight of sample. Relative plant growth was calculated as the ratio of the dry weight of an individual plant in a particular treatment group to the corresponding control group.

To improve the application of PMs compost in a wide range of soils with different microbial activities, future studies should determine the duration of the plant growth inhibition period to ascertain the best time to start planting after mixing PMs compost. In addition, the inorganic N dynamics that occur during the mineralization process after mixing PMs compost into the soil should be clarified to determine the amount and timing of PM addition that is required for various soil types.

5. Conclusion

Because PMs compost contain a large amount of mineralizable C, its application to the soil can readily cause N starvation due to the increased microbial activity. During the early stage of PMs compost application, the addition of 7.5% and 15% PMs compost effectively promoted plant growth in fertile soils (total N content > 4.6 g/kg). However, the application of 30% PMs compost to fertile soils led to an excessive consumption of N, resulting in no promotion, and even the inhibition, of plant growth. Furthermore, the application of any PM to infertile soil (total N < 4.6 g/kg) inhibited plant growth during the early stage, with this effect becoming more pronounced with increasing amounts of PMs compost. Therefore, further research is required to determine the duration of N starvation as a result of PMs compost application to determine the optimal amount and timing.

Acknowledgments

We would like to thank the Agora Landscape Architecture and Weihai Weinong Biotechnology Corporation for providing PMs compost and support. We would like to acknowledge the professional support of Professor Tatsuaki Kobayashi and Akira Kato of Chiba University for their advice. We are also thankful to Mr. Hitomi Takoya, for his assistance during the experiment. The experiments comply with the current laws of the country in which they were performed.

Reference

[1] Zimmerman A R, Gao B and Ahn M Y 2011. Positive and negative c mineralization priming effects among a variety of biochar-amended soils. *Soil Biol. Biochem.* **43**(6): 1169–79. doi: doi.org/10.1016/j.soilbio.2011.02.005

[2] Benito M A, Masaguer R De Antonio and Moliner A 2005. Use of pruning waste compost as a component in soilless growing media. *Bioresour. Technol.* **96**(5): 597–603. doi: doi.org/10.1016/j.biortech.2004.06.006

[3] Takahashi T, Kanbara D, Ishii M, Ogino J, Harada H, Yairo H, Yamada H and Torigoe A 2014. Estimation of carbon dynamics in decomposition of pruning material. *Landscape Res. Japan* Online **7**(0): 17–19. doi: http://doi.org/10.5632/jilaonline7.17 (in Japanese)

[4] Boldrin A, Andersen J K, Moller J, Christensen T H and Favoino E 2009. Composting and compost utilization: accounting of greenhouse gases and global warming contributions. *Waste Manag. Res.*: The Journal of the International Solid Wastes and Public Cleansing Association, ISWA **27** (8): 800–812. doi: 10.1177/0734242x09345275

[5] Takahashi T, Iizumi K, Hirano M, Hirano Y and Matsuda H 2009. Soil improvement by mixing with woody compost in crop land. *J. Japan. Soc. Reveget. Tech.* **35**(1): 194–197. doi: 10.7211/jjsrt.35.194 (in Japanese)

[6] Liu E, Takahashi T and Liu C 2016. The effect of compost made from pruning materials and traditional compost on soil chemical properties. *Forest. Environ. Sci.* **32**(2): 68–72. doi: 10.3969/j.issn.1006-4427.2016.02.013. (in Chinese)

[7] Zimmerman A R, Gao B and Ahn M Y 2011. Positive and negative c mineralization priming effects among a variety of biochar-amended soils. *Soil Biol. Biochem.* **43** (6): 1169–1179. doi: doi.org/10.1016/j.soilbio.2011.02.005

[8] Dempster D, Gleeson D, Solaiman Z I, Jones D and Murphy D 2012. Decreased soil microbial biomass and n mineralisation with eucalyptus biochar addition to a coarse textured soil. Plant Soil **354**(1-2): 311–324. doi: 10.1007/s11104-011-1067-5

[9] Benito M A, Masaguer R De Antonio and Moliner A 2005. Use of pruning waste compost as a component in soilless growing media. *Bioresour. Technol.* **96** (5): 597–603. doi: doi.org/10.1016/j.biortech.2004.06.006

[10] Thiessen S, Gleixner G, Wutzler T and Reichstein M 2013. Both priming and temperature sensitivity of soil organic matter decomposition depend on microbial biomass–an incubation study. *Soil Biol. Biochem.* **57**: 739–748. doi: 10.1016/j.soilbio.2012.10.029

[11] Cambardella C A, Richard T and Russell A 2003. Compost mineralization in soil as a function of composting process conditions. *Europ. J. Soil Biol.* **39** (3): 117–127. doi: 10.1016/s1164-5563(03)00027-x

[12] Henriksen T and Breland T 1999. Nitrogen availability effects on c mineralization, fungal and bacterial growth, and enzyme activities during decomposition of wheat straw in soil. *Soil Biol. Biochem.* **31** (8): 1121–1134. doi: 10.1097/00010694-200002000-00007

[13] Riffaldi R, Saviozzi A and Levi-minzi R 1996. C Mineralization kinetics as influenced by soil properties. *Biol. Fertil. Soils* **22** (4): 293–298. doi: 10.1007/s003740050114

[14] Toda H and Haibara K 1999. Effects of C properties on characteristics of nitrogen mineralization in forest soil of kanto region, Japan. *Japanese J. Forest Environ.* **41** (2): 59–66. doi: https://doi.org/10.18922/jjfe.41.2_59

[15] Ichikawa A and Urakwa R 2008. A simple method of measuring of microbial avtivities in organic matter using a FDA (Fluorescein Diacetate) Hydrolysis. *Japan. Soc. Forest Environ.* **50** (2): 175–177. doi: 10.7211/jjsrt.26.337 (in Japanses)

[16] López-González J A, Suárez-Estrella F, Vargas-García M C, .López M M J, Jurado M and Moreno J 2015. Dynamics of bacterial microbiota during lignocellulosic waste composting: studies upon its structure, functionality and biodiversity. Bioresour. Technol. **175**: 406–416. Doi: 10.1016/j.biortech.2014.10.123

[17] LóPez-GonzáLez J.A, L lóPez M.J, Vargas-García M.C, Suáez-Estrella F, Jurado M, Moreno J .2013. Tracking organic matter and microbiota dynamics during the stages of lignocellulosic waste composting. Bioresource Technology 146:574-584. doi: 10.1016/j.biortech.2013.07.122

[18] Kaye J P and Hart S C 1997. Competition for nitrogen between plants and soil microorganisms. *Trends Ecol. Evol.* **12** (4): 139–143.doi： 10.1016/S0169-5347(97)01001-X

[19] He Z, Alva A, Yan P, Li Y, Calvert D, Stoffella P and Banks D 2000. Nitrogen mineralization and transformation from composts and biosolids during field incubation in a sandy soil. *Soil Sci.* **165** (2): 161–169. doi: 10.1097/00010694-200002000-00007

[20] Smith J, Papendick R, Bezdicek D and Lynch J 1993. Soil organic matter dynamics and crop residue management. *Soil Microb. Ecol.*, 65–95. New York: Marcel Dekker.

IOP Conf. Series: Earth and Environmental Science **346** (2019) 012080 doi:10.1088/1755-1315/346/1/012080

Optimization of process conditions for the development of rice milk by using response surface methodology

P Plengsaengsri[1], T Pimsuwan[2], T Wiriya-Aree[2], J Luecha[2], S Nualkaekul[1] and P Deetae[1, 3]

[1]Faculty of Agro-industry, King Mongkut's Institute of Technology Ladkrabang, Bangkok 10520, Thailand
[2]Food Technology Program, Kanchanaburi Campus, Mahidol University Kanchanaburi 71150, Thailand

E-mail: pawinee_deetae@hotmail.com

Abstract. Rice milk is a dairy product alternative for who are allergenic milk protein, lactose and concern about cholesterol. Unlike cow's milk, rice milk contains little protein but has a higher carbohydrate content. The purpose of this research was to study the optimal condition to develop rice milk nutritionally equivalent to cow milk with consumer acceptability. In this study, we used rice protein concentrate, rice flour, rice bran oil, sugar and calcium lactate as ingredients. Firstly, the selection of enzymes for rice protein hydrolysis was carried out and neutrase had high degree of hydrolysis with the highest consumer acceptability ($p<0.05$). Response surface methodology (RSM) base on central composite design was then employed to investigate the optimum condition with three independent variables: enzyme: protein ratios, speeds and times of homogenization. Our results reveal that the experimental variables significantly affected the color and sensory attributes. The optimum condition for rice milk production was found to be 16000 rpm,20 minutes of homogenization and 1:40 (w/w). The milk produced from this condition had protein 3.14 %, fat 3.28%, fiber 0.45%, ash 0.44% moisture 86.85% and carbohydrate 5.84%, which was similar to the nutritional value of cow's milk. The developed product showed good nutritional and sensory characteristics and this could be used for the development of commercial rice milk beverage after conducting appropriate scale up and pilot plant study.

1. Introduction

Cow's milk is one of the most consumed food relished by human population since ancient time due to its nutritional value and its convenience in satiating appetite. Nowadays, its consumption has raise concerned about the health awareness. Clinical researches have demonstrated that some components of cow's milk are allergy, hypercholesterolemia, lactose intolerance or coronary heart accomplice with health problems including milk protein diseases [1,2,3]. For these reasons, consumers have become more interested in non-dairy alternatives over cow's milk. In recent years, plant-based milks products such as soy milk, almond milk, coconut milk, rice milk and cashew milk have been lanched on markets by various food industries worldwide and rice milk is one of the most popular dairy free alternatives in Asian countries such as China, Japan, Vietnam and Thailand [4]. This beverage can be prepared by using milled broken rice or rice flour with water and communition to a size range of 5-20 μm through homogenization [5]. Heating is applied during the milk process, mostly to ensure food safety and extend the shelf life of the product [6]. Rice based milk generally present inferior nutritional substitutes of

Content from this work may be used under the terms of the Creative Commons Attribution 3.0 licence. Any further distribution of this work must maintain attribution to the author(s) and the title of the work, journal citation and DOI.
Published under licence by IOP Publishing Ltd

bovine milk, in which it is very rich in carbohydrate (9.4-14.2 g 100 ml^{-1}) but contains small amounts of protein (0.1-0.8 g 100 ml^{-1}) and fat (0.8-0.2 g 100 ml^{-1}) also lacks of calcium and vitamins [7]. Consumption of rice milk as an alternative's without suitable care can result in malnutrition, especially in the case of infants and toddlers [8].

Rice base milk alternative contains blend of rice protein concentrate, rice bran oil, sugar and calcium lactate can increase its nutritional value and could be equivalent to cow's milk nutrition. Rice protein is hypoallergenic and suitable for people allergic to cow or soy milks [9]. The use of rice protein ingredient in food and sports nutrition as an alternative to the more commonly used casein, whey and soy protein, is increasing but limit its applications to products that do not require high solubility of the protein ingredient such as meat analogues, baked goods, protein bars because rice protein is low solubility in water [10]. The functionality of rice protein may be improved by using enzyme hydrolysis. According to the study of Humiski *et al.* [11] protease enzymes such as an alcalase, flavourzyme, trypsin can be used to increase the solubility of pea proteins. The effect of enzymatic hydrolysis is dependent on various factors such as type of enzyme [12,13] treatment conditions (enzyme-substrate ratio, reaction temperature, time of enzyme action, characteristics of substrate [14]. Moreover, it has been found that using combination between homogenization and high temperature during milk production could increase protein solubility [15,16]. Response surface methodology (RSM) is a widely used mathematical and statistical method for modeling and optimizing a process, which requires minimum experimentation [17]. The purpose of this research was to evaluate the effect of enzymatic hydrolysis of rice protein concentrates on the solubility of the resulting hydrolysates and study the optimal condition to develop rice milk nutritionally equivalent to cow's milk with consumer acceptability.

2. Materials and methods

2.1. Materials

Rice flour was purchased from a local supplier (Thaiflour industry company limited, Thailand). Rice protein concentrate was obtained by Fenchem co. ltd and rice bran oil was obtained by Thai edible oil co., ltd (Thailand). Calcium lactate was purchased from Bkkchemi Co., Ltd. Rennet enzyme 2000 U/g was obtained by Tinnakorn Chemical and Supply Co., Ltd. Flavourzyme1000LAPU/g, Alcalase 2.4 AU-A/g and Neutrase 0.8AU-Nu/g were supplied by Brenntag Ingredients (Thailand) Co., Ltd. All other chemicals and reagents were of analytical grade.

2.2. Enzymatic screening for improved protein solubility

2.2.1. Hydrolysis of rice protein concentrate. Enzymatic hydrolysis of rice protein concentrate was studied by using the modified method of Barać *et al.* [18]. Four protease enzymes including rennet, flavourzyme, alcalase and neutrase were explored as the part of enzymatic screening. The factors affecting the degree of hydrolysis such as pH (5, 6, 7) and reaction time (0, 15, 30, 60, 120 minutes) were also investigated. Protein concentrate solution was prepared at 4% (w/w) then the pH was adjusted to 5, 6, 7 with 1 M NaOH or 1M HCl. The enzyme to substrate (E/S) ratio was 1:200 was later added to the solution. Incubated at 50 °C then inactivate the enzyme immediately at 80 °C for 10 min.The hydrolyzed proteins were immediately used for the determination of the degree of hydrolysis.

2.2.2. Degree of hydrolysis. Degree of hydrolysis (DH) was determined according to the method of Barać *et al.* [18]. The protein hydrolysate solution was centrifuged at 12000×g for 15 minutes. Then 10 ml of supernatant was taken to mix with 10 ml of 20% w/v trichloroacetic acid (TCA) and centrifuged at 12,000 g for 15 minutes. The soluble protein content of the supernatant was determined by the Lowry method [19] using bovine serum albumin as the standard. Total nitrogen was determined from 10 ml of suspension prepared in the same way as for enzymatic hydrolysis, but without the enzyme [20].

DH(%) = (soluble protein content in 20 %w/v of TCA (mg) / total protein content (mg)) x 100

2.3. *Manufacture of rice milk*

First, preparing rice protein hydrolysates using 4 % (w/v) rice protein with 0.4-0.04 g of enzymes at the reaction temperature of 50°C then inactivating at 80°C for 10 minutes. Then the rice protein hydrolysate was mixed with the following ingredients: 6.1 g rice four gelatinized (12 %w/v of rice flour gelatinized at 70 °C 5 minutes) according to Coda *et al.* [21], 4 g rice bran oil, 4.5 g sugar and 1g calcium lactate. Then homogenized at speed varying from 8000 rpm to 16000 rpm for 5 to 20 minutes. Each rice milk formulation was suggested by the mixture design based on a central composite design as shown in **Table 1.**

2.4. *Experimental design*

The optimum condition for development of rice milk formulation was determined by Central Composite Design with response surface methodology (RSM) using 3-level-3-factors in the Design-Expert (7.00) software. In the experimental design, it showed twenty formulations with six replications of the central point. Independent variable parameters were speed of homogenize (X_1), time of homogenize (X_2), enzyme content (X_3) and varied from -1 to +1 **(Table1)**. Dependent variable parameters were apparent viscosity, color expressed as L*, a*, b* values and sensory analysis. The regression model was used to generate contour maps for optimization and explanation of parameters effecting rice milk properties.

Table 1. Independent variables and their levels used for Central Composite Rotatable Design.

Independent variables	Symbol	Coded variable levels		
		-1	0	+1
Speed of homogenize (rpm)	X_1	8000	12000	16000
Time of homogenize (min)	X_2	5.00	12.50	20.00
Content of enzyme (g)	X_3	0.04	0.22	0.40

The regression model equation for viscosity (Y_1), color L* (Y_2), color a* (Y_3), color b* (Y_4), bitterness (Y_5), smoothness (Y_6), roughness (Y_7), aftertaste(Y_8) could be speculated as follows:

$$Y_n = \beta_0 + \beta_1 X_1 + \beta_2 X_2 + \beta_{12} X_1 X_2 + \beta_{13} X_1 X_3 + \beta_{23} X_2 X_3 + \beta_{11} X^2_1 + \beta_{22} X^2_2 + \beta_{33} X^2_3$$

Where a dependent variable Y expressed response viscosity, color L*, color a*, color b*, bitterness, smoothness, roughness, aftertaste. The β_0, β_1 and β_2 expressed constant coefficients of the intercept, linear, quadratic and interaction terms, respectively. After study the optimum condition validation was carried out to check the validity of the predicted value suggested from RSM.

2.5. *Apparent viscosity measurement*

The apparent viscosity of rice milk was measured by viscometer (Brookfield DV-III Ultra programmable viscometer, MA, USA) with small sample adapter (spindle sc4-31) at sample size of 9 ml. Five different shear rates was used (6.60 13.2 19.8 26.4 and 33.0 S^{-1}). The results are presented in pascal-seconds (Pa.s) [22].

$$\tau = K\dot{\gamma}^n$$

τ is shear stress (Pa), k is consistency coefficient in (Pa/s), $\gamma^{\cdot}$ is shear rate (s^{-1}) and n is flow-behaviour index.

2.6. *Color measurement*

Color was measured using a hunter colorimeter (Color Quest II, Reston, USA). Samples were contained in optical glass cells 3.8 cm high and 6 cm diameter. Results were given in CIELAB system for illuminant D 65 and a 10° angle of vision. Registered parameters were: L* (brightness), a* (red component), b* (yellow component) [23].

2.7. Sensory analysis
Sensory analysis of the hydrolytic rice protein concentrate solution as obtained from 2.2.1 was evaluated using 9 points hedonic scale (1 = "liked extremely", 5 = "neither liked nor disliked" and 9 = "disliked extremely") with 30 untrained panelists for bitterness, roughness and overall preference according to the method of Pokora M *et al.* [24]. For analysis each formulas of rice milk products from experiment design using intensity analysis modified method of Christine Wilson *et al* [25] by using ten-point scale (rating from 1 to 10) to rate each descriptor's intensity with 30 untrained panelists for bitterness, roughness, smoothness and aftertaste.

2.8. Proximate analysis
Calcium content and proximate analysis procedures for moisture, protein (N*6.25), fat, ash, fiber, carbohydrate of rice milk product was performed according to AOAC 2000 [26].

2.9. Statistical analysis
Analysis of variance (ANOVA) was carried out to analyze the differences between groups. Data was indicated as mean ± SD and statistically significant differences were determined at $p \leq 0.05$. MINITAB release 16 (Minitab Inc., State college, USA) was used to carry out all statistical analyses.

3. Result and discussion

3.1. Enzymatic screening for improved rice protein solubility

3.1.1. Effect of pH on enzymatic hydrolysis. In this experiment, the enzyme to protein ratio was fixed at 1:200 g and the reaction time was set at 60 minutes. It was found that pH significantly influenced on enzymatic hydrolysis ($p \leq 0.05$) as shown in **Table 2**. Among the enzymes studied, alcalase showed the highest degree of rice protein hydrolysis, followed by neutrase, flavourzyme and rennet, respectively and pH 7 was the optimum condition. The obtained results were compatible with those reported by Zhang *et al.* [27] who studied the solubility of soy protein and rice protein at various pH with enzyme acalase. They found that rice proteins at pH more than 6 had higher degree of hydrolysis and Amiza *et al.* [28] found that neutrase to hydrolysis mixture of angle wing clam flesh at pH 7 and long reaction time caused higher % Degree of hydrolysis

Table 2. Degree of rice protein hydrolysis from different enzymes at various pH.

pH	%DH			
	Alcalase	Neutrease	Flavourzyme	Rennet
pH 5	0.9^{C}±0.01	1.01^{C}±0.003	1.19^{A}±0.04	0.38^{A}±0.02
pH 6	5.68^{B}±0.37	3.62^{B}±0.07	0.88^{C}±0.03	0.30^{B}±0.03
pH 7	8.79^{A}±0.05	5.74^{A}±0.14	1.02 ^{B}±0.02	0.29^{B}±0.01

Values that are not followed by the same letter within the same column are significantly different ($P \leq 0.05$).

3.1.2. Effect of reaction time on enzymatic hydrolysis. In this experiment, the ratio of enzyme to protein was fixed at 1:200 and at pH 6. **Table 3** showed that when the reaction time increased, the % degree of hydrolysis increased significantly ($p \leq 0.05$). The enzyme reaction time of 120 minutes had the highest % Degree of hydrolysis. Haslaniza *et al.* [29] reported that a longer reaction time would allow enzyme to act more extensively on the protein, thus resulting in more % degree of hydrolysis.

Table 3. Degree of rice protein hydrolysis from different enzymes at various reaction time.

Time	%DH			
	Alcalase	Neutrease	Flavourzyme	Rennet
0	2.84^{E}±0.12	0.94^{D}±0.02	0.58^{C}±0.02	0.24^{C}±0.00
15	3.49^{D}±0.02	3.07^{C}±0.01	0.67^{B}±0.02	0.29^{C}±0.01
30	4.57^{C}±0.02	3.63^{B}±0.04	0.7^{B}±0.02	0.33^{B}±0.01
60	5.68^{B}±0.37	3.64^{B}±0.07	0.88^{A}±0.03	0.30^{B}±0.03
120	6.33^{A}±0.18	5.84^{A}±0.12	0.92^{A}±0.00	0.37^{A}±0.01

Values that are not followed by the same letter within the same column are significantly different ($P \leq 0.05$).

3.1.3. Sensory analysis for enzyme selection. Since enzyme alcalase and netrease had a higher DH than rennet and flavourzyme from the previous experiments therefore we selected alcalase and netrease enzymes for sensory analysis. neutrase enzymes had higher score in aroma, sweetness, bitterness, roughness and overall acceptance than alcalasse. Therefore, neutrase enzyme was selected for the study of the optimum conditions for rice milk production.

3.2. Conditions for optimum responses

The experimental results using RSM with CCD were presented in **Table 4** and **5**. Three independent variables used were speed of homogenization (X1), time of homogenization (X2), and enzyme content (X3) with three levels for each variable. While the dependent variables were the apparent viscosity, color expressed as L*, a*, b* values and sensory analysis

Table 4. Central composite design for optimizing condition to rice milk production and responses of physical properties.

Run number	Variable levels			Response variables			
	X_1	X_2	X_3	Y_1	Y_2	Y_3	Y_4
1	0	0	0	7.37	3.52	0.1	-0.03
2	+1	+1	+1	6.95	3.73	0.19	0.12
3	-1	+1	-1	6.75	3.02	0.46	0.25
4	0	0	-1	7.56	2.91	0.06	0.48
5	-1	+1	+1	7.75	2.43	0.64	0.14
6	0	-1	0	6.90	2.79	0.31	0.34
7	0	0	0	7.77	2.86	0.08	0.26
8	+1	0	0	7.93	1.8	0.18	-0.02
9	0	0	0	7.44	2.71	0.31	0.06
10	0	+1	0	5.83	2.56	0.07	0.13
11	-1	-1	-1	7.13	1.98	1.37	0.3
12	0	0	0	5.72	2.11	0.7	0.2
13	-1	-1	+1	6.83	2.09	1.4	-0.18
14	0	0	-1	7.76	1.68	0.68	-0.08
15	-1	0	0	7.96	1.96	1.24	-0.18
16	+1	-1	+1	6.95	1.88	0.73	-0.44
17	0	0	0	6.61	1.71	0.69	-0.57
18	0	0	0	7.22	1.78	0.93	-0.23
19	+1	-1	-1	7.84	1.51	0.71	-0.32
20	+1	+1	-1	7.65	1.91	0.05	0.05

X_1= Speed of homogenization (rpm), X_2=Time of homogenization (minutes), X_3= enzyme content (g)
Y_1=Viscosity (pa.s), Y_2=Color L*, Y_3= Color a*, Y_4=Color b*

Table 5. Central composite design for optimizing condition to rice milk production and responses of sensory analysis.

	Variable levels			Response variables			
Run number	X_1	X_2	X_3	Y_5	Y_6	Y_7	Y_8
1	0	0	0	4.45	6.59	3.82	4.3
2	+1	+1	+1	6.49	6.58	3.26	3.78
3	-1	+1	-1	4.05	6.73	4.27	4.1
4	0	0	-1	5.95	6.5	3.16	3.94
5	-1	+1	+1	6.55	6.87	3.49	4.01
6	0	-1	0	3.86	6.73	3.47	4.05
7	0	0	0	4.39	6.76	3.62	4.16
8	+1	0	0	4.41	6.53	3.68	4.09
9	0	0	0	4.47	6.26	3.72	3.91
10	0	+1	0	4.85	6.27	3.88	4.56
11	-1	-1	-1	2.13	6.58	4.19	4.25
12	0	0	0	4.67	6.64	3.91	4.16
13	-1	-1	+1	6.54	6.28	3.25	4.01
14	0	0	-1	2.65	6.59	4.12	4.39
15	-1	0	0	4.59	6.59	3.79	4.18
16	+1	-1	+1	6.75	6.53	2.98	3.54
17	0	0	0	4.33	6.39	3.72	4.06
18	0	0	0	4.38	6.63	3.85	4
19	+1	-1	-1	2.24	6.08	4.05	4.06
20	+1	+1	-1	2.19	6.29	3.71	4.09

X_1= Speed of homogenization (rpm), X_2=Time of homogenization (minutes), X_3= enzyme content(g), Y_5= Bitterness, Y_6=Smoothness, Y_7=Roughness, Y_8=Aftertaste

The quadratic model suggested from program and used for optimization and explanation of parameters effecting rice milk properties. Three independent variables did not affect the physical properties of the milk, in terms of the viscosity and the color L * b * values and the sensory analysis of smoothness and aftertaste. In contrast, three factors affected color as a * value and sensory analysis of bitterness and roughness significantly ($p \leq 0.05$). From the results of regression analysis, the mathematical models for Color a* value (Y_3), bitterness(Y_5) and roughness(Y_7) as functions of speed of homogenize (rpm) (X_1), time of homogenize (min) (X_2) and content of enzyme (g) (X_3) can be expressed by the following equation.

$$Y_3 = 0.43-0.32x_1-0.31x_2+0.062x_3+ 0.059x_1 x_2 - 60250E\text{-}003x_1 x_3 + 0.034x_2 x_3 + 0.35x^2_1 -0.17x^2_2 +0.072x^2_3$$

$$Y_5 = 4.67-0.18x_1+0.26x_2+1.88x_3-0.28x_1 x_2 +0.24x_1 x_3-0.27x_2 x_3 - 0.5x^2_1 -0.65x^2_2 +1.18x^2_3$$

$$Y_7 = 3.70-0.13x_1+0.067x_2-0.38x_3-0.047x_1 x_2 +0.025x_1 x_3-0.098x_2 x_3 +0.15x^2_1 +0.092x^2_2 -0.32 x^2_3$$

3.3. Analysis of response surface graphs

Factors affecting a * values were speed and time of homogenization. From the figure (1A), it showed that when the time and speed of homogenization increased, color a * value decreased. Consistent of Park *et al.* [30] who found high pressure homogenized decreased both the a* and b* values of the natural plant-based model emulsion and homogenization reduced the fat granules and helped dispersion. When the amount of fat pellets increased, it caused more light reflection, resulting in lighter color [31]. The color of the sterilized milk can occur from Maillard reaction between free amino acids and reducing sugar at high temperature.

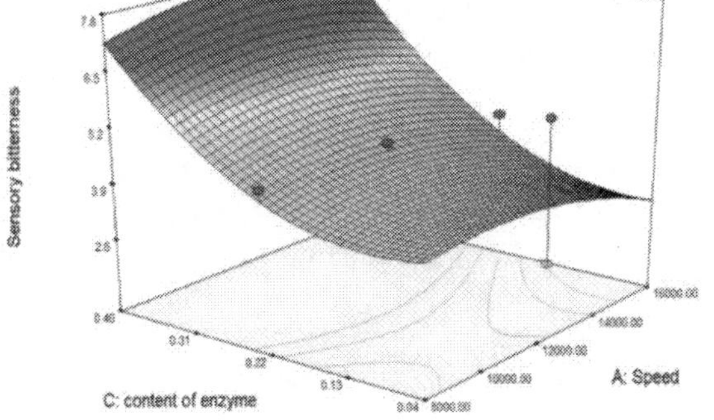

(A)

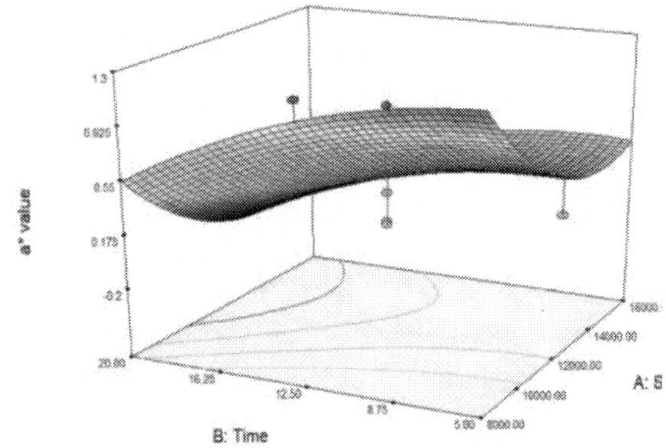

(B)

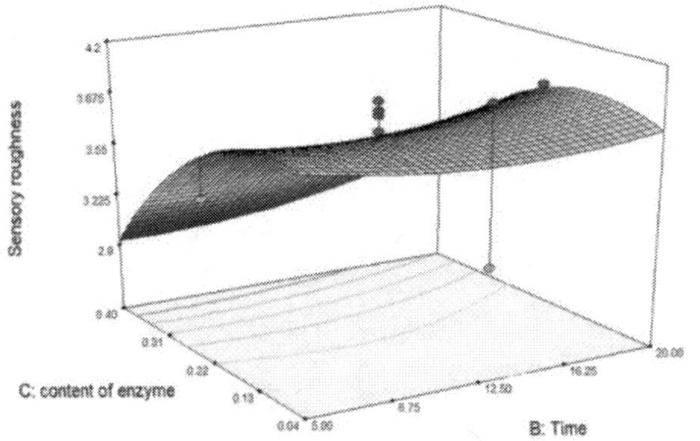

(C)

Figure 1. Response surface plots. (A) Speed of homogenize and Time of homogenize on color a *values, (B) Speed of homogenize and content of enzyme on sensory evaluation of bitterness, (C) Time of homogenize and content of enzyme on sensory evaluation of roughness.

The content of enzymes and the speed of homogenization affected to bitterness. Figure (1B) showed that as enzyme content increased, bitterness increased. In contrast, when the speed of homogenization increased, the bitterness was decreased. This can be due to that the neutrase is an endoprotease and can produce free amino acids or short peptides that have hydrophobic amino acid residues. Therefore, nonpolar amino acid residues at the C- terminus of the resulting peptides remain and cause a relatively high bitterness[32,33,34].Zhang *et al.* and Wang *et al.*[35,36] found a higher DH at higher content of enzyme to substrate is due to the increase of the contact chance of enzyme and protein and enhanced concentration of peptide bonds from hydrolysis by the proteases. Hence more content of enzyme, more hydrophobic amino acid residues relatively with bitterness of the product.

Enzyme content and the time of homogenization affected roughness. Figure (1C) showed that when the content of enzyme and the time of homogenization was increase, the roughness decreased. This could be that the enzyme could break the peptide bond to a free amino acid or short chain peptide, which would increase interactions between the hydrophilic amino acids and water molecules also can induce changes in the balance of net charges that lead to enhanced solubility [37]. Therefore, the high of enzyme content effect more enzymatic hydrolysis resulting in protein more solubility and less roughness. Jingqi Yang *et al.* [38] found the homogenization was demonstrated to modulate the hydrophobic interactions among protein aggregates, leading to dissociation of large protein aggregates into soluble supramolecular aggregates that due to can decrease aggregate size of protein, the solubility of protein increase. Hence, selection of speed and time of homogenization and content of enzyme is importance to produce rice milk with good acceptability. The optimum condition for sterilized rice milk production was found to be at 16000 rpm, 20 minutes of homogenization and 1:40 (w/w) of enzyme to protein.

3.4. Validation of the optimized conditions
In order to confirm the optimized conditions. This optimum conditions from RSM optimization approach was used to validate experimentally shown in **Table 6**. All predicted and experimental values corresponded very well showed the adequacy of selected models. After the validation process, optimized conditions use to produce rice milk product.

Table 6. Predicted and experimental values of responses for rice milk production using optimum process parameter.

Response variable	Predicted values	Experimental values
a* value	-0.130956	-0.12333
Bitterness	2.61882	2.78±2.07
Roughness	3.80658	3.92±1.37

3.5. Proximate analysis
The proximate analysis of rice milk product produced from the optimum condition had 3.14 % protein, 3.28% fat, 0.45% fiber, 0.44% ash, and 5.84%carbohydrate. It also contained 157.3 mg of calcium. The nutritional value of rice milk formulated in this study was like that of cow's milk and better than rice milk beverages commercial in Thailand as shown in **Table 7.**

Table 7. Chemical composition of cow's milk, rice milk and rice milk commercial in Thailand (g/100ml).

Chemical composition	Cow's milk[a]	Rice milk formulated	Rice milk beverages commercialized in Thailand[b]
Protein	2.9-5.0	3.14 ± 0.02	0.4-2.0
Fat	2.5-6.0	3.28 ± 0.13	1.0-1.8
Fiber	-	0.45 ± 0.03	0.0-4.4
Ash	0.6-0.9	0.44 ± 0.02	-
Moisture	85.5-89.5	86.85 ± 0.04	52.0-86.0
Carbohydrate	3.6-5.5	5.84 ± 0.04	2.28-13.6
Calcium (mg)	113-125	157.3	0.0-80.0

[a] source nutrition value of cow's milk [39].
[b] Proximate analysis of rice milk.
[c] rice milk commercial was explored from the market in Thailand.

4. Conclusion
The optimum condition for rice milk production was found to be at 16000 rpm, 20 minutes of homogenization and 1:40 (w/w) of enzyme to protein. The rice milk produced from this condition had nutrition value similar to cow's milk and better than rice milk beverages commercial in Thailand

Acknowledgements
This work was supported by Faculty of Agro-industry research fund (Grant number A118-61-010), King Mongkut's Institute of Technology Ladkrabang, Thailand.

Reference

[1] Hochwallner H, Schulmeister U, Swoboda I, Spitzauer S and Valenta R 2014 Cow's milk allergy: From allergens to new forms of diagnosis, therapy and prevention. *Methods* **66** 22–33
[2] Luthy K E, Larimer S G and Freeborn D S 2017 Differentiating between lactose intolerance, celiac disease, and irritable bowel syndrome-diarrhea. *J Nurse Practition.* **13**(5) 348–53
[3] Kumar B V, Vijayendra S V and Reddy S 2015 Trends in dairy and non-dairy probiotic product - a review *J. Food Sci. Tech.* **52**(10) 6112–24
[4] Mintel's Dairy Milk US 2017 Report and Non-Dairy Milk US 2017 report, (January 04th, 2018) Retrieved July 10, 2019, from https://www.mintel.com/press-centre/food-and-drink/us-non-dairy-milk-sales-grow-61-over-the-last-five-years
[5] Sethi S, Tyagi S K and Anurag R K 2016 Plant-based milk alternatives an emerging segment of functional beverages: a review *J. Food Sci. Tech.* **53**(9) 3408–23
[6] Deeth H 2017 Optimum thermal processing for extended shelf-life (ESL) milk *Foods* **6** 102
[7] Vitoria Miñana I. 2017 The nutritional limitations of plant-based beverages in infancy and childhood *Nutrición Hospitalaria ARAN Ediciones* **34**(5) 1205-14
[8] Chalupa-Krebzdak S, Long C J and Bohrer B M 2018 Nutrient density and nutritional value of milk and plant-based milk alternatives *Int. Dairy. J.* **87** 84–92
[9] Day L 2013 Proteins from land plants – potential resources for human nutrition and food security *Trends Food Sci. Technol.* **32** 25–42
[10] Amagliani L, O'Regan J, Kelly A L and O'Mahony J A 2017 The composition, extraction, functionality and applications of rice proteins: A review *Trends Food Sci. Technol* **64** 1–12
[11] Humiski L M and Aluko R E 2007 Physicochemical and bitterness properties of enzymatic pea protein hydrolysates *J. Food Sci.* **72**(8) 605–11
[12] Kim S Y, Park P S W and Rhee K C 1990 Functional properties of proteolytic enzyme modified soy protein isolate *J. Agric. Food Chem.* **38** 651–6
[13] Van der Ven C, Gruppen H, De Bont D B A and Voragen A G J 2002 Correlations between biochemical characteristics and foam-forming and stabilizing ability of whey and casein hydrolysates *J. Agric. Food Chem.* **50** 2938–46
[14] Betancur-Ancona D, Martínez-Rosado R, Corona-Cruz A, Caellanos-Ruelas A, Jaramillo-Flores M E and Chel-Guerrero L 2009 Functional properties of hydrolysates from Phaseolus lunatus seeds *Int. J. Food Sci. Technol.* **44** 128–37
[15] Park S H, Min S G, Jo Y J, Chun J Y 2015 Effect of high pressure homogenization on the physicochemical properties of natural plant-based model emulsion applicable for dairy products *Korean J. Food Sci. Anim. Resour.* **35**(5) 630–710
[16] Sikand V, Tong P, Vink S and Walker J 2012 Effect of powder source and processing conditions on the solubility of milk protein concentrates 80 *Milchwissenschaft* **67** 300–3
[17] Morshedi A and Akbarian M 2014 Application of responese surface methodology: design of experiments and optimization: a mini review *Indian J. Fund. Applied Life Sci.* **4** 2434-9
[18] Barać M, Čabrilo S, Pešić M, Stanojević S, Pavlićević M, Maćej O and Ristić N 2011 Functional properties of pea (Pisum sativum, L.) Protein isolates modified with chymosin *J. Molecular Sci.* **12** 8372-87
[19] Lowry O H, Rosebrough N J, Farr A L and Randall R J 1951 Protein measurement with the Folin phenol reagent *J. Biol. Chem.* **193** 265
[20] Silvestre M P C, Morais H A, Silva V D M and Silva M R 2013 Degree of hydrolysis and peptide profile of whey proteins using pancreatin *J. Brazilian Soc. Food Nutr.* **38**(3) 278–90
[21] Coda R, Lanera A, Trani A, Gobbetti M and Di Cagno R 2012 Yogurt-like beverages made of a

mixture of cereals, soy and grape must:Microbiology, texture, nutritional and sensory properties. *Int. J. Food Micr* **155**(3) 120–7

[22] Pushpadass H A, Emerald F M E, Balasubramanyam B V and Patel S S 2019 Rheological properties of milk-based beverages *Sci. Bev.* **9** 373-96

[23] Pandalaneni K, Amamcharla J K, Marella C and Metzger L E 2018 Influence of milk protein concentrates with modified calcium content on enteral dairy beverage formulations: Physicochemical properties *J. Dairy Sci.* **101**(11) 9714–24

[24] Pokora M, Eckert E, Zambrowicz A, Bobak Ł, Szołtysik M and Dąbrowska A 2103 Biological and functional properties of proteolytic enzyme-modified egg protein by-products. *Food Sci. Nutri.* **14** 184-95

[25] Christine W, Jeanne B, Wessel du T and Astrid B 2019.Matrix effects influencing the perception of 3-mercaptohexan-1-ol (3MH) and 3-mercaptohexyl acetate (3MHA) in different Chenin Blanc wines by Projective Mapping (PM) with Ultra Flash profiling (UFP) intensity ratings *Food Res. Int.* **121** 633–40

[26] Association of Official Analytical Chemists (AOAC). *Official Methods of Analysis of AOAC International* 17th ed AOAC International: Gaithersburg, MD, USA, 2000.

[27] Zhang H, Zhang J, Wang L, Guo XN. 2012.Preparation and functional properties of rice bran proteins from heat-stabilized defatted rice bran. *Food Res. Internation.* **4** 359-63.

[28] Amiza A M,Liyana H A and Zaliha H 2016 Optimization of enzymatic protein hydrolysis conditions to obtain maximum angiotensin-iconverting enzyme (ACE) inhibitory activity from Angel Wing Clam (Pholas orientalis) meat *J. Food Tech.* **2** 65-73

[29] Haslaniza H,Maskat M Y,Wan A and Mamot A 2010 The effects of enzyme concentration temperature and incubation time on nitrogen content and degree of hydrolysis of protein precipitate from cockle (Anadara granosa) meat wash water *Int. Food Res. J.* **17** (1) 147–52

[30] Park SH,Min S-G,Jo Y-J and Chun J-Y 2015 Effect of High Pressure Homogenization on the Physicochemical Properties of Natural Plant-based Model Emulsion Applicable for Dairy Products *Korean J. for Food Sci. of Ani. Res. Korean* **35** (5) 630–7

[31] Walstra P,Geurts T J, and Wouters J T M 2006 *Dairy Science and Technology* CRC Press Boca Raton, FL pp 325-8

[32] Adler-Nissen J 1986 *Enzymic hydrolysis of food proteins* Applied Science Publishers , New York pp 110 – 69

[33] Ishibashi N,Ono I,Kato K,Shigenaga T,Shinoda I,Okai H and Sakuzo F 1988 Role of the hydrophobic amino- acid residue in the bitterness of peptides *Agric. Biol. Chem* **52** 91 – 4

[34] Saha BC and Hayashi K 2001 Debittering of protein hydrolyzates *Biotech. Adv.* **19** 355 – 70

[35] Zhang Y and Wang Q 2007 Preparation of peanut functional oligopeptides by two-step-hydrolysis *Trans. Chin. Soc. Agric. Eng.* **23** 258–63

[36] Wang C,Feng B and Dong Y 2008 Optimization of the enzymatic hydrolysis of oat bran by Protamexprotease response surface method *Food Sci. Technol.* **5** 134–37

[37] Lee Jookyeong 2011 Soy protein hydrolysate; solubility, thermal stability, bioactivity, and sensory acceptability in a tea beverage A thesis submitted to the facculty of the graduate school of the univerity of minnesota p 12

[38] Yang J, Liu G, Zeng H and Chen L 2018 Effects of high pressure homogenization on faba bean protein aggregation in relation to solubility and interfacial properties *Food Hydrocoll.* **83** 275–86

[39] Gösta Bylund, M.Sc. 1994. Composition of cows' milk. *Dairy processing handbook* Teknotext Sweden : Tetra Pak Processing Systems pp 15-8

IOP Conf. Series: Earth and Environmental Science **346** (2019) 012081 doi:10.1088/1755-1315/346/1/012081

Responses of Thai recommended rice cultivars to high temperature in *PSII* efficiency at panicle initiation stage

N Sumthonglang[1,2], B Jongdee[1] and A Dongsansuk[1,2,3]

[1]Department of Agronomy, Faculty of Agriculture, Khon Kaen University, Khon Kaen 40002, Thailand
[2]Salt-Tolerant Rice Research Group, Khon Kaen University, Khon Kaen 40002, Thailand

E-mail: Danoma@kku.ac.th

Abstract. Photosynthesis is a major process for determining plant growth and development. The temperature plays important roles in plant photosynthesis. Thus, this research aimed to study the temperature response of *PSII* efficiency in different Thai recommended rice cultivars at panicle initiation (*PI*) stage. Youngest fully expanded leaves of fifteen rice cultivars namely, N22 (heat tolerance), IR64 (heat sensitive), CN1, Dular, PSL2, PTT1, Riceberry, RD29, RD31, RD41, RD49, RD57, RD61, RD63 and SPT1 at *PI* stage were exposed to temperature ranged from 25-55°C. During the exposure experiment to the temperature, coinciding measurements on the effective quantum yield of *PSII* efficiency ($\Delta F/F_m'$) and maximum quantum yield of *PSII* efficiency (F_v/F_m) were carried out. The experiment was designed in *CRD* with 4 replications. Temperature response to $\Delta F/F_m'$ and F_v/F_m showed a parabolic curve in all rice cultivars but F_v/F_m showed a wider parabolic curve than $\Delta F/F_m'$. Optimum temperature of $\Delta F/F_m'$ and F_v/F_m was 31.83-35.64°C and 31.82-34.10°C, respectively. The ranges of supraoptimal temperature of $\Delta F/F_m'$ and F_v/F_m were 47.15-50.60°C and 46.93-48.65°C, respectively. On the other hand, the high temperature stress of $\Delta F/F_m'$ and F_v/F_m was 50.80-54.70°C and 50.93-52.50°C, respectively. Optimum temperature of $\Delta F/F_m'$ showed high correlation with supraoptimal temperature especially in *cv.* PTT1. Broad sense heritability of $\Delta F/F_m'$ under high temperature stress was the highest which could be used as selection criteria for heat tolerance in rice breeding programme.

1. Introduction

Photosynthesis is an important process for plant growth and it is sensitive to temperature. Thus temperature has been considered as a major factor for photosynthesis and high temperature can result in a decline in plant leaf photosynthesis [1]. The higher than normal temperature causes plant heat injury and photosynthetic damage [2]. Haldimann and Feller [3] reported that when temperature was increased rapidly or gradually photosynthesis decreased with increasing leaf temperature. This kind of effect reduced photosynthesis more than 90% at 45°C compared to 25°C. The photosynthetic fluorescence parameters such as F_v/F_m also decreased with increasing temperature [4]. The F_v/F_m declined markedly at 45°C within 30 min after the beginning of treatment. It is an important parameter for assessing *PSII* efficiency whereby, decreasing F_v/F_m indicated loss of *PSII* efficiency. Therefore, changes in chlorophyll fluorescence can be used as plant screening method for heat tolerance. *PSII* efficiency was inhibited by high temperature and plant response to high temperature depended on the extent of increasing temperature, temperature duration and plant type [5]. When light falls on the

Content from this work may be used under the terms of the Creative Commons Attribution 3.0 licence. Any further distribution of this work must maintain attribution to the author(s) and the title of the work, journal citation and DOI.
Published under licence by IOP Publishing Ltd

leaves resulted in increasing in leaf temperature. Therefore, leaf temperatures can be used as indicators of the optimum temperature for plant photosynthesis under different conditions [6]. Measurements of $\Delta F/F_m'$ and F_v/F_m in response to temperature determined the whole thermal adaptation of photosynthesis [7] which used parabolic function of $\Delta F/F_m'$ for indicating optimum temperature, supraoptimal temperature and high temperature stress in tropical plant species. The purpose of the present research was to study the temperature response on photosynthetic efficiency of *PSII* in Thai recommended rice cultivars at panicle initiation stage and to indicate optimum temperature, optimum temperature range, supraoptimal temperature and high temperature stress by *PSII* efficiency response to temperature.

2. Materials and methods

Plant materials: Rice seeds of fifteen recommended rice cultivars such as N22 (heat tolerance), IR64 (heat sensitive), CN1, Dular, PSL2, PTT1, Riceberry, RD29, RD31, RD41, RD49, RD57, RD61, RD63 and SPT1 were used as experimental plant materials. Seeds were sterilized first by soaking in 70% alcohol for 1 minute followed by rinsing with tap water for 3 times. After that, the seeds were soaked in 10% Clorox for 15 minutes and rinsed further with tap water for 3 times. The seeds were soaked in tap water for 24 hours, and placed on wet paper until seeds germinated. Germinated seeds were transferred to pots of 20 cm in diameter containing paddy. Rice seedling was grown until panicle initiation (*PI*) stage in an open air greenhouse at the Agronomy Field Station, Faculty of Agriculture, Khon Kaen University, Thailand. At the *PI* stage photosynthetic efficiency of *PSII* in response to temperature were measured according to the method of Hacker and Neuner [8] and Dongsansuk [7]. Determination of *PSII* efficiency was carried out on youngest fully expanded leaf by using a chlorophyll fluorometer (Mini-PAM; Walz, Effeltrich, Germany). For preparing leaf sample, each leaf was placed on wet paper for keeping the leaf humidity under normal condition. The samples were then put in a temperature controlled cabinet (VRV. Corp. Ltd, Thailand) set with a relative humidity of 70%. All the samples were exposed to a temperature ranged from 25°C reached to 55°C coincided recorded fluorescence parameters and generated fluorescence response curve. The following fluorescence parameters were measured: effective quantum yield of *PSII* efficiency; $\Delta F/F_m'$ under light intensity at 61.80 μmol m^{-2} s^{-1} and maximum quantum yield of *PSII* efficiency; F_v/F_m under dark 30 min condition according to Schreiber [9]. $\Delta F/F_m'$ and F_v/F_m response curve was calculated for optimum temperature (T_{opt}), optimum temperature range, supraoptimal temperature and high temperature stress. Broad sense heritability (h_b^2) was calculated as following: $h_b^2 = \delta^2_G / \delta^2_P$, according to Holland *et al* [10] (δ^2_G; genotypic variance, δ^2_P; phenotypic variance, $\delta^2_P = \delta^2_G + \delta^2_E$ and δ^2_E; error variance).

This experiment was designed in Completely Randomized Design (*CRD*) with 4 replications. Data were analyzed by one-way ANOVA and Duncan Multiple Range Test at $p \leq 0.01$ by software SPSS for windows version 16 to compare the significant differences among different rice cultivars.

3. Results and discussion

Temperature response of effective quantum yield of *PSII* efficiency; $\Delta F/F_m'$ and maximum quantum yield of *PSII* efficiency; F_v/F_m exposed to temperature range of 25-55°C in each leaf of fifteen rice cultivars (N22, IR64, CN1, Dular, PSL2, PTT1, Riceberry, RD29, RD31, RD41, RD49, RD57, RD61, RD63 and SPT1) at *PI* stage are shown in figure 1. Temperature response curves of $\Delta F/F_m'$ and F_v/F_m for fifteen rice cultivars fitted in a parabolic curve (figure 1). The result is in consistent with that obtained by Dongsansuk and Neuner [11] which showed parabolic response curve (from -5 to +60°C) of $\Delta F/F_m'$ and F_v/F_m in tropical and temperate plants. The shape of parabolic function of $\Delta F/F_m'$ in fifteen rice cultivars showed unchanged value by the temperature between 12 and 21*K* (figure 1A) but curves of F_v/F_m maintained value by temperatures in the range of 18-21*K* (from+25°C) (figure 1B) (data not shown). The curve range of F_v/F_m was wider than the $\Delta F/F_m'$. The reduction of F_v/F_m indicates damages of *PSII* function [12]. With increasing leaf temperature F_v/F_m was declining, this indicated plants were under stress.

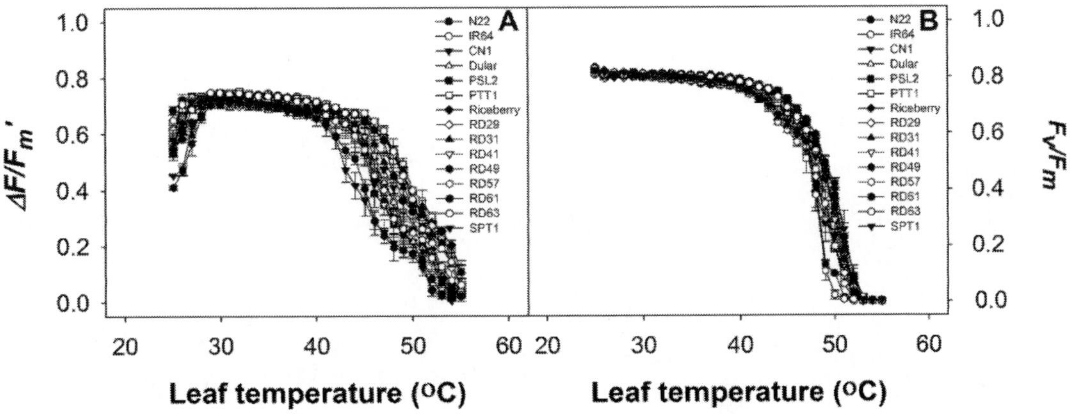

Figure 1. Effect of temperature exposure between 25°C and 55°C on $\Delta F/F_m'$ (A) and F_v/F_m (B) in fifteen rice cultivars (N22, IR64, CN1, Dular, PSL2, PTT1, Riceberry, RD29, RD31, RD41, RD49, RD57, RD61, RD63 and SPT1) at *PI* stage. The values were mean ± *SE* (n=3-4).

From the temperature response curves of $\Delta F/F_m'$ and F_v/F_m in 15 rice cultivars at *PI* stage, optimum temperature, optimum temperature range, supraoptimal temperature and high temperature stress could be inferred as shown in table 1. The optimum temperature of $\Delta F/F_m'$ and F_v/F_m was defined from parabolic curve according to Battaglia *et al* [13]. The optimum temperature of $\Delta F/F_m'$ and F_v/F_m in 15 rice cultivars was found in the range of 31.83-35.64°C and 31.82-34.10°C, respectively. The highest optimum temperature of $\Delta F/F_m'$ and F_v/F_m were found in *cvs.* CN1 and RD61, respectively (table 1). Whereas, the lowest optimum temperature of $\Delta F/F_m'$ and F_v/F_m were found in *cvs.* SPT1 and RD57, respectively. The optimum temperature in 15 rice cultivars was related to the highest of $\Delta F/F_m'$ and F_v/F_m as defined by Battaglia *et al* [13]. As a result, temperature increased up to optimum temperature resulted in an increasing photosynthesis and then photosynthesis tended to decline with increasing upper optimum level [14].

In the present research, the optimum temperature range of $\Delta F/F_m'$ in 15 rice cultivars at *PI* stage were in the range between 25°C and 45°C (table 1). Relating this, Dongsansuk *et al* [15] suggested that the optimal temperature of healthy structure was between 30°C and 45°C and for inactive structure was higher than 60°C. For their study, the optimum temperature range of F_v/F_m was found only maximum value (approximately 42°C) but the minimum of optimum temperature range of F_v/F_m could not be defined from F_v/F_m curve. For other plant species, Dongsansuk *et al* [15] suggested that optimum temperature range of $\Delta F/F_m'$ in *Musa* sp. and *A. ursinum* exposed to a wide range of temperature (-15 and 60°C) showed a range of 27-36°C and 18-27°C, respectively. In the present investigation, the supraoptimal temperature of $\Delta F/F_m'$ and F_v/F_m ranged 47.15-50.60°C and 46.93-48.65°C, respectively (table 1). The highest supraoptimal temperature of $\Delta F/F_m'$ and F_v/F_m was found in *cvs.* PTT1 and RD49, respectively. On the other hand, the lowest supraoptimal temperature of $\Delta F/F_m'$ and F_v/F_m was found in *cvs.* SPT1 and N22, respectively. High temperature stress of $\Delta F/F_m'$ and F_v/F_m in 15 rice cultivars at *PI* stage was in the range of 50.80-54.70°C and 50.93-52.50°C, respectively. The highest and the lowest high temperature stress of $\Delta F/F_m'$ were found in *cvs.* PTT1 and SPT1, respectively. For F_v/F_m, the highest and the lowest high temperature stress were found in PSL2 and RD57, respectively (table 1). As a result, each species showed their critical and lethal temperature at which it could not survive and caused damage in cell structure resulting cell death [14].

Table 1. Optimum temperature, optimum temperature range, supraoptimal temperature and high temperature stress of $\Delta F/F_m'$ and F_v/F_m between 25°C and 55°C in fifteen rice cultivars at *PI* stage. The values were mean ± *SE* (n=3-4). The different small letters indicated significantly different by Duncan's New Multiple Range Test at $p \leq 0.01$. $^+$ indicated calculation by fitting a parabolic function. ** is significant different at $p \leq 0.01$. N22 is heat tolerance rice cultivar and IR64 is heat sensitive rice cultivar

Rice cultivars	$\Delta F/F_m'$						F_v/F_m				
	optimum temperature (°C±SE)$^+$	T_{opt} range (°C)		T_{opt} span	supraoptimal temperature (°C±SE)$^+$	high temperature stress (°C±SE)$^+$	optimum temperature (°C±SE)$^+$	T_{opt} range (°C)		supraoptimal temperature (°C±SE)$^+$	high temperature stress (°C±SE)$^+$
		Min	Max					Min	Max		
N22	34.26±0.42^{a-d}	29	42	13	47.90±0.10^{def}	51.51±0.04^{fgh}	32.64±0.05^{bcd}	-	32	47.22±0.10^{bc}	51.24±0.04^{bc}
IR64	33.44±0.20^{a-e}	26	41	15	48.92±0.12^{b-e}	53.02±0.14^{b-e}	33.26±0.17^{abc}	-	38	47.95±0.07^{ab}	51.84±0.09^{abc}
CN1	35.64±0.41^{a}	27	45	18	49.42±0.67^{a-d}	52.67±0.67^{c-f}	33.26±0.30^{abc}	-	39	48.34±0.10^{a}	52.03±0.33^{ab}
Dular	33.89±0.87^{a-e}	26	43	17	48.23±0.37^{def}	52.30±0.19^{d-g}	32.76±0.11^{bcd}	-	31	48.06±0.06^{ab}	52.12±0.09^{ab}
PSL2	32.93±0.63^{cde}	26	42	16	47.92±0.48^{def}	51.89±0.45^{e-h}	33.68±0.27^{ab}	-	40	48.52±0.27^{a}	52.45±0.28^{a}
PTT1	35.11±0.22^{abc}	28	41	13	50.60±0.20^{a}	54.70±0.30^{a}	32.55±0.26^{cd}	-	40	47.67±0.13^{abc}	51.68±0.10^{abc}
Riceberry	35.21±0.25^{ab}	27	41	14	50.32±0.34^{ab}	54.33±0.38^{ab}	32.64±0.26^{bcd}	-	42	47.72±0.48^{abc}	51.88±0.45^{abc}
RD29	34.61±0.05^{a-d}	27	44	17	50.11±0.25^{ab}	54.36±0.21^{ab}	32.84±0.25^{bcd}	-	39	47.73±0.17^{abc}	51.67±0.16^{abc}
RD31	35.17±0.17^{abc}	28	41	13	50.17±0.25^{ab}	52.14±0.28^{ab}	33.58±0.23^{abc}	-	37	47.98±0.34^{ab}	51.96±0.26^{abc}
RD41	33.34±0.42^{b-e}	26	42	16	48.06±0.38^{def}	52.08±0.39^{d-h}	33.38±0.36^{abc}	-	36	48.27±0.10^{a}	52.22±0.10^{ab}
RD49	33.44±0.89^{a-e}	28	41	13	47.46±0.08^{ef}	51.17±0.16^{gh}	34.20±0.38^{a}	-	40	48.64±0.03^{a}	52.13±0.35^{ab}
RD57	32.63±0.18^{de}	28	37	9	48.36±0.10^{c-f}	52.53±0.08^{c-g}	31.82±0.18^{d}	-	39	46.93±0.16^{c}	50.93±0.16^{c}
RD61	33.95±0.51^{a-e}	25	45	20	48.81±0.38^{b-e}	53.38±0.49^{a-d}	34.10±0.18^{a}	-	41	48.65±0.22^{a}	52.50±0.30^{a}
RD63	34.53±0.30^{a-d}	26	42	16	49.84±0.20^{adc}	53.90±0.20^{abc}	33.16±0.21^{abc}	-	37	47.83±0.17^{abc}	51.71±0.16^{abc}
SPT1	31.83±1.09^{e}	26	41	15	47.15±0.80^{f}	50.80±0.33^{h}	32.92±0.30^{bc}	-	40	48.05±0.41^{ab}	52.26±0.34^{ab}
Mean	34.00	26.87	41.87	15.00	48.88	52.85	33.12		38.07	47.97	51.91
F-test	**				**	**	**			**	**

For 15 rice cultivars at *PI* stage, significant positive relationships were found between optimum temperature and supraoptimal temperature ($\Delta F/F_m'$; r=0.82 and F_v/F_m; r=0.89), between optimum temperature and high temperature stress ($\Delta F/F_m'$; r=0.71 and F_v/F_m; r=0.77) and between supraoptimal temperature and high temperature stress ($\Delta F/F_m'$; r=0.97 and F_v/F_m; r=0.93) as shown in figures 2-4, respectively. The highest positive of all relationships of $\Delta F/F_m'$ was found in *cv.* PTT1 and the lowest positive of all correlation of $\Delta F/F_m'$ and F_v/F_m was found in *cvs.* SPT1 and RD57, respectively. This positive relationship suggested that the increasing optimum temperature resulted in increasing trend in supraoptimal temperature and high temperature stress in 15 rice cultivars at *PI* stage.

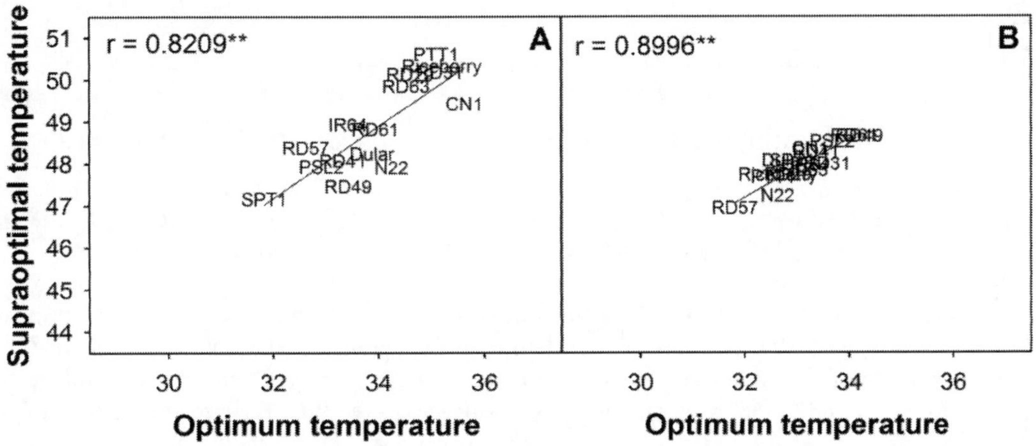

Figure 2. Relationships between optimum and supraoptimal temperature of $\Delta F/F_m'$ (A) and F_v/F_m (B) for fifteen rice cultivars at panicle initiation (*PI*) stage; N22, IR64, CN1, Dular, PSL2, PTT1, Riceberry, RD29, RD31, RD41, RD49, RD57, RD61, RD63 and SPT1.

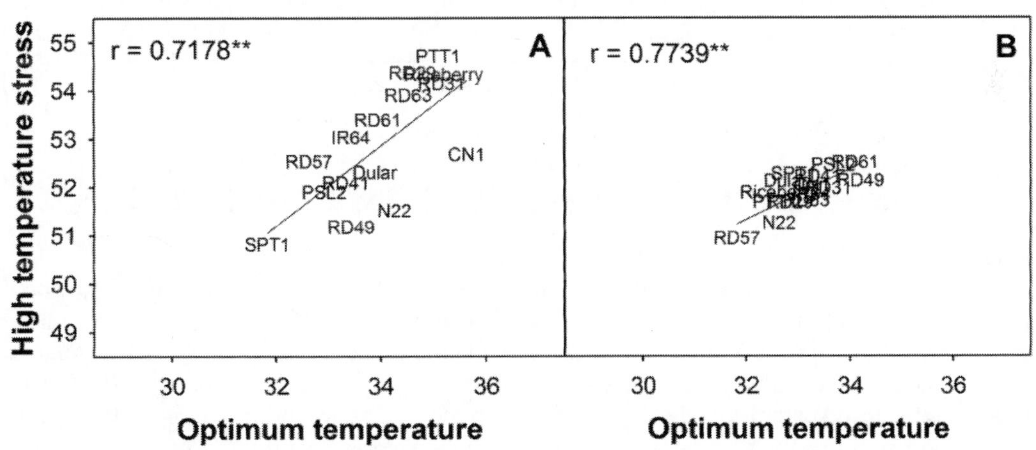

Figure 3. Relationships between optimum temperature and high temperature stress of $\Delta F/F_m'$ (A) and F_v/F_m (B) for fifteen rice cultivars at panicle initiation (*PI*) stage; N22, IR64, CN1, Dular, PSL2, PTT1, Riceberry, RD29, RD31, RD41, RD49, RD57, RD61, RD63 and SPT1.

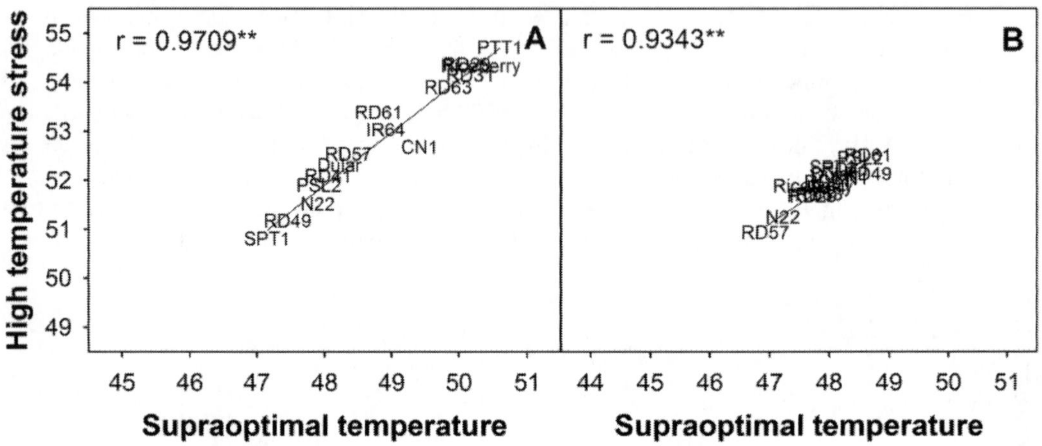

Figure 4. Relationships between supraoptimal temperature and high temperature stress of $\Delta F/F_m'$ (A) and F_v/F_m (B) for fifteen rice cultivars at panicle initiation (*PI*) stage; N22, IR64, CN1, Dular, PSL2, PTT1, Riceberry, RD29, RD31, RD41, RD49, RD57, RD61, RD63 and SPT1.

Broad sense heritability (h_b^2) of $\Delta F/F_m'$ under supraoptimal temperature and high temperature stress for 15-rice cultivars at *PI* stage showed higher than of F_v/F_m (table 2). The highest h_b^2 of $\Delta F/F_m'$ was found in high temperature stress (h_b^2=0.760) followed by supraoptimal temperature and optimum temperature as 0.693 and 0.466, respectively. This result suggested that, for rice cultivars at *PI* stage, screening for high temperature stress tolerance by measuring $\Delta F/F_m'$ should be performed under supraoptimal temperature and high temperature stress.

Table 2. Broad sense heritability of optimum temperature, supraoptimal temperature and high temperature stress of $\Delta F/F_m'$ and F_v/F_m in fifteen rice cultivars at *PI* stage.

Parameter	$\Delta F/F_m'$			F_v/F_m		
	δ^2_G	δ^2_E	h_b^2	δ^2_G	δ^2_E	h_b^2
Optimum temperature	0.789	0.905	0.466	0.305	0.215	0.586
Supraoptimal temperature	1.134	0.502	0.693	0.173	0.215	0.445
High temperature stress	1.367	0.431	0.760	0.120	0.241	0.332

4. Conclusion

From the results it could be concluded that temperature response of $\Delta F/F_m'$ and F_v/F_m showed a parabolic function in all 15 rice cultivars at *PI* stage. However, the F_v/F_m showed a wider parabolic curve. Moreover, the *PSII* efficiency was limited by high temperature stress. High temperature significantly reduced the leaf *PSII* efficiency. Therefore, the temperature response of *PSII* efficiency could be used as indicators of the optimum temperature, supraoptimal temperature and high temperature stress in different rice cultivars and also indicated temperature tolerance in rice. There was the highest positive relationship of $\Delta F/F_m'$ between optimum temperature and supraoptimal temperature. The high temperature stress of $\Delta F/F_m'$ could be considered as a useful parameter to perform screening of heat tolerance rice in breeding programme.

References

[1] Krishnan P, Ramakrishnan B, Raja Reddy K and Reddy V R 2011 *High-Temperature Effects on Rice Growth, Yield, and Grain Quality Advances in Agronomy* vol 111 ed D L Sparks (Burlington: Academic Press) chapter 3 pp 87-206
[2] Schuster W S and Monson R K 1990 *J. Plant, Cell Environ.* **13** 903-12
[3] Haldimann P and Feller U 2004 *J. Plant, Cell Environ.* **27** 1169-83

[4] Yamada M, Hidaka T and Fukamachi H 1996 *J. Scientia Horticulturae* **67** 39-48
[5] Mittler R 2006 *Trends Plant Sci.* **1** 15-9
[6] Salvucci M E and Crafts-Brandner S J 2004 *J. Physiologia Plantarum* **120** 179-86
[7] Dongsansuk A 2012 Photosynthesis of plants from various biomes-temperature dependency and thermal limits of photosynthetic light reactions (Austria: University of Innsbruck)
[8] Hacker J and Neuner G 2007 *J. Tree Physiology* **27** 1661-70
[9] Schreiber U 2004 *Pulse-Amplitude-Modulation (PAM) Fluorometry and Satuartion Pulse Method: An Overview Chlorophyll a fluorescence. A Signature of Photosynthesis Advances in Photosynthesis and Respiration* ed G Papageorgiou and C Govindjee (Dordrecht: Springer) chapter 6 pp 279-319
[10] Holland J B 2003 *J. Plant Breeding* **22** 9-112
[11] Dongsansuk A and Neuner G 2013 *J. Trends Photochem. Photobiol.* **15** 77-87
[12] Han F, Chen H, Li X J, Yang M F, Liu G S and Shen S H 2009 *J. Biochim Biophys Acta*. **1794** 1625-34
[13] Battaglia M, Beadle C and Loughhead S 1996 *J. Tree Physiol.* **16** 81-9
[14] Zróbek-sokolnik A 2012 *Temperature Stress and Responses of Plants Environmental Adaptations and Stress Tolerance of Plants in the Era of Climate Change* ed P Ahmad and M N V Prasad (New York: Springer) pp 113-34
[15] Dongsansuk A, LüTZ C and Neuner G 2013 *J. Photosynthetica* **51** 13-21

IOP Conf. Series: Earth and Environmental Science **346** (2019) 012082 doi:10.1088/1755-1315/346/1/012082

Research on the improvement of the core competitiveness of tea industry in famous tea town: A case study of Huangshan City, Anhui Province

N Jiang[1,2]

[1]School of Economics, Shanghai University, 99 Shangda Road, 200444, China

E-mail: 234117612@qq.com

Abstract. Huangshan City, Anhui Province, is one of the most suitable tea producing areas in China, with a long history of tea making. There are many varieties of tea in Huangshan, such as Maofeng, Taiping Houkui, Qimen black tea and so on. Tea industry is an important economic pillar of Huangshan City. However, statistics in recent years show that there are obvious shortcomings in the quantity and quality of tea production in Huangshan City, and the development in domestic sales and export is slow. In this paper, the diamond model proposed by Professor Michael Porter is used to analyze the competitiveness of tea industry in Huangshan City from six aspects: production factors, demand situation, related and supporting industries, enterprise strategy and structure, as well as inter-industry competition, government and opportunity. It is suggested to try to promote the operation mode of "Leading Enterprises + Production Cooperatives + Farmers + Traditional Channels + Internet +", so as to form a more complete local tea industry chain and help more local tea farmers to increase income and become rich. Furthermore, relying on the rich tourism and cultural resources in the region, this paper tries to explore a unique development path that combines tea garden experience, tea culture and local tourism, highlights local characteristics and advantages, and seeks an optimized model suitable for the development of local tea industry.

1. Introduction

China is the hometown of tea and tea drinking is very common in China. Chinese tea culture has a long history, extensive and profound, not only contains the level of material culture, but also contains a profound level of spiritual civilization. Tea has the curative effect of cure disease, rich appreciate taste, which can cultivate sentiment. Tea tasting and hospitality are elegant entertainments and social activities of Chinese people. Tea is China's main cash crop and traditional bulk export products, playing an important role in national economy.

Huangshan City in Anhui Province is the main production area of Chinese tea, known as the "home of famous tea in China" reputation. Huangshan City is mild and rainy, rich in land and forest, which is one of the most suitable tea producing areas in China. Huangshan City has a long history of tea making and is the breeding base of Chinese famous tea, among which Huangshan Maofeng, Qimen Black Tea and Taiping Houkui are among the "top ten famous tea in China" and enjoy a good reputation both at home and abroad. There are hundreds of tea enterprises, hundreds of thousands of tea farmers and more than 700,000 mu of tea garden in Huangshan City (more than 98% of them have passed the certification of "Three Certificates").

Tea industry is one of the economic pillar industries in Huangshan City, which is of vital importance

Content from this work may be used under the terms of the Creative Commons Attribution 3.0 licence. Any further distribution of this work must maintain attribution to the author(s) and the title of the work, journal citation and DOI.
Published under licence by IOP Publishing Ltd

in earning foreign exchange through export, developing rural economy and increasing farmers' income. However, in recent years, tea enterprises in Huangshan City have also exposed some problems, such as hidden dangers of tea quality and safety, insufficient deep processing, weak brand marketing, etc., which restrict the improvement of competitiveness of tea industry in Huangshan City. Tea is the key product of Huangshan City. Exploring an operation mode and characteristic path suitable for the development of local tea industry not only has practical significance to help tea farmers increase income and expand local employment, but also has reference significance for tea enterprises in other provinces and cities. With its unique natural geographical environment, extensive and profound tea culture and rich tourism resources, Huangshan City will certainly take a new step in the production and development of tea.

2. Literature review

There are two kinds of literatures closely related to this paper. One is the research on the competitiveness of tea industry in China, the other is the research on the development status and influencing factors of tea industry in Huangshan City.

2.1. Research on competitiveness of Chinese tea industry

Xu [1] studied the factors affecting the competitiveness of China's tea industry from the perspective of tea industry, adopted the index of dominant comparative advantage (RCA) and total labor productivity (OLP) as the proxy indexes of the international competitiveness of China's tea industry, and proposed targeted countermeasures and suggestions on this basis. Bao and Fang [2] used Porter's diamond theory to analyze the four elements of competitiveness of Anhui tea industry, held that cluster development is the strategic choice for the development of Anhui tea industry, and proposed the ideas and relevant measures of cluster development of Anhui tea industry. Xu and Chen [3] quantified the gap between China and developed countries in the international competitiveness of tea by calculating the index of revealed comparative advantage (RCA), and analyzed the reasons for the gap in competitiveness by using Porter's "national diamond" model theory, and put forward policy suggestions to improve the international competitiveness of China's tea. Liu and Zhang [4] used comprehensive advantage index method, trade competition index method, revealed comparative advantage method, and combined with the international market share index to make an empirical analysis on the competitiveness of tea industry in Anhui Province. Compared with other tea-producing provinces, the factors restricting the development of tea industry in Anhui Province are obtained, and the corresponding countermeasures and suggestions are put forward. Guan [5] used the classified data of the trade database of the United Nations Statistical Office to explore the comparative advantages of different categories and packages of tea exported from China. The results showed that, among the tea products exported from China at the present stage, the large package green tea had the most comparative advantage over other products, the large package fermented tea had the least comparative advantage, and the small package fermented tea and small package fermented tea had the fastest increase.

2.2. Research on tea industry in Huangshan City

Wu *et al* [6] analyzed the impact of the geological background of Huangshan tea on the element combination and soil type in the soil from the geological perspective through the field investigation of Huangshan tea, discussed the relationship between the main quality of tea and the geological background of the producing area, and analyzed the correlation between the excellent quality of tea and the geological background conditions. Zhan and Chen [7] found that consumers have a relatively low cognition of Huangshan tea geographical indication, but they have a strong purchase intention of Huangshan tea and are willing to pay a certain price premium. At the same time, consumers' understanding of the knowledge of Huangshan tea geographical indication, trust degree of Huangshan tea geographical indication label, consumers' preference of tea and the main purpose of buying tea are the important factors influencing consumers to buy Huangshan geographical indication tea. Zhu *et al* [8] discussed the current situation and development countermeasures of tea culture tourism in Huangshan City, holding that it is necessary to promote the development of tea culture tourism in

Huangshan City from such aspects as guaranteeing the quality of tea products, strengthening the role of tea culture, propagating and popularizing tea culture knowledge, distributing benefits well and developing related products. Wang *et al* [9] took 76 tea processing enterprises with geographical indications in Huangshan City as research objects, adopted the methods of questionnaire survey and field interview of middle-level and above managers, analyzed the current situation and development difficulties of the geographical indication tea industry in Huangshan City with diamond model, and proposed corresponding countermeasures.

The existing literature provides abundant information and theoretical origin for the research of this paper. This paper will further deepen the analysis of the core competitiveness of the local tea industry based on the local characteristics of Huangshan City and through its deep integration with regional culture and local tourism, so as to provide some models for reference for the discussion of the development path of characteristic agriculture in less developed areas of China.

3. The development status of tea industry in Huangshan City

3.1. Tea cultivation in Huangshan City

According to the Statistical Bulletin of National Economic and Social Development of Huangshan City over the years, the annual output of tea was 28,000 tons in 2013, up 4.0% over the previous year. At the end of the year, there were 205 agricultural products processing enterprises above the scale in the city, 20 more than the previous year, and 217 specialized farmers cooperatives were newly established, totaling 1,198. In 2015, the annual output of tea was 33,700 tons, an increase of 2.4% over the previous year. At the end of the year, there were 218 agricultural products processing enterprises above the scale in the city, and the annual output value of agricultural products processing was 19.08 billion yuan, an increase of 0.1%. Moreover, 167 newly registered farmers' cooperatives were established, and the total number of farmers' cooperatives reached 1,654. In 2017, the output of tea reached 34,300 tons, with an increase of 3.3%. The city has won the title of 9 top 100 tea enterprises in China, ranking the first in the number of prefecture-level cities in China. By the end of the year, the city had 2.45 million mu of characteristic industrial bases, including 761,000 mu of tea plantation area. By the end of the year, 28 national and provincial agricultural standard bases had been built, 10 well-known agricultural products trademarks in China, and 1,905 farmer cooperatives, among which 84 were newly registered. The data of annual tea yield and growth rate in Huangshan City from 2013 to 2017 are shown in table 1.

Table 1. The change of tea yield (ten thousand tons) and growth rate (%) in Huangshan City from 2013 to 2017.

Year	2013	2014	2015	2016	2017
Annual tea yield	2.8	3.3	3.37	3.3	3.43
Growth rate	4	5.8	2.4	-2.1	3.3

Source: Huangshan statistical bulletin on national economic and social development

3.2. Huangshan tea varieties

There are many famous teas in Huangshan, including Maofeng, Houkui and Qihong. In addition, there are Zhulanhua Tea, Huangshan Green Peony, Laozhu Dafang Tea, Songluo Tea, Xinanyuan Yinhao Tea and Huihongqing Tea, etc. The peak of the hair is like a finch's tongue. It is picked from the tender part of the tea, and there are many tiny hairs on the surface. The more hairs, the better the quality of the tea. Houkui is flatter, leaf slant big, has machine press and manual suppress, the tea that manual suppress drinks very mellow sweet nature. Qihong is warmer, more suitable for women to drink, if milk is added to drink, it has a good effect of tonifying blood gas. Different kinds of tea taste different, you can choose according to your own preferences.

3.3. Tea trade in Huangshan City

IOP Conf. Series: Earth and Environmental Science **346** (2019) 012082 doi:10.1088/1755-1315/346/1/012082

Tea is the traditional bulk agricultural products in Huangshan City, and tea industry is the pillar of Huangshan's agricultural economy. Tea trade plays an important role in the city's agricultural income and export earnings. In 2003, the total output of tea in the city was 17,800 tons, of which more than 4,800 tons were exported by itself and the foreign exchange was 7.805 million dollars. In 2013, the city's tea export volume reached 25,300 tons, earning $101.58 million in foreign exchange, and the tea export volume exceeded us $100 million for the first time. At the end of 2015, Huangshan's tea exports amounted to about $140 million, accounting for about 30% of the city's total exports and about 65% of Anhui's tea exports. The data of tea output, export volume and total export amount in some years of Huangshan City are shown in table 2.

Table 2. Tea export volume (tons), tea output (tons) and total export amount (US$10,000) in Huangshan City.

Year	Export volume of tea	Tea yield	Total export amount
2003	4,800	17,192	780.5
2008	6,000	22,956	1,000
2013	25,300	29,151	10,158

Source: Huangshan statistical bulletin on national economic and social development

4. International competitiveness analysis of Huangshan tea industry: Based on diamond model

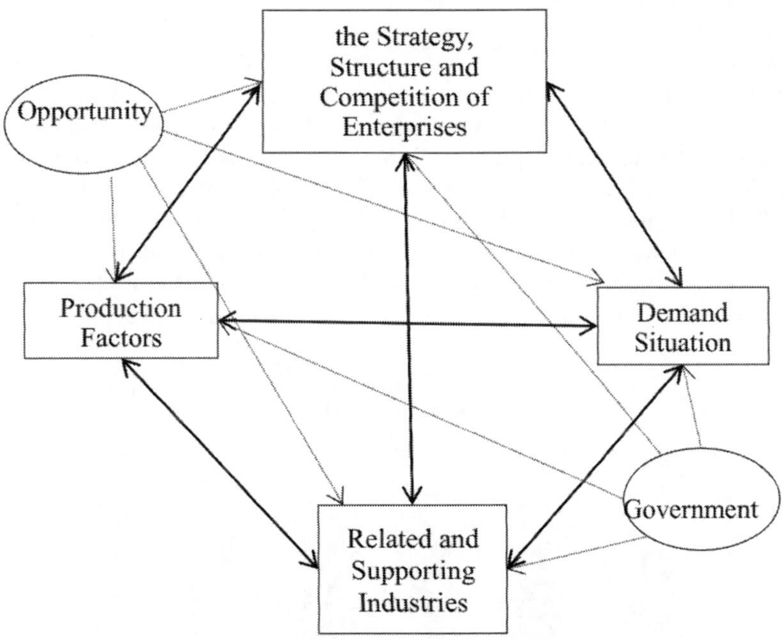

Figure 1. Diamond model.

The "diamond model" was proposed by Michael Porter, a famous strategic management scientist at Harvard Business School. Porter believes there are four factors that determine the competitiveness of a country's industry: (1) factors of production: including human resources, natural resources, knowledge resources, capital resources and infrastructure. (2) demand conditions: mainly the demand of the domestic market. (3) The performance of related industries and supporting industries: whether these industries and related upstream industries have international competitiveness. (4) The strategy, structure and performance of competitors. These four elements play a two-way role in forming a diamond system. Besides the four elements, there are two variables: government and opportunity. Opportunity cannot be controlled, and the influence of government policies cannot be ignored. The relationship is shown in figure 1.

4.1. Production factors

Primary factors of production refer to the geographical location, natural resources, climate and non-technical labor of the area where the enterprise is located, which can be obtained through passive inheritance or simple investment. Huangshan City is a typical subtropical humid monsoon climate, located on the mysterious golden line of 30° north latitude, with forest coverage rate of 82.9%. Tea trees grow in the altitude of 300-1000 meters between the hillside and deep valley land, surrounded by trees shade, rivers and streams and moisture, deep soil layer, loose texture, good permeability, strong water retention, rich in organic matter and phosphorus and potash fertilizer, which is very suitable for their growth.

Advanced production factors refer to senior talents, scientific research institutes, higher education system and modern communication infrastructure. In the aspect of tea research, Anhui Province has a high human capital and has many professional tea research institutions, such as Qimen Tea Research Institute of Anhui Agricultural Academy, tea department of Anhui Agricultural University, Huangshan Tea School and so on. Although the natural conditions of tea production in Huangshan City are advantageous, the development of tea industry in Huangshan City has been very slow due to a series of factors, such as the weak innovation ability of tea science and technology, the lack of large-scale promotion of tea technology, the poor ability of tea export, the lack of management technology and the low professional quality of tea farmers.

4.2. Demand situation

In recent years, with the continuous improvement of living standards, consumers' demand for tea has gradually become more focused on famous tea, paying more attention to the brand and packaging of tea. As is shown in figure 2, the tea consumption in China is on the rise, but it has not completely reached saturation, and there is still a lot of room for growth. From the perspective of foreign trade, according to the database of Chinese Institute of Business and Industry, the export volume of tea in 2018 was 364,742 tons, with a year-on-year growth of 2.7%, and the export amount of tea was 1,777.86 million dollars, with a year-on-year growth of 10.4%. Huangshan tea is a small-scale production for the global market, but the huge market demand will bring opportunities for its rapid development and growth.

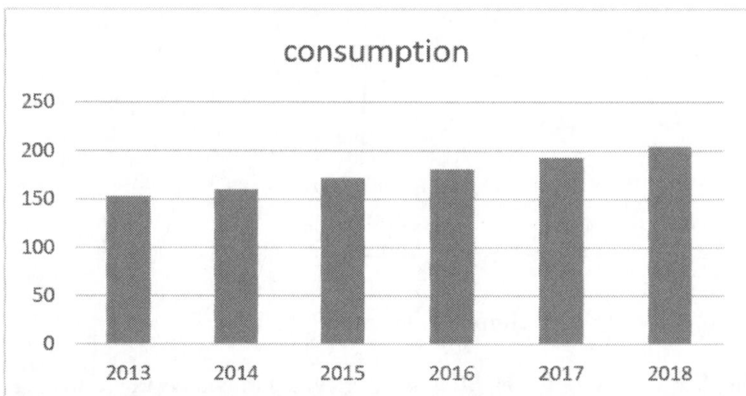

Figure 2. Statistics of tea consumption (10,000 ton) in China from 2013 to 2018.
Data source: publicly available data collation. http://www.chyxx.com/industry/201809/676082.html

4.3. Related and supporting industries

The upstream materials of tea industry chain include tea planting and picking, fertilizers and pesticides, packaging materials, etc. The middle reaches are tea production, while the downstream channels are sales, such as supermarkets, tea houses, catering industry, e-commerce platforms, etc. Huangshan City has basically realized the mechanization of tea production and processing. By 2015, Huangshan City has 61,495 tea trimming machines, 4,891 tea picking machines, 6,000 solar insecticidal lamps, 11,000 electrostatic sprayers, 219,377 tea processing machines and 80 tea production and processing lines. The mechanical processing of tea leaves in Huangshan City is not limited to the production of tea leaves, but gradually extends to the whole process of tea plantation cultivation, plant protection, pruning, picking and keeping fresh. However, it is difficult for a single enterprise or industry to maintain its competitive advantage. Only when the upstream and downstream industries interact with each other and form an effective "industrial cluster", can the competitive advantage of the industry be sustained developed.

4.4. *The strategy, structure and competition of enterprises*
From the perspective of enterprise strategy, tea producers in Huangshan City generally lack clear strategic objectives and strategic planning. The decentralized operation of the tea farmers market has poor information, poor circulation channels, high transaction costs to enter the market, and the low cost strategy cannot be realized. From the organizational structure of tea production, small-scale family management of farmers is the main mode of tea garden management in Huangshan City. This mode of production leads to low productivity of tea and the quality of tea is difficult to guarantee, unable to keep up with changes in market demand, and individual tea farmers bear greater market risks. From the perspective of industry competition, excessive dispersion and narrow production of tea is not only unfavorable to the improvement of competitiveness, but also hindering the development of the industry. If the competitive mode of Huangshan tea still stays at the low level of price competition, it will seriously reduce the market value of famous and excellent tea.

4.5. *Government and opportunity*
In the past two years, with the launch of the National Poverty Alleviation Plan, the municipal and county governments in major tea-producing areas have incorporated the development of tea industry into the local economic and social development plan and poverty alleviation plan and increased their support. The government summarized and promoted the effective model of tea industry poverty alleviation, and improved the effect of tea industry in poverty alleviation. Huangshan Municipal Government has done a lot of measures in the development of tea industry. Firstly, based on the local resource advantages, the local industrial development plan of Huangshan City is made, and the policies conducive to the development of tea industrialization are formulated and promulgated. Secondly, standardize the order of tea market and create a fair competition environment for tea merchants. Thirdly, through typical demonstration, the successful experiences of various regions are summarized and promoted to guide the healthy development of tea industrialization. Moreover, with the rapid development of the economy, people's income has generally increased, and the demand for drinks is increasing day by day. Tea's unique health effect makes people pay more and more attention to tea beverage. Green tea has gradually become the most popular beverage in the world, which is undoubtedly an excellent opportunity for the expansion of the tea market in Huangshan City at home and abroad.

5. Exploration of new operational modes
The first is to set up production cooperatives. There are many tea enterprises in Huangshan City, but there is a lack of unified standards and scientific guidance from the picking and production of tea gardens to the packaging and sales, which leads to chaos in the tea industry in Huangshan City. In order to overcome this problem, tea production cooperatives can be set up, and the government can provide technical guidance and financial support in terms of pest management, soil composition testing and other aspects for tea farmers and tea gardens. The cooperative will be built into the raw material base of leading enterprises within the territory.

The second is the introduction of high-tech tea-making technology and technological innovation. In the process of tea planting, tea enterprises in Huangshan should actively introduce modern and intelligent tea planting technologies, such as Internet of Things technology, mechanized picking integration technology, etc. In the process of tea processing, using Microwave drying machine for tea processing can not only improve the quality of tea but also improve the safety factor of tea products. In addition, attention should be paid to the development of tea deep-processing products. With the improvement of people's living standard and the acceleration of life rhythm, the demand for tea products will gradually tend to tea drinks, refreshments and instant tea products. They can try to cooperate with local tea schools or scientific research institutions to develop a series of tea deep-processing products.

Thirdly, offline and online marketing promotes the circulation of tea products. In terms of traditional offline sales channels, expand the market scale of the existing "tea city", improve the taste of the tea city, and accelerate the construction of the tea wholesale market of Huangshan Agricultural Products Logistics Park in the City Economic Development Zone. In addition, explore the online operation mode of "Internet +", so as to vigorously expand sales channels. Give full play to the platform advantages of e-commerce, it is not only the distribution center of commodities, but also the distribution center of production information and sales information. It can not only output products, but also output information, producers and consumers closely linked, achieve the production and marketing docking.

Through the above efforts, a new operation mode of "leading enterprises + production cooperatives + farmers + traditional channels + Internet +" for the development of tea industry in Huangshan City is finally formed.

6. The way to develop local characteristics

The development of tea industry in Huangshan City is faced with both opportunities and challenges. However, by virtue of its unique natural geographical environment, famous tea varieties at home and abroad and extensive and profound tea culture, Huangshan City can try to explore a development path integrating local characteristics and advantages in the tea industry.

6.1. Developing tea-picking activities and experience tea culture

Huangshan tea culture has a long history, extensive and profound. When people taste Huangshan tea, the combination of tea and tea culture can not only enhance the connotation of tea products, but also make Huizhou culture better go to the world. Therefore, we should formulate the most acceptable integration scheme of tea and tea culture for different consumer markets, so that the Huangshan tea culture can be understood by people all over the world. Deep and rich culture and beautiful natural scenery are undoubtedly one of Huangshan's great advantages in publicizing local industries. Along with the rise of cultural tourism boom, tea culture tourism has been more and more popular. Develop tourism projects with tea garden as background.

6.2. Establishing tea museum to promote brand effect

There are many famous teas in Huangshan City, but having famous teas does not mean having famous brands. Although "Xie Yuda" and "Xin'anyuan" and other leading tea enterprises in Huangshan City have been listed among the top 100 tea industry in China, so far, the world reputation of Huangshan tea brand is still not high. The establishment of the Xie Yuda Tea Museum will help promote tea brands and enhance the influence of local tea brands. Only by gradually creating a tea brand with international popularity, can Huangshan tea be better sold abroad and become "world famous tea".

6.3. Deep integration with local tourism

Tourism is an important carrier and bridge of cultural exchange, through which people can contact and experience different cultures. Huangshan is a tourist mecca, having its own unique culture - tea culture. The development of tourism will undoubtedly promote the spread and exchange of Huizhou tea culture, and it is also a kind of inheritance and protection of huizhou tea culture. Combining with the current consumption pattern of the public and the development of tourism, we will develop tea tourism commodities with Huizhou characteristics, such as tea food, crafts, tea utensils, etc., so as to realize the tea way of making friends with tea and promoting the development with culture, and let more people know and like Huangshan tea. Increase the development of tourism and promote the increase of tea trade, so as to realize the common development of famous mountains and famous teas.

7. Conclusion

Based on the current situation of tea industry development in Huangshan City, this paper analyzes the competitiveness of tea industry in Huangshan City with diamond model, and then discusses the new operation mode and the development path of local tea industry characteristics. Based on the above analysis, this paper thinks that the following three aspects can be considered to enhance the competitiveness of Huangshan tea industry. First, cluster development is the strategic choice for the development of tea industry in Huangshan City. The diamond model analysis shows that there are some defects in the four factors of the competitiveness of Huangshan tea industry, which makes it difficult to form an organic whole to strengthen the overall competitive advantage. Industrial clusters are characterized by specialized division of labor, geographical proximity, organizational proximity, intra-group organization correlation, synergy and spillover effects, and thus produce regional competitive advantages, so as to promote and strengthen the development of local economy. Second, actively explore new operating modes and optimize the efficiency of resource allocation. "Leading Enterprises + Production Cooperatives + Farmers" will build cooperatives into the raw material base of leading enterprises in China, and make the production of farmers more efficient. The introduction and innovation of new technologies are crucial to the improvement of tea quality. The creation of "traditional channels + Internet +" will realize the interaction between offline and online sales and the smooth flow of production and marketing information. Third, to seek the characteristic development of local tea industry. By carrying out tea picking and tea tasting activities, experiencing tea culture, establishing Huizhou Tea Expo Garden, expanding brand effect, and combining tea industry with local tourism based on abundant local tourism resources, highlighting regional culture and characteristics, Huangshan tea can be better promoted to the whole country and the world.

References

[1] Xu Y M 2005 An empirical study on the influencing factors of international competitiveness of Chinese tea-making industry *China Rural Obs.* **3** 19-24+50

[2] Bao J H and Fang S J 2006 Anhui tea industry cluster development: Analysis based on diamond model *Agric. Econ. Probl.* **6** 65-8

[3] Xu G Q and Chen C G 2007 Analysis of China's tea international competitiveness based on RCA and national diamond model *J. Int. Trade.* **11** 38-43

[4] Liu Y N and Zhang J Q 2009 Analysis on the competitiveness of tea industry in Anhui Province *Anhui J. Agric Sci.* **4** 1818-20

[5] Guan X 2010 A study on the comparative advantage of China's export tea products based on the analysis of different types and packaging *Rural Econ. China.* **1** 28-34

[6] Wu Y D *et al* 2010 Discussion on the relation between the quality of Huangshan tea and the geological background of its origin *Resour. Surv. Environ.* **1** 39-49

[7] Zhan H B and Chen C 2011 Econometric analysis of consumers' cognition and purchasing behavior of Huangshan geographical mark tea *Agric. Syst. Sci. Compr. Res.* **2** 181-5

[8] Zhu S D, Zhao L and Dong H J 2014 Present situation and development countermeasure of tea culture tourism in Huangshan City *J. Yunnan Agric. Univ. (Soc Sci Ed.)* **1** 20-4

[9] Wang Y R, Wang D D and Gao S Y 2019 Current situation and countermeasure of the development of characteristic industry in the Yangtze River economic belt: A case study of the development of Huangshan tea industry *J. Changjiang Univ. (Soc Sci Ed.)* **1** 65-69+113

IOP Conf. Series: Earth and Environmental Science **346** (2019) 012083 doi:10.1088/1755-1315/346/1/012083

Impact of short-term exposure to elevated temperatures on physiology of Thai rice (cv. Riceberry)

W Paethaisong[1,2], W Lontom[1,2] and A Dongsansuk[2,3,4]

[1]Department of Biology, Faculty of Science, Khon Kaen University, Khon Kaen, 40002, Thailand

[2]Salt-tolerant Rice Research Group, Khon Kaen University, Khon Kaen, 40002, Thailand

[3]Department of Agronomy, Faculty of Agriculture, Khon Kaen University, Khon Kaen, 40002, Thailand

E-mail: Danoma@kku.ac.th

Abstract. High temperature is an important limiting factor for the growth and development of rice cultivation worldwide. The current research aimed to investigate the efficiency of *PSII* system, photosynthetic changes in photosynthetic pigments and membrane stability in Thai seedlings of Thai rice cv. Riceberry after exposure to different temperatures *viz*., 25, 30, 35, 40, 45, 50 and 55°C for 30 min. The experiment was designed in CRD with four replications. The results showed that the highest F_v/F_m was realized when seedlings were exposed to 30°C and 35°C and decreased when exposed to 40°C for 30 min. The content of photosynthetic pigments such as total chlorophyll (*TC*) and chlorophyll a (*Chl a*) were the highest after short-term exposure to a temperature of 25°C. However, the contents of all photosynthetic pigment such as *TC, Chl a, Chl b* and carotenoids were the lowest when exposed to 55°C for 30 min. The membrane stability of Riceberry seedlings affected on short-term exposure to temperatures of 50°C and 55°C. Thus, F_v/F_m and Hill reaction were the first mechanisms to be significantly affected by the short-term exposure to a temperature of 40°C followed by membrane stability and contents of photosynthetic pigments which were affected by short-term exposure to temperature of 50°C -55°C and 55°C, respectively.

1. Introduction

Rice (*Oryza sativa* L.) is an essential food for a large part of the world's population. It can adapt and grow well in many climatic zones. However, its production is severely affected by various factors such as pests, weeds, drought, high and low temperature, salinity and environmental pollution [1]. In recent years, the global warming and climate change have seriously affected the cultivation of rice and other crops. The Intergovernmental Panel on Climate Change (IPCC) [2] has predicted that the air temperature will increase by more than 4.0°C in the 21st century. High temperature affects the photosynthesis system especially the *PSII* efficiency because the *PSII* is the most heat-sensitive among the two photosynthetic systems [3]. Mathur *et al* [4] reported that, in wheat exposure to temperatures of 25°C, 35°C, 40°C and 45°C for 15 min showed a decreased in *PSII* efficiency with increasing temperature. High temperature decreased chlorophyll synthesis by inhibiting the chloroplast function [5] and also reduction in the Hill reaction rate [6]. Cui *et al* [7] reported that in tall fescue (*Festuca arundinacea* S.) seedlings after treatment with 35°C/30°C (day/night) for 20 days showed a reduction in chlorophyll a+b and chlorophyll/carotenoid ratio. Heat stress also damaged the cell membrane by

Content from this work may be used under the terms of the Creative Commons Attribution 3.0 licence. Any further distribution of this work must maintain attribution to the author(s) and the title of the work, journal citation and DOI.
Published under licence by IOP Publishing Ltd

lipid peroxidation causing an increase in leaked electrolytes [8]. Wilson *et al* [9] reported that in 4-day-old *Brassica juncea* seedlings, after exposure to a temperature of 45°C for 4.5 h showed an increase in electrolyte leakage. Therefore, this research aimed to investigate the physiological changes in Thai rice cv. Riceberry seedlings exposed to varied temperature for short term.

2. Material and methods

Thai rice cv. Riceberry was used in this study. The seeds were sterilized and germinated on wet tissue papers until the root length was approximately 3 cm. Then, the germinated seeds were transplanted to the pots containing paddy soil in open air greenhouse at the Agronomy Field Station, Department of Agronomy, Khon Kaen University, Khon Kaen, Thailand. Further, 21-day-old seedlings were subjected to different temperatures in the temperature control cabinet (VRV. Corp. Ltd, Thailand). The initial temperature was 25°C which was further increased to 30°C, 35°C, 40°C, 45°C, 50°C, 55°C, and maintained for 30 min. Relative humidity and light intensity were 70% and 400 µmol (photons) $m^{-2}s^{-1}$, respectively. For the temperature was first increased to the target temperature, maintained for 30 min and then cooled down to 25°C [10] before measuring the *PSII* efficiency, Hill reaction, photosynthetic pigments and membrane stability.

The *PSII* efficiency, as indicated by F_v/F_m is the maximal quantum yield of *PSII* efficiency in dark-adapted leaves, as described by Schreiber [11]. The Hill reaction was determined according to Giebel [12]. Fresh leaves (1 g) were extracted in sucrose-phosphate buffer pH 7.0 using a homogenizer (Vinayak Industries Thane, Maharashtra, India). Chloroplast (1 ml) was extracted in sucrose-phosphate buffer pH 7.0 (1.5 ml) and 0.28 mM 2,6-Dichlorophenolindophenol (*DPCIC*) (0.5 ml). The Hill reaction was indicated by *DCPIP* reduction at 620 nm. It was expressed in $\mu M.mg.Chl^{-1}.hr^{-1}$. Arnon's [13] procedure was used to determine photosynthetic pigment contents such as total chlorophyll (*TC*), chlorophyll a (*Chl a*), chlorophyll b (*Chl b*) and carotenoid (*Car*). Fresh leaves (0.1 g) were collected, cut in small pieces and put in a glass test tube and then 80% acetone was added (10 ml) and kept in the dark for 72 h. For the total extraction volume (*V*) was recorded and the optical density of solution was measured and recorded at 440, 645 and 663 nm using by UV-VIS spectrophotometer (Model i3, Hanon, China). The contents of *TC*, *Chl a*, *Chl b* and *Car* were calculated as described by Arnon [13] and the photosynthetic parameters were expressed as $mg.gFW^{-1}$. Membrane stability was indicated by electrolyte leakage (*EL*) as described by Bajji *et al* [14]. Fresh leaves (0.1 g) were soaked in deionized water (10 ml) for 24 h. The electrical conductivity (EC_1) of the suspension was measured. The suspension was incubated at 100°C for 15 min and cooled down a temperature to 25°C on ice and then the final electrical conductivity (EC_2) was measured. For *EL* was calculated as follows: $EL\ (\%) = (EC_1/EC_2) \times 100$.

3. Results and discussion

3.1. *Effect of short-term exposure to elevated temperature on PSII efficiency*

In rice (cv. Riceberry), the efficiency of *PSII* was influenced by short-term exposure to elevated temperatures, as indicated by decreased F_v/F_m values as shown in figure 1. The highest F_v/F_m was observed after short-term exposure to temperatures of 30°C and 35°C. A significant reduction in F_v/F_m was found when the temperature increased from 40°C to 55°C (figure 1). These results suggested that F_v/F_m in rice seedling was significantly affected by heat exposure at 55°C for 30 min (table 1), which was in agreement with the effect on rice cv. Pathumthani 60 at heading stage in which F_v/F_m decreased after exposure to 45°C for 30 min [10]. Moreover, Tang *et al* [15] also showed a decrease in F_v/F_m with increasing temperature from 45°C to 50°C in spinach seedlings (*Spinacia oleracea*) and also in rice cv. 9311 exposed to 35°C for 1 h [16].

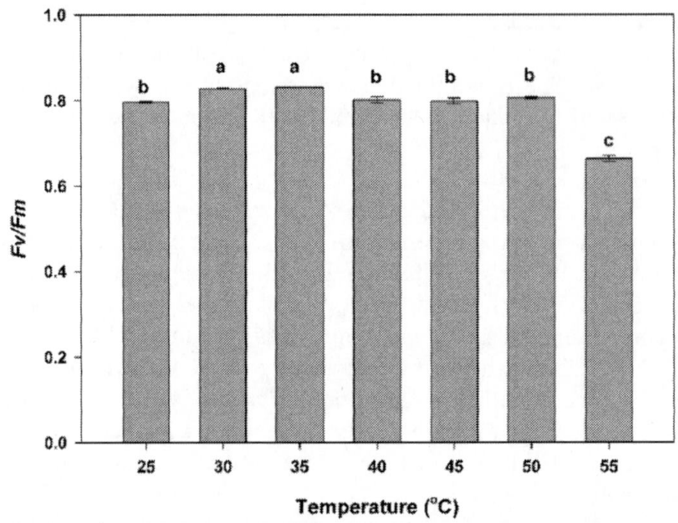

Figure 1. Effect of short-term treatment at different temperatures on F_v/F_m in seedlings of rice cv. Riceberry. Data were expressed as means ± *SE* (*n*=3-4). The different letters showed significant differences among different temperatures by *DMRT* at $P \leq 0.05$.

3.2. Effect of short-term exposure to elevated temperature on Hill reaction and photosynthetic pigments

The effect of short-term exposure to elevated temperatures on Hill reaction was indicated by a reduction of DCPIP in rice cv. Riceberry as shown in figure 2A. The highest rate of Hill reaction was observed after short-term exposure to temperature of 35°C (approximately 62.18±3.07μ M.mg.Chl^{-1}.hr^{-1}) and the decreasing trend of Hill reaction rates was found on increasing temperature from 40°C to 55°C (figure 2A). According to Misr and Meena [17] reported that Hill reaction dropped in rice (*Oryza sativa* L. cv. Mousouri) when 12-day-old seedlings were exposed to high temperature (40°C) [17]. The photosynthetic pigment contents, namely *TC* and *Chl a,* were the highest after short-term exposure to a temperature of 25°C (approximately 1.853±0.129 and 1.391±0.096 mg.gFW^{-1}, respectively); the contents decreased when exposed to temperatures of 30°C to 55°C (figures 2B and 2C). However, *Chl b* and *Car* contents were the highest after short-term exposure to a temperature of 45°C, which then significantly decreased at 55°C (for *Chl b*) and 50°C -55°C (for *Car*) (figures 2D and 2E). The *Chl b* and *Car* contents were the lowest after treatment of 55°C for 30 min. This suggested that short-term exposure to elevated temperatures (50°C -55°C) affected the Hill reaction and the contents of *TC*, *Chl a*, *Chl b* and *Car* (table 1). Earlier studies showed a decrease in the contents of *TC*, *Chl a* and *Chl b* in five rice cultivars (N22, Dular, IR64, KDML105 and PTT1) at dough grain stage when they were exposed to 42°C for 30 min [18].

Table 1. Analysis of variance of the effects of short-term exposure to different temperatures on F_v/F_m, Hill reaction and contents of *TC*, *Chl a*, *Chl b* and *Car* in seedlings of rice cv. Rice berry (*n*=3-4).

Parameters	Mean Square						
	F_v/F_m	*Hill reaction*	*TC*	*Chl a*	*Chl b*	*Car*	*EL*
Temperature	0.013*	800.095*	0.110*	0.074*	0.004*	0.007*	880.9*
Error	9.54×10^{-5}	13.844	0.024	0.013	0.001	0.001	46.435

* indicated significant differences at $P \leq 0.05$ by *DMRT*.

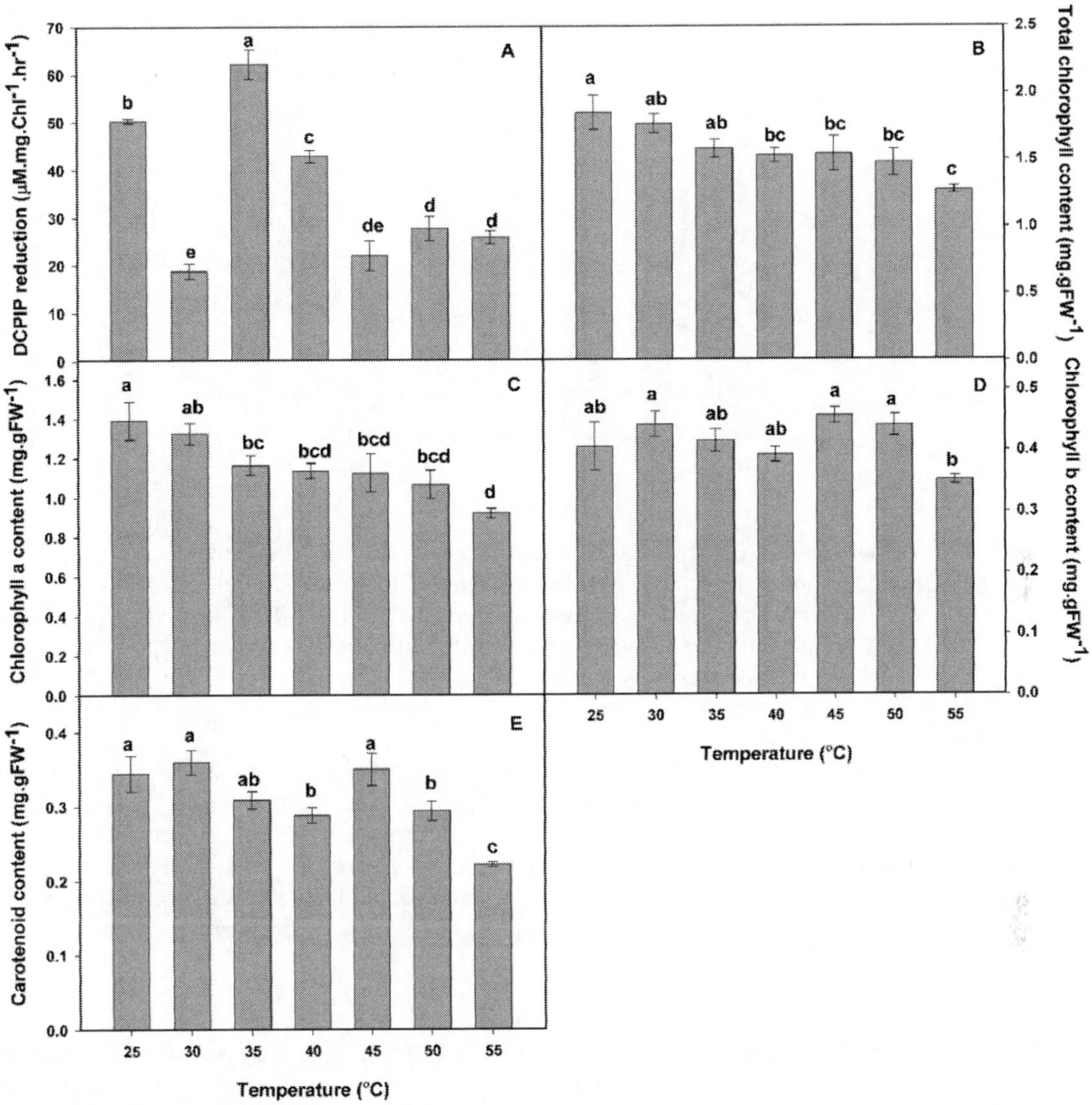

Figure 2. Effect of short-term exposure to different temperatures on Hill reaction (A) and contents of total chlorophyll (B), chlorophyll a (C), chlorophyll b (D) and carotenoid (E) in seedlings of rice cv. Riceberry. Data were expressed as means ± *SE* (n=3-4). The different letters showed significant differences among different temperatures by *DMRT* at $P \leq 0.05$.

3.3. Effect short-term exposure to elevated temperature on membrane stability

The effect of short-term exposure to elevated temperatures on electrolyte leakage (*EL*) in seedlings of rice cv. Riceberry is shown in figure 3. The highest *EL* value was observed after short-term exposure to temperatures of 50°C -55°C (figure 3) This suggested that rice seedlings were affected by short-term exposure to elevated temperatures (table 1), resulting in increased cell membrane damage and high leaked solute to outside the cell which probably induced the death of rice cell [19]. The lowest of *EL* was found when rice was exposed to a temperature of 35°C (figure 3). This suggested that a temperature of 35°C was the optimum temperature for rice [20]. Ali *et al* [21] showed the highest *EL* in rice seedlings on exposure to 42°C for 72 h. Other studies also reported that the membrane leakage significantly increased in wheat [22] and lentil (*Lens culinaris* Medik.) [23] after exposure to high temperatures.

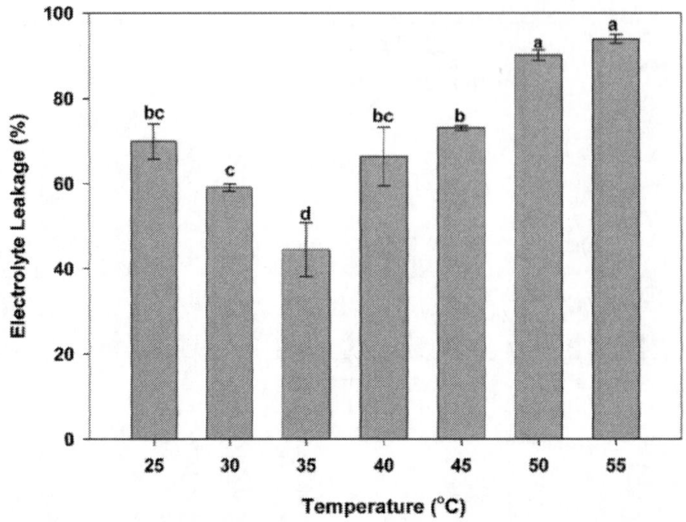

Figure 3. Effect of short-term exposure to different temperatures on electrolyte leakage in seedlings of rice cv. Riceberry. Data were expressed as means ± *SE* (*n*=3-4). The different letters showed significant differences among different temperatures by *DMRT* at $P \leq 0.05$.

4. Conclusions

Based on the results of the effect of short-term temperature on the physiology of rice seedlings (cv. Riceberry) indicated that short-term exposure to temperatures of 40°C - 55°C affected the efficiency of *PSII* by decreasing F_v/F_m values. Also the short-term heat stress (at 55°C) severely affected Hill reaction and the contents of photosynthetic pigments (*TC*, *Chl a*, *Chl b* and *Car*) leading to low membrane stability as indicated by increased membrane damage and high *EL*. Short-term temperature of 35°C could be the optimum temperature for growing seedling rice (cv. Riceberry).

References

[1] Al-Whaibi M H 2011 *Saud. Univ. Sci.* **23** 139-50
[2] Intergovernmental Panel on Climate Change (IPCC) 2007 Climate change 2007: The Physical Science Basis. *Contribution of Working Group I to the Fourth Assessment Report of the Intergovernmental Panel on Climate Change ed S Solomon et al* , Cambridge: Cambridge University Press, Cambridge, United Kingdom
[3] Al-Khatib K and Paulsen G M 1999 *Crop Sci.* **31** 119-25
[4] Mathur S, Allakhverdiev S I and Jajoo J 2011 *Biochim. Biophys. Acta* **1807** 22-9
[5] Xu Q, Paulsen A Q, Guikema J A and Paulsen G M 1995 *Environ. Exp. Bot.* **35** 43-54
[6] Vijayalakskmi D and Bangarusamy U 2014 *Univers. J. Agric. Res.* **2** 131-4
[7] Cui L, Li J, Fan Y, Xu S and Zhang Z 2005 *Bot. Stud.* **47** 61-9
[8] Asthir B 2015 *Biol. Plant.* **59** 620-8
[9] Wilson R A, Sangha M K, Banga S S, Atwal A K and Gupta S 2014 J. *Environ. Biol.* **35** 383-7
[10] Dongsansuk A, Theerakulpisut P and Pongdontri P 2017 *Pertanika J. Trop. Agric. Sci.* **40** 621-8
[11] Schreiber U 2004 Pulse-amplitude-modulation (PAM) fluorometry and saturation Pulse method: an overview Chlorophyll a fluorescence. Advances in photosynthesis and respiration vol 19 ed G C Papageorgiou and Govindjee (Dordrecht: Springer) chapter 6 pp 279-319
[12] Giebel P E 2009 Extraction of chloroplasts from plant tissue and their use in demonstrating the hill reaction (Richmond: Virginia Commonwealth University)
[13] Arnon D I 1949 *Plant Physiol.* **24** 1-15

[14] Bajji M, Lutts S and Kinet J M 2001 *Plant Growth Regul.* **36** 61-70
[15] Tang Y, Wen X, Lu Q, Yang Z, Chen Z and Lu C 2007 *Plant Physiol.* **143** 629-38
[16] Yin Y, Li S, Liao W, Lu Q, Wen X and Lu C 2010 *J. Plant Physiol.* **167** 959-66
[17] Misr A N and Meena M 1986 *Plant Sci.* **46** 1-4
[18] Pansarakham P, Pongdontri P, Theerakulpisut P and Dongsansuk A 2018 *Acta Physiol. Plant.* **40** 173
[19] Ferreira S, Hjernø K, Larsen M, Wingsle G, Larsen P, Fey S, Roepstorff P and Salomé Pais M 2006 *Ann. Bot.* **98** 361-77
[20] Yoshida S 1981 Fundamentals of Rice Crop Science (Los Baños: IRRI)
[21] Ali K, Azhar A and Galani S 2013 *RJOAS* **8** 11-9
[22] Almeselmani M, Deshmukh P S, Sairam R K, Kushwaha S R and Singh T P 2006 *Plant Sci.* **171** 382-8
[23] Chakraborty U and Pradhan D 2011 *J. Plant Interac.* **6** 43-52

IOP Conf. Series: Earth and Environmental Science **346** (2019) 012084 doi:10.1088/1755-1315/346/1/012084

Egg freshness recognition based on a fuzzy radial-basis-function neural network technology

H B Zhao[1, 2, 4] and X H Zhou[3]

[1]Engineering Technology Research Center of Optoelectronic Appliance, Anhui Province, Tongling, Anhui, 244061, China
[2]College of Electrical Engineering, Tongling University, Tongling Anhui, 244061, China
[3]NO. 43 Research Institute, China Electronic Science and Technology Group Company, Hefei, Anhui, 230088, China

E-mail: happyzhaohaibo@126.com

Abstract. There are often mobile, unconscientious vendors in China who sell stale or even rotten eggs in produce markets, earning money by dishonest means. To prevent the entry of substandard eggs into the market that endanger the health of consumers, we designed an egg freshness recognition system based on a fuzzy radial-basis-function (RBF) neural network. This system acquires the color characteristic parameters red (R), green (G), and blue (B) of the transmitted light of eggs through a computer vision device. The system converts the RGB values into HIS (hue, intensity, and saturation) values, uses the egg transmitted light color characteristic parameter HIS as an input value and employs the Huff value coding as an output. A test sample was used to verify the identification system. Our experimental results show that when using a fuzzy RBF neural network and simple RBF neural network algorithm, the average recognition accuracy of the system is 96.35% and 92.72%, respectively, both of which are higher than the average recognition accuracy of 88.14% when using the back propagation neural network algorithm. The feasibility and superiority of the identification system proposed in this paper were verified; therefore, this system may serve as a reference for future research on egg freshness recognition.

1. Introduction

Eggs contain a variety of nutrients that are beneficial to the human body and are one of the most common foods consumed by society. However, some mobile and unconscientious vendors sell stale or even rotten eggs in grocery markets, thereby threatening the health of consumers. Figure 1 is a comparison of various eggs, including the shape of the yolk of a fresh egg. If the yolk is loose, then the egg is not fresh. If the egg yolk sticks to the interior of the eggshell or if the interior of the eggshell is black in color, the egg is rotten. Therefore, it is necessary to study egg freshness recognition.

When we purchase eggs, we typically hold the eggs close to our ear and forcefully shake them up and down. Hearing a sound while shaking the egg indicates that it is not fresh, whereas the absence of sound reveals egg freshness. In general, the recognition accuracy of this method is poor, and vendors often prevent consumers from shaking the eggs they sell, as the vendors are concerned that shaking further loosens the yolk inside the egg. In recent years, several scholars worldwide have studied egg freshness recognition. The main methods used include near-infrared spectroscopy [1-3], electronic nose detection [4], hyperspectral detection [5,6], microwave spectroscopy [7], and gas

Content from this work may be used under the terms of the Creative Commons Attribution 3.0 licence. Any further distribution of this work must maintain attribution to the author(s) and the title of the work, journal citation and DOI.
Published under licence by IOP Publishing Ltd

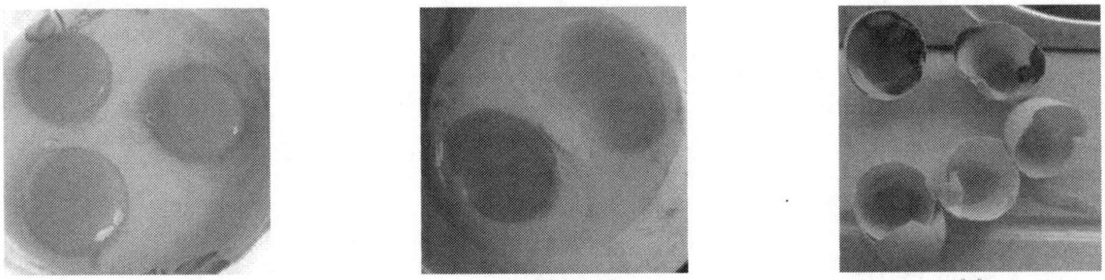
(a) (b) (c)

Figure 1. Eggs of different degrees of freshness. (a) Fresh egg, (b) Loose yolk eggs and (c) Rotten eggs.

chromatography detection [8]. The applicability of near-infrared spectroscopy is relatively limited, as the position and number of selected points are relatively random and one-sided. Electronic nose detection is suitable for sampling detection but is relatively time-consuming when compared to the optical method. Hyperspectral detection and microwave spectroscopy are susceptible to instrument noise and are affected by the uneven surfaces of eggs, and gas chromatography detection cannot directly generate qualitative analytical results as it requires derivation. Computer vision technology has been successfully applied to agricultural product testing [9-11], including the integration of back propagation (BP) neural network and computer vision technology to identify egg freshness [9]. However, the established system does not have a high recognition accuracy for egg freshness, which is mainly due to slow convergence speed, long training time, and low approximation accuracy of the BP neural network; there are also local optimization problems. Relative to the numerous shortcomings of BP neural network, RBF neural network can approximate arbitrary nonlinear functions with arbitrary precision and display a global approximation ability, which fundamentally solves the local optimization problem inherent to BP network. RBF neural network also have rapid convergence speeds and a shorter training time [12,13]. However, studies on the application of combined RBF neural network and computer vision technology for egg freshness recognition are rare. According to the optical properties of eggs, the transmittance of fresh eggs to light decreases with storage time [9,14]. Changes in the transmittance of light indicate alterations in the internal egg quality, which can be reflected as modifications of color inside the egg. Here, a fuzzy RBF neural network and computer vision technology were combined to construct an automatic system for the recognition of egg freshness. The experimental device used in this paper has been successfully applied in the recognition of artificial ripening tomatoes [15], it is a low cost and easy-to-use device, and the RBF neural network algorithm itself is mature, so any person can apply it immediately.

2. Material and methods

2.1. Materials
600 fresh eggs, provided by xiantao food corporation in China's Hubei province on March 1, 2019.

2.2. Methods formulation
There are multiple color models for describing the color of an object. In practical applications, the commonly used models include the RGB and HIS color models. Based on the requirements of our test, the HIS color model was selected [15]. The conversion formula from the RGB model to the HIS model is [15]:

$$\begin{cases} H=\cos^{-1}\left(\dfrac{1/2((R-G)+(R-B))}{\sqrt{(R-G)^2+(R-B)(G-B)}}\right) \\ I=(R+G+B)/3 \\ S=1-3\min(R,G,B)/(R+G+B) \end{cases} \tag{1}$$

where R is red, G is green, B is blue, H is hue, I is intensity, and S is saturation.

The test device consists of a light source, an egg tray, a video camera, an image acquisition card and a computer, as shown in figure 2. During the test, it is critical to ensure that there is no gap between the egg and the light-transmitting hole to prevent light from entering the light box directly from the gap. Both sides of the egg are equipped with two mirrors at an angle of 45° to the horizontal plane, which ensures that the charge-coupled device (CCD) camera can capture the transmitted light image information of three sides of each egg at a time, which satisfies the comprehensive requirements of color detection.

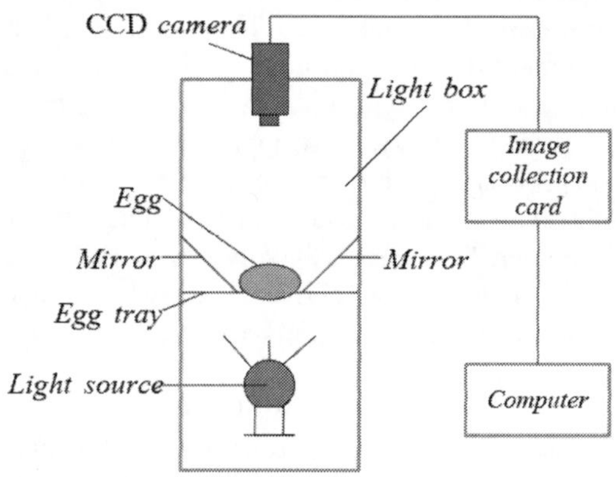

Figure 2. Schematic diagram of test drive for egg freshness recognition.

The operation process of system identification is as follows: (1) select 200 pieces each of brown- and white-shell eggs, and label all eggs to facilitate test data collection; (2) place each egg in the egg holder of the test device, collect the images (R, G, B) of the transmitted light of the eggs in the light box and transfer the images to a computer, in order to process the image through a test procedure to obtain color information of the egg content (H, I, S); (3) weigh each egg on an electronic balance after the images are obtained; and (4) break each egg after the images are obtained and measure the Huff value as an indicator of the freshness of the egg. The formula for calculating the Huff value is as follows [9]:

$$H_a = 100\lg(h + 7.57 - 1.7w^{0.37}) \tag{2}$$

In the formula, H_a represents the Huff value. The Huff value is a test specified by the U.S. Department of Agriculture as the egg standard and serves as a factor indicating the freshness of eggs. The Huff value of a fresh egg is usually between 75 and 82, with relatively high values up to 90; edible eggs must have a Huff value above 72. According to the requirements of this test, the Huff value is divided into three grades according to international standards and are coded as different codes (table 1). In equation (2), w (g) indicates the mass of the egg and h (mm) indicates the height of the albumen, which can be measured using the device in figure 3 [14].

Table 1. Huff value coding.

Egg freshness grade	Huff value coding
Special ($H_a \geq 72$)	[1,0,0]
Grade A($60 \leq H_a < 72$)	[0,1,0]
Grade B($30 < H_a < 60$)	[0,0,1]

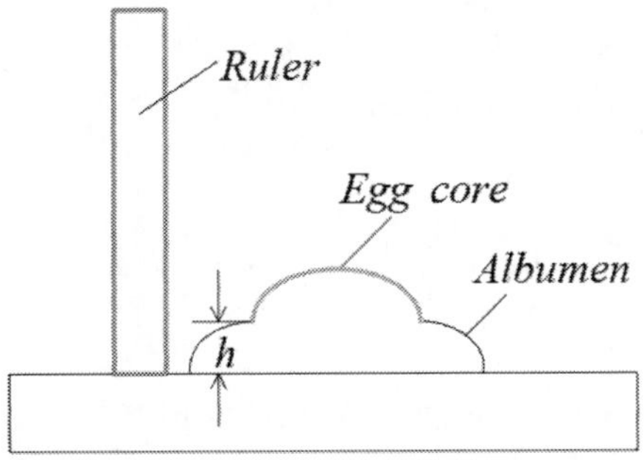

Figure 3. Measurement of albumen height.

The method of measuring albumen height involves placing a cracked egg on a horizontal glass plate, identifying the height values of four different parts of the egg with a ruler, namely, h_1, h_2, h_3 and h_4, and computing the average value, which is designated as albumen height h. In the experiments, we respectively obtained 196 and 190 groups of brown- and white-shell egg color data (H, I, S). The corresponding Huff value data indicated freshness grade, which we used as sample data to train the fuzzy RBF neural network. Finally, we identified each test sample by employing the trained network model.

3. Construction of fuzzy RBF neural network

The RBF neural network has a strong self-learning function and self-adaptive ability, and can effectively learn and adjust to the sample; however, its internal mapping rules are invisible and difficult to understand. Fuzzy control has the characteristics of strong reasoning ability, and the control rules of fuzzy control are summarized and determined based on the experience of experts or skilled professionals, so the rules are easy to understand and intuitive. Therefore, we combined fuzzy control and a RBF neural network to form a fuzzy RBF neural network.

3.1. Neural network structure

The structure of the fuzzy RBF neural network is shown in figure 4 [16]. The network consists of an input layer, a fuzzification layer, a fuzzy inference layer, and an output layer. The functions of each layer are as follows:

- The first layer is the input layer. This layer has three nodes, each node is directly connected to each component of the input quantity, and the input quantity is passed to the next layer. The input and output of each node of the layer is expressed as follows:

$$f_1(i) = \mathrm{X} = [x_1, x_2, x_3] = [\mathrm{H}, \mathrm{I}, \mathrm{S}] \tag{3}$$

where the input and output of the layer are equal and represent the color data H, I, and S.

- The second layer is the fuzzification layer. According to the principle of least membership function [16], three nodes are selected in this layer to divide the three inputs into three fuzzy sets, and the Gaussian function is used as the membership function. c_{ij} and b_{ij} are the mean and standard deviation of the membership function of the i th input variable and the j th fuzzy set, respectively:

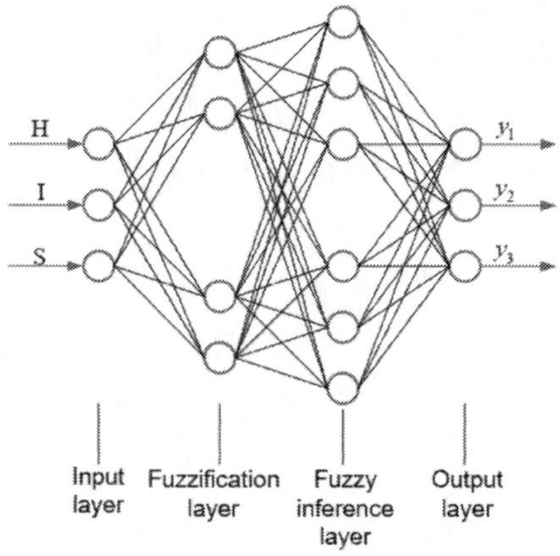

Figure 4. Fuzzy RBF neural network structure.

$$f_2(i,j) = \exp\left\{-\frac{(f_1(i)-c_{ij})^2}{(b_{ij})^2}\right\} \tag{4}$$

where $i = 1,2,3$ and $j = 1,2,3$.

- The third layer is the fuzzy inference layer. The number of nodes in this layer is $3^3 \times 3$ = 81 [17], and the fuzzy operation is implemented between each node. Specifically, the corresponding ignition intensity is obtained by the combination of each fuzzy node. The output of each node j is the product of all input signals of the node, namely:

$$f_3(j) = \prod_{j=1}^{N} f_2(i,j) \tag{5}$$

In the formula, $N = \prod_{i=1}^{n} N_i$, where N_i is the fuzzy segmentation number of the i th input.

- The fourth layer is the output layer. This layer has three nodes and the output is a three-dimensional Huff value coding $[y_1, y_2, y_3]$, i.e.:

$$f_4(i) = W \cdot f_3 = \sum_{j=1}^{N} W(i,j) \cdot f_3(j)(i = 1,2,3) \tag{6}$$

where $W_{ij}(i = 1,2,3)$ is the connection weight matrix of the output node and each node of the third layer.

The fields of input quantities, H, I, and S, are all [−1, 1], and the fuzzy sets are all {NB, ZO, PB}, indicating that the input is {small, moderate, and large}; the output values are all 0 or 1. The membership functions of the input quantities (H, I, and S) are shown in figure 5. After several trials, the fuzzy control rules were developed as shown in table 2. Figure 6 shows the fuzzy relationship between the inputs (H, I, S) and output y_2 of the system (the fuzzy relationships between the three inputs of the system and the other two outputs are not listed).

IOP Conf. Series: Earth and Environmental Science **346** (2019) 012084 doi:10.1088/1755-1315/346/1/012084

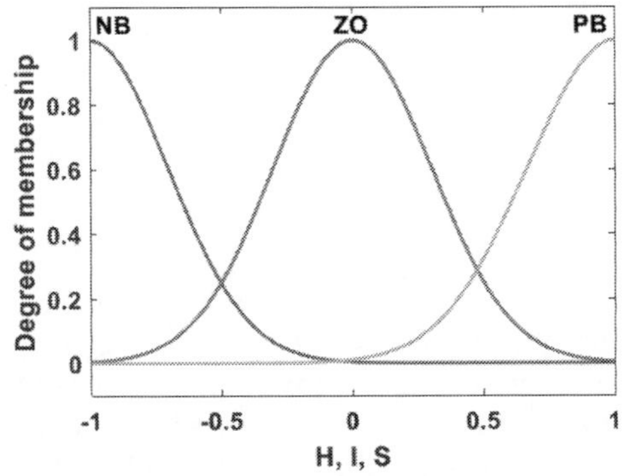

Figure 5. Membership function.

Table 2. Fuzzy control rules.

H	I	S	Huff value
① NB	NB	NB	①[0,0,1] ②[0,0,1] ③[0,0,1]
	NB	ZO	①[0,0,1] ②[0,1,0] ③[1,0,0]
	NB	PB	①[0,0,1] ②[1,0,0] ③[0,0,1]
② ZO →	ZO	NB	①[0,0,1] ②[0,0,1] ③[1,0,0]
	ZO	ZO	①[0,1,0] ②[0,0,1] ③[0,1,0]
	ZO	PB	①[1,0,1] ②[1,0,0] ③[1,0,0]
	PB	NB	①[0,1,0] ②[0,1,0] ③[0,1,0]
③ PB	PB	ZO	①[0,1,0] ②[1,0,0] ③[0,0,1]
	PB	PB	①[0,1,0] ②[0,0,1] ③[1,0,0]

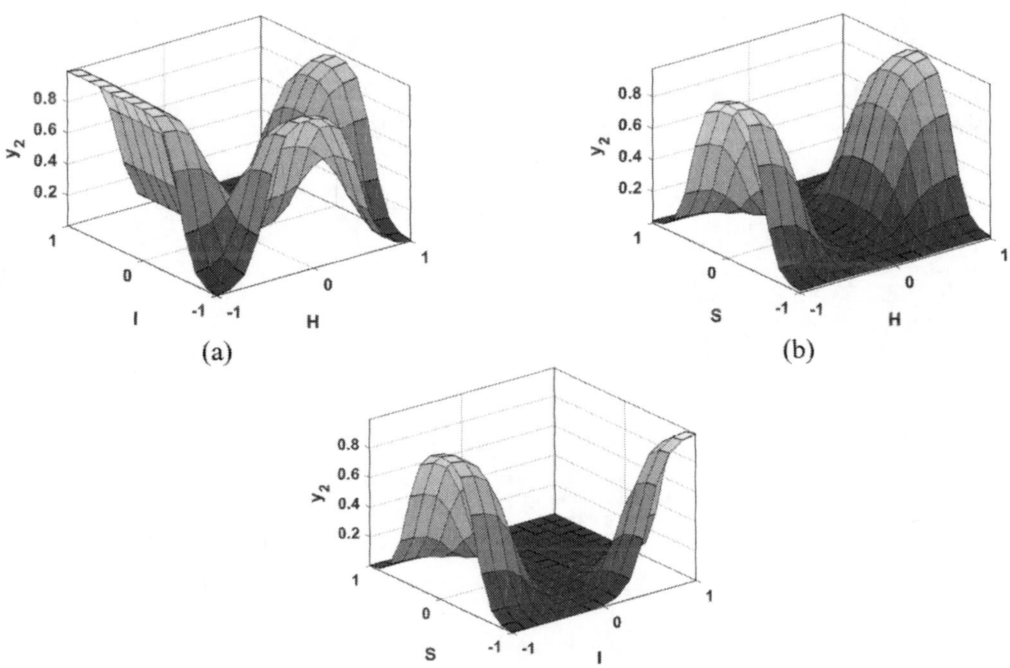

Figure 6. Fuzzy relationships between input and output. (a) Fuzzy relationship between H, I, and output y_2, (b) Fuzzy relationship between H, S, and output y_2 and (c) Fuzzy relationship between

I, S, and output y_2.

3.2. Network training

Using Matlab (2017a) as network training software. Figures 7-9 illustrate the network training errors when using a BP neural network, simple RBF neural network, and fuzzy RBF neural network, respectively. Figure 7 shows that the network training error is 0.01193 when the network training is iterated to 40 steps and the training is stopped at 238 steps. Figure 8 shows that when the network training is iterated to 11 steps, the network training error is 0.01193; the training is stopped when the network training is iterated to 78 steps. Figure 9 shows network training iterated only to the 4th step, and the network training error reached 0.01424; the network training was stopped only when it was iterated to the 21st step. Therefore, compared to the BP and simple RBF neural network, when the fuzzy RBF neural network is adopted, the system has a faster convergence speed, shorter training time, and higher approximation accuracy.

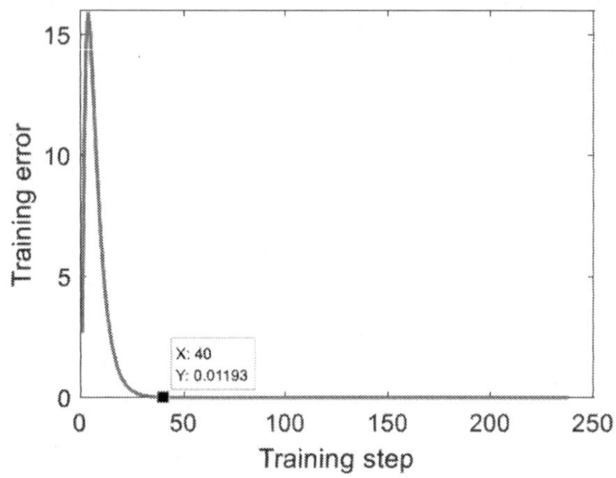

Figure 7. Network training error when using BP neural network.

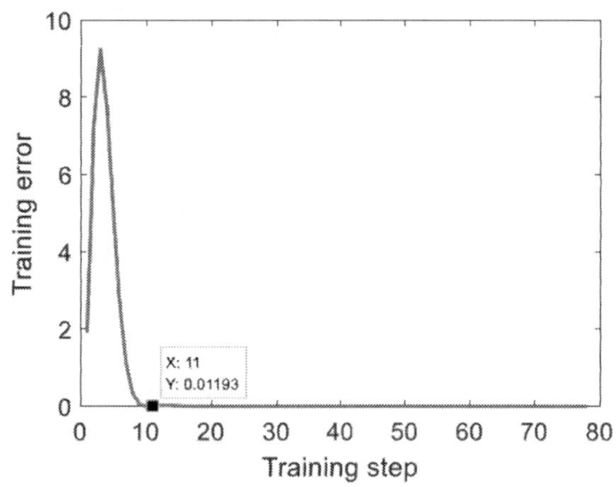

Figure 8. Network training error when using simple RBF neural network.

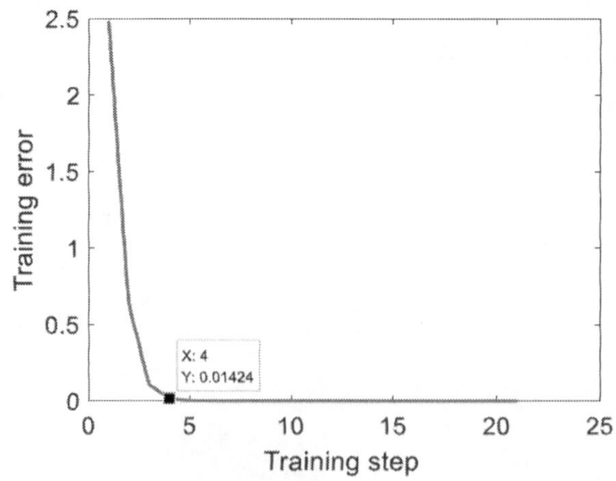

Figure 9. Network training error when using fuzzy RBF neural network.

4. System design

Matlab (2017a) was used as the development software in the Win7 environment in the system, and detection was conducted using a computer vision device. Eggs with different shell colors and the same freshness have different image color characteristics, whereas eggs of various shell colors and freshness sometimes have similar image color characteristics. Therefore, to avoid the influence of shell color on egg freshness detection, we detected freshness according to different shell color classifications to be efficient and accurate. Since the color of the egg shell is related to the color information of the yolk area (H, I, S), the fuzzy identification is conducted using the I value with greater influence. The membership functions A(I) and B(I) of the fuzzy set of brown-shell and white-shell eggs was established, and the I value of the tested egg was substituted into A(I) and B(I), respectively. According to the principle of maximum membership, if A(I) > B(I), then the egg is considered to have a brown-shell, and vice versa [9]. The system identification flow chart is shown in figure 10.

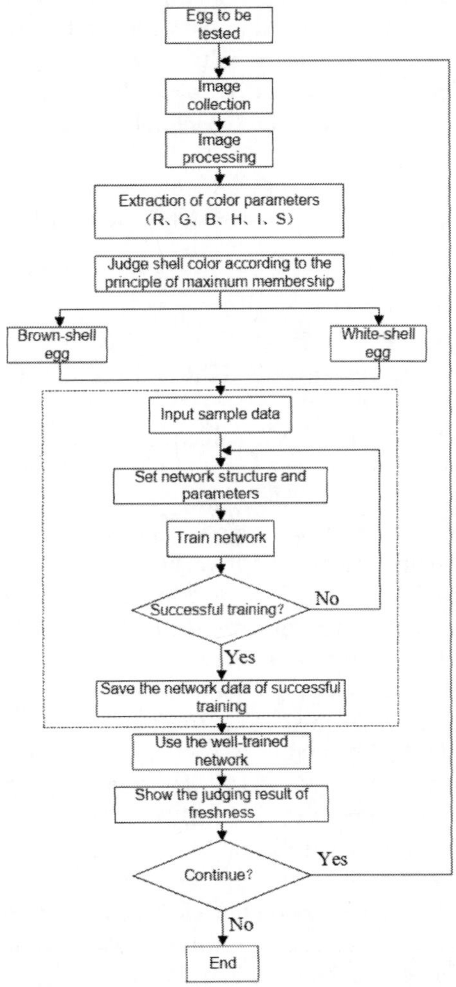

Figure 10. System identification flow chart.

5. Results

To detect the performance and efficiency of the network, several batches of brown- and white-shell eggs of the same variety as the previous test were randomly selected, and the test conditions were kept the same. First, we extracted color information from the egg to be tested on the test device. Next, we broke the egg, measured and calculated the true freshness of the egg, and then input the color information parameters (H, I, S) of the egg into the network. Then, we assessed the freshness grade of the egg using the network and recorded the test and network results of each egg separately. The system detection results as shown in table 3.

Table 3. Recognition accuracy.

Algorithm	Brown-shell egg	White-shell egg
BP neural network	87.258%	89.029%
Simple RBF neural network	91.682%	93.758%
Fuzzy RBF neural network	95.175%	97.525%

6. Conclusions

We developed a novel fuzzy RBF neural network egg freshness recognition system to address the problem of poor recognition accuracy of egg freshness using a BP neural network. To avoid the influence of shell color on egg freshness detection, we detected egg freshness according to different shell color classifications, and our experimental results show that the average recognition accuracy of the fuzzy RBF neural network detection system is 96.35%, which is higher than that of the simple RBF neural network detection system (92.72%) and the BP neural network detection system (88.14%).

Acknowledgments

The authors thank the Engineering Technology Research Center of Optoelectronic Appliance, Anhui Province of China for support.

References

[1] Duan Y F, Wang Q H, Ma M H, Lu X and Wang C Y 2016 Study on non-destructive detection method for egg freshness based on LLE-SVR and visible/near-infrared spectrum *Spectrosc. Spectr. Anal.* **34** 981-5

[2] Aboonajmi M, Saberi A, Najafabadi T A and Kondo N 2016 Quality assessment of poultry egg based on visible-near infrared spectroscopy and radial basis function networks *Int. J. Food Prop.* **19** 1163-72

[3] Li X M 2017 Study on on-line detection of egg freshness based on visible-nearinfrared spectrum (Wuhan, China: Huazhong Agriculture University)

[4] Li J T, Zhu S S, Jiang S and Wang J 2017 Prediction of egg storage time and yolk index based on electronic nose combined with chemometric methods *LWT* **82** 369-76

[5] Fu D D and Wang Q H 2016 Predictive models for the detection of egg freshness, acidity and viscosity using hyper-spectral imaging *Food Sc.* **37** 173-9

[6] Wang Q H, Zhou K, Wu L L and Wang C Y 2016 Egg freshness detection based on hyper spectr *Spectrosc. Spectr. Anal.* **36** 2596-600

[7] Akbarzadeh N, Mireei S A and Askari G 2019 Microwave spectroscopy based on the waveguide technique for the nondestructive freshness evaluation of egg *Food Chem.* **277** 558-65

[8] Cavanna D, Zanardi S, Dall'Asta C and Suman M 2019 Ion mobility spectrometry coupled to gas chromatography: A rapid tool to assess eggs freshness *Food Chem.* **271** 691-6

[9] Wang Q, Ren Y and Wen Y 2006 Study on non-destructive detection method for fresh degree of eggs based on BP neural network *Trans. Chin. Soc. Agricul. Machinery* **37** 104-6

[10] Pan L, Tu K, Zhan G, Liu M and Zou X 2010 Eggshell crack detection based on information fusion between computer vision and acoustic response *Trans. CSAE* **26** 332-7

[11] Zhao H and Zhou X 2013 Recognition of artificial ripening tomato and nature mature tomato based on the double parallel genetic neural network *Adv. J. Food Sci. Technol.* **5** 482-7

[12] Song Q, Meng G J, Yang L, Du D Q and Mao X F 2014 Comparison between BP and RBF neural network pattern recognition process applied in the droplet analyzer *International Conference on Vehicle and Mechanical Engineering and Information Technology (VMEIT), Beijing, China* pp 2333-6

[13] Wang H, Kong C, Li D, Qin N, Fan H, Hong H and Luo Y 2015 Modeling quality changes in brined bream (*Megalobrama amblycephala*) fillets during storage: comparison of the arrhenius model, BP, and RBF neural network *Food. Bioprocess Technol.* **8** 2429-43

[14] Wang Q 2004 The study of recognition of egg fresh degree by BP neural network (Wuhan, China: Huazhong Agricultural University)
[15] Zhao H B and Zhou X H 2013 Recognition of artificial ripening tomato and nature mature tomato based on the double parallel genetic neural network *Adv. J. Food Sci. Technol.* **5** 482-7
[16] Zhang Y and Lv H 2016 Study on PID controller based on fuzzy RBF neural network in rolling mill hydraulic AGC system *Proceedings of the 28th Chinese Control and Decision Conference* (Yinchuan, China) pp 4616-20
[17] Cao H, Sun B, Duan J, Pan D and Li T 2000 Principle of determining the minimum reasoning rules in fuzzy control *Veh. Engin.* **5** 24-7

IOP Conf. Series: Earth and Environmental Science **346** (2019) 012085 doi:10.1088/1755-1315/346/1/012085

Chemical composition, antioxidant and anticancer activity of the essential oil from myric rubra leaves

J Yin[1], X Li[1], F F Huang[1], M H Lu[1], J Yang[1, 3] and L Y Zhu[2, 3]

[1]China tobacco Zhejiang industrial.co., ltd., Hangzhou 310024, Zhejiang, China

[2]Key Laboratory of Marine Food Quality and Hazard Controlling Technology of Zhejiang Province, China Jiliang University, Hangzhou 310018, Zhejiang, China

E-mail: yangjun@zjtobacco.com(J Yang); 1059443655@qq.com (L Y Zhu)

Abstract. The Myric rubra leaves essential oil (MEO) was extracted by steam distillation and its major component was analyzed by gas chromatography mass spectrum (GC-MS). The result showed that 20 kinds of essential oils were identified. with humulene (26.03%), caryophyllene (19.74%), ledene oxide-(II) (16.37%) and γ-himachalene (7.7%) being the major constituents. The antioxidant activity of MEO were analyzed by 2,2-diphenyl-1-picrylhydrazylassay (DPPH·) and oxygen radical absorbance capacity (ORAC) assay respectively, Analogues of vitamin E (Trolox) was positive control. The results showed that the IC50 of DPPH radical scavenging is 0.0576 mg/mL and the ORAC value of fluorescence spectrometry is 1.32mmol TE/g. The anticancer effect of obtained MEO was tested in A549 lung cancer cells by 3-(4,5-dimethylthiazol-2-yl)-2,5-diphenyl tetrazolium bromide (MTT) assay. MEO IC50 value of 163 μ g/mL was obtained with significantly inhibited cell proliferation in concentration-dependant manner. MEO could be useful for pharmaceutical, food and cosmetic industry, because of its strong antioxidant and anticancer activity.

1. Introduction

Chinese bayberry belongs to Myricaceae, with crown spherical, single leaf alternate, and widely grown in Zhejiang Province. The kind of Myrica rubra cv. DongKui Orient Pearl originated in Zhejiang Huangyan, is famous for its largest fruit size in the world, known as "Bayberry king". The fruit can help produce saliva, slake thirst and digestion, having high medicinal and food value. Their roots, stems, leaves have also been used externally for dysentery, diarrhea, rectal bleeding, rheumatism pain, stomach pain, and cardiovascular disease in Chinese traditional medicine [1,2]. There are many active components in Chinese bayberry leaves, such as procyanidins, tannins, quercetin, bayberry, coumarin derivatives, triterpenes, etc., which have the functions of anti-depression, regulating lipid and glucose metabolism, anti-oxidation, anti-inflammatory, analgesic, tumor inhibition and anticancer. Chinese bayberry leaf essential oil is a kind of water insoluble volatile liquid extracted by steam distillation, supercritical extraction, simultaneous distillation and extraction, etc. The composition is complex and diverse, mainly terpenes. Significant anti-oxidation [3], anti-viral [4], anti-microbial [5,6], anti-tumor [7], anti-allergic activity of MEO has been reported in many studies. The anti-proliferation effect of MEO was tested in human colon cancer cell lines SW480, SW620, CaCO2, HCT8 and HT29. Concentration dependent proliferation of cells in all cell lines could be inhibited by MEO, and Caco2 is the most sensitive [8]. The antioxidant activity of essential oils of Myrica rubra Var. astropurea Tsen was reported on [9].

Content from this work may be used under the terms of the Creative Commons Attribution 3.0 licence. Any further distribution of this work must maintain attribution to the author(s) and the title of the work, journal citation and DOI.
Published under licence by IOP Publishing Ltd

IOP Conf. Series: Earth and Environmental Science **346** (2019) 012085 doi:10.1088/1755-1315/346/1/012085

The antioxidant activity of the non-volatle compounds of Myrica rubra Var has been further reported [10]. One of the aims of the current study was to investigate EMO anti-oxidant activity by DPPH and ORAC methods and anticancer activity by MTT method for their future application in either pharmaceuticals or cosmetics.

2. Materials and methods

2.1. Plant materials

The naturally dried leaves of Myrica rubra cv. DongKui Orient Pearl was collected from Xianju, Zhejiang Province, China in Jun-2014.

1,1-Diphenyl-2-picryl-hydrazyl (DPPH) and Trolox was from Sigma–Aldrich (USA), Stock solutions of DPPH were prepared in methanol.

2.2. Extraction of the essential oils

MEO was isolated by steam distillation. Air-dried Myrica rubra (500g, at room temperature, 6d) was crushed into powder with a mixer (HF-6300, Biaohang Co., Ltd., Hangzhou, China). The powder was mixed with water (1: 6), and refluxed for 6-8 hours at 160 oC to extract the essential oils. The obtained oil was dried with anhydrous sodium sulphate and stored in a 4°C sealed vial in the refrigerator before analysis.

2.3. Gas chromatography-mass spectrometry (GC-MS)

GC–MS analysis was performed on a combined GC-MS instrument (Agilent 6890N GC with 5973 inert MSD) fitting with a HP-5 fused silica capillary column (30 m length, 0.25 mm diameter, 0.25 μm film thickness). GC-MS conditions: injection volume was 0.5 μl, solvent delay time was 2 min, split ratio was 1:40, injector temperature was 250 ℃, carrier gas was helium at 1.0 ml/min constant flow mode, the initial column temperature was from 50 to 280 ℃ at 8 ℃/min. Mass spectra: electron impact (EI+) mode, 70 eV and ion source temperature was 200 ℃. Mass spectra was recorded over 50–500 amu range.

The relative percentage of MEO components was expressed as percentage by peak area normalization method. According to the gc-ms spectra from NIST'98 literature data of the United States, the composition of MEO was identified according to the retention time.

2.4. Antioxidant activity

DPPH radical scavenging activity [11,12] has been detected by the following method:

Stock solutions of DPPH (25mg/ml) were prepared in methanol. Various concentrations of oil sample in methanol (0.001, 0.005, 0.01, 0.025, 0.05, 0.075 and 0.10mg/mL) were added to 2 mL (25mg/ml) DPPH and the volume was made up to 2.2mL with methanol. The mixture was shaken thoroughly and left at room temperature for 30 min. The mixture's absorbance was then measured on a spectrophotometer at 517 nm. Trolox was used as a positive control, and all determinations were performed in triplicate (n = 3).

The percentage inhibition was calculated by the following equation:

DPPH radical scavenging (%) = [(Ac- As)/Ac) × 100%]

where, Ac is the absorbance of the control (only DPPH in methanol) and As is the absorbance of the sample.

2.5. ORAC assay

The ORAC assay for essential oils was carried out using an automated multimode reader with 96-well plates. Trolox standard was formulated into 6.25, 12.5, 25, 50μmol/L with phosphate buffer (pH7.4). Fluorescein (0.12 μmol/L) was used as the substrate and 2,2'-azobis (2-methylpropion amidine) dihydrochloride (AAPH) (40 mmol/L) was used as radical generator. The trolox standard or test compound solution (20μl, used as the control phosphate buffer) was added to each of the 96-well black microporous plates and 120μl of fluorescein was added to the phosphate buffer solution. The plate was incubated at 37℃ for 20 minutes, and the machine was programmed to inject 60 μl of AAPH into each well. Fluorescence conditions were excitation at 485nm and emission at 538nm. A standard curve was drawn and the ORAC value of the tested compounds was obtained as Trolox equivalent (TE).

2.6. Inhibition of A549 lung cancer cells

A549 lung cancer cells (3000 cells/well) were inoculated into a 96-well plate and preincubated for 3 hours for cell adherence. Test samples of 10 μl EMO for various concentrations (100, 150, 200, 250, 300, 350 and 400 μg/mL) and the control cells without MEO were added into the cultures and incubated at 37℃ for 72 h under humidified air containing 5% CO2 respectively.

After removing the medium from the wells, 10 μL of tetrazolium salt solution (MTT) (5mg/ml) was added. After been incubated at 37℃ for 4h, the medium was removed and added into 150 μl of DMSO and the formazan crystals were dissolved in the shaker for 10 minutes. Absorbance was measured in a multimode reader at 570 nm. The cell viability ratio (%) was calculated from the following equation: viability % = (absorbance of test sample/absorbance of control) × 100%.

3. Results

MEO was obtained as a light-yellow oil, looked clear and transparent. The yield was 0.129% of the dried plant material.

Table 1. Identified chemical composition and the relative content of Myric rubra leaves essential oil.

NO	RT (min)	Compound	Molecular formula	Composition (%)
1	15.681	Caryophyllene	C15H24	19.74
2	15.951	Alloaromadendrene	C15H24	0.56
3	16.216	Humulene	C15H24	26.03
4	16.299	Calarene	C15H24	0.18
5	16.501	γ-Muurolene	C15H24	1.65
6	16.703	(-)-a-Selinene	C15H24	3.68
7	16.792	β-copaene	C15H24	0.13
8	17.092	(-)-g-Cadinene	C15H24	4.72
9	17.259	Simvastatin	C25H38O5	1.07
10	17.44	α-Muurolene	C15H24	0.34
11	18.41	Caryophyllene oxide	C15H24O	1.64
12	18.566	Ledene oxide-(II)	C15H24O	16.37
13	18.794	1,2,3,4,4a,7-hexahydro-1,6-dimethyl-4-(1-methylethyl)- Naphthalene	C15H24	0.1
14	18.94	4,4-Dimethyl-3-(3-methylbut-3-enylidene)-2-methylenebicyclo[4.1.0]heptane	C15H22	6.85
15	19.033	γ-Himachalene	C15H24	7.7
16	19.204	β-Guaiene	C15H24	2.31
17	19.272	Retinol, acetate	C22H32O2	0.25
18	19.458	Cadalin	C15H18	0.73
19	21.56	2-Pentadecanone, 6,10,14-trimethyl-	C18H36O	1.44
20	24.808	Dehydroisophytol	C20H40O	0.77

3.1. Chemical composition of MEO

Chemical constituents of Myrica rubra cv. DongKui Orient Pearl essential oil analyzed by GC-MS is shown in Table 1, and Figure 1 shows the total ion flow of the essential oil. 20 chemical components were identified by GC/MS in MEO, covering 96.25% of the total peak area. The major constituents of the essential oil were Humulene (26.03%), Caryophyllene (19.74%), Ledene oxide-(II) (16.37%), γ-Himachalene (7.7%), 4,4-Dimethyl-3-(3-methylbut-3-enylidene)-2-methylenebicyclo[4.1.0]heptane (6.85%), (-)-g-Cadinene (4.72%), (-)-a-Selinene (3.68%),β-Guaiene (2.31%), γ-Muurolene (1.65%), Caryophyllene oxide (1.64%).

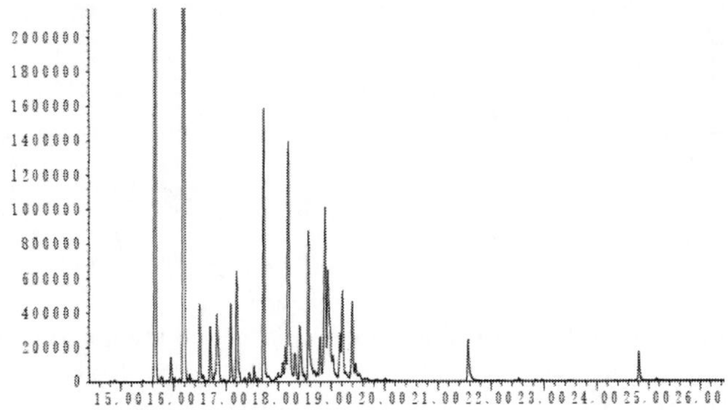

Figure 1. Identified chemical composition and the relative content of Myric rubra leaves essential oil.

3.2. Antioxidant activity of MEO

Free radical scavenging activities of MEO and trolox measured by DPPH assay are reported in Figure 2. It was observed that the radical scavenging activities enhanced with increasing concentration. The linear function obtained after plotting Radical scavenging versus the concentration were y = 633.76x + 35.926, R2 = 0.9543 and y = 734.08x + 7.6931, R2 = 0.9738 within 0.005-0.1mg / mL. IC50 of Trolox and essential oils (IC50) were 0.0222 mg / mL and 0.0576 mg / mL from the fitted curve, the radical scavenging activities of trolox was stronger than MEO's.

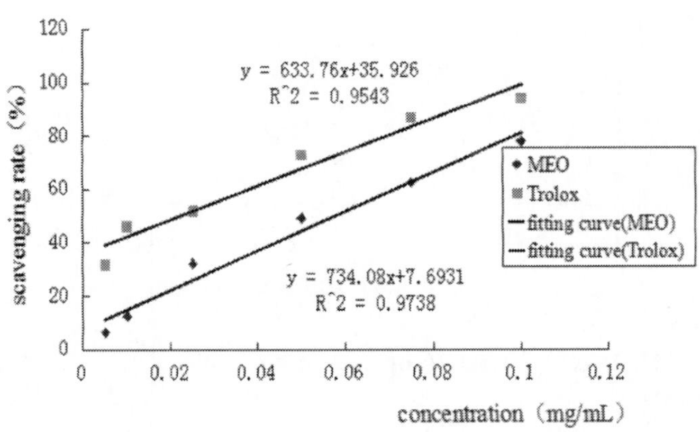

Figure 2. The fitting curve of DPPH radical scavenging activity of EMO and Trolox.

The antioxidant capacity of MEO and trolox determined by ORAC assay is reported in Figure 3. The linear function obtained after plotting net AUC versus the concentration was y = 4.8589x + 11.296, R2= 0.9973 within 6.25-50 μ mol/L. The ORAC value of MEO, 1.32mmol TE/g from the fitted curve, showed good oxygen radical scavenging ability. Seen from component analysis of essential oils, the reason why MEO had the radical scavenging activity was MEO had high content of antioxidant ingredients, such as Himachalene, Caryophyllene, Ledene oxide-(II), γ-Himachalene having strongly radical scavenging activities.

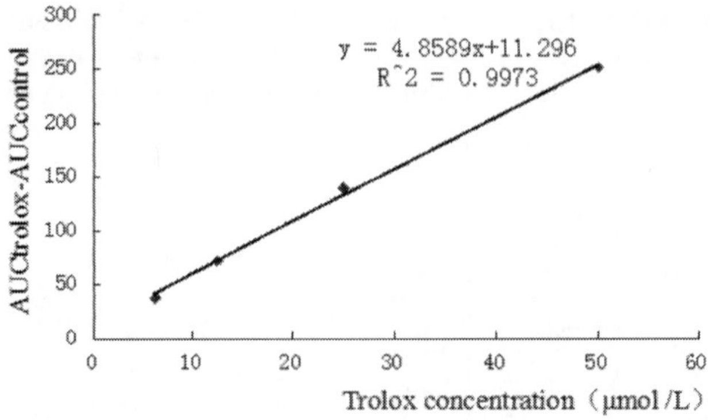

Figure 3. Standard curve for the trolox concentration and the net area under the FL.

3.3. Anticancer activity of MEO

The yellow MTT (3- (4, 5-dimethythiazole-2-group) -2, 5-diphenyltetrazolium bromide) can be reduced to purple formazan by the living cell mitochondria. Anti-proliferation effects of MEO at various concentrations (100-400 mug/mL) were tested in A549 lung cancer cells. The survival rates after 48h exposure shown in Figure 4 showed MEO inhibits cell line proliferation in a concentration-dependent manner. Essential oils were found to be very active at a concentration of 250 mug/mL, where the survival rate is 15%. The linear function was obtained after plotting survival rates versus the concentration: y=-0.3215x+102.43, R2=0.9757 within 0-300μg/mLThe values of IC50 is found to be 163μg/mL.

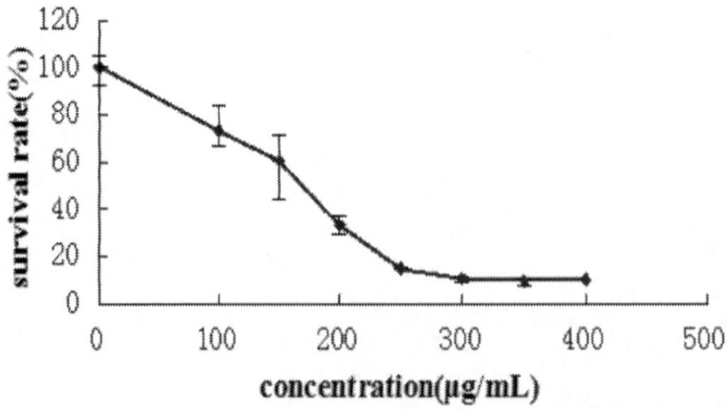

Figure 4. Effect of MEO on proliferation of human lung cancer cell A549. Values are mean ±SD of three experiments.

4. Discussion

The results obtained were 20 components in MEO, with α -Humulene (26.03%), β-Caryophyllene (19.74%), Ledene oxide-(II) (16.37%) and γ-Himachalene (7.7%) determined as the most abundant molecules among the principal compounds. While our team reported that the main components of MEO from Chaoshan carbon variety are Caryophyllene (44.01%), Humulene (34.40%), Ledene oxide-(II) (4.81%), α-cis-Himachalene (2.83%) [13]. Composition differences are due to different varieties of Myric rubra leavies and seasonal variability.

In the last few years, an increasing interest on the antioxidant and anticancer activity of the different essential oils has been reported and in many cases these bioactivities are due to the presence of active constituents such as α-Humulene, caryophyllene oxide and β-caryophyllene [14,15]. β-Caryophyllene is reported to possess antioxidant [16], antimicrobial [17, 18], anti-inflammatory [19-21] and anticancer activities [22]. β-Caryophyllene induces apoptosis in tumor cells with the catalytic activity of DNA ladder and caspase-3. Intraperitoneally administered caryophyllene oxide exhibits anti-injury sensitivity and might play a role on a variety of mechanisms that involve central and peripheral pathways [23]. The essential oils with high content of caryophyllene and caryophyllene oxide have a strong anti-tumor activity. Therefore, Antitumor activity of MEO possibly associated with the major components of Humulene, caryophyllene and caryophyllene oxide, this result needs further study yet.

Free radical scavenging activities of MEO and trolox measured by DPPH assay have been researched above. It was observed that the radical scavenging activities enhanced with increasing concentration. IC50 of Trolox and essential oils (IC50) were 0.0222 mg / mL and 0.0576 mg / mL from the fitted curve, the radical scavenging activities of trolox was stronger than MEO's.

The antioxidant capacity of MEO and trolox determined by ORAC assay has been researched above. The ORAC value of MEO, 1.32mmol TE/g from the fitted curve, showed good oxygen radical scavenging ability. Seen from component analysis of essential oils, the reason why MEO had the radical scavenging activity was MEO had high content of antioxidant ingredients, such as Himachalene, Caryophyllene, Ledene oxide-(II), γ -Himachalene having strongly radical scavenging activities.

The yellow MTT (3- (4, 5-dimethythiazole-2-group) -2, 5-diphenyltetrazolium bromide) can be reduced to purple formazan by the living cell mitochondria. Anti-proliferation effects of MEO at various concentrations (100-400 mug/mL) were tested in A549 lung cancer cells. The survival rates after 48h exposure showed MEO inhibits cell line proliferation in a concentration-dependent manner. Essential oils were found to be very active at a concentration of 250 mug/mL, where the survival rate is 15%.

References

[1] Jing T, Toshio M, Iwao T, Shin A, Hisashi M and Masayuki Y 2002 Inhibitors of nitric oxide production from the bark of myrica rubra: structures of new biphenyl type diaryl heptanoid glycosides and taraxerane type triterpene. *Bioorg. Medic. Chem.*. **10**: 4005-12.

[2] Han H K, Myeong H O, Kwang J P, Jun H H and Min W L 2014 Anti-inflammatory activity of sulfate-containing phenolic compounds isolated from the leaves of Myrica rubra. *Fitoterapia*. **92**(1):188-93.

[3] Takahashi Y, Inaba N, Kuwahara S and Kuki W 2003. Antioxidative effect of Citrus essential oil components on human lowdensity lipoprotein in vitro. *Biosci Biotechnol Biochem*. **67**:195-7.

[4] Ramy M, Sayed A and Mahmoud I 2010 Chemical composition, antivirus and antioxidant activities of seven essential oils. *J. Appl. Sci. Res.* **6**: 50-62.

[5] Gkogkaa E, Hazelegera W C, Posthumusb M A and Beumera R R 2013. The antimicrobial activity of the essential oil of pistacia lentiscus var. chia. *J. Essent. Oil Bearing Plants*. **16**(6):714-29.

[6] Josefa R, Francisca S, Isidora S B, Amparo C and Santamarinaa M P 2015 Antifungal activity and potential use of essential oils against fusarium culmorum and fusarium verticillioides. *J. Essent. Oil Bearing Plants*. **18**(2):359-67.

[7] Amr E E 2007 Pharmaceutical and therapeutic potentials of essential oils and their individual volatile constituents: a review. *Phytoth. Res.* **21**(4): 308-23.

[8] Lenka L, Veronika H, Jan R, Barbora S, Martin A, Vera K, Tomas V, Lou J D, Zhu L Y, Yang J and Lenka S 2014. Essential oil from Myrica rubra leaves inhibits cancer

cell proliferationand induces apoptosis in several human intestinal lines. *Industr. Crops Prod.* **59**:20-6.
[9] Zhong R M, Zhang Z M, Wang Y M, Zeng Q X and Peng H G 2006 Chemical composition and antioxidant components of oils from leaves, bark and root of Myrica rubra. *Chem. Indust. Forest Products.* **26**(1):1-5.
[10] Yang J, Zhu L Y, Wu J Q, Yin J, Lenka L, Tomas V, Gao Y S and Lou J D 2014 Antioxidant, antimicrobial and antisepsis properties of the water-soluble extracts of Myric rubra leaves. *China Food Addit.* **123**(2):198-203.
[11] Liu C J, Zhang S Q, Zhang J S, Liang Q and Li D S 2012. Chemical composition and antioxidant activity of the essential oil from berries of Schisandra chinensis (Turcz.) Baill. *Nat. Prod. Res.* **26**(23): 2199-203.
[12] Om P S and Tej K B 2009 DPPH antioxidant assay revisited. *Food Chem.* **113**:1202-5
[13] Yang J, Wu J Q, Lenka L, Jiang J, Yin J, Zhu L Y, Gao Y S and Zhang Y J 2014. Chemical composition, antioxidant and antimicrobial properties of the essential oils of myric rubra charcoal variety leaves. *Food Sci. Tech.* **39**(1):286-90.
[14] Noudogbessi J P, Delort L, Chalard P, Billard H, Figueredo G, Ruiz N, Chalchat J C, Sohounhloue D and Chezet F C 2013 Antiproliferative activity of four aromatic plants of Benin. *J. Nat. Prod.* **76**(6):123-31.
[15] Keawsaard S and Kongtaweelert S 2012 Antioxidant, antibacterial, anticancer activities and chemical constituents of the essential oil from Mesua ferrea leaves. *Chiang Mai J. Sci.* **39**(3):455-63.
[16] Kubo I, Chaudhuri S K, Kubo Y, Sanchez Y, Oqura T, Ishikawa H and Haraguchi H 1996 Cytotoxic and antioxidative sesquiterpenoids from Heterotheca nuloides. *Planta Med.* **62**(5):427-30.
[17] Huang M S, Sanchez-Moreiras A M, Abel C, Sohrabi R, Lee S, Gershenzon J and Tholl D 2012 The major volatile organic compound emitted from Arabidopsis thaliana flowers, the sesquiterpene (E)-β-caryophyllene, is a defense against a bacterial pathogen. *New Phytol.* **193**(4):997-1008.
[18] Sabulal B, Dan M, J A J, Kurup R, Pradeep N S, Valsamma R K and George V 2006. Caryophyllene-rich rhizome oil of Zingiber nimmonii from South India: Chemical characterization and antimicrobial activity. *Phytochemistry.* **67**(22):2469-73.
[19] Medeiros R, Passos G F, Vitor C E, Koepp J, Mazzuco T L, Pianowski L F, Campos M M and Calixto J B 2007 Effect of two active compounds obtained from the essential oil of Cordia verbenacea on the acute inflammatory responses elicited by LPS in the rat paw. *Br. J. Pharmacol.* **151**(5):618-27.
[20] Al-Reza S M, Yoon J I, Kim H J and Kang S C 2010 Anti-inflammatory activity of seed essential oil from Zizyphus jujuba. *Food Chem. Toxicol.* **48**(2):639-643.
[21] Bento A F, Marcon R, Dutra R C, Claudino R F, Cola M, Leite D F, Calixto J B 2011 β-caryophyllene inhibits dextran sulfate sodium-induced colitis in mice through CB2 receptor activation and PPARγ pathway. *Am. J. Pathol.* **178**(3):1153-66.
[22] Eitan A, Rivka O, Nativ D, Elaine S, Tatiana R and Shimon R 2012. β-caryophyllene, a compound isolated from the biblical balm of gilead (Commiphora gileadensis), Is a selective apoptosis inducer for tumor cell lines. *Evidence-Based Complementary and Alternative Medicine*. doi: 10.1155/2012/872394
[23] Chavana M J, Wakteb P S and Shinde D B 2010. Analgesic and anti-inflammatory activity of Caryophyllene oxide from Annona squamosa L. bark. *Phytomedicine.* **17**:149-51.

IOP Conf. Series: Earth and Environmental Science **346** (2019) 012086 doi:10.1088/1755-1315/346/1/012086

Analysis on the incentive effect of abolishment of agricultural tax in China staple grain production input

W Fang[1]

Institute of Agricultural Economics and Rural Development, Guangdong Academy of Agricultural Sciences, Guangzhou, 510640, China

E-mail:184347068@qq.com

Abstract: By integrating the process of abolishing the agricultural tax into the empirical analysis of regression discontinuity, this paper studies the incentive effect of the policy on different input types and the production of different food crops. According to research findings, mechanical operating cost is positively stimulated by the abolishment of agricultural tax, while the positive stimulation on fertilizer and labor costs degrades with time. From the perspective of different crops, the abolishment of agricultural tax has incentive effect on the fertilizer input to rice, corn and wheat, the labor input to rice and wheat, and the machinery input to corn. The method of regression discontinuity can determine whether the abolishment of agricultural tax has any incentive effect on crop production, but cannot identify the incentive scale. Therefore, appropriate instrumental variables are introduced in further study to identify the scale of incentives.

1. Introduction

China started agricultural tax and fee reform in 2000 [1-2] when Anhui was determined as the pilot province for rural tax and fee reform and the agricultural tax rate was set no more than 7%. However, the problem of rising agricultural specialty tax rate was derived from the implementation of the reform [3]. In 2001, the Chinese Government took some mitigation measures, including increasing transfer payment and properly dealing with debts, incorporated Jiangsu to the pilot provinces, but slowed down the pilot work in other provinces of China and determined to stop the expansion of provinces with local pilot work. In 2002, the pilot agricultural tax reform was comprehensively implemented in 20 provinces of China on the basis of experience from Anhui and Jiangsu, which drove the rural population in pilot regions up to 620 million and benefited 3/4 of the national rural population. In 2003, China got through the pilot phase of rural tax and fee reform and started to implement it comprehensively across the country. In 2004, the Chinese Government continued the rural tax and fee

Content from this work may be used under the terms of the Creative Commons Attribution 3.0 licence. Any further distribution of this work must maintain attribution to the author(s) and the title of the work, journal citation and DOI.
Published under licence by IOP Publishing Ltd

reform, by which the agricultural specialty tax was abolished, Heilongjiang and Jilin led in realizing complete exemption of agricultural tax, and the agricultural tax rate was decreased by 3% in 11 provinces including Hebei and by 1% in other provinces. In 2005, the Chinese Government further expanded the scope of agricultural tax exemption and abolished the animal husbandry tax. At that time, there were 28 provinces where agricultural tax was exempted, except Shandong, Hebei and Yunnan. In 2006, China abolished many taxes on agricultural production throughout the country, including agricultural tax, agricultural specialty tax and animal husbandry tax.

2. Status of domestic and foreign research

The research on agricultural tax in European and American countries shows that effective agricultural tax policy is favorable to the protection for national environment, health and land productivity [4-5]. Soliwoda and Paw [6] deem that most EU member states (such as Netherlands, Germany and Austria) take taxes as the policy instrument for influencing sustainable agricultural development. As found by research, taxes have neutral or positive effect on the environmental sustainability. Berendse [7] has proposed that EU should levy progressive tax on such agricultural production activities affecting health and environment, such as farmers' purchase of pesticides and antibiotics, and import of animal feeds including soybean. Munoz-Zamponi and Mardones-Poblete [8] have simulated the economic and environmental impacts of taxes on CO_2 emissions of the industry by applying the Leontief Price Model. Researches show that the levy of environmental tax on agriculture and animal husbandry will not reduce CO_2 emissions obviously. Both positive and negative effects are reflected in domestic literatures regarding the abolishment of agricultural tax. Yang [9] deem that the abolishment of agricultural tax has positive effect on the change of rural powers, mainly represented by the fact that the initiative of "farmers no longer providing for the government" forces the overall transformation of "rural political" system. However, Zhang [10], Ni *et al.* [11] deem that the abolishment of agricultural tax has material negative effect on the finance in some regions and eventually makes the supply of rural public goods difficult. Zuo *et al.*[12] have studied the panel data of 108 county-level administrative units in Henan during 2001-2008 by applying the double difference method, which shows that the abolishment of agricultural tax causes such behavior as allocating more fiscal expenditure by the local government to policy objectives easy to assess, such as infrastructure, ignoring capital input to public services and manpower. Some scholars deem that the abolishment of agricultural tax has positive effect on agricultural production input since it provides real benefits for producers. Wang *et al.* [13] based on the investigation data of 340 farmer households in 5 counties of Hebei, Henan and Shandong, the abolishment of agricultural tax has incentive effect on the increase of food output.

3. Comparison of food production inputs before and after abolishment of agricultural tax in china

The original intention of reduction and exemption in agricultural tax by Chinese Government is to increase farmers' confidence in national support for "agriculture, farmer and rural area", arouse the enthusiasm of farmers for food growing, guide farmers to input more resources to production and therefore promote food production. To verify the authenticity of such incentive effect, the process of abolishment of agricultural tax is divided into three stages (see Table 1): I- before abolishment of

IOP Conf. Series: Earth and Environmental Science 346 (2019) 012086 doi:10.1088/1755-1315/346/1/012086

agricultural tax (1995-2000), II- during abolishment of agricultural tax (2000-2005) and III- after abolishment of agricultural tax (2005-2010), with attention to various agricultural production inputs and the average annual growth rate of food output at different stages.

Table 1. Comparison of agricultural production factors and average annual growth rate of food product output before and after abolishment of agricultural tax in china.

Stage	Year	Cultivated area (1,000 hectares)	Total sowing area of crops (1,000 hectares)	Gross power of agricultural machinery (10,000 KW)	Irrigated area (1,000 hectares)	Converted application of fertilizer (10,000t)
I before abolishment of agricultural tax	1995-2000	0.86%	0.70%	6.46%	1.48%	2.41%
II during abolishment of agricultural tax	2000-2005	0.81%	-0.09%	4.48%	0.37%	2.35%
III after abolishment of agricultural tax	2005-2010	2.92%	0.55%	5.21%	1.55%	2.61%
Stage	**Year**	**Food (10,000t)**	**Cereal (10,000t)**	**Rice (10,000t)**	**Wheat (10,000t)**	**Corn (10,000t)**
I before abolishment of agricultural tax	1995-2000	-0.16%	-0.44%	0.24%	-0.42%	-0.91%
II during abolishment of agricultural tax	2000-2005	0.77%	0.91%	-0.66%	-0.37%	4.67%
III after abolishment of agricultural tax	2005-2010	2.04%	2.51%	1.35%	2.83%	4.09%

Data source: Hexun macro data (http://mac.hexun.com/)

From the perspective of production factors, the growth rate was in recession during transition from stage I to stage II, which was specifically reflected by the great decrease in the average annual growth rate of agricultural land scale and mechanization, and mainly caused by ecological restoration and construction occupation [2]. However, such situation was reversed from stage II to stage III when the growth rate of several production factors, including cultivated area, sowing area of crops, gross power of machinery, irrigated area and fertilizer application, was increased compared with that at stage II. At the early stage of reduction and exemption in agricultural tax during 2000-2003, China carried out pilot tax reduction and exemption gradually, but local governments put focus on settlement of owing taxes since they were under the time pressure regarding to the agricultural tax exemption imposed by the Central Government (China Taxation Yearbook 2005). This explains, to some degree, why the agricultural tax reached the peak-peak in 2002 and 2003 as shown in Table 1, and such situation was gradually mitigated in 2004 and 2005. In other words, during implementation of the tax reduction at stage II, the policy effect of "alleviating burden" had not been fully developed, and the effect of abolishment of agricultural tax was observed after 2004.

From the perspective of food output, the average annual growth rate declined slightly at stage I, but such declining was restrained at stage II, and then the average annual growth rate rose greatly at stage III. The average annual growth rate of rice and wheat was negative at stage II and positive at stage III, while that of corn was maintained above 4.0% at both stage II and stage III. It can be preliminarily inferred from above that the incentive effect of abolishment of agricultural tax in China on food production cannot be underestimated.

4. Empirical analysis on incentive effect of abolishment of agricultural tax in china on food production input

4.1. Research idea and method

It is the most reasonable to take 2004, 2005 and 2006 as the observation nodes since the abolishment of agricultural tax is a process and policy effect is with hysteretic nature in general. In addition, such incentive may vary depending on different food crops. If the years 2003, 2004, 2005, 2006 and 2007 are taken as the breakpoints to observe the time of showing policy effect, the policy effect on different products can be observed more clearly by using the data of production inputs to such four food varieties as rice, corn and wheat. It is necessary to compare the agricultural situation at stages with and without effect of agricultural tax and take the time (year) as grouping variable in order to reflect the effects of abolishment of agricultural tax on agriculture in a more objective and clearer way. To obtain the estimates by applying the accurate regression discontinuity, this study applies the local linear regression, namely to obtain the required estimates by minimizing the following objective function:

$$\min_{|\alpha,\beta,\gamma,\delta|} \sum_{i=1}^{n} K\left[\frac{x_i - c}{h}\right][y_i - \alpha - \beta(x_i - c) - \delta D_i - \gamma(x_i - c)D_i]^2$$

$$D_i = \begin{cases} 1 & if x_i \geq 2006 \\ 0 & if x_i < 2006 \end{cases}$$

In the formula, K represents kernel function, h represents bandwidth, y_i represents mechanical operating cost, x_i represents year and c is 2006 (2006 as the breakpoint). The essence of local linear regression is to estimate by weighted least square method in a small neighborhood, where the weight is calculated by the kernel function and the weight of the point closer to the point is larger. With hysteretic nature between agricultural production decision and actual production, food farmers may have not made the decision to increase production input in the year of tax exemption, and farmers always make fixed asset input until the year after they obtain increased income. Therefore, it is necessary to take 2004, 2005 and 2006 as the potential jump points and the discontinuity occurring in these three years can be regarded as the incentive effect of abolishment of agricultural tax.

4.2. Data description

The data used in this study is from the Data Compilation of Costs and Benefits of Agricultural Products in China 1980-2016 and China Statistical Yearbook 2016, covering the material and manpower inputs to three food crops (rice, corn and wheat) in 31 provinces of China during 1979-2016. This study puts emphasis on analyzing whether the abolishment of agricultural tax is incentive to food farmers' material and labor inputs, where the representative variables are fertilizer cost per mu, mechanical operating cost per mu and labor cost per mu. Control variables involved in the calculation include seed and seedling cost, farmyard manure cost and pesticide cost per mu (Table 2), which are all converted to the constant price in 1978 based on the CPI over the years.

Table 2. Key variables and control variables.

variable	mean	sd	p50
Key variable			
X3 fertilizer cost	272.4	281.7	197.0
X7 mechanical operating cost	174.6	287.6	40.63
X13 labor cost	144	169.7	86.12
Other control variables			
X1 seed and seedling cost	123.9	507.1	60.37
X2 farmyard manure cost	37.07	56.22	19.87
X5 pesticide cost	59.53	100.4	20.25
X8 irrigation and drainage cost	62.83	98.75	21.61
X9 fuel and power cost	14.29	45.54	3.720
X11 other direct costs	9.800	25.76	2.480
X12 indirect production costs	39.25	48.21	23.94

4.3. Calculation results

STATA15 is used for accurate regression discontinuity analysis. 3 key variables are taken as dependent variables in sequence and the years 2004, 2005 and 2006 are taken as independent variables for operation (Table 3).

Table 3. Regression discontinuity results of different inputs in years.

	(1)	(2)	(3)	(4)	(5)
Variables	-1	2004	2005	2006	+1
Fertilizer cost	5.978**	15.37***	-3.360	-10.35*	6.713
	(2.792)	(3.464)	(4.304)	(5.649)	(4.622)
Mechanical operating cost	1.273	2.416	4.430*	5.036	3.260
	(2.876)	(2.738)	(2.646)	(3.286)	(4.060)
Labor cost	-2.006	27.15**	0.0962	-13.98	-23.89**
	(11.17)	(10.96)	(11.33)	(12.14)	(10.37)

IOP Conf. Series: Earth and Environmental Science **346** (2019) 012086 doi:10.1088/1755-1315/346/1/012086

Notes: 1. The optimum bandwidth is selected for regression and we try to use 75% and 125% of the optimum bandwidth for stability test. Similar results are obtained.

2. The table shows the results without control variable, which are slightly different from the results obtained by adding control variables for regression, but they still back up the opinions in this paper.

3. The content in the bracket is standard error, *** p<0.01, ** p<0.05, * p<0.1.

The incentive effect of abolishment of agricultural tax on the production inputs to 3 food crops is observed for the year when a positive jump occurs as mentioned above. The results are shown in Table 4:

Table 4. Regression discontinuity results of different material inputs to three food crops.

Variables	(1) Rice	(2) Corn	(3) Wheat
Fertilizer cost (2004)	16.15***	20.68**	14.89***
	(4.541)	(8.536)	(5.506)
Mechanical operating cost (2005)	6.430	13.37**	4.866
	(5.201)	(6.040)	(4.543)
Labor cost (2004)	30.63**	21.83	25.67*
	(11.94)	(17.94)	(19.20)

Notes: 1. The optimum bandwidth is selected for regression and we try to use 75% and 125% of the optimum bandwidth for stability test. Similar results are obtained.

2. The table shows the results without control variable, which are slightly different from the results obtained by adding control variables for regression, but they still back up the opinions in this paper.

3. The content in the bracket is standard error, *** p<0.01, ** p<0.05, * p<0.1.

4.4. Test

A series of tests are required after regression discontinuity: the first is to test whether the covariate condition at the breakpoint shows a jumpn [2], and the second is to test whether the density function of grouping variable is discontinuous at the breakpoint. The above statistical significant results are verified respectively in Table 6 to observe whether the covariates show a jump at the breakpoint. The results are not reliable if there are many jumps. It can be seen that only a few covariates in 5 regression functions show a jump, which means the results are acceptable.

Table 5. Test of covariate condition and density.

	Fertilizer cost (2004)	Fertilizer cost (2006)	Mechanical operating cost (2005)	Labor cost (2004)	Labor cost (2007)
X1	-2.8712	-4.2739	-4.5369	-0.1849	-5.5709
	(6.1054)	(6.6948)	(5.4224)	(10.3256)	(6.5942)
X2	0.7019	-0.1753	0.2714	0.2934	-0.1601

	(0.4591)	(0.7796)	(0.4485)	(0.4280)	(0.3485)
X3			2.3589	3.2879	1.9550
			(1.3120)	(1.7649)	(1.3818)
X4	-0.1236	-0.2964	-0.0208	0.0997	-0.3423
	(0.3773)	(0.5822)	(0.4079)	(0.6538)	(0.6240)
X5	0.2232	0.1095	0.6298	0.3248	0.4979
	(0.3438)	(0.6123)	(0.3976)	(0.4223)	(0.5070)
X6	0.0634	-0.1212	0.7221	-0.1615	0.7009
	(0.7062)	(0.9007)	(0.5425)	(0.4499)	(0.4079)
X7	0.5441	0.2239		1.0051	1.2883
	(0.5751)	(0.9842)		(0.7618)	(0.7980)
X8	-0.0836	0.4119	0.1504	-0.0971	-0.0800
	(0.6098)	(0.7438)	(0.5607)	(0.7528)	(0.6203)
X9	-14.6908*	0.9792	-0.3718	-24.6926**	-0.1837
	(7.3827)	(0.9966)	(0.4150)	(8.0468)	(0.5152)
X11	-0.3861**	0.1056	-0.0866	-0.4222*	0.1872
	(0.1374)	(0.0863)	(0.0879)	(0.1780)	(0.1791)
X13	6.8707	-1.5857	2.7314		
	(8.1440)	(11.5174)	(8.9612)		
N	3130	3130	3053	3162	3162

5. Main study conclusions

By integrating the process of abolishment of agricultural tax into the empirical analysis of regression discontinuity, this paper studies the incentive effect of the policy on different input types and the production of different food crops. According to research findings, mechanical operating cost is positively stimulated by the abolishment of agricultural tax, while the positive stimulation on fertilizer and labor costs degrades with time. From the perspective of different crops, the abolishment of agricultural tax has incentive effect on the fertilizer input to rice, corn and wheat, the labor input to rice and wheat, and only the machinery input to corn. The method of regression discontinuity can determine whether the abolishment of agricultural tax has any incentive effect on crop production, but cannot identify the incentive scale. Therefore, appropriate instrumental variables are introduced in further study to identify the scale of incentives.

References

[1] He X F 2007 Effects of Abolishment of Agricultural Tax on State-Farmer Relationship. Gansu Soc. Sci. [Chinese]. **02**: 1-3.

[2] Chen X W 2010 Accelerating the construction of new socialist countryside. *Qiushi J.* [Chinese]. **21**: 40-3.

[3] Hu Z H 2014 Transition of Agricultural Tax Reform and Chinese Farmers since the Reform and Opening up . *Nankai Univ.* [Chinese].

[4] Butcher B and Xu Y 2017 Chinese tax policy and the promotion of agricultural cooperatives and environmental protection. *Social Sci. Electr. Publishing.* **31**(4): 649-69.

[5] Miller S R 2015 Tax Policies, Agriculture and the Environment. Staff Papers.

[6] Soliwoda M and Pawłowska-Tyszko J 2015 Tax policy tools vs. sustainable development of agriculture. The case of Poland. International Academic Conferences. International Institute of Social and Eco. Sci.

[7] Berendse F 2017 Health and environment:Add a tax to the EU. Agricultural Policy. Nature. **543**(7645):315.

[8] Munoz-Zamponi T B and Mardones-Poblete C A 2016 Simulation of a CO2e tax to mitigate impacts from Chilean agriculture and livestock sector on climate change. *Agrociencia.* **50**(3):271-85.

[9] Yang M H 2006 Pressure and innovation: analysis on China's rural governance structure reform after agricultural tax exemption. *Nanjing Soc. Sci.* [Chinese]. **05**(05):17-22.

[10] Zhang G 2006 Abolishment of agricultural tax: unbalance of fiscal impacts and adjustment of transfer payment policy. *The World of Survey and Research.* [Chinese]. **03**(13):38-41.

[11] Ni Z L 2017 How fiscal and cultural expenditure affects subjective well-being of residents - based on the analysis of CGSS2013 survey data. *J. Shanxi Uni. Finance Econom.* [Chinese]. **07**(39):1-13.

[12] Zuo X, Yin X M and Pan X T 2011 Does Centralizati on of Fiscal Revenue Increase Grassroot Government Expenditure on Public Services? Taking the Agricultural Tax Exemption in Henan as an Example. CEQ. [Chinese]. **07**(10):1350-70.

[13] Wang J and Xiao H F 2007 Analysis on effects of subsidies for superior crop varieties and agricultural machinery and policies on agricultural tax reduction and exemption in China. Issues in Agricultural Economy. [Chinese]. **02**:24-2.

IOP Conf. Series: Earth and Environmental Science **346** (2019) 012087 doi:10.1088/1755-1315/346/1/012087

Research on dynamic monitoring of cultivated land quality in mountainous regions around the Sichuan Basin

J H He[1,2]

[1]Sichuan University of Arts and Science, Dazhou, China

E-mail: drifter626@163.com

Abstract. The mountainous regions around the Sichuan Basin are characterized by complicated topography, large population and little land, and the per capita cultivated land resources are seriously insufficient. In order to realize the goal of protecting cultivated land from quantity to quality, dynamic monitoring of cultivated land quality was carried out to accurately understand the cultivated land quality. By establishing the cultivated land quality monitoring system and forming the regional monitoring network in the form of coordinate grid, the monitoring indexes are measured regularly. It is found that the change of cultivated land quality in the region can be divided into five types and their causes are analyzed thoroughly. The dynamic monitoring of cultivated land quality is conducive to the timely improvement and correction of the control measures, so as to improve the quality of cultivated land at the source and achieve the purpose of protecting cultivated land.

1. Introduction

To implement the requirements of the CPC central committee and the state council on the control of cultivated land quantity, quality management and ecological management, further enriching the connotation of the cultivated land and basic farmland protection, enhance the level of cultivated land and basic farmland protection, strengthening the construction of cultivated land quality in an all-round way and the management work. It is necessary to monitor the cultivated land quality and improve the management level of cultivated land quality. Do a good job in the annual basic farmland daily management and protection work.

At home and abroad, a lot of research has been done on the quality of cultivated land and its monitoring methods. With the help of its national resource inventory program, the United States completed national resource monitoring sites covering all 50 states from a statistical perspective [1]. The monitoring sample points of soil and land use types covering the whole eu member states were constructed by using grid method [2]. In recent years, on the basis of the achievements of agricultural land classification, domestic scholars have conducted in-depth exploration on the monitoring of cultivated land quality. Xiangbin Kong, Kening Wu, Jian Peng *et al* introduced the application principle and method of standard sample land in the monitoring of cultivated land quality grade in detail [3-5]. Yi Liu, Lina Guo, He *et al* discussed the selection method of farmland quality monitoring counties at the national scale and the layout method of monitoring sample belts at the provincial scale [6-8]. The existing researches on the monitoring of cultivated land quality mainly focus on the distribution and selection of monitoring points, and lack of the system and indicators for further dynamic monitoring, as well as the monitoring of dynamic changes of cultivated land quality from the indicators.

Content from this work may be used under the terms of the Creative Commons Attribution 3.0 licence. Any further distribution of this work must maintain attribution to the author(s) and the title of the work, journal citation and DOI.
Published under licence by IOP Publishing Ltd

With the ecological "trinity" protection of cultivated land quantity and quality, the importance of the investigation, evaluation and monitoring of cultivated land quality has become increasingly apparent. Because of its unique landform and complex and varied land use patterns, the mountainous areas in the basin have an all-round influence on the quantity, quality and even ecological environment of cultivated land. Although the mountainous areas in the basin have achieved national cultivated land quality and other achievements based on the first detailed land survey and change survey, After a comprehensive investigation of the quality of cultivated land within the region, the basic map of the land change survey results was used to improve the supplementary work on the quality of cultivated land, and the latest results on the quality of cultivated land were formed based on the secondary survey and the change survey [9]. However, the quantity, quality and ownership of cultivated land are changing at any time due to construction occupation, disaster damage, agricultural structural adjustment, ecological conversion and land improvement. Only through the annual land change survey and remote sensing monitoring, master only cultivated land quantity, the change of the ownership situation, can more accurately reflect the mountainous regions around the characteristics and change trend of cultivated land use, such as a comprehensive grasp of cultivated land quality don't change the situation, don't establish cultivated land quality monitoring and evaluation of long-term mechanism, to improve the quality of cultivated land, the implementation of divination balance of arable land and realize the target of cultivated land protection is of great significance.

2. Methodology- study area

2.1. Confirm the gradient type of farmland

To organize technical experts in agriculture, meteorology, hydrology and other fields to put forward and establish different gradual types of regional cultivated land and so on, on the basis of fully considering factors such as climate, topography, soil type, spatial distribution of water resources and so on, and combining with the actual situation of studying regional land use change and food production. According to the investigation of the research area, the gradual change types of cultivated land can be preliminarily determined as: gradually drought type, gradually waterlogging type, soil erosion type, fertility enhancement type, fertility decline type, desertification type and soil erosion type.

2.2. Confirm the dominant factors of gradient farmland

According to periods don't data such as cultivated land, the comprehensive analysis of regional distribution of cultivated land resource background, revealing the regional cultivated land quality distribution and variation characteristics, combined with agricultural meteorological data, land use planning, soil census data and agricultural research, combined with the county don't gradient type, such as initial set up arable land in preliminary sure don't gradient dominant factors such as cultivated land in the area. Detailed data are shown in table 1.

Table 1. Comparison table of cultivated land quality gradient type and dominant factors.

Gradient type	Gradient type code	Dominant factor	Dominant factor code
progressive drought	ZBGHX	probability of irrigation irrigation	GGBZL
progressive waterlogging waterlogging	ZBZLX	drainage condition	PSTJ
soil erosion	STLSX	effective soil thickness	YXTCHD
fertility enhancement	FLTSX	soil organic matter content	TRYJZHL
declining fertility	FLSTX		
acidification of type	SHX	surface soil texture	BCTRZD

2.3. Confirm the distribution range of arable land and other gradient types

According to the classification and renewal of regional agricultural land and the leading factors of gradual change of cultivated land, the distribution range of gradual change of cultivated land is determined. The gradation type distribution range of cultivated land and so on can include the dominant factor, uniform farmland and so on different agricultural land gradation units. Gradually changing type should reflect the main trend of regional cultivated land quality change, not too fine. Each gradient type can be centrally distributed in several areas that are not adjacent; The monitoring of this study is mainly to evaluate the changing rules of gradual change range, which should be considered on the basis of watershed, geomorphic unit, soil type and cultivated land use type, and not restricted by administrative boundaries. When a gradient type contains more than one grade, the grade that is greatly affected by the dominant factor is selected for monitoring, and the grade that has no influence or has a small area may not be monitored.

2.4. Confirm different gradation monitoring units such as farmland

The spatial location of the monitoring unit within the scope of gradual change type is determined, and field sampling of the monitoring unit is conducted through field investigation to form the monitoring sample point information within the monitoring range. Fixed monitoring unit and random monitoring unit were selected for each grade of cultivated land within the grade gradient type area of cultivated land quality. There should be at least one fixed monitoring unit in each grade within the distribution range of cultivated land quality grade gradation [8]. Fixed monitoring units should be distributed in the distribution range of the type of other typical parts, away from the city, road. Based on the change of the dominant factor attribute of the fixed monitoring unit, the change trend of the quality and productivity of the cultivated land in the county was evaluated. Based on the random monitoring unit, the distribution range of cultivated land quality and other gradation type areas was adjusted year by year. The driving factor, leading factor and monitoring index of farmland quality gradual change of fixed monitoring unit were monitored for a long time, and the tendency evaluation of farmland quality gradual change caused by fixed monitoring unit was made.

3. Results of dynamic monitoring of cultivated land quality

According to the special topography and geomorphology of the research area combined with the coordinate grid, it is divided into 6 regions for the identification of cultivated land, distribution and determination of gradual change types, establishment of monitoring units and evaluation of monitoring data. The total monitoring area is 446848.3 hectares, with 158 monitoring units and 6 categories of gradual change types. The total gradual change area is 89456.6 hectares, accounting for 64.67% at most. Detailed data are shown in table 2.

Table 2. Arable land quality and other gradient types and monitoring data table.

	cultivated land	paddy field	dry land	Irrigable land	monitoring units	gradient types	gradient type	gradient area	variable proportion
A	31473.50	18069.26	13347.37	56.87	3	1	Soil erosion	5146.24	35.90%
B	103618.66	50506.32	53003.73	108.61	10	2	progressive drought	3815.7	3.40%
							Soil erosion	5737.85	4.60%
C	117208.13	58042.07	59077.3	88.77	11	6	Acidification	8927.65	7.62%
							Soil erosion	4594.62	3.92%
D	40994.72	21310.11	19675.76	8.85	6	2	Declining fertility	14484.03	35.33%
							Acidification	26510.69	64.67%
E	96128.47	41153.92	54974.55	0	120	2	Soil erosion	1798.48	1.87%
							Acidification	4152.02	4.32%
F	57424.77	22535.43	34889.34	0	8	2	Fertility enhancement	7851.951	13.67%
							Soil erosion	6437.37	11.21%

4. Discussion

4.1. Analysis of the gradient type

Through comprehensive analysis of basic data such as farmland quality update and agricultural land productivity calculation results by experts from land, agriculture, meteorology, water conservancy and other departments, and collection of effective soil layer thickness changes by monitoring unit sample points, the reasons for the gradual change of various cultivated land types are finally obtained.

4.1.1. Causes of soil erosion. The fragile mountain natural environment is the potential factor of soil erosion and water loss. Along with the population growth, a large number of slope farmland reclamation, forest vegetation deforestation, various degrees of development and construction, and the concentration of heavy rainfall runoff caused serious erosion [10].

4.1.2. Causes of soil acidification. The natural acidification rate of soil is relatively slow, while under the large-scale influence of modern human activities, atmospheric acid deposition brings more H^+ into the soil. Excessive use of chemical fertilizer and unreasonable fertilization structure in the study area are the main reasons for soil acidification. Taking field fertilization as an example, after long-term application of nitrogen fertilizer, NH^{4+} in the soil produces acid through nitrification (1 mol of NH^{4+}produces 2 mol of H^+), thus reducing the pH of the soil, increase the activation of soil exchangeable and soluble aluminum [11].

4.1.3. Causes of declining soil fertility. Heavy application of chemical fertilizer results in lower soil organic matter content, no significant improvement of soil structure, poor soil water and fertilizer retention ability, and low improvement of cultivated land quality. Due to the decrease of rural labor force and land wastage, it is difficult for farmers to change their traditional planting habits, so they only apply chemical fertilizer, and less farm manure with high organic matter content. Due to the influence of climate and environment, soil and water loss, loose soil quality, sparse vegetation, destruction of vegetation and excessive reclamation also cause the loss of soil organic matter and decline of soil fertility.

4.1.4. Causes of soil fertility enhancement. Along the river basin and the punching groove, the surface runoff from organic matter such as precipitation, high degree of cultivation, long-term application of organic manure, straw returned, can improve the physical properties of the soil, promote the formation of soil aggregate structure, etc., as well as reasonable rotation farming methods of utilizing land while maintaining it, increase the leguminous crops planting area, leguminous crops can increase the nitrogen content in the soil, improving soil physical and chemical properties, improve the soil fertility.

4.1.5. Causes of gradual soil drought. Cause of gradual drought in the study area is mainly multi-cropping index increased, soil erosion, vegetation destruction, engineering lack or water infrastructure is fragile, no water conservation, irrigation, low reliability, not adapt to floods and drought flood rule, do when flood water conservation, the law of the drought water diversion to promote the dynamic balance of water resources.

4.2. The effects of gradient type

4.2.1. The effects of soil fertility enhancement. In the region of increasing fertility, the soil organic matter content of cultivated land in the region has been gradually increased due to the high degree of cultivation, good hydrological and climatic conditions in the small region, convenient irrigation and the application of organic fertilizer. The content of soil organic matter is closely related to the level of soil fertility. Organic matter is the main source of plant nutrients. Organic matter contains various nutrient elements needed by plant growth and development, especially nitrogen in soil, where more than 95% nitrogen exists in an organic state. Therefore, soil with more organic matter has more nutrient content, and fertilizer can be applied with less. Organic matter can promote the growth and development of crops, among which humic acid can enhance plant respiration, improve the permeability of cell membrane and enhance the absorption of nutrients. Humus in organic matter is the main cementing agent of soil aggregates, which can promote the formation of good structure. Soil organic matter is an organic colloid with a strong adsorption capacity, which can absorb a large amount of nutrients and water and improve the soil's ability to retain water and fertilizer. Soil organic matter provides energy and nutrients needed by soil microorganisms, which is beneficial to their activities.

4.2.2. The effects of soil erosion. Topsoil plays a vital role in crop growth. Rain washes away topsoil and thins the available soil layer. Will bring many serious harm: destroy land and farmland， soil erosion and fertility loss, ecological imbalance, frequent droughts and floods, the deposition in the reservoir, blocking river, river bed sedimentation, aggravating floods cut, sediment deposition in reservoirs, lakes, reduce its comprehensive utilization function, reduce soil development, intensify drought, soil and water loss and poverty vicious circle, synchronous development.

4.2.3. The effects of soil acidification. Soil acidification will affect the growth of crops. When the acidity of the soil is reduced below the threshold of the ability and tolerance of crops, crops will grow badly or even fail to grow. Soil acidification will lead to the activation of heavy metal ions and the increase of acid-root ions in the soil. Acid-root ions can react with most of the heavy metals and change the precipitated heavy metals into dissolved ones, which will be absorbed by plant roots and enter into the plant body for enrichment. Soil acidification will lead to the decline of soil ion exchange ability, and the substitution of saline-base ions adsorbed on acidic soil colloids into soil solution will be leached, resulting in the reduction of soil exchangeable base content, exchangeable cation content and saline-base saturation, and soil exchange performance will be affected.

5. Conclusion

The dynamic monitoring and evaluation of cultivated land quality reflects the quality change trend of various cultivated land types with gradual change in a wide range through monitoring the leading factors of gradual change in the monitoring unit. In combination with the potential status of resources, the key areas of land improvement can be reasonably determined and the direction of land improvement in each region can be clearly defined. It not only provides accurate technical support for the reconstruction of low and medium yield fields in the region and the improvement of other restrictions on cultivated land, improves the quality of cultivated land from the source, but also realizes the feedback of implementation effect on an annual basis, which is conducive to the timely improvement and correction of the control measures. The results of dynamic monitoring of cultivated land quality can also directly serve the comprehensive updating of agricultural land classification results [12]. Based on the monitoring results of the change of cultivated land use mode, cultivated land soil condition, land use level and benefit level, the correlation values of land use coefficient, land economic coefficient, standard farming system, light and temperature production potential and climate production potential in agricultural land classification are reasonably adjusted.

According to the influence of the indexes that monitor the change of cultivated land quality, the weights of the indexes in the regulations are adjusted reasonably, so as to enhance the scientific rationality of agricultural land grading results. In addition, the dynamic monitoring work makes up for the deficiency that the annual update evaluation of cultivated land quality and other categories only evaluates the cultivated land with sudden changes in quality. The evaluation results of the two forms a complete local cultivated land quality system, which has reference and argumentation significance for the comprehensive update results of agricultural land classification results.

References

[1] Nusser S M, Breidt F J and Fuller W A 1998 Designandestimation for investigating the dynamics of natural resources *J Eco. Appl.* **8** 234-45

[2] Jones R J A, Spoor G and Thomasson A J 2003 Assessing the vulnerability of subsoils in Europe to compaction: A preliminary analysis *Soil Till. Res.* **73** 131-43
[3] Kong X B 2014 Distribution characteristics and representativeness of national standard sample plots in western *China J. Agric. Eng.* **30** 227-35
[4] Wu K N 2008 Research on the framework of dynamic monitoring points of cultivated land quality based on national summary of standard sample plots *J. Agric. Eng.* **24** 74-9
[5] Peng J 2004 The application of standard sample plots in the summary of agricultural land classification results *Land Sci. China* **18** 126-7
[6] Liu Y 2013 Research on selection method of cultivated land quality monitoring counties in western *Resources Sci. China* **35** 2248-54
[7] Guo L N 2008 Discussion on the layout method of agricultural land grade and quality monitoring belt *Resources Sci.***30** 1199-205
[8] He J H 2010 Research on farmland quality monitoring based on standard sample plots *Hebei Agric. Sci.* **14** 125-7
[9] Mi C L and Po C Y. 2017 Research on quality monitoring and evaluation of cultivated land in Gansu province based on 3S technology *Glac. Per.* **39** 1374-80
[10] Liang P W 2018 Annual monitoring and evaluation of cultivated land quality in yuan district *J. Agric. Sci. Anhui* **24** 87-90
[11] Du M 2017 Causes analysis and countermeasures of farmland acidification *Agric. Dev. Equip.* **3** 32-4
[12] Chen Z and Tan Y S 2017 Research on dynamic monitoring of cultivated land quality *J. Agric. Sci. Anhui* **45** 193-5

IOP Conf. Series: Earth and Environmental Science **346** (2019) 012088 doi:10.1088/1755-1315/346/1/012088

Correlations between circulating ghrelin concentrations and growth performance, carcass traits, meat quality indices in finishing bulls fed high-concentrate diets

C Y Geng[1,3], Q X Meng[2] and M Zhang[1]

[1]Agricultural University College, Yanbian University, Yanji 133000, China
[2]Beef Cattle Research Center, China Agricultural University, Beijing 100094, China

E-mail: cygeng1011@163.com

Abstract. Circulating ghrelin concentrations could potentially be used as a predictor of production traits in cattle. This study aimed to clarify the correlations between circulating total ghrelin concentrations and growth performance, carcass traits, meat quality indices in finishing bulls fed high-concentrate diets. Fifty-seven Simmental × Luxi F1 crossbred bulls (approximately 24 month of age, mean body weight ± SD: 514 ± 27 kg) were used for 112-days finishing experiment. Circulating total ghrelin concentrations and relevant indices of growth performance, carcass traits and meat quality were measured during or after finishing trial. The results indicated that the total ghrelin concentrations were positively correlated with dry matter intake (DMI) ($r = 0.278$, $P = 0.046$) and were not correlated to other growth indices including initial body weight (BW), final body weight, average daily gain (ADG) and feed efficiency (feed : gain) ($P > 0.05$). The total ghrelin concentrations were positively correlated with meat-bone ratio ($r = 0.269$, $P = 0.045$) and tended to be positively correlated to highrib weight ($r = 0.234$, $P = 0.083$). Moreover, total ghrelin concentrations were not correlated to other carcass traits indices such as hot carcass weight, dressing percentage, Longissimus dorsi muscle (LM) area and 12th rib fat thickness ($P > 0.05$). Total ghrelin concentrations tended to be negatively correlated to the shear force of beef ($r = -0.299$, $P = 0.064$), and were not correlated to other meat quality indices such as moisture, pH, drip loss, cooking loss, meat color conventional chemical composition and fatty acid composition ($P > 0.05$). In conclusion, circulating total ghrelin concentrations were associated with DMI, meat-bone ratio and beef tenderness in finishing bulls fed high concentrated diet, and it could be used as a potential biomarker for evaluation of animal production traits such as DMI and beef tenderness in finishing bulls fed high concentrate diets.

1. Introduction

Ghrelin is peptide composed of 27 amino acid peptide in ruminants [1], and it is mainly synthesized by the abomasal and ruminal tissues of cattle [2]. Ghrelin is thought to be able to stimulate growth hormone release and appetite by binding to the growth hormone secretagogue receptor (GHSR1a) [3,4]. In addition, it's reported that ghrelin plays a prominent role in the regulation of fat metabolism in ruminants [5]. Given the potential functions of ghrelin on growth, appetite regulation and body composition, it is possible that circulating ghrelin concentrations could be used as a biomarker for evaluation of animal production traits [6].

Up till now, although the correlations analysis of circulating ghrelin concentrations with dry matter intake, growth performance, and carcass characteristics of finishing beef cattle have been carried out

Content from this work may be used under the terms of the Creative Commons Attribution 3.0 licence. Any further distribution of this work must maintain attribution to the author(s) and the title of the work, journal citation and DOI.
Published under licence by IOP Publishing Ltd

[6,7], relationship of circulating ghrelin concentrations with meat quality traits is still unclear in finishing cattle. Therefore, The objective of this study was to clarify the associations of circulating total ghrelin concentrations with animal production traits including growth performance, carcass traits and meat quality in finishing bulls.

2. Materials and methods

All procedures involving animal care were under the approval of the Yanbian University Institutional Animal Care and Use Committee.

2.1. Animals feeding and management

Fifty-seven crossbred bulls (Simmental × Luxi, approximately 24 month of age, mean body weight ± SD: 514 ± 27 kg) were used in this study. All bulls were ear tagged and dewormed before finishing trial. All bulls were individually fed the same high-concentrate basal diets (table 1). The bulls were fed twice a day. The finishing trial lasted 112 days, and dietary dry matter intake (DMI), average daily gain (ADG) and feed efficiency (feed: gain, F: G) were recorded and calculated during the trial.

Table 1. Ingredient and nutritional composition of basal diets.

Ingredient composition	% of DM	Nutritional composition [b]	% of DM
Corn meal	51.14	Dry matter, DM	71.45
Corn silage	30.61	Crude protein, CP	12.23
Cottonseed meal	7.79	Ether extract, EE	2.98
Soybean meal	5.79	Calcium, Ca	0.63
Salt	0.40	Phosphorus, P	0.37
Sodium bicarbonate	1.04	Sulphur, S	0.16
Compound premix [a]	3.45	Net energy for growth, NEg (Mcal/kg DM)	1.29
Total （%）	100.00	Net energy for maintenance, NEm (Mcal/kg DM)	1.94

Abbreviation: DM, dry matter.
[a]Compound premix, supplied per kilogram of product. Ca: 160 g; P: 30 g; Cu: 450 mg; Zn: 1600 mg; Mn: 800 mg; I: 10 mg; Co: 10 mg; Se: 5 mg; vitamin A : 120 000 IU; vitamin D: 55000 IU; vitamin E: 400mg; vitamin B3: 600 mg; vitamin B5: 200 mg; Monesin: 1000 mg.
[b]The value for nutritional composition of diets including DM, CP, EE, Ca and P was calculated based on the nutrient analysis from ingredient samples. S, NEg and NEm was estimated from NRC (2000) values.

2.2. Sample collection, processing, and analyses

Blood samples were collected on the last day of the study prior to feeding in the early morning by 10 Ml vacuum blood collection tubes. Tubes were centrifuged for 15 min (3200 rpm, 4°C) and the separated plasma was used for measurement of concentration of total ghrelin concentration. The plasma total ghrelin concentration was analyzed by a r-911 full automatic radio-immune counter (University of Science and Technology of China, Co. Ltd, China) with radioimmunoassay methods in a service corporation (Beijing Sino-Uk of Biological technology, Beijing, China).

The dressing percentage of each bull was calculated according to Live weight and hot carcass weight (HCW). Other carcass characteristics were measured during Carcass segmentation (24-h chill), including 1) subcutaneous fat thickness between the 12th and 13th ribs (12th-rib fat thickness), 2) LM area between the 12th and 13th ribs (LM area), meat weight in carcass (net meat weight), meat percentage in carcass (net meat percentage), bone weight in carcass (bone weight), meat-bone ratio in carcass, weight and proportion of high quality beef (high rib, ribeye, sirloin and tenderloin) in carcass. Briefly, 12th-rib fat thickness was directly measured with vernier calipers, and LM area was measured by a planimeter (QCJ-2000, Harbin optical instrument factory, China) after drawing the cross-sectional figure of LM with transparent sulfuric acid paper. Other indices including live weight, hot carcass weight, weight of bone and high quality beef were all from plant personnel.

Thirty nine samples of LM (sirloin) were randomly selected from fifty-seven samples for meat quality evaluation, including water content, pH, cooking loss, drip loss, shear force, color indices, crude ash, intramuscular fat, protein and cholesterol content. The water content, protein and ash contents of beef were determined according to standard procedures [8], and the details of determination of drip loss, cooking loss percentage, shear force and cholesterol content of beef refer to method of Geng *et al* [9].

Fatty acid methyl ester (FAME) synthesis was conducted according to method of O'Fallon [10]. Saturated fatty acids (SFA) were the sum of C10:0, C12:0, C14:0, C15:0, C16:0, C17:0, C18:0 , C20:0, C21:0 and C22:0. Monounsaturated fatty acids (MUFA) were the sum of C14:1n5, C16:1n7, C17:1n7, C18:1n9c and C:20:1n9. Polyunsaturated fatty acids (PUFA) were the sum of C18:2n6, C:18:2n6c, C:18:3n6c, C:18:3n3, C20:3n6, C20:3n3, C20:2, C20:4n6, C22:2 and C20:5n3. N-6 fatty acids were the sum of C18:2n6t, C:18:2n6c, C:18:3n6c, C20:3n6 and C20:4n6. N-3 fatty acids were the sum of C:18:3n3, C20:3n3 and C20:5n3.

2.3. Statistical analysis

All production traits indices were sorted into three equal groups according to descending order of total ghrelin concentrations, and data were statistically described by GLM model of SPSS 18.0 (SPSS Inc., Chicago, IL, USA), and multiple comparisons were conducted by one-way with method of Duncan. The correlation analysis were performed by the Pearson option (2-tailed). Significance for correlation was declared at $P < 0.05$, and tendencies were declared at $P < 0.10$.

3. Results

3.1. Association of circulating total ghrelin concentrations with growth performance traits

The statistical description and correlation analysis between circulating total ghrelin concentrations and growth performance indices of cattle were showed in table 2. There was no significant difference in growth traits indices among three groups ($P > 0.05$), and the final body weight (BW), average daily gain (ADG) and feed conversion (FC) improved with the increase of total ghrelin concentrations ($P > 0.05$; table 2). The total ghrelin concentrations showed a positive correlation with dry matter intake (DMI) (r = 0.278, $P = 0.046$), and were not correlated to other indices including initial BW, final BW, ADG and FC ($P > 0.05$; table 2).

Table 2. The description of the growth performance and correlations analysis with circulating total ghrelin concentrations in finishing bulls fed high-concentrate diets.

Items	Data description [d]					Correlation analysis	
	LC ghrelin (n=19)	MC ghrelin (n=19)	HC ghrelin (n=19)	SEM [e]	*P* value	Correlation coefficient (r)	*P* value
Ghrelin, ng/mL	73.18[c]	94.50[b]	121.40[a]	1.981	<0.001	/	/
Initial BW, kg	513.17	517.20	513.00	6.757	0.892	-0.010	0.945
Final BW, kg	584.88	592.48	594.53	11.07	0.811	0.131	0.338
ADG, kg/day	0.73	0.84	0.88	0.078	0.392	0.206	0.139
Average DMI, kg/day	9.35	9.70	9.98	0.353	0.454	0.278**	0.046
Feed conversion (feed: gain)	14.07	13.29	13.57	1.170	0.897	-0.059	0.675

Abbreviation: BW, body weight; ADG, average daily gain; DMI, dry matter intake; LC ghrelin, low concentration of ghrelin; MC ghrelin, medium concentration of ghrelin; HC ghrelin, high concentration of ghrelin.
[a, b, c] Values within row with differing superscripts differ $P < 0.05$.
[d] Data were blocked into three equal groups according to descending order of measured circulating total ghrelin concentrations.
[e] Standard error of the mean (n = 57).
**, Significant correlation ($P < 0.05$); *, Correlation tendency ($0.05 < P < 0.1$).

3.2. Association of circulating total ghrelin concentrations with carcass traits

There was no significant difference in carcass traits indices but proportion of tenderloin in carcass among three groups ($P > 0.05$; table 3). The proportion of tenderloin in carcass was significantly decreased with the increase of total ghrelin concentrations ($P = 0.038$; table 3). The total ghrelin concentrations were positively correlated with meat-bone ratio (r = 0.269, $P = 0.045$), and tended to be positively correlated to high rib weight (r = 0.234, $P = 0.083$; table 3). In addition, total ghrelin concentrations tended to be negatively correlated to proportion of tenderloin in carcass (r = - 0.259, $P = 0.057$; table 3). Moreover, total ghrelin concentrations did not show association with other carcass traits indices including hot carcass weight, dressing percentage, LM area, 12th rib fat thickness, bone weight, net meat percentage, ribeye weight, sirloin weight, tenderloin weight, proportion of high rib in carcass, proportion of ribeye in carcass and proportion of sirloin in carcass ($P > 0.05$; table 3).

Table 3. The description of the carcass traits and correlations analysis with circulating total ghrelin concentrations in finishing bulls fed high-concentrate diets.

Items	Data description[d]					Correlation analysis	
	LC ghrelin (n=19)	MC ghrelin (n=19)	HC ghrelin (n=19)	SEM[e]	*P* value	Correlation coefficient (r)	*P* value
Ghrelin, ng/mL	73.18[c]	94.50[b]	121.40[a]	1.981	<0.001	/	/
Hot carcass weight, kg	319.16	331.11	328.11	6.346	0.396	0.126	0.355
Dressing percentage, %	55.15	56.47	55.87	0.674	0.399	0.008	0.954
LM area, cm^2	69.11	72.36	70.33	2.178	0.579	0.113	0.408
12th rib fat thickness, cm	0.77	0.80	0.84	0.080	0.835	0.053	0.699
Net meat weight, kg	271.06	281.70	281.29	5.738	0.343	0.167	0.218
Bone weight, kg	48.10	49.41	46.82	1.190	0.323	-0.138	0.311
Net meat percentage, %	46.83	48.05	47.88	0.655	0.373	0.080	0.559
Meat-bone ratio	5.69	5.75	6.03	0.142	0.201	0.269**	0.045
High quality beef[f], kg	53.82	55.63	55.67	1.157	0.440	0.182	0.180
Highrib, kg	22.25	23.19	23.57	0.552	0.232	0.234*	0.083
Ribeye, kg	11.97	11.98	12.17	0.356	0.904	0.081	0.551
Sirloin, kg	12.79	13.52	13.25	0.342	0.323	0.168	0.217
Tenderloin, kg	6.81	6.95	6.68	0.159	0.515	-0.042	0.758
Proportion of high quality beef in carcass, %	16.86	16.83	16.99	0.214	0.863	0.114	0.404
Highrib, %	6.97	7.02	7.20	0.143	0.497	0.114	0.404
Ribeye, %	3.75	3.63	3.73	0.091	0.591	0.009	0.949
Sirloin, %	4.00	4.08	4.02	0.072	0.705	0.096	0.486
Tenderloin, %	2.14[a]	2.10[ab]	2.03[b]	0.028	0.038	-0.259*	0.057

Abbreviation: LM, Longissimus dorsi muscle; LC ghrelin, low concentration of ghrelin; MC ghrelin, medium concentration of ghrelin; HC ghrelin, high concentration of ghrelin.
[a, b, c] Values within row with differing superscripts differ $P < 0.05$.
[d] Data were blocked into three equal groups according to descending order of measured circulating total ghrelin concentrations.
[e] Standard error of the mean (n = 57).
[f] High quality beef included highrib, ribeye, sirloin and tenderloin.
**, Significant correlation ($P < 0.05$); *, Correlation tendency ($0.05 < P < 0.1$).

3.3. *Association of circulating total ghrelin concentrations with conventional indices of meat quality and chemical composition*

There was no significant difference in *conventional indices of meat quality and chemical composition* among three groups ($P > 0.05$; table 4). The total ghrelin concentrations tended to be negatively correlated to shear force of beef (r = - 0.299, $P = 0.064$; table 4). Moreover, total ghrelin concentrations were not correlated to other meat quality indices including moisture, pH, drip loss, cooking loss, meat color and chemical composition including content of ash, protein, intramuscular fat and cholesterol (*P* > 0.05; table 4).

Table 4. The description of the conventional indices of meat quality and chemical composition and correlations analysis with circulating total ghrelin concentrations in finishing bulls fed high-concentrate diets.

Items	Data description[d]					Correlation analysis	
	LC ghrelin (n=13)	MC ghrelin (n=13)	HC ghrelin (n=13)	SEM[e]	*P* value	Correlation coefficient (r)	*P value*
Ghrelin, ng/mL	74.82[c]	99.47[b]	125.37[a]	3.614	<0.001	/	/
Conventional indices							
Moisture, %	72.40	72.24	73.27	0.280	0.273	0.134	0.415
pH	5.62	5.54	5.70	0.057	0.534	0.094	0.570
Drip loss, %	35.79	33.58	29.04	1.629	0.231	-0.185	0.260
Cooking loss, %	43.74	42.78	41.14	1.30	0.724	-0.079	0.631
Shear force, kg	7.35	6.82	6.36	0.268	0.327	-0.299*	0.064
Meat color							
L	35.59	35.53	35.36	0.661	0.990	-0.042	0.800
a	21.82	20.51	20.60	0.596	0.619	-0.103	0.534
b	10.61	10.07	9.48	0.451	0.603	-0.135	0.414
Chemical composition							
Ash, % of DM	3.97	3.97	4.02	0.051	0.904	0.022	0.896
Protein, % of DM	84.42	85.75	86.45	0.650	0.435	0.123	0.460
Intramuscular fat, % of DM	10.75	10.43	8.09	0.805	0.348	-0.143	0.386
Cholesterol, mg/100g DM	118.98	115.04	118.14	1.606	0.586	0.031	0.854

Abbreviation: DM, dry matter of beef; LC ghrelin, low concentration of ghrelin; MC ghrelin, medium concentration of ghrelin; HC ghrelin, high concentration of ghrelin.
[a, b, c] Values within row with differing superscripts differ $P < 0.05$.
[d] Data were blocked into three equal groups according to descending order of measured circulating total ghrelin concentrations.
[e] Standard error of the mean (n = 39).
**, Significant correlation ($P < 0.05$); *, Correlation tendency ($0.05 < P < 0.1$).

3.4. Association of circulating total ghrelin concentrations with fatty acid composition

There was no significant difference in proportions and concentrations of SFA, MUFA and PUFA fatty acids and some nutritional values including n-6/n-3, PUFA/SFA and C18:2n6/C:18:3n3 among three groups (*P* > 0.05; table 5). Moreover, total ghrelin concentrations did not show significant correlation with these indices above (*P* > 0.05; table 5).

Table 5. The description of the proportions (% of total fatty acids) and concentrations (mg/100g fresh muscle tissue) of saturated (SFA), monounsaturated (MUFA) and polyunsaturated (PUFA) fatty acids and some nutritional values and correlations analysis with circulating total ghrelin concentrations in finishing bulls fed high-concentrate diets.

Items	Data description[d]					Correlation analysis	
	LC ghrelin (n=13)	MC ghrelin (n=13)	HC ghrelin (n=13)	SEM[e]	*P* value	Correlation coefficient (r)	*P* value
Ghrelin, ng/mL	74.82[c]	99.47[b]	125.37[a]	3.614	<0.001	/	/
% of total fatty acids							
SFA	46.78	47.46	45.75	0.425	0.261	-0.235	0.151

MUFA	43.70	42.74	45.44	0.614	0.193	0.145	0.379
PUFA	9.52	9.80	8.80	0.497	0.713	0.013	0.939
n-6	9.26	9.53	8.54	0.492	0.708	0.013	0.940
n-3	0.189	0.20	0.20	0.007	0.825	-0.005	0.975
mg/100 g fresh muscle tissue							
SFA	1333.92	1337.83	1320.12	75.770	0.995	0.126	0.444
MUFA	1288.96	1239.91	1322.26	78.302	0.915	0.180	0.273
PUFA	235.77	286.71	230.96	18.944	0.423	0.120	0.465
n-6	228.75	279.22	223.71	18.787	0.422	0.118	0.474
n-3	5.12	5.40	5.20	0.179	0.817	0.203	0.216
Nutritionally important values							
n-6/n-3	44.72	50.03	43.46	1.953	0.354	0.031	0.850
PUFA/ SFA	0.19	0.22	0.19	0.009	0.364	0.038	0.820
C18:2n6/C:18:3n3	46.73	46.17	46.09	1.204	0.974	-0.103	0.534

Abbreviation: SFA = C10:0 + C12:0 + C14:0 + C15:0 + C16:0 + C17:0 + C18:0 + C20:0 + C21:0 + C22:0; MUFA = C14:1n5 + C16:1n7 + C17:1n7 + C18:1n9c + C:20:1n9; PUFA = C18:2n6 + C:18:2n6c + C:18:3n6c + C:18:3n3 + C20:3n6 + C20:3n3 + C20:2 + C20:4n6 + C22:2 + C20:5n3; n-6 = C18:2n6t + C:18:2n6c + C:18:3n6c + C20:3n6 + C20:4n6; n-3 = C:18:3n3 + C20:3n3 + C20:5n3. LC ghrelin, low concentration of ghrelin; MC ghrelin, medium concentration of ghrelin; HC ghrelin, high concentration of ghrelin.

[a, b, c] Values within row with differing superscripts differ $P < 0.05$.

[d] Data were blocked into three equal groups according to descending order of measured circulating total ghrelin concentrations.

[e] Standard error of the mean (n = 39).

**, Significant correlation ($P < 0.05$); *, Correlation tendency ($0.05 < P < 0.1$).

4. Discussion

Ghrelin is a 28 amino acid peptide [11]. In the circulation, ghrelin is presented as acylated (AG) and un-acylated (UAG) forms which play opposing action on appetite [12]. AG was considered as the bioactive form of ghrelin on appetite, and UAG can be transformed into AG by the enzyme ghrelin o-acyltransferase (GOAT) [13]. It was reported that exogenous AG can increase time spent eating and DMI in beef cattle [4]. Presently, for finishing bulls fed high-concentrated diets, the report of relationship of circulating total ghrelin concentration with production traits is limited. Our study indicated that circulating total ghrelin concentrations showed a positive correlation with DMI in finishing bulls. However, in other studies, total ghrelin concentrations have no significant association with DMI in finishing cattle even tended to be negatively associated with DMI in finishing beef steers and heifers [6,7].

It was reported that the ratio of AG to total ghrelin or AG concentrations rather total ghrelin concentrations showed a positive correlation with dry matter intake (DMI) in finishing cattle [7]. It is likely to be more rational for AG or active: total ghrelin ratio to show a positive association with DMI in cattle because only AG can bind the GHSR for stimulating appetite. However, it is worth noting that circulating active and total ghrelin concentrations and their ratio in plasma may alter with different physiological stages, sex and diets composition in ruminants [5,14,15]. In the present study, the intact finishing bulls and high concentrate diets were used, which may cause a disparity for relationship of circulating total ghrelin concentration and feed intake in different studies. Given this, it is interesting and necessary to further research into the relationship of changes of AG, UAG and GOAT concentrations with DMI in finishing bulls fed high concentrate diets.

It was reported that administration of ghrelin can cause an increased BW gain and improved feed efficiency in rats and mice [16,17]. Nevertheless, in cattle, published literature regarding ghrelin did not show consistent benefit to ADG and feed efficiency. For ADG, a previous study reported that total ghrelin was negatively correlated to ADG [7], however, the data presented here or reported by another study indicated that total ghrelin did not show a significant association with ADG [6]. For feed efficiency, total circulating ghrelin concentrations showed a negative correlation to feed efficiency or tended to be positively associated with feed efficiency in cattle [6,7], and our data did not show any association of total ghrelin concentrations with feed efficiency ($P > 0.01$; table 2).

There are very limited reports for relationship between ghrelin and carcass characteristics. A previous study reported that total ghrelin concentrations did not show significant correlation with carcass characteristics including HCW, LM area, 12th rib fat, marbling score and yield grade ($P > 0.10$) [7]. In present study, we also found that circulating total ghrelin concentrations have no significant relationship with these carcass indices above. However, total ghrelin concentrations showed a positive correlation meat-bone ratio ($r = 0.269$, $P = 0.045$; table 3). In addition, total ghrelin concentrations tended to be positively correlated with high rib weight high-quality beef ($r = 0.234$, $P = 0.083$; table 3) and tended to be negatively correlated with tenderloin proportion of high-quality beef in carcass ($r = -0.259$, $P = 0.057$; table 3).

It was reported that sheep with high levels of adiposity have higher ghrelin concentrations compared with lean animals after overnight fasting [18], which may lead to increased meat-bone ratio with ghrelin concentration increased. Moreover, ghrelin can increase time spent eating and DMI of beef cattle [4], which may lead to difference in weight of related muscle tissue.

It was reported that ghrelin has a certain effect on cattle accreting fat [5], which hinted that circulating ghrelin concentrations may have a relationship with meat quality of cattle. For the first time, we revealed the relationship between circulating total ghrelin concentrations and meat quality traits in finishing bulls fed high-concentrate diets. The results indicated that total ghrelin concentrations tended to be negatively correlated to shear force of beef ($r = -0.299$, $P = 0.064$; table 4). Moreover, total ghrelin concentrations was not correlated to other meat quality indices such as moisture, pH, drip loss, cooking loss, meat color, chemical composition and fatty acid composition ($P > 0.05$; tables 4 and 5).

The muscle characteristics depend on many factors such as feeding management, age, sex, and breed, which is related to the variability of beef tenderness [19,20]. In this study, intact bulls were used and a negative correlation tendency between total ghrelin concentrations and shear force of beef was found which indicating that ghrelin may contribute to regulation of beef tenderness in finishing bulls fed concentrate diets. It was reported that the degradability of muscle connective tissue protein is a key factor for meat tenderness [19]. Available evidences have indicated that increased intracellular Ca^{2+} concentrations play an important role in improving the beef tenderness by regulating degradability of connective tissue protein [21-23], and increased endogenous ghrelin concentrations can increase intracellular Ca^{2+} concentration in muscle cells [24]. These results suggest that ghrelin may potentially contribute to regulation of beef tenderness by increasing Ca^{2+} concentrations in muscle, and further research is necessary for effect of ghrelin on beef quality.

5. Conclusion

In conclusion, circulating total ghrelin concentrations are associated with DMI, meat bone ratio and beef tenderness in finishing bulls fed high concentrate diet. It is probable that circulating total ghrelin concentrations could be used as a potential biomarker for evaluation of animal production traits such as DMI and beef tenderness in finishing bulls fed high concentrate diets.

Acknowledgments

This work was supported by the National Natural Science Foundation of China (grant no. 31660669, 2016), Jilin scientific research planning project of Jilin Province in 13th Five-Year (grant no. JJKH20180904KJ, 2018).

References

[1] Dickin J C, Thue T D and Buchanan F C 2004 An alternative splice site in ghrelin is missing in ruminants *Anim. Genet.* **35** 411-2

[2] Hayashida T *et al* 2001 Ghrelin in domestic animals: distribution in stomach and its possible role *Domest. Anim. Endocrinol.* **21** 17-24

[3] Kojima M *et al* 1999 Ghrelin is a growth-hormone-releasing acylated peptide from stomach. *Nature* **402** 656–60

[4] Wertz-Lutz A E *et al* 2006 Circulating ghrelin concentrations fluctuate relative to nutritional status and influence feeding behavior in cattle *J. Anim. Sci.* **84** 3285-300

[5] Jennings J S *et al* 2011 Circulating ghrelin and leptin concentrations and growth hormone secretagogue receptor abundance in liver, muscle, and adipose tissue of beef cattle exhibiting differences in composition of gain *J. Anim. Sci.* **89** 3954-72

[6] Foote A P, Hales K E and Freetly H C 2016 Changes in acyl and total ghrelin concentrations and their association with dry matter intake, average daily gain, and feed efficiency of finishing beef steers and heifers *Domest. Anim. Endocrinol.* **57** 100-7

[7] Foote A P *et al* 2014 Association of circulating active and total ghrelin concentrations with dry matter intake, growth, and carcass characteristics of finishing beef cattle *J. Anim. Sci.* **92** 5651-8

[8] AOAC 2000 *Official Methods of Analysis* 17th ed. (Gaithersburg, MD: Association of Official Analytical Chemists)

[9] Geng C Y *et al* 2016 Comparison of active dry yeast (Saccharomyces cerevisiae) and yeast culture for growth performance, carcass traits, meat quality and blood indexes in finishing bulls *Anim. Sci. J.* **87** 982-8

[10] O'Fallon J V *et al* 2007 A direct method for fatty acid methyl ester synthesis: Application to wet meat tissues, oils, and feedstuffs *J. Anim. Sci.* **85** 1511-21

[11] Kojima M and Kangawa K 2002 Ghrelin, an orexigenic signalling molecule from the gastrointestinal tract *Curr. Opin. Pharmacol.* **2** 665-8

[12] Asakawa A *et al* 2005 Stomach regulates energy balance via acylated ghrelin and desacyl ghrelin *Gut* **54** 18-24

[13] Yang J *et al* 2008 Identification of the acyltransferase that octanoylates ghrelin, an appetite stimulating peptide hormone *Cell* **132** 387-96

[14] ThidarMyint H *et al* 2006 Dose-dependent response of plasma ghrelin and growth hormone concentrations to bovine ghrelin in Holstein heifers *J. Endocrinol.* **189** 655-64

[15] Fukumori R *et al* 2013 Ingestion of medium chain fatty acids by lactating dairy cows increases concentrations of plasma ghrelin *Domest. Anim. Endocrinol.* **45** 216-23

[16] Tschop M, Smiley D L and Heiman M L 2000 Ghrelin induces adiposity in rodents *Nature* **407** 908-13

[17] Nakazato M *et al* 2001 A role for ghrelin in the central regulation of feeding *Nature* **409** 194-8

[18] Sugino T *et al* 2002 A transient surge of ghrelin secretion before feeding is modified by different feeding regimens in sheep. *Biochem. Biophys. Res. Commun.* **15** 785–8.

[19] Koohmaraie M and Geesink G H 2006 Contribution of postmortem muscle biochemistry to the delivery of consistent meat quality with particular focus on the calpain system *Meat. Sci.* **74** 34-43

[20] Chriki S *et al* 2013 Meta-analysis of the relationships between beef tenderness and muscle characteristics *Livest. Sci.* **155** 424-34

[21] Foote M R *et al* 2004 The use of vitamin D3 and its metabolites to improve beef tenderness *J. Anim. Sci.* **82** 242-9

[22] Rider Sell N, Mikel W B, Xiong Y L and Behrends J M 2004 Vitamin D3 supplementation of cull cows: effects on longissimus and semitendinosus muscle tenderness *J. Anim. Sci.* **282** 225-30

[23] Mohrhauser D A *et al* 2015 The influence of maternal energy status during mid-gestation on beef offspring tenderness, muscle characteristics, and gene expression *Meat. Sci.* **110** 201-11

[24] Fang H *et al* 2012 Effects of ghrelin on the intracellular calcium concentration in rat aorta vascular smooth muscle cells. *Cell Physiol. Biochem.* **30** 1299–309

The study on mechanical properties of Phytagel medium

Y Zhou[1], J Yan[1, 3], B Y Xu[1] and B C Wang[2]

[1]Chongqing Normal University, Chongqing 401331, China
[2]Chongqing University, Chongqing 400030, China

Email: yanjie_cqu@163.com

Abstract. Background: The mechanical properties of plant culture medium such as Phytagel affect plant growth and development. Given that the mechanical properties of Phytagel medium are vital for biomechanical experiments designing, a systematic study on mechanical properties of Phytagel medium with different concentrations were carried out here to better understand the response of plant to mechanical stimulation. **Results:** Uniaxial compression test was conducted for the mechanical strength and Young's modulus. The variation of concentrations of media results in different mechanical strength. The linear-regression analysis of the breaking load shows that there is a lack of fit of the linear regression model to the observed data points for all these Phytagel concentrations ($R^2 = 0.9708$). The spline regression model, however, fits well to the Young's modulus for Phytagel medium data. The rheological measurements from the oscillation tests (frequency sweep from 0.1 to 20 Hz at 1% strain) indicate that increasing the Phytagel concentrations results in a stiffer structure at 0.5%-1.2%. **Conclusions:** Although the biological effects of Phytagel on plant along with mechanical power are expected to check, this contribution provides a useful reference in biomechanical experiments to choose the best Phytagel concentration for the culture of plants and tissues.

1. Introduction

Physical laws and processes have profoundly impacted plant morphogenesis [1], mechanical stimulus can influence plant architecture by triggering organ initiation and causing alterations in the stem or root growth direction. Plants have evolved various responsive behaviors to adapt to these stimuli [2-4]. In plant biomechanical experiments, the mechanical properties of plant growth medium such as the stiffness, elasticity and surface roughness, influence the root system architecture [5, 6]. Varying amounts of Phytagel have been used to regulate the mechanical strength of plant growth medium [7]. Murashige and Skoog medium (MS) is a plant growth medium used in the laboratories for cultivation of plant cells, tissues and organs. It was originally developed for the in vitro culture of tobacco tissues. Since MS is a mixture of inorganic salts, it is mixed into the Agar or Phytagel as the culture medium to

Content from this work may be used under the terms of the Creative Commons Attribution 3.0 licence. Any further distribution of this work must maintain attribution to the author(s) and the title of the work, journal citation and DOI.
Published under licence by IOP Publishing Ltd

provide macro elements, trace elements and vitamins for the plant growth [8]. Phytagel concentrations of 0.5%-2.0% in half strength MS medium were commonly used in the research of mechanical response of root behavior [9]. Composition of plant culture medium can change its rheological properties and then the growth behavior of plant living in it. In many cases the rheological character is the determinant of growth medium performance. However, the lack of mechanical properties of Phytagel medium is one of the main barriers in applying quantitative study of the effect of stiffness on root development.

The application of high stability, high transparency Phytagel as a curing agent in plant culture medium has received an upsurge of interest. Research in plant development has advanced greatly by application of gelling agent in plant growth medium [10]. In recent decades, Phytagel has been mainly used in root cultivation because soil particles bound to roots are difficult to remove without injury to the roots [9]. Previously, Phytagel has been used to improve the food quality [11], and it is fabricated from a bacterial substrate and as an agar substitute composed of glucuronic acid, rhamnose and glucose [11]. Phytagel has been considered as green composites and an economical alternative to agar as a gelling agent. Half strength MS medium, sucrose and Phytagel were blended to form an interpenetrating network-like cross-linked complex.

Successful plant tissue culture depends on properties of growth medium, especially the mechanical property. Phytagel belongs to a complex fluid. The accurate measurement on the rheological characteristics of Phytagel growth medium can make the application in plant or tissue culture to be perfect. Therefore, it is essential to study mechanical properties of Phytagel in depth. In this paper, the mechanical properties of different composites of culture medium containing a range of Phytagel concentrations were investigated through tests and comparison. A systematic study of Phytagel were performed using uniaxial compression testing, together with rotational rheometer to provide a better understanding of the effect of Phytagel concentration on the mechanical properties of culture medium. Uniaxial compression test was performed to determine the mechanical strength of Phytagel media. Dynamic mechanical properties of low Phytagel concentration (0.5%-1.2%) were detected by frequency sweep tests. Quantitative measures were preliminarily reported for different concentrations of Phytagel. The results indicate that Phytagel should be regarded as the best regulator for the mechanical property of plant growth medium.

2. Materials and methods

2.1. Phytagel medium preparation

The standard plant growth medium consists of 1/2 MS basal salt (phytotechnology laboratories M519) and 15 g/l sucrose. Phytagel was added to 1/2 MS medium in various ratios from 0.5% to 2.0% w/v of fluid medium, and then they were blended at room temperature on high speed to eliminate any lumps before microwave heating. The pH of all media was adjusted to 5.8 with KOH 1 mM. The melted Phytagel media were sterilized by high pressure steam (1.05 kg/cm, 121.3 °C for 20 minutes). Homogeneous Phytagel media with a series of concentration from 0.5% to 2.0% were poured into the petri dishes (diameter = 9 cm, thickness = 12 mm). The Phytagel specimens were cut to dimensions (length = 20 mm, width = 40 mm and thickness = 12 mm). These rectangle specimens were under the condition of 23 °C and saturation moisture content.

IOP Conf. Series: Earth and Environmental Science **346** (2019) 012089 doi:10.1088/1755-1315/346/1/012089

2.2. *Uniaxial compression test*

Uniaxial compression experiment was conducted using Electron Puls E1000 machine (capacity, 1000 newtons) with Console Bulehill 3 software, and began when the load force touches the sample top area and reached 0.001 N. Loading rate was 1 mm per min and load force was increasing at the beginning, reaching the maximum, then declining. When the load force begins to decline, the test was finished. In this study, various amounts of Phytagel have been used to regulate the stiffness of plant growth medium. Phytagel media were tested from 0.5% to 2.0%. The minimum of three specimens with each Phytagel concentration was tested to obtain average values for Young's modulus and breaking load. The tensile properties of Phytagel mediums could not be measured due to their low stretch ability.

2.3. *Dynamic oscillatory tests*

The Phytagel specimens were prepared in the same way as described above. Homogeneous Phytagel media with concentration from 0.5% to 1.2% were poured into the petri dishes (diameter = 9 cm, thickness = 2 mm). A hole puncher with 2 cm diameter was pushed vertically into the Phytagel layer to make round Phytagel. The rheological properties were measured using a rotational rheometer (Haake MARS, Schwerte, Germany) and 20 mm parallel plates. Because Phytagel media with high concentrations exhibited high hardness from uniaxial compression test and rotational rheometer can't test high-hardness materials, rheological behavior of 0.5%-1.2% Phytagel media were carefully documented to determine complex modulus, elastic modulus and viscous modulus from dynamic oscillatory tests. All rheology measurements were performed at 25 °C. The round sample was placed between the plates and aligned with plates, and the gap was adjusted to 2 mm. A standard loading sequence was used to ensure both samples were subjected to a consistent and controllable loading protocol. The specimens were scanned from low frequency (0.1 Hz) to high frequency (20 Hz). The number of measuring points is 21. The data of frequency sweeps for each sample at different concentration are plotted as storage and loss moduli, viscous modulus and complex viscosity.

2.4. *Statistical analysis*

Multiple samples with various ratios of Phytagel were used in each type of experiment. Each experiment was repeated independently for three times. The results of all experiments were subjected to statistical analysis. Means and standard deviations were determined and confidence intervals were evaluated. Statistical differences were analyzed by one-way analysis of variance (ANOVA) followed by the Student's t-test, and P-value ≤ 0.05 was significant.

3. Results and discussion

3.1. *Breaking strength*

Gelling agent concentrations were varied to study the effect on the breaking strength properties. Their breaking load was analyzed by load-displacement curves (Figure 1). There was no evidence of any kind of physical phase separation based on the visual observation.

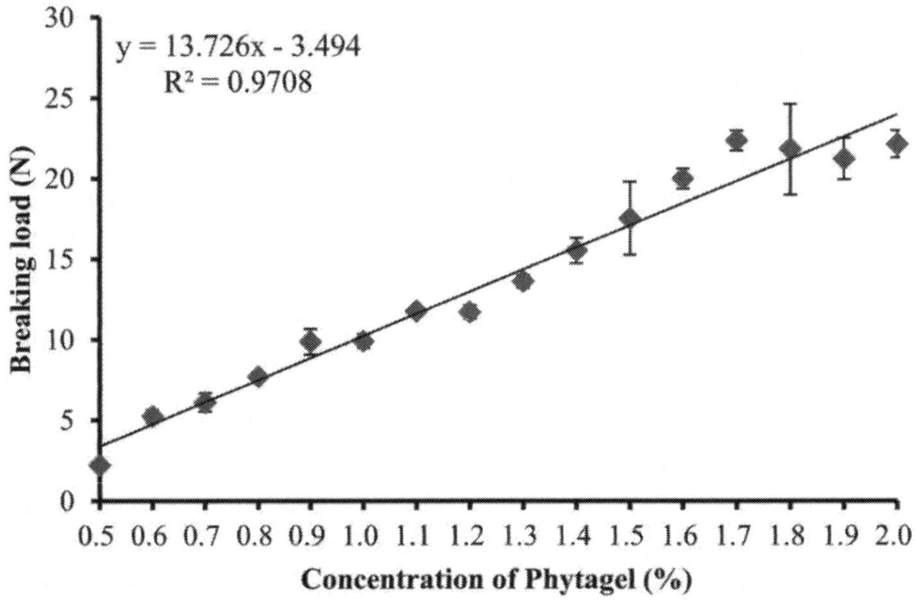

Figure 1. Breaking load of Phytagel media with concentration from 0.5%-2.0%.
The R^2 and the slope of the linear regression (black solid line) was calculated from data points in the breaking load.

Load-displacement data indicated that Phytagel can increase the mechanical strength of culture medium. As the Phytagel concentration (x) increased, value of breaking load (y) increased linearly: $y = 13.726x - 3.49$ (R-squared value of 97.08%). In terms of breaking load, the highest value was measured for 1.7% concentration, indicating a more brittle structure for concentrations higher than 1.7%. The lowest Phytagel concentration has the lowest value, suggesting a comparatively more ductile structure. These results seem to indicate that the incorporation of Phytagel in 1/2 MS liquid medium was regarded as good mechanical structure, and the best level for designing plant biomechanical experiments is from 0.5% to 1.7%. All tested Phytagel media demonstrated that high concentration of Phytagel enhanced the resistance to fracture and damage tolerance and dosing with water did not affect the breaking force for all samples.

3.2. Young's modulus

Because the breaking load reflects the maximum force loading on surface of samples when the samples are broken, it was expected that the Young's modulus would provide more information for the mechanical strength of sample.

Young's modulus (E) describes compression elasticity or the tendency of an object to deform along an axis when compressional forces are applied along that axis. It is defined as the ratio of positive stress and the positive strain of elastic material under stress.

$$E = \frac{\text{STRESS}}{\text{STRAIN}} = \frac{\sigma}{\varepsilon} = \frac{F/A_0}{\Delta L/L_0} = \frac{FL_0}{A_0 \Delta L},$$

where

F/A_0: the force per unit area, A_0: cross-sectional area;

E: Young's modulus;
F: the force loaded on a sample under compression;
A_0: original cross-sectional area through which the force is applied;
ΔL : the amount by which the length of the object changes;
L_0: original length of the object.

As Table 1 shows, when the percentage of Phytagel increased from 0.5 % to 2.0%, Young's modulus increased from 0.01530 to 0.12251 MPa. With the same medium composition, Phytagel at 1.7% generally tended to result in a higher mechanical strength than others.

Table 1. Effect of Phytagel on the mechanical properties of media with phytagel concentration from 0.5% to 2.0%.

The concentration of Phytagel media	Young's modulus (MPa)
0.5%	0.01530 ±0.00057
0.6%	0.02063 ±0.00111
0.7%	0.02653 ±0.00179
0.8%	0.04017 ±0.00151
0.9%	0.05000 ±0.00681
1.0%	0.04093 ±0.00114
1.1%	0.04150 ±0.00125
1.2%	0.04838 ±0.00144
1.3%	0.05755 ±0.00321
1.4%	0.06244 ±0.00079
1.5%	0.07601 ±0.00844
1.6%	0.08685 ±0.00268
1.7%	0.12251 ±0.01596
1.8%	0.10214 ±0.01265
1.9%	0.09657 ±0.00978
2.0%	0.08869 ±0.01597

3.3. Further modelling

An analysis of repeated measures design was done using the data provided. As a result, Phytagel with different concentration have different influence on the stiffness of the plant culture media with a *P*-value of 0.0007 (< 0.05). To establish the relationship between Phytagel concentration level and stiffness of the culture media, further modeling using the spline regression is subsequently applied to the data for Phytagel (Figure 2). With Phytagel at 5%-1.7 % Young's modulus also showed a steady increase and that was consistent with the result of breaking load. It should be noted that the slop of modeling curve from 0.5%-1.2% was small, and the slop from 1.2%-1.7% was steep. In other words, stiffness increased slowly when Phytagel concentration incresed to 1.2%, and stiffness increased rapidly when Phytagel concentration incresed from 1.3% to 1.7%. It could be concluded that mechanical strength of Phytagel medium have a transition point at 1.2%. Generally speaking, high concentration of Phytagel would enhance extrusion force for plant tissues when they were living in it. Phytagel of 1.0%-1.2% have been widely used in plant tissue culture, and the reason for that may be those Phytagel level provide a suitable mechanical stress for plant development. Plant morphogenesis depends on the integration of both mechanical and biochemical signals. For the published literatures of plant morphogenesis, the cause of formation of phenotype could be estimated from these data.

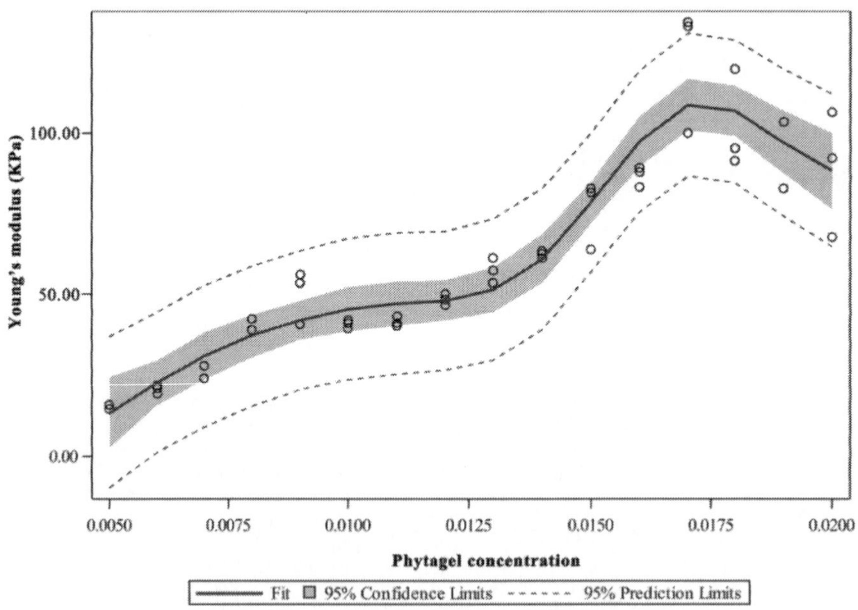

Figure 2. Summary of the model fitting to the Young's modulus for Phytagel medium.

$y = 0.01341804 + 0.03376900X + 0.05079903X^2 + 0.04575657X^3 + 0.13323091(X - 0.011)^3 + 0.09362697(X - 0.014)^3 + 0.08838223(X - 0.017)^3$ where y denotes the predicted stiffness of Phytagel medium and X represents the concentration level of Phytagel medium. The above model is statistically significant with *P*-value < 0.0001 and adjusted R-squared value of 89.6%.

3.4. Rheological properties

Phytagel properties were evaluated from dynamic oscillatory tests for a frequency sweep from 0.1 to 20 Hz [12]. The data of frequency sweeps for each media with different concentration are plotted as G* (Complex modulus), G_s (Elastic modulus, Storage modulus, Pa), G_l (Viscous modulus, Loss modulus,

IOP Conf. Series: Earth and Environmental Science **346** (2019) 012089 doi:10.1088/1755-1315/346/1/012089

Pa). G_s measures the stored energy, representing the elastic portion, and G_l measures the energy dissipated as heat, representing the viscous portion. Mechanical strength was expressed as complex modulus. The mathematical forms of these equations are given as below:

Storage Modulus: $G_s = \frac{\sigma_0}{\varepsilon_0}\cos\delta$

Loss Modulus: $G_l = \frac{\sigma_0}{\varepsilon_0}\sin\delta$

stress and strain are represented by σ_0 and ε_0, respectively, and phase lag between stress and strain is

represented by a.

Complex Modulus: $G^* = G_s + i\,G_l$

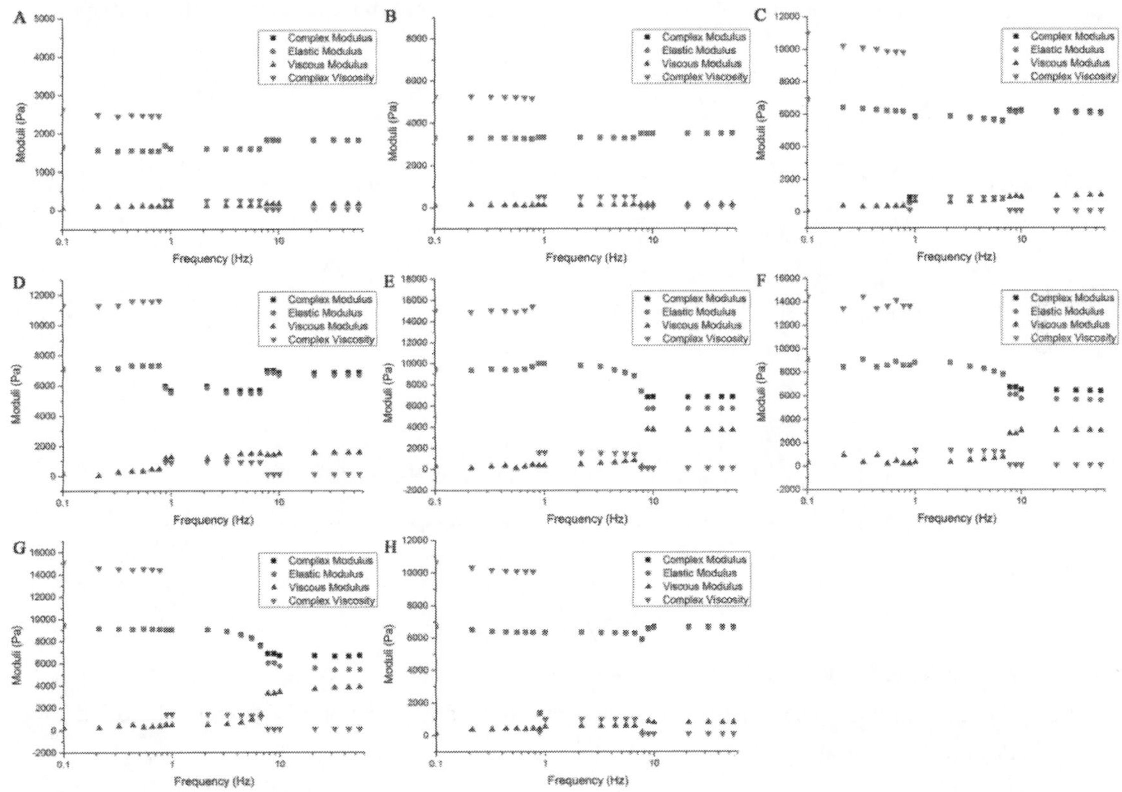

Figure 3. The variation of moduli (Storage and loss moduli, viscous modulus, and complex viscosity) for all the Phytagel medium with different concentration among various frequency (0.1-20 Hz). Master curves are constructed by using time – concentration superposition, referenced to the labeled concentrations, respectively. The reference concentrations are 0.5% for A, 0.6% for B, 0.7% for C and 0.8% for D. 0.5% -1.2% from left to right, up to down.

Phytagel medium combine the characteristics of elastic solids and Newtonian fluids based on the results of dynamic mechanical properties of all the Phytagel concentration. Complex modulus with higher value indicated a stiffer structure. Of the samples A-H representing media with 0.5%-1.2% Phytagel concentration with almost increasing complex modulus from 3101.22 ± 1308.73 Pa to 8607.205 ± 1357.059 Pa, sample A of 0.5% exhibited the lowest gel strength and sample F of 1.0% exhibited the highest gel strength (Figure 3). Rheological results showed that the mechanical response of Phytagel media was viscoelastic and well structured. The storage modulus is higher than the loss modulus. Whether structural integrated Phytagel media presented elasticity or viscidity depended on which modulus was predominant under the specific frequency. Almost each modulus exhibited frequency dependence during the dynamic oscillatory tests, indicating of structural relaxation.

With the increase of Phytagel concentration, the relative increase of complex modulus was observed. It was also observed for all Phytagel concentrations that elastic modulus varies very little with frequency, suggesting little structural relaxation is occurring with time. 0.8%-1.2% exhibited increase of viscous modulus at high frequency (10-20 Hz), especially 0.9%-1.1%. It is generally accepted that the chain relaxation governs the response at low frequencies while segmental dynamics dominate the response at high frequencies.

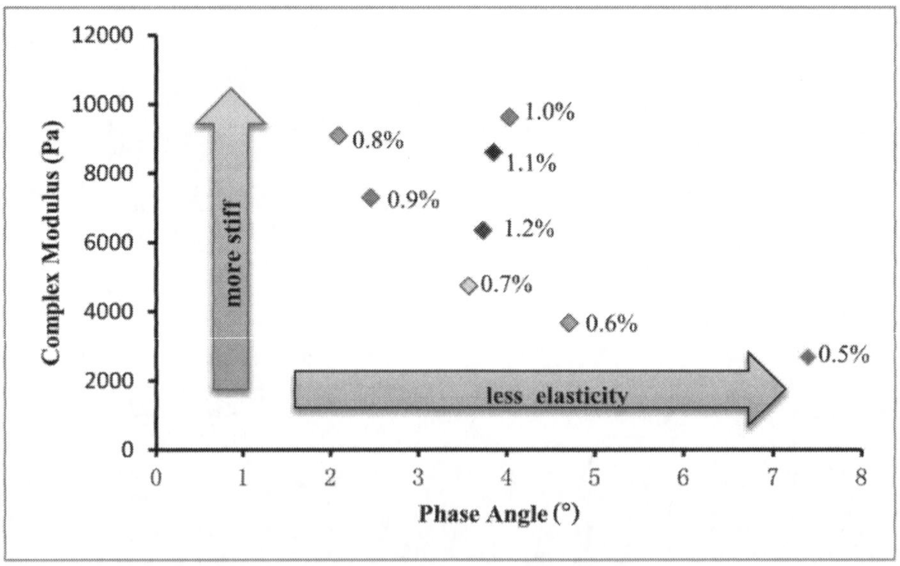

Figure 4. A schematic representation of relationship of phase angle and complex modulus. When oscillatory frequency was 0.1, viscoelastic properties of Phytagel media from 0.5% to 1.2% were reflected by phase angle.

Another characteristic of viscoelastic properties is reflected by phase angle. As the amount of Phytagel increased from 0.5% to 1.2%, the structure of media changed irregularly. In frequency of 0.1 Phase angle for all the samples were < 10°, indicating all samples could be considered as ideal gel. Phase angle of zero presented a perfect gel while the value between 0 and 45° suggested some degree of viscous damping, which could facilitate structural relaxation. Of the samples labeled 0.5%-1.2% in

Figure 4, Phytagel concentration increased from 0.5% to 0.8%, and the media exhibit increasing elasticity and stiffness. It was interesting that Phytagel medium of 0.9% was less stiff but more elastic than that of 0.8%. As Phytagel concentration increased from 1.0% to 1.2%, elasticity increased very little and stiffness decreased.

4. Discussion

Uniaxial tension and compression experiment was a common test for mechanical properties of materials [13]. Only compression test could be carried out for Phytagel because of its poor tensile properties. Previous studies raised concerns about the mechanical strength of Phytagel, for instance, when Phytagel was mixed in Soy Protein, its tensile and moisture properties were significantly improved [11, 14]. Measurement of rheological properties can provide microscopic structure for better understanding of the application features. There are various methods for determining rheological characteristics of gel materials. However, for the gel material one of the most sensitive methods is the dynamic oscillatory test. Oscillation testing included variable amplitude and variable frequency tests [15]. Amplitude sweep measurement was taken before frequency sweep measurement to make sure the response was under critical strain and within the linear viscoelastic region (LVER). Subsequent frequency sweep tests were performed between 0.1 and 20 Hz using a constant strain within the LVER. Oscillatory rheological experiments can systematically investigate nonlinear viscoelastic material responses. This study employed a combination of two types of mechanical experiments to characterize the properties of Phytagel medium. There are two factors determining the application of high concentration of Phytagel: medium moisture content, stiffness or hardness. Phytagel media were cross-linked complex solid which could congeal the liquid medium.

If the measurement is only made through compression test, elastic relaxation time scales will be lost. In addition, yield stress and strain were evaluated from these two kinds of tests. The results demonstrated how such an approach can be used to quantify and compare the properties of different gel systems.

Linear viscoelasticity theory is only valid when the total deformation is quite small [16]. Single shear rate can only provide little information and cannot reach the relaxation time scales. In contrast, frequency sweep test was useful for a broad class of complex fluids and soft matter because strain amplitude and frequency can be varied independently allowing a broad spectrum of conditions to be attained.

Previous studies considered Basal medium, plant growth regulators, and Phytagel concentration as factors affecting initiation frequency[17]. When the new biomechanical experiments are designed, it is essential to investigate the properties of the materials and chose the best Phytagel concentration to optimize the experiments.

Acknowledgments

We would love to acknowledge Dr. Jie Chen (Augusta University) for her valuable suggestions to improve this work. This research is funded by Chongqing Research Program of Basic Research and Frontier Technology (cstc2015jcyjBX0142, cstc2016jcyjA0375).

References

[1] Niklas K J 2013 Biophysical and size-dependent perspectives on plant evolution *J Exp Bot.* **64**

4817-27

[2] Valentine T A, Hallett P D, Binnie K, Young M W, Squire G R, Hawes C and Bengough A G 2012 Soil strength and macropore volume limit root elongation rates in many UK agricultural soils *Ann Bot.* **110** 259-70

[3] Acuña T L B and Wade L J 2013 Use of genotype × environment interactions to understand rooting depth and the ability of wheat to penetrate hard soils *Ann Bot.* **112** 359-68

[4] Jin K, Shen J, Ashton R W, White R P, Dodd I C, Phillips A L, Parry M A and Whalley W R 2015 The effect of impedance to root growth on plant architecture in wheat *Plant Soil* **392** 323-32

[5] Zhu L, Wang B, Wang Y, Liu J, Yang X and Fu X 2014 A micromechanics model for turgor pressure of *arabidopsis thaliana* protoplast *J Plant Growth Regul.* **33** 751-6

[6] Massa G D and Gilroy S 2003 Touch modulates gravity sensing to regulate the growth of primary roots of *Arabidopsis thaliana Plant J: Cell Molecul. Biol.* **33** 435-45

[7] Buer C S, Masle J and Wasteneys G O 2000 Growth conditions modulate root-wave phenotypes in *Arabidopsis Plant Cell Physiol.* **41** 1164-70

[8] Murashige T and Skoog F 1962 A revised medium for rapid growth and bioassay with tobacco tissue culture *Physiol. Plant* **15** 473-97

[9] Yamamoto C, Sakata Y, Taji T, Baba T and Tanaka S 2008 Unique ethylene-regulated touch responses of *Arabidopsis thaliana* roots to physical hardness *J. Plant Res.* **121** 509-19

[10] Stolz A, Bender P L, Faller J E, Silverberg E C, Mulholland J D, Shelus P J, Williams J G, Carter W E, Currie D G and Kaula W M 1976 Earth rotation measured by lunar laser ranging *Science* **193** 997-9

[11] Lodha P and Netravali A N 2005 Characterization of Phytagel® modified soy protein isolate resin and unidirectional flax yarn reinforced "green" composites. *Polym. Compos.* **26** 647-59

[12] Sivaramakrishnan H P, Senge B and Chattopadhyay P 2004 Rheological properties of rice dough for making rice bread *J. Food Eng.* **62** 37-45

[13] Oliver W C and Pharr G M 1992 An improved technique for determining hardness and elastic modulus using load and displacement sensing indentation experiments. *J. Mater Res.* **7** 1564-83

[14] Huang X and Netravali A N 2006 Characterization of nano-clay reinforced phytagel-modified soy protein concentrate resin *Biomacromolecules* **7** 2783-9

[15] Hyun K, Wilhelm M, Klein C O, Cho K S, Nam J G, Ahn K H, Lee S J, Ewoldt R H and McKinley G H 2011 A review of nonlinear oscillatory shear tests: Analysis and application of large amplitude oscillatory shear (LAOS). *Progr. Polym. Sci.* **36** 1697-753

[16] Dealy J M and Wissbrun K F 2012 Melt rheology and its role in plastics processing: theory and applications Springer Science & Business Media

[17] Li X, Huang F and Gbur Jr E 1998 Effect of basal medium, growth regulators and Phytagel concentration on initiation of embryogenic cultures from immature zygotic embryos of loblolly pine (Pinus taeda L.) *Plant Cell Rep.* **17** 298-301

Determination of lipophilic shellfish toxins in bivalve mollusks by ultra-performance liquid chromatography quadrupole linear ion trap mass spectrometry

G Y Liu[1], Z H Wang[1, 3], K Ge[2], L Song[1], J Du[1], J H Wu[1], K Wang and A Li[1]

[1]Liaoning Ocean and Fisheries Science Research Institute, No. 50 Heishijiao Street, Shahekou District, Dalian, China
[2]Dalian Ocean School, No.40 Lingshui Street, Shahekou District, Dalian, China

Email: 37765762@qq.com

Abstract. In the study, a method developed for the determination of lipophilic shellfish toxins in muscular tissues of *Ruditapes philippinarum* was presented. Lipophilic shellfish toxins contain Dinophysistoxin-1 (DTX_1), dinophysistoxin-2 (DTX_2) and homo-yessotoxins (h-YTX). We used microwave-assisted extraction (MAE) method with methanol as extractant to perform the sample preparation for muscular tissues of *Ruditapes philippinarum* and analyzed the crude extracts by ultra-performance liquid chromatography quadrupole linear ion trap mass spectrometry (UPLC-MS/MS). The microwave-assisted extraction operational parameters, including extraction temperature and volume, the solvent type and time, were selected detailed about extraction efficiency of lipophilic shellfish toxins from *Ruditapes philippinarum* muscular tissues. The method recoveries at three different spiked levels proved to be in the range from 50.47% to 94.98%. (<10%). The limits of detection (LOD) of three compounds were between 0.001–0.05μg/ kg, and it could meet the requirements for the actual samples. Then, this developed method was applied to actual samples which demonstrated that the method was a useful tool for the rapid quantitative detection lipophilic shellfish toxins. All in all, this determined method shows its good selectivity and sensitivity for the analysis of lipophilic shellfish toxins in tissues of *Ruditapes philippinarum*.

1. Experimental

Lipophilic shellfish toxins (LSTs) are produced by several microalage species and frequently bioaccumulated in filter-feeding molluscan shellfish. Although some toxins have no adverse

Content from this work may be used under the terms of the Creative Commons Attribution 3.0 licence. Any further distribution of this work must maintain attribution to the author(s) and the title of the work, journal citation and DOI.
Published under licence by IOP Publishing Ltd

IOP Conf. Series: Earth and Environmental Science **346** (2019) 012090 doi:10.1088/1755-1315/346/1/012090

effects on shellfish themselves, severe intoxication may occur when the contaminated shellfish is consumed by human, with diarrhea, nausea and emesis as the common symptoms [1]. Dinophysistoxin-1(DTX_1), dinophysistoxin-2 (DTX_2) and homo-yessotoxins (h-YTX) were found in bivalves or algae. They are lipophilic marine biotoxins with similarity in structure to lipid-soluble cyclic polyether compounds and are now regarded as typical LSTs. Their chemical structures are shown in Figure 1. These toxins are easy to accumulate in shellfish fat and may influence human beings by trophic transfer. It is reported that DTX_1 and DTX_2 are potent phosphatase inhibitors [2] and theycan cause inflammation of the intestinal tract and diarrhea [3]. YTXs have been reported to modulate the calcium homeostasis in human lymphocytes through studies [4]. In addition, YTX also induces a selective disruption of the E-cadherin-catenin system in epithelial cells [5]. In this case, maximum levels of several toxins that should not be exceeded when placing shellfish products on the shelf has been established in the European Union (EU) regulation, for protecting shellfish consumers, [6]. But the mouse bioassay analysis shows some obvious drawbacks, for example, low sensitivity and reproducibility, and lack of specificity and information on toxin composition [7].

Therefore, some methods for simultaneous analysis of LSTs in different matrix were reported, for example, liquid chromatograph [8], chromatography-tandem mass spectrometry (LC-MS) [9-15], Enzyme-linked immunosorbent assay (ELISA) [16], potentiometric chemical sensors assay [17]. Some analytical determination methods for the LSTs in seawater and sediment have been developed with LC-MS [18-20].

However, DTX_1, DTX_2 and h-YTX used as lipophilic toxins in bivalve molluscsby have not been detected at the same time. Because the similar potential hazards to people, simultaneous analysis of DTX1, DTX2 and h-YTX in shellfish is critical.

In this paper, based on the ultra-high performance liquid chromatography and microwave assisted extraction (MAE) device, we have established a method to detect DTX_1 and DTX_2 and hYTX simultaneously. The developed method was applied to separate and test DTX_1, DTX_2 and h-YTX in bivalve molluscsby though extraction and separation. Figure 1 showed the chemical structure of the target.

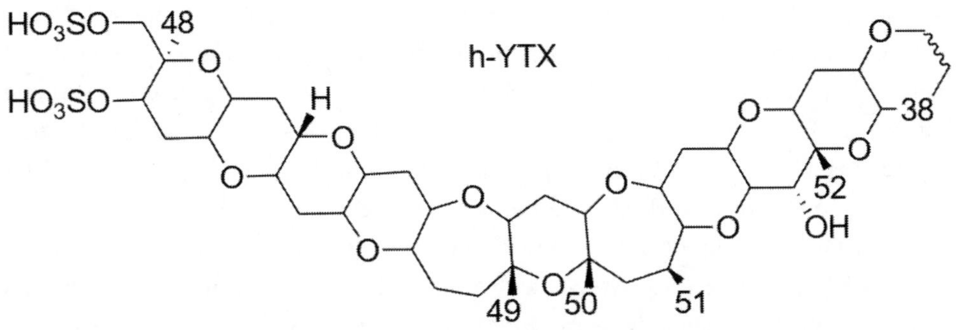

Figure 1. Chemical structures of three LSTs.

1.1. Reagents and materials
Methanol, acetonitrile, ammonium hydroxide (absolute hypergrade for LC-MS) were bought from Fisher Scientific (Pittsburgh, U.S.A). Water was deionized and passed through a Milli-Q water purification system (Millipore, Billerica, MA, USA). DTX_1 (15.1±1.1 μg/mL), DTX_2 (7.8±0.4 μg/mL), h-YTX (5.8g±0.3 μg/mL) standard were bought from the National Research Council. Institute for Marine Biosciences (Halifax, Canada). Individual DTX_1, DTX_2, h-YTX stock solutions were prepared in methanol with different concentration and they were stored at -18°C for three months. The different concentration was prepared in methanol with 1.0, 2.0, 5.0, 20.0, 50.0 μg /L. They were used to prepare calibration curves.

1.2. Apparatuses
The extraction of LSTs from molluscan shellfish tissue was extracted from a CEM MARS microwave sample preparation system, which equipped with 24 teflon vessels (Matthews, North Carolina). A Waters ACQUITY UPLC system (Waters, Milford, USA) was equipped with a vacuum degasser, an autosampler and empower workstation (Waters Milford, USA). The chromatographic separation was carried out on an ACQUITY UPLC BEH C18 (2.1×50mm i.d., 1.7μm, Waters, USA). A XEVO-TQD Mass Spectrometer (Waters, Milford, USA) coupled with electrospray ionization (ESI) source was tested in both positive and negative ionization mode and interfaced to a computer for running. Analyst Masslynx Version 4.1 software was applied for instrument control, data acquisition and the processing. MS/MS data acquisition was implemented in the multiple-reaction monitoring (MRM). Quantification was applied by using Targetlynx.

1.3. Sample preparation
Ruditapes philippinarum which was obtained from the food market was distinguished by Prof. Lun Song. The samples were crushed completely. The sample was divided into four parts after homogenization and stored at -18°C. One part was blank. Another three parts were used for spiking. The concentration were 2.0 μg/kg, 10 μg/kg and 20 μg/kg. They were prepared by adding an appropriate volume of spiking solution of three LSTs (DTX_1, DTX_2 and h-YTX) to the sample, then homogenize carefully. All samples were cryopreserved no more than three months.

1.4. Extraction

We weighed accurately 1.00±0.01 g of Ruditapes philippinarum tissue sample and placed them in a MAE extraction vessel with Methanol (15 mL). Every vessel was tightened and rocked violently for a few seconds. The vessels were put into the microwave sample preparation system. Extraction was implemented at 90 ℃ for 10 minutes. The time of ramp to temperature was 5 minutes. When the parameters are set, the MAE was opened. After extraction, the vessels were cooled down to the room temperature for 15 minutes. Then the vessels were taken off from the microwave sample preparation system. Then, this extract was filtered. The residue was washed three times with 5mL methanol at each time. These aliquots of methanol were collected together with methanol extract after filtration. Subsequently, each collected solution was distilled to dry at low pressure using a rotary evaporator at 60◦C. The residues were eluted using 3 mL methanol. Then, the aliquot of methanol was purified by Prime Oasis HLB (Waters). The purified liquid was blown by nitrogen blowing instrument and reserved to 1ml. Finally, the analyte solution was filtered by 0.22 μm filter membrane, then filter solution was used for UPLC-MS/MS analysis.

1.5. Conditions of UPLC

The reversed-phase BEH C-18 column was used for the UPLC method. Temperature of the column was set at 25◦C. Sample injection volume was 5μL. The flow rate was set at 0.4 mL/min. The mobile phase consisted of acetonitrile (solvent B) and water (solvent A) with 0.1% ammonium hydroxide, the initial proportion in volume of solvent B was 30%, which was increased to 80% within 1 minutes, kept constant for 3.5 minutes, and reduced to 30% in 0.5 minutes. The total run time was 5 minutes. Compared to HPLC, UPLC supplied improved resolution and sensitivity, higher efficiency, a fast run time, and reduced solvent use.

1.6. Conditions of MS

The mass spectrometer equipped with an ESI source and operated in negative polarity. MRM experiments were carried out. Nitrogen (99.99%) was used for desolvation gas and argon (99.99%) was used for cone gas. In order to obtain maximum sensitivity for identification and detection of LSTs, the ion source temperature (TEM) was set at 150℃, and the desolvation TEM was set at 500℃.The flow rate of the desolvation gas was 800 L/hr. The flow rate of the cone gas was 50 L/hr. Table 1 showed that the analyte specific parameters (cone voltage, collision energy, etc) for three compounds.

Table 1. The parameter of MS/MS for the tested three compounds.

Compound	Ionisation	Parention	MS/MS transition (m/z)	Dwell (s)	Cone Voltage (V)	Collision energy (eV)
hYTX	ESI^-	$[M-H]^-$	577.3>474.2	0.038	72	32
			577.3>403.2		72	33
DTX_2	ESI^-	$[M-H]^-$	803.3>255.1	0.038	80	60
			803.3>113.1		80	45
DTX_1	ESI^-	$[M-H]^-$	817.3>255.0	0.038	80	60
			817.3>112.9		80	45

2. Results and discussion

2.1. Selection of solvent

Methanol, 80% methanol, acetonitrile, 80% acetonitrile were selected as the extractants of LSTs from *Ruditapes philippinarum* tissue. The extractions were detected at 90°C with 15 minutes using 15 mL solvent and tissue spiked with LSTs at 10 μg/kg level. Figure 2 shows the extraction recoveries of LSTs for detected extractants. When 80% methanol was used for the extraction solvent, Three LSTs contents was lower. So under this condition, the recoveries of DTX_1, DTX_2 and h-YTX were lower. The recoveries nearly to 90% of three LSTs were the highest inmethanol. However, the recoveries of h-YTX and DTX_1 were comparatively lower, especially that of DTX_1 which is only about 35%, lower than other solvent, although DTX_2 is higher. The data presented in Figure 2, methanol can be used as the best solvent for extraction.

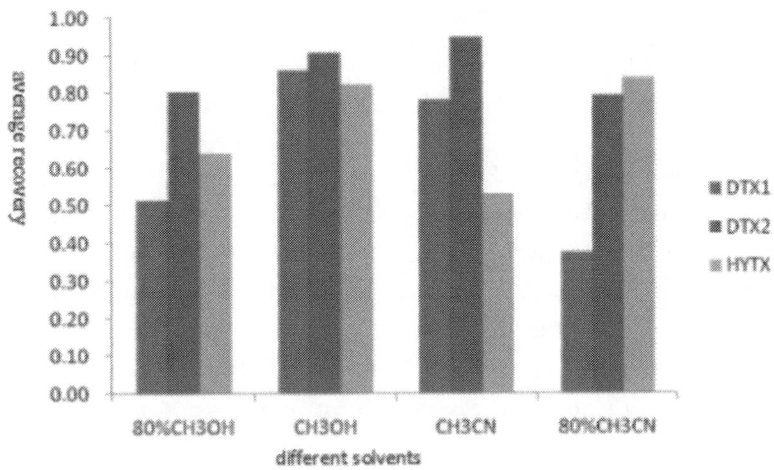

Figure 2. The influence of solvent types ($n = 3$).

2.2. Selection of extraction temperature

The interactions, equilibrium rate and control partition of analytes between sample and solvent could be influenced by the temperature of microwave irradiation. In order to extraction temperature, extractions were tested at 60, 80, 85, 90 and 100 °C, respectively. Figure 3. Showed the extraction recoveries. For DTX_1, the recoveries increased from 60 °C to 90 °C and the recovery decreased at 100 °C, it is possible that DTX_1 was decomposition and transferred into other matter. So, the recoveries of DTX_1 reach at a pinnacle and the irradiation temperature hardly influenced the recoveries in 100 °C. when temperature reached 80 °C, the extraction meets its maximum and further temperature increasing to 90 °C did not ultimately influence the recovery of DTX_2. At 100 °C, the recovery decreased distinctly. For h-YTX, the recovery was highest at 90 °C than other temperatures. On the basis of data indicating a different degree of thermal sensitivity for each solute presented in Figure 3, the 90 °C extraction temperature was selected for further experiment.

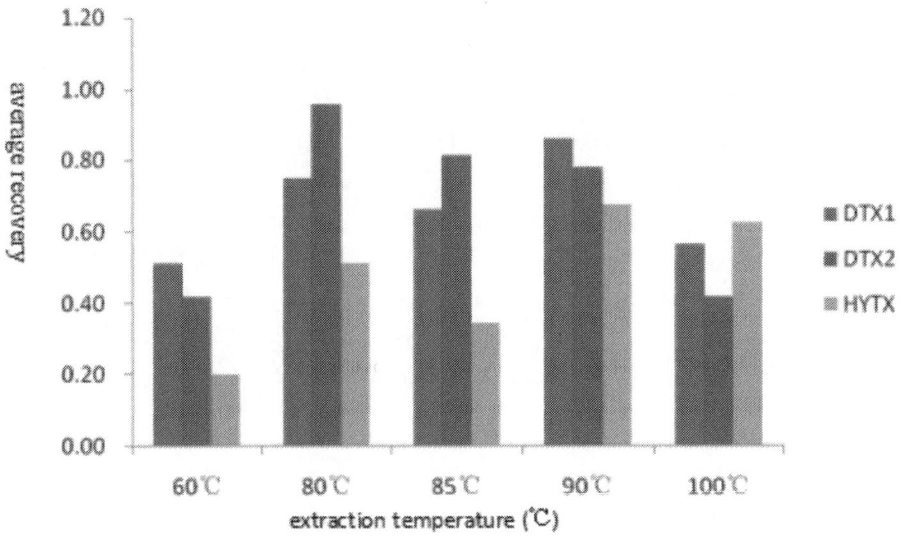

Figure 3. The influence of temperature (*n*=3).

2.3. Selection of solvent volume

Four kinds of solvent volume were selected, Figure 4. showed the influence of solvent volume on the extraction efficiency by MAE. The volume was important. If the volume of solvent is small, the extraction of target material will be incomplete, and the volume of solvent is too large, resulting in the waste of solvent and the increase of workload. After repeated tests, the volume of the solvent is determined. The recoveries of DTX_1, DTX_2 and h-YTX were the highest with 15mL. So, the solvent volume was set at 15 mL.

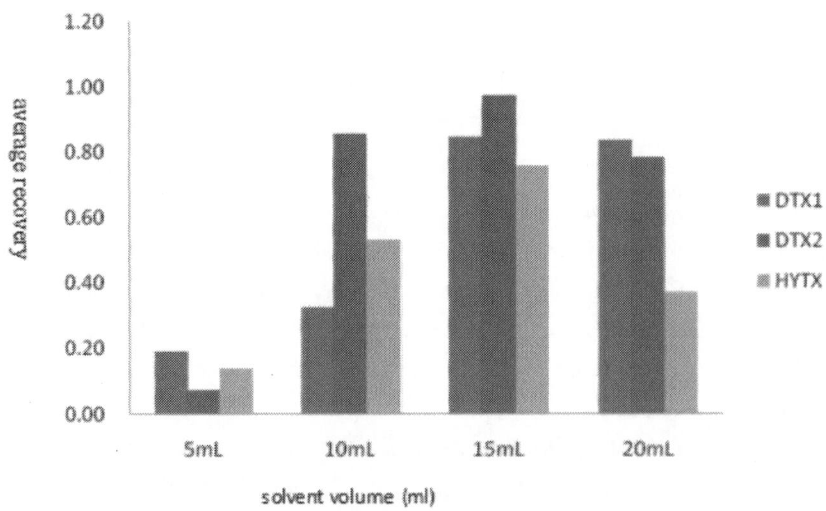

Figure 4. The influence of extactant volume (n = 3).

2.4. Selection of extraction time

The speed of extraction time affects the whole analysis process. The analysis time is short and the analysis efficiency is high. It is an important part of the analysis work. According to the microwave extraction device and the actual sample requirements, we chose5, 10, 15 and 20 minutes. So, extractions were operated at 90 °C for 5, 10, 15 and 20 min with using 15 mL methanol as the extraction solvent for optimizing extraction time. The results were suggested in Figure 5 under different extraction time. From 5 minutes to 20 minutes, the recoveries of all LSTs increased Obviously, the recovery of LSTs 5 minutes was lower than 10 minutes. However, the recovery of LSTs did not changed significantly after the extraction time which was prolonged from 10 to 20 minutes. Further increase in extraction time did not lead to the significant recovery improvement. In generally, the optimum extraction time was 10 minutes for giving to its comparatively better performance.

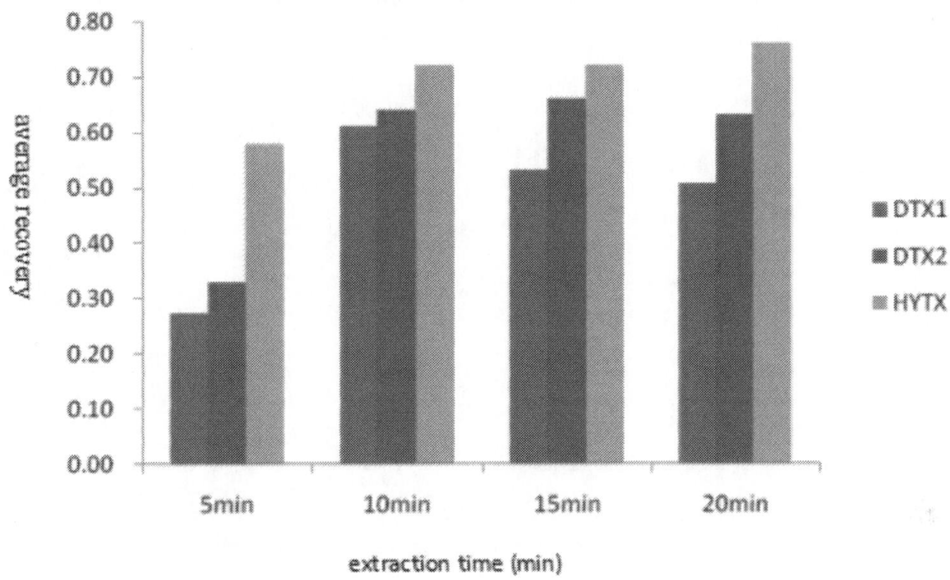

Figure 5. The influence of extraction time (n = 3).

Table 2. Calibration curve and linear coefficient of three LSTs (Y: peak area, X: concentration of the anlalytes).

LSTs	Calibration curve	Linear coefficient （r）
DTX_1	Y=155.23X+103.93	0.9942
DTX_2	Y=198.42X+136.6	0.9942
h-YTX	Y=63.65X-38.20	0.9995

2.5. Determination

From table 2, the relationships between the analyte concentration and measured signal for three LSTs was showed. The linearity of the calibration curves was verified by the correlation coefficients. Calibration curves were set up at concentration range from 1.0 μg/L to 50.0 μg/L by UPLC-MS. Table 2 illustrated the results. The LODs of DTX , DTX and h-YTX were

0.05 μg/kg, 0.01 μg/kg and 0.01 μg/kg. The LOQs of DTX_1, DTX_2 and h-YTX were 0.16 μg/kg, 0.04 μg/ kg and 0.03 μg/kg.

2.6. Accuracy and precision

Three level spiked samples (2μg/kg, 10μg/kg and 20μg/kg) were selected for continuous injection 5 times, respectively. The method validation studies for spiked samples suggested that the developed method provides good accuracy and high sensitivity for LSTs of three levels (2μg/kg, 10μg/kg and 20μg/kg). The recoveries were 50.47–94.98% and the relative standard deviation. was between 3.45%-7.86% in table 3. Figure 6 showed the TIC chromatograms of LSTs (A) and linear ion trap MS spectra of LSTs (B). The results showed that the method can separate the three LSTs well. In addition, these results illustrated the applicability of the method to extract DTX_1, DTX_2 and h-YTX from Ruditapes philippinarum tissue.

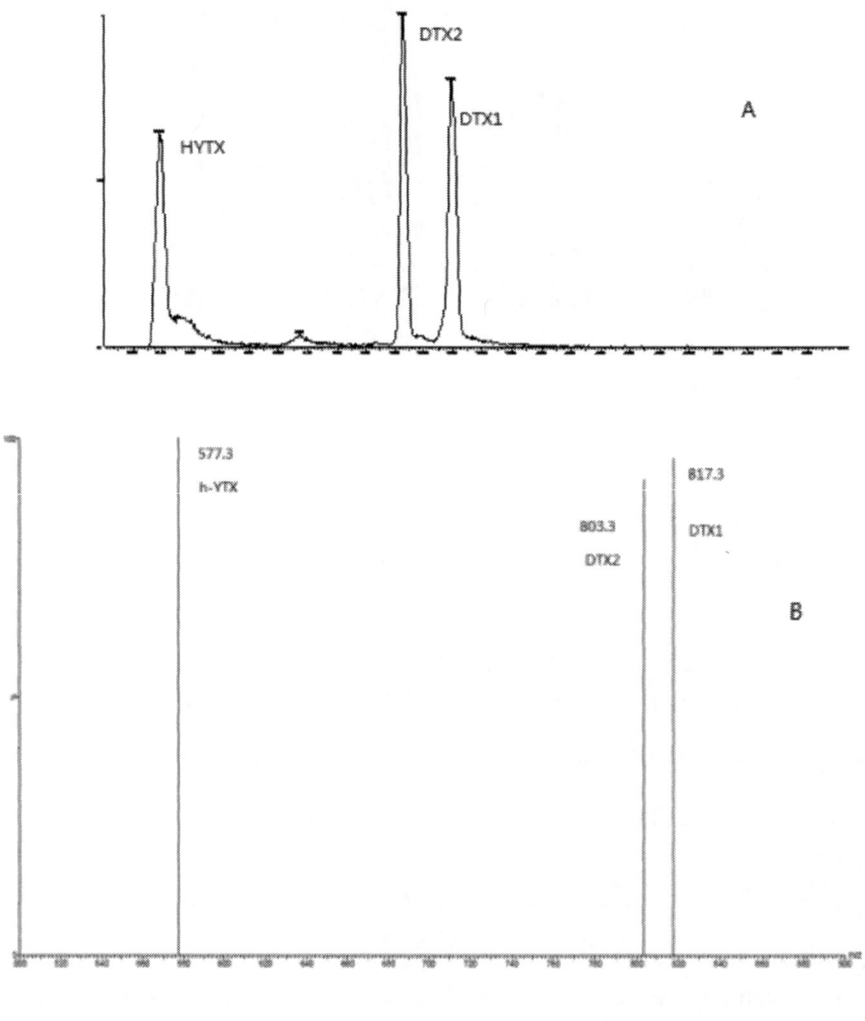

Figure 6. The TIC chromatograms of LSTs(A) and Linear ion trap MS spectra of LSTs (B).

Table 3. Recoveries and precision of extracted procedure of the samples (n = 5).

LSTs	Added (μg/kg)	Average Recovery (%)	R. S. D. (%)
	2.0	50.47	4.19
DTX_1	10.0	86.09	3.45
	20.0	94.98	5.62
	2.0	52.2	7.86
DTX_2	10.0	79.34	5.78
	20.0	82.69	3.56
	2.0	57.62	6.89
h-YTX	10.0	75.76	4.35
	20.0	64.11	5.23

2.7. Application of the method

We applied the method to detect the LSTs in the different *Ruditapes philippinarum* samples. Table 4 presented the result. DTX_1, DTX_2 and h-YTX were analysed in the visceral mass and muscle tissue, but none was detected in the. muscle tissue. Because the internal viscera have detoxification function, the toxin can be detected. People try not to eat internal organs of bivalve mollusks.

Table 4. Results of the LSTs in the different *Ruditapes philippinarum* samples.

LSTs	Muscle tissue (μg/kg)	Visceral mass (μg/kg)
DTX_1	-	1.2
DTX_2	-	-
h-YTX	-	1.3

"-" not detected.

3. Conclusion

In this study, a method developed for the analysis of lipophilic shellfish toxins in muscular tissues from *Ruditapes philippinarum* was presented. The parameters of microwave-assisted extraction were determined by recovery rate, including solvent types and volume, extraction temperature and time. So, it is optimization that methanol was used as extractant at 90 °C for 10 min in 15mL. This study demonstrates that the microwave-assisted extraction with the UPLC-MS/MS method is applicable to determine DTX_1, DTX_2 and h-YTX in shellfish simultaneously. Compared with the reported method [21], the results presented that this extraction procedure was efficient and accurate. In the process of extraction, we only used small amounts of organic solvent, which was greee and was harmful to operators, under final optimized conditions, the total analysis time was only 20 minutes, including sample extraction, injection, chromatographic separation and data acquisition. The method has the advantages of simple operation, fast, high efficiency, accuracy and reliability. So, the validated method can be applied successfully to determine the presence of DTX_1, DTX_2 and h-YTX in *Ruditapes philippinarum* tissue. It can be concluded that the determined method provided good selectivity and reproducibility for the quantitative analysis of LSTs in bivalve molluscs.

Acknowledgments
This study was funded by the Scientific Research Public Foundation of Liaoning Province (Grant No. 2014004019 and 20170036).

References

[1] Pfannkuchen M, Godrijan J and Pfannkuchen D 2012 *Environ. Sci. Technol.* **46**(10): 5574-82
[2] Lucia S, Francesc X and Jorge D 2015 *Toxicol. in Vitro*, **29**: 59-62
[3] Miguel C, Zhong T and Ricardo T 2017 *Toxicon.* 126: 23-31
[4] De la Rosa L A, Alfonso A, Vilarino N, *et al.* 2001 *Biochem. Pharmacol.*, **61**: 827-33
[5] Ronzitti G, Callegari F, Malaguti C, *et al.* 2004 *Br. J. Cancer*, **90**: 1100-7
[6] European Commission 2004 Regulation (EC) No 853/2004 of the European Parliament and of the Council of 29 April 2004 laying down specific hygiene rules for food of animal origin. *Official J. Europ. Commun.*, L226, 22-82
[7] Hess P. *Anal.* 2010 *Bioanal. Chem.* **397**: 1683-94
[8] Lee J S, Yanagi T, Kenma R, *et al.* 1987 *Agric. Biol. Chem.* **51**: 877-81
[9] Shen H H, Chen J H, Xu X L, *et al.* 2018 *Chin. J. Analyt. Chem.* **46**(6): 985-92
[10] Wu H Y, Guo M M and Tan Z J 2014 J. *Chromatogr. A.* **1358**: 172-80
[11] Gabriel O, Lieven V and Maarten D 2017 *Harmful Algae*, **64**: 30-41
[12] Liu Y, Yu R and Kong F 2017 Chemosphere, **183**: 380-8
[13] Liu Y, Yu R and Kong F 2019 *Environ. Pollut.*, **249**: 171-80
[14] Wang L, Shi X, Zhao Q, et al, *Food Chem.*, 2019 **272**: 427-33
[15] Choonshik S J,Yun H J, Hong Y S, et al, 2018 Food Contr., **91**: 365-71
[16] Zhao R, Liu L and Liu L. 2015 *Chin. J. Anal. Lab.* **34**(8): 882-5
[17] Ferreira N, Cruz M and Gomes M. 2018 *Toxins*, **181**: 380-4
[18] Li X, Li Z Y and Chen J H 2014 *Chemosphere*, **111**: 560-7
[19] Lane J Q, Roddam C, Langlois GW and Kudela R M 2010 *Limnol Oceanogr. Meth.* **8**: 645-60
[20] Li Z, Guo M, Yang S, et al, 2010 *Mar. Drugs*. **8**: 1263-72
[21] Yao J, Tan Z and Zhou D, 2010 *Chin. J. Chromatogr.* **28**(4):363-7

IOP Conf. Series: Earth and Environmental Science **346** (2019) 012091 doi:10.1088/1755-1315/346/1/012091

The border between species with and without feelings

J van Rooijen[1]

Former poultry behaviour and welfare researcher, Centre for Applied Poultry Research "Het Spelderholt" Beekbergen, The Netherlands

E-mail: Jeroenvrooijenakira@gmail.com

Abstract. Animal protection makes only sense when animal suffering and welfare exist. Suffering and welfare require the existence of feelings in a species. The question "Which species possess feelings?" is, therefore, important. Scientists draw the line between species with and without feelings at different places in the animal kingdom. Reasons for laymen to assume or discard the existence of feelings in different species are considered. It is mentioned that science is fundamentally unable to draw this line. (A view often attacked under the influence of psychology that belongs to the humanities). It is also mentioned that science is able to make the presence of feelings in an animal species more, or less, convincing. The views of the different scientists are discussed. It is argued why, in my opinion, all animal species possess feelings. Although not all scientists share this view, many scientists are of opinion that at least farm animals possess feelings. Laypeople who are not influenced by economic or other interests mostly do not question this opinion. The protection of farm animals is justified to avoid suffering in case these farm animals possess feelings.

1. Introduction

Animal protection is based on the presence of animal suffering and welfare. "Animal welfare research includes the existence of feelings. Otherwise it would be comparable with: "How to handle a machine in order to make it neatly running" or "What is wrong with a machine behaving in this way" [1]. A definition of "feelings" is not necessary. Everybody knows by introspection what a feeling is. I believe that the distinction between affective and non-affective feelings is artificial. The presence of all feelings indicates the presence of a mind. The capability to feel a neutral touch implies, in my opinion, also the capability to feel a negatively experienced touch ("pain"). Also, a bacteria swims away of a puncture needle. An important question in relation with animal welfare is "Which species have feelings?" In other words: where do we draw the line between animal species which do have and species which do not have feelings? This seems at first sight a matter of personal taste.

2. Where do different scientists draw the line?

Different scientists draw the line between species with and without feelings on different places. I give some examples:

In 1992, Kennedy drew this line between man and all other animal species [2]. This author took an "uncompromisingly negative view of animal consciousness." (p.32). Under the heading "Uniqueness of Homo sapiens sapiens" he attacked the most widely held view that there must be some consciousness in animals because of the Darwinian principle that evolution has been a continuous process (p.15). In his opinion are quantitative differences with the emergence of Homo sapiens sapiens "translated" into

Content from this work may be used under the terms of the Creative Commons Attribution 3.0 licence. Any further distribution of this work must maintain attribution to the author(s) and the title of the work, journal citation and DOI.
Published under licence by IOP Publishing Ltd

a qualitative effect. A phenomenon described by Geertz [3] (cited Kennedy 1992 [2]) "as when water reduced degree by degree without any loss of fluidity suddenly freezes at 0°C."

Bermond drew the line between human beings, anthropoïd apes and all other animal species, possibly with an exception for dolphins (Delphinidae sp.) etc [4]. He was of the opinion that the presence of a well-developed prefrontal cortex and a right neocortical hemisphere is necessary for a species to be able to experience suffering.

Wiepkema (1997), and others, drew the line between vertebrates and invertebrates, sometimes an exception was made for the octopus (e.g. Mather 1991) [5]. Wiepkema claimed that "apart from performing species-specific goal directed behaviour vertebrates show so-called emotional behaviours or expressions" [6]. In Wiepkema's opinion, vertebrates have emotions derived from the difference between what is and what should be and emotions corresponding to how well an organism trusts its means to restore or maintain its homeostasis [5].

Griffin (1976) seemed apt to draw the line lower than the insects [7]. In his chapter "The versatility of animal communication" he described the bee dances and the communication between fireflies and between spiders, suggesting that insects and spiders could have awareness.

Van Rooijen (2001) draws the line lower than the insects [8].

3. Reasons for people to assume or discard the existence of feelings in different species

Scientists are human beings. Normal human feelings will influence their judgement. Therefore, I will first discuss some reasons why human beings in general deny or accept the presence of feelings in different animal species.

3.1. Reasons to assume that a particular species has feelings

3.1.1. Simulation. We may imagine ourselves in the position of another organism and register the feelings that we should have. For instance, "I would feel cold when I, like that animal, was outside in the winter." Or" I should miss my freedom when I was in a cage as that animal." This method seems to tell us more about human beings than about animals. It may be misleading because of all the differences in sense organs and nervous system between us and the other organisms [9]. For instance, Icelandic ponies (Equus caballus) are adapted to a cold climate. They often prefer to stay outside in the winter.

3.1.2. Empathy. Other individuals may induce particular feelings in us

We may spontaneously get the feeling that we understand what goes on in the mind of another human being or animal. Certain brain regions are not only active when a person experiences a feeling (disgust, happiness, pain, etc.) but also when he sees another person (or animal) showing signs that seem to indicate similar feelings (e.g. [10]).

This spontaneous feeling may be less reliable, especially towards individuals of other species. For instance, Heinroth described a fight between a peacock (Pavo cristatus) and a turkey (Meleagris gallopavo) [11]. He also described the misunderstanding between both species during this fight. For us fish stimuli may be misleading because fish do not possess facial expressions for pain.

A special form of empathy is the empathy we feel for children. Stimuli that resemble the child stimuli of our own species are especially effective in our species to elicit a tendency to protect and take care [12]. This phenomenon is important in relation with our wish to protect animals. Many young animals of other species are able to elicit such feelings in individuals of our species. Even young scorpions (Scorpio sp.) are, in some degree, able to do so. Of course, not all species are equally effective in this respect. Piglets (Sus scrofa) resemble human babies in such a way that in some culture's piglets are mouth to mouth fed as is done with the own babies [13]. Also, adult individuals of particular species may be attractive to us because they resemble human babies, for instance the round forms and relatively large head of brown bears (Ursus arctos) are very effective to make us want to pamper them. Not surprisingly, Teddy bears are able to elicit parental care even in very young children. A round head silhouette, as in short-nosed dogs (Canis familiaris), is attractive to many of us; even in birds as parrots (Psittacus sp.), budgerigars (Melopsittacus undulatus), love birds (Agapornis sp.) and puffins (Fratercula sp.). Even the round form of turtles (Testudines sp.) and ladybirds (Coccinellinea sp.) induces positive feelings in many human beings.

Babies have relatively large eyes. Because of the black surrounding of the eyes of panda bears they seem even more baby-like than other bears. Most species have eyes at both sides of the head. Species with eyes in the front side have a "face", as in owls (Strigidae sp.). Such animals are more human-like, and, therefore, more appealing to us.

We are fond of smiling babies. Therefore, we like the smiling face of the dolphin (Delphinidae sp.). We get enthusiastic when our babies develop hair on the top of their head. This explains the attractiveness of longer hair on the mammalian head, as in horses (Equus caballus) and male African lions (Panthera leo), or longer feathers on the head of birds, as in cacatoos (Cacattua sp.), Nicobar pigeons (Caloenas nicobarica), grey crowned cranes (Balearica regulorum), plumed guineafowl (Guttera plumifera), emperor penguins (Aptenodytes forsteri) and some breeds of domestic poultry (Gallus gallus). Also hanging ears in domestic goats (Capra aegagrus hircus), domestic rabbits (Oryctolagus cuniculus) and dogs (Canis familiaris) may give us the appearance of hanging hair.

Before babies start to crawl, they have a stage in which they lay on their belly with their head up. The normal posture of seals (Phoca vitulina), California sea lions (Zalophus californianus), etc. is very similar to this baby posture. This makes seals attractive for us, especially young seals. Young seals combine this attractive posture with large eyes and a crying that resembles the crying of a human baby. Many species with four legs move in a way that resembles the crawling of a young human infant. We are enthusiastic when a child starts to walk upright. This explains our enthusiastic reaction on species that adopt an upright posture like squirrels (Sciurus vulgaris), prairie dogs (Cynomys ludovicianus), meerkats (Suricata suricatta), kangaroos (Macropodidae sp.) and even seahorses (Hippocampus brevirostris). Especially the clumsy walking movements of penguins (Spheniscidae sp.) are very attractive to us.

We are also enthusiastic when a child starts to talk. Talking birds like parrots (Psittaciformes sp.) and Myna Birds (Gracula religiosa) charm us.

3.1.3. Familiarity. Personal experience with a species provides thorough knowledge of this species. Familiarity with individuals makes it easier for us to understand the needs and the behavioural and morphological stimuli of a species. When individuals of different species live close together they have a tendency for mutual adaptation. For example, domestic house cats (Felis catus) frequently make sounds that resemble sounds made by a human baby. The high frequency of these baby-like sounds is an adaptation to man. Among each other cats make these sounds much less frequently. Not surprisingly the frequency of these sounds lowers in cats that become feral. Wild cats hardly ever make such sounds [14]. Students considered the sounds of house cats as more friendly than the sounds of wild cats, suggesting an adaptation by house cats to human wishes [15]. Also dogs probably have developed barking as an adaptation to human talking. The wolf and wild dogs species bark much less than domestic dogs. Human beings, even people that did not live with a dog understood the meaning of different types of barking [16]. Industrial poultry keepers are less familiar with individual hens than fancy fowl keepers. Fancy fowl keepers generally understand hens (Gallus gallus) better than industrial farmers. This holds also for dog keepers in comparison to scientists: in general dog owners understand dogs (Canis vulgaris) better than behaviour scientists do [17]. Only recently the amazing learning abilities of dogs, already known for a long time among dog keepers, are recognised by science, see, for instance, [18]. Another example is the remarkable ability of dogs to react to human pointing [19]. These authors interpreted this finding by the assumption that dogs possess a 'theory of mind' This last assumption is questioned by [20].

Still another example is an experiment in which a dog was trained to lower a hanging stick with its paw in order to obtain a reward. Eighty-three percent of the dogs that, after observing this act, were placed in the same situation, also used their paw to obtain the reward, and not the easier technique of grabbing the stick with its mouth. Another group of dogs watched the same act. But now the demonstrator dog had a ball in his mouth. In this case only 21 % of the dogs used their paws. In a control group, that did not observe any act, 15 % used their paws. This showed that most dogs of the second group had not learned the behavioural response but that they "understood" why the demonstrator dog used his paw and not his mouth [21].

In zoos we may become familiar with more exotic species. A male chimpanzee always threw stones to the public during his dominance display. During this display he was too aroused to collect stones. It turned out that he started his day with the stone caching and the manufacture of discs from concrete. He used these objects for his display, a few hours later. He kept his stone collection always at the visitor's side. He did not collect stones outside the visitor's season [22]. This behaviour showed that this chimpanzee found a, probably, individual solution to anticipate on his own future needs.

An increasing body of refined experiments reveals that animals are less different from human beings then until recently assumed in science. For instance, New Caledonian crows (Corvus moneduloides) use tools and are also able to manufacture them and show even an understanding of how tools can be combined to reach a goal [23]. Also, the Kea (Nestor notabilis) turned out to be able to solve a problem step by step with the help of tools [24].

Most people are most familiar with vertebrates. This is one of the reasons why most people take the presence of feelings in vertebrates for granted (however, see III.2.1 and IV). It must also be stressed that the presence of cognition is not a prerequisite for the presence of feelings. However, the presence of cognition makes it easier for us to assume that also feelings are present. But then we use the degree of similarity.

3.1.4. The degree of similarity. Similarity is important because stimuli may be misleading. Examples are already mentioned under II.3.1. Another example that shows the unreliability of stimuli is an experiment done by Krach (Aachen University) and Hegel [25]. These researchers made experimental persons play a prisoner's dilemma game against a computer, a machinelike robot, a human like robot and a human. In all plays the answers were already determined on forehand.

The researchers established in a fMRI-scanner the activity of particular brain areas associated with a "theory of mind." They found that these areas were least active in humans while playing a prisoner's dilemma game against a computer. They became more active while playing against a machinelike robot and even more active while playing against a human like robot. However, they were most active while playing against a human being. Most fun was experienced in the play against the human being. The human's play was considered to be most intelligent. Although this play, as in the other ones, the answers were already determined on forehand.

This experiment shows the importance of stimuli in our experience of the presence of a mind in our opponent.

They also show how misleading stimuli may be. Only in the case of the human player a genuine similarity with the experimental person was present. In all three other opponents (computer, machinelike robot, human like robot) no genuine similarity with us was present.

The more similar a species is in relevant respects with our own species the more reason we have to assume a similarity in feelings.

3.1.5. Evolutionary biology. Sometimes it is suggested that feelings are emerged during the evolution because they have a function [26]. However not everything that emerged during the evolution has a function [27].

3.2. Reasons to deny that a particular species has feelings

3.2.1. Unfamiliarity. A reason to deny that a particular species has feelings may be the unfamiliarity with a species. For instance, many people are unfamiliar with the fact that ear wigs (Dermaptera sp.)

are caring mothers or that some species of cockroaches (Cryptocercus sp.) may stay together as a family for three years [28]. Also, behaviour biologists of the hunters' type are unfamiliar with individual animals. Therefore, they are no animal protectionists but species protectionists. Further, they are accustomed to the cruelty frequently found in nature [29].

3.2.2. The absence of appealing stimuli. A reason to deny the existence of feelings may be that a species hardly has sign stimuli that resemble our own. Hobby fishermen often deny the presence of feelings in fish. Apart from their own interest in this denial it may play a role that fish have hardly sign stimuli that resemble the stimuli of our species: they have no legs, facial expression or gestures like our own. An exception is the seahorse (Hippocampus brevirostris) because of its upright position and parental care. Invertebrates hardly possess sign stimuli that resemble the stimuli of our own species. This absence of stimuli that resemble the stimuli of our species in invertebrates is an important reason for the denial of the existence of feelings in invertebrates.

3.2.3. The absence of stimuli that stimulate help. Some species have developed the capacity to help each other. They have also developed stimuli that may elicit the tendency to protect and take care in other individuals. Such species mostly more easily elicit such feelings in our own species. In many social species, however, it increases the fitness of an individual when it does not ask for help but is able to mask its bad condition as long as possible to prevent expulsion of the group, or worse. In hens a bad condition may result in cannibalism by flock mates.

3.2.4. Aversive stimuli. Great apes (Pongidae sp.) have much in common with us. Nevertheless, they often appear to be "ugly" to us. Their unattractiveness is perhaps the result of an inhibition by morphological and behavioural traits developed to prevent sexual contact. Ten million years ago the ancestors of man and chimpanzee (Pan troglodytes) belonged to the same species [30]. When two species adapt further to each own niche individuals resulting from crosses between two species in almost all cases get an increasingly lower fitness to the parental niches than the individuals of each original species in their own niche. Therefore, signals develop in such situations to prevent further gene exchange. Probably such signals have developed between Homo sapiens and great apes.

Other species possess incidentally stimuli that resemble stimuli that signal disease in our own species, e.g. toad (Bufonidae sp.) skin, bat (Chiroptera sp.) wing bones and skin. Still other species may pose a threat to our health. For instance, snakes, spiders and insects. As adaptation to these threats the stimuli of such species have become aversive, even when these stimuli are present in a completely harmless species. During our evolution also darkness has been an indication of danger. Therefore, we have a tendency to consider nocturnal species, like bats and rats, as aversive.

Aversive stimuli make a species fit into an enemy image. We are apt to deny the existence of feelings in our enemies.

3.2.5. Economic advantage. The main reason for the denial of feelings in another species seems to be that it silences our conscience. Economic advantage seems a strong drive in this respect. It may be found among hunters, fishermen, industrial farmers and animal experimenters. However, this attitude seems largely absent among traditional hunters and farmers.

3.2.6. Lack of similarity. Another reason to deny the existence of feelings in a particular species may be that we are of opinion that the similarity with our own species is too limited. However, this may be a rationalization of an opinion based on emotional arguments.

Science may help us to establish the exact degree of similarity in morphology and behaviour. We may consider behaviour as a "part" of the body. Further I will sometimes refer to this combination as [body and behaviour]. Is science also able to establish the similarity in feelings between different species?

4. Is science able to help us to draw the line?

Van Rooijen pointed out that from epistemology we know that it is impossible to prove that a species has feelings and, thus, to prove that a species is capable of having feelings [27, 31]. Within science only models are acceptable as explanations for behaviour [32]. Opposition against this insight is mostly the result of the psychological influences on applied behaviour science [17]. However, this insight is strongly supported by the fact that until now nobody ever successfully performed an experiment that proved that a particular species has feelings. Otherwise, the majority of the scientific community would already agree on the presence of feelings in the investigated species.

However, scientific research is able to establish the relation between the feelings and bodily structures and/or behaviour in our own species (feelings/body and feelings/behaviour relation). Scientific research is also able to study the bodily structures and behaviour in other species. Van Rooijen suggested that the combination of both studies made [body and/or behaviour]/feelings speculations possible [33]. Similarities in relevant aspects could make the assumption that a particular species has feelings more convincing. Differences in relevant aspects could make the assumption that a particular species has feelings less convincing. This approach stimulates research.

For instance, in humans a bleeding wound is mostly connected with pain, a red painted spot mostly not (feelings/body relation). With the scientific method we are able to establish whether the red spot on the throat of a bleeding-heart pigeon (Gallicolumba luzonica) is caused by blood out of a wound or by red feathers. When a wound is present the presence of pain is likely, when no wound is present pain is unlikely (body/feelings speculation). The same holds for the throat of the cut throat finch (Amadina fasciata). It was a long time believed that the Dalmatian pelican (Pelecanus crispus) wounded itself to feed the young with its own blood. Scientifc research has shown that these pelicans do not possess a wound but red breast feather and a red coloured bill tip. Therefore, it is assumed that they are not in pain (body/feelings speculation).

In humans a broken arm is mostly connected with pain (feelings/body relation). The ringed plover (Charaderius maticula) sometimes behaves as having a broken wing. This species sometimes performs a "broken wing display" to lure predators away from the eggs or chicks. By using the scientific method, we are also able to establish whether a bird really has a broken wing or is acting a broken wing display. Only when the bird really has a broken wing it seems reasonable to assume that these birds indeed experience a feeling of pain (body/feelings speculation).

Another example of such research is the application of choice and preference tests [34]. We make a choice for the alternative that seems the more rewarding or the less punishing to us (feelings/behaviour relation). Scientific research has indicated that pigs are able to make similar choices. The finding that pigs are able to make such choices makes the assumption that they possess feelings more convincing (behaviour/feelings speculation).

Humans are prepared to work to obtain a rewarding or to avoid a punishing factor, and to work more to obtain a more rewarding or to avoid a more punishing factor (feelings/behaviour relation). Also, pigs are prepared to work and to work harder for some items than for other [34]. This seems connected with the reward or punishment they expect to receive from this particular factor (behaviour/feelings speculation). This finding that pigs are able to show preferences also makes the assumption that pigs possess feelings more convincing

Still another example of such research is described by Bateson [35]. From human psychology and psychiatry, it is known that a depressed person will classify a glass as half empty, whereas a happy person will classify the same glass as half full (feelings/behaviour relation).

It was shown that starlings (Sturnus vulgaris) that were recently moved from enriched cages

to standard cages were more apt to interpret an ambiguous stimulus as non-rewarding, while starlings that remained housed in enriched cages were more apt to interpret these same ambiguous stimuli as rewarding.

The similarity between the behaviour of depressed people and the behaviour of the birds that recently experienced a decline in the quality of their housing conditions suggests that these birds also were depressed (behaviour/feelings speculation). The results of these scientific experiments also makes the assumption that starlings possess feelings more convincing.

Mendl performed similar comparisons between human feelings/behaviour relations and rat (Rattus norvegicus) behavior [36]. The resulting behaviour/feelings speculations made the assumption that rats possess feelings more convincing.

Braithwaite described examples of research on the feelings-body relation in humans and the body-feelings relation in fish [37]. In the human brain the amygdala is a structure which processes emotional information. A similar structure is found in the fish brain. Lesions in this last structure impair escape- and avoidance learning in fish. We may assume that these lesions influence feelings in fish too ([body and behaviour]/feelings speculation).

In the human brain the hippocampus is a structure associated with learning and memory (feelings/[body and behaviour] relation). A structure comparable to this hippocampus is also found in the fish brain. Lesions in this last structure impair also in the fish spatial and temporal learning and memory tasks ([body and behaviour]/feelings speculation).

In human beings specialised receptors registrate pain (feelings/body relation). Similar receptors are found in the rainbow trout (Oncoryhnchus mykiss) [38]. These scientific findings suggest that at least particular fish species know pain (body/feelings speculation). This makes the presence of feelings more convincing.

In humans beings a local anaesthetic suppresses the pain (feelings/body relation). Elwood treated one antenna of glass prawns (Dendrobranchiata sp.) with a local anaesthetic, in the control prawns one antenna was treated with water [39]. The same antenna was then treated with a chemical abuse or water. Those receiving chemical abuse groomed the treated antenna more and rubbed it on the side of the tank. However, this response was reduced in the antenna that were pre-treated with a local anaesthetic instead of water. These results make the presence of pain perception in glass prawns more convincing (behaviour/feelings speculation).

These examples show that science is able to help us to draw the line. However, the title of Elwood's contribution was "Pain and stress in crustaceans?" (Braithwaite's contribution was called "Do fish feel pain?") This implied that the question marks indeed remained. The decision whether a species has feelings or not remains in the end a matter of personal taste.

Although it is, within science, only possible to approach the proof that particular animal species have feelings (and other species not) we have to talk about feelings of farm animals for ethical reasons [9]. The protection of farm animals is justified to avoid suffering in case these farm animals possess feelings. I have to mention that this view is able to convince scientists that believe that no other animal species than man possesses feelings. Kennedy takes, according to himself, an "uncompromisingly negative view of animal consciousness." [2]. Nevertheless, he described my opinion that it is impossible to prove that species possess feelings but that it nevertheless is possible to perform welfare research as "This would seem to be a realistic position to adopt" (p.115).

5. A short discussion of the views of different scientists

Kennedy is of opinion that man is a unique species [2]. Evidence for the assumption of the existence of feelings in other species is already elaborately mentioned in this article. Kennedy is probably right that Homo sapiens sapiens is "unique" because of its variety of rich cultures [2]. However, even culture did not emerge out of the blue. Bonner defined culture as the transfer of information by behavioural means [40]. Following this definition culture is common throughout the animal kingdom. Different groups of killer whales (Orcinus orca) developed different vocal and behavioural cultures of a high complexity [41]. Field studies have shown that also groups of wild chimpanzees (pan troglodytes) spontaneously developed locally different cultures of a high complexity [42]. Lycett et al. made a phylogenetic tree of chimpanzee cultures [43]. This tree did not coïncide with the genetic relation between the different groups of chimpanzees. The reason for this distinction is that young adult females that migrate from one group to an adjacent group transfer their genes to this new group but not their culture: They conform themselves to the habits and traditions of their new social environment.

Whiten et al. were able to create experimentally two different traditions in two different groups of laboratory chimpanzees (pan troglodytes) [44]. In each of two groups of chimpanzees only the dominant animal was trained to obtain food. Both dominant animals were trained to use another method to obtain the food. The other members of each group copied these strategies. After two months both groups still used only its own solution. In a third, control, group, without any trained animal, nobody found any solution.

Wild chimpanzees use sticks to "fish" termites out of their holes. Only wild chimpanzees in the Goualougo Traingle, Replublic of Congo, applied, prior to fishing, a set of deliberately, distinguishable actions to modify herb stems into a brush-tipped probe. Experiments showed that such brush-tipped probes collect more insects [45].

Fiorito and Scotto presented two identical balls, but differing in colour, to octopuses (Octopus vulgaris) [46]. The octopuses were trained to attack only the ball with a particular colour. After training such attacks were observed by naive octopuses. After isolation these observer octopuses attacked the same coloured ball. The result of this experiment was independent of the used colour. This could be called culture (in the sense of transfer of information by behavioural means).

Another example of culture is honey bees (Apis mellifera) scouts that with the "waggle dance" provide information about the location of potential nest sites. After some time, some dancers give up their own dance and join the dance of another dancer (they become "convinced" by this other dancer that she dances toward a better site). Within one-hour unanimity is reached and the group heads for the "democratically" chosen site [47].

However, the most important question, in this context, is not whether species possess culture but whether or not man is unique in the sense that man is the only species with feelings as suffering and pain. I do not belief that this is the case.

I disagree also with Bermond [4] that complex bodily structures as prefrontal cortex and a right neocortical hemisphere are necessary to be able to experience feelings as suffering or pain. Damasio (2010) mentioned that in humans sometimes children are born without a cerebral cortex, they only have a brain stem. Nevertheless, these children react with joy when they are tickled and they clearly prefer one piece of music over another one. According to Damasio [48] poses such children a "self".

Tomer et al. already found a precursor of the cerebral cortex in bristle worms (Platynereis dumerilii) [49]. These worms possess mushroom-like brain structures, the corpora pedunculata, that process sensory information. These structures are probably developed to find food with help of smell. They are supposed to exist for about 600.000.000 years. Tomer et al. used "cellular profiling by image registration" to establish genexpression [49]. The expression of essential genes in corpora pedunculata resembles very much the expression in the pallium (the embryonic precursor of the cerebral cortex in vertebrates).

I consider suffering and pain as such basic feelings that complex bodily structures are not necessary. Locomotion in lower animals is also connected with other body structures than locomotion in man [50]. Although snails have no structures comparable with our legs, they are still able to move autonomously.

In humans' feelings are processed in the limbic system. In a part of this limbic system, the amygdala, incoming information elicits especially negative feelings. This amygdala system is evolutionary much older than the prefrontal cortex. A precursor of this structure, as already mentioned above, is even present in fish. The prefrontal cortex has an inhibitory top-down control of amygdala function, resulting in, for humans, contextually appropriate "emotional" behaviour.

Subhuman animals perform predominantly behaviour that in humans is linked with feelings (emotions). For instance, parental, aggressive, flight and sexual behaviour. Behaviour that is linked with thinking (ratio) is rarer in subhuman animals, but not absent. For instance, Yerkes and Coburn stated that domestic pigs had "an approach to free ideas" that these researchers had not in any way expected [50]. Hunters ascribe to wild swine an "almost humanlike logical ability" [51]. Above is mentioned that naïve dogs that had observed a dog obtaining a reward by lowering a stick with its paw imitated this behaviour. However, when the demonstrator dog lowering the stick with his paw had a ball in its mouth the naïve dogs used their mouth. Then the naïve observer dogs apparently "understood" why the demonstrator dog used his paw [21]. This is an example of thinking.

Kaminski et al. described a case of reasoning by exclusion [18]. When a particular owner asked his border collie collect to an object while using a novel word the dog collected the novel toy, instead of one of the familiar ones. Lorenz described animals, in terms derived from daily life, as "emotional people with little cognitive abilities" [52].

After 35 hours of sleep deprivation the prefrontal cortex does no longer function properly in experimental persons. Such people react more "emotional": that means stimuli elicit more easily crying or attack behaviour [53]. Their behaviour has become more animal-like. This implies that the prefrontal cortex is less important for subhuman animals, especially not as substrate for feelings.

Morphological structures connected with the experience of feelings seem already present in species lower on the evolutionary scale than fish. In humans sexual behaviour is connected with a pleasant feeling (feelings/body relation). Snails (Helix sp.) learned to touch a rod in order to electrically stimulate an area in their own brain. This area is normally active during their sexual behaviour, and probably gives a pleasant feeling ([body and behaviour]/feelings speculation) [54].

In Wiepkema's opinion, vertebrates have emotions derived from the difference between what is and what should be and emotions corresponding to how well an organism trusts its means to restore or maintain its homeostasis [5].

For a critique on Wiepkema's ideas [5] about the relation between homeostasis mechanisms and feelings see Van Rooijen [55]. More important, within this context, is that also invertebrates possess homeostasis mechanisms.

Wiepkema claimed that vertebrates have emotions because they show so-called emotional behaviours or expressions [5]. This is an example of circular reasoning. When behaviours (expressions are also behaviours) in animals are labelled as "emotional" this does not prove the existence of animal emotions. However, one could argue that in humans particular behaviours are connected with emotions (feelings/behaviour relation). Therefore, one could speculate that similar behaviours in animals are also connected with emotions (behaviour/feelings speculation).

Further Wiepkema claimed that vertebrates perform species-specific goal directed behavior [5]. However, invertebrates also perform goal-directed behaviours. A well-known example of unicellular organisms that direct themselves towards a goal are sperm cells: they direct themselves toward egg cells. This implies that the presence of such goal-directed behaviours in vertebrates are no reason to restrict the presence of feelings to vertebrates only.

The decision whether a species has a feeling or not remains in the end a matter of personal taste. Because of my research with crab spiders and with fruit flies I have some familiarity with spiders and insects. Above (III.1.3) I have mentioned the importance of familiarity for our tendency to accept the presence of feelings in a particular species. My familiarity is one of the reasons why I have the conviction that spiders and insects have feelings. I have defended this conviction at a workshop at the International Ethological Conference in 1999 in Bangalore, India. Probably to test this view Sherwin performed an elaborate study of publications on invertebrates as cockroaches, flies and slugs [56]. These species turned out to have short- and long-term memory, age effects on memory, complex spatial, associative and social learning and were able to perform appropriately in preference tests and demand studies, and to exhibit behavioural and physiological responses indicative of pain, and to experience learned helplessness. This author concluded:" This indicates that we should either be more cautious when using argument-by-analogy, or remain open-minded in the possibility that invertebrates are capable of suffering in a similar way as vertebrates."

Even Lady Slippers (paramecium caudatum) are able, by using a voltage as a reinforcement, to learn to discriminate between different brightness levels [57]. This means that scientific results strengthened my conviction that all animal species possess feelings. One has to realize that even in the case of primates the assumption that they possess feelings is in the end a matter of personal taste.

6. Conclusion.

Scientists vary in their opinion about the border between species with and without feelings. In my opinion it is reasonable to assume that all animal species possess feelings. Further research is not able to prove this opinion. It is only able to approach, or leave, this proof by making this opinion more, or less, convincing. Although not all scientists share the view that all animal species possess feelings, many scientists are of the opinion that at least farm animals possess feelings. Laypeople that are not influenced by economic or other interests mostly do not question this opinion. The protection of farm animals is justified to avoid suffering because the possibility exist that these animals possess feelings.

References

[1] Van Rooijen J 1981 Are feelings adaptations? The basis for modern applied ethology. *Appl. Anim. Ethol.*, **7**, 187-9.

[2] Kennedy J S 1992 The New Anthropomorphism. Cambridge, United Kingdom: University Press.

[3] Geertz C 1975 The growth of culture and the evolution of mind. 55-83. In C Geertz, (Editor). The interpretation of cultures: selected essays. London, UK: Hutchinson.

[4] Bermond B 1997 The myth of animal suffering. In: M Dol, S Kasanmoentalib, S Lijmbach, E Rivas and R van den Bos (Editor). Animal Consciousness and Animal Ethics. Perspectives from the Netherlands. Animals in Philosophy and Science, **I**, (125-143). Assen, The Netherlands: Van Gorcum.

[5] Wiepkema P R 1997 The emotional vertebrate. In M Dol, S Kasanmoentalib, S Lijmbach, E Rivas and R van den Bos (Editor). Animal Consciousness and Animal Ethics. Perspectives from the Netherlands. Animals in Philosophy and Science Volume **I**, (93-102). Assen, The Netherlands: Van Gorcum.

[6] Mather J A 1991 Personalities of octopuses. Abstracts 22nd International Ethological Conference, (138) 22-29 August 1991. Kyoto, Japan: Otani University.

[7] Griffin D R 1976 The Question of Animal Awareness. The Rockefeller University Press. New York.

[8] Van Rooijen J 2001 Possibilities and limitations of the study of the animal mind. *Anim. Welf.* 2001, **10**, 248-9.

[9] Van Rooijen J 1985 Du-Evidenz, applied ethology and animal welfare. *Brit. Vet. Journ.*, **141**, 245-8.

[10] Botvinick M, Jha A, Bylsma L M, Fabian S A, Solomon P E and Prkachin K M 2005 Viewing facial expressions of pain engages cortical areas involved in the direct experience of pain. *NeuroImage*, **25**, 312-9.

[11] Heinroth O 1941 Pfauen und Truthahnbalz. *Zeitschr. f. Tierpsych.*: vergleichende Verhaltensforschung, **4**, 330.

[12] Lorenz K 1950 Ganzheit und Teil in die tierischen und menschlichen Gesellschaft. Reprinted in K Lorenz (Editor) (1965). Über tierisches und menschliches Verhalten. Gesammelte Abhandlungen. Band II, 114-95. München, Germany: Piper & Co.

[13] Schultz H 1962. Hombu: Urwaldleben der brasilianischer Indianer. Stuttgart, Germany: Belser.

[14] Nicastro N and. Owren M 2003 Classification of domestic cats (Felis catus) vocalizations by naïve and experienced human listeners. *Journ. Comp. Psych.*, **117**, nr 1, 44-52.

[15] Nicastro N 2004 Perceptual and acoustic evidence for species-level differences in meow vocalizations by domestic cats (Felis Catus} and african wild cats (Felis silvestris lybica). *Journ. Comp. Psych.*, **118**(3), 287-96. http://dx.doi.org/10.1037/0735-7036.118.3.287

[16] Pongracz P, Molnar C, Miklosi A and Csanyi V 2005 Human listeners are able to classify dog (Canis familiaris) barks recorded in different situations. *J. Comp. Psych.,* **119**, nr 2, 136-44.

[17] Van Rooijen J 1991 The value of a philosophical fundament for animal welfare research. *Appl. Anim. Behav. Sci.*, **28**, 387-8.

[18] Kaminski J, Call J and Fisher J 2004 Word learning in a domestic dog: evidence for "fast mapping". *Science*, **304**, 1682-3. DOI: 10.126/science1097859

[19] Hare B and Tomasello M 2005. Human-like social skills in dogs? *Trends Cogn. Sci.*, **9**, 439-44.

[20] Van Rooijen J 2009 Do dogs and bees possess a 'theory of mind'? *Anim. Behav.*, 79, 2, e7-e8. doi:10.1016/j.anbehav.2009.11.016434

[21] Range F, Viranyi Z and Huber L 2007 Selective imitation in domestic dogs. *Curr. Biol.*, **17**, Issue

10, 868-72.
[22] Osvath M 2009 Spontaneous planning for future stone throwing by a male chimpanzee. *Curr. Biol.,* **19**, 5, R190-R191. doi: 10.1016/j.cub.2009.01.010
[23] Taylor A H, Hunt G R, Holzhaider J C and Gray R D 2007. Spo7.
[24] Auersperg A M F, Gajdon G K and Huber L 2010 Kea, Nestor notabilis, produce dynamic relationships between objects in a second-order tool use task. *Anim. Behav.*, **80**(5), 783-9.
[25] Krach S and Hegel F 2008 Third International Conferences of leading philosophers. **III**, Part 7, (62). Routledge.
[26] Costa J T 2006 The other insect societies. Harvard, USA: Universities Press.
[27] Van Rooijen J 1990 Backgrounds of students of behaviour in relation with their attitude toward the issue of animal well-being. *J. Agric. Eth.*, **2**, 235-40.
[28] Patterson N, Richter D J Gnerre S, Lander E S and Reich D 2006 Genetic evidence for complex speciation of humans and chimpanzees. *Nature*, June 29, **441**, 1103-8.
[29] Van Rooijen J 1983 Awareness and Griffin's circular reasoning. *Anim. Behav.* **31**, 613-4.
[30] Van Rooijen J 1983/1984 Preference tests, motivations, models and welfare. (Guest Editorial) *Appl. Anim. Ethol.*, **11**, 1-6.
[31] Van Rooijen J 1997. Suffering and Well-being and the Study of Behaviour. In M Dol, S. Kasanmoentalib, S Lijmbach, E Rivas and R van den Bos (Editor). Animal Consciousness and Animal Ethics. Perspectives from the Netherlands. Animals in Philosophy and Science Volume **I**, (114-24). Assen, The Netherlands: Van Gorcum.
[32] Van Rooijen J 1982 Operant preference tests with pigs. *Appl. Anim. Ethol.*, **9**, 87-8.
[33] Bateson M 2007 Optimism and pessimism in captive European starlings: Interpretative biases as novel indicators of animal welfare. Animal Suffering and Well-being. International Symposium on the State of the Science. 21st IGN-meeting, September 20-21, 2007, (47). Giessen, Germany: International Society of livestock Husbandry.
[34] Mendl M 2007 Cognitive measures of emotional states. Animal Suffering and Well-being. International Symposium on the State of the Science. 21st IGN-meeting, September 20-21, 2007, (45-6). Giessen, Germany: International Society of livestock Husbandry.
[35] Braithwait V 2007 Do fish feel pain? Animal Suffering and Well-being. International Symposium on the State of the Science. 21st IGN-meeting, September 20-21, 2007, (37). Giessen, Germany: International Society of livestock Husbandry.
[36] Sneddon L U, Braithwaite V A and Gentle M J 2003 Do fishes have nociceptors: evidence for the evolution of a vertebrate sensory system. London: *Proc. Royal Soc. B*, On line, 02PB1068.1 - 02PB1068.6
[37] Elwood R W 2007 Pain and stress in crustaceans? Animal Suffering and Well-being. International Symposium on the State of the Science. 21st IGN-meeting, September 20-21, 2007, (39). Giessen, Germany: International Society of livestock Husbandry.
[38] Bonner J 1980 The Evolution of Culture in Animals. Princeton University Press, Princeton, NJ.
[39] Rendell L and Whitehead H 2001 Culture in Whales and Dolphins. *Behav. Brain Sci.*, **24**, 2, 302-82.
[40] Whiten A, J Goodall, W C McGrew, T Nishida, V Reynolds, Y Sugiyama, C E G Tutin, R Wrangham W and. Boesch C 1999 Cultures in Chimpanzees. *Nature*, **399**, 682-5.
[41] Lycett S J, Collard, M and McGrew, M C 2007. From the Cover: Phylogenetic analyses of behavior support existence of culture among wild chimpanzees. *Proc. Nat. Acad. Sci. USA.*, **104** (45), 17588-92.
[42] Whiten A, Horner V and de Waal F B M 2005 Conformity to cultural norms of tool use in chimpanzees (letter) *Nature* (29 september), **437**, 737-40.
[43] Sanz C, Call J and Morgan D 2009 Design complexity in termite-fishing tools of chimpanzees (Pan troglodytes). *Biol. Letters.* Published online before print. DOI: 10.1098/rsbl/.2008.0786
[44] Fiorito G and Scotto P 1992. Observational learning in Octopus vulgaris. *Science,* **256**, 545-7.
[45] Seeley T D and Buhrman S C 1999 Group decision making in swarms of honey bees. *Behav.*

Ecol. and Sociobiol., **45**, 19-31.
[46] Damasio A 2010 Self Comes to Mind — Constructing the Conscious Brain: The Evolution of Consciousness, Heinemann.
[47] Tomer R, Denes A, Tessmar-Raible K and Arendt D 2010 Profiling by image registration reveals common origin of annelid mushroom bodies and vertebrate pallium. *Cell,* **142**(5), 800-9.
[48] Yerkes R M and Coburn, C A 1915 A study of the behavior of the pig by the mutiple choice method. *J. Anim. Behav.* **5**, 185-225.
[49] Snetlage K 1957 Das Schwarzwild. Hamburg, Germany: Parey.
[50] Lorenz K 1980 Tiere sind Gefühlsmenschen. *Der Spiegel,* **47**, 251-264.
[51] Yoo S-S, Gujar N, Hu P, Jolesz F A and Walker M P 2007 The human emotional brain without sleep - a prefrontal amygdala disconnect. *Curr. Biol.*, **20** (17), 25 October, R877-8
[52] Balaban P and Maksimova O A 1993 Positive and negative brain zones in the snail. *Europ. J. Neurosci.*, **5**, 768-74.
[53] Van Rooijen J 1984 (Guest editorial) Impoverished environments and welfare. *Appl. Anim. Behav. Sci.*, 12, 3-13.
[54] Sherwin C M 2001 Can invertebrates suffer? Or how robust is argument-by-analogy? *Anim. Welf.* **10**, 103-18.
[55] Armus H, Montgomery A R and Jellison J L 2006 Discrimination learning in paramecia (P. caudatum), *Psychol. Rec.*, **56**, 489-98.

IOP Conf. Series: Earth and Environmental Science **346** (2019) 012092 doi:10.1088/1755-1315/346/1/012092

Physiological and ethylene accumulation responses of cassava under drought stress

P Ogaddee[1]and T Girdthai[1, 2]

[1]School of Crop Production Technology, Institute of Agricultural Technology, Suranaree University of Technology, Thailand

E-mail : teerayoot@sut.ac.th

Abstract. Drought is a major production constraint of cassava worldwide. The objective of this study was to evaluate the chemo-physiological and phytohormone responses of cassava under water stress conditions. Five cassava (Manihot esculenta) varieties were grown in greenhouse conditions under different water regimes (field capacity (FC), 2/3 and 1/3 available water (AW)). The physiological responses to drought were monitored. Plant height, relative water content (RWC) and SPAD Chlorophyll meter reading (SCMR) were recorded at 3, 4 and 5 months after planting. Ethylene accumulate were recorded at 1, 2 and 3 weeks after withholding water at 2 months after planting. The result revealed that plant height and RWC decreased significant difference under water deficit. Plant height and RWC under 1/3 AW had lowest followed by 2/3 AW and FC conditions. Under 1/3 AW, ethylene and SCMR increased significant difference followed by 2/3 AW and FC conditions. Thus, plant height, RWC, SCMR and ethylene accumulation could be used as a criterion in breeding programs to improve drought tolerance in cassava.

1. Introduction

Cassava scientific name is Manihot esculenta. Origin from South America [1]. Cassava an important industrial crop for starch, biofuel production and animal feed [2]. Cassava is classified as a drought-resistant crop, extensively cultivated as an annual crop in tropical and subtropical regions. The planted area for cassava is mainly under rain-fed conditions. An indirect effect of global warming can higher plant water demand due to increased transpiration at water limitation making drought conditions more severe. [3, 4]. Cassava as resistance to drought tolerance and low nutrient soil. It can grow in the area receiving 300 mm of rain annually [5]. Abiotic stress, such as drought, heat, cold and salinity are the cause of reducing plant growth and crop yields quality. Drought is serious threats to plant, in addition to physiological hormones are rapidly occurring responses to drought stress [6]. Drought is multidimensional stress, affect cell organs until the whole plant [7].

Several hormones, such as ethylene has been reported to associate with stress signaling [8]. Ethylene gas is a plant hormone produced by plant tissue and effect to physiological metabolism including fruit ripening leaf senescence and senescence [9, 10]. Under water deficit also increases ethylene accumulation [11] and leaf abscission occurs as the result of hormone signaling by ethylene consequently reduced water loss [12].

Measuring leaf water content is a simple way to evaluate the water status of leaf and important parameter to the determination of plant drought tolerance [13]. Leaf water content is an attribute for discriminating tolerance and sensitive genotypes. Hight relative water content is a resistant mechanism

Content from this work may be used under the terms of the Creative Commons Attribution 3.0 licence. Any further distribution of this work must maintain attribution to the author(s) and the title of the work, journal citation and DOI.
Published under licence by IOP Publishing Ltd

IOP Conf. Series: Earth and Environmental Science **346** (2019) 012092 doi:10.1088/1755-1315/346/1/012092

to drought stress and more osmotic regulation [14]. RWC has been proposed as a selection criterion for drought tolerance in many crops such as wheat [15] and potato [16].

The chlorophyll content is closely linked with photosynthetic systems and drought is one of the most important factors constraining plant photosynthesis, stomatal and metabolic limitations [17]. Recent research has shown that the chlorophyll content can be measured using the SPAD chlorophyll meter reading [18]. SPAD-502 chlorophyll meter reading is a tool that measures the greenness of leaves by the leaf transmission of red light and infrared light [19]. SCMR is easy to use, stable and low cost [20].

Therefore, this study aimed to verify the effect of genotypes, water regimes and plant age on parameters related to cassava breeding program. Understanding of tolerance mechanisms under drought stress in cassava is a key in developing tolerance varieties and increased yield.

2. Materials and methods

2.1. Plant materials experimental design

The experiment was arranged in 3 X 5 Factorial in randomized complete block design (RCBD) with four replications. Five cassava varieties including Rayong 9, Rayong 90, Kasetsart 50, Huay Bong 80 and Hanatee were assigned as factor A. 3 different water levels [field capacity (FC), 2/3 Available Water (AW) and 1/3 AW] were arranged as factor B. The experiment was conducted under greenhouse conditions at Suranaree University of Technology Farm, Nakhon Ratchasima province, Thailand. The Cassava were grown in 20-L plastic plots. The plant spacing was 1m x 1m between plots. The drought conditions in the greenhouse were created by soil moisture content. In FC treatment, fully irrigated conditions were applied throughout the experiment for maintaining soil moisture levels at 28.06 %. In 2/3 and 1/3 AW conditions, soil moisture content was maintained at 16.84% and 11.22 %, respectively.

2.2. Leaf retention

Leaf retention was monitored by counting the number of leaves on the ground. Data collected at 1, 2 and 3 weeks after withholding water at 2 months after planting.

2.3. Ethylene accumulation

Ethylene gas on fresh leaves was collected weighted and placed into 70 ml. tube. After incubating at room temperature for 48 h. 1 ml. of the gas sample was injected into gas chromatography (GC). Data collected at 1, 2 and 3 weeks after withholding water at 2 months after planting.

2.4. Relative Water Content (RWC)

Relative water content (RWC) using the fully expanded leaf from the top of the plant, RWC was accorded by calculated of leaf fresh weight, dry weight, and turgid weight. Mature leave (FW) sample was recorded and soaked overnight in distilled water at room temperature, turgid weight (TW) was recorded and oven dried at 75°C for 24 h (DW) Collected at 3,4 and 5 months after planting.

RWC (%) was calculated as $[(FW - DW) / (TW - DW)] \times 100$

2.5. SCMR

Collected at 3,4 and 5 months after planting using SPAD-502 meter (Minolta Konica Co.Ltd., Japan). The SCMR measurement was recorded three times at different positions from the expended leaf and average for each plot at 09:00 – 10:00 AM

2.6. Statistical analysis

The physiological and biochemical parameters were statistically analysed for Analysis of variance (ANOVA). The significance of difference among the treatment was performed by Tukey's HSD (Honestly Significant Difference) test. ($p = 0.05$) using statistic 8.

3. Results

The averages of soil moisture content field during the experiment at 0–10 cm depth shown in figure 1. It was expected that the soil water content at FC conditions has a higher compared to the 2/3 and 1/3 AW conditions. The water content started to differentiate after no irrigation after 2 months after planting. The FC, 2/3 and 1/3 AW conditions of soil moisture were 26.02, 22.30 and 19.30% respectively.

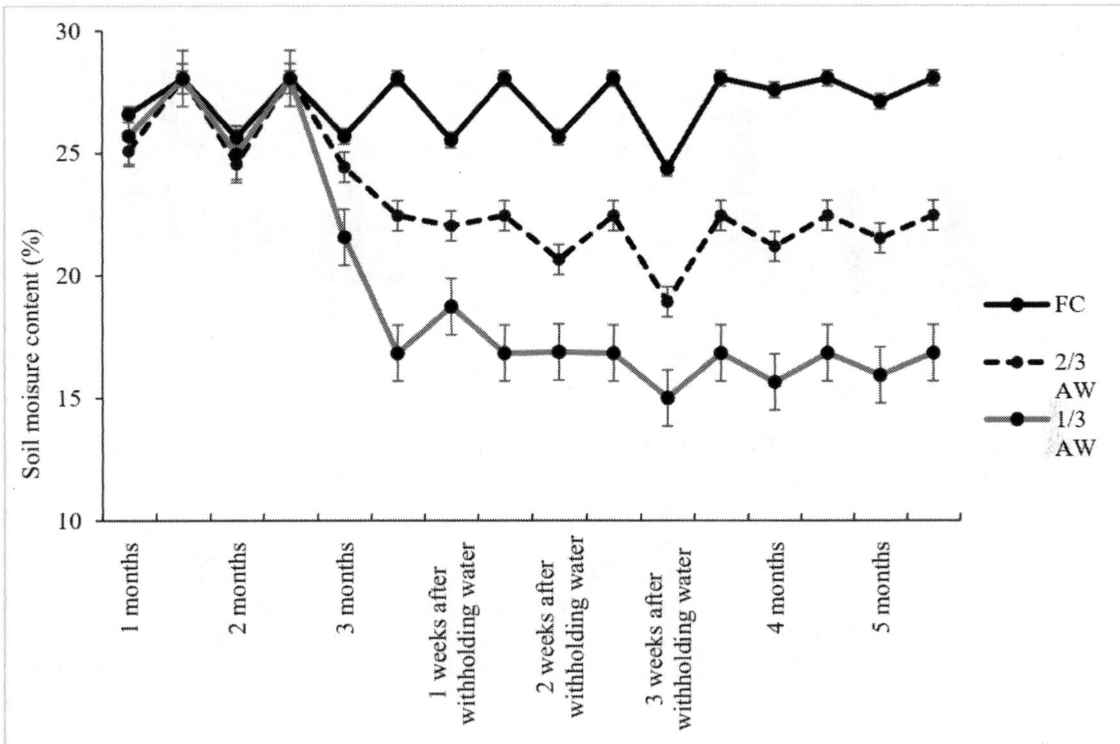

Figure 1. The average value of soil moisture at the FC, 2/3 and 1/3 AW conditions by gravimetric methods

3.1. Ethylene accumulation

Water stress also affected to ethylene accumulation under 1/3 AW was an increase when compared to 2/3 AW and FC conditions. Ethylene accumulation under FC conditions was low ranging between 0.75 – 1.30 µmol $g^{-1}FW\ h^{-1}$ at 1, 2 and 3 weeks after withholding water. Ethylene accumulation increased to a maximum rate at 3 weeks after withholding water reached 4.56 µmol $g^{-1}FW\ h^{-1}$ at 3 weeks after withholding water (figure 2a). The highest ethylene accumulation of 3.62 µmol $g^{-1}FW\ h^{-1}$ was recorded for KU50 and the lowest ethylene accumulation of 1.73 µmol $g^{-1}FW\ h^{-1}$ for HB50 at 3 weeks after withholding water, but was not a significant difference between Rayong 9, Rayong 90 and Hanatee (figure 2b)

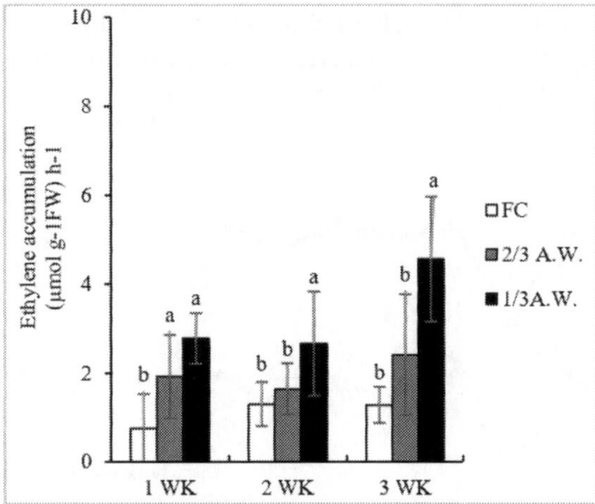

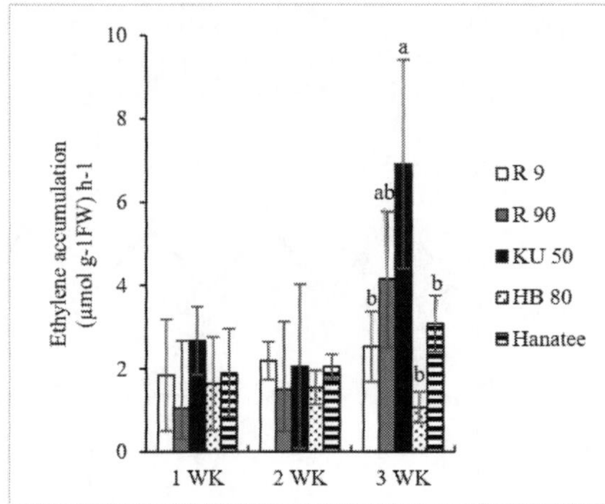

(a) (b)

Figure 2. Effect of water on ethylene accumulation at 1, 2 and 3 weeks after withholding water at 2 months after planting.

3.2. Leaf retention

At 1, 2 and 3 weeks after withholding water leaf retention of the five-cassava variety was significantly affected by water deficit with the hight (7 leaves) and the lowest (3 leaves) were obtained at 3 weeks after withholding water (figure 3a). Rayong 90 had the highest leaf retention (8 leaves) followed by Huay Bong 80, Rayong 9, Kasetsart 50 and Hanatee, but no significant difference between them. (figure 3b)

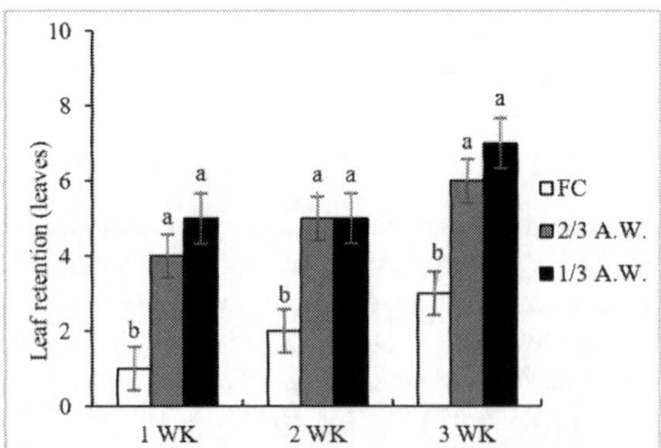

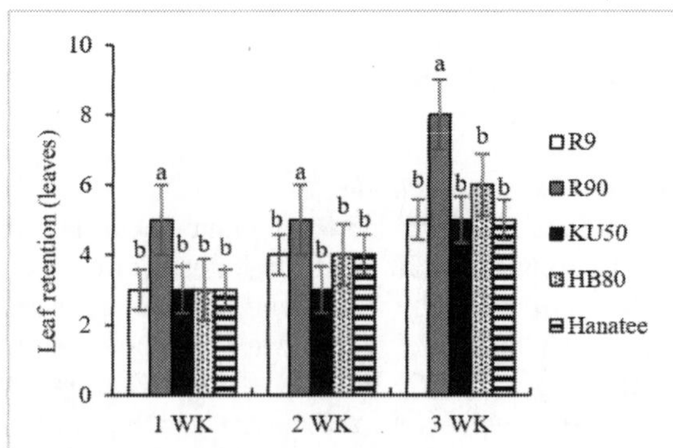

(a) (b)

Figure 3. Effect of water on Leaf retention at 1, 2 and 3 weeks after withholding water at 2 months after planting.

3.3. Relative Water Content (RWC)

A significant difference was obtained for RWC among the water regimes. Under the drought stress groups induced significant decreases in RWC in all varieties, RWC at 3,4 and 5 months after planting under 1/3 AW were higher than 2/3 AW and FC conditions (figure 4a). 3 months after planting was no significant difference in all variety. however, this difference became significant at 4 and 5 months after planting. Rayong 9 had the highest RWC (86.18 and 87.26 %) follow with Rayong 90 (83.93 and 84.91 %) figure 4b

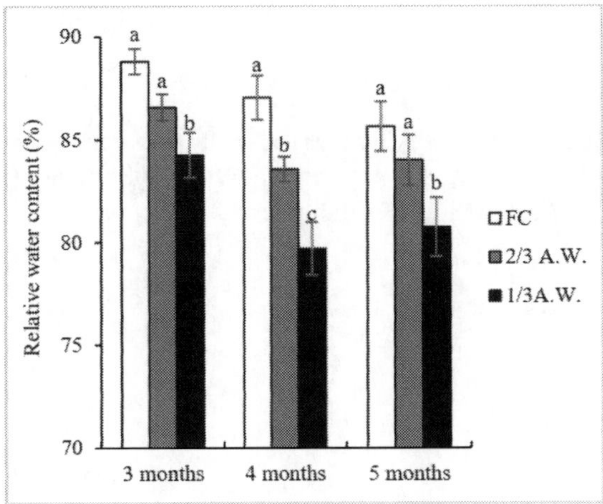

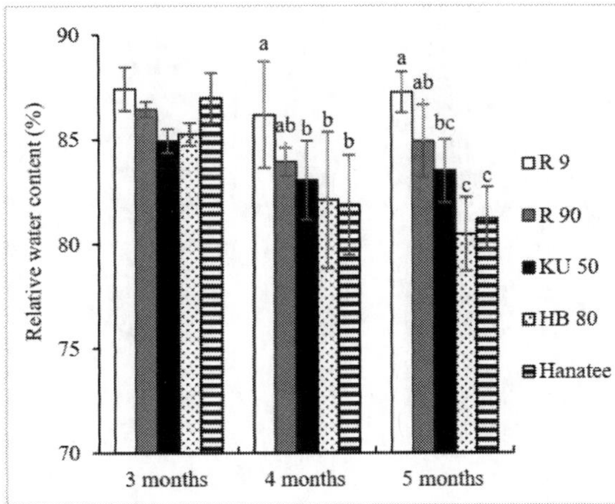

(a) (b)

Figure 4. Effect of water on Relative water content (RWC) at 3, 4 and 5 months after planting.

3.4. SPAD chlorophyll meter reading (SCMR)

Analysis of variance (ANOVA) showed that under drought stress regimes had a significant effect on trits. Results showed SPAD was increased by increasing water restriction and at 3, 4 and 5 months after planting, SCMR under 1/3 AW conditions have a highest (figure 5a) Rayong 9 has a hightest SCMR follow by HB80 and Hanatee has the lowes. Hanatee had a light green leaf color, but Rayong 9 had a dark green leaf color. (figure 6))

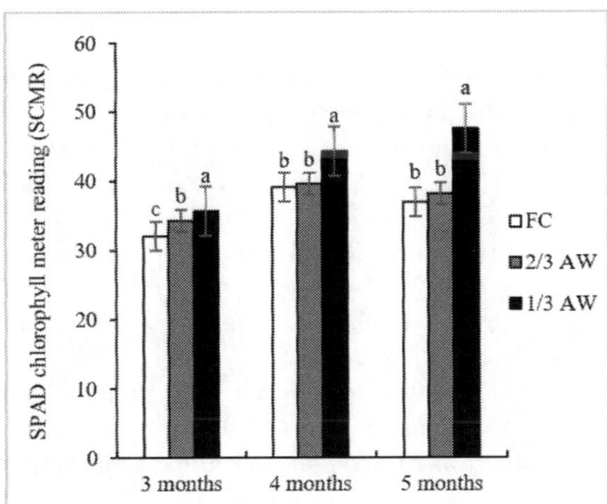

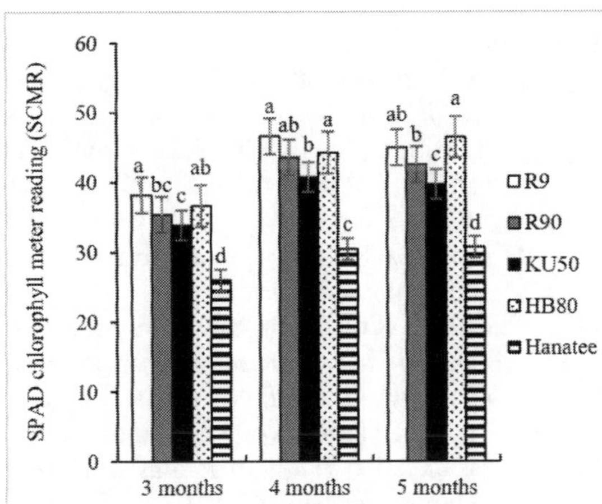

(a) (b)

Figure 5. Effect of water on SCMR at 3, 4 and 5 months after planting.

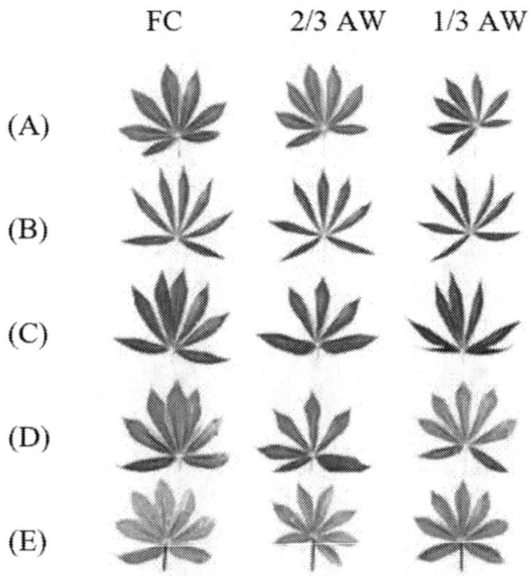

Figure 6. Five cassava leaves under 3 water regimes. (A) Rayong 9, (B) Rayong 90, (C) Huaybong 80, (D) Kasetsart 50 and (E) Hanatee

4. Conclusion

In a review of the hormonal regulation of abscission [21] concluded that auxin is the only one that has been shown to function in preventing leaf abscission. Drought stress is greatly affected by leaf fall. There was also the possibility that leaf fall was related to an ethylene accumulation. However, water deficit also affected to the SPAD and RWC within the first 3 months after planting under water deficit conditions, Therefore, this study supported the screening period of 5 months for cassava under water deficit

Acknowledgements

The authors gratefully thank to Suranaree University of Technology, Thailand for providing the funding and research facilities support and Thai Tapioca Development Institute (TTDI) (Huay Bong) at Huaybong, Dankhuntod district, Nakornratchasima province, Thailand for supporting all cassava varieties.

References

[1] Kuete V 2014 *Physical hematological and histopathological signs of toxicity induced by african medicinal plants* Toxicological Survey of African Medicinal Plants. pp 635-57

[2] Fu L, Ding Z, Han B, Hu W, Li Y and Zhanh J 2016 Physiological investigation and transcriptome analysis of polyethylene glycol (PEG)-induced dehydration stress in cassava *Int. J. Mol. Sci.* **17** 1-18

[3] Howeler H R 2002 Cassava mineral nutrition and fertilization. CIAT retinol office in Asia, department of agriculture

[4] Pareek A, Sopory S K and Bohnert H J 2010 *Abiotic stress adaptation in plants: physiological, molecular and genomic foundation* springer, Dordrecht p 526

[5] Olubanjo O O, Ollifayo A A and Oguntunde P G 2011 Water use, growth, and yield of drip irrigated cassava in a humid tropical environment *Soil Water Res.* **6** 10-20

[6] Sara I Z 2016 *Plant strategies to deal with a combination of drought and high temperatures* Universitat Jaume I p 195

[7] Vurukonda S S, Vardharajula S, Shrivastava M and SkZ A 2016 Enhancement of drought stress tolerance in crops by plant growth promoting rhizobacteria *Microbiol. Res.* **184** 13-24

[8] Müller M and Munné-Bosch S 2015 Ethylene response factors: a key regulatory hub in hormone and stress signaling *Plant Physiol.* **169** 32-41

[9] Iqbal N, Khan N A, Ferrante A, Trivellini A, Francini A, Khan MIR 2017 Ethylene role in plant

growth, development and senescence: interaction with other phytohormones *Front. Plant Sci.* 1-19

[10] Arraes F B, Beneventi M A, Lisei de Sa M E, Paixao J F, Albuquerque E V, Marin S R, Purgatto E, Nepomuceno A L and Grossi-de-Sa M F 2015 Implications of ethylene biosynthesis and signaling in soybean drought stress tolerance *BMC Plant Biol.* **15** 1-20

[11] Wilkinson S and Davies W 2009 Drought, ozone, ABA and ethylene: New insights from cell to plant to community *Plant Cell Environ.* **33** 510-525

[12] Liao W, Wang G, Li Y, Wang B, Zhang P and Peng M 2016 Reactive oxygen species regulate leaf pulvinus abscission zone cell separation in response to water-deficit stress in cassava *Sci. Rep.* **6** 1-17

[13] Jin X, Shi C, Yu CY, Yamada T and Sacks E J 2017 Determination of leaf water content by visible and near-infrared spectrometry and multivariate calibration in miscanthus *Plant Sci.* **8** 1-8

[14] Canavar Ö, Götz K P, Ellmer F, Chmielewski F M and Ali Kaynak M 2014 Determination of the relationship between water use efficiency, carbon isotope discrimination and proline in sunflower genotypes under drought stress *Plant Sci.* **8** 232-242

[15] Arjenaki F G, Morshedi A and Jabbari R 2012 Evaluation of drought stress on relative water content, chlorophyll content and mineral elements of wheat (*Triticum aestivum* L.) varieties *Int. J. Agricul. Crop Sci.* **4** 726-729

[16] Soltys-Kalina D, Plich J, Strzelczyk-Żyta D, Śliwka J and Marczewski W 2016 The effect of drought stress on the leaf relative water content and tuber yield of a half-sib family of 'Katahdin'-derived potato cultivars *Breed Sci.* **66** 328-331

[17] Xu Z and Zhou G 2008 Responses of leaf stomatal density to water status and its relationship with photosynthesis in a grass *Botany.* **59** 3317-25

[18] Arunyanark A, Jogloy S, Akkasaeng C, Vorasoot N, Kesmala T, Nageswara Rao R C, Wright G C and Patanothai A 2008 Chlorophyll stability is an indicator of drought tolerance in peanut *J. Agron. Crop Sci.* **194** 113-125

[19] Jiang C, Johkan M, Hohjo M, Tsukagoshi S and Maruo T 2017. A correlation analysis on chlorophyll content and SPAD value in tomato leaves *Hort. Res.* **71** 37-42

[20] Nigam S N and Aruna R 2007 Stability of soil plant analytical development (SPAD) chlorophyll meter reading (SCMR) and specific leaf area (SLA) and their association across varying soil moisture stress conditions in groundnut (*Arachis hypogaea* L.) *Euphytica.* **160** 111-117

[21] Jacobs W P 1968 Hormonal regulation of leaf abscission *Plant Physiol.* **43** 1480–95

AUTHOR INDEX

AUTHOR INDEX

AUTHOR INDEX

AUTHOR INDEX

Institute of Physics
Dirac House, Temple Back
Bristol BS1 6BE UK

ISSN: 1755-1307
ISBN 978-1-5108-9670-3

CURRAN ASSOCIATES INC.
proceedings.com

2nd International Conference on Computing and Applied Informatics (ICCAI 2017)

Journal of Physics: Conference Series Volume 978

Medan, Indonesia
28 – 30 November 2017

IOP
Institute of Physics